Environmental Endocrine Disruptors

Environmental Endocrine Disruptors

A Handbook of Property Data

LAWRENCE H. KEITH, PhD

A Wiley-Interscience Publication

JOHN WILEY & SONS, INC.

New York • Chichester • Weinheim • Brisbane • Singapore • Toronto

Library of Congress Cataloging in Publication Data:

Keith, Lawrence H., 1938–
    Environmental endocrine disruptors : a handbook of property data /
Lawrence H. Keith.
        p.   cm
    Includes bibliographical (p.   ) references and index.
    ISBN 0-471-19126-4 (alk. paper). — ISBN 0-471-24114-8 (set : alk.
paper)
        1. Environmental toxicology—Handbooks, manuals, etc.
    2. Environmental toxicology—Handbooks, manuals, etc.   I. Title.
RC649.K45   1997
616.4—dc21   97-17406
                                                        CIP

Printed in the United States of America

10 9 8 7 6 5 4 3 2 1

About the Author

Dr. Lawrence H. Keith is a Corporate Fellow at Radian International LLC in Austin, Texas. At Radian he helps to develop new analytical reference materials and manages contracts involving environmental sampling and analysis, quality assurance, and quality control (QA/QC). Before joining Radian in 1977, he was a research chemist with the U.S. Environmental Protection Agency Laboratory in Athens, Georgia. A pioneer in environmental sampling and analysis, methods development, and handling hazardous chemicals, Dr. Keith has published numerous books and technical articles on these subjects. Recent publications have also used electronic books and expert systems as a medium rather than traditional paper publications like this one. In 1986 Dr. Keith and his wife, Virginia, founded Instant Reference Sources, Inc. as a company dedicated to the electronic publication of technical databases, books, and expert systems. He also teaches short courses on environmental sampling and analysis, quality assurance and quality control, and the use of data quality objectives (DQOs) to facilitate efficient, economical generation of data of known quality for making informed decisions. He has lectured extensively on these subjects throughout the U.S. and abroad. Dr. Keith has long been active in the American Chemical Society and has served as Chairman, Program Chair, Secretary, and Alternate Councilor of its Division of Environmental Chemistry. He currently is the Editor of that Division's newsletter, *EnvirofACS,* and the Division's Internet home page. He is also Chairman of the Environmental Monitoring Subcommittee for the ACS Committee on Environmental Improvement. Dr. Keith serves on several academic and industrial advisory boards and is a past member of the EPA Advisory Board and the ACS Environmental Science & Technology Advisory Board.

Acknowledgment

Chemical structures were downloaded from the ChemFinder WebServer (http://chem-finder.camsoft.com) and reproduced with permission of CambridgeSoft Corp., 875 Mass-achusetts Ave., Cambridge, MA 02139 USA (Fax: 617-491-0555; URL http://www.cam-soft.com/).

Electronic Versions of This Database

Electronic versions of the data compiled in this publication are also available on CD-ROM and via Wiley's Worldwide-Web site (www.wiley.com) from John Wiley & Sons, Inc. They contain the same information as this printed book. However, the electronic publications are fully searchable using and/or/not/near type Boolean searches. In addition, all the terms in the three Appendices are hyperlinked to their respective definitions throughout the text and all of the compounds are hyperlinked to their primary sources of information from Instant Reference Sources, Inc. (these publications, *Instant Pesticide Facts, Instant EPA's IRIS, Instant EPA's Air Toxics,* and *Instant Tox-Base,* must be purchased separately on diskette or on CD-ROM in order to complete the hyperlink access to them).

Contents

BENZO(K)FLUORANTHENE   237

BETA-BHC   250

BISPHENOL A   261

BUTYL BENZYL PHTHALATE   271

Introduction

The purpose of this publication is to provide background information on chemical, physical, and toxicological properties of suspect environmental endocrine disruptors in a convenient easy-to-use format. However, this publication does not include in-depth summaries of endocrine disrupting effects of these chemicals. This is an intensely developing field with a great deal of research currently being performed so new publications will be rapidly forthcoming and would quickly make a publication of this kind obsolete. However, a field named "Endocrine Disruptor Information" provides a brief overview of the important endocrine disrupting information for each chemical which lead to it's selection as a endocrine disrupting chemical.

Endocrine Disrupting Chemicals

The endocrine system refers to the complex system that involves the brain and associated organs and tissues of the body. These include the pituitary, thyroid, and adrenal glands and the male and female reproductive systems, all of which release hormones into the bloodstream. These glands and their hormones regulate the growth, development and functions of various tissues and coordinate many of the metabolic processes that occur. In particular the sex hormones include estrogens in females and androgens in males.

Endocrine disrupting chemicals (EDCs) consist of synthetic and naturally occurring chemicals that affect the balance of normal hormonal functions in animals. Depending on their activity they may be characterized as estrogen modulators or androgen modulators. They may mimic the sex hormones estrogen or androgen (thereby producing similar responses to them) or they may block the activities of estrogen or androgen. (i.e., be anti-estrogens or anti-androgens).

One of the first recognized synthetic EDCs was diethylstilbestrol (DES), a pharmaceutical product that was given to pregnant women from 1948 to 1972 to help prevent miscarriages. It caused clear-cell carcinoma in the vagina and reproductive abnormalities in female offspring and a much higher than normal rate of genital defects in male babies. The latter included undescended testicles and stunted testicles. [826]

Another source of EDCs includes foods such as soybeans, apples, cherries, wheat, and peas. There are also many other plants that produce polycyclic substances such as isoflavonoids, phytosterols, and coumestans that have estrogenic activity. These are collectively called "phyto-estrogens" and the degree to which they are important causes of endocrine modifying activity in animals is not yet clear although in ancient Rome the now extinct giant fennel called "silphium" was used for birth control. [826]

Environmental Endocrine Disruptors

A third group of endocrine modifying chemicals are those which are synthetic chemicals that end up as environmental pollutants. These *environmental endocrine disruptors* (EEDs) are the subject of this publication. Attention to this group of chemicals was focused in early 1996 by the publication of the book, "*Our Stolen Future*" by Colborn, Dumanoski and Myers with a foreword by Vice President Gore [810].

What are the EEDs? Indeed, this is one of the basic problems at hand. *Our Stolen Future* lists about 50 suspect EEDs. Another expanded list of about 70 suspect EEDs is published on the Internet by the World Wildlife Fund Canada. [813]. In addition, a list of about 50 suspect EEDs from the Centers for Disease Control & Prevention (CDC) in Atlanta [812] and another list of about 50 suspect EEDs from the U.S. Environmental Protection Agency [811] were obtained by private communication with scientists there. Many of the same chemicals are on all of these lists but there are also significant differences in additions and omissions of some of these chemicals. This highlights the fact that we really do not have good enough information on this group of chemicals to even agree on the basic question of their definition. The 67 chemicals in this publication were selected on the basis of their inclusion in one or more of the priority lists from the above sources and the availability of a reasonable amount of information on their chemical, physical and/or toxicological properties. Given that there are about 600 registered pesticides and 80,000 industrial chemicals in commercial use today it is likely that the lists of suspect EEDs will continue to grow and vary over the next several years as the results of current research begin to throw more light on this subject.

The information on the chemicals in this publication was drawn from a large number of references including several electronic publications and Internet sources. Especially important resources include three volumes of *the National Toxicology Program's Chemical Database* [822], [823], [824] and four members of the *Professional PC References / Windows Series* [818], [819], [820], and [821].

Unresolved Issues

Much of the research on EEDs has focused on the effects of these chemicals on wildlife. There are many studies involving certain pesticides, polychlorobiphenyls (PCBs) and polychlorodibenzo-p-dioxins (e.g., 2,3,7,8-TCDD) that link them to birth defects and aberrant sexual behavior. Birth defects include deformed reproductive organs in male alligators from Lake Apopka, FL associated with the pesticide Dicofol [827] to abnormally small and sometimes undescended testicles in species as diverse as river otters and Florida panthers. Laboratory tests have also produced genital defects, reduced testicular weights and low sperm counts in rats fed with DDE, PCBs, Vinclozolin, and 2,3,7,8-TCDD. [826]

It is also believed by some researchers that EEDs may be the cause of similar types of recent observations in humans. The CDC in Atlanta has performed surveys that show that the average US resident has hundreds of chemicals accumulated in their fat tissues including polychlorodibenzo-p-dioxin and polychlorodibenzofuran isomers ("dioxins" and "furans"). [813]

People in industrial countries typically carry 30-50 ppt of dioxin-like pollutants in their bodies (as measured by "toxic equivalency factors"). [883], [882] This is the same range of exposure believed to be affecting some wildlife populations. Our continuing exposure is in the order of 3-10 pg/kg/day, which translates to about 0.2-0.6 ppt (<u>parts per trillion</u>) per day for a 140-pound person. However, nursing infants may have 10-20 times higher daily exposure, relative to their weight. [882] These are still fantastically low concentrations, but if they are indeed causing the kinds of problems described below then we have a very challenging job to detect and measure them in complex matrices such as environmental or food samples. Knowledge of their chemical, physical and toxicological properties will help make this job easier.

Some chemicals, including many EEDs, can "bioaccumulate" or build up in animals. Once they are incorporated into the tissues and fat of animals and humans, they can remain there for long periods of time until they are ultimately metabolized. Thus, an embryo, the most sensitive stage of life, can be damaged by chemicals the mother was exposed to weeks or years earlier. [882]

The most significant source of exposure by the general public to environmental pollutants is probably from consuming contaminated fish. Any time a contaminated organism is eaten, the contaminants may be passed on to the consumer. Thus, organisms at higher levels of the food chain, such as humans, are typically exposed to greater concentrations of contaminants than the lower level organisms that they eat. The result is that they often develop higher accumulations in their tissues than animals at lower levels of the food chain. [882]

Women should be more prone to higher levels of accumulation of pollutants than men since they generally have a higher percentage of body fat, but everyone who consumes contaminated fish is potentially at risk. Pregnant and lactating women are more likely to mobilize contaminants stored in fatty tissues, which can result in substantial fetal and neonatal (infant) exposure. [882]

Another factor that is important with EEDs is that timing of exposure to them can be critical. In fact, timing of exposure may be more important than the dose or concentration of their exposure. A single exposure at a vulnerable moment for a developing embryo has the potential to cause damage and, of course, long term exposure to relatively small amounts of them also could cause damage. [882]

It has also been discovered that some combinations of two or three EEDs can be many times more potent than any one of them by themselves. Thus, in addition to the timing of exposure, it appears that the combinations of EEDs that one is exposed to can cause a synergistic effect that can magnify the damage they can cause. These are all unresolved issues that will have to be worked out with time and additional research.

Mechanisms of Endocrine Disrupting Chemical Effects

A particularly good summary of the mechanisms and effects of EEDs was written by Wayne A. Schmidt for the Great Lakes Natural Resource Center, National Wildlife Federation [882] and portions of it are provided in the following subsections.

There are at least five mechanisms [882] by which environmental contaminants are able to disrupt vital functions of the endocrine system:

1 Some contaminants are similar enough in structure to hormones that they are able to bind to cellular receptors designed to be targets for natural hormones. This causes unpredictable and abnormal cell activity.

2 Other contaminants appear to block these binding sites so that hormones are unable to bind to them, thus impairing normal cell activity.

3 Some contaminants induce the creation of extra receptor sites in the cell. The consequence can be an amplification of the impact of hormones on cell activity.

4 Contaminants can interact directly and indirectly with natural hormones, changing the hormones' messages and thus altering cell activity.

5 The natural pattern of hormone synthesis can be disrupted by contaminants, resulting in an improper balance or quantity of circulating hormones.

Apparently some environmental contaminants can trigger the mixed function oxidase system (MFO), which is responsible for the production of enzymes involved in the biosynthesis of sex steroids. In addition, the MFO produces enzymes that influence the synthesis of new hormones and the breakdown of hormones. These three enzyme activities control the level of sex hormones circulating in the body. By inducing the MFO to produce the controlling enzymes, contaminants are able to drastically alter the level of sex hormones in the body. [882]

The complexities of potential effects and mechanisms for action are considerable. [882] For example, natural hormones and EEDs can have:

- different effects on different parts of the body;
- different effects at different stages of development;
- different effects depending on nutritional health, age, genetic disposition and even time of the year or the day that exposure occurs;
- variance in their effects with the presence of other chemicals;
- and variance in effects among different species.

PCBs, for example, can produce either estrogenic or antiestrogenic effects.[889], [882] These common EEDs can affect reproductive organs directly, or they can indirectly affect reproductive function through changing pituitary gland secretions. Cumulative small PCB doses can be more damaging than single, larger doses. Predicting effects is confounded further by the complexity of PCB mixtures in the environment. [882]

Effects of Endocrine Disrupting Chemicals

The exact effect of hormone exposure, both natural and unnatural, is greatly dependent on factors such as species, age and gender. Generally, the offspring of exposed adults are the most vulnerable to these effects. Fetuses and newborns are especially susceptible to environmental contaminants. At these early stages in life their protective systems are not fully developed and exposure to estrogenic chemicals early in embryonic or fetal life can lead to the development of major structural changes in the genital tract, including tumors and other abnormal cell growth. Damage done at this stage can be permanent and irreversible. Often the cause of this damage is difficult to determine because the parents may show no obvious adverse effect, and the damage to the offspring may not reveal itself until maturity. [882]

Effects on Sexual Development in General

Early in their existence, vertebrate embryos exhibit neither male nor female traits. The presence or absence of a Y-chromosome generally signals whether an embryo is to be male or female. However, sexual differentiation of all other aspects of masculinization and feminization results from the effects of androgens (male hormones) or estrogens on the embryo.[885], [882]

In mammals, males typically begin differentiation of the sex organs earlier than females. Shortly after differentiation, male embryos secrete testosterone at a high rate, which serves to initiate the process of masculinization (and defeminization) of the accessory reproductive organs, external genitalia and enzyme systems in other tissues such as liver, kidney and brain. In the absence of a Y-chromosome and testosterone, normal female development occurs. [882]

In addition to regulating sexual differentiation during fetal development, sex hormones play a role in the organization of specific areas of the brain. Less is known about this action, but studies have shown a correlation between levels of estrogen and brain morphology, as well as with sexual behavior in male rats and mice. The brain and central nervous system continue development throughout the fetal stage and early natal period making them particularly susceptible to chemical exposure. [882]

Depending on the level and timing of exposure of the fetus to EEDs, impairments caused by them may not even be recognizable at birth. For example, tumors or other abnormal cell growth, may not manifest themselves until adolescence, adulthood or even late in life. The detrimental effects seen in the children of women exposed to the estrogenic drug DES, for example, included decreased fertility, malformed genital ducts, rare reproductive cancers, depressed immune function and possibly effects on behavior and intelligence.[886], [887], [882] In addition, some studies suggested that prenatal DES exposure increased the likelihood of major depressive disorders and bisexual activity and interest in adult women.[888] One study, for example, compared 12 pairs of sisters; 42% of the DES-exposed group indicated a life-long bisexual orientation versus 8% of the unexposed sisters. [882]

Effects on Humans

EEDs affect humans as well as wildlife. There are many documented effects on humans and even more suspected effects. These include decrease in male fertility, defects in male sexual development, increases in prostate cancer, female reproductive problems, increases in breast cancer, endometriosis, immune system damage, increased incidence of goiters, and behavioral and developmental problems in children. Some examples of these are described below.

Male Fertility

The most likely effect of endocrine disruption in men may be a reduction in sperm production and also in the sperm's ability to fertilize an egg. The reproductive system of male rats is exceptionally sensitive to dioxin, leading researchers to believe that human males may be susceptible as well. In normal human males, the number of sperm produced per ejaculate is normally close to the level required for fertility. Thus, even a small reduction in daily sperm production can lead to infertility. [882]

Sperm production by the average man in western countries, including the U.S., today is reported by some to be half of what it was in 1940. One report indicates that average sperm count has declined 42% and average volume of semen diminished by 20%. [891], [882] Another report showed an increase in infertility in the last twenty years concluding that one in twenty men are either subfertile or infertile.[892], [882] However, these may be an over simplification and later reports have questioned these kind of conclusions.

The mechanism by which environmental contaminants could be reducing sperm production is not fully understood. Alterations in hormone activity rates in developing male embryos or infants can affect Sertoli cells, which are the nursery cells to produce sperm. Sertoli cell secretions regulate when testicles descend in baby boys, as well as cell division in the testes and development of the urethra. In addition, the number of Sertoli cells are fixed during gestation; the more Sertoli cells, the greater the sperm production in adult males. Thus, these cells play a very important part in the development of male babies. [882]

Studies of men occupationally exposed to dioxin revealed a decreased level of circulating testosterone. This same effect was seen in laboratory rats exposed to dioxin; in addition, the dioxin-exposed male rats found it difficult to successfully copulate, even after mounting the female. Reduced levels of testosterone are caused by two hormonal actions: decreased secretion from testes and increased metabolism due to induction of the MFO. The similar results in rats and occupationally exposed men are consistent with the effects expected from contaminant-caused disruption at the cellular level. Thus, the theory of disrupted hormone levels is supported in men exposed to relatively higher doses of dioxin but it is not supported yet at levels encountered in the environment. [882]

Male Sexual Development Defects and Cancer

There appears to be an increase in sexual development defects in recent rears and some of these may be linked to exposure to EEDs. For example, more baby boys have to undergo operations to correct undescended testicles ("cryptorchidism") now than 30 years ago; the rate appears to have increased 2- to 3-fold during the past 30 years. A birth defect called "hypospadias," in which the male urinary canal is open on the underside of the penis, also is increasing. "Inter-sex" features in baby boys, where the penis is covered with a layer of fat and genitals have a cleft resembling female features, also appear to be increasing. In some cases where pregnant mothers were exposed to very high levels of toxic chemicals, the mothers' boys have shorter than normal penises, similar to Lake Apopka's alligators. Boys born to women who were exposed to PCB-poisoned rice bran cooking oil in 1978-79 in central Taiwan, the so-called "Yucheng" boys, were found to have significantly shorter penis lengths at ages 11 to 14.[893], [882]

These effects are similar to what happens to male rat pups when their mother is exposed to minute levels of dioxin (64 ng/kg).[894] When the pups reached puberty, decreased sperm count, altered sexual behavior and shortened penises were observed. (Female pups showed malformations of the urogenital organs, including, in some cases, no vaginal openings). [882]

In addition, the rate of sexual development defects and disease in men is reportedly sharply increasing, which also could be linked to effects of environmental hormone exposure.[895], [896], [882] Studies in some industrialized western nations show that cancer of the testicles, relatively more common in young men than older men, has increased at least 3-fold in the past 30 years. Another possible effect of exposure to estrogen-like contaminants is prostate enlargement in older men. This condition affects 80% of men 70 years and older. The exact cause of prostate enlargement, however, is often unknown. Prostate cancer in men also has increased by 80% in the last 20 years.[897], [882]

Female Reproductive Effects

Women normally are exposed to estrogen, but the effects of EEDs on females are more difficult to track due to the estrous cycle and the resulting huge differences in circulating hormone concentrations at different stages of the cycle. Studies that have been done on rodents indicate that estrogenic substances produce accelerated sexual maturation, irregular estrous cycles and prolonged estrous. Some female mice also displayed masculine behavior in conjunction with the other effects. The presence of estrogen mimicking compounds in adult women can impair reproductive capacity by interfering with natural hormone cycles, potentially rendering women unable to conceive or to maintain pregnancy. [882]

Female Breast Cancer

Schmidt speculates that breast cancer may also have links to the estrogenic contaminants. [882] Women in the US and Canada who live to age 85 have a one in nine risk of contracting breast cancer in their lifetime, double the risk in 1940.[898] Furthermore, breast cancer mortality since the 1940s has increased by 1% per year.[899] Two leading theories of the primary risk factors for breast cancer are exposure to estrogen and high fat diets. [882]

Diet can affect women's circulating estrogen level, thus indicating a link between these two factors, as well as to the possibility of a link to EEDs. Fat and fiber are known to alter the intestinal resorption of estrogen. High fat diets increase circulating estrogen levels while high fiber diets are known to decrease estrogen levels. Furthermore, vegetable consumption is associated with the levels of estrogen-binding proteins. A reduction in vegetable consumption results in a decrease in the proteins, increasing the level of circulating estrogen. When exposure to estrogen-mimicking compounds is considered, the potential for appreciably increased estrogen levels is amplified. [882]

In a recent study of 15,000 New York City women, the women with the highest DDE levels had a 4-times higher risk of breast cancer, compared to women with the lowest DDE levels. [900], [882] In this study other risk factors such as age at first menstruation, age at first pregnancy and family history of breast cancer, were controlled and a positive relationships between breast cancer and DDE was still observed. [882]

It also may be possible that some of these chemicals are promoters or inducers of cancer rather than being direct carcinogens. This theory is supported by the findings that some EEDs have estrogenic properties, and that estrogen is known to promote abnormal cell growth. If estrogen exposure after maturation plays a role in the full expression of early developmental changes then this could provide an explanation for both the increased risk of breast cancer to women exposed to estrogens in utero and the rare cancers initiated at maturation in the women whose mothers took DES. [882]

Endometriosis

Recent animal studies strongly suggest that human exposure to dioxin, an endocrine-disrupting chemical, may be linked to endometriosis, a painful disease currently affecting 10% of reproductive-age women. The disease appears to becoming more common and afflicting women at younger ages. [882]

Endometriosis causes bits of uterine lining to migrate generally to other pelvic organs and can cause infertility, internal bleeding and other serious problems. Recently, scientists discovered that four out of five rhesus monkeys used for long-term dioxin research developed endometriosis (compared to one out of three "control" monkeys).[901], [882] At just 5 ppt dioxin in the monkeys' food, 71% of the monkeys developed the disease. These monkeys had dioxin levels in their bodies only ten times more than the average level of dioxins currently found in humans. [882] Other laboratory animal studies also link PCBs with the disease. And a German study indicated that women with endometriosis had significantly higher levels of PCBs in their blood than those without this disease.[902], [882]

Immune System Damages

The interactions between the immune system, by which white blood cells produce hormonal secretions that help ward off diseases, and the endocrine and central nervous systems are complex. However, the associations between endocrine-disrupting pollutants and immune system damages in wildlife are well established.[903] Similar associations are being discovered in humans. [882]

The best evidence may be from humans living closest to the top of the food chain, and hence, most vulnerable to the bioaccumulation of environmental contaminants. For example, the Inuit from Arctic Quebec rely heavily on meat and blubber from seal and whale. These marine mammals are highly contaminated with PCBs and other toxic chemicals so that, as a result, the total PCB concentration in the milk fat of Inuit women was similar to that found in beluga whale blubber. [904], [882] Inuit infants have a very high rate of infectious diseases, including respiratory and ear infections (otitis) and vaccinations are less effective than in the general population, suggesting impaired immune systems. [882]

Goiters

Another effect of endocrine disruption in both adult males and females may be thyroid gland enlargement, more commonly known as goiter. The thyroid gland controls growth hormones and the hormones that regulate metabolism and enlargement of the thyroid gland can disrupt metabolism. The exact effects of environmental contaminants on the thyroid are not well understood, but thyroid dysfunction may be contributing to the "wasting syndrome" seen in wildlife in polluted areas. The EEDs that have been implicated in this syndrome include PCBs, dioxin, DDT, toxaphene and lead. [882] The properties of these chemicals are all included in this publication.

Most cases of goiter are attributed to dietary iodine deficiency, but the addition of iodine to table salt has largely reduced this source of disease. So, in the US, cases of goiter cannot be attributed to lack of dietary iodine. Also, Great Lakes salmon have enlarged thyroid and those problems are not iodine-related. [882] Some environmental factor is suspected to be affecting the endocrine systems of Great Lakes salmon and it also could be affecting the people living there. There are reports of the prevalence of endemic goiter in the US, specifically in the state of Michigan, that are not attributable to iodine deficiency.[890], [882]

Hyperactivity, Learning, and Attention Problems With Children

Developing embryos and fetuses, infants and children are most vulnerable to the effects of EEDs. Recent studies found a dose-response relationship between the quantity of contaminated Great Lakes fish consumed by the mother and such measures in newborn infants as abnormally weak reflexes, reduced responsiveness, motor coordination and muscle tone. [905], [906], [882] Further studies indicated that more highly exposed children had slower reaction times to visual stimuli, made more errors on a memory test and took longer to solve problems. Performance with short-term memory tests also depends on selective and sustained attention suggesting that attentional deficits may have been contributing to the children's poor performance. Hyperactivity and learning deficits are among the likely effects in children exposed in utero to endocrine-disrupting chemicals, based on many related studies. If only a small part of the learning and behavioral problems of children can be attributed to endocrine, immune or nervous system damages caused by maternal or childhood exposure to EEDs, the implications for our culture, society and survival are profound. [882]

Fields of Information Included With Each Chemical

Specific "fields" of information have been gathered and organized for each chemical. The purpose is to provide a consistent organizational structure so that information of interest can be easily located. In general, the chemical's name and molecular structure (or elemental symbol) and introductory background information are followed by chemical "identifying" data (the unique CAS Registry Number, synonyms, and, when available, NIOSH Registry Number). Next in importance is a field describing what Endocrine Disruptor Information is available to understand why each chemical is included in this publication. This is followed by fields of information organized into five main groups: Chemical and Physical Properties, sources of Environmental Reference Materials (needed to conduct analyses and toxicological studies), Medical Symptoms of Exposure, Toxicological Properties, and, for pesticides only, Production, Use, and Pesticide Labeling Information. Each of these are described in detail below.

Introduction

General facts on manufacture, uses, regulatory limits, and dispersion of each chemical in the environment are provided with other miscellaneous background information. The information in this field serves to acquaint the reader with the relative importance of each chemical in our society by describing their useful features and detrimental effects.

CAS Registry Number

This is a unique nine digit number assigned by the American Chemical Society Chemical Abstracts Service. The last digit in a CAS number is the first number to the left of the decimal in the sum which is derived from (n+1) times each of the CAS number in sequence from right to left beginning with the first CAS numbers to the left of the last hyphen. This is illustrated below:

CAS No.   123-72-8

1 x	2	=	2
2 x	7	=	14
3 x	3	=	9
4 x	2	=	8
5 x	1	=	5

               Last Digit: 8

Synonyms

Most compounds are known by more than one chemical name. All available synonyms are listed in this section for each preferred name. In cases where some compounds are formulated as commercial preparations, even these commercial "trade" names are listed. This category is very useful for finding compound's entries which can be searched by keywords.

Endocrine Disruptor Information

Each chemical in this publication is here because it is suspected of causing endocrine disrupting effects as a result of its dispersion in the environment. The sources of these environmental endocrine disruptors are lists from the U. S. Environmental Protection Agency, the Center for Disease Control and Prevention, and the World Wildlife Fund. Dr. Theo Colborn works with the latter organization and all of the approximately 50 chemicals discussed in her book, "*Our Stolen Future*," plus some additional ones, were considered. Unfortunately, the reasons for selecting some of these chemicals by these organizations are not all well documented. But, where documentation was able to be found, or deduced from literature searches by the Editor, it is listed. Thus, the purpose of this section is to give the reader some sense of the background reasoning as to why various organizations have included each chemical in their prioritized lists. With the intense interest now focused on EEDs it is expected that the effects of endocrine disruption will become more clearly defined and documented in future publications by environmental toxicologists. These will provide important sources for updated information in future editions of this publication.

Chemical Formula

Empirical formulas show the number of each type of atom present in the molecules.

Molecular Weight

The molecular weights given are formula weights, or molar masses in daltons, usually to the nearest one hundredth of an atomic mass unit (AMU) and are based on multiples of average atomic mass units.

WLN

The Wiswesser Line Notation (WLN) uses numbers to express lengths of alkyl chains and the sizes of rings. These symbols then are cited in connected order from one end of the molecule to the other.

Physical Description

The physical description of each chemical is drawn from the literature. It includes whether the material is liquid, solid or a gas and whether it is colored or colorless when this information is available.

Specific Gravity

The specific gravity of a chemical is a ratio of the mass of the chemical, solid or liquid, to the mass of an equal volume of water at a specified temperature. The usual convention is to use the mass of water at 4°C, which is the temperature at which it is at its greatest density. The first temperature listed is that at which the mass of the chemical was measured; the second temperature listed is that at which the mass of the water was measured and it is separated from the first temperature by a slash.

Density

Density is the mass of a substance per unit volume. In all cases in these volumes it is the mass in grams per milliliter (cubic centimeter). Density measurements were drawn from the available literature sources or were experimentally determined at Radian Corporation in Austin, Texas.

Melting Point

The melting points are drawn from available literature sources or were experimentally determined at Radian Corporation. They are reported as a range in degrees Centigrade. Where two or more melting points are recorded the highest, or the one considered to be most probable, is listed in this column. The second melting point is listed in the column Other Physical Data.

Boiling Point

The boiling points are drawn from available literature sources and are reported in degrees Centigrade. If no pressure notations are included with the boiling point it is assumed that the boiling point recorded was at standard pressure (760 mm Hg). If the boiling point was obtained at a reduced pressure it is so noted directly after the temperature. When two or more boiling points at a given pressure are noted in the literature the highest temperature or the one considered most probable is listed.

Solubility

Common solvents used in toxicity testing include, water, DMSO (dimethyl sulfoxide), 95% ethanol, acetone, methanol, and toluene.   When available, solubility data is presented for each of these solvents over a specific weight per volume range using units of milligrams per liter. To be considered soluble at the referenced range all of the subject material must have dissolved in the given amount of solvent.   All solubility measurements in this data field were conducted by Radian Corporation and the temperatures at which the measurement was made is included.

Other Solvents

Solubility data in this field is drawn from literature sources.   While it sometimes significantly increases the solubility information from additional, and sometimes more diverse solvents, it also has the disadvantage of being less precise than the experimentally determined values in the above field.   Literature information involving chemical solubilities is often poorly documented, highly variable, and subjective.   In addition, inconsistent terms such as "insoluble," "slightly soluble," "poorly.

Other Physical Data

A large variety of miscellaneous chemical and physical information is included in this section.   Data involving refractive index, boiling points under reduced pressure, conflicting information on any of the above physical or chemical properties, multiple crystalline or other physical states, vapor pressure, vapor density, odor, pK, optical rotation, etc., are all listed in this data field.

HAP Weighting Factor

The hazardous information characteristics for each Hazardous Air Pollutant (HAP) in this publication includes EPA's proposed toxicity weighting factor, shown as "HAP Weighting Factor." The Clean Air Act Amendments of 1990 establish a strategy to reduce and control emissions of Hazardous Air Pollutants (HAP's) from major sources (i.e., those which emit 10 tons per year of a single HAP or 25 tons per year for any combination of HAP's). It required the application of Maximum Achievable Control Technology (MACT) to specific source categories. Proposed regulations in 56 FR 27338-27374 and EPA's "Enabling Document for Regulations Governing Compliance Extensions for Early Reductions of Hazardous Air Pollutants" of July 1991 set forth proposed guidance for early reduction demonstrations.

Volatility

Information on volatility is provided in two forms whenever possible: vapor pressure and vapor density. Data on vapor pressure is usually provided in units of millimeters (mm) of mercury at a given temperature. The higher the vapor pressure the more volatile, and potentially dangerous, a chemical is. The vapor density is expressed in units relative to air. Thus a chemical with a vapor density of 2.0 is twice as dense as air and will tend to "flow" or "pool" around ground level and low places whereas chemicals with a vapor density less than 1.0 will tend to rise and more easily become dispersed in air.

Fire Hazard

This data field describes whether a compound is known or believed to be flammable (according to NFPA definitions) and thus to present a potential serious fire hazard or just to be combustible and therefore present a less potential fire hazard. Where flammability data is listed by class identifiers such as IA, IB, IC, II, etc., OSHA criteria (29 CFR 1910.106) have been used. Appendix 2 summarizes several code systems. Flash points are provided when they could be found. In some cases flash points of liquids were measured by chemists at Radian Corporation in Austin, Texas. In addition, when available information permitted, recommended fire extinguishing methods are provided.

LEL and UEL

These are fields for Lower Explosive Limits and Upper Explosive Limits (also known as lower flammability limits and upper flammability limits -- LFL and UFL), respectively. These numbers represent the lowest and highest percentages of these compounds in air that, in the presence of a spark or flame, can cause an explosive combustion.

Reactivity

This data field contains information on incompatibility of chemicals with water, air, or other chemicals.

Stability

This data field provides information on known or believed stability of chemicals. If protection from light, air, or heat is known or believed to be required or prudent for long term storage, these recommendations are provided. And, since many of the compounds were prepared in water or other solvents for toxicological dosing studies with animals, information on their stability in one or more solutions is often provided. In some cases stability measurements were made by chemists at Radian Corporation. In other cases an evaluation was made from data in the literature and/or from a review of the hazardous properties data. In the latter case, the chemical stability information is qualified by the words "...should be stable...".

Acute/Chronic Hazards

Acute hazards have a rapid onset of symptoms of exposure and a short duration. Examples may include irritation of exposed tissues (a reversible reaction that may vary from mild spots that itch to more severe rashes and welts), corrosion (irreversible destruction) of tissues, lachrymation (tearing of the eyes) onset of coma, or death. Other acute hazards may result from heating chemicals to the point of decomposition, either in a laboratory or in an uncontrolled fire. Hazards from thermally decomposing chemicals may vary from acrid, noxious fumes to toxic gases such as phosgene or carbon monoxide.

Chronic hazards may include sensitization, carcinogenicity (cause cancer) or reproductive and developmental effects (including teratogenicity), among other hazards.

Medical Symptoms of Exposure

Symptoms of exposure are tremendously variable and range from irritation of the eyes, nose, and other exposed tissues to coma and, ultimately, death. Because of the thousands of different combinations and medically related grouping of symptoms, the electronic version of this publication is especially useful for searching for compounds that have various symptoms or combinations of symptoms. In the electronic version a glossary of commonly found symptoms are hyperlinked to many of the terms used for symptoms of exposure in order to help make these more understandable to people who lack medical training.

Sources of Information

In addition to literature searches and searches of the Internet, primary sources of information came from several major publications containing the selected fields of information discussed in the following section.

Chemical and Physical Properties data were obtained from:

- *Agrochemical Desk Reference Environmental Data,* CRC Press, Inc./Lewis Publishers;
- *Instant EPA's Air Toxics*, Instant Reference Sources, Inc.;
- *Instant EPA's Pesticide Facts*, Instant Reference Sources, Inc.;
- *The National Toxicology Program's Chemical Database, Volume 2,* CRC Press, Inc./Lewis Publishers; and
- *The National Toxicology Program's Chemical Compendium, Volume 2,* CRC Press, Inc./Lewis Publishers.

Chemical Structures were downloaded from the ChemFinder WebServer (http://chemfinder.camsoft.com) and reproduced with permission of CambridgeSoft Corp, Cambridge, MA.

Sources of Environmental Reference Materials were obtained from catalogs of three companies that provide high purity analytical reference materials:

- *Analytical Reference Materials*, 1997, Radian International LLC;
- *Chemical Standards*, 1996, Cambridge Isotope Labs, Inc., and
- *Chemical Standards*, 1996, Crescent Chemical Co.

There are other companies besides these, of course, but between them they covered all of the chemicals in this publication. Thus, convenience for ordering these materials and the author's knowledge of the high quality of their products played a significant role in choosing these three sources.

Information on **Hazardous Properties** and **Medical Symptoms of Exposure** was obtained primarily from:

- *Agrochemicals Desk Reference Environmental Data,* CRC Press, Inc./Lewis Publishers;
- *Instant EPA's Air Toxics,* Instant Reference Sources, Inc.;
- *The National Toxicology Program's Chemical Database, Volume 4,* CRC Press, Inc./Lewis Publishers; and
- *The National Toxicology Program's Chemical Compendium, Volume 4,* CRC Press, Inc./Lewis Publishers.

Information on **Toxicological Properties** was from several sources including two national databases:

- *Integrated Risk Information System* (IRIS) U. S. EPA
- *The National Toxicology Program's Chemical Database, Volume 7,* CRC Press, Inc./Lewis Publishers;
- *The National Toxicology Program's Chemical Compendium, Volume 7,* CRC Press, Inc./Lewis Publishers; and

These databases contain a wealth of information on "traditional" toxicological data which includes LD50, LC50, and LCLo data, data on carcinogenicity (e.g., tumorigenic data), teratogenicity (reproductive effects data), plus information on chronic health hazard assessments, carcinogenicity assessments, Drinking Water Health Advisories and other EPA regulatory actions. EPA's IRIS database, obtainable as a series of over 500 ASCII files, is annually republished in Microsoft® Windows™ format by Instant Reference Sources. This searchable hyperlinked publication, *Instant EPA's IRIS,* was the actual publication from which IRIS data was abstracted.

Production, Use, and Pesticide Labeling Information was obtained from *Instant EPA's Pesticide Facts,* a database republished from an on-line EPA database in Microsoft® Windows™ format by Instant Reference Sources.

Miscellaneous environmental impact information found in the **Introduction** was often found in a database at the website of Spectrum Laboratories, Inc.

Reference Source Addresses

The addresses, telephone or fax numbers, and Internet URLs for the principal reference sources listed in this publication are provided below in alphabetical order.

Cambridge Isotope Laboratories, 50 Frontage Rd., Andover, MA 01810-5413 (Fax: 508-749-2768; Internet URL, http://www.isotope.com).

CambridgeSoft Corp., 875 Massachusetts Ave., Cambridge, MA 02139 (Fax: 617-491-0555; URL http://www.camsoft.com/).

CRC Press, Inc./Lewis Publishers, 2000 Corporate Blvd., Boca Raton, FL 33431 (Fax: 407-998-9114; Internet URL, http://www.crcpress.com/).

Crescent Chemical Co., 1324 Motor Parkway, Hauppauge, NY 11788
(Fax: 516-348-0913; Internet URL, http://iscpubs.com/manu0351.html).

Instant Reference Sources, Inc., 7605 Rockpoint Drive, Austin, TX 78731
(Fax: 512-345-2386; Internet URL, http://www.instantref.com/inst-ref.htm).

Radian International LLC, PO Box 201088, 8501 N. Mopac Blvd., Austin, TX
78720 (Fax 512-454-0268; Internet URL, http://www.radian.com/standards);

Spectrum Laboratories, Inc., Ft Lauderdale, FL and Savannah, GA
(Tel: 1-800-262 5983; Internet URL, http://www.speclab.com)

Literature Citations

Reference numbers from [001] through [700] were incorporated from the *National Toxicology Program's Chemical Database* and it's companion printed publication, the *National Toxicology Program's Chemical Compendium.* There are gaps in the reference number sequence and this was done intentionally so that when new editions of the same reference were procured they could be given a new number that was sequential to earlier editions. In other cases spaces were left for potential future references of a like nature. Reference numbers higher than [700] have come from other sources; for example, reference numbers [801] through [806] come from *Instant EPA's Air Toxics,* and reference numbers starting from [810] are those specifically added for this publication. In the Microsoft® Windows™ electronic version of this publication these are all hyperlinked to their reference citations throughout.

Microsoft® and Windows™ are registered trademarks of the Microsoft Corporation.

Acenaphthene

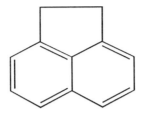

Introduction

Acenaphthene is polynuclear aromatic hydrocarbon (PAH) that is derived from coal tar and petroleum refining. It is used in the synthesis of many chemical products such as dye intermediates, plastics, pharmaceuticals and pesticides. Not surprisingly it is a commonly found environmental pollutant in water, air and soil due to its wide usage as a chemical intermediate in the synthesis of many useful products.

Emissions from petroleum, coal tar distillation and diesel fueled engines are major contributors of acenaphthene to the environment. It is expected to exist entirely in the vapor phase in ambient air and reaction with photochemically produced hydroxyl radicals may decompose it within days. Non-occupational exposures would most likely occur from urban atmospheres, contaminated drinking water supplies and recreational contaminated waterways. Over the years it has been found as a contaminant in many wastes, groundwater and drinking water samples from all over the world [828].

Acenaphthene is rapidly decomposed in domestic wastewater [816] and, as a relatively small hydrocarbon molecule, would be expected to rapidly decompose under many environmental conditions where ambient temperatures are representative of temperate climates. It is a planar (flat) aromatic molecule, a characteristic of many of the compounds that are suspected EEDs.

Identifying Information

CAS NUMBER: 83-32-9

NIOSH Registry Number: AB1000000

SYNONYMS: ETHYLENENAPHTHALENE, 1,2-DIHYDROACE-NAPHTHYLENE, 1,8-DIHYDROACENAPHTHYLENE, PERI-ETHYLENENAPHTHALENE, 1,8-ETHYLENENAPHTHALENE, NAPHTHYLENEETHYLENE

Endocrine Disruptor Information

Acenaphthene is on EPA's list of suspected EEDs [811] and it is an aromatic coplanar compound. However, no additional information has been located with respect to its endocrine modifying properties. It is very prevalent as an environmental pollutant and is widely found in drinking water, wastewater, groundwater, air and landfill wastes [828] so this may also have contributed to its inclusion on EPA's list.

Chemical and Physical Properties

CHEMICAL FORMULA: $C_{12}H_{10}$

MOLECULAR WEIGHT: 154.21

WLN: L566 1Λ LT&&J

PHYSICAL DESCRIPTION: White crystals

SPECIFIC GRAVITY: 1.024 @ 99/4°C [041], [062], [071]

DENSITY: 1.02 g/mL @ 25°C [430]

MELTING POINT: 95°C [031], [038], [041]

BOILING POINT: 279°C [017], [031], [058], [205], [274]

SOLUBILITY: Water: <1 mg/mL @ 20°C [700]; DMSO: 10-50 mg/mL @ 20°C [700]; 95% Ethanol: 1-5 mg/mL @ 20°C [700]; Acetone: 50-100 mg/mL @ 20°C [700]; Methanol: 1 g/56 mL [031]; Toluene: 1 g/5 mL [031]

OTHER SOLVENTS: Chloroform: 1 g/2.5 mL [031]; Ether: Soluble [430]; Benzene: 1 g/5 mL [031]; Propanol: 1 g/25 mL [031]; Glacial acetic acid: 1 g/100 mL [031]

OTHER PHYSICAL DATA: Specific gravity: 1.069 @ 95/95°C [205]; 1.0242 @ 90/4°C [017]; Odor threshold: 0.02-0.22 ppm; Refractive index: 1.6048 @ 95°C

Source: *The National Toxicology Program's Chemical Database, Volume 2*, 1992 CRC Press, Inc./Lewis Publishers, 2000 Corporate Blvd., Boca Raton, FL 33431 (Fax: 407-998-9114).

Environmental Reference Materials

A source of EED environmental reference materials for acenaphthene is Radian International LLC, Austin, TX (Fax 512-454-0268; Internet URL, http://www.radian.com/standards); Catalog No. ERA-009.

Hazardous Properties

ACUTE/CHRONIC HAZARDS: This compound may be harmful by inhalation, ingestion or skin absorption. It may be an irritant of the skin, eyes, mucous membranes and upper respiratory tract. When heated to decomposition it emits toxic fumes of carbon monoxide and carbon dioxide [058], [269].

VOLATILITY: Vapor pressure: 0.001-0.01 mm Hg @ 20°C [051], [071]; 5 mm Hg @ 114.8°C [038]; Vapor density: 5.32

FIRE HAZARD: Flash point data for Acenaphthene are not available. It is probably combustible. Fires involving this material can be controlled with a dry chemical, carbon dioxide or Halon extinguisher. A water spray may also be used [058], [269].

LEL: 0.6% [430]   UEL: Not available

REACTIVITY: Acenaphthene is incompatible with strong oxidizing agents [058], [071], [269]. It is also incompatible with ozone and chlorinating agents [071]. It forms crystalline complexes with desoxycholic acid [031].

STABILITY: Acenaphthene is stable under normal laboratory conditions. Solutions of Acenaphthene in water, DMSO, 95% ethanol or acetone should be stable for 24 hours under normal laboratory conditions [700].

Source: *The National Toxicology Program's Chemical Database, Volume 7*, Copyright 1992 by CRC Press, Inc./Lewis Publishers, 2000 Corporate Blvd., Boca Raton, FL 33431 (Fax: 407-998-9114).

Medical Symptoms of Exposure

Symptoms of exposure to this compound may include irritation of the skin, eyes, mucous membranes and upper respiratory tract [058], [269]. If ingested, it may cause vomiting [041], [051], [071], [346]. Chronic exposure may result in kidney and liver damage [301].

Source: *The National Toxicology Program's Chemical Database, Volume 4*, 1992, CRC Press, Inc./Lewis Publishers, 2000 Corporate Blvd., Boca Raton, FL 33431 (Fax: 407-998-9114).

Toxicological Information

The toxicological information below is gathered from several sources including two national databases: one from the *National Toxicology Program Chemical Repository Database* and another from *EPA's Integrated Risk Information System (IRIS)*.

Toxicology Information from the National Toxicology Program

CARCINOGENICITY: Not available

TERATOGENICITY: Not available

MUTATION DATA:

Test	Lowest dose		Test	Lowest dose
mmo-omi	3			

TOXICITY: Not available

OTHER TOXICITY DATA:
Status:

> EPA Genetox Program 1986, Inconclusive: D melanogaster-nondisjunction
> EPA TSCA Chemical Inventory, 1986
> EPA TSCA Test Submission (TSCATS) Data Base, December 1986
> NIOSH Analytical Methods: see Polynuclear Aromatic Hydrocarbons
> (HPLC), 5506; (GC), 5515

SAX TOXICITY EVALUATION:

> THR: An irritant to the skin and mucous membranes. An experimental
> neoplastigen.

Source: ***Instant Tox-Base***, 1995, Instant Reference Sources, Inc., 7605 Rockpoint Drive, Austin, TX 78731 (Fax: 512-345-2386; Internet URL, http://www.instantref.com/inst-ref.htm).

Toxicology Information from EPA's Integrated Risk Information System (IRIS)

I.   CHRONIC HEALTH HAZARD ASSESSMENTS FOR NONCARCINOGENIC EFFECTS

A. REFERENCE DOSE FOR CHRONIC ORAL EXPOSURE (RfD)

The Reference Dose (RfD) is based on the assumption that thresholds exist for certain toxic effects such as cellular necrosis, but may not exist for other toxic effects such as carcinogenicity. In general, the RfD is an estimate (with uncertainty spanning perhaps an order of magnitude) of a daily exposure to the human population (including sensitive subgroups) that is likely to be without an appreciable risk of deleterious effects during a lifetime. Please refer to Background Document 1 in Service Code 5 for an elaboration of these concepts. RfDs can also be derived for the noncarcinogenic health effects of compounds which are also carcinogens. Therefore, it is essential to refer to other sources of information concerning the carcinogenicity of this substance. If the U.S. EPA has evaluated this substance for potential human carcinogenicity, a summary of that evaluation will be contained in Section II of this file when a review of that evaluation is completed.

I.A.1. ORAL RfD SUMMARY

Critical Effect	Experimental Doses*	UF	MF	RfD
Hepatotoxicity	NOAEL: 175 mg/kg/day	3000	1	6E-2 mg/kg/day
	LOAEL: 350 mg/kg/day			

*Conversion Factors: None

Study Mouse Oral Subchronic, U.S. EPA, 1989

I.A.2. PRINCIPAL AND SUPPORTING STUDIES (ORAL RfD)

U.S. EPA. 1989. Mouse oral subchronic study with acenaphthene. Study conducted by Hazelton Laboratories, Inc., for the Office of Solid Waste, Washington, DC.

Four groups of CD-1 mice (20/sex/group) were gavaged daily with 0, 175, 350, or 700 mg/kg/day acenaphthene for 90 days. The toxicological evaluations of this study included body weight changes, food consumption, mortality, clinical pathological evaluations (including hematology and clinical chemistry), organ weights and histopathological evaluations of target organs. The results of this study indicated no treatment-related effects on survival, clinical signs, body weight changes, total food intake, and ophthalmological alterations. Liver weight changes accompanied by microscopic alterations (cellular hypertrophy) were noted in both mid-and high-dose animals and seemed to be dose-dependent. Additionally, high-dose males and mid-and high-dose females showed significant increases in cholesterol levels. Although increased liver weights, without accompanying microscopic alterations or increased cholesterol levels, were also observed at the low dose, this change was considered to be adaptive and was not considered adverse. The LOAEL is 350 mg/kg/day based on hepatotoxicity); the NOAEL is 175 mg/kg/day.

I.A.3. UNCERTAINTY AND MODIFYING FACTORS (ORAL RfD)

UF -- An uncertainty factor of 3000 reflects 10 each for inter- and intraspecies variability, 10 for the use of a subchronic study for chronic RfD derivation, and 3 for the lack of adequate data in a second species and reproductive/developmental data.

MF -- None

I.A.4. ADDITIONAL COMMENTS (ORAL RfD)

Reshetyuk et al. (1970) examined the comparative toxicity of acenaphthene and acenaphthylene with respect to naphthalene. On intraperitoneal administration in rats (species/number/sex unspecified), naphthalene was more toxic than acenaphthene and acenaphthylene. Two LD50 values (0.6 and 1.7 g/kg) were reported, but it is unclear to which of the three chemicals these values belonged. Intraperitoneal and intratracheal administration of naphthalene, acenaphthene, and acenaphthylene produced monotypic effects in the form of vascular disorders, and degeneration in the internal organs and central nervous system. Inflammatory changes were also observed in the lungs; the degree was the same for all three substances. Splenic degeneration was noted among the unscheduled deaths in this study. Reshetyuk et al. (1970) concluded that chronic inhalation of acenaphthene and acenaphthylene had more pronounced toxic effects than naphthalene.

Gershbein (1975) exposed partially hepatectomized rats to 15 mg/kg acenaphthene in the diet for 7 days. The only parameters used to assess toxicity were body weight, absolute liver weight, and liver regeneration. Information on histopathologic alterations and food intake is needed to evaluate the adversity of decreased body weight gain and increased liver weight observed in this study. Increased liver regeneration was reported. Because of its inherent deficiencies, this study is not considered adequate for RfD derivation.

Knobloch et al. (1969) administered 2 g/kg acenaphthene orally to rats and mice for 32 days. Weight loss and mild histopathological alterations in the liver and kidney were observed. It is unclear whether experimental controls were used.

I.A.5. CONFIDENCE IN THE ORAL RfD

Study -- LowData Base -- LowRfD -- Low

Confidence in the study is low, because the observed effects were adaptive and not considered adverse. Confidence in the data base is low because of the lack of supporting chronic toxicity and developmental/reproductive studies. Low confidence in the RfD follows.

I.A.6. EPA DOCUMENTATION AND REVIEW OF THE ORAL RfD

Source Document -- This assessment is not presented in any existing U.S. EPA document.

Other EPA Documentation -- U.S. EPA, 1980

Agency Work Group Review -- 11/15/89

Verification Date -- 11/15/89

I.A.7. EPA CONTACTS (ORAL RfD)

Harlal Choudhury / NCEA -- (513)569-7536

Kenneth A. Poirier / OHEA -- (513)569-7453

I.B. REFERENCE CONCENTRATION FOR CHRONIC INHALATION EXPOSURE (RfC)

[Ed. Note: EPA does not have information yet in this section of IRIS].

II. CARCINOGENICITY ASSESSMENT FOR LIFETIME EXPOSURE

This substance/agent has been evaluated by the U.S. EPA for evidence of human carcinogenic potential. This does not imply that this agent is necessarily a carcinogen. The evaluation for this chemical is under review by an inter-office Agency work group. A risk assessment summary will be included on IRIS when the review has been completed.

III. HEALTH HAZARD ASSESSMENTS FOR VARIED EXPOSURE DURATIONS

III.A. DRINKING WATER HEALTH ADVISORIES

[Ed. Note: EPA does not have information yet in this section of IRIS].

IV. U.S. EPA REGULATORY ACTIONS

[Ed. Note: EPA does not have information yet in this section of IRIS].

VI. BIBLIOGRAPHY

VI.A. ORAL RfD REFERENCES

Gershbein, L.L. 1975. Liver regeneration as influenced by the structure of aromatic and heterocyclic compounds. Res. Commun. Chem. Pathol. Pharmacol. 11: 445.

Knobloch, K., S. Szendzikowski and A. Slusarczyk-Zalobona. 1969. Acute and subacute toxicity of acenaphthene and acenaphthylene. Med. Pracy. 20: 210-222. (Pol.) (Cited in U.S. EPA, 1980)

Reshetyuk, A.L, E.I. Talakina and P.A. En'yakova. 1970. Toxicological evaluation of acenaphthene and acenaphthylene. Gig. Tr. Prof. Zabol. 14:46-47.

U.S. EPA. 1980. Ambient Water Quality Criteria Document for Acenaphthene. Prepared by the Office of Health and Environmental Assessment, Environmental Criteria and Assessment Office, Cincinnati, OH for the Office of Water Regulation and Standards, Washington, DC. EPA-440/5-80-015. NTIS PB81-117269.

U.S. EPA. 1989. Mouse Oral Subchronic Study with Acenaphthene. Study conducted by Hazelton Laboratories, Inc., for the Office of Solid Waste, Washington, DC.

VI.B. INHALATION RfC REFERENCES

None

VI.C. CARCINOGENICITY ASSESSMENT REFERENCES

None

VI.D. DRINKING WATER HA REFERENCES

None

VII. REVISION HISTORY

Date	Section	Description
11/01/90	I.A.	Oral RfD summary on-line
11/01/90	VI.	Bibliography on-line
01/01/92	IV.	Regulatory Action section on-line
07/01/92	IV.	Wrong information removed
05/01/93	II.	Carcinogenicity assessment under review

SYNONYMS

Acenaphthylene,1,2-dihydro-
Acenaphthene
HSDB 2659
Naphthyleneethylene

NSC 7657
PERI-ETHYLENENAPHTHALENE
1,2-DIHYDROACENAPHTHYLENE
1,8-ETHYLENENAPHTHALENE

Source: ***Instant EPA's IRIS***, 1997, Instant Reference Sources, Inc., 7605 Rockpoint Drive, Austin, TX 78731 (Fax: 512-345-2386; Internet URL, http://www.instantref.com/inst-ref.htm).

Alachlor

Introduction

Alachlor is a commonly used herbicide for preemergence and early postemergence control of many annual broad-leaved weeds and most grasses. It is used on crops such as corn, peas, soybeans, peanuts, beans, cotton, milo, and sunflowers. It is usually sold in the form of emulsifiable concentrates, microscopic capsules and granules for application to the soil or weeds where the above types of crops are grown. It's persistence in soil ranges from about 1 to 3 months and it is degraded by hydrolysis, photolysis and microorganisms to a large number of metabolites containing functional groups of dihydroindoles, tetrahydroquinones, acetanilides, and anilines.

The release of alachlor in the environment occurs particularly as a result of its application in the field. In soil, alachlor is transformed to its metabolites primarily by biodegradation. The half-life of alachlor disappearance from soil is about 15 days, although very little mineralization has been observed. Alachlor is highly to moderately mobile in soil and the mobilization decreases with an increase in organic carbon and clay content in soil. In water, both photolysis and biodegradation are important for the loss of alachlor, although the role of photolysis becomes important in shallow clean water, particularly in the presence of sensitizers. The bioconcentration of alachlor in aquatic organisms is not important. [828]

Identifying Information

CAS NUMBER: 15972-60-8

SYNONYMS: 2-Chloro-2',6'-diethyl-N-(methoxymethyl)acetanilide, 2-Chloro-2',6'-diethyl-N-methoxymethyl)acetanilide, 2-Chloro-N-(2,6-diethyl)phenyl-N-ethoxymethylacetamide, 2-Chloro-N-(2,6-diethylphenyl)-N-(methoxymethyl)acetamide, Alanex, Alochlor, Lasso, Lazo, Metachlor, Methachlor, Pillarzo

Source: *Instant EPA's IRIS*, 1997, Instant Reference Sources, Inc., 7605 Rockpoint Drive, Austin, TX 78731 (Fax: 512-345-2386; Internet URL, http://www.instantref.com/inst-ref.htm).

Endocrine Disruptor Information

Alachlor is a strongly suspected EED in that it is listed by EPA [811], the Centers for Disease Control & Prevention [812], and the World Wildlife Fund [813] as a potential endocrine modifying chemical. Although no biologically relevant fetal anomalies were noted in feeding studies with rats and rabbits, there were pathological gross and microscopic findings in the thyroids, prostates and ovaries in rats [819]. Thyroid follicular tumors were also reported in male rats and a NOEL of 10 mg/kg/day has been established for reproductive effects of kidneys to offspring [821].

Chemical and Physical Properties

CHEMICAL FORMULA: $C_{14}H_{20}ClNO_2$

MOLECULAR WEIGHT: 269.77

PHYSICAL DESCRIPTION: White or cream colored solid or crystals

SPECIFIC GRAVITY: 1.133 @ 25/15.6 °C

DENSITY: Not available

MELTING POINT: 40 - 41.5 °C

BOILING POINT: 100 °C at 0.02 mm Hg (decomposes at 105 °C)

SOLUBILITY: Soluble in ether, acetone, benzene, alcohol (unspecified) and ethyl acetate; slightly soluble in hexane; solubility in water - 240 ppm

OTHER PHYSICAL DATA: Octanol/Water Partition Coefficient: 434

Source: *Instant EPA's Pesticide Facts*, 1994, Instant Reference Sources, Inc., 7605 Rockpoint Drive, Austin, TX 78731 (Fax: 512-345-2386; Internet URL, http://www.instantref.com/inst-ref.htm).

Environmental Reference Materials

A source of EED environmental reference materials for Alachlor is Radian International LLC, Austin, TX (Fax 512-454-0268; Internet URL, http://www.radian.com/standards); Catalog No. ERA-053.

Hazardous Properties

ACUTE/CHRONIC HAZARDS: Acute Toxicity: Technical alachlor is not an acutely toxic product by any route of exposure and Alachlor does not cause significant eye or skin irritation in rabbits. However, in terms of chronic toxicity, alachlor is oncogenic in both mice and rats. In mice, alachlor caused a statistically significant increase in lung bronchioalveolar tumors in females at 260 mg/kg/day (Highest Dose Tested). In rats, alachlor causes statistically significant increases at 42 mg/kg/day and above in nasal turbinate and stomach tumors in both sexes and thyroid follicular tumors in males.

HAP WEIGHTING FACTOR: Not applicable

VOLATILITY: Not available

FIRE HAZARD: Not available

LEL: Not available   UEL: Not available

REACTIVITY: Alachlor hydrolyzes in strongly acidic or alkaline solutions to give methanol, chloroacetic acid, formaldehyde, and 2,6-diethylaniline and it emits toxic fumes of nitrogen oxides and chlorine when heated to decomposition [817].

STABILITY: Alachlor degrades in soil, plants, and water to a large variety of chemical products. In addition, it is photodegraded by sunlight undergoing dechlorination, N-dealkylation, N-deacylation, and cyclization of N-dealkylated products [817].

Source: *Instant EPA's Pesticide Facts*, 1994, Instant Reference Sources, Inc., 7605 Rockpoint Drive, Austin, TX 78731 (Fax: 512-345-2386; Internet URL, http://www.instantref.com/inst-ref.htm); and *Agrochemicals Desk Reference Environmental Data*, 1993, CRC Press, Inc./Lewis Publishers, 2000 Corporate Blvd., Boca Raton, FL 33431 (Fax: 407-998-9114).

Medical Symptoms of Exposure

Beagle dogs administered alachlor had hemosiderosis seen in the liver and this was correlated with hematologic findings showing red cell destruction and consequent red cell replenishment. In addition, liver weights (absolute and relative-to-body-weight) were significantly ($p < 0.05$) higher in high-dose males. High-dose females also showed dose-related increases, but they were not statistically significant. Rats fed alachlor exhibited kidney discoloration and significant increases in kidney weights and kidney-to-body weight ratios. These adverse effects were further confirmed microscopically by the presence of chronic nephritis. Lower ovary weights were also noted in the high-dose females of each parental generation and in pups. Other symptoms included soft stools, red matter around the nose and mouth, hair loss, and anogenital staining. Degenerative ocular and hepatic changes as well as other pathological gross and microscopic findings in the thyroid, kidneys, brain, spleen, heart, prostate, and ovaries were noted. Discoloration of the liver, fatty degenerations and billiary hyperplasia were also noted

Source: *Instant EPA's IRIS*, 1997, Instant Reference Sources, Inc., 7605 Rockpoint Drive, Austin, TX 78731 (Fax: 512-345-2386; Internet URL, http://www.instantref.com/inst-ref.htm).

Toxicological Information

The toxicological information below is gathered from several sources including two national databases: one from the *National Toxicology Program Chemical Repository Database* and another from *EPA's Integrated Risk Information System (IRIS)*.

Toxicology Information from the National Toxicology Program

Alachlor is not currently listed in NTP's Chemical Repository database nor is it scheduled for addition in the future.

Toxicology Information from EPA's Integrated Risk Information System (IRIS)

I.   CHRONIC HEALTH HAZARD ASSESSMENTS FOR NONCARCINOGENIC EFFECTS

I.A. REFERENCE DOSE FOR CHRONIC ORAL EXPOSURE (RfD)

The Reference Dose (RfD) is based on the assumption that thresholds exist for certain toxic effects such as cellular necrosis, but may not exist for other toxic effects such as carcinogenicity. In general, the RfD is an estimate (with uncertainty spanning perhaps an order of magnitude) of a daily exposure to the human population (including sensitive subgroups) that is likely to be without an appreciable risk of deleterious effects during a lifetime. Please refer to Background Document 1 in Service Code 5 for an elaboration of these concepts. RfDs can also be derived for the noncarcinogenic health effects of compounds which are also carcinogens. Therefore, it is essential to refer to other sources of information concerning the carcinogenicity of this substance. If the U.S. EPA has evaluated this substance for potential human carcinogenicity, a summary of that evaluation will be contained in Section II of this file when a review of that evaluation is completed.

I.A.1. ORAL RfD SUMMARY

Critical Effect	Experimental Doses*	UF	MF	RfD
Hemosiderosis, hemolytic anemia	NOAEL: 1 mg/kg/day	100	1	1E-2 mg/kg/day
	LOAEL: 3 mg/kg/day			

*Conversion Factors and Assumptions: Actual dose tested

1-Year Dog Feeding Study Monsanto Co., 1984a

I.A.2. PRINCIPAL AND SUPPORTING STUDIES (ORAL RfD)

Monsanto Company. 1984a. MRID No. 00148923; HED Doc No. 004660. Available from EPA. Write to FOI, EPA, Washington, DC 20460.

Beagle dogs (6/sex/dose) were administered alachlor at levels of 0, 1, 3, and 10 mg/kg-day for 1 year. The test material was given neat (with no dilution) in gelatin capsules and administered daily at least 1 hour after feeding a ration of pellated dog chow. Control dogs were given empty gelatin capsules. Animals were observed twice daily during the study for signs of toxicity and mortality. Body weight and food consumption were measured, and clinical chemistry, hematology, and urinalysis testing were performed. All surviving animals were sacrificed after 12 months and examined for gross pathological changes.

The LEL for systemic toxicity is 3 mg/kg-day based on hemosiderosis seen in the kidney (1/6) and spleen (1/6). No hemosiderosis was associated with the liver in the mid-dose group. In the high-dose group, the hemosiderosis was seen in the liver (3/6) and correlated with hematologic findings showing red cell destruction and consequent red cell replenishment. The hematologic findings noted in the high-dose males were also seen as a trend (except for the increased reticulocyte count) in the mid-dose dogs. In addition, liver weights (absolute and relative-to-body-weight) were significantly (p<0.05) higher in high-dose males. High-dose females also showed dose-related increases, but they were not statistically significant. Therefore, the NOEL for systemic toxicity is 1 mg/kg-day.

I.A.3. UNCERTAINTY AND MODIFYING FACTORS (ORAL RfD)

UF -- The uncertainty factor of 100 reflects 10 for interspecies extrapolation and 10 for intraspecies-variability.
MF -- None

I.A.4. ADDITIONAL STUDIES / COMMENTS (ORAL RfD)

1) 1-Year Feeding - dog: Principal study -- see previous description; core grade guideline (Monsanto Co., 1984a)

2) 2-Year Feeding/Oncogenicity - rat: Dietary levels tested: 0, 0.5, 2.5, and 15 mg/kg-day; Groups of Long-Evans rat (50/sex/dose) were administered alachlor in the diet for 2 years. The LEL for systemic toxicity is 15 mg/kgday based on molting of retinal pigmentation and increased mortality rate in females and abnormal disseminated foci in the liver of males. The NOEL for systemic toxicity is therefore 2.5 mg/kg-day. Core grade minimum (Monsanto Co., 1984b)

3) 3-Generation Reproduction - rat: Dietary levels tested: 0, 3, 10, and 30 mg/kg-day; Groups of Charles River Sprague-Dawley CD rats were fed diets containing alachlor over three generations. No significant adverse effects on the reproduction of adult rats were observed at any dose tested. Therefore the NOEL for reproductive toxicity is equal to or greater than 30 mg/kg-day. Compound-related effects were not observed in litter size, survival and lactation indices, or the pup body weights. However, gross pathology examinations indicated compound-related effects on kidneys (in parents and progeny) in the high-dose males and females especially in the F2 parent generation and F3b pups. These effects were kidney discoloration and significant increases (5-18%, p<0.05) in kidney weights and kidney-to-body weight ratios. These adverse effects were further confirmed microscopically by the presence of chronic nephritis in the F2 high-dose males (8/10 compared with 1/10 animals in the control groups) and a healing infarct in 1/10 F3b male pups in addition to hydronephrosis in another F3b male pup (total of 2/10 animals as compared with none in the control group). Lower ovary weights were also

noted in the high-dose females of each parental generation and in F3b pups. This decrease was maximal in the F0 generation in which a 17% decrease (p<0.05) in ovary weights was noted. This decrease was also associated with 17% decrease (p<0.05) in the ovaries-to-body-weight ratio in the F0 high-dose females. This effect on kidneys is more remarkable in male. Based on kidney effects in both F2 adults and F3b pups, the LEL for systemic toxicity is 30 mg/kg-day. The NOEL for systemic toxicity is 10 mg/kg-day. Core grade minimum (Monsanto Co., 1981a)

4) Developmental Toxicity - rat: Dose levels tested: 0, 50, 150, and 400 mg/kg-day; Groups of pregnant Charles River COBS CD rats (25/dose) were administered alachlor in a corn oil vehicle by gavage on days 6 though 19 of gestation. High-dose dams exhibited soft stools, red matter around the nose and mouth, hair loss, and anogenital staining. Four high-dose dams died during the last 5 days of gestation. The cause of death was not apparent at necropsy. Mean body weight gains were moderately reduced in the high-dose group throughout the treatment period. Uterine examinations indicated that for each group of 25 females, 0, 5, 2, and 2 in each of the control, low-, mid-, and high-dose groups, respectively, were non-gravid. The four high-dose dams which died were gravid. All were viable fetuses in the treated groups. There were no statistically significant differences in mean numbers of viable fetuses, resorptions, post-implantation losses, total implantations, corpora lutea, sex distribution of pups, or mean fetal body weights in any of the treated groups when compared to the control group. In the high-dose group, a slight increase in the mean numbers of early and late resorptions resulted in a slight increase in mean post-implantation loss and a slight decrease in the mean number of viable fetuses. Based on soft stools, hair loss, anogenital staining, and maternal death, the NOEL and LEL for maternal toxicity are 150 and 400 mg/kg-day, respectively. Based on a slight decrease in mean fetal body weight and a slight increase in mean post-implantation loss, the NOEL and LEL for developmental toxicity are 150 and 400 mg/kg-day, respectively. Core grade guideline (Monsanto Co., 1980)

5) Developmental Toxicity - rabbit: Dose levels tested: 0, 50, 100, and 150 mg/kg-day; Groups of pregnant New Zealand White rabbits (18/dose) were administered alachlor by gastric intubation from gestation days 7 through 19, inclusive. Due to the death of two females in the control group during the treatment period, two additional females were mated and added to this group. Maternal toxicity was noted at the high dose in the form of reduced body weight gain during the dosing period with a rebound increase in body weight gain in the period following dosing. No other maternal toxicity was noted. No biologically relevant fetal external, visceral, or skeletal anomalies were noted. The NOEL and LEL for maternal toxicity are 100 and 150 mg/kg-day, respectively. The NOEL for developmental toxicity is equal to or greater than 150 mg/kg-day. Core grade minimum (Monsanto Co., 1988)

Other Data Reviewed:

1) 2-Year Feeding/Oncogenicity - rat: Dietary levels tested: 0, 100, 300, and 1000 ppm (0, 14, 42, and 126 mg/kg-day); Groups of Long-Evans rats (50/sex/dose with a additional 10/sex/dose for clinical studies) were fed diets containing alachlor for 2 years. Degenerative ocular and hepatic changes as well as other pathological gross and microscopic findings in the thyroid, kidneys, brain, spleen, heart, prostate, and ovaries were noted at all dose levels tested. The LEL for systemic toxicity is therefore 100 ppm

(14 mg/kg-day), the lowest dose tested. A NOEL for systemic toxicity could not be established. Core grade minimum (Monsanto Co., 1981b)

2) 6-Month Feeding - dog: Dietary levels tested: 0, 5, 25, 50, and 75/100 mg/kg-day; Groups of beagle dogs (6/sex/dose) were administered alachlor daily in capsules for 6 months. The high-dose level was 100 mg/kg-day for weeks 1 through 3 but due to the severe toxicity the dose was reduced to 75 mg/kg-day for the remainder of the study. Liver weight values (absolute, relative-to body-weight, and relative-to-brain) increased in males at 5 mg/kg-day and in both sexes at 25 mg/kg-day and above. The increases were often statistically significant ($p < 0.05$). Discoloration of the liver, fatty degenerations and billiary hyperplasia were also noted in both sexes at 25 mg/kg-day and above. Significant increases in SAP and LDH activity were also indicative of liver pathology. Other organ weight changes were noted at 5 mg/kg-day. Dose related emaciation and mortality were noted in both sexes at 25 mg/kg-day and above. Mortality in the mid- and high-dose groups was high; all but 1 high dose female died during the first 2-3 months of the study, and 4/6 males and 3/6 females of the mid-dose group were sacrificed in extremis during the study. A slight reduction in body weight gain was also noted at 5 mg/kg-day as compared with the control group. Based on the effects noted at the all dose levels, the LEL for systemic toxicity is equal to 5 mg/kg-day. A NOEL for systemic toxicity was not established. Core grade minimum (Monsanto Co., 1981c)

Data Gap(s): None

I.A.5. CONFIDENCE IN THE ORAL RfD

Study -- High
Data Base -- High
RfD -- High

The principal study is of good quality and is given a high confidence rating. In addition, there are generally good toxicologic studies available on alachlor which, overall provide high confidence in the data base. High confidence in the RfD follows.

I.A.6. EPA DOCUMENTATION AND REVIEW OF THE ORAL RfD

Source Document -- This assessment is not presented in any existing U.S. EPA document.
Other EPA Documentation -- None
Agency Work Group Review -- 03/11/86, 03/27/91
Verification Date -- 03/27/91

I.A.7. EPA CONTACTS (ORAL RfD)

George Ghali / OPP -- (703)305-7490

William Burnam / OPP -- (703)305-7491

I.B. REFERENCE CONCENTRATION FOR CHRONIC INHALATION EXPOSURE (RfC)

II. CARCINOGENICITY ASSESSMENT FOR LIFETIME EXPOSURE

This substance/agent has been evaluated by the U.S. EPA for evidence of human carcinogenic potential. This does not imply that this agent is necessarily a carcinogen. The evaluation for this chemical is under review by an inter-office Agency work group. A risk assessment summary will be included on IRIS when the review has been completed.

III. HEALTH HAZARD ASSESSMENTS FOR VARIED EXPOSURE DURATIONS

III.A. DRINKING WATER HEALTH ADVISORIES

The Office of Drinking Water provides Drinking Water Health Advisories (HAs) as technical guidance for the protection of public health. HAs are not enforceable Federal standards. HAs are concentrations of a substance in drinking water estimated to have negligible deleterious effects in humans, when ingested, for a specified period of time. Exposure to the substance from other media is considered only in the derivation of the lifetime HA. Given the absence of chemical-specific data, the assumed fraction of total intake from drinking water is 20%. The lifetime HA is calculated from the Drinking Water Equivalent Level (DWEL) which, in turn, is based on the Oral Chronic Reference Dose. Lifetime HAs are not derived for compounds which are potentially carcinogenic for humans because of the difference in assumptions concerning toxic threshold for carcinogenic and noncarcinogenic effects.   A more detailed description of the assumptions and methods used in the derivation of HAs is provided in Background Document 3 in Service Code 5.

III.A.1. ONE-DAY HEALTH ADVISORY FOR A CHILD

No appropriate data are available to derive a One-day HA; therefore, it is recommended that the Ten-day HA of 0.1 mg/L be used as the One-day HA.

III.A.2. TEN-DAY HEALTH ADVISORY FOR A CHILD

Ten-day HA -- 1E-1 mg/L
NOAEL -- 1 mg/kg/day

UF -- 100 (allows for interspecies and intrahuman variability with the use of a NOAEL from an animal study)

Assumptions -- 1 L/day water consumption for a 10-kg child

Principal Study -- Monsanto Company, 1984

Data from a 6-month dog feeding study (Monsanto, 1981) using 5, 25, 50 and 75 mg/kg/day dose levels was considered for the Ten-day HA calculations. This study reflect 92, 58, and 17% mortality at 75, 50 and 25 mg/kg/day, respectively. The lowest dose tested (LDT) in this study, 5 mg/kg/day, reflected mild hepatotoxic responses (i.e., increase in liver weight) that were intensified at the higher doses. However, a more recent 1-year dog feeding study (Monsanto, 1984) at 1, 3, and 10 mg/kg/day dose levels reflected a NOAEL at 1 mg/kg/day for this endpoint. Therefore, the Monsanto (1984) study is used in the calculation of the Ten-day HA.

III.A.3. LONGER-TERM HEALTH ADVISORY FOR A CHILD

A Longer-term Health Advisory has not been determined for alachlor because it has been shown to produce cancer in less than 5.6 months in rats (at the same rate as did the lifetime exposure).

III.A.4. LONGER-TERM HEALTH ADVISORY FOR AN ADULT

A Longer-term Health Advisory has not been determined for alachlor because it has been shown to produce cancer in less than 5.6 months in rats (at the same rate as did the lifetime exposure).

III.A.5. DRINKING WATER EQUIVALENT LEVEL / LIFETIME HEALTH ADVISORY

DWEL -- 3.5E-1 mg/L
Assumptions -- 2 L/day water consumption for a 70-kg adult

RfD Verification Date = 03/11/86 (see Section I.A. of this file)

Lifetime Health Advisory:

The assessment for the potential human carcinogenicity of alachlor is currently under review. Until this review is completed a Lifetime HA is not recommended.

Principal Study (DWEL) -- Monsanto Company, 1984 (This study was used in the derivation of the chronic oral RfD; see Section I.A.2.)

III.A.6. ORGANOLEPTIC PROPERTIES

[Ed. Note: EPA does not have information yet in this section of IRIS].

III.A.7. ANALYTICAL METHODS FOR DETECTION IN DRINKING WATER

Determination of alachlor is by a liquid-liquid extraction gas chromatographic procedure.

III.A.8. WATER TREATMENT

Data are available on the removal of alachlor from potable water using conventional treatment and adsorption. The use of air stripping has also been considered.

III.A.9. DOCUMENTATION AND REVIEW OF HAs

Monsanto Company, 1981, 1984; U.S. EPA, 1984
EPA review of HAs in 1985.
Public review of HAs following notification of availability in October, 1985.
Scientific Advisory Panel review of HAs in January, 1986.
Preparation date of this IRIS summary --   06/17/87

III.A.10. EPA CONTACTS

Amal Mahfouz / OST -- (202)260-9568
Edward V. Ohanian / OST -- (202)260-7571

III.B. OTHER ASSESSMENTS

Content to be determined.

IV. U.S. EPA REGULATORY ACTIONS

EPA risk assessments may be updated as new data are published and as assessment
methodologies evolve. Regulatory actions are frequently not updated at the same time.
Compare the dates for the regulatory actions in this section with the verification dates for
the risk assessments in sections I and II, as this may explain inconsistencies. Also note
that some regulatory actions consider factors not related to health risk, such as technical
or economic feasibility. Such considerations are indicated for each action. In addition,
not all of the regulatory actions listed in this section involve enforceable federal
standards. Please direct any questions you may have concerning these regulatory actions
to the U.S. EPA contact listed for that particular action. Users are strongly urged to read
the background information on each regulatory action in Background Document 4 in
Service Code 5.

IV.A. CLEAN AIR ACT (CAA)

No data available

IV.B. SAFE DRINKING WATER ACT (SDWA)

IV.B.1. MAXIMUM CONTAMINANT LEVEL GOAL (MCLG) for Drinking Water

Value (status) -- 0 mg/L (Final, 1991)

Considers technological or economic feasibility? -- NO

Discussion -- An MCLG of 0 mg/L for alachlor is promulgated based upon carcinogenic
potential (B2).

Reference --   56 FR 3526 (01/30/91)

EPA Contact --   Health and Ecological Criteria Division / OST / (202) 260-7571 / FTS 260-7571; or Safe Drinking Water Hotline / (800) 426-4791

IV.B.2. MAXIMUM CONTAMINANT LEVEL (MCL) for Drinking Water

Value -- 0.002 mg/L (Final, 1991)

Considers technological or economic feasibility? -- YES

Discussion -- EPA has set an MCL equal to the PQL, which is associated with a lifetime individual risk in the range of 10E-6.

Monitoring requirements -- All systems initially monitored for four consecutive quarters every three years; repeat monitoring dependent upon detection, vulnerability status and system size.

Analytical methodology -- Microextraction/gas chromatography (EPA 505); nitrogen-phosphorus detector/gas chromatography (EPA 507); gas chromatographic/mass spectrometry (EPA 525): PQL= 0.002 mg/L.

Best available technology -- Granular activated carbon

Reference --   56 FR 3526 (01/30/91)

EPA Contact --   Drinking Water Standards Division / OGWDW / (202) 260-7575 / FTS 260-7575; or Safe Drinking Water Hotline / (800) 426-4791

IV.B.3. SECONDARY MAXIMUM CONTAMINANT LEVEL (SMCL) for Drinking Water

[Ed. Note: EPA does not have information yet in this section of IRIS].

IV.B.4. REQUIRED MONITORING OF "UNREGULATED" CONTAMINANTS

[Ed. Note: EPA does not have information yet in this section of IRIS].

IV.C. CLEAN WATER ACT (CWA)

[Ed. Note: EPA does not have information yet in this section of IRIS].

IV.D. FEDERAL INSECTICIDE, FUNGICIDE, AND RODENTICIDE ACT (FIFRA)

IV.D.1. PESTICIDE ACTIVE INGREDIENT, Registration Standard

Status -- Issued (1984)

Reference --   Alachlor Pesticide Registration Standard. November, 1984 (NTIS No. PB86-179835).

EPA Contact --   Registration Branch / OPP (703)557-7760 / FTS 557-7760

IV.D.2. PESTICIDE ACTIVE INGREDIENT, Special Review

Action -- Final regulatory decision - PD4 on dietary and applicator risk (1987) (PD4 for ground water deferred until 1991).

Considers technological or economic feasibility? -- NO

Summary of regulatory action --   Required restricted use label warning and closed systems for large-scale mixer/loaders; criterion of concern: oncogenicity.

Reference --   51 FR 36166 (10/08/86)

EPA Contact --   Special Review Branch / OPP (703)557-7400 / FTS 557-7400

IV.E. TOXIC SUBSTANCES CONTROL ACT (TSCA)

> [Ed. Note: EPA does not have information yet in this section of IRIS].

IV.F. RESOURCE CONSERVATION AND RECOVERY ACT (RCRA)

> [Ed. Note: EPA does not have information yet in this section of IRIS].

IV.G. SUPERFUND (CERCLA)

> [Ed. Note: EPA does not have information yet in this section of IRIS].

VI. BIBLIOGRAPHY

VI.A. ORAL RfD REFERENCES

Monsanto Company. 1980. MRID No. 00043645; HED Doc. No. 000399, 001021. Available from EPA. Write to FOI, EPA, Washington, DC 20460.

Monsanto Company. 1981a. MRID No. 0075062; HED Doc. No. 001338. Available from EPA. Write to FOI, EPA, Washington, DC 20460.

Monsanto Company. 1981b. MRID No. 00091050, 00109319; HED Doc. 001991. Available from EPA. Write to FOI, EPA, Washington, DC 20460.

Monsanto Company. 1981c. MRID No. 00100659; HED Doc. No. 002483. Available from EPA. Write to FOI, EPA, Washington, DC 20460.

Monsanto Company. 1984a. MRID No. 00148923; HED Doc No. 004660. Available from EPA. Write to FOI, EPA, Washington, DC 20460.

Monsanto Company. 1984b. MRID No. 00091050, 00109319, 40284001; HED Doc. No. 003753, 004091, 004855. Available from EPA. Write to FOI, EPA, Washington, DC 20460.

Monsanto Company. 1988. MRID No. 40579402; HED Doc. No. 006886. Available from EPA. Write to FOI, EPA, Washington, DC 20460.

VI.B. INHALATION RfD REFERENCES

None

VI.C. CARCINOGENICITY ASSESSMENT REFERENCES

None

VI.D. DRINKING WATER HA REFERENCES

Monsanto Company. 1981. MRID No. 00100659. Available from EPA. Write to FOI, EPA, Washington, DC 20460.

Monsanto Company. 1984. MRID No. 00148923. Available from EPA. Write to FOI, EPA, Washington, DC 20460.

U.S. EPA. 1984. Special Review: Position Document 1. Office of Pesticide Programs.

VII. REVISION HISTORY

Date	Section	Description
03/01/88	III.A.	Health Advisory added
12/01/88	I.A.4.	Core graded added to studies 1, 3 and 4
11/01/89	VI.	Bibliography on-line
01/01/92	IV.	Regulatory actions updated
09/01/93	I.A.	Oral RfD revised; same study and number
09/01/93	VI.A.	Oral RfD references revised
11/01/93	III.A.2.	Ten-day Health Advisory corrected
11/01/93	III.A.9	Documentation revised
11/01/93	IV.B.2.	Discussion corrected
11/01/93	VI.D.	Health Advisory references revised

SYNONYMS

ACETAMIDE, 2-CHLORO-N-(2,6-DIETHYLPHENYL)-N-(METHOXYMETHYL)-
ACETANILIDE, 2-CHLORO-2',6'-DIETHYL-N-(METHOXYMETHYL)-
Alachlor
ALANEX
ALOCHLOR
CHLORESSIGSAEURE-N-(METHOXYMETHYL)-2,6-DIAETHYLANILID
2-CHLORO-2',6'-DIETHYL-N-(METHOXYMETHYL)ACETANILIDE
2-CHLORO-N-(2,6-DIETHYL)PHENYL-N-METHOXYMETHYLACETAMIDE
CP 50144
LASSO
LAZO
METACHLOR
METHACHLOR
PILLARZO

Source: *Instant EPA's IRIS*, 1997, Instant Reference Sources, Inc., 7605 Rockpoint
Drive,   Austin,   TX   78731   (Fax: 512-345-2386;   Internet   URL,
http://www.instantref.com/inst-ref.htm).

Production, Use, and Pesticide Labeling Information

1. Description of the Chemical

> Chemical Name: 2-chloro-2',6'-diethyl-N-(methoxymethyl)-acetanilide

> Common Name: Alachlor

> Trade Names: Lasso, Pillarzo, Alanex

> EPA Shaughnessy Number: 090501

> Chemical Abstracts Service (CAS) Number: 15972-60-8

> Year of Initial Registration: 1969

> Pesticide Type: Herbicide

> Chemical Family: Acetanilide

2. Use Patterns and Formulations

> Application Sites: Pre-emergent use on field corn (including sweet corn,
> popcorn), soybeans, peanuts, dry beans, lima beans, green peas, cotton, grain
> sorghum, sunflowers, and ornamental plants.

> Types of Formulations: Emulsifiable concentrate, granular, microencapsulate

Types and Methods of Application: Ground or aerial methods

Application Rates: Rate and frequency vary according to site application; typically 1 to 4 pounds of active ingredient per acre.

3. Science Findings

Physical and Chemical Characteristics

Color: White

Melting Point: 40 to 41 °C

Specific Gravity: 1.133 25/15.6 °C

Solubility: Soluble in ether, acetone, benzene, alcohol (unspecified) and ethyl acetate; slightly soluble in hexane; solubility in water - 240 ppm

Octanol/Water Partition Coefficient: 434

Stability: Stable (first detectable heat evolution at 105 °C)

Appearance at room Temperature: White, crystalline solid at 23 °C

Toxicological Characteristics

- Acute Toxicity: Technical alachlor is not an acutely toxic product by any route of exposure. The acute oral LD50 in rats is 0.93 g/kg

(Category III), the acute dermal LD50 in rabbits is 13.3 g/kg (Category III), and Alachlor does not cause significant eye or skin irritation in rabbits (Category IV).

- Chronic Toxicity: Alachlor is oncogenic in both mice and rats. In mice, alachlor causes a statistically significant increase in lung bronchioalveolar tumors in females at 260 mg/kg/day (Highest Dose Tested). In rats, alachlor causes statistically significant increases at 42 mg/kg/day and above in nasal turbinate and stomach tumors in both sexes and thyroid follicular tumors in males. The following No Observed Effect Levels (NOELS) have been established for non-oncogenic effects: 1 mg/kg/day for liver and kidney effects; 2.5 mg/kg/day for uveal degeneration syndrome (UDS) of the eyes; 10 mg/kg/day for reproductive effects to kidneys of offspring. No birth defects were seen in highest dose tested for rats (400 mg/kg/day), but an additional teratogenicity study in rabbits is pending completion in 1988.

Tolerance Assessment

- Tolerances were established in 40 CFR 180.249 for residues of alachlor and its metabolites.

4. Summary of Regulatory Position and Rationale

- The Agency position is that current registrations for the use of alachlor on agricultural and nonfood crops provided the following conditions and label modifications are met to reduce applicator exposure:

- a. Reclassification as restricted use pesticide: The application of alachlor is restricted to use by certified applicators or persons under their direct supervision.

- b. Continue use of tumor warning on label: The following tumor warning statement imposed in the alachlor Registration Standard must remain on the label: "The use of this product may be hazardous to your health. This product contains alachlor which has been determined to cause tumors in laboratory animals."

- c. Require the use of mechanical transfer devices by all mixer/loaders and/or applicators who treat 300 acres or more annually with alachlor.

- d. Reinstate aerial application on the alachlor label with the following additional label restriction: "Human flaggers prohibited. Aerial application may be performed using mechanical flaggers only."

- The Agency has determined that the upper bound U.S. dietary risk from alachlor residues on food crops is in the range of 10^{-6} (approximately one increased tumor case per million persons exposed) based on the actual percent of crops treated with alachlor. The Agency believes this risk is reasonable given the benefits of continued use.

- The Agency has determined that the risks associated with alachlor exposure through ground water cannot be adequately assessed at this time. The true extent of alachlor occurrence in ground water is not known and cannot be properly estimated for most areas of the country. Further monitoring will be necessary considering the large volume, multi-regional use of alachlor, and the lack of statistically representative data from the available studies. The Agency's final evaluation of the potential for alachlor contamination of ground water is being deferred, pending receipt of required monitoring data being generated by the registrant under an EPA-approved protocol. This study is scheduled for completion in late 1989, and the results will be evaluated by EPA in early 1990.

- Existing data on alachlor residues in surface water indicate that the risk from drinking water sources supplied by surface water will generally not exceed a range of 10^{-6}. The Agency believes this level of risk is reasonable given the benefits of continued use of alachlor products, and is not proposing regulatory action under FIFRA on alachlor residues in surface water. The Agency plans to promulgate regulations establishing a Maximum Contaminant Level (MCL) for alachlor under the Safe Drinking Water Act in the near future. These regulations would require the treatment of drinking water which contains alachlor residues in excess of the MCL, thereby maintaining the level of risk from exposure at reasonable levels.

5. Summary of Major Data Gaps

> None

6. Contact person at EPA

> James V. Roelofs
> Review Manager
> Special Review Branch
> Registration Division (TS-767C)
> Office of Pesticide Programs
> U.S. Environmental Protection Agency
> Washington, DC 20460
>
> Phone: (703) 557-0064

Source: *Instant EPA's Pesticide Facts*, 1994, Instant Reference Sources, Inc., 7605 Rockpoint Drive, Austin, TX 78731 (Fax: 512-345-2386; Internet URL, http://www.instantref.com/inst-ref.htm).

Aldicarb

Introduction

Aldicarb is an extremely toxic systemic carbamate insecticide. It is one of the most acutely poisonous pesticides known. [173], [346] It is a soil-applied insecticide, acaricide and nematocide for use on cotton, sugar beets, potatoes, peanuts, ornamentals, yams, oranges, pecans, dry beans, soybeans and sugarcane. It is also a systemic nematocide and insecticide used against millipedes, eelworms and many insect pests of root crops. Highly publicized incidents involving contaminated cucumbers and watermelons occurred in the mid 1980s. In these cases, misapplication led to adverse effects in people. In 1990 Rhone-Poulenc Ag Company, the manufacturer of aldicarb, announced a voluntary halt on the sale of aldicarb for use on potatoes because of concerns about groundwater contamination. Aldicarb is a Restricted Use Pesticide (RUP) in the United States as determined by the US Environmental Protection Agency. Restricted Use Pesticides may be purchased and used only by certified applicators [829]. Release of aldicarb to the environment will occur due to its manufacture and use as a systemic insecticide, acaricide and nematocide for soil use. If aldicarb is released to the soil it should not bind to the soil. It will be susceptible to chemical and possibly biological oxidation to form the sulfoxide and sulfone. Hydrolysis is both acid and base catalyzed If aldicarb is released to water it should not adsorb to sediments or bioconcentrate in aquatic organisms. [828]

Identifying Information

CAS NUMBER: 116-06-3

NIOSH Registry Number: UE2275000

SYNONYMS: 2-METHYL-2-(METHYLTHIO)PROPIONALDEHYDE O-(METHYLCARBAMOYL)OXIME, 2-METHYL-2-(METHYLTHIO)PROPANAL, O-((METHYLAMINO)CARBONYL))-OXIME, CARBAMIC ACID, METHYL-, O-((2-METHYL-2-(METHYLTHIO)PROPYLIDENE)AMINO) DERIV., ALDECARB, TEMIC, AMBUSH, TEMIK G10, TEMIK 10 G, TEMIK, NCI-C08640, ENT 27,093, UC-21149, OMS-771, UNION CARBIDE 21149, UNION CARBIDE UC-21149, RCRA WASTE NUMBER P070, PROPIONALDEHYDE, 2-METHYL-2-(METHYLTHIO)-, O-(METHYLCARBAMOYL)OXIME, PROPANAL, 2-METHYL-2-(METHYLTHIO)-, O-((METHYLAMINO)CARBONYL)OXIME

42

Endocrine Disruptor Information

Aldicarb is a strongly suspected EED that it is listed by the Centers for Disease Control & Prevention [812], and the World Wildlife Fund [813] as a potential endocrine modifying chemical. It is member of the carbamate class of pesticides which are toxic to immune systems. When administered to pregnant rats at very low levels (0.001 to 0.1 mg/kg) it depressed acetylcholinesterase activity more in the fetuses than in the mothers [829] and there may be a link between low-level exposure and immunological abnormalities [829], [869]. Aldicarb is still widely used in the United States with an annual U.S. consumption of 2, 110 tons. [825]

Chemical and Physical Properties

CHEMICAL FORMULA: $C_7H_{14}N_2O_2S$

MOLECULAR WEIGHT: 190.27

WLN: 1SX1&1&1UNOVM1

PHYSICAL DESCRIPTION: White crystalline solid

SPECIFIC GRAVITY: 1.1950 @ 25/20°C [051], [055], [071]

DENSITY: Not available

MELTING POINT: 98-100°C [029], [172], [346]

BOILING POINT: Decomposes

SOLUBILITY: Water: 0.1-1.0 mg/mL @ 22°C [700]; DMSO: >=100 mg/mL @ 21°C [700]; 95% Ethanol: >=100 mg/mL @ 21°C [700]; Acetone: >=100 mg/mL @ 21° [700]; Toluene: 10% [173]

OTHER SOLVENTS: Chloroform: 35% [173]; Isopropane: 20% [173]; Chlorobenzene: 15% [173]; Xylene: 5% @ 25° [033]; Methylene chloride: 30% @ 25° [033]; Heptane: Insoluble [169], [172]; Ether: 20% [173]; Mineral oils: Insoluble [169]; Benzene: 15% @ 25°C [033]; Absolute ethanol: Soluble [052]; Most organic solvents: Soluble [169], [172]

OTHER PHYSICAL DATA: Odorless [051], [071], [173]; Crystals from isopropyl ether [033]

Source: *The National Toxicology Program's Chemical Database, Volume 2*, 1992, CRC Press, Inc./Lewis Publishers, 2000 Corporate Blvd., Boca Raton, FL 33431 (Fax: 407-998-9114).

Environmental Reference Materials

A source of EED environmental reference materials for Aldicarb is Radian International LLC, Austin, TX (Fax 512-454-0268; Internet URL, http://www.radian.com/standards); Catalog No. ERA-054.

Hazardous Properties

ACUTE/CHRONIC HAZARDS: Aldicarb may be fatal by ingestion, inhalation or skin absorption It may be absorbed through the skin [107]. It is a cholinesterase inhibitor [051], [071], [107], [186]. When heated to decomposition it emits toxic fumes of nitrogen oxides and sulfur oxides [043] and it may also emit fumes of carbon dioxide and methylamine. It is extremely toxic through both oral and dermal routes and exhibits cholinesterase inhibition [829].

VOLATILITY: Vapor pressure: 0.001 mm Hg @ 25°C [055]

FIRE HAZARD: Literature sources indicate that Aldicarb is nonflammable [051], [071], [172]. Fires involving this material can be controlled with a dry chemical, carbon dioxide or Halon extinguisher.

REACTIVITY: Aldicarb is incompatible with highly alkaline substances [051], [071], [107], [173]. It is rapidly converted by oxidizing agents [169].

STABILITY: Aldicarb decomposes at temperatures greater than 100°C. It is stable in neutral, acidic and weakly alkaline media [169]. Solutions of aldicarb in water, DMSO, 95% ethanol or acetone should be stable for 24 hours under normal laboratory conditions [700].

Source: *The National Toxicology Program's Chemical Database, Volume 7*, 1992, CRC Press, Inc./Lewis Publishers, 2000 Corporate Blvd., Boca Raton, FL 33431 (Fax: 407-998-9114).

Medical Symptoms of Exposure

Symptoms of exposure may include pupillary constriction, nausea, vomiting, diarrhea, muscle twitching and convulsions [051], [071], [173]. It may also cause anorexia, salivation, bronchoconstriction and respiratory failure [051], [071]. Other symptoms may include involuntary urination, weakness, difficulty breathing, abdominal pain, excessive perspiration, blurred vision, headache, temporary paralysis of the extremities, dyspnea, excessive secretions, spontaneous evacuation of the bladder and bowel and cyanosis [173]. It may also cause abdominal cramps, visual disturbances, angina pectoris, nervousness, lacrimation, drooling, frothing of the mouth and nose, and central nervous system depression [107]. Unconsciousness, irregular heartbeat, marked pulmonary edema, sinus brachy cardia with recurrent heart failure, acute renal failure and death have also been reported [295]. Exposure may cause tremors. Eye contact may cause redness. Skin irritation may be aggravated in persons with existing skin lesions. Breathing of dust may aggravate asthma and inflammatory or fibrotic pulmonary disease. Overexposure may aggravate existing chronic cardiovascular or respiratory disease leading to respiratory difficulty and cyanosis.

Source: *The National Toxicology Program's Chemical Database, Volume 4*, 1992, CRC Press, Inc./Lewis Publishers, 2000 Corporate Blvd., Boca Raton, FL 33431 (Fax: 407-998-9114).

Toxicological Information

The toxicological information below is gathered from several sources including two national databases: one from the *National Toxicology Program Chemical Repository Database* and another from *EPA's Integrated Risk Information System (IRIS)*.

Toxicology Information from the National Toxicology Program

CARCINOGENICITY:

> Status:
>> NCI Carcinogenesis Bioassay (Feed); Negative: Male and Female Rat; Male and Female Mouse

TERATOGENICITY: Not available

MUTATION DATA: See RETCS.

TOXICITY:

Typ.dose	Mode	Specie	Amount	Units	Other
LC50	ihl	rat	200	mg/mg/5H	
LD50	orl	rat	650	ug/kg	
LD50	skn	rbt	1400	mg/kg	
LD50	skn	rat	2500	ug/kg	
LD50	scu	rat	666	ug/kg	
LD50	unr	rat	930	ug/kg	
LD50	orl	mus	300	ug/kg	
LD50	skn	gpg	2400	mg/kg	
LD50	orl	dck	3400	ug/kg	
LD50	orl	ckn	8	mg/kg	
LD50	orl	bwd	750	ug/kg	
LD50	orl	pgn	3160	ug/kg	
LD50	orl	qal	2	mg/kg	
LD50	scu	mus	250	ug/kg	

OTHER TOXICITY DATA:

Review:
        Toxicology Review

        Status:
                EPA TSCA Test Submission (TSCATS) Data Base, September 1989
                EPA TSCA Chemical Inventory, 1986
                EPA Genetox Program 1988, Negative: Carcinogenicity-mouse/rat

SAX TOXICITY EVALUATION:

        THR: Deadly poison by ingestion, skin contact, subcutaneous and possibly other
        routes. Human mutagenic data. A powerful systemic poison, nematocide and
        acaricide.

Source: *Instant Tox-Base*, 1995, Instant Reference Sources, Inc., 7605 Rockpoint Drive,
Austin, TX 78731 (Fax: 512-345-2386; Internet URL, http://www.instantref.com/inst-
ref.htm).

Toxicology Information from EPA's Integrated Risk Information System (IRIS)

I. CHRONIC HEALTH HAZARD ASSESSMENTS FOR NONCARCINOGENIC
EFFECTS

I.A. REFERENCE DOSE FOR CHRONIC ORAL EXPOSURE (RfD)

The Reference Dose (RfD) is based on the assumption that thresholds exist for certain
toxic effects such as cellular necrosis, but may not exist for other toxic effects such as
carcinogenicity. In general, the RfD is an estimate (with uncertainty spanning perhaps
an order of magnitude) of a daily exposure to the human population (including sensitive
subgroups) that is likely to be without an appreciable risk of deleterious effects during a
lifetime. Please refer to Background Document 1 in Service Code 5 for an elaboration
of these concepts. RfDs can also be derived for the noncarcinogenic health effects of
compounds which are also carcinogens. Therefore, it is essential to refer to other
sources of information concerning the carcinogenicity of this substance. If the U.S. EPA
has evaluated this substance for potential human carcinogenicity, a summary of that
evaluation will be contained in Section II of this file when a review of that evaluation is
completed.

I.A.1. ORAL RfD SUMMARY

Critical Effect	Experimental Doses*	UF	MF	RfD
Critical Effect -- Sweating as clinical sign of AChE inhibition	Critical Dose -- 0.01 mg/kg-day {1Na}	10	1	1E-3 mg/kg/day
Critical Study 1 Study Type -- Acute Human Oral Exposure Study	NOAEL -- 0.01 mg/kg			
	NOAEL(ADJ) -- 0.01 mg/kg-day			
Reference -- Rhone-Poulenc, 1992	LOAEL -- 0.025 mg/kg			
	LOAEL(ADJ) -- 0.025 mg/kg-day			
	Conversion Factors and Assumptions -- Oral doses administered in 200 mL orange juice ingested over 15 minutes with a light breakfast.			
Critical Study 2	NOAEL -- None			
Critical Effect -- Clinical signs and symptoms of acetylcholinesterase inhibition including sweating, pinpoint pupils, leg weakness, and other effects.	LOAEL -- (FEL) 0.1 mg/kg-day			
	Conversion Factors and Assumptions -- Oral doses administered neat in 100 mL water. Dose administered affected 4 out of 4 exposed men.			
Study Type -- Acute Human Oral Exposure Study				
Reference -- Union Carbide, 1971				
Critical Study 3	NOAEL -- None			
Critical Effect -- Nausea, diarrhea, and other signs and symptoms	LOAEL -- (FEL) 0.01 mg/kg-day			
	Conversion Factors and Assumptions -- Oral doses estimated based on self reports of amount of commodities consumed, measured residue levels in commodities, and average body weights for given age and sex. Level listed is median of 41 cases.			
Study Type -- Acute Human Oral Poisoning Episodes				
References -- Goldman et al., 1990a,b; Hirsch et al., 1987				

I.A

.2. PRINCIPAL AND SUPPORTING STUDIES (ORAL RfD)

Rhone-Poulenc Ag Company. 1992. A Safety and Tolerability Study of Aldicarb at Various Dose Levels in Healthy Male and Female Volunteers. Inveresk Clinical Research Report No. 7786. MRID No. 423730-01. HED Doc. No. 0010459. Available from EPA. Write to FOI, EPA, Washington, DC 20460.

Union Carbide Corporation. 1971. R. Haines, J.B. Dernehl, and J.B. Block, supervising physicians. Ingestion of Aldicarb by Human Volunteers: A Controlled Study of the Effects of Aldicarb on Man. ALD-03-77-2215. February 11, 1971. MRID No. 00101911. HED Doc. No. 010450. Available from EPA. Write to FOI, EPA, Washington, DC 20460.

Goldman, L.R., M. Beller and R.J. Jackson. 1990a. Aldicarb food poisonings in California, 1985-1988: Toxicity estimates for humans. Arch. Environ. Health. 45(3): 141-147. HED Doc. No. 010451, 010455, 010458. Available from EPA. Write to FOI, EPA, Washington, DC 20460.

Goldman L.R., D.F. Smith, R.R. Neutra, et al. 1990b. Pesticide Food Poisoning from Contaminated Watermelons in California, 1985. Archives of Environmental Health. 45(4): 229-236.

Hirsch, G.H., B.T. Mori, G.B. Morgan, P.R. Bennett, and B.C. Williams. 1987. Report of Illnesses Caused by Aldicarb-Contaminated Cucumbers. Food Additives and Contaminants. 5(2): 155-60. HED Doc. No. 010455, 010458. Available from EPA. Write to FOI, EPA, Washington, DC 20460.

The double blind, placebo controlled study included 38 men and 9 women, with 6 men and 5 women receiving both a dose and a placebo exposure (Rhone- Poulenc Ag Company, 1992). Men were exposed to doses of 0, 0.01, 0.025, 0.05, 0.06, or 0.075 mg/kg of aldicarb, while women received 0, 0.025, or 0.05 mg/kg. Subjects were given a light breakfast on the day of the study, including a drink of orange juice containing one of the doses of aldicarb or the placebo to be consumed over 15-30 minutes of the breakfast period. Subjects remained generally seated or recumbent for the first 4 hours after dosing. A number of biological parameters known to be affected by cholinesterase inhibitors were monitored before dosing, hourly for the first 6 hours after dosing, and at 24 hours after dosing. These measures included recording of signs and symptoms (e.g., sweating), measurements of pulse and blood pressure, evaluation of pulmonary functions (FEV-1 and FVC), saliva and urine output, pupil diameter measurements, and measurement of plasma and red blood cell cholinesterase activity. All study subjects were evaluated with respect to the above consequences after dosing with aldicarb or placebo. Emphasis was placed on the first 6 hours after exposure, because it is known that the effects and cholinesterase inhibition caused by aldicarb are acute and readily reversible. The major endpoints seen in the study and discussed as potentially treatment-related were effects on red blood cell and plasma cholinesterases, sweating, light-headedness, headaches, salivation, and supine diastolic blood pressure.

Aldicarb treatment of both males and females resulted in statistically significant inhibition of both red blood cell and plasma cholinesterase at all dose levels. Peak effects were noted at 1 hour after the dose, and the degree and duration of effect increased with increasing doses. One male in the 0.075 mg/kg group who had mistakenly received 0.06 mg/kg, developed diffuse and profuse sweating that began within 2 hours and abated within 6 hours of dosing. Two other treated males, one given 0.05 mg/kg and another given 0.025 mg/kg, developed localized and mild sweating with onset within the first 2 hours of dosing and abated within 6 hours of dosing. One male given 0.075 mg/kg reported that he was light-headed within 1 hour of dosing. Three men in the 0.01 mg/kg group reported headaches, two with onset within 6 hours of dosing, and one within 8 hours. This long time between dosing and onset is beyond the peak of cholinesterase inhibition and the other effects seen here and in both the Union Carbide study and the poisoning episodes. None of the females developed any clinical signs or symptoms consistent with cholinesterase inhibition or treatment.

Females given 0.05 mg/kg showed higher saliva output than controls, with marginal statistical significance. Observed changes in blood pressure were generally small in magnitude, limited to supine diastolic pressure, and statistically significant in some, but not other analyses. There were no treatment-related changes in standing or supine pulse, pupil size, or urine volume in either males or females. As expected, there were no changes in hematology and clinical chemistry parameters.

There were statistically significant increases in FVC in men at the 0.01 and 0.075 mg/kg doses, but these were not considered to be treatment-related based on a one way analysis of variance and on the observation that the statistically significant findings were likely a result of a drop in control values during the session.

A number of questions arose during the review of this study that made it more difficult to fully interpret the results. Some of these were related to incomplete reporting by the study authors. Others are simply limitations in the design and conduct of the study.

The paucity of clearly and statistically significant chemical-related effects other than inhibition of plasma and red blood cell cholinesterase activity, and the limitations noted, make a definitive judgment of toxic and non-toxic doses difficult. Effects noted tended not to demonstrate statistical significance, dose-response or dose effect relationships, or correspondence between males and females. Diffuse and profuse sweating in one man given 0.06 mg/kg was the most clear-cut sign of toxicity; the appearance of localized sweating in one man at 0.05 and another at 0.025 mg/kg suggests some dose-related response, especially in light of the Union Carbide (1971) study, where all males receiving 0.1 mg/kg showed sweating. Some other effects consistent with cholinesterase inhibition were noted in the range of 0.025-0.075 mg/kg.

In conclusion, the NOAEL for this study was considered to be 0.01 mg/kg- day and the LOAEL is 0.025 mg/kg-day, based on the sweating seen in males. This NOAEL serves as the operational basis for the RfD derivation. In another human study (Union Carbide Corporation, 1971), 12 adult male volunteers were weighed and assigned to different treatment groups based on nearly equal average weights. None of the subjects had known exposure to aldicarb or other cholinesterase inhibitors for a week prior to the study. Subjects were divided into three test groups (4/group) and administered aldicarb at 0.025, 0.05, or 0.1 mg/kg. A stock solution of 1 mg/mL of aldicarb was prepared by dissolving 0.2 g of analytical grade aldicarb in 200 mL of distilled water. Dosages were prepared by diluting the appropriate amount of aldicarb solution into 100 mL of distilled water, which was then ingested in one draft. Subjects were given their doses between 9:00 and 9:15 a.m. and engaged in normal business activities except during blood and urine sampling and clinical observations. Liquids were provided ad libitum during the post-exposure period. Observations were reported 1, 2, 3, 4, and 6 hours following the dose. These observations included measurement of pulse, blood pressure, observation of pupil size, and subjects' complaints.

All three groups experienced significant cholinesterase inhibition in whole blood, with the peak inhibition between 1-2 hours and almost complete recovery in 6 hours. The 0.1 mg/kg dose elicited clinical signs in all four subjects, predominantly sweating and leg weakness, while most subjects given the two lower doses had no signs or symptoms. At 0.025 mg/kg, one subject reported apprehension. The method of analysis of cholinesterase in blood was valid and appropriate for this carbamate. The range of cholinesterase inhibition at this dose(0.025 mg/kg) was 30-57%. Therefore, an FEL of 0.1 mg/kg-day can be established for this study based on clinical signs and symptoms of acetylcholinesterase inhibition including sweating, pinpoint pupils, leg weakness, and other effects.

Dosage estimates for 28 cases of alleged aldicarb poisoning were derived from average body weights by age and sex (from standard tables), self-reported symptoms and estimated consumption, and aldicarb sulfoxide residues from watermelons and cucumbers (Goldman et al., 1990a). Estimates for 13 additional cases were provided by Hirsch et al. (1987), also based on estimates of body weights and consumption, and measurements of residues of total aldicarb, believed to be primarily sulfoxide. This total population (N=41) had a median of 0.01 mg/kg (for total aldicarb), a first quartile of 0.06 mg/kg, and a third quartile of 0.029 mg/kg. The description of cases used for estimates was limited in terms of onset, duration, and severity, and many of the reported symptoms of cholinesterase inhibition, (i.e., nausea, vomiting, and diarrhea) are nonspecific. The analytical methodology was valid, although the limit of detection of 0.2 ppm (Goldman et al., 1990b) was somewhat higher than in other reports. As a result, some misclassification errors due to these factors (i.e., some false positives and false negatives), may have occurred among the over 1000 reported cases of illness. Further, the use of sex and age averages for body weights and self-reported food consumption values are also subject to estimation errors, but these are expected to include both under and overestimates. Nevertheless, these effects were consistent with the expected syndrome, the analytical techniques were considered valid and these dosage estimates are regarded as reasonable general estimates of effects.

In conclusion, an FEL of 0.01 mg/kg-day can be established from these three studies based on nausea, diarrhea, and other signs and symptoms.

I.A.3. UNCERTAINTY AND MODIFYING FACTORS (ORAL RfD)

UF -- An uncertainty factor of 10 is proposed based on the NOAEL in the Rhone-Poulenc human study to account for variation in sensitivity among persons in the population. Several considerations went into the choice of this uncertainty factor. Based on a relatively complete data base for systemic toxicity, effects from repeated exposures have not been seen in laboratory animals at levels comparable to those seen in humans exposed acutely, and so would not support a lower RfD. While human data on the adverse consequences of repeated aldicarb exposure is lacking, available evidence both in experimental animals and in humans exposed to aldicarb suggest that neurobehavioral effects are short lived with no accumulation of effects over time. Thus, the doses producing effects following repeated daily exposure are comparable to those following a single dose. Also, comparable degrees of cholinesterase inhibition are seen from the same dose levels, whether delivered in one acute dose or following subchronic or chronic dosing (Hazelton et al., 1988; Rhone-Poulenc, 1992). Since aldicarb does not appear to produce neurobehavioral effects at doses below those producing inhibition of cholinesterase, an acute human experimental study is expected to reasonably evaluate the potential neurobehavioral consequences of repeated human exposure. Compared with the controlled studies, the human poisoning episodes document effects in a group of individuals probably self-selected as a sensitive population and more heterogeneous than the healthy adults chosen for the controlled studies. The dosage estimates from these reports, however, were derived from estimates of the body weights of the people involved, based on tables of average weights for a given age and sex. They also were derived from self-reported estimates of the amount consumed. Thus, while these dosage estimates are based on reasonable estimates of body weight and amount of watermelon consumed for each population sample, they are not as precise as those derived in the controlled studies, where each subject was weighed and received a known dose. It is more reasonable to examine the distribution of estimated doses over which effects were reported, rather than to regard each individual estimate as precise. All the estimates for the 41 cases from the poisoning episodes are subsumed within this RfD. The proposed RfD provides for a margin of exposure of 10 from that recorded in the controlled studies or from the median poisoning estimate. It also subsumes the entire range over which effects have been reported in the poisoning episodes which empirically define, to some extent, the sensitivity of a more sensitive heterogeneous population.

MF -- None

I.A.4. ADDITIONAL STUDIES / COMMENTS (ORAL RfD)

There is a rich data base bearing on the toxicity of aldicarb. The compound is a potent cholinesterase inhibitor that rapidly induces adverse effects that are rapidly reversed. Chronic toxicity in laboratory animals is manifest at doses comparable to those that produce acute toxicity. Essentially all hazards from aldicarb exposure are associated with cholinesterase inhibition (ChEI). Both human and laboratory animal data can be used to determine levels of aldicarb exposure that are probably not associated with significant risk. An overall view of the data indicates that various species show toxic effects at similar doses. Because of the availability of human studies, it is logical to place primary reliance on these studies in determining the RfD.

These four studies of humans provide the best information on the toxic effects of aldicarb: the recent experiment conducted on behalf of Rhone- Poulenc (1992), the Haines experimental study conducted by Union Carbide (1971), and the evaluations of some pesticide misuse poisoning incidents developed by Goldman and Hirsch (Goldman et al., 1990a,b; Hirsch et al., 1987). In developing the RfD, primary emphasis has been placed upon the Rhone-Poulenc study. The Union Carbide study provides some correlative evidence of a frank effect level and sweating, while the poisoning incidents provides important general estimates of potential population sensitivity. The weight of the evidence from all of the available human data cited here were considered critical in the estimation of this reference dose. Each of these sources has limitations that raise concerns about sole reliance on any one as a basis for this estimate. However, it was considered equally important to consider use of all of the available evidence to estimate potential risks.

The Union Carbide study (1971) helps define a dose (0.1 mg/kg) that is clearly associated with adverse effects in humans. At least three of four subjects showed sweating, pupillary constriction, muscle weakness, and increased salivation, while fewer demonstrated slurred speech, malaise, nausea, gastrointestinal cramping, and vomiting. No confirmation of these signs was noted in groups receiving 0.05 or 0.025 mg/kg of aldicarb. Significant reductions in blood cholinesterase activity, which returned to normal within a few hours, was noted in all dosed individuals.

In the Rhone-Poulenc study (1992), where groups received doses of aldicarb between 0.01 and 0.075 mg/kg, there were less obvious indications of toxicity. Some male subjects manifested sweating, headache, or light-headedness that could have been due to aldicarb exposure. Only the sweating was a manifestation in common with the subjects in the Union Carbide study. The most obvious sign in the new study was the finding of diffuse and profuse sweating in one male who had received 0.06 mg/kg of aldicarb; the other two cases were males demonstrating localized sweating of the palms with or without sweating of the soles (0.05 and 0.025 mg/kg, respectively). One of the control males also developed sweating of the palms and forehead. Diffuse body sweating is a well characterized cholinergic sign, whereas localized sweating is mediated by sympathetic fibers that synapse at cholinergic ganglia and communicate with the sweat glands of the palms and soles by norepinephrine. Cholinesterase inhibition at the ganglia may sensitize these neurons to other potential stimuli, like emotional factors. At 0.1 mg/kg, all four of the treated males in the Union Carbide study demonstrated sweating, and two of them had sweating localized to the palms and forehead. As in the Union Carbide study (1971), dosed subjects in the Rhone-Poulenc study (1992) showed depressions in blood cholinesterase that generally began to recover within a few hours.

The relative drop in supine diastolic blood pressure during the first hour post-dosing in the Rhone-Poulenc (1992) study was significantly greater in the male 0.075 mg/kg group than in the controls when an unweighted analysis was performed but not when a weighted analysis was performed. Females in the 0.025 mg/kg group showed a decrease in the relative diastolic pressure at 1 hour post-dosing, which was significant in a weighted analysis but not in an unweighted analysis. Higher-dosed females (0.05 mg/kg) showed no significant differences. In contrast, the mean supine diastolic pressure of the male 0.05 mg/kg group in the Rhone-Poulenc (1992) study was statistically significantly higher than that in the placebo group. There were no differences in the supine systolic pressures or the standing blood pressures in treated males and females. Also of interest is the absence of blood pressure changes in the males in the Union Carbide (1971) study who received doses of 0.025, 0.05, or 0.1 mg/kg of aldicarb.

As would be expected from ChEI, salivation was statistically significantly increased in the 0.05 mg/kg female dose group (highest dose tested in females) as compared with controls in the Rhone-Poulenc study. In contrast, males in the 0.01 and 0.05 mg/kg groups and the male who received 0.06 mg/kg showed a significant decrease in salivation; males receiving 0.075 mg/kg showed no difference from the placebo group. Increased salivation was noted in 3 of the 4 treated males in the Union Carbide study that received 0.1 mg/kg aldicarb, but not in those receiving 0.025 or 0.05 mg/kg. Headache was reported by three males in the 0.01 mg/kg dose group in the Rhone-Poulenc study (1992), but this symptom was not declared among males in any of the other groups or in any of the females. Headache was not reported in the males in the Union Carbide study.

In sum, there is very good information that 0.1 mg/kg of aldicarb is toxic to humans and produces multiple effects including such things as sweating, muscular weakness, and pinpoint pupils. Diffuse or localized sweating was noted in 4 of 4 subjects at 0.1 mg/kg and in one subject each who had received the 0.06, 0.05, and 0.025 mg/kg doses; no sweating was reported at 0.01 mg/kg. Other indications of a potential cholinergic response in these studies are less reliable: they occur at some doses but not at doses higher or doses lower; there is a failure across sexes to confirm the presence of effects or there is an opposite effect; there is no statistical significance and there is no dose response. These limitations apply to the evaluation of all the effects in these studies.

A reasoned course is to place major emphasis on sweating, the only sign of cholinergic response that was noted in both experimental studies. Therefore based on sweating, a LOAEL of 0.025 mg/kg and a NOAEL of 0.010 mg/kg were identified. These considerations include the bulk of other potential effects such as headaches that were noted in the Rhone-Poulenc study. Recognizing that these judgments are based upon a limited number of observations in humans, an uncertainty factor of 10 is included to account for potential human variability. The estimates of potential exposure from five pesticide misuse poisoning incidents (Goldman et al., 1990a,b; Hirsch et al., 1987) were all higher than the proposed RfD. Based upon all of the data on humans from these sources, exposure to 0.001 mg/kg of aldicarb, the present RfD, is expected to be without significant adverse effect.

Two other general comments related to the population sensitivity and the nature and extent of effects should be noted. First, the broad range of exposure levels over which effects in humans have been seen in these studies (0.002-0.1 mg/kg) suggests that some portion of population sensitivity is accounted for in the data. Second, the effects seen are acute, relatively small in magnitude in many cases, and of relatively low incidence at all but the highest dose levels.

Other Data Reviewed:

1) 1-Year Feeding - dog: Core grade supplementary (Rhone-Poulenc, 1988a; Rhone-Poulenc, 1991a).

Groups of beagle dogs (5/sex/dose) were administered Aldicarb, technical grade, in the diet for 52 weeks at doses of 0, 0.028, 0.054, 0.132, and 0.231 mg/kg-day for males and 0, 0.027, 0.055, 0.131, and 0.251 mg/kg-day for females. It was initially concluded that the lowest dose tested, 1 ppm (0.028 mg/kg-day) was a LOAEL for plasma cholinesterase inhibition in males and that the next higher dose level (0.055 mg/kg) and above produced signs consistent with cholinesterase inhibition, including diarrhea and mucoid and/or soft stool. According to the study report, group brain cholinesterase was significantly inhibited only at the highest dose in males compared with controls. A NOAEL for ChE inhibition in the study was not established. Review of an addendum submitted by the registrant(Rhone-Poulenc, 1991a) which provided additional pre-exposure data on clinical signs seen in the 1-year dog study, and after extensive statistical analyses by EPA, it was concluded that the available data and evaluations on clinical signs in dogs of diarrhea and soft and mucoid stool were insufficient to support the conclusion that this was an effect of treatment.

2) Subchronic Feeding - dog: (Rhone-Poulenc, 1991b).

This study established a clear NOAEL for plasma cholinesterase between 0.35 and 0.7 ppm in the diet (0.012-0.025 mg/kg) for dogs fed these levels in the diet for 5 weeks. No clinical signs were noted as effects of treatment. Levels of red blood cell cholinesterase inhibition may have been underestimated, however.

3) 2-Year Feeding - rat: Core grade supplementary (Union Carbide, 1966c).

Four groups of rats (20/sex/dose) were maintained on diets containing 0, 0.005, 0.025, 0.05, or 0.1 mg/kg-day of aldicarb for 2 years. Based upon measurements of food consumption; mortality and lifespan; incidence of infection; liver and kidney weights as percentage of body weight; body weight gain; hematology; incidence of neoplasms; incidence of pathological lesions; and brain, plasma, and erythrocyte cholinesterase levels, the animals were found not to differ significantly from controls for any of these parameters. Therefore, the NOAEL for systemic toxicity is greater than or equal to 0.1 mg/kg-day.

4) 2-Year Feeding (carcinogenicity) - rat: Core grade minimum (Union Carbide, 1972).

Groups of 20 Greenacres laboratory controlled flora rats of each sex were fed 0 or 0.3 mg/kg-day of aldicarb in the diet for 2 years. There were no differences from controls in mortality, growth, hematological characteristics, or other histological abnormalities. The NOAEL for systemic toxicity is therefore greater than or equal to 0.3 mg/kg-day.

5) 3-Generation Reproduction - rat: Core grade minimum (Union Carbide, 1974a).

Rats were fed aldicarb at dose levels of 0, 0.2, 0.3, or 0.7 mg/kg-day for 90 to 100 days and mated to produce the respective F1, F2, and F3 generations. All animals were maintained continuously on diets containing aldicarb. The F3 animals were histologically examined either at weaning or at 90 days of age. No reproductive effects were noted at any dose tested. Decreased body weight of F2 pups was observed at 0.7 mg/kg-day. Therefore, the NOAEL and LOAEL for fetotoxicity are 0.3 and 0.7 mg/kg-day, respectively.

6) 2-Generation Reproduction - rat: core grade minimum (Rhone Poulenc, 1991c).

Twenty-six males and females/dose were administered aldicarb at dose levels of 0, 0.1, 0.4, 0.7-9, and 1.4-1.7 mg/kg-day for 70 days prior to mating and then bred to obtain the F1A litters. These progeny were raised until weaning (day 21). The F0 rats were then bred producing the F1B litters.   Aldicarb had no adverse effects on reproductive capacity during either mating. At 1.4-1.7 mg/kg-day dose, there were decrease pup weights and reduced pup viability observed at lactation day 4. At 0.7-0.9 mg/kg-day, in parents decreased body weights and decreased RBC and plasma cholinesterase levels were seen. The NOEL for parental systemic toxicity was 0.4 mg/kg-day and the fetotoxic NOEL was 0.7-0.9 mg/kg-day.

7) Developmental toxicity - rat: Core grade minimum (Union Carbide, 1966a).

Pregnant rats were administered aldicarb in the diet at dose levels of 0, 0.04, 0.2, and 1 mg/kg-day. The rats were further divided into three groups: Group 1 rats were fed aldicarb in the diet throughout pregnancy or until the pups were weaned; Group 2 rats were administered aldicarb from the day the vaginal plug first appeared through the seventh day; and Group 3 received aldicarb from days 5 through 15 of gestation. No congenital malformations were reported for any of the treated groups, and body weights of both the mothers and pups were normal. No significant effects were observed on fertility, gestation, viability of offspring, or lactation. Therefore, the NOAEL for systemic and developmental effects is equal to or greater than 1 mg/kg-day (HDT).

8) Developmental toxicity - rat: Core grade minimum (Rhone-Poulenc, 1988b).

Groups of CD rats were administered aldicarb by gavage at dose levels of 0, 0.125, 0.25 and 0.5 mg/kg-day. Effects on pregnant dams were found at 0.25 and 0.5 mg/kg-day and consisted of decreased body weight gain and food consumption. There were some maternal deaths at 0.5 mg/kg-day. The NOEL for maternal toxicity was 0.125 mg/kg-day. In offspring, at the 0.5 mg/kg-day dose level, there were significant increases in the dilation of the lateral ventricles of the brain, poor ossification of the sixth sternebra, and significant decreases in fetal body weight. In addition, ecchymosis (small hemorrhages) of the trunk were significantly increased ($p<0.05$) at the 0.25 and 0.5 mg/kg-day dose levels. The incidence of this finding at the low-dose group was not statistically significant and was within the historical control range. The NOEL for developmental effects was 0.125 mg/kg-day.

9) Developmental toxicity - rabbit: Core grade guideline (Union Carbide, 1983).

Groups of pregnant Dutch Belted rabbits (16/group) were administered aldicarb by gavage at dose levels of 0, 0.1, 0.25, and 0.5 mg/kg-day from days 7 through 27 of gestation. On the first day of dosing (day 7), 8 animals/group were misdosed with 3 mL/kg instead of 1 mL/kg. Consequently, 5/8 rabbits died in the 0.5 mg/kg-day group. Remaining misdosed animals in all groups were sacrificed and replaced. Survival, other than noted above, was comparable among all groups. General observations for toxic signs were unremarkable, except for pale kidneys and hydroceles on the oviducts at doses greater than or equal to 0.25 mg/kg-day. There were compound-related decreases in body weight for dams at the two highest doses. Based on decreased body weight, pale kidneys, and hydroceles on the oviducts, the NOEL and LOEL for maternal toxicity are 0.25 and 0.5 mg/kg-day, respectively. The number of viable fetuses/doe was significantly reduced in all treatment groups, as was the total number of implantations/doe. Group mean post-implantation loss was significantly reduced in all treatment groups, as was the total number of implantations/doe. These observations (which were statistically significant only at the lowest dose tested) were considered to be due to the unusually large number of corpora lutea/dam and the low rate of pre-implantation loss in the control group, both of which contributed to a higher number of viable fetuses and implantations in the control group. The historical control data supported this conclusion. Therefore, the NOEL for fetotoxicity was 0.5 mg/kg-day, and the overall NOEL for developmental toxicity is 0.25 mg/kg-day. 10) 2-Year Feeding - dog: Core grade minimum (Union Carbide, 1966b).

Four groups of dogs (3/sex/dose) were fed aldicarb in the diet at dose levels of 0, 0.025, 0.05, and 0.1 mg/kg-day for 2 years. Based upon observations of body weight changes, appetite, mortality, histopathology, hematology, biochemistry, and terminal liver and kidney weights, there were no statistically significant effects found at any dose tested. Therefore, the NOAEL for systemic toxicity is greater than or equal to 0.1 mg/kg-day.

11) 3-Month Feeding - dog: Core grade minimum (Union Carbide, 1974b).

Dogs were fed aldicarb in the diet at dose levels of 0, 0.2, 0.3, and 0.7 mg/kg-day for 90 days. The only effects observed were the slightly decreased weight of the testes and the slightly increased weight of the adrenal glands in males at 0.7 mg/kg-day. Therefore, the NOAEL and LOAEL for systemic toxicity are 0.3 and 0.7 mg/kg-day, respectively.

12) 14-Day Feeding - dog: Core grade supplementary (Union Carbide, 1987).

One dog/sex/dose received aldicarb in the diet at dose levels of 0, 0.1, 0.3, 1, 3, and 10 ppm (Male: 0, 0.003, 0.008, 0.029, 0.08, and 0.269 mg/kg-day; Female: 0, 0.003, 0.008, 0.029, 0.114, and 0.294 mg/kg-day) for 2 weeks. Treatment-related effects were observed for inhibition of red blood cell (RBC) acetylcholinesterase (AChE) and plasma butyrlcholinesterase (BuChE) which occurred at or about 3 ppm. Three weekly pre-dose determinations for both RBC and plasma were utilized for comparison of an individual to its own baseline cholinesterase levels. Serial measurements performed every 2 hours for the first 8 hours post-dosing in Group 6 (10 ppm dose) clearly demonstrated inhibition of RBC AChE to 45% of normal and plasma BuChE to 34% of normal. Inhibition was not fully recovered at 8 hours. Based on plasma and RBC ChE inhibition, the NOAEL and LOAEL for systemic toxicity are 1 (0.029 mg/kg-day) and 3 ppm (Male: 0.08 mg/kg-day; Female: 0.114 mg/kg-day).

13) Immunotoxicity - humans: Fiore et al., 1988; Mirkin et al., 1990.

These two published studies suggest that data from women exposed to aldicarb in their drinking water indicated immunomodulatory effects on T Cell subsets. The second, follow-up study notes evidence of lymphocyte (CD8+ T-cell increases) in peripheral blood without clinical signs in five women. U.S. EPA, Office of Pesticide

Programs reviews conclude that for a number of methodological, statistical, and other reasons, that immunological hazards due to Aldicarb have not been demonstrated in these studies.

Data Gap(s): The Agency is preparing a Data Call In for aldicarb that is expected to call for further neurotoxicity studies, and a rat dominant lethal study.

I.A.5. CONFIDENCE IN THE ORAL RfD

The principal studies are given a medium to low confidence rating because none establish a definitive state of the science NOAEL for adverse effects and are limited in the ways described above. Their corroboration of one another in some ways provide additional support. The data base consists of numerous studies and is given a medium confidence rating for completion due to the lack of definitive neurotoxicity studies for a chemical which is a potent neurotoxicant. Confidence in the RfD can be considered medium.

I.A.6. EPA DOCUMENTATION AND REVIEW OF THE ORAL RfD

Source Document -- This assessment is not presented in any existing U.S. EPA document.

Other EPA Documentation -- U.S. EPA, 1984, 1988, 1991

Agency Work Group Review -- 12/02/85, 02/05/86, 05/15/86, 06/20/90, 07/25/90, 09/22/92, 10/15/92

Verification Date -- 10/15/92

I.A.7. EPA CONTACTS (ORAL RfD)

George Ghali / OPP -- (703)305-7490

William Sette / OPP -- (703)305-6375

I.B. REFERENCE CONCENTRATION FOR CHRONIC INHALATION EXPOSURE (RfC)

[Ed. Note: EPA does not have information yet in this section of IRIS].

II. CARCINOGENICITY ASSESSMENT FOR LIFETIME EXPOSURE

Section II provides information on three aspects of the carcinogenic risk assessment for the agent in question; the U.S. EPA classification, and quantitative estimates of risk from oral exposure and from inhalation exposure. The classification reflects a weight-of-evidence judgment of the likelihood that the agent is a human carcinogen. The quantitative risk estimates are presented in three ways. The slope factor is the result of application of a low-dose extrapolation procedure and is presented as the risk per (mg/kg)/day. The unit risk is the quantitative estimate in terms of either risk per ug/L drinking water or risk per ug/cu.m air breathed. The third form in which risk is presented is a drinking water or air concentration providing cancer risks of 1 in 10,000, 1 in 100,000 or 1 in 1,000,000. Background Document 2 (Service Code 5) provides details on the rationale and methods used to derive the carcinogenicity values found in IRIS. Users are referred to Section I for information on long-term toxic effects other than carcinogenicity.

II.A. EVIDENCE FOR CLASSIFICATION AS TO HUMAN CARCINOGENICITY

II.A.1. WEIGHT-OF-EVIDENCE CLASSIFICATION

Classification -- D; not classifiable as to human carcinogenicity.

Basis -- Aldicarb was not found to induce statistically significant increases in tumor incidence in mice or rats in feeding studies or mice in a skin painting study. In the feeding studies there were, however, significant trends in pituitary tumors in female rats and fibrosarcomas in the male mouse. This evidence, together with the fact that less than maximum tolerated doses were used, indicates that the available assays are inadequate to assess the carcinogenic potential of aldicarb.

II.A.2. HUMAN CARCINOGENICITY DATA

None.

II.A.3. ANIMAL CARCINOGENICITY DATA

Inadequate. The NCI (1979) conducted a bioassay of aldicarb for possible carcinogenic effects in rats and mice. Fifty male and 50 female F344 rats and the same numbers of male and female B6C3F1 mice were administered doses of 2 or 6 ppm in the diet for 103 weeks. The animals were then observed for an additional 0 to 2 weeks before terminal sacrifice. Matched controls were composed of 25 untreated rats and 25 untreated mice of each sex. There was a dose-related trend in incidence of pituitary adenomas or carcinomas in the female rats for which the Cochran-Armitage test was statistically significant but the Fisher exact test was not significant. There were occurrences of pancreatic islet-cell adenomas in both treated male and female rats. The incidences were not statistically significant but were nonetheless regarded as compound-related by NCI since there were no pancreatic tumors in the concurrent controls. A dose-related trend in incidence of fibrosarcoma or sarcoma of the subcutaneous tissue was observed in treated male mice for which the Cochran-Armitage test was statistically significant. The Fisher exact test, however was not significant. NCI acknowledged that maximum tolerated doses were not achieved in this study, and it was concluded that none of the tumors could be clearly attributed to the administration of aldicarb.

In a 2-year feeding study, Weil and Carpenter (1965) administered 0.005, 0.025, 0.05 or 0.1 mg aldicarb/kg/day in the diet (0.1, 0.5, 1.0 or 2.0 ppm, assuming food consumption of 5% of body weight per day) to an unspecified strain of rats. The tumor incidences were not significantly greater than those of the control animals. Weil and Carpenter (1972) reported similar results in Greenacres Laboratory Controlled Flora rats fed 0.3 mg/kg/day (6 ppm) for 2 years. No adverse effects were observed due to the aldicarb administration.

Weil (1973) conducted a skin painting study using male C3H/HEJ mice. Female mice were not used for the study due to the high incidence of spontaneous mammary tumors. Mice were administered the aldicarb in the form of applications of 0.125% concentration to hair-free skin on the backs of the animals twice a week for up to 28 months or until death. When compared to controls (positive control group painted with cholanthrene), aldicarb was determined to be noncarcinogenic under the conditions of the experiment.

II.A.4. SUPPORTING DATA FOR CARCINOGENICITY

Ercegovich and Rashid (1973) found aldicarb to be weakly mutagenic in five strains of Salmonella typhimurium in the absence of liver microsomal enzymes.

II.B. QUANTITATIVE ESTIMATE OF CARCINOGENIC RISK FROM ORAL EXPOSURE

[Ed. Note: EPA does not have information yet in this section of IRIS].

II.C. QUANTITATIVE ESTIMATE OF CARCINOGENIC RISK FROM INHALATION EXPOSURE

[Ed. Note: EPA does not have information yet in this section of IRIS].

II.D. EPA DOCUMENTATION, REVIEW, AND CONTACTS (CARCINOGENICITY ASSESSMENT)

II.D.1. EPA DOCUMENTATION

Source Document -- U.S. EPA, 1987

The 1987 Drinking Water Criteria Document for Aldicarb has received Agency peer and administrative review.

II.D.2. REVIEW (CARCINOGENICITY ASSESSMENT)

Agency Work Group Review -- 08/05/87, 08/26/87

Verification Date -- 08/26/87

II.D.3. U.S. EPA CONTACTS (CARCINOGENICITY ASSESSMENT)

Amal Mahfouz / OST -- (202)260-9568

Edward V. Ohanian / OST -- (202)260-7571

III. HEALTH HAZARD ASSESSMENTS FOR VARIED EXPOSURE DURATIONS

III.A. DRINKING WATER HEALTH ADVISORIES

The Health Advisory for aldicarb has been withdrawn on 11/01/93. A revised Health Advisory is in preparation by the Office of Water. For further details contact Amal Mahfouz / OST -- (202)260-9568.

III.B. OTHER ASSESSMENTS

[Ed. Note: EPA does not have information yet in this section of IRIS].

IV. U.S. EPA REGULATORY ACTIONS

EPA risk assessments may be updated as new data are published and as assessment methodologies evolve. Regulatory actions are frequently not updated at the same time. Compare the dates for the regulatory actions in this section with the verification dates for the risk assessments in sections I and II, as this may explain inconsistencies. Also note that some regulatory actions consider factors not related to health risk, such as technical or economic feasibility. Such considerations are indicated for each action. In addition, not all of the regulatory actions listed in this section involve enforceable federal standards. Please direct any questions you may have concerning these regulatory actions to the U.S. EPA contact listed for that particular action. Users are strongly urged to read the background information on each regulatory action in Background Document 4 in Service Code 5.

IV.A. CLEAN AIR ACT (CAA)

[Ed. Note: EPA does not have information yet in this section of IRIS].

IV.B. SAFE DRINKING WATER ACT (SDWA)

IV.B.1. MAXIMUM CONTAMINANT LEVEL GOAL (MCLG) for Drinking Water

Value (status) -- 0.001 mg/L (Final, 1991)

Considers technological or economic feasibility? -- NO

Discussion -- An MCLG of 0.001 mg/L for aldicarb is based on potential adverse effects (cholinesterase inhibition) reported in humans.

Reference -- 56 FR 3600 (01/30/91); 56 FR 30266 (07/01/91)

EPA Contact -- Health and Ecological Criteria Division / OST / (202) 260-7571 / FTS 260-7571; or Safe Drinking Water Hotline / (800) 426-4791

IV.B.2. MAXIMUM CONTAMINANT LEVEL (MCL) for Drinking Water

Value -- 0.003 mg/L (Final, 1991)

Considers technological or economic feasibility? -- YES

Discussion -- The EPA has set an MCL equal to the PQL of 0.003 mg/L.

Monitoring requirements -- All systems monitored for four consecutive quarters every 3 years; repeat monitoring dependent upon detection, vulnerability status and system size.

Analytical methodology -- Derivatization/gas chromatography (EPA 531.1): PQL= 0.003 mg/L.

Best available technology -- Granular activated carbon

Reference -- 56 FR 3526 (01/30/91); 56 FR 3600 (01/30/91); 56 FR 30266 (01/07/91).

EPA Contact -- Drinking Water Standards Division / OGWDW / (202) 260-7575 / FTS 260-7575; or Safe Drinking Water Hotline / (800) 426-4791

IV.B.3. SECONDARY MAXIMUM CONTAMINANT LEVEL (SMCL) for Drinking Water

[Ed. Note: EPA does not have information yet in this section of IRIS].

IV.B.4. REQUIRED MONITORING OF "UNREGULATED" CONTAMINANTS

[Ed. Note: EPA does not have information yet in this section of IRIS].

IV.C. CLEAN WATER ACT (CWA)

[Ed. Note: EPA does not have information yet in this section of IRIS].

IV.D. FEDERAL INSECTICIDE, FUNGICIDE, AND RODENTICIDE ACT (FIFRA)

IV.D.1. PESTICIDE ACTIVE INGREDIENT, Registration Standard

Status -- Issued (1984)

Reference -- Aldicarb Pesticide Registration Standard. March, 1984 (NTIS No. PB84-207653)

EPA Contact -- Registration Branch / OPP (703)557-7760 / FTS 557-7760

IV.D.2. PESTICIDE ACTIVE INGREDIENT, Special Review

Action -- Risk Benefit Analysis - PD2/3 (1988)

Considers technological or economic feasibility? -- NO

Summary of regulatory action -- Based on concerns for acute toxicity and ground water contamination.

Reference -- 53 FR 24630 (06/29/88)

EPA Contact -- Special Review Branch / OPP (703)557-7400 / FTS 557-7400

IV.E. TOXIC SUBSTANCES CONTROL ACT (TSCA)

> [Ed. Note: EPA does not have information yet in this section of IRIS].

IV.F. RESOURCE CONSERVATION AND RECOVERY ACT (RCRA)

> [Ed. Note: EPA does not have information yet in this section of IRIS].

IV.G. SUPERFUND (CERCLA)

IV.G.1. REPORTABLE QUANTITY (RQ) for Release into the Environment

Value (status) -- 1 pound (Final, 1985)

Considers technological or economic feasibility? -- NO

Discussion -- The RQ for aldicarb is based on aquatic toxicity. Available data indicate the aquatic 96-hour Median Threshold Limit for aldicarb is 0.05 ppm, which corresponds to an RQ of 1 pound.

Reference -- 50 FR 13456 (04/04/85); 54 FR 33418 (08/14/89)

EPA Contact -- RCRA/Superfund Hotline (800)424-9346 / (202)260-3000 / FTS 260-3000

VI. BIBLIOGRAPHY

VI.A. ORAL RfD REFERENCES

Fiore M.C., H.A. Anderson, R. Hong, et al. 1986. Chronic Exposure to Aldicarb Contaminated Groundwater and Human Immune Function. Environ. Res. 41: 633-645.

Goldman L.R., M. Beller, and R.J. Jackson. 1990a. Aldicarb Food Poisonings in California, 1985-1988: Toxicity Estimates for Humans. Arch. Environ. Health. 45(3): 141-147.

Goldman L.R., D.F. Smith, R.R. Neutra, et al. 1990b. Pesticide Food Poisoning from Contaminated Watermelons in California, 1985. Arch. Environ. Health. 45(4): 229-236.

Hirsch G.H., B.T. Mori, G.B. Morgan, P.R. Bennett, and B.C. Williams. 1987. Report of Illnesses Caused by Aldicarb-Contaminated Cucumbers. Food Add. Contam. 5(2): 155-60.

Mirkin I.R., H.A. Anderson, L. Hanrahan, R. Hong, R. Golubjatnikov, and D. Belluck. 1990. Changes in T-Lymphocyte Distribution Associated with Ingestion of Aldicarb Contaminated Drinking Water: A Follow-up Study. Environ. Res. 51: 35-50.

Rhone-Poulenc Ag Company. 1988a. MRID No. 40695901; HED Doc Nos. 007058, 010449, 010456. Available from EPA. Write to FOI, EPA, Washington, DC 20460.

Rhone-Poulenc Ag Company. 1988b. MRID No. 41004501; HED Doc. Nos. 007254, 010453. Available from EPA. Write to FOI, EPA, Washington, DC 20460.

Rhone-Poulenc Ag Company. 1991a. MRID No. 42191501; HED Doc. No. 010456. Available from EPA. Write to FOI, EPA, Washington, DC 20460.

Rhone-Poulenc Ag Company. 1991b. MRID No. 41919901, 41956101; HED Doc. Nos. 008388, 010454. Available from EPA. Write to FOI, EPA, Washington, DC 20460.

Rhone-Poulenc Ag Company. 1991c. MRID No. 421484-01; HED Doc. No. 010457. Available from EPA. Write to FOI, EPA, Washington, DC 20460.

Rhone-Poulenc Ag Company. 1992. MRID No. 423730-01; HED Doc. No. 0010459. Available from EPA. Write to FOI, EPA, Washington, DC 20460.

Union Carbide Corporation. 1966a. MRID No. 00058631, 00085456; HED Doc. No. 004022. Available from EPA. Write to FOI, EPA, Washington, DC 20460.

Union Carbide Corporation. 1966b. MRID No. 00085458; HED Doc. No. 004022. Available from EPA. Write to FOI, EPA, Washington, DC 20460.

Union Carbide Corporation. 1966c. MRID No. 00085460; HED Doc. No. 004022. Available from EPA. Write to FOI, EPA, Washington, DC 20460.

Union Carbide Corporation. 1971. MRID No. 00101911; HED Doc. No. 010450. Available from EPA. Write to FOI, EPA, Washington, DC 20460.

Union Carbide Corporation. 1972. MRID No. 00029943; HED Doc No. 004022. Available from EPA. Write to FOI, EPA, Washington, DC 20460.

Union Carbide Corporation. 1974a. MRID No. 00044736, 00069918; HED Doc. No. 004022. Available from EPA. Write to FOI, EPA, Washington, DC 20460.

Union Carbide Corporation. 1974b. MRID No. 00044737; HED Doc. No. 004022. Available from EPA. Write to FOI, EPA, Washington, DC 20460.

Union Carbide Agricultural Product Company, Inc. 1983. MRID No. 00131661, 00132668; HED Doc No. 003466, 010460. Available from EPA. Write to FOI, EPA, Washington, DC 20460.

Union Carbide Agricultural Product Company, Inc. 1987. MRID No. 40166601; HED Doc. No. 005925. Available from EPA. Write to FOI, EPA, Washington, DC 20460.

U.S. EPA. 1984. Requirements for Interim Registration of Pesticide Products containing Aldicarb as the Active Ingredient. Office of Pesticides and Toxic Substances, Washington DC. March 30. NTIS PB84-207653.

U.S. EPA. 1988. Aldicarb Special Review Technical Support Document. Office of Pesticides and Toxic Substances, Washington DC. June. NTIS PB88-236856.

U.S. EPA. 1991. Drinking Water Criteria Document for Aldicarb. Office of Health and Environmental Assessment, Environmental Criteria and Assessment Office, Cincinnati, OH for the Office of Drinking Water, Washington, DC.

VI.B. INHALATION RfC REFERENCES

None

VI.C. CARCINOGENICITY ASSESSMENT REFERENCES

Ercegovich, C.D. and K.A. Rashid. 1973. Mutagenesis induced in mutant strains of Salmonella typhimurium by pesticides. Abstracts of papers. Am. Chem. Soc. Abstract No. 43.

NCI (National Cancer Institute). 1979. Bioassay of aldicarb for possible carcinogenicity. NCI Report No. 136. DHEW Publ. No. (NIH) 79-1391.

U.S. EPA. 1987. Drinking Water Criteria Document for Aldicarb. Prepared by the Office of Health and Environmental Assessment, Cincinnati, OH, for the Office of Drinking Water, Washington, DC. ECAO-CIN-420.

Weil, C.S. 1973. Miscellaneous toxicity studies. Mellon Institute Report 35-41. EPA Pesticide Petition No. 3F1414.

Weil, C.S. and C.P. Carpenter. 1965. Two-year feeding of Compound 21149 in the diet of rats. Mellon Institute Report No. 28-123. EPA Pesticide Petition No. 9F0798.

Weil, C.S. and C.P. Carpenter. 1972. Aldicarb (A), aldicarb sulfoxide (ASO), aldicarb sulfone (ASO2) and a 1:1 mixture of ASO:ASO2. Two-year feeding in the diet of rats. Mellon Institute Report No. 35-82. EPA Pesticide Petition No. 9F0798.

VI.D. DRINKING WATER HA REFERENCES

Not available at this time

VII. REVISION HISTORY

```
--------  -----------  --------------------------------------------------------
Date      Section      Description
--------  -----------  --------------------------------------------------------
```

Date	Section	Description
03/01/88	I.A.5.	Confidence levels revised
03/01/88	I.A.7.	Primary contact changed
03/01/88	III.A.	Health Advisory added
08/22/88	II.	Carcinogen summary on-line
05/01/89	I.A.5.	Confidence statement revised
05/01/89	I.A.7.	Secondary contact changed
06/01/89	II.D.2.	Work group review date added
08/01/89	VI.	Bibliography on-line
07/01/90	I.A.	Oral RfD summary noted as pending change
08/01/90	I.A.	Withdrawn; new Oral RfD verified (in preparation)
08/01/90	III.A.5.	DWEL & Lifetime HA withdrawn (RfD withdrawn)
08/01/90	VI.A.	Oral RfD references withdrawn
03/01/91	II.D.3.	Secondary contact changed
07/01/91	I.A.	Oral RfD summary replaced; RfD changed
07/01/91	III.A.5.	DWEL and Lifetime HA replaced
07/01/91	VI.A.	Oral RfD references replaced
07/01/91	III.A.	Section III.A.1-4 replaced
08/01/91	I.A.	General edit
01/01/92	IV.	Regulatory actions updated
10/01/92	I.A.	Oral RfD summary noted as pending change
10/01/92	I.A.6.	Work group review date added
12/01/92	I.A.6.	Work group review date added
11/01/93	I.A.	Oral RfD summary replaced; new RfD
11/01/93	III.A.	Health Advisory withdrawn
11/01/93	VI.A.	Oral RfD references replaced
11/01/93	VI.D.	Health Advisory references withdrawn
01/01/94	III.A.	Message revised to include contact's name and number

SYNONYMS

Aldecarb
Aldicarb
Ambush
Carbamyl
Carbanolate
ENT 27,093
2-Methyl-2-(Methylthio)Propanal, O-((Methylamino)Carbonyl) Oxime
2-Methyl-2-(Methylthio)Propionaldehyde O-(Methylcarbamoyl)Oxime
NCI-C08640
OMS 771
Propanal, 2-Methyl-2-(Methylthio)-, O-((Methylamino)Carbonyl)Oxime
Propionaldehyde, 2-Methyl-2-(Methylthio)-, O-(Methyl-carbamoyl)Oxime
Sulfone aldoxycarb
Temic
Temik
Temik 10 G
Temik G 10
Temik TSK
UC 21149
Union Carbide 21149
Union Carbide UC-21149

Source: *Instant EPA's IRIS*, 1997, Instant Reference Sources, Inc., 7605 Rockpoint Drive, Austin, TX 78731 (Fax: 512-345-2386; Internet URL, http://www.instantref.com/inst-ref.htm).

Production, Use, and Pesticide Labeling Information

1. Description of Chemical

   Common Name: Aldicarb

   Generic Name: 2-methyl-2-(methylthio)proprionaldehyde O-(methylcarbamoyl)oxime

   Trade Name: Temik

   EPA Shaughnessy Code: 098301

   Chemical Abstracts Service (CAS) Number: 116-06-3

   Year of Initial Registration: 1970

   Pesticide Type: Insecticide, acaricide, nematicide

   Chemical Family: Carbamate

U.S. and Foreign Producers: Rhone-Poulenc (formerly Union Carbide Agricultural Chemical Co.)

2. Use Patterns and Formulations

- Aldicarb is currently registered for use only on cotton, potatoes, citrus, peanuts, soybeans, sugar beets, pecans, tobacco, sweet potatoes, ornamentals, seed alfalfa, grain sorghum, dry beans, and sugar cane.

- Types and Methods of Application: Soil incorporated.

- Application Rates: 0.3 - 10.0 lbs. active ingredient.

- Types of Formulation: Granular formulation (15%, 10%, and 5%). Also as a granular in a mixture with the fungicides pentachloronitrobenzene and 5-ethoxy-3-(trichloromethyl)-1,2,4-thiadiazole.

3. Science Findings

Chemical Characteristics

- Technical aldicarb is a white crystalline solid with a melting point of 98-100 °C (pure material). Under normal conditions, aldicarb is a heat-sensitive, inherently unstable chemical and must be stabilized to obtain a practical shelf-life.

Toxicological Characteristics

- Aldicarb is a carbamate insecticide which causes cholinesterase inhibition (ChE) at very low exposure levels. It is highly toxic by the oral, dermal and inhalation routes of exposure (Toxicity Category I).

The oral LD50 value for technical aldicarb is 0.9 mg/kg and 1.0 mg/kg for male and female rats, respectively. The acute dermal LD50 for aldicarb in rats is 3.0 mg for males and 2.5 mg for females. Rats, mice and guinea pigs were exposed to aldicarb, finely ground, mixed with talc, and dispersed in the air at a concentration of 200 mg/m^3 for five minutes; all animals died. At a lower concentration (6.7 mg/m^3), a 15 minute exposure was not lethal; however, 5 of 6 animals died during a 30 minute exposure. Exposure of rats for eight hours to air that had passed over technical aldicarb or granular aldicarb produced no mortality. Aldicarb applied to the eyes of rabbits at 100 mg of dry powder caused ChE effects and lethality.

- The toxicity database for aldicarb is complete. The toxicity database includes a 2-year rat feeding/oncogenicity study which was negative for oncogenic effects at the no-observed effect level (NOEL) of 0.3 mg/kg bw/day; a 100-day dog feeding study and a 2-year dog feeding study with NOELs of 0.7 and 0.1 mg/kg bw/day, respectively, for effects other than cholinesterase inhibition (highest levels tested (HLT)); an 18-month mouse feeding/oncogenicity study with a NOEL of 0.7 mg/kg bw/day and was negative for oncogenic effects at the levels tested (0.1, 0.3 and 0.7 mg/kg bw/day);a 2-year mouse oncogenicity study which was negative for oncogenic effects; a 6-month rat feeding study using aldicarb sulfoxide with a NOEL of 0.125 mg/kg bw/day for ChE inhibition; a 3-generation rat reproduction study with a 0.7 mg/kg bw/day NOEL; a rat teratology study, which was negative for teratogenic effects at 0.5 mg/kg bw/day (HLT); a hen neurotoxicity study which was negative at up to 4.5 mg/kg bw/day; a mutagenicity study utilizing the rat hepatocyte primary culture/ DNA repair test which was negative for mutagenic effects at 10,000 ug/well; and a mutagenicity test utilizing an in vivo chromosome aberration analysis in Chinese hamster ovary cells which was negative for mutagenic effects at 500 ug/mL.

Physiological and Biochemical Behavioral Characteristics

- Aldicarb and its metabolites are absorbed by plants from the soil and translocated into the roots, stems, leaves, and fruit. The available data indicate that the metabolism of aldicarb in plants and small animals is similar.

- Aldicarb is metabolized rapidly by oxidation to the sulfoxide metabolite, followed by slower oxidation to the sulfone metabolite, which is 25 times less acutely toxic than aldicarb. Both metabolites are subsequently hydrolyzed and degraded further to yield less toxic entities. Available studies demonstrate that the administration of aldicarb to a lactating ruminant results in the rapid metabolism and elimination of the material. No residues of the parent compound and little, if any, residues of aldicarb sulfoxide or aldicarb sulfone are found in the tissues and milk. the predominant residue detected in tissues and milk is aldicarb sulfone nitrile.

Environmental Characteristics

- Sufficient data are available to fully assess the environmental fate of aldicarb. From the available data, aldicarb has been determined to be mobile in fine to coarse textured soils, even including those soils with high organic matter content, and has been found to reach ground water. Aldicarb is not expected to move horizontally from a bare, sloping field. Therefore, accumulation of aldicarb in aquatic nontarget organisms is expected to be minimal. This is further supported by an octanol/water partition coefficient of 5 and an ecological magnification value of 42.

Ecological Characteristics

- Aldicarb is highly toxic to mammals, birds, estuarine/marine organisms, and freshwater organisms. LC50 values for the bluegill sunfish and rainbow trout have been reported as 50 ug/liter and 560 ug/liter, respectively. An LC50 of 410.7 ug/liter was reported for the Daphnia magna. Studies on the toxicity of aldicarb to the mallard duck and bobwhite quail indicate LD50 values of 1.0 and 2.0 mg/kg, respectively.

- Limited exposure to mammals is expected from a dietary standpoint. However, data from field studies and the use history of aldicarb provide sufficient information to suggest that application of this pesticide may result in some mortality, with possible local population reductions of some avian species. Whether these effects are excessive, long-lasting, or likely to diminish wildlife resources cannot be stated with any degree of certainty. Therefore, additional field studies have been required to further quantify the impact on avian and small mammal populations. Field study results will be submitted in April 1988.

- Aldicarb has also been found to pose a threat to the endangered Attwater's Greater Prairie Chicken, living in or near aldicarb-treated fields. Accordingly, all aldicarb products are required to bear labeling restrictions prohibiting the use of the product in the Texas counties of Aransas, Austin, Brazoria, Colorado, Galveston, Goliad, Harris, Refugio, and Victoria if this species is located in or immediately adjacent to the treatment area.

Tolerance Assessment

- The Agency is in the process of reassessing the existing tolerances for aldicarb. Processing studies for coffee and potatoes have been submitted and are acceptable. A large animal metabolism study has been submitted to the Agency and satisfies the data requirement. A completed study of aldicarb residues on soybean processing fractions is to be submitted by August, 1988. The requirement to submit a study analyzing aldicarb residues on treated cotton forage has been satisfied with a label restriction prohibiting the feeding of treated forage to livestock.

Problems known to Have Occurred with Use of the Chemical

- In 1979 aldicarb residues were found in drinking water wells located near aldicarb treated potato fields in Suffolk County, Long Island, New York, at levels greater than 200 parts per billion (ppb). Subsequently, aldicarb has been detected in ground water in 48 counties within 15 other States at levels up to 515 ppb. In all, the Agency has evaluated over 35,000 ground water samples of which 32% were positive for residues of aldicarb. The Agency's Office of Drinking Water (ODW) has established a Health Advisory level (HA) of 10 ppb for residues of aldicarb in drinking water.

- The Pesticide Incident Monitoring System (PIMS) reports on aldicarb from 1966 through 1982 contained 165 incidents associated with human injury. Most of the human incidents alleged that aldicarb was the cause of the problem, but there was insufficient evidence to support such a conclusion. Those incidents involving confirmed aldicarb poisonings appeared to be the result of failure to use label recommended safety equipment while applying aldicarb. Other incidents resulted from accidental spillage, ingestion of aldicarb, or consumption of food commodities improperly treated with aldicarb.

- The largest document episode of food-borne pesticide poisoning in North American history occurred in July 1985 from aldicarb-contaminated California watermelons. More than a thousand probable causes were reported from California, Oregon, Washington, Alaska, Idaho, Nevada, Arizona and Canada. The spectrum of illness attributed to aldicarb ranged from mild to severe and included cases of grand mal seizures, cardiac arrhythmias, severe dehydration, bronchospasms, and at least two stillbirths occurring shortly after maternal illness. The prompt embargo of watermelons on July 4, 1985 abruptly terminated the major portion of the outbreak and reported illnesses occurring after the implementation of the watermelon certification program were far fewer and milder in comparison to earlier cases. Contamination of the watermelons ranged up to 3.3 ppm of aldicarb sulfoxide (ASO), a metabolite of aldicarb. Clinical signs occurred from exposures to dosages estimated to be as low as 0.0026 mg/kg ASO.

4. Summary of Regulatory Position and Rationale:

- Dietary Exposure to Treated Food Commodities

- The Agency has recently received the final results of a National Food Survey which monitored raw agricultural commodities for residues of aldicarb in the market place. After these data have been evaluated, the dietary exposure from consuming treated food commodities will be estimated, and a risk assessment will be conducted. The Agency may propose further regulatory action depending on the results of this study.

- Dietary Exposure to Contaminated Ground Water

- The Agency has concluded that there are unacceptable risks to persons consuming drinking water that is contaminated with aldicarb at levels greater than the HA of 10 ppb due to a reduced margin of safety for ChE inhibition.

- The Agency cannot identify all specific areas of the nation where aldicarb residues exceed the HA, or the number of people who would be exposed to these high levels of contamination. However, the Agency can predict certain areas of the nation where the ground water supplies have a relatively high vulnerability to aldicarb contamination due to the hydrogeology and/or agronomic practices found in that area. Additionally, the Agency can predict certain areas which would have a medium vulnerability to contamination, although the vulnerability within some of these areas could vary greatly with some areas being much more vulnerable.

- It is the Agency's presumption that the risks posed by aldicarb contamination of ground water above the HA in current or potential drinking waters will likely be more significant, in almost all cases, than any local benefit derived from aldicarb's continued use. Consequently, the Agency is proposing to regulate the use of aldicarb in order to eliminate or prevent contamination of ground water at levels above the HA. As a basic level of protection for all areas where aldicarb is used, the Agency is proposing a number of restrictions on the label. Specifically, no use of aldicarb would be permitted within 300 feet of a drinking water well, and aldicarb would be classified as a restricted use pesticide due to ground water concerns. (Aldicarb is already classified as a restricted use pesticide due to its acute toxicity.) Additionally, the Agency is seeking public comment as to what, if any, additional measures should be considered regarding the use of aldicarb and site-conditional restrictions.

- The Agency will also require monitoring in those areas classified as having a medium tendency to leach. The data generated will be used to determine whether further regulatory action is required in these areas.

- Finally, for those areas where there is the greatest likelihood of ground water contamination, states will need to implement, either for the entire state or for a county or counties within the state, State Pesticide Ground Water Management Plans (MPs). Briefly, MPs are comprehensive plans which describe the measures states will impose to prevent ground water contamination. The Agency believes that MPs provide the best method of protection ground water pesticide contamination.

- The Agency is soliciting public comment on a number of issues regarding its preliminary determination for aldicarb. Included are questions regarding the components of an MP, which assessment (hydrogeologic region or county) should be used in identifying those areas where contamination is most likely to occur, how should a localized risk/benefit analysis be performed and who should conduct it, and who is responsible for the costs associated with cleaning up ground water contamination.

5. Summary of Major Data Gaps

      None

6. Contact Person at EPA

Bruce Kapner Special Review Branch,
Registration Division
Office of Pesticide Programs (TS-767C)
401 M St., SW.
Washington, DC 20460

Telephone: (703) 557-1170

Source: *Instant EPA's Pesticide Facts*, 1994, Instant Reference Sources, Inc., 7605 Rockpoint Drive, Austin, TX 78731 (Fax: 512-345-2386;   Internet URL, http://www.instantref.com/inst-ref.htm).

Aldrin

Introduction

Aldrin is an effective insecticide that has been used primarily against soil insects that attack field, forage, vegetable, and fruit crops. It was used to kill cotton insects, turf pests, white grubs, and corn rootworms. It is also effective against termites and was used for wood preservation and to combat ant infestations. [828] Aldrin is a chlorinated cyclodiene compound with high acute toxicity. It is extremely toxic to aquatic organisms and birds. It is persistent in the environment and bioaccumulates. It was formerly used as an insecticide and fumigant in the United States but its manufacture and use have been discontinued because of its adverse effects on the environment. It usually degrades in the environment to dieldrin, another commercial pesticide. Biodegradation of aldrin is also relatively slow and it mostly resides in water, soil or plants but when it is in the vapor state it may be degraded within hours by photochemical reactions with hydroxyl radicals in the atmosphere. Aldrin has been found frequently in environmental samples of all kinds in the past, especially in field runoff to surface waters such as lakes and rivers and also in groundwaters. Today in the U.S. it may mostly be found as its decomposed product, dieldrin.

Aldrin residues in soil and plants will volatilize from soil surfaces or be slowly transformed to dieldrin in soil. Biodegradation is expected to be slow and aldrin is not expected to leach. Aldrin residues in water will volatilize from the water surface and photooxidization is expected to be significant. Photolysis has been observed in water, although the absorption characteristics of aldrin indicate it should not extensively directly photolyze in the environment. Bioconcentration will be significant. Adsorption to sediments is expected and biodegradation is expected to be slow. [828]

Identifying Information

CAS NUMBER: 309-00-2

NIOSH Registry Number: IO2100000

SYNONYMS: ALDREX, ALDRITE, ALDROSOL, DRINOX, HEXACHLOROHEXAHYDRO-ENDO-EXO-DIMETHANO-NAPTH-ALENE, 1,2,3,4,10,10-HEXACHLORO-1,4,4A,5,8,8A-HEXAHYDRO-1,4,5,8-DIMETHANONAPTHALENE, HHDN, NCI-C00044, OCTALENE, SEEDRIN, OCTALENE COMPOUND 118

Endocrine Disruptor Information

Aldrin is on EPA's list of suspect EEDs [811] but little information has been located to determine the Agency's reason for its inclusion. More information should be forthcoming after EPA's studies are concluded.

Chemical and Physical Properties

CHEMICAL FORMULA: $C_{12}H_8Cl_6$

MOLECULAR WEIGHT: 364.93

WLN: L D5 C55 A D- EU JUTJ AG AG BG IG JG KG

PHYSICAL DESCRIPTION: Brown to white crystalline solid

SPECIFIC GRAVITY: Not available

DENSITY: 1.70 g/cm^3 @ 20°C [173]

MELTING POINT: 104°C [017], [031], [047], [346]

BOILING POINT: 145°C @ 2 mm Hg [169], [173]

SOLUBILITY: Water: <1 mg/mL @ 24°C [700]; DMSO: 1-5 mg/mL @ 24°C [700]; 95% Ethanol: <1 mg/mL @ 24°C [700]; Acetone: 5-10 mg/mL @ 24°C [700]

OTHER SOLVENTS: Most organic solvents: Very soluble [031], [395]; Aromatics: Soluble [043]; Esters: Soluble [043]; Ketones: Soluble [043]; Paraffins: Soluble [043]; Halogenated solvents: Soluble [043]; Petroleum ether: Moderately soluble [029]; Alcohol: Soluble [017]; Benzene: >600 g/L [172]; Ether: Soluble [017], [047]; Xylene: >600 g/L [172]

OTHER PHYSICAL DATA: Technical grade: Tan with melting point of 49-60°C [055], [172], [346], [395]

Source: *The National Toxicology Program's Chemical Database, Volume 2*, 1992, CRC Press, Inc./Lewis Publishers, 2000 Corporate Blvd., Boca Raton, FL 33431 (Fax: 407-998-9114).

Environmental Reference Materials

A source of EED environmental reference materials for aldrin is Radian International LLC, Austin, TX (Fax 512-454-0268; Internet URL, http://www.radian.com/standards); Catalog No. ERA-006.

Hazardous Properties

ACUTE/CHRONIC HAZARDS: Aldrin may be toxic by ingestion and skin contact [043].

FIRE HAZARD: Literature sources indicate that aldrin is nonflammable [430]. Fires involving this material can be controlled with a dry chemical, carbon dioxide or Halon extinguisher.

REACTIVITY: Aldrin reacts with concentrated acids and phenols in the presence of oxidizing agents. It can be corrosive to metals [169], [172]. It can react with acid catalysts, acid oxidizing agents and active metals. [430]

STABILITY: Aldrin may be sensitive to prolonged exposure to light. Aldrin is stable to heat and in the presence of inorganic and organic bases. It is stable to hydrated metal chlorides and mild acids [395]. Aldrin is thermally stable up to 200°C and it is stable between pH 4 and 8 [169], [172]. Solutions of aldrin in water, DMSO, 95% ethanol or acetone should be stable for 24 hours under normal laboratory conditions. [700]

Source: *The National Toxicology Program's Chemical Database, Volume 7*, 1992, CRC Press, Inc./Lewis Publishers, 2000 Corporate Blvd., Boca Raton, FL 33431 (Fax: 407-998-9114).

Medical Symptoms of Exposure

Symptoms of exposure to aldrin may include excitement, tremors, nausea, vomiting and liver damage [043]. Other symptoms may include renal damage, ataxia, convulsions by central nervous system depression, respiratory failure and death [031]. Headache, dizziness, malaise, coma and myoclonic jerks of the limbs, hematuria and azotemia have also been reported [346]. Diarrhea, paresthesia, giddiness, fatigue, pumonary edema, loss of appetite and muscular weakness may also occur [295]. Other symptoms may include hyper excitability and ventricular fibrillation [371].

Source: *The National Toxicology Program's Chemical Database, Volume 4*, 1992, CRC Press, Inc./Lewis Publishers, 2000 Corporate Blvd., Boca Raton, FL 33431 (Fax: 407-998-9114).

Toxicological Information

The toxicological information below is gathered from several sources including two national databases: one from the *National Toxicology Program Chemical Repository Database* and another from *EPA's Integrated Risk Information System (IRIS)*.

Toxicology Information from the National Toxicology Program

CARCINOGENICITY:
>   Tumorigenic Data:
>
>>   TDLo: orl-rat 200 mg/kg/2Y-C
>>   TDLo: orl-mus 270 mg/kg/80W-I
>>   TD : orl-mus 540 mg/kg/80W-I
>>   TD : orl-rat 188 mg/kg/2Y-C

>   Review:
>
>>   IARC Cancer Review: Animal Limited Evidence
>>   IARC Cancer Review: Human Inadequate Evidence
>>   IARC: Not classifiable as a human carcinogen (Group 3)

>   Status:
>
>>   NCI Carcinogenesis Bioassay (Feed): Equivocal: Male and
>>   Female Rat
>>   NCI Carcinogenesis Bioassay (Feed): Positive: Male Mouse
>>   NCI Carcinogenesis Bioassay (Feed): Negative: Female Mouse
>>   EPA Carcinogen Assessment Group

TERATOGENICITY:

>   Reproductive Effects Data:
>
>>   TDLo: orl-mus 25 mg/kg (9D preg)
>>   TDLo: orl-ham 50 mg/kg (7D preg)
>>   TDLo: scu-rat 10 mg/kg (2D pre)
>>   TDLo: orl-dog 73 mg/kg (44W pre/1-8W preg)

MUTATION DATA:

Test	Lowest dose	Test	Lowest dose
dns-mus-orl	11 umol/kg	dns-hmn:fbr	1 umol/L
cyt-mus-ipr	9560 ug/kg	cyt-hmn:leu	19125 ug/L
cyt-mus-orl	13 mg/kg	cyt-rat-ipr	9560 ug/kg
mmo-smc	5 ppm	dnd-rat:lvr	300 umol/L
cyt-hmn:lym	19125 ug/kg		

TOXICITY:

Typ. Dose	Mode	Specie	Amount	Units	Other
TDLo	orl	hmn	14	mg/kg	
LDLo	orl	orl chd	1250	ug/kg	
LD50	orl	rat	39	mg/kg	
LD50	skn	rat	98	mg/kg	
LD50	ipr	rat	150	mg/kg	
LCLo	ihl	rat	5800	ug/m3/4H	
LD50	scu	rat	62	mg/kg	
LD50	orl	mus	44	mg/kg	
LD50	ipr	mus	50	mg/kg	
LD50	ivn	mus	21	mg/kg	
LD50	orl	dog	65	mg/kg	
LD50	ipr	dog	1999	mg/kg	
LDLo	unr	dog	95	mg/kg	
LDLo	skn	cat	75	mg/kg	
LD50	orl	rbt	50	mg/kg	
LD50	skn	rbt	15	mg/kg	
LDLo	scu	rbt	100	mg/kg	
LD50	orl	gpg	33	mg/kg	
LD50	orl	ham	100	mg/kg	
LD50	orl	pgn	56200	ug/kg	
LD50	orl	ckn	10	mg/kg	
LD50	orl	qal	42100	ug/kg	
LD50	orl	dck	520	mg/kg	
LD50	orl	mam	39	mg/kg	
LD50	unr	mam	40	mg/kg	
LD50	orl	bwd	7200	ug/kg	
LDLo	orl	cat	15	mg/kg	

OTHER TOXICITY DATA:

Review:

Toxicology Review-11

Status:

NIOSH Analytical Methods: see Aldrin and Lindane

Meets criteria for proposed OSHA Medical Records Rule

EPA Genetox Program 1988, Positive/limited: Carcinogenicity-mouse/rat

EPA Genetox Program 1988, Negative: Histidine reversion-Ames test S cerevisiae-homozygosis

EPA Genetox Program 1988, Inconclusive: D melanogaster Sex-linked lethal

EPA TSCA Test Submission (TSCATS) Data Base, January 1989

Standards and Regulations:
          DOT-Hazard: Poison B; Label: Poison
          DOT-Hazard: ORM-A; Label: None, solid
          IDLH value: 100 mg/m3

SAX TOXICITY EVALUATION:

THR: Poison by ingestion, skin contact, intravenous, intraperitoneal and other routes. An experimental tumorigen, neoplastigen, carcinogen and teratogen. Human systemic effects by ingestion. Experimental reproductive effects. Human mutagenic data.

Source: *Instant Tox-Base*, 1995, Instant Reference Sources, Inc., 7605 Rockpoint Drive, Austin, TX 78731 (Fax: 512-345-2386; Internet URL, http://www.instantref.com/inst-ref.htm).

Toxicology Information from EPA's Integrated Risk Information System (IRIS)

I. CHRONIC HEALTH HAZARD ASSESSMENTS FOR NONCARCINOGENIC EFFECTS

I.A. REFERENCE DOSE FOR CHRONIC ORAL EXPOSURE (RfD)

The Reference Dose (RfD) is based on the assumption that thresholds exist for certain toxic effects such as cellular necrosis, but may not exist for other toxic effects such as carcinogenicity. In general, the RfD is an estimate (with uncertainty spanning perhaps an order of magnitude) of a daily exposure to the human population (including sensitive subgroups) that is likely to be without an appreciable risk of deleterious effects during a lifetime. Please refer to Background Document 1 in Service Code 5 for an elaboration of these concepts. RfDs can also be derived for the noncarcinogenic health effects of compounds which are also carcinogens. Therefore, it is essential to refer to other sources of information concerning the carcinogenicity of this substance. If the U.S. EPA has evaluated this substance for potential human carcinogenicity, a summary of that evaluation will be contained in Section II of this file when a review of that evaluation is completed.

I.A.1. ORAL RfD SUMMARY

Critical Effect	Experimental Doses*	UF	MF	RfD
Liver toxicity	NOAEL: none	1000	1	3E-5 mg/kg/day
	LOAEL: 0.5 ppm (0.025 mg/kg/day)			

*Conversion Factors: 1 ppm = 0.05 mg/kg/day (assumed rat food consumption)

Rat Chronic Feeding diet Study, Fitzhugh et al., 1964

I.A.2. PRINCIPAL AND SUPPORTING STUDIES (ORAL RfD)

Fitzhugh, O.G., A.A. Nelson, and M.L. Quaife. 1964. Chronic oral toxicity of aldrin and dieldrin in rats and dogs. Food Cosmet. Toxicol. 2: 551-562.

Groups of 24 rats (12/sex) were fed aldrin in the diet at levels of 0, 0.5, 2, 10, 50, 100, or 150 ppm for 2 years. Liver lesions characteristic of chlorinated insecticide poisoning were observed at dose levels of 0.5 ppm and greater. These lesions were characterized by enlarged centrilobular hepatic cells, with increased cytoplasmic oxyphilia, and peripheral migration of basophilic granules. A statistically significant increase in liver-to-body weight ratio was observed at all dose levels. Kidney lesions occurred at the highest dose levels. Survival was markedly decreased at dose levels of 50 ppm and greater.

Additional data are fairly supportive. Effect and no-effect levels are similar (to those found for rats) for liver effects in dogs after 15 months' exposure to aldrin in the diet. Liver effects were observed at slightly higher doses in several other subchronic-to-chronic rat and dog studies. Short-term exposure to higher doses resulted in mortality for a number of species.

I.A.3. UNCERTAINTY AND MODIFYING FACTORS (ORAL RfD)

UF -- The composite UF of 1000 encompasses the uncertainty of extrapolation from animals to humans, the uncertainty in the range of human sensitivities, and an additional uncertainty because the RfD is based on a LOAEL rather than a NOAEL.

MF -- None

I.A.4. ADDITIONAL COMMENTS (ORAL RfD)

None.

I.A.5. CONFIDENCE IN THE ORAL RfD

Study -- MediumData Base -- MediumRfD -- Medium

The principal study, designed as a carcinogenesis bioassay, is strong in histopathologic analysis but lacks other toxicologic parameters, and is therefore rated medium. The data base is fairly extensive, and generally supportive, but is rated medium because of the lack of NOELs for some studies. Also, no chronic data exist for the dog, which may be a more sensitive species than the rat. Medium confidence in the RfD follows.

I.A.6. EPA DOCUMENTATION AND REVIEW OF THE ORAL RfD

U.S. EPA. 1982. Toxicity-Based Protective Ambient Water Levels for Various Carcinogens. Environmental Criteria and Assessment Office, Cincinnati, OH. ECAO-CIN-431. Internal review draft.

The RfD has been reviewed internally by ECAO-Cin.

Agency Work Group Review -- 12/18/85

Verification Date -- 12/18/85

I.A.7. EPA CONTACTS (ORAL RfD)

Michael L. Dourson / OHEA -- (513)569-7533

I.B. REFERENCE CONCENTRATION FOR CHRONIC INHALATION EXPOSURE (RfC)

> [Ed. Note: EPA does not have information yet in this section of IRIS].

II. CARCINOGENICITY ASSESSMENT FOR LIFETIME EXPOSURE

Section II provides information on three aspects of the carcinogenic risk assessment for the agent in question; the U.S. EPA classification, and quantitative estimates of risk from oral exposure and from inhalation exposure. The classification reflects a weight-of-evidence judgment of the likelihood that the agent is a human carcinogen. The quantitative risk estimates are presented in three ways. The slope factor is the result of application of a low-dose extrapolation procedure and is presented as the risk per (mg/kg)/day. The unit risk is the quantitative estimate in terms of either risk per ug/L drinking water or risk per ug/cu.m air breathed. The third form in which risk is presented is a drinking water or air concentration providing cancer risks of 1 in 10,000, 1 in 100,000 or 1 in 1,000,000. Background Document 2 (Service Code 5) provides details on the rationale and methods used to derive the carcinogenicity values found in IRIS. Users are referred to Section I for information on long-term toxic effects other than carcinogenicity.

II.A. EVIDENCE FOR CLASSIFICATION AS TO HUMAN CARCINOGENICITY

II.A.1. WEIGHT-OF-EVIDENCE CLASSIFICATION

Classification -- B2; probable human carcinogen

Basis -- Orally administered aldrin produced significant increases in tumor responses in three different strains of mice in both males and females. Tumor induction has been observed for structurally related chemicals, including dieldrin, a metabolite.

II.A.2. HUMAN CARCINOGENICITY DATA

Inadequate. Two studies of workers exposed to aldrin and dieldrin (a metabolite of aldrin) did not find these workers to have an excess risk of cancer. Both studies, however, were limited in their ability to detect an excess of deaths from cancer. Van Raalte (1977) observed two cases of cancer (gastric and lymphosarcoma) among 166 pesticide manufacturing workers exposed 4 to 19 years and followed from 15 to 20 years. Exposure was not quantified, and workers were also exposed to other organochlorine pesticides (endrin and telodrin). A small number of workers was studied, the mean age of the cohort (47.7 years) was low, the number of expected deaths was not calculated, and the duration of exposure and of latency was relatively short.

In a retrospective mortality study, Ditraglia et al. (1981) reported no increased incidence of deaths from cancer among 1155 organochlorine pesticide manufacturing workers (31 observed vs. 37.8 expected, SMR=82). This result was not statistically significant. Workers were employed for 6 or more months and followed for 13 or more years (24,939 person-years). Workers with no exposure (for example, office workers) were included in the cohort. Vital status was not known for 112 (10%) of the workers, and these workers were assumed to be alive; therefore, additional deaths may have occurred but were not observed. Exposure was not quantified and workers were also exposed to other chemicals and pesticides (including endrin). An increased incidence of deaths from cancer was seen at several specific sites: esophagus (2 deaths observed, SMR=235), rectum (3, SMR=242); liver (2, SMR=225), and lymphatic and hematopoietic system (6, SMR=147); but these site-specific incidences were not statistically significant.

II.A.3. ANIMAL CARCINOGENICITY DATA

Sufficient. Davis and Fitzhugh (1962) fed a group of 215 male and female C3HeB/Fe mice a dietary mixture containing 10 ppm aldrin for up to 2 years. The control group consisted of 217 mice. The aldrin-treated mice died 2 months earlier than controls. Intercurrent disease, pneumonia, and intestinal parasitism may have influenced the long-term survival rate. A statistically significant increase of hepatomas was reported in the treated animals as compared with controls. An independent reevaluation of the liver lesions showed most of the hepatomas to be liver carcinomas (Epstein, 1975). In a follow-up study, Davis (1965) administered aldrin at 0 or 10 ppm in the diet to 100 male and 100 female C3H mice for 2 years. The incidence of hepatic hyperplasia and benign hepatomas in the aldrin group was approximately double that of controls, whereas the number of hepatic carcinomas was about the same. Neither study provided a detailed pathologic examination or data separated by sex.

Aldrin (95% pure) was administered in the diet to 50 male and 50 female B6C3FI mice at TWA doses of 4 and 8 ppm or 3 and 6 ppm. Treatment was for 80 weeks, and animals were observed for an additional 10 to 13 weeks (NCI, 1978). In male mice, there was a significant dose-related increase in hepatocellular carcinomas when compared with matched or pooled controls.

Treon and Cleveland (1955) administered aldrin in the diet to 40 Carworth rats/sex at concentrations of 2.5, 12.5, or 25 ppm for a period of 2 years. Forty animals/sex served as controls. Mortality of the treated rats was greater than controls, with 50% surviving in the 2.5 and 12.5 ppm groups and 40% surviving in the 25 ppm group at the end of the experiment. Cleveland (1966) reported that no apparent treatment-related tumors were present in the above study. Deichmann et al. (1970) fed 50 male and 50 female Osborne-Mendel rats aldrin (95% pure) at final concentrations of 20, 30, or 50 ppm for 31 months. Controls consisted of 100 rats/sex. There was no evidence of carcinogenic response in male or female rats fed aldrin. The NCI (1978) fed 50 Osborne-Mendel rats/sex aldrin (95% pure) at 30 or 60 ppm. Male rats were treated 111 to 113 weeks and followed for 37 to 38 weeks of observation, and female rats were treated for 80 weeks and followed for 32 to 33 weeks of observation. Aldrin produced no significant effect on the mortality of the rats of either sex. The tumors observed were randomly distributed, with no apparent relationship to aldrin treatment. Four additional bioassays observed no carcinogenic effect of aldrin in rats, but were considered inadequate for carcinogenicity assessment.

II.A.4. SUPPORTING DATA FOR CARCINOGENICITY

Aldrin causes chromosomal aberrations in mouse, rat, and human cells (Georgian, 1974) and unscheduled DNA synthesis in rats (Probst et al., 1981) and humans (Rocchi et al., 1980) cells. Aldrin does not cause reverse mutations in S. typhimurium, E. coli, or S. marcesans, or mitotic gene conversion in S. cerevisiae (Fahrig, 1974).

Five compounds structurally related to aldrin--dieldrin, chlordane, heptachlor, heptachlor epoxide, and chlorendic acid--have induced malignant liver tumors in mice. Chlorendic acid has also induced liver tumors in rats.

II.B. QUANTITATIVE ESTIMATE OF CARCINOGENIC RISK FROM ORAL EXPOSURE

II.B.1. SUMMARY OF RISK ESTIMATES

Oral Slope Factor -- 1.7E+1 per (mg/kg)/day

Drinking Water Unit Risk -- 4.9E-4 per (ug/L)

Extrapolation Method -- Linearized multistage procedure, extra risk

Drinking Water Concentrations at Specified Risk Levels:

Risk Level	Concentration
E-4 (1 in 10,000)	2E-1 ug/L
E-5 (1 in 100,000)	2E-2 ug/L
E-6 (1 in 1,000,000)	2E-3 ug/L

II.B.2. DOSE-RESPONSE DATA (CARCINOGENICITY, ORAL EXPOSURE)

Tumor Type -- liver carcinoma
Test Animals -- mouse/C3H (Davis); mouse/B6C3F1, male (NCI)
Route -- diet
Reference -- Davis, 1965 (see table); NCI, 1978

Administered Dose (ppm)	Human Equivalent Dose	(mg/kg-day)	Tumor Incidence	Reference
females	0	0	2/53	Davis, 1965
	10	0.104	72/85	reevaluated males
				by Reuber
	0	0	22/73	(cited in
	10	0.104	75/91	Epstein, 1975)
	0	0	3/20	NCI, 1978
	4	0.04	16/49	
	8	0.08	25/45	

II.B.3. ADDITIONAL COMMENTS (CARCINOGENICITY, ORAL EXPOSURE)

Body weights for mice were assumed to be 0.03 kg for purposes of dose conversion. The above data sets were used for calculation of the following slope factors: 2.3E+1 per (mg/kg)/day for female C3H mice, 1.8E+1 per (mg/kg)/day for male C3H mice, and 1.2E+1 per (mg/kg)/day for male B6C3F1 mice. No strain or sex specificity was noted in the studies, since aldrin treatment induced liver tumors in all mouse strains tested. A geometric mean of 1.7E+1 per (mg/kg)/day was thus chosen for the quantitative estimate, since all three slope factors were very similar.

The unit risk should not be used if the water concentration exceeds 20 ug/L, since above this concentration the unit risk may not be appropriate.

II.B.4. DISCUSSION OF CONFIDENCE (CARCINOGENICITY, ORAL EXPOSURE)

Adequate numbers of animals were treated for a large proportion of their lifetime. The route of treatment was appropriate. Slope factors calculated from three data sets from two independent assays were within a factor of 2. A slope factor for dieldrin, a major metabolite of aldrin, was determined to be 1.6E+1, essentially identical to that of aldrin.

II.C. QUANTITATIVE ESTIMATE OF CARCINOGENIC RISK FROM INHALATION EXPOSURE

II.C.1. SUMMARY OF RISK ESTIMATES

Inhalation Unit Risk -- 4.9E-3 per (ug/cu.m)

Extrapolation Method -- Linearized multistage procedure, extra risk

Air Concentrations at Specified Risk Levels:

Risk Level	Concentration
E-4 (1 in 10,000)	2E-2 ug/cu.m
E-5 (1 in 100,000)	2E-3 ug/cu.m
E-6 (1 in 1,000,000)	2E-4 ug/cu.m

II.C.2. DOSE-RESPONSE DATA FOR CARCINOGENICITY, INHALATION
EXPOSURE

The unit risk was calculated from the oral data presented in II.B.2.

II.C.3. ADDITIONAL COMMENTS (CARCINOGENICITY, INHALATION
EXPOSURE)

The unit risk should not be used if the air concentration exceeds 2 ug/cu.m, since above
this concentration the unit risk may not be appropriate.

II.C.4. DISCUSSION OF CONFIDENCE (CARCINOGENICITY, INHALATION
EXPOSURE)

See II.B.4.

II.D. EPA DOCUMENTATION, REVIEW, AND CONTACTS (CARCINOGENICITY
ASSESSMENT)

II.D.1. EPA DOCUMENTATION

Source Document -- U.S. EPA, 1986

The values in the 1986 Carcinogenicity Assessment for Aldrin/Dieldrin have been
reviewed by the Carcinogen Assessment Group.

II.D.2. REVIEW (CARCINOGENICITY ASSESSMENT)

Agency Work Group Review -- 03/22/87

Verification Date -- 03/22/87

II.D.3. U.S. EPA CONTACTS (CARCINOGENICITY ASSESSMENT)

Dharm V. Singh / NCEA -- (202)260-5889

Jim Cogliano / NCEA -- (202)260-3814

III. HEALTH HAZARD ASSESSMENTS FOR VARIED EXPOSURE DURATIONS

[Ed. Note: EPA does not have information yet in this section of IRIS].

IV. U.S. EPA REGULATORY ACTIONS

EPA risk assessments may be updated as new data are published and as assessment
methodologies evolve. Regulatory actions are frequently not updated at the same time.
Compare the dates for the regulatory actions in this section with the verification dates for
the risk assessments in sections I and II, as this may explain inconsistencies. Also note

that some regulatory actions consider factors not related to health risk, such as technical or economic feasibility. Such considerations are indicated for each action. In addition, not all of the regulatory actions listed in this section involve enforceable federal standards. Please direct any questions you may have concerning these regulatory actions to the U.S. EPA contact listed for that particular action. Users are strongly urged to read the background information on each regulatory action in Background Document 4 in Service Code 5.

IV.A. CLEAN AIR ACT (CAA)

[Ed. Note: EPA does not have information yet in this section of IRIS].

IV.B. SAFE DRINKING WATER ACT (SDWA)

IV.B.1. MAXIMUM CONTAMINANT LEVEL GOAL (MCLG) for Drinking Water

[Ed. Note: EPA does not have information yet in this section of IRIS].

IV.B.2. MAXIMUM CONTAMINANT LEVEL (MCL) for Drinking Water

[Ed. Note: EPA does not have information yet in this section of IRIS].

IV.B.3. SECONDARY MAXIMUM CONTAMINANT LEVEL (SMCL) for Drinking Water

[Ed. Note: EPA does not have information yet in this section of IRIS].

IV.B.4. REQUIRED MONITORING OF "UNREGULATED" CONTAMINANTS

Status -- Listed (Final, 1991)

Discussion -- "Unregulated" contaminants are those contaminants for which EPA establishes a monitoring requirement but which do not have an associated final MCLG, MCL, or treatment technique. EPA may regulate these contaminants in the future.

Monitoring requirement -- All systems to be monitored unless a vulnerability assessment determines the system is not vulnerable.

Analytical methodology -- Microextraction/gas chromatography (EPA 505);electron-capture/gas chromatography (EPA 508); gas chromatographic/mass spectrometry (EPA 525).

Reference -- 56 FR 3526 (01/30/91)

EPA Contact -- Drinking Water Standards Division / OGWDW /(202) 260-7575 / FTS 260-7575; or Safe Drinking Water Hotline / (800) 426-4791

IV.C. CLEAN WATER ACT (CWA)

IV.C.1. AMBIENT WATER QUALITY CRITERIA, Human Health

Water and Fish Consumption -- 7.4E-5 ug/L

Fish Consumption Only -- 7.9E-5 ug/L

Considers technological or economic feasibility? -- NO

Discussion -- For the maximum protection from the potential carcinogenic properties of this chemical, the ambient water concentration should be zero. However, zero may not be attainable at this time, so the recommended criteria represents a E-6 estimated incremental increase of cancer risk over a lifetime.

Reference -- 45 FR 79318 (11/28/80)

EPA Contact -- Criteria and Standards Division / OWRS(202)260-1315 / FTS 260-1315

IV.C.2. AMBIENT WATER QUALITY CRITERIA, Aquatic Organisms

Freshwater:

   Acute -- 3.0E+0 ug/L   Chronic -- None

Marine:

   Acute -- 1.3E+0 ug/L   Chronic -- None

Considers technological or economic feasibility? -- NO

Discussion -- Criteria were derived from a minimum data base consisting of acute tests on a variety of species. Requirements and methods are covered in the reference to the Federal Register.

Reference -- 45 FR 79318 (11/28/80)

EPA Contact -- Criteria and Standards Division / OWRS(202)260-1315 / FTS 260-1315

IV.D. FEDERAL INSECTICIDE, FUNGICIDE, AND RODENTICIDE ACT (FIFRA)

IV.D.1. PESTICIDE ACTIVE INGREDIENT, Registration Standard

Status -- Issued (1986)

Reference -- Aldrin Pesticide Registration Standard. December, 1986(NTIS No. PB-87-183778).

EPA Contact -- Registration Branch / OPP (703)557-7760 / FTS 557-7760

IV.D.2. PESTICIDE ACTIVE INGREDIENT, Special Review

Action -- Cancellations issued prior to RPAR/special review process (1974)

Considers technological or economic feasibility? -- YES

Summary of regulatory action -- All uses canceled except those in the following list: 1) subsurface ground insertion for termite control, 2) dipping of nonfood roots and tops, 3) moth-proofing by manufacturing processes in a closed system. Accelerated Decision of the Chief Administrative Law Judge (5/27/75) and the order Declining Review of the Accelerated Decision of the Administrative Law Judge issued by the Chief Judicial Officer (6/30/75);criterion of concern: carcinogenicity, bioaccumulation, wildlife hazard and other chronic effects.

Reference -- 39 FR 37246 (10/18/74)

EPA Contact -- Special Review Branch / OPP (703)557-7400 / FTS 557-7400

IV.E. TOXIC SUBSTANCES CONTROL ACT (TSCA)

[Ed. Note: EPA does not have information yet in this section of IRIS].

IV.F. RESOURCE CONSERVATION AND RECOVERY ACT (RCRA)

IV.F.1. RCRA APPENDIX IX, for Ground Water Monitoring

Status -- Listed

Reference -- 52 FR 25942 (07/09/87)

EPA Contact -- RCRA/Superfund Hotline(800)424-9346 / (202)260-3000 / FTS 260-3000

IV.G. SUPERFUND (CERCLA)

IV.G.1. REPORTABLE QUANTITY (RQ) for Release into the Environment

Value (status) -- 1 pound (Final, 1989)

Considers technological or economic feasibility? -- NO

Discussion -- The RQ for aldrin is 1 pound, based on its aquatic toxicity and its potential carcinogenicity. The available data, as established under the CWA Section 311 (40 CFR 117.3), indicate the aquatic 96-hour Median Threshold Limit for aldrin is less than 0.1 ppm. This corresponds to an RQ of 1 pound. In addition, aldrin has been identified as a potential carcinogen and assigned a hazard ranking of high, based on a potency factor of 180.00/mg/kg/day and weight-of-evidence group B2, which also corresponds to an RQ of 1 pound.

Reference -- 54 FR 33418 (08/14/89)

EPA Contact -- RCRA/Superfund Hotline (800)424-9346 / (202)260-3000 / FTS 260-3000

VI. BIBLIOGRAPHY

Substance Name – Aldrin CASRN -- 309-00-2Last Revised -- 09/01/89

VI.A. ORAL RfD REFERENCES

Fitzhugh, O.G., A.A. Nelson, and M.L. Quaife. 1964. Chronic oral toxicity of aldrin and dieldrin in rats and dogs. Food Cosmet. Toxicol. 2: 551-562.

U.S. EPA. 1982. Toxicity-Based Protective Ambient Water Levels for Various Carcinogens. Environmental Criteria and Assessment Office, Cincinnati, OH. ECAO-CIN-431. Internal review draft.

VI.B. INHALATION RfD REFERENCES

None

VI.C. CARCINOGENICITY ASSESSMENT REFERENCES

Cleveland, F.P. 1966. A summary of work on aldrin and dieldrin toxicity at the Kettering Laboratory. Arch. Environ. Health. 13: 195.

Davis, K.J. 1965. Pathology report on mice fed dieldrin, aldrin, heptachlor, or heptachlor epoxide for two years. Internal FDA memorandum to Dr. A.J. Lehrman, July 19.

Davis, K.J. and O.G. Fitzhugh. 1962. Tumorigenic potential of aldrin and dieldrin for mice. Toxicol. Appl. Pharmacol. 4: 187-189.

Deichmann, W.B., W.E. McDonald, E. Blum, et al. 1970. Tumorigenicity of aldrin, dieldrin and endrin in the albino rat. Ind. Med. 39(10): 426-434.

Ditraglia, D., D.P. Brown, T. Namekata and N. Iverson. 1981. Mortality study of workers employed at organochlorine pesticide manufacturing plants. Scand. J. Environ. Health. 7(suppl 4): 140-146.

Epstein, S.S. 1975. The carcinogenicity of dieldrin. Part 1. Sci. Total Environ. 4: 1-52.

Fahrig, R. 1974. Comparative mutagenicity with pesticides. IARC Publ. (U.N.) 10: 161-181.

Georgian, L. 1975. The comparative cytogenic effects of aldrin and phosphamidon. Mutat. Res. 31: 103-108. NCI (National Cancer Institute). 1978. Bioassays of aldrin and dieldrin for possible carcinogenicity. DHEW Publication No. (NIH) 78-821. NCI Carcinogenesis Tech. Rep. Ser. No. 21. NCI-C6-TR-21.

Probst, G.S., R.E. McMahon, L.W. Hill, D.Z. Thompson, J.K. Epp and S.B. Neal. 1981. Chemically-induced unscheduled DNA synthesis in primary rat hepatocyte cultures: A comparison with bacterial mutagenicity using 218 chemicals. Environ. Mutagen. 3: 11-32.

Rocchi, P., P. Perocco, W. Alberghini, A. Fini and G. Prodi. 1980. Effect of pesticides on scheduled and unscheduled DNA synthesis of rat thymocytes and human lymphocytes. Arch. Toxicol. 45: 101-108.

Treon, J.F. and F.P. Cleveland. 1955. Toxicity of certain chlorinated hydrocarbon insecticides for laboratory animals, with special reference to aldrin and dieldrin. Agric. Food Chem. 3: 402-408.

U.S. EPA. 1986. Carcinogenicity Assessment of Aldrin and Dieldrin. Prepared by the Office of Health and Environmental Assessment, Carcinogen Assessment Group, Washington, DC, for the Hazard Evaluation Division, Office of Pesticides and Toxic Substances, Office of Pesticide Programs, Washington, DC.

Van Raalte, H.G.S. 1977. Human experience with dieldrin in perspective. Ecotoxicol. Environ. Safety. 1: 203-210.

VI.D. DRINKING WATER HA REFERENCES

None

VII. REVISION HISTORY

Date	Section	Description
09/30/87	II.	Carcinogenicity section added
03/01/88	II.B.4.	Confidence statement revised
12/01/88	II.B.4.	Corrected slope factor in text
09/01/89	II.A.2.	Ditraglia reference changed to Ditraglia et al.
09/01/89	II.A.3.	Deichmann reference changed to Deichmann et al.
09/01/89	II.B.3.	Body weight for mice corrected to kg
09/01/89	VI.	Bibliography on-line
01/01/91	II.	Text edited
01/01/91	II.C.1.	Inhalation slope factor removed (global change)
01/01/92	I.A.7.	Secondary contact changed
01/01/92	IV.	Regulatory actions updated
07/01/93	II.D.3.	Secondary contact's phone number changed

SYNONYMS

Aldrex
Aldrin
Aldrite
Aldrosol
1,4:5,8-Dimethanonaphthalene
1,2,3,4,10,10-Hexachloro-1,4,4a,5,8,8a-
  Hexahydro-, (1 alpha, 4 alpha, 4a beta, 5
alpha, 8 alpha, 8a beta)-
1,4:5,8-Dimethanonaphthalene,
1,2,3,4,10,10-Hexachloro-1,4,4a,5,8,8a-
Hexahydro-
Drinox
ENT 15,949
1,2,3,4,10,10-Hexachloro-1,4,4a,5,8,8a-
Hexahydro-1,4,5,8-Dimethanonaphthalene

1,2,3,4,10,10-Hexachloro-1,4,4a,5,8,8a-
Hexahydro-1,4-endo-exo-5,8-
Dimethanonaphthalene
1,2,3,4,10,10-Hexachloro-1,4,4a,5,8,8a-
Hexahydro-exo-1,4-endo-5,8-
Dimethanonaphthalene
Hexachlorohexahydro-endo-exo-
Dimethanonaphthalene
HHDN
NCI-C00044
Octalene
Seedrin

Source: *Instant EPA's IRIS*, 1997, Instant Reference Sources, Inc., 7605 Rockpoint
Drive, Austin, TX 78731 (Fax: 512-345-2386; Internet URL,
http://www.instantref.com/inst-ref.htm).

Production, Use, and Pesticide Labeling Information

1. Description of Chemical

Generic Name:1,2,3,4,10,10-hexachloro-1,4,4a,5,8,8a-hexahydro-exo-1,4-endo-
5,8-dimethanonaphthalene

Common Name: Aldrin

Trade Names: Aldrine, HHDN, Aldres™, Aldrex 30™, Aldrite™, Aldrosol™,
Altox, Bangald™, Drinox™, Octalene™, Rasayaldrin™, Seedrin™, Liquid,
Entoma 15949, and Compound 118.

EPA Shaughnessy Code: 045101

Chemical Abstracts Service (CAS) Number: 309-00-2

Year of Initial Registration: 1949

Pesticide Type: Insecticide

Chemical Family: Chlorinated cyclodiene

U.S. and Foreign Producers: Shell International Corp.

2. Use Patterns and Formulations

　Application Sites: soil surrounding wooden structures for termite control

　Types of Formulations: 2 and 4 lb/gal emulsifiable concentrates

　Types and Methods of Application: trenching, rodding, subslab injection, low pressure spray

　Application Rates: 0.25 to 0.5% emulsion

3. Science Findings

　Chemical Characteristics

　　- Physical state: crystalline solid

　　- Color: tan to dark brown

　　- Odor: mild chemical odor

　　- Molecular weight and formula: 364.93 - $C_{12}H_8C_{16}$

　　- Melting point: 104 to 104.5 °C

　　- Boiling Point: Decomposes at 1 atm.

　　- Vapor pressure: 6.6×10^{-6} mm Hg at 25 °C

　　- Solubility in various solvents: Very soluble in most organic solvents; practically insoluble in water.

　　- Stability: Stable with alkali and alkaline-oxidizing agents; not stable with phenols, concentrated mineral acids; acid catalysts, acid-oxidizing agents, or active metals.

　Toxicology characteristics

　　- Acute oral: data gap
　　- Acute dermal: data gap
　　- Primary dermal irritation: data gap
　　- Skin sensitization: not a sensitizer
　　- Acute inhalation: data gap
　　- Major routes of exposure: Inhalation exposure to occupants of treated structures; dermal and respiratory exposure to termiticide applicators.
　　- Delayed neurotoxicity: does not cause delayed neurotoxic effects.
　　- Oncogenicity:
　　- This chemical is classified as a Group B2 (probable human oncogen).

Rat oncogenicity study is a data gap. There are three long-term carcinogenesis bioassays of aldrin in mice which were independently conducted by investigators affiliated with the National Cancer Institute and the Food and Drug Administration. These studies were found to produce significant tumor responses in three different strains of mice (C3H, CF1, and B6C3F1) in males and females with a dose-related increase in the proportion of tumors that were malignant.

- Available data from seven existing carcinogenesis bioassays in rats are inadequate and inconclusive and a well-designed study in rats is needed to determine the carcinogenic potential of aldrin in this species.

- Chronic Feeding: Based on a rat chronic feeding study, a Lowest Effect Level (LEL) of 0.025 mg/kg/day has been calculated.

- Metabolism: In biological systems, aldrin is readily epimerized to dieldrin.

- Teratogenicity: data gap

- Reproduction: data gap

- Mutagenicity: Aldrin does not possess mutagenic activity in bacteria. Further testing is required to assess the mutagenic potential of aldrin in eukaryotes.

Physiological and Biochemical Characteristics

- The precise mode of action in biological systems is not known. In humans, signs of acute intoxication are primarily related to the central nervous system (CNS), including hyperexcitability, convulsions, depression and death.

Environmental characteristics

- Available data are insufficient to fully assess the environmental fate of aldrin. Data gaps exist for all applicable studies. However, available supplementary data indicate general trends of aldrin behavior in the environment. Aldrin degrades readily to dieldrin, which is persistent in the environment. Reports on leaching and field studies suggest that aldrin/dieldrin would be unlikely to leach to underground aquifers. However, additional data are necessary to assess the potential for ground-water contamination as a result of the termiticide use of aldrin.

Ecological characteristics

- Avian oral toxicity: 6.59 mg/kg in bobwhite quail; 52 mg/kg in mallard ducks.

- Avian dietary toxicity: 34 and 155 ppm in Japanese quail and mallard duck, respectively.

- Freshwater fish acute toxicity: 5 and 53 ppb in largemouth bass and channel catfish, respectively (warm water species), 2.6 and 8.2 ppb in rainbow trout and chinook salmon, respectively (cold water species).

- Freshwater invertebrate toxicity: 18 ppb in a species of seed shrimp; 32 ppb in a species of water flea.

Tolerance Reassessment

- No tolerance reassessment for Aldrin is necessary since there are no food or feed uses.

Summary Science Statement

- Aldrin is a chlorinated cyclodiene with high acute toxicity. The chemical has demonstrated adverse chronic effects in mice (causing liver tumors). Aldrin may pose a significant health risk of chronic liver effects to occupants of structures treated with aldrin. The Agency is continuing to evaluate the potential risk from the termiticide use of aldrin to determine whether further regulatory action may be warranted. Aldrin is extremely toxic to aquatic organisms and birds. Aldrin is persistent and bioaccumulates. Aldrin may have a potential for contaminating surface water; thus, a special study is required to delineate this potential. Applicator exposure studies are required to determine whether exposure to applicators may be posing health risks. Special subacute inhalation testing is required to evaluate the respiratory hazards to humans in structures treated with aldrin. Data available to the Agency show a pattern of misuse and misapplication of aldrin. The Agency is requiring restricted use classification of all end-use products containing aldrin. Application must be made either in the actual physical presence of the Certified Applicator, or if the Certified Applicator is not physically present at the site, each uncertified applicator must have completed a State approved training course and be registered in the State in which the uncertified applicator is working.

4. Summary of Regulatory Position and Rationale

- EPA is currently evaluating the potential human health risks of 1) non-oncogenic chronic liver effects, and 2) oncogenic effects from exposure to aldrin. Following the completion of Registration Standards on alternative chlorinated cyclodiene termiticides (chlordane and heptachlor) EPA will determine whether the risks posed by the termiticide use of aldrin warrant further regulatory action.

- In order to meet the statutory standard for continued registration, retail sale and use of all end-use products containing Aldrin must be restricted to Certified Applicators or persons under their direct supervision. For purposes of Aldrin use, direct supervision by a Certified Applicator means 1) the actual physical presence of a Certified Applicator at the application site during application, or 2) if the Certified Applicator is not physically present at the site, each uncertified applicator must have completed a State approved training course and be registered in the State in which the uncertified applicator is working; the Certified Applicator must be available if and when needed.

- In order to meet the statutory standard for continued registration, Aldrin product labels must be revised to provide specified Aldrin disposal procedures, and to provide fish and wildlife toxicity warnings.

- The Agency is requiring a special monitoring study to evaluate whether and to what extent surface water contamination may be resulting from the use of Aldrin as a termiticide.

- Special product-specific subacute inhalation testing is required to evaluate the respiratory hazards to humans in structures treated with termiticide products containing Aldrin.

- The Agency is requiring the submission of applicator exposure data from dermal and respiratory routes of exposure.

- While data gaps are being filled, currently registered manufacturing use products and end use products containing Aldrin may be sold, distributed, formulated, and used, subject to the terms and conditions specified in the Registration Standard for Aldrin. Registrants must provide or agree to develop additional data in order to maintain existing registrations.

5. Summary of Major Data Gaps

- Hydrolysis
- Photodegradation in water
- Aerobic soil metabolism
- Anaerobic soil metabolism
- Leaching and adsorption/desorption
- Aerobic aquatic metabolism
- Soil dissipation
- Reproductive effects in rats
- Rat oncogenicity study
- Mutagenicity studies
- Teratology studies
- Battery of acute toxicity studies
- Special surface water monitoring studies

- Applicator exposure studies
- Special guinea pig inhalation study
- All product chemistry studies

6. Contact Person at EPA

George LaRocca
Product Manager No. 15 Insecticide- Rodenticide Branch
Registration Division (TS-767C), Office of Pesticide Programs
Environmental Protection Agency
401 M Street, SW
Washington, DC 20460

- Office location and telephone number:

- Room 204, Crystal Mall #2
- 1921 Jefferson Davis Highway
- Arlington, VA 22202

Telephone: (703) 557-2400

Source: *Instant EPA's Pesticide Facts*, 1994, Instant Reference Sources, Inc., 7605 Rockpoint Drive, Austin, TX 78731 (Fax: 512-345-2386; Internet URL, http://www.instantref.com/inst-ref.htm).

Allethrin

Introduction

Allethrin is used almost exclusively in homes and gardens for control of flies and mosquitoes, and in combination with other pesticides to control flying or crawling insects. The purified d-trans-isomer of allethrin is more toxic to insects and is used for control of crawling insects in homes and restaurants. Allethrin is a synthetic duplicate of a component of pyrethrum, a botanical insecticide extracted from chrysanthemum flowers. Allethrin, the first synthetic pyrethroid, was introduced in 1949, and is a mixture of several isomeric forms. The most common form is a 4:1 mixture of the trans- and cis-isomers. It is available in aerosol, coil, mat, dust and oil formulations. Aerosol and spray formulations of the purified d-trans-isomer of allethrin are also available. D-trans allethrin is usually combined with synergists such as piperonyl-butoxide [829], [870].

Identifying Information

CAS NUMBER: 584-79-2

SYNONYMS: (2-METHYL-1-PROPENYL-2-METHYL-4-OXO-3-(2-PROPENYL)-2-CYCLOPENTEN-1-YL ESTER OF CIS AND TRANS DL-2-ALLYL-4-HYDROXY-3-METHYL-2-CYCLOPENTEN-1-ONE; PYNAMIN; BIOALLETHRIN; CINERIN I, ALLYL HOMOLOG

Endocrine Disruptor Information

Allethrin is on EPA's list of suspect EEDs [811] but little information has been located to determine the Agency's reason for its inclusion. More information should be forthcoming after EPA's studies are concluded.

Chemical and Physical Properties

CHEMICAL FORMULA: $C_{19}H_{26}O_3$

MOLECULAR WEIGHT: 302.41

PHYSICAL DESCRIPTION: Clear, amber-colored viscous liquid [829]

SPECIFIC GRAVITY: 1.01 [828]

DENSITY: Not available

MELTING POINT: Not available

BOILING POINT: 160 °C [828], [829]

SOLUBILITY: Insoluble in water [829], [870]

OTHER SOLVENTS: Miscible with most organic solvents [829], [870] and with petroleum oils and it is soluble in paraffinic and aromatic hydrocarbons [829], [871].

OTHER PHYSICAL DATA: Not available

Environmental Reference Materials

A source of EED environmental reference materials for allethrin is Crescent Chemical Co., Hauppauge, NY (Fax: 516-348-0913); Catalog No. 45317.

Hazardous Properties

ACUTE/CHRONIC HAZARDS: Allethrin is slightly to moderately toxic by dermal absorption and ingestion . Short-term dermal exposure to allethrin may cause itching, burning, tingling, numbness, a feeling of warmth, with no dermatitis. Persons sensitive to ragweed pollen are at increased risk from exposure to allethrin [829], [872]. The toxicity of allethrin varies with the amounts of different isomers present. The oral LD50 for allethrin in male rats is 1,100 mg/kg, in female rats is 685 mg/kg [829], [870], 370 mg/kg in mice, and 4,290 mg/kg in rabbits [829], [872]. The oral LD50 for d-trans allethrin in rats is 860 mg/kg [829], [870]. The dermal LD50 in rabbits is 11,332 mg/kg [829], [872]. A dosage of 50 mg/kg/day for 2 years produced no detectable effect in the dog [829], [872].

Medical Symptoms of Exposure

Allethrin is a central nervous system stimulant. Exposure to large doses by any route may lead to nausea, vomiting, diarrhea, hyperexcitability, incoordination, tremors, convulsive twitching, convulsions, bloody tears, incontinence, muscular paralysis, prostration and coma. Heavy respiratory exposure caused incoordination and urinary incontinence in mice and rats [829], [872].

Toxicological Information

The toxicological information below is gathered from several sources including two national databases: one from the *National Toxicology Program Chemical Repository Database* and another from *EPA's Integrated Risk Information System (IRIS).*

Toxicology Information from the National Toxicology Program

Allethrin is not currently listed in NTP's Chemical Repository database nor is it scheduled for addition in the future.

Toxicology Information from EPA's Integrated Risk Information System (IRIS)

Allethrin is not currently listed in EPA's IRIS database nor is it listed as scheduled for addition in the near future.

Production, Use, and Pesticide Labeling Information

1. Description of Chemicals

> - Common Name: Allethrin

> - Generic Name: (2-methyl-1-propenyl)-2-methyl-4-oxo-3-(2-propenyl)-2-cyclopenten-1-yl ester of cis and trans dl-2-allyl-4-hydroxy-3-methyl-2-cyclopenten-1-one

> - Trade Name: Pynamin

> - EPA Shaughnessy Code: 004001 and 004002 (Allethrin coil)

> - Chemical Abstracts Service (CAS) Number: 584-79-2

> - Producers: McLaughlin Gormley King, Sumitomo Chemical Company, Fairfield American

> - Common Name: d-trans Allethrin, Bioallethrin

> - Generic Name: d-trans-chrysanthemum monocarboxylic ester of dl-2-allyl-4-hydroxy-3-methyl-2-cyclopenten-1-one

> - Trade Name: Bioallethrin

> - EPA Shaughnessy Code: 004003

> - Chemical Abstracts Service (CAS) Number: 584-79-2

> - Producers: McLaughlin Gormley King, Roussel Uclaf

- Common Name: S-bioallethrin; Esbiol

    - Generic Name: d-trans-chrysanthemum monocarboxylic acid ester of dl-2-allyl-4-hydroxy-3-methyl-2-cyclopenten-1-one.

    - Trade Name: Esbiol

    - EPA Shaughnessy Code: 004004

    - Chemical Abstracts Service (CAS) Number: 28434-00-6

    - Producers: McLaughlin Gormley King, Roussel Uclaf

- Common Name: d-cis/trans allethrin; Pynamin Forte

    - Generic Name: dl-3-allyl-2-methyl-4-oxo-2-cyclopentenyl d-cis/trans chrysanthemate

    - Trade Name: Pynamin Forte

    - EPA Shaughnessy Code: 004005

    - Chemical Abstracts Service (CAS) Number 42534-61-2

    - Producers: Sumitomo Chemical Co., Ltd

2. Use Patterns and Formulations

- Application Sites: Broad spectrum insecticides and acaricides registered for use on terrestrial food crops (vegetables, citrus fruits, and orchard crops); terrestrial nonfood uses (ornamental plants, turf, recreational areas, and forest trees); greenhouse food and nonfood crops (ornamentals and vegetables); indoor and outdoor domestic dwellings; post-harvest use on fruit, vegetables and grains, and stored food; commercial and industrial uses (food handling establishments).

- Types of Formulations: Pressurized liquids, mosquito coils, dusts, emulsifiable concentrates, soluble concentrate liquids, and ready-to-use liquids. Almost always formulated with a synergist and one or more additional active ingredients.

- Predominant Uses and Methods of Application: Primarily indoor and outdoor use around the home as foggers, plant, carpet and general purpose aerosols, and mosquito coils to control common pests including, but not limited to, ants, bedbugs, carpet beetles, cockroaches, fleas, ticks, moths, wasps and bees. Applied to crops foliarly by aerial or ground equipment. Post-harvest applications made as an emulsive dip.

3. Science Findings

    Chemical Characteristics of the Technical Material

        - Physical State: Viscous oil; liquid, clear oil

        - Color: Pale yellow, yellow-orange, slightly brownish

        - Odor: Mild to slightly aromatic

        - Molecular weight and empirical formula: 302 - $C_{19}H_{26}O_3$

        - Solubility: Insoluble in water; miscible with petroleum oils, and soluble in paraffinic and aromatic hydrocarbons.

    Toxicological Characteristics

        - Acute Toxicity: The acute oral toxicity of bioallethrin and also of s-bioallethrin is low to moderate. Adequate data to discern other acute effects of these compounds are not available.

        - Subchronic Toxicity: In a 90-day feeding study on bioallethrin, rats were administered 0, 500, 1500, 5000, and 10,000 ppm bioallethrin in the diet. A no-observed-effect (NOEL) was established at 1500 ppm based upon a decrease in body weight gain and increased levels of serum liver enzymes in females and increased liver weights in both sexes. This study, however, is presently classified as only supplementary, but may be upgraded upon submission of additional information.

        - Chronic Toxicity: In a 6-month oral feeding study using beagle dogs, the animals were administered 0, 200, 1000, and 5000 ppm bioallethrin in the diet. The NOEL was determined to be 200 ppm based on effects on the liver.

        - One rodent chronic feeding/oncogenicity study is available for d-cis/trans allethrin. In this study, rats were fed 0, 125, 500, and 2000 ppm of the test substance in the diet for 2 years. No oncogenic effects were observed. For systemic toxicity, the NOEL was determined to be 125 ppm based on decreased body weight gain and the presence of liver effects.

        - Teratogenicity: One teratology study conducted with bioallethrin is available. In this study, rats were dosed with 50, 125, and 195 mg/kg/day bioallethrin in the diet. The test compound did not induce developmental effects at the dose levels tested.

        - Mutagenicity: Two mutagenicity studies (DNA damage and reverse mutation) conducted with bioallethrin are negative for genetic damage.

Environmental Characteristics

- No data on the allethrin stereoisomers are available to assess the environmental fate and transport, and the potential exposure of humans and nontarget organisms. The potential of these compounds to contaminate ground water is unknown. Because the allethrins are thought to degrade rapidly in the environment, environmental fate data are being required on a "tiered" basis. This approach will permit the Agency to make a preliminary assessment of the persistence of these compounds. The requirement for additional testing will be deferred until evaluation of all data submitted under Tier I.

Ecological Characteristics

- Avian Acute Oral Toxicity:

Species	Stereoisomer	LD50 or LC50
Mallard Duck	Technical allethrin	>2000 mg/kg
Mallard Duck	D-cis/trans allethrin	5620 ppm
Bobwhite Quail	Bioallethrin	2030 ppm
Bobwhite Quail	D-cis/trans allethrin	5620 ppm

- These data show that the allethrins are practically nontoxic to birds on both an acute and subacute exposure basis.

- Freshwater Fish Acute Toxicity: Twenty-seven toxicity tests conducted with cold water and warm water fish species indicate that the allethrins are highly toxic to fish. The LC50 values ranged from 2.6 ppb (coho salmon -- bioallethrin) to 80 ppb (fathead minnow -- S-bioallethrin).

- Toxicity to Aquatic Invertebrates: Data show that allethrin is highly toxic to aquatic invertebrates with LC50 values of 5.6 ppb for stoneflies and 56 ppb for blackflies.

- Toxicity to Non-Target Insects: Although technical allethrin is moderately toxic to honey bees, the outdoor application rates are so low that even a direct application to bees is not likely to result in significant mortality.

Tolerance Assessment

The available data reviewed are insufficient to evaluate the adequacy of the established tolerances (covering post-harvest use) for residues of allethrin in or on food/feed items (40 CFR 180.113). Allethrin is the only stereoisomer with established tolerances.

- Because of insufficient residue chemistry and toxicity data for all of the allethrin stereoisomers, the Agency is unable to calculate an acceptable daily intake under the Tolerance Assessment System.

- There are no Canadian or Mexican tolerances or Codex Maximum Residue Limits for residues of the allethrins in or on any plant commodity. Therefore, no compatibility questions exist.

Summary Science Statement

- The Agency has very little acceptable toxicity data for the allethrin stereoisomers. There are no data available to assess the environmental fate characteristics of these compounds, including their potential to contaminate ground water. There are ecological effects data which show that the stereoisomers are highly toxic to fish and aquatic invertebrates, and essentially non-toxic to avian species. There are no acceptable residue data available to assess the adequacy of the current tolerances for allethrin.

4. Summary of Regulatory Position and Rationale

- The Agency is not starting a special review the allethrins.

- Since EPA believes that the allethrins may degrade rapidly in the environment, the Agency is requiring environmental fate data on a tiered basis. Additional data may be required upon evaluation of the tier I studies.

- The Agency is permitting registrants to use the technical product Esbiothrin as a representative test material for chronic studies on Bioallethrin and S-bioallethrin since it is a mixture of the two compounds, and they are of similar toxicity. Separate chronic studies are being required for Allethrin and D-cis/trans allethrin.

- The Agency is not requiring any ground water advisory labeling, on reentry, spray drift, or protective clothing restrictions at this time.

- The Agency is not imposing restricted use classification on the allethrins.

- While the required data are under development all currently registered products containing the allethrins may be sold, distributed, formulated and used, provided that they are in compliance with all other terms specified in the Registration Standard.

- REQUIRED UNIQUE LABELING

> - The registration standard for the allethrins contains no unique labeling requirements. It requires only updated environmental precautionary and disposal statements and a statement for outdoor use products that the product is highly toxic to fish.

5. Summary of Major Data Gaps

- Toxicology

- Acute Toxicity

- Acute oral LD50 toxicity (Allethrin, Pynamin-forte)
- Acute dermal LD50 toxicity (Allethrin, Pynamin-forte, S-bioallethrin, Bioallethrin)
- Acute inhalation LC50 Toxicity (all allethrins)
- Eye irritation (all allethrins except Esbiothrin)
- Dermal irritation (all allethrins except Esbiothrin)
- Dermal sensitization (all allethrins)

- Subchronic Toxicity
- 90-day feeding
- Rodent (all allethrins except Pynamin-forte)
- Nonrodent (all allethrins except Bioallethrin)
- 21-day dermal (all allethrins)
- 90-day inhalation (reserved for all allethrins)

- Chronic Toxicity
- Rodent feeding (all allethrins except Pynamin-forte)
- Nonrodent feeding (all allethrins except Bioallethrin)
- Rat oncogenicity (all allethrins except Pynamin-forte)
- Mouse oncogenicity (all allethrins)
- Rat teratogenicity (all allethrins except Bioallethrin)
- Rabbit teratogenicity (all allethrins)

- Reproduction (all allethrins)

- Mutagenicity
- Gene mutation (Allethrin, Pynamin-forte, S-bioallethrin)
- Chromosomal aberration (all allethrins)
- Other mechanisms of mutagenicity (all allethrins except Bioallethrin)

- Special Testing

- Metabolism (all allethrins)

- Ecological Effects

    - Avian reproduction
    - Field testing - mammals and birds (reserved pending reproduction data)
    - Freshwater fish LC50 (typical EP)
    - Freshwater aquatic invertebrate LC50 (typical EP)
    - Acute estuarine and marine LC50 (fish, shrimp, oyster)
    - Fish early life stage and invertebrate life cycle (freshwater, estuarine)
    - Fish life cycle
    - Field testing (aquatic organisms)

- Environmental Fate

    - Tier 1

        - Degradation Studies - Lab
              - Hydrolysis
              - Photodegradation - water, soil, and air

        - Metabolism Studies - Lab
              - Aerobic metabolism (soil and aquatic)
              - Anaerobic metabolism in soil

        - Mobility Studies

              - Leaching/aged leaching
              - Volatility (lab)

    - Tier II

        - Anaerobic aquatic metabolism   - Reserved
        - Volatility (field)
        - Field dissipation (soil)   - Reserved
        - Field dissipation (aquatic, sediment)   - Reserved
        - Field dissipation (soil, long-term)   - Reserved
        - Accumulation studies on rotational crops (confined)-
              Reserved
        - Accumulation studies on rotationalcrops (field)   -
              Reserved
        - Accumulation studies on irrigated crops   - Reserved
        - Accumulation studies in fish   - Reserved
        - Accumulation studies in aquatic nontarget organisms -
              Reserved
        - Reentry   - Reserved
        - Spray drift   - Reserved
        - Exposure   - Reserved

- Product Chemistry

  - Product Identity and Composition
  - Analysis and Certification of Product Ingredients
  - Physical and Chemical Characteristics

- Residue Chemistry

  - Nature of the Residue (Metabolism) in Plant and Livestock
  - Residue Analytical Methods (may be required if additional metabolites of toxicological concern are identified)

  - Stability Data

  - Magnitude of Residue
    - Crop field trials
    - Post-harvest treatment of fruits and vegetables
    - Stored commodities
    - Processing studies
    - Meat/milk/poultry/eggs
    - Food handling

6. Contact Person at EPA

Philip O. Hutton,
Product Manager 17, Registration Division (TS-767C)
Office of Pesticide Programs
Environmental Protection Agency
401 M Street, S.W.
Washington, DC 20460

Telephone: (703) 557-2600

Source: *Instant EPA's Pesticide Facts*, 1994, Instant Reference Sources, Inc., 7605 Rockpoint Drive, Austin, TX 78731 (Fax: 512-345-2386; Internet URL, http://www.instantref.com/inst-ref.htm).

Amitrole

Introduction

Amitrole is used as a defoliant, a herbicide, a reagent in photography and a plant growth regulator. As an herbicide it has a wide spectrum of activity and appears to act by inhibiting the formation of chlorophyll. It is commonly used around orchard trees, on fallow land, along roadsides and railway lines, or for pond weed control. It is also used in non-selective weed control. It is manufactured by the condensation of formic acid with aminoguanidine bicarbonate in an inert solvent at 100-200 °C.

Identifying Information

CAS NUMBER: 61-82-5

NIOSH Registry Number: XZ3850000

SYNONYMS: 3-AMINO-1,2,4-TRIAZOLE TRIAZOLE, 3-AMINO-1,2,4-, AMEROL, AMINOTRIAZOLE, 3-AMINO-S-TRIAZOLE, 2-AMINO-1,3,4-TRIAZOLE, 3-AMINO-1H-1,2,4-TRIAZOLE, AMITOL, AMITRIL, AMITROL, AMIZOL, ENT25445, NCI-C00373, AZOLE, CYTROLE, AMITROLE, DIUROL 5030, DOMATOL, EMISOL, TRIAZOLAMINE, VOROX, USAF XR-22, RAMIZOL, CYTROL, 3-AMINOTRIAZOLE, 3,A-T, ATA, AZAPLANT, FENAMINE, WEEDAR ADS, WEEDAZIN, WEEDOCLOR, 2-AMINOTRIAZOLE, AMINO TRIAZOLE WEEDKILLER 90, AMINOTRIZOL-SPRITZPULVER, AT, AZAPLANT KOMBI, AZOLAN, CAMPAPRIM A 1544, ELMASIL, HERBIDAL TOTAL, HERBIZOLE, KLEER-LOT, ORGA-414, RADOXONE TL, SIMAZOL, WEEDAZOL, WEEDEX GRANULAT, X-ALL LIQUID

Endocrine Disruptor Information

Triazine herbicides are listed as a class of endocrine disrupters in *Our Stolen Future* [810] Amitrole, one of the triazine herbicides, is specifically listed by the World Wildlife Fund Canada [813], and the Center for Disease Control and Prevention [812]. Amitrole was found to affect the thyroid after single, short-term and long-term exposures. Amitrole is goitrogenic; it causes thyroid hypertrophy and hyperplasia, depletion of colloid and increased vascularity. In long-term experiments these changes precede the development of thyroid neoplasia in rats. The carcinogenic effect of amitrole on the thyroid is thought to be related to the continuous stimulation of the gland by increased thyroid stimulating hormone (TSH) levels, which are caused by the interference of amitrole with thyroid hormone synthesis. Equivocal results have been reported in some

106

studies on the genotoxic potential of amitrole. In carcinogenicity testing in rats, amitrole did not induce tumors in organs other than the thyroid. However, high doses of amitrole caused liver tumors in mice. Several criteria have been used to assess the early effects of amitrole on the thyroid. The lowest no-observed-adverse-effect level (NAOEL) derived from these studies was 2 mg/kg in the diet of rats and was assessed on the basis of thyroid hyperplasia [873].

Chemical and Physical Properties

CHEMICAL FORMULA: $C_2H_4N_4$

MOLECULAR WEIGHT: 84.08

WLN: T5MN DNJ CZ

PHYSICAL DESCRIPTION: White crystals

SPECIFIC GRAVITY: Not available

DENSITY: 1.138 g/cm^3 @ 20°C [169]

MELTING POINT: 159°C [016], [058], [395], [430]

SOLUBILITY: Water: >=100 mg/mL @ 17.5°C [700]; DMSO: >=100 mg/mL @ 17.5°C [700]; 95% Ethanol: 5-10 mg/mL @ 17.5°C [700]; Acetone: 1-5 mg/mL @ 17.5°C [700]; Methanol: Soluble [033]

OTHER SOLVENTS: Alcohol: Soluble [016], [062], [421], [430]; Non-polar solvents: Insoluble [172], [173]; Chloroform: Soluble [016], [033], [421], [430]; Ethyl acetate: Sparingly soluble [033], [395]; Ether: Insoluble [033], [172], [395], [421]

OTHER PHYSICAL DATA: Sublimes undecomposed @ reduced pressure [071]; Bitter taste [430]; Odorless [173]; pH (10% solution): 6.5-7.5 [058]

Source: *The National Toxicology Program's Chemical Database, Volume 2*, 1992, CRC Press, Inc./Lewis Publishers, 2000 Corporate Blvd., Boca Raton, FL 33431 (Fax: 407-998-9114).

Environmental Reference Materials

A source of EED environmental reference materials for amitrole is Crescent Chemical Co., Hauppauge, NY (Fax: 516-348-0913); Catalog No. 45324.

Hazardous Properties

ACUTE/CHRONIC HAZARDS: When heated to decomposition this compound emits toxic fumes of carbon monoxide, carbon dioxide and oxides of nitrogen [043], [058], [269]. It may be highly toxic by intraperitoneal and moderately toxic by ingestion [043]. It may be toxic by skin absorption and inhalation [269].

FIRE HAZARD: Literature sources indicate that Amitrole is non-combustible [058]. Fires involving this material can be controlled with a dry chemical, carbon dioxide or Halon extinguisher. A water spray may also be used [269].

REACTIVITY: Amitrole forms chelates with some metals. It is corrosive to iron, copper and aluminum [071], [169], [172]. Forms salts with most acids and alkalis [169], [172]. Amitrole is incompatible with strong oxidizers, strong acids, acid chlorides and acid anhydrides [269].

STABILITY: Amitrole is sensitive to moisture [058]. UV spectrophotometric stability screening indicates that solutions of Amitrole in 95% ethanol are stable for at least 24 hours [700].

Source: *The National Toxicology Program's Chemical Database, Volume 7,* 1992, CRC Press, Inc./Lewis Publishers, 2000 Corporate Blvd., Boca Raton, FL 33431 (Fax: 407-998-9114).

Medical Symptoms of Exposure

SYMPTOMS: Symptoms of exposure to Amitrole may include irritation of the eyes, skin and upper respiratory tract, depression, dyspnea, labored respiration, diarrhea, ataxia, vomiting, convulsions, coma, lung injury and, possibly, thyroid effects [058]. It may also enhance yellowing or browning of lenses with ultraviolet exposure [099].

Source: *The National Toxicology Program's Chemical Database, Volume 4,* 1992, CRC Press, Inc./Lewis Publishers, 2000 Corporate Blvd., Boca Raton, FL 33431 (Fax: 407-998-9114).

Toxicological Information

The toxicological information below is gathered from several sources including two national databases: one from the *National Toxicology Program Chemical Repository Database* and another from *EPA's Integrated Risk Information System (IRIS).*

Toxicology Information from the National Toxicology Program

CARCINOGENICITY:

Tumorigenic Data:

Type of Effect	Route/Animal	Amount Dosed (Notes)
TDLo:	orl-rat	4595 mg/kg/2.5Y-C
TDLo:	orl-mus	113 gm/kg/3W-I
TD :	orl-rat	3670 mg/kg/2Y-C
TD :	orl-rat	122 gm/kg/70W-C
TD :	orl-rat	105 gm/kg/60W-C
TD :	orl-mus	366 gm/kg/26W-C

Review:
> IARC Cancer Review: Animal Sufficient Evidence
> IARC Cancer Review: Human Inadequate Evidence
> IARC possible human carcinogen (Group 2B)

Status:
> NTP Fourth Annual Report on Carcinogens, 1984
> NTP anticipated human carcinogen
> EPA Carcinogen Assessment Group

TERATOGENICITY: See RTECS printout for data

MUTATION DATA: See RTECS printout for data

TOXICITY:

Typ.dose	Mode	Specie	Amount	Units	Other
LD50	orl	mus	14700	mg/kg	
LD50	orl	rat	1100	mg/kg	
LD50	ipr	mus	200	mg/kg	

OTHER TOXICITY DATA:

> Review: Toxicology Review-2

> Status:
> EPA Genetox program 1988, Positive: Carcinogenicity-mouse/rat; SHE-clonal assay
> EPA Genetox program 1988, Positive: Cell transform.-RLV F344 rat embryo; Host-mediated assay
> EPA Genetox program 1988, Negative: D melanogaster-whole sex chrom. loss
> EPA Genetox program 1988, Negative: D melanogaster-nondisjunction; N crassa-aneuploidy
> EPA Genetox program 1988, Negative: E coli polA without S9; E coli polA with S9
> EPA Genetox program 1988, Negative: Histidine reversion-Ames test; Sperm morphology-mouse
> EPA Genetox program 1988, Negative: In vitro UDS-human fibroblast; S cerevisiae-homozygosis
> EPA Genetox program 1988, Inconclusive: Mammalian micronucleus
> EPA Genetox program 1988, Inconclusive: D melanogaster Sex-linked lethal
> EPA TSCA Chemical Inventory, 1986

SAX TOXICITY EVALUATION:

> THR: Poison by intraperitoneal. Moderately toxic by ingestion. An experimental carcinogen, tumorigen, neoplastigen and teratogen. Other experimental reproductive effects. Mutagenic data.

Source: *Instant Tox-Base*, 1995, Instant Reference Sources, Inc., 7605 Rockpoint Drive, Austin, TX 78731 (Fax: 512-345-2386; Internet URL, http://www.instantref.com/inst-ref.htm).

Toxicology Information from EPA's Integrated Risk Information System (IRIS)

Amitrole is not currently listed in EPA's IRIS database nor is it listed as scheduled for addition in the near future.

Production, Use, and Pesticide Labeling Information

1. Description of Chemical

- Generic Name: 3-amino-1,2,4-triazole

- Common Name: amitrole

- Trade Names: Weedazole, Amino Triazole Weed Killer, Cytrol, Amitrol T, Domatol, Vorox, Amizole, X-All, Ustinex, AT, ATA, Aminotriazole 90 and Chempar Amitrole

- EPA Shaughnessy Code: 004401

- Chemical Abstracts Service (CAS) Number: 61-82-5

- Year of Initial Registration: 1948

- Pesticide Type: Herbicide

- Chemical Family:

- U.S. and Foreign Producers: Not produced in U.S.; major importers are Union Carbide, American Cyanamid, and Aceto Chemical

2. Use Patterns and Formulations

- Application sites: Noncrop sites including rights-of-way, marshes, drainage ditches, ornamentals, and around commercial, industrial, agricultural, domestic, and recreational premises.

- Types of formulations: Technical (90%, 95%); wettable powder (15%, 25%); flowable concentrate (0.33 lb/gal, 0.44 lb/gal, 1%); soluble concentrate/liquid (0.3 lb/gal, 2 lb/gal); soluble concentrate/solid (50%, 90%) and pressurized liquid (0.36%, 1%).

- Types and methods of application: Applied as a spray for broadcast, spot, or directed treatments, using aerial or ground equipment.

- Application rates: 0.9 to 20 lbs a.i./A, depending upon weed species and method of application.

- Usual carriers: water

3. Science Findings

- Chemical Characteristics

- Physical state: crystalline powder

- Color: transparent, colorless

- Odor: odorless

- Melting point: 159 °C

- Solubility: 28g/100g water, soluble in some polar solvents

- Stability: Stable in heat to 100 °C. Amitrole sublimes under reduced pressure.

- pH: aqueous solutions are neutral

- Unusual handling characteristics: none

- Toxicological Characteristics

- Acute toxicology results

- Acute oral LD50 (rat) > 4.08 gm/kg, Toxicity Category III

- Acute dermal LD50 (rabbit): No mortalities reported, Toxicity Category III

- Primary eye irritation (rabbit): Amitrole is slightly irritating, additional testing is required.

- Chronic toxicology results

- Feeding/Oncogenicity: Amitrole has an anti-thyroid effect in laboratory rats. Dogs fed Amitrole exhibited thyroid and pituitary changes.

- Reproduction: Amitrole does not cause reproductive effects.

- Teratology: Additional testing required.

- Mutagenicity: Amitrole does not cause mutagenic effects.

- Metabolism: Amitrole is rapidly eliminated from the body.

- Major routes of exposure: Mixers, loaders, and applicators would be expected to receive the most exposure via skin contact and inhalation.

- Physiological and Biochemical Behavioral Characteristics

- Absorption and translocation: It is readily absorbed and rapidly translocated in the roots and leaves of higher plants.

- Mechanism of pesticidal action: Amitrole interferes with the metabolism of nucleic acid precursors, disrupts chloroplast development and regrowth from buds.

- Environmental Characteristics

- Adsorption and leaching in basic soil types: Amitrole exhibits intermediate soil mobility.

- Microbial breakdown: Microbial metabolism is the expected major route of degradation.

- Resultant average persistence: Amitrole residues degrade with a half-life of <1 to 56 days in non-sterile aerobic soils. The soil dissipation rate is affected by moisture, temperature, cation exchange capacity, and clay content, but is unaffected by soil pH. Amitrole is persistent in pond water and hydrosoil.

- Ecological Characteristics

- Hazards to fish and wildlife:

- Avian dietary LC50: Mallard duck > 5,000 ppm; ring-neck pheasant > 5,000 ppm

- Freshwater fish LC50: Rainbow trout > 180 mg/l; bluegill sunfish > 180 mg/l

- Aquatic invertebrate LC50 > 10 ppm

- Tolerance Assessments

Temporary maximum residue limits for amitrole of 0.02 ppm have been established by FAO/WHO for those crops where residues are likely to occur. There are no established tolerances for amitrole in the

U.S., Canada, and Mexico. There are no food or feed uses in the U.S., and residues are not permitted on any food or water intended for irrigation, drinking, or other domestic purposes.

- Problems Known to Have Occurred From Use

The Pesticide Incident Monitoring System (PIMS) listed eight incidents resulting from the use of amitrole alone from 1972 to 1977. One incident involved illegal residue on apples, and two others involved plant injury resulting from soil residues. The remaining five incidents involved pesticide applicators receiving medical attention after exposure. Symptoms included skin rash, vomiting, diarrhea, and nosebleed. There were no reported fatalities. PIMS incidents are voluntarily reported, do not include detailed follow-ups, and are not validated in any way.

- Summary Science Statement

Extensive data gaps exist for amitrole in product chemistry, toxicology, ecological effects, and environmental fate. Amitrole has demonstrated oncogenic potential and is a candidate for Special Review. Because of this oncogenic risk, all use patterns and application techniques (except for homeowner uses) are classified as restricted.

4. Summary of Regulatory Position and Rationale

- Use classification: Restricted (for all uses except for homeowner uses)

- Use, formulation, or geographical restrictions: Noncrop land areas only

- Unique label warning statements:

- Manufacturing-Use Products: Products intended for formulation into end-use products must bear the following statements:

- For formulation only into end-use herbicide products intended for noncrop land, outdoor use.

- The use of this product may be hazardous to your health. This product contains amitrole, which has been determined to cause cancer in laboratory animals. Products intended for formulation into restricted-use pesticides must require on their labeling that a respirator be worn during mixing and loading. Lightweight waterproof clothing (jumpsuit [or coverall], boots [or shoes], gloves, and a wide-brimmed plastic hard-hat) must be worn when mixing and loading all products and when applying all products to control dense, tall vegetation. Workers applying this product in all other situations must wear lightweight waterproof gloves and boots (or shoes).

Products intended for formulation into general- use pesticides must require on their labeling that waterproof gloves be worn while handling the product.

- All products must bear the following statements:

> - Each formulator is responsible for obtaining EPA registrations for its formulated product(s).

> - Do not discharge into lakes, streams, ponds, or public waters unless in accordance with NPDES permit. For guidance, contact your Regional Office of the EPA.

- End-Use Products: All restricted-use products must bear the following statements:

> - Restricted Use Pesticide

> - For retail sale to and application only by certified applicators or personnel under their direct supervision.

> - The use of this product may be hazardous to your health. This product contains amitrole, which has been determined to cause cancer in laboratory animals. Wear a respirator during mixing and loading of all products. Wear lightweight, waterproof clothing (jumpsuit [or coverall], boots [or shoes], gloves, and a wide- brimmed plastic hard-hat) when applying all products to control dense, tall vegetation. Workers applying this product in all other situations must wear lightweight waterproof gloves and boots (or shoes).

- All homeowner products must bear the following statement:

> - The use of this product may be hazardous to your health. This product contains amitrole, which has been determined to cause cancer in laboratory animals. Wear waterproof gloves when using this product.

- All products intended for nonaquatic uses must bear the following statement on the label:

> - Do not apply directly to water or wetlands. Do not contaminate water by cleaning of equipment or disposal of wastes.

- All products intended for aquatic uses must bear the following statement on the label:

- Consult your state Fish and Game Agency before applying this product to public waters. Permits may be required before treating such waters.

- All products must bear the following statements, regardless of classification:

    - Do not allow spray or spray drift to contaminate edible crops or water intended for irrigation, drinking, or other domestic purposes.

    - Do not allow livestock to graze or feed in treated noncrop areas.

- Summary of preliminary risk/benefit review:

- Risks:

    - Amitrole is not used on food crops, and there is no dietary exposure to amitrole. Dermal exposure is the major source of exposure, with inhalation furnishing only a minor contribution to the total body burden. Human exposure, in some circumstances, occurs at doses which resulted in antithyroid effects in laboratory animals.

    - Conservatively assuming 100% dermal penetration, the oncogenic risk associated with some use patterns and application techniques is high. Lightweight, waterproof clothing and a respirator are expected to reduce exposure and risk for all uses except the power wagon application.

Benefits:

    The largest use site by production volume, the highway rights-of- way site was selected for this limited analysis. Amitrole is not produced in the United States, with under 800 thousand pounds being imported by Union Carbide, American Cyanamid, and Aceto Chemical. Amitrole, in combination with other chemicals, offers low cost, broad spectrum control of both newly emerged or established broadleaf weeds, as well as seasonal control by residual chemicals with which it is mixed. Alternatives include contact herbicides and mechanical cutting.

5. Summary of Major Data Gaps

- Generic data requirements:
    - Product chemistry: data due 6 months after receipt of Standard
        - Statement of composition
        - Discussion of formation of unintentional ingredients
        - Preliminary analysis
        - Density, bulk density, or specific gravity
        - Solubility
        - Vapor pressure
        - Dissociation constant
        - Octanol/Water partition coefficient
        - Submittal of samples

- Toxicology:
    - Acute testing: data due 6 months after receipt of Standard
    - Primary eye irritation
    - Primary skin irritation
    - Dermal sensitization

- Subchronic testing: data due 6 months after receipt of Standard
    - 90-day dermal
    - 90-day inhalation

- Chronic testing: data due 24 months after receipt of Standard
    - Teratogenicity - 2 species

- Special testing: data due 6 months after receipt of Standard
    - Dermal absorption study

- Wildlife and aquatic organisms: data due 24 months after receipt of Standard
    - Avian oral LD50
    - Freshwater fish LC50
    - Acute LC50 freshwater invertebrates
    - Acute LC50 estuarine and marine organisms

- Environmental fate:
    - Data due 6 months after receipt of Standard
    - Hydrolysis studies
    - Photodegradation studies in water
    - Photodegradation studies on soil
    - Leaching and adsorption/desorption
    - Special exposure study - protective clothing effectiveness

- Data due 24 months after receipt of Standard
    - Aerobic soil metabolism study
    - Anaerobic aquatic metabolism study
    - Aerobic aquatic metabolism study
    - Soil dissipation study - field

- Aquatic (sediment) dissipation study - field
- Forestry dissipation study - field
- Soil, long-term dissipation study (field) - reserved, depending upon results of field dissipation study
- Accumulation studies - irrigated crops

- Product specific data requirements for manufacturing-use products containing amitrole:
  - Product chemistry: data due 6 months after receipt of Standard
    - Statement of composition
    - Discussion of formation of unintentional ingredients
    - Preliminary analysis
    - Certification of limits
    - Analytical methods for enforcement of limits
    - Density, bulk density, or specific gravity
    - pH
    - Oxidizing or reducing action
    - Flammability
    - Explodability
    - Storage stability

- Toxicology:
  - Acute testing: data due 6 months after receipt of Standard - Primary eye irritation - rabbit
    - Primary dermal irritation
    - Dermal sensitization

6. Contact Person at EPA

Robert J. Taylor
Product Manager (25), TS-767C
Environmental Protection Agency
401 M Street
SW Washington, DC 20460.
Telephone: (703) 557-1800

Source: *Instant EPA's Pesticide Facts*, 1994, Instant Reference Sources, Inc., 7605 Rockpoint Drive, Austin, TX 78731 (Fax: 512-345-2386; Internet URL, http://www.instantref.com/inst-ref.htm).

Anthracene

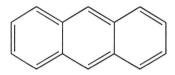

Introduction

Anthracene is an important source of dyestuffs and it is used in the manufacture of anthraquinone, alizarin dyes, insecticides, wood preservatives and as scintillation counter crystals. Anthracene's release to the environment is quite general since it is a ubiquitous product of incomplete combustion and has extensive natural and man made sources. It is largely associated with particulate matter, soils, and sediments when it occurs as an environmental pollutant. Human exposure may occur from inhalation of contaminated air and consumption of contaminated food and water. Especially high exposure will occur through the smoking of cigarettes and the ingestion of certain foods (eg smoked and charcoal broiled [828].

Identifying Information

CAS NUMBER: 120-12-7

NIOSH Registry Number: CA9350000

SYNONYMS: ANTHRACEN; ANTHRACIN; GREEN OIL; HSDB 702; NSC 7958; PARANAPHTHALENE; TETRA OLIVE N2G

Endocrine Disruptor Information

Anthracene is on EPA's list of suspected EEDs [811] and it is an aromatic coplanar compound. Chronic toxic effects may include shortened lifespan, reproductive problems, lower fertility, and changes in appearance or behavior [830].

Chemical and Physical Properties

CHEMICAL FORMULA: $C_{14}H_{10}$

MOLECULAR WEIGHT: 178.24 [703]

PHYSICAL DESCRIPTION: Colorless crystals, violet fluorescence [703]

DENSITY: 1.24 @ 27°C/4°C [703]

MELTING POINT: 217°C [703]

BOILING POINT: 339.9°C [703]

SOLUBILITY: Water: Insoluble [703]; Alcohol: Soluble in alcohol @ 1.9/100 @ 20°C [703]; Ether: Soluble in ether @ 12.2/100 @ 20°C [703]

OTHER PHYSICAL DATA: Vapor Pressure: 1 mm @ 145.0°C (sublimes) [703]; Vapor Density: 6.15 [703]

Source: *Instant EPA's Air Toxics*, 1994, Instant Reference Sources, Inc., 7605 Rockpoint Drive, Austin, TX 78731 (Fax: 512-345-2386; Internet URL, http://www.instantref.com/inst-ref.htm).

Environmental Reference Materials

A source of EED environmental reference materials for Anthracene is Radian International LLC, Austin, TX (Fax 512-454-0268; Internet URL, http://www.radian.com/standards); Catalog No. ERA-010.

Hazardous Properties

ACUTE/CHRONIC HAZARDS: Anthracene is an irritant and may be toxic by ingestion.

VOLATILITY:
    Vapor pressure: 1 mm Hg @ 145°C (sublimes)
    Vapor density: 6.15

FIRE HAZARD: The flash point for anthracene is 121°C (250°F). It is combustible. Fires involving this material should be controlled using a dry chemical, carbon dioxide, foam or Halon extinguisher.
The autoignition temperature is 538°C (1004°F).

LEL: 0.6%   UEL: Not available

REACTIVITY: Anthracene darkens in sunlight and reacts with oxidizers. It reacts explosively with flame, $Ca(OCl)_2$ and chromic acid.

STABILITY: Anthracene is sensitive to prolonged exposure to air and light.

Source: *Instant EPA's Air Toxics*, 1994, Instant Reference Sources, Inc., 7605 Rockpoint Drive, Austin, TX 78731 (Fax: 512-345-2386; Internet URL, http://www.instantref.com/inst-ref.htm).

Medical Symptoms of Exposure

Exposure may cause irritation of the eyes and respiratory tract and gastrointestinal irritation if swallowed. Long-term contact may result in pigmentation or carcinogenesis on the skin.

Source: *Instant EPA's Air Toxics*, 1994, Instant Reference Sources, Inc., 7605 Rockpoint Drive, Austin, TX 78731 (Fax: 512-345-2386; Internet URL, http://www.instantref.com/inst-ref.htm).

Toxicological Information

The toxicological information below is gathered from several sources including two national databases: one from the *National Toxicology Program Chemical Repository Database* and another from *EPA's Integrated Risk Information System (IRIS)*.

Toxicology Information from the National Toxicology Program

CARCINOGENICITY:

        Tumorigenic Data:
                TDLo: orl-rat   20 gm/kg/79W-I
                TDLo: scu-rat 3300 mg/kg/33W-I
                TD : scu-rat   660 mg/kg/33W-I

      Review:
                IARC Cancer Review: Animal Inadequate Evidence
                IARC: Not classifiable as a human carcinogen (Group 3)

TERATOGENICITY: Not available

MUTATION DATA:

Test	Lowest Dose		Test	Lowest Dose
dns-hmn:fbr	10 mg/L		hma-mus:sat	125 mg/kg
dnd-mam:lym	100 umol		mma-sat	100 ug/plate

TOXICITY:
    Not available

OTHER TOXICITY DATA:
    Skin and Eye Irritation Data: skn-mus 118 ug MLD

    Status:
        "NIOSH Manual of Analytical Methods" Vol 1 206
        Reported in EPA TSCA Inventory, 1980
        Meets criteria for proposed OSHA Medical Records Rule

SAX TOXICITY EVALUATION:

    THR: Allergen and mild irritant. A recognized carcinogen of the skin, hands, forearms and scrotum. An experimental carcinogen of the bladder.

Source: ***Instant Tox-Base***, 1995, Instant Reference Sources, Inc., 7605 Rockpoint Drive, Austin, TX 78731 (Fax: 512-345-2386; Internet URL, http://www.instantref.com/inst-ref.htm).

Toxicology Information from EPA's Integrated Risk Information System (IRIS)

I. CHRONIC HEALTH HAZARD ASSESSMENTS FOR NONCARCINOGENIC EFFECTS

I.A. REFERENCE DOSE FOR CHRONIC ORAL EXPOSURE (RfD)

The Reference Dose (RfD) is based on the assumption that thresholds exist for certain toxic effects such as cellular necrosis, but may not exist for other toxic effects such as carcinogenicity. In general, the RfD is an estimate (with uncertainty spanning perhaps an order of magnitude) of a daily exposure to the human population (including sensitive subgroups) that is likely to be without an appreciable risk of deleterious effects during a lifetime. Please refer to Background Document 1 in Service Code 5 for an elaboration of these concepts. RfDs can also be derived for the noncarcinogenic health effects of compounds which are also carcinogens. Therefore, it is essential to refer to other sources of information concerning the carcinogenicity of this substance. If the U.S. EPA has evaluated this substance for potential human carcinogen-icity, a summary of that evaluation will be contained in Section II of this file when a review of that evaluation is completed.

I.A.1. ORAL RfD SUMMARY

Critical Effect	Experimental Doses*	UF	MF	RfD
No observed effects	NOEL: 1000 mg/kg/day	3000	1	3E-1 mg/kg/day
	LOAEL: none			

*Conversion Factors: none

Subchronic Toxicity Study in Mice U.S. EPA, 1989

I.A.2. PRINCIPAL AND SUPPORTING STUDIES (ORAL RfD)

U.S. EPA. 1989. Subchronic toxicity in mice with anthracene. Final Report. Hazelton Laboratories America, Inc. Prepared for the Office of Solid Waste, Washington, DC.

Anthracene was administered to groups of 20 male and female CD-1 (ICR)BR mice by oral gavage at doses of 0, 250, 500, and 1000 mg/kg/day for at least 90 days. Mortality, clinical signs, body weights, food consumption, opthalmology findings, hematology and clinical chemistry results, organ weights, organ-to-body weight ratios, gross pathology, and histopathology findings were evaluated. No treatment-related effects were noted. The no-observed-effect level (NOEL) is the highest dose tested (1000 mg/kg/day).

I.A.3. UNCERTAINTY AND MODIFYING FACTORS (ORAL RfD)

UF -- An uncertainty factor of 3000 was used: 10 to account for interspecies extrapolation, 10 for intraspecies variability and 30 for both the use of a subchronic study for chronic RfD derivation and for lack of reproductive/developmental data and adequate toxicity data in a second species.

MF -- None

I.A.4. ADDITIONAL COMMENTS (ORAL RfD)

In a chronic bioassay (Schmahl, 1955), a group of 28 BD I and BD III rats received anthracene in the diet, starting when the rats were approximately 100 days old. The daily dosage was 5 to 15 mg/rat, and the experiment was terminated when a total dose of 4.5 g/rat was achieved, on the 550th experimental day. The rats were observed until they died, with some living more than 1000 days. No treatment-related effects on lifespan or gross and histological appearance of tissues were observed. Body weights were not mentioned, and hematological parameters were not measured. No chronic LOAEL could be determined from this study.

I.A.5. CONFIDENCE IN THE ORAL RfD

Study -- LowData Base -- LowRfD -- Low

Confidence in the study is low. It was a well-designed experiment examining a variety of toxicological endpoints; however, failure to identify a LOAEL precludes a higher level of confidence. Confidence in the data base is low, because of the lack of adequate toxicity data in a second species and developmental/reproductive studies. Low confidence in the RfD follows.

I.A.6. EPA DOCUMENTATION AND REVIEW OF THE ORAL RfD

Source Document -- U.S. EPA, 1987

ECAO-CIN Internal Review and Limited Agency Review.

Other EPA Documentation -- U.S. EPA,1989

Agency Work Group Review -- 10/19/89, 11/15/89

Verification Date -- 11/15/89

I.A.7. EPA CONTACTS (ORAL RfD)

Harlal Choudhury / NCEA -- (513)569-7536

Kenneth A. Poirier / OHEA -- (513)569-7553

I.B. REFERENCE CONCENTRATION FOR CHRONIC INHALATION EXPOSURE (RfC)

A risk assessment for this substance/agent is under review by an EPA work group.

II. CARCINOGENICITY ASSESSMENT FOR LIFETIME EXPOSURE

Section II provides information on three aspects of the carcinogenic risk assessment for the agent in question; the U.S. EPA classification, and quantitative estimates of risk from oral exposure and from inhalation exposure. The classification reflects a weight-of-evidence judgment of the likelihood that the agent is a human carcinogen. The quantitative risk estimates are presented in three ways. The slope factor is the result of application of a low-dose extrapolation procedure and is presented as the risk per (mg/kg)/day. The unit risk is the quantitative estimate in terms of either risk per ug/L drinking water or risk per ug/cu.m air breathed. The third form in which risk is presented is a drinking water or air concentration providing cancer risks of 1 in 10,000, 1 in 100,000 or 1 in 1,000,000. Background Document 2 (Service Code 5) provides details on the rationale and methods used to derive the carcinogenicity values found in IRIS. Users are referred to Section I for information on long-term toxic effects other than carcinogenicity.

II.A. EVIDENCE FOR CLASSIFICATION AS TO HUMAN CARCINOGENICITY

II.A.1. WEIGHT-OF-EVIDENCE CLASSIFICATION

Classification -- D, not classifiable as to human carcinogenicity

Basis -- Based on no human data and inadequate data from animal bioassays.

II.A.2. HUMAN CARCINOGENICITY DATA

None.

II.A.3. ANIMAL CARCINOGENICITY DATA

Inadequate. A group of 28 BDI or BDIII rats (sex not specified) were fed a diet containing anthracene in oil, 6 days/week for 78 weeks, and observed until natural death, approximately 700 days (Schmahl, 1955). The total dose was 4.5 g anthracene/rat (approximately 28 mg/kg/day). No concurrent controls were used. No tumors were observed.

Groups of 60 female 3-to 6-month-old Osborne-Mendel rats were observed for 55-81 weeks after receiving a single lung implant of anthracene (0.5 mg/rat, approximately 2 mg/kg) dissolved in a 1:1 (v:v) mixture of beeswax and trioctanoin (0.1 mL) (Stanton et al., 1972). Controls received an implant of the vehicle. No tumors were observed.

Tests for complete carcinogenicity and initiating activity in mouse skin-painting assays have not shown positive results. No tumors were observed in an assay of initiating activity in which Crl:CD/1 (ICR)BR female albino mice were exposed to 1 mg anthracene in acetone, and then treated with 12-o-tetradecanoyl-phorbol-13 acetate as the promoting agent 3 times/week for 20 weeks (LaVoie et al., 1985).

A single dermal application of 10 um anthracene (purity not stated) in benzene was administered to 30 female CD-1 mice; this initial application was followed 7 days later by twice-weekly applications of 5 um 12-0-tetradecanoyl phorbol-13-acetate (TPA) for 35 weeks. Survival in the group was 93% after 35 weeks. By week 20 of the test, 2/28 mice had developed skin tumors; this increased to 4/28 by week 35. In the control group, in which 30 mice received only the TPA applications, a mouse developed a skin tumor at week 25 (Scribner, 1973).

Kennaway (1924) administered 40% anthracene (purity unknown) either in lanolin or as an ether-extract to two groups of 100 mice each (sex and strain not stated). In the lanolin-group, 44% of the mice survived 131 days and in the ether-extract group only 6% survived until day 160. In the lanolin-group 1/44 surviving mice had developed a papilloma by day 131; no mice had developed tumors in the ether-extract by day 160. No information pertaining to the use of a control group was given.

Druckrey and Schmahl (1955) administered a diet containing anthracene in oil 6 days/week to 28 BDI or BDIII rats (sex not stated) for 78 weeks. The total dose was 4.5 g anthracene/rat. No treatment-related tumors were found, and no control groups appear to have been utilized.

II.A.4. SUPPORTING DATA FOR CARCINOGENICITY

Tests for DNA damage and gene mutations in prokaryotes have generally shown negative results. Negative results were observed in tests for DNA damage in Escherichia coli at concentrations up to 250 ug/mL and Bacillus subtilis at 62 ug/mL (Rosenkrantz and Poirier, 1979; McCarroll et al., 1981; DeFlora et al., 1984). Negative results were obtained in tests for reverse mutation in six strains of Salmonella typhimurium, at concentrations up to 1000 ug/plate (McCann et al., 1975; Simmon, 1979a; LaVoie et al., 1979; Salamone et al., 1979; Ho et al., 1981; DeFlora et al., 1984; Bos et al., 1988). Tests for forward mutation at 40 ug/mL were negative (Kaden et al., 1979). Positive results for reverse mutation in Salmonella typhimurium (TA97) at 10 ug/plate were reported (Sakai et al., 1985). Anthracene was tested in bacterial assays in 20 laboratories as part of an international collaborative study. One lab reported a positive in TA100 without activation, one lab reported a positive in TA98 and TA100 but only with S9 and all other labs reported negative results (Bridges et al., 1981).

Anthracene has consistently been negative in yeast test systems measuring mitotic recombination (Simmon, 1979b; de Serres and Hoffman, 1981), gene conversion, mutation and chromosome loss (de Serres and Hoffman, 1981).

Tests for DNA damage, mutation, chromosome effects and cell transformation in a variety of eukaryotic cell preparations have shown negative results. Anthracene showed negative results in tests for DNA damage (DNA synthesis) in primary rat hepatocytes (1 ug/mL), Chinese hamster ovary cells (1000 ug/mL), or HeLa cells (100 ug/mL) (Williams, 1977; Probst et al., 1981; Garrett and Lewtas, 1983; Martin et al., 1978; Martin and McDermid, 1981). It yielded negative results in tests for forward mutation in Chinese hamster V79 cells (125 ug/mL), mouse lymphoma L5178Y cells (18 ug/mL) and human lymphoblastoid cells (36 ug/mL) (Knapp et al., 1981; Amacher and Turner, 1980; Amacher et al., 1980; Barfknecht et al., 1981). Results obtained in tests for sister-chromatid exchange and chromosome breaks in Chinese hamster D6 cells and rat liver epithelial ARL-18 cells at 178 ug/mL were negative (Abe and Sasaki, 1977; Tong et al.,

1981). Results reported in tests for cell transformation (morphological changes) at concentrations up to 30 ug/mL in mouse BALB/3T3 cells, guinea pig fetal cells, Syrian hamster embryo cells and mouse embryo C3H10T1/2 cells (DiPaolo et al., 1972; Evans and DiPaolo, 1975; Pienta et al., 1977; Lubet et al., 1983) were negative. In the international collaborative study negative results were reported with in vitro assays measuring unscheduled DNA synthesis, sister chromatid exchange, chromosome aberrations, and gene mutations (Brookes and Preston, 1981).

II.B. QUANTITATIVE ESTIMATE OF CARCINOGENIC RISK FROM ORAL EXPOSURE

None.

II.C. QUANTITATIVE ESTIMATE OF CARCINOGENIC RISK FROM INHALATION EXPOSURE

None.

II.D. EPA DOCUMENTATION, REVIEW, AND CONTACTS (CARCINOGENICITY ASSESSMENT)

II.D.1. EPA DOCUMENTATION

Source Document -- U.S. EPA, 1990

The 1990 Drinking Water Criteria Document for Polycyclic Aromatic Hydrocarbons has received Agency and external review.

II.D.2. REVIEW (CARCINOGENICITY ASSESSMENT)

Agency Work Group Review -- 02/07/90

Verification Date -- 02/07/90

II.D.3. U.S. EPA CONTACTS (CARCINOGENICITY ASSESSMENT)

Rita Schoeny / NCEA -- (513)569-7544

Robert McGaughy / NCEA -- (202)260-5889

III. HEALTH HAZARD ASSESSMENTS FOR VARIED EXPOSURE DURATIONS

III.A. DRINKING WATER HEALTH ADVISORIES

Not available at this time.

III.B. OTHER ASSESSMENTS

Content to be determined.

IV. U.S. EPA REGULATORY ACTIONS

EPA risk assessments may be updated as new data are published and as assessment methodologies evolve. Regulatory actions are frequently not updated at the same time. Compare the dates for the regulatory actions in this section with the verification dates for the risk assessments in sections I and II, as this may explain inconsistencies. Also note that some regulatory actions consider factors not related to health risk, such as technical or economic feasibility. Such considerations are indicated for each action. In addition, not all of the regulatory actions listed in this section involve enforceable federal standards. Please direct any questions you may have concerning these regulatory actions to the U.S. EPA contact listed for that particular action. Users are strongly urged to read the background information on each regulatory action in Background Document 4 in Service Code 5.

IV.A. CLEAN AIR ACT (CAA)

[Ed. Note: EPA does not have information yet in this section of IRIS].

IV.B. SAFE DRINKING WATER ACT (SDWA)

[Ed. Note: EPA does not have information yet in this section of IRIS].

IV.C. CLEAN WATER ACT (CWA)

IV.C.1. AMBIENT WATER QUALITY CRITERIA, Human Health

Water and Fish Consumption: 2.8E-3 ug/L

Fish Consumption Only: 3.11E-2 ug/L

Considers technological or economic feasibility? -- NO

Discussion -- For the maximum protection from the potential carcinogenic properties of this chemical, the ambient water concentration should be zero. However, zero may not be obtainable at this time, so the recommended criteria presents a E-6 estimated incremental increase of cancer over a lifetime. The values given represent polynuclear aromatic hydrocarbons as a class.   Reference -- 45 FR 79318 (11/28/80)

EPA Contact -- Criteria and Standards Division / OWRS(202)260-1315 / FTS 260-1315

IV.C.2. AMBIENT WATER QUALITY CRITERIA, Aquatic Organisms

Freshwater:
    Acute LEC -- none
    Chronic LEC -- none

Marine:
    Acute LEC --   3.0E+2 ug/L
    Chronic LEC -- none

Considers technological or economic feasibility? -- NO

Discussion -- The values that are indicated as "LEC" are not criteria, but are the lowest effect levels found in the literature. LEC's are given when the minimum data required to derive water quality criteria are not available. The values given represent polynuclear aromatic hydrocarbons as a class.

Reference -- 45 FR 79318 (11/28/80)
EPA Contact -- Criteria and Standards Division / OWRS (202)260-1315 / FTS 260-1315

IV.D. FEDERAL INSECTICIDE, FUNGICIDE, AND RODENTICIDE ACT (FIFRA)

[Ed. Note: EPA does not have information yet in this section of IRIS].

IV.E. TOXIC SUBSTANCES CONTROL ACT (TSCA)

[Ed. Note: EPA does not have information yet in this section of IRIS].

IV.F. RESOURCE CONSERVATION AND RECOVERY ACT (RCRA)

IV.F.1. RCRA APPENDIX IX, for Ground Water Monitoring

Status -- Listed
Reference -- 52 FR 25942 (07/09/87)
EPA Contact -- RCRA/Superfund Hotline (800)424-9346 / (202)260-3000 / FTS 260-3000

IV.G. SUPERFUND (CERCLA)

IV.G.1. REPORTABLE QUANTITY (RQ) for Release into the Environment

Value (status) -- 5000 pounds (Final, 1989)
Considers technological or economic feasibility? -- NO

Discussion -- No data have been found to permit the ranking of this hazardous substance. The available data for acute hazards may lie above the upper limit for the 5000-pound RQ, but since it is a designated hazardous substance, the largest assignable RQ is 5000 pounds. This chemical has been assessed for carcinogenicity and is not, at this time, considered to be a potential carcinogen.

Reference -- 54 FR 33418 (08/14/89)
EPA Contact -- RCRA/Superfund Hotline (800)424-9346 / (202)260-3000 / FTS 260-3000

VI. BIBLIOGRAPHY

VI.A. ORAL RfD REFERENCES

Schmahl, D. 1955. Testing of naphthalene and anthracene as carcinogenic agents in the rat. Krebsforsch. 60: 697-710. (Ger.)

U.S. EPA. 1987. Health and Environmental Effects Profile for Anthracene. Prepared by the Office of Health and Environmental Assessment, Environmental Criteria and Assessment Office, Cincinnati, OH for the Office of Solid Waste and Emergency Response, Washington, DC.

U.S. EPA. 1989. Subchronic toxicity in mice with anthracene. Final Report. Hazelton Laboratories America, Inc. Prepared for the Office of Solid Waste, Washington, DC.

VI.B. INHALATION RfC REFERENCES

None

VI.C. CARCINOGENICITY ASSESSMENT REFERENCES

Abe, S. and M. Sasaki. 1977. Studies on chromosomal aberrations and sister chromatid exchanges induced by chemicals. Proc. Jap. Acad. Sci. 53(1):46-49.

Amacher, D.E. and G.N. Turner. 1980. Promutagen activation by rodent-liver postmitochondrial fractions in the L5178Y/TK cell mutation assay. Mutat. Res. 74: 485-501.

Amacher, D.E., S.C. Paillet, G.N. Turner, V.A. Ray and D.S. Salsburg. 1980. Point mutations at the thymidine kinase locus in L5178Y mouse lymphoma cells. II. Test validation and interpretation. Mutat. Res. 72: 447-474.

Barfknecht, T.R., B.M. Andon, W.G. Thilly and R.A. Hites. 1981. Soot and mutation in bacteria and human cells. In: Chemical Analysis and Biological Fate: Polynuclear Aromatic Hydrocarbons. 5th Int. Symp., M. Cooke and A.J. Dennis, Ed. Battelle Press, Columbus, OH. p. 231-242.

Bos, R.P., J.L.G. Theuws, F.J. Jongeneelen and P.T. Henderson. 1988. Mutagenicity of bi-, tri- and tetra-cyclic aromatic hydrocarbons in the "taped-plate assay" and in the conventional Salmonella mutagenicity assay. Mutat. Res. 204: 203-206.

Bridges, B.A., D.B. McGregor, E. Zeiger, et al . 1981. Summary report on the performance of bacterial mutation assays. In: Evaluation of Short-term Tests for Carcinogens. Report of the International Collaborative Program. Progress in Mutation Research, Vol. 1, F.J. de Serres and J. Ashby, Ed. Amsterdam, Elsevier, North Holland. p. 49-67.

Brookes, P. and R.J. Preston. 1981. Summary report on the performance of in vitro mammalian assays. In: Evaluation of Short-term Tests for Carcinogens. Report of the

International Collaborative Program. Progress in Mutation Research, Vol. 1, F.J. de Serres and J. Ashby, Ed. Amsterdam, Elsevier, North Holland. p. 77-85.

DeFlora, S., P. Zanacchi, A. Camoirano, C. Bennicelli and G.S. Badolati. 1984. Genotoxic activity and potency of 35 compounds in the Ames reversion test and in a bacterial DNA-repair test. Mutat. Res. 133(3): 161-198.

de Serres, F.J., G.R. Hoffman, et al. 1981. Summary report on the performance of yeast assays. In: Evaluation of Short-term Tests for Carcinogens. Report of the International Collaborative Program. Progress in Mutation Research, Vol. 1, F.J. de Serres and J. Ashby, Ed. Amsterdam, Elsevier, North Holland. p. 68-76.

DiPaolo, J.A., K. Takano and N.C. Popescu. 1972. Quantitation of chemically induced neoplastic transformation of BALB/3T3 cloned cell lines. Cancer Res. 32: 2686-2695.

Druckrey, H. and D. Schmahl. 1955. Carcinogenic effect of anthracene. Naturwissenschaften. 42: 159-160.

Evans, C.H. and J.A. DiPaolo. 1975. Neoplastic transformation of guinea pig fetal cells in culture induced by chemical carcinogens. Cancer Res. 35: 1035-1044.

Garrett, N.E. and J. Lewtas. 1983. Cellular toxicity in Chinese hamster ovary cell cultures. I. Analysis of cytotoxicity endpoints for twenty-nine priority pollutants. Environ. Res. 32(2): 455-465.

Ho, C-H., B.R. Clark, M.R. Guerin, B.D. Barkenhus, T.K. Rao and J.L. Epier. 1981. Analytical and biological analyses of test materials from the synthetic fuel technologies. IV. Studies of chemical structure-mutagenic activity relationships of aromatic nitrogen compounds relevant to synfuels. Mutat. Res. 85: 335-345.

Kaden, D.A., R.A. Hites and W.G. Thilly. 1979. Mutagenicity of soot and associated polycyclic aromatic hydrocarbons to Salmonella typhimurium. Cancer Res. 39: 4152-4159.

Kennaway, E.L. 1924. On cancer-producing tars and tar-fractions. J. Ind. Hyg. 5(12): 462-488.

Knapp, A., C. Goze and J. Simons. 1981. Mutagenic activity of seven coded samples of V79 Chinese hamster cells. In: Evaluation of Short-term Tests for Carcinogens. Report of the International Collaborative Program. Progress in Mutation Research, Vol. 1, F.J. de Serres and J. Ashby, Ed. Amsterdam, Elsevier, North Holland. p. 608-613.

LaVoie, E.J., E.V. Bedenko, N. Hirota, S.S. Hecht and D. Hoffmann. 1979. A comparison of the mutagenicity, tumor-initiating activity and complete carcinogenicity of polynuclear aromatic hydrocarbons. In: Polynuclear Aromatic Hydrocarbons, P.W. Jones and P. Leber, Ed. Ann Arbor Science Publishers, Ann Arbor, MI. p. 705-721.

LaVoie, E.J., D.T. Coleman, J.E. Rice, N.G. Geddie and D. Hoffmann. 1985. Tumor-initiating activity, mutagenicity, and metabolism of methylated anthracenes. Carcinogenesis. 6(10): 1483-1488.

Lubet, R.A., E. Kiss, M.M. Gallagher, C. Dively, R.E. Kouri and L.M. Schectman. 1983. Induction of neoplastic transformation and DNA single- strand breaks in C3H/10T1/2 clone 8 cells by polycyclic hydrocarbons and alkylating agents. J. Natl. Cancer Inst. 71(5): 991-997.

Martin, C.N. and A.C. McDermid. 1981. Testing of 42 coded compounds for their ability to induce unscheduled DNA repair synthesis in HeLa cells. In: Evaluation of Short-term Tests for Carcinogens. Report of the International Collaborative Program. Progress in Mutation Research, Vol. 1, F.J. de Serres and J. Ashby, Ed. Amsterdam, Elsevier, North Holland. p. 533-537.

Martin, C.N., A.C. McDermid and R.C. Garner. 1978. Testing of known carcinogens and noncarcinogens for their ability to induce unscheduled DNA synthesis in HeLa cells. Cancer Res. 38: 2621-2627.

McCann, J.E., E. Choi, E. Yamasaki and B.N. Ames. 1975. Detection of carcinogens as mutagens in the Salmonella/microsome test: Assay of 300 chemicals. Proc. Natl. Acad. Sci. USA. 72(12): 5135-5139.

McCarroll, N.E., B.H. Keech and C.E. Piper. 1981. A microsuspension adaptation of the Bacillus subtilis 'rec' assay. Environ. Mutagen. 3:607-616.

Pienta, R.J., J.A. Poiley and W.B. Libherz, III. 1977. Morphological transformation of early passage golden Syrian hamster embryo cells derived from cryopreserved primary cultures as a reliable in vitro bioassay for identifying diverse carcinogens. Int. J. Cancer. 19: 642-655.

Probst, G.S., R.E. McMahon, L.E. Hill, C.Z. Thompson, J.K. Epp and S.B. Neal. 1981. Chemically-induced unscheduled DNA synthesis in primary rat hepatocyte cultures: A comparison with bacterial mutagenicity using 218 compounds. Environ. Mutagen. 3: 11-32.

Rosenkrantz, H.S. and L.A. Poirier. 1979. Evaluation of the mutagenicity and DNA-modifying activity of carcinogens and noncarcinogens in microbial systems. J. Natl. Cancer Inst. 62(4): 873-892.

Sakai, M., D. Yoshida and S. Mizusdki. 1985. Mutagenicity of polycyclic aromatic hydrocarbons and quinones on Salmonella typhimurium TA97. Mutat. Res. 156: 61-67.

Salamone, M.F., J.A. Heddle and M. Katz. 1979. The mutagenic activity of thirty polycyclic aromatic hydrocarbons (PAH) and oxides in urban airborne particulates. Environ. Int. 2: 37-43.

Schmahl, D. 1955. Examination of the carcinogenic action of naphthalene and anthracene in rats. Z. Krebsforsch. 60: 697-710.

Scribner, J.D. 1973. Brief communication: Tumor initiation by apparently noncarcinogenic polycyclic aromatic hydrocarbons. J. Natl. Cancer Inst. 50: 1717-1719.

Simmon, V.F. 1979a. In vitro mutagenicity assays of chemical carcinogens and related compounds with Salmonella typhimurium. J. Natl. Cancer Inst. 62(4): 893-899.

Simmon, V.F. 1979b. In vitro assays of recombinogenic activity of chemical carcinogens and related compounds with Saccharomyces cerevisiae D3. J. Natl. Cancer Inst. 62(4): 901-909.

Stanton, M.F., E. Miller, C. Wrench and R. Blackwell. 1972. Experimental induction of epidermoid carcinoma in the lungs of rats by cigarette smoke condensate. J. Natl. Cancer Inst. 49(3): 867-877.

Tong, C., S.V. Brat and G.M., Williams. 1981. Sister-chromatid exchange induction by polycyclic aromatic hydrocarbons in an intact cell system of adult rat-lever epithelial cells. Mutat. Res. 91: 467-473.

U.S. EPA. 1990. Drinking Water Criteria Document for Polycyclic Aromatic Hydrocarbons (PAHs). Prepared by the Office of Health and Environmental Assessment, Environmental Criteria and Assessment Office, Cincinnati, OH for the Office of Drinking Water, Washington, DC. ECAO-CIN-D010, September, 1990. (Final Draft) Williams, G.M. 1977. Detection of chemical carcinogens by unscheduled DNA synthesis in rat liver primary cell cultures. Cancer Res. 37: 1845-1851.

VI.D. DRINKING WATER HA REFERENCES

None

VII. REVISION HISTORY

Date	Section	Description
09/01/90	I.A.	Oral RfD summary on-line
09/01/90	VI.	Bibliography on-line
01/01/91	II.	Carcinogen assessment on-line
01/01/91	VI.C.	Carcinogen assessment references added
07/01/91	I.A.7.	Primary and secondary contacts changed
01/01/92	IV.	Regulatory Action section on-line
07/01/93	I.A.6.	Source Document and Other EPA Documentation clarified
09/01/94	I.B.	Inhalation RfD now under review

SYNONYMS

ANTHRACEN [GERMAN]	HSDB 702
ANTHRACENE	NSC 7958
ANTHRACIN	PARANAPHTHALENE
GREEN OIL	TETRA OLIVE N2G

Source: *Instant EPA's IRIS*, 1997, Instant Reference Sources, Inc., 7605 Rockpoint Drive, Austin, TX 78731 (Fax: 512-345-2386; Internet URL, http://www.instantref.com/inst-ref.htm).

Arsenic

As

Introduction

Arsenic is a silver-gray brittle, crystalline solid. It also exists in black and yellow amorphous forms. It is used as an alloying agent for heavy metals, in special solders and in medicine [830]. It is also used in bronzing, pyrotechny, and for hardening and improving the sphericity of shot, and as a doping agent in solid-state devices such as transistors. Calcium arsenate, lead arsenate, and others have been used as agricultural insecticides and poisons. Gallium arsenide is used as a laser material [706]. Arsenic is highly persistent in water, with a half-life of more than 200 days. Some arsenic substances increase in concentration, or bioaccumulate, in living organisms as they breathe contaminated air, drink contaminated water, or eat contaminated food. These chemicals can become concentrated in the tissues and internal organs of animals and humans. The concentration of arsenic found in fish tissues is expected to be somewhat higher than the average concentration of arsenic in the water from which the fish was taken. [830]

Identifying Information

CAS NUMBER: 7440-38-2

NIOSH Registry Number: CG0525000

SYNONYMS: GRAY-ARSENIC; ARSENIC, INORGANIC

Endocrine Disruptor Information

Arsenic is on EPA's priority list of suspect EEDs [811]. In support of this importance, arsenic is considered a metal similar in suspected EED properties to lead. Evidence has been obtained from rats, which includes lead exposure *in utero*, "suggesting a dual site of lead action: (a) at the level of the hypothalamic pituitary unit, and (b) directly at the level of gonadal steroid biosynthesis." Blood-serum testosterone levels (in males) and estradiol levels (in females) were most severely suppressed in those animals where exposure began *in utero*. Both sexes in this group showed a 25% suppression in prepubertal growth (as measured by weight). [868], [811] Arsenic is also reported as being able to damage the developing fetus and it should be handled as a potential teratogenic agent since some arsenic compounds are known teratogens. [830]

Chemical and Physical Properties

ATOMIC FORMULA: As

ATOMIC WEIGHT: 74.92 [703]

PHYSICAL DESCRIPTION: Silvery to black, brittle, crystalline and amorphous metalloid [703]

SPECIFIC GRAVITY: 5.73 @ 20°C/4°C [704]

MELTING POINT: 814°C @ 36 atm [703]

BOILING POINT: Sublimes @ 612°C [703]

SOLUBILITY: Water: Insoluble [703] HNO_3: Soluble [703]

OTHER PHYSICAL DATA: Vapor Pressure: 1 mm @ 372°C (sublimes) [703]; Density: black crystals 5.724 @ 14°C [703]

Source: *Instant EPA's Air Toxics*, 1994, Instant Reference Sources, Inc., 7605 Rockpoint Drive, Austin, TX 78731 (Fax: 512-345-2386; Internet URL, http://www.instantref.com/inst-ref.htm).

Environmental Reference Materials

A source of EED environmental reference materials for arsenic is Radian International LLC, Austin, TX (Fax 512-454-0268; Internet URL, http://www.radian.com/standards); Catalog No. ICA-006L

Hazardous Properties

ACUTE/CHRONIC HAZARDS: Arsenic is a human carcinogen. It is a poison by subcutaneous, intramuscular, and intraperitoneal routes. Human systemic skin and gastrointestinal effects occur by ingestion. It is an experimental teratogen and tumorigen. Mutagenic data exists. When heated or on contact with acid or acid fumes, it emits highly toxic fumes. [703] Target organs include the liver, kidneys, skin, lungs and lymphatic system. [704]

VOLATILITY: Vapor Pressure: 1 mm @ 372°C (sublimes) [703]

FIRE HAZARD: Flash Point: Not available. Flammable in the form of dust when exposed to heat or flame or by chemical reaction with powerful oxidizers. Slightly explosive in the form of dust when exposed to flame. [703]

REACTIVITY: Flammable by chemical reaction with powerful oxidizers. Emits highly toxic fumes on contact with acid or acid fumes; can react vigorously on contact with oxidizing materials. Incompatible with bromine azide; dirubidium acetylide; halogens; palladium; zinc; platinum; NCl_3; $AgNO_3$; CrO_3; Na_2O_2; hexafluoro isopropylideneamino lithium. [703] Arsenical materials are incompatible with any reducing agent [706].

STABILITY: When heated, emits highly toxic fumes [703].

Source: *Instant EPA's Air Toxics*, 1994, Instant Reference Sources, Inc., 7605 Rockpoint Drive, Austin, TX 78731 (Fax: 512-345-2386; Internet URL, http://www.instantref.com/inst-ref.htm).

Medical Symptoms of Exposure

Symptoms of exposure to arsenic may include ulceration of the nasal septum, dermatitis, gastrointestinal disturbances, peripheral neuropathy, respiratory irritation, and hyperpigmentation of the skin [704]. Repeated exposure can also damage the liver, cause narrowing of the blood vessels, or interfere with the bone marrow's ability to make red blood cells [830].

Source: *Instant EPA's Air Toxics*, 1994, Instant Reference Sources, Inc., 7605 Rockpoint Drive, Austin, TX 78731 (Fax: 512-345-2386; Internet URL, http://www.instantref.com/inst-ref.htm).

Toxicological Information

The toxicological information below is gathered from several sources including two national databases: one from the *National Toxicology Program Chemical Repository Database* and another from *EPA's Integrated Risk Information System (IRIS).*

Toxicology Information from the National Toxicology Program

Arsenic is not currently listed in NTP's Chemical Repository database nor is it scheduled for addition in the future.

Source: *Instant Tox-Base*, 1995, Instant Reference Sources, Inc., 7605 Rockpoint Drive, Austin, TX 78731 (Fax: 512-345-2386; Internet URL, http://www.instantref.com/inst-ref.htm).

Toxicology Information from EPA's Integrated Risk Information System (IRIS)

I. CHRONIC HEALTH HAZARD ASSESSMENTS FOR NONCARCINOGENIC EFFECTS

I.A. REFERENCE DOSE FOR CHRONIC ORAL EXPOSURE (RfD)

The Reference Dose (RfD) is based on the assumption that thresholds exist for certain toxic effects such as cellular necrosis, but may not exist for other toxic effects such as carcinogenicity. In general, the RfD is an estimate (with uncertainty spanning perhaps an order of magnitude) of a daily exposure to the human population (including sensitive subgroups) that is likely to be without an appreciable risk of deleterious effects during a lifetime. Please refer to Background Document 1 in Service Code 5 for an elaboration of these concepts. RfDs can also be derived for the noncarcinogenic health effects of compounds which are also carcinogens. Therefore, it is essential to refer to other sources of information concerning the carcinogenicity of this substance. If the U.S. EPA has evaluated this substance for potential human carcinogenicity, a summary of that evaluation will be contained in Section II of this file when a review of that evaluation is completed.

NOTE: There was not a clear consensus among Agency scientists on the oral RfD. Applying the Agency's RfD methodology, strong scientific arguments can be made for various values within a factor of 2 or 3 of the currently recommended RfD value, i.e., 0.1 to 0.8 ug/kg/day. It should be noted, however, that the RfD methodology, by definition, yields a number with inherent uncertainty spanning perhaps an order of magnitude. New data that possibly impact on the recommended RfD for arsenic will be evaluated by the Work Group as it becomes available. Risk managers should recognize the considerable flexibility afforded them in formulating regulatory decisions when uncertainty and lack of clear consensus are taken into account.

I.A.1. ORAL RfD SUMMARY

Critical Effect	Experimental Doses*	UF	MF	RfD
Hyperpigmentation, keratosis and possible vascular complications	NOAEL: 0.009 mg/L converted to 0.0008 mg/kg-day LOAEL: 0.17 mg/L converted to 0.014 mg/kg-day oral exposure	3	1	3E-4 mg/kg-day

*Conversion Factors: NOAEL was based on an arithmetic mean of 0.009 mg/L in a range of arsenic concentration of 0.001 to 0.017 mg/L. This NOAEL also included estimation of arsenic from food. Since experimental data were missing, arsenic concentrations in sweet potatoes and rice were estimated as 0.002 mg/day. Other assumptions included consumption of 4.5 L water/day and 55 kg bw (Abernathy et al., 1989). NOAEL = [(0.009 mg/L x 4.5 L/day) + 0.002 mg/day] / 55 kg = 0.0008 mg/kg-day. The LOAEL dose was estimated using the same assumptions as the NOAEL starting with an arithmetic mean water concentration from Tseng (1977) of 0.17 mg/L. LOAEL = [(0.17 mg/L x 4.5 L/day) + 0.002 mg/day] / 55 kg = 0.014 mg/kg-day.

Human chronic Tseng, 1977; Tseng et al., 1968

I.A.2. PRINCIPAL AND SUPPORTING STUDIES (ORAL RfD)

Tseng, W.P. 1977. Effects and dose-response relationships of skin cancer and blackfoot disease with arsenic. Environ. Health Perspect. 19: 109-119. Tseng, W.P., H.M. Chu, S.W. How, J.M. Fong, C.S. Lin and S. Yeh. 1968. Prevalence of skin cancer in an endemic area of chronic arsenicism in Taiwan. J. Natl. Cancer Inst. 40: 453-463.

The data reported in Tseng (1977) show an increased incidence of blackfoot disease that increases with age and dose. Blackfoot disease is a significant adverse effect. The prevalences (males and females combined) at the low dose are 4.6 per 1000 for the 20-39 year group, 10.5 per 1000 for the 40-59 year group, and 20.3 per 1000 for the >60 year group. Moreover, the prevalence of blackfoot disease in each age group increases with increasing dose. However, a recent report indicates that it may not be strictly due to arsenic exposure (Lu, 1990). The data in Tseng et al. (1968) also show increased incidences of hyperpigmentation and keratosis with age. The overall prevalences of hyperpigmentation and keratosis in the exposed groups are 184 and 71 per 1000, respectively. The text states that the incidence increases with dose, but data for the individual doses are not shown. These data show that the skin lesions are the more sensitive endpoint. The low dose in the Tseng (1977) study is considered a LOAEL.

The control group described in Tseng et al. (1968; Table 3) shows no evidence of skin lesions and presumably blackfoot disease, although this latter point is not explicitly stated. This group is considered a NOAEL.

The arithmetic mean of the arsenic concentration in the wells used by the individuals in the NOAEL group is 9 ug/L (range: 1-17 ug/L) (Abernathy et al., 1989). The arithmetic mean of the arsenic concentration in the wells used by the individuals in the LOAEL group is 170 ug/L (Tseng, 1977; Figure 4). Using estimates provided by Abernathy et al. (1989), the NOAEL and LOAEL doses for both food and water are as follows: LOAEL - [170 ug/L x 4.5 L/day + 2 ug/day (contribution of food)] x (1/55 kg) = 14 ug/kg/day; NOAEL - [9 ug/L x 4.5 L/day + 2 ug/day (contribution of food)] x (1/55 kg) = 0.8 ug/kg/day.

Although the control group contained 2552 individuals, only 957 (approximately 38%) were older than 20, and only 431 (approximately 17%) were older than 40. The incidence of skin lesions increases sharply in individuals above 20; the incidence of blackfoot disease increases sharply in individuals above 40 (Tseng, 1968; Figures 5, 6 and 7). This study is less powerful than it appears at first glance. However, it is certainly the most powerful study available on arsenic exposure to people.

This study shows an increase in skin lesions, 22% (64/296) at the high dose vs. 2.2% (7/318) at the low dose. The average arsenic concentration in the wells at the high dose is 410 ug/L and at the low dose is 5 ug/L (Cebrian et al., 1983; Figure 2 and Table 1) or 7 ug/L (cited in the abstract). The average water consumption is 3.5 L/day for males and 2.5 L/day for females. There were about an equal number of males and females in the study. For the dose estimates given below we therefore assume an average of 3 L/day. No data are given on the arsenic exposure from food or the body weight of the participants (we therefore assume 55 kg). The paper states that exposure times are directly related to chronological age in 75% of the cases. Approximately 35% of the participants in the study are more than 20 years old (Figure 1).

Exposure estimates (water only) are: high dose - 410 ug/L x 3 L/day x (1/55 kg) = 22 ug/kg/day; low dose - 5-7 ug/L x 3 L/day x (1/55 kg) = 0.3-0.4 ug/kg/day.

The high-dose group shows a clear increase in skin lesions and is therefore designated a LOAEL. There is some question whether the low dose is a NOAEL or a LOAEL since there is no way of knowing what the incidence of skin lesions would be in a group where the exposure to arsenic is zero. The 2.2% incidence of skin lesions in the low-dose group is higher than that reported in the Tseng et al. (1968) control group, but the dose is lower (0.4 vs. 0.8 ug/kg/day).

The Southwick et al. (1983) study shows a marginally increased incidence of a variety of skin lesions (palmar and plantar keratosis, diffuse palmar or plantar hyperkeratosis, diffuse pigmentation, and arterial insufficiency) in the individuals exposed to arsenic. The incidences are 2.9% (3/105) in the control group and 6.3% (9/144) in the exposed group. There is a slight, but not statistically significant increase in the percent of exposed individuals that have abnormal nerve conduction (8/67 vs. 13/83, or 12% vs. 16% (Southwick et al., 1983; Table 8). The investigators excluded all individuals older than 47 from the nerve conduction portion of the study. These are the individuals most likely to have the longest exposure to arsenic.

Although neither the increased incidence of skin lesions nor the increase in abnormal nerve conduction is statistically significant, these effects may be biologically significant because the same abnormalities occur at higher doses in other studies. The number of subjects in this study was insufficient to establish statistical significance.

Table 3 (Southwick et al., 1983) shows the annual arsenic exposure from drinking water. No data are given on arsenic exposure from food or the body weight (assume 70 kg). Exposure times are not clearly defined, but are >5 years, and dose groups are ranges of exposure.

Exposure estimates (water only) are: dosed group - 152.4 mg/year x 1 year/365 days x (1/70) kg = 6 ug/kg/day; control group - 24.2 mg/year x year/365 days x (1/70) kg = 0.9 ug/kg/day.

Again because there are no data for a group not exposed to arsenic, there is some question if the control group is a NOAEL or a LOAEL. The incidence of skin lesions in this group is about the same as in the low-dose group from the Cebrian et al. (1983) study; the incidence of abnormal nerve conduction in the control group is higher than that from the low-dose group in the Hindmarsh et al. (1977) study described below. The control dose is comparable to the dose to the control group in the Tseng et al. (1968) and Hindmarsh et al. (1977) studies. The dosed group may or may not be a LOAEL, since it is does not report statistically significant effects when compared to the control.

This study shows an increased incidence of abnormal clinical findings and abnormal electromyographic findings with increasing dose of arsenic (Hindmarsh et al., 1977; Tables III and VI). However, the sample size is extremely small. Percentages of abnormal clinical signs possibly attributed to As were 10, 16, and 40% at the low, mid and high doses, respectively. Abnormal EMG were 0, 17 and 53% in the same three groups.

The exact doses are not given in the Hindmarsh et al. (1977) paper; however, some well data are reported in Table V. The arithmetic mean of the arsenic concentration in the high-dose and mid-dose wells is 680 and 70 ug/L, respectively. Figure 1 (Hindmarsh et al., 1977) shows that the average arsenic concentration of the low-dose wells is about 25 ug/L. No data are given on arsenic exposure from food. We assume daily water consumption of 2 liters and body weight of 70 kg. Exposure times are not clearly stated.

Exposure estimates (water only) are: low - 25 ug/L x 2 L/day x (1/70) kg = 0.7 ug/kg/day; mid - 70 ug/L x 2 L/day x (1/70) kg = 2 ug/kg/day; high - 680 ug/L x 2 L/day x (1/70) kg = 19 ug/kg/day.

The low dose is a no-effect level for abnormal EMG findings. However, because there is no information on the background incidence of abnormal clinical findings in a population with zero exposure to arsenic, there is no way of knowing if the low dose is a no-effect level or another marginal effect level for abnormal clinical findings. The low dose is comparable to the dose received by the control group in the Tseng (1977) and Southwick et al. (1983) studies.

The responses at the mid dose do not show a statistically significant increase but are part of a statistically significant trend and are biologically significant. This dose is an equivocal NOAEL/LOAEL. The high dose is a clear LOAEL for both responses.

As discussed previously there is no way of knowing whether the low doses in the Cebrian et al. (1983), Southwick et al. (1983) and Hindmarsh et al. (1977) studies are NOAELs for skin lesions and/or abnormal nerve conduction. However, because the next higher dose in the Southwick and Hindmarsh studies only shows marginal effects at doses 3-7 times higher, the Agency feels comfortable in assigning the low doses in these studies as NOAELs.

The Tseng (1977) and Tseng et al. (1968) studies are therefore considered superior for the purposes of developing an RfD and show a NOAEL for a sensitive endpoint. Even discounting the people <20 years of age, the control group consisted of 957 people that had a lengthy exposure to arsenic with no evidence of skin lesions.

The following is a summary of the defined doses in mg/kg-day from the principal and supporting studies:

1) Tseng (1977): NOAEL = 8E-4; LOAEL = 1.4E-2

2) Cebrian et al. (1983): NOAEL = 4E-4; LOAEL = 2.2E-2

3) Southwick et al. (1983): NOAEL = 9E-4; LOAEL = none (equivocal effects at 6E-3)

4) Hindmarsh et al., 1977: NOAEL = 7E-4; LOAEL = 1.9E-2 (equivocal effects at 2E-3)

I.A.3. UNCERTAINTY AND MODIFYING FACTORS (ORAL RfD)

UF -- The UF of 3 is to account for both the lack of data to preclude reproductive toxicity as a critical effect and to account for some uncertainty in whether the NOAEL of the critical study accounts for all sensitive individuals.

MF -- None

I.A.4. ADDITIONAL STUDIES / COMMENTS (ORAL RfD)

Ferm and Carpenter (1968) produced malformations in 15-day hamster fetuses via intravenous injections of sodium arsenate into pregnant dams on day 8 of gestation at dose levels of 15, 17.5, or 20 mg/kg bw. Exencephaly, encephaloceles, skeletal defects and genitourinary systems defects were produced. These and other terata were produced in mice and rats all at levels around 20 mg/kg bw. Minimal effects or no effects on fetal development have been observed in studies on chronic oral exposure of pregnant rats or mice to relatively low levels of arsenic via drinking water (Schroeder and Mitchner, 1971). Nadeenko et al. (1978) reported that intubation of rats with arsenic solution at a dose level of 25 ug/kg/day for a period of 7 months, including pregnancy, produced no significant embryotoxic effects and only infrequent slight expansion of ventricles of the cerebrum, renal pelves and urinary bladder. Hood et al. (1977) reported that very high single oral doses of arsenate solutions (120 mg/kg) to pregnant mice were necessary to cause prenatal fetal toxicity, while multiple doses of 60 mg/kg on 3 days had little effect.

Extensive human pharmacokinetic, metabolic, enzymic and long-term information is known about arsenic and its metabolism. Valentine et al. (1987) established that human blood arsenic levels did not increase until daily water ingestion of arsenic exceeded approximately 250 ug/day (approximately 120 ug of arsenic/L. Methylated species of arsenic are successively 1 order of magnitude less toxic and less teratogenic (Marcus and Rispin, 1988). Some evidence suggests that inorganic arsenic is an essential nutrient in goats, chicks, minipigs and rats (NRC, 1989). No comparable data are available for humans.

I.A.5. CONFIDENCE IN THE ORAL RfD

Study -- MediumData Base -- MediumRfD -- Medium

Confidence in the chosen study is considered medium. An extremely large number of people were included in the assessment (>40,000) but the doses were not well-characterized and other contaminants were present. The supporting human toxicity data base is extensive but somewhat flawed. Problems exist with all of the epidemiological studies. For example, the Tseng studies do not look at potential exposure from food or other source. A similar criticism can be made of the Cebrian et al. (1983) study. The U.S. studies are too small in number to resolve several issues. However, the data base does support the choice of NOAEL. It garners medium confidence. Medium confidence in the RfD follows.

I.A.6. EPA DOCUMENTATION AND REVIEW OF THE ORAL RfD

Source Document -- This assessment is not presented in any existing U.S. EPA document.

This analysis has been reviewed by EPA's Risk Assessment Council on 11/15/90. This assessment was discussed by the Risk Assessment Council of EPA on 11/15/90 and verified through a series of meetings during the 1st, 2nd and 3rd quarters of FY91.

Other EPA Documentation -- U.S. EPA, 1984, 1988

Agency Work Group Review -- 03/24/88, 05/25/88, 03/21/89, 09/19/89, 08/22/90, 09/20/90

Verification Date -- 11/15/90

I.A.7. EPA CONTACTS (ORAL RfD)

Charles Abernathy / OST -- (202)260-5374

I.B. REFERENCE CONCENTRATION FOR CHRONIC INHALATION EXPOSURE (RfC)

[Ed. Note: EPA does not have information yet in this section of IRIS].

II. CARCINOGENICITY ASSESSMENT FOR LIFETIME EXPOSURE

Section II provides information on three aspects of the carcinogenic risk assessment for the agent in question; the U.S. EPA classification, and quantitative estimates of risk from oral exposure and from inhalation exposure. The classification reflects a weight-of-evidence judgment of the likelihood that the agent is a human carcinogen. The quantitative risk estimates are presented in three ways. The slope factor is the result of application of a low-dose extrapolation procedure and is presented as the risk per (mg/kg)/day. The unit risk is the quantitative estimate in terms of either risk per ug/L drinking water or risk per ug/cu.m air breathed. The third form in which risk is presented is a drinking water or air concentration providing cancer risks of 1 in 10,000, 1 in 100,000 or 1 in 1,000,000. Background Document 2 (Service Code 5) provides details on the rationale and methods used to derive the carcinogenicity values found in IRIS. Users are referred to Section I for information on long-term toxic effects other than carcinogenicity.

II.A. EVIDENCE FOR CLASSIFICATION AS TO HUMAN CARCINOGENICITY

II.A.1. WEIGHT-OF-EVIDENCE CLASSIFICATION

Classification -- A; human carcinogen

Basis -- based on sufficient evidence from human data. An increased lung cancer mortality was observed in multiple human populations exposed primarily through inhalation. Also, increased mortality from multiple internal organ cancers (liver, kidney, lung, and bladder) and an increased incidence of skin cancer were observed in populations consuming drinking water high in inorganic arsenic.

II.A.2. HUMAN CARCINOGENICITY DATA

Sufficient.   Studies of smelter worker populations (Tacoma, WA; Magma, UT; Anaconda, MT; Ronnskar, Sweden; Saganoseki-Machii, Japan) have all found an association between occupational arsenic exposure and lung cancer mortality (Enterline and Marsh, 1982; Lee-Feldstein, 1983; Axelson et al., 1978; Tokudome and Kuratsune, 1976; Rencher et al., 1977). Both proportionate mortality and cohort studies of pesticide manufacturing workers have shown an excess of lung cancer deaths among exposed persons (Ott et al., 1974; Mabuchi et al., 1979). One study of a population residing near a pesticide manufacturing plant revealed that these residents were also at an excess risk of lung cancer (Matanoski et al., 1981). Case reports of arsenical pesticide applicators have also corroborated an association between arsenic exposure and lung cancer (Roth, 1958).

A cross-sectional study of 40,000 Taiwanese exposed to arsenic in drinking water found significant excess skin cancer prevalence by comparison to 7500 residents of Taiwan and Matsu who consumed relatively arsenic-free water (Tseng et al., 1968; Tseng, 1977). Although this study demonstrated an association between arsenic exposure and development of skin cancer, it has several weaknesses and uncertainties, including poor nutritional status of the exposed populations, their genetic susceptibility, and their exposure to inorganic arsenic from non-water sources, that limit the study's usefulness in risk estimation. Dietary inorganic arsenic was not considered nor was the potential confounding by contaminants other than arsenic in drinking water. There may have been bias of examiners in the original study since no skin cancer or preneoplastic lesions were seen in 7500 controls; prevalence rates rather than mortality rates are the endpoint; and furthermore there is concern of the applicability of extrapolating data from Taiwanese to the U.S. population because of different background rates of cancer, possibly genetically determined, and differences in diet other than arsenic (e.g., low protein and fat and high carbohydrate) (U.S. EPA, 1988).

A prevalence study of skin lesions was conducted in two towns in Mexico, one with 296 persons exposed to drinking water with 0.4 mg/L arsenic and a similar group with exposure at 0.005 mg/L. The more exposed group had an increased incidence of palmar keratosis, skin hyperpigmentation and hypopigmentation, and four skin cancers (histologically unconfirmed) (Cebrian et al. (1983). The association between skin cancer and arsenic is weak because of the small number of cases, small cohort size, and short duration follow-up; also there was no unexposed group in either town. No excess skin cancer incidence has been observed in U.S. residents consuming relatively high levels of arsenic in drinking water but the numbers of exposed persons were low (Morton et al., 1976; Southwick et al., 1981). Therapeutic use of Fowler's solution (potassium arsenite) has also been associated with development of skin cancer and hyperkeratosis (Sommers and McManus, 1953; Fierz, 1965); several case reports implicate exposure to Fowler's solution in skin cancer development (U.S. EPA, 1988).

Several follow-up studies of the Taiwanese population exposed to inorganic arsenic in drinking water showed an increase in fatal internal organ cancers as well as an increase in skin cancer. Chen et al. (1985) found that the standard mortality ratios (SMR) and cumulative mortality rates for cancers of the bladder, kidney, skin, lung and liver were significantly greater in the Blackfoot disease endemic area of Taiwan when compared with the age adjusted rates for the general population of Taiwan. Blackfoot disease (BFD, an endemic peripheral artery disease) and these cancers were all associated with high levels of arsenic in drinking water. In the endemic area, SMRs were greater in villages that used only artesian well water (high in arsenic) compared with villages that partially or completely used surface well water (low in arsenic). However, dose-response data were not developed (Chen et al. 1985).

A retrospective case-control study showed a significant association between duration of consuming high-arsenic well water and cancers of the liver, lung and bladder (Chen et al., 1986). In this study, cancer deaths in the Blackfoot disease endemic area between January 1980 and December 1982 were chosen for the case group. About 90% of the 86 lung cancers and 95 bladder cancers in the registry were histologically or cytologically confirmed and over 70% of the liver cancers were confirmed by biopsy or _-fetoprotein presence with a positive liver x-ray image. Only confirmed cancer cases were included in the study. A control group of 400 persons living in the same area was frequency-matched with cases by age and sex. Standardized questionnaires of the cases (by proxy) and controls determined the history of artesian well water use, socioeconomic variables, disease history, dietary habits, and lifestyle. For the cancer cases, the age-sex adjusted odds ratios were increased for bladder (3.90), lung (3.39), and liver (2.67) cancer for persons who had used artesian well water for 40 or more years when compared with controls who had never used artesian well water. Similarly, in a 15-year study of a cohort of 789 patients of Blackfoot disease, an increased mortality from cancers of the liver, lung, bladder and kidney was seen among BFD patients when compared with the general population in the endemic area or when compared with the general population of Taiwan. Multiple logistic regression analysis to adjust for other risk factors including cigarette smoking did not markedly affect the exposure-response relationships or odds ratios (Chen et al., 1988).

A significant dose-response relationship was found between arsenic levels in artesian well water in 42 villages in the southwestern Taiwan and age-adjusted mortality rates from cancers at all sites, cancers of the bladder, kidney, skin, lung, liver and prostate (Wu et al., 1989). An ecological study of cancer mortality rates and arsenic levels in drinking water in 314 townships in Taiwan also corroborated the association between arsenic levels and mortality from the internal cancers (Chen and Wang, 1990).

Chen et al.(1992) conducted a recent analysis of cancer mortality data from the arsenic-exposed population to compare risk of various internal cancers and compare risk between males and females. The study area and population have been described by Wu et al. (1989). It is limited to 42 southwestern coastal villages where residents have used water high in arsenic from deep artesian wells for more than 70 years. Arsenic levels in drinking water ranged from 0.010 to 1.752 ppm. The study population had 898,806 person-years of observation and 202 liver cancer, 304 lung cancer, 202 bladder cancer and 64 kidney cancer deaths. The study population was stratified into four groups according to median arsenic level in well water (<0.10 ppm, 0.10-0.29 ppm, 0.30-0.59 ppm and 60+ ppm), and also stratified into four age groups (<30 years, 30-49 years, 50-69 years and 70+ years). Mortality rates were found to increase significantly with age for all cancers and significant dose-response relationships were observed between arsenic level and mortality from cancer of the liver, lung, bladder and kidney in most age groups of both males and females. The data generated by Chen et al. (1992) provide evidence for an association of the levels of arsenic in drinking water and duration of exposure with the rate of mortality from cancers of the liver, lung, bladder, and kidney. Dose-response relationships are clearly shown by the tabulated data (Tables II-V of Chen et al., 1992). Previous studies summarized in U.S. EPA (1988) showed a similar association in the same Taiwanese population with the prevalence of skin cancers (which are often non-fatal). Bates et al. (1992) and Smith et al. (1992) have recently reviewed and evaluated the evidence for arsenic ingestion and internal cancers.

II.A.3. ANIMAL CARCINOGENICITY DATA

Inadequate.   There has not been consistent demonstration of carcinogenicity in test animals for various chemical forms of arsenic administered by different routes to several species (IARC, 1980). Furst (1983) has cited or reviewed animal carcinogenicity testing studies of nine inorganic arsenic compounds in over nine strains of mice, five strains of rats, in dogs, rabbits, swine and chickens. Testing was by the oral, dermal, inhalation, and parenteral routes. All oxidation states of arsenic were tested. No study demonstrated that inorganic arsenic was carcinogenic in animals. Dimethylarsonic acid (DMA), the end metabolite predominant in humans and animals, has been tested for carcinogenicity in two strains of mice and was not found positive (Innes et al., 1969); however, this was a screening study and no data were provided. The meaning of non-positive data for carcinogenicity of inorganic arsenic is uncertain, the mechanism of action in causing human cancer is not known, and rodents may not be a good model for arsenic carcinogenicity testing. There are some data to indicate that arsenic may produce animal lung tumors if retention time in the lung can be increased (Pershagen et al., 1982, 1984).

II.A.4. SUPPORTING DATA FOR CARCINOGENICITY

A retrospective cohort mortality study was conducted on 478 British patients treated between 1945-1969 with Fowler's solution (potassium arsenite). The mean duration of treatment was 8.9 months and the average total oral consumption of arsenic was about 1890 mg (daily dose x duration). In 1980, 139 deaths had occurred. No excess deaths from internal cancers were seen after this 20-year follow-up. Three bladder cancer deaths were observed (1.19 expected, SMR 2.5) (Cuzick et al., 1982). A recent follow-up (Cuzick et al., 1992) indicated no increased mortality from all cancers but a significant excess from bladder cancer (5 cases observed/1.6 expected; SMR of 3.07). A subset of the original cohort (143 persons) had been examined by a dermatologist in 1970 for signs of arsenicism (palmar keratosis). In 1990, there were 80 deaths in the subcohort and 11 deaths from internal cancers. All 11 subjects had skin signs (keratosis-10, hyperpigmentation-5 and skin cancer-3). A case-control study of the prevalence of palmar keratoses in 69 bladder cancer patients, 66 lung cancer patients and 218 hospital controls (Cuzick et al., 1984), indicated an association between skin keratosis (as an indicator of arsenic exposure) and lung and bladder cancer. Above the age of 50, 87% of bladder cancer patients and 71% of lung cancer patients but only 36% of controls had one or more keratoses. Several case reports implicate internal cancers with arsenic ingestion or specifically with use of Fowler's solution but the associations are tentative (U.S. EPA, 1988).

Sodium arsenate has been shown to transform Syrian hamster embryo cells (Dipaolo and Casto, 1979) and to produce sister chromatid-exchange in DON cells, CHO cells, and human peripheral lymphocytes exposed in vitro (Wan et al., 1982; Ohno et al., 1982; Larramendy et al., 1981; Andersen, 1983; Crossen, 1983).   Jacobson-Kram and Montalbano (1985) have reviewed the mutagenicity of inorganic arsenic and concluded that inorganic arsenic is inactive or very weak for induction of gene mutations in vitro but it is clastogenic with trivalent arsenic being an order of magnitude more potent than pentavalent arsenic.

Both the pentavalent and trivalent forms of inorganic arsenic are found in drinking water. In both animals and humans, arsenate (As+5) is reduced to arsenite (As+3) and the trivalent form is methylated to give the metabolites mononomethylarsinic acid (MMA) and dimethylarsonic acid (DMA) (Vahter and Marafante, 1988). The genotoxicity of arsenate (As+5) and arsenite (As+3) and the two methylated metabolites, MMA and DMA were compared in the thymidine kinase forward mutation assay in mouse lymphoma cells (Harrington-Brock et al. 1993; Moore et al., 1995, in press). Sodium arsenite (+3) and sodium arsenate (+5) were mutagenic at concentration of 1-2 ug/mL and 10-14 ug/mL, respectively, whereas MMA and DMA were significantly less potent, requiring 2.5-5 mg/mL and 10 mg/mL, respectively, to induce a genotoxic response. Based on small colony size the mutations induced were judged chromosomal rather than point mutations. The authors have previously shown that for chemicals having clastogenic activity (i.e., cause chromosomal mutations), the mutated cells grow more slowly than cells with single gene mutations and this results in small colony size. In the mouse lymphoma assay, chromosomal abberations were seen at approximately the same arsenic levels as TK forward mutations. Arsenate, arsenite and MMA were considered clastogenic but the abberation response with DMA was insufficient to consider it a clastogen. Since arsenic exerts its genotoxicity by causing chromosomal mutations, it has been suggested by the above authors that it may act in a latter stage of carcinogenesis as a progressor, rather than as a classical initiator or promotor (Moore et al., 1994). A finding which supports this process is that arsenate (8-16 uM) and arsenite (3 uM) have been shown to induce 2-10 fold amplification of the dihydrofolate reductase gene in culture in methotrexate resistant 3T6 mouse cells (Lee et al., 1988). Although the mechanism of induction in rodent cells is not known, gene amplification of oncogenes is observed in many human tumors. Inorganic arsenic has not been shown to mutate bacterial strains, it produces preferential killing of repair deficient strains (Rossman, 1981). Sodium arsenite (As+3) induces DNA-strand breaks which are associated with DNA-protein crosslinks in cultured human fibroblasts at 3 mM but not 10 mM (Dong and Luo, 1993) and it appears that arsenite inhibits the DNA repair process by inhibiting both excision and ligation (Jha et al., 1992; Lee-Chen et al., 1993).

The inhibitory effect of arsenite on strand-break rejoining during DNA repair was found to be reduced by adding glutathione to cell cultures (Huang et al., 1993). The cytotoxic effects of sodium arsenite in Chinese hamster ovary cells also has also found to correlate with the intracellular glutathione levels (Lee et al., 1989).

In vivo studies in rodents have shown that oral exposure of rats to arsenate (As+5) for 2-3 weeks resulted in major chromosomal abnormalities in bone marrow (Datta et al., 1986) and exposure of mice to As (+3) in drinking water for 4 weeks (250 mg As/L as arsenic trioxide) caused chromosomal aberrations in bone marrow cells but not spermatogonia (Poma et al., 1987); micronuclei in bone marrow cells were also induced by intraperitoneal dosing of mice with arsenate (DeKnudt et al., 1986; Tinwell et al., 1991). Chromosomal aberrations and sister chromatid exchange have been seen in patients exposed to arsenic from treatment with Fowler's solution (Burgdorf et al., 1977) and subjects exposed occupationally (Beckman et al., 1977) but no increase in either endpoint was seen in lymphocytes of subjects exposed to arsenic in drinking water (Vig et al., 1984).

II.B. QUANTITATIVE ESTIMATE OF CARCINOGENIC RISK FROM ORAL EXPOSURE

II.B.1. SUMMARY OF RISK ESTIMATES

Oral Slope Factor -- 1.5E+0 per (mg/kg)/day

Drinking Water Unit Risk -- 5E-5 per (ug/L)

Extrapolation Method -- Time- and dose-related formulation of the multistagemodel (U.S. EPA, 1988)

Drinking Water Concentrations at Specified Risk Levels:

Risk Level	Concentration
E-4 (1 in 10,000)	2E+0 ug/L
E-5 (1 in 100,000)	2E-1 ug/L
E-6 (1 in 100,000)	2E-2 ug/L

II.B.2. DOSE-RESPONSE DATA (CARCINOGENICITY, ORAL EXPOSURE)

The Risk Assessment Forum has completed a reassessment of the carcinogenicity risk associated with ingestion of inorganic arsenic (U.S. EPA, 1988). The data provided in Tseng et al., 1968 and Tseng, 1977 on about 40,000 persons exposed to arsenic in drinking water and 7500 relatively unexposed controls were used to develop dose-response data. The number of persons at risk over three dose intervals and four exposure durations, for males and females separately, were estimated from the reported prevalence rates as percentages. It was assumed that the Taiwanese persons had a constant exposure from birth, and that males consumed 3.5 L drinking water/day and females consumed 2.0 L/day. Doses were converted to equivalent doses for U.S. males and females based on differences in body weights and differences in water consumption and it was assumed that skin cancer risk in the U.S. population would be similar to the Taiwanese population. The multistage model with time was used to predict dose-specific and age-specific skin cancer prevalence rates associated with ingestion of inorganic arsenic; both linear and quadratic model fitting of the data were conducted. The maximum likelihood estimate (MLE) of skin cancer risk for a 70 kg person drinking 2 L of water per day ranged from 1E-3 to 2E-3 for an arsenic intake of 1 ug/kg/day. Expressed as a single value, the cancer unit risk for drinking water is 5E-5 per (ug/L). Details of the assessment are in U.S. EPA (1988).

Dose response data have not been developed for internal cancers for the Taiwanese population. The data of Chen et al. (1992) are considered inadequate at present.

II.B.3. ADDITIONAL COMMENTS (CARCINOGENICITY, ORAL EXPOSURE)

None.

II.B.4. DISCUSSION OF CONFIDENCE (CARCINOGENICITY, ORAL EXPOSURE)

This assessment is based on prevalence of skin cancer rather than mortality because the types of skin cancer studied are not normally fatal. However, competing mortality from Blackfoot disease in the endemic area of Taiwan would cause the risk of skin cancer to be underestimated. Other sources of inorganic arsenic, in particular those in food sources have not been considered because of lack of reliable information. There is also uncertainty on the amount of water consumed/day by Taiwanese males (3.5 L or 4.5 L) and the temporal variability of arsenic concentrations in specific wells was not known. The concentrations of arsenic in the wells was measured in the early 1960s and varied between 0.01 and 1.82 ppm. For many villages 2 to 5 analyses were conducted on well water and for other villages only one analysis was performed; ranges of values were not provided. Since tap water was supplied to many areas after 1966, the arsenic-containing wells were only used in dry periods. Because of the study design, particular wells used by those developing skin cancer could not be identified and arsenic intake could not be assigned except by village. Several uncertainties in exposure measurement reliability existed and subsequent analysis of drinking water found fluorescent substances in water that are possible confounders or caused synergistic effects. Uncertainties have been discussed in detail in U.S. EPA (1988). Uncertainties in exposure measurement can affect the outcome of dose-response estimation.

II.C.   QUANTITATIVE ESTIMATE OF CARCINOGENIC RISK FROM INHALATION EXPOSURE

II.C.1. SUMMARY OF RISK ESTIMATES

Inhalation Unit Risk -- 4.3E-3 per (ug/cu.m)

Extrapolation Method -- absolute-risk linear model

Air Concentrations at Specified Risk Levels:

Risk Level	Concentration
E-4 (1 in 10,000)	2E-2 per (ug/cu.m)
E-5 (1 in 100,000)	2E-3 per (ug/cu.m)
E-6 (1 in 1,000,000)	2E-4 per (ug/cu.m)

II.C.2.   DOSE-RESPONSE DATA FOR CARCINOGENICITY, INHALATION EXPOSURE

Tumor Type -- lung cancer
Test Animals -- human, male
Route -- inhalation, occupational exposure
Reference -- Brown and Chu, 1983a,b,c; Lee-Feldstein, 1983; Higgins, 1982; Enterline and Marsh, 1982

Ambient Unit Risk Estimates (per (ug/cu.m))

Exposure Source	Unit Study	Geometric Risk	Mean Unit Risk	Final Estimates Unit Risk
Anaconda	Brown and Chu Lee-Feldstein,	1.25E-3	smelter	1983a,b,c
	1983	2.80E-3	2.56E-3	
	Higgins, 1982; Higgins et al., 1982; Welch et al., 1982	4.90E-3	4.29E-3	
ASARCO smelter	Enterline and Marsh, 1982	7.6E-3	6.81E-3	7.19E-3

II.C.3. ADDITIONAL COMMENTS (CARCINOGENICITY, INHALATION EXPOSURE)

A geometric mean was obtained for data sets obtained with distinct exposed populations (U.S. EPA, 1984). The final estimate is the geometric mean of those two values. It was assumed that the increase in age-specific mortality rate of lung cancer was a function only of cumulative exposures.

The unit risk should not be used if the air concentration exceeds 2 ug/cu.m, since above this concentration the unit risk may not be appropriate.

II.C.4. DISCUSSION OF CONFIDENCE (CARCINOGENICITY, INHALATION EXPOSURE)

Overall a large study population was observed. Exposure assessments included air measurements for the Anaconda smelter and both air measurements and urinary arsenic for the ASARCO smelter. Observed lung cancer incidence was significantly increased over expected values. The range of the estimates derived from data from two different exposure areas was within a factor of 6.

II.D. EPA DOCUMENTATION, REVIEW, AND CONTACTS (CARCINOGENICITY ASSESSMENT)

II.D.1. EPA DOCUMENTATION

U.S. EPA. 1984, 1988, 1993

A draft of the 1984 Health Assessment Document for Inorganic Arsenic was independently reviewed in public session by the Environmental Health Committee of the U.S. EPA Science Advisory Board on September 22-23, 1983. A draft of the 1988 Special Report on Ingested Inorganic Arsenic; Skin Cancer; Nutritional Essentiality was externally peer reviewed at a two-day workshop of scientific experts on December 2-3, 1986. A draft of the Drinking Water Criteria Document for Arsenic was reviewed by the Drinking Water Committee of the U.S. EPA Science Advisory Board on March 10, 1993. The comments from these reviews were evaluated and considered in the revision and finalization of these reports.

II.D.2. REVIEW (CARCINOGENICITY ASSESSMENT)

Agency Work Group Review -- 01/13/88, 12/07/89, 02/03/94

Verification Date -- 02/03/94

II.D.3. U.S. EPA CONTACTS (CARCINOGENICITY ASSESSMENT)

Herman Gibb / NCEA -- (202)260-7315

Charles Abernathy / OST -- (202)260-5374

III. HEALTH HAZARD ASSESSMENTS FOR VARIED EXPOSURE DURATIONS

[Ed. Note: EPA does not have information yet in this section of IRIS].

IV. U.S. EPA REGULATORY ACTIONS

EPA risk assessments may be updated as new data are published and as assessment methodologies evolve. Regulatory actions are frequently not updated at the same time. Compare the dates for the regulatory actions in this section with the verification dates for the risk assessments in sections I and II, as this may explain inconsistencies. Also note that some regulatory actions consider factors not related to health risk, such as technical or economic feasibility. Such considerations are indicated for each action. In addition, not all of the regulatory actions listed in this section involve enforceable federal standards. Please direct any questions you may have concerning these regulatory actions to the U.S. EPA contact listed for that particular action. Users are strongly urged to read the background information on each regulatory action in Background Document 4 in Service Code 5.

IV.A. CLEAN AIR ACT (CAA)

[Ed. Note: EPA does not have information yet in this section of IRIS].

IV.B. SAFE DRINKING WATER ACT (SDWA)

IV.B.1. MAXIMUM CONTAMINANT LEVEL GOAL (MCLG) for Drinking Water

Value (status) -- 0.05 mg/L (Proposed, 1985)

Considers technological or economic feasibility? -- NO

Discussion -- An MCLG of 0.05 mg/L for arsenic is proposed based on the current MCL of 0.05 mg/L. Even though arsenic is potentially carcinogenic in humans by inhalation and ingestion, its potential essential nutrient value was considered in determination of an MCLG. The basis for this evaluation is nutritional requirements by NAS (NAS, 1983, Vol. 5, Drinking Water and Health, National Academy of Sciences Press, Washington, DC.)
Reference -- 50 FR 46936 (11/13/85)

EPA Contact -- Health and Ecological Criteria Division / OST /(202) 260-7571 / FTS 260-7571; or Safe Drinking Water Hotline / (800) 426-4791

IV.B.2. MAXIMUM CONTAMINANT LEVEL (MCL) for Drinking Water

Value (status) -- 0.05 mg/L (Interim, 1980)

Considers technological or economic feasibility? -- YES

Discussion -- As an interim measure the U.S. EPA is using the value previously derived by the Public Health Service.

Monitoring requirements -- Ground water systems every three years; surface water systems annually.

Analytical methodology -- Atomic absorption/furnace technique (EPA 206.2; SM 304); atomic absorption/gaseous hydride (EPA 206.3; SM 303E; ASTM D-2972-78B)

Best available technology -- No data available.

Reference -- 45 FR 57332 (08/27/80); 50 FR 46936 (11/13/85)

EPA Contact -- Drinking Water Standards Division / OGWDW /(202) 260-7575 / FTS 260-7575; or Safe Drinking Water Hotline / (800) 426-4791

IV.B.3. SECONDARY MAXIMUM CONTAMINANT LEVEL (SMCL) for Drinking Water

[Ed. Note: EPA does not have information yet in this section of IRIS].

IV.B.4. REQUIRED MONITORING OF "UNREGULATED" CONTAMINANTS

[Ed. Note: EPA does not have information yet in this section of IRIS].

IV.C. CLEAN WATER ACT (CWA)

IV.C.1. AMBIENT WATER QUALITY CRITERIA, Human Health

Water and Fish Consumption -- 2.2E-3 ug/L

Fish Consumption Only -- 1.75E-2 ug/L

Considers technological or economic feasibility? -- NO

Discussion -- For the maximum protection from the potential carcinogenic properties of this chemical, the ambient water concentration should be zero. However, zero may not be attainable at this time, so the recommended criteria represents a E-6 estimated incremental increase of cancer risk over a lifetime.

Reference -- 45 FR 79318 (11/28/80)

EPA Contact -- Criteria and Standards Division / OWRS(202)260-1315 / FTS 260-1315

IV.C.2. AMBIENT WATER QUALITY CRITERIA, Aquatic Organisms

Freshwater:
    Acute -- 3.6E+2 ug/L (Arsenic III)
    Chronic -- 1.9E+2 ug/L (Arsenic III)

Marine:
    Acute -- 6.9E+1 ug/L (Arsenic III)
    Chronic -- 3.6E+1 ug/L (Arsenic III)

Considers technological or economic feasibility? -- NO

Discussion -- The criteria given are for Arsenic III. Much less data are available on the effects of Arsenic V to aquatic organisms, but the toxicity seems to be less. A complete discussion may be found in the referenced notice.

Reference -- 50 FR 30784 (07/29/85)

EPA Contact -- Criteria and Standards Division / OWRS(202)260-1315 / FTS 260-1315

IV.D. FEDERAL INSECTICIDE, FUNGICIDE, AND RODENTICIDE ACT (FIFRA)

IV.D.1. PESTICIDE ACTIVE INGREDIENT, Registration Standard

Status -- Issued (1988)

Reference -- Arsenic, Chromium and Chromated Arsenical Compounds Pesticide Registration Standard. June, 1988. [NTIS# PB89-102842]

EPA Contact -- Registration Branch / OPP(703)557-7760 / FTS 557-7760

IV.D.2. PESTICIDE ACTIVE INGREDIENT, Special Review

Action -- Final regulatory decision - PD4 (1988)

Considers technological or economic feasibility? -- NO

Summary of regulatory action -- Cancellation of specified non-wood uses. Registrant of lead arsenate voluntarily canceled 09/87. Registrant of calcium arsenate voluntarily canceled 02/14/89. Use of sodium arsenate as ant bait canceled on 07/26/89. Criterion of concern: oncogenicity, mutagenicity and teratogenicity. Previous actions: 1) Voluntary cancellation of sodium arsenite (1978). Voluntary cancellation of two products. Criterion of concern: oncogenicity, mutagenicity and teratogenicity; 2) PD4 (1984). Requires label changes for wood use including a restricted use classification. Criterion of concern: oncogenicity, mutagenicity and teratogenicity; 3) Voluntary cancellation of copper arsenate (1977). Criterion of concern: oncogenicity.

Reference -- 53 FR 24787 (06/30/88); 43 FR 48267 (10/18/78); 42 FR 18422 (04/07/77); 49 FR 28666 (07/13/84) [NTIS# PB84-241538]; 49 FR 43772 (10/31/84); 50 FR 4269 (01/30/85)

EPA Contact -- Special Review Branch / OPP(703)557-7400 / FTS 557-7400

IV.E. TOXIC SUBSTANCES CONTROL ACT (TSCA)

[Ed. Note: EPA does not have information yet in this section of IRIS].

IV.F. RESOURCE CONSERVATION AND RECOVERY ACT (RCRA)

IV.F.1. RCRA APPENDIX IX, for Ground Water Monitoring

Status -- Listed

Reference -- 52 FR 25942 (07/09/87)

EPA Contact --   RCRA/Superfund Hotline(800)424-9346 / (202)260-3000 / FTS 260-3000

IV.G. SUPERFUND (CERCLA)

IV.G.1. REPORTABLE QUANTITY (RQ) for Release into the Environment

Value (status) -- 1 pound (Final, 1989)

Considers technological or economic feasibility? -- NO

Discussion -- The 1-pound RQ for arsenic is based on its potential carcinogenicity. Available data indicate a hazard ranking of high based on a potency factor of 142.31/mg/kg/day and a weight-of-evidence group A, which corresponds to an RQ of 1 pound.   Evidence found in "Water-Related Environmental Fate of 129 Priority Pollutants" (EPA 440/4-79-029a) also indicates that this material, or a constituent of this material, is bioaccumulated to toxic levels in the tissue of aquatic and marine organisms, and has the potential to concentrate in the food chain. Reporting of releases of massive forms of this hazardous substance is not required if the diameter of the pieces released exceeds 100 micrometers (0.004 inches).

Reference -- 54 FR 33418 (08/14/89)

EPA Contact -- RCRA/Superfund Hotline (800)424-9346 / (202)260-3000 / FTS 260-3000

VI. BIBLIOGRAPHY

VI.A. ORAL RfD REFERENCES

Abernathy, C.O., W. Marcus, C. Chen, H. Gibb and P. White. 1989. Office of Drinking Water, Office of Research and Development, U.S. EPA. Memorandum to P. Cook, Office of Drinking Water, U.S. EPA and P. Preuss, Office of Regulatory Support and Scientific Management, U.S. EPA. Report on Arsenic (As) Work Group Meetings. February 23.

Cebrian, M.E., A. Albores, M. Aguilar and E. Blakely. 1983. Chronic arsenic poisoning in the north of Mexico. Human Toxicol. 2: 121-133.

Ferm, V.H. and S.J. Carpenter. 1968. Malformations induced by sodium arsenate. J. Reprod. Fert. 17: 199-201.

Hindmarsh, J.T., O.R. McLetchie, L.P.M. Heffernan et al. 1977. Electromyographic abnormalities in chronic environmental arsenicalism. J. Analyt. Toxicol. 1: 270-276.

Hood, R.D., G.T. Thacker and B.L. Patterson. 1977. Effects in the mouse and rat of prenatal exposure to arsenic. Environ. Health Perspect. 19: 219-222.

Lu, F.J. 1990. Blackfoot disease: Arsenic or humic acid? The Lancet. 336: 115-116.

Marcus, W.L. and A.S. Rispin. 1988. Threshold carcinogenicity using arsenic as an example. In: Advances in Modern Environmental Toxicology, Vol. XV. Risk Assessment and Risk Management of Industrial and Evironmental Chemicals, C.R. Cothern, M.A. Mehlman and W.L. Marcus, Ed. Princeton Scientific Publishing Company, Princeton, NJ. p. 133-158.

Nadeenko, V.G., V. Lenchenko, S.B. Genkina and T.A. Arkhipenko. 1978. The influence of tungsten, molibdenum, copper and arsenic on the intrauterine development of the fetus. TR-79-0353. Farmakologiya i Toksikologiya. 41: 620-623.

NRC (National Research Council). 1989. Recommended Dietary Allowances, 10th ed. Report of the Food and Nutrition Board, National Academy of Sciences, Washington, National Academy Press, Washington, DC. 285 p.

Schroeder, H.A. and M. Mitchner. 1971. Toxic effects of trace elements on the reproduction of mice and rats. Arch. Environ. Health. 23(2): 102-106.

Southwick, J.W., A.E. Western, M.M. Beck, et al. 1983. An epidemiological study of arsenic in drinking water in Millard County, Utah. In: Arsenic: Industrial, Biomedical,

Environmental Perspectives, W.H. Lederer and R.J. Fensterheim, Ed. Van Nostrand Reinhold Co., New York. p. 210-225.

Tseng, W.P. 1977. Effects and dose-response relationships of skin cancer and blackfoot disease with arsenic. Environ. Health Perspect. 19: 109-119.

Tseng, W.P., H.M. Chu, S.W. How, J.M. Fong, C.S. Lin and S. Yeh. 1968. Prevalence of skin cancer in an endemic area of chronic arsenicism in Taiwan. J. Natl. Cancer. Inst. 40(3): 453-463.

Valentine, J.L., L.S. Reisbord, H.K. Kang and M.D Schluchter. 1987. Arsenic effects on population health histories. In: Trace Elements in Man and Animals - TEMA 5, C.F. Mills, I. Bremner and J.K. Chesters, eds. Commonwealth Agricultural Bureaux, Aberdeen, Scotland.

U.S. EPA. 1984. Health Assessment Document for Inorganic Arsenic. Prepared by the Office of Health and Environmental Assessment, Environmental Criteria and Assessment Office, Research Triangle Park, NC. EPA 600/8-83-021F.

U.S. EPA. 1988. Quantitative Toxicological Evaluation of Ingested Arsenic. Office of Drinking Water, Washington, DC. (Draft)

VI.B. INHALATION RfD REFERENCES

None

VI.C. CARCINOGENICITY ASSESSMENT REFERENCES

Axelson, O., E. Dahlgren, C.D. Jansson and S.O. Rehnlund. 1978. Arsenic exposure and mortality: A case referent study from a Swedish copper smelter. Br. J. Ind. Med. 35: 8-15.

Andersen, O. 1983. Effects of coal combustion products and metal compounds on sister chromatid exchange (SCE) in a macrophage like cell line. Environ. Health Perspect. 47: 239-253.

Bates, M.N., A.H. Smith and C. Hopenhayn-Rich. 1992. Arsenic ingestion and internal cancers: A review. Am. J. Epidemiol. 135(5): 462-476.

Beckman, G., L. Beckman and I. Nordenson. 1977. Chromosome aberrations in workers exposed to arsenic. Environ. Health Perspect. 19: 145-146.

Brown, C.C. and K.C. Chu. 1983a. Approaches to epidemiologic analysis of prospective and retrospective studies: Example of lung cancer and exposure to arsenic. In: Risk Assessment Proc. SIMS Conf. on Environ. Epidemiol. June 28-July 2, 1982, Alta, VT. SIAM Publications.

Brown, C.C. and K.C. Chu. 1983b. Implications of the multistage theory of carcinogenesis applied to occupational arsenic exposure. J. Natl. Cancer Inst. 70(3): 455-463.

Brown, C.C. and K.C. Chu. 1983c. A new method for the analysis of cohort studies: Implications of the multistage theory of carcinogenesis appled to occupational arsenic exposure. Environ. Health Perspect. 50: 293-308.

Burgdorf, W., K. Kurvink and J. Cervenka. 1977. Elevated sister chromatid exchange rate in lymphocytes of subjects treated with arsenic. Hum. Genet. 36(1): 69-72.

Cebrian, M.E., A. Albores, M. Aquilar and E. Blakely. 1983. Chronic arsenic poisoning in the north of Mexico. Human Toxicol. 2: 121-133.

Chen, C-J. and C-J. Wang. 1990. Ecological correlation between arsenic level in well water and age-adjusted mortality from malignant neoplasms. Cancer Res. 50(17): 5470-5474.

Chen, C-J., Y-C. Chuang, T-M. Lin and H-Y. Wu. 1985. Malignant neoplasms among residents of a Blackfoot disease-endemic area in Taiwan: High-arsenic artesian well water and cancers. Cancer Res. 45: 5895-5899.

Chen, C-J., Y-C. Chuang, S-L. You, T-M. Lin and H-Y. Wu. 1986. A retrospective study on malignant neoplasms of bladder, lung, and liver in blackfoot disease endemic area in Taiwan. Br. J. Cancer. 53: 399-405.

Chen, C-J., M-M. Wu, S-S. Lee, J-D. Wang, S-H. Cheng and H-Y. Wu. 1988. Atherogenicity and carcinogenicity of high-arsenic artesian well water. Multiple risk factors and related malignant neoplasms of Blackfoot disease. Arteriosclerosis. 8(5): 452-460.

Chen, C-J., CW. Chen, M-M. Wu and T-L. Kuo. 1992. Cancer potential in liver, lung bladder and kidney due to ingested inorganic arsenic in drinking water. Br. J. Cancer. 66(5): 888-892.

Crossen, P.E. 1983. Arsenic and SCE in human lymphocytes. Mutat. Res. 119: 415-419.

Cuzick, J., S. Evans, M. Gillman and D.A. Price Evans. 1982. Medicinal arsenic and internal malignancies. Br. J. Cancer. 45(6): 904-911.

Cuzick, J., R. Harris, P.S. Mortimer. 1984. Palmar keratoses and cancers of the bladder and lung. March 10. The Lancet. 1(8376): 530-533.

Cuzick, J., P. Sasieni and S. Evans. 1992. Ingested arsenic, keratoses, and bladder cancer. Am. J. Epidemiol. 136(4): 417-421.

Datta, S., G. Talukder and A. Sharma. 1986. Cytotoxic effects of arsenic in dietary oil primed rats. Sci. Culture. 52: 196-198.

DeKnudt, G., A. Leonard, A. Arany, G.D. Buisson and E. Delavignette. 1986. In vivo studies in male mice on the mutagenic effects of inorganic arsenic. Mutagenesis. 1(1): 33-34.

DiPaolo, J.A. and B.C. Casto. 1979. Qantitative studies of in vitro morphological transformation of Syrian hamster cells by inorganic metal salts. Cancer Res. 39: 1008-1013.

Dong, J-T., X-M. Luo. 1993. Arsenic-induced DNA-strand breaks associated with DNA-protein crosslinks in human fetal lung fibroblasts. Mutat. Res. 302(2): 97-102.

Enterline, P.E. and G.M. Marsh. 1982. Cancer among workers exposed to arsenic and other substances in a copper smelter. Am. J. Epidemiol. 116(6): 895-911.

Fierz, U. 1965. Catamnestic investigations of the side effects of therapy of skin diseases with inorganic arsenic. Dermatologica. 131: 41-58.

Furst, A. 1983. A new look at arsenic carcinogenesis. In: Arsenic: Industrial, Biomedical, and Environmental Perspectives, W. Lederer and R. Fensterheim, Ed. Van Nostrand Reinhold, New York. p. 151-163.

Harrington-Brock, K., T.W. Smith, C.L. Doerr, and M.M. Moore. 1993. Mutagenicity of the human carcinogen arsenic and its methylated metabolites monomethylarsonic and dimethylarsenic acids in L5178Y TK+/- mouse lymphoma cells (abstract). Environ. Mol. Mutagen. 21(Supplement 22): 27.

Higgins, I. 1982. Arsenic and respiratory cancer among a sample of Anaconda smelter workers. Report submitted to the Occupatinal Safety and Health Administration in the comments of the Kennecott Minerals Company on the inorganic arsenic rulemaking. (Exhibit 203-5)

Higgins, I., K. Welch and C. Burchfield. 1982. Mortality of Anaconda smelter workers in relation to arsenic and other exposures. University of Michigan, Dept. Epidemiology, Ann Arbor, MI.

Huang, H., C.F. Huang, D.R. Wu, C.M. Jinn and K.Y. Jan. 1993. Glutathione as a cellular defence against arsenite toxicity in cultured Chinese hamster ovary cells. Toxicology. 79(3): 195-204.

IARC (International Agency for Research on Cancer). 1980. IARC Monographs on the Evaluation of Carcinogenic Risk of Chemicals to Man, Vol. 23. Some metals and metallic compounds. World Health Organization, Lyon, France.

Innes, J.R.M., B.M. Ulland, M.G. Valerio, et al. 1969. Bioassay of pesticides and industrial chemicals for tumorigenicity in mice: A preliminary note. JNCI. 42: 1101-1114.

Jacobson-Kram, D. and D. Montalbano. 1985. The reproductive effects assessment group's report on the mutagenicity of inorganic arsenic. Environ. Mutagen. 7(5): 787-804.

Jha, A.N., M. Noditi, R. Nilsson and A.T. Natarajan. 1992. Genotoxic effects of sodium arsenite on human cells. Mutat. Res. 284(2): 215-221.

Larramendy, M.L., N.C. Popescu and J.A. DiPaolo. 1981. Induction by inorganic metal salts of sister chromatid exchanges and chromosome aberrations in human and Syrian hamster strains. Environ. Mutagen. 3: 597-606.

Lee, Te-Chang, J. Ko and K.Y. Jan. 1989. Differential cytotoxocity of sodium arsenite in human fibroblasts and chinese hamster ovary cells. Toxicology. 56(3): 289-300.

Lee, Te-Chang, N. Tanaka, P.W. Lamb, T.M. Gilmer and J.C. Barrett. 1988. Induction of gene amplification by arsenic. Science. 241(4861): 79-81.

Lee-Chen, S.F., J.R. Gurr, I.B. Lin and K.Y. Jan. 1993. Arsenite enhancesDNA double-strand breaks and cell killing of methyl methanesulfonate-treatedcells by inhibiting the excision of alkali-labile sites. Mutat. Res. 294(1): 21-28.

Lee-Feldstein, A. 1983. Arsenic and respiratory cancer in man: Follow-up of an occupational study. In: Arsenic: Industrial, Biomedical, and Environmental Perspectives, W. Lederer and R. Fensterheim, Ed. Van Nostrand Reinhold, New York.

Mabuchi, K., A. Lilienfeld and L. Snell. 1979. Lung cancer among pesticide workers expsoed to inorganic arsenicals. Arch. Environ. Health. 34: 312-319.

Matanoski, G., E. Landau, J. Tonascia, et al. 1981. Cancer mortality in an industrial area of Baltimore. Environ. Res. 25: 8-28.

Moore, M.M., K. Harrington-Brock and C.L. Doerr. 1995. Genotoxicity of arsenic and its methylated metabolites. Mutagenesis and Cellular Toxicology Brance, Genetic Toxicology Dividion, Health Effects Research Laboratory, U.S. Environmental Protection Agency, Research Triangle Park, NC. (In press).

Morton, W., G. Starr, D. Pohl, J. Stoner, S. Wagner, and P. Weswig. 1976. Skin cancer and water arsenic in Lane County, Oregon. Cancer. 37: 2523-2532.

Ohno, H., F. Hanaoka and M. Yamada. 1982. Inductibility of sister chromatid exchanges by heavy-metal ions. Mutat. Res. 104(1-3): 141-146.

Ott, M.G., B.B. Holder and H.L. Gordon. 1974. Respiratory cancer and occupational exposure to arsenicals. Arch. Environ. Health. 29: 250-255.

Pershagen, G., B. Lind and N.E. Bjorklund. 1982. Lung retention and toxicityof some inorganic arsenic compounds. Environ. Res. 29: 425-434.

Pershagen, G., G. Norberg and N.E. Bjorklund. 1984. Carcinomas of the respiratory tract in hamsters given arsenic trioxide and/or benxo(a)pyrene by the pulmonary route. Environ. Res. 34: 227-241.

Poma, K., N. Degraeve and C. Susanne. 1987. Cytogenic effects in mice after chronic exposure to arsenic followed by a single dose of ethyl methanesulfonate. Cytologia. 52(3): 445-450.

Rencher, A.C., M.W. Carter and D.W. McKee. 1977. A retrospective epidemiological study of mortality at a large western copper smelter. J. Occup. Med. 19(11): 754-758.

Rossman, T.G. 1981. Enhancement of UV-mutagenesis by low concentrations ofarsenite in Escherichia coli. Mutat. Res. 91: 207-211.

Roth, F. 1958. Uber den Bronchialkrebs Arsengeschadigter Winzer. Virchows Arch. 331: 119-137.

Smith, A.H., C. Hopenhayn-Rich, M.N. Bates, et al. 1992. Cancer risks from arsenic in drinking water. Environ. Health Perspect. 97: 259-267.

Sommers, S.C. and R.G. McManus. 1953. Multible arsenical cancers of the skin and internal organs. Cancer. 6: 347-359.

Southwick, J., A. Western, M. Beck, et al. 1981. Community health associated with arsenic in drinking water in Millard County, Utah. Health Effects Research Laboratory, Cincinnati, OH. EPA-600/1-81-064.

Tinwell, H., S.C. Stephens, J. Ashby. 1991. Arsenite as the probable active species in the human carcinogenicity of arsenic: Mouse micronucleus assays on Na and K arsenite, orpiment, and Fowler's solution. Environ. Health Perspec. 95: 205-210.

Tokudome, S. and M. Kuratsune. 1976. A cohort study on mortality from cancer and other causes among workers at a metal refinery. Int. J. Cancer. 17:310-317.

Tseng W.P., H.M. Chu, S.W. How, J.M. Fong, C.S. Lin, and S. Yen. 1968. Prevalence of skin cancer in an endemic area of chronic arsenicism in Taiwan. J. Natl. Cancer Inst. 40(3): 453-463.

Tseng W.P. 1977. Effects and dose-response relationships of skin cancer and Blackfoot disease with arsenic. Environ. Health Perspect. 19: 109-119.

U.S. EPA. 1984. Health Assessment Document for Inorganic Arsenic. Prepared by the Office of Research and Development, Environmental Criteria and Assessment Office, Reasearch Triangle Park, NC.

U.S. EPA. 1988. Special Report on Ingested Inorganic Arsenic; Skin Cancer; Nutritional Essentiality Risk Assessment Forum. July 1988. EPA/625/3-87/013.

U.S. EPA. 1993. Drinking Water Criteria Document for Arsenic. Office of Water, Washington, DC. Draft.

Vahter, M. and E. Marafante. 1988. In vivo methylation and detoxification of arsenic. Royal Soc. Chem. 66: 105-119.

Vig, B.K., M.L. Figueroa, M.N. Cornforth, S.H. Jenkins. 1984. Chromosome studies in human subjects chronically exposed to arsenic in drinking water. Am. J. Ind. Med. 6(5): 325-338.

Wan, B., R.T. Christian and S.W. Soukup. 1982. Studies of cytogenetic effects of sodium arsenicals on mammalian cells in vitro. Environ. Mutag. 4: 493-498.

Welch, K., I. Higgins, M. Oh and C. Burchfield. 1982. Arsenic exposure, smoking, and respiratory cancer in copper smelter workers. Arch. Environ. Health. 37(6): 325-335.

Wu, M-M., T-L. Kuo, Y-H Hwang and C-J. Chen. 1989. Dose-response relation between arsenic concentration in well water and mortality from cancers and vascular diseases. Am. J. Epidemiol. 130(6): 1123-1132.

VI.D. DRINKING WATER HA REFERENCES

None

VII. REVISION HISTORY

Date	Section	Description
06/30/88	II.B.	Revised last paragraph
06/30/88	II.C.1.	Inhalation slope factor changed
06/30/88	II.C.3.	Paragraph 2 added
09/07/88	II.B.	Major text changes
12/01/88	II.A.2.	Mabuchi et al. citation year corrected
12/01/88	II.A.3.	Pershagen et al. citation year corrected
09/01/89	II.C.2.	Citations added to anacondor smelter
09/01/89	VI.	Bibliography on-line
06/01/90	II.A.2.	2nd & 3rd paragraph - Text revised
06/01/90	II.A.4.	Text corrected
06/01/90	II.C.1.	Inhalation slope factor removed (format change)
06/01/90	IV.F.1.	EPA contact changed
06/01/90	VI.C.	References added
12/01/90	II.B.	Changed slope factor to "unit risk", 2nd para, 1st sen
02/01/91	II.C.3.	Text edited
09/01/91	I.A.	Oral RfD summary now on-line
09/01/91	I.A.	Oral RfD bibliography added
10/01/91	I.A.1.	Conversion factor text clarified
10/01/91	IV.B.1.	MCLG noted as pending change
01/01/92	IV.	Regulatory actions updated

08/01/92	II.	Note added to indicate text in oral quant. estimate
10/01/92	VI.C.	Missing reference added to bibliography
02/01/93	I.A.4.	Citations added to second paragraph
02/01/93	VI.A.	References added to bibliography
03/01/93	VI.A.	Corrections to references
03/01/94	II.D.2.	Work group review date added
06/01/94	II.	Carcinogen assessment noted as pending change
01/01/95	II.	Pending change note revised
01/01/95	II.B.	Dates and document no. added to oral quant. estimate
06/01/95	II.	Carcinogenicity assessment replaced
06/01/95	VI.C.	Carcinogenicity references replaced
07/01/95	II.D.1.	Documentation year corrected; review statement revised
07/01/95	VI.C.	U.S. EPA, 1994 corrected to 1993

SYNONYMS

Arsenic   Arsenic, inorganic
gray-arsenic

Source: *Instant EPA's IRIS*, 1997, Instant Reference Sources, Inc., 7605 Rockpoint Drive, Austin, TX 78731 (Fax: 512-345-2386; Internet URL, http://www.instantref.com/inst-ref.htm).

Atrazine

Introduction

Atrazine is a member of the s- triazine herbicides and is the most widely used herbicide in the world [811].]. Production in the U.S. alone is some 100 million pounds annually. [882] It is used as a selective herbicide for weed control in agriculture. Atrazine may be released into the environment via effluents at manufacturing sites and at points of application where it is employed as a herbicide. The s- triazine ring of atrazine is fairly resistant to degradation. 2-Chloro-4-ethyl-amino-6-amino-s-triazine, 2-chloro-4-amino-6-isopropylamino-s-triazine, 2-hydroxy-4-ethylamino-6-isopropyl-amino-s-triazine and 2-hydroxy-4-ethylamino-6-amino-s-triazine have been identified as microbial transformation products of atrazine. Chemical degradation of atrazine may be more important environmentally than biodegradation. Atrazine may hydrolyze fairly rapidly in either acidic or basic environments, yet is fairly resistant to hydrolysis at neutral pHs. Furthermore the rate of hydrolysis was found to drastically increase upon small additions of humic materials, indicating atrazine hydrolysis could be catalyzed. The most probable exposure would be occupational exposure, which may occur through dermal contact or inhalation at places where atrazine is produced or used as a herbicide [828].

Identifying Information

CAS NUMBER: 1912-24-9

NIOSH Registry Number: XY5600000

SYNONYMS: 6-CHLORO-N-ETHYL-N'-(1-METHYLETHYL)-1,3,5-TRIAZINE-2,4-DIAMINE, 2-CHLORO-4-ETHYLAMINO-6-ISOPROPYLAMINO-S-TRIAZINE, 2-CHLORO-4-ETHYLAMINO-6-ISOPROPYLAMINO-1,3,5-TRIAZINE, 2-CHLORO-4-(ETHYLAMINO)-6-(ISOPROPYLAMINO)TRIAZINE, 1,3,5-TRIAZINE-2,4-DIAMINE, 6-CHLORO-N-ETHYL-N'-(1-METHYLETHYL)-, S-TRIAZINE, 2-CHLORO-4-(ETHYLAMINO)-6-(ISOPROPYLAMINO)-, A 361, G 30027, AKTINIT A, HUNGAZIN PK, TRIAZINE A 1294, GESAPRIM, ZEAZIN, HUNGAZIN, ARGEZIN, AKTIKON, AKTIKON PK, ATRAZIN, AKTINIT PK, CYAZIN, WONUK, PRIMATOL A, AATREX, OLEOGESAPRIM, ZEAZINE, CHROMOZIN, PITEZIN, ACTINITE PK, RADAZIN, GESAPRIM 50, AKTICON, ATRATAF, ZEAPOS, CET, OLEOGESAPRIM 200, GESAPRIM 500, MAIZINA, AATREX 4L, AATREX NINE-O, AATREX 80W, ATAZINAX, ATRANEX, ATRASINE, ATRATOL A, ATRED, ATREX, CANDEX, CEKUZINA-T, 2-CHLORO-4-

ETHYLAMINEISOPROPYLAMINE-S-TRIAZINE, 1-CHLORO-3-ETHYLAMINO-5-
ISOPROPYLAMINO-S-TRIAZINE, 1-CHLORO-3-ETHYLAMINO-5-
ISOPROPYLAMINO-2,4,6-TRIAZINE, 2-CHLORO-4-(2-PROPYLAMINO)-6-
ETHYLAMINO-S-TRIAZINE, CRISATRINA, CRISAZINE, FARMCO ATRAZINE,
FENAMIN, FENAMINE, FENATROL, GEIGY 30,027, GESOPRIM, GRIFFEX,
INAKOR, PRIMATOL, PRIMAZE, RADIZINE, SHELL ATRAZINE HERBICIDE,
STRAZINE, VECTAL, VECTAL SC, WEEDEX A, ZEAPHOS

Endocrine Disruptor Information

Atrazine is discussed in *Our Stolen Future* [810] and it is on the EED list of EPA [811],
the Centers for Disease Control & Prevention [812], and the World Wildlife Fund [813].
Exposure of breast cells in tissue culture leads to increased production of a metabolite of
estradiol, 16-alpha- hydroxyestrone (16aOHE) which has been implicated in increasing
the risk of breast cancer. [811], [839] Atrazine has been shown to produce breast cancer
in laboratory rats. [811], [840] and chronic toxicity studies noted morphlogical changes
in the brain heart liver lungs ovaries and endocrine glands. [811], [841] Atrazine has also
exhibited immunotoxic effects. [811], [826]

Chemical and Physical Properties

CHEMICAL FORMULA: $C_8H_{14}ClN_5$

MOLECULAR WEIGHT: 215.72

WLN: T6N CN ENJ BMY1&1 DM2 FG

PHYSICAL DESCRIPTION: Colorless crystals

SPECIFIC GRAVITY: Not available

DENSITY: 1.187 g/cm^3 @ 20°C [172]

MELTING POINT: 175-177°C [029], [172], [173], [346]

SOLUBILITY: Water: <1 mg/mL @ 19.5°C [700]; DMSO: <1 mg/mL @ 19.5°C [700];
95% Ethanol: <1 mg/mL @ 19.5°C [700]; Acetone: <1 mg/mL @ 19.5°C [700];
Methanol: <1 mg/mL @ 18°C [700]; Toluene: <1 mg/mL @ 18°C [700]

OTHER SOLVENTS: Chloroform: 52 g/kg @ 20°C [043], [169], [172], [173]; Ether:
12000 ppm @ 25°C [033], [043], [421]; Ethyl acetate: 18 g/kg @ 20°C [169], [172];
Octan-1-ol: 10 g/kg @ 20°C [172]; n-Pentane: 0.36 g/kg @ 20°C [169]

OTHER PHYSICAL DATA: pKa: 1.7 [172]

Source: *The National Toxicology Program's Chemical Database, Volume 2*, 1992,
CRC Press, Inc./Lewis Publishers, 2000 Corporate Blvd., Boca Raton, FL 33431 (Fax:
407-998-9114).

Environmental Reference Materials

A source of EED environmental reference materials for Atrazine is Radian International LLC, Austin, TX (Fax 512-454-0268; Internet URL, http://www.radian.com/standards); Catalog No. ERA-055.

Hazardous Properties

ACUTE/CHRONIC HAZARDS: The inhalation hazard is low in humans. There is no apparent skin irritation or other toxic manifestations in humans [033], [151]. Atrazine may be toxic and a local irritant. When heated to decomposition this compound emits toxic fumes of chlorine and nitrogen oxides [043].

FIRE HAZARD: Literature sources indicate that Atrazine is nonflammable [107], [421], [430]. Fires involving this material can be controlled with a dry chemical, carbon dioxide or Halon extinguisher.

REACTIVITY: Atrazine undergoes slow hydrolysis at 70°C under neutral conditions. Hydrolysis is more rapid in acidic or alkaline conditions [033], [169], [172], [173]. It forms salts with acids [172].

STABILITY: Atrazine is stable under normal laboratory conditions. Solutions of Atrazine in water, DMSO, 95% ethanol or acetone should be stable for 24 hours under normal laboratory conditions [700].

Source: *The National Toxicology Program's Chemical Database, Volume 7*, 1992, CRC Press, Inc./Lewis Publishers, 2000 Corporate Blvd., Boca Raton, FL 33431 (Fax: 407-998-9114).

Medical Symptoms of Exposure

Symptoms of exposure to atrazine may include irritation of the eyes and mucous membranes, erythema, conjunctivitis, dermatitis, muscular weakness, lethargy, ptosis, dyspnea, prostration, ataxia, respiratory distress, paralysis, anorexia, anemia, gastroenteritis and convulsions [107]. Skin irritation may occur [043], [107]. Other symptoms may include convulsions, coma, liver damage and kidney damage [301].

Source: *The National Toxicology Program's Chemical Database, Volume 4*, 1992, CRC Press, Inc./Lewis Publishers, 2000 Corporate Blvd., Boca Raton, FL 33431 (Fax: 407-998-9114).

Toxicological Information

The toxicological information below is gathered from several sources including two national databases: one from the *National Toxicology Program Chemical Repository Database* and another from *EPA's Integrated Risk Information System (IRIS)*.

Toxicology Information from the National Toxicology Program

TOXICITY:

Typ. Dose	Mode	Specie	Amount	Units	Other
LD50	orl	rat	672	mg/kg	
LD50	ipr	rat	235	mg/kg	
LD50	orl	mus	850	mg/kg	
LD50	orl	rbt	750	mg/kg	
LD50	skn	rbt	7500	mg/kg	
LD50	orl	ham	1000	mg/kg	
LD50	unr	mam	1400	mg/kg	
LC50	ihl	rat	5200	mg/m3/4H	
LD50	ipr	mus	626	mg/kg	

OTHER TOXICITY DATA:

Skin and Eye Irritation Data:
skn-rbt   38 mg open MLD
skn-mam   500 mg MLD
eye-rbt 6320 ug SEV
eye-mam   100 mg SEV

Review: Toxicology Review-2

Status:
EPA Genetox Program 1988, Positive: Aspergillus-reversion; S cerevisiae gene conversion
EPA Genetox Program 1988, Negative: N crassa-aneuploidy
EPA Genetox Program 1988, Inconclusive: D melanogaster-whole sex chrom. loss
EPA Genetox Program 1988, Inconclusive: D melanogaster-n ondisjunction
EPA Genetox Program 1988, Inconclusive: D melanogaster Sex-linked lethal
EPA TSCA Chemical Inventory, 1986
EPA TSCA Test Submission (TSCATS) Data Base, September 1989

CARCINOGENICITY:

Tumorigenic Data:

Type of Effect	Route/Animal	Amount Dosed (Notes)
TDLo	orl-mus	9000 mg/kg/78W-I

Atrazine   165

TERATOGENICITY: See RTECS printout for data

MUTATION DATA: See RTECS printout for data

SAX TOXICITY EVALUATION:

> THR: Poison by intraperitoneal route. Moderately toxic by ingestion. Mildly toxic by inhalation and skin contact. An experimental tumorigen. Human mutagenic data. Experimental reproductive effects. A skin and severe eye irritant.

Source: *Instant Tox-Base*, 1995, Instant Reference Sources, Inc., 7605 Rockpoint Drive, Austin, TX 78731 (Fax: 512-345-2386; Internet URL, http://www.instantref.com/inst-ref.htm).

Toxicology Information from EPA's Integrated Risk Information System (IRIS)

I.   CHRONIC HEALTH HAZARD ASSESSMENTS FOR NONCARCINOGENIC EFFECTS

I.A. REFERENCE DOSE FOR CHRONIC ORAL EXPOSURE (RfD)

The Reference Dose (RfD) is based on the assumption that thresholds exist for certain toxic effects such as cellular necrosis, but may not exist for other toxic effects such as carcinogenicity. In general, the RfD is an estimate (with uncertainty spanning perhaps an order of magnitude) of a daily exposure to the human population (including sensitive subgroups) that is likely to be without an appreciable risk of deleterious effects during a lifetime. Please refer to Background Document 1 in Service Code 5 for an elaboration of these concepts. RfDs can also be derived for the noncarcinogenic health effects of compounds which are also carcinogens. Therefore, it is essential to refer to other sources of information concerning the carcinogenicity of this substance. If the U.S. EPA has evaluated this substance for potential human carcinogenicity, a summary of that evaluation will be contained in Section II of this file when a review of that evaluation is completed.

I.A.1. ORAL RfD SUMMARY

Critical Effect	Experimental Doses*	UF	MF	RfD
Decreased body weight gain Cardiac toxicity and moderate-to-severe dilation of the right atrium	NOAEL: 70 ppm (3.5 mg/kg-day) LOAEL: 500 ppm (25 mg/kg-day) NOAEL: 150 ppm (4.97 mg/kg-day) LOAEL: 1000 ppm (33.65 mg/kg/-day, male 33.8 mg/kg-day, female)	100	1	3.5E-2 mg/kg-day

2-Year Rat Feeding Study Ciba-Geigy Corp., 1986

1-Year Dog Feeding Study Ciba-Geigy Corp., 1987a

*Conversion Factors and Assumptions -- 1 ppm = 0.05 mg/kg-day (assumed rat food consumption)

I.A.2. PRINCIPAL AND SUPPORTING STUDIES (ORAL RfD)

Ciba-Geigy Corporation. 1986. MRID No. 00141874, 00157875, 00158930, 40629302. HED Doc. No. 005940, 006937. Available from EPA. Write to FOI, EPA, Washington, DC 20460.

Ciba-Geigy Corporation, Agricultural Division. 1987a. MRID No. 40431301, 41293801. HED Doc. No. 006718, 006937, 007647. Available from EPA. Write to FOI, EPA, Washington, DC 20460.

Groups of Sprague-Dawley rats (20/sex/dose for the chronic study, 50/sex/dose for the oncogenicity study) were administered atrazine in the diet for 2 years at dietary concentrations of 0, 10, 70, 500, and 1000 ppm (0, 0.5, 3.5, 25 and 50 mg/kg-day) (Ciba-Geigy Corp., 1986). An additional 10 rats/sex were placed on control and high-dose (1000 ppm) diets for a 12-month interim sacrifice and 10/sex (control and high-dose) for a 13-month sacrifice (the 1000 ppm group was placed on control diet for 1 month prior to sacrifice). Animals were caged individually and received food and water ad libitum.

In males, survival was increased in a dose-related manner (p<0.003, using the Cox Tarone test) and was significantly higher in males receiving 1000 ppm when compared with controls (p=0.0055, using pairwise comparison). In contrast, survival in females was decreased in a dose-related manner (p value for a negative trend = 0.0016) and was significantly lower (p=0.0042) in high-dose females when compared with controls.

Mean body weights were significantly depressed ($p<0.01$) in males and females receiving 500 (M: 8%; F: 19%) and 1000 (M: 19%; F: 27%) ppm with the exception of mean weights for males in the last 2 months of the study. The 24-month weight gain in the high-dose animals was 76% of control for males and 64.5% of control for females. In the recovery groups, the weight gain for month 13 for males previously receiving 1000 ppm was 63 +/- 14.6 g, compared with 20 +/- 17.9 g for controls ($p<0.01$), and for females previously receiving 1000 ppm it was 56 +/- 31.6 g, compared with 16 +/- 11.9 g for controls ($p<0.01$). However, the mean weight at 13 months in these males was still significantly ($p<0.01$) lower than controls.

In females receiving 1000 ppm, statistically significantly ($p<0.05$) lower mean red cell count, hemoglobin and hematocrit were noted at 6, 12, and 18 months when compared with controls. The values were somewhat depressed in high-dose females at 24 months; however, only four females were used for clinical studies. Red cell count, hemoglobin and hematocrit were also decreased in the 10 females receiving 1000 ppm and scheduled for sacrifice at 12 months. The values approached control levels at 13 months in the 1000 ppm recovery group. All values for parameters in dosed males were similar to those in the control groups with the exception of an increased mean platelet count at 6 months in rats receiving 1000 ppm. Increase platelet counts were also seen at 6 and 12 months in females receiving 1000 ppm. The level of serum triglycerides in high-dose males (56.55 +/- 18.12 mg/dL) was, in general, lower than control values (139.26ñ88.35 mg/dL) throughout the study; however, the decrease was significant ($p<0.05$) only at 6 months. In groups scheduled for the 12-month sacrifice, the level in high-dose males (103.1 +/- 47.81 mg/dL) was significantly lower ($p<0.01$) than in control males (228.3 +/- 95.44 mg/dL). At the 13-month sacrifice, the triglyceride level was similar in control males and those that had previously received 1000 ppm. In females, glucose levels were decreased ($p<0.01$) in the high-dose group at 3, 6 and 12 months when compared with controls.

The absolute weight of liver and kidney in high-dose males sacrificed at 12 months was significantly ($p<0.05$) lower than controls. The mean weight of the liver was 14.71 +/- 2.81 g for the high-dose groups and 19.7 +/- 3.46 g for the controls; the mean weight of the kidneys was 3.67 +/- 0.4 g for high-dose males, compared with 4.39 +/- 0.58 g for controls.

At 24 months, the mean absolute weights of liver and kidney in high-dose males were lower than those of controls, but the decrease was not statistically significant. There were no other changes in absolute organ weights of males and females. There were several increases in organ-to-body weight ratios in high-dose animals that were significant ($p<0.05$) when compared with controls but they were not accompanied by changes in absolute organ weights. These changes were the result of decreased body weights.

Acinar hyperplasia of the mammary gland and epithelial hyperplasia of the prostate were increased in males receiving 1000 ppm when compared to controls. In females receiving 500 or 1000 ppm, there was an increased myeloid hyperplasia in the bone marrow of both the femur and sternum. It was reported that the bone marrow changes, as well as an increase in extramedullary hematopoiesis in the spleen, were sequellae related to mammary fibroadenomas and adenocarcinomas. The myeloid hyperplasia was characterized by a decrease in the number of fat cells in the marrow and an increase in the hematopoietic tissue, particularly cells of the granulocytic series. Muscle degeneration (femoral muscle) was found in both high-dose males and females. Retinal degeneration was increased in both males and females, the incidence being significantly ($p<0.05$) higher in high-dose females than in controls. In high-dose females there was an increase in coagulative centrolobular necrosis in the liver.

Based on decreased body weight gain, the LEL for systemic toxicity is 500 ppm (25 mg/kg-day). The NOEL for systemic toxicity is 70 ppm (3.5 mg/kg-day).

Groups of 5-month-old beagle dogs (6/sex control, 4/sex low- and mid-dose, and 6/sex high-dose) were administered atrazine for 1 year at dietary levels of 0, 15, 150 and 1000 ppm (Male: 0, 0.48, 4.97 and 33.65 mg/kg-day; Female: 0, 0.48, 4.97 and 33.8 mg/kg-day) (Ciba-Geigy Corp., Agricultural Div., 1987a). Animals received food and water ad libitum.

The most significant effect of atrazine administration was the syndrome of cardiopathy, featuring discrete myocardial degeneration, which was most prominently found in animals receiving 1000 ppm. Clinical signs referable to cardiac toxicity, such as ascites, cachexia, labored/shallow breathing, and abnormal EKG (irregular heart beat and increased heart rate, decreased P-II values, atrial premature complexes, atrial fibrillation) were first observed as early as 17 weeks into the study. Gross pathological examination revealed moderate-to-severe dilation of the right atrium (and occasionally the left atrium), microscopically manifest as atrophy and myelosis (degeneration of the atrial myocardium).

Three animals had to be sacrificed during the study in moribund condition: one 150 ppm male on day 75; one 1000 ppm female on day 113; and one 1000 ppm male on day 250. Control and 15 ppm animals survived the entire study period without incident. The study authors concluded that only the death of the high-dose female was compound related.

No effects were observed in the low- and mid-dose groups. Therefore, based on the effects observed in the high-dose group, the LEL for systemic toxicity is 1000 ppm (Male: 33.65 mg/kg-day; Female: 33.8 mg/kg-day). The NOEL for systemic toxicity is 150 ppm (4.97 mg/kg-day).

One hundred twenty male and 120 female Charles River CD rats were randomly distributed into four treatment groups, 0, 10, 50 and 500 ppm (Male: 0, 0.69, 3.5 and 34.97 mg/kg-day; Female: 0, 0.76, 3.78 and 37.45 mg/kg-day) (Ciba-Geigy Corp., Agricultural Div., 1987b). Male rats were placed on the control and test diets at 47 days of age and females at 48 days of age. They were maintained on these diets for a period of 10 weeks prior to mating. Males and females were housed together in a 1:1 ratio for mating. One litter was produced in each generation. After weaning of the first generation, 30 males and 30 females were selected for the second parental generation. The remaining male parental animals were sacrificed on days 133-134 of the study. Animals selected for the second generation were exposed to test diets for 12 weeks prior to mating. Mating was conducted in the same manner as for the first generation. Parental males were sacrificed on day 138 of the study and parental females on days 138, 139 and 152 after weaning of their litters.

Body weights were statistically significantly (p<0.05) lower for high-dose animals (14% and 16% for the F0 and F1 males, respectively; 14% and 13% for the F0 and F1 females, respectively) throughout the study. Body weight gains were also statistically significantly depressed (p<0.05) at the high-dose (27% and 17% for the F0 and F1 males, respectively; 28% and 15% for the F0 and F1 females, respectively). At the mid-dose, sporadic decreases in body weight gain were also noted. These changes were not considered to be treatment-related since they were occasional and sporadic. Food consumption was statistically significantly reduced for males and females during the premating period for both parental generations and for F1 females on days 0-7 of gestation. No other effects were noted.

Based on the effects observed at the high-dose, the LEL for parental toxicity is 500 ppm (Male: 34.97 mg/kg-day; Female: 37.45 mg/kg-day). The NOEL for parental toxicity is 50 ppm (Male: 3.5 mg/kg-day; Female: 3.78 mg/kg-day).

In the initial Office of Pesticide Programs (OPP) review of this study for reproductive effects, a NOEL and LEL of 10 and 50 ppm were selected based on a statistically significant (p<0.05) decrease in the F2 generation male pup body weights at day 21. The registrant subsequently submitted the original and revised "Healy analyses" for the F1 and F2 male and female body weights. This method of analysis was proposed by M.J.R. Healy, who observed that "experiments using animal litters as experimental units usually require a weighted analysis to allow for variations in litter size" and described "methods for assessing appropriate weights" (Healy, 1972). The registrant contended that a review of the original Healy analysis procedure determined that pairwise comparisons should not have been conducted in the absence of a statistically significant F-statistic. The correct procedure, as the registrant pointed out, is to carry out pairwise comparisons only if the F-Test is significant, thereby controlling the Type I (false-positive) error rate. Therefore, the revised Healy analysis should be considered correct as far as determining which differences are statistically significant. Therefore, the registrant concluded that the NOEL for reproductive toxicity should be greater than 10 ppm.

Upon reviewing this additional information, the OPP toxicologists and statisticians concluded that the NOEL and LEL for reproductive toxicity be changed to 50 and 500 ppm respectively, based on reduced body weight in the F2 generation male pups at lactation days 7 and 14. Further rationale for this change is described below.

Body weights of the F2 male pups in the 10 and 50 ppm groups at day 7 (13.39 g and 13.66 g, respectively) and day 14 (28.26 g and 28.33 g, respectively) are essentially the same even though they are less than the control group value (14.01 g and 29.32 g for days 7 and 14, respectively). The day 4 post-culling data should be used as a control time point to determine the lactational effect endpoint. The added week of additional compound intake, especially at 500 ppm, is expected to further increase the body weight reduction changes already seen. The only change noted at 500 ppm was a consistent, though not statistically significant, fall in F2 male pup body weights at 500 ppm (9.02, 13.28 and 28.06 g, for days 4, 7 and 14, respectively) when compared to the post culling controls (9.29, 14.01 and 29.32 g for days 4, 7 and 14, respectively). Therefore, the decreases in body weights of the F2 male pups are considered to be an equivocal biological effect.

Based on an equivocal decrease in body weights of the F2 male pups, the LEL for reproductive toxicity is 500 ppm (Male: 34.97 mg/kg-day; Female: 37.45 mg/kg-day). The NOEL for reproductive toxicity is 50 ppm (Male: 3.5 mg/kg-day; Female: 3.78 mg/kg-day).

I.A.3. UNCERTAINTY AND MODIFYING FACTORS (ORAL RfD)

UF -- The uncertainty factor of 100 reflects 10 for interspecies extrapolation and 10 for intraspecies-variability.

MF -- None

I.A.4. ADDITIONAL STUDIES / COMMENTS (ORAL RfD)

In the previous review of atrazine, a core grade supplementary 2-year feeding study in dogs (Ciba-Geigy Corp., 1964) with a systemic NOEL and LEL of 0.35 and 3.54 mg/kg-day, respectively, was included under other data reviewed. Since the stated NOEL for this study was approximately 10-fold lower than the NOEL established in a newer 1-year dog study (Ciba-Geigy Corp., 1987a), a reevaluation of the older study by the Office of Pesticide Programs (OPP) was considered necessary. OPP's reevaluation noted the following deficiencies: 1) purity of test material was not reported; 2) only 3/dogs/sex were used, animals were obtained from different suppliers and weights of individual animals were not provided; 3) animals were given double portions of food on Saturdays and none on the following day; 4) individual animal feeding and observation data were not submitted; 5) the study reported that tetrachloroethylene, an antihelmenthic, was used several times a day. This chemical can cause effects on certain blood parameters. Since hematological effects were noted in the study, the possibility that this antihelmenthic chemical caused or contributed to these effects cannot be ruled out.; 6) limited clinical chemistries determinations; and 7) urinalysis determinations were carried out on cage collected urine. Based on the many deficiencies and reporting omissions in this study, which have not been satisfactorily addressed by the sponsor, the study can no longer be classified as core supplementary and has been down-graded to invalid.

1) 2-Year Feeding/Oncogenicity - rat: Principal study -- see previous description; core grade minimum (Ciba-Geigy Corp., 1986).

2) 1-Year Feeding - dog: Co-principal study -- see previous description; core grade minimum (Ciba-Geigy Corp., 1987a).

3) 2-Generation Reproduction - rat: See previous description; core grade minimum (Ciba-Geigy Corp., Agricultural Div., 1987b).

4) Developmental toxicity - rat: Core grade minimum (Ciba-Geigy Corp., 1984a).

Groups of pregnant Charles River rats (27/dose) were administered atrazine by gastric intubation at dose levels of 0, 10, 70 and 700 mg/kg-day. Dosing occurred daily and was performed on days 6 through 15 of presumed gestation. The control group received 10 ml/kg-day of 3% corn starch containing 0.5% Tween 80, which was the volume equivalent to that received by treated rats.

Maternal toxicity was observed during and after the treatment period at the 700 mg/kg-day dose level (HDT). Signs of toxicity in the HDT included death (21 of 27 dams), reduced food consumption, reduced weight gain, salivation, ptosis, swollen abdomen, oral/nasal discharge and bloody vulva. Maternal toxicity was also observed at the 70 mg/kg-day dose level. Toxicity signs in this group included reduced food consumption, reduced body weight and reduced weight gain. No maternal toxicity was observed in the 10 mg/kg-day or control groups. Based on the above effects, the NOEL and LEL for maternal toxicity are 10 and 70 mg/kg-day, respectively.

At 70 mg/kg-day, there were statistically significant increases by both fetal and litter incidence in skeletal variations indicating delayed ossification. These included: skull not completely ossified, presphenoid not ossified, teeth not ossified, metacarpals not ossified, metacarpals bipartite, and distal phalanx not ossified. The incidences of these effects in the control and low-dose groups were comparable. Based on these effects, the NOEL and LEL for developmental toxicity are 10 and 70 mg/kg-day.

5) Developmental toxicity - rabbit: Core grade minimum (Ciba-Geigy Corp., 1984b)

Groups of pregnant New Zealand White rabbits (19/dose) were administered atrazine at dose levels of 0, 1, 5 and 75 mg/kg-day on days 7 through 19 of gestation. The control group (vehicle control) received 5 ml/kg-day of 3% corn starch with Tween 80, which was a volume equivalent to that received by rabbits treated with atrazine.

Maternal toxicity was noted in the high-dose group only. From the body weight gain data, it is not apparent that there was a dose related decrease in body weight gain at all dose levels as originally reported. A treatment-related effect was noted in the body weight gain in the high-dose group during the dosing period, with a rebound increase in body weight gain following the dosing period. Combined-dosing period and post-dosing period body weight gain for the high-dose group also shows a decrease, as does the corrected body weight gain for the high-dose group. Food consumption data for these periods also support the observation of a high-dose group effect. Low food efficiency was noted in the high-dose group during the dosing period and for the combined period including the dosing period and post-dosing period. There appears to be increased food efficiency in the post-dosing period, which supports what was noted for food consumption and the rebound in body weight gain in the post dosing period. Other signs of maternal toxicity related to clinical observations included an increase in "stool: little, none and/or soft" in the high-dose group, along with the evidence of "blood on vulva/in cage" in the high-dose group. No treatment-related signs were noted in the low- or mid-dose groups. Cesarean section data showed no additional maternal toxicity data; however, developmental toxicity was noted at the high-dose in the form of increased total resorptions and resorptions per dam, decreased total live fetuses and live fetuses per dam (litter size), an increased post-implantation loss and a decrease in mean fetal weight. Therefore, based on the effects observed at the highest dose tested, the LEL for maternal and developmental toxicity is 75 mg/kg-day. The NOEL for maternal and developmental toxicity is 5 mg/kg-day.

Other Data Reviewed:

1) 91-Week Feeding/Oncogenicity - mouse: Core grade guideline (Ciba-Geigy Corp., Agricultural Division, 1987c).

Groups of CD-1 mice (60/sex/dose) were administered atrazine for 91 weeks at dietary levels of 0, 10, 300, 1500, and 3000 ppm (Male: 0, 1.4, 38.4, 194.0 and 385.7 mg/kg-day; Female: 0, 1.6, 47.9, 246.9 and 482.7 mg/kg-day). Animals received food and water ad libitum.

Dose-related reductions were observed in mean body weight gain in both sexes of mice receiving 1500 or 3000 ppm. In mice receiving 1500 ppm, the decrease in mean body weight gain at weeks 12 and 91 were 33% and 23% for males, respectively, and 14.3% and 11% for females, respectively. In mice receiving 3000 ppm, the decrease in mean body weight gain at weeks 12 and 91 were 40.6% and 24.4% for males, respectively, and 12.1% and 48.5% for females, respectively.

At the termination of the study, statistically significant ($p < 0.05$) reductions in mean erythroid variables (erythrocyte count, hematocrit and hemoglobin) were observed in mid- and high-dose males and high-dose females. High-dose females also had increased neutrophil percentage ($p < 0.05$) and decreased lymphocyte percentage ($p < 0.05$) when compared with controls.

Dose-related cardiac thrombi (primarily in the atria) were seen in male and female mice. The incidence for males was 3/59, 6/60, 3/60, 7/60 and 9/58 for the 0, 10, 300, 1500 and 3000 ppm dose groups, respectively. The incidence for females was 3/60, 4/59, 2/60, 11/60 and 26/60 for the 0, 10, 300, 1500 and 3000 ppm dose groups, respectively. This effect was observed primarily in those animals who had died or were killed in the course of the study.

Based on the above effects, the LEL for systemic toxicity is 1500 ppm (Male: 194.0 mg/kg-day; Female: 246.9 mg/kg-day). The NOEL for systemic toxicity is 300 ppm (Male: 38.4 mg/kg-day; Female: 47.9 mg/kg-day).

2) Developmental toxicity - rat: Core grade minimum (Ciba-Geigy, Corp., 1971).

Groups of pregnant rats were administered atrazine by gavage at dose levels of 0, 100, 500 and 1000 mg/kg-day on days 6 through 15 of gestation. In the high-dose group, 7 out of the 30 animals treated died. Slight weight losses in females were observed at mid-dose. A reduction in mean fetal weights and an increase in the number of embryonic and fetal resorptions were observed in the mid- and high-dose groups. Based on the above effects, the NOEL and LEL for maternal toxicity and fetotoxicity are 100 and 500 mg/kg-day, respectively.

Data Gap(s): None

I.A.5. CONFIDENCE IN THE ORAL RfD

Study -- HighData Base -- HighRfD -- High

The principal studies are of good quality and are given high confidence ratings. Additional studies are supportive of the principal studies and of good quality. Therefore, the data base is given a high confidence rating. High confidence in the RfD follows.

I.A.6. EPA DOCUMENTATION AND REVIEW OF THE ORAL RfD

Source Document -- This assessment is not presented in any existing U.S. EPA document.

Other EPA Documentation -- None

Agency Work Group Review -- 07/08/86, 12/09/86, 05/20/87, 06/22/88, 02/21/90, 11/04/92, 12/16/92, 09/23/93

Verification Date -- 09/23/93

I.A.7. EPA CONTACTS (ORAL RfD)

George Ghali / OPP -- (703)305-7490

William Burnam / OPP -- (703)305-7491

I.B. REFERENCE CONCENTRATION FOR CHRONIC INHALATION EXPOSURE (RfC)

[Ed. Note: EPA does not have information yet in this section of IRIS].

II. CARCINOGENICITY ASSESSMENT FOR LIFETIME EXPOSURE

This substance/agent has been evaluated by the U.S. EPA for evidence of human carcinogenic potential. This does not imply that this agent is necessarily a carcinogen. The evaluation for this chemical is under review by an inter-office Agency work group. A risk assessment summary will be included on IRIS when the review has been completed.

III. HEALTH HAZARD ASSESSMENTS FOR VARIED EXPOSURE DURATIONS

[Ed. Note: EPA does not have information yet in this section of IRIS].

IV. U.S. EPA REGULATORY ACTIONS

EPA risk assessments may be updated as new data are published and as assessment methodologies evolve. Regulatory actions are frequently not updated at the same time. Compare the dates for the regulatory actions in this section with the verification dates for the risk assessments in sections I and II, as this may explain inconsistencies. Also note that some regulatory actions consider factors not related to health risk, such as technical or economic feasibility. Such considerations are indicated for each action. In addition, not all of the regulatory actions listed in this section involve enforceable federal standards. Please direct any questions you may have concerning these regulatory actions to the U.S. EPA contact listed for that particular action. Users are strongly urged to read the background information on each regulatory action in Background Document 4 in Service Code 5.

IV.A. CLEAN AIR ACT (CAA)

No data available

IV.B. SAFE DRINKING WATER ACT (SDWA)

IV.B.1. MAXIMUM CONTAMINANT LEVEL GOAL (MCLG) for Drinking Water

Value -- 0.003 mg/L (Final, 1991)

Considers technological or economic feasibility? -- NO

Discussion -- A MCLG of 0.003 mg/L is set based on its potential adverse effects (liver and kidney damage) reported in dog and rat studies. The MCL Gis based upon a DWEL of 0.2 mg/L and an assumed drinking water contribution of 20 percent.

Reference -- 56 FR 3526 (01/30/91)

EPA Contact -- Health and Ecological Criteria Division / OST /(202) 260-7571 / FTS 260-7571; or Safe Drinking Water Hotline / (800) 426-4791

IV.B.2. MAXIMUM CONTAMINANT LEVEL (MCL) for Drinking Water

Value -- 0.003 mg/L (Final, 1991)

Considers technological or economic feasibility? -- YES

Discussion -- EPA has set a MCL equal to the MCLG of 0.003 mg/L.

Monitoring requirements -- All systems initially monitored for four consecutive quarters every three years; repeat monitoring dependent upon detection, vulnerability status and size.

Analytical methodology -- Microextraction/gas chromatography (EPA505);liquid-solid extraction/capillary column gas chromatography/mass spectrometry (EPA 525); (EPA 507); PQL=0.001 mg/L.

Best available technology -- Granular activated carbon.

Reference -- 56 FR 3526 (01/30/91)

EPA Contact -- Drinking Water Standards Division / OGWDW /(202) 260-7575 / FTS 260-7575; or Safe Drinking Water Hotline / (800) 426-4791

IV.B.3. SECONDARY MAXIMUM CONTAMINANT LEVEL (SMCL) for Drinking Water

[Ed. Note: EPA does not have information yet in this section of IRIS].

IV.B.4. REQUIRED MONITORING OF "UNREGULATED" CONTAMINANTS

[Ed. Note: EPA does not have information yet in this section of IRIS].

IV.C. CLEAN WATER ACT (CWA)

[Ed. Note: EPA does not have information yet in this section of IRIS].

IV.D. FEDERAL INSECTICIDE, FUNGICIDE, AND RODENTICIDE ACT (FIFRA)

IV.D.1. PESTICIDE ACTIVE INGREDIENT, Registration Standard

Status -- Issued (1983). Second Round Review still in progress.

Reference -- Atrazine Pesticide Registration Standard. September, 1983(NTIS No. PB84-149541)

EPA Contact -- Registration Branch / OPP(703)557-7760 / FTS 557-7760

IV.D.2. PESTICIDE ACTIVE INGREDIENT, Special Review

[Ed. Note: EPA does not have information yet in this section of IRIS].

IV.E. TOXIC SUBSTANCES CONTROL ACT (TSCA)

[Ed. Note: EPA does not have information yet in this section of IRIS].

IV.F. RESOURCE CONSERVATION AND RECOVERY ACT (RCRA)

[Ed. Note: EPA does not have information yet in this section of IRIS].

IV.G. SUPERFUND (CERCLA)

[Ed. Note: EPA does not have information yet in this section of IRIS].

VI. BIBLIOGRAPHY

VI.A. ORAL RfD REFERENCES

Ciba-Geigy Corporation. 1964. MRID No. 00059213. HED Doc. No. 000525, 002917, 006605. Available from EPA. Write to FOI, EPA, Washington, DC 20460.

Ciba-Geigy Corporation. 1971. MRID No. 00038041. HED Doc. No. 002917. Available from EPA. Write to FOI, EPA, Washington, DC 20460.

Ciba-Geigy Corporation. 1984a. MRID No. 00143008, 40566302. HED Doc. No. 006131, 006761, 006937. Available from EPA. Write to FOI, EPA, Washington, DC 20460.

Ciba-Geigy Corporation. 1984b. MRID No. 00143006, 40566301. HED Doc. No. 006131, 006761, 006937. Available from EPA. Write to FOI, EPA, Washington, DC 20460.
Ciba-Geigy Corporation. 1986. MRID No. 00141874, 00157875, 00158930, 40629302. HED Doc. No. 005940, 006937. Available from EPA. Write to FOI, EPA, Washington, DC 20460.

Ciba-Geigy Corporation, Agricultural Division. 1987a. MRID No. 40431301, 41293801. HED Doc. No. 006718, 006937, 007647. Available from EPA. Write to FOI, EPA, Washington, DC 20460.

Ciba-Geigy Corporation. 1987b. MRID No. 40431303, 41986801. HED Doc. No. 006718, 06937. Available from EPA. Write to FOI, EPA, Washington, DC 20460.

Ciba-Geigy Corporation, Agricultural Division. 1987c. MRID No. 40431302, 40629301. HED Doc. No. 006718, 006765, 006937. Available from EPA. Write to FOI, EPA, Washington, DC 20460.

Healy, M.R.C. 1972. Animal Litters as Experimental Units. J. Royal Stat. Soc. (Series C), Applied Statistics. 21(2): 155-159.

VI.B. INHALATION RfD REFERENCES

None

VI.C. CARCINOGENICITY ASSESSMENT REFERENCES

None

VI.D. DRINKING WATER HA REFERENCES

None

VII. REVISION HISTORY

Date	Section	Description
03/01/88	I.A.5	Confidence levels revised
08/22/88	I.A.	Withdrawn; new oral RfD in preparation
09/07/88	I.A.	Revised oral RfD summary added
10/01/89	VI.	Bibliography on-line
03/01/90	I.A.	Withdrawn; new oral RfD in preparation
03/01/90	VI.A.	Oral RfD references withdrawn
05/01/90	I.A.	Oral RfD summary replaced; RfD unchanged
05/01/90	VI.A.	Oral RfD references replaced
01/01/91	I.A.	Text edited
01/01/91	II.	Carcinogen assessment now under review
01/01/92	IV.	Regulatory Action section on-line
12/01/92	I.A.	Oral RfD summary noted as pending change
12/01/92	I.A.6.	Work group review date added
01/01/93	I.A.	Withdrawn; new oral RfD in preparation
01/01/93	I.A.	Work group review date added
01/01/93	VI.A.	Oral RfD references withdrawn
10/01/93	I.A.	Oral RfD summary replaced; RfD changed
10/01/93	VI.A.	Oral RfD references replaced

SYNONYMS

A 361
AATREX
AATREX 4L
AATREX 80W
AATREX NINE-O
2-AETHYLAMINO-4-CHLOR-6-
ISOPROPYLAMINO-1,3,5-TRIAZIN
2-AETHYLAMINO-4-
ISOPROPYLAMINO-6-CHLOR-1,3,5-
TRIAZIN
AKTIKON
AKTIKON PK
AKTINIT A
AKTINIT PK
ARGEZIN
ATAZINAX
ATRANEX
ATRASINE
ATRATOL A
ATRAZIN
Atrazine
ATRED
ATREX

TRIAZINE-2,4-DIAMINE
2-CHLORO-4-(2-PROPYLAMINO)-6-
ETHYLAMINO-s-TRIAZINE
CRISATRINA
CRISAZINE
CYAZIN
FARMCO ATRAZINE
FENAMIN
FENAMINE
FENATROL
G 30027
GEIGY 30,027
GESAPRIM
GESOPRIM
GRIFFEX
HUNGAZIN
HUNGAZIN PK
INAKOR
OLEOGESAPRIM
PRIMATOL
PRIMATOL A
PRIMAZE
RADAZIN

CANDEX
CEKUZINA-T
2-CHLORO-4-
ETHYLAMINEISOPROPYLAMINE-s-
TRIAZINE
1-CHLORO-3-ETHYLAMINO-5-
ISOPROPYLAMINO-2,4,6-TRIAZINE
1-CHLORO-3-ETHYLAMINO-5-
ISOPROPYLAMINO-s-TRIAZINE
2-CHLORO-4-ETHYLAMINO-6-
ISOPROPYLAMINO-1,3,5-TRIAZINE
2-CHLORO-4-ETHYLAMINO-6-
ISOPROPYLAMINO-s-TRIAZINE
6-CHLORO-N-ETHYL-N'-(1-
METHYLETHYL)-1,3,5-

RADIZINE
STRAZINE
TRIAZINE A 1294
s-TRIAZINE, 2-CHLORO-4-
ETHYLAMINO-6-
ISOPROPYLAMINO-
1,3,5-TRIAZINE-2,4-DIAMINE, 6-
CHLORO-N-ETHYL-N'-(1-
METHYLETHYL)-
VECTAL
VECTAL SC
WEEDEX A
WONUK
ZEAZIN
ZEAZINE

Source: *Instant EPA's IRIS*, 1997, Instant Reference Sources, Inc., 7605 Rockpoint Drive, Austin, TX 78731 (Fax: 512-345-2386; Internet URL, http://www.instantref.com/inst-ref.htm).

Production, Use, and Pesticide Labeling Information

Atrazine is not currently listed in EPA's Pesticide Factsheet database and there is no indication of whether it will be added or not in the future.

Benomyl

Introduction

Benomyl is a fungicide used in the control of a wide range of diseases of fruit, nuts, vegetables, mushrooms, field crops, ornamentals, turf and trees. It is also used as an acaricide to slow the growth of mites. It is used as a veterinary anthelmintic and as an oxidizer in sewage treatment. When it is used as a protective and eradicant fungicide and is effective against a wide range of fungi. Benomyl released to soil will not tend to leach but volatilization of benomyl from soil may be significant. Hydrolysis of benomyl in soil is probably the most significant removal process although biodegradation may also be significant, especially for the benomyl hydrolysis products. Benomyl released to water will have a low to moderate tendency to sorb to sediments, suspended sediments and biota and will not tend to bioconcentrate to any significant extent. Volatilization of benomyl from water is probably insignificant. Hydrolysis will probably be the most significant removal process for benomyl in water (t 1/2 <1 week) although biodegradation and photolysis may also be important. Humans may be exposed to benomyl through dermal contact where it is mixed and used, through the inhalation of dust particles to which it has sorbed in fields, and from dermal contact from picking fruits and vegetables that have been sprayed with benomyl [828].

Identifying Information

CAS NUMBER: 17804-35-2

NIOSH Registry Number: DD6475000

SYNONYMS: 1-(BUTYLCARBAMOYL)-2-BENZIMIDAZOLECARBAMIC ACID, METHYL ESTER, ARILATE, BENOSAN, ARBORTRINE , BBC, BENLAT, BENLATE, BENLATE 50, BENLATE 50 W, BENOMYL 50 W, BENEX, FIBENZOL, 1-(N-BUTYLCARBAMOYL)-2-(METHOXYCARBOXAMIDO)-BENZIMIDAZOL, METHYL 1-(BUTYLCARBAMOYL)-2-BENZIMIDAZOLYLCARBAMATE, METHYL 1-((BUTYLAMINO)CARBONYL)-1H-BENZIMIDAZOL-2-YL CARBAMATE, METHYL 1-(BUTYLCARBAMOYL)BENZIMIDAZOL-2-YLCARBAMATE, METHYL 1-(BUTYLCARBAMOYL)-2-BENZIMIDAZOLECARBAMATE, BNM, D 1991, DU PONT 1991, F1991,

FUNDASOL, FUNDAZOL, FUNGICIDE 1991, MBC, TERSAN 1991, TERSAN, UZGEN, NS O2,

Endocrine Disruptor Information

Benomyl is a strongly suspected EED that it is listed by the Centers for Disease Control & Prevention [812], and the World Wildlife Fund [813] as a potential endocrine modifying chemical.

Chemical and Physical Properties

CHEMICAL FORMULA: $C_{14}H_{18}N_4O_3$

MOLECULAR WEIGHT: 290.36

WLN: T56 BN DNJ BVM4 CMVO1

PHYSICAL DESCRIPTION: Colorless or white crystals

MELTING POINT: Decomposes before melting [025], [169], [346], [430]

BOILING POINT: Decomposes [107]

SOLUBILITY: Water: <1 mg/mL @ 20°C [700]; DMSO: 10-50 mg/mL @ 21°C [700]; 95% Ethanol: 1-5 mg/mL @ 21°C [700]; Acetone: 1-5 mg/mL @ 21°C [700]; Methanol: 1-10 mg/mL @ 23°C [700]; Toluene: <1 mg/mL @ 23°C [700]

OTHER SOLVENTS: Chloroform: 94 g/kg @ 25°C [169], [172]; Dimethylformamide: 53 g/kg @ 25° [169], [172]; Xylene: 10 g/kg @ 25°C [169], [172]; Ethanol: 4 g/kg @ 25°C [169], [172]; Heptane: 400 g/kg @ 25°C [172]; Oils: Insoluble [031], [430]; Most organic solvents: Slightly soluble [025], [173]

OTHER PHYSICAL DATA: Technical grade: Tan solid [058]; In some solvents; dissociation occurs [172]; Very soluble in water @ pH 1 [172]; Decomposes in water @ pH 13 [172]; Faint acrid odor [107], [173], [346]; Odorless [058]

Source: *The National Toxicology Program's Chemical Database, Volume 2*, 1992, CRC Press, Inc./Lewis Publishers, 2000 Corporate Blvd., Boca Raton, FL 33431 (Fax: 407-998-9114).

Environmental Reference Materials

A source of EED environmental reference materials for benomyl is Crescent Chemical Co., Hauppauge, NY (Fax: 516-348-0913); Catalog No. 45339.

Hazardous Properties

ACUTE/CHRONIC HAZARDS: Benomyl is a cholinesterase inhibitor [107], [301]. It may be highly toxic and an irritant [058] and when heated to decomposition it emits toxic fumes of nitrogen oxides [043].

VOLATILITY: Vapor pressure: Negligible @ room temperature [058], [169], [172], [421]

FIRE HAZARD: Literature sources indicate that benomyl is nonflammable [107]. Fires involving this material can be controlled with a dry chemical, carbon dioxide or Halon extinguisher. A water spray may also be used [058].

LEL: 0.05 g/L [058]   UEL: Not available

REACTIVITY: Benomyl is incompatible with strong acids, peroxides and strong oxidizers [107]. It is decomposed by strong alkalis [169]. It also decomposes on storage with water [172].

STABILITY: Benomyl decomposes slowly in the presence of moisture [169], [172], [430]. It also decomposes with heat [058]. UV spectrophotometric stability screening indicates that solutions of benomyl in 95% ethanol are stable for at least 24 hours [700].

Source: *The National Toxicology Program's Chemical Database, Volume 7*, 1992, CRC Press, Inc./Lewis Publishers, 2000 Corporate Blvd., Boca Raton, FL 33431 (Fax: 407-998-9114).

Medical Symptoms of Exposure

Symptoms of exposure may include skin irritation, erythema, salivation, sweating, lassitude, reproductive effects, muscular incoordination, nausea, vomiting, abdominal cramps, angina pectoris, central nervous system depression and keratitis [107]. Cholinesterase inhibition may occur [107], [301]. Skin contact may result in contact dermatitis, redness and edema [173]. It may also cause irritation of the eyes, nose and throat and lacrimation [058]. This type of compound may cause visual disturbances, respiratory difficulty, gastrointestinal hyperactivity, anorexia, headache, dizziness, weakness, anxiety, substernal discomfort, tremors of the tongue and eyelids, miosis, impairment of visual acuity, slow pulse, muscular fasciculations, diarrhea, pinpoint and nonreactive pupils, pulmonary edema, cyanosis, loss of sphincter control, convulsions, coma, heart block, hyperglycemia and possible acute pancreatitis [301].

Source: *The National Toxicology Program's Chemical Database, Volume 4*, 1992, CRC Press, Inc./Lewis Publishers, 2000 Corporate Blvd., Boca Raton, FL 33431 (Fax: 407-998-9114)

Toxicological Information

The toxicological information below is gathered from several sources including two national databases: one from the *National Toxicology Program Chemical Repository Database* and another from *EPA's Integrated Risk Information System (IRIS)*.

Toxicology Information from the National Toxicology Program

CARCINOGENICITY: Not available

TERATOGENICITY: See RTECS printout for data

MUTATION DATA: See RTECS printout for data

TOXICITY:

Typ. Dose	Mode	Specie	Amount	Units	Other
LD50	orl	rat	10	gm/kg	
LD50	unr	rat	9920	mg/kg	
LD50	orl	bwd	100	mg/kg	
LD50	unr	mam	10	gm/kg	
LD50	orl	mus	5600	mg/kg	

OTHER TOXICITY DATA:
     Skin and Eye Irritation Data:
       skn-man   0.1% MLD

Review: Toxicology Review-2
     Status:
          EPA Genetox Program 1988, Positive: Aspergillus-aneuploidy; TRP reversion
          EPA Genetox Program 1988, Negative: Aspergillus-recombination; Host-mediated assay
          EPA Genetox Program 1988, Negative: S cerevisiae gene conversion
          EPA Genetox Program 1988, Inconclusive: TRP reversion
          EPA TSCA Chemical Inventory, 1986
          EPA TSCA Test Submission (TSCATS) Data Base, January 1990

SAX TOXICITY EVALUATION:

THR: Poison by ingestion. Mildly toxic by inhalation. An experimental teratogen. Experimental reproductive effects. Human mutagenic data. A human skin irritant. An agricultural chemical and pesticide, pharmaceutical and veterinary drug.

Source: *Instant Tox-Base*, 1995, Instant Reference Sources, Inc., 7605 Rockpoint Drive, Austin, TX 78731 (Fax: 512-345-2386; Internet URL, http://www.instantref.com/inst-ref.htm).

Toxicology Information from EPA's Integrated Risk Information System (IRIS)

I. CHRONIC HEALTH HAZARD ASSESSMENTS FOR NONCARCINOGENIC EFFECTS

I.A. REFERENCE DOSE FOR CHRONIC ORAL EXPOSURE (RfD)

The Reference Dose (RfD) is based on the assumption that thresholds exist for certain toxic effects such as cellular necrosis, but may not exist for other toxic effects such as carcinogenicity. In general, the RfD is an estimate (with uncertainty spanning perhaps an order of magnitude) of a daily exposure to the human population (including sensitive subgroups) that is likely to be without an appreciable risk of deleterious effects during a lifetime. Please refer to Background Document 1 in Service Code 5 for an elaboration of these concepts. RfDs can also be derived for the noncarcinogenic health effects of compounds which are also carcinogens. Therefore, it is essential to refer to other sources of information concerning the carcinogenicity of this substance. If the U.S. EPA has evaluated this substance for potential human carcinogenicity, a summary of that evaluation will be contained in Section II of this file when a review of that evaluation is completed.

I.A.1. ORAL RfD SUMMARY

Critical Effect	Experimental Doses*	UF	MF	RfD
Decreased pup weanling weights	NOEL: 100 ppm diet (5 mg/kg/day) LEL: 500 ppm (25 mg/kg/day)	100	1	5E-2mg/kg/day

*Conversion Factors: 1 ppm = 0.05 mg/kg/day (assumed rat food consumption)

3-Generation Reproduction Rat Study du Pont, 1968a

I.A.2. PRINCIPAL AND SUPPORTING STUDIES (ORAL RfD)

E.I. duPont de Nemours and Co. 1968a. MRID No. 00066773. Available from EPA. Write to FOI, EPA, Washington DC. 20460.

Benomyl, 50 or 70% wettable powder (dose based on % active ingredient), was administered in the diet at 0, 100, 500, and 2500 ppm (0, 5, 25, and 125 mg/kg/day) to male and female ChR-CD rats for 3 generations (7 litters). Six males and females were mated for the first generation, 12 males and females for the second generation, and 20 males and females for the third generation. Histology was performed on F3bn weanlings. F3c pups were used for a post weaning growth curve study. No treatment-related effects were seen with the exception of pup weanling weights in F2b, F3b, and F3c litters at 500 and 2500 ppm as compared with control values. The NOEL was 100 ppm (5 mg/kg/day) and the LEL, based on decreased pup weanling weights, was 500 ppm (25 mg/kg/day).

I.A.3. UNCERTAINTY AND MODIFYING FACTORS (ORAL RfD)

UF -- A UF of 100 includes uncertainties in extrapolation from laboratory animals to humans. The extrapolation from the teratology data was considered to be sufficiently covered by this UF, since the NOEL for teratogenic effects is 30 mg/kg/day, that is, 6 times higher than the NOEL of 5 mg/kg/day used to establish the RfD. Thus, there is an overall 600-fold margin between the teratogenic NOEL and the RfD.

MF -- None

I.A.4. ADDITIONAL COMMENTS (ORAL RfD)

Data Considered for Establishing the RfD

1) 3-Generation Reproduction - rat: Principal study - see previous description; core grade minimum

2) 2-Year Feeding (oncogenic) - rat: Systemic NOEL=2500 ppm (125 mg/kg/day) (HDT); core grade minimum (E.I. du Pont de Nemours and Co., 1969)

3) 2-Year Feeding - dog: NOEL=500 ppm (12.5 mg/kg/day); LEL=2500 ppm (62.5 mg/kg/day) (biochemical alterations, hepatic cirrhosis, decreased weight gain and lower food consumption); core grade minimum (E.I. du Pont de Nemours and Co., 1970)

4) Teratology - rat: NOEL=30 mg/kg/day; LEL=62.5 mg/kg/day (microphthalmia); core grade minimum (E.I. du Pont de Nemours and Co., 1982)

5) Teratology - rat: Fetotoxic NOEL=30 mg/kg/day; Fetotoxic LEL=62.5 mg/kg/day (decreased fetal weight); Maternal NOEL=125 mg/kg/day (HDT); core grade minimum (E.I. du Pont de Nemours and Co., 1980a)

6) Teratology - mouse: Teratogenic NOEL=50 mg/kg/day; Teratogenic LEL=100 mg/kg/day (supra occipital scars, subnormal vertebral centrum, supernumary ribs, cleft palate); core grade minimum (Kavlock et al., 1982)

Other Data Reviewed:

1) Ocogenic - mice: CD-1 mice were given 0, 500, 1500, and 7500 (reduced to 5000 ppm after 37 weeks) ppm (0, 75, 225, 1125/750 mg/kg/day). High dose males had microscopic evidence of hepatocellular and testicular (and epididymal) degeneration; core grade minimum (E.I. du Pont de Nemours and Co., 1980b)

2) 90-Day Feeding - rat: NOEL=500 ppm (25 mg/kg/day); LEL=2500 ppm (125 mg/kg/day) (increased relative and absolute liver weight in females and increased SGPT values in males); core grade minimum (E.I. du Pont de Nemours and Co., 1967)

3) 90-Day Feeding - dog: NOEL=500 ppm (12.5 mg/kg/day); LEL=2500 ppm (62.5 mg/kg/day) (depressed albumen/globulin, A/G ratio, and increased SGPT in males); core grade minimum (E.I. du Pont de Nemours and Co., 1968b)

Data Gap(s): None

I.A.5. CONFIDENCE IN THE ORAL RfD

Study -- Medium
Data Base -- High
RfD -- High

The principal study is of adequate quality and is therefore given medium confidence. Since additional studies are of adequate quality, the data base is given high confidence. High confidence in the RfD follows.

I.A.6. EPA DOCUMENTATION AND REVIEW OF THE ORAL RfD

Source Document -- This assessment is not presented in any existing U.S. EPA document.

Other EPA Documentation -- Pesticide Registration Standard April, 1986; Special Review Position Document; Pesticide Registration Files

Agency Work Group Review -- 03/26/86

Verification Date -- 03/26/86

I.A.7. EPA CONTACTS (ORAL RfD)

George Ghali / OPP -- (703)557-7490

William Burnam / OPP -- (703)557-7491

I.B. REFERENCE CONCENTRATION FOR CHRONIC INHALATION EXPOSURE (RfC)

II. CARCINOGENICITY ASSESSMENT FOR LIFETIME EXPOSURE

This substance/agent has been evaluated by the U.S. EPA for evidence of human carcinogenic potential. This does not imply that this agent is necessarily a carcinogen. The evaluation for this chemical will be reviewed at a later date by an inter-office Agency work group. A risk assessment summary will be included on IRIS when the review has been completed.

III. HEALTH HAZARD ASSESSMENTS FOR VARIED EXPOSURE DURATIONS

[Ed. Note: EPA does not have information yet in this section of IRIS].

IV. U.S. EPA REGULATORY ACTIONS

EPA risk assessments may be updated as new data are published and as assessment methodologies evolve. Regulatory actions are frequently not updated at the same time. Compare the dates for the regulatory actions in this section with the verification dates for the risk assessments in sections I and II, as this may explain inconsistencies. Also note that some regulatory actions consider factors not related to health risk, such as technical or economic feasibility. Such considerations are indicated for each action. In addition, not all of the regulatory actions listed in this section involve enforceable federal standards. Please direct any questions you may have concerning these regulatory actions to the U.S. EPA contact listed for that particular action. Users are strongly urged to read the background information on each regulatory action in Background Document 4 in Service Code 5.

IV.A. CLEAN AIR ACT (CAA)

[Ed. Note: EPA does not have information yet in this section of IRIS].

IV.B. SAFE DRINKING WATER ACT (SDWA)

[Ed. Note: EPA does not have information yet in this section of IRIS].

IV.C. CLEAN WATER ACT (CWA)

[Ed. Note: EPA does not have information yet in this section of IRIS].

IV.D. FEDERAL INSECTICIDE, FUNGICIDE, AND RODENTICIDE ACT (FIFRA)

IV.D.1. PESTICIDE ACTIVE INGREDIENT, Registration Standard

Status -- Issued (1987)

Reference -- Benomyl Pesticide Registration Standard. June, 1987 (NTIS No. PB89-100150).

EPA Contact -- Registration Branch / OPP (703)557-7760 / FTS 557-7760

IV.D.2. PESTICIDE ACTIVE INGREDIENT, Special Review

Action -- Final regulatory decision - PD4 (1982)

Considers technological or economic feasibility? -- YES

Summary of regulatory action -- Requires use of either cloth or commercially available disposable dust masks by mixer/loaders of benomyl intended for aerial application and that registrants of benomyl products conduct field monitoring studies to identify residues that may enter aquatic sites after use on rice; criterion of concern: mutagenicity, teratogenicity, reproductive effects, wildlife hazard and reduction in non-target organisms.

Reference -- 47 FR 46747 (10/20/82)

EPA Contact -- Special Review Branch / OPP (703)557-7400 / FTS 557-7400

IV.E. TOXIC SUBSTANCES CONTROL ACT (TSCA)

[Ed. Note: EPA does not have information yet in this section of IRIS].

IV.F. RESOURCE CONSERVATION AND RECOVERY ACT (RCRA)

[Ed. Note: EPA does not have information yet in this section of IRIS].

IV.G. SUPERFUND (CERCLA)

[Ed. Note: EPA does not have information yet in this section of IRIS].

VI. BIBLIOGRAPHY

VI.A. ORAL RfD REFERENCES

E.I. du Pont de Nemours & Company. 1967. MRID No. 00066771. Available from EPA. Write to FOI, EPA, Washington D.C. 20460.

E.I. du Pont de Nemours & Company. 1968a. MRID No. 00066773. Available from EPA. Write to FOI, EPA, Washington D.C. 20460.

E.I. du Pont de Nemours & Company. 1968b. MRID No. 00066785. Available from EPA. Write to FOI, EPA, Washington D.C. 20460.

E.I. du Pont de Nemours & Company. 1969. MRID No. 00097284. Available from EPA. Write to FOI, EPA, Washington D.C. 20460.

E.I. du Pont de Nemours & Company. 1970. MRID No. 00097305, 00097318, 00097326. Available from EPA. Write to FOI, EPA, Washington D.C. 20460.

E.I. du Pont de Nemours & Company. 1980a. EPA Accession No. 256575. Available from EPA. Write to FOI, EPA, Washington D.C. 20460.

E.I. du Pont de Nemours & Company. 1980b. MRID No. 00096514. Available from EPA. Write to FOI, EPA, Washington D.C. 20460.

E.I. du Pont de Nemours & Company. 1982. MRID No. 00115674. Available from EPA. Write to FOI, EPA, Washington D.C. 20460.

Kavlock, R.J., N. Chernoff, L.E. Gray, Jr., J.A. Gray and D. Whitehouse. 1982. Teratogenic effects of benomyl in the Wistar rat and CD-1 mouse, with emphasis on the route of administration. Toxicol. Appl. Pharmacol. 62:
44-54.

VI.B. INHALATION RfC REFERENCES

None

VI.C. CARCINOGENICITY ASSESSMENT REFERENCES

None

VI.D. DRINKING WATER HA REFERENCES

None

VII. REVISION HISTORY

-------- -------- ---
Date   Section   Description
-------- -------- ---
03/01/88 I.A.5.   Confidence levels revised
03/01/89 I.A.   Major text revisions
07/01/89 VI.   Bibliography on-line
01/01/92 IV.   Regulatory action updated

SYNONYMS

ARILATE
BBC
BENLAT
BENLATE
BENLATE 50
BENLATE 50 W
Benomyl
BENOMYL 50W
2-BENZIMIDAZOLECARBAMIC ACID,
1-(BUTYLCARBAMOYL)-, METHYL
ESTER
BNM
1-(BUTYLCARBAMOYL)-2-
BENZIMIDAZOLECARBAMIC ACID,
METHYL ESTER
1-(BUTYLCARBAMOYL)-2-
BENZIMIDAZOL-METHYLCARBAMAT

CARBAMIC ACID, METHYL-, 1-
(BUTYLCARBAMOYL)-2-
BENZIMIDAZOLE ESTER
D 1991
DU PONT 1991
F1991
FUNDASOL
FUNDAZOL
FUNGICIDE 1991
MBC
METHYL 1-(BUTYLCARBAMOYL)-2-
BENZIMIDAZOLYLCARBAMATE
1-(N-BUTYLCARBAMOYL)-2-
(METHOXY-CARBOXAMIDO)-
BENZIMIDAZOL
TERSAN 1991

Source: *Instant EPA's IRIS*, 1997, Instant Reference Sources, Inc., 7605 Rockpoint Drive, Austin, TX 78731 (Fax: 512-345-2386; Internet URL, http://www.instantref.com/inst-ref.htm).

Production, Use, and Pesticide Labeling Information

Benomyl is not currently listed in EPA's Pesticide Factsheet database and there is no indication of whether it will be added or not in the future.

Benz(a)anthracene

Introduction

Benz(a)anthracene is a polycyclic aromatic hydrocarbon (PAH) and is found in petroleum, wax, and smoke. The pattern of benz(a)anthracene (BA) release into air and water is quite general since it is a universal product of combustion of organic matter. Both in air and water it is largely associated with particulate matter. When released into water it rapidly becomes adsorbed to sediment or particulate matter in the water column, and bioconcentrates into aquatic organisms. In the unadsorbed state, it will degrade by photolysis in a matter of hours to days. Its slow desorption from sediment and particulate matter will maintain a low concentration of BA in the water. Because it is strongly adsorbed to soil it will usually remain in the upper few centimeters of soil and may not leach into groundwater. BA will very slowly biodegrade when colonies of microorganisms are acclimated but this is too slow a process to be significant since the half life is about 1 year. Benz(a)anthracene in the atmosphere can be transported long distances and will probably be subject to photolysis and photooxidation although there is little documentation about the rate of these processes in the literature. Humans can be exposed to benz(a)anthracene in ambient air, particularly in industrial areas, from stoves, cigarette smoke, food (particularly when smoked or charcoal broiled), and drinking water [828].

Identifying Information

CAS NUMBER: 56-55-3

NIOSH Registry Number: CV9275000

SYNONYMS: BA, 2,3-BENZOPHENANTHRENE, 1,2-BENZANTHRACENE, 1,2-BENZ(A)ANTHRACENE, 2,3-BENZPHENANTHRENE, TETRAPHENE, BENZANTHRENE, 1,2-BENZANTHRENE, NAPHTHANTHRACENE, BENZO(A)ANTHRACENE, BENZO(A)PHENANTHRENE, B(A)A, BAA, NAPHTHANTHRACENE, RCRA WASTE NUMBER U018

Endocrine Disruptor Information

Benzo(a)anthracene is on EPA's list of suspected EEDs [811] and it is an aromatic coplanar compound. However, no additional information has been located with respect to its endocrine modifying properties.

190

Chemical and Physical Properties

CHEMICAL FORMULA: $C_{18}H_{12}$

MOLECULAR WEIGHT: 228.28

WLN: L D6 B666J

PHYSICAL DESCRIPTION: Colorless leaflets or plates

MELTING POINT: 157-159°C

BOILING POINT: 435°C (sublimes)

SOLUBILITY: Water: <1 mg/mL @ 20°C [700]; DMSO: 10-50 mg/mL @ 20°C [700]; 95% Ethanol: <1 mg/mL @ 20°C [700]; Acetone: 10-50 mg/mL @ 20° [700]

OTHER SOLVENTS: Acetic acid: Slightly soluble; Toluene: Soluble; Most organic solvents: Soluble; Hot ethanol: Soluble; Ether: Soluble; Benzene: Very soluble

OTHER PHYSICAL DATA: Plates from glacial acetic acid or alcohol; greenish-yellow fluorescence

Source: *The National Toxicology Program's Chemical Database, Volume 2*, 1992, CRC Press, Inc./Lewis Publishers, 2000 Corporate Blvd., Boca Raton, FL 33431 (Fax: 407-998-9114).

Environmental Reference Materials

A source of EED environmental reference materials for benz(a)anthracene is Radian International LLC, Austin, TX (Fax 512-454-0268; Internet URL, http://www.radian.com/standards); Catalog No. ERB-006.

Hazardous Properties

ACUTE/CHRONIC HAZARDS: When heated to decomposition this compound emits acrid smoke and irritating fumes.

FIRE HAZARD: Flash point data for this chemical are not available. It is probably combustible. Fires involving benz(a)anthracene can be controlled using a dry chemical, carbon dioxide or Halon extinguisher.

STABILITY: This compound is stable under normal laboratory conditions. Solutions of benz(a)anthracene should be stable for 24 hours under normal lab conditions [700].

Source: *The National Toxicology Program's Chemical Database, Volume 7*, 1992, CRC Press, Inc./Lewis Publishers, 2000 Corporate Blvd., Boca Raton, FL 33431 (Fax: 407-998-9114).

Medical Symptoms of Exposure

Symptoms of exposure to benz(a)anthracene have not been found. They may be expected to be similar to symptoms recorded for some of the other PAHs such as anthracene, phenantharene and pyrene.

Toxicological Information

The toxicological information below is gathered from several sources including two national databases: one from the *National Toxicology Program Chemical Repository Database* and another from *EPA's Integrated Risk Information System (IRIS)*.

Toxicology Information from the National Toxicology Program

MUTATION DATA:

Test	Lowest dose	Test	Lowest dose
mma-sat	4 ug/plate	dnd-esc	10 umol/L
pic-esc	50 mg/L	dni-omi	200 ug/L
slt-dmg-par	5 mmol/L	sln-dmg-par	5 mmol/L
dnd-sal:tes	5 ug/1H-C	dns-hmn:hla	100 umol/L
dni-hmn:oth	10 umol/L	msc-hmn:lym	9 umol/L
otr-rat-orl	180 mg/kg	otr-rat:emb	3 mg/L
dns-rat:lvr	100 umol/L	sce-rat:lvr	10 umol/L
otr-mus:fbr	4 mg/L	otr-mus:emb	12500 ug/L
dnd-mus:lvr	60 umol/L	dnd-mus:oth	1 mg/L
dnd-mus-skn	192 umol/kg	dni-mus:oth	10 umol/L
msc-mus:lym	10 mg/L	hma-mus/sat	1600 mg/kg
mnt-ham-ipr	1800 mg/kg/24H	mma-ham:lng	109 umol/L
otr-ham:emb	6 mg/L/7D-C	otr-ham:kdy	80 ug/L
otr-ham:lng	100 ug/L	dnd-ham:fbr	5 mg/L
dnd-ham:kdy	5 mg/L	dnd-ham:lng	1 mg/L
oms-ham:lng	1 mg/L	cyt-ham-ipr	1800 mg/kg/24H
sce-ham-ipr	900 mg/kg/24H	sce-ham:ovr	100 umol/L
msc-ham:lng	1 mg/L/3H	spm-ham-orl	900 mg/kg/24H
dnd-mam:lym	100 umol		

TOXICITY:

Typ. Dose	Mode	Specie	Amount	Unit	Other
LDLo	ivn	mus	10	mg/kg	

CARCINOGENICITY:

    Tumorigenic Data:
        TDLo: skn-mus   18 mg/kg
        TDLo: scu-mus   2 mg/kg
        TDLo: imp-mus   80 mg/kg
        TD : skn-mus   18 mg/kg
        TD : skn-mus   360 mg/kg/56W-I
        TD : skn-mus   240 mg/kg/1W-I

    Review:
        IARC Cancer Review: Animal Sufficient Evidence
        IARC probable human carcinogen (Group 2A)

    Status:
        NTP Third Annual Report on Carcinogens, 1982
        EPA Carcinogen Assessment Group
        NTP anticipated human carcinogen

TERATOGENICITY: Not available

OTHER TOXICITY DATA:

    Review:
        Toxicology Review-3

    Status:
        Reported in EPA TSCA Inventory, 1983
        EPA Genetic Toxicology Program, January 1984
        Meets criteria for proposed OSHA medical records rule

SAX TOXICITY EVALUATION:

THR: MUTATION data. Experimental neoplastigen, equivocal tumorigenic agent, carcinogen. It is found in oils, waxes, smoke, food, drugs. HIGH acute intravenous.

Source: *Instant Tox-Base*, 1995, Instant Reference Sources, Inc., 7605 Rockpoint Drive, Austin, TX 78731 (Fax: 512-345-2386; Internet URL, http://www.instantref.com/inst-ref.htm).

Toxicology Information from EPA's Integrated Risk Information System (IRIS)

I. CHRONIC HEALTH HAZARD ASSESSMENTS FOR NONCARCINOGENIC EFFECTS

I.A. REFERENCE DOSE FOR CHRONIC ORAL EXPOSURE (RfD)

[Ed. Note: EPA does not have information yet in this section of IRIS].

I.B. REFERENCE CONCENTRATION FOR CHRONIC INHALATION EXPOSURE (RfC)

A risk assessment for this substance/agent is under review by an EPA work group.

II. CARCINOGENICITY ASSESSMENT FOR LIFETIME EXPOSURE

Section II provides information on three aspects of the carcinogenic risk assessment for the agent in question; the U.S. EPA classification, and quantitative estimates of risk from oral exposure and from inhalation exposure. The classification reflects a weight-of-evidence judgment of the likelihood that the agent is a human carcinogen. The quantitative risk estimates are presented in three ways. The slope factor is the result of application of a low-dose extrapolation procedure and is presented as the risk per (mg/kg)/day. The unit risk is the quantitative estimate in terms of either risk per ug/L drinking water or risk per ug/cu.m air breathed. The third form in which risk is presented is a drinking water or air concentration providing cancer risks of 1 in 10,000, 1 in 100,000 or 1 in 1,000,000. Background Document 2 (Service Code 5) provides details on the rationale and methods used to derive the carcinogenicity values found in IRIS. Users are referred to Section I for information on long-term toxic effects other than carcinogenicity.

II.A. EVIDENCE FOR CLASSIFICATION AS TO HUMAN CARCINOGENICITY

II.A.1. WEIGHT-OF-EVIDENCE CLASSIFICATION

Classification -- B2; probable human carcinogen

Basis -- Based on no human data and sufficient data from animal bioassays. Benz[a]anthracene produced tumors in mice exposed by gavage; intraperitoneal, subcutaneous or intramuscular injection; and topical application. Benz[a]anthracene produced mutations in bacteria and in mammalian cells, and transformed mammalian cells in culture.

II.A.2. HUMAN CARCINOGENICITY DATA

None. Although there are no human data that specifically link exposure to benz[a]anthracene to human cancers, benz[a]anthracene is a component of mixtures that have been associated with human cancer. These include coal tar, soots, coke oven emissions and cigarette smoke (U.S. EPA, 1984, 1990; IARC, 1984; Lee et al., 1976; Brockhaus and Tomingas, 1976).

II.A.3. ANIMAL CARCINOGENICITY DATA

Sufficient. Benz[a]anthracene administration caused an increase in the incidence of tumors by gavage (Klein, 1963); dermal application (IARC, 1973); and both subcutaneous injection (Steiner and Faulk, 1951; Steiner and Edgecomb, 1952) and intraperitoneal injection (Wislocki et al., 1986) assays. A group of male B6AF1/J mice was exposed to gavage solutions containing 3% benz[a]anthracene in Methocel-Aerosol O.T. (dioctyl ester of sodium sulfo-succinic acid), 3 doses/week for 5 weeks (total dose of approximately 225 mg/mouse, 500 mg/kg/day) or the vehicle (Klein, 1963). Mice were evaluated for tumors on days 437-444 and 547 after treatment was initiated. A statistical analysis was not reported. Increased incidences of pulmonary adenoma and hepatoma in treated vs. control mice were reported by the authors at both observation times. The incidence of pulmonary adenoma at 437-444 days was 37/39 (95%) in treated animals vs. 10/38 (26%) in controls; whereas at 547 days, 19/20 (95%) treated animals and 7/20 (35%) controls had pulmonary adenomas. The incidence of hepatomas at 437 to 440 days was 18/39 (46%) in treated animals compared with 0/38 among the vehicle controls. After 547 days, the hepatoma incidences increased to 20/20 for the treated animals versus 2/20 (10%) for vehicle controls.

Mice (strain and sex not specified) were exposed to a single gavage dose of 0.5 mg benz[a]anthracene in mineral oil (approximately 17 mg/kg). No tumors were reported in 13 mice examined 16 months after exposure. In another part of the study, multiple gavage treatments, 8 or 16 treatments at 3-7 day intervals over a 16-month period, resulted in forestomach papillomas in 2/27 treated mice compared with 0/16 in vehicle controls (Bock and King, 1959).

Groups of male and female CD-1 mice (n=90-100) received intraperitoneal injections of benz[a]anthracene in DMSO on days 1, 8, and 15 of age (total dose = 638 ug/mouse) (Wislocki et al., 1986). Tumors were evaluated in animals that died spontaneously after weaning and in all remaining animals at 1 year after exposure. In treated male mice, a statistically significant increase in the incidence of liver adenomas or carcinomas (31/39 treated vs. 2/28 controls) occurred; 25/39 had carcinomas. Female mice did not develop liver tumors. The incidence of pulmonary adenomas or carcinomas in benz[a]anthracene-treated males (6/39, with a majority of adenomas) was increased but not statistically significantly relative to the vehicle controls (1/28). In the female mice, however, the incidence of pulmonary adenomas was significantly elevated in the treated group (6/32) when compared with vehicle controls (0/31).

Benz[a]anthracene yielded positive results in tests for complete carcinogenicity and initiating activity in skin painting assays in C3H/He, CAF1 and ICR/Ha mouse strains. These studies are reviewed in IARC (1973).

Subcutaneous injection of benz[a]anthracene in tricaprylin into C57Bl mice (40-50/group) produced injection site sarcomas 9 months after treatment (Steiner and Falk, 1951; Steiner and Edgecomb, 1952). The sarcoma incidences were: uninjected controls,

0/76; tricaprylin controls, 3/28 (11%); 0.05 mg, 5/43 (12%); 0.2 mg, 11/43 (26%); 1.0 mg, 15/31 (48%); 5.0 mg, 49/145 (34%); and 10 mg, 5/16 (31%). The results of similar experiments in this series were combined (Steiner and Edgecomb, 1952). A statistical analysis of the results was not reported. Survival was roughly equivalent in all groups (70%).

Klein (1952) showed that an intramuscular injection of benz[a]anthracene in combination with 1 or 3% croton oil produced injection site fibrosarcomas and hemangioendotheliomas in Strain A-derived albino mice; 3/24 mice injected with benz[a]anthracene and 1% croton oil and 1/26 mice injected with benz[a]anthracene and 3% croton oil developed tumors. None of the 30 mice injected with benz[a]anthracene and 0.1% croton oil and none of the 30 mice injected with benz[a]anthracene and 5% croton oil developed tumors. In the control groups none of the 35 mice injected only with 1% croton oil and none of the 32 mice injected only with benz[a]anthracene developed tumors. The survival rate for all groups was roughly equivalent (74%).

II.A.4. SUPPORTING DATA FOR CARCINOGENICITY

The results of tests for DNA damage in Escherichia coli have not been positive at concentrations of benz[a]anthracene up to 250 ug/mL and 1000 ug/well (Rosenkrantz and Poirier, 1979; DeFlora et al., 1984). Positive results were obtained in tests for reverse mutation in five different strains of Salmonella typhimurium and for forward mutation in one strain (McCann et al., 1975; Coombs et al., 1976; Simmon, 1979; Salamone et al., 1979; Bartsch et al., 1980; DeFlora et al., 1984; Norpoth et al., 1984; Utesch et al., 1987; Bos et al., 1988; Kaden et al. 1979).

Benz[a]anthracene produced positive results in an assay for mutations in Drosophila melongaster (Fahmy and Fahmy, 1973).

Tests for DNA damage, mutation, chromosomal effects and cell transformation in a variety of eukaryotic cell preparations have yielded mostly positive results. Benz[a]anthracene tested positive for DNA damage in primary rat hepatocytes and HeLa cells (Probst et al., 1981; Martin et al., 1978). It also tested positive for forward mutation in Chinese hamster cells, V79 cells, mouse lymphoma L5178Y cells and rat liver epithelial cells (Slaga et al., 1978; Krahn and Heidelberger, 1977; Amacher et al., 1980; Amacher and Turner, 1980; Tong et al., 1981). Benz[a]anthracene tested positive for chromosomal affects in Chinese hamster ovary cells (Pal, 1981). Tests for cell transformation (cell morphology) have yielded positive results in Syrian hamster embryo cells and mouse prostate C3HG23 cells (Pienta et al., 1977; DiPaolo et al., 1969, 1971; Marquardt and Heidelberger, 1972).

Current theories on mechanisms of metabolic activation of polycyclic aromatic hydrocarbons are consistent with a carcinogenic potential for benz[a]anthracene. Benz[a]anthracene has a "bay-region" structure (Jerina et al., 1978). It is metabolized by mixed function oxidases to reactive "bay-region" diol epoxides that are mutagenic in bacteria and tumorigenic in mouse skin painting assays (Booth and Sims, 1974; Wood et al., 1977a,b).

II.B.   QUANTITATIVE ESTIMATE OF CARCINOGENIC RISK FROM ORAL
EXPOSURE

[Ed. Note: EPA does not have information yet in this section of IRIS].

II.C.   QUANTITATIVE   ESTIMATE   OF   CARCINOGENIC   RISK   FROM
INHALATION EXPOSURE

[Ed. Note: EPA does not have information yet in this section of IRIS].

II.D. EPA DOCUMENTATION, REVIEW, AND CONTACTS (CARCINOGENICITY
ASSESSMENT)

II.D.1. EPA DOCUMENTATION

Source Document -- U.S. EPA, 1984

The 1990 Drinking Water Criteria Document for Polycyclic Aromatic Hydrocarbons has
received Agency and external review.

II.D.2. REVIEW (CARCINOGENICITY ASSESSMENT)

Agency Work Group Review -- 02/07/90, 08/05/93, 09/21/93, 02/02/94

Verification Date -- 02/07/90

II.D.3. U.S. EPA CONTACTS (CARCINOGENICITY ASSESSMENT)

Rita S. Schoeny / NCEA -- (513)569-7544

Robert E. McGaughy / NCEA -- (202)260-5889

III. HEALTH HAZARD ASSESSMENTS FOR VARIED EXPOSURE DURATIONS

III.A. DRINKING WATER HEALTH ADVISORIES

[Ed. Note: EPA does not have information yet in this section of IRIS].

IV. U.S. EPA REGULATORY ACTIONS

EPA risk assessments may be updated as new data are published and as assessment
methodologies evolve. Regulatory actions are frequently not updated at the same time.
Compare the dates for the regulatory actions in this section with the verification dates for
the risk assessments in sections I and II, as this may explain inconsistencies. Also note
that some regulatory actions consider factors not related to health risk, such as technical
or economic feasibility. Such considerations are indicated for each action. In addition,
not all of the regulatory actions listed in this section involve enforceable federal
standards. Please direct any questions you may have concerning these regulatory actions
to the U.S. EPA contact listed for that particular action. Users are strongly urged to read

the background information on each regulatory action in Background Document 4 in Service Code 5.

IV.A. CLEAN AIR ACT (CAA)

[Ed. Note: EPA does not have information yet in this section of IRIS].

IV.B. SAFE DRINKING WATER ACT (SDWA)

IV.B.1. MAXIMUM CONTAMINANT LEVEL GOAL (MCLG) for Drinking Water

Value -- 0 mg/L (Proposed, 1990)

Considers technological or economic feasibility? -- NO

Discussion -- The proposed MCLG for benz[a]anthracene is zero based on theevidence of carcinogenic potential (B2).

Reference -- 55 FR 30370 (07/25/90)

EPA Contact -- Health and Ecological Criteria Division / OST /(202) 260-7571 / FTS 260-7571; or Safe Drinking Water Hotline / (800) 426-4791

IV.B.2. MAXIMUM CONTAMINANT LEVEL (MCL) for Drinking Water

Value -- 0.0002 mg/L (Proposed, 1990)

Considers technological or economic feasibility? -- YES

Discussion -- The proposed MCL is equal to the PQL and is associated with a maximum lifetime individual risk of 1 E-4.

Monitoring requirements -- Community and non-transient water system monitoring based on state vulnerability assessment; vulnerable systems to be monitored quarterly for one year; repeat monitoring dependent upon detection and size of system.

Analytical methodology -- High pressure liquid chromatography (EPA 550, 550.1); gas chromatographic/mass spectrometry (EPA 525): PQL= 0.0001 mg/L.

Best available technology -- Granular activated carbon.

Reference -- 55 FR 30370 (07/25/90)

EPA Contact -- Drinking Water Standards Division / OGWDW /(202) 260-7575 / FTS 260-7575; or Safe Drinking Water Hotline / (800) 426-4791

IV.B.3. SECONDARY MAXIMUM CONTAMINANT LEVEL (SMCL) for Drinking Water

[Ed. Note: EPA does not have information yet in this section of IRIS].

IV.B.4. REQUIRED MONITORING OF "UNREGULATED" CONTAMINANTS

Status -- Listed (Proposed, 1991)

Discussion --   "Unregulated" contaminants are those contaminants for which EPA establishes a monitoring requirement but which do not have an associated final MCLG, MCL, or treatment technique. EPA may regulate these contaminants in the future.

Monitoring requirement -- All systems to be monitored unless a vulnerability assessment determines the system is not vulnerable.

Analytical methodology --   Gas chromatography/mass spectrometry (EPA 525); high pressure liquid chromatography (EPA 550, 550.1).

Reference -- 56 FR 3526 (01/30/91)

EPA Contact -- Drinking Water Standards Division / OGWDW /(202) 260-7575 / FTS 260-7575; or Safe Drinking Water Hotline / (800) 426-4791

IV.C. CLEAN WATER ACT (CWA)

IV.C.1. AMBIENT WATER QUALITY CRITERIA, Human Health

Water and Fish Consumption: 2.8E-3 ug/L

Fish Consumption Only: 3.11E-2 ug/L

Considers technological or economic feasibility? -- NO

Discussion -- For the maximum protection from the potential carcinogenic properties of this chemical, the ambient water concentration should be zero. However, zero may not be obtainable at this time, so the recommended criteria presents a E-6 estimated incremental increase of cancer over a lifetime. The values given represent polynuclear aromatic hydrocarbons as a class.   Reference -- 45 FR 79318 (11/28/80)

EPA Contact -- Criteria and Standards Division / OWRS(202)260-1315 / FTS 260-1315

IV.C.2. AMBIENT WATER QUALITY CRITERIA, Aquatic Organisms

Freshwater:
    Acute LEC -- none
    Chronic LEC -- none

Marine:
    Acute LEC --   3.0E+2 ug/L
    Chronic LEC -- none

Considers technological or economic feasibility? -- NO

Discussion -- The values that are indicated as "LEC" are not criteria, but are the lowest effect levels found in the literature. LEC's are given when the minimum data required to derive water quality criteria are not available. The values given represent polynuclear aromatic hydrocarbons as a class.

Reference -- 45 FR 79318 (11/28/80)

EPA Contact -- Criteria and Standards Division / OWRS(202)260-1315 / FTS 260-1315

IV.D. FEDERAL INSECTICIDE, FUNGICIDE, AND RODENTICIDE ACT (FIFRA)

        [Ed. Note: EPA does not have information yet in this section of IRIS].

IV.E. TOXIC SUBSTANCES CONTROL ACT (TSCA)

        [Ed. Note: EPA does not have information yet in this section of IRIS].

IV.F. RESOURCE CONSERVATION AND RECOVERY ACT (RCRA)

IV.F.1. RCRA APPENDIX IX, for Ground Water Monitoring

Status -- Listed

Reference -- 52 FR 25942 (07/09/87)

EPA Contact -- RCRA/Superfund Hotline(800)424-9346 / (202)260-3000 / FTS 260-3000

IV.G. SUPERFUND (CERCLA)

IV.G.1. REPORTABLE QUANTITY (RQ) for Release into the Environment

Value (status) -- 10 pounds (Final, 1989)

Considers technological or economic feasibility? -- NO

Discussion -- The final RQ for benz[a]anthracene is based on potential carcinogenicity. Available data indicate a hazard ranking of medium and a weight of evidence classification of Group B2, which corresponds to an RQ of 10 pounds.

Reference -- 54 FR 33418 (08/14/89)

EPA Contact -- RCRA/Superfund Hotline(800)424-9346 / (202)260-3000 / FTS 260-3000

VI. BIBLIOGRAPHY

Substance Name -- Benz[a]anthracene CASRN -- 56-55-3Last Revised -- 12/01/90

VI.A. ORAL RfD REFERENCES

None

VI.B. INHALATION RfC REFERENCES

None

VI.C. CARCINOGENICITY ASSESSMENT REFERENCES

Amacher, D.E. and G.N. Turner. 1980. Promutagen activation by rodent-liver post mitochondrial fractions in the L5178Y/TK cell mutation assay. Mutat. Res. 74: 485-501.

Amacher, D.E., S.C. Paillet, G.N. Turner and D.S. Salsburg. 1980. Point mutations at the thymidine kinase locus in L5178Y mouse lymphoma cells. II. Test validation and interpretation. Mutat. Res. 72: 447-474.

Bartsch, H., C. Malaveille, A.M. Camus, et. al. 1980. Validation and comparative studies on 180 chemicals with S. typhimurium strains and V79 Chinese hamster cells in the presence of various metabolizing systems. Mutat. Res. 76: 1-50.

Bock, F.G. and D.W. King. 1959. A study of the sensitivity of the mouse forestomach toward certain polycyclic hydrocarbons. J. Natl. Cancer Inst. 23(4): 833-839.

Booth, J. and P. Sims. 1974. 8,9-Dihydro-8,9-dihydroxybenz[a]anthracene 10,11-oxide: A new type of polycyclic aromatic hydrocarbon metabolite. FEBS Lett. 47(1): 30-33.

Bos, R.P., J.L.G. Theuws, F.J. Jongeneelen and P.Th. Henderson. 1988. Mutagenicity of bi-, tri and tetra-cyclic aromatic hydrocarbons in the "taped-plate assay" and in the conventional Salmonella mutagenicity assay. Mutat. Res. 204: 203-206.

Brockhaus, A. and R. Tomingas, 1976. Emission of polycyclic hydrocarbons from combustion processes in small heating units and their concentration in the atmosphere. Staub-Reinhalt. Luft. 36(3): 96-101.

Coombs, M.M., C. Dixon and A.M. Kissonerghis. 1976. Evaluation of the mutagenicity of compounds of known carcinogenicity, belonging to the benz[a]anthracene, chrysene, and cyclopenta[a]phenanthrene series, using Ame's test. Cancer Res. 36: 4525-4529.

DeFlora, S., P. Zanacchi, A. Camoirano, C. Bennicelli and G.S. Badolati. 1984. Genotoxic activity and potency of 35 compounds in the Ames reversion test and in a bacterial DNA-repair test. Mutat. Res. 133(3): 161-198.

DiPaolo, J.A., J.P. Donovan and R.L. Nelson. 1969. Quantitative studies of in vitro transformation by chemical carcinogens. J. Natl. Cancer Inst. 42(5): 867-874.

DiPaolo, J.A., P.J. Donovan and R.L. Nelson. 1971. Transformation of hamster cells in vitro by polycyclic hydrocarbons without cytotoxicity. Proc. Natl. Acad. Sci. USA. 68(12): 2958-2961.

Fahmy, O.G. and M.J. Fahmy. 1973. Oxidative activation of benz(a)anthracene and methylated derivatives in mutagenesis and carcinogenesis. Cancer Res. 33: 2354-2361.

IARC (International Agency for Research on Cancer). 1973. Certain Polycyclic Aromatic Hydrocarbons and Heterocyclic Compounds. Monographs on the Evaluation of the Carcinogenic Risk of Chemicals to Man. Polynuclear Aromatic Compounds. Vol. 3. Lyon, France.

IARC (International Agency for Research on Cancer). 1984. Monographs on the Evaluation of the Carcinogenic Risk of Chemicals to Man. Polynuclear Aromatic Compounds. Part 3. Industrial Exposures in Aluminum Production, Coal Gasification, Coke Production, and Iron and Steel Founding. Vol. 34. World Health Organization.

Jerina, D.M., H. Yagi, R.E. Lehr, et al. 1978. The bay-region theory of carcinogenesis by polycyclic aromatic hydrocarbons. In: Polycyclic Hydrocarbons and Cancer: Vol. 1, Environment, Chemistry, and Metabolism, H.V. Gelboin and P.O.P. Ts'O, Ed. Academic Press, NY. p. 173-188.

Kaden, D.A., R.A. Hites and W.G. Thilly. 1979. Mutagenicity of soot and associated polycyclic aromatic hydrocarbons to Salmonella typhimurium. Cancer Res. 39: 4152-4159.

Klein, M. 1952. Effect of croton oil on induction of tumors by 1,2-benzanthracene, desoxycholic acid, or low doses of 20-methylcholanthrene in mice. J. Natl. Cancer Inst. 13: 333-341.

Klein, M. 1963. Susceptibility of strain B6AF/J hybrid infant mice to tumorigenesis with 1,2-benzanthracene, deoxycholic acid, and 3-methylcholanthrene. Cancer. Res. 23: 1701-1707.

Krahn, D.F. and C. Heidelberger. 1977. Liver homogenate-mediated mutagenesis in Chinese hamster V79 cells by polycyclic aromatic hydrocarbons and aflatoxins. Mutat. Res. 46: 27-44.

Lee, M.L., M. Novotny and K.D. Bartle. 1976. Gas chromatography/mass spectrometric and nuclear magnetic resonance studies of carcinogenic polynuclear aromatic hydrocarbons in tobacco and marijuana smoke condensates. Anal. Chem. 48(2): 405-416.

Marquardt, H. and C. Heidelberger. 1972. Influence of "feeder cells" and inducers and inhibitors of microsomal mixed-function oxidases on hydrocarbon-induced malignant transformation of cells derived from C3H mouse prostate. Cancer Res. 32: 721-725.

Martin, C.N., A.C. McDermid and R.C. Garner. 1978. Testing of known carcinogens and noncarcinogens for their ability to induce unscheduled DNA synthesis in HeLa cells. Cancer Res. 38: 2621-2627.

McCann, J.E., E. Choi, E. Yamasaki and B.N. Ames. 1975. Detection of carcinogens as mutagens in the Salmonella/microsome test: Assay of 300 chemicals. Proc. Natl. Acad. Sci. USA. 72(12): 5135-5139.

Norpoth, K., A. Kemena, J. Jacob and C. Schumann. 1984. The influence of 18 environmentally relevant polycyclic aromatic hydrocarbons and Clophen A50, as liver monooxygenase inducers, on the mutagenic activity of benz[a]anthracene in the Ames test. Carcinogenesis. 5(6): 747-752.

Pal, K. 1981. The induction of sister-chromatid exchanges in Chinese hamster ovary cells by K-region epoxides and some dihydrodiols derived from benz[a]anthracene, dibenz[a,c]anthracene and dibenz[a,h]anthracene. Mutat. Res. 84: 389-398.

Pienta, R.J., J.A. Poiley and W.B. Lebherz, III. 1977. Morphological transformation of early passage golden Syrian hamster embryo cells derived from cryopreserved primary cultures as a reliable in vitro bioassay for identifying diverse carcinogens. Int. J. Cancer. 19: 642-655.

Probst, G.S., R.E. McMahon, L.E. Hill, C.Z. Thompson, J.K. Epp and S.B. Neal. 1981. Chemically-induced unscheduled DNA synthesis in primary rat hepatocyte cultures: A comparison with bacterial mutagenicity using 218 compounds. Environ. Mutagen. 3: 11-32.

Rosenkrantz, H.S. and L.A. Poirier. 1979. Evaluation of the mutagenicity and DNA-modifying activity of carcinogens and noncarcinogens in microbial systems. J. Natl. Cancer Inst. 62(4): 873-892.

Salamone, M.F., J.A. Heddle and M. Katz. 1979. The mutagenic activity of thirty polycyclic aromatic hydrocarbons (PAH) and oxides in urban airborne particulates. Environ. Int. 2: 37-43.

Simmon, V.F. 1979. In vitro mutagenicity assays of chemical carcinogens and related compounds with Salmonella typhimurium. J. Natl. Cancer Inst. 62(4): 893-899.

Slaga, T.J., E. Huberman, J.K. Selkirk, R.G. Harvey and W.M. Braken. 1978. Carcinogenicity and mutagenicity of benz[a]anthracene diols and diol-epoxides. Cancer Res. 38: 1699-1704.

Steiner, P.E. and J.H. Edgecomb. 1952. Carcinogenicity of 1,2-benzanthracene. Cancer Res. 12: 657-659.

Steiner, P.E. and H.L. Falk. 1951. Summation and inhibition effects of weak and strong carcinogenic hydrocarbons: 1,2-Benzanthracene, chrysene, 1:2:5:6-dibenzanthracene and 20-methylcholanthrene. Cancer Res. 11: 56-63.

Tong, C., M.F. Laspia, S. Telang and G.M. Williams. 1981. The use of adult rat liver cultures in the detection of the genotoxicity of various polycyclic aromatic hydrocarbons. Environ. Mutagen. 3: 477-487.

U.S. EPA. 1984. Carcinogen Assessment of Coke Oven Emissions. Office of Health and Environmental Assessment, Washington, DC. EPA 600/6-82-003F. NTIS PB 84-170181.

U.S. EPA. 1990. Drinking Water Criteria Document for Polycyclic Aromatic Hydrocarbons (PAHs). Prepared by the Office of Health and Environmental Assessment, Environmental Criteria and Assessment Office, Cincinnati, OH for the Office of Drinking Water, Washington, DC. Final Draft. ECAO-CIN-D010, September, 1990.

Utesch, D., H. Glatt and F. Oesch. 1987. Rat hepatocyte-mediated bacterial mutagenicity in relation to the carcinogenic potency of benz(a)anthracene, benzo(a)pyrene and twenty-five methylated derivatives. Cancer Res. 47: 1509-1515.

Wislocki, P.G., E.S. Bagan, A.Y.H. Lu, et al. 1986. Tumorigenicity of nitrated derivatives of pyrene, benz[a]anthracene, chrysene and benzo[a]pyrene in the newborn mouse assay. Carcinogenesis. 7(8): 1317-1322.

Wood, A.W., R.L. Chang, W. Levin, et al. 1977a. Mutagenicity and cytotoxicity of benz[a]anthracene diol epoxides and tetrahydro-epoxides: Exceptional activity of the bay region 1,2-epoxides. Proc. Natl. Acad. Sci. USA. 74(7): 2746-2750.

Wood, A.W., W. Leven. R.L. Chang, et al. 1977b. Tumorigenicity of five dihydrodiols of benz[a]anthracene on mouse skin: Exceptional activity of benz[a]anthracene 3,4-dihydrodiol. Proc. Natl. Acad. Sci. USA. 74(8):3137-3179.

VI.D. DRINKING WATER HA REFERENCES

None

VII. REVISION HISTORY

Date	Section	Description
12/01/90	II.	Carcinogen assessment on-line
12/01/90	VI.	Bibliography on-line
01/01/92	IV.	Regulatory Action section on-line
09/01/93	II.	Carcinogenicity assessment noted as pending change
09/01/93	II.D.2.	Work group review date added
11/01/93	II.D.2.	Work group review date added
03/01/94	II.	Pending change note removed; no change
03/01/94	II.D.2.	Work group review date added
09/01/94	I.B.	Inhalation RfC now under review

SYNONYMS

Benz(a)anthracene
benz(a)anthracene
Benzanthracene
Benzanthrene
BENZO(a)ANTHRACENE
BENZO(b)PHENANTHRENE
Benzoanthracene
HSDB 4003
NSC 30970

RCRA WASTE NUMBER U018
Tetraphene
1,2-BENZ(a)ANTHRACENE
1,2-Benzanthracene
1,2-BENZANTHRAZEN [German]
1,2-BENZANTHRENE
1,2-BENZOANTHRACENE
2,3-Benzophenanthrene

Source: ***Instant EPA's IRIS***, 1997, Instant Reference Sources, Inc., 7605 Rockpoint Drive, Austin, TX 78731 (Fax: 512-345-2386; Internet URL, http://www.instantref.com/inst-ref.htm).

Benzo(a)pyrene

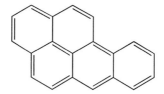

Introduction

Benzo(a)pyrene (BaP) is a polycyclic aromatic hydrocarbon (PAH) and is found in petroleum, wax, and smoke. Its release to the environment is quite wide spread since it is an ubiquitous product of incomplete combustion. It is largely associated with particulate matter, soils, and sediments. Although environmental concentrations are highest near sources, its presence in places distant from primary sources indicates that it is reasonably stable in the atmosphere and capable of long distance transport. When released to air it may be subject to direct photolysis, although adsorption to particulates apparently can retard this process. If released to water, it will adsorb very strongly to sediments and particulate matter and bioconcentrate in aquatic organisms which can not metabolize it. It may be subject to significant biodegradation, and direct photolysis may be important near the surface of waters; adsorption, however, may significantly retard these two processes. If released to soil it will be expected to adsorb very strongly to the soil and will not be expected to appreciably leach to the groundwater, although its presence in some samples of groundwater illustrates that it can be transported there. It will not be expected to hydrolyze or significantly evaporate from soils and surfaces. It may be subject to appreciable biodegradation in soils. Human exposure can be from inhalation of contaminated air and consumption of contaminated food and water. Especially high exposure will occur through the smoking of cigarettes and the ingestion of certain foods (e.g., smoked and charcoal broiled meats and fish) [828].

Identifying Information

CAS NUMBER: 50-32-8

NIOSH Registry Number: DJ3675000

SYNONYMS: 3,4-Benzopyrene, 3,4-Benzpyrene, 3,4-Benzylpyrene, Benzo(D,E,F)chrysene, 6,7-Benzopyrene, 3,4-Benz(a)pyrene, 3,4-Benzypyrene

Endocrine Disruptor Information

Benzo(a)pyrene is a strongly suspected EED that it is listed by EPA [811] and the World Wildlife Fund [813] as a potential endocrine modifying chemical. It also is an aromatic coplanar compound. BaP has been implicated in at least one study as having potential estrogenic effects. [867], [811]

Chemical and Physical Properties

CHEMICAL FORMULA: $C_{20}H_{12}$

MOLECULAR WEIGHT: 252.32 [703]

PHYSICAL DESCRIPTION: Yellow crystals [703]

SPECIFIC GRAVITY: Not available

MELTING POINT: 179°C [703]

BOILING POINT: 312°C @ 10mm [703]

SOLUBILITY: Water: Insoluble [703]; Benzene: Soluble [703]; Toluene: Soluble [703]; Xylene: Soluble [703]

Source: *Instant EPA's Air Toxics*, 1994, Instant Reference Sources, Inc., 7605 Rockpoint Drive, Austin, TX 78731 (Fax: 512-345-2386; Internet URL, http://www.instantref.com/inst-ref.htm).

Environmental Reference Materials

A source of EED environmental reference materials for benzo(a)pyrene is Radian International LLC, Austin, TX (Fax 512-454-0268; Internet URL, http://www.radian.com/standards); Catalog No. ERB-007.

Hazardous Properties

ACUTE/CHRONIC HAZARDS: Benzo(a)pyrene may be harmful by ingestion or inhalation [071], [269]. It may be an irritant and when heated to decomposition it emits acrid smoke and toxic fumes of carbon monoxide and carbon dioxide [043], [269].

FIRE HAZARD: Benzo(a)pyrene is combustible. Fires involving it can be controlled with a dry chemical, carbon dioxide or Halon extinguisher. A water spray may also be used [269].

REACTIVITY: Benzo(a)pyrene is incompatible with strong oxidizers [051], [102], [107], [269]. It readily undergoes nitration and halogenation [395]. Ozone, chromic acid and chlorinating agents oxidize this compound. Benzo(a)pyrene may react with organic and inorganic oxidants including various electrophiles, peroxides, nitrogen oxides and sulfur oxides [051], [071]. Hydrogenation occurs with platinum oxide [395].

STABILITY: Benzo(a)pyrene undergoes photo-oxidation after irradiation in indoor sunlight or by fluorescent light in organic solvents. Solutions of it in benzene oxidize under the influence of light and air [395]. Solutions of it in water, DMSO, 95% ethanol or acetone should be stable for 24 hours under normal lab conditions [700].

Source: *Instant EPA's Air Toxics*, 1994, Instant Reference Sources, Inc., 7605 Rockpoint Drive, Austin, TX 78731 (Fax: 512-345-2386; Internet URL, http://www.instantref.com/inst-ref.htm).

Medical Symptoms of Exposure

Symptoms of exposure may include irritation of mucous membranes, dermatitis, bronchitis, coughing, dyspnea, conjunctivitis, photosensitization, pulmonary edema, reproductive effects and leukemia [107]. Contact with the skin may result in erythema, pigmentation, desquamation, formation of verrucae and infiltration [395]. It may also cause keratosis which are relatively small, heaped up, scaling, brown plaques on the skin, some of which may be fissured and may itch [401]. Exposure also may cause reddening and squamous eczema of the eye lid margins with only small erosion of the corneal epithelium and superficial changes in the stroma which disappear a month following exposure. Chronic exposure to the fumes and dust of this type of compound may cause discoloration of the cornea and epithelioma of the eye lid margin [102]. Repeated exposure may cause sunlight to have more severe effects on a skin including allergic skin rash. Aplastic anemia may also occur.

Source: *Instant EPA's Air Toxics*, 1994, Instant Reference Sources, Inc., 7605 Rockpoint Drive, Austin, TX 78731 (Fax: 512-345-2386; Internet URL, http://www.instantref.com/inst-ref.htm).

Toxicological Information

The toxicological information below is gathered from several sources including two national databases: one from the *National Toxicology Program Chemical Repository Database* and another from *EPA's Integrated Risk Information System (IRIS)*.

Toxicology Information from the National Toxicology Program

TERATOGENICITY: See RTECS printout for data

MUTATION DATA: See RTECS printout for data

TOXICITY:

Typ. Dose	Mode	Specie	Amount	Units	Other
LD50	scu	rat	50	mg/kg	
LDLo	ipr	mus	500	mg/kg	
LDLo	irn	frg	11	mg/kg	

OTHER TOXICITY DATA:
  Skin and Eye Irritation Data:
    skn-mus   14 ug MLD

  Review: Toxicology Review-11
  Status:
    NIOSH Analytical Methods: see polynuclear aromatic hydrocarbons
    (HPLC), 5506; (GC), 5515
    EPA Genetox Program 1988, Positive: Cell transform.-SA7/F344 rat;
    SHE-focus assay
    EPA Genetox Program 1988, Positive: Cell transform.-mouse embryo
    EPA Genetox Program 1988, Positive: Cell transform.-mouse prostate
    EPA Genetox Program 1988, Positive: Cell transform.-RLV F344 rat
    embryo
    EPA Genetox Program 1988, Positive: Cell transform.-SA7/SHE;
    Host-mediated assay
    EPA Genetox Program 1988, Positive: L5178Y cells in vitro-TK test;
    Mammalian micronucleus
    EPA Genetox Program 1988, Positive: Mouse spot test; E coli polA
    with S9
    EPA Genetox Program 1988, Positive: Histidine reversion-Ames test
    EPA Genetox Program 1988, Positive: In vitro SCE-human
    lymphocytes; In vitro SCE-human
    EPA Genetox Program 1988, Positive: Sperm morphology-mouse; D
    melanogaster Sex-linked lethal
    EPA Genetox Program 1988, Positive: In vitro UDS in rat liver; V79
    cell culture-gene mutation
    EPA Genetox Program 1988, Positive/dose response: Cell transform.-
    BALB/c-3T3; SHE-clonal assay
    EPA Genetox Program 1988, Positive/dose response: Cell transform.-
    C3H/10T1/2
    EPA Genetox Program 1988, Positive/dose response: In vitro SCE-
    nonhuman; in vivo SCE-nonhuman
    EPA Genetox Program 1988, Negative: D melanogaster-nondisjunction
    EPA Genetox Program 1988, Negative: Rodent heritable translocation;
    Mouse specific locus
    EPA Genetox Program 1988, Negative: UDS in mouse germ cells; S
    cerevisiae-homozygosis
    EPA Genetox Program 1988, Inconclusive: E coli polA without S9;
    Invitro UDS-human fibroblast
    EPA Genetox Program 1988, Positive: Body fluid assay;
    Carcinogenicity- mouse/rat
    EPA Genetox Program 1988, Positive: CHO gene mutation
    EPA TSCA Chemical Inventory, 1989
    EPA TSCA Test Submission (TSCATS) Data Base, April 1990
    OSHA Analytical Method #ID-58

CARCINOGENICITY:
Tumorigenic Data:

Type of Effect	Route/Animal	Amount Dosed (Notes)
TDLo	orl-rat	15 mg/kg
TCLo	ihl-mus	200 ng/m3/6H/13W-I
TDLo	ipr-rat	16 mg/kg
TDLo	scu-rat	455 ug/kg/60D-I
TDLo	ivn-rat	39 mg/kg/6D-I
TDLo	ims-rat	2400 ug/kg
TDLo	ice-rat	22 mg/kg
TDLo	itr-rat	68 mg/kg/15W-I
TDLo	imp-rat	150 ug/kg
TDLo	orl-mus	700 mg/kg/75W-I
TDLo	skn-mus	120 mg/kg (MGN)
TDLo	skn-mus	28500 ug/kg/19W-I
TD	skn-mus	12 mg/kg/20D-I
TDLo	skn-mus	25 ng/kg/110W-I
TDLo	ipr-mus	10 mg/kg
TDLo	scu-mus	9 mg/kg
TDLo	ivn-mus	10 mg/kg
TDLo	itr-mus	200 mg/kg/10W-I
TDLo	imp-mus	200 mg/kg
TDLo	unr-mus	80 mg/kg/8D-I
TDLo	rec-mus	200 mg/kg
TDLo	par-dog	819 mg/kg/26W-I
TDLo	imp-dog	651 mg/kg/21W-C
TDLo	scu-mky	40 mg/kg
TDLo	skn-rbt	17 mg/kg/57W-I
TDLo	itr-rbt	145 mg/kg/2Y-I
TDLo	orl-ham	420 mg/kg/21W-I
TCLo	ihl-ham	9500 ug/m3/4H/96W-I
TDLo	scu-ham	4000 ug/kg
TDLo	itr-ham	64 mg/kg
TDLo	imp-frg	45 mg/kg
TD	imp-rat	500 ug/kg
TD	itr-rat	200 mg/kg/15W-I
TD	skn-mus	26 mg/kg/65W-I
TD	itr-ham	360 mg/kg/36W-I
TD	ims-rat	3150 ug/kg
TD	skn-mus	18 mg/kg/73W-I
TD	scu-mus	12 mg/kg
TDLo	ipr-mus	300 mg/kg (16-18D preg)
TDLo	scu-mus	480 mg/kg (11-15D preg)
TD	rec-mus	560 mg/kg/14W-I
TD	scu-mus	8 mg/kg
TDLo	ivn-rbt	30 mg/kg (25D preg)
TDLo	itr-ham	120 mg/kg/17W-I
TDLo	imp-ham	5 mg/kg/22W-C

Review:

  IARC Cancer Review: Animal Sufficient Evidence
  IARC probable human carcinogen (Group 2A)
  ACGIH suspected human carcinogen

Status:

  NTP Fifth Annual Report on Carcinogens, 1989; anticipated to be
  carcinogen
  EPA Carcinogen Assessment Group

SAX TOXICITY EVALUATION:

  THR: A poison via subcutaneous, intraperitoneal and intrarenal routes. An
  experimental carcinogen, tumorigen, neoplastigen and teratogen. Other
  experimental reproductive effects. Human mutagenic data. A common
  aircontaminant of water, food and smoke.

Source: ***Instant Tox-Base***, 1995, Instant Reference Sources, Inc., 7605 Rockpoint Drive,
Austin, TX 78731 (Fax: 512-345-2386; Internet URL, http://www.instantref.com/inst-
ref.htm).

Toxicology Information from EPA's Integrated Risk Information System (IRIS)

I.   CHRONIC HEALTH HAZARD ASSESSMENTS FOR NONCARCINOGENIC
EFFECTS

I.A. REFERENCE DOSE FOR CHRONIC ORAL EXPOSURE (RfD)

  [Ed. Note: EPA does not have information yet in this section of IRIS].

I.B. REFERENCE CONCENTRATION FOR CHRONIC INHALATION EXPOSURE
(RfC)

  [Ed. Note: EPA does not have information yet in this section of IRIS].

II. CARCINOGENICITY ASSESSMENT FOR LIFETIME EXPOSURE

Section II provides information on three aspects of the carcinogenic risk assessment for
the agent in question; the US EPA classification, and quantitative estimates of risk from
oral exposure and from inhalation exposure. The classification reflects a weight-of-
evidence judgment of the likelihood that the agent is a human carcinogen. The
quantitative risk estimates are presented in three ways. The slope factor is the result of
application of a low-dose extrapolation procedure and is presented as the risk per
(mg/kg)/day. The unit risk is the quantitative estimate in terms of either risk per ug/L
drinking water or risk per ug/cu.m air breathed. The third form in which risk is presented
is a drinking water or air concentration providing cancer risks of 1 in 10,000, 1 in

100,000 or 1 in 1,000,000. Background Document 2 (Service Code 5) provides details on the rationale and methods used to derive the carcinogenicity values found in IRIS. Users are referred to Section I for information on long-term toxic effects other than carcinogenicity.

NOTE: At the June 1992 CRAVE Work Group meeting, a revised risk estimate for benzo[a]pyrene was verified (see Additional Comments for Oral Exposure). This section provides information on three aspects of the carcinogenic risk assessment for the agent in question; the US EPA classification, and quantitative estimates of risk from oral exposure and from inhalation exposure. The classification reflects a weight-of-evidence judgment of the likelihood that the agent is a human carcinogen. The quantitative risk estimates are presented in three ways. The slope factor is the result of application of a low-dose extrapolation procedure and is presented as the risk per (mg/kg)/day. The unit risk is the quantitative estimate in terms of either risk per ug/L drinking water or risk per ug/cu.m air breathed. The third form in which risk is presented is a drinking water or air concentration providing cancer risks of 1 in 10,000 or 1 in 1,000,000. The Carcinogenicity Background Document provides details on the rationale and methods used to derive the carcinogenicity values found in IRIS. Users are referred to the Oral Reference Dose (RfD) and Reference Concentration (RfC) sections for information on long-term toxic effects other than carcinogenicity.

II.A. EVIDENCE FOR CLASSIFICATION AS TO HUMAN CARCINOGENICITY

II.A.1. WEIGHT-OF-EVIDENCE CLASSIFICATION

Classification -- B2; probable human carcinogen

Basis -- Human data specifically linking benzo[a]pyrene (BAP) to a carcinogenic effect are lacking. There are, however, multiple animal studies in many species demonstrating BAP to be carcinogenic following administration by numerous routes. BAP has produced positive results in numerous genotoxicity assays.

II.A.2. HUMAN CARCINOGENICITY DATA

Inadequate. Lung cancer has been shown to be induced in humans by various mixtures of polycyclic aromatic hydrocarbons known to contain BAP including cigarette smoke, roofing tar and coke oven emissions. It is not possible, however, to conclude from this information that BAP is the responsible agent.

II.A.3. ANIMAL CARCINOGENICITY DATA

Sufficient.   The animal data consist of dietary, gavage, inhalation, intratracheal instillation, dermal and subcutaneous studies in numerous strains of at least four species of rodents and several primates. Repeated BAP administration has been associated with increased incidences of total tumors and of tumors at the site of exposure. Distant site tumors have also been observed after BAP administration by various routes. BAP is frequently used as a positive control in carcinogenicity bioassays.

BAP administered in the diet or by gavage to mice, rats and hamsters has produced increased incidences of stomach tumors. Neal and Rigdon (1967) fed BAP (purity not reported) at concentrations of 0, 1, 10, 20, 30, 40, 45, 50, 100 and 250 ppm in the diets of male and female CFW-Swiss mice. The age of the mice ranged from 17-180 days old and the treatment time from 1-197 days; the size of the treated groups ranged from 9 to 73. There were 289 mice (number of mice/sex not stated) in the control group. No forestomach tumors were reported in the 0-, 1- and 10-ppm dose groups. The incidence of forestomach tumors in the 20-, 30-, 40-, 45-, 50-, 100- and 250-ppm dose groups were 1/23, 0/37, 1/40, 4/40, 23/34, 19/23 and 66/73, respectively. The authors felt that the increasing tumor incidences were related to both the concentration and the number of doses administered. Historical control forestomach tumor data are not available for CFW-Swiss strain mice. In historical control data from a related mouse strain, SWR/J Swill, the forestomach tumor incidence rate was 2/268 and 1/402 for males and females, respectively (Rabstein et al., 1973).

Brune et al., (1981) fed 0.15 mg/kg BAP (reported to be "highly pure") in the diet of 32 Sprague-Dawley rats/sex/group either every 9th day or 5 times/week. These treatments resulted in annual average doses of 6 or 39 mg/kg, respectively. An untreated group of 32 rats/sex served as the control. Rats were treated until moribund or dead; survival was similar in all groups. Histologic examinations were performed on each rat. The combined incidence of tumors of the forestomach, esophagus and larynx was 3/64, 3/64 and 10/64 in the control group, the group fed BAP every 9th day and the group fed BAP 5 times/week, respectively. A trend analysis showed a statistically significant tendency for the proportion of animals with tumors of the forestomach, esophagus or larynx to increase steadily with dose (Knauf and Rice, 1992).

As part of the same study, Brune et al. (1981) administered BAP ("highly pure") orally to Sprague-Dawley rats by caffeine gavage. The rats were treated until moribund or dead; all rats were subjected to terminal histopathologic examination. Gavaged rats were divided into 3 dose groups of 32 rats/sex/group; the groups received 0.15 mg/kg per gavage either every 9th day (Group A), every 3rd day (Group B) or 5 times per week (Group C); these treatments resulted in annual average doses of 6, 18 or 39 mg/kg, respectively. Untreated and gavage (5 times/week) controls (32 rats/sex/group) were included. The median survival times for the untreated control group; the gavage control group; and groups A, B and C were 129, 102, 112, 113 and 87 weeks, respectively. The survival time of Group C was short compared with controls and may have precluded tumor formation (Knauf and Rice, 1992). The combined tumor incidence in the forestomach, esophagus and larynx was 3/64, 6/64, 13/64, 26/64 and 14/64 for the untreated control group, gavage control group, group A, group B and group C, respectively. There was a statistically significant association between the dose and the proportions of rats with tumors of the forestomach, esophagus or larynx. This association is not characterized by a linear trend. The linearity was affected by the apparently reduced tumor incidence that is seen in the high-dose group (Knauf and Rice, 1992).

Intratracheal instillation and inhalation studies in guinea pigs, hamsters and rats have resulted in elevated incidences of respiratory tract and upper digestive tract tumors (US EPA, 1991a). Male Syrian golden hamsters (24/group) were exposed by inhalation to 0, 2.2, 9.5 or 46.5 mg BAP/cu.m in a sodium chloride aerosol (Thyssen et al., 1981).

(Greater than 99% of the particles had diameters between 0.2 and 0.5 um.) For the first 10 weeks of the study, the hamsters were exposed to BAP daily for 4.5 hours/day; thereafter, daily for 3 hours/day. Animals dying within the first year of the study were replaced; the effective number of hamsters in the control, low-, mid- and high-dose groups was 27, 27, 26 and 25, respectively. (The total time of treatment, although over 60 weeks, was not stated.) During the first 10 weeks, animals in the 3 dose groups reportedly lost weight. After week 10, however, the body weights in all groups were similar until week 60 when the body weights of hamsters in the high-dose group decreased and the mortality increased significantly. The incidence of respiratory tract tumors (including tumors of the nasal cavity, larynx and trachea) in the control, low-, mid- and high-dose groups was 0/27, 0/27, 9/26 and 13/25, respectively; the incidences of upper digestive tract tumors (including tumors of the pharynx, esophagus and forestomach) were 0/27, 0/27, 7/26 and 14/25, respectively. Trend analysis for incidences of both respiratory tract tumors and upper gastrointestinal tract tumors showed a statistically significant tendency for the proportion of animals with either tumor type to increase steadily with increased dose (Knauf and Rice, 1992).

Intraperitoneal BAP injections have caused increases in the number of injection site tumors in mice and rats (reviewed in US EPA, 1991a). Subcutaneous BAP injections have caused increases in the number of injection site tumors in mice, rats, guinea pigs, hamsters and some primates (IARC, 1983; US EPA, 1991a). BAP is commonly used as a positive control in many dermal application bioassays and has been shown to cause skin tumors in mice, rats, rabbits and guinea pigs. BAP is both an initiator and a complete carcinogen in mouse skin (IARC, 1983). Increased incidences of distant site tumors have also been reported in animals as a consequence of dermal BAP exposure (reviewed in US EPA, 1991a).

BAP has also been reported to be carcinogenic in animals when administered by the following routes: i.v.; transplacentally; implantation in the stomach wall, lung, renal parenchyma and brain; injection into the renal pelvis; and vaginal painting (US EPA, 1991a).

II.A.4. SUPPORTING DATA FOR CARCINOGENICITY

Benzo[a]pyrene has been shown to cause genotoxic effects in a broad range of prokaryotic and mammalian cell assay systems (US EPA, 1991a). In prokaryotes, BAP tested positive in DNA damage assays and in both reverse and forward mutation assays. In mammalian cell culture assays, BAP tested positive in DNA damage assays, forward mutation assays, chromosomal effects assays and cell transformation assays.

II.B. QUANTITATIVE ESTIMATE OF CARCINOGENIC RISK FROM ORAL EXPOSURE

NOTE: The range of oral slope factors calculated was: 4.5E+0 to 11.7E+0 per (mg/kg)/day.

II.B.1. SUMMARY OF RISK ESTIMATES

Oral Slope Factor -- 7.3E+0 per (mg/kg)/day
Drinking Water Unit Risk -- 2.1E-4 per (ug/L)

Extrapolation Method -- Risk estimate based on a geometric mean of four slope factors obtained by differing modeling procedures. Derived from the combination of multiple data sets from two different reports using more than one sex and species.

Drinking Water Concentrations at Specified Risk Levels:

Risk Level	Concentration
E-4 (1 in 10,000)	5E-1 ug/L
E-5 (1 in 100,000)	5E-2 ug/L
E-6 (1 in 1,000,000)	5E-3 ug/L

II.B.2. DOSE-RESPONSE DATA (CARCINOGENICITY, ORAL EXPOSURE)

Tumor Type -- forestomach, squamous cell papillomas and carcinomas
Test Animals -- CFW mice, sex unknown Route -- oral, diet
Reference -- Neal and Rigdon, 1967

a) Conditional upper bound two-stage model with terms for promotion (modification of Moolgavkar-Venson-Knudson, generalized forms of two-stage model)

Administered Dose (ppm)	Tumor Incidence
0	0/289
1	0/25
10	0/24
20	1/23
30	0/37
40	1/40
45	4/40
50	24/34
100	19/23
250	66/73

Tumor Type -- squamous cell carcinoma of the forestomach
Test Animals -- SWR/J Swill mice   Route -- oral, diet
Reference -- Rabstein et al., 1973

Administered Dose (ppm)	Tumor Incidence
0	2/268* male
0	1/402* female

*See additional comments concerning the use of control data from other studiesthat utilized similar mouse strains.

b) Same data as above. Upper bound estimate by extrapolation from 10% response point to background of empirically fitted dose-response curve. (Procedure using two-stage model described in (a)).

c) Same data as above except the additional 2 control groups (Rabstein et al.,1973) were excluded. Generalized Weibull-type dose-response model.

d) Tumor Type -- forestomach, larynx and esophagus, papillomas and carcinomas(combined). Linearized Multistage Model, Extra Risk.

Test Animals -- Sprague-Dawley rats, males and female Route -- oral, diet Reference -- Brune et al., 1981

Dose (mg/kg diet/year)	Tumor Incidence
0	3/64
6	3/64
39	10/64

II.B.3. ADDITIONAL COMMENTS (CARCINOGENICITY, ORAL EXPOSURE)

At the June 1992 CRAVE Work Group meeting, it was noted that an error had been made in the 1991 document "Dose-Response Analysis of Ingested Benzo[a]pyrene" which is quoted in the Drinking Water Criteria Document for PAH. In the calculation of the doses in the Brune et al. (1981) study it was erroneously concluded that doses were given in units of mg/year, whereas it was in fact mg/kg/year. When the doses are corrected the slope factor is correctly calculated as 11.7 per (mg/kg)/day, as opposed to 4.7 per (mg/kg)/day as reported in the Drinking Water Criteria Document. The correct range of slope factors is 4.5 to 11.7 per (mg/kg)/day, with a geometric mean of 7.3 per (mg/kg)/day. A drinking water unit risk based on the revised slope factor is 2.1E-4 per (ug/L). Therefore, these values have been changed on IRIS and an Erratum to the Drinking Water Criteria Document is being prepared.

Risk estimates were calculated from two different studies in two species of outbred rodents (Neal and Rigdon, 1967; Brune et al., 1981). These studies have several commonalities including mode of administration, tumor sites, tumor types and the presumed mechanisms of action. The data sets were not combined prior to modeling (the preferred approach) because they employed significantly dissimilar protocols.

The geometric mean from several slope factors, each considered to be of equal merit, was used to calculate a single unit risk. These four slope factor estimates span less than a factor of three and each is based on an acceptable, but less-than-optimal, data set. Each estimate is based on a low-dose extrapolation procedure which entails the use of multiple assumptions and default procedures.

Clement Associates (1990) fit the Neal and Rigdon (1967) data to a two-stage dose response model. In this model the transition rates and the growth rate of preneoplastic cells were both considered to be exposure-dependent. (The functional form for the dose-dependence of preneoplastic cell growth rate was simple saturation.) A term to permit the modeling of BAP as its own promoter was also included. Historical control stomach tumor data from a related, but not identical, mouse strain, SWR/J Swill (Rabstein et al., 1973) and the CFW Texas colony (Neal and Rigdon, 1967) were used in the modeling. In calculating the lifetime unit risk for humans several standard assumptions were made: mouse food consumption was 13% of its body weight/day; human body weight was assumed to be 70 kg and the assumed body weight of the mouse 0.034 kg. The standard assumption of surface area equivalence between mice and humans was the cube root of 70/0.034. A conditional upper bound estimate was calculated to be 5.9 per (mg/kg)/day (US EPA, 1991a).

A US EPA report (1991b) argued that the upper-bound estimate calculated in Clement Associates (1990) involved the use of unrealistic conditions placed on certain parameters of the equation. Other objections to this slope factor were also raised. The authors of this report used the Neal and Rigdon (1967) data to generate an upper-bound estimate extrapolated linearly from the 10% response point to the background of an empirically fitted dose-response curve (Clement Associates, 1990). Other results, from similar concepts and approaches used for other compounds, suggest that the potency slopes calculated in this manner are comparable to those obtained from a linearized multistage procedure for the majority of the other compounds. The upper bound estimate calculated in US EPA (1991b) is 9.0 per (mg/kg)/day.

The authors of US EPA (1991b) selected a model to reflect the partial lifetime exposure pattern over different parts of the animals' lifetimes. The authors thought that this approach more closely reflected the Neal and Rigdon (1967) regimen. A Weibull-type dose-response model was selected to accommodate the partial lifetime exposure; the upper-bound slope factor calculated from this method was 4.5 per (mg/kg)/day.

Using the dietary portion of the Brune et al. (1981) rat data, a linearized multistage procedure was used to calculate an upper bound slope factor for humans. In the interspecies conversion the assumed human body weight was 70 kg and the rat 0.4 kg. The slope factor calculated by this method was 11.7 per (mg/kg)/day.

II.B.4. DISCUSSION OF CONFIDENCE (CARCINOGENICITY, ORAL EXPOSURE)

The data are considered to be less than optimal, but acceptable. There are precedents for using multiple data sets from different studies using more than one sex, strain and species; the use of the geometric mean of four slope factors is preferred because it makes use of more of the available data. The use of the geometric means was based on arguments presented in a personal communication (Stiteler, 1991).

II.C.   QUANTITATIVE   ESTIMATE   OF   CARCINOGENIC   RISK   FROM
INHALATION EXPOSURE

[Ed. Note: EPA does not have information yet in this section of IRIS].

II.D. EPA DOCUMENTATION, REVIEW, AND CONTACTS (CARCINOGENICITY
ASSESSMENT)

II.D.1. EPA DOCUMENTATION

Source Document -- US EPA, 1991a,b

The 1991 Drinking Water Criteria Document for the polycyclic aromatic hydrocarbons
has received agency review.

II.D.2. REVIEW (CARCINOGENICITY ASSESSMENT)

Agency Work Group Review -- 01/07/87, 12/04/91, 06/03/92, 08/05/93, 02/02/94,
06/09/94

Verification Date -- 12/04/91

II.D.3. US EPA CONTACTS (CARCINOGENICITY ASSESSMENT)

Robert E. McGaughy / NCEA -- (202)260-5889

Rita Schoeny / NCEA -- (513)569-7544

III. HEALTH HAZARD ASSESSMENTS FOR VARIED EXPOSURE DURATIONS

[Ed. Note: EPA does not have information yet in this section of IRIS].

IV. US EPA REGULATORY ACTIONS

EPA risk assessments may be updated as new data are published and as assessment
methodologies evolve. Regulatory actions are frequently not updated at the same time.
Compare the dates for the regulatory actions in this section with the verification dates for
the risk assessments in sections I and II, as this may explain inconsistencies. Also note
that some regulatory actions consider factors not related to health risk, such as technical
or economic feasibility. Such considerations are indicated for each action. In addition,
not all of the regulatory actions listed in this section involve enforceable federal
standards. Please direct any questions you may have concerning these regulatory actions
to the US EPA contact listed for that particular action. Users are strongly urged to read
the background information on each regulatory action in Background Document 4 in
Service Code 5.

IV.A. CLEAN AIR ACT (CAA)

[Ed. Note: EPA does not have information yet in this section of IRIS].

IV.B. SAFE DRINKING WATER ACT (SDWA)

IV.B.1. MAXIMUM CONTAMINANT LEVEL GOAL (MCLG) for Drinking Water

Value -- 0 mg/L (Proposed,1990)

Considers technological or economic feasibility? -- NO

Discussion -- The proposed MCLG for benzo(a)pyrene is zero based on the evidence of carcinogenic potential (B2).

Reference -- 55 FR 30370 (07/25/90)

EPA Contact -- Health and Ecological Criteria Division / OST /(202) 260-7571 / FTS 260-7571; or Safe Drinking Water Hotline / (800) 426-4791

IV.B.2. MAXIMUM CONTAMINANT LEVEL (MCL) for Drinking Water

Value -- 0.0002 mg/L (Proposed, 1990)

Considers technological or economic feasibility? -- YES

Discussion -- The proposed MCL is equal to the PQL and is associated with a maximum lifetime individual risk of 1 E-4.

Monitoring requirements -- Community and non-transient water system monitoring based on state vulnerability assessment; vulnerable systems to be monitored quarterly for one year; repeat monitoring dependent upon detection and size of system.

Analytical methodology -- High pressure liquid chromatography (EPA 550, 550.1); gas chromatographic/mass spectrometry (EPA 525): PQL= 0.0002 mg/L.

Best available technology -- Granular activated carbon

Reference -- 55 FR 30370 (07/25/90)

EPA Contact -- Drinking Water Standards Division / OGWDW /(202) 260-7575 / FTS 260-7575; or Safe Drinking Water Hotline / (800) 426-4791

IV.B.3. SECONDARY MAXIMUM CONTAMINANT LEVEL (SMCL) for Drinking Water

[Ed. Note: EPA does not have information yet in this section of IRIS].

IV.B.4. REQUIRED MONITORING OF "UNREGULATED" CONTAMINANTS

Status -- Listed (Proposed, 1991)

Discussion -- "Unregulated" contaminants are those contaminants for which EPA establishes a monitoring requirement but which do not have an associated final MCLG, MCL, or treatment technique. EPA may regulate these contaminants in the future.

Monitoring requirement -- All systems to be monitored unless a vulnerability assessment determines the system is not vulnerable.

Analytical methodology -- Gas chromatography/mass spectrometry (EPA 525); high pressure liquid chromatography (EPA 550, 550.1).

Reference -- 56 FR 3526 (01/30/91)

EPA Contact -- Drinking Water Standards Division / OGWDW /(202) 260-7575 / FTS 260-7575; or Safe Drinking Water Hotline / (800) 426-4791

IV.C. CLEAN WATER ACT (CWA)

IV.C.1. AMBIENT WATER QUALITY CRITERIA, Human Health

Water and Fish Consumption: 2.8E-3 ug/L

Fish Consumption Only: 3.11E-2 ug/L

Considers technological or economic feasibility? -- NO

Discussion -- For the maximum protection from the potential carcinogenic properties of this chemical, the ambient water concentration should be zero. However, zero may not be obtainable at this time, so the recommended criteria represents a E-6 estimated incremental increase of cancer over a lifetime. The values given represent polynuclear aromatic hydrocarbons as a class.   Reference -- 45 FR 79318 (11/28/80)

EPA Contact -- Criteria and Standards Division / OWRS(202)260-1315 / FTS 260-1315

IV.C.2. AMBIENT WATER QUALITY CRITERIA, Aquatic Organisms

Freshwater:
   Acute LEC -- none
   Chronic LEC -- none

Marine:
   Acute LEC --   3.0E+2 ug/L
   Chronic LEC -- none

Considers technological or economic feasibility? -- NO

Discussion -- The values that are indicated as "LEC" are not criteria, but are the lowest effect levels found in the literature. LEC's are given when the minimum data required to derive water quality criteria are not available. The values given represent polynuclear aromatic hydrocarbons as a class.

Reference -- 45 FR 79318 (11/28/80)

EPA Contact -- Criteria and Standards Division / OWRS(202)260-1315 / FTS 260-1315

IV.D. FEDERAL INSECTICIDE, FUNGICIDE, AND RODENTICIDE ACT (FIFRA)

[Ed. Note: EPA does not have information yet in this section of IRIS].

IV.E. TOXIC SUBSTANCES CONTROL ACT (TSCA)

[Ed. Note: EPA does not have information yet in this section of IRIS].

IV.F. RESOURCE CONSERVATION AND RECOVERY ACT (RCRA)

IV.F.1. RCRA APPENDIX IX, for Ground Water Monitoring

Status -- Listed
Reference -- 52 FR 25942 (07/09/87)
EPA Contact -- RCRA/Superfund Hotline(800)424-9346 / (202)260-3000 / FTS 260-3000

IV.G. SUPERFUND (CERCLA)

IV.G.1. REPORTABLE QUANTITY (RQ) for Release into the Environment

Value -- 1 pound (Final, 1989)
Considers technological or economic feasibility? -- NO

Discussion -- The RQ for benzo(a)pyrene is based on potential carcinogenicity (group B2). This chemical is currently under assessment for carcinogenicity and chronic toxicity and the RQ is subject to change in future rule making.

Reference -- 54 FR 33418 (08/14/89)
EPA Contact -- RCRA/Superfund Hotline(800) 424-9346 / (202) 260-3000 / FTS 260-3000

VI. BIBLIOGRAPHY

VI.A. ORAL RfD REFERENCES

None

VI.B. INHALATION RfD REFERENCES

None

VI.C. CARCINOGENICITY ASSESSMENT REFERENCES

Brune, H., R.P. Deutsch-Wenzel, M. Habs, S. Ivankovic and D. Schmahl. 1981. Investigation of the tumorigenic response to benzo[a]pyrene in aqueous caffeine solution applied orally to Sprague-Dawley rats. J. Cancer Res. Clin. Oncol. 102(2): 153-157.

Clement Associates. 1990. Ingestion dose-response model to benzo(a)pyrene. EPA Control No. 68-02-4601.

IARC (International Agency for Research on Cancer). 1983. Certain Polycyclic Aromatic Hydrocarbons and Heterocyclic Compounds. Monographs on the Evaluation of Carcinogenic Risk of the Chemical to Man, Vol. 3. Lyon, France.

Knauf, L. and G. Rice. 1992. Statistical Evaluation of Several Benzo[a]pyrene Bioassays. Memorandum to R. Schoeny, US EPA, Cincinnati, OH. January 2.

Neal, J. and R.H. Rigdon. 1967. Gastric tumors in mice fed benzo[a]pyrene --A quantitative study. Tex. Rep. Biol. Med. 25(4): 553-557.

Rabstein, L.S., R.L. Peters and G.J. Spahn. 1973. Spontaneous tumors and pathologic lesions in SWR/J mice. J. Natl. Cancer Inst. 50: 751-758.

Stiteler, W. 1991. Syracuse Research Corporation, Syracuse, NY. Personal communication with R. Schoeny, US EPA, Cincinnati, OH.

Thyssen, J., J. Althoff, G. Kimmerle and U. Mohr. 1981. Inhalation studies with benzo[a]pyrene in Syrian golden hamsters. J. Natl. Cancer Inst. 66: 575-577.

US EPA. 1991a. Drinking Water Criteria Document for PAH. Prepared by the Office of Health and Environmental Assessment, Environmental Criteria and Assessment Office, Cincinnati, OH for the Office of Water Regulations and Standards, Washington, DC.

US EPA. 1991b. Dose-Response Analysis of Ingested Benzo[a]pyrene (CAS No. 50-32-8). Human Health Assessment Group, Office of Health and Environmental Assessment, Washington, DC. EPA/600/R-92/045.

VI.D. DRINKING WATER HA REFERENCES

None

VII. REVISION HISTORY

Date	Section	Description
08/01/89	VI.	Bibliography on-line
01/01/92	II.	Carcinogen assessment noted as pending change
01/01/92	IV.	Regulatory actions updated
04/01/92	II.	Summary revised; oral quantitative section added
04/01/92	VI.C.	Carcinogen assessment references revised
05/01/92	II.D.2.	Work group review and verification date corrected
07/01/92	II.	Text revised in NOTE
07/01/92	II.B.	Range of slope factors corrected
07/01/92	II.B.1.	Slope factor and risks corrected
07/01/92	II.B.2.	Data table heading corrected
07/01/92	II.B.3	Slope factor corrected; last paragraph
07/01/92	II.D.3.	Secondary contact changed
09/01/93	II.	Carcinogenicity assessment noted as pending change
09/01/93	II.D.2.	Work group review date added
12/01/93	VI.C.	Reference revised - US EPA, 1991b
02/01/94	II.D.3.	Primary contact's phone number changed
03/01/94	II.	Pending change note removed; no change
03/01/94	II.D.2.	Work group review date added
07/01/94	II.D.2.	Work group review date added
11/01/94	II.B.1.	Slope factor clarified; changed O to "0"

SYNONYMS

BaP
Benzo[a]pyrene
BENZO(d,e,f)CHRYSENE
3,4-BENZOPIRENE
3,4-BENZOPYRENE
6,7-BENZOPYRENE
BENZO(a)PYRENE
3,4-BENZPYREN

3,4-BENZPYRENE
3,4-BENZ(a)PYRENE
BENZ(a)PYRENE
3,4-BENZYPYRENE
BP
3,4-BP
B(a)P
RCRA WASTE NUMBER U022

Source: *Instant EPA's IRIS*, 1997, Instant Reference Sources, Inc., 7605 Rockpoint Drive, Austin, TX 78731 (Fax: 512-345-2386; Internet URL, http://www.instantref.com/inst-ref.htm).

Benzo(b)fluoranthene

Introduction

Benzo(b)fluoranthene is a polycyclic aromatic hydrocarbon (PAH) and is found in petroleum, wax, and smoke. Release of benzo(b)fluoranthene is most likely to result from the incomplete combustion of a variety of fuels including wood and fossil fuels. Plants, however, also produce benzo(b)fluoranthene. When released to water, adsorption to suspended sediments usually removes most of the benzo(b)fluoranthene from solution. Photolysis and photo-oxidation of the benzo(b)fluoranthene which remains in solution may occur but adsorbed benzo(b)fluoranthene is expected to resist these processes. Volatilization and biodegradation of dissolved benzo(b)fluoranthene may also occur. Bioconcentration in fish may occur; however microsomal oxidase, an enzyme capable of rapidly metabolizing polynuclear aromatic hydrocarbons, is present in most fish. Release to the soil may result in some biodegradation but due to typical strong adsorption of benzo(b)fluoranthene to soil, volatilization, photolysis and leaching to groundwater are not usually significant. Benzo(b)fluoranthene in the atmosphere is likely to be adsorbed to particulate matter, and will be subject to wet and dry deposition. Benzo(b)fluoranthene in the vapor phase can react with photochemically generated atmospheric hydroxyl radicals resulting in an estimated half-life of 1 day. Photolysis of vapor phase benzo(b)fluoranthene can be rapid, but the adsorbed compound may not photolyze significantly. Benzo(b)fluoranthene is a contaminant in air, water, sediment, soil, fish, and other aquatic organisms and food. Human exposure results primarily from air [828].

Identifying Information

CAS NUMBER: 205-99-2

NIOSH Registry Number: CU1400000

SYNONYMS: 3,4-BENZ(E)ACEPHENANTHRYLENE, 2,3-BENZFLUORANTHENE, 2,3-BENZOFLUORANTHENE, 2,3-BENZOFLUORANTHENE, BENZO(B)FLUORANTHENE, BENZO(B) FLUORANTHENE, 3,4-BENZOFLUORANTHENE, BENZ(E)ACEPHENANTHRYLENE, B (B) F, BENZO(E)FLUORANTHENE, B(B)F, B B F, BBF, B(E)F, B E F, BEF

Endocrine Disruptor Information

Benzo(b)fluoranthene is on EPA's list of suspected EEDs [811] and it is an aromatic coplanar compound. However, no additional information has been located with respect to its endocrine modifying properties.

Chemical and Physical Properties

CHEMICAL FORMULA: $C_{20}H_{12}$

MOLECULAR WEIGHT: 252.32

WLN: L C65 K666 1A TJ

PHYSICAL DESCRIPTION: Needles

SPECIFIC GRAVITY: Not available

DENSITY: Not available

MELTING POINT: 163-165°C [275]

BOILING POINT: Not available

SOLUBILITY: Water: <1 mg/mL @ 19°C [700]; DMSO: 10-50 mg/mL @ 19°C [700]; 95% Ethanol: <1 mg/mL @ 19° [700]; Acetone: 10-50 mg/mL @ 19° [700]

OTHER SOLVENTS: Benzene: Slightly soluble [395]

Source: *The National Toxicology Program's Chemical Database, Volume 2*, 1992, CRC Press, Inc./Lewis Publishers, 2000 Corporate Blvd., Boca Raton, FL 33431 (Fax: 407-998-9114).

Environmental Reference Materials

A source of EED environmental reference materials for benzo(b)fluoranthene is Radian International LLC, Austin, TX (Fax 512-454-0268; Internet URL, http://www.radian.com/standards); Catalog No. ERB-002.

Hazardous Properties

ACUTE/CHRONIC HAZARDS: When heated to decomposition, this compound emits acrid smoke and irritating
 fumes [042]. There is evidence it may be also an animal carcinogen.

FIRE HAZARD: Flash point data for benzo(b)fluoranthene are not available. It is probably combustible. Fires involving benzo(a)fluoranthene can be controlled with a dry chemical, carbon dioxide or Halon extinguisher.

REACTIVITY: This compound can react with strong oxidizers [107]. Ozone and chlorinating agents oxidize this type of compound [071]. It may also react with various electrophiles, peroxides, nitrogen oxides and sulfur oxides [051], [071].

STABILITY: This compound is stable under normal laboratory conditions. Solutions of benzo(a)fluoranthene in water, DMSO, 95% ethanol or acetone should be stable for 24 hours under normal lab conditions [700].

Source: *The National Toxicology Program's Chemical Database, Volume 7*, 1992, CRC Press, Inc./Lewis Publishers, 2000 Corporate Blvd., Boca Raton, FL 33431 (Fax: 407-998-9114).

Medical Symptoms of Exposure

Symptoms of exposure may include skin tumors in laboratory animals [107].

Source: *The National Toxicology Program's Chemical Database, Volume 4*, 1992, CRC Press, Inc./Lewis Publishers, 2000 Corporate Blvd., Boca Raton, FL 33431 (Fax: 407-998-9114).

Toxicological Information

The toxicological information below is gathered from several sources including two national databases: one from the *National Toxicology Program Chemical Repository Database* and another from *EPA's Integrated Risk Information System (IRIS)*.

Toxicology Information from the National Toxicology Program

CARCINOGENICITY:

Tumorigenic Data:

Type of Effect	Route/Animal	Amount dosed (Notes)
TDLo:	imp-rat	5 mg/kg
TD:	imp-rat	5 mg/kg
TDLo:	ipr-mus	5046 ug/kg/15D-I
TDLo:	scu-mus	72 mg/kg/9W-I
TDLo:	skn-mus	88 ng/kg/120W-I
TD:	skn-mus	72 mg/kg/60W-I
TD:	skn-mus	4037 ug/kg/20D-I

Review:
　　　　　IARC Cancer Review: Animal Sufficient Evidence
　　　　　IARC possible human carcinogen (Group 2B)

Status:

    EPA Carcinogen Assessment Group
    NTP Fourth Annual Report on Carcinogens, 1984
    NTP anticipated human carcinogen

TERATOGENICITY: Not available

MUTATION DATA:

Test	Lowest dose	Test	Lowest dose
mma-sat	31 nmol/plate	otr-ham:lng	100 ug/L
sce-ham-ipr	900 mg/kg/24H		

TOXICITY:

    Not available

OTHER TOXICITY DATA:

Review:
    Toxicology Review-2

Status:

    EPA Genetox Program 1986, Positive: Carcinogenicity-mouse/rat
    EPA Genetox Program 1986, Inconclusive: In vivo SCE-nonhuman
    EPA TSCA Test Submission (TSCATS) Data Base, June 1987
    NIOSH Analytical Methods: see Polynuclear aromatic hydrocarbons
    Meets criteria for proposed OSHA Medical Records Rule

SAX TOXICITY EVALUATION:

    THR: MUTATION Data. An experimental carcinogen and equivocal
    tumorigenic agent.

Source: *Instant Tox-Base*, 1995, Instant Reference Sources, Inc., 7605 Rockpoint Drive, Austin, TX 78731 (Fax: 512-345-2386; Internet URL, http://www.instantref.com/inst-ref.htm).

Toxicology Information from EPA's Integrated Risk Information System (IRIS)

I. CHRONIC HEALTH HAZARD ASSESSMENTS FOR NONCARCINOGENIC EFFECTS

I.A. REFERENCE DOSE FOR CHRONIC ORAL EXPOSURE (RfD)

[Ed. Note: EPA does not have information yet in this section of IRIS].

I.B. REFERENCE CONCENTRATION FOR CHRONIC INHALATION EXPOSURE (RfC)

[Ed. Note: EPA does not have information yet in this section of IRIS].

II. CARCINOGENICITY ASSESSMENT FOR LIFETIME EXPOSURE

Section II provides information on three aspects of the carcinogenic risk assessment for the agent in question; the US. EPA classification, and quantitative estimates of risk from oral exposure and from inhalation exposure. The classification reflects a weight-of-evidence judgment of the likelihood that the agent is a human carcinogen. The quantitative risk estimates are presented in three ways. The slope factor is the result of application of a low-dose extrapolation procedure and is presented as the risk per (mg/kg)/day. The unit risk is the quantitative estimate in terms of either risk per ug/L drinking water or risk per ug/cu.m air breathed. The third form in which risk is presented is a drinking water or air concentration providing cancer risks of 1 in 10,000, 1 in 100,000 or 1 in 1,000,000. Background Document 2 (Service Code 5) provides details on the rationale and methods used to derive the carcinogenicity values found in IRIS. Users are referred to Section I for information on long-term toxic effects other than carcinogenicity.

II.A. EVIDENCE FOR CLASSIFICATION AS TO HUMAN CARCINOGENICITY

II.A.1. WEIGHT-OF-EVIDENCE CLASSIFICATION

Classification -- B2; probable human carcinogen

Basis -- Based on no human data and sufficient data from animal bioassays. Benzo[b]fluoranthene produced tumors in mice after lung implantation, intraperitoneal (i.p.) or subcutaneous (s.c.) injection, and skin painting.

II.A.2. HUMAN CARCINOGENICITY DATA

None.   Although there are no human data that specifically link exposure to benzo[b]fluoranthene to human cancers, benzo[b]fluoranthene is a component of mixtures that have been associated with human cancer. These include coal tar, soots, coke oven emissions and cigarette smoke (US. EPA, 1984, 1990; IARC, 1984).

II.A.3. ANIMAL CARCINOGENICITY DATA

Sufficient.   In a lifetime implant study, 3-month-old female Osborne-Mendel rats (35/group) received a single lung implant of either 0.1 mg (0.4 mg/kg), 0.3 mg (1.2 mg/kg) or 1 mg (4.1 mg/kg) benzo[b]fluoranthene in 0.05 mL of a 1:1 (v:v) mixture of beeswax and trioctanoin (Deutsch-Wenzel et al., 1983).   Controls consisted of an untreated group and a group receiving an implant of the vehicle.   The median survival times were: 118, 104, 110, 113 and 112 weeks, for the untreated, vehicle control, low-, mid- and high-dose groups, respectively.   The incidences of epidermoid carcinomas and pleomorphic sarcomas in the lung and thorax (combined) were: untreated controls, 0/35; vehicle controls, 0/35; low-dose group, 1/35; mid-dose group, 3/35; and high-dose group, 13/35. These incidences showed a statistically significant dose-response relationship.

Groups of 15-17 male and 17-18 female CD-1 mice received i.p. injections of benzo[b]fluoranthene in DMSO on days 1, 8 and 15 after birth (total dose was approximately 126 ug/mouse) and were sacrificed at 52 weeks of age (LaVoie et al., 1987).   A statistically significant increase in the incidence of liver adenomas and hepatomas (combined) occurred in treated males (8/15) relative to vehicle controls (1/17), but not in females. Lung adenomas (2/15 males, 3/17 females) were reported in treated animals, whereas none were found in controls.

Injection site sarcomas occurred in 18/24 survivors of a total of 16 male and 14 female XVIInc/Z mice that received three s.c. injections of benzo[b]fluoranthene (total dose = 2.6 mg) over a period of 2 months (Lacassagne et al., 1963).

Benzo[b]fluoranthene has yielded positive results for complete carcinogenic activity and initiating activity in mouse skin-painting assays. In skin-painting assays groups of 20 female Swiss mice were treated 3 times/week with 0.01, 0.1 or 0.5% solutions of benzo[b]fluoranthene in acetone (Wynder and Hoffmann, 1959).   The high dose produced papillomas in 100% of the mice and carcinomas in 90% of the mice within 8 months. The middle dose produced papillomas in 65% and carcinomas in 85% within 12 months, while the low dose produced a papilloma in only 1 animal among 10 survivors at 14 months.   No concurrent controls were observed.   LaVoie et al. (1982) applied solutions of 0, 10, 30 or 100 ug benzo[b]fluoranthene in 0.1 mL acetone (10 doses, one every other day) to the skins of groups of 20 Crl:CD-1 mice. This regimen was followed by treatment with 2.5 ug 12-0-tetradecanoyl-phorbol-13-acetone (TPA) (a tumor promoter), 3 times/week for 20 weeks. Increases in the percentage of tumor-bearing animals (0, 45, 60, 80) as well as the number of skin tumors/animal (0, 0.9, 2.3, 7.1) appeared to be dose-related.   Similar studies by Amin et al. (1985a,b) resulted in comparable elevations of tumor incidence.

II.A.4. SUPPORTING DATA FOR CARCINOGENICITY

Positive results have been reported for a reverse mutation assay in Salmonella TA98 and the results for Salmonella TA100 have been positive and not positive (Mossanda et al., 1979; LaVoie et al., 1979; Hermann, 1981; Amin et al., 1985a,b).

Current theories on mechanisms of metabolic activation of polycyclic aromatic hydrocarbons are consistent with a carcinogenic potential for benzo[b]fluoranthene. Benzo[b]fluoranthene does not have a "classic bay-region" structure (Jerina et al., 1978). It is metabolized by mixed function oxidases to dihydrodiols (Amin et al., 1982). The 9,10-dihydrodiol is tumorigenic in mouse skin-painting assays, suggesting the possible formation of a reactive diol-epoxide (LaVoie et al., 1982).

II.B.   QUANTITATIVE ESTIMATE OF CARCINOGENIC RISK FROM ORAL EXPOSURE

[Ed. Note: EPA does not have information yet in this section of IRIS].

II.C.   QUANTITATIVE ESTIMATE OF CARCINOGENIC RISK FROM INHALATION EXPOSURE

[Ed. Note: EPA does not have information yet in this section of IRIS].

II.D. EPA DOCUMENTATION, REVIEW, AND CONTACTS (CARCINOGENICITY ASSESSMENT)

II.D.1. EPA DOCUMENTATION

Source Document -- US. EPA, 1984, 1990

The 1990 Drinking Water Criteria Document for Polycyclic Aromatic Hydrocarbons has received Agency and external review.

II.D.2. REVIEW (CARCINOGENICITY ASSESSMENT)

Agency Work Group Review -- 02/07/90, 08/05/93, 09/21/93, 02/02/94

Verification Date -- 02/07/90II.D.3. US. EPA CONTACTS (CARCINOGENICITY ASSESSMENT)

Rita S. Schoeny / NCEA -- (513)569-7544

Robert E. McGaughy / NCEA -- (202)260-5889

III. HEALTH HAZARD ASSESSMENTS FOR VARIED EXPOSURE DURATIONS

[Ed. Note: EPA does not have information yet in this section of IRIS].

IV. US. EPA REGULATORY ACTIONS

EPA risk assessments may be updated as new data are published and as assessment methodologies evolve. Regulatory actions are frequently not updated at the same time. Compare the dates for the regulatory actions in this section with the verification dates for the risk assessments in sections I and II, as this may explain inconsistencies. Also note that some regulatory actions consider factors not related to health risk, such as technical or economic feasibility. Such considerations are indicated for each action. In addition, not all of the regulatory actions listed in this section involve enforceable federal standards. Please direct any questions you may have concerning these regulatory actions to the US. EPA contact listed for that particular action. Users are strongly urged to read the background information on each regulatory action in Background Document 4 in Service Code 5.

IV.A. CLEAN AIR ACT (CAA)

[Ed. Note: EPA does not have information yet in this section of IRIS].

IV.B. SAFE DRINKING WATER ACT (SDWA)

IV.B.1. MAXIMUM CONTAMINANT LEVEL GOAL (MCLG) for Drinking Water

Value -- 0 mg/L (Proposed, 1990)

Considers technological or economic feasibility? -- NO

Discussion -- The proposed MCLG is zero. This value is based on carcinogenic PAH's as a class.

Reference -- 55 FR 30370 (07/25/90)

EPA Contact -- Health and Ecological Criteria Division / OST /(202) 260-7571 / FTS 260-7571; or Safe Drinking Water Hotline / (800) 426-4791

IV.B.2. MAXIMUM CONTAMINANT LEVEL (MCL) for Drinking Water

Value -- 0.0002 mg/L (Proposed, 1990)

Considers technological or economic feasibility? -- YES

Discussion -- The proposed MCL is equal to the PQL and is associated with a maximum lifetime individual risk of 1 E-4.

Monitoring requirements -- Community and non-transient water system monitoring based on state vulnerability assessment; vulnerable systems to be monitored quarterly for one year; repeat monitoring dependent upon detection and size of system.

Analytical methodology -- High pressure liquid chromatography (EPA 550, 550.1); gas chromatographic/mass spectrometry (EPA 525): PQL= 0.0002 mg/L.

Best available technology -- Granular activated carbon.

Reference -- 55 FR 30370 (07/25/90)

EPA Contact -- Drinking Water Standards Division / OGWDW /(202) 260-7575 / FTS 260-7575; or Safe Drinking Water Hotline / (800) 426-4791

IV.B.3. SECONDARY MAXIMUM CONTAMINANT LEVEL (SMCL) for Drinking Water

[Ed. Note: EPA does not have information yet in this section of IRIS].

IV.B.4. REQUIRED MONITORING OF "UNREGULATED" CONTAMINANTS

Status -- Listed (Proposed, 1991)

Discussion -- "Unregulated" contaminants are those contaminants for which EPA establishes a monitoring requirement but which do not have an associated final MCLG, MCL, or treatment technique. EPA may regulate these contaminants in the future.

Monitoring requirement -- All systems to be monitored unless a vulnerability assessment determines the system is not vulnerable.

Analytical methodology -- Gas chromatography/mass spectrometry (EPA 525); high pressure liquid chromatography (EPA 550, 550.1).

Reference -- 56 FR 3526 (01/30/91)

EPA Contact -- Drinking Water Standards Division / OGWDW /(202) 260-7575 / FTS 260-7575; or Safe Drinking Water Hotline / (800) 426-4791

IV.C. CLEAN WATER ACT (CWA)

IV.C.1. AMBIENT WATER QUALITY CRITERIA, Human Health

Water and Fish Consumption: 2.8E-3 ug/L

Fish Consumption Only: 3.11E-2 ug/L

Considers technological or economic feasibility? -- NO

Discussion -- For the maximum protection from the potential carcinogenic properties of this chemical, the ambient water concentration should be zero. However, zero may not be obtainable at this time, so the recommended criteria presents a E-6 estimated incremental increase of cancer over a lifetime. The values given represent polynuclear aromatic hydrocarbons as a class.

Reference -- 45 FR 79318 (11/28/80)

EPA Contact -- Criteria and Standards Division / OWRS(202)260-1315 / FTS 260-1315

IV.C.2. AMBIENT WATER QUALITY CRITERIA, Aquatic Organisms

Freshwater:
    Acute LEC -- none
    Chronic LEC -- none

Marine:
    Acute LEC --   3.0E+2 ug/L
    Chronic LEC -- none

Considers technological or economic feasibility? -- NO

Discussion -- The values that are indicated as "LEC" are not criteria, but are the lowest effect levels found in the literature. LEC's are given when the minimum data required to derive water quality criteria are not available. The values given represent polynuclear aromatic hydrocarbons as a class.

Reference -- 45 FR 79318 (11/28/80)

EPA Contact -- Criteria and Standards Division / OWRS(202)260-1315 / FTS 260-1315

IV.D. FEDERAL INSECTICIDE, FUNGICIDE, AND RODENTICIDE ACT (FIFRA)

        [Ed. Note: EPA does not have information yet in this section of IRIS].

IV.E. TOXIC SUBSTANCES CONTROL ACT (TSCA)

        [Ed. Note: EPA does not have information yet in this section of IRIS].

IV.F. RESOURCE CONSERVATION AND RECOVERY ACT (RCRA)

IV.F.1. RCRA APPENDIX IX, for Ground Water Monitoring

Status -- Listed

Reference -- 52 FR 25942 (07/09/87)

EPA Contact -- RCRA/Superfund Hotline(800)424-9346 / (202)260-3000 / FTS 260-3000

IV.G. SUPERFUND (CERCLA)

IV.G.1. REPORTABLE QUANTITY (RQ) for Release into the Environment

Value (status) -- 1 pound (Final, 1989)

Considers technological or economic feasibility? -- NO

Discussion --   The final RQ for benzo[b]fluoranthene is based on potential carcinogenicity.   Available data indicate a hazard ranking of high and a weight of evidence classification of Group B2, which corresponds to an RQ of 1 pound.

Reference -- 54 FR 33418 (08/14/89)

EPA Contact -- RCRA/Superfund Hotline(800)424-9346 / (202)260-3000 / FTS 260-3000

VI. BIBLIOGRAPHY

VI.A. ORAL RfD REFERENCES

None

VI.B. INHALATION RfC REFERENCES

None

VI.C. CARCINOGENICITY ASSESSMENT REFERENCES

Amin, S., E.J. LaVoie and S.S. Hecht.   1982.   Identification of metabolites of benzo[b]fluoranthene. Carcinogenesis. 3(2): 171-174.

Amin, S., K. Huie and S.S. Hecht. 1985a. Mutagenicity and tumor initiating activity of methylated benzo[b]fluoranthene. Carcinogenesis. 6(7): 1023-1025.

Amin, S., N. Hussain, G. Balanikas, K. Huie and S.S. Hecht. 1985b. Mutagenicity and tumor initiating activity of methylated benzo[k]fluoranthenes. Cancer Lett. 26: 343-347.

Deutsch-Wenzel, R., H. Brune, G. Grimmer, G. Dettbarn and J. Misfeld.   1983. Experimental studies in rat lungs on the carcinogenicity and dose-response relationships of eight frequently occurring environmental polycyclic aromatic hydrocarbons. J. Natl. Cancer Inst. 71(3): 539-543.

Hermann, M. 1981. Synergistic effects of individual polycyclic aromatic hydrocarbons on the mutagenicity of their mixtures. Mutat. Res. 90: 399-409.

IARC (International Agency for Research on Cancer).   1984.   Monographs on the Evaluation of the Carcinogenic Risk of the Chemical to Man. Polynuclear Aromatic Hydrocarbons. Part 3. Industrial Exposures in Aluminum Production, Coal Gasification, Coke Production, and Iron and Steel Founding. Vol. 34. World Health Organization.

Jerina, D.M., H. Yagi, R.E. Lehr, et al. 1978. The Bay-region theory of carcinogenisis by polycyclic aromatic hydrocarbons. In: Polycyclic Hydrocarbons and Cancer, Vol. 1. Environment, Chemistry and Metabolism, H.V. Gelboin and P.O.P. Ts'o, Ed. Academic Press, NY.

Lacassagne, A., N.P. Buu-Hoi, F. Zajdela, D. Lavit-Lamy and O. Chalvet. 1963. Activite cancerogene d'hydrocarbures aromatiques polycycliques a noyau fluoranthene. Un. Int. Cancer Acta. 19(3-4): 490-496. (Fre.)

LaVoie, E.J., E.V. Bedenko, N. Hirota, S.S. Hecht and D. Hoffmann. 1979. A comparison of the mutagenicity, tumor-initiating activity and complete carcinogenicity of polynuclear aromatic hydrocarbons. In: Polynuclear Aromatic Hydrocarbons, P.W. Jones and P. Leber, Ed. Ann Arbor Science Publishers, Ann Arbor, MI. p. 705-721.

LaVoie, E.J., S. Amin., S.S. Hecht, K. Furuya and D. Hoffmann. 1982. Tumor initiating activity of dihydrodiols of benzo[b]fluoranthene, benzo[j]fluoranthene and benzo[k]fluoranthene. Carcinogenesis. 3(1): 49-52.

LaVoie, E.J., J. Braley, J.E. Rice and A. Rivenson. 1987. Tumorigenic activity for non-alternant polynuclear aromatic hydrocarbons in newborn mice. Cancer Lett. 34: 15-20.

Mossanda, K., F. Poncelet, A. Fouassin and M. Mercier. 1979. Detection of mutagenic polycyclic aromatic hydrocarbons in African smoked fish. Food Cosmet. Toxicol. 17: 141-143.

US. EPA. 1984. Carcinogen Assessment of Coke Oven Emissions. Office of Health and Environmental Assessment, Washington, DC. EPA 600/6-82-003F. NTIS PB 84-170181.

US. EPA. 1990. Drinking Water Criteria Document for Polycyclic Aromatic Hydrocarbons (PAHs). Prepared by the Office of Health and Environmental Assessment, Environmental Criteria and Assessment Office, Cincinnati, OH for the Office of Drinking Water, Washington, DC. Final Draft. ECAO-CIN-D010, September, 1990.

Wynder, E.L. and D. Hoffmann. 1959. A study of tobacco carcinogenesis. VII. The role of higher polycyclic hydrocarbons. Cancer. 12: 1079-1086.

VI.D. DRINKING WATER HA REFERENCES

None

VII. REVISION HISTORY

```
--------  --------  ------------------------------------------------------------
Date      Section   Description
--------  --------  ------------------------------------------------------------
12/01/90  II.       Carcinogen assessment on-line
12/01/90  VI.       Bibliography on-line
01/01/92  IV.       Regulatory Action section on-line
09/01/93  II.       Carcinogenicity assessment noted as pending change
09/01/93  II.D.2.   Work group review date added
11/01/93  II.D.2.   Work group review date added
03/01/94  II.       Pending change note removed; no change
03/01/94  II.D.2.   Work group review date added
```

SYNONYMS

Benz(e)acephenanthrylene
B(b)F
BENZ(e)ACEPHENANTHRYLENE
Benzo(b)fluoranthene
Benzo(e)fluoranthene
HSDB 4035
NSC 89265

2,3-BENZFLUORANTHENE
2,3-BENZOFLUORANTHENE
2,3-BENZOFLUORANTHRENE
3,4-BENZ(e)ACEPHENANTHRYLENE
3,4-BENZFLUORANTHENE
3,4-Benzofluoranthene

Source: ***Instant EPA's IRIS***, 1997, Instant Reference Sources, Inc., 7605 Rockpoint Drive, Austin, TX 78731 (Fax: 512-345-2386; Internet URL, http://www.instantref.com/inst-ref.htm).

Benzo(k)fluoranthene

Introduction

Benzo(k)fluoranthene (BkF) is a polycyclic aromatic hydrocarbon (PAH) and is found in petroleum, wax, and smoke. Its release into air and water is quite general since it is a ubiquitous product of incomplete combustion. Both in air and water it is largely associated with particulate matter. Although environmental concentrations are greatest near sources, BkFs presence in distant places indicates that it is reasonably stable in the atmosphere and capable of long distant transport. Atmospheric losses are caused by gravitational settling and washing out during rain events. On land it is strongly adsorbed to soil and remains in the upper soil layers and usually does not leach into groundwater. Biodegradation may occur but will usually be very slow (its half-life is estimated to be about 2 years with acclimated microorganisms). It can get into surface water from dust and precipitation in addition to runoff and effluents. In the water it will sorb to sediment and particulate matter in the water column and it can bioconcentrate in fish and seafood. Human exposure is from smoking, inhalation of polluted air and eating food contaminated with products of combustion or prepared in such a way (smoked, charcoal broiled) that PAHs are generated. Since water treatment such as filtration, chlorination and ozonolysis removes benzo(k)fluoranthene, exposure from drinking water should be minor [828].

Identifying Information

CAS NUMBER: 207-08-9

NIOSH Registry Number: DF6350000

SYNONYMS: 8,9-BENZOFLUORANTHENE, 11,12-BENZOFLURANTHENE, 11,12-BENZO(K)FLUOR-ANTHENE, 2,3,1',8'-BINAPTHYLENE, BENZO(K)FLUORANTHENE, BENZO(K) FLUORANTHENE, DIBENZO(B,JK)FLUORENE, B (K) F, B(K)F, B K F, BKF

Endocrine Disruptor Information

Benzo(k)fluoranthene is on EPA's list of suspected EEDs [811] and it is an aromatic coplanar compound.

Chemical and Physical Properties

CHEMICAL FORMULA: $C_{20}H_{12}$

MOLECULAR WEIGHT: 252.32

WLN: L E6 C6566 1A TJ

PHYSICAL DESCRIPTION: Pale yellow needles

SPECIFIC GRAVITY: Not available

DENSITY: Not available

MELTING POINT: 217°C [017],[029],[051],[395]

BOILING POINT: 480°C [017],[029],[047],[395]

SOLUBILITY: Water: <1 mg/mL @ 20°C [700]; DMSO: <1 mg/mL @ 20°C [700]; 95% Ethanol: <1 mg/mL @ 20°C [700]; Acetone: 1-10 mg/mL @ 20°C [700]; Methanol: <1 mg/mL @ 20°C [700]; Toluene: 5-10 mg/mL @ 20°C [700]

OTHER SOLVENTS: Benzene: Soluble [017],[047],[395]; Acetic acid: Soluble [017],[047],[395]

Source: *The National Toxicology Program's Chemical Database, Volume 2*, 1992, CRC Press, Inc./Lewis Publishers, 2000 Corporate Blvd., Boca Raton, FL 33431 (Fax: 407-998-9114).

Environmental Reference Materials

A source of EED environmental reference materials for benzo(k)fluoranthene is Radian International LLC, Austin, TX (Fax 512-454-0268; Internet URL, http://www.radian.com/standards); Catalog No. ERB-001.

Hazardous Properties

ACUTE/CHRONIC HAZARDS: When heated to decomposition this compound emits acrid smoke and irritating fumes [042].

FIRE HAZARD: Flash point data for benzo(k)fluoranthene are not available. It is probably combustible. Fires involving this material can be controlled with a dry chemical, carbon dioxide or Halon extinguisher.

REACTIVITY: Ozone and chlorinating agents oxidize this type of compound. It may also react with various electrophiles, peroxides, nitrogen oxides and sulfur oxides [051],[071]. Benzo(k)fluoranthene can react with strong oxidizers [107].

STABILITY: This compound is stable under normal laboratory conditions. Solutions of benzo(k)fluoranthene in water, DMSO, 95% ethanol or acetone should be stable for 24 hours under normal lab conditions [700].

Source: *The National Toxicology Program's Chemical Database, Volume 7*, 1992, CRC Press, Inc./Lewis Publishers, 2000 Corporate Blvd., Boca Raton, FL 33431 (Fax: 407-998-9114).

Medical Symptoms of Exposure

Symptoms of exposure may include irritation to the eyes, skin and mucous membranes.

Source: *The National Toxicology Program's Chemical Database, Volume 4*, 1992, CRC Press, Inc./Lewis Publishers, 2000 Corporate Blvd., Boca Raton, FL 33431 (Fax: 407-998-9114).

Toxicological Information

The toxicological information below is gathered from several sources including two national databases: one from the *National Toxicology Program Chemical Repository Database* and another from *EPA's Integrated Risk Information System (IRIS)*.

Toxicology Information from the National Toxicology Program

CARCINOGENICITY:

Tumorigenic Data:

Type of Effect	Route/Animal	Amount Dosed (Notes)
TDLo:	imp-rat	5 mg/kg
TDLo:	scu-mus	72 mg/kg/9W-I
TDLo:	skn-mus	2820 mg/kg/47W-I

Review:
> IARC Cancer Review: Animal Sufficient Evidence
> IARC possible human carcinogen (Group 2B)
> Status: NTP anticipated human carcinogen

TERATOGENICITY: Not available

MUTATION DATA:

Test	Lowest dose		Test	Lowest dose
mma-sat	10 ug/plate			

TOXICITY:

Not available

OTHER TOXICITY DATA:

Status:
> EPA TSCA Test Submission (TSCATS) Database, June 1987
> NIOSH Analytical Methods: see Polynuclear Aromatic Hydrocarbons
> Meets criteria for proposed OSHA Medical Records Rule
> EPA Genetox Program 1986, Positive: Carcinogenicity-mouse/rat

SAX TOXICITY EVALUATION:

THR: An experimental equivocal tumorigenic agent.

Source: *Instant Tox-Base*, 1995, Instant Reference Sources, Inc., 7605 Rockpoint Drive, Austin, TX 78731 (Fax: 512-345-2386; Internet URL, http://www.instantref.com/inst-ref.htm).

Toxicology Information from EPA's Integrated Risk Information System (IRIS)

I.   CHRONIC HEALTH HAZARD ASSESSMENTS FOR NONCARCINOGENIC EFFECTS

I.A. REFERENCE DOSE FOR CHRONIC ORAL EXPOSURE (RfD)

[Ed. Note: EPA does not have information yet in this section of IRIS].

I.B. REFERENCE CONCENTRATION FOR CHRONIC INHALATION EXPOSURE (RfC)

[Ed. Note: EPA does not have information yet in this section of IRIS].

II. CARCINOGENICITY ASSESSMENT FOR LIFETIME EXPOSURE

Section II provides information on three aspects of the carcinogenic risk assessment for the agent in question; the US. EPA classification, and quantitative estimates of risk from oral exposure and from inhalation exposure. The classification reflects a weight-of-evidence judgment of the likelihood that the agent is a human carcinogen. The quantitative risk estimates are presented in three ways. The slope factor is the result of application of a low-dose extrapolation procedure and is presented as the risk per (mg/kg)/day. The unit risk is the quantitative estimate in terms of either risk per ug/L drinking water or risk per ug/cu.m air breathed. The third form in which risk is presented is a drinking water or air concentration providing cancer risks of 1 in 10,000, 1 in 100,000 or 1 in 1,000,000. Background Document 2 (Service Code 5) provides details on the rationale and methods used to derive the carcinogenicity values found in IRIS. Users are referred to Section I for information on long-term toxic effects other than carcinogenicity.

II.A. EVIDENCE FOR CLASSIFICATION AS TO HUMAN CARCINOGENICITY

II.A.1. WEIGHT-OF-EVIDENCE CLASSIFICATION

Classification -- B2; probable human carcinogen

Basis -- Based on no human data and sufficient data from animal bioassays. Benzo[k]fluoranthene produced tumors after lung implantation in mice and when administered with a promoting agent in skin-painting studies. Equivocal results have been found in a lung adenoma assay in mice. Benzo[k]fluoranthene is mutagenic in bacteria.

II.A.2. HUMAN CARCINOGENICITY DATA

None. Although there are no human data that specifically link exposure to benzo[k]fluoranthene to human cancers, benzo[k]fluoranthene is a component of mixtures that have been associated with human cancer. These include coal tar, soots, coke oven emissions and cigarette smoke (US. EPA, 1984, 1990; IARC, 1984).

II.A.3. ANIMAL CARCINOGENICITY DATA

Sufficient. In a lifetime implant study, female Osborne-Mendel rats (27-35/group) received lung implants of 0.16 mg (0.65 mg/kg), 0.83 mg (3.4 mg/kg) or 4.15 mg (17 mg/kg) benzo[k]fluoranthene in 0.05 mL of a 1:1 (v:v) mixture of beeswax and trioctanoin (Deutsch-Wenzel et al., 1983). Controls consisted of an untreated group and a group receiving an implant of the vehicle. Median survival times (weeks) were: 118 (untreated controls), 104 (vehicle controls); 114 (0.16 mg dose); 95 (0.83 mg dose); 98 (4.15 mg dose). The incidences of epidermoid carcinomas in the lung and thorax (combined) showed a statistically significant dose-related increase. The observed incidences were: untreated controls, 0/35; vehicle controls, 0/35; low-dose, 0/35; mid-dose, 3/31; high-dose, 12/27.

Groups of 16-17 male and 18 female newborn CD-1 mice received intraperitoneal injections of benzo[k]fluoranthene in DMSO on days 1, 8 and 15 after birth (total dose approximately 126 ug/mouse) and were sacrificed at 52 weeks of age (LaVoie et al., 1987). The incidence of hepatic adenomas and hepatomas was increased in treated male mice (3/16) relative to vehicle controls (1/17), although this increase was not statistically significant. No liver tumors were found in females. Lung adenomas were found in treated male (1/16) and female (3/18) mice, whereas none were reported for the controls. This assay is considered to be a short-term, in vivo, lung tumor assay.

Benzo[k]fluoranthene has yielded positive results for initiating activity in several mouse skin-painting assays. A single dermal application of 11 mg benzo[k]fluoranthene to 20 Swiss mice in a 63-week study did not induce tumors (Van Duuren et al., 1966). However, when the same dose was followed by promoting treatments with croton resin, 18/20 animals developed papillomas and 5/20 developed carcinomas. LaVoie et al. (1982) applied doses of 0, 30, 100 or 1000 ug benzo[k]fluoranthene (10 doses each, every other day, in 0.1 mL acetone) to the skin of groups of 20 Crl:CD-1 mice. This regimen was followed by treatment with 2.5 ug 12-O-tetradecanoyl phorbol-13-acetate (TPA) (a tumor promoter), 3 times/week for 20 weeks. Increases in the percentage of tumor- bearing animals (0, 5, 25, 75), as well as the number of tumors per animal (0, 0.1, 0.4, 2.8), appeared to be dose-related. These results were corroborated by reports of Amin et al. (1985a,b).

II.A.4. SUPPORTING DATA FOR CARCINOGENICITY

Tests for mutagenicity in prokaryotic cells have produced positive results. Tests for reverse mutation in Salmonella typhimurium strain TA100 and TA98 yielded positive results for benzo[k]fluoranthene in the presence of a metabolic activation system (rat liver S9) (LaVoie et al., 1980; Hermann et al., 1980).

II.B. QUANTITATIVE ESTIMATE OF CARCINOGENIC RISK FROM ORAL EXPOSURE

[Ed. Note: EPA does not have information yet in this section of IRIS].

II.C. QUANTITATIVE ESTIMATE OF CARCINOGENIC RISK FROM INHALATION EXPOSURE

[Ed. Note: EPA does not have information yet in this section of IRIS].

II.D. EPA DOCUMENTATION, REVIEW, AND CONTACTS (CARCINOGENICITY ASSESSMENT)

II.D.1. EPA DOCUMENTATION

Source Document -- US. EPA, 1984, 1990

The 1990 Drinking Water Criteria Document for Polycyclic Aromatic Hydrocarbons has received Agency and external review.

II.D.2. REVIEW (CARCINOGENICITY ASSESSMENT)

Agency Work Group Review -- 02/07/90, 08/05/93, 09/21/93, 02/02/94

Verification Date -- 02/07/90

II.D.3. US. EPA CONTACTS (CARCINOGENICITY ASSESSMENT)

Rita S. Schoeny / NCEA -- (513)569-7544

Robert E. McGaughy / NCEA -- (202)260-5889

III. HEALTH HAZARD ASSESSMENTS FOR VARIED EXPOSURE DURATIONS

[Ed. Note: EPA does not have information yet in this section of IRIS].

IV. US. EPA REGULATORY ACTIONS

EPA risk assessments may be updated as new data are published and as assessment methodologies evolve. Regulatory actions are frequently not updated at the same time. Compare the dates for the regulatory actions in this section with the verification dates for the risk assessments in sections I and II, as this may explain inconsistencies. Also note that some regulatory actions consider factors not related to health risk, such as technical or economic feasibility. Such considerations are indicated for each action. In addition, not all of the regulatory actions listed in this section involve enforceable federal standards. Please direct any questions you may have concerning these regulatory actions to the US. EPA contact listed for that particular action. Users are strongly urged to read the background information on each regulatory action in Background Document 4 in Service Code 5.

IV.A. CLEAN AIR ACT (CAA)

[Ed. Note: EPA does not have information yet in this section of IRIS].

IV.B. SAFE DRINKING WATER ACT (SDWA)

IV.B.1. MAXIMUM CONTAMINANT LEVEL GOAL (MCLG) for Drinking Water

Value -- 0 mg/L (Proposed, 1990)

Considers technological or economic feasibility? -- NO

Discussion -- The proposed MCLG is zero. This value is based on carcinogenic PAH's as a class.

Reference -- 55 FR 30370 (07/25/90)

EPA Contact -- Health and Ecological Criteria Division / OST / (202) 260-7571 / FTS 260-7571; or Safe Drinking Water Hotline / (800) 426-4791

IV.B.2. MAXIMUM CONTAMINANT LEVEL (MCL) for Drinking Water

Value -- 0.0002 mg/l (Proposed, 1990)

Considers technological or economic feasibility? -- YES

Discussion -- The proposed MCL is equal to the PQL and is associated with a maximum lifetime individual risk of 1 E-4.

Monitoring requirements -- Community and non-transient water system monitoring based on state vulnerability assessment; vulnerable systems to be monitored quarterly for one year; repeat monitoring dependent upon detection and size of system.

Analytical methodology -- High pressure liquid chromatography (EPA 550, 550.1); gas chromatographic/mass spectrometry (EPA 525): PQL= 0.0002 mg/L.

Best available technology -- Granular activated carbon.

Reference -- 55 FR 30370 (07/25/90)

EPA Contact -- Drinking Water Standards Division / OGWDW / (202) 260-7575 / FTS 260-7575; or Safe Drinking Water Hotline / (800) 426-4791

IV.B.3. SECONDARY MAXIMUM CONTAMINANT LEVEL (SMCL) for Drinking Water

No data available

IV.B.4. REQUIRED MONITORING OF "UNREGULATED" CONTAMINANTS

Status -- Listed (Proposed, 1991)

Discussion -- "Unregulated" contaminants are those contaminants for which EPA establishes a monitoring requirement but which do not have an associated final MCLG, MCL, or treatment technique. EPA may regulate these contaminants in the future.

Monitoring requirement -- All systems to be monitored unless a vulnerability assessment determines the system is not vulnerable.

Analytical methodology -- Gas chromatography/mass spectrometry (EPA 525); high pressure liquid chromatography (EPA 550, 550.1).

Reference -- 56 FR 3526 (01/30/91)

EPA Contact -- Drinking Water Standards Division / OGWDW / (202) 260-7575 / FTS 260-7575; or Safe Drinking Water Hotline / (800) 426-4791

IV.C. CLEAN WATER ACT (CWA)

IV.C.1. AMBIENT WATER QUALITY CRITERIA, Human Health

Water and Fish Consumption: 2.8E-3 ug/L

Fish Consumption Only: 3.11E-2 ug/L

Considers technological or economic feasibility? -- NO

Discussion -- For the maximum protection from the potential carcinogenic properties of this chemical, the ambient water concentration should be zero. However, zero may not be obtainable at this time, so the recommended criteria represents a E-6 estimated incremental increase of cancer over a lifetime. The values given represent polynuclear aromatic hydrocarbons as a class.   Reference -- 45 FR 79318 (11/28/80)

EPA Contact -- Criteria and Standards Division / OWRS (202)260-1315 / FTS 260-1315

IV.C.2. AMBIENT WATER QUALITY CRITERIA, Aquatic Organisms

Freshwater:
   Acute LEC -- none
   Chronic LEC -- none

Marine:
  Acute LEC --   3.0E+2 ug/L
  Chronic LEC -- none

Considers technological or economic feasibility? -- NO

Discussion -- The values that are indicated as "LEC" are not criteria, but are the lowest effect levels found in the literature. LEC's are given when the minimum data required to derive water quality criteria are not available. The values given represent polynuclear aromatic hydrocarbons as a class.

Reference -- 45 FR 79318 (11/28/80)

EPA Contact -- Criteria and Standards Division / OWRS (202)260-1315 / FTS 260-1315

IV.D. FEDERAL INSECTICIDE, FUNGICIDE, AND RODENTICIDE ACT (FIFRA)

    [Ed. Note: EPA does not have information yet in this section of IRIS].

IV.E. TOXIC SUBSTANCES CONTROL ACT (TSCA)

    [Ed. Note: EPA does not have information yet in this section of IRIS].

IV.F. RESOURCE CONSERVATION AND RECOVERY ACT (RCRA)

IV.F.1. RCRA APPENDIX IX, for Ground Water Monitoring

Status -- Listed

Reference -- 52 FR 25942 (07/09/87)

EPA Contact -- RCRA/Superfund Hotline (800)424-9346 / (202)260-3000 / FTS 260-3000

IV.G. SUPERFUND (CERCLA)

IV.G.1. REPORTABLE QUANTITY (RQ) for Release into the Environment

Value (status) -- 5000 pounds (Final, 1989)

Considers technological or economic feasibility? -- NO

Discussion -- No data have been found to permit the ranking of this hazardous substance. The available data for acute hazards may lie above the upper limit for the 5000-pound RQ, but since it is a designated hazardous substance, the largest assignable RQ is 5000 pounds. This chemical has been assessed for carcinogenicity and is not, at this time, considered to be a potential carcinogen.

Reference -- 54 FR 33418 (08/14/89)

EPA Contact -- RCRA/Superfund Hotline (800)424-9346 / (202)260-3000 / FTS 260-3000

VI. BIBLIOGRAPHY

VI.A. ORAL RfD REFERENCES

None

VI.B. INHALATION RfC REFERENCES

None

VI.C. CARCINOGENICITY ASSESSMENT REFERENCES

Amin, S., K. Huie and S.S. Hecht. 1985a. Mutagenicity and tumor initiating activity of methylated benzo[b]fluoranthenes. Carcinogenesis. 6(7): 1023-1025.

Amin, S., N. Hussain, G. Balanikas, K. Huie and S.S. Hecht. 1985b. Mutagenicity and tumor initiating activity of methylated benzo[k]fluoranthenes. Cancer Lett. 26: 343-347.

Deutsch-Wenzel, R., H. Brune, G. Grimmer, G. Dettbarn and J. Misfeld. 1983. Experimental studies in rat lungs on the carcinogenicity and dose-response relationships of eight frequently occurring environmental polycyclic aromatic hydrocarbons. J. Natl. Cancer Inst. 71(3): 539-543.

Hermann, M., J.P. Durand, J.M. Charpentier, et al. 1980. Correlations of mutagenic activity with polynuclear aromatic hydrocarbon content of various mineral oils. In: Polynuclear Aromatic Hydrocarbons: Chemistry and Biological Effects, 4th Int. Symp., A. Bjorseth and A.J. Dennis, Ed. Battelle Press, Columbus, OH. p. 899-916.

IARC (International Agency for Research on Cancer). 1984. Monographs on the Evaluation of the Carcinogenic Risk of Chemicals to Man. Polynuclear Aromatic Hydrocarbons. Part 3. Industrial Exposures in Aluminum Production, Coal Gasification, Coke Production, and Iron and Steel Founding. Vol. 34. World Health Organization.

LaVoie, E.J., S.S. Hecht, S. Amin, V. Bedenko and D. Hoffmann. 1980. Identification of mutagenic dihydrodiols as metabolites of benzo(j)fluoranthene and benzo(k)fluoranthene. Cancer Res. 40: 4528-4532.

LaVoie, E.J., S. Amin., S.S. Hecht, K. Furuya and D. Hoffmann. 1982. Tumor initiating activity of dihydrodiols of benzo[b]fluoranthene, benzo[j]fluoranthene and benzo[k]fluoranthene. Carcinogenesis. 3(1): 49-52.

LaVoie, E.J., J. Braley, J.E. Rice and A. Rivenson. 1987. Tumorigenic activity for non-alternant polynuclear aromatic hydrocarbons in newborn mice. Cancer Lett. 34: 15-20.

US. EPA. 1984. Carcinogen Assessment of Coke Oven Emissions. Office of Health and Environmental Assessment, Washington, DC. EPA 600/6-82-003F. NTIS PB84-170181.

US. EPA. 1990. Drinking Water Criteria Document for Polycyclic Aromatic Hydrocarbons (PAHs). Prepared by the Office of Health and Environmental Assessment, Environmental Criteria and Assessment Office, Cincinnati, OH for the Office of Drinking Water, Washington, DC. Final Draft. ECAO-CIN-D010, September, 1990.

Van Duuren, B.L., A. Sivak, A. Segal, L. Orris and L. Langseth. 1966. The tumor-promoting agents of tobacco leaf and tobacco smoke condensate. J. Natl. Cancer Inst. 37(4): 519-526.

VI.D. DRINKING WATER HA REFERENCES

None

VII. REVISION HISTORY

```
--------  --------  ------------------------------------------------------
```
Date   Section Description
```
--------  --------  ------------------------------------------------------
```
11/01/90 II.   Carcinogen assessment on-line
11/01/90 VI.   Bibliography on-line
01/01/92 IV.   Regulatory action section on-line
09/01/93 II.   Carcinogenicity assessment noted as pending change
09/01/93 II.D.2. Work group review date added
11/01/93 II.D.2. Work group review date added
02/01/94 II.D.3. Secondary contact's phone number changed
03/01/94 II.   Pending change note removed; no change
03/01/94 II.D.2. Work group review date added

SYNONYMS

Benzo(k)fluoranthene   11,12-Benzofluoranthene
Dibenzo(b,jk)fluorene   2,3,1',8'-Binaphthylene
HSDB 6012   8,9-BENZOFLUORANTHENE
11,12-BENZO(k)FLUORANTHENE

Source: *Instant EPA's IRIS*, 1997, Instant Reference Sources, Inc., 7605 Rockpoint Drive, Austin, TX 78731 (Fax: 512-345-2386; Internet URL, http://www.instantref.com/inst-ref.htm).

beta-BHC

Introduction

Hexachlorobenzenes are created as byproducts of the manufacture of chlorobenzenes, pentachlorophenol, vinyl chloride, tetrachloroethylene, many pesticides (including atrazine and simazine) and other useful synthetic chemicals. It is also formed in the electrolytic production of chlorine and in the incineration of chlorinated wastes. [813] In addition to being formed as a byproduct, it is also manufactured for use as an insecticide and is produced in numerous convenient formulations such as emulsifiable concentrates, wettable powders, fumigants, suspension concentrates, granules, and dustable powders. [817] beta-BHC is one of several isomers of benzenehexachloride, also commonly called beta-hexachlorobenzene and beta-hexachlorocyclohexane (beta-HCH).

Identifying Information

CAS NUMBER: 319-85-7

NIOSH Registry Number: GV4375000

SYNONYMS: B-BHC, BETA-BENZENEHEXACHLORIDE, BETA-HEXACHLOROCYCLOHEXANE, TRANS-ALPHA-BENZENEHEXACHLORIDE, BETA-ISOMER, 1A,2B,3A,4B,5A,6B-HEXACHLORO-CYCLOHEXANE, BETA-LINDANE, BETA-1,2,3,4,5,6-HEXACHLOROCYCLOHEXANE, BENZENE-CIS-HEXACHLORIDE, BETA-HCH

Endocrine Disruptor Information

beta-BHC is a strongly suspected EED that it is listed by EPA [811], the Centers for Disease Control & Prevention [812], and the World Wildlife Fund [813] as a potential endocrine modifying chemical. Levels of BHC in water and fish pose a significant threat to human health if consumed. This is especially problematic for mothers who wish to breast feed because nursing infants ingest 200 to 300 times the adult intake based on bodyweight. Acute exposure to BHC in utero has been linked to increased stillbirths and death before the age of two. [882]

Chemical and Physical Properties

CHEMICAL FORMULA: $C_6H_6C_{l6}$

MOLECULAR WEIGHT: 290.82

250

WLN: L6TJ AG BG CG DG EG FG (BETA)

PHYSICAL DESCRIPTION: white crystalline solid

SPECIFIC GRAVITY: 1.89 g/cm^3 (19°C)

DENSITY: Not available

MELTING POINT: 312°C

OTHER SOLVENTS: Water: Insoluble; Chloroform: Slightly soluble; Benzene: Slightly soluble

Source: *The National Toxicology Program's Chemical Database, Volume 2*, 1992, CRC Press, Inc./Lewis Publishers, 2000 Corporate Blvd., Boca Raton, FL 33431 (Fax: 407-998-9114).

Environmental Reference Materials

A source of EED environmental reference materials for beta-BHC is Radian International LLC, Austin, TX (Fax 512-454-0268; Internet URL, http://www.radian.com/standards); Catalog No. ERB-013.

Hazardous Properties

ACUTE/CHRONIC HAZARDS: Beta-BHC may be toxic and an irritant. There is evidence it may be a carcinogen.

STABILITY: beta-BHC is stable under normal laboratory conditions.

Source: *The National Toxicology Program's Chemical Database, Volume 7*, 1992, CRC Press, Inc./Lewis Publishers, 2000 Corporate Blvd., Boca Raton, FL 33431 (Fax: 407-998-9114).

Medical Symptoms of Exposure

Exposure to beta-BHC may cause irritation to the eyes, skin and mucous membranes.

Source: *The National Toxicology Program's Chemical Database, Volume 4*, 1992, CRC Press, Inc./Lewis Publishers, 2000 Corporate Blvd., Boca Raton, FL 33431 (Fax: 407-998-9114).

Toxicological Information

The toxicological information below is gathered from several sources including two national databases: one from the *National Toxicology Program Chemical Repository Database* and another from *EPA's Integrated Risk Information System (IRIS)*. EPA regulatory actions.

Toxicology Information from the National Toxicology Program

CARCINOGENICITY:

Tumorigenic Data:

Type of Effect	Route/Animal	Amount Dosed (Notes)
TDLo:	orl-mus	18 gm/kg/2Y-C

Review:

       IARC Cancer Review: Animal Limited Evidence
       IARC Note: Although IARC has assigned an overall evaluation to hexachlorocyclohexanes, it has not assigned an overall evaluation to all substances within this group

Status:

       NTP anticipated human carcinogen

TERATOGENICITY: Not available

MUTATION DATA: Not available

TOXICITY:

Typ.dose	Mode	Specie	Amount	Unit	Other
LD50	orl	rat	6000	mg/kg	

OTHER TOXICITY DATA:

Status:

       Reported in EPA TSCA Inventory, 1980

SAX TOXICITY EVALUATION:

       THR: An experimental (+,-) carcinogen.

**Source: *Instant Tox-Base*, 1995, Instant Reference Sources, Inc., 7605 Rockpoint Drive, Austin, TX 78731 (Fax: 512-345-2386; Internet URL, http://www.instantref.com/inst-ref.htm).

Toxicology Information from EPA's Integrated Risk Information System (IRIS)

I. CHRONIC HEALTH HAZARD ASSESSMENTS FOR NONCARCINOGENIC EFFECTS

I.A. REFERENCE DOSE FOR CHRONIC ORAL EXPOSURE (RfD)

I.B. REFERENCE CONCENTRATION FOR CHRONIC INHALATION EXPOSURE (RfC)

[Ed. Note: EPA does not have information yet in this section of IRIS].

II. CARCINOGENICITY ASSESSMENT FOR LIFETIME EXPOSURE

Section II provides information on three aspects of the carcinogenic risk assessment for the agent in question; the US. EPA classification, and quantitative estimates of risk from oral exposure and from inhalation exposure. The classification reflects a weight-of-evidence judgment of the likelihood that the agent is a human carcinogen. The quantitative risk estimates are presented in three ways. The slope factor is the result of application of a low-dose extrapolation procedure and is presented as the risk per (mg/kg)/day. The unit risk is the quantitative estimate in terms of either risk per ug/L drinking water or risk per ug/cu.m air breathed. The third form in which risk is presented is a drinking water or air concentration providing cancer risks of 1 in 10,000, 1 in 100,000 or 1 in 1,000,000. Background Document 2 (Service Code 5) provides details on the rationale and methods used to derive the carcinogenicity values found in IRIS. Users are referred to Section I for information on long-term toxic effects other than carcinogenicity.

II.A. EVIDENCE FOR CLASSIFICATION AS TO HUMAN CARCINOGENICITY

II.A.1. WEIGHT-OF-EVIDENCE CLASSIFICATION

Classification -- C; possible human carcinogen

Basis -- Increases in benign liver tumors in CF1 mice fed beta-HCH

II.A.2. HUMAN CARCINOGENICITY DATA

Inadequate. One case report of a Japanese sanitation employee with acute leukemia was associated with occupational exposure to HCH and DDT (Hoshizaki et al., 1969).

II.A.3. ANIMAL CARCINOGENICITY DATA

Positive or marginally positive tumorigenic responses, characterized as benign hepatomas or hepatocellular carcinomas, have been observed in two strains of mice. The studies are limited in that small numbers of animals were used, no dose-response data are available, not all of the animals were examined histologically, or the duration of exposure was less than lifetime.

Thorpe and Walker (1973) fed 30 each male and female CF1 mice dietary beta-HCH at 200 ppm for 110 weeks. This resulted in 12% mortality of males and 25% of females during the first 3 months. A significantly increased incidence of liver tumors was observed in treated males and females.

No statistically significant evidence of increased tumor incidence as a consequence of beta-HCH feeding was seen in several small (5-20/group) studies with male and female dd mice fed 0-600 ppm for 24-32 weeks (Nagasaki et al., 1972; Hanada et al. 1973; Ito et al., 1973) or in male Wistar rats (Ito et al., 1975; Fitzhugh et al., 1950) fed 0-1000 ppm for >72 weeks.

Goto et al. (1972) maintained male ICR-JCL mice on a diet containing 600 ppm beta-HCH for 26 weeks. Relative liver weight was increased in the treated animals, and there was histologic evidence of benign neoplasms at an unspecified incidence.

II.A.4. SUPPORTING DATA FOR CARCINOGENICITY

No data on genetic toxicology of beta-HCH are available.

II.B. QUANTITATIVE ESTIMATE OF CARCINOGENIC RISK FROM ORAL EXPOSURE

II.B.1. SUMMARY OF RISK ESTIMATES

Oral Slope Factor -- 1.8E+0 per (mg/kg)/day

Drinking Water Unit Risk -- 5.3E-5 per (ug/L)

Extrapolation Method -- Linearized multistage procedure, extra risk

Drinking Water Concentrations at Specified Risk Levels:

Risk Level	Concentration
E-4 (1 in 10,000)	2E+0 ug/L
E-5 (1 in 100,000)	2E-1 ug/L
E-6 (1 in 1,000,000)	2E-2 ug/L

II.B.2. DOSE-RESPONSE DATA (CARCINOGENICITY, ORAL EXPOSURE)

Tumor Type -- hepatic nodules and hepatocellular carcinomas
Test Animals -- Mouse/CF1, male
Route -- diet
Reference -- Thorpe and Walker, 1973

Administered Dose (ppm)	Human Equivalent Dose (mg/kg)/day	Tumor Incidence
0	0	11/45
200	1.96	22/24

II.B.3. ADDITIONAL COMMENTS (CARCINOGENICITY, ORAL EXPOSURE)

Animal TWA dose (26 mg/kg/day) was calculated based on exposure and lifetime of 110 weeks (reported). This unit risk should not be used if water concentration exceeds 200 ug/L, since above this concentration the slope factor may differ from that stated.

II.B.4. DISCUSSION OF CONFIDENCE (CARCINOGENICITY, ORAL EXPOSURE)

The risk estimate was calculated on data from only one non-zero dose group. Relatively few animals and limited number of doses were used. A slope factor based on the data of Nagasaki et al. (1972) was calculated to be 4.7 per (mg/kg)/day and one based on Ito et al. (1973) to be 6.3 per (mg/kg)/day. These are generally supportive of the slope factor derived from Thorpe and Walker, which is the only available chronic study.

II.C.   QUANTITATIVE ESTIMATE OF CARCINOGENIC RISK FROM INHALATION EXPOSURE

II.C.1. SUMMARY OF RISK ESTIMATES

Inhalation Unit Risk -- 5.3E-4 per (ug/cu.m)

Extrapolation Method -- Linearized multistage procedure, extra risk

Air Concentrations at Specified Risk Levels:

Risk Level	Concentration
E-4 (1 in 10,000)	2E-1 ug/cu.m
E-5 (1 in 100,000)	2E-2 ug/cu.m
E-6 (1 in 1,000,000)	2E-3 ug/cu.m

II.C.2.   DOSE-RESPONSE DATA FOR CARCINOGENICITY, INHALATION EXPOSURE

The inhalation risk estimates were calculated from the oral data presented in II.B.2.

II.C.3.   ADDITIONAL COMMENTS (CARCINOGENICITY, INHALATION EXPOSURE)

The unit risk should not be used if the air concentration exceeds 20 ug/cu.m, since above this concentration the unit risk may not be appropriate.

II.C.4.   DISCUSSION OF CONFIDENCE (CARCINOGENICITY, INHALATION EXPOSURE)

See II.B.4.

II.D. EPA DOCUMENTATION, REVIEW, AND CONTACTS (CARCINOGENICITY ASSESSMENT)

II.D.1. EPA DOCUMENTATION

Source Document -- US. EPA, 1986

The 1986 HEEP has received Agency review.

II.D.2. REVIEW (CARCINOGENICITY ASSESSMENT)

Agency Work Group Review -- 10/29/86, 12/17/86

Verification Date -- 12/17/86

II.D.3. US. EPA CONTACTS (CARCINOGENICITY ASSESSMENT)

James W. Holder / NCEA -- (202)260-5721

Jim Cogliano / NCEA -- (202)260-3814

III. HEALTH HAZARD ASSESSMENTS FOR VARIED EXPOSURE DURATIONS

[Ed. Note: EPA does not have information yet in this section of IRIS].

IV. US. EPA REGULATORY ACTIONS

EPA risk assessments may be updated as new data are published and as assessment methodologies evolve. Regulatory actions are frequently not updated at the same time. Compare the dates for the regulatory actions in this section with the verification dates for the risk assessments in sections I and II, as this may explain inconsistencies. Also note that some regulatory actions consider factors not related to health risk, such as technical or economic feasibility. Such considerations are indicated for each action. In addition, not all of the regulatory actions listed in this section involve enforceable federal standards. Please direct any questions you may have concerning these regulatory actions to the US. EPA contact listed for that particular action. Users are strongly urged to read the background information on each regulatory action in Background Document 4 in Service Code 5.

IV.A. CLEAN AIR ACT (CAA)

[Ed. Note: EPA does not have information yet in this section of IRIS].

IV.B. SAFE DRINKING WATER ACT (SDWA)

[Ed. Note: EPA does not have information yet in this section of IRIS].

IV.C. CLEAN WATER ACT (CWA)

IV.C.1. AMBIENT WATER QUALITY CRITERIA, Human Health

Water and Fish Consumption -- 1.63E-2 ug/L

Fish Consumption Only -- 5.47E-2 ug/L

Considers technological or economic feasibility? -- NO

Discussion -- For the maximum protection from the potential carcinogenic properties of this chemical, the ambient water concentration should be zero. However, zero may not be attainable at this time, so the recommended criteria represents a E-6 estimated incremental increase of cancer risk over a lifetime.

Reference -- 45 FR 79318 (11/28/80)

EPA Contact -- Criteria and Standards Division / OWRS (202)260-1315 / FTS 260-1315

IV.C.2. AMBIENT WATER QUALITY CRITERIA, Aquatic Organisms

Freshwater:
    Acute LEC -- 1.0E+2 ug/L
    Chronic LEC -- None

Marine:
    Acute LEC -- 3.4E-1 ug/L
    Chronic LEC -- None

Considers technological or economic feasibility? -- NO

Discussion -- The values that are indicated as "LEC" are not criteria, but are the lowest effect levels found in the literature. LECs are given when the minimum data required to derive water quality criteria are not available. The values given are for a mixture of isomers. Criteria for lindane (gamma isomer) are also available.

Reference -- 45 FR 79318 (11/28/80)

EPA Contact -- Criteria and Standards Division / OWRS (202)260-1315 / FTS 260-1315

IV.D. FEDERAL INSECTICIDE, FUNGICIDE, AND RODENTICIDE ACT (FIFRA)

No data available

IV.E. TOXIC SUBSTANCES CONTROL ACT (TSCA)

[Ed. Note: EPA does not have information yet in this section of IRIS].

IV.F. RESOURCE CONSERVATION AND RECOVERY ACT (RCRA)

IV.F.1. RCRA APPENDIX IX, for Ground Water Monitoring

Status -- Listed

Reference -- 52 FR 25942 (07/09/87)

EPA Contact -- RCRA/Superfund Hotline (800)424-9346 / (202)260-3000 / FTS 260-3000

IV.G. SUPERFUND (CERCLA)

IV.G.1. REPORTABLE QUANTITY (RQ) for Release into the Environment

Value (status) -- 1 pound (Statutory, 1989)

Considers technological or economic feasibility? -- NO

Discussion -- The statutory RQ for all isomers of hexachlorocyclohexane is 1 pound, based on the chemical structural similarity to the gamma isomer of hexachlorocyclohexane, commonly known as lindane. An RQ of 1 pound, based on aquatic toxicity, was established for lindane under the Clean Water Act Section 311 (40 CFR 117.3). The available data indicate the aquatic 96-hour Median Threshold Limit for lindane (LC50) is less than 0.1 ppm, which corresponds to an RQ of 1 pound (50 FR 13456).

Reference -- 54 FR 33418 (08/14/89)

EPA Contact -- RCRA/Superfund Hotline (800)424-9346 / (202)260-3000 / FTS 260-3000

VI. BIBLIOGRAPHY

VI.A. ORAL RfD REFERENCES

None

VI.B. INHALATION RfD REFERENCES

None

VI.C. CARCINOGENICITY ASSESSMENT REFERENCES

Goto, M., M. Hattori and T. Miyagawa. 1972. Contribution on ecological chemistry. II. Formation of hepatoma in mice after ingestion of HCH isomers in high doses. Chemosphere. No. 6. p. 279-282.

Fitzhugh, O.G., A.A. Nelson, and J.P. Frawley. 1950. The chronic toxicities of technical benzene hexachloride and its alpha, beta, and gamma isomers. J. Pharmacol. Exp. Therap. 100: 59-66.

Hanada, M., C. Yutani and T. Miyaji. 1973. Induction of hepatoma in mice by benzene hexachloride., Gann. 64: 511-513.

Hoshizaki, H., Y. Niki, H. Tajima, Y. Terada and A. Kasahara. 1969. A case of leukemia following exposure of insecticide. Nippon Ketsueki Gakkai Zasshi. 32(4): 672-677.

Ito, N., H. Nagasaki, M. Arai, S. Makiura, S. Sugihara and K. Hirao. 1973. Histopathologic studies on liver tumorigenesis induced in mice by technical polychlorinated biphenyls and its promoting effect on liver tumors induced by benzene hexachloride. J. Natl. Cancer Inst. 51(5): 1637-1646.

Ito, N., H. Nagasaki, H. Aoe, et al. 1975. Brief communication: Development of hepatocellular carcinomas in rats treated with benzene hexachloride. J. Natl. Cancer Inst. 54(3): 801-805.

Nagasaki, H., S. Tomii, T. Mega, M. Marugami and N. Ito. 1972. Carcinogenicity of benzene hexachloride. In: Proc. 2nd International Symposium of the Princess Takamatsu Cancer Research Fund. Topics in Chemical Carcinogenesis, W. Nakahara, S. Takayama, T. Sugimura and S. Odashima, Ed. University Park Press, Baltimore, MD. p. 343-353.

Thorpe, E. and A.I.T. Walker. 1973. Toxicology of dieldrin (HEOD). II. Comparative long-term oral toxicity studies in mice with dieldrin, DDT, phenobarbitone, beta-HCH and gamma-HCH. Food Cosmet. Toxicol. 11: 433-442.

US. EPA. 1986. Health and Environmental Effects Profile for Hexachloro-cyclohexanes. Prepared by the Office of Health and Environmental Assessment, Environmental Criteria and Assessment Office, Cincinnati, OH for the Office of Solid Waste and Emergency Response, Washington, DC.

VI.D. DRINKING WATER HA REFERENCES

None

VII. REVISION HISTORY

```
--------   --------   ----------------------------------------------------------
Date       Section    Description
--------   --------   ----------------------------------------------------------
01/01/90   VI.        Bibliography on-line
01/01/91   II.        Text edited
01/01/91   II.C.1.    Inhalation slope factor removed (global change)
01/01/92   IV.        Regulatory actions updated
07/01/93   II.D.3.    Secondary contact's phone number changed
```

SYNONYMS

BENZENEHEXACHLORIDE, trans-alpha-BETA-ISOMER
beta-BHC
CYCLOHEXANE, 1,2,3,4,5,6-HEXACHLORO-, beta-
CYCLOHEXANE, 1,2,3,4,5,6-HEXACHLORO-, trans-
CYCLOHEXANE, beta-1,2,3,4,5,6-HEXACHLORO-
CYCLOHEXANE, 1,2,3,4,5,6-HEXACHLORO-, beta-isomer
ENT 9,233

beta-HCH
beta-HEXACHLOROBENZENE
1-alpha,2-beta,3-alpha,4-beta,5-alpha,6-beta-HEXACHLOROCYCLOHEXANE
Hexachlorocyclohexane, beta-
beta-1,2,3,4,5,6-HEXACHLOROCYCLOHEXANE
beta-LINDANE
trans-alpha-BENZENEHEXACHLORIDE

Source: *Instant EPA's IRIS*, 1997, Instant Reference Sources, Inc., 7605 Rockpoint Drive, Austin, TX 78731 (Fax: 512-345-2386; Internet URL, http://www.instantref.com/inst-ref.htm).

Production, Use, and Pesticide Labeling Information

beta-BHC is not currently listed in EPA's Pesticide Factsheet database and there is no indication of whether it will be added or not in the future.

Bisphenol A

Introduction

Bisphenol A is an intermediate in the manufacture of polymers, epoxy resins, polycarbonates, fungicides, antioxidants, dyes, phenoxy, polysulfone and certain polyester resins, flame retardant and rubber chemicals. It may leach from plastic due to incomplete polymerization or breakdown of the polymer during heating, e.g., during sterilization by autoclaving. Bisphenol A is also used in plastic dental fillings and polycarbonate plastics are used to coat teeth, especially children's teeth. [813]

The primary sources of environmental release of bisphenol A are expected to be effluents and emissions from its manufacturing facilities and facilities which manufacture epoxy, polycarbonate, and polysulfone resins. The most probable routes of human exposure to bisphenol A are inhalation and dermal contact of workers involved in the manufacture, use, transport or packaging of this compound or use of epoxy powder paints. If released to soil bisphenol A is expected to have moderate to low mobility. It may biodegrade under aerobic conditions following acclimation. If released to the atmosphere, bisphenol A is expected to exist almost entirely in the particulate phase. Bisphenol A in particulate form may be removed from the atmosphere by dry deposition or photolysis. Photodegradation products of bisphenol A vapor are phenol, 4-isopropylphenol, and a semiquinone derivative of bisphenol A. [828]

Identifying Information

CAS NUMBER: 80-05-7

NIOSH Registry Number: SL6300000

SYNONYMS: 4,4'-ISOPROPYLIDENEDIPHENOL, BIS(4-HYDROXYPHENYL)DIMETHYLMETHANE, BIS(4-HYDROXYPHENYL)PROPANE, 2,2-BIS(P-HYDROXYPHENYL)PROPANE, 2,2-BIS(4-HYDROXYPHENYL)PROPANE, BISPHENOL, 4,4'-BISPHENOL A, DIAN, P,P'-DIHYDROXYDIPHENYLDIMETHYLMETHANE, 4,4'-DIHYDROXYDIPHENYLDIMETHYLMETHANE, P,P'-DIHYDROXYDIPHENYLPROPANE, 2,2-(4,4-DIHYDROXYDIPHENYL)PROPANE, 4,4'-DIHYDROXDIPHENYLPROPANE, 4,4'-DIHYDROXYDIPHENYL-2,2-PROPANE, 4,4'-DIHYDROXY-2,2-DIPHENYLPROPANE, DIMETHYLMETHYLENE-P,P'-DIPHENOL, BETA-DI-P-HYDROXYPHENYLPROPANE, 2,2-DI(4-HYDROXYPHENYL)PROPANE,

DIMETHYL BIS(P-HYDROXYPHENYL)METHANE, DIPHENYLOLPROPANE, 2,2-DI(4-PHENYLOL)PROPANE, P,P'-ISOPROPYLIDENEBISPHENOL, 4,4'-ISOPROPYLIDENEBISPHENOL, P,P'-ISOPROPYLIDENEDIPHENOL, NCI-C50635, 4,4'-DIMETHYLMETHYLENEDIPHENOL

Endocrine Disruptor Information

Bisphenol A has been known to exert estrogenic effects since 1938. [907] It is listed by EPA [811], the Centers for Disease Control & Prevention [812], and the World Wildlife Fund [813] as an endocrine modifying chemical. Bisphenol A has been shown to leach from polycarbonate, a common plastic used for many consumer products including jugs used to bottle drinking water. It binds to estrogen and two to five parts per billion of bisphenol A was enough to produce an estrogenic response in cell cultures. [810] Polycarbonates also are commonly used to line food cans and recent studies found bisphenol A in half the canned foods samples where the foods contained up to 80 parts per billion (0.080 parts per million), an amount 27 times more than the amount reported to make breast cancer cells proliferate. [813] A 1993 study showed it to be estrogenic using a MCF-7 human breast cancer cell culture assay. [908] Also recently, liquor obtained from tinned vegetables that contained bisphenol-A exhibited estrogenic effects with human breast cancer cells [909] and when bisphenol A was fed to pregnant rats the male pups had smaller testes and reduced sperm counts.

Chemical and Physical Properties

CHEMICAL FORMULA: $C_{15}H_{16}O_2$

MOLECULAR WEIGHT: 228.29

WLN: QR DX1&1&R DQ

PHYSICAL DESCRIPTION: White or tan crystals or flakes

SPECIFIC GRAVITY: 1.195 g/cm^3 @ 25/25°C [058], [062]

DENSITY: Not available

MELTING POINT: 153-156°C [047], [205], [269], [275]

BOILING POINT: 220°C @ 4 mm Hg [047], [062], [269], [275]

SOLUBILITY: Water: <1 mg/mL @ 21.5°C [700]; DMSO: >=100 mg/mL @ 25°C [700]; 95% Ethanol: >=100 mg/mL @ 25°C [700]; Acetone: >=100 mg/mL @ 25°C [700]

OTHER SOLVENTS: Carbon tetrachloride: Slightly soluble [043], [062]; Dilute alkalis: Soluble [043], [062]; Acetic acid: Soluble [017]; Ether: Soluble [017]; Benzene: Soluble [017]; Alcohol: Soluble [062]; Acids: Soluble [047]

OTHER PHYSICAL DATA: Boiling point: 250-252°C @ 13 mm Hg [025]; 222°C @ 3 mm Hg [047]; Decomposes @ pressures above 8 mm Hg when heated above 220°C

[058]; Mild phenolic odor [043], [062], [346]; Lambda max (95% ethanol): 285 nm (shoulder), 278 nm, 227 nm (epsilon:; 0.00308, 0.00359,0.01390) [052]

Source: *The National Toxicology Program's Chemical Database, Volume 2*, 1992, CRC Press, Inc./Lewis Publishers, 2000 Corporate Blvd., Boca Raton, FL 33431 (Fax: 407-998-9114).

Environmental Reference Materials

A source of EED environmental reference materials for Bisphenol A is Radian International LLC, Austin, TX (Fax 512-454-0268; Internet URL, http://www.radian.com/standards); Catalog No. ERB-061.

Hazardous Properties

ACUTE/CHRONIC HAZARDS: Bisphenol A may be toxic by ingestion, inhalation or skin absorption [269]. It may be an irritant and when heated to decomposition it emits toxic fumes of carbon monoxide and carbon dioxide [043], [269].

VOLATILITY: Vapor pressure: 0.2 mm Hg @ 170°C [058]; 1 mm Hg @ 193.0°C [038]

FIRE HAZARD: Bisphenol A has a flash point of 79.4°C (175°F) [062]. It is combustible. Fires involving this material can be controlled with a dry chemical, carbon dioxide or Halon extinguisher. A water spray may also be used [058], [371]. This compound may form explosive dust clouds [058]. Static electricity can cause its dust to explode[052]

REACTIVITY: Bisphenol A is incompatible with strong oxidizers [058], [269]. It is also incompatible with strong bases, acid chlorides and acid anhydrides. [269]

STABILITY: Bisphenol A is stable under normal laboratory conditions [052]. Solutions of bisphenol A in water, DMSO, 95% ethanol or acetone should be stable for 24 hours under normal lab conditions. [700]

Source: *The National Toxicology Program's Chemical Database, Volume 7*, 1992, CRC Press, Inc./Lewis Publishers, 2000 Corporate Blvd., Boca Raton, FL 33431 (Fax: 407-998-9114).

Medical Symptoms of Exposure

Symptoms of exposure may include irritation of the eyes, nose, throat, skin, mucous membranes and upper respiratory tract [269], [371]. It may cause sneezing [371]. It may cause contact dermatitis, redness and edema with weeping followed by crusting and scaling of the skin [346]. Skin sensitization may also occur [058]. Eye contact may lead to severe eye injury [099].

Source: *The National Toxicology Program's Chemical Database, Volume 4*, 1992, CRC Press, Inc./Lewis Publishers, 2000 Corporate Blvd., Boca Raton, FL 33431 (Fax: 407-998-9114).

Toxicological Information

The toxicological information below is gathered from several sources including two national databases: one from the *National Toxicology Program Chemical Repository Database* and another from *EPA's Integrated Risk Information System (IRIS)*.

Toxicology Information from the National Toxicology Program

CARCINOGENICITY:

    Status:

        NTP Carcinogenesis Bioassay (Feed); Equivocal: Male Rat
        NTP Carcinogenesis Bioassay (Feed); Negative: Female Rat, Male and Female Mouse

TERATOGENICITY: See RTECS printout for most current data

    Reproductive Effects Data:

Type of Effect	Route/Aminal	Amount Dosed (Notes)
TDLo	ipr-rat	1275 mg/kg (1-15D preg)
TDLo	ipr-rat	1875 mg/kg (1-15D preg)
TDLo	orl-mus	12500 mg/kg (6-15D preg)
TDLo	orl-mus	7500 mg/kg (6-15D preg)
TDLo	orl-rat	10 gm/kg (6-15D preg)
TDLo	orl-rat	15 gm/kg (6-15D preg)

MUTATION DATA: Not available

TOXICITY:

Typ. Dose	Mode	Specie	Amount	Units	Other
LD50	orl	rat	3250	mg/kg	
LC50	ihl	rat	200	ppm	
LD50	orl	mus	2500	mg/kg	
LD50	ipr	mus	150	mg/kg	
LD50	orl	rbt	2230	mg/kg	
LD50	skn	rbt	3000	mg/kg	
LD50	orl	mam	6500	mg/kg	

OTHER TOXICITY DATA:

> Skin and Eye Irritation Data:
> > skn-rbt 250 mg open MLD
> > eye-rbt 250 ug/24H SEV
> > skn-rbt 500 mg/24H MLD

> Status:
> > EPA TSCA Chemical Inventory, 1989
> > EPA TSCA Test Submission (TSCATS) Data Base, April 1990

SAX TOXICITY EVALUATION:

> THR: Poison by intraperitoneal route. Moderately toxic by ingestion and skin contact. An experimental teratogen. Other experimental reproductive effects. A skin and eye irritant.

Source: *Instant Tox-Base*, 1995, Instant Reference Sources, Inc., 7605 Rockpoint Drive, Austin, TX 78731 (Fax: 512-345-2386; Internet URL, http://www.instantref.com/inst-ref.htm).

Toxicology Information from EPA's Integrated Risk Information System (IRIS)

I. CHRONIC HEALTH HAZARD ASSESSMENTS FOR NONCARCINOGENIC EFFECTS

I.A. REFERENCE DOSE FOR CHRONIC ORAL EXPOSURE (RfD)

The Reference Dose (RfD) is based on the assumption that thresholds exist for certain toxic effects such as cellular necrosis, but may not exist for other toxic effects such as carcinogenicity. In general, the RfD is an estimate (with uncertainty spanning perhaps an order of magnitude) of a daily exposure to the human population (including sensitive subgroups) that is likely to be without an appreciable risk of deleterious effects during a lifetime. Please refer to Background Document 1 in Service Code 5 for an elaboration of these concepts. RfDs can also be derived for the noncarcinogenic health effects of compounds which are also carcinogens. Therefore, it is essential to refer to other sources of information concerning the carcinogenicity of this substance. If the U.S. EPA has evaluated this substance for potential human carcinogenicity, a summary of that evaluation will be contained in Section II of this file when a review of that evaluation is completed.

I.A.1. ORAL RfD SUMMARY

Critical Effect	Experimental Doses*	UF	MF	RfD
Reduced mean body weight	NOEL: None LOAEL: 1000 ppm diet (50 mg/kg/day)	1000	1	5E-2 mg/kg/day

*Conversion Factors: Assumed food consumption equivalent to 5% of body weight/day

Rat Chronic Oral Bioassay NTP, 1982

I.A.2. PRINCIPAL AND SUPPORTING STUDIES (ORAL RfD)

NTP (National Toxicology Program). 1982. NTP Technical report on the carcinogenesis bioassay of bisphenol A (CAS No. 80-05-7) in F344 rats and B6C3F1 mice (feed study). NTP-80-35. NIH Publ. No. 82-1771.

In this 103-week dietary study, groups of 50 rats/sex were fed diets containing 0, 1000, or 2000 ppm bisphenol A. All treated groups of rats had reduced body weights, compared with controls, evident from the 5th week of exposure. Food consumption was also reduced, compared with controls, but this effect was not observed until the 12th week of treatment. Reduced body weights in rats, therefore, was considered a direct adverse effect of exposure to bisphenol A.

In the same study (NTP, 1982), male mice (50/group) were fed diets containing 0, 1000, or 5000 ppm bisphenol A and female mice (50/group) were fed 0, 5000, or 10,000 ppm bisphenol A. Male mice at 5000 ppm and female mice at 5000 and 10,000 had reduced body weights. At 1000 and 5000 ppm, there was an increase in the number of multinucleated giant hepatocytes in male mice. This effect was not considered to be adverse, and this level is a NOAEL in mice. Assuming a food factor for mice of 0.13, this dietary concentration corresponds to a dosage of 130 mg/kg/day. Because the LOAEL of 50 mg/kg/day in rats is less than the NOAEL of 130 mg/kg/day in mice, the NOAEL in mice cannot be chosen as a basis for the RfD. The LOAEL of 50 mg/kg/day in rats, the lowest dosage used in either species in the chronic studies, is chosen as the basis for a chronic oral RfD.

I.A.3. UNCERTAINTY AND MODIFYING FACTORS (ORAL RfD)

UF -- The UF of 1000 includes 10 for uncertainty in the extrapolation of dose levels for animals to humans, 10 for uncertainty in the threshold for sensitive humans, and 10 for uncertainty in the effects of duration on toxicity when extrapolating for subchronic to chronic exposure.

MF -- None

I.A.4. ADDITIONAL COMMENTS (ORAL RfD)

Three subchronic oral toxicity studies of bisphenol A have been considered using dogs, rats and mice (U.S. EPA, 1984a,b,c; NTP, 1982). The only toxic effect seen in beagle dogs fed 1000-9000 ppm bisphenol A in the diet for 90 days was an increase in group mean liver weight in the high-dose group (U.S. EPA, 1984a). The only effect seen in 2-generation bisphenol A feeding studies (100-9000 ppm) conducted with Charles River rats (U.S. EPA, 1984b,c) were decreases in body weight in the F0 generation at 9000 ppm and F1 generation at greater than or equal to 1000 ppm. Rats and mice of both sexes were fed bisphenol A (250 to 4000 ppm rats; 5000to 25,000 ppm mice) in the diet for 90 days (NTP, 1982). Doses >1000 ppm produced decreased body weight in both sexes of rats with no alteration in food consumption. Male mice receiving >15,000 ppm and all treated females had decreased body weight gain compared with controls. A dose-related increase in severity of multinucleated giant hepatocytes was found in the treated male mice.

In mice, a dosage of 1250 mg/kg/day was associated with fetotoxicity and maternal toxicity, but did not cause a significant increase in the incidence of malformations at any dose level (NTP, 1985a). In rats, dosages of less than or equal to 1280 mg/kg/day were not toxic and did not cause malformations to the fetus (NTP, 1985b).

I.A.5. CONFIDENCE IN THE ORAL RfD

Study -- Medium
Data Base -- High
RfD -- High

Confidence in the key study is medium because this study, although well controlled and performed, failed to identify a chronic NOAEL for reduced body weight, the critical effect, in rats, the most sensitive species. Confidence in the data base is high, however, because the subchronic studies in rats indicate that the NOAEL for reduced body weight in rats is probably not far below the LOAEL of 1000 ppm of the diet and the uncertainty factor of 10 to estimate a NOAEL from the LOAEL is probably conservative. The developmental toxicity of bisphenol A has been adequately investigated. Confidence in the RfD, therefore, is high.

I.A.6. EPA DOCUMENTATION AND REVIEW OF THE ORAL RfD

Source Document -- U.S. EPA, 1987

Other EPA Documentation -- U.S. EPA, 1984a,b,c

Agency Work Group Review -- 09/16/87, 03/24/88, 04/20/88

Verification Date -- 04/20/88

I.A.7. EPA CONTACTS (ORAL RfD)

Harlal Choudhury / NCEA -- (513)569-7536

Chris Cubbison / NCEA -- (513)569-7599

I.B. REFERENCE CONCENTRATION FOR CHRONIC INHALATION EXPOSURE (RfC)

Not available at this time.

II. CARCINOGENICITY ASSESSMENT FOR LIFETIME EXPOSURE

This substance/agent has not been evaluated by the U.S. EPA for evidence of human carcinogenic potential.

III. HEALTH HAZARD ASSESSMENTS FOR VARIED EXPOSURE DURATIONS

[Ed. Note: EPA does not have information yet in this section of IRIS].

IV. U.S. EPA REGULATORY ACTIONS

[Ed. Note: EPA does not have information yet in this section of IRIS].

VI. BIBLIOGRAPHY

VI.A. ORAL RfD REFERENCES

NTP (National Toxicology Program). 1982. NTP Technical Report on the carcinogenesis bioassay of bisphenol A (CAS No. 80-05-7) in F344 rats and B6C3F1 mice (feed study). NTP-80-35. NIH Publ. No. 82-1771.

NTP (National Toxicology Program). 1985a. Teratologic evaluation of bisphenol A (CAS No. 80-05-7) administered to CD-1 mice on gestational days6-15. NTP, NIEHS, Research Triangle Park, NC.

NTP (National Toxicology Program). 1986a. Teratologic evaluation of bisphenol A (CAS No. 80-05-7) administered to CD(R) rats on gestational days 6-15. NTP, NIEHS, Research Triangle Park, NC.

U.S. EPA. 1984a. Ninety-day oral toxicity study in dogs. Office of Pesticides and Toxic Substances. Fiche No. OTS0509954.

U.S. EPA. 1984b. Reproduction and ninety-day oral toxicity study in rats. Office of Pesticides and Toxic Substances. Fiche No. OTS0509954.

U.S. EPA. 1984c. Fourteen-day range finding study in rats. Office of Pesticides and Toxic Substances. Fiche No. OTS0509954.

U.S. EPA. 1987. Health and Environmental Effects Document on Bisphenol A. Prepared by the Office of Health and Environmental Assessment, Environmental Criteria and Assessment Office, Cincinnati, OH for the Office of Solid Waste and Emergency Response, Washington, DC.

VI.B. INHALATION RfD REFERENCES

None

VI.C. CARCINOGENICITY ASSESSMENT REFERENCES

None

VI.D. DRINKING WATER HA REFERENCES

None

VII. REVISION HISTORY

```
--------  --------  ---------------------------------------------------------
Date      Section   Description
--------  --------  ---------------------------------------------------------
09/26/88  I.A.      Oral RfD summary on-line
07/01/92  I.A.2.    Principal study clarified
07/01/92  I.A.4.    Citations clarified
07/01/92  VI.A.     Oral RfD references on-line
07/01/93  I.A.6.    Other EPA Documentation clarified
```

SYNONYMS

bisferol A
Bishpenol A.
2,2-bis-4'-hydroxyfenylpropan
bis(4-hydroxyphenyl) dimethylmethane
2,2-bis(4-hydroxyphenyl)propane
2,2-bis(p-hydroxyphenyl)propane
bis(4-hydroxyphenyl)propane
bisphenol
bisphenol A
Bisphenol A.
4,4'-bisphenol A
dian
4,4'-dihydroxydiphenyldimethylmethane
p,p'-dihydroxydiphenyldimethylmethane
2,2-(4,4'-dihydroxydiphenyl)propane
4,4'-dihydroxydiphenylpropane

4,4'-dihydroxydiphenyl-2,2-propane
p,p'-dihydroxydiphenylpropane
2,2-di(4-hydroxyphenyl)propane
beta-di-p-hydroxyphenylpropane
dimethyl bis(p-hydroxyphenyl)methane
dimethylmethylene-p,p'-diphenol
diphenylolpropane
2,2-di(4-phenylol)propane
4,4'-isopropylidenebisphenol
p,p'-isopropylidenebisphenol
p,p'-isopropylidenediphenol
NCI-C50635
phenol, 4,4'-dimethylmethylenedi-
phenol, 4,4'-isopropylidenedi-
propane, 2,2-bis(p-hydroxyphenyl)-

Source: *Instant EPA's IRIS*, 1997, Instant Reference Sources, Inc., 7605 Rockpoint Drive, Austin, TX 78731 (Fax: 512-345-2386; Internet URL, http://www.instantref.com/inst-ref.htm).

Butyl benzyl phthalate

Introduction

Phthalates, best known for their ability to make plastics flexible are the most abundant industrial contaminants in the environment. [811] Butyl benzyl phthalate is used as a plasticizer for cellulosic resins, polyvinyl acetates, polyurethanes and polysulfides and in regenerated cellulose films for packaging. It is also used in vinyl products such as synthetic leathers, floor tiles, acrylic caulking, adhesive for medical devices and in the cosmetic industry as well as a dispersent and carrier for insecticides and repellents. [813] In addition it is used as an organic intermediate, a solvent and a fixative in perfume.

Possible sources of butyl benzyl phthalate release to the environment are from its manufacture, distribution, and PVC blending operations; however, release from consumer products is expected to be minimal. Most butyl benzyl phthalate releases will be to soil and water and not to the air. Butyl benzyl phthalate released to soil is expected to adsorb and not to leach extensively although it has been detected in groundwater. Butyl benzyl phthalate released to aquatic systems will adsorb to sediments and biota but will not volatilize significantly except under windy conditions or from shallow rivers. Biodegradation appears to be the primary fate mechanism for butyl benzyl phthalate. It is readily biodegraded in activated sludge, semicontinous activated sludge, salt water, lake water, and under anaerobic conditions. Over 66,000 US workers are potentially exposed to butyl benzyl phthalate while a larger population of the US may be exposed to lower concentrations of it found in some drinking waters. [828]

Identifying Information

CAS NUMBER: 85-68-7

SYNONYMS: BENZYL BUTYL PHTHALATE, BENZYL N-BUTYL PHTHALATE, N-BUTYL BENZYL PHTHALATE, 1,2-BENZENEDICARBOXYLIC ACID, BUTYL PHENYLMETHYL ESTER, BUTYL PHENYLMETHYL 1,2-BENZENEDICARBOXYLATE, PHTHALIC ACID, BENZYL BUTYL ESTER, PHTHALIC ACID, BUTYL BENZYL ESTER, BENZYL BUTYL-1,2-PHTHALATE, SANTICIZER 160, BBP, PALATINOL BB, SICOL 160, UNIMOLL BB, NCI-C54375

Endocrine Disruptor Information

Butyl benzyl phthalate is a strongly suspected EED that it is listed by EPA [811], the Centers for Disease Control & Prevention [812], and the World Wildlife Fund [813] as a potential endocrine modifying chemical. Susan Jobling of Brunel University in Uxbridge, England has studied butyl benzyl phthalate (BBP) and di-n-butyl phthalate (DBP) for their estrogenicity. It seems that not only do they appear estrogenic, but they also stimulate the growth of breast cancer cells in culture. [864], [811]

Chemical and Physical Properties

CHEMICAL FORMULA: $C_{19}H_{20}O_4$

MOLECULAR WEIGHT: 312.39

WLN: 4OVR BVO1R

PHYSICAL DESCRIPTION: Clear oily liquid

SPECIFIC GRAVITY: 1.116 @ 25/25°C [043], [051], [053]

DENSITY: Not available

MELTING POINT: <-35°C [043], [051], [055]

BOILING POINT: 370°C [043], [051], [052], [055]

SOLUBILITY: Water: <0.1 mg/mL @ 22.5°C [700]; DMSO: >=100 mg/mL @ 23°C [700]; 95% Ethanol: >=100 mg/mL @ 23°C [700]; Acetone: >=100 mg/mL @ 23°C [700]

OTHER SOLVENTS: Most organic solvents: Soluble [058]

OTHER PHYSICAL DATA: Refractive index: 1.535-1.540 @ 25°C [395]; Slight odor [053], [058], [062], [371]; Heat of combustion: -8090 cal/g [371]; Boiling point: 240°C @ 10 mm Hg [058]; Vapor pressure: 0.16 mm Hg @ 150°C [052], [058]; 14.4 mm Hg @ 250°C [058]; Surface tension: 39.9 dynes/cm @ 25°C [058]; Lambda max (nm, in methanol): 281 (shoulder), 274.5, 268, 263.8, 257; (shoulder), 230 (shoulder) [052]

Source: *The National Toxicology Program's Chemical Database, Volume 2*, 1992, CRC Press, Inc./Lewis Publishers, 2000 Corporate Blvd., Boca Raton, FL 33431 (Fax: 407-998-9114).

Environmental Reference Materials

A source of EED environmental reference materials for butyl benzyl phthalate is Radian International LLC, Austin, TX (Fax 512-454-0268; Internet URL, http://www.radian.com/standards); Catalog No. ERB-063.

Hazardous Properties

ACUTE/CHRONIC HAZARDS: Butyl benzyl phthalate may be an irritant of the mucous membranes [151]. When heated to decomposition it emits acrid smoke and irritating fumes of carbon monoxide and carbon dioxide [043], [058].

FIRE HAZARD: Butyl benzyl phthalate has a flash point of 198°C (390°F) [043], [051], [062], [371]. It is combustible. Fires involving this material can be controlled with a dry chemical, carbon dioxide or Halon extinguisher.

The autoignition temperature is 233°C (451°F) [051].

LEL: 1.2% @ 233°C [058]   UEL: Not available

REACTIVITY: Butyl benzyl phthalate is incompatible with oxidizing materials [043], [051], [053], [058]. It can also react with oxygen, peroxides and nitrates [107].

STABILITY: Butyl benzyl phthalate may be sensitive to light. NMR stability screening indicates that solutions of butyl benzyl phthalate in DMSO are stable for at least 24 hours [700].

Butyl benzyl phthalate is not currently listed in NTP's Chemical Repository database nor is it scheduled for addition in the future.

Source: *The National Toxicology Program's Chemical Database, Volume 7*, 1992, CRC Press, Inc./Lewis Publishers, 2000 Corporate Blvd., Boca Raton, FL 33431 (Fax: 407-998-9114).

Medical Symptoms of Exposure

Symptoms of exposure may include irritation of the skin, eyes and mucous membranes [151], [371]. It may cause central nervous system depression [151]. It may also cause respiratory distress, lightheadedness, nausea, sore throat, sleepiness and erythema [107].

Source: *The National Toxicology Program's Chemical Database, Volume 4*, 1992, CRC Press, Inc./Lewis Publishers, 2000 Corporate Blvd., Boca Raton, FL 33431 (Fax: 407-998-9114).

Toxicological Information

The toxicological information below is gathered from several sources including two national databases: one from the *National Toxicology Program Chemical Repository Database* and another from *EPA's Integrated Risk Information System (IRIS)*.

Toxicology Information from the National Toxicology Program

CARCINOGENICITY:

Tumorigenic Data:

Type of Effect	Route/Animal	Amount Dosed (Notes)
TDLo	orl-rat	433 gm/kg/2Y-C
TD	orl-rat	437 gm/kg/2Y-C

Review:

IARC Cancer Review: Animal Inadequate Evidence
IARC: Not classifiable as a human carcinogen (Group 3)

Status:

NTP Carcinogenesis Bioassay (Feed); Inadequate Study: Male Rat
NTP Carcinogenesis Bioassay (Feed); Positive: Female Rat
NTP Carcinogenesis Bioassay (Feed); Negative: Male and Female Mouse

TERATOGENICITY: See RTECS printout for most current data

Reproductive Effects Data:

Type of Effect	Route/Animal	Amount Dosed (Notes)
TDLo	orl-rat	21 gm/kg (14D male)

MUTATION DATA: Not available

TOXICITY:

Typ. Dose	Mode	Specie	Amount	Units	Other
LD50	ipr	mus	3160	mg/kg	
LD50	orl	gpg	13750	mg/kg	
LD50	orl	rat	2330	mg/kg	
LD50	orl	mus	4170	mg/kg	

OTHER TOXICITY DATA:

Status:

EPA TSCA Chemical Inventory, 1989
EPA TSCA 8(a) Preliminary Assessment Information, Final Rule
EPA TSCA Test Submission (TSCATS) Data Base, April 1990

SAX TOXICITY EVALUATION:

> THR: Moderately toxic by ingestion and intraperitoneal routes. An experimental carcinogen. Experimental reproductive effects.

Source: *Instant Tox-Base*, 1995, Instant Reference Sources, Inc., 7605 Rockpoint Drive, Austin, TX 78731 (Fax: 512-345-2386; Internet URL, http://www.instantref.com/inst-ref.htm).

Toxicology Information from EPA's Integrated Risk Information System (IRIS)

I.   CHRONIC HEALTH HAZARD ASSESSMENTS FOR NONCARCINOGENIC EFFECTS

I.A. REFERENCE DOSE FOR CHRONIC ORAL EXPOSURE (RfD)

The Reference Dose (RfD) is based on the assumption that thresholds exist for certain toxic effects such as cellular necrosis, but may not exist for other toxic effects such as carcinogenicity. In general, the RfD is an estimate (with uncertainty spanning perhaps an order of magnitude) of a daily exposure to the human population (including sensitive subgroups) that is likely to be without an appreciable risk of deleterious effects during a lifetime. Please refer to Background Document 1 in Service Code 5 for an elaboration of these concepts. RfDs can also be derived for the noncarcinogenic health effects of compounds which are also carcinogens. Therefore, it is essential to refer to other sources of information concerning the carcinogenicity of this substance. If the US. EPA has evaluated this substance for potential human carcinogenicity, a summary of that evaluation will be contained in Section II of this file when a review of that evaluation is completed.

I.A.1. ORAL RfD SUMMARY

Critical Effect	Experimental Doses	UF	MF	RfD
Significantly increased liver-to-body weight and liver-to-brain weight ratios	NOAEL: 2800 ppm (159 mg/kg/day) LOAEL: 8300 ppm (470 mg/kg/day)	1000	1	2E-1 mg/kg/day

*Conversion Factors: approximately 300 g bw and 17 g of food consumption/dayfrom data presented in the report

6-Month Rat Study Oral Exposure (diet) NTP, 1985

I.A.2. PRINCIPAL AND SUPPORTING STUDIES (ORAL RfD)

NTP (National Toxicology Program). 1985. Twenty-six week subchronic study and modified mating trial in F344 rats. Butyl benzyl phthalate. Final Report. Project No. 12307-02, -03. Hazelton Laboratories America, Inc. Unpublished study.

NTP (1985) conducted a toxicity study in F344 rats in which 15 males/group were administered concentrations of either 0, 0.03, 0.09, 0.28, 0.83, or 2.5% BBP in the diet for 26 weeks. Using body weight and food consumption data presented in the report these dietary levels correspond to 0, 17, 51, 159, 470, and 1417 mg/kg/day, respectively. In this study powdered rodent meal was provided in such a way that measured food consumption at the highest dose level could include significant waste and spillage rather than true food intake. For this reason a standard food consumption rate of 5% rat body weight was used in the 2.5% dose conversion. Throughout the study body weight gain was significantly depressed at the 2.5% BBP level when compared with the controls. There were no deaths attributed to BBP toxicity. All the rats given 2.5% BBP had small testes upon gross necropsy; 5/11 had soft testes and 1/11 had a small prostate and seminal vesicle. In the 0.03, 0.09, 0.28, and 0.83% dose groups there were no grossly observable effects on male reproductive organs. Terminal mean organ weight values were significantly decreased ($p<0.05$) for the heart, kidney, lungs, seminal vesicles and testes in the 2.5% group. Hematological effects at 2.5% BBP included decreased red cell mass (which the authors state is indicative of deficient hemoglobin synthesis), reduced values for hemoglobin, total RBC and hematocrit. The kidneys of six animals in the 2.5% group contained focal cortical areas of infarct-like atrophy. In addition, testicular lesions were also observed at the 2.5% dose level. Lesions were characterized by atrophy of seminiferous tubules and aspermia. At 0.83% the effects noted were significantly ($p<0.05$) increased absolute liver weight, increased liver-to-body weight and liver-to-brain weight ratios and increases in mean corpuscular hemoglobin. The 0.03, 0.09, 0.28, and 0.83% treatment groups showed no evidence of abnormal morphology in any organ. No adverse effects were observed at the 0.28% treatment level or below.

The only other information on subchronic effects is reported by Krauskopf (1973) from an unpublished study by Monsanto (1972). Rats fed diets containing 0.25% (125 mg/kg/day) and 0.5% (250 mg/kg/day) for 90 days showed no toxic effects. Liver weights were increased in animals fed diets containing 1.0, 1.5, or 2.0% (500, 750, or 1000 mg/kg/day, respectively) for 90 days, and a mild decrease in growth rate was reported for the 1.5 and 2.0% groups. No other hematologic, histopathologic or urinalysis effects were observed. When dogs were administered gelatin capsules containing doses equivalent to 1.0, 2.0, or 5.0% of the daily diet (10,000 20,000 and 50,000 ppm) for 90 days, no effect on hematological parameters, urinalysis or liver and kidney functions were observed. No further details of this study were available for review.

Similar LOAELs of 470 and 500 mg/kg/day for increased liver weight were identified in both the NTP (1985) and Monsanto (1972) studies, respectively. NOAELs differ slightly: 159 (NTP, 1985) versus 250 mg/kg/day (Monsanto, 972). It is recommended that the NOAEL of 159 mg/kg/day from the NTP (1985) study be used to derive the RfD for two reasons: 1) the NTP (1985) study is of longer duration, and 2) The Monsanto (1972) study provides an incomplete description of methods comparing study design and clinical analysis. Treatment-related effects across similar dose ranges including liver effects in both studies support the use of 159 mg/kg/day as a NOAEL.

I.A.3. UNCERTAINTY AND MODIFYING FACTORS (ORAL RfD)

UF -- 10 for intraspecies sensitivity, 10 for interspecies variability and 10 for extrapolating from subchronic to chronic NOAELs.

MF -- None

I.A.4. ADDITIONAL COMMENTS (ORAL RfD)

Two 14-day studies support the selection of the NTP (1985) bioassay for deriving the oral RfD. Agarwal et al. (1985) administered BBP to male F344 rats in the diet for 14 consecutive days at dose levels of 0.625, 1.25, 2.5, and 5.0%. Significant increases in liver and kidney weights and kidney pathology (proximal tubular regeneration) was observed at 0.625% (375 mg/kg/day), which represents a LOAEL.

In male Sprague-Dawley rats administered 160, 480, or 1600 mg/kg/day BBP for 14 days by gastric intubation, biochemical or morphological changes in the liver as well as effects on testes weights were not observed in the 160 mg/kg/day dose group (Lake et al., 1978). However, at 480 mg/kg/day activities of ethyl morphine N-demethylase and cytochrome oxidase were significantly increased and testicular atrophy was observed in one-third Sprague-Dawley rats in the first portion of this experiment. In the second position, the 480 mg/kg/day dose induced testicular atrophy in one-sixth Sprague-Dawley rats, whereas the Wistar albino strain revealed no such effects. A NOAEL for this study would be 160 mg/kg/day based on the absence of liver and testicular effects.

In an addendum to the NTP (1985) final report, evaluation of the data revealed a significantly reduced total marrow cell count in the 2.5% dose group (NTP, 1986). The change in total cell count was comprised primarily of significant decreases in neutrophil, metamyelocytes, bands, segmeters, lymphocytes and leasophilic rubricytes. The total marrow cell counts, metamyelocyte, and leasophilic rubricyte counts were also significantly decreased in the lowest dose group, 0.03%. No statistically significant differences were noted in the middle dose groups (0.09, 0.28, or 0.83%) when compared with controls. The addendum states that decreased total marrow cell count in the 0.03 and 2.5% dose group represent change of uncertain meaning in light of the systemic effects noted in the middle dose groups. Trend analysis by the Terpstra-Jonckheere test revealed significantly (p<0.05%) decreasing trends in all of the previously mentioned parameters as well as an increasing trend for monocytes at 0.03 and 2.5%.

NTP (1985) also conducted a male mating trial study concomitantly with the toxicity study. Testicular atrophy was observed in male F344 rats after 10 weeks of exposure to 2.5% (2875 mg/kg/day) BBP. Throughout the study, body weight gain was significantly depressed at the 2.5% BBP level when compared with the controls.

I.A.5. CONFIDENCE IN THE ORAL RfD

Study -- Medium
Data Base -- Low
RfD -- Low

The critical study is of adequate quality and is given a medium confidence rating. Since the critical study used only male rats and there are no adequate supporting studies of chronic duration, the data base is given a low confidence rating. Low confidence in the RfD follows.

I.A.6. EPA DOCUMENTATION AND REVIEW OF THE ORAL RfD

Source Document -- US. EPA, 1987a,b; 1988

US. EPA (1987a, 1988) have been both OHEA reviewed and Agency reviewed. US. EPA (1987b) has been OHEA reviewed.

Other EPA Documentation -- None

Agency Work Group Review -- 06/15/89

Verification Date -- 06/15/89

I.A.7. EPA CONTACTS (ORAL RfD)

Brian J. Commons / OST -- (202)260-7589

Linda R. Papa / NCEA -- (513)569-7587

I.B. REFERENCE CONCENTRATION FOR CHRONIC INHALATION EXPOSURE (RfC)

[Ed. Note: EPA does not have information yet in this section of IRIS].

II. CARCINOGENICITY ASSESSMENT FOR LIFETIME EXPOSURE

Section II provides information on three aspects of the carcinogenic risk assessment for the agent in question; the US. EPA classification, and quantitative estimates of risk from oral exposure and from inhalation exposure. The classification reflects a weight-of-evidence judgment of the likelihood that the agent is a human carcinogen. The quantitative risk estimates are presented in three ways. The slope factor is the result of application of a low-dose extrapolation procedure and is presented as the risk per (mg/kg)/day. The unit risk is the quantitative estimate in terms of either risk per ug/L drinking water or risk per ug/cu.m air breathed. The third form in which risk is presented is a drinking water or air concentration providing cancer risks of 1 in 10,000, 1 in 100,000 or 1 in 1,000,000. Background Document 2 (Service Code 5) provides details on the rationale and methods used to derive the carcinogenicity values found in IRIS. Users are referred to Section I for information on long-term toxic effects other than carcinogenicity.

II.A. EVIDENCE FOR CLASSIFICATION AS TO HUMAN CARCINOGENICITY

II.A.1. WEIGHT-OF-EVIDENCE CLASSIFICATION

Classification -- C; possible human carcinogen.

Basis -- Based on statistically significant increase in mononuclear cell leukemia in female rats; the response in male rats was inconclusive and there was no such response in mice.

II.A.2. HUMAN CARCINOGENICITY DATA

None.

II.A.3. ANIMAL CARCINOGENICITY DATA

Limited. A bioassay was performed by the NTP (1982) to evaluate the carcinogenic potential of orally administered butyl benzyl phthalate (BBP) to both rats and mice. Dietary levels of 0, 6000, and 12,000 ppm BBP were fed to groups of 50 male and 50 female F344 rats and 50 male and 50 female B6C3F1 mice for 103 weeks. The male rats at both dose levels experienced high mortality within the first 30 weeks of the study due to apparent internal hemorrhaging; all male rats were, thus, terminated at 30 weeks. No chronic toxicity or carcinogenic effects were observed in male or female mice. Among female rats a statistically significant increase in mononuclear cell leukemia (MCL) or lymphoma (p=0.007) was observed at the high dose level compared with controls with an increasing trend at p=0.006. The time to first tumor was 83 weeks in control as well as in the high-dose group. NTP indicated that BBP was "probably" carcinogenic in female rats. The tumor incidence was 7/49 (14%) for controls, 7/49 (14%) in low dose and 19/50 (38%) in the high dose as compared with historical control incidence in the laboratory of 19% (12-24% range).

Given the similarity of the MCL pathology in the control and the dosed female rats as well as the absence of a reduction in time to first tumor, the response is judged to be an acceleration of an old age tumor in the F344 rats. This weakens somewhat the interpretive value of the MCL response. The NTP has initiated a retest in the rats.

BBP did not induce lung adenomas in strain A mice administered 24 intraperitoneal injections of 160, 400 or 800 mg/kg (Theiss et al., 1977).

II.A.4. SUPPORTING DATA FOR CARCINOGENICITY

Studies indicate that BBP is not a direct acting mutagen in the reverse mutation assay in Salmonella typhimurium (Rubin et al., 1979; Kozumbo et al., 1982; Zeiger et al., 1982) or in E. coli (NTP, 1982). Mammalian cytogenicity studies using Chinese hamster ovary cells were also negative (NTP, 1982). NTP (1982) noted that additional studies on metabolites, benzyl alcohol and butanol was important.

II.B. QUANTITATIVE ESTIMATE OF CARCINOGENIC RISK FROM ORAL EXPOSURE

Not available. The qualitative weaknesses of the MCL response does not provide a compelling basis to model the dose-response data.

II.C. QUANTITATIVE ESTIMATE OF CARCINOGENIC RISK FROM INHALATION EXPOSURE

[Ed. Note: EPA does not have information yet in this section of IRIS].

II.D. EPA DOCUMENTATION, REVIEW, AND CONTACTS (CARCINOGENICITY ASSESSMENT)

II.D.1. EPA DOCUMENTATION

Source Document -- US. EPA, 1987

The 1987 Draft Drinking Water Quality Criteria Document received Agency review.

II.D.2. REVIEW (CARCINOGENICITY ASSESSMENT)

Agency Work Group Review -- 08/26/87

Verification Date -- 08/26/87

II.D.3. US. EPA CONTACTS (CARCINOGENICITY ASSESSMENT)

Brian J. Commons / OST -- (202)260-7589

Linda R. Papa / NCEA -- (513)569-7587

III. HEALTH HAZARD ASSESSMENTS FOR VARIED EXPOSURE DURATIONS

III.A. DRINKING WATER HEALTH ADVISORIES

[Ed. Note: EPA does not have information yet in this section of IRIS].

IV. US. EPA REGULATORY ACTIONS

EPA risk assessments may be updated as new data are published and as assessment methodologies evolve. Regulatory actions are frequently not updated at the same time. Compare the dates for the regulatory actions in this section with the verification dates for the risk assessments in sections I and II, as this may explain inconsistencies. Also note that some regulatory actions consider factors not related to health risk, such as technical or economic feasibility. Such considerations are indicated for each action. In addition, not all of the regulatory actions listed in this section involve enforceable federal standards. Please direct any questions you may have concerning these regulatory actions to the US. EPA contact listed for that particular action. Users are strongly urged to read

the background information on each regulatory action in Background Document 4 in Service Code 5.

IV.A. CLEAN AIR ACT (CAA)

[Ed. Note: EPA does not have information yet in this section of IRIS].

IV.B. SAFE DRINKING WATER ACT (SDWA)

IV.B.1. MAXIMUM CONTAMINANT LEVEL GOAL (MCLG) for Drinking Water

Value -- 0.1 mg/L (Proposed, 1990)

Considers technological or economic feasibility? -- NO

Discussion -- A MCLG of 0.1 mg/L for butyl benzyl phthalate is proposed basedon its potential adverse effects (hepatic toxicity) reported in a 26-weekfeeding study in rats. The MCLG is based upon a DWEL of 6.0 mg/L and an assumed drinking water contribution of 20 percent.

Reference -- 55 FR 30370 (07/25/90)

EPA Contact -- Health and Ecological Criteria Division / OST /(202) 260-7571 / FTS 260-7571; or Safe Drinking Water Hotline / (800) 426-4791

IV.B.2. MAXIMUM CONTAMINANT LEVEL (MCL) for Drinking Water

Value -- 0.1 mg/L (Proposed, 1990)

Considers technological or economic feasibility? -- YES

Discussion -- EPA is proposing an MCL equal to the proposed MCLG of 0.1 mg/L.

Monitoring requirements -- Community and non-transient water system monitoring based on state vulnerability assessment; vulnerable systems to be monitored quarterly for one year; repeat monitoring dependent upon detection and system size.

Analytical methodology -- Photoionization/gas chromatography (EPA 506); gas chromatography/mass spectrometry (EPA 525). PQL= 0.002 mg/L.

Best available technology -- Granular activated carbon

Reference -- 55 FR 30370 (07/25/90)

EPA Contact -- Drinking Water Standards Division / OGWDW /(202) 260-7575 / FTS 260-7575; or Safe Drinking Water Hotline / (800) 426-4791

IV.B.3. SECONDARY MAXIMUM CONTAMINANT LEVEL (SMCL) for Drinking Water

[Ed. Note: EPA does not have information yet in this section of IRIS].

IV.B.4. REQUIRED MONITORING OF "UNREGULATED" CONTAMINANTS

[Ed. Note: EPA does not have information yet in this section of IRIS].

IV.C. CLEAN WATER ACT (CWA)

IV.C.1. AMBIENT WATER QUALITY CRITERIA, Human Health

[Ed. Note: EPA does not have information yet in this section of IRIS].

IV.C.2. AMBIENT WATER QUALITY CRITERIA, Aquatic Organisms

Freshwater:
    Acute LEC -- 9.4E+2 ug/L
    Chronic LEC -- 3.0E+0 ug/L

Marine:
    Acute LEC -- 2.944E+3 ug/L
    Chronic LEC -- None

Considers technological or economic feasibility? -- NO

Discussion -- The values that are indicated as "LEC" are not criteria, but are the lowest effect levels found in the literature. LECs are given when the minimum data required to derive water quality criteria are not available. The criteria are based on phthalate esters as a class.

Reference -- 45 FR 79318 (11/28/80)

EPA Contact -- Criteria and Standards Division / OWRS(202)260-1315 / FTS 260-1315

IV.D. FEDERAL INSECTICIDE, FUNGICIDE, AND RODENTICIDE ACT (FIFRA)

[Ed. Note: EPA does not have information yet in this section of IRIS].

IV.E. TOXIC SUBSTANCES CONTROL ACT (TSCA)

[Ed. Note: EPA does not have information yet in this section of IRIS].

IV.F. RESOURCE CONSERVATION AND RECOVERY ACT (RCRA)

IV.F.1. RCRA APPENDIX IX, for Ground Water Monitoring

Status -- Listed

Reference -- 52 FR 25942 (07/09/87)

EPA Contact -- RCRA/Superfund Hotline (800)424-9346 / (202)260-3000 / FTS 260-3000

IV.G. SUPERFUND (CERCLA)

IV.G.1. REPORTABLE QUANTITY (RQ) for Release into the Environment

Value (status) -- 100 pounds (Final, 1989)

Considers technological or economic feasibility? -- NO

Discussion -- The final RQ for butyl benzyl phthalate is based on aquatic toxicity. The available data indicate that the aquatic 96-Hour Median Threshold Limit is 2-40 ppm, which corresponds to an RQ of 100 pounds.

Reference -- 54 FR 33418 (08/14/89)

EPA Contact -- RCRA/Superfund Hotline (800)424-9346 / (202)260-3000 / FTS 260-3000

VI. BIBLIOGRAPHY

VI.A. ORAL RfD REFERENCES

Agarwal, D.K., R.R. Maronpot, J.C. Lamb, IV and W.M. Kluwe. 1985. Adverse effects of butyl benzyl phthalate on the reproductive and hematopoietic systems of male rats. Toxicology. 35: 189-206.

Krauskopf, L.G. 1973. Studies on the toxicity of phthalates via ingestion. Environ. Health Perspect. 3: 61-72.

Lake, B.G., R.A. Harris, P. Grasso and S.D. Gangollia. 1978. Studies on the metabolism and biological effects of n-butyl benzyl phthalate in the rat. Prepared by British Industrial Biological Research Association for Monsanto, Report No. 232/78, June.

Monsanto Chemical Company. 1972. Unpublished work. (Cited in Krauskopf, 1973)

NTP (National Toxicology Program). 1985. Twenty-six week subchronic study and modified mating trial in F344 rats. Butyl benzyl phthalate. Final Report. Project No. 12307-02, -03. Hazelton Laboratories America, Inc. Unpublished study.

NTP (National Toxicology Program). 1986. Addendum to Final Report. Bone marrow differential results - 26 Week study. LBI/HLA Project No. 12307-02. Hazelton Laboratories America, Inc. Unpublished report.

US. EPA. 1987a. Health and Environmental Profile for Phthalic Acid Alkyl, Aryl and Alkyl/Aryl Esters. Prepared by the Office of Health and Environmental Assessment, Environmental Criteria and Assessment Office, Cincinnati, OH for the Office of Solid Waste and Emergency Response, Washington, DC.

US. EPA. 1987b. Health Effects Assessment for Selected Phthalic Acid Esters. Prepared by the Office of Health and Environmental Assessment, Environmental Criteria and Assessment Office, Cincinnati, OH for the Office of Emergency and Remedial Response, Washington, DC.

US. EPA. 1988. Drinking Water Criteria Document for Phthalic Acid Esters (PAEs). Prepared by the Office of Health and Environmental Assessment Environmental Criteria and Assessment Office, Cincinnati, OH for the Office of Drinking Water, Washington, DC. Final Draft.

VI.B. INHALATION RfD REFERENCES
None

VI.C. CARCINOGENICITY ASSESSMENT REFERENCES

Kozumbo, W.J., R. Kroll and R.J. Rubin. 1982. Assessment of the mutagenicity of phthalate esters. Environ. Health Perspect. 45: 103-109.

NTP (National Toxicology Program). 1982. Carcinogenesis Bioassay of Butyl Benzyl Phthalate (CAS No. 85-68-7) in F344 Rats and B6C3F1 Mice (Feed Study). NTP Tech. Rep. Ser. TR No. 213, NTP, Research Triangle Park, NC. p. 98

Rubin, R.J., W. Kozumbo and R. Kroll. 1979. Ames mutagenic assay of a series of phthalic acid esters: Positive response of the dimethyl and diethyl esters in TA100. Soc. Toxicol. Ann. Meet., New Orleans, March 11-15. p. 11. (Abstract)

Theiss, J.C., G.D. Stoner, M.B. Shimkin and E.K. Weisburger. 1977. Test for carcinogenicity of organic contaminants of United States drinking waters by pulmonary tumor response in strain A mice. Cancer Res. 37: 2717-2720.

US. EPA. 1987. Drinking Water Criteria Document for Phthalic Acid Esters. Prepared by the Office of Health and Environmental Assessment, Environmental Criteria and Assessment Office, Cincinnati, OH for the Office of Drinking Water, Washington, DC. External Review Draft.

Zeiger, E., S. Haworth, W. Speck and V. Mortlemans. 1982. Phthalate ester testing in the National Toxicology Program's environmental mutagenesis test development program. Environ. Health Perspect. 45: 99-101.

VI.D. DRINKING WATER HA REFERENCES
None

VII. REVISION HISTORY

---------- ---------- --
Date   Section   Description
---------- ---------- --
08/22/88 II.A.   Carcinogen summary on-line
02/01/89 II.D.3. Primary contact's phone number corrected
07/01/89 I.A.   Oral RfD now under review
09/01/89 I.A.   Oral RfD summary on-line
09/01/89 VI.   Bibliography on-line
08/01/91 I.A.7.   Primary and secondary contacts changed
08/01/91 II.D.3. Primary and secondary contacts changed
01/01/92 IV.   Regulatory Action section on-line
02/01/93 I.A.7.   Primary contact changed
02/01/93 II.D.3. Primary contact changed

SYNONYMS

BBP
1,2-BENZENEDICARBOXYLIC ACID,
BUTYL PHENYLMETHYL ESTER
BENZYL-BUTYLESTER KYSELINY
FTALOVE
BENZYL BUTYL PHTHALATE
BENZYL n-BUTYL PHTHALATE
Butyl benzyl phthalate
n-BUTYL BENZYL PHTHALATE

BUTYL PHENYLMETHYL 1,2-
BENZENEDICARBOXYLATE
NCI-C54375
PALATINOL BB
PHTHALIC ACID, BENZYL BUTYL
ESTER
SANTICIZER 160
SICOL 160
UNIMOLL BB

**Source: *Instant EPA's IRIS*, 1997, Instant Reference Sources, Inc., 7605 Rockpoint Drive, Austin, TX 78731 (Fax: 512-345-2386; Internet URL, http://www.instantref.com/inst-ref.htm).

Cadmium

Cd

Introduction

Cadmium is used as a constituent of easily fusible alloys; soft solder and solder for aluminum. Electroplating is its major use. It is also used as a deoxidizer in nickel plating, in process engraving, in electrodes for cadmium vapor lamps, in photoelectric cells, with photometry of ultraviolet sun-rays, and in Ni-Cd storage batteries. The powder is also used as an amalgam (1 Cd : 4 Hg) in dentistry.

Identifying Information

CAS NUMBER: 7440-43-9

NIOSH Registry Number: EU9800000

SYNONYMS: KADMIUM

Endocrine Disruptor Information

Cadmium is a strongly suspected EED that it is listed by EPA [811], the Centers for Disease Control & Prevention [812], and the World Wildlife Fund [813] as a potential endocrine modifying chemical. In support of this importance, cadmium is considered a metal similar in suspected EED properties to lead. Evidence has been obtained from rats, which includes lead exposure *in utero*, "suggesting a dual site of lead action: (a) at the level of the hypothalamic pituitary unit, and (b) directly at the level of gonadal steroid biosynthesis." Blood-serum testosterone levels (in males) and estradiol levels (in females) were most severely suppressed in those animals where exposure began *in utero*. Both sexes in this group showed a 25% suppression in prepubertal growth (as measured by weight). [868], [811]

Chemical and Physical Properties

ATOMIC FORMULA: Cd

ATOMIC WEIGHT: 112.41

WLN: CD

PHYSICAL DESCRIPTION: Silver-white, blue tinged and lustrous metal

SPECIFIC GRAVITY: 8.642

DENSITY: Not available

MELTING POINT: 320.9°C

BOILING POINT: 765°C

SOLUBILITY: Water: Insoluble

OTHER SOLVENTS: Acid: Soluble; NH_4NO_3: Soluble; Sulfuric Acid (hot): Soluble

OTHER PHYSICAL DATA: Refractive index: 1.13 Specific gravity: 8.642 Hexagonal crystals, silver-white, malleable metal [703].

Source: *Instant EPA's Air Toxics*, 1994, Instant Reference Sources, Inc., 7605 Rockpoint Drive, Austin, TX 78731 (Fax: 512-345-2386; Internet URL, http://www.instantref.com/inst-ref.htm).

Environmental Reference Materials

A source of EED environmental reference materials for cadmium is Radian International LLC, Austin, TX (Fax 512-454-0268; Internet URL, http://www.radian.com/standards); Catalog No. ICC-002L.

Hazardous Properties

ACUTE/CHRONIC HAZARDS: Cadmium and its salts may be highly toxic and there is evidence that it is a suspected carcinogen.

VOLATILITY:
   Vapor pressure: 1 mm Hg @ 394°C

FIRE HAZARD: Cadmium is flammable in powder form.

LEL: Not available   UEL: Not available

REACTIVITY: Cadmium reacts readily with dilute nitric acid and reacts slowly with hot hydrochloric acid but it does not react with alkalis. Cadmium dust can react vigorously with oxidizing materials.

STABILITY: Cadmium slowly oxidizes by moist air to form cadmium oxide (CdO).

Source: *Instant EPA's Air Toxics*, 1994, Instant Reference Sources, Inc., 7605 Rockpoint Drive, Austin, TX 78731 (Fax: 512-345-2386;   Internet URL, http://www.instantref.com/inst-ref.htm).

Medical Symptoms of Exposure

Ingestion of cadmium metal and solvated compounds may cause increased salivation, choking, vomiting, abdominal pain, anemia, renal dysfunction, diarrhea and tenesmus. Inhalation (dust or fumes) may cause throat dryness, cough, headache, vomiting, chest pain, extreme restlessness and irritability, pneumonitis and possibly bronchopneumonia.

Source: *Instant EPA's Air Toxics*, 1994, Instant Reference Sources, Inc., 7605 Rockpoint Drive, Austin, TX 78731 (Fax: 512-345-2386; Internet URL, http://www.instantref.com/inst-ref.htm).

Toxicological Information

The toxicological information below is gathered from several sources including two national databases: one from the *National Toxicology Program Chemical Repository Database* and another from *EPA's Integrated Risk Information System (IRIS).*

Toxicology Information from the National Toxicology Program

CARCINOGENICITY:

    Review:

        IARC Cancer Review: Human Limited Evidence
        IARC Cancer Review: Animal Sufficient Evidence
        IARC probable human carcinogen (Group 2A)

    Status:

        EPA Carcinogen Assessment Group

TERATOGENICITY: Not available

MUTATION DATA:

Test	Lowest dose
cyt-ham:ovr	1 mol/L

TOXICITY:

Typ. Dose	Mode	Specie	Amount	Units	Other
TCLo	ihl	man	88	ug/m3/8.6Y	
LCLo	ihl	hmn	39	mg/m3/20m	
LDLo	unk	man	15	mg/kg	
LD50	orl	rat	4	mg/kg	
LD50	ipr	rat	4	mg/kg	
LD50	scu	rat	9	mg/kg	

OTHER TOXICITY DATA:
    Review: Toxicology Review-19

SAX TOXICITY EVALUATION:

    THR: The oral toxicity of Cd and its compounds is high. However, when these materials are ingested, the irritant and emitic action is so violent that little of the Cd is adsorbed and fatal poisoning does not as a rule ensue.

Source: *Instant Tox-Base*, 1995, Instant Reference Sources, Inc., 7605 Rockpoint Drive, Austin, TX 78731 (Fax: 512-345-2386; Internet URL, http://www.instantref.com/inst-ref.htm).

Toxicology Information from EPA's Integrated Risk Information System (IRIS)

I. CHRONIC HEALTH HAZARD ASSESSMENTS FOR NONCARCINOGENIC EFFECTS

I.A. REFERENCE DOSE FOR CHRONIC ORAL EXPOSURE (RfD)

The Reference Dose (RfD) is based on the assumption that thresholds exist for certain toxic effects such as cellular necrosis, but may not exist for other toxic effects such as carcinogenicity. In general, the RfD is an estimate (with uncertainty spanning perhaps an order of magnitude) of a daily exposure to the human population (including sensitive subgroups) that is likely to be without an appreciable risk of deleterious effects during a lifetime. Please refer to Background Document 1 in Service Code 5 for an elaboration of these concepts. RfDs can also be derived for the noncarcinogenic health effects of compounds which are also carcinogens. Therefore, it is essential to refer to other sources of information concerning the carcinogenicity of this substance. If the US. EPA has evaluated this substance for potential human carcinogenicity, a summary of that evaluation will be contained in Section II of this file when a review of that evaluation is completed.

I.A.1. ORAL RfD SUMMARY

Critical Effect	Experimental Doses*	UF	MF	RfD
Significant proteinuria	NOAEL (water): 0.005 mg/kg/day	10	1	5E-4 mg/kg/day (water)
Human studies involving chronic exposures	NOAEL (food): 0.01 mg/kg/day	10	1	1E-3 mg/kg/day (food)

*Conversion Factors: See text for discussion

US. EPA, 1985

I.A.2. PRINCIPAL AND SUPPORTING STUDIES (ORAL RfD)

US. EPA. 1985. Drinking Water Criteria Document on Cadmium. Office of Drinking Water, Washington, DC. (Final draft)

A concentration of 200 ug cadmium (Cd)/gm wet human renal cortex is the highest renal level not associated with significant proteinuria (US. EPA, 1985). A toxicokinetic model is available to determine the level of chronic human oral exposure (NOAEL) which results in 200 ug Cd/gm wet human renal cortex; the model assumes that 0.01% day of the Cd body burden is eliminated per day (US. EPA, 1985). Assuming 2.5% absorption of Cd from food or 5% from water, the toxicokinetic model predicts that the NOAEL for chronic Cd exposure is 0.005 and 0.01 mg Cd/kg/day from water and food, respectively (i.e., levels which would result in 200 ug Cd/gm wet weight human renal cortex). Thus, based on an estimated NOAEL of 0.005 mg Cd/kg/day for Cd in drinking water and a UF of 10, an RfD of 0.0005 mg Cd/kg/day (water) was calculated; an equivalent RfD for Cd in food is 0.001 mg Cd/kg/day (see Section VI.A. for references).

I.A.3. UNCERTAINTY AND MODIFYING FACTORS (ORAL RfD)

UF -- This uncertainty factor is used to account for intrahuman variability to the toxicity of this chemical in the absence of specific data on sensitive individuals.

MF -- None

I.A.4. ADDITIONAL COMMENTS (ORAL RfD)

Cd is unusual in relation to most, if not all, of the substances for which an oral RfD has been determined in that a vast quantity of both human and animal toxicity data are available. The RfD is based on the highest level of Cd in the human renal cortex (i.e., the critical level) not associated with significant proteinuria (i.e., the critical effect). A toxicokinetic model has been used to determine the highest level of exposure associated with the lack of a critical effect. Since the fraction of ingested Cd that is absorbed appears to vary with the source (e.g., food vs. drinking water), it is necessary to allow for this difference in absorption when using the toxicokinetic model to determine an RfD.

I.A.5. CONFIDENCE IN THE ORAL RfD

Study -- Not applicable
Data Base -- High
RfD -- High

The choice of NOAEL does not reflect the information from any single study. Rather, it reflects the data obtained from many studies on the toxicity of cadmium in both humans and animals. These data also permit calculation of pharmacokinetic parameters of cadmium absorption, distribution, metabolism and elimination. All of this information considered together gives high confidence in the data base. High confidence in either RfD follows as well.

I.A.6. EPA DOCUMENTATION AND REVIEW OF THE ORAL RfD

Source Document -- US. EPA, 1985
Other EPA Documentation -- None
Agency Work Group Review -- 05/15/86, 08/19/86, 09/17/87, 12/15/87, 01/20/88, 05/25/88
Verification Date -- 05/25/88

I.A.7. EPA CONTACTS (ORAL RfD)

Ken Bailey / OST -- (202)260-5535
Yogi Patel / OST -- (202)260-5849

I.B. REFERENCE CONCENTRATION FOR CHRONIC INHALATION EXPOSURE (RfC)

A risk assessment for this substance/agent is under review by an EPA work group.

II. CARCINOGENICITY ASSESSMENT FOR LIFETIME EXPOSURE

Section II provides information on three aspects of the carcinogenic risk assessment for the agent in question; the US. EPA classification, and quantitative estimates of risk from oral exposure and from inhalation exposure. The classification reflects a weight-of-evidence judgment of the likelihood that the agent is a human carcinogen. The quantitative risk estimates are presented in three ways. The slope factor is the result of application of a low-dose extrapolation procedure and is presented as the risk per (mg/kg)/day. The unit risk is the quantitative estimate in terms of either risk per ug/L drinking water or risk per ug/cu.m air breathed. The third form in which risk is presented is a drinking water or air concentration providing cancer risks of 1 in 10,000, 1 in 100,000 or 1 in 1,000,000. Background Document 2 (Service Code 5) provides details on the rationale and methods used to derive the carcinogenicity values found in IRIS. Users are referred to Section I for information on long-term toxic effects other than carcinogenicity.

II.A. EVIDENCE FOR CLASSIFICATION AS TO HUMAN CARCINOGENICITY

II.A.1. WEIGHT-OF-EVIDENCE CLASSIFICATION

Classification -- B1; probable human carcinogen

Basis -- Limited evidence from occupational epidemiological studies of cadmium is consistent across investigators and study populations. There is sufficient evidence of carcinogenicity in rats and mice by inhalation and intramuscular and subcutaneous injection. Seven studies in rats and mice wherein cadmium salts (acetate, sulfate, chloride) were administered orally have shown no evidence of carcinogenic response.

II.A.2. HUMAN CARCINOGENICITY DATA

Limited. A 2-fold excess risk of lung cancer was observed in cadmium smelter workers. The cohort consisted of 602 white males who had been employed in production work a minimum of 6 months during the years 1940-1969. The population was followed to the end of 1978. Urine cadmium data available for 261 workers employed after 1960 suggested a highly exposed population. The authors were able to ascertain that the increased lung cancer risk was probably not due to the presence of arsenic or to smoking (Thun et al., 1985). An evaluation by the Carcinogen Assessment Group of these possible confounding factors has indicated that the assumptions and methods used in accounting for them appear to be valid. As the SMRs observed were low and there is a lack of clear cut evidence of a causal relationship of the cadmium exposure only, this study is considered to supply limited evidence of human carcinogenicity.

An excess lung cancer risk was also observed in three other studies which were, however, compromised by the presence of other carcinogens (arsenic, smoking) in the exposure or by a small population (Varner, 1983; Sorahan and Waterhouse, 1983; Armstrong and Kazantzis, 1983).

Four studies of workers exposed to cadmium dust or fumes provided evidence of a statistically significant positive association with prostate cancer (Kipling and Waterhouse, 1967; Lemen et al., 1976; Holden, 1980; Sorahan and Waterhouse, 1983), but the total number of cases was small in each study. The Thun et al. (1985) study is an update of an earlier study (Lemen et al., 1976) and does not show excess prostate cancer risk in these workers. Studies of human ingestion of cadmium are inadequate to assess carcinogenicity.

II.A.3. ANIMAL CARCINOGENICITY DATA

Exposure of Wistar rats by inhalation to cadmium as cadmium chloride at concentrations of 12.5, 25 and 50 ug/cu.m for 18 months, with an additional 13-month observation period, resulted in significant increases in lung tumors (Takenaka et al., 1983). Intratracheal instillation of cadmium oxide did not produce lung tumors in Fischer 344 rats but rather mammary tumors in males and tumors at multiple sites in males (Sanders and Mahaffey, 1984). Injection site tumors and distant site tumors (for example, testicular) have been reported by a number of authors as a consequence of intramuscular or subcutaneous administration of cadmium metal and chloride, sulfate, oxide and sulfide compounds of cadmium to rats and mice (US. EPA, 1985). Seven studies in rats and mice where cadmium salts (acetate, sulfate, chloride) were administered orally have shown no evidence of a carcinogenic response.

II.A.4. SUPPORTING DATA FOR CARCINOGENICITY

Results of mutagenicity tests in bacteria and yeast have been inconclusive. Positive responses have been obtained in mutation assays in Chinese hamster cells (Dom and V79 lines) and in mouse lymphoma cells (Casto, 1976; Ochi and Ohsawa, 1983; 0berly et al., 1982).

Conflicting results have been obtained in assays of chromosomal aberrations in human lymphocytes treated in vitro or obtained from exposed workers. Cadmium treatment in vivo or in vitro appears to interfere with spindle formation and to result in aneuploidy in germ cells of mice and hamsters (Shimada et al., 1976; Watanabe et al., 1979; Gilliavod and Leonard, 1975).

II.B. QUANTITATIVE ESTIMATE OF CARCINOGENIC RISK FROM ORAL EXPOSURE

Not available. There are no positive studies of orally ingested cadmium suitable for quantitation.

II.C.   QUANTITATIVE ESTIMATE OF CARCINOGENIC RISK FROM INHALATION EXPOSURE

II.C.1. SUMMARY OF RISK ESTIMATES

Inhalation Unit Risk -- 1.8E-3 per (ug/cu.m)

Extrapolation Method -- Two stage; only first affected by exposure; extra risk

Air Concentrations at Specified Risk Levels:

Risk Level	Concentration
E-4 (1 in 10,000)	6E-2 ug/cu.m
E-5 (1 in 100,000)	6E-3 ug/cu.m
E-6 (1 in 1,000,000)	6E-4 ug/cu.m

II.C.2.   DOSE-RESPONSE DATA FOR CARCINOGENICITY, INHALATION EXPOSURE

Tumor Type -- lung, trachea, bronchus cancer deaths
Test Animals -- human/white male
Route -- inhalation, exposure in the workplace
Reference -- Thun et al., 1985

Cumulative Exposure (mg/day/cu.m)	Median Observation	24 hour/ug/cu.m Equivalent	No. of Expected Lung, Trachea and Bronchus Cancers Assuming No Cadmium Effect	Observed No. of Deaths (lung, trachea, bronchus cancers)
Less than or equal to 584	280	168	3.77	2
585-2920	1210	727	4.61	7
greater than or equal to 2921	4200	2522	2.50	7

The 24-hour equivalent = median observation x 1E+3 x 8/24 x 1/365 x 240/365.

II.C.3.   ADDITIONAL COMMENTS (CARCINOGENICITY, INHALATION EXPOSURE)

The unit risk should not be used if the air concentration exceeds 6 ug/cu.m, since above this concentration the unit risk may not be appropriate.

II.C.4.   DISCUSSION OF CONFIDENCE (CARCINOGENICITY, INHALATION EXPOSURE)

The data were derived from a relatively large cohort. Effects of arsenic and smoking were accounted for in the quantitative analysis for cadmium effects.

An inhalation unit risk for cadmium based on the Takenaka et al. (1983) analysis is 9.2E-2 per (ug/cu.m). While this estimate is higher than that derived from human data [1.8E-3 per (ug/cu.m)] and thus more conservative, it was felt that the use of available human data was more reliable because of species variations in response and the type of exposure (cadmium salt vs. cadmium fume and cadmium oxide).

II.D.   EPA DOCUMENTATION, REVIEW, AND CONTACTS (CARCINOGENICITY ASSESSMENT)

II.D.1. EPA DOCUMENTATION

Source Document -- US. EPA, 1985

The Addendum to the Cadmium Health Assessment has received both Agency and external review.

II.D.2. REVIEW (CARCINOGENICITY ASSESSMENT)

Agency Work Group Review -- 11/12/86

Verification Date -- 11/12/86

II.D.3. US. EPA CONTACTS (CARCINOGENICITY ASSESSMENT)

William E. Pepelko / NCEA -- (202)260-5904

David Bayliss / NCEA -- (202)260-5726

III. HEALTH HAZARD ASSESSMENTS FOR VARIED EXPOSURE DURATIONS

[Ed. Note: EPA does not have information yet in this section of IRIS].

IV. US. EPA REGULATORY ACTIONS

EPA risk assessments may be updated as new data are published and as assessment methodologies evolve. Regulatory actions are frequently not updated at the same time. Compare the dates for the regulatory actions in this section with the verification dates for the risk assessments in sections I and II, as this may explain inconsistencies. Also note that some regulatory actions consider factors not related to health risk, such as technical or economic feasibility. Such considerations are indicated for each action. In addition, not all of the regulatory actions listed in this section involve enforceable federal standards. Please direct any questions you may have concerning these regulatory actions to the US. EPA contact listed for that particular action. Users are strongly urged to read

the background information on each regulatory action in Background Document 4 in Service Code 5.

IV.A. CLEAN AIR ACT (CAA)

[Ed. Note: EPA does not have information yet in this section of IRIS].

IV.B. SAFE DRINKING WATER ACT (SDWA)

IV.B.1. MAXIMUM CONTAMINANT LEVEL GOAL (MCLG) for Drinking Water

Value (status) -- 0.005 mg/L (Final, 1991)

Considers technological or economic feasibility? -- NO

Discussion -- Cadmium has been classed as a Category III contaminant with an MCLG of 0.005 mg/L based upon reports of renal toxicity in humans. The MCLG is based upon a DWEL of 0.018 mg/L and an assumed drinking water contribution (plus aquatic organisms) of 25 percent. An uncertainty factor of 10 was also applied.

Reference -- 56 FR 3526 (01/30/91)

EPA Contact -- Health and Ecological Criteria Division / OST /(202) 260-7571 / FTS 260-7571; or Safe Drinking Water Hotline / (800) 426-4791

IV.B.2. MAXIMUM CONTAMINANT LEVEL (MCL) for Drinking Water

Value (status) -- 0.005 mg/L (Final, 1991)

Considers technological or economic feasibility? -- YES

Discussion -- EPA has promulgated an MCL equal to the established MCLG of 0.005 mg/L.

Monitoring requirements -- Ground water systems monitored every three years surface water systems monitored annually; systems out of compliance must begin monitoring quarterly until system is reliably and consistently below MCL.

Analytical methodology -- Atomic absorption/ furnace technique (EPA 213.2;SM 304); inductively coupled plasma (200.7): PQL= 0.002 mg/L.

Best available technology -- Coagulation/filtration; ion exchange; lime softening; and reverse osmosis.

Reference -- 56 FR 3526 (01/30/91)

EPA Contact -- Drinking Water Standards Division / OGWDW /(202) 260-7575 / FTS 260-7575; or Safe Drinking Water Hotline / (800) 426-4791

IV.B.3. SECONDARY MAXIMUM CONTAMINANT LEVEL (SMCL) for Drinking Water

[Ed. Note: EPA does not have information yet in this section of IRIS].

IV.B.4. REQUIRED MONITORING OF "UNREGULATED" CONTAMINANTS

[Ed. Note: EPA does not have information yet in this section of IRIS].

IV.C. CLEAN WATER ACT (CWA)

IV.C.1. AMBIENT WATER QUALITY CRITERIA, Human Health

Water and Fish Consumption: 1E+1 ug/L

Fish Consumption Only: None

Considers technological or economic feasibility? -- NO

Discussion -- The criteria is the same as the existing standard for drinking water.

Reference -- 45 FR 79318 (11/28/80)

EPA Contact -- Criteria and Standards Division / OWRS (202)260-1315 / FTS 260-1315

IV.C.2. AMBIENT WATER QUALITY CRITERIA, Aquatic Organisms

Freshwater:
    Acute -- 3.9E+0 ug/L (1-hour average)
    Chronic -- 1.1E+0 ug/L (4-day average)

Marine:
    Acute -- 4.3E+1 ug/L (1-hour average)
    Chronic -- 9.3E+0 ug/L (4-day average)

Considers technological or economic feasibility? -- NO

Discussion -- Criteria were derived from a minimum data base consisting of acute and chronic tests on a variety of species. The freshwater criteria are hardness dependent. Values given here are calculated at a hardness of 100 mg/L CaCO3. A complete discussion can be found in the referenced notice.

Reference -- 50 FR 30784 (07/29/85)

EPA Contact -- Criteria and Standards Division / OWRS (202)260-1315 / FTS 260-1315

IV.D. FEDERAL INSECTICIDE, FUNGICIDE, AND RODENTICIDE ACT (FIFRA)

IV.D.1. PESTICIDE ACTIVE INGREDIENT, Registration Standard

Status -- Voluntary Cancellation [cadmium chloride] (1990)

Reference -- 55 FR 31227 (08/01/90)

EPA Contact -- Registration Branch / OPP (703)557-7760 / FTS 557-7760

IV.D.2. PESTICIDE ACTIVE INGREDIENT, Special Review

Action -- Termination of Special Review (1991)

Considers technological or economic feasibility? -- YES

Summary of regulatory action -- All uses of cadmium pesticides canceled. Criterion of concern: oncogenicity, mutagenicity, teratogenicity, and fetotoxicity.

Reference -- 56 FR 14522 (04/10/91)

EPA Contact -- Special Review Branch / OPP (703)557-7400 / FTS 557-7400

IV.E. TOXIC SUBSTANCES CONTROL ACT (TSCA)

No data available

IV.F. RESOURCE CONSERVATION AND RECOVERY ACT (RCRA)

IV.F.1. RCRA APPENDIX IX, for Ground Water Monitoring

Status -- Listed

Reference -- 52 FR 25942 (07/09/87)

EPA Contact -- RCRA/Superfund Hotline (800)424-9346 / (202)260-3000 / FTS 260-3000

IV.G. SUPERFUND (CERCLA)

IV.G.1. REPORTABLE QUANTITY (RQ) for Release into the Environment

Value (status) -- 10 pounds (Final, 1989)

Considers technological or economic feasibility? -- NO

Discussion -- The RQ for cadmium is 10 pounds, based on potential carcinogenicity. Available data indicate a hazard ranking of medium, based on a potency factor of 57.87/mg/kg/day and weight-of-evidence group B1, which corresponds to an RQ of 10 pounds. Cadmium has also been found to bioaccumulate in the tissues of aquatic and marine organisms, and has the potential to concentrate in the food chain. Reporting of releases of massive forms of this hazardous substance is not required if the diameter of the pieces released exceeds 100 micrometers (0.004 inches).

Reference -- 54 FR 33418 (08/14/89)

EPA Contact -- RCRA/Superfund Hotline (800)424-9346 / (202)260-3000 / FTS 260-3000

VI. BIBLIOGRAPHY

VI.A. ORAL RfD REFERENCES

Foulkes, E.C.   1986.   Absorption of cadmium.   In: Handbook of Experimental Pharmacology, E.C. Foulkes, Ed. Springer Verlag, Berlin. Vol. 80, p. 75-100.

Friberg, L., M. Piscator, G.F. Nordberg and T. Kjellstrom.   1974.   Cadmium in the environment, 2nd ed. CRC Press, Inc., Boca Raton, FL.

Shaikh, Z.A. and J.C. Smith. 1980. Metabolism of orally ingested cadmium in humans. In: Mechanisms of Toxicity and Hazard Evaluation, B. Holmstedt et al., Ed. Elsevier Publishing Co., Amsterdam. p. 569-574.

US. EPA. 1985. Drinking Water Criteria Document on Cadmium. Office of Drinking Water, Washington, DC. (Final draft)

WHO (World Health Organization). 1972. Evaluation of certain food additives and the contaminants mercury, lead, and cadmium. Sixteenth Report of the Joint FAO/WHO Expert Committee on Food Additives. WHO Technical Report Series No. 505, FAO Nutrition Meetings Report Series No. 51. Geneva, Switzerland.

WHO (World Health Organization). 1984. Guidelines for drinking water quality -- recommendations. Vol. 1. Geneva, Switzerland.

VI.B. INHALATION RfD REFERENCES

None

VI.C. CARCINOGENICITY ASSESSMENT REFERENCES

Armstrong, B.G. and G. Kazantzis. 1983. The mortality of cadmium workers. Lancet. June 25, 1983: 1425-1427.

Casto, B. 1976. Letter to Richard Troast, US. EPA. Enclosing mutagenicity data on cadmium chloride and cadmium acetate.

Gilliavod, N. and A. Leonard. 1975. Mutagenicity tests with cadmium in the mouse. Toxicology. 5: 43-47.

Holden, H. 1980. Further mortality studies on workers exposed to cadmium fumes. Presented at Seminar on Occupational Exposure to Cadmium, March 20, 1980, London, England.

Kipling, M.D. and J.A.H. Waterhouse. 1967. Cadmium and prostatic carcinoma. Lancet. 1: 730.

Lemen, R.A., J.S. Lee, J.K. Wagoner and H.P. Blejer. 1976. Cancer mortality among cadmium production workers. Ann. N.Y. Acad. Sci. 271: 273.

Oberly, T., C.E. Piper and D.S. McDonald. 1982. Mutagenicty of metal salts in the L5178 Y mouse lymphoma assay. J. Toxicol. Environ. Health. 9:367-376.

Ochi, T. and M. Ohsawa. 1983. Induction of 6-thioguanine-resistant mutants and single-strand scission DNA by cadmium chloride in cultured Chinese hamster cells. Mutat. Res. 111: 69-78.

Sanders, C.L. and J.A. Mahaffey. 1984. Carcinogenicity of single and multiple intratracheal instillations of cadmium oxide in the rat. Environ. Res. 33: 227-233.

Shimada, T., T. Watanabe and A. Endo. 1976. Potential mutagenicity of cadmium in mammalian oocytes. Mutat. Res. 40: 389-396.

Sorahan, T. and J.A.H. Waterhouse. 1983. Mortality study of nickel-cadmiumbattery workers by the method of regression models in life tables. Br. J.Ind. Med. 40: 293-300.

Takenaka, S., H. Oldiges, H. Konig, D. Hochrainer and G. Oberdoerster. 1983. Carcinogenicity of cadmium aerosols in Wistar rats. J. Natl. Cancer Inst. 70: 367-373.

Thun, M.J., T.M. Schnorr, A.B. Smith and W.E. Halperin. 1985. Mortalityamong a cohort of US. cadmium production workers: An update. J. Natl. Cancer Inst. 74(2): 325-333.

US. EPA. 1985. Updated Mutagenicity and Carcinogenicity Assessment of Cadmium. Addendum to the Health Assessment Document for Cadmium (EPA 600/B-B1-023). EPA 600/B-83-025F.

Varner, M.O. 1983. Updated epidemiologic study of cadmium smelter workers. Presented at the Fourth International Cadmium Conference. Unpublished.

Watanabe, T., T. Shimada and A. Endo. 1979. Mutagenic effects of cadmium on mammalian oocyte chromosomes. Mutat. Res. 67: 349-356.

VI.D. DRINKING WATER HA REFERENCES

None

VII. REVISION HISTORY

```
--------  ----------   ---------------------------------------------------------
Date      Section      Description
--------  ----------   ---------------------------------------------------------
```

Date	Section	Description
05/21/87	II.C.	Slope factor corrected
03/01/88	II.A.1.	Text added
03/01/88	II.C.3.	Text revised
03/01/88	II.C.4.	Confidence statement revised
03/01/88	II.D.3.	Secondary contact changed
01/01/89	IV.C.1.	Water quality human health criteria added
01/01/89	IV.C.2.	Corrected marine acute criterion
08/01/89	VI.	Bibliography on-line
10/01/89	I.A.	Oral RfD summary on-line
10/01/89	VI.A.	Oral RfD references added
12/01/89	I.B.	Inhalation RfD now under review
06/01/90	IV.A.1.	Area code for EPA contact corrected
06/01/90	IV.F.1.	EPA contact changed
08/01/90	II.A.1.	Basis statement revised
08/01/90	II.A.2.	Text revised, paragraph 1
08/01/90	II.B.	Text revised
01/01/91	II.	Text edited
01/01/91	II.C.1.	Inhalation slope factor removed (global change)
03/01/91	II.A.1.	Text revised
03/01/91	II.B.	Text revised
01/01/92	IV.	Regulatory actions updated
04/01/92	IV.A.1.	CAA regulatory action withdrawn
05/01/92	II.C.2.	Number correction in data table
06/01/92	II.A.2.	Text revised, paragraph 1
06/01/92	II.A.3.	Text clarified
02/01/94	I.A.7.	Secondary contact changed

SYNONYMS

C.I. 77180   KADMIUM
Cadmium

Source: *Instant EPA's IRIS*, 1997, Instant Reference Sources, Inc., 7605 Rockpoint Drive, Austin, TX 78731 (Fax: 512-345-2386; Internet URL, http://www.instantref.com/inst-ref.htm).

Carbaryl

Introduction

Carbaryl is a widely used contact type insecticide. It is used to control many different kinds of insects on fruits, vegetables, and ornamentals and it is also used in flea powders for domestic animals because of its relatively low acute toxicity. It is manufactured in many convenient forms including wettable powders, emulsifiable concentrates, suspension concentrates, powders, emulsifiable concentrates and oil-miscible concentrates. One of the most common trade names used in the US is "Sevin."

Carbaryl release to the environment results from its uses as a molluscicide and an insecticide and acaricide on a variety of crops. Carbaryl has been used to treat oyster grounds and has become a controversial issue in the Pacific northwest. Oyster growers maintain carbaryl is needed to kill two types of burrowing shrimp which render Pacific oyster beds soft and silty, preventing oyster culture. Release to soil will result in photolysis at the soil surface. Carbaryl will hydrolyze relatively rapidly in moist alkaline soil, but only slowly in acidic soil. Carbaryl may leach to groundwater based on its moderate soil sorption coefficient. Release to water will result in rapid hydrolysis at pH values of 7 and above. In acidic water, hydrolysis will be slow. At lower pH values, biodegradation may be significant. Adsorption to high organic content sediments has been demonstrated to be important. Bioconcentration is not expected to be significant. Release to the atmosphere may result in direct photolysis as carbaryl absorbs light wavelengths above 290 nm. Monitoring data indicate that carbaryl is a contaminant in food and a minor contaminant in drinking and surface water. Human exposure to carbaryl is expected to result mainly from ingestion of contaminated food and occupational exposure in farm workers is expected. [828]

Identifying Information

CAS NUMBER: 63-25-2

NIOSH Registry Number: FC5950000

SYNONYMS: ARYLAM; CARBAMINE; CARBARIL; CARBATOX; CARBATOX-60; CARBATOX 75; CARPOLIN; CARYLDERM; CEKUBARYL; CRAG SEVIN; DENAPON; DEVICARB; DICARBAM; GAMONIL; GERMAIN'S; HEXAVIN; KARBARYL; KARBASPRAY; KARBATOX; KARBOSEP; METHYLCARBAMATE 1-NAPHTHALENOL; METHYLCARBAMATE 1-NAPHTHOL;

METHYLCARBAMIC ACID; alpha-NAFTYL-N-METHYLKARBAMAT; 1-
NAPHTHOL N-METHYLCARBAMATE; 1-NAPHTHYL ESTER; 1-NAPHTHYL
METHYLCARBAMATE; 1-NAPHTHYL N-METHYLCARBAMATE; alpha-
NAPHTHYL N-METHYLCARBAMATE; 1-NAPHTHYL-N-METHYL-KARBAMAT;
N-METHYLCARBAMATE DE 1-NAPHTYLE; N-METHYL-1-NAFTYL-
CARBAMAAT; N-METHYL-1-NAPHTHYL-CARBAMAT;
N-METHYL-1-NAPHTHYL CARBAMATE; N-METHYL-alpha-
NAPHTHYLCARBAMATE; N-METHYL-alpha-NAPHTHYLURETHAN; N-METIL-
1-NAFTIL-CARBAMMATO; PANAM; RAVYON; RYLAM; SEFFEIN; SEPTENE;
SEVIMOL; SEVIN; SOK; TERCYL; TOXAN; TRICARNAM; UNION CARBIDE
7,744

Endocrine Disruptor Information

Carbaryl is a strongly suspected EED that it is listed by the Centers for Disease Control
& Prevention [812], and the World Wildlife Fund [813] as a potential endocrine
modifying chemical.

Chemical and Physical Properties

CHEMICAL FORMULA: $C_{12}H_{11}NO_2$

MOLECULAR WEIGHT: 201.24 [702]

PHYSICAL DESCRIPTION: White crystalline solid, essentially odorless [705]

SPECIFIC GRAVITY: 1.23 @ 20°C/4°C [704]

MELTING POINT: 142°C [705]

BOILING POINT: Decomposes [702]

SOLUBILITY: Water: 40 mg/L [702]; Water: 40 ppm @ 30°C [709]; Soluble in most
organic solvents [709]

OTHER PHYSICAL DATA: Vapor pressure: <0.005 mm Hg @ 26°C [705]; Vapor
pressure: 0.002 mm Hg @ 40°C [709]; Melting Point: 145°C [702]

Source: ***Instant EPA's Air Toxics***, 1994, Instant Reference Sources, Inc., 7605
Rockpoint Drive, Austin, TX 78731 (Fax: 512-345-2386; Internet URL,
http://www.instantref.com/inst-ref.htm).

Environmental Reference Materials

A source of EED environmental reference materials for Carbaryl is Radian International
LLC, Austin, TX (Fax 512-454-0268; Internet URL, http://www.radian.com/standards);
Catalog No. ERC-030.

Hazardous Properties

ACUTE/CHRONIC HAZARDS: A poison by ingestion, intravenous, intraperitoneal, and possibly other routes; human mutagenic data exists. An experimental carcinogen, teratogen, and tumorigen; produces experimental reproductive effects. An eye and severe skin irritant; absorbed by all routes; skin absorption is slow with no accumulation in tissue. A reversible cholinesterase inhibitor. [703] Targeted organs include the respiratory system, central nervous system, cardiovascular system, and skin. [704] When heated to decomposition, emits toxic fumes of NOx. [703]

HAP WEIGHTING FACTOR: 1 [713]

VOLATILITY: Vapor Pressure: 0.002 mm Hg @ 40°C [709]

FIRE HAZARD: Flammability: Noncombustible solid, but it may be dissolved in flammable liquids [704].

REACTIVITY: Incompatible or reacts with strong oxidizers such as chlorates, bromates, and nitrates [704].

STABILITY: When heated to decomposition it emits toxic fumes of NOx [703]; carbaryl hydrolyzes rapidly in alkaline solutions [709].

Source: *Instant EPA's Air Toxics*, 1994, Instant Reference Sources, Inc., 7605 Rockpoint Drive, Austin, TX 78731 (Fax: 512-345-2386; Internet URL, http://www.instantref.com/inst-ref.htm).

Medical Symptoms of Exposure

Symptoms of exposure to carbaryl (trade name "Sevin") may include miosis, blurred vision, tearing; nasal discharge, salivation; sweat; abdominal cramps, nausea, vomiting, diarrhea; tremor; cyanosis; convulsions; and skin irritation. [704] Symptoms include blurred vision, headache, stomach ache, and vomiting. Symptoms similar to but less severe than those due to parathion. [703]

Source: *Instant EPA's Air Toxics*, 1994, Instant Reference Sources, Inc., 7605 Rockpoint Drive, Austin, TX 78731 (Fax: 512-345-2386; Internet URL, http://www.instantref.com/inst-ref.htm).

Toxicological Information

The toxicological information below is gathered from several sources including two national databases: one from the *National Toxicology Program Chemical Repository Database* and another from *EPA's Integrated Risk Information System (IRIS).*

Toxicology Information from the National Toxicology Program

Carbaryl is not currently listed in NTP's Chemical Repository database nor is it scheduled for addition in the future.

Toxicology Information from EPA's Integrated Risk Information System (IRIS)

I. CHRONIC HEALTH HAZARD ASSESSMENTS FOR NONCARCINOGENIC EFFECTS

I.A. REFERENCE DOSE FOR CHRONIC ORAL EXPOSURE (RfD)

The Reference Dose (RfD) is based on the assumption that thresholds exist for certain toxic effects such as cellular necrosis, but may not exist for other toxic effects such as carcinogenicity. In general, the RfD is an estimate (with uncertainty spanning perhaps an order of magnitude) of a daily exposure to the human population (including sensitive subgroups) that is likely to be without an appreciable risk of deleterious effects during a lifetime. Please refer to Background Document 1 in Service Code 5 for an elaboration of these concepts. RfDs can also be derived for the noncarcinogenic health effects of compounds which are also carcinogens. Therefore, it is essential to refer to other sources of information concerning the carcinogenicity of this substance. If the US. EPA has evaluated this substance for potential human carcinogen-icity, a summary of that evaluation will be contained in Section II of this file when a review of that evaluation is completed.

I.A.1. ORAL RfD SUMMARY

Critical Effect	Experimental Doses*	UF	MF	RfD
Kidney and liver	NOAEL: 200 ppm of diet (9.6 mg/kg/day)	100	1	1E-1 mg/kg/day
	LOAEL: 400 ppm of diet (15.6 mg/kg/day)			

*Conversion Factors: Dose conversion based on body weight and foodconsumption data reported by the authors.

Rat Chronic Feeding Study, Carpenter et al., 1961

I.A.2. PRINCIPAL AND SUPPORTING STUDIES (ORAL RfD)

Carpenter, C.P., C.W. Weil, P.E. Polin, et al. 1961. Mammalian toxicity of 1-naphthayl-N-methylcarbamate (Sevin insecticide). J. Agric. Food Chem. 9: 30-39.

Groups of 20 CF-N rats/sex were fed carbaryl at 0, 50, 100, 200, or 400 ppm of diet for 2 years. Food consumption and body weight records were maintained. Interim sacrifices (4-8 animals) from concurrent auxiliary groups were performed at 6, 9, and 12 months for organ weight comparisons and histopathologic analysis. Hematologic analyses were done at irregular intervals throughout the study. Surviving animals were sacrificed at 2 years with gross and histopathologic examinations performed. The only noteworthy effects reported were slight histopathologic changes in the kidneys and liver at the high-dose level. Diffuse cloudy swelling of renal tubules was observed at 1 and 2 years. A statistically significant increase in cloudy swelling of the hepatic cords was also observed after 2 years. Based on body weight and food consumption data, the LOAEL of 400 ppm was equivalent to a dose of 15.6 mg/kg bw/day. The NOAEL established was 9.6 mg/kg bw/day.

I.A.3. UNCERTAINTY AND MODIFYING FACTORS (ORAL RfD)

UF -- UF = 10a x 10h. The UF of 100 includes uncertainties in interspecies and intrahuman variability.

MF -- None

I.A.4. ADDITIONAL COMMENTS (ORAL RfD)

Effect and no-effect levels (14 and 7 mg/kg/day, respectively) similar to those found in the critical study were observed for rat body weight reduction and cholinesterase inhibition in a 1-year study. In subchronic rat studies, higher dose levels (85-200 mg/kg/day) caused kidney toxicity and biochemical changes. Kidney lesions were observed in dogs fed carbaryl at 5 mg/kg/day for 1 year; however, the effect was not clearly associated with treatment, since the lesions appeared in control animals but not in lower dose groups.

Carbaryl was teratogenic for several species, with widely varying NOELs. The lowest effect levels of 5-6 mg/kg were observed for dogs, with NOELs of 2-3 mg/kg. Other LOELs were higher than the established chronic LOAEL of 15.6 mg/kg/day. Carbaryl was not teratogenic for monkeys at 20 mg/kg. The dog studies were judged inappropriate for human health risk assessment because of differences in the metabolism of carbaryl between dogs and humans.

I.A.5. CONFIDENCE IN THE ORAL RfD

Study -- HighData Base -- MediumRfD -- Medium

The principal study was well designed and clearly reported with unequivocal effect levels established. The data base is moderately supportive of the nature of the critical effect, if somewhat sparse. The principal problem is the observation of teratogenicity in dogs at lower doses. Because the significance of these data cannot be discounted entirely, confidence in the RfD should be considered medium to low.

I.A.6. EPA DOCUMENTATION AND REVIEW OF THE ORAL RfD

Source Document -- US. EPA, 1984

Limited Agency review of 1984 Health and Environmental Effects Profile with the help of two external scientists.

Other EPA Documentation -- None

Agency Work Group Review -- 05/31/85
Verification Date -- 05/31/85

I.A.7. EPA CONTACTS (ORAL RfD)

Joan Dollarhide / NCEA -- (513)569-7539

I.B. REFERENCE CONCENTRATION FOR CHRONIC INHALATION EXPOSURE
(RfC)

The health effects data for carbaryl were reviewed by the US. EPA RfD/RfC Work
Group and determined to be inadequate for derivation of an inhalation RfC. The
verification status of this chemical is currently not verifiable. For additional information
on the health effects of this chemical, interested parties are referred to the EPA
documentation listed below.

US. EPA. 1984. Health and Environmental Effects Profile for Carbaryl. Prepared by the
Office of Health and Environmental Assessment, Environmental Criteria and Assessment
Office, Cincinnati, OH for the Office of Solid Waste, Washington, DC. EPA/600/X-
84/155.

Agency Work Group Review -- 08/15/91

EPA Contacts:
Jeffrey S. Gift / NCEA -- (919)541-4828
Annie M. Jarabek / NCEA -- (919)541-4847

II. CARCINOGENICITY ASSESSMENT FOR LIFETIME EXPOSURE

This substance/agent has not been evaluated by the US. EPA for evidence of human
carcinogenic potential.

III. HEALTH HAZARD ASSESSMENTS FOR VARIED EXPOSURE DURATIONS

[Ed. Note: EPA does not have information yet in this section of IRIS].

IV. US. EPA REGULATORY ACTIONS

EPA risk assessments may be updated as new data are published and as assessment
methodologies evolve. Regulatory actions are frequently not updated at the same time.
Compare the dates for the regulatory actions in this section with the verification dates for
the risk assessments in sections I and II, as this may explain inconsistencies. Also note
that some regulatory actions consider factors not related to health risk, such as technical
or economic feasibility. Such considerations are indicated for each action. In addition,
not all of the regulatory actions listed in this section involve enforceable federal
standards. Please direct any questions you may have concerning these regulatory actions
to the US. EPA contact listed for that particular action. Users are strongly urged to read
the background information on each regulatory action in Background Document 4 in
Service Code 5.

IV.A. CLEAN AIR ACT (CAA)

[Ed. Note: EPA does not have information yet in this section of IRIS].

IV.B. SAFE DRINKING WATER ACT (SDWA)

IV.B.1. MAXIMUM CONTAMINANT LEVEL GOAL (MCLG) for Drinking Water

[Ed. Note: EPA does not have information yet in this section of IRIS].

IV.B.2. MAXIMUM CONTAMINANT LEVEL (MCL) for Drinking Water

[Ed. Note: EPA does not have information yet in this section of IRIS].

IV.B.3. SECONDARY MAXIMUM CONTAMINANT LEVEL (SMCL) for Drinking Water

[Ed. Note: EPA does not have information yet in this section of IRIS].

IV.B.4. REQUIRED MONITORING OF "UNREGULATED" CONTAMINANTS

Status -- Listed (Final, 1991)

Discussion -- "Unregulated" contaminants are those contaminants for which EPA establishes a monitoring requirement but which do not have an associated final MCLG, MCL, or treatment technique. EPA may regulate these contaminants in the future.

Monitoring requirement -- All systems to be monitored unless a vulnerability assessment determines the system is not vulnerable.

Analytical methodology -- Derivatization/gas chromatography (EPA 531.1).

Reference -- 56 FR 3526 (01/30/91)

EPA Contact -- Drinking Water Standards Division / OGWDW /(202) 260-7575 / FTS 260-7575; or Safe Drinking Water Hotline / (800) 426-4791

IV.C. CLEAN WATER ACT (CWA)

[Ed. Note: EPA does not have information yet in this section of IRIS].

IV.D. FEDERAL INSECTICIDE, FUNGICIDE, AND RODENTICIDE ACT (FIFRA)

IV.D.1. PESTICIDE ACTIVE INGREDIENT, Registration Standard

Status -- Issued (1984)
Reference -- Carbaryl Pesticide Registration Standard. March, 1984(NTIS No. PB84-232602).
EPA Contact -- Registration Branch / OPP (703)557-7760 / FTS 557-7760

IV.D.2. PESTICIDE ACTIVE INGREDIENT, Special Review

Action -- Final determination on pre-special review (1980)

Considers technological or economic feasibility? -- NO

Summary of regulatory action -- The agency has returned this compound to the registration process with the stipulation that the following measures be considered: (1) a section 3(c)(2)(b) action that registrants remedy data gaps identified in pre-special review and (2) appropriate label changes be made to minimize exposure.

Reference -- 45 FR 81869 (12/12/80)

EPA Contact -- Special Review Branch / OPP (703)557-7400 / FTS 557-7400

IV.E. TOXIC SUBSTANCES CONTROL ACT (TSCA)

[Ed. Note: EPA does not have information yet in this section of IRIS].

IV.F. RESOURCE CONSERVATION AND RECOVERY ACT (RCRA)

[Ed. Note: EPA does not have information yet in this section of IRIS].

IV.G. SUPERFUND (CERCLA)

IV.G.1. REPORTABLE QUANTITY (RQ) for Release into the Environment

Value (status) -- 100 pounds (Final, 1985)

Considers technological or economic feasibility? -- NO

Discussion -- The final RQ is based on aquatic toxicity, as established under CWA Section 311(b)(4). The available data indicate that carbaryl has an aquatic 96-Hour Median Threshold Limit between 1 and 10 ppm.

Reference -- 50 FR 13456 (04/04/85); 54 FR 33418 (08/14/89)

EPA Contact -- RCRA/Superfund Hotline (800)424-9346 / (202)260-3000 / FTS 260-3000

VI. BIBLIOGRAPHY

VI.A. ORAL RfD REFERENCES

Carpenter, C.P., C.W. Weil, P.E. Polin, et al. 1961. Mammalian toxicity of 1-naphthayl-N-methylcarbamate (Sevin insecticide). J. Agric. Food Chem.9: 30-39.

US. EPA. 1984. Health and Environmental Effects Profile for Carbaryl. Prepared by the Office of Health and Environmental Assessment, Environmental Criteria and Assessment Office Cincinnati, OH for the Office of Solid Waste, Washington, DC.

VI.B. INHALATION RfC REFERENCES

US. EPA. 1984. Health and Environmental Effects Profile for Carbaryl. Prepared by the Office of Health and Environmental Assessment, Environmental Criteria and Assessment Office, Cincinnati, OH for the Office of Solid Waste, Washington, DC. EPA/600/X-84/155.

VI.C. CARCINOGENICITY ASSESSMENT REFERENCES

None

VI.D. DRINKING WATER HA REFERENCES

None

VII. REVISION HISTORY

```
--------    --------    ---------------------------------------------------------
Date        Section     Description
--------    --------    ---------------------------------------------------------
03/01/88    I.A.1.      Dose conversion clarified
08/01/89    VI.         Bibliography on-line
09/01/91    I.B.        Inhalation RfC now under review
11/01/91    I.B.        Inhalation RfC message on-line
11/01/91    VI.B.       Inhalation references added
01/01/92    I.A.7.      Secondary contact changed
01/01/92    IV.         Regulatory actions updated
```

SYNONYMS

ARYLAM
CARBAMINE
CARBARIL
Carbaryl
CARBATOX
CARBATOX-60
CARBATOX 75
CARPOLIN
CARYLDERM
CEKUBARYL
CRAG SEVIN
DENAPON
DEVICARB
DICARBAM
ENT 23,969
GAMONIL
GERMAIN'S
HEXAVIN
KARBARYL
KARBASPRAY
KARBATOX
KARBOSEP
METHYLCARBAMATE 1-
NAPHTHALENOL
METHYLCARBAMATE 1-NAPHTHOL
METHYLCARBAMIC ACID
NA 2757
NAC
alpha-NAFTYL-N-METHYLKARBAMAT
1-NAPHTHOL N-
METHYLCARBAMATE
1-NAPHTHYL ESTER

1-NAPHTHYL METHYLCARBAMATE
1-NAPHTHYL N-
METHYLCARBAMATE
alpha-NAPHTHYL N-
METHYLCARBAMATE
1-NAPHTHYL-N-METHYL-
KARBAMAT
N-METHYLCARBAMATE DE 1-
NAPHTYLE
N-METHYL-1-NAFTYL-
CARBAMAAT
N-METHYL-1-NAPHTHYL-
CARBAMAT
N-METHYL-1-NAPHTHYL
CARBAMATE
N-METHYL-alpha-
NAPHTHYLCARBAMATE
N-METHYL-alpha-
NAPHTHYLURETHAN
N-METIL-1-NAFTIL-CARBAMMATO
OMS-29
PANAM
RAVYON
RYLAM
SEFFEIN
SEPTENE
SEVIMOL
SEVIN
SOK
TERCYL
TOXAN
TRICARNAM
UC 7744
UNION CARBIDE 7,744

Source: *Instant EPA's IRIS*, 1997, Instant Reference Sources, Inc., 7605 Rockpoint Drive, Austin, TX 78731 (Fax: 512-345-2386; Internet URL, http://www.instantref.com/inst-ref.htm).

Production, Use, and Pesticide Labeling Information

1. Description of Chemical

   Generic Name: 1-napthyl N-methylcarbamate

   Common Name: carbaryl

   Trade Name: Sevin

   EPA Shaughnessy Code: 056801

   Year of Initial Registration: 1958

   Pesticide Type: Insecticide

   Chemical Family: Carbamates

   US. and Foreign Producers: Union Carbide, Makteshim Chemical Works, Inc.

2. Use Patterns and Formulations

   - Application sites: citrus, pome, stone and berry fruits, forage, field and vegetable crops, nuts, lawns, forests, ornamental plants, rangeland, shade trees, poultry and pets, indoor use

   - Types of formulations: baits, dusts, granules, wettable powders, flowables, and aqueous dispersions - Types and methods of application: ground and aerial - Application rates: range from 0.53 lbs. a.i./A to 6.4 lbs. a.i./A - Usual carriers: synthetic clays, talc, various solvents

3. Science Findings

   - Toxicological Characteristics
         - Acute Oral LD50: 225 mg/kg, Toxicity Category II
         - Acute Dermal LD50: > 2 g/kg, Toxicity Category III
         - Primary Dermal Irritation: no irritation, Toxicity Category IV
         - Primary Eye Irritation: Conjunctival irritation at 24 hours. Cleared at
         48 hours. Toxicity   Category III

     - Acute Inhalation LC50: data gap

   - Oncogenicity: Ten studies. Each study classified as supplemental. Collectively these studies provide sufficient evidence that carbaryl is not oncogenic in experimental animals. Eighteen-month mouse study was negative at 400 ppm. A 2-year rat feeding and oncogenicity study was negative at 200 ppm.

- Teratogenicity: Twenty-four studies have been evaluated to determine the teratogenic potential of carbaryl. In evaluating these studies, some were found to be flawed. Other studies demonstrated no teratogenicity or maternal toxicity. There are studies which demonstrate teratogenic effects, although the doses also caused maternal toxicity. Two studies produced teratogenic effects in the beagle dog. These two studies are the primary reason carbaryl was made a candidate for RPAR in 1976.

- The Agency has concluded (45 FR 81869) that carbaryl does not constitute a potential human teratogen or reproductive hazard under proper environmental usage. However, the Agency is requesting that the teratology study in the beagle dog be repeated, because the results of the dog studies continue to be a concern that has never been fully resolved. A repeat dog study will provide the basis for any appropriate regulatory action. In the interim, the Agency has determined that a label precaution stating not to use carbaryl on pregnant dogs is warranted until a new dog teratology study is submitted and reviewed.

- There have been proposals that there are differences in the metabolism of carbaryl between the dog and man. These differences, however, have never been demonstrated. Therefore, a metabolism study in the beagle dog versus the rat is being required. This metabolism study should allow us to determine if there are meaningful differences between the dog and other mammalian species.

- Reproduction: A rat 3-generation study was negative at 200 mg/kg.

- Mutagenicity: Carbaryl is characterized as a weak mutagen. The Agency has determined that carbaryl does not pose a mutagenic risk. No additional data are being requested.

- 1-year dog feeding study:
    - 2-year rat feeding study: Demonstrated an apparent effect on renal function. A kidney effect was also noted in a short-term human study. A 1-year dog feeding study using carbaryl is being requested in order to determine the effects of carbaryl on kidney dysfunction. The results of these may necessitate a re-evaluation of the ADI for carbaryl.

- Physiological and Biochemical Behavioral Characteristics
    - Mechanism of pesticidal action: A contact insecticide which causes reversible carbamylation of the acetylcholinesterase enzyme of tissues, allowing accumulation of acetylcholine at cholinergic neuroeffector junctions (muscarinic effects), and at skeletal muscle myoneural junctions and autonomic ganglia. Poisoning also impairs the central nervous system function.

- Metabolism and persistence in plants and animals:
    Carbaryl is rapidly excreted in animals, mainly in the urine. Residues in animals are carbaryl, 1-naphthol, and hydroxycarbaryl. The hydroxy metabolites are found mainly as glucuronide and sulfate conjugates. Carbaryl is slowly taken up into plants, after which it is metabolized.

The disappearance of carbaryl residue from plant surfaces is attributed to mechanical attribution, volatilization and uptake into plant. Photochemical degradation does not appear to be a factor. 1-naphthol is the major metabolite.

- Environmental Characteristics
    - Available data are insufficient to fully assess the environmental fate of carbaryl.
    - Adsorption and leaching in basic soil types: The Agency is requesting data to determine if carbaryl will contaminate groundwater.

- Microbial breakdown:
    Carbaryl is degraded by fungi. The soil fungi attack carbaryl by hydroxylation of the side chain and ring structure.
    - Loss from photodecomposition: Data gaps. Data are required.
    - Bioaccumulation: Preliminary data indicate that there may be a potential for carbaryl and its residue(s) to accumulate in catfish, crayfish, snail, duckweed, and algae. Additional data are requested.

- Resultant average persistence:
    Carbaryl is metabolized by pure and mixed cultures of bacteria, fungi, and to some extent by other soil and water organisms. The half-life appears to range from 7 to 28 days in aerobic and anaerobic soils, respectively.

- Ecological Characteristics
    - Avian oral LD50: Mallard duck, > 2179 mg/kg; Ring-necked pheasant, > 2000 mg/kg
    - Avian dietary LC50: Mallard duck, > 5000 ppm, Ring-necked pheasant, > 5000 ppm, Bobwhite quail, > 5000 ppm
    - Freshwater fish LC50: cold water fish, rainbow trout, 1.95 ppm; warm water fish, bluegill sunfish, 6.76 ppm
    - Acute LC50 freshwater invertebrates - Daphnia pulex - 6.4 ppb
    - Acute LC50 estuarine and marine organisms: Data gap. Data being requested.
    - Freshwater fish early life-stage: Fathead minnow, Maximum Acceptable Theoretical Concentration (MATC) - >0.21 <0.68 ppb
    - No precautionary language is required for birds or fish. However, carbaryl is highly toxic to aquatic invertebrates. There is insufficient information to characterize the chronic toxicity of carbaryl to aquatic invertebrates.

- Tolerance Assessments
    - The Agency is unable to complete a full tolerance reassessment because of certain residue chemistry and toxicology data gaps, namely

a 1-year dog feeding study and the need for residue data on various processed food commodities.

Tolerances:

Commodity	Parts Per Million
Alfalfa	100
Almonds	1
Almonds, hulls	40
Apples	10
Apricots	10
Asparagus	10
Bananas	10
Barley, grain	0
Barley, green fodder	100
Barley, straw	100
Beans	10
Beans, forage	100
Beans, hay	100
Beets, garden (roots)	5
Beets, garden (tops)	12
Birdsfoot trefoil, forage	100
Birdsfoot trefoil, hay	100
Blackberries	12
Blueberries	10
Boysenberries	12
Broccoli	10
Brussels sprouts	10
Cabbage	10
Carrots	10
Cauliflower	10
Celery	10
Cherries	10
Chestnuts	1
Chinese cabbage	10
Citrus fruits	10
Clover	100
Clover, hay	100
Collards	12
Corn, fresh (including sweet) Kernel (K) 5 + Corn with husk removed (CWHR) Corn, fodder	100
Corn, forage	100
Cotton, forage	100
Cottonseed	0
Cowpeas	5
Cowpeas, forage	100
Cowpeas, hay	100
Cranberries	10
Cucumbers	10
Dandelions	12
Dewberries	12
Eggplants	10

Table Continued	
Commodity	**Parts Per Million**
Endive (escarole)	10
Filberts (hazelnuts)	1
Flax, seed	5
Flax, straw	100
Grapes	10
Grass	100
Grass, hay	100
Horseradish	5
Kale	12
Kohlrabi	10
Lentils	10
Lettuce1	0
Loganberries	12
Maple sap	0.5
Melons	10
Millet, proso, grain	3
Millet, proso, straw	100
Mustard greens	12
Nectarines	10
Oats, fodder, green	100
Oats, grain	0
Oats, straw	100
Okra	10
Olives	10
Oysters	0.25
Parsley	12
Parsnips	5
Peaches	10
Peanuts	5
Peanuts, hay	100
Pears	10
Peas (with pods)	10
Peavines	100
Pecans	1
Peppers	10
Pistachio nuts	1
Plums (fresh prunes)	10
Poultry, fat	5
Poultry, meat	5
Potatoes	0.2 (N)
Prickly pear cactus, fruit	12.0
Prickly pear cactus, pads	12.0
Pumpkins	10
Radishes	5
Raspberries	12
Rice	5
Rice, straw	100
Rutabagas	5
Rye, fodder, green	100
Rye, grain	0

Table Continued	
Commodity	Parts Per Million
Rye, straw	100
Salsify (roots)	5
Salsify (tops)	10
Sorghum, forage	100
Sorghum, grain	10
Soybeans	5
Soybeans, forage	100
Soybeans, hay	100
Spinach	12
Squash, summer	10
Squash, winter	10
Strawberries	10
Sugar beets, tops	100
Sunflower seeds	1
Sweet potatoes	0.2
Swiss chard	12
Tomatoes	10
Turnips, roots	5
Turnips, tops	12
Walnuts	1
Wheat, fodder, green	100
Wheat (grain)	3
Wheat, straw	100
Cattle, fat	0.1
Cattle, kidney	1
Cattle, liver	1
Cattle, meat	0.1
Cattle, (mbyp)	0.1
Goats, fat	0.1
Goats, kidney	1
Goats, liver	1
Goats, meat	0.1
Goats, (mbyp)	0.1
Horses, fat	0.1
Horses, kidney	1
Horses, liver	1
Horses, meat	0.1
Horses, (mbyp)	0.1
Sheep, fat	0.1
Sheep, kidney	1
Sheep, liver	1
Sheep, meat	0.1
Sheep (mbyp)	0.1
Swine, fat	0.1
Swine, kidney	1
Swine, liver	1
Swine, meat	0.1

- Based on established tolerances, the theoretical maximum residue contribution (TMRC) for carbaryl residues in the human diet is calculated to be 5.48 mg/day.

The acceptable daily intake (ADI) of carbaryl is 0.1 mg/kg/day. The maximum permissible intake (MPI) is 6 mg/day. To provide for conformity between US. tolerances for carbaryl and tolerances established by the Codex Alimentarius, Canada and Mexico, the expression of the US. tolerances for carbaryl will be changed to omit reference to 1-naphthol.

- A 1-year dog feeding study is being requested in order to determine the effects of carbaryl on kidney dysfunction. The results of these data may require that the ADI for carbaryl be recalculated.

- US. tolerances for most raw agricultural commodities are supported by current residue chemistry data. In some cases, however, more data are required.

- Summary Science Statement

- Carbaryl has moderate to low mammalian toxicity. It is not considered to be an oncogen. It is a weak mutagen. Available data indicates that carbaryl has only low teratogenic potential. Long term dietary studies in rats and dogs and a short term study in humans (highest dose only) demonstrate an apparent effect on renal function.

- No reentry interval is necessary for carbaryl. Carbaryl is not expected to contaminate groundwater. Data are insufficient to assess the environmental fate of carbaryl.

- Carbaryl is extremely toxic to aquatic invertebrates and certain estuarine organisms. It is extremely toxic to honeybees. It is moderately toxic to both warm water and cold water fishes and has only low toxicity to birds.

- A full tolerance reassessment cannot be completed. A 1-year dog feeding study is required, as well as residue data on numerous processed commodities.

4. Summary of Regulatory Position and Rationale

- The Agency has determined that it should continue to allow the registration of carbaryl. Adequate studies are available to assess the acute toxicological effects of carbaryl to humans. None of the criteria listed in section 162.11(a) of Title 40 of the US. Code of Federal Regulations have been met or exceeded. However, because of gaps in the data base, a full risk assessment of carbaryl cannot be completed.

- A full tolerance reassessment cannot be completed because of certain residue chemistry and toxicology data gaps, namely a 1-year dog feeding study and the need for residue data on various processed commodities.

- No federal or state reentry intervals have been established for carbaryl or will be established.

- Available data are insufficient to fully assess the environmental fate of carbaryl. The Agency is requesting data to determine if carbaryl will contaminate groundwater.

5. Summary of Major Data Gaps

- Residue data on various processed commodities
- 1-year dog feeding study
- Teratology study in beagle dog
- Hydrolysis study
- Photodegradation studies
- Soil metabolism studies
- Mobility studies
- Dissipation studies
- Accumulation studies
- Acute LC50 freshwater invertebrates study
- Metabolism study in dog versus rat

6. Contact Person at EPA

Jay S. Ellenberger
Product Manager (12)
Insecticide- Rodenticide Branch Registration Division (TS-767C)
Office of Pesticide Programs
Environmental Protection Agency
401 M Street, SW
Washington, DC 20460

Office location and telephone number:
Room 202, Crystal Mall #2
1921 Jefferson Davis Highway
Arlington, VA 22202

Phone: (703) 557-2386

Source: *Instant EPA's Pesticide Facts*, 1994, Instant Reference Sources, Inc., 7605 Rockpoint Drive, Austin, TX 78731 (Fax: 512-345-2386; Internet URL, http://www.instantref.com/inst-ref.htm).

Chlordane

Introduction

Chlordane is used as a non-systemic contact and stomach insecticide with some fumigant action. It is also used as an acaricide, a pesticide and a wood preservative. In addition, it is used in termite control and as a protective treatment for underground cables. Chlordane is an extremely persistent organochlorine insecticide that was in wide use prior to 1983, when it was banned from general use except for termite control and was banned from all use in 1988. [811] There are two isomeric forms of chlordane: cis-chlordane and trans-chlordane. These vary by whether the two chlorine atoms at the far left in the structure above are both in equivalent positions (cis-chlordane) or whether one is equatorial and the other axial to each other (trans-chlordane). Commercial insecticides are mixtures of the two isomers and the primary CAS Number used for referencing this material is the number for the mixture but CAS Numbers for the two isomers are also included and, when available, properties of the pure isomers are also provided.

Chlordane has been released into the environment primarily from its application as an insecticide. Currently, there are no approved uses for chlordane in the US. If released to soil, chlordane may persist for long periods of time. Chlordane is expected to be generally immobile or only slightly mobile in soil; however, its detection in various groundwaters indicates that movement to groundwater can occur. Chlordane can volatilize significantly from soil surfaces on which it has been sprayed, particularly moist soil surfaces; however, shallow incorporation into soil will greatly restrict volatile losses. [828]

Although sufficient biodegradation data are not available, it has been suggested that chlordane is very slowly biotransformed in the environment which is consistent with the long persistence periods observed under field conditions. If released to water, chlordane is not expected to undergo significant hydrolysis, oxidation or direct photolysis. Adsorption to sediment is expected to be a major fate process. Bioconcentration is expected to be important. Sensitized photolysis in the water column may be possible. [828]

If released to the atmosphere chlordane will be expected to exist predominately in the vapor phase. The detection of chlordane in remote atmospheres (Pacific and Atlantic Oceans; the Arctic) indicates that long range transport occurs. It has been estimated that

96% of the airborne reservoir of chlordane exists in the sorbed state which may explain why its long range transport is possible without chemical transformation. The detection of chlordane in rainwater and its observed dry deposition at various rural locations indicates that physical removal via wet and dry deposition occurs in the environment. [828]

Major general population exposure to chlordane can occur through oral consumption of contaminated food and inhalation of contaminated air as well as through contact with treated soil. Occupational exposure by dermal and inhalation routes related to the use of chlordane as an insecticide may have been significant. [828]

Identifying Information

CAS NUMBER: 57-74-9
          cis-Chlordane CAS NUMBER: 5103-74-2
          trans Chlordane CAS NUMBER: 5103-71-9

NIOSH Registry Number: PB9800000

SYNONYMS: 1,2,4,5,6,7,8,8-octachloro-2,3,3a,4,7,7a-hexahydro-4,7-methano-indene, Octachlorodihydrodicyclopentadiene

Endocrine Disruptor Information

Chlordane is a strongly suspected EED that it is listed by EPA [811], the Centers for Disease Control & Prevention [812], and the World Wildlife Fund [813] as a potential endocrine modifying chemical. It is also listed as one of the commonly found pesticides in wildlife exhibiting symptoms of EDCs. [810] EPA has ruled that chlordane is a probable human carcinogen and hepatic tumors have been produced in test animals [853] as has acute leukemia. [854] Prenatal mice exposed to chlordane exhibited immunotoxic effects. [855]

Chemical and Physical Properties

CHEMICAL FORMULA: $C_{10}H_6C_{l8}$

MOLECULAR WEIGHT: 409.80

WLN: L C555 A IUTJ AG AG BG DG EG HG IG JG

PHYSICAL DESCRIPTION: Off-white powder (pure cis- or trans- isomers); amber solid or a yellow-brown liquid (impure mixture)

SPECIFIC GRAVITY: 1.57-1.63 @ 15.5/15.5°C [043]

DENSITY: 1.59-1.63 g/mL @ 25°C [033], [172], [395], [430]

MELTING POINT: 106-107°C (cis); 104-105°C (trans) [051], [172], [173], [395]

BOILING POINT: 175°C @ 2 mm Hg (decomposes) [102]

SOLUBILITY: Water: <1 mg/mL @ 23°C [700]; DMSO: >=100 mg/mL @ 23°C [700]; 95% Ethanol: 50-100 mg/mL @ 23°C [700]; Acetone: >=100 mg/mL @ 23°C [700]

OTHER SOLVENTS: Aliphatic hydrocarbon solvents: Miscible [033], [169], [430]; Aromatic hydrocarbon solvents: Miscible [033], [169], [430]; Deodorized kerosene: Miscible [033], [062], [169], [430]; Petroleum solvents: Soluble [151]; Petroleum hydrocarbons: Soluble [395]; Cyclohexanone: Miscible [172]; Propan-2-ol: Miscible [172]; Trichloroethylene: Miscible [172]; Most organic solvents: Soluble [062], [173], [395], [421]

OTHER PHYSICAL DATA: Chlorine-like odor [102], [371]; Viscosity: 69 poises @ 25°C [033]; Refractive index: 1.56-1.57 @ 25°C [062], [172], [395], [430]; Liquid surface tension (estimated): 0.025 N/m @ 20° [051], [371]; Heat of combustion (estimated): -2200 cal/g [051], [371]

Source: *Instant EPA's Air Toxics*, 1994, Instant Reference Sources, Inc., 7605 Rockpoint Drive, Austin, TX 78731 (Fax: 512-345-2386; Internet URL, http://www.instantref.com/inst-ref.htm).

Environmental Reference Materials

A source of EED environmental reference materials for Chlordane is Radian International LLC, Austin, TX (Fax 512-454-0268; Internet URL, http://www.radian.com/standards); Catalog No. ERC-005.

Hazardous Properties

ACUTE/CHRONIC HAZARDS: Chlordane may be toxic by ingestion, inhalation or skin absorption [033], [062]. It is an irritant and may be absorbed through the skin [051], [102], [346], [371]. Effects at higher dosage levels may be cumulative [421]. When heated to decomposition it emits toxic fumes of carbon monoxide, hydrogen chloride gas, chlorine and phosgene [043], [051], [102], [371].

HAP WEIGHTING FACTOR: 10 [713]

FIRE HAZARD: Chlordane is noncombustible [102], [421]. However, the technical grade material has a flash point of 56°C (133°F) [051]. Fires involving this material can be controlled with a dry chemical, carbon dioxide or Halon extinguisher.

The autoignition temperature of technical grade material is 210°C (410°F) [051], [371].

LEL: 0.7% [051], [371];   UEL: 5% [051], [371]

REACTIVITY: Chlordane is incompatible with strong oxidizers [051], [102], [346]. It decomposes in weak alkalis [033], [062], [421], [430]. It is also decomposed by sodium in isopropyl alcohol [186]. It is corrosive to iron, zinc and various protective coatings [169] and it attacks some forms of plastics and rubber [102].

STABILITY: Chlordane is stable under normal laboratory conditions. Solutions of it in water, DMSO, 95% ethanol or acetone should be stable for 24 hours under normal lab conditions [700].

Source: *Instant EPA's Air Toxics*, 1994, Instant Reference Sources, Inc., 7605 Rockpoint Drive, Austin, TX 78731 (Fax: 512-345-2386;   Internet URL, http://www.instantref.com/inst-ref.htm).

Medical Symptoms of Exposure

Symptoms of exposure may include skin irritation, irritability, convulsions, deep depression and degenerative changes in the liver [033]. Tremors, excitement, ataxia, gastritis, central nervous system stimulation, respiratory failure, fatty degeneration and death may also occur [043]. Other symptoms may include nervousness, loss of coordination, blood dyscrasias and acute leukemia [395]. Exposure to this compound may lead to diplopia, blurring of vision, twitching of the extremities, nausea and vomiting [099]. Additional symptoms may include headache, dizziness and mild clonic jerking [406]. Symptoms may also include severe cough, dyserythropoiesis, eosinophilia, megaloblastosis, aplastic anemia, neuroblastoma, hyperexcitability of the central nervous system, hyperactive reflexes, muscle twitching, coma, anorexia, weight loss and severe gastroenteritis [151]. Exposure to this compound may also cause confusion, oliguria, proteinuria, hematuria, mild hypertension, albuminuria, paresthesia, unconsciousness, weakness, enteritis, diffuse pneumonia, lower nephron syndrome and lateral nystagmus [173]. Delirium, abdominal pain, diarrhea and anuria may also occur [102], [346]. Other symptoms may include scattered petechial hemorrhages in the lungs, kidneys and brain, damage to tubular cells of the kidneys, degenerative changes in hepatic cells and renal tubules, hyperexcitability, central nervous system depression, cognitive and emotional deterioration, impairment of memory and impairment of visual motor coordination [301]. Symptoms may also include giddiness, fatigue, pulmonary edema, liver, kidney and myocardial toxicity, hypothermia, accelerated respiration followed by depressed respiration, loss of appetite, muscular weakness and apprehensive mental state [295]. Symptoms may also include light headedness, tightness of the chest, arthralgias, sore throat, eye irritation, neurological symptoms and dry and red skin [051]. Shaking, staggering and mania may also occur [102]. Symptoms may also include gastrointestinal tract irritation [371].

Source: *Instant EPA's Air Toxics*, 1994, Instant Reference Sources, Inc., 7605 Rockpoint Drive, Austin, TX 78731 (Fax: 512-345-2386;   Internet URL, http://www.instantref.com/inst-ref.htm).

Toxicological Information

The toxicological information below is gathered from several sources including two national databases: one from the *National Toxicology Program Chemical Repository Database* and another from *EPA's Integrated Risk Information System (IRIS)*.

Toxicology Information from the National Toxicology Program

TOXICITY:

Typ. Dose	Mode	Specie	Amount	Units	Other
LC50	ihl	cat	100	mg/m3/4H	
LDLo	ipr	mus	240	mg/kg	
LD50	ipr	rat	343	mg/kg	
LD50	ivn	mus	100	mg/kg	
LDLo	ivn	rbt	10	mg/kg	
LD50	orl	ckn	220	mg/kg	
LD50	orl	dck	1200	mg/kg	
LD50	orl	dom	50	mg/kg	
LD50	orl	ham	1720	mg/kg	
LDLo	orl	hmn	29	mg/kg	
LD50	orl	mam	180	mg/kg	
LD50	orl	mus	145	mg/kg	
LD50	orl	rat	200	mg/kg	
LD50	orl	rbt	100	mg/kg	
LDLo	orl	wmn	120	ug/kg	
LDLo	skn	hmn	428	mg/kg	
LD50	skn	rat	690	mg/kg	
LD50	skn	rbt	780	mg/kg	
LDLo	unr	man	118	mg/kg	
TDLo	orl	man	3071	ug/kg	
LD50	orl	qal	83	mg/kg	

OTHER TOXICITY DATA:

Review: Toxicology Review-4

Standards and Regulations:
DOT-Hazard: Flammable liquid; Label: Flammable liquid
DOT-Hazard: Combustible liquid; Label: None

Status:
EPA TSCA Test Submission (TSCATS) Data Base, September 1989
Human fatal dose (estimated): 6 to 60 grams
IDLH level: 500 mg/m3

CARCINOGENICITY:

Tumorigenic Data:

Type of Effect	Route/Animal	Amount Dosed (Notes)
TDLo	orl-mus	2020 mg/kg/80W-C
TD	orl-mus	3780 mg/kg/80W-C

Review:

IARC Cancer Review: Animal Limited Evidence
IARC Cancer Review: Human Inadequate Evidence
IARC: Not classifiable as a human carcinogen (Group 3)

Status:

NCI Carcinogenesis Bioassay (Feed); Positive: Male and Female Mouse
NCI Carcinogenesis Bioassay (Feed); Negative: Male and Female Rat
EPA Carcinogen Assessment Group

TERATOGENICITY: See RTECS printout for data

MUTATION DATA: See RTECS printout for data

SAX TOXICITY EVALUATION:

THR: Poison to humans by ingestion and possibly other routes. An experimental poison by ingestion, inhalation, intravenous and intraperitoneal routes. Moderately toxic by skin contact. Human systemic effects by ingestion or skin contact: tremors, convulsions, excitement, ataxia and gastritis. A suspected human carcinogen. An experimental carcinogen and teratogen. Other experimental reproductive effects. Human mutagenic data.

Source: *Instant Tox-Base*, 1995, Instant Reference Sources, Inc., 7605 Rockpoint Drive, Austin, TX 78731 (Fax: 512-345-2386; Internet URL, http://www.instantref.com/inst-ref.htm).

Toxicology Information from EPA's Integrated Risk Information System (IRIS)

I. CHRONIC HEALTH HAZARD ASSESSMENTS FOR NONCARCINOGENIC EFFECTS

I.A. REFERENCE DOSE FOR CHRONIC ORAL EXPOSURE (RfD)

The Reference Dose (RfD) is based on the assumption that thresholds exist for certain toxic effects such as cellular necrosis, but may not exist for other toxic effects such as carcinogenicity. In general, the RfD is an estimate (with uncertainty spanning perhaps an order of magnitude) of a daily exposure to the human population (including sensitive subgroups) that is likely to be without an appreciable risk of deleterious effects during a lifetime. Please refer to Background Document 1 in Service Code 5 for an elaboration of these concepts. RfDs can also be derived for the noncarcinogenic health effects of compounds which are also carcinogens. Therefore, it is essential to refer to other sources of information concerning the carcinogenicity of this substance. If the US EPA has evaluated this substance for potential human carcinogenicity, a summary of that evaluation will be contained in Section II of this file when a review of that evaluation is completed.

I.A.1. ORAL RfD SUMMARY

Critical Effect	Experimental Doses*	UF	MF	RfD
Regional liver hypertrophy in females 30-Month Rat Feeding Study	NOEL: 1 pm (0.055 mg/kg/day) LEL: 5 ppm (0.273 mg/kg/day)	1000	1	6E-5 mg/kg/day

*Conversion Factors: Actual dose tested

Velsicol Chemical Co., 1983a

I.A.2. PRINCIPAL AND SUPPORTING STUDIES (ORAL RfD)

Velsicol Chemical Company. 1983a. MRID No. 00138591, 00144313. Available from EPA. Write to FOI, EPA, Washington, DC 20460.

Charles River Fischer 344 rats (80/sex/dose) were fed technical chlordane at dietary levels of 0, 1, 5, and 25 ppm for 130 weeks. Body weight, food consumption, and water uptake were monitored at regular intervals. Clinical laboratory studies were performed and organ weights measured on eight animals/sex/group at weeks 26 and 52, and on all survivors at week 130. Gross and microscopic pathology were performed on all tissues. Daily dose level of 0.045, 0.229, and 1.175 mg/kg/day for males and 0.055, 0.273, and 1.409 mg/kg/day for females for the 1, 5, and 25 ppm treatment groups, respectively, were calculated from food consumption and body weight data.

Following the submission of a 30-month chronic feeding/oncogenicity study in Fischer 344 rats, the Agency reviews by the Office of Pesticides Programs and the Cancer Assessment Group of these data indicated that male rats at the highest dosage exhibited an increase in liver tumors (ICF Clement, 1987). The registrant, Velsicol Chemical Company, subsequently convened the Pathology Working Group to reevaluate the slides of livers of the chlordane-treated rats reported in MRID No. 00138591. It was concluded that liver lesions had not occurred in male rats and that 25 ppm (0.1175 mg/kg/day) was the NOEL for males. Liver lesions (hypertrophy), however, had occurred in female rats at 5 ppm (0.273 mg/kg/day), which was considered an LEL. Therefore an NOEL of 1 ppm (0.055 mg/kg/day) (LDT) was established for female rats.

I.A.3. UNCERTAINTY AND MODIFYING FACTORS (ORAL RfD)

UF -- An uncertainty factor of 100 was used to account for the inter- and intraspecies differences. An additional UF of 10 was used to account for the lack of an adequate reproduction study and adequate chronic study in a second mammalian species, and the generally inadequate sensitive endpoints studied in existing studies, particularly since chlordane is known to bioaccumulate over a chronic duration.

MF -- None

I.A.4. ADDITIONAL COMMENTS (ORAL RfD)

Data Considered for Establishing the RfD

1) 30-Month Feeding (oncogenic) - rat: Principal study - see previous description; core grade minimum

2) 24-Month Chronic Toxicity - mouse: NOEL=1 ppm (0.15 mg/kg/day); LEL=5 ppm (0.75 mg/kg/day) (hepatocellular swelling and necrosis in males; hepatocyte swelling in males, and increased live weight in males and females); At 12.5 ppm (1.875 mg/kg/day) (HDT); core grade minimum (Velsicol Chemical Co., 1983b)

Data Gap(s): Chronic Dog Feeding Study, Rat Reproduction Study, Rat Teratology Study, Rabbit Teratology Study

I.A.5. CONFIDENCE IN THE ORAL RfD

Study -- Medium
Data Base -- Low
RfD -- Low

The critical study is of adequate quality and is given a medium rating. The data base is given a low confidence rating because of 1) the lack of an adequate reproduction study and adequate chronic study in a second mammalian species and 2) inadequate sensitive endpoints studied in existing studies, particularly since chlordane is known to bioaccumulate over a chronic duration. Low confidence in the RfD follows.

I.A.6. EPA DOCUMENTATION AND REVIEW OF THE ORAL RfD

Source Document -- This assessment is not presented in any existing US EPA document.

Other EPA Documentation -- Pesticide Registration Standard, November 1986; Pesticide Registration Files

Agency Work Group Review -- 12/18/85, 03/22/89

Verification Date -- 03/22/89

I.A.7. EPA CONTACTS (ORAL RfD)

George Ghali / OPP -- (703)557-7490

William Burnam / OPP -- (703)557-7491

I.B. REFERENCE CONCENTRATION FOR CHRONIC INHALATION EXPOSURE (RfC)

A risk assessment for this substance/agent is under review by an EPA work group.

II. CARCINOGENICITY ASSESSMENT FOR LIFETIME EXPOSURE

Section II provides information on three aspects of the carcinogenic risk assessment for the agent in question; the US EPA classification, and quantitative estimates of risk from oral exposure and from inhalation exposure. The classification reflects a weight-of-evidence judgment of the likelihood that the agent is a human carcinogen. The quantitative risk estimates are presented in three ways. The slope factor is the result of application of a low-dose extrapolation procedure and is presented as the risk per (mg/kg)/day. The unit risk is the quantitative estimate in terms of either risk per ug/L drinking water or risk per ug/cu.m air breathed. The third form in which risk is presented is a drinking water or air concentration providing cancer risks of 1 in 10,000, 1 in 100,000 or 1 in 1,000,000. Background Document 2 (Service Code 5) provides details on the rationale and methods used to derive the carcinogenicity values found in IRIS. Users are referred to Section I for information on long-term toxic effects other than carcinogenicity.

II.A. EVIDENCE FOR CLASSIFICATION AS TO HUMAN CARCINOGENICITY

II.A.1. WEIGHT-OF-EVIDENCE CLASSIFICATION

Classification -- B2; probable human carcinogen

Basis -- Sufficient evidence in studies in which benign and malignant liver tumors were induced in four strains of mice of both sexes and in F344 male rats; structurally related to other liver carcinogens

II.A.2. HUMAN CARCINOGENICITY DATA

Inadequate. There were 11 case reports involving central nervous system effects, blood dyscrasias and neuroblastomas in children with pre-/postnatal exposure to chlordane and heptachlor (Infante et al., 1978). As no other information was available, no conclusions can be drawn.

There were three epidemiological studies of workers exposed to chlordane and/or heptachlor. One study of pesticide applicators was considered inadequate in sample size and duration of follow-up. This study showed marginal statistically significant increased mortality from bladder cancer (3 observed) (Wang and McMahon, 1979a). The other two studies were of pesticide manufacturing workers. Neither of them showed any statistically significantly increased cancer mortality (Wang and McMahon, 1979b; Ditraglia et al., 1981). Both these populations also had confounding exposures from other chemicals.

II.A.3. ANIMAL CARCINOGENICITY DATA

Sufficient.   Chlordane has been studied in four mouse and four rat long-term carcinogenesis bioassays.   Dose-related incidences of liver carcinoma constitute the major finding in mice.   Becker and Sell (1979) tested chlordane (90:10 mixture of chlordane to heptachlor) in C57B1/6N mice, a strain historically known not to develop spontaneous liver tumors. An unspecified number of mice were fed chlordane at 0, 25 and 50 ppm (0, 3.57, 7.14 mg/kg bw) for 18 months. None of the controls developed tumors or nodular lesions of the liver. Twenty-seven percent (16 mice) of the surviving treated mice developed primary hepatocellular carcinomas. Velsicol (1973) fed groups of 100 male and 100 female CD-1 mice diets with 0, 5, 25 or 50 ppm analytical grade chlordane for 18 months.   A significant (p<0.01) dose-related increase in nodular hyperplasias in the liver of male and female mice was reported at the two highest dose levels.   A histological review by Reuber (US EPA, 1985) reported a high incidence (p<0.01) of hepatic carcinomas instead of hyperplastic nodules at 25 and 50 ppm.

A dose-related increase (p<0.001 after lifetable adjustment) of hepatocellular carcinomas was also observed in both sexes of B6C3F1 mice (NCI, 1977). Male and female mice were fed technical-grade chlordane (purity = 94.8%) at TWA concentrations (TWAC) of 29.9 and 56.2 ppm and 30.1 and 63.8 ppm, respectively, for 80 weeks. In this study there were individual matched controls for the low and high dose groups. ICR male mice developed hepatocellular adenomas and hemangiomas when fed 12.5 ppm chlordane for 24 months. No tumors were observed in the female mice when tested at the same concentrations: 0, 1, 5, and 12.5 ppm (Velsicol, 1983a).

Velsicol (1983b) reported a long-term (130 weeks) carcinogenesis bioassay on 80 male and 80 female F344 rats fed concentrations of 0, 1, 5, and 25 ppm chlordane.   A significant increase in adenomas of the liver was observed in male rats receiving 25 ppm. Although no tumors were observed in female rats, hepatocellular swelling was significantly increased at 25 ppm.   The NCI (1977) reported a significant increase (p<0.05) of neoplastic nodules of the liver in low-dose Osborne-Mendel female rats (TWAC of 120.8 ppm) but not in the high-dose group (TWAC of 241.5 ppm). No tumor incidence was reported for the males fed TWAC of 203.5 and 407 ppm. Loss of body weight and a dose-related increase in mortality was observed in all treated groups. High mortality and reduced growth rates in Osborne-Mendel rats was also observed by Ingle (1952) when the rats were exposed to 150 and 300 ppm chlordane but not at 5, 10, and 30 ppm. No treatment-related incidence of tumors was reported. Significantly enlarged livers and liver lesions were found in male and female albino rats fed chlordane at greater than or equal to 80 ppm (Ambrose et al., 1953a,b). No treatment-related increase in tumors was found, but the study duration (400 days) was short.

II.A.4. SUPPORTING DATA FOR CARCINOGENICITY

Gene mutation assays indicate that chlordane is not mutagenic in bacteria (Wildeman and Nazar, 1982; Probst et al., 1981; Gentile et al., 1982). Positive results have been reported in Chinese hamster lung V79 cells and mouse lymphoma L5178Y cells with and without exogenous metabolism, as well as in plant assays. Chlordane did not induce DNA repair in bacteria, rodent hepatocytes (Maslansky and Williams, 1981), or human lymphoid cells (Sobti et al., 1983). It is a genotoxicant in yeast (Gentile et al., 1982; Chambers and Dutta, 1976), human fibroblasts (Ahmed et al., 1977), and fish (Vigfusson et al., 1983).

Five compounds structurally related to chlordane (aldrin, dieldrin, heptachlor, heptachlor epoxide, and chlorendic acid) have produced liver tumors in mice. Chlorendic acid has also produced liver tumors in rats.

II.B. QUANTITATIVE ESTIMATE OF CARCINOGENIC RISK FROM ORAL EXPOSURE

II.B.1. SUMMARY OF RISK ESTIMATES

Oral Slope Factor -- 1.3E+0 per (mg/kg)/day
Drinking Water Unit Risk -- 3.7E-5 per (ug/L)
Extrapolation Method -- Linearized multistage procedure, extra risk

Drinking Water Concentrations at Specified Risk Levels:

Risk Level	Concentration
E-4 (1 in 10,000)	3E+0 ug/L
E-5 (1 in 100,000)	3E-1 ug/L
E-6 (1 in 1,000,000)	3E-2 ug/L

II.B.2. DOSE-RESPONSE DATA (CARCINOGENICITY, ORAL EXPOSURE)

Tumor Type -- hepatocellular carcinoma
Test Animals -- mouse/CD-1 (Velsicol); mouse/B6C3F1 (NCI)
Route -- diet
Reference -- Velsicol, 1973; NCI, 1977

Administered Dose (ppm)	Human Equivalent Dose (mg/kg-day)	Tumor Incidence	Reference
female	0.000	0/45	Velsicol, 1973
0	0.052	0/61	
5	0.260	32/50	
25	0.520	26/37	
50			
male	0.000	3/33	Velsicol, 1973
0	0.052	5/55	
5	0.260	41/5232/30	
25	0.520		
50			
male	0.00	2/18	NCI, 1977
0	0.31	16/48	
29.9	0.58	43/49	
56.2			
female	0.00	0.19	NCI, 1977
0	0.31	3/47	
30.1	0.66	34/49	
63.8			

II.B.3. ADDITIONAL COMMENTS (CARCINOGENICITY, ORAL EXPOSURE)

Four data sets for mice and one data set for rats showed a significant increase in liver tumors; namely hepatocellular carcinomas in mice (NCI, 1977; Velsicol, 1973) and hepatocellular adenomas in rats (Velsicol, 1983a). The quantitative estimate is based on the geometric mean from the four mouse data sets as mice were the more sensitive species tested and as risk estimates for a similar compound (heptachlor) were similarly derived from mouse tumor data. The slope factors for the data sets are these: 2.98 per (mg/kg)/day for CD-1 female mice, 4.74 per (mg/kg)/day for CD-1 male mice, 0.76 per (mg/kg)/day for B6C3F1 male mice, and 0.25 per (mg/kg)/day for B6C3F1 female mice. Low and high dose groups in the NCI (1977) study had individual matched controls.

The unit risk should not be used if the water concentration exceeds 300 ug/L, since above this concentration the unit risk may not be appropriate.

II.B.4. DISCUSSION OF CONFIDENCE (CARCINOGENICITY, ORAL EXPOSURE)

Liver carcinomas were induced in mice of both sexes in two studies. An adequate number of animals was observed, and dose-response effects were reported in all studies. The geometric mean of slope factors (0.25 to 4.74 per (mg/kg)/day for the most sensitive species is consistent with that derived from rat data (1.11/mg/kg/day).

II.C.   QUANTITATIVE   ESTIMATE   OF   CARCINOGENIC   RISK   FROM
INHALATION EXPOSURE

II.C.1. SUMMARY OF RISK ESTIMATES

Inhalation Unit Risk -- 3.7E-4 per (ug/cu.m)

Extrapolation Method -- Linearized multistage procedure, extra risk

Air Concentrations at Specified Risk Levels:

```
-------------------------------------------------
  Risk Level              Concentration
------------------------   ------------------
  E-4 (1 in 10,000)        3E-1 ug/cu.m
  E-5 (1 in 100,000)       3E-2 ug/cu.m
  E-6 (1 in 1,000,000)     3E-3 ug/cu.m


-------------------------------------------------
```

II.C.2.   DOSE-RESPONSE   DATA   FOR   CARCINOGENICITY,   INHALATION
EXPOSURE

The inhalation risk estimates were calculated from the oral data presented in II.B.2.

II.C.3.   ADDITIONAL   COMMENTS   (CARCINOGENICITY,   INHALATION
EXPOSURE)

The unit risk should not be used if the air concentration exceeds 30 ug/cu.m, above this
concentration the unit risk may not be appropriate.

II.C.4.   DISCUSSION   OF   CONFIDENCE   (CARCINOGENICITY,   INHALATION
EXPOSURE)

See II.B.4.

II.D. EPA DOCUMENTATION, REVIEW, AND CONTACTS (CARCINOGENICITY
ASSESSMENT)

II.D.1. EPA DOCUMENTATION

Source Document -- US EPA, 1986, 1985

The   values   in   the   1986   Carcinogenicity   Assessment   for   Chlordane   and
Heptachlor/Heptachlor Epoxide have been reviewed by the Carcinogen Assessment
Group.

II.D.2. REVIEW (CARCINOGENICITY ASSESSMENT)

Agency Work Group Review -- 04/01/87

Verification Date -- 04/01/87

II.D.3. US EPA CONTACTS (CARCINOGENICITY ASSESSMENT)

Dharm V. Singh / NCEA -- (202)260-5958

Jim Cogliano / NCEA -- (202)260-3814

III. HEALTH HAZARD ASSESSMENTS FOR VARIED EXPOSURE DURATIONS

III.A. DRINKING WATER HEALTH ADVISORIES

The Office of Drinking Water provides Drinking Water Health Advisories (HAs) as technical guidance for the protection of public health. HAs are not enforceable Federal standards. HAs are concentrations of a substance in drinking water estimated to have negligible deleterious effects in humans, when ingested, for a specified period of time. Exposure to the substance from other media is considered only in the derivation of the lifetime HA. Given the absence of chemical-specific data, the assumed fraction of total intake from drinking water is 20%. The lifetime HA is calculated from the Drinking Water Equivalent Level (DWEL) which, in turn, is based on the Oral Chronic Reference Dose. Lifetime HAs are not derived for compounds which are potentially carcinogenic for humans because of the difference in assumptions concerning toxic threshold for carcinogenic and noncarcinogenic effects.   A more detailed description of the assumptions and methods used in the derivation of HAs is provided in Background Document 3 in Service Code 5.

III.A.1. ONE-DAY HEALTH ADVISORY FOR A CHILD

Appropriate data for calculating a One-day HA are not available. It is recommended that the Ten-day HA of 0.06 mg/L be used as the One-day HA. III.A.2. TEN-DAY HEALTH ADVISORY FOR A CHILD

Ten-day HA -- 6E-2 mg/L

LOAEL -- 6.25 mg/kg/day
UF -- 1000 (allows for interspecies and intrahuman variability with the use of a LOAEL from an animal study)
Assumptions -- 1 L/day water consumption for a 10-kg child

Principal Study -- Ambrose et al., 1953

The toxic effects in rats resulting from daily gastric intubation of chlordane at doses of 6.25, 12.5, 25.0, 50.0, 100.0, or 200 mg/kg for 15 days were histologic changes in the liver of the treated animals at all dose levels and central nervous system effects at higher dose levels. Only minimal histopathologic changes characterized by the presence of abnormal intra-cytoplasmic bodies of various diameters were evident at the lowest dose level (6.25 mg/kg). That dose level was identified as the LOAEL in this study.

III.A.3. LONGER-TERM HEALTH ADVISORY FOR A CHILD

Appropriate data for calculating a Longer-term HA are not available. It is recommended that the modified DWEL (adjusted for a 10-kg child) of 0.5 ug/L be used as the Longer-term HA.

III.A.4. LONGER-TERM HEALTH ADVISORY FOR AN ADULT

Appropriate data for calculating a Longer-term HA are not available. It is recommended that the DWEL of 2 ug/L be used as the Longer-term HA for the 70-kg adult.

III.A.5. DRINKING WATER EQUIVALENT LEVEL / LIFETIME HEALTH ADVISORY

DWEL -- 2E-3 mg/L

Assumptions -- 2 L/day water consumption for a 70-kg adult

RfD Verification Date = 03/22/89 (see Section I.A. of this file)

Lifetime HA -- None

Chlordane is considered to be a probable human carcinogen. Refer to Section II of this file for information on the carcinogenicity of this substance.

Principal Study (DWEL) -- Velsicol Chemical Corporation, 1983 (This study was used in the derivation of the chronic oral RfD; see Section I.A.2.)

III.A.6. ORGANOLEPTIC PROPERTIES

[Ed. Note: EPA does not have information yet in this section of IRIS].

III.A.7. ANALYTICAL METHODS FOR DETECTION IN DRINKING WATER

Determination of chlordane is by a liquid-liquid extraction gas chromatographic procedure.

III.A.8. WATER TREATMENT

Treatment technologies which are capable of removing chlordane from drinking water include adsorption by granular or powdered activated carbon and air stripping.

III.A.9. DOCUMENTATION AND REVIEW OF HAs

US EPA. 1985. Final Draft of the Drinking Water Criteria Document on Chlordane. Office of Drinking Water, Washington, DC.

EPA review of HAs in 1985.

Public review of HAs following notification of availability in October, 1985.

Scientific Advisory Panel review of HAs in January, 1986.

Preparation date of this IRIS summary -- 06/17/87

III.A.10. EPA CONTACTS

Jennifer Orme Zavaleta / OST -- (202)260-7586

Edward V. Ohanian / OST -- (202)260-7571

III.B. OTHER ASSESSMENTS

Content to be determined.

IV. US EPA REGULATORY ACTIONS

EPA risk assessments may be updated as new data are published and as assessment methodologies evolve. Regulatory actions are frequently not updated at the same time. Compare the dates for the regulatory actions in this section with the verification dates for the risk assessments in sections I and II, as this may explain inconsistencies. Also note that some regulatory actions consider factors not related to health risk, such as technical or economic feasibility. Such considerations are indicated for each action. In addition, not all of the regulatory actions listed in this section involve enforceable federal standards. Please direct any questions you may have concerning these regulatory actions to the US EPA contact listed for that particular action. Users are strongly urged to read the background information on each regulatory action in Background Document 4 in Service Code 5.

IV.A. CLEAN AIR ACT (CAA)

[Ed. Note: EPA does not have information yet in this section of IRIS].

IV.B. SAFE DRINKING WATER ACT (SDWA)

IV.B.1. MAXIMUM CONTAMINANT LEVEL GOAL (MCLG) for Drinking Water

Value (status) -- 0 mg/L (Final, 1991)

Considers technological or economic feasibility? -- NO

Discussion -- An MCLG of 0 mg/L for chlordane is promulgated based upon carcinogenic effects (B2).

Reference -- 56 FR 3526 (01/30/91)

EPA Contact -- Health and Ecological Criteria Division / OST /(202) 260-7571 / FTS 260-7571; or Safe Drinking Water Hotline / (800) 426-4791

IV.B.2. MAXIMUM CONTAMINANT LEVEL (MCL) for Drinking Water

Value -- 0.002 mg/L (Final, 1991)

Considers technological or economic feasibility? -- YES

Discussion -- EPA has set a MCL equal to the PQL of 0.002, which is associated with a lifetime individual risk of 1.5 E-4.

Monitoring requirements -- All systems monitored for four consecutive quarters every three years; repeat monitoring dependent upon detection, vulnerability status and system size.

Analytical methodology -- Microextraction/gas chromatography (EPA 505):electron-capture/gas chromatography (EPA 508); gas chromatography/mass spectrometry (EPA 525): PQL= 0.002 mg/L.

Best available technology -- Granular activated carbon

Reference -- 56 FR 3526 (01/30/91)

EPA Contact -- Drinking Water Standards Division / OGWDW /(202) 260-7575 / FTS 260-7575; or Safe Drinking Water Hotline / (800) 426-4791

IV.B.3. SECONDARY MAXIMUM CONTAMINANT LEVEL (SMCL) for Drinking Water

[Ed. Note: EPA does not have information yet in this section of IRIS].

IV.B.4. REQUIRED MONITORING OF "UNREGULATED" CONTAMINANTS

[Ed. Note: EPA does not have information yet in this section of IRIS].

IV.C. CLEAN WATER ACT (CWA)

IV.C.1. AMBIENT WATER QUALITY CRITERIA, Human Health

Water and Fish Consumption: 4.6E-4 ug/L

Fish Consumption Only: 4.8E-4 ug/L

Considers technological or economic feasibility? -- NO

Discussion -- For the maximum protection from the potential carcinogenic properties of this chemical, the ambient water concentration should be zero. However, zero may not be obtainable at this time, so the recommended criteria represents an E-6 estimated incremental increase of cancer risk over a lifetime.

Reference -- 45 FR 79318 (11/28/80)

EPA Contact -- Criteria and Standards Division / OWRS (202)260-1315 / FTS 260-1315

IV.C.2. AMBIENT WATER QUALITY CRITERIA, Aquatic Organisms

Freshwater:
    Acute -- 2.4 E+ 0 ug/L (at any time)
    Chronic -- 4.3 E- 3 ug/L (24-hour average)

Marine:
    Acute -- 9.0 E-2 ug/L (at any time)
    Chronic -- 4.0 E-3 ug/L (24-hour average)

Considers technological or economic feasibility? -- NO

Discussion -- Criteria were derived from a minimum data base consisting of acute and chronic tests on a variety of species. Requirements and methods are covered in the reference to the Federal Register.

Reference -- 45 FR 79318 (11/28/80)

EPA Contact -- Criteria and Standards Division / OWRS (202)260-1315 / FTS 260-1315

IV.D. FEDERAL INSECTICIDE, FUNGICIDE, AND RODENTICIDE ACT (FIFRA)

IV.D.1. PESTICIDE ACTIVE INGREDIENT, Registration Standard

Status -- Issued (1986)

Reference -- Chlordane Pesticide Registration Standard. December, 1986 (NTIS No. PB87-175816).

EPA Contact -- Registration Branch, OPP / (703)557-7760 / FTS 557-7760

IV.D.2. PESTICIDE ACTIVE INGREDIENT, Special Review

Action -- Cancellation of many termiticide products (1988)

Considers technological or economic feasibility? -- YES

Summary of regulatory action -- 43 FR 12372 (03/24/78) - Cancellation of all but termiticide use; under the provisions of the Administrator's acceptance of the settlement plan to phase out certain uses of chlordane, most registered products containing chlordane were effectively canceled or the applications for registration were denied by 12/31/80. A summary of those uses not affected by this settlement, or a previous suspension, follows: 1) subsurface ground insertion for termite control (clarified by 40 FR 30522, July 21, 1975, to apply to the use of emulsifiable or oil concentrate formulations for controlling subterranean termites on structural sites such as buildings, houses, barns, and sheds, using current control practices), 2) dipping of nonfood roots and tops./52 FR 42145 (11/03/87) - Negotiated agreement ontermiticide use. The agreement (order) accepted voluntary cancellations of the registration of certain pesticide products and imposed limitations on the continued sale, distribution, and use of existing stocks of such products/criterion of concern: oncogenicity,

Reference -- 43 FR 12372 (03/24/78); 52 FR 42145 (11/03/87); 53 FR 11798 (04/08/88)

EPA Contact -- Special Review Branch / OPP (703)557-7400 / FTS 557-7400

IV.E. TOXIC SUBSTANCES CONTROL ACT (TSCA)

[Ed. Note: EPA does not have information yet in this section of IRIS].

IV.F. RESOURCE CONSERVATION AND RECOVERY ACT (RCRA)

IV.F.1. RCRA APPENDIX IX, for Ground Water Monitoring

Status -- Listed

Reference -- 52 FR 25942 (07/09/87)

EPA Contact -- RCRA/Superfund Hotline (800)424-9346 / (202)260-3000 / FTS 260-3000

IV.G. SUPERFUND (CERCLA)

IV.G.1. REPORTABLE QUANTITY (RQ) for Release into the Environment

Value (status) -- 1 pound (Final, 1989)

Considers technological or economic feasibility? -- NO

Discussion -- The RQ for chlordane is 1 pound, based on aquatic toxicity, as established under CWA Section 311 (40 CFR 117.3). Available data indicate the aquatic 96-hour Median Threshold Limit for chlordane is less than 0.1 ppm. This corresponds to an RQ of 1 pound. Chlordane has also been found to bioaccumulate in the tissues of aquatic and marine organisms, and has the potential to concentrate in the food chain.

Reference -- 54 FR 33418 (08/14/89)

EPA Contact -- RCRA/Superfund Hotline (800)424-9346 / (202)260-3000 / FTS 260-3000

VI. BIBLIOGRAPHY

VI.A. ORAL RfD REFERENCES

ICF-Clement. 1987. MRID No. 40433701. Available from EPA. Write to FOI, EPA, Washington DC 20460.

Velsicol Chemical Co. 1983a. MRID No. 00138591, 00144313. Available from EPA. Write to FOI, EPA, Washington DC 20460.

Velsicol Chemical Co. 1983b. MRID No. 00144312. Available from EPA. Write to FOI, EPA, Washington DC 20460.

VI.B. INHALATION RfD REFERENCES

None

VI.C. CARCINOGENICITY ASSESSMENT REFERENCES

Ahmed, F.E., R.W. Hart and N.J. Lewis. 1977. Pesticide induced DNA damage and its repair in cultured human cells. Mutat. Res. 42: 161-174.

Ambrose, A.M., H.E. Christensen, D.J. Robbins and L.J. Rather. 1953a. Toxicological and pharmacological studies on chlordane. Arch. Ind. Hyg. Occup. Med. 7: 197-210.

Ambrose, A.M., H.E. Christensen and D.J. Robbins. 1953b. Pharmacological observations on chlordane. Fed. Proceed. 12: 298. (Abstract #982)

Becker, F.F. and S. Sell. 1979. Fetoprotein levels and hepatic alterations during chemical carcinogenesis in C57BL/6N mice. Cancer Res. 39: 3491-3494.

Chambers, D. and S.K. Dutta. 1976. Mutagenic tests of chlordane on different microbial tester strains. Genetics. 83: s13. (Abstract)

Ditraglia, D., D.P. Brown, T. Namekata and N. Iverson. 1981. Mortality study of workers employed at organochlorine pesticide manufacturing plants. Scand. J. Work Environ. Health. 7(4): 140-146.

Gentile, J.M., G.J. Gentile, J. Bultman, R. Sechriest, E.D. Wagner and M.J. Plewa. 1982. An evaluation of the genotoxic properties of insecticides following plant and animal activation. Mutat. Res. 101: 19-29.

Infante, P.F., S.S. Epstein and W.A. Newton. 1978. Blooddyscrasis and childhood tumors and exposure to chlordane and heptachlor. Scand. J. Work Environ. Health. 4: 137-150.

Ingle, L. 1952. Chronic oral toxicity of chlordan to rats. Arch. Ind. Hyg. Occup. Med. 6: 357-367.

Maslansky, C.J. and G.M. Williams. 1981. Evidence for an epigenetic mode of action in organochlorine pesticide hepatocarcinogenicity: A lack of genotoxicity in rat, mouse, and hamster hepatocytes. J. Toxicol. Environ. Health. 8: 121-130.

NCI (National Cancer Institute). 1977. Bioassay of Chlordane for possible Carcinogenicity. NCI Carcinogenesis Tech. Rep. Ser. No. 8. US DHEW Publ. No. (NIH) 77-808. Bethesda, MD.

Probst, G.S., R.E. McMahon, L.E. Hill, C.Z. Thompson, J.K. Epp and S.B. Neal. 1981. Chemically-induced unscheduled DNA synthesis in primary rat hepatocyte cultures: A comparison with bacterial mutagenicity using 218 compounds. Environ. Mutagen. 3: 11-31.

Sobti, R.C., A. Krishan and J. Davies. 1983. Cytokinetic and cytogenetic effect of agricultural chemicals on human lymphoid cells in vitro. Arch. Toxicol. 52: 221-231.

US EPA. 1985. Hearing Files on Chlordane, Heptachlor Suspension (unpublished draft). Available for inspection at US EPA, Washington, DC.

US EPA. 1986. Carcinogenicity Assessment of Chlordane and Heptachlor/ Heptachlor Epoxide. Prepared by the Office of Health and Environmental Assessment, Carcinogen Assessment Group, Washington, DC.

Velsicol Chemical Corporation. 1973. MRID No. 00067568. Available from EPA. Write to FOI, EPA, Washington, DC. 20460.

Velsicol Chemical Corporation. 1983a. MRID No. 00144312, 00132566. Available from EPA. Write to FOI, EPA, Washington, DC. 20460.

Velsicol Chemical Corporation. 1983b. MRID No. 00138591. Available from EPA. Write to FOI, EPA, Washington, DC. 20460.

Vigfusson, N.V., E.R. Vyse, C.A. Pernsteiner and R.J. Dawson. 1983. In vivo induction of sister-chromatid exchange in Umbra limi by the insecticides endrin, chlordane, diazinon and guthion. Mutat. Res. 118: 61-68.

Wang, H.H. and B. MacMahon. 1979a. Mortality of workers employed in the manufacture of chlordane and heptachlor. J. Occup. Med. 21(11): 745-748.

Wang, H.H. and B. MacMahon. 1979b. Mortality of pesticide applicators. J. Occup. Med. 21(11): 741-744.

Wildeman, A.G. and R.N. Nazar. 1982. Significance of plant metabolism in the mutagenicity and toxicity of pesticides. Can. J. Genet. Cytol. 24: 437-449.

VI.D. DRINKING WATER HA REFERENCES

Ambrose, A.M., H.E. Christensen, D.J. Robbins and L.J. Rather. 1953. Toxicological and pharmacological studies on chlordane. Arch. Ind. Hyg. Occup. Med. 7: 197-210.

US EPA. 1985. Final Draft of the Drinking Water Criteria Document on Chlordane. Office of Drinking Water, Washington, DC.

Velsicol Chemical Corp. 1983. MRID No. 00138591. Available from EPA Write to FOI, EPA, Washington, DC 20460.

VII. REVISION HISTORY

Date	Section	Description
09/30/87	II.	Carcinogenicity section added
03/01/88	I.A.1.	Dose conversion clarified
03/01/88	I.A.2.	Text clarified in paragraph 3
03/01/88	II.A.1.	Basis for classification clarified
03/01/88	III.A.	Health Advisory added
04/01/89	I.A.	Withdrawn; new RfD verified (in preparation)
06/01/89	I.A.	Revised oral RfD summary added
06/01/89	VI.	Bibliography on-line
07/01/89	I.A.2.	Reference clarified in paragraph 2
07/01/89	II.	Velsicol (1983) references clarified
07/01/89	VI.C	Carcinogen references added
03/01/90	I.B.	Inhalation RfD now under review
08/01/90	III.A.5.	DWEL changed reflecting change in RfD
08/01/90	III.A.10	Primary contact changed
08/01/90	IV.F.1.	EPA contact changed
01/01/91	II.	Text edited
01/01/91	II.C.1.	Inhalation slope factor removed (global change)
01/01/92	IV.	Regulatory actions updated
07/01/93	II.D.3.	Secondary contact's phone number changed

SYNONYMS

Belt	Niran
CD 68	Octachlorodihydrodicyclopentadiene
Chlordane	1,2,4,5,6,7,8,8-Octachloro-2,3,3a,4,7,7a-
Chlorindan	Hexahydro-4,7-Methano-indene
Chlor Kil	1,2,4,5,6,7,8,8-Octachloro-3a,4,7,7a-
Corodan	Hexahydro-4,7-Methylene Indane
Dowchlor	Octachloro-4,7-Methanohydroindane
ENT 9,932	Octachloro-4,7-Methanotetrahydroindane
HCS 3260	Octa-Klor
Kypchlor	Oktaterr
M 140	Ortho-Klor
M 410	Synklor
4,7-Methanoindan, 1,2,4,5,6,7,8,8-	TAT Chlor 4
Octachloro-3a,4,7,7a-Tetrahydro-	Topiclor
4,7-Methano-1H-Indene, 1,2,4,5,6,7,8,8-	Toxichlor
Octachloro-2,3,3a,4,7,7a-Hexahydro-	Velsicol 1068
NCI-C00099	

Source: *Instant EPA's IRIS*, 1997, Instant Reference Sources, Inc., 7605 Rockpoint Drive, Austin, TX 78731 (Fax: 512-345-2386; Internet URL, http://www.instantref.com/inst-ref.htm).

Production, Use, and Pesticide Labeling Information

1.Description of Chemical

Generic Name: 1,2,4,5,6,7,8,8-octachloro-2,3,3a,4,7,7a-hexahydro-4,7-methanoindene

Common Name: Chlordane

Trade Names: 1,2,4,5,6,7,8,8-octachloro-3a,4,7,7a-tetrahydro-4,7-methanoindan; Velsicol 1068; Velsicol 168; M-410; Belt; Chlor-Kil; Chlortox; Corodane; Gold Crest C-100; Gold Crest C-50; Kilex; Kypchlor; Niran; Octachlor; Synchlor; Termi-Ded; Topiclor 20; Chlordan; Prentox; and Penticklor.

EPA Shaughnessy Code: 058201

Chemical Abstracts Service (CAS) Number: 57-47-9

Year of Initial Registration: 1948

Pesticide Type: Insecticide

Chemical Family: Chlorinated cyclodiene

US and Foreign Producers: Velsicol Chemical Corporation

2. Use Patterns and Formulations

- Application Sites: Subsurface soil treatment for termite control; underground cables for termite control; above ground structural application for control of termites and other wood-destroying insects.

- Types of Formulations: Emulsifiable concentrates; granular; soluble concentrates

- Types and Methods of Application: trenching, rodding, subslab injection, low pressure spray for subsurface termite control; brush, spray, or dip for applying to structural wood

- Application Rates: 0.5 to 2.0% emulsion for termite control; 3.0 to 4.25% solution for above ground structural wood treatment. [EDITOR'S NOTE: See the fact sheet for heptachlor regarding EPA cancellation proceedings for both chlordane and heptachlor.]

3. Science Findings

Chemical Characteristics

- Physical State: Crystalline solid

- Color: White

- Odor: Chlorine odor

- Molecular weight and formula: 409.8 - $C_{10}H_6C_{18}$

- Melting Point: 95 to 96 °C

- Boiling Point: 118 C at 0.66 mm Hg (technical)

- Density: 1.59 - 1.63 at 25 °C

- Vapor Pressure: 0.00001 mm Hg at 25 °C (technical)

- Solubility in various solvents: Miscible with aliphatic and aromatic hydrocarbon solvents, including deodorized kerosene, insoluble in water.

- Stability: Loses its chlorine in presence of alkaline reagents and should not be formulated with any solvent, carrier, diluent or emulsifier which has an alkaline reaction (technical)

Toxicological Characteristics

- Acute oral: data gap

- Acute dermal: data Gap

- Primary dermal irritation: data gap

- Primary eye irritation: data gap (except for a 72% technical formulation)

- Skin sensitization: not a sensitizer

- Acute inhalation: data gap

- Subchronic inhalation (2-week duration) using rats or guinea pigs: data gap

- Subchronic inhalation (1-year duration) using rats: data gap

- Major routes of exposure: inhalation exposure to occupants of treated structures; dermal and respiratory exposure to termiticide applicators

- Delayed neurotoxicity: does not cause delayed neurotoxic effects

- Oncogenicity: This chemical is classified as a Group B2 oncogen (probable human oncogen).

- There are three long-term carcinogenesis bioassays of chlordane in mice which were independently conducted by investigators affiliated with the National Cancer Institute, the International Research and Development Corporation, and the Research Institute for Animal Science in Biochemistry and Toxicology, Japan. Reported in these studies were significant tumor responses in three different strains of mice (IRC, CF1, and B6C3F1) in males and females with a dose-related increase in the proportion of tumors that were malignant. In Fischer 344 rats, significant tumor responses were reported in a study conducted by the Research Institute for Animal Science in Biochemistry and Toxicology.

- Chronic feeding: Based on a rat chronic feeding study with chlordane, a Lowest Effect Level (LEL) of 0.05 mg/kg/day for liver effects has been calculated.

- Teratogenicity: data gap

- Reproduction: data gap

- Mutagenicity: Data gap. Further testing is required in all three categories (gene mutation, structural chromosome aberrations and other genotoxic effects.

Physiological and Biochemical Characteristics

- The precise mode of action in biological systems is not known. In humans, signs of acute intoxication are primarily related to the central nervous system (CNS), including hyperexcitability, convulsions, depression and death.

- Metabolism: Chlordane's major metabolite is oxychlordane. Oxychlordane has been found to be a major fat tissue residue in rats. Human fat samples frequently contain trans-nonachlor, a contaminant found in technical chlordane, as a major residue.

Environmental Characteristics

- Available data are insufficient to fully assess the environmental fate of chlordane. Data gaps exist for all applicable studies. However, available supplementary data indicate general trends of chlordane behavior in the environment. Chlordane is persistent and bioaccumulates. Chlordane is not expected to leach, since it is insoluble in water and should adsorb to the soil surface; thus it should not reach underground aquifers. However, additional data are necessary to fully assess the potential for ground-water contamination as a result of the termiticide use of chlordane.

Ecological Characteristics

- Avian acute toxicity: LD50 of 83.0 mg/kg in bobwhite quail

- Avian dietary toxicity (8-day): 858 ppm in mallard duck; 331 ppm in bobwhite quail; and 430 ppm in pheasant.

- Freshwater fish acute toxicity (96-Hr. LC50): 57 to 74.8 ug/L for bluegill; 42 to 90 ug/L for rainbow trout.

- Freshwater invertebrate toxicity: 15 to 590 ug/L for Pteronarcys and Daphnia, respectively.

Tolerance Assessment

- No tolerance reassessment for chlordane is necessary, since there are no food or feed uses. The Agency is proceeding to revoke all tolerances and replace them with action levels. The final rule is scheduled for publication in the Federal Register in early 1987.

Summary Science Statement

- Chlordane is a chlorinated cyclodiene with moderate acute toxicity. the chemical has demonstrated adverse chronic effects in mice (causing liver tumors). Chlordane may pose a significant health risk of chronic liver effects to occupants of structures treated with chlordane for termite control. This risk may be determined to be of regulatory concern, pending further evaluation. Chlordane is highly toxic to aquatic organisms and birds. Chlordane is persistent and bioaccumulates. Chlordane may have a potential for contaminating surface water; thus, a special study is required to delineate this potential. Applicator exposure studies are required to determine whether exposure to applicators may be posing health risks. special product-specific subacute inhalation testing is required to evaluate the short-term respiratory hazards to humans in structures treated with chlordane. An inhalation study of one-year duration using rats is required to assess potential hazards to humans in treated residences from this route of exposure. The Agency has been apprised of reported cases of optic neuritis associated with termiticide treatment of homes. To determine whether this is a significant health effect, the registrant must have eye tissue from the latest two-year rat oncogenicity study analyzed by neuropathologists specializing in optic tissue pathology. Data available to the Agency show an occurrence of misuse and misapplication of chlordane. The Agency is requiring restricted use classification of all end-use products containing chlordane. Application must be made either in the actual physical presence of a Certified Applicator, or if the Certified Applicator is not physically present at the site, each uncertified applicator must have completed a State approved training course in termiticide application meeting minimal EPA training requirements and be registered in the State in which the uncertified applicator is working.

4. Summary of Regulatory Position and Rationale

- EPA is currently evaluating the potential human health risks of 1) non-oncogenic chronic liver effects, and 2) oncogenic effects to determine whether additional regulatory action on chlordane may be warranted.

- In order to meet the statutory standard for continued registration, retail sale and use of all end-use products containing chlordane must be restricted to Certified Applicators or persons under their direct supervision. for purposes of chlordane use, direct supervision by a Certified Applicator means 1) the actual physical presence of a Certified Applicator at the application site during application, or 2) if the Certified Applicator is not physically present at the site, each uncertified applicator must have completed a State approved training course in termiticide application meeting minimal EPA training requirements and be registered in the State in which the uncertified applicator is working; the Certified Applicator must be available if and when needed.

- In order to meet the statutory standard for continued registration, chlordane product labels must be revised to provide specific chlordane disposal procedures, and to provide fish and wildlife toxicity warnings.

- The Agency is requiring a special monitoring study to evaluate whether and to what extent surface water contamination may be resulting from the use of chlordane as a termiticide.

- Special product-specific subacute inhalation testing is required to evaluate the respiratory hazards to humans in structures treated with termiticide products containing chlordane.

- Evaluation of eye tissue from the latest two-year rat oncogenicity study is required to determine whether chlordane's termiticide use may be causing optic neuritis in humans.

- The Agency is requiring the submission of applicator exposure data from dermal and respiratory routes of exposure.

- While data gaps are being filled, currently registered manufacturing use products and end use products containing chlordane may be sold, distributed, formulated, and used, subject to the terms and conditions specified in the Registration Standard for chlordane, and any additional regulatory action taken by the Agency. Registrants must provide or agree to develop additional data in order to maintain existing registrations.

5. Summary of Major Data Gaps

- Hydrolysis
- Photodegradation in water
- Aerobic soil metabolism
- Anaerobic soil metabolism
- Leaching and adsorption/desorption
- Aerobic aquatic metabolism
- Soil dissipation
- Chronic toxicity studies - rodents and non-rodents
- Teratogenicity
- Mutagenicity studies
- Acute toxicity studies
- Optic tissue pathology
- Special surface water monitoring studies

- Applicator exposure studies
- Indoor air exposure studies
- Special product-specific subchronic inhalation study (two-week duration using guinea pigs or rats)
- Subchronic inhalation study (one-year duration using rats)
- All product chemistry studies

6. Contact Person at EPA

> George LaRocca
> Product Manager No. 15
> Insecticide- Rodenticide Branch Registration Division (TS-767C)
> Office of Pesticide Programs
> Environmental Protection Agency
> 401 M Street, SW
> Washington, DC 20460

> Office location and telephone number:

> Room 204, Crystal Mall #2
> 1921 Jefferson Davis Highway
> Arlington, VA 22202

> Phone: (703) 557-2386

Source: *Instant EPA's Pesticide Facts*, 1994, Instant Reference Sources, Inc., 7605 Rockpoint Drive, Austin, TX 78731 (Fax: 512-345-2386;   Internet URL, http://www.instantref.com/inst-ref.htm).

Chlorpyrifos

Introduction

Chlorpyrifos is an acaricide and has a broad range of insecticidal activity and is effective by contact, ingestion and vapor action, but is not systemic. It is used for control of flies, household pests, mosquitoes (larvae and adults) and of various crop pests in soil and on foliage; and it is also used for control of ectoparasites on cattle and sheep. [828] It is also used to treat poultry, dogs, livestock premise treatment, domestic dwellings, terrestrial structures, and direct application to stagnant water. [821] It is one of the largest use urban insecticides in the US and is a member of the organophosphate family of pesticides with an annual US usage of 9 million kg/yr. [811]

In addition to treating animals, it is used to control insects on a wide variety of crops including grain crops, nut crops, bananas, cole crops, citrus, pome and strawberry fruits, forage, field and vegetable crops, lawns and ornamental plants. Chlorpyrifos is extremely toxic to fish, birds, and other wildlife and it is highly toxic to honeybees. [821]

It is manufactured in several convenient forms including emulsifiable concentrates, wettable powders, granules, suspension concentrates, dustable powder and pellets. The acceptable daily intake of chlorpyrifos is 0.003 mg/kg/day and the maximum permissible intake is 0.18 mg/day. [821]

Chlorpyrifos is released into the environment primarily from its application as an insecticide. If released to soil, chlorpyrifos can degrade by a combination of chemical hydrolysis and microbial degradation. Volatilization from soil surfaces, is expected to contribute to its loss from soil. Chlorpyrifos is tightly absorbed by soil and not expected to leach significantly. If released to water, chlorpyrifos partitions significantly from the water column to sediments. The photolysis half-life at the water surface in the US during the mid summer is about 3 to 4 weeks, however, photolysis is not expected to be a very significant removal mechanism in relatively deep waters, in the winter-time, or in any natural waters containing sufficient light attenuating material. Microbial degradation may contribute to removal in some natural waters. If released to air, chlorpyrifos will react in the vapor-phase with photochemically produced hydroxyl but it is not expected to react with ozone. Photolysis in air may contribute to its transformation. Major general population exposure to chlorpyrifos will occur through consumption of contaminated food and inhalation of contaminated air. Occupational exposure by dermal and inhalation routes may be significant. [828]

348

Identifying Information

CAS NUMBER: 2921-88-2

SYNONYMS: BRODAN; CHLORPYRIFOS-ETHYL; DETMOL U.A.; DOWCO 179; DURSBAN; DURSBAN F; ENT 27311; ERADEX; ETHION, dry; LORSBAN; O,O-DIAETHYL-O-3,5,6-TRICHLOR-2-PYRIDYLMONOTHIOPHOSPHAT; O,O-DIETHYL O-3,5,6-TRICHLORO-2-PYRIDYL PHOSPHOROTHIOATE; PHOSPHOROTHIOIC ACID, O,O-DIETHYL O-(3,5,6-TRICHLORO-2-PYRIDYL) ESTER; PYRINEX

Endocrine Disruptor Information

Chlorpyrifos is on EPA's list of suspect EEDs [811] and it is highly toxic to birds and fish as a cholinesterase inhibitor [842] and thus must be held suspect in having an impact upon the neuroendocrine system. [811] Immunologic abnormalities have been noted in the peripheral immune system of humans exposed to chlorpyrifos. [843]

Chemical and Physical Properties

CHEMICAL FORMULA: $C_9H_{11}Cl_3NO_3PS$

MOLECULAR WEIGHT: 350.59

PHYSICAL DESCRIPTION: colorless to white crystals

DENSITY: 1.398 g/cm^3 at 43.5/4°C [816], [055]

MELTING POINT: 41.5-43.5 °C [817]

BOILING POINT: decomposes at 160°C [817]

SOLUBILITY: water: ~1 mg/L at 25°C; acetone: 6.5 g/mL [816], [878]; benzene: 7.9 g/mL [816], [878]; chloroform 6.3 g/mL [816], [878]; methanol: 0.45 g/mL [816], [878]

OTHER PHYSICAL DATA: Vapor pressure: 1.87×10^{-5} mm Hg at 20 °C [821]

Environmental Reference Materials

A source of EED environmental reference materials for chlorpyrifos is Crescent Chemical Co., Hauppauge, NY (Fax: 516-348-0913); Catalog No. 45830.

Hazardous Properties

ACUTE/CHRONIC HAZARDS: Chlorpyrifos has moderate mammalian toxicity. Poisoning from chlorpyrifos impairs the central nervous system function. Studies available to EPA do not suggest oncogenicity or teratogenicity potential. The Agency has determined that this chemical is not teratogenic at levels up to 25 mg/kg/day. However, the oncogenicity and mutagenicity studies used to draw these conclusions are not up to current Agency standards. [821]

Medical Symptoms of Exposure

Symptoms include headache, dizziness, extreme weakness, ataxia, tiny pupils, twitching, tremor, nausea, slow heartbeat, pulmonary edema, and sweating. Continual absorption at intermediate dosages may cause influenza-like illness which includes symptoms like weakness, anorexia, and malaise. Conjunctival irritation in eyes may occur but noneurotoxic responses were noted at doses up to 100 mg/kg (highest dose tested). No reproductive effects were demonstrated at dose levels up to 1.2 mg/kg/day. [821]

Toxicological Information

The toxicological information below is gathered from several sources including two national databases: one from the *National Toxicology Program Chemical Repository Database* and another from *EPA's Integrated Risk Information System (IRIS)*.

Toxicology Information from the National Toxicology Program

Chlorpyrifos is not currently listed in NTP's Chemical Repository database nor is it scheduled for addition in the future.

Toxicology Information from EPA's Integrated Risk Information System (IRIS)

I.A. REFERENCE DOSE FOR CHRONIC ORAL EXPOSURE (RfD)

The Reference Dose (RfD) is based on the assumption that thresholds exist for certain toxic effects such as cellular necrosis, but may not exist for other toxic effects such as carcinogenicity. In general, the RfD is an estimate (with uncertainty spanning perhaps an order of magnitude) of a daily exposure to the human population (including sensitive subgroups) that is likely to be without an appreciable risk of deleterious effects during a lifetime. Please refer to Background Document 1 in Service Code 5 for an elaboration of these concepts. RfDs can also be derived for the noncarcinogenic health effects of compounds which are also carcinogens. Therefore, it is essential to refer to other sources of information concerning the carcinogenicity of this substance. If the US EPA has evaluated this substance for potential human carcinogenicity, a summary of that evaluation will be contained in Section II of this file when a review of that evaluation is completed.

I.A.1. ORAL RfD SUMMARY

Critical Effect	Experimental Doses*	UF	MF	RfD
Decreased plasma ChE activity after 9 days	NOEL: 0.03 mg/kg/day	10	1	3E-3mg/kg/day
	LEL: 0.10 mg/kg/day			

*Conversion Factors: none

20-Day Human Cholinesterase Inhibition Study, Dow Chemical, 1972

I.A.2. PRINCIPAL AND SUPPORTING STUDIES (ORAL RfD)

Dow Chemical Company. 1972. Accession No. 112118. Available from EPA. Write to FOI, EPA, Washington DC 20460.

Sixteen human male volunteers were treated (4/dose) with 0, 0.014, 0.03, or 0.10 mg/kg/day of chlorpyrifos by capsule for a total of 20 days at the low and mid dose, and for 9 days at the high dose. Treatment of the high-dose group (0.1 mg/kg/day) was discontinued after 9 days due to a runny nose and blurred vision in one individual. Mean plasma ChE in this group was inhibited by about 65% compared to the control average. No effect on RBC ChE activity was apparent at any dose.

I.A.3. UNCERTAINTY AND MODIFYING FACTORS (ORAL RfD)

UF -- The UF of 10 is the standard factor allowing for the range of human sensitivity for cholinesterase inhibition.

MF -- None

I.A.4. ADDITIONAL COMMENTS (ORAL RfD) Data Considered for Establishing the RfD

1) 20-day ChE inhibition study - humans: Principal study - see previous description

2) Chronic feeding - dog: Plasma ChE NOEL=0.01 mg/kg/day (LDT); NOEL (other effects)=3.0 mg/kg/day (HDT); core grade minimum (Dow Chemical, 1971a)

3) Chronic feeding - rat: Plasma ChE NOEL=0.1 mg/kg/day; NOEL (other effects)=3.0 mg/kg/day (HDT); few animals used (Dow Chemical, 1971b)

4) Reproduction - rat: ChE NOEL=0.1 mg/kg/day; Reproductive NOEL=1.0 mg/kg/day (HDT); core grade minimum (Dow Chemical, 1971c)

5) Teratology - rat: Maternal (ChE) NOEL=0.1 mg/kg/day; Developmental NOEL=15 mg/kg/day (HDT); core grade minimum (Dow Chemical, 1983)

6) Teratology - mouse: Maternal (ChE) NOEL=0.1 mg/kg/day; Developmental NOEL=10 mg/kg/day (decreased fetal length, increased skeletal variations at 25 mg/kg/day); core grade minimum (Dow Chemical, 1979)

Data Gap(s): Chronic feeding/oncogenicity - rat.

I.A.5. CONFIDENCE IN THE ORAL RfD

Study -- MediumData Base -- MediumRfD -- Medium

The principal human study was limited because only 4 males/dose were studied, but several doses were used; thus, the study is given a medium to low confidence rating. The available experimental data in rats, dogs, rabbits, and mice appears to be similar to the NOEL for ChE inhibition in humans; thus, confidence in the data base is rated medium. Medium confidence in the RfD follows.

I.A.6. EPA DOCUMENTATION AND REVIEW OF THE ORAL RfD

Office of Pesticide Programs files

Agency Work Group Review -- 03/11/86

Verification Date -- 03/11/86

I.A.7. EPA CONTACTS (ORAL RfD)

William Burnam / OPP -- (703)557-7491

George Ghali / OPP -- (703)557-7490

I.B. REFERENCE CONCENTRATION FOR CHRONIC INHALATION EXPOSURE (RfC)

[Ed. Note: EPA does not have information yet in this section of IRIS].

II. CARCINOGENICITY ASSESSMENT FOR LIFETIME EXPOSURE

This substance/agent has not been evaluated by the US EPA for evidence of human carcinogenic potential.

III. HEALTH HAZARD ASSESSMENTS FOR VARIED EXPOSURE DURATIONS

[Ed. Note: EPA does not have information yet in this section of IRIS].

IV. US EPA REGULATORY ACTIONS

EPA risk assessments may be updated as new data are published and as assessment methodologies evolve. Regulatory actions are frequently not updated at the same time. Compare the dates for the regulatory actions in this section with the verification dates for the risk assessments in sections I and II, as this may explain inconsistencies. Also note that some regulatory actions consider factors not related to health risk, such as technical or economic feasibility. Such considerations are indicated for each action. In addition, not all of the regulatory actions listed in this section involve enforceable federal standards. Please direct any questions you may have concerning these regulatory actions to the US EPA contact listed for that particular action. Users are strongly urged to read the background information on each regulatory action in Background Document 4 in Service Code 5.

IV.A. CLEAN AIR ACT (CAA)

[Ed. Note: EPA does not have information yet in this section of IRIS].

IV.B. SAFE DRINKING WATER ACT (SDWA)

[Ed. Note: EPA does not have information yet in this section of IRIS].

IV.C. CLEAN WATER ACT (CWA)

IV.C.1. AMBIENT WATER QUALITY CRITERIA, Human Health

[Ed. Note: EPA does not have information yet in this section of IRIS].

IV.C.2. AMBIENT WATER QUALITY CRITERIA, Aquatic Organisms

Freshwater:
    Acute -- 8.3E-2 ug/L (1-hour average)
    Chronic -- 4.1E-2 ug/L (4-day average)

Marine:
    Acute -- 1.1E-2 ug/L (1-hour average)
    Chronic -- 5.6E-3 ug/L (4-day average)

Considers technological or economic feasibility? -- NO

Discussion -- Criteria were derived from a minimum data base consisting of acute and chronic tests on a variety of species. Requirements and methods are covered in the reference to the Federal Register.

Reference -- 51 FR 43665 (12/03/86)

EPA Contact -- Criteria and Standards Division / OWRS(202)260-1315 / FTS 260-1315

IV.D. FEDERAL INSECTICIDE, FUNGICIDE, AND RODENTICIDE ACT (FIFRA)

IV.D.1. PESTICIDE ACTIVE INGREDIENT, Registration Standard

Status -- Issued (1984)

Reference -- Chlorpyrifos Pesticide Registration Standard. September, 1984(NTIS No. PB87-110284).

EPA Contact -- Registration Branch / OPP (703)557-7760 / FTS 557-7760

IV.D.2. PESTICIDE ACTIVE INGREDIENT, Special Review

[Ed. Note: EPA does not have information yet in this section of IRIS].

IV.E. TOXIC SUBSTANCES CONTROL ACT (TSCA)

[Ed. Note: EPA does not have information yet in this section of IRIS].

IV.F. RESOURCE CONSERVATION AND RECOVERY ACT (RCRA)

[Ed. Note: EPA does not have information yet in this section of IRIS].

IV.G. SUPERFUND (CERCLA)

IV.G.1. REPORTABLE QUANTITY (RQ) for Release into the Environment

Value (status) -- 1 pound (Final, 1985)

Considers technological or economic feasibility? -- NO

Discussion -- The final RQ is based on aquatic toxicity as established under CWA Section 311 (40 CFR 117.3). The available data indicate that the aquatic 96-Hour Median Threshold Limit for chlorpyrifos is <0.1 ppm.

Reference -- 50 FR 13456 (04/04/85); 54 FR 33418 (08/14/89)

EPA Contact -- RCRA/Superfund Hotline (800)424-9346 / (202)260-3000 / FTS 260-3000

VI. BIBLIOGRAPHY

VI.A. ORAL RfD REFERENCES

Dow Chemical Company. 1972. Accession No. 112118. Available from EPA. Write to FOI, EPA, Washington DC 20460.

Dow Chemical USA. 1971a. MRID No. 00029063, 00064933, 00146519. Available from EPA. Write to FOI, EPA, Washington DC 20460.

Dow Chemical USA. 1971b. MRID No. 00081270, 00095178. Available from EPA. Write to FOI, EPA, Washington DC 20460.

Dow Chemical USA. 1971c. MRID No. 00029064, 00064934. Available from EPA. Write to FOI, EPA, Washington DC 20460.

Dow Chemical USA. 1979. MRID No. 00095268. Available from EPA. Write to FOI, EPA, Washington DC 20460.

Dow Chemical USA. 1983. MRID No. 00130400. Available from EPA. Write to FOI, EPA, Washington DC 20460.

VI.B. INHALATION RfC REFERENCES

None

VI.C. CARCINOGENICITY ASSESSMENT REFERENCES

None

VI.D. DRINKING WATER HA REFERENCES

None

VII. REVISION HISTORY

```
--------  --------  ------------------------------------------------------
Date      Section   Description
--------  --------  ------------------------------------------------------
03/01/88  I.A.5.    Core grades added
11/01/89  VI.       Bibliography on-line
01/01/92  IV.       Regulatory actions updated
```

SYNONYMS

BRODAN	LORSBAN
Chlorpyrifos	NA 2783
CHLORPYRIFOS-ETHYL	OMS-0971
DETMOL U.A.	O,O-DIAETHYL-O-3,5,6-TRICHLOR-2-
DOWCO 179	PYRIDYLMONOTHIOPHOSPHAT
DURSBAN	O,O-DIETHYL O-3,5,6-TRICHLORO-2-PYRIDYL
DURSBAN F	PHOSPHOROTHIOATE
ENT 27311	PHOSPHOROTHIOIC ACID, O,O-DIETHYL O-
ERADEX	(3,5,6-TRICHLORO-2-PYRIDYL) ESTER
ETHION, dry	PYRINEX

Source: *Instant EPA's IRIS*, 1997, Instant Reference Sources, Inc., 7605 Rockpoint Drive, Austin, TX 78731 (Fax: 512-345-2386; Internet URL, http://www.instantref.com/inst-ref.htm).

Production, Use, and Pesticide Labeling Information

1. Description of Chemical

> Generic Name: O,O-diethyl O-(3,5,6-trichloro-2-pyridyl) phosphorothioate

> Common Name: chlorpyrifos

Trade Name: Dursban for household products, Lorsban for agricultural products

EPA Shaughnessy Code: 059101

Chemical Abstracts Service (CAS) Number: 2921-88-2

Year of Initial Registration: 1965

Pesticide Type: Insecticide

Chemical Family: Organophosphate

US and Foreign Producers: Dow Chemical USA., Makhteshim-Beer Shiva, All India Medical Corp., Planters Products, Inc.

2. Use Patterns and Formulations

  - Application sites: Grain crops, nut crops, bananas, cole crops, citrus, pome and strawberry fruits, forage, field and vegetable crops, lawns and ornamental plants, poultry, beef cattle, sheep and dogs, livestock premise treatment, domestic dwellings, terrestrial structures, and direct application to stagnant water, etc.

  - Types of formulations: Baits, dusts, granules, wettable powders, flowables, impregnated plastics, and pressurized liquids.

  - Types and methods of application: Ground and aerial, sprays and dust applications

  - Application rates: Range from 0.5 lbs. a.i./A to 3 lbs. a.i./A, and crack and crevice treatment to broadcast treatment for indoor uses.

  - Usual carriers: Synthetic clays, talc, various solvents

3. Science Findings

  - Chemical Characteristics

    - Physical state: crystalline solid

    - Color: white to tan

    - Odor: mild mercaptan

    - Melting point: 41.5-43.5 °C

    - Vapor pressure: 1.87×10^{-5} mm Hg at 20 °C

    - Flash point: none

- Toxicological Characteristics

- Acute Oral: 163 mg/kg, Toxicity Category II

- Acute Dermal: 1505 mg/kg, Toxicity Category II

- Primary Dermal Irritation: No irritation, Toxicity Category III

- Primary Eye Irritation: Conjunctival irritation at 24 hours. Cleared at 48 hours. Toxicity Category III

- Acute Inhalation: Data gap

- Neurotoxicity: Not an acute delayed neurotoxic agent at doses up to 100 mg/kg (highest dose tested).

- Oncogenicity: Two studies submitted, but neither meet Agency standards. Neither suggest oncogenicity potential.

- Teratogenicity: Three studies have been evaluated to determine the teratogenic potential of chlorpyrifos. The Agency has determined that this chemical is not teratogenic at levels up to 25 mg/kg/day.

- Reproduction-2 generation: Two studies adequately demonstrate that chlorpyrifos does not produce reproductive effects. No effects were demonstrated at dose levels up to 1.2 mg/kg/day.

- Metabolism: The submitted studies suggest that chlorpyrifos is rapidly absorbed and metabolized to 3,5,6-trichloro-2-pyridinal (TCP). The parent compound and metabolite are rapidly excreted in the urine. The submitted studies do not meet Agency standards.

- Mutagenicity: Data gap

- Physiological and Biochemical Behavioral Characteristics

   - Mechanism of pesticidal action: An insecticide which is active by contact, ingestion, and vapor action and almost irreversibly causes phosphorylation of the acetylcholinesterase enzyme of tissues, allowing accumulation of acetylcholine at cholinergic neuroeffect or junctions (muscarinic effects), and at skeletal muscle myoneural junctions and autonomic ganglia. Poisoning also impairs the central nervous system function.

   - Symptoms of poisoning include: headache, dizziness, extreme weakness, ataxia, tiny pupils, twitching, tremor, nausea, slow heartbeat, pulmonary edema, and sweating. Continual absorption at intermediate dosages may cause influenza-like illness which includes symptoms like weakness, anorexia, and malaise.

- The metabolism of chlorpyrifos in plants and animals is not adequately understood. The major metabolite is 3,5,6-trichloro-2-pyridinol (TCP). The Agency does not have adequate data on TCP to determine if this metabolite should continue to be a part of the tolerance expression.

- Environmental Characteristics

- Available data are insufficient to fully assess the environmental fate of chlorpyrifos. Data gaps exist on all required studies except for aerobic and anaerobic soil studies.

- Adsorption and leaching in basic soil types: The Agency is requesting data to determine if chlorpyrifos will contaminate groundwater.

- Microbial breakdown: Depending on the soil type, microbial metabolism of chlorpyrifos may have a half-life of up to 279 days.

- Ecological Characteristics

- Avian oral: Mallard duck -- 76.6 mg/kg; Ring-necked pheasant -- 17.7 mg/kg

- Avian dietary: Mallard duck -- 136 ppm; Bobwhite quail -- 721 ppm

- Freshwater fish: Cold water fish (rainbow trout) -- 3.0 ppm; Warm water fish (bluegill sunfish) -- 2.4 ppm

- Acute freshwater invertebrates: Daphnia -- 0.176 ppb

- Acute estuarine and marine organisms: Oyster -- 0.27 ppm; Grass shrimp -- 1.5 ppm; Killifish -- 3.2 ppm

- Precautionary language is being required for hazards to birds, fish, and aquatic organisms. Chronic effects to non-target aquatic invertebrate species are not adequately characterized, and therefore appropriate studies are required.

- Tolerance Assessment

- The Agency is unable to complete a full tolerance reassessment because of certain residue chemistry and toxicology data gaps.

Tolerances:

Commodity	Parts Per Million
Alfalfa, green forage	4.0
Alfalfa, hay	15.5
Almonds	0.05
Almonds, hull	0.05
Apples	1.5
Bananas (whole)	0.25
Bananas, pulp with peel removed	0.05
Bean forage	1.0
Beans, lima	0.05
Beans, lima, forage	1.0
Beans, snap	0.05
Beans, snap, forage	1.0
Beets, sugar, roots	1.0
Beets, sugar, tops	8.0
Broccoli	2.0
Brussels sprouts	2.0
Cabbage	2.0
Cattle, fat	2.0
Cattle, meat by-products (mbyp)	2.0
Cattle, meat	2.0
Cauliflower	2.0
Cherries	2.0
Citrus fruits	1.0
Corn, field, grain	0.1
Corn, fresh (including sweet corn; kernel plus cob with husk removed)	0.1
Corn, fodder	10.0
Corn, forage	10.0
Cottonseed	0.5
Cranberries	1.0
Cucumbers	0.1
Eggs	0.1
Figs	0.1
Goats, fat	1.0
Goats, mbyp	1.0
Goats, meat	1.0
Grapes	0.5
Hogs, fat	0.5
Hogs, mbyp	0.5
Hogs, meat	0.5
Horses, fat	1.0
Horses, mbyp	1.0
Horses, meat	1.0
Milk, fat (reflecting 0.02 ppm in whole milk)	0.5

(Table Continued)

Commodity	Parts Per Million
Mint, hay	1.0
Nectarines	0.05
Onions (dry bulb)	0.5
Pea forage	1.0
Peaches	0.05
Peanuts	0.5
Peanut hulls	15.0
Pears	0.05
Peppers	1.0
Plums (fresh prunes	0.05
Poultry, fat (including turkeys)	0.5
Poultry, mbyp (including turkeys)	0.5
Poultry, meat (including turkeys)	0.5
Pumpkins	0.1
Radishes	3.0
Seed and pod vegetables	0.1
Sheep, fat	1.0
Sheep, mbyp	1.0
Sheep, meat	1.0
Sorghum, fodder	6.0
Sorghum, forage	1.5
Sorghum, grain	0.75
Soybeans	0.5
Soybeans, forage	8.0
Soybeans, straw	15.0
Strawberries	0.5
Sunflower, seeds	0.25
Sweet potatoes	0.1
Tomatoes	0.5
Turnips (roots)	3.0
Turnips (greens)	1.0

- Based on established tolerances, the theoretical maximum residue contribution (TMRC) for chlorpyrifos residues in the human diet is calculated to be 0.5637 mg/day. The acceptable daily intake (ADI) of chlorpyrifos is 0.003 mg/kg/day. The maximum permissible intake (MPI) is 0.18 mg/day. The percent utilized ADI is 313%. To provide for conformity between US tolerances for chlorpyrifos and tolerances established by the Codex Alimentarius, Canada and Mexico, the expression of the US tolerances for chlorpyrifos would have to exclude the major metabolite TCP, but the Agency is not recommending this now.

- US tolerances for most raw agricultural commodities are supported by current residue chemistry data. In some cases, however, more data are required.

- Summary Science Statement

- Chlorpyrifos has moderate mammalian toxicity. It is not considered to be oncogenic, mutagenic, or teratogenic. However, the oncogenicity and mutagenicity studies used to draw these conclusions are not up to current Agency standards. Additional information from these studies is required.

- The Agency is imposing a 24-hour reentry restriction for crop uses until appropriate reentry studies are submitted and evaluated and a decision is reached whether a different time interval is more appropriate. The 24-hour interval also coincides with the requirements of California.

- Data are insufficient to fully assess the environmental fate of chlorpyrifos. The Agency is requesting necessary data to make this assessment and also to specifically assess whether or not chlorpyrifos has a potential to leach into groundwater. Data are also insufficient to measure human exposure in outdoor and indoor applications.

- Chlorpyrifos is extremely toxic to fish, birds, and other wildlife. It is highly toxic to honeybees. Use precautions and restrictions are being imposed to reduce potential hazards.

- A full tolerance reassessment cannot be completed. The previous ADI (at 94% of the TMRC) was established based on a 2-year fat feeding study. The present ADI (313% of the TMRC) was calculated using a human study. Chronic feeding studies are required, as well as metabolism and residue data on numerous commodities.

4. Summary of Regulatory Position and Rationale

- The Agency has determined that it should continue to allow the registration of chlorpyrifos. Adequate studies are available to assess the acute toxicological effects of chlorpyrifos to humans. None of the criteria for unreasonable adverse effects listed in section 162.11(a) of Title 40 of the US Code of Federal regulations have been met or exceeded. However, because of certain gaps in the data base, a full risk assessment of chlorpyrifos cannot be completed.

- Also, a full tolerance reassessment cannot be completed because of certain residue chemistry and toxicology data gaps.

- The Agency is concerned whether or not the potential total human exposure to chlorpyrifos and its metabolites, from its widespread use and its ADI being exceeded three-fold, poses any unacceptable hazards. To resolve this concern, additional residue, metabolism, and exposure data are required, and until it is resolved no significant new tolerances or uses will be granted.

- A federal 24-hour reentry interval is established for treated crop areas until reentry data are submitted, as required, and the Agency decides on the most appropriate time interval.

- Available data are insufficient to fully assess the environmental fate of chlorpyrifos. The Agency is requesting data to determine if chlorpyrifos will contaminate groundwater.

5. Summary of Major Data Gaps

- Additional residue data on various processed commodities are being required. Also, additional chronic toxicity, oncogenicity, and mutagenicity testing is needed to better define the long-term effects of this chemical. Plant, animal, and exposure data are required to better qualify and quantify human exposure to residues from dietary and nondietary sources.

- Other requirements:

> - Acute inhalation
> - General metabolism
> - Hydrolysis study
> - Photodegradation studies
> - Soil metabolism studies
> - Mobility studies
> - Dissipation studies
> - Accumulation studies
> - Fish embryo-larvae study
> - Large scale field testing
> - Monitoring for crop runoff
> - Phytotoxic effects on algae and other aquatic plants
> - Indoor monitoring

6. Contact Person at EPA

Jay S. Ellenberger
Product Manager (12), Insecticide- Rodenticide Branch
Registration Division (TS-767C)
Office of Pesticide Programs
Environmental Protection Agency
401 M Street, SW
Washington, DC 20460

Office location and telephone number:
Room 202, Crystal Mall #2
1921 Jefferson Davis Highway
Arlington, VA 22202

Phone: (703) 557-2386

Source: ***Instant EPA's Pesticide Facts***, 1994, Instant Reference Sources, Inc., 7605 Rockpoint Drive, Austin, TX 78731 (Fax: 512-345-2386;   Internet URL, http://www.instantref.com/inst-ref.htm).

Chrysene

Introduction

Chrysene occurs in coal tar and it is formed during distillation of coal. It is also formed in very small amounts during distillation or pyrolysis of many fats and oils. Thus, it is very wide spread and is commonly found in the environment, especially in conjunction with particulate matter from smoke stacks.

Chrysene's release to the environment is largely associated with particulate matter, soils, and sediments. If released to soil it will be expected to adsorb very strongly to the soil and will not be expected to leach appreciably to groundwater. It will not hydrolyze or appreciably evaporate from soils or surfaces, and it may be subject to biodegradation in soils. If released to water, it will adsorb very strongly to sediments and particulate matter, but will not hydrolyze or appreciably evaporate. It will bioconcentrate in species which lack microsomal oxidase. [828]

The small amount of information available suggests that chrysene may be subject to biodegradation in water systems. Adsorption to various materials may affect the rate of these processes. If released to air, chrysene will be subject to direct photolysis, although adsorption to particulates may affect the rate of this process. The estimated half-life of any gas phase chrysene in the atmosphere is 1.25 hours as a result of reaction with photochemically produced hydroxyl radicals. Human exposure will be from inhalation of contaminated air and consumption of contaminated food and water. Especially high exposure will occur through the smoking of cigarettes and ingestion of certain foods (eg smoked and charcoal broiled meats and fish). [828]

Identifying Information

CAS NUMBER: 218-01-9

NIOSH Registry Number: GC0700000

SYNONYMS: 1,2-Benzophenanthrene; 1,2,5,6-Dibenzonaphthalene; Benzo(a)phenanthrene

Endocrine Disruptor Information

Chrysene is on EPA's list of suspect EEDs [811] but little information has been located to determine the Agency's reason for its inclusion. It is a co-planar molecule like many of the suspect EEDs and a polycyclic aromatic hydrocarbon (PAH). More information should be forthcoming after EPA's studies are concluded.

Chemical and Physical Properties

CHEMICAL FORMULA: $C_{18}H_{12}$

MOLECULAR WEIGHT: 228.30 [703]

PHYSICAL DESCRIPTION: Orthorhombic bipyramidal plates from benzene. [703]

DENSITY: 1.274 [703]

MELTING POINT: 254°C [703]

BOILING POINT: 448°C [703]

SOLUBILITY: Water: Insoluble [703]; Alcohol: Slightly soluble [703]; Ether: Slightly soluble [703]; Carbon bisulfide: Slightly soluble [703]; Glacial acetic acid: Slightly soluble [703]; Boiling benzene: Moderately soluble [703]

NOTE: Chrysene is generally only slightly soluble in cold organic solvents, but fairly soluble in these solvents when hot, including glacial acetic acid. [703]

OTHER PHYSICAL DATA: Sublimes easily in vacuo [703]

Source: *Instant EPA's Air Toxics*, 1994, Instant Reference Sources, Inc., 7605 Rockpoint Drive, Austin, TX 78731 (Fax: 512-345-2386; Internet URL, http://www.instantref.com/inst-ref.htm).

Environmental Reference Materials

A source of EED environmental reference materials for Chrysene is Radian International LLC, Austin, TX (Fax 512-454-0268; Internet URL, http://www.radian.com/standards); Catalog No. ERC-001.

Hazardous Properties

ACUTE/CHRONIC HAZARDS: An experimental carcinogen, neoplastigen and tumorigen by skin contact; human mutagenic data exists. When heated to decomposition, emits acrid smoke and fumes. [703] Has been implicated as an etiologic determinant of chemical carcinogenesis. Able to induce aryl hydrocarbon hydroxylase in cultured human lymphocytes. A significant increase in enzyme induction occurred in chrysene-induced cultures compared with controls. No data are available in humans. Limited evidence of carcinogenicity in animals. [723] Chrysene may be a mild irritant.

VOLATILITY: Not volatile

FIRE HAZARD: Chrysene is combustible. Fires involving this compound may be controlled using a dry chemical, carbon dioxide or Halon extinguisher. A water fog may also be used.

STABILITY: Chrysene is stable under normal laboratory conditions. When heated to decomposition it emits acrid smoke and fumes [703].

Source: *Instant EPA's Air Toxics*, 1994, Instant Reference Sources, Inc., 7605 Rockpoint Drive, Austin, TX 78731 (Fax: 512-345-2386; Internet URL, http://www.instantref.com/inst-ref.htm).

Medical Symptoms of Exposure

Symptoms of exposure to chrysene have not been found. They may be expected to be similar to symptoms recorded for some of the other PAHs such as anthracene, phenantharene and pyrene.

Toxicological Information

The toxicological information below is gathered from several sources including two national databases: one from the *National Toxicology Program Chemical Repository Database* and another from *EPA's Integrated Risk Information System (IRIS)*.

Toxicology Information from the National Toxicology Program

CARCINOGENICITY:

Tumorigenic Data:

Type of Effect	Route/Animal	Amount Dosed (Notes)
TDLo:	scu-mus	200 mg/kg
TDLo:	skn-mus	99 mg/kg/31W-I

Review: IARC Cancer Review: Animal Limited Evidence
        IARC: Not classifiable as a human carcinogen (Group 3)

Status:
        EPA Carcinogen Assessment Group
        ACGIH suspected human carcinogen

TERATOGENICITY: Not available

MUTATION DATA:

Test	Lowest dose	Test	Lowest dose
mma-sat	10 ug/plate	msc-mus-orl	450 mg/kg
otr-ham:kdy	25 ug/L	sce-ham-ipr	900 mg/kg/24H

TOXICITY: Not available

OTHER TOXICITY DATA:

Review:
Toxicology Review-2

Status:
"NIOSH Manual of Analytical Methods" Vol 1 183-4, 206
Reported in EPA TSCA Inventory, 1980

SAX TOXICITY EVALUATION:

THR: High via sc and dermal and probably inhale routes. An experimental (+) neoplasm and carcinogen. A polycyclic hydrocarbon air pollutant.

Source: *Instant Tox-Base*, 1995, Instant Reference Sources, Inc., 7605 Rockpoint Drive, Austin, TX 78731 (Fax: 512-345-2386; Internet URL, http://www.instantref.com/inst-ref.htm).

Toxicology Information from EPA's Integrated Risk Information System (IRIS)

I. CHRONIC HEALTH HAZARD ASSESSMENTS FOR NONCARCINOGENIC EFFECTS

I.A. REFERENCE DOSE FOR CHRONIC ORAL EXPOSURE (RfD)

[Ed. Note: EPA does not have information yet in this section of IRIS].

I.B. REFERENCE CONCENTRATION FOR CHRONIC INHALATION EXPOSURE (RfC)

[Ed. Note: EPA does not have information yet in this section of IRIS].

II. CARCINOGENICITY ASSESSMENT FOR LIFETIME EXPOSURE

Section II provides information on three aspects of the carcinogenic risk assessment for the agent in question; the US EPA classification, and quantitative estimates of risk from oral exposure and from inhalation exposure. The classification reflects a weight-of-evidence judgment of the likelihood that the agent is a human carcinogen. The quantitative risk estimates are presented in three ways. The slope factor is the result of application of a low-dose extrapolation procedure and is presented as the risk per (mg/kg)/day. The unit risk is the quantitative estimate in terms of either risk per ug/L drinking water or risk per ug/cu.m air breathed. The third form in which risk is presented is a drinking water or air concentration providing cancer risks of 1 in 10,000, 1 in 100,000 or 1 in 1,000,000. Background Document 2 (Service Code 5) provides details on the rationale and methods used to derive the carcinogenicity values found in IRIS. Users are referred to Section I for information on long-term toxic effects other than carcinogenicity.

II.A. EVIDENCE FOR CLASSIFICATION AS TO HUMAN CARCINOGENICITY

II.A.1. WEIGHT-OF-EVIDENCE CLASSIFICATION

Classification -- B2; probable human carcinogen

Basis -- No human data and sufficient data from animal bioassays. Chrysene produced carcinomas and malignant lymphoma in mice after intraperitoneal injection and skin carcinomas in mice following dermal exposure. Chrysene produced chromosomal abnormalities in hamsters and mouse germ cells after gavage exposure, positive responses in bacterial gene mutation assays and transformed mammalian cells exposed in culture.

II.A.2. HUMAN CARCINOGENICITY DATA

None. Although there are no human data that specifically link exposure to chrysene to human cancers, chrysene is a component of mixtures that have been associated with human cancer. These include coal tar, soots, coke oven emissions and cigarette smoke (US EPA, 1984, 1990; IARC, 1983, 1984).

II.A.3. ANIMAL CARCINOGENICITY DATA

Sufficient. Intraperitoneal chrysene injections in male mice caused an increased incidence of liver tumors (Wislocki et al., 1986; Buening et al., 1979) and increased incidences of malignant lymphoma and lung tumors (Wislocki et al., 1986). In mouse skin painting assays chrysene tested positive in both initiation and complete carcinogen studies (Wynder and Hoffman, 1959).

On days 1, 8, and 15 of age, groups of male (28 to 35/group) and female (24 to 34/group) CD-1 mice received intraperitoneal injections of chrysene in dimethyl sulfoxide (DMSO) (total dose = 0, 160 ug or 640 ug/mouse) (Wislocki et al., 1986). The low-dose and high-dose experiments were initiated 10 weeks apart and had separate concurrent vehicle controls. Tumors were evaluated in animals that died spontaneously after weaning and in all remaining animals at 1 year after exposure. A statistically significant increase in the incidence of liver adenomas or carcinomas occurred in treated male mice relative to their respective controls: 10/35 (29%) and 5/45 (11%) in the low-dose mice and controls, respectively; and 14/34 (41%) and 2/28 (7%) in the high-dose mice and controls, respectively. The majority of the liver tumors in the high-dose males were carcinomas and the incidence was statistically significantly greater than in its respective control group, whereas the majority of tumors in the low-dose males were adenomas. Liver adenonas, but no carcinomas were observed in the control groups. In female mice no tumors were observed. The incidence of lung adenomas or carcinomas in the low-dose male mice was 6/35 (17%) (one of which was a carcinoma) and 4/45 (9%) (two of which were carcinomas) in their control group. The incidence of lung adenomas was statistically elevated in high-dose males 7/34 (21%) when compared with their control group (1/28, 4%). The incidence of malignant lymphoma was significantly elevated (3/35, 9%) in low-dose males relative to the controls (0/45), but not in the high-dose males (1/34) relative to their controls (1/28). In females, there was no statistically significant increase in lung tumors or lymphoma. This is generally regarded as a short-term exposure study with a less-than-lifetime (1 year) experiment.

Male and female Swiss Webster BLU/Ha(ICR) mice received intraperitoneal injections of chrysene in DMSO (total dose = 320 ug/mouse) or DMSO alone on days 1, 8 and 15 after birth (Buening et al., 1979). Mice were killed at 38-42 weeks of age. The incidences of lung tumors in the treated group appeared to be elevated (5/24 (21%) and 1/11 (9%) in males and females, respectively), although not statistically significantly, when compared with the control groups (2/21 (10%) and 7/38 (18%) in males and females, respectively). The incidence of hepatic tumors in the treated males was statistically significantly greater (6/24, 25%) than in control males (0/21), whereas no hepatic tumors were found in the females. In a replication of this study, lung tumor incidence was not increased; however, the incidence of hepatic tumors in treated male mice was significantly elevated (6/27, 22%) over the incidence in the control group (0/52) (Chang et al., 1983). No liver tumors were reported in the females. These studies are regarded as short-term exposure, less-than-lifetime experiments.

Chrysene has been tested for complete carcinogenic activity and initiating activity in mouse skin painting assays. It was shown to be a complete carcinogen (Wynder and Hoffmann, 1959). Chrysene has produced positive results for initiating activity in several mouse strains (C3H, ICR/Ha Swiss, Ha/ICR/Mil Swiss, CD-1, Sencar) when applied in combination with various promoting agents (decahydronaphthalene, croton oil, TPA) producing skin papillomas and carcinomas (Van Duuren et al., 1966; Scribner, 1973; Horton and Christian, 1974; Hecht et al., 1974; Levin et al., 1978; Wood et al., 1979, 1980; Slaga et al., 1980; Rice et al., 1985).

II.A.4. SUPPORTING DATA FOR CARCINOGENICITY

Chrysene produced positive results in tests for reverse mutation in three strains of Salmonella typhimurium and positive results for forward mutation in one strain (McCann et al., 1975; Tokiwa et al., 1977; Wood et al., 1977; LaVoie et al., 1979; Dunkel and Simmon, 1980; Sakai et al., 1985; Kaden et al., 1979).

Chromosomal effects were observed in Chinese hamster cells, mouse oocytes and hamster spermatogonia following gavage doses of 450 or 900 mg/kg (Basler et al., 1977; Roszinsky-Kocher et al., 1979). Positive results were obtained (10 ug/mL) in tests for cell transformation in Syrian hamster embryo cells and negative results in mouse prostrate C3HG23 cells (Marquardt and Heidelberger, 1972; Pienta et al., 1977).

Current theories on mechanisms of metabolic activation of polycyclic aromatic hydrocarbons are consistent with a carcinogenic potential for chrysene. Chrysene has a "bay-region" in structure (Jerina et al., 1978). It is metabolized by mixed function oxidases to reactive "bay-region" diol epoxides (Nordqvist et al., 1981; Vyas et al., 1982) that are mutagenic in bacteria and tumorigenic in mouse skin painting assays and when injected into newborn mice (Levin et al., 1978; Wood et al., 1977, 1979; Slaga et al., 1980; Chang et al., 1983).

II.B.   QUANTITATIVE ESTIMATE OF CARCINOGENIC RISK FROM ORAL EXPOSURE

[Ed. Note: EPA does not have information yet in this section of IRIS].

II.C.   QUANTITATIVE ESTIMATE OF CARCINOGENIC RISK FROM INHALATION EXPOSURE

[Ed. Note: EPA does not have information yet in this section of IRIS].

II.D. EPA DOCUMENTATION, REVIEW, AND CONTACTS (CARCINOGENICITY ASSESSMENT)

II.D.1. EPA DOCUMENTATION

Source Document -- US EPA, 1984, 1990

The 1990 Drinking Water Criteria Document for Polychlorinated Aromatic Hydrocarbons has received Agency and external review.

II.D.2. REVIEW (CARCINOGENICITY ASSESSMENT)

Agency Work Group Review -- 02/07/90, 08/05/93, 09/21/93, 02/02/94

Verification Date -- 02/07/90

II.D.3. US EPA CONTACTS (CARCINOGENICITY ASSESSMENT)

Rita S. Schoeny / NCEA -- (513)569-7544

Robert E. McGaughy / NCEA -- (202)260-5889

III. HEALTH HAZARD ASSESSMENTS FOR VARIED EXPOSURE DURATIONS

[Ed. Note: EPA does not have information yet in this section of IRIS].

IV. US EPA REGULATORY ACTIONS

EPA risk assessments may be updated as new data are published and as assessment methodologies evolve. Regulatory actions are frequently not updated at the same time. Compare the dates for the regulatory actions in this section with the verification dates for the risk assessments in sections I and II, as this may explain inconsistencies. Also note that some regulatory actions consider factors not related to health risk, such as technical or economic feasibility. Such considerations are indicated for each action. In addition, not all of the regulatory actions listed in this section involve enforceable federal standards. Please direct any questions you may have concerning these regulatory actions to the US EPA contact listed for that particular action. Users are strongly urged to read the background information on each regulatory action in Background Document 4 in Service Code 5.

IV.A. CLEAN AIR ACT (CAA)

[Ed. Note: EPA does not have information yet in this section of IRIS].

IV.B. SAFE DRINKING WATER ACT (SDWA)

IV.B.1. MAXIMUM CONTAMINANT LEVEL GOAL (MCLG) for Drinking Water

Value -- 0 mg/L (Proposed, 1990)

Considers technological or economic feasibility? -- NO

Discussion -- The proposed MCLG is zero. This value is based on carcinogenic PAH's as a class.

Reference -- 55 FR 30370 (07/25/90)

EPA Contact -- Health and Ecological Criteria Division / OST /(202) 260-7571 / FTS 260-7571; or Safe Drinking Water Hotline / (800) 426-4791

IV.B.2. MAXIMUM CONTAMINANT LEVEL (MCL) for Drinking Water

Value -- 0.0002 mg/L (Proposed, 1990)

Considers technological or economic feasibility? -- YES

Discussion -- The proposed MCL is equal to the PQL and is associated with a maximum lifetime individual risk of 1 E-4.

Monitoring requirements -- Community and non-transient water system monitoring based on state vulnerability assessment; vulnerable systems to be monitored quarterly for one year; repeat monitoring dependent upon detection and size of system.

Analytical methodology -- High pressure liquid chromatography (EPA 550, 550.1); gas chromatographic/mass spectrometry (EPA 525): PQL= 0.0002 mg/L.

Best available technology -- Granular activated carbon.

Reference -- 55 FR 30370 (07/25/90)

EPA Contact -- Drinking Water Standards Division / OGWDW /(202) 260-7575 / FTS 260-7575; or Safe Drinking Water Hotline / (800) 426-4791

IV.B.3. SECONDARY MAXIMUM CONTAMINANT LEVEL (SMCL) for Drinking Water

[Ed. Note: EPA does not have information yet in this section of IRIS].

IV.B.4. REQUIRED MONITORING OF "UNREGULATED" CONTAMINANTS

Status -- Listed (Proposed, 1991)

Discussion -- "Unregulated" contaminants are those contaminants for which EPA establishes a monitoring requirement but which do not have an associated final MCLG, MCL, or treatment technique. EPA may regulate these contaminants in the future.

Monitoring requirement -- All systems to be monitored unless a vulnerability assessment determines the system is not vulnerable.

Analytical methodology -- Gas chromatography/mass spectrometry (EPA 525); high pressure liquid chromatography (EPA 550, 550.1).

Reference -- 56 FR 3526 (01/30/91)

EPA Contact -- Drinking Water Standards Division / OGWDW /(202) 260-7575 / FTS 260-7575; or Safe Drinking Water Hotline / (800) 426-4791

IV.C. CLEAN WATER ACT (CWA)

IV.C.1. AMBIENT WATER QUALITY CRITERIA, Human Health

Water and Fish Consumption: 2.8E-3 ug/L

Fish Consumption Only: 3.11E-2 ug/L

Considers technological or economic feasibility? -- NO

Discussion -- For the maximum protection from the potential carcinogenic properties of this chemical, the ambient water concentration should be zero. However, zero may not be obtainable at this time, so the recommended criteria are presents a E-6 estimated incremental increase of cancer over a lifetime. The values given represent polynuclear aromatic hydrocarbons as a class.　　Reference -- 45 FR 79318 (11/28/80)

EPA Contact -- Criteria and Standards Division / OWRS(202)260-1315 / FTS 260-1315

IV.C.2. AMBIENT WATER QUALITY CRITERIA, Aquatic Organisms

Freshwater:
  Acute LEC -- none
  Chronic LEC -- none

Marine:
  Acute LEC --   3.0E+2 ug/L
  Chronic LEC -- none

Considers technological or economic feasibility? -- NO

Discussion -- The values that are indicated as "LEC" are not criteria, but are the lowest effect levels found in the literature. LEC's are given when the minimum data required to derive water quality criteria are not available. The values given represent polynuclear aromatic hydrocarbons as a class.

Reference -- 45 FR 79318 (11/28/80)

EPA Contact -- Criteria and Standards Division / OWRS(202)260-1315 / FTS 260-1315

IV.D. FEDERAL INSECTICIDE, FUNGICIDE, AND RODENTICIDE ACT (FIFRA)

[Ed. Note: EPA does not have information yet in this section of IRIS].

IV.E. TOXIC SUBSTANCES CONTROL ACT (TSCA)

[Ed. Note: EPA does not have information yet in this section of IRIS].

IV.F. RESOURCE CONSERVATION AND RECOVERY ACT (RCRA)

IV.F.1. RCRA APPENDIX IX, for Ground Water Monitoring

Status -- Listed

Reference -- 52 FR 25942 (07/09/87)

EPA Contact -- RCRA/Superfund Hotline(800)424-9346 / (202)260-3000 / FTS 260-3000

IV.G. SUPERFUND (CERCLA)

IV.G.1. REPORTABLE QUANTITY (RQ) for Release into the Environment

Value (status) --   100 pounds (Final, 1989)

Considers technological or economic feasibility? -- NO

Discussion -- The final RQ for chrysene is based on potential carcinogenicity. Available data indicate a hazard ranking of low and a weight of evidence classification of Group B2, which corresponds to an RQ of 100 pounds.

Reference -- 54 FR 33418 (08/14/89)

EPA Contact -- RCRA/Superfund Hotline(800)424-9346 / (202)260-3000 / FTS 260-3000

VI. BIBLIOGRAPHY

VI.A. ORAL RfD REFERENCES

None

VI.B. INHALATION RfC REFERENCES

None

VI.C. CARCINOGENICITY ASSESSMENT REFERENCES

Basler, A., B. Herbold, S. Peter and G. Rohrborn. 1977. Mutagenicity of polycyclic hydrocarbons. II. Monitoring genetical hazards of chrysene in vitro and in vivo. Mutat. Res. 48: 249-254.

Buening, M.K., W. Levin, J.M. Karle, H. Yagi, D.M. Jerina and A.H. Conney. 1979. Tumorigenicity of bay-region epoxides and other derivatives of chrysene and phenanthrene in newborn mice. Cancer Res. 39: 5063-5068.

Chang, R.L., W. Levin, A.W. Wood, et al. 1983. Tumorigenicity of enantiomers of chrysene 1,2-dihydrodiol and of the diastereomeric bay-region chrysene 1,2-diol-3,4-epoxides on mouse skin and in newborn mice. Cancer Res. 43: 192-196.

Dunkel, V.C. and V.F. Simmon. 1980. Mutagenic activity of chemicals previously tested for carcinogenicity in the National Cancer Institute Bioassay Program. In: Molecular and Cellular Aspects of Carcinogenic Screening Tests, R. Montesano, H. Bartsch and L. Tomatis, Ed. IARC Sci. Publ. No. 27, International Agency for Research on Cancer, Lyon, France. p. 283-301.

Hecht, S.S., W.E. Bondinell and D. Hoffmann. 1974. Chrysene and methyl chrysenes: Presence in tobacco smoke and carcinogenicity. J. Natl. Cancer Inst. 53: 1121-1133.

Horton, A.W. and G. M. Christian. 1974. Carcinogenic versus incomplete carcinogenic activity among aromatic hydrocarbons: Contrast between chrysenes and benzo[b]triphenylene. J. Natl. Cancer Inst. 53(4): 1017-1020.

IARC (International Agency for Research on Cancer). 1983. Monographs on the Evaluation of the Carcinogenic Risk of Chemicals for Humans. Polynuclear Aromatic

Compounds. Part 1. Chemical, Environmental and Experimental Data. Vol. 32. World Health Organization.

IARC (International Agency for Research on Cancer). 1984. Monographs on the Evaluation of the Carcinogenic Risk of the Chemical to Man. Polynuclear Aromatic Hydrocarbons. Part 3. Industrial Exposures in Aluminum Production, Coal Gasification, Coke Production, and Iron and Steel Founding. Vol. 34. World Health Organization.

Jerina, D.M., H. Yagi, R.E. Lehr, et al., 1978. The bay-region theory of carcinogenesis by polycyclic aromatic hydrocarbons. In: Polycyclic Hydrocarbons and Cancer, Vol. 1, Environment, Chemistry and Metabolism, H.V. Gelboin and P.O.P. Ts'o, Ed. Academic Press, NY. p. 173-188.

Kaden, D.A., R.A. Hites and W.G. Thilly. 1979. Mutagenicity of soot and associated polycyclic aromatic hydrocarbons to Salmonella typhimurium. Cancer Res. 39: 4152-4159.

LaVoie, E.J., E.V. Bedenko, N. Hirota, S.S. Hecht and D. Hoffmann. 1979. A comparison of the mutagenicity, tumor-initiating activity and complete carcinogenicity of polynuclear aromatic hydrocarbons. In: Polynuclear Aromatic Hydrocarbons, P.W. Jones and P. Leber, Ed. Ann Arbor Science Publishers, Ann Arbor, MI. p. 705-721.

Levin, W., A.W. Wood, R.L. Chang, et al. 1978. Evidence for bay region activation of chrysene 1,2-dihydrodiol to an ultimate carcinogen. Cancer Res. 38: 1831-1834.

Marquardt, H. and C. Heidelberger. 1972. Influence of "feeder cells" and inducers and inhibitors of microsomal mixed-function oxidases on hydrocarbon-induced malignant transformation of cells derived from C3H mouse prostate. Cancer Res. 32: 721-725.

McCann, J.E., E. Choi, E. Yamasaki and B.N. Ames. 1975. Detection of carcinogens as mutagens in the Salmonella/microsome test: Assay of 300 chemicals. Proc. Natl. Acad. Sci. USA. 72(12): 5135-5139.

Nordqvist, M., D.R. Thakker, K.P. Vyas, et al. 1981. Metabolism of chrysene and phenanthrene to bay-region diol epoxides by rat liver enzymes. Mol. Pharmacol. 19: 168-178.

Pienta, R.J., J.A. Poiley and W.B. Lebherz, III. 1977. Morphological transformation of early passage golden Syrian hamster embryo cells derived from cryopreserved primary cultures as a reliable in vitro bioassay for identifying diverse carcinogens. Int. J. Cancer. 19: 642-655.

Rice, J.E., G.S. Mokowski, T.J. Hosted, Jr. and E. LaVoie. 1985. Methylene-bridged bay region chrysene and phenanthrene derivatives and their keto-analogs: Mutagenicity in Salmonella typhimurium and tumor-initiating activity on mouse skin. Cancer Lett. 27: 199-206.

Roszinsky-Kocher, G., A. Basler and G. Rohrborn. 1979. Mutagenicity of polycyclic hydrocarbons. V. Induction of sister-chromatid exchanges in vivo. Mutat. Res. 66: 65-67.

Sakai, M., D. Yoshida and S. Mizusaki. 1985. Mutagenicity of polycyclic aromatic hydrocarbons and quinones on Salmonella typhimurium TA97. Mutat. Res. 156: 61-67.

Scribner, J.D. 1973. Brief Communication: Tumor initiation by apparently noncarcinogenic polycyclic aromatic hydrocarbons. J. Natl. Cancer Inst. 50(6): 1717-1719.

Slaga, T.J., G.L. Gleason, G. Mills, et al. 1980. Comparison of the skin tumor-initiating activities of dihydrodiols and diol-epoxides of various polycyclic aromatic hydrocarbons. Cancer Res. 40: 1981-1984.

Tokiwa, H., K. Morita, H. Takeyoshi, K. Takahashi and Y. Ohnishi. 1977. Detection of mutagenic activity in particulate air pollutants. Mutat. Res. 48: 237-248.

US EPA. 1984. Carcinogen Assessment of Coke Oven Emissions. Office of Health and Environmental Assessment, Washington, DC. EPA 600/6-82-003F. NTIS PB 84-170181.

US EPA. 1990. Drinking Water Criteria Document for Polycyclic Aromatic Hydrocarbons (PAHs). Prepared by the Office of Health and Environmental Assessment, Environmental Criteria and Assessment Office, Cincinnati, OH for the Office of Drinking Water, Washington, DC. Final Draft. ECAO-CIN-D010, September, 1990.

Van Duuren, B.L., A. Sivak, A. Segal, L. Orris and L. Langseth. 1966. The tumor promoting agents of tobacco leaf and tobacco smoke condensate. J. Natl. Cancer Inst. 37: 519-526.

Vyas, K.P., W. Levin, H. Yagi, et al. 1982. Stereoselective metabolism of the (+)- and (-)-enantiomers of trans-1,2-dihydroxy-1,2-dihydrochrysene to bay-region 1,2-diol-3,4-epoxide diastereomers by rat liver enzymes. Mol. Pharmacol. 22: 182-189.

Wislocki, P.G., E.S. Bagan, A.Y.H. Lu, et al. 1986. Tumorigenicity of nitrated derivatives of pyrene, benz[a]anthracene, chrysene and benzo[a]pyrene in the newborn mouse assay. Carcinogenesis. 7(8): 1317-1322.

Wood, A.W., W. Levin, D. Ryan, et al. 1977. High mutagenicity of metabolically activated chrysene 1,2-dihydrodiol: Evidence for bay region activation of chrysene. Biochem. Biophys. Res. Commun. 78(3): 847-854.

Wood, A.W., R.L. Chang, W. Levin, et al. 1979. Mutagenicity and tumorigenicity of phenanthrene and chrysene epoxides and diol epoxides. Cancer Res. 39: 4069-4077.

Wood, A.W., W. Levin, R.L. Chang, et al. 1980. Mutagenicity and tumor-initiating activity of cyclopenta(c,d)pyrene and structurally related compounds. Cancer Res. 40: 642-649.

Wynder, E.L. and D. Hoffmann. 1959. A study of tobacco carcinogenesis. VII. The role of higher polycyclic hydrocarbons. Cancer. 12: 1079-1086.

VI.D. DRINKING WATER HA REFERENCES

None

VII. REVISION HISTORY

Date	Section	Description
12/01/90	II.	Carcinogen assessment on-line
12/01/90	VI.	Bibliography on-line
01/01/92	IV.	Regulatory Action section on-line
09/01/93	II.	Carcinogenicity assessment noted as pending change
09/01/93	II.D.2.	Work group review date added
11/01/93	II.D.2.	Work group review date added
03/01/94	II.	Pending change note removed; no change
03/01/94	II.D.2.	Work group review date added

SYNONYMS

Chrysene	NSC 6175
BENZ(a)PHENANTHRENE	RCRA WASTE NUMBER U050
BENZO(a)PHENANTHRENE	1,2-BENZOPHENANTHRENE
Chrysene	1,2-BENZPHENANTHRENE
HSDB 2810	1,2,5,6-DIBENZONAPHTHALENE

Source: *Instant EPA's IRIS*, 1997, Instant Reference Sources, Inc., 7605 Rockpoint Drive, Austin, TX 78731 (Fax: 512-345-2386; Internet URL, http://www.instantref.com/inst-ref.htm).

Cypermethrin

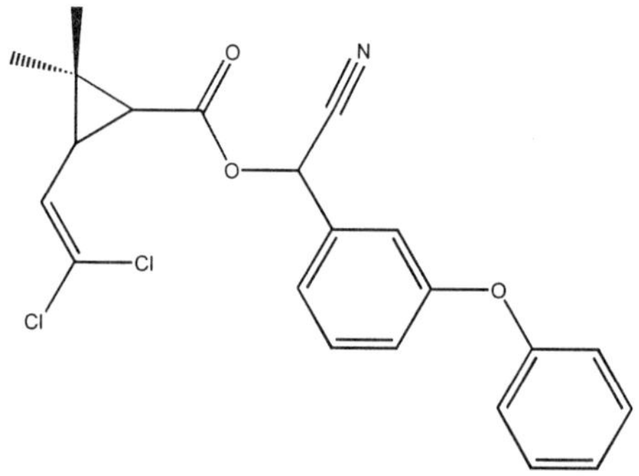

Introduction

Cypermethrin is a pyrethroid type insecticide used on cotton, lettuce, and pecans. It may also be applied into overhead sprinkler irrigation water and by pest control operators as a crack, crevice, and spot spray treatment in and around areas including, but not limited to, stores, warehouses, industrial buildings, houses, apartment buildings, greenhouses, laboratories, and on vessels, railcars, buses, trucks, trailers, and aircraft. Also it may be used in nonfood areas of schools, nursing homes, hospitals, restaurants, and hotels; and food manufacturing, processing, and servicing establishments; as barrier treatments; and as an insect repellent for horses and ponies. It is extremely toxic to fish and highly toxic to bees exposed to direct treatment on blooming crops or weeds but it has a low toxicity to mammals

Identifying Information

CAS NUMBER: 5235-07-8

SYNONYMS: Ammo™; Cymbush®; Demon®; Cynoff™; Barricade; CCN52; Cymperator; Cyperkill; Folcord; Kafil; Super; NRDC 149; Siperin; Ripcord

Endocrine Disruptor Information

Cypermethrin is a suspected EED that is listed by the World Wildlife Foundation, Canada [813]. Synthetic pyrethroids are listed as suspected EDCs in *Our Stolen Future* [810] but there is no discussion of their endocrine disrupting effects

Chemical and Physical Properties

CHEMICAL FORMULA: $C_{22}H_{19}Cl_2O_3N$

MOLECULAR WEIGHT: 416.3

PHYSICAL DESCRIPTION: Colorless, odorless crystals when pure [821] and a viscous yellow-brown liquid when impure [817]

378

DENSITY: 1.249 g/cm^3 at 20°C [821]

MELTING POINT: 60 to 80°C [821]

BOILING POINT: 170 to 195°C [821]

SOLUBILITY: Water: 4 ug/L; propylene glycol: insoluble (less than 0.5%); methanol: very soluble (approx. 75%); acetone: completely miscible; cyclohexanone: completely miscible; hexane: slightly soluble (7%); xylene: completely miscible; methylene chloride: completely miscible [821]

Environmental Reference Materials

A source of EED environmental reference materials for cypermethrin is Radian International LLC, Austin, TX (Fax 512-454-0268; Internet URL, http://www.radian.com/standards); Catalog No. ERC-032.

Hazardous Properties

ACUTE/CHRONIC HAZARDS: Cypermethrin has a low toxicity to mammals. [821] EPA data provides limited evidence of oncogenicity in female mice and is is classified as a weak Category C oncogen (possible human carcinogen with limited evidence of carcinogenicity in animals). [821]

STABILITY: Cypermethrin slowly hydrolyzes at pH 7 and below: It hydrolyzes more rapidly at pH 9. Slow photodegradation in sterile solution in sunlight (< 10% in 32 days). [821]

Medical Symptoms of Exposure

Cypermethrin produced benign lung adenomas in female mice.

Toxicological Information

The toxicological information below is gathered from several sources including two national databases: one from the *National Toxicology Program Chemical Repository Database* and another from *EPA's Integrated Risk Information System (IRIS)*.

Toxicology Information from the National Toxicology Program

Cypermethrin is not currently listed in NTP's Chemical Repository database nor is it scheduled for addition in the future.

Toxicology Information from EPA's Integrated Risk Information System (IRIS)

I. CHRONIC HEALTH HAZARD ASSESSMENTS FOR NONCARCINOGENIC EFFECTS

I.A. REFERENCE DOSE FOR CHRONIC ORAL EXPOSURE (RfD)

The Reference Dose (RfD) is based on the assumption that thresholds exist for certain toxic effects such as cellular necrosis, but may not exist for other toxic effects such as carcinogenicity. In general, the RfD is an estimate (with uncertainty spanning perhaps an order of magnitude) of a daily exposure to the human population (including sensitive subgroups) that is likely to be without an appreciable risk of deleterious effects during a lifetime. Please refer to Background Document 1 in Service Code 5 for an elaboration of these concepts. RfDs can also be derived for the noncarcinogenic health effects of compounds which are also carcinogens. Therefore, it is essential to refer to other sources of information concerning the carcinogenicity of this substance. If the US EPA has evaluated this substance for potential human carcinogenicity, a summary of that evaluation will be contained in Section II of this file when a review of that evaluation is completed.

I.A.1. ORAL RfD SUMMARY

Critical Effect	Experimental Doses*	UF	MF	RfD
G.I. tract disturbances	NOEL: 1 mg/kg/day	100	1	1E-2mg/kg/day
	LEL: 5 mg/kg/day			

*Conversion Factors: Actual dose tested

1-Year Dog Feeding Study, ICI Americas, Inc., 1982a

I.A.2. PRINCIPAL AND SUPPORTING STUDIES (ORAL RfD)

ICI Americas, Inc. 1982a. MRID No. 00112909. Available from EPA. Write to FOI, EPA, Washington, DC 20460.

Groups of beagle dogs, 6/sex, were dosed with 0, 1, 5, or 15 mg/kg/day of cypermethrin in corn oil for 52 weeks. The test material was administered by gelatin capsule and the amount administered was based on the current weight of the dog. Males and females in the high dose group, 15 mg/kg/day, displayed signs of nervous system stimulation in the form of body tremors, gait abnormalities and in coordination, disorientation, and hypersensitivity to noise. At all doses, the dogs showed increases in vomiting during the first week and the passing of liquid feces throughout the study. The increased incidence of liquid feces was 10-fold for groups dosed with 5 mg/kg/day and 30-fold for groups dosed with 15 mg/kg/day. A NOEL for systemic effects is 1 mg/kg/day based on the increased incidence of liquid feces observed at 5 mg/kg/day.

I.A.3. UNCERTAINTY AND MODIFYING FACTORS (ORAL RfD)

UF -- An uncertainty factor of 100 was used to account for the inter- and intra species differences.

MF -- None

I.A.4. ADDITIONAL COMMENTS (ORAL RfD)

Data Considered for Establishing the RfD:

1) 1-Year Feeding - dog: Principal study - see previous description;

2) 3-Generation Reproduction - rat: Systemic NOEL=2.5 mg/kg/day; Systemic LEL=7.5 mg/kg/day (decreased body weight gain in maturing pups); core grade guideline (ICI Americas, Inc., 1982b)

3) 3-Generation Reproduction - rat: Systemic NOEL=0.5 mg/kg/day; Systemic LEL=5 mg/kg/day (decrease in pup weight); core grade minimum (ICI Americas, Inc., 1979)

4) 2-Year Feeding (oncogenic) - rat: NOEL=7.5 mg/kg/day; LEL=75 mg/kg/day (weight loss, general change in blood elements and cholesterol); core grade guideline (ICI Americas, Inc., 1982c)

5) Teratology - rat: Maternal NOEL=17.5 mg/kg/day; Maternal LEL=35 mg/kg/day (decreases in weight); core grade minimum (ICI Americas, Inc., 1978a)

6) Teratology - rabbit: Teratogenic and Maternal NOEL=30 mg/kg/day (HDT); core grade minimum (ICI Americas, Inc., 1978b)

Other Data Reviewed:

1) 90-Day Feeding - dog: NOEL=500 ppm (12.5 mg/kg/day); LEL=1500 ppm (37.5 mg/kg/day) (diarrhea, anorexia, behavioral signs of nervous system effects); core grade minimum (ICI Americas, Inc., 1977)

2) 90-Day Feeding - rat: NOEL=150 ppm (7.5 mg/kg/day); LEL=1500 ppm (75 mg/kg/day) (decreased body weight and possible nerve damage); core grade minimum (ICI Americas, Inc., 1980)

Data Gap(s): None

I.A.5. CONFIDENCE IN THE ORAL RfD

Study -- HighData Base -- HighRfD -- High

The critical studies are of good quality and are given high confidence ratings. Additional studies are also of good quality; therefore, the data base is given a high confidence rating. High confidence in the RfD follows.

I.A.6. EPA DOCUMENTATION AND REVIEW OF THE ORAL RfD

Source Document -- This assessment is not presented in any existing US EPA document.

Other EPA Documentation -- Pesticide Registration Files

Agency Work Group Review -- 10/12/88, 01/18/89

Verification Date -- 01/18/89

I.A.7. EPA CONTACTS (ORAL RfD)

George Ghali / OPP -- (703)557-7490

William Burnam / OPP -- (703)557-7491

I.B. REFERENCE CONCENTRATION FOR CHRONIC INHALATION EXPOSURE (RfC)

[Ed. Note: EPA does not have information yet in this section of IRIS].

II. CARCINOGENICITY ASSESSMENT FOR LIFETIME EXPOSURE

This substance/agent has not been evaluated by the US EPA for evidence of human carcinogenic potential.

III. HEALTH HAZARD ASSESSMENTS FOR VARIED EXPOSURE DURATIONS

[Ed. Note: EPA does not have information yet in this section of IRIS].

IV. US EPA REGULATORY ACTIONS

EPA risk assessments may be updated as new data are published and as assessment methodologies evolve. Regulatory actions are frequently not updated at the same time. Compare the dates for the regulatory actions in this section with the verification dates for the risk assessments in sections I and II, as this may explain inconsistencies. Also note that some regulatory actions consider factors not related to health risk, such as technical or economic feasibility. Such considerations are indicated for each action. In addition, not all of the regulatory actions listed in this section involve enforceable federal standards. Please direct any questions you may have concerning these regulatory actions to the US EPA contact listed for that particular action. Users are strongly urged to read the background information on each regulatory action in Background Document 4 in Service Code 5.

IV.A. CLEAN AIR ACT (CAA)

[Ed. Note: EPA does not have information yet in this section of IRIS].

IV.B. SAFE DRINKING WATER ACT (SDWA)

[Ed. Note: EPA does not have information yet in this section of IRIS].

IV.C. CLEAN WATER ACT (CWA)

[Ed. Note: EPA does not have information yet in this section of IRIS].

IV.D. FEDERAL INSECTICIDE, FUNGICIDE, AND RODENTICIDE ACT (FIFRA)

IV.D.1. PESTICIDE ACTIVE INGREDIENT, Registration Standard

Status -- List "B" Pesticide

Reference -- 54 FR 22706 (05/25/89)

EPA Contact -- Registration Branch / OPP(703)557-7760 / FTS 557-7760

IV.D.2. PESTICIDE ACTIVE INGREDIENT, Special Review

[Ed. Note: EPA does not have information yet in this section of IRIS].

IV.E. TOXIC SUBSTANCES CONTROL ACT (TSCA)

[Ed. Note: EPA does not have information yet in this section of IRIS].

IV.F. RESOURCE CONSERVATION AND RECOVERY ACT (RCRA)

[Ed. Note: EPA does not have information yet in this section of IRIS].

IV.G. SUPERFUND (CERCLA)

[Ed. Note: EPA does not have information yet in this section of IRIS].

VI. BIBLIOGRAPHY

VI.A. ORAL RfD REFERENCES

ICI Americas, Inc. 1977. MRID No. 00112929. Available from EPA. Write to FOI, EPA, Washington D.C. 20460.

ICI Americas, Inc. 1978a. Accession No. 099855. Available from EPA. Write to FOI, EPA, Washington D.C. 20460.

ICI Americas, Inc. 1978b. Accession No. 099855. Available from EPA. Write to FOI, EPA, Washington D.C. 20460.

ICI Americas, Inc. 1979. MRID No. 00090040. Available from EPA. Write to FOI, EPA, Washington D.C. 20460.

ICI Americas, Inc. 1980. Accession No. 099855. Available from EPA. Write to FOI, EPA, Washington D.C. 20460.

ICI Americas, Inc. 1982a. MRID No. 00112909. Available from EPA. Write to FOI, EPA, Washington D.C. 20460.

ICI Americas, Inc. 1982b. MRID No. 00112912. Available from EPA. Write to FOI, EPA, Washington D.C. 20460.

ICI Americas, Inc. 1982c. MRID No. 00112910. Available from EPA. Write to FOI, EPA, Washington D.C. 20460.

VI.B. INHALATION RfD REFERENCES

None

VI.C. CARCINOGENICITY ASSESSMENT REFERENCES

None

VI.D. DRINKING WATER HA REFERENCES

None

VII. REVISION HISTORY

Date	Section	Description
03/01/89	I.A.	Oral RfD summary on-line
07/01/89	VI.	Bibliography on-line
01/01/90	I.A.2.	Text edited
01/01/92	IV.	Regulatory Action section on-line

SYNONYMS

Agrothrin
Ambush CY
Ammo
Antiborer 3767
Ardap
Barricade
CCN 52
alpha-Cyano-m-phenoxybenzyl 3-(2,2-dichlorovinyl)-2,2-
dimethylcyclopropane
   carboxylate
Cymbush
Cyperil
Cyperkill
Cypermethrin
3-(2,2-dichloroethenyl)-2,2-dimethyl-Cyclopropanecarboxylic acid,
cyano(3-phenoxyphenyl)methyl ester
EXP 5598

Fendona
FMC 30980
FMC 45497
FMC 45806
JF 5705F
NRDC 149
NRDC 160
NRDC 166
Nurele
Ripcord
RU 27998
Sherpa
SF 06646
WL 8517
WL 43467

Source: *Instant EPA's IRIS*, 1997, Instant Reference Sources, Inc., 7605 Rockpoint Drive, Austin, TX 78731 (Fax: 512-345-2386; Internet URL, http://www.instantref.com/inst-ref.htm).

Production, Use, and Pesticide Labeling Information

1. Description of Chemical

Generic Name: Cyclopropanecarboxylic acid, 3-(2,2-dichloroethenyl)-2,2-dimethyl, cyano(3-phenoxyphenyl)methyl ester

Common Name: Cypermethrin

Trade Names: Ammo™ Cymbush®; Demon®; Cynoff™

Other Names: Barricade; CCN52; Cymperator; Cyperkill; Folcord; Kafil; Super; NRDC 149; Siperin; Ripcord.

EPA Shaughnessy Code: 109702

Chemical Abstracts Service (CAS) Number: 52315-07-8

Year of Initial Registration: 1984

Pesticide Type: Pyrethroid-like; Insecticide/Miticide

Chemical Family: Pyrethroid

Manufacturers: FMC Corporation; ICI Americas, Inc.; Shell International Chemical Company, Ltd. (London).

2. Use Patterns and Formulations

- Application Rates:

- Applied to cotton, lettuce (head) and pecans at a rate up to 0.1 pounds active ingredient per acre. It may also be applied into overhead sprinkler irrigation water. Cypermethrin is also applied by Pest Control Operators as a crack, crevice, and spot spray treatment in and around areas including, but not limited to, stores, warehouses, industrial buildings, houses, apartment buildings, greenhouses, laboratories, and on vessels, railcars, buses, trucks, trailers, and aircraft.

- Also it may be used in nonfood areas of schools, nursing homes, hospitals, restaurants, and hotels; and food manufacturing, processing, and servicing establishments; as barrier treatments; and as an insect repellent for horses and ponies.

- Usual Carrier: Water and oil.

- Type of Formulations: 30.6%, 25.3%, and 36.6% emulsifiable concentrate and 88% technical.

3. Science Findings

Chemical/Physical Characteristics of the Technical Grade

- Physical State: Solid

- Color: Colorless crystals

- Odor: Odorless

- Molecular Weight: 416.3

- Molecular Formula: $C_{22}H_{19}O_3NCl_2$

- Melting Point: 60 to 80 °C

- Boiling Point: 170 to 195 °C

- Density: 1.249 q CM^3 at 20 °C

- Vapor Pressure: 8×10^{-4} at 80 °C; 1×10^{-7} at 20 °C

- Solubility in Various Solvents:

- Water - 4 ppb
- Base - as water
- Acid - as water
- Propylene glycol - insoluble (less than 0.5%)
- Methanol - very soluble (approx. 75%)
- Acetone - completely miscible
- Cyclohexanone - completely miscible
- Hexane - slightly soluble (7%)
- Xylene - completely miscible
- Methylene dichloride - completely miscible

- Stability - as neat material:

    - No detectable decomposition at normal ambient temperatures and for at least 3 months at 50 °C to date.

- Stability - as dilute aqueous solution:

    - Slowly hydrolyzes at pH 7 and below: Hydrolyzes more rapidly at pH 9. Slow photodegradation in sterile solution in sunlight (< 10% in 32 days).

Toxicological Characteristics of the Technical Grade

- Acute Oral LD50 - Rat: LD50 = 247 (187-326) mg/kg (males); LD50 = 309 (150-500) mg/kg (females).

- Acute Dermal LD50 - Rabbit: LD50 > 2460 mg/kg.

- Primary Dermal Irritation - Rabbit: PIS = 0.71 (not irritating).

- Primary Eye Irritation - Rabbit: Mild irritation.

- Skin Sensitization - Guinea Pig: May cause allergic skin reactions.

- Subchronic Oral - Rat: NOEL of 75 ppm for pharmacological effects. NOEL of 150 ppm for toxic effects.

- Chronic Toxicity - Rat: NOEL = 150 ppm; LEL = 1500 ppm (HDT).

- Oncogenicity - 24-Month Mouse: Positive neoplastic response in lung tissue. Increased incidence of benign adenomas in females (only), statistically significant at 1600 ppm (HDT). No evidence of oncogenicity in the rat at up to 1500 ppm (HDT).

- Teratogenicity

    - Rabbit: Not teratogenic at 30 mg/kg/day (HDT).

    - Rat: Not teratogenic at 70 mg/kg/day (HDT).

    - Reproduction - 3-Generation Rat: NOEL for adverse reproductive effects = 750 ppm (HDT), NOEL for systemic effects = 50 ppm, LEL = 150 ppm (decreased body weight gain in maturing pups).

    - Mutagenicity - Ames Test: Not mutagenic.

    - Mutagenesis - Host-Mediated Assay: Not mutagenic at 50 mg/kg.

    - Mutagenesis - Dominant Lethal: Not mutagenic at 25 mg/kg.

Physiological and Biochemical Characteristics

    - The mode of action in biological systems is stomach and contact exhibiting neurotoxicological characteristics typical of pyrethroid insecticides. Slight repellent effect.

Environmental Characteristics

    - Adequate data are sufficient to define the fate of cypermethrin in the environment. Cypermethrin is stable to hydrolysis, with an estimated T 1/2 exceeding 50 days at environmental expected temperatures and pH values. Cypermethrin is extremely stable to photolysis in water with an estimated T 1/2 exceeding 100 days at environmentally expected temperatures and pH values. Photoproducts produced included DCVA, 3-phenoxybenzaldehyde, and 3-phenoxybenzoic acid. Cypermethrin photodegrades rapidly on soil surfaces (T 1/2: 8 to 16 days) to many photoproducts, the major ones identified as 3-phenoxybenzoic acid and compound XIV. Cypermethrin degrades in soil under laboratory conditions. The rate is more rapid on sandy clay and sandy loam soils than on clay soils and more rapid on soils lower in organic matter content and cation exchange capacity under all aerobic conditions.

    The T 1/2 in aerobic soils ranged from 2 to 8 weeks. In sterile aerobic soils, cypermethrin degraded with a T 1/2 of 20 to 25 weeks indicating that microbes play a significant role in soil degradation. Cypermethrin degraded more slowly under anaerobic or waterlogged conditions with the major metabolite 3-phenoxybenzoic acid produced. Under aerobic conditions the major metabolites produced were DCVA and 3-phenoxybenzoic acid. Cypermethrin does not leach significantly in soil. It has a low solubility in water (0.2 ppm) and, consequently, high adsorption characteristics. The leaching potential for its degrades may be higher. Under field conditions, runoff of cypermethrin has been shown to occur to some degree and was probably due to physical transport of the solid particles via erosion.

Cypermethrin itself degrades rapidly in the field with a T 1/2 of 4 to 12 days. The persistence of the major aerobic soil metabolites is not known. Both accumulation and depuration of cypermethrin residues will occur in trout and catfish. Bioconcentration factors of approximately 1200X were calculated in rainbow trout in a flowthrough study. The data requirement for accumulation of cypermethrin in rotational crops has not been satisfactorily completed.

Ecological Characteristics

- Avian acute oral LD50 - Mallard Duck: >4640 mg/kg
- Avian dietary LC50 - Mallard Duck and Bobwhite Quail: LC50 >
    20,000 ppm.
- Avian reproduction - Mallard Duck and Bobwhite Quail: NOEL > 50
    ppm (HDT).
- Fish acute 96-hour LC50 - Rainbow Trout = 0.82 ppb.
- Fish acute 96-hour LC50 - Bluegill Sunfish = 1.78 ppm.
- Aquatic invertebrate acute LC50 - Daphnia magna = 0.26 ppb.

Tolerance Assessments

- Section 408 tolerances under the Federal Food, Drug, and Cosmetic Act are established until December 31, 1989 for residues of the insecticide cypermethrin (+/=/-) alpha-cyano-(3-phenoxyphenyl) methyl (+/=-cis,trans-3-(2,2-dichloroethanyl)-2-panecarboxylate and its metabolites* 3-PB Acid and DCVA in or on the following raw agricultural commodities:

Commodity	ppm
Cattle, fat	0.05
Cattle, meat	0.05
Cattle, meat byproducts	0.05
Cottonseed	0.5
Goats, fat	0.05
Goats, meat	0.05
Goats, meat byproducts	0.05
Hogs, fat	0.05
Hogs, meat	0.05
Hogs, meat byproducts	0.05
Horses, fat	0.05
Horses, meat	0.05
Horses, meat byproducts	0.05
Lettuce (head)	10.00
Milk	0.05
Pecans*	0.05
Sheep, fat	0.05
Sheep, meat	0.05
Sheep, meat byproducts	0.05

- The acceptable daily intake (ADI) is calculated to be 0.01 mg/kg/day based on a dog study with a NOEL of 1.0 mg/kg/day and using a safety factor of 100. The maximum permissible intake (MPI) is calculated to be 0.60 mg/kg/day for a 60-kg person. Published tolerances result in a theoretical maximum residue contribution (TMRC) of 0.002773 mg/kg bwt/day. The existing TMRC is equivalent to 27.7 percent of the ADI. No additional data are required to support the current crop tolerances listed in 40 CFR 180.418.

Summary Science Statement

- Cypermethrin, a pyrethroid, is extremely toxic to fish. It is highly toxic to bees exposed to direct treatment on blooming crops or weeds. Cypermethrin has a low toxicity to mammals. EPA's review of the cypermethrin pond study indicated that the data, as presented, were not adequate or scientifically complete to allow the Agency to evaluate the actual impact that the use of cypermethrin would have on aquatic life forms. Therefore, ICI/FMC must repeat this study. The EPA Peer Review Committee completed its evaluation of cypermethrin with respect to its oncogenic potential and concluded that the data available for cypermethrin provide limited evidence of oncogenicity for the chemical in female mice. According to EPA Guidelines for Carcinogen Risk Assessment (Federal Register September 24, 1986), the Committee classified cypermethrin as a weak Category C oncogen (possible human carcinogen with limited evidence of carcinogenicity in animals). That is, cypermethrin produced benign lung adenomas at the highest dose level in only one sex and species of animal (female mice) and was not considered strong enough to warrant a "quantitative estimation of human risk."

4. Summary of Regulatory Position and Rationale

- Adequate data are available to assess the acute and chronic toxicological effects of cypermethrin to humans. The Agency's review of the field study (Part 72-7) found that the data, as presented, were not scientifically adequate nor sufficiently complete to allow the Agency to evaluate the actual impact that the use of cypermethrin would have on aquatic life forms. ICI/FMC disagreed with the Agency's conclusion and will submit additional information necessary in order for the Agency to complete the risk assessment.

- On the basis of this information, EPA seriously considered whether to issue new conditional registrations for cypermethrin. Based on the information submitted, on January 3, 1989 the Agency issued new conditional registrations of cypermethrin which will expire on June 15, 1989.

- The Agency issued these new conditional registrations for this short period of time to ICI and FMC in light of their agreement to: - Submit all data generated in the course of the cypermethrin Alabama pond study pertaining to runoff and residues in water and sediment to the Agency by January 1989.

- Submit all other data now in existence and not previously submitted to the Agency, as well as all other data generated in the future in the course of the cypermethrin Alabama pond study, as soon as is practical.

- Conduct an aquatic mesocosm study (a simulated 2-year field study) for which EPA will develop and provide the protocol no later than April 15, 1989, in case ICI/FMC do not persuade EPA that the current cypermethrin pond study is acceptable.

- Within 30 days of receipt of an EPA protocol for an aquatic mesocosm study, provide the Agency with written unconditional acceptance of the protocol and an unconditional commitment to conduct the study through completion. For any modification of the protocol to be valid, it must be agreed to by EPA within the above-mentioned 30-day period.

- EPA has concluded that the continual use of cypermethrin for this short period of time will not cause a significant increase in the risk of adverse effects to the environment.

- The Delaney Clause in section 409 of the Federal Food, Drug, and Cosmetic Act bars the establishment of food additive regulations for substances which induce cancer in man or test animals. Since cypermethrin has been found to produce an oncogenic response in test animals, no 409 tolerances will be granted. Cypermethrin as parent does not leach significantly in soil. At this time, there are no concerns for ground water contamination.

- Limitations:

> - Registrations are being extended with an expiration date of June 15, 1989. Tolerances expire December 31, 1989.
>
> - RESTRICTED USE PESTICIDE - Extremely toxic to fish. For retail sale to and use only by Certified Applicators, or persons under their direct supervision, and only for those uses covered by the Certified Applicator's Certification.
>
> - REENTRY STATEMENTS - Do not treat areas while unprotected humans or domestic animals are present in the treatment areas.
>
> - Do not allow entry into treated areas without protective clothing until sprays have dried.
>
> - Do not apply within 21 days of harvest.
>
> - Do not graze livestock in treated orchards or cut treated cover crops for feed.
>
> - CROP ROTATION RESTRICTION - Do not plant rotational crops within 30 days of last application.

5. Summary of Major Data Gaps

> - Simulated and/or actual field study (Part 72-7).

6. Contact Person at EPA

> George T. LaRocca Product Manager 15
> Insecticide- Rodenticide Branch, Registration Division (H7504C)
> Office of Pesticide Programs
> Environmental Protection Agency
> 401 M Street SW
> Washington, DC 20460
>
> Office location and telephone number:
>
> Room 211, Crystal Mall #2
> 1921 Jefferson Davis Highway
> Arlington, VA 22202
>
> Phone: (703) 557-2400

Source: ***Instant EPA's Pesticide Facts***, 1994, Instant Reference Sources, Inc., 7605 Rockpoint Drive, Austin, TX 78731 (Fax: 512-345-2386; Internet URL, http://www.instantref.com/inst-ref.htm).

2,4-D

Introduction

2,4-D is a defoliant and herbicide [703]. In the US it is used as a herbicide for control of broadleaf plants and also as a plant growth regulator. It is typically applied to grasses, wheat, barley, oats, sorghum, corn, sugarcane as well as to noncrop plants such as pasture and range land, lawns and turf. It controls post-emergent weeds such as thistle, dandelions, annual mustards, ragweed, etc. It is used on tomatoes to cause all fruits to ripen at the same time for machine harvesting.. It is used in forest management for brush control and conifer release. [828] It is also used to increase latex output of old rubber trees. [715]

2,4-D is released into the environment through its use in herbicide formulations and as a hydrolysis product of 2,4-D esters or from spills. If released on land, it will probably readily biodegrade (typical half-lives <1 day to several weeks). Its adsorption to soils will depend upon organic content and pH of the soil (2,4-D pKa= 2.64). Leaching to groundwater will likely be a significant process in coarse-grained sandy soils with low organic content or with very basic soils. If released to water, it will be lost primarily due to biodegradation (typical half-lives 10 to >50 days). It will be more persistent in oligotrophic waters and where high concentrations are released. Degradation will be rapid in sediments (half-life <1 day). It will not bioconcentrate in aquatic organisms or appreciably adsorb to sediments, especially at basic pH's. If released in air, it will be subject to photooxidation (estimated half-life of 1 day) and rainout. Human exposure will be primarily to workers involved in the making and using 2,4-D compounds as herbicides and those who work in and live near fields sprayed and treated with 2,4-D compounds. Exposure may also occur through ingestion of contaminated food products and drinking water. [828]

Identifying Information

CAS NUMBER: 94-75-7

NIOSH Registry Number: AG6825000

SYNONYMS: 2,4-DICHLOROPHENOXYACETIC ACID; AGROTECT; AMIDOX; AMOXONE; AQUA-KLEEN; BRUSH-RHAP; B-SELEKTONON; CHLOROXONE; CROP RIDER; CROTILIN; D 50; DACAMINE; 2,4-D ACID; DEBROUSSAILLANT 600; DECAMINE; DED-WEED LV-69; DESORMONE; (2,4-DICHLOOR-FENOXY)-AZIJNZUUR; (2,4-DICHLOR-PHENOXY)-ESSIGSAEURE; DICOPUR; DICOTOX; DINOXOL; DMA-4; DORMONE; EMULSAMINE BK; EMULSAMINE E-3; ENT 8,538; ENVERT 171; ENVERT DT; ESTERON; ESTERONE FOUR; ESTONE;

FARMCO; FERNESTA; FERNIMINE; FERNOXONE FERXONE; FOREDEX 75;
FORMULA 40; HEDONAL; HERBIDAL; IPANER; KROTILINE; LAWN-KEEP;
MACRONDRAY; MIRACLE; MONOSAN; MOXONE; NA 2765; NETAGRONE;
NETAGRONE 600; NSC 423; PENNAMINE; PENNAMINE D; PHENOX; PIELIK;
PLANOTOX; PLANTGARD; RCRA WASTE NUMBER U240; RHODIA; SALVO;
SPRITZ-HORMIN/2,4-D; SPRITZ-HORMIT/2,4-D; TRANSAMINE; TRIBUTON;
TRINOXOL; U 46DP; U-5043; VERGEMASTER; VIDON 638; VISKO-RHAP;
WEED-AG-BAR; WEEDAR-64; WEEDATUL; WEED-B-GON; WEEDEZ WONDER
BAR; WEEDONE LV4; WEED-RHAP; WEED TOX; WEEDTROL

Endocrine Disruptor Information

2,4-D is a strongly suspected EED that it is listed by EPA [811], the Centers for Disease
Control & Prevention [812], and the World Wildlife Fund [813] as a potential endocrine
modifying chemical.

Chemical and Physical Properties

CHEMICAL FORMULA: $C_8H_6Cl_2O_3$

MOLECULAR WEIGHT: 221.04 [703]

PHYSICAL DESCRIPTION: White to yellow crystalline, odorless powder [704]

SPECIFIC GRAVITY: (86 °F) 1.57 [704]

MELTING POINT: 141°C [703]

BOILING POINT: 160°C @ 0.4 mm [703]

SOLUBILITY: Water: 540 ppm [702]

OTHER PHYSICAL DATA: Vapor Density: 7.63 [703] Vapor Pressure: 0.4 mm Hg @
160°C [705]

Source: *Instant EPA's Air Toxics*, 1994, Instant Reference Sources, Inc., 7605
Rockpoint Drive, Austin, TX 78731 (Fax: 512-345-2386; Internet URL,
http://www.instantref.com/inst-ref.htm).

Environmental Reference Materials

A source of EED environmental reference materials for 2,4-D is Radian International
LLC, Austin, TX (Fax 512-454-0268; Internet URL, http://www.radian.com/standards);
Catalog No. ERD-076.

Hazardous Properties

ACUTE/CHRONIC HAZARDS: Poison by ingestion, intravenous, and intraperitoneal
routes. Moderately toxic by skin contact; experimental carcinogen and teratogen;
suspected human carcinogen. Human systemic effects by ingestion: somnolence,
convulsions, coma, nausea or vomiting. Can cause liver and kidney injury. A skin and

severe eye irritant. Human mutagenic data exists; produces experimental reproductive effects. When heated to decomposition, emits toxic fumes of Cl^-. [703]

HAP WEIGHTING FACTOR: 1 [713]

VOLATILITY:
   Vapor Pressure: Low [704]
   Vapor Density: 7.63 [703]

FIRE HAZARD: Flash Point: 88°C [702]
   Noncombustible solid, but may be dissolved in flammable liquids [704].
   Autoignition temperature: Not available

REACTIVITY: Incompatible or reacts with strong oxidizers [704].

STABILITY: When heated to decomposition it emits toxic fumes of Cl^- [703].

Source: *Instant EPA's Air Toxics*, 1994, Instant Reference Sources, Inc., 7605 Rockpoint Drive, Austin, TX 78731 (Fax: 512-345-2386; Internet URL, http://www.instantref.com/inst-ref.htm).

Medical Symptoms of Exposure

Symptoms of exposure may include weakness, stupor, hyporeflexia, muscle twitch; convulsions; dermatitis; in animals: liver and kidney damage. [704] It has also caused somnolence, convulsions, coma, and nausea or vomiting. It is a skin and severe eye irritant; it can cause liver and kidney injury. [703] Dust may irritate eyes. Acute eye or skin irritation has been reported in agricultural and forestry workers following occupational exposure. [828]

Source: *Instant EPA's Air Toxics*, 1994, Instant Reference Sources, Inc., 7605 Rockpoint Drive, Austin, TX 78731 (Fax: 512-345-2386; Internet URL, http://www.instantref.com/inst-ref.htm).

Toxicological Information

The toxicological information below is gathered from several sources including two national databases: one from the *National Toxicology Program Chemical Repository Database* and another from *EPA's Integrated Risk Information System (IRIS)*.

Toxicology Information from the National Toxicology Program

2,4-DICHLOROPHENOXYACETIC ACID

TOXICITY:

Typ.dose	Mode	Specie	Amount	Units	Other
LDLo	orl	hmn	80	mg/kg	
LDLo	orl	man	93	mg/kg	
LD50	orl	rat	370	mg/kg	
LD50	skn	rat	1500	mg/kg	
LDLo	ipr	rat	666	mg/kg	
LD50	orl	mus	347	mg/kg	
LDLo	ipr	mus	125	mg/kg	
LDLo	orl	rbt	800	mg/kg	
LD50	orl	dog	100	mg/kg	
LD50	skn	rbt	1400	mg/kg	
LDLo	ipr	rbt	400	mg/kg	
LDLo	ivn	rbt	400	mg/kg	
LDLo	ipr	gpg	666	mg/kg	
LD50	orl	ham	500	mg/kg	
LD50	orl	ckn	541	mg/kg	
LD50	orl	mam	375	mg/kg	

TERATOGENICITY:

Reproductive Effects Data:

Type of Effect	Route/Animal	Amount Dosed (Notes)
TDLo:	orl-rat	1 gm/kg (6-15D preg)
TDLo:	orl-rat	125 mg/kg (6-15D preg)
TDLo:	orl-rat	500 mg/kg (6-15D preg)
TDLo:	orl-mus	707 mg/kg (11-14D preg)
TDLo:	orl-mus	900 mg/kg (6-14D preg)
TDLo:	scu-mus	882 mg/kg (6-14D preg)
TDLo:	scu-mus	900 mg/kg (6-14D preg)
TDLo:	orl-ham	200 mg/kg (7-11D preg)
TDLo:	orl-rat	220 ug/kg (1-22D preg)
TDLo:	orl-mus	438 mg/kg (8-12D preg)

MUTATION DATA:

Test	Lowestdose		Test	Lowestdose
dnr-esc	5 mg/disc	|	mmo-omi	1 gm/L
dnd-esc	20 umol/L	|	cyt-rat-ipr	100 ug/kg
dnr-bcs	5 mg/disc	|	dni-ham:ovr	1 mmol/L
sln-dmg-orl	25 ppm	|	cyt-ham:ovr	2400 mg/L
sln-dmg-unr	1000 ppm/15D	|	dnd-sal:spr	1 mmol/L
mmo-smc	150 mg/L	|	sce-hmn:lym	10 mg/L
mrc-asn	4 umol/L	|	msc-ham:lng	10 umol/L
mma-sat	250 ug/plate	|	cyt-ctl:kdy	1 ppm
dns-hmn:fbr	1 umol/L	|	dnd-mam:lym	1 mmol/L
cyt-hmn:lym	20 ug/L	|	dni-mus-orl	200 mg/kg
cyt-mus-orl	100 mg/kg	|	sce-ham:ovr	167 mg/L

OTHER TOXICITY DATA:

    Skin and Eye Irritation Data:
        skn-rbt 500 mg/24H MLD
        eye-rbt 750 ug/24H SEV
    Review: Toxicology Review-6

    Standards and Regulations: DOT-Hazard: ORM-A; Label: None

    Status: EPA Genetox Program 1988, Positive: In vivo cytogenetics-nonhuman bone marrow
    EPA Genetox Program 1988, Positive: In vitro cytogenetics-human lymphocyte
    EPA Genetox Program 1988, Positive: B subtilis rec assay; E coli polA without S9
    EPA Genetox Program 1988, Positive: V79 cell culture-gene mutation
    EPA Genetox Program 1988, Positive: S cerevisiae gene conversion
    EPA Genetox Program 1988, Negative: D melanogaster-whole sex chrom. loss
    EPA Genetox Program 1988, Negative: D melanogaster-nondisjunction
    EPA Genetox Program 1988, Negative: Histidine reversion-Ames test
    EPA Genetox Program 1988, Negative: D melanogaster Sex-linked lethal
    EPA Genetox Program 1988, Negative: In vitro UDS-human fibroblast;

TRP reversion

> EPA Genetox Program 1988, Negative: S cerevisiae-homozygosis
> EPA Genetox Program 1988, Inconclusive: Carcinogenecity-mouse/rat;
> Mammalian micronucleus
> EPA TSCA Chemical Inventory, 1986
> EPA TSCA Test Submission (TSCATS) Data Base, January 1989
> NIOSH Analytical Methods: see 2,4-D and 2,4,5-T, 5001
> Meets criteria for proposed OSHA Medical Records Rule
>> Lethal dose: 700 mg/kg [051]
>> Acceptable daily intake for man: 0-0.3 mg/kg [395]

SAX TOXICITY EVALUATION:

THR: Poison by ingestion, intravenous and intraperitoneal routes. Moderately toxic by skin contact. An experimental carcinogen and teratogen. A suspected human carcinogen. Human systemic effects by ingestion. Human mutation data. A defoliant and herbicide.

Source: *Instant Tox-Base*, 1995, Instant Reference Sources, Inc., 7605 Rockpoint Drive, Austin, TX 78731 (Fax: 512-345-2386; Internet URL, http://www.instantref.com/inst-ref.htm).

Toxicology Information from EPA's Integrated Risk Information System (IRIS)

I. CHRONIC HEALTH HAZARD ASSESSMENTS FOR NONCARCINOGENIC EFFECTS

I.A. REFERENCE DOSE FOR CHRONIC ORAL EXPOSURE (RfD)

The Reference Dose (RfD) is based on the assumption that thresholds exist for certain toxic effects such as cellular necrosis, but may not exist for other toxic effects such as carcinogenicity. In general, the RfD is an estimate (with uncertainty spanning perhaps an order of magnitude) of a daily exposure to the human population (including sensitive subgroups) that is likely to be without an appreciable risk of deleterious effects during a lifetime. Please refer to Background Document 1 in Service Code 5 for an elaboration of these concepts. RfDs can also be derived for the noncarcinogenic health effects of compounds which are also carcinogens. Therefore, it is essential to refer to other sources of information concerning the carcinogenicity of this substance. If the U.S. EPA has evaluated this substance for potential human carcinogenicity, a summary of that evaluation will be contained in Section II of this file when a review of that evaluation is completed.

NOTE -- The Oral RfD for 2,4-D may change in the near future pending the outcome of a further review now being conducted by the Oral RfD Workgroup.

I.A.1. ORAL RfD SUMMARY

Critical Effect	Experimental Doses*	UF	MF	RfD
Hematologic, hepatic and renal toxicity.	NOAEL: 1 mg/kg/day	100	1	1E-2 mg/kg/day
Reduced neonatal survival	LOAEL: 50 mg/kg/day			

90-Day Rat Oral Bioassay and 1-Year Interim Report from a Year Rat Oral Bioassay Dow Chemical Co., 1983

*Conversion Factors: none

I.A.2. PRINCIPAL AND SUPPORTING STUDIES (ORAL RfD)

Dow Chemical Co. 1983. Acc. No. 251473. Available from EPA. Write to FOI, EPA, Washington, DC 20460.

Hematologic, hepatic, and renal toxicity were demonstrated in a study in Fischer rats (strain 344) during subchronic feeding performed at the Hazleton Laboratories (1983). 2,4-D (97.5% pure) was added to the diet chow and fed to the rats for 91 days at doses calculated to be 0.0 (controls), 1.0, 5.0, 15.0, or 45.0 mg/kg/day. In each of the five groups there were 20 animals/sex and 40 animals/treatment group, for a total of 200 animals. Criteria examined to determine toxicity were survival, daily examination for clinical symptomatology, weekly change in body weights, growth rates, food intake, ophthalmologic changes, changes in organ weights, and clinical, gross and histopathologic alterations. The results of the study demonstrated statistically significant reductions in mean hemoglobin (both sexes), mean hematocrit and red blood cell levels (both sexes), and mean reticulocyte levels (males only) at the 5.0 mg/kg/day dose or higher after 7 weeks. There were also statistically significant reductions in liver enzymes LDH, SGOT, SGPT, and alkaline phosphatase at week 14 in animals treated at the 15.0 mg/kg/day or higher doses. Kidney weights (absolute and relative) showed statistically significant increases in all animals at the 15.0 mg/kg/day dose or higher at the end of the experimental protocol. Histopathologic examinations correlated well with kidney organ weight changes showing cortical and subcortical pathology. Increases in ovarian weights, T-4 levels, and a decrease in BUN were reported, but were not considered to be treatmentrelated.

In a second part of this study (Dow Chemical Co., 1983), B6C3F1 mice (20/sex/group) were fed the diet chow mixed with 97.5% pure 2,4-D at 0.0, 5.0, 15.0, 45.0 or 90.0 mg/kg bw/day (calculated doses) for 91 days. Criteria used to determine toxicity were the same as for rats. The only effect reported at 5 mg/kg/day was increased weight of adrenals in females. Effects at 15 mg/kg/day included altered organ weights and hematologic effects. Kidney weights were not affected below 45 mg/kg/day.

I.A.3. UNCERTAINTY AND MODIFYING FACTORS (ORAL RfD)

UF -- The 100-fold uncertainty factor accounts for both interspecies and interhuman variability in the toxicity of this chemical in lieu of specific data. Because an analysis of the 90-day and 1-year interim results suggests that the NOAEL would hold for the full 2-year duration, inclusion of the subchronic-to-chronic UF is not warranted.

MF -- None

I.A.4. ADDITIONAL COMMENTS (ORAL RfD)

The subchronic studies previously discussed provide a more sensitive basis for the RfD than available chronic or reproduction studies. Chronic toxicity and reproduction studies of 2,4-D indicated no adverse effects at dietary levels up to 500 ppm in dogs (approximately 14.5 mg/kg bw/day), up to 1250 ppm in rats (approximately 62.5 mg/kg bw/day) (Hansen et al., 1971), or at levels of 1000 ppm in drinking water (50-100 mg/kg bw/day) in pregnant rats (exposed through gestation and for 10 months following parturition) or their offspring (exposed for up to 2 years after weaning) (Bjorklund and Erne, 1966). A secondary reference to another study reported an increase in mortality among rats whose dams received approximately 50 mg/kg bw/day of 2,4-D in the diet for 3 months before mating and throughout gestation and lactation (Gaines and Kimbrough, 1970).

I.A.5. CONFIDENCE IN THE ORAL RfD

Study -- Medium
Data Base -- Medium
RfD -- Medium

Confidence in the principal study is medium because a fair number of animals of each sex was used, four doses were given, and a good number of parameters were measured. Confidence in the data base is medium because several studies support both the observation of critical toxic effects and the levels at which they occur. Medium confidence in the RfD follows.

I.A.6. EPA DOCUMENTATION AND REVIEW OF THE ORAL RfD

Source Document -- U.S. EPA, 1984
This document has received several internal EPA reviews, reviews by outside expert scientists and a public review.
Other EPA Documentation -- None
Agency Work Group Review -- 05/20/85, 02/05/86
Verification Date -- 02/05/86

I.A.7. EPA CONTACTS (ORAL RfD)

Krishan Kahnna / OST -- (202)260-7588

I.B. REFERENCE CONCENTRATION FOR CHRONIC INHALATION EXPOSURE (RfC)

[Ed. Note: EPA does not have information yet in this section of IRIS].

II. CARCINOGENICITY ASSESSMENT FOR LIFETIME EXPOSURE

This substance/agent has not been evaluated by the U.S. EPA for evidence of human carcinogenic potential.

III. HEALTH HAZARD ASSESSMENTS FOR VARIED EXPOSURE DURATIONS

[Ed. Note: EPA does not have information yet in this section of IRIS].

IV. U.S. EPA REGULATORY ACTIONS

EPA risk assessments may be updated as new data are published and as assessment methodologies evolve. Regulatory actions are frequently not updated at the same time. Compare the dates for the regulatory actions in this section with the verification dates for the risk assessments in sections I and II, as this may explain inconsistencies. Also note that some regulatory actions consider factors not related to health risk, such as technical or economic feasibility. Such considerations are indicated for each action. In addition, not all of the regulatory actions listed in this section involve enforceable federal standards. Please direct any questions you may have concerning these regulatory actions to the U.S. EPA contact listed for that particular action. Users are strongly urged to read the background information on each regulatory action in Background Document 4 in Service Code 5.

IV.A. CLEAN AIR ACT (CAA)

[Ed. Note: EPA does not have information yet in this section of IRIS].

IV.B. SAFE DRINKING WATER ACT (SDWA)

IV.B.1. MAXIMUM CONTAMINANT LEVEL GOAL (MCLG) for Drinking Water

Value (status) -- 0.07 mg/L (Final, 1991)

Considers technological or economic feasibility? -- NO

Discussion -- EPA has set a MCLG for 2,4-D based on potential adverse effects (hepatic and renal toxicity) reported in a 90-day dietary study in rats. The MCLG based upon a DWEL of 0.35 mg/L and an assumed drinking water contribution of 20 percent.

Reference -- 56 FR 3526 (01/30/91)

EPA Contact -- Health and Ecological Criteria Division / OST / (202) 260-7571 / FTS 260-7571; or Safe Drinking Water Hotline / (800) 426-4791

IV.B.2. MAXIMUM CONTAMINANT LEVEL (MCL) for Drinking Water

Value -- 0.07 mg/L (Final, 1991)

Considers technological or economic feasibility? -- YES

Discussion -- EPA has set a MCL equal to the MCLG of 0.07 mg/L.

Monitoring requirements -- All systems monitored for four consecutive quarters every 3 years; repeat monitoring dependent upon detection, vulnerability status and system size.

Analytical methodology -- Electron-capture/gas chromatography (EPA 515.1): PQL= 0.001 mg/L

Best available technology -- Granular activated carbon.

Reference --   56 FR 3526 (01/30/91)

EPA Contact --   Drinking Water Standards Division / OGWDW / (202) 260-7575 / FTS 260-7575; or Safe Drinking Water Hotline / (800) 426-4791

IV.B.3. SECONDARY MAXIMUM CONTAMINANT LEVEL (SMCL) for Drinking Water

        [Ed. Note: EPA does not have information yet in this section of IRIS].

IV.B.4. REQUIRED MONITORING OF "UNREGULATED" CONTAMINANTS

        [Ed. Note: EPA does not have information yet in this section of IRIS].

IV.C. CLEAN WATER ACT (CWA)

IV.C.1. AMBIENT WATER QUALITY CRITERIA, Human Health

Water and Fish Consumption: 1.0E+3 ug/L

Fish Consumption Only: none

Considers technological or economic feasibility? -- NO

Discussion -- This value is the same as the drinking water standard and approximates a safe level assuming consumption of contaminated organisms and water.

Reference --   Quality Criteria for Water, EPA 440/9-76-023 (7/76), [NTIS No. PB-263943].

EPA Contact --   Criteria and Standards Division / OWRS (202)260-1315 / FTS 260-1315

IV.C.2. AMBIENT WATER QUALITY CRITERIA, Aquatic Organisms

[Ed. Note: EPA does not have information yet in this section of IRIS].

IV.D. FEDERAL INSECTICIDE, FUNGICIDE, AND RODENTICIDE ACT (FIFRA)

IV.D.1. PESTICIDE ACTIVE INGREDIENT, Registration Standard

Status -- Issued (1988)
Reference --   2,4-D Pesticide Registration Standard.September, 1988 (NTIS No. PB89-102396).
EPA Contact --   Registration Branch / OPP (703)557-7760 / FTS 557-7760

IV.D.2. PESTICIDE ACTIVE INGREDIENT, Special Review

Action -- Proposed decision not to initiate a special review (1988)

Considers technological or economic feasibility? -- NO

Summary of regulatory action --   Criterion of concern: carcinogenicity/final decision to be issued in FY 91 upon completion and review of one epidemiologic NCI study.

Reference --   53 FR 9590 (03/23/88)

EPA Contact --   Special Review Branch / OPP (703)557-7400 / FTS 557-7400

IV.E. TOXIC SUBSTANCES CONTROL ACT (TSCA)

[Ed. Note: EPA does not have information yet in this section of IRIS].

IV.F. RESOURCE CONSERVATION AND RECOVERY ACT (RCRA)

IV.F.1. RCRA APPENDIX IX, for Ground Water Monitoring

Status -- Listed
Reference -- 52 FR 25942 (07/09/87)
EPA Contact --   RCRA/Superfund Hotline (800)424-9346 / (202)260-3000 / FTS 260-3000

IV.G. SUPERFUND (CERCLA)

IV.G.1. REPORTABLE QUANTITY (RQ) for Release into the Environment

Value (status) -- 100 pounds (Final, 1985)

Considers technological or economic feasibility? -- NO

Discussion -- The RQ for 2,4-D acid is based on the aquatic toxicity RQ established by Section 311(b)(4) of the Clean Water Act (40 CFR 117.3). Available data indicate a 96-

hour Median Threshold Limit between 1 and 10 ppm, corresponding to an RQ of 100 pounds.

Reference --   50 FR 13456 (04/04/85); 54 FR 33418 (08/14/89)

EPA Contact --   RCRA/Superfund Hotline (800)424-9346 / (202)260-3000 / FTS 260-3000

VI. BIBLIOGRAPHY

Last Revised --   08/01/89

VI.A. ORAL RfD REFERENCES

Bjorklund, N. and K. Erne. 1966. Toxicological studies of phenoxyacetic herbicides in animals. Acta. Vet. Scand. 7: 364-390.

Dow Chemical Company. 1983. Accession No. 251473. Available from EPA. Write to FOI, EPA, Washington, DC 20460.

Gaines, T. and R. Kimbrough. 1970. Personal Communication. (Cited in Hansen et al., 1971).

Hansen, W.H., M.L. Quaife, R.T. Habermann and O.G. Fitzhugh. 1971. Chronic toxicity of 2,4-dichlorophenoxyacetic acid in rats and dogs. Toxicol. Appl. Pharmacol. 20(1): 122-129.

U.S. EPA. 1984. Drinking Water Criteria Document for 2,4- Dichlorophenoxyacetic Acid (2,4-D). Prepared by the Office of Health and Environmental Assessment, Environmental Criteria and Assessment Office, Cincinnati, OH for the Office of Drinking Water, Washington, DC.

VI.B. INHALATION RfD REFERENCES

None

VI.C. CARCINOGENICITY ASSESSMENT REFERENCES

None

VI.D. DRINKING WATER HA REFERENCES

None

VII. REVISION HISTORY

Date	Section	Description
09/30/87	IV.	Regulatory Action section on-line
06/30/88	I.A.	RfD under review - change possible
08/01/89	VI.	Bibliography on-line
02/01/90	STATUS	Carcinogenicity assessment review status changed
01/01/92	IV.	Regulatory actions updated

SYNONYMS

ACETIC ACID, (2,4-
DICHLOROPHENOXY)-
ACIDE 2,4-DICHLORO
PHENOXYACETIQUE
ACIDO(2,4-DICLORO-FENOSSI)-
ACETICO
AGROTECT
AMIDOX
AMOXONE
AQUA-KLEEN
BH 2,4-D
BRUSH-RHAP
B-SELEKTONON
CHLOROXONE
CROP RIDER
CROTILIN
2,4-D
D 50
DACAMINE
2,4-D ACID
DEBROUSSAILLANT 600
DECAMINE
DED-WEED LV-69
DESORMONE
(2,4-DICHLOOR-FENOXY)-AZIJNZUUR
DICHLOROPHENOXYACETIC ACID
2,4-Dichlorophenoxyacetic acid
Dichlorophenoxyacetic acid, 2,4-
2,4-DICHLORPHENOXYACETIC ACID
(2,4-DICHLOR-PHENOXY)-
ESSIGSAEURE
DICOPUR
DICOTOX
DINOXOL
DMA-4
DORMONE

FERNESTA
FERNIMINE
FERNOXONE FERXONE
FOREDEX 75
FORMULA 40
HEDONAL
HERBIDAL
IPANER
KROTILINE
LAWN-KEEP
MACRONDRAY
MIRACLE
MONOSAN
MOXONE
NA 2765
NETAGRONE
NETAGRONE 600
NSC 423
PENNAMINE
PENNAMINE D
PHENOX
PIELIK
PLANOTOX
PLANTGARD
RCRA WASTE NUMBER U240
RHODIA
SALVO
SPRITZ-HORMIN/2,4-D
SPRITZ-HORMIT/2,4-D
TRANSAMINE
TRIBUTON
TRINOXOL
U 46DP
U-5043
VERGEMASTER
VIDON 638

EMULSAMINE BK   VISKO-RHAP
EMULSAMINE E-3   WEED-AG-BAR
ENT 8,538   WEEDAR-64
ENVERT 171   WEEDATUL
ENVERT DT   WEED-B-GON
ESTERON   WEEDEZ WONDER BAR
ESTERONE FOUR   WEEDONE LV4
ESTONE   WEED-RHAP
FARMCO   WEED TOX
                                WEEDTROL

Source: *Instant EPA's IRIS*, 1997, Instant Reference Sources, Inc., 7605 Rockpoint Drive, Austin, TX 78731 (Fax: 512-345-2386; Internet URL, http://www.instantref.com/inst-ref.htm).

Production, Use, and Pesticide Labeling Information

1. Description of Chemical

   Generic Name: 2,4-dichlorophenoxyacetic acid

   Common Name: 2,4-D (includes parent acid as well as salt, amine and ester derivatives).

   Trade Name: 2,4-D is available under a large selection of trade names, most often formulated as an inorganic salt, amine or ester.

   EPA Chemical Code: 030001 (acid)

   Chemical Abstracts Service (CAS) Number: 94-75-7 (acid)

   Year of Initial Registration: 1948

   Pesticide Type: Herbicide, plant growth regulator

   Chemical Family: Chlorinated phenoxy

   U.S. and Foreign Producers: 2,4-D technical products are manufactured by a large number of companies, both U.S. and foreign.

2. Use Patterns and Formulations

   - Registered Uses: Terrestrial, food and nonfood; aquatic, food and nonfood; domestic; and forestry.

   - Predominant Uses: Postemergent weed control in agricultural crops (approximately 57 percent of total usage; over 45 percent of total usage is on wheat and corn; 20 percent of total usage on pastures and rangelands; other major crops are sorghum, other small grains, rice and sugarcane); the remainder is used on noncrop areas, with a small amount used as a plant growth regulator (in filberts, citrus and potatoes).

- Formulation Types Registered: Granular; amine and ester liquids; aerosol spray (foam).

- Methods of Application: Aerial and ground equipment, knapsack sprayers, pressure and hose-end applicators, and lawn spreaders.

3. Science Findings

Chemical Characteristics (Acid)

- Physical State: Flakes, powder, and crystalline powder and solid.

- Color: White to light tan.

- Odor: Phenolic to odorless.

- Melting Point: 135-142 C.

- Boiling Point: 160 C at 0.4 mm Hg.

- Solubility: Soluble in acetone, ethanol, aqueous alkali, alcohols, diethyl ether, ethyl ether, isopropanol, methyl isobutyl ketone, most organic solvents; insoluble in benzene, petroleum oils.

- Vapor Pressure: 0.4 mm Hg at 160 C.

- Stability: Stable to melting point.

Toxicological Characteristics (Acid, except as noted)

- Acute Toxicity

- 2,4-D Acid:

- Oral (rat): 639 mg/kg (males); 764 mg/kg (females); Toxicity Category III.

- Inhalation (rat): 1.79 mg/L; Toxicity Category III.

- Dermal Sensitization (guinea pig): Not a sensitizer.

- 2,4-D Sodium Salt:

  - Oral (rat): 876 mg/kg (males); 975 mg/kg (females); Toxicity Category III.

  - Dermal (rat): >2000 mg/kg; Toxicity Category III.

- Diethanolamine Salt (Manufacturing-Use Product):

  - Oral (rat): >2000 mg/kg (males); 1605 mg/kg (females); Toxicity Category III.

  - Dermal (rabbit): >2000 mg/kg (males and females); Toxicity Category III.

  - Inhalation (rat): >3.8 mg/L; Toxicity Category III.

  - Primary Eye (rabbit): Severe irritation and corneal ulcer not resolved 21 days post-treatment; Toxicity Category I.

  - Primary Dermal (rabbit): No signs of dermal irritation; Toxicity Category IV.

  - Dermal Sensitization (guinea pig): Not a dermal sensitizer.

- Butoxyethyl Ester:

  - Oral (rat): 866 mg/kg; Toxicity Category III.

  - Dermal (rabbit): >2000 mg/kg (females); 1829 mg/kg (males); Toxicity Category III.

  - Inhalation (rat): >4.6 mg/L; Toxicity Category III.

  - Primary Eye (rabbit): Very mild eye irritation resolved in 72 hours; Toxicity Category III.

  - Primary Dermal (rabbit): Very slightly erythema cleared in 72 hours; Toxicity Category III.

  - Dermal Sensitization (guinea pig): Was a sensitizer in two tests and not a sensitizer in a third test.

- Isooctyl Ester:

  - Oral (rat): 982 mg/kg (males); >720 <864 mg/kg (females); Toxicity Category III.

  - Dermal (rabbit): >2000 mg/kg; Toxicity Category III.

- Isobutyl Ester:

> - Oral (rat): 700 mg/kg (males); 553 mg/kg (females); Toxicity Category III.

> - Dermal (rabbit): >2000 mg/kg; Toxicity Category III.

- Isopropyl Ester:

> - Oral (rat): 640 mg/kg (males); 440 mg/kg (females); Toxicity Category II.

> - Dermal (rabbit): >2000 mg/kg; Toxicity Category III.

> - Inhalation (rat): >4.97 mg/L; Toxicity Category III.

> - Primary Eye (rabbit): All irritation cleared at 4 days; Toxicity Category III.

> - Primary Dermal (rabbit): No irritation at 72 hours; Toxicity Category IV.

> - Dermal Sensitization (guinea pig): Nonsensitizer

- Subchronic Toxicity: No acceptable data are available on 2,4-D. The requirement for subchronic oral studies on the acid is waived because chronic studies are required; a subchronic dermal study is required. Subchronic studies are required for the esters and amines.

- Chronic Toxicity:

> - Oncogenicity (rats): No observed effects level (NOEL) for systemic effects - 1 mg/kg/day; lowest observed effects level (LOEL) for
>
> systemic effects - 5 mg/kg/day; further evaluation needed to determine if maximum tolerated dose was reached.

> - Oncogenicity (mice): NOEL for systemic effects - 1 mg/kg/day; LOEL for systemic effects - 15 mg/kg/day; further evaluation needed to determine if maximum tolerated dose was reached.

> - Teratology (rats): Fetotoxicity (delayed ossification) LOEL 75 mg/kg/day and NOEL 25 mg/kg/day; Maternal toxicity NOEL 75 mg/kg/day (highest dose tested).

- Reproduction (rats): NOEL 5 mg/kg/day.

- Major Routes of Exposure: The major route of exposure is dermal; respiratory exposure is negligible.

Physiological and Behavioral Characteristics:

>   - Foliar Absorption: 2,4-D is absorbed through the roots and/or leaves depending upon the type of formulation. A rain-free period of 4 to 6 hours usually is adequate for uptake.

>   - Translocation: Following foliar absorption, 2,4-D translocates within the phloem, probably moving with the food material. Following root absorption, it may move upward in the transpiration stream.

>>   Translocation rate is influenced by the growth rate of the plant.

>>   Accumulation occurs principally at the rapid growth regions of shoots and roots.

- Mechanism of Pesticide Action: 2,4-D acid stimulates nucleic acid and protein synthesis effecting the activity of enzymes, respiration and cell division. Broadleaf plants exhibit malformed leaves, stems and roots.

Environmental Characteristics:

>   - Absorption and Leaching: 2,4-D is mobile to highly mobile in five soil types. Based on available data, aged 2,4-D residues are only slightly mobile.

>   - Microbial Breakdown: 2,4-D degrades rapidly in aerobic silty clay and loam soil systems.

>   - Bioaccumulation: Available data indicate a low potential for 2,4-D to accumulate in fish.

>   - Resultant Average Persistence: In aerobic silty clay and loam soils, 1.9-2.2 percent of applied 2,4-D remained at 51 days post-treatment; in four other soils, only 0.7-2.5 percent remained at 150 days post-treatment.

>   - Environmental Fate and Surface and Groundwater Contamination Concerns:

>>   Although laboratory data demonstrate that 2,4-D is mobile in soils, its potential to contaminate groundwater is limited by its rapid rate of degradation and uptake by target plants. However, residues of 2,4-D have been detected in

groundwater, mostly from point sources, such as mixing, loading and disposal.

- Exposure of Humans and Nontarget Organisms: Accidental human poisoning with 2,4-D, which resulted in severe neurotoxicity, has been reported. Reports have been received concerning off-target movement of 2,4-D resulting in damage to crops or other desirable plants.

- Exposure during Reentry Operations: Based on available data, 2,4-D products are of low toxicity (Toxicity Categories III and IV). Because of these low levels of toxicity, reentry is not a concern.

Ecological Characteristics

- Avian Toxicity:

> - Acceptable data indicate that 2,4-D acid can be characterized as moderately toxic to practically nontoxic to avian species on an acute basis. Butyl ester can be characterized as practically nontoxic on an acute and chronic basis.

- Fish Toxicity:

> - Acceptable data indicate that 2,4-D acid and certain of its salts, esters and amines can be characterized in the range of moderately toxic to practically nontoxic to fish. However, the compounds N-oleyl-1,3-propylenediamine salt, N,N-dimethyloleyl-linoleylamine, butyl ester, butoxyethanol ester and propylene glycol butyl ether ester can be characterized as highly toxic to fish, based on the following toxicity values:

N-oleyl-1,3-propylenediamine salt	0.3 ppm (bluegill sunfish)
	0.8 ppm (channel catfish)
N,N-dimethyloleyl-linoleylamine	0.64 ppm (rainbow trout)

Butyl ester

0.49-2.82 ppm (cutthroat trout)
0.5-2.8 ppm (lake trout)
0.4-0.96 ppm (rainbow trout)
0.29-0.3 ppm (bluegill sunfish)

Butoxyethanol ester
>   0.65 ppm (rainbow trout)
>   0.76-1.2 ppm (bluegill sunfish)
>   3.3 ppm (fathead minnow)
>   0.78-1.35 ppm (channel catfish)

Butoxypropyl ester
>   5.4 ppm (rainbow trout)

Propylene glycol butyl ether ester
>   0.33-2.8 ppm (cutthroat trout)
>   0.39-2.93 ppm (lake trout)
>   0.95-1.44 ppm (rainbow trout)
>   0.56-0.67 ppm (bluegill sunfish)

- Freshwater Invertebrates Toxicity:

Of those compounds for which the Agency has data, reported toxicity values indicate that the compounds can be characterized as slightly toxic to practically nontoxic, excepted as noted below. The compounds set forth below have toxicity values which characterize them as highly toxic to aquatic invertebrates.

Dimethylamine   0.15 ppm (grass shrimp)

Isooctyl ester   0.5 ppm (waterflea)

Butoxyethanol ester
>   1.7-6.4 ppm (waterflea)
>   2.2 ppm (seed shrimp)
>   2.6 ppm (sow bug)
>   0.44-6.1 ppm (side swimmer)
>   0.39--0.79 ppm (midge)

Propylene glycol butyl ether ester
>   0.1-14 ppm (waterflea)
>   0.42 ppm (seed shrimp)

Estuarine and Marine Organisms Toxicity:

- Acceptable data are available only for the butoxyethanol ester which report toxicity values of 5.0 mg/L (longnose killifish), 2.6 mg/L (Eastern oyster) and 5.6 mg/L (brown shrimp), which indicate that the material is moderately toxic to estuarine and marine organisms.

- Effects on Plants:

  - Limited plant protection studies are available. In a spray drift study, two application methods were compared as to quantity and pattern of deposition. No difference was found between the amine derivatives (diethanolamine and dimethylamine). With these amines, drift was observed beyond 225 feet from the site of application. No residues, attributable to drift, were found when applied postemergent to wheat or corn.

  - The toxicity of butoxyethanol ester was tested on four species of algae. Toxicity values ranged from 75 mg/L to 150 mg/L.

- Nontarget Insects:

  - There is sufficient information to characterize 2,4-D as relatively nontoxic to honey bees, when bees are exposed to direct treatment.

- Potential Problems Related to Endangered Species:

  - The Office of Endangered Species has determined that certain uses of 2,4-D may jeopardize the continued existence of endangered species or critical habitat of certain endangered species.

- Tolerance Assessments

  - Tolerances Established: Tolerances and food and feed additive regulations have been established for residues of 2,4-D in a variety of raw agricultural commodities and meat byproducts (40 CFR 180.142), and in processed food (40 CFR 185.1450) and feed (40 CFR 186.1450).

  - Results of Tolerance Assessment: A provisional acceptable daily intake (PADI) of 0.003 mg/kg/day for 2,4-D acid has been established based on a two-year rat feeding study. Compound related effects were observed in the kidneys of both male and female rats. The LOEL was 5 mg/kg/day and the NOEL was 1 mg/kg/day. An uncertainty factor of 100 was used to account for the inter- and intraspecies differences. An additional uncertainty factor of 3 was used since there is no dog study available and no information available that indicates the dog is less sensitive than the rat.

Reported Pesticide Incidents

  - Based on the Pesticide Incidence Monitoring System files, covering the period 1966 to 1979, reports were received concerning the off-target movement of 2,4-D in unspecified formulations, esters and amines. The incidents involved drift from aerial (173 reports) and ground (104 reports) applications, as well as volatilization and drift (35 reports) and resulted in damage to off-target crops or other desirable plants.

Summary Science Statement

- The Agency's Office of Pesticide Programs (OPP) has classified 2,4-D as a Group D oncogen (not classifiable as to human carcinogenicity) because existing data are inadequate to assess the carcinogenic potential of 2,4-D. Accidental human poisoning with 2,4-D, which resulted in severe neurotoxicity, has been reported; adequate neurotoxicity studies are not available. While published data indicate that 2,4-D may be teratogenic, an acceptable rat teratology study is negative; a study in rabbits is needed.

- 2,4-D is often formulated as various esters and amines. These formulations may affect the physical characteristics, biological activity and environmental fate of the parent compound. Data are needed on each ester and amine before the Agency can completely assess 2,4-D.

- Although laboratory data demonstrate that 2,4-D is mobile in soils, its potential to contaminate groundwater is limited by its rapid rate of degradation and uptake by target plants. However, residues of 2,4-D have been detected in groundwater, mostly from point sources, such as mixing, loading and disposal.

- Certain formulations of 2,4-D are highly toxic to fish and/or aquatic invertebrates. Other formulations, for which the Agency has data, are in the range of moderately toxic to practically nontoxic to nontarget organisms. The Office of Endangered Species has issued biological opinions indicating that certain endangered species may be in jeopardy from the use of 2,4-D.

4. Summary of Regulatory Position and Rationale

- OPP has classified 2,4-D as a Group D oncogen (not classifiable as to human carcinogenicity). EPA is, however, requiring additional data, including additional information on oncogenicity and teratogenicity and neurotoxicity studies, for further evaluation of 2,4-D. Data are being required on the ester and amine formulations of 2,4-D as well as on the acid. EPA will not establish any significant new food use tolerances or register any significant new uses at this time.

- Additional data are needed to thoroughly evaluate the ecological effects of 2,4-D and its potential to contaminate groundwater.

- EPA is developing a program to reduce or eliminate exposure to endangered species from the use of 2,4-D to a point where use does not result in jeopardy, and will issue notice of any labeling revisions when the program is developed. Endangered species labeling is not required at this time.

- Unique Warning Statements Required on Labels:

  - Manufacturing-Use Products:

  "Do not discharge effluent containing this product into lakes, streams, ponds, estuaries, oceans, or public waters unless this product is specifically identified and addressed in an NPDES permit. Do not discharge effluent containing this product to sewer systems without previously notifying the sewage treatment plant authority. For guidance, contact your State Water Board or Regional Office of the EPA."

- End-Use Products:

  - Aquatic Uses:

    "Drift or runoff may adversely affect nontarget plants. Do not apply directly to water except as specified on this label. Do not contaminate water when disposing of equipment washwaters."

  - Nonaquatic Uses:

    "Drift or runoff may adversely affect nontarget plants. Do not apply directly to water or wetlands (swamps, bogs, marshes, and potholes). Do not contaminate water when disposing of equipment washwaters."

    - End-Use Products - Certain Formulations: End-use products containing the following formulations must contain the above environmental precautions modified to indicate that the product is toxic either to fish or aquatic invertebrates:

- Toxic to Fish -   N-Oleyl-1,3-Propylenediamine salt
                                                N,N-Dimethyloleyl-Linoleylamine
                                                 Butyl ester
                                                 Butoxylethanol ester
                                                 Propylene glycol butyl ether ester

- Toxic to Aquatic Invertebrates -   Dimethylamine
                                                 Isooctyl ester

- All End-Use Products: The following statements are required in the use directions for all end-use products.

- Liquid Formulations:

> - "This product can reach groundwater from mixing and loading. To minimize groundwater contamination from spills during mixing, loading and cleaning of equipment, take the following steps:
>
>> - "Mixing and Loading: When mixing, loading or applying this product, wear chemical resistant gloves. Wash non-disposable gloves thoroughly with soap and water before removing.
>>
>> - "The mixing and loading of spray mixtures into the spray equipment must be carried out on an impervious pad (i.e., concrete slab, plastic sheeting) large enough to catch any spilled material. If spills occur, contain the spill by using an absorbent material (e.g., sand, earth or synthetic absorbent). Dispose of the contaminated absorbent material by placing in a plastic bag and following disposal instructions on this label.
>>
>> - "Triple rinse empty containers and add the rinsate to the mixing tank.
>>
>> - "Cleaning of Equipment: When cleaning equipment, do not pour the washwater on the ground; spray or drain over a large area away from wells and other water sources."

- Granular Formulations:

> - "This product can reach groundwater from improper handling. To minimize groundwater contamination from spills during loading and cleaning of equipment, take the following steps:
>
>> - "Handling: When handling this product, wear chemical resistant gloves. Wash non-disposable gloves thoroughly with soap and water before removing. If spills occur, collect the material and dispose of by following disposal instructions on this label.
>>
>> - "Cleaning of Equipment: When cleaning equipment, do not pour the washwater on the ground; spray or drain over a large area away from wells and other water sources."
>>
>> - End-Use Products - Certain Food/Feed Uses. Labels for products registered for certain food/feed uses must contain revised use directions pertaining to appropriate preharvest, pregrazing and preslaughter intervals; allowable range of diluent; and/or maximum seasonal application rate and/or number of applications.

5. Summary of Major Data Gaps

- The following data are required for 2,4-D acid. The Agency is also requiring data on each individual ester and amine of 2,4-D. Specific requirements are detailed in the Data Tables, Appendix I of the Registration Standard, which can be obtained from the Product Manager listed below.

Due Date
From Study   Date of Standard
------------------------   ----------------------
- Product Chemistry   6-15 Months
- Residue Chemistry: 18-24 Months

- Plant and animal metabolism
- Analytical methods
- Residue studies

Due Date
From Study   Date of Standard
------------------------   ----------------------
- Toxicology: 9-50 Months

- Primary Eye and Dermal Irritation
        - 21-Day Dermal
- Chronic Toxicity (nonrodent)
- Teratogenicity (rabbit)
- Mutagenicity
- Metabolism
- Special Dermal (Neurotoxicity)
- Reserved: Oncogenicity (two species)

Due Date
From Study   Date of Standard
------------------------   ----------------------
- Ecological Effects: 9-18 Months

- Avian Dietary
- Aquatic Organism (freshwater fish and invertebrates; estuarine and marine organisms; accumulation)
- Phytotoxicity (Tier II)

```
--------------------------------------------------------
Due Date
  From Study                   Date of Standard
  --------------------------   ----------------------
- Environmental Fate:          9-50 Months
```

- Hydrolysis
- Photodegradation (water, soil, air)
- Metabolism (anaerobic soil; aerobic and anaerobic aquatic)
- Leaching and Adsorption/Desorption
- Volatility (lab and field)
- Dissipation (soil, aquatic and forestry)
- Accumulation (confined rotational crops; irrigated crops; fish and aquatic nontarget organisms)
- Spray drift

```
--------------------------------------------------------
```

6. Contact Person at EPA

Mr. Richard Mountfort
Product Manager (Team 23)
Fungicide-Herbicide Branch
Registration Division (TS-787C)
Office of Pesticide Programs, EPA
Washington, DC 20460

Telephone: (703) 557-1830

Source: *Instant EPA's Pesticide Facts*, 1994, Instant Reference Sources, Inc., 7605 Rockpoint Drive, Austin, TX 78731 (Fax: 512-345-2386; Internet URL, http://www.instantref.com/inst-ref.htm).

DDD

Introduction

DDD has been used as an insecticide for contract control of leaf rollers and other insects on vegetables and tobacco [817]. It is also a degredation product of DDT. There are two isomers of DDD: o,p'-DDD and p,p'-DDD. The most common isomer is p,p'-DDD and most of the information herein pertains to this isomer. When information was not available on p,p'-DDD but was available for o,p'-DDD then the latter is cited with notes that it is for the o,p'- isomer. The structure and CAS Number herein is for p,p'-DDD.

DDD was released to the environment through its use as a non-systemic contact and stomach insecticide and as a biodegradation product of DDT. Its use in the US has been banned since the early 1970's. If released to soil it will adsorb very strongly to the soil and will not be expected to appreciably leach to the groundwater, although its presence in certain groundwater samples illustrates that it can be transported there. It will not hydrolyze in the soil and biodegradation is expected to be slow. Indirect photolysis many be substantial based upon the behavior of the related compound DDT. If released in water it will be expected to strongly adsorb to sediment and to bioconcentrate in aquatic organisms. It will not hydrolyze or directly photodegrade and biodegradation is expected to be slow. If released to the atmosphere, it will not be expected to directly photolyze. The estimated vapor phase half-life in the atmosphere is 1.71 days as a result of reaction with photochemically produced hydroxyl radicals. Fallout and washout will be the major removal mechanisms from the air since DDD is expected to adsorb to particulate matter. General human exposure will mainly be from consumption of contaminated food especially contaminated fish and seafoods. [828]

Identifying Information

CAS NUMBER: 72-54-8

SYNONYMS: 1,1-bis(4-chlorophenyl)-2,2-dichloroethane; 1,1-bis(p-chlorophenyl)-2,2-dichloroethane; p,p'-TDE; 2,2-bis(p-chlorophenyl)-1,1-dichloroethane; 4,4'-DDD; p,p'-DDD; 1,1-dichloro-2,2-bis(p-chlorophenyl)ethane; dilene; rothane; TDE; p,p'-dichlorodiphenyl dichloroethane

Endocrine Disruptor Information

DDD is a strongly suspected EED that it is listed by EPA [811], the Centers for Disease Control & Prevention [812], and the World Wildlife Fund [813] as a potential endocrine modifying chemical. It also is listed as one of the commonly found pesticides in wildlife exhibiting symptoms of EDCs. [810] DDD and its analogs DDT and DDD along with their structural isomers have long been implicated as being endocrine disruptors. [845] It also is listed as one of the commonly found pesticides in wildlife exhibiting symptoms of EDCs. DDD and its congeners have reproductive endocrine effects and also have a major toxic effect on the adrenal glands, DDE being utilized therapeutically in the treatment of Cushing's syndrome. [846] Deformaties in birds including clubbed feet, crossed bills, and missing eyes are also suspected of having links with exposures to DDT. [810] These organochlorine insecticides have been banned from use in this country, except for public health emergencies, for almost 20 years but due to these chemicals' extraordinary persistence, residues still appear in approximately 10% of the hazardous waste sites sampled. [811]

Chemical and Physical Properties

CHEMICAL FORMULA: $C_{14}H_{10}Cl_4$

MOLECULAR WEIGHT: 320.04

PHYSICAL DESCRIPTION: Colorless powder

DENSITY: 1.476 g/cm^3 at 20/4°C [816], [874]

MELTING POINT: 109-112°C [817]

BOILING POINT: 193°C [817]

SOLUBILITY: Water: 90 ug/L at 25°C [816], [875]

Environmental Reference Materials

A source of EED environmental reference materials for DDD is Radian International LLC, Austin, TX (Fax 512-454-0268; Internet URL, http://www.radian.com/standards); Catalog No. ERD-011.

Hazardous Properties

ACUTE/CHRONIC HAZARDS: The o,p'- isomer of DDD is toxic by ingestion, inhalation, and skin absorption and, when heated to decomposition it emits toxic fumes. [815] The p,p'- isomer is expected to pose similar hazards.

STABILITY: DDD is stable under normal laboratory conditions. [815]

Medical Symptoms of Exposure

Contact with eyes causes irritation and ingestsion causes vomiting [817]. The o,p'- isomer has been shown to be an inhibitor of adrenal steroid biosynthesis. [815]

Toxicological Information

The toxicological information below is gathered from several sources including two national databases: one from the *National Toxicology Program Chemical Repository Database* and another from *EPA's Integrated Risk Information System (IRIS)*.

Toxicology Information from the National Toxicology Program

O,P'-DDD

CARCINOGENICITY: Not available

TERATOGENICITY: Not available

MUTATION DATA: Not available

TOXICITY:

Typ. Dose	Mode	Specie	Amount	Units	Other
TDLo	orl	man	17	gm/kg/35W	
TDLo	orl	wmn	800	mg/kg/4D	
TDLo	orl	wmn	11	gm/kg/15W	
TDLo	orl	wmn	14	gm/kg/22W	

OTHER TOXICITY DATA: Not available

SAX TOXICITY EVALUATION:

    THR: Not available

Source: *Instant Tox-Base*, 1995, Instant Reference Sources, Inc., 7605 Rockpoint Drive, Austin, TX 78731 (Fax: 512-345-2386; Internet URL, http://www.instantref.com/inst-ref.htm).

Toxicology Information from EPA's Integrated Risk Information System (IRIS)

I.   CHRONIC HEALTH HAZARD ASSESSMENTS FOR NONCARCINOGENIC EFFECTS

I.A.   REFERENCE DOSE FOR CHRONIC ORAL EXPOSURE (RfD)

    [Ed. Note: EPA does not have information yet in this section of IRIS].

I.B.   REFERENCE CONCENTRATION FOR CHRONIC INHALATION EXPOSURE (RfC)

[Ed. Note: EPA does not have information yet in this section of IRIS].

II. CARCINOGENICITY ASSESSMENT FOR LIFETIME EXPOSURE

Section II provides information on three aspects of the carcinogenic risk assessment for the agent in question; the US EPA classification, and quantitative estimates of risk from oral exposure and from inhalation exposure. The classification reflects a weight-of-evidence judgment of the likelihood that the agent is a human carcinogen. The quantitative risk estimates are presented in three ways. The slope factor is the result of application of a low-dose extrapolation procedure and is presented as the risk per (mg/kg)/day. The unit risk is the quantitative estimate in terms of either risk per ug/L drinking water or risk per ug/cu.m air breathed. The third form in which risk is presented is a drinking water or air concentration providing cancer risks of 1 in 10,000, 1 in 100,000 or 1 in 1,000,000. Background Document 2 (Service Code 5) provides details on the rationale and methods used to derive the carcinogenicity values found in IRIS. Users are referred to Section I for information on long-term toxic effects other than carcinogenicity.

II.A. EVIDENCE FOR CLASSIFICATION AS TO HUMAN CARCINOGENICITY

II.A.1. WEIGHT-OF-EVIDENCE CLASSIFICATION

Classification -- B2; probable human carcinogen

Basis -- based on an increased incidence of lung tumors in male and female mice, liver tumors in male mice and thyroid tumors in male rats. DDD is structurally similar to, and is a known metabolite of DDT, a probable human carcinogen.

II.A.2. HUMAN CARCINOGENICITY DATA

None. Human epidemiological data are not available for DDD. Evidence for the carcinogenicity in humans of DDT, a structural analog, is based on autopsy studies relating tissue levels of DDT to cancer incidence. These studies have yielded conflicting results. Three studies reported that tissue levels of DDT and DDE were higher in cancer victims than in those dying of other diseases (Casarett et al., 1968; Dacre and Jennings, 1970; Wasserman et al., 1976). In other studies no such relationship was seen (Maier-Bode, 1960; Robinson et al., 1965; Hoffman et al., 1967). Studies of occupationally exposed workers and volunteers have been of insufficient duration to determine the carcinogenicity of DDT to humans.

II.A.3. ANIMAL CARCINOGENICITY DATA

Sufficient. Tomatis et al. (1974) fed DDD for 130 weeks at 250 ppm (TWA) to 60 CF-1 mice/sex. A statistically significant increase in incidence of lung tumors was seen in both sexes compared with controls. In males, a statistically significant increase in incidence of liver tumors was also seen.

NCI (1978) fed DDD at 411 and 822 ppm (TWA) to 50 B6C3F1 mice/sex/dose for 78 weeks. Actual doses were 350 or 630 ppm for 5 weeks, 375 or 750 ppm for 11 weeks, and 425 or 850 ppm for the next 62 weeks. After an additional 15 weeks, an increased incidence of hepatocellular carcinomas was seen in both sexes by comparison to controls, but the increase was not statistically significant.

NCI (1978) also fed DDD at 1647 and 3294 ppm TWA for males and 850 and 1700 ppm TWA for females for 78 weeks to 50 Osborne-Mendel rats/sex/dose. Males were fed 1400 or 2800 ppm for 23 weeks followed by 1750 or 3500 ppm for 55 weeks. Females were fed 850 or 1700 ppm for the entire 78 weeks. After an additional 35 weeks, an increased incidence of thyroid tumors (follicular cell adenomas and carcinomas) was observed in males. Due to a wide variation in incidence of these tumors in the control groups for DDD, DDE and DDT, the increased incidence was not statistically significant by comparison to concurrent controls. Although tumor incidence did not appear to be dose-related, the increase was significant at the low dose by comparison to historical controls. Thus, the pathologists' judgment and statistical results suggest a possible carcinogenic effect of DDD in male rats. NCI concluded that a definitive interpretation of the data was not possible.

II.A.4. SUPPORTING DATA FOR CARCINOGENICITY

DDD is structurally similar to, and is a metabolite of, DDT, a probable human carcinogen, in rats (Peterson and Robinson, 1964), mice (Gingell and Wallcave, 1976), and humans (Morgan and Roan, 1977).

Positive effects were found with DDD in mammalian cytogenetic assays and a host-mediated assay (ICPEMC, 1984).

II.B.   QUANTITATIVE ESTIMATE OF CARCINOGENIC RISK FROM ORAL EXPOSURE

II.B.1. SUMMARY OF RISK ESTIMATES

Oral Slope Factor -- 2.4E-1/mg/kg/day

Drinking Water Unit Risk -- 6.9E-6/ug/L

Extrapolation Method -- Linearized multistage procedure, extra risk

Drinking Water Concentrations at Specified Risk Levels:

Risk Level	Concentration
E-4 (1 in 10,000)	1E+1 ug/L
E-5 (1 in 100,000)	1 ug/L
E-6 (1 in 1,000,000)	1E-1 ug/L

II.B.2. DOSE-RESPONSE DATA (CARCINOGENICITY, ORAL EXPOSURE)

Tumor Type -- liver
Test Animals -- mouse/CF-1, males
Route -- diet
Reference -- Tomatis et al., 1974

Administered Dose (ppm)	Human Equivalent Dose (mg/kg)/day	Tumor Incidence
0	0	33/98
250	245	31/59

II.B.3. ADDITIONAL COMMENTS (CARCINOGENICITY, ORAL EXPOSURE)

DDD used in the Tomatis study was 99% pure p,p'-isomer. In the NCI bioassay, technical grade DDD was used, in which 60% of the material consisted of the p,p'-isomer. The composition of the remaining 40% was unspecified, but it was stated that analysis by gas chromatography revealed at least 19 impurities.

The unit risk should not be used if the water concentration exceeds 1E+3 ug/L, since above this concentration the slope factor may differ from that stated.

II.B.4. DISCUSSION OF CONFIDENCE (CARCINOGENICITY, ORAL EXPOSURE)

An adequate number of animals was tested. The slope factor was calculated using tumor incidence data from only one dose. The slope factor was similar to, and within a factor of 2, of the slope factors for this same site of three other structurally similar compounds: DDT, 3.4E-1/mg/kg/day; DDE, 3.4E-1/mg/kg/day; and dicofol, 4.4E-1/mg/kg/day.

II.C.   QUANTITATIVE ESTIMATE OF CARCINOGENIC RISK FROM INHALATION EXPOSURE

[Ed. Note: EPA does not have information yet in this section of IRIS].

II.D. EPA DOCUMENTATION, REVIEW, AND CONTACTS (CARCINOGENICITY ASSESSMENT)

II.D.1. EPA DOCUMENTATION

Source Document -- US EPA, 1980, 1985

The 1985 Carcinogen Assessment Group's report has received Agency review.

The 1980 Hazard Assessment Report has received peer review.

II.D.2. REVIEW (CARCINOGENICITY ASSESSMENT)

Agency Work Group Review -- 06/03/87, 06/24/87

Verification Date -- 06/24/87

II.D.3. US EPA CONTACTS (CARCINOGENICITY ASSESSMENT)

James H. Holder / NCEA -- (202)260-5721

Chao W. Chen / NCEA -- (202)260-5898

III. HEALTH HAZARD ASSESSMENTS FOR VARIED EXPOSURE DURATIONS

[Ed. Note: EPA does not have information yet in this section of IRIS].

IV. US EPA REGULATORY ACTIONS

EPA risk assessments may be updated as new data are published and as assessment methodologies evolve. Regulatory actions are frequently not updated at the same time. Compare the dates for the regulatory actions in this section with the verification dates for the risk assessments in sections I and II, as this may explain inconsistencies. Also note that some regulatory actions consider factors not related to health risk, such as technical or economic feasibility. Such considerations are indicated for each action. In addition, not all of the regulatory actions listed in this section involve enforceable federal standards. Please direct any questions you may have concerning these regulatory actions to the US EPA contact listed for that particular action. Users are strongly urged to read the background information on each regulatory action in Background Document 4 in Service Code 5.

IV.A. CLEAN AIR ACT (CAA)

[Ed. Note: EPA does not have information yet in this section of IRIS].

IV.B. SAFE DRINKING WATER ACT (SDWA)

[Ed. Note: EPA does not have information yet in this section of IRIS].

IV.C. CLEAN WATER ACT (CWA)

[Ed. Note: EPA does not have information yet in this section of IRIS].

IV.D. FEDERAL INSECTICIDE, FUNGICIDE, AND RODENTICIDE ACT (FIFRA)

IV.D.1. PESTICIDE ACTIVE INGREDIENT, Registration Standard

[Ed. Note: EPA does not have information yet in this section of IRIS].

IV.D.2. PESTICIDE ACTIVE INGREDIENT, Special Review

Action -- Registration canceled (1971)

Considers technological or economic feasibility? -- NO

Summary of regulatory action -- Criteria of concern: carcinogenicity, bio-accumulation, hazard to wildlife and other chronic effects.

Reference -- 36 FR 5254 (03/18/71)

EPA Contact -- Special Review Branch / OPP(703)557-7400 / FTS 557-7400

IV.E. TOXIC SUBSTANCES CONTROL ACT (TSCA)

> [Ed. Note: EPA does not have information yet in this section of IRIS].

IV.F. RESOURCE CONSERVATION AND RECOVERY ACT (RCRA)

IV.F.1. RCRA APPENDIX IX, for Ground Water Monitoring

Status -- Listed

Reference -- 52 FR 25942 (07/09/87)

EPA Contact -- RCRA/Superfund Hotline (800)424-9346 / (202)260-3000 / FTS 260-3000

IV.G. SUPERFUND (CERCLA)

IV.G.1. REPORTABLE QUANTITY (RQ) for Release into the Environment

Value (status) -- 1 pound (Final, 1989)

Considers technological or economic feasibility? -- NO

Discussion -- The final RQ for p,p'-dichlorodiphenyl dichloroethaneis based on aquatic toxicity as established under CWA Section 311(40 CFR 117.3). The available data indicate that the aquatic 96-Hour Median Threshold Limit is less than 0.1 ppm, which corresponds to an RQ of 1 pound.

Reference -- 54 FR 33418 (08/14/89)

EPA Contact -- RCRA/Superfund Hotline (800)424-9346 / (202)260-3000 / FTS 260-3000

VI. BIBLIOGRAPHY

VI.A. ORAL RfD REFERENCES

None

VI.B. INHALATION RfD REFERENCES

None

VI.C. CARCINOGENICITY ASSESSMENT REFERENCES

Casarett, L.J., G.C. Fryer, W.L. Yauger, Jr. and H. Klemmer. 1968. Organochlorine pesticide residues in human tissue. Hawaii. Arch. Environ. Health. 17: 306-311.

Dacre, J.C. and R.W. Jennings. 1970. Organochlorine insecticides in normal and carcinogenic human lung tissues. Toxicol. Appl. Pharmacol. 17: 277.

Gingell, R. and L. Wallcave. 1976. Metabolism of 14C-DDT in the mouse and hamster. Xenobiotica. 6: 15.

Hoffman, W.S., H. Adler, W.I. Fishbein and F.C. Bauer. 1967. Relation of pesticide concentrations in fat to pathological changes in tissues. Arch. Environ. Health. 15: 758-765.

ICPEMC (International Commission for Protection Against Environmental Mutagens and Carcinogens). 1984. Report of ICPEMC task group 5 on the differentiation between genotoxic and nongenotoxic carcinogens. ICPEMC Publication No. 9. Mutat. Res. 133: 1-49.

Maier-Bode, H. 1960. DDT in Koperfett des Menschen. Med. Exp. 1: 132-137. (Russian)

Morgan, D.P. and C.C. Roan. 1977. The metabolism of DDT in man. Essays Toxicol. 5: 39.

NCI (National Cancer Institute). 1978. Bioassay of DDT, TDE and p,p'-DDE for possible carcinogenicity. NCI Report No. 131. DHEW Publ. No. (NIH) 78-1386.

Peterson, J.R. and W.H. Robinson. 1964. Metabolic products of p.p'-DDT in the rat. Toxicol. Appl. Pharmacol. 6: 321.

Robinson, J., A. Richardson, C.G. Hunter, A.N. Crabtree and H.J. Rees. 1965. Organochlorine insecticide content of human adipose tissue in south-eastern England. Br. J. Ind. Med. 22: 220-224.

Tomatis, L., V. Turusov, R.T. Charles and M. Boicchi. 1974. Effect of long-term exposure to 1,1-dichloro-2,2-bis(p-chlorophenyl)ethylene, to 1,1-dichloro-2,2-bis(p-chlorophenyl)-ethane, and to the two chemicals combined on CF-1 mice. J. Natl. Cancer Inst. 52(3): 883-891.

US EPA. 1980. Hazard Assessment Report on DDT, DDD, DDE. Prepared by the Office of Health and Environmental Assessment, Environmental Criteria and Assessment Office, Cincinnati, OH.

US EPA. 1985. The Carcinogenic Assessment Group's Calculation of the Carcinogenicity of Dicofol (Kelthane), DDT, DDE and DDD (TDE). Prepared by the Office of Health and Environmental Assessment, carcinogen Assessment Group, Washington, DC, for the Hazard Evaluation Division, Office of Toxic Substances, Washington, DC. (Internal Report) EPA-600/X-85-097.

Wasserman, M., D.P. Nogueira, L. Tomatis, et al. 1976. Organochlorine compounds in neoplastic and adjacent apparently normal breast tissue. Bull. Environ. Contam. Toxicol. 15: 478-484.

VI.D. DRINKING WATER HA REFERENCES

None

VII. REVISION HISTORY

```
---------- ------------ --------------------------------------------------------
Date     Section   Description
---------- ------------ --------------------------------------------------------
08/22/88  II.      Carcinogen summary on-line
08/01/89  VI.      Bibliography on-line
01/01/92  IV.      Regulatory Action section on-line
```

SYNONYMS

1,1-bis(4-chlorophenyl)-2,2-dichloroethane	dichlorodiphenyl dichloroethane
1,1-bis(p-chlorophenyl)-2,2-dichloroethane	Dichlorodiphenyl dichloroethane, p,p'-
2,2-bis(p-chlorophenyl)-1,1-dichloroethane	dilene
DDD	rothane
4,4'-DDD	TDE
p,p'-DDD	p,p'-TDE
1,1-dichloro-2,2-bis(p-chlorophenyl)ethane	

Source: *Instant EPA's IRIS*, 1997, Instant Reference Sources, Inc., 7605 Rockpoint Drive, Austin, TX 78731 (Fax: 512-345-2386; Internet URL, http://www.instantref.com/inst-ref.htm).

Production, Use, and Pesticide Labeling Information

DDD is not currently listed in EPA's Pesticide Factsheet database and there is no indication of whether it will be added or not in the future.

DDE

Introduction

DDE is an impurity in DDT as well as a biodegradation product of DDT and therefore occurs in the environment as a result of the use of DDT as an insecticide. General human exposure will mainly be from consumption of contaminated food. Minor exposure may also occur through the manufacture and use of DDD as an insecticide. [828] Concentrations of DDT in humans appear to be decreasing, but DDE is not. DDT does not convert to DDE in humans, so it is assumed that the primary source of DDE in humans is from food, with fish consumption the most likely source. [882]

If released to soil it will adsorb very strongly to the soil and will not be expected to leach through soil to groundwater. It will not hydrolyze under normal environmental conditions and will probably not significantly biodegrade. Evaporation from the surface of soils with low organic content (such as sandy soils) may be significant, but adsorption of DDE to soils may reduce the rate of evaporation. [828]

If released to water DDE will adsorb very strongly to sediment, bioconcentrate in aquatic organisms, and be subject to photolysis with half-lives of 15-26 hrs for photodegradation by environmentally significant wavelengths of light. It will not appreciably hydrolyze or biodegrade in water. Evaporation from water may be important with half-lives of 5.6-6.4 hrs predicted for evaporation from a river 1 m deep, flowing at 1 m/sec with a wind velocity of 3 m/sec; however, the expected adsorption of DDE to sediments may retard the evaporation process. [828]

If DDE is released to the atmosphere it may be subject to direct photolysis. The estimated vapor phase half-life in the atmosphere is 4.63 hrs as a result of reaction with photochemically produced hydroxyl radicals. The major translocation mechanisms for DDE in air is fallout and washout since DDE should adsorb to particulate matter. [828]

Identifying Information

CAS NUMBER: 72-55-9

NIOSH Registry Number: KV9450000

SYNONYMS: 2,2-bis(p-Chlorophenyl)-1,1-, dichloroethene 2,2-bis(4-Chlorophenyl)-1,1-dichloroethene, 2,2-bis(p-Chlorophenyl)-1,1-dichloroethylene, 2,2-bis(4-Chlorophenyl)-1,1-dichloroethylene, P,P'-DDE

Endocrine Disruptor Information

DDE is a strongly suspected EED that it is listed by EPA [811], the Centers for Disease Control & Prevention [812], and the World Wildlife Fund [813] as a potential endocrine modifying chemical. It also is listed as one of the commonly found pesticides in wildlife exhibiting symptoms of EDCs. p,p'-DDE is a potent androgen blocker. Poor reproduction of bald eagles has been linked to levels of DDE found in their body fat. [810] DDE and its analogs DDT and DDD along with their structural isomers have long been implicated as being endocrine disruptors. [845] DDE and its congeners have reproductive endocrine effects and also have a major toxic effect on the adrenal glands, DDE being utilized therapeutically in the treatment of Cushing's syndrome. [846] Deformities in birds including clubbed feet, crossed bills, and missing eyes are also suspected of having links with exposures to DDT. [810] These organochlorine insecticides have been banned from use in this country, except for public health emergencies, for almost 20 years but due to these chemicals' extraordinary persistence, residues still appear in approximately 10% of the hazardous waste sites sampled. [811]

Chemical and Physical Properties

CHEMICAL FORMULA: $C_{14}H_8Cl_4$

MOLECULAR WEIGHT: 318.03

WLN: GYGUYR XG&R XG

PHYSICAL DESCRIPTION: White crystalline solid

DENSITY: Not available

MELTING POINT: 88-90°C [269], [275]

BOILING POINT: 316.5°C [025]

SOLUBILITY: Water: <0.1 mg/mL @ 22°C [700]; DMSO: 50-100 mg/mL @ 21°C [700]; 95% Ethanol: 10-50 mg/mL @ 21°C [700]; Acetone: 50-100 mg/mL @ 21°C [700]

OTHER SOLVENTS: Fats: Soluble [395]; Most organic solvents: Soluble [395]

OTHER PHYSICAL DATA: Adsorption capacity: 232 mg/g

Source: *Instant EPA's Air Toxics*, 1994, Instant Reference Sources, Inc., 7605 Rockpoint Drive, Austin, TX 78731 (Fax: 512-345-2386; Internet URL, http://www.instantref.com/inst-ref.htm).

Environmental Reference Materials

A source of EED environmental reference materials for DDE is Radian International LLC, Austin, TX (Fax 512-454-0268; Internet URL, http://www.radian.com/standards); Catalog No. ERD-007.

Hazardous Properties

ACUTE/CHRONIC HAZARDS: DDE is harmful if ingested, inhaled or absorbed through the skin and is an irritant. There is evidence that this compound is an animal carcinogen [015]. When heated to decomposition it emits toxic fumes of carbon monoxide and carbon dioxide [269]. It may also emit toxic fumes of hydrogen chloride gas [042], [269].

HAP WEIGHTING FACTOR: 1 [713]

FIRE HAZARD: Flash point data for DDE are not available. It is probably combustible. Fires involving DDE may be controlled with a dry chemical, carbon dioxide or Halon extinguisher.

REACTIVITY: DDE is incompatible with strong oxidizing agents and strong bases [269]. Oxidation is catalyzed by UV radiation [395].

STABILITY: DDE is sensitive to exposure to light. Solutions of it in water, DMSO, 95% ethanol or acetone should be stable for 24 hours under normal lab conditions [700].

Source: *Instant EPA's Air Toxics*, 1994, Instant Reference Sources, Inc., 7605 Rockpoint Drive, Austin, TX 78731 (Fax: 512-345-2386; Internet URL, http://www.instantref.com/inst-ref.htm).

Medical Symptoms of Exposure

Symptoms of exposure to DDE may include liver and kidney damage [052]. Based on data for a similar compound, symptoms may also include vomiting, headache, fatigue, malaise, numbness and partial paralysis of the extremities, moderate ataxia, exaggeration of part of the reflexes, mild convulsions, loss of proprioception and vibratory sensation of the extremities, hyperactive knee-jerk reflexes, excitement, confusion and increased respiration [215]. It may also cause nausea and diarrhea [042]. Other symptoms may include tremors of the head and neck muscles, cardiac and respiratory failure and even death [031]. It may also cause paresthesias of the tongue, lips and face, irritability and dizziness [406]. It may cause tonic and clonic convulsions [031], [406]. Other symptoms include apprehension and hyperesthesia of the mouth and face [215], [406]. It may also cause "yellow vision" [099].

Source: *Instant EPA's Air Toxics*, 1994, Instant Reference Sources, Inc., 7605 Rockpoint Drive, Austin, TX 78731 (Fax: 512-345-2386; Internet URL, http://www.instantref.com/inst-ref.htm).

Toxicological Information

The toxicological information below is gathered from several sources including two national databases: one from the *National Toxicology Program Chemical Repository Database* and another from *EPA's Integrated Risk Information System (IRIS)*.

Toxicology Information from the National Toxicology Program

CARCINOGENICITY:

Tumorigenic Data:

Type of Effect	Route/Animal	Amount Dosed (Notes)
TDLo	orl-mus	9700 mg/kg/78W-C
TDLo	orl-ham	36 gm/kg/86W-C
TD	orl-mus	28 gm/kg/80W-C
TD	orl-mus	17 gm/kg/78W-C
TD	orl-ham	57 gm/kg/68W-C
TD	orl-ham	41 gm/kg/97W-C
TD	orl-ham	81 gm/kg/97W-C

Status:
NCI Carcinogenesis Bioassay (Feed); Negative: Male and Female Rat
NCI Carcinogenesis Bioassay (Feed); Positive: Male and Female
Mouse

TERATOGENICITY:
Reproductive Effects Data:

Type of Effect	Route/Animal	Amount Dosed (Notes)
TDLo	ipr-rat	3500 ug/kg (7D pre)

MUTATION DATA:

Test	Lowest dose	Test	Lowest dose
sln-dmg-orl	1 pph	dnd-rat:lvr	300 umol/L
cyt-rat:oth	10 ug/L	otr-mus:emb	42600 nmol/L
dni-mus-orl	50 mg/kg	msc-mus:lym	40 mg/L/4H
msc-ham:ovr	20 mg/L		

TOXICITY:

Typ. Dose	Mode	Specie	Amount	Unit	Other
LD50	orl	rat	880	mg/kg	

OTHER TOXICITY DATA:

Review: Toxicology Review

Status:
EPA Genetox Program 1986, Positive: In vitro cytogenetics-nonhuman
EPA Genetox Program 1986, Positive: V79 cell culture-gene mutation
EPA Genetox Program 1986, Positive/limited: Carcinogenicity-mouse/rat
EPA Genetox Program 1986, Weakly positive: L5178Y cells in vitro-TK test
EPA Genetox Program 1986, Negative: Host-mediated assay; Histidine reversion-Ames test
EPA Genetox Program 1986, Negative: S cerevisiae-homozygosis
EPA Genetox Program 1986, Inconclusive: E coli polA without S9
EPA Genetox Program 1986, Inconclusive: D melanogaster Sex-linked lethal
EPA TSCA Test Submission (TSCATS) Data Base, June 1987
Meets criteria for proposed OSHA Medical Records Rule

SAX TOXICITY EVALUATION:

THR: An experimental carcinogen and neoplastigen. MODERATE via oral route. MUTATION data.

Source: *Instant Tox-Base*, 1995, Instant Reference Sources, Inc., 7605 Rockpoint Drive, Austin, TX 78731 (Fax: 512-345-2386; Internet URL, http://www.instantref.com/inst-ref.htm).

Toxicology Information from EPA's Integrated Risk Information System (IRIS)

I. CHRONIC HEALTH HAZARD ASSESSMENTS FOR NONCARCINOGENIC EFFECTS

I.A. REFERENCE DOSE FOR CHRONIC ORAL EXPOSURE (RfD)

[Ed. Note: EPA does not have information yet in this section of IRIS].

I.B. REFERENCE CONCENTRATION FOR CHRONIC INHALATION EXPOSURE (RfC)

[Ed. Note: EPA does not have information yet in this section of IRIS].

II. CARCINOGENICITY ASSESSMENT FOR LIFETIME EXPOSURE

Section II provides information on three aspects of the carcinogenic risk assessment for the agent in question; the US EPA classification, and quantitative estimates of risk from oral exposure and from inhalation exposure.   The classification reflects a weight-of-evidence judgment of the likelihood that the agent is a human carcinogen.   The quantitative risk estimates are presented in three ways. The slope factor is the result of application of a   low-dose extrapolation procedure and is presented as the risk per (mg/kg)/day.   The unit risk is the quantitative estimate in terms of either risk per ug/L drinking water or risk per ug/cu.m air breathed. The third form in which risk is presented is a drinking water or air concentration providing cancer risks of 1 in 10,000, 1 in 100,000 or 1 in 1,000,000. Background Document 2 (Service Code 5) provides details on the rationale and methods used to derive the carcinogenicity values found in IRIS. Users are referred to Section I for information on long-term toxic effects other than carcinogenicity.

II.A. EVIDENCE FOR CLASSIFICATION AS TO HUMAN CARCINOGENICITY

II.A.1. WEIGHT-OF-EVIDENCE CLASSIFICATION

Classification -- B2; probable human carcinogen

Basis -- increased incidence of liver tumors including carcinomas in two strains of mice and in hamsters and of thyroid tumors in female rats by diet.

II.A.2. HUMAN CARCINOGENICITY DATA

Human epidemiological data are not available for DDE.   Evidence for the carcinogenicity in humans of DDT, a structural analog, is based on autopsy studies relating tissue levels of DDT to cancer incidence. These studies have yielded conflicting results. Three studies reported that tissue levels of DDT and DDE were higher in cancer victims than in those dying of other diseases (Casarett et al., 1968; Dacre and Jennings, 1970; Wasserman et al., 1976). In other studies no such relationship was seen (Maier-Bode, 1960; Robinson et al., 1965; Hoffman et al., 1967). Studies of volunteers and workers occupationally exposed to DDT have been of insufficient duration to determine the carcinogenicity of DDT to humans.

II.A.3. ANIMAL CARCINOGENICITY DATA

Sufficient. NCI (1978) administered DDE in feed at TWA doses of 148 and 261 ppm to 50 B6C3F1 mice/sex/dose for 78 weeks. After an additional 15 weeks, a dose-dependent and statistically significant increase in incidence of hepatocellular carcinomas was observed in males and females in comparison with controls. Increased weight loss and mortality was observed in females.

Tomatis et al. (1974) administered 250 ppm DDE in feed for lifetime (130 weeks) to 60 CF-1 mice/sex. A statistically significant increase in incidence of hepatomas was observed in both males and females in comparison
with controls. In females, 98% of the 55 surviving exposed animals developed hepatomas, compared to 1% of the surviving controls.

Rossi et al. (1983) administered DDE in feed for 128 weeks to 40-46 Syrian Golden hamsters/sex/dose at doses of 500 and 1000 ppm. After 76 weeks, a statistically significant increase in incidence of neoplastic nodules of the liver were observed in both sexes in comparison with vehicle-treated controls.

NCI (1978) also fed DDE at TWA doses of 437 and 839 ppm for males and 242 and 462 ppm for females for 78 weeks to 50 Osborne-Mendel rats/sex/ dose, with an additional 35 week observation period. A dose-dependent trend in incidence of thyroid tumors was observed in females which was statistically significant by the Cochran Armitage trend test after adjustment for survival. The Fischer Exact test, however, was not statistically significant. Overall, the results of the bioassay were not considered by NCI to provide convincing evidence for carcinogenicity.

II.A.4. SUPPORTING DATA FOR CARCINOGENICITY

DDE was mutagenic in mouse lymphoma (L5178Y) cells and chinese hamster (V79) cells, but not in Salmonella (ICPEMC, 1984). DDE is structurally similar to and a metabolite of DDT (Peterson and Robinson, 1964; Gingell and Wallcave, 1976; Morgan and Roan, 1977) which is a probable human carcinogen.

II.B. QUANTITATIVE ESTIMATE OF CARCINOGENIC RISK FROM ORAL EXPOSURE

II.B.1. SUMMARY OF RISK ESTIMATES

Oral Slope Factor -- 3.4E-1/mg/kg/day
Drinking Water Unit Risk -- 9.7E-6/ug/L

Extrapolation Method -- Linearized multistage procedure, extra risk

Drinking Water Concentrations at Specified Risk Levels:

Risk Level	Concentration
E-4 (1 in 10,000)	1E+1 ug/L
E-5 (1 in 100,000)	1 ug/L
E-6 (1 in 1,000,000)	1E-1 ug/L

II.B.2. DOSE-RESPONSE DATA (CARCINOGENICITY, ORAL EXPOSURE)

Tumor Type -- hepatocellular carcinomas, hepatomas
Test Animals -- mouse/B6C3F1; mouse/CF-1; hamsters/Syrian Golden
Route -- diet
Reference -- NCI, 1978; Tomatis et al., 1974; Rossi et al., 1983

Administered Dose (ppm)	Human Equivalent Dose (mg/kg)/day	Tumor Incidence female	male	Reference
Mouse/B6C3F1; hepatocellular carcinomas				
0	0.0	0/19	0/19	NCI, 1978
148	0.90	19/47	7/41	
261	1.584	34/48	17/47	
Mouse/CF-1; hepatomas				
0	0	1/90	33/98	Tomatis et
250	2.45	54/55	39/53	al., 1974
Hamsters/Syrian Golden; neoplastic nodules (hepatomas)				
0	0	0/31	0/42	Rossi et
500	4.79	7/30	4/39	al., 1983
1000	9.57	8/39	6/39	

II.B.3. ADDITIONAL COMMENTS (CARCINOGENICITY, ORAL EXPOSURE)

NCI (1978) used DDE of about 95% purity, while that used by Tomatis et al. (1974) and Rossi et al. (1983) was 99% pure. In the hamster study, Rossi et al. described the observed lesions as neoplastic liver nodules or hepatocellular tumors, using these terms interchangeably. The oral quantitative estimate is a geometric mean of six slope factors computed from incidence data by sex from the studies cited in Section II.A.3.

The unit risk should not be used if the water concentration exceeds 1E+3 ug/L, since above this concentration the slope factor may differ from that stated.

II.B.4. DISCUSSION OF CONFIDENCE (CARCINOGENICITY, ORAL EXPOSURE)

An adequate number of animals was observed. The geometric mean obtained using the slope factors from the mouse studies alone is 7.8E-1/mg/kg/day. This is within a factor of 2 of that derived from the mouse and hamster studies combined. In addition, the slope factor for DDE was within a factor of 2 of the slope factors for liver tumors for three structurally similar compounds: DDT, 3.4E-1/mg/kg/day; DDD, 2.4E-1/mg/kg/day; and Dicofol, 4.4E-1/mg/kg/day.

II.C.   QUANTITATIVE ESTIMATE OF CARCINOGENIC RISK FROM INHALATION EXPOSURE

[Ed. Note: EPA does not have information yet in this section of IRIS].

II.D. EPA DOCUMENTATION, REVIEW, AND CONTACTS (CARCINOGENICITY ASSESSMENT)

II.D.1. EPA DOCUMENTATION

Source Document -- US EPA, 1980, 1985

The 1985 Carcinogen Assessment Group's report has received Agency review. The 1980 Hazard Assessment Report has received peer review.

II.D.2. REVIEW (CARCINOGENICITY ASSESSMENT)

Agency Work Group Review -- 06/24/87

Verification Date -- 06/24/87

II.D.3. US EPA CONTACTS (CARCINOGENICITY ASSESSMENT)

James Holder / NCEA -- (202)260-5721

Chao Chen / NCEA -- (202)260-5719

III. HEALTH HAZARD ASSESSMENTS FOR VARIED EXPOSURE DURATIONS

[Ed. Note: EPA does not have information yet in this section of IRIS].

IV. US EPA REGULATORY ACTIONS

EPA risk assessments may be updated as new data are published and as assessment methodologies evolve. Regulatory actions are frequently not updated at the same time. Compare the dates for the regulatory actions in this section with the verification dates for the risk assessments in sections I and II, as this may explain inconsistencies. Also note that some regulatory actions consider factors not related to health risk, such as technical or economic feasibility. Such considerations are indicated for each action. In addition, not all of the regulatory actions listed in this section involve enforceable federal standards. Please direct any questions you may have concerning these regulatory actions to the US EPA contact listed for that particular action. Users are strongly urged to read the background information on each regulatory action in Background Document 4 in Service Code 5.

IV.A. CLEAN AIR ACT (CAA)

[Ed. Note: EPA does not have information yet in this section of IRIS].

IV.B. SAFE DRINKING WATER ACT (SDWA)

[Ed. Note: EPA does not have information yet in this section of IRIS].

IV.C. CLEAN WATER ACT (CWA)

IV.C.1. AMBIENT WATER QUALITY CRITERIA, Human Health

[Ed. Note: EPA does not have information yet in this section of IRIS].

IV.C.2. AMBIENT WATER QUALITY CRITERIA, Aquatic Organisms

Freshwater:
    Acute LEC -- 1.05E+3 ug/L
    Chronic -- None

Marine:
    Acute LEC -- 1.4E+1 ug/L
    Chronic -- None

Considers technological or economic feasibility? -- NO

Discussion -- The values that are indicated as "LEC" are not criteria, but are the lowest effect levels found in the literature. LECs are given when the minimum data required to derive water quality criteria are not available.

Reference -- 45 FR 79318 (11/28/80)

EPA Contact -- Criteria and Standards Division / OWRS (202)260-1315 / FTS 260-1315

IV.D. FEDERAL INSECTICIDE, FUNGICIDE, AND RODENTICIDE ACT (FIFRA)

[Ed. Note: EPA does not have information yet in this section of IRIS].

IV.E. TOXIC SUBSTANCES CONTROL ACT (TSCA)

[Ed. Note: EPA does not have information yet in this section of IRIS].

IV.F. RESOURCE CONSERVATION AND RECOVERY ACT (RCRA)

IV.F.1. RCRA APPENDIX IX, for Ground Water Monitoring

Status -- Listed

Reference -- 52 FR 25942 (07/09/87)

EPA Contact -- RCRA/Superfund Hotline (800)424-9346 / (202)260-3000 / FTS 260-3000

IV.G. SUPERFUND (CERCLA)

IV.G.1. REPORTABLE QUANTITY (RQ) for Release into the Environment

Value (status) -- 1 pound (Final, 19)

Considers technological or economic feasibility? -- NO

Discussion -- The final RQ for p,p'-dichlorodiphenyldichloroethylene is based on aquatic toxicity. The available data indicate that the aquatic 96-Hour Median Threshold Limit is less than 0.1 ppm, which corresponds to an RQ of 1 pound.

Reference -- 54 FR 33418 (08/14/89)

EPA Contact -- RCRA/Superfund Hotline (800)424-9346 / (202)260-3000 / FTS 260-3000

VI. BIBLIOGRAPHY

VI.A. ORAL RfD REFERENCES

None

VI.B. INHALATION RfD REFERENCES

None

VI.C. CARCINOGENICITY ASSESSMENT REFERENCES

Casarett, L.J., G.C. Fryer, W.L. Yauger, Jr. and H. Klemmer. 1968. Organochlorine pesticide residues in human tissue. Hawaii. Arch. Environ. Health. 17: 306-311.

Dacre, J.C. and R.W. Jennings. 1970. Organochlorine insecticides in normal and carcinogenic human lung tissues. Toxicol. Appl. Pharmacol. 17: 277.

Gingell, R. and L. Wallcave. 1976. Species differences in the acute toxicity and tissue distribution of DDT in mice and hamsters. Toxicol. Appl. Pharmacol. 28: 385.

Hoffman, W.S., H. Adler, W.I. Fishbein and F.C. Bauer. 1967. Relation of pesticide concentrations in fat to pathological changes in tissues. Arch. Environ. Health. 15: 758-765.

ICPEMC (International Commision for Protection Against Environmental Mutagens and Carcinogens). 1984. Report of ICPEMC Task Group 5 on the differentiation between genotoxic and nongenotoxic carcinogens. ICPEMC Publication No. 9. Mutat. Res. 133: 1-49.

Maier-Bode, H. 1960. DDT im Korperfett des Menschen. Med. Exp. 1: 146-152.

Morgan, D.P. and C.C. Roan. 1977. The metabolism of DDT in man. Essays Toxicol. 5: 39.

NCI (National Cancer Institute). 1978. Bioassay of DDT, TDE and p,p'-DDE for possible carcinogenicity. NCI Report No. 131. DHEW Publ. No. (NIH) 78-1386.

Peterson, J.E. and W.H. Robinson. 1964. Metabolic products of p,p'-DDT in the rat. Toxicol. Appl. Pharmacol. 6: 321-327.

Robinson, J., A. Richardson, C.G. Hunter, A.N. Crabtree and H.J. Rees. 1965. Organochlorine insecticide content of human adipose tissue in south-eastern England. Br. J. Ind. Med. 22: 220-224.

Rossi, L., O. Barbieri, M. Sanguineti, J.R.P. Cabral, P. Bruzzi and L. Santi. 1983. Carcinogenicity study with technical-grade DDT and DDE in hamsters. Cancer Res. 43: 776-781.

Tomatis, L., V. Turusov, R.t. Charles and M. Boicchi. 1974. Effect of long-term exposure to 1,1-dichloro-2,2-bis(p-chlorophenyl)ethylene, to 1,1-dichloro-2,2-bis(p-chlorophenyl)ethane, and the two chemicals combined on CF-1 mice. J. Natl. Cancer Inst. 52: 883-891.

US EPA. 1980. Hazard Assessment Report on DDT, DDD, DDE. Prepared by the Office of Health and Environmental Assessment, Environmental Criteria and Assessment Office, Cincinnati, OH.

US EPA. 1985. The Carcinogen Assessment Group's Calculation of the Carcinogenicity of Dicofol (Kelthane), DDT, DDE and DDD (TDE). Prepared by the Office of Health and Environmental Assessment, Carcinogen Assessment Group, Washington, DC for the Hazard Evaluation Division, Office of Toxic Substances, Washington, DC.

Wasserman, M., D.P. Nogueira, L. Tomatis, et al. 1976. Organochlorine compounds in neoplastic and adjacent apparently normal breast tissue. Bull. Environ. Contam. Toxicol. 15: 478-484.

VI.D. DRINKING WATER HA REFERENCES

None

VII. REVISION HISTORY

Date	Section	Description
08/22/88	II.	Carcinogen summary on-line
08/01/89	VI.	Bibliography on-line
01/01/92	IV.	Regulatory Action section on-line

SYNONYMS

2,2-BIS(4-CHLOROPHENYL)-
1,1-DICHLOROETHENE
2,2-BIS(p-CHLOROPHENYL)-
1,1-DICHLOROETHYLENE
DDE
p,p'-DDE
DDT DEHYDROCHLORIDE
1,1-DICHLORO-2,2-BIS(p-
CHLOROPHENYL)ETHYLENE

DICHLORODIPHENYLDICHLOROETHYLENE
Dichlorodiphenyldichloroethylene, p,p'-
1,1'-DICHLOROETHENYLIDENE)BIS(4-
CHLOROBENZENE)
ETHYLENE, 1,1-DICHLORO-2,2-BIS(p-
CHLOROPHENYL)-
NCI-C00555

Source: *Instant EPA's IRIS*, 1997, Instant Reference Sources, Inc., 7605 Rockpoint Drive, Austin, TX 78731 (Fax: 512-345-2386; Internet URL, http://www.instantref.com/inst-ref.htm).

Production, Use, and Pesticide Labeling Information

DDE is not currently listed in EPA's Pesticide Factsheet database and there is no indication of whether it will be added or not in the future.

DDT

Introduction

DDT was originally produced in 1938 and was used extensively in North America as an insecticide until it was prohibited in the United States in 1973. However, it is still manufactured in the US for export and is also still produced and used in developing countries. [813] If sprayed on land DDT will adsorb very strongly to soil and be subject to evaporation and photodegradation at the surface of soils. It will not leach appreciably to groundwater or hydrolyze but may be subject to biodegradation in flooded soils or under anaerobic conditions. If released to water it will adsorb very strongly to sediments and be subject to evaporation and photooxidation near the surface. It will not hydrolyze and will not significantly biodegrade in most waters. Biodegradation may be significant in sediments. If released to the air it will be subject to direct photodegradation and reaction with photochemically produced hydroxyl radicals. Wet and dry deposition will be major removal mechanisms from the atmospheric compartment. General population exposure will occur mainly through ingestion of contaminated food, especially contaminated fish and human milk. [828] Concentrations of DDT in humans appear to be decreasing, but DDE is not. DDT does not convert to DDE in humans, so it is assumed that the primary source of DDE in humans is from food, with fish consumption the most likely source. [882]

Identifying Information

CAS NUMBER: 50-29-3

NIOSH Registry Number: KJ3325000

SYNONYMS: ALPHA,ALPHA-BIS(P-CHLOROPHENYL)-BETA,BETA,BETA-TRICHLORETHANE, 2,2-BIS(P-CHLOROPHENYL)-1,1,1-TRICHLOROETHANE, CHLOROPHENOTHANE, DICHLORODIPHENYL-TRICHLOROETHANE, P,P'-DICHLORODIPHENYLTRICHLOROETHANE, 4,4'-DICHLORODIPHENYL-TRICHLOROETHANE, 4,4'-DDT, 1,1-BIS(P-CHLOROPHENYL)-2,2,2-TRICHLOROETHANE, DIPHENYLTRICHLOROETHANE, 1,1,1-TRICHLORO-2,2-DI(4-CHLOROPHENYL)ETHANE, ANOFEX, P,P'-DDT, DICOPHANE, DIDIGAM, DIDIMAC, ENT 1,506, ESTONATE, GENITOX, GESAROL, GYRON, IXODEX, NCI-C00464, NEOCID, PENTACHLORIN, SANTOBANE, TRICHLOROBIS(4-CHLOROPHENYL)ETHANE, 1,1,1-TRICHLORO-2,2-BIS(P-

CHLOROPHENYL)ETHANE, ZEIDANE, ZERDANE, AGRITAN, ARKOTINE, AZOTOX, 1,1'-(2,2,2-TRICHLOROETHYLIDENE)BIS(4-CHLOROBENZENE), 1,1-BIS-(P-CHLOROPHENYL)-2,2,2-TRICHLOROETHANE, BOSAN SUPRA, BOVIDERMOL, CHLORPHENOTHAN, CHLOROPHENOTOXUM, CITOX, CLOFENOTANE, DEDELO, DEOVAL, DETOX, DETOXAN, DIBOVAN, DODAT, DYKOL, GESAFID, GESAPON, GESAREX, GUESAPON, GUESAROL, HAVERO-EXTRA, HILDIT, IVORAN, KOPSOL, MICRO DDT 75, MUTOXIN, NA 2761, OMS 16, PARACHLOROCIDUM, PEB1, PENTECH, PPZEIDAN, R50, RCRA WASTE NUMBER U061, RUKSEAM, TECH DDT, TRICHLOROBIS(4-CHLOROPHENYL)ETHANE

Endocrine Disruptor Information

DDT is a strongly suspected EED that it is listed by EPA [811], the Centers for Disease Control & Prevention [812], and the World Wildlife Fund [813] as a potential endocrine modifying chemical. DDT and its analogs DDE and DDD along with their structural isomers have long been implicated as being endocrine disruptors. [845] It also is listed as one of the commonly found pesticides in wildlife exhibiting symptoms of EDCs. DDT and its congeners have reproductive endocrine effects and also have a major toxic effect on the adrenal glands, DDE being utilized therapeutically in the treatment of Cushing's syndrome. [846] Deformaties in birds including clubbed feet, crossed bills, and missing eyes are also suspected of having links with exposures to DDT. [810] These organochlorine insecticides have been banned from use in this country, except for public health emergencies, for almost 20 years but due to these chemicals' extraordinary persistence, residues still appear in approximately 10% of the hazardous waste sites sampled. [811] In spite of its long time ban in the United States, 96 tons of DDT were exported from the US in 1991. [810]

Chemical and Physical Properties

CHEMICAL FORMULA: $C_{14}H_9Cl_5$

MOLECULAR WEIGHT: 354.49

WLN: GXGGYR DG&R DG

PHYSICAL DESCRIPTION: White powder

SPECIFIC GRAVITY: 1.56 @ 15°C [371]

DENSITY: Not available

MELTING POINT: 108.5-109°C [025], [169], [173], [430]

BOILING POINT: 260°C [017], [047], [430]

SOLUBILITY: Water: <1 mg/mL @ 21°C [700]; DMSO: >=100 mg/mL @ 21°C [700]; 95% Ethanol: 5-10 mg/mL @ 21°C [700]; Acetone: 10-50 mg/mL @ 21°C [700]; Methanol: 40 g/L @ 27°C [169]

OTHER SOLVENTS: Alkalies: Insoluble [031], [430]; Carbon tetrachloride: 45 g/100 mL [031], [430]; Chlorobenzene: 74 g/100 mL [031]; Cyclohexanone: 116 g/100 mL [031], [430]; Dilute acids: Insoluble [031], [430]; Dioxane: Freely soluble [031], [062]; Isopropanol: 3 g/100 mL [031]; Tetralin: 61 g/100 mL [031]; Pyridine: Freely soluble [031], [062]; Ether: 28 g/100 mL [031]; Benzene: 78 g/100 mL [031], [430]; Tributyl phosphate: 50 g/100 mL [031]; Peanut oil: 11 g/100 mL [031]; Pine oil: 10-16 g/100 mL [031]; Chloroform: Soluble [017], [047]; Kerosene: 8-10 g/100 mL [031]; Petroleum ether: Soluble [017]; Xylene: 600 g/L @ 27°C [169]; Benzyl benzoate: 42 g/100 mL [031]; Morpholine: 75 g/100 mL [031]; Gasoline: 10 g/100 mL [031]; Alcohol: 1 in 5 [295], [455]; Dichloromethane: 850 g/L @ 27°C [169]; Trichloroethylene: 720 g/L @ 27°C [169]; Dimethyl phthalate: 1 in 4 [295]; Aromatic and chlorinated solvents: Readily soluble [169], [172]; Polar organic solvents: Moderately soluble [169], [172]; Hydroxylic solvents: Moderately soluble [172]

OTHER PHYSICAL DATA: Odorless or with slight aromatic odor [062], [295], [421], [455]; UV max (in 95% alcohol): 236 nm [031]; Crystallizing point is not less than 89°C [295]; Boiling point: 185-187°C @ 0.05 mm Hg (with decomposition) [169]; Tasteless [173]; Odor threshold: 0.2 ppm [051]

Source: *The National Toxicology Program's Chemical Database, Volume 2*, 1992, CRC Press, Inc./Lewis Publishers, 2000 Corporate Blvd., Boca Raton, FL 33431 (Fax: 407-998-9114).

Environmental Reference Materials

A source of EED environmental reference materials for xxx is Radian International LLC, Austin, TX (Fax 512-454-0268; Internet URL, http://www.radian.com/standards); Catalog No. ERD-005.

Hazardous Properties

ACUTE/CHRONIC HAZARDS: DDT is toxic by ingestion, inhalation and skin absorption, especially in solution [029], [033], [062]. Poisoning may occur by absorption through the respiratory tract [033]. Absorption is facilitated by oily substances [295]. It is also an irritant [173] and when heated to decomposition it emits toxic fumes of chloride [051].

FIRE HAZARD: DDT has a flash point of 162-171°C (324-340°F) [051], [371]. It is combustible. Fires involving DDT can be controlled with a dry chemical, carbon dioxide or Halon extinguisher. A water spray may also be used [371].

REACTIVITY: DDT may react with iron, aluminum, aluminum and iron salts, and alkalis [033]. It is incompatible with ferric chloride and aluminum chloride [172], [395]. It can also react with strong oxidizing materials [051], [346].

STABILITY: DDT is sensitive to exposure to ultraviolet light [172], [395]. It is also sensitive to high temperatures [033]. Solutions of DDT in water, DMSO, 95% ethanol or acetone should be stable for 24 hours under normal lab conditions [700].

Source: *The National Toxicology Program's Chemical Database, Volume 7*, 1992, CRC Press, Inc./Lewis Publishers, 2000 Corporate Blvd., Boca Raton, FL 33431 (Fax: 407-998-9114).

Medical Symptoms of Exposure

Acute symptoms of exposure to DDT include tremors of head and neck muscles, tonic and clonic convulsions, cardiac or respiratory failure and death. Chronic symptoms of exposure include hepatic damage, central nervous system degeneration, agranulocytosis, dermatitis, weakness, convulsions and coma [033]. It may cause retrobulbar neuritis and "yellow vision" [099]. It may also cause perspiration, headache, nausea, vomiting, decrease in hemoglobin, moderate leukocytosis, variable hyperesthesia of the mouth, disturbance of sensitivity of the lower part of the face, uncertain gait, malaise, cold and moist skin, hypersensitivity to contact, fatigue, difficulty in seeing or hearing, prickling of the tongue and around the mouth and nose, disturbance of equilibrium, dizziness, confusion, tremor of extremities, aching of the limbs, anxiety, irritability and irritation of the nose, throat, eyes and skin. Other symptoms of exposure may include ataxia of the hands, restlessness, weak and slow heartbeat, giddiness, incoordination, palpitations, numbness of the hands, urge to defecate, wide and nonreactive pupils, dysarthria, facial weakness, reduced reflexes, positive Romberg test, slightly lowered blood pressure, persistent and irregular heart action, slight jaundice, abdominal pain, diarrhea, increased salivation, tachycardia and cyanosis of the lips [173]. It may cause paresthesia (abnormal sensation) and hyperactive knee-jerk reflexes [051]. It may also cause abnormal EKG patterns, aplastic anemia, thrombocytopenia, hyperglycemia, glycosuria, decrease in liver glycogen, sore throat and apprehension [151]. Other symptoms include paresthesia of the tongue, lips and face [346] and may cause ataxia, confusion and partial paralysis [371].

Source: *The National Toxicology Program's Chemical Database, Volume 4*, 1992, CRC Press, Inc./Lewis Publishers, 2000 Corporate Blvd., Boca Raton, FL 33431 (Fax: 407-998-9114).

Toxicological Information

The toxicological information below is gathered from several sources including two national databases: one from the *National Toxicology Program Chemical Repository Database* and another from *EPA's Integrated Risk Information System (IRIS)*.

Toxicology Information from the National Toxicology Program

CARCINOGENICITY:

Tumorigenic Data:

Type of Effect	Route/Animal	Amount Dosed (Notes)
TDLo	orl-rat	1225 mg/kg/7W-C
TDLo	orl-mus	24 mg/kg (MGN)
TDLo	orl-mus	73 mg/kg/26W-C
TDLo	scu-mus	370 mg/kg/80W-I
TDLo	orl-ham	21280 mg/kg/38W-I
TD	orl-rat	12096 mg/kg/3Y-C
TD	orl-mus	7560 mg/kg/90W-C
TD	orl-mus	5600 mg/kg/80W-I
TD	orl-rat	8100 mg/kg/2Y-C
TD	orl-mus	3150 mg/kg/15W-C
TD	orl-mus	3408 mg/kg (MGN)
TD	orl-rat	19 gm/kg/2Y-C
TD	orl-rat	438 mg/kg/2Y-C
TD	orl-rat	17976 mg/kg/2Y-C
TD	orl-rat	24192 mg/kg/3Y-C

Review:
IARC Cancer Review: Animal Sufficient Evidence
IARC Cancer Review: Human Inadequate Evidence
IARC possible human carcinogen (Group 2B)

Status:
NCI Carcinogenesis Bioassay (Feed); Negative: Male and Female Rat, Male and Female Mouse
NTP Fifth Annual Report on Carcinogens, 1989; anticipated to be carcinogen
EPA Carcinogen Assessment Group

TERATOGENICITY: See RTECS printout for data

MUTATION DATA: See RTECS printout for data

TOXICITY:

Typ. Dose	Mode	Specie	Amount	Units	Other
LDLo	orl	inf	150	mg/kg	
LDLo	unr	man	221	mg/kg	
LD50	orl	rat	87	mg/kg	
LD50	skn	rat	1931	mg/kg	
LD50	ipr	rat	9100	ug/kg	
LDLo	orl	hmn	500	mg/kg	
LD50	scu	rat	1500	mg/kg	
LD50	ivn	rat	68	mg/kg	
LD50	orl	mus	135	mg/kg	
LD50	ipr	mus	32	mg/kg	
LD50	ivn	mus	68500	ug/kg	
LD50	orl	dog	150	mg/kg	
LDLo	ivn	dog	75	mg/kg	
LD50	orl	mky	200	mg/kg	
LDLo	ivn	mky	50	mg/kg	
LDLo	orl	cat	250	mg/kg	
LDLo	ivn	cat	40	mg/kg	
LD50	orl	rbt	250	mg/kg	
LD50	skn	rbt	300	mg/kg	
LD50	scu	rbt	250	mg/kg	
LDLo	ivn	rbt	50	mg/kg	
LD50	orl	gpg	150	mg/kg	
LD50	skn	gpg	1000	mg/kg	
LD50	scu	gpg	900	mg/kg	
LDLo	orl	ckn	300	mg/kg	
LD50	orl	frg	7600	ug/kg	
LD50	scu	frg	35	mg/kg	
LDLo	orl	dom	300	mg/kg	
TDLo	orl	man	6	mg/kg	
TDLo	orl	hmn	16	mg/kg	
TDLo	orl	hmn	5	mg/kg	

OTHER TOXICITY DATA:

   Review: Toxicology Review-29

   Standards and Regulations:
          DOT-Hazard: ORM-A, Label: None

Status:
          EPA TSCA Test Submission (TSCATS) Data Base, April 1990
          EPA Genetox Program 1988, Positive: Carcinogenicity-mouse/rat; In
          vitro cytogenetics-nonhuman
          EPA TSCA Chemical Inventory, 1989

EPA Genetox Program 1988, Positive: D melanogaster-partial sex chrom. loss
EPA Genetox Program 1988, Positive: V79 cell culture-gene mutation
EPA Genetox Program 1988, Negative: Host-mediated assay; Sperm morphology-mouse
EPA Genetox Program 1988, Negative: D melanogaster Sex-linked lethal; S cerevisiae-homozygosis
EPA Genetox Program 1988, Inconclusive: D melanogaster-whole sex chrom. loss
EPA Genetox Program 1988, Inconclusive: D melanogaster-nondisjunction; Rodent dominant lethal
EPA Genetox Program 1988, Inconclusive: Mammalian micronucleus; E coli polA without S9 Lethal dose for humans: 500 mg/kg

SAX TOXICITY EVALUATION:

THR: Human poison by ingestion and possibly other routes. Experimental poison by ingestion, skin contact, subcutaneous, intravenous and intraperitoneal routes. A suspected human carcinogen. An experimental carcinogen, neoplastigen, tumorigen and teratogen. Experimental reproductive effects. Human systemic effects by ingestion.

Experimental reproductive effects. Human mutagenic data. A dose of 20 grams has proved highly dangerous though not fatal to man. This dose was taken by 5 persons who vomited an unknown portion of the material and, even so, recovered only incompletely after 5 weeks. Smaller doses produced less important symptoms with relatively rapid recovery. Experimental ingestion of 1.5 g resulted in great discomfort and moderate neurological changes. Recovery was complete on the following day. The fatal dose for humans is not known. Judging from the literature, no one has ever been killed in the absence of other insecticides and/or a variety of toxic solvents. However, these common solvent formulations are highly fatal when taken in small doses, partly because of the toxicity of the solvent, and perhaps because of the increased absorbability of the compound; several fatal cases in humans have been reported. Little is known of the hazard of chronic poisoning. Human volunteers have ingested up to 35 mg/day for 21 months with no ill effects.

This compound and some of its degradation products, particularly DDE, are stored in fat. This storage effect leads to a concentration of this compound at higher levels of the food chain. DDT stored in the fat is at least largely inactive since a greater total dose may be stored in an experimental animal than is sufficient as a lethal dose for that same animal if given at one time. A study based on 75 human cases reported an average of 5.3 ppm of DDT stored in the fat. A higher content of DDT and its derivatives (up to 434 ppm of DDE and 648 ppm of DDT) was found in workers who had very extensive exposure. Without exception, the samples were taken from persons who were either asymptomatic or suffering from some disease completely unrelated to DDT. Careful hospital examination of workers, who had been very extensively exposed and who had volunteered for examination revealed no abnormality which could be attributed to DDT. Much higher levels have been found in humans than have been observed in the fat of experimental animals which were apparently asymptomatic. DDT stored in the fat is eliminated only very gradually when further dosage is discontinued. However, weight loss can speed the release of this stored DDT (and DDE) into the blood. After a single dose, the secretion of DDT in the milk and its excretion in the urine reach their height with in a day or two and continue at a lower level thereafter.

Source: *Instant Tox-Base*, 1995, Instant Reference Sources, Inc., 7605 Rockpoint Drive, Austin, TX 78731 (Fax: 512-345-2386; Internet URL, http://www.instantref.com/inst-ref.htm).

Toxicology Information from EPA's Integrated Risk Information System (IRIS)

I.A. REFERENCE DOSE FOR CHRONIC ORAL EXPOSURE (RfD)

The Reference Dose (RfD) is based on the assumption that thresholds exist for certain toxic effects such as cellular necrosis, but may not exist for other toxic effects such as carcinogenicity. In general, the RfD is an estimate (with uncertainty spanning perhaps an order of magnitude) of a daily exposure to the human population (including sensitive subgroups) that is likely to be without an appreciable risk of deleterious effects during a lifetime. Please refer to Background Document 1 in Service Code 5 for an elaboration of these concepts. RfDs can also be derived for the noncarcinogenic health effects of compounds which are also carcinogens. Therefore, it is essential to refer to other sources of information concerning the carcinogenicity of this substance. If the US EPA has evaluated this substance for potential human carcinogenicity, a summary of that evaluation will be contained in Section II of this file when a review of that evaluation is completed.

I.A.1. ORAL RfD SUMMARY

Critical Effect	Experimental Doses*	UF	MF	RfD
Liver lesions	NOEL: 1 ppm diet (0.05 mg/kg bw/day) LOAEL: 5 ppm	100	1	5E-4 mg/kg/day

*Conversion Factors: Food consumption = 5% bw/day

Laug et al., 1950

I.A.2. PRINCIPAL AND SUPPORTING STUDIES (ORAL RfD)

Laug, E.P., A.A. Nelson, O.G. Fitzhugh and F.M. Kunze. 1950. Liver cell alteration and DDT storage in the fat of the rat induced by dietary levels of 1-50 ppm DDT. J. Pharmacol. Exp. Therap. 98: 268-273.

Weanling rats (25/sex/group) were fed commercial DDT (81% P,P isomer and 19% O,P isomer) at levels of 0, 1, 5, 10 or 50 ppm for 15-27 weeks. The diet was prepared by mixing appropriate amounts of DDT in corn oil solution with powdered chow. No interference with growth was noted at any level. Females stored more DDT in peripheral fat than did males, but pathologic changes were seen to a greater degree in males. Increasing hepatocellular hypertrophy, especially centrilobularly, increased cytoplasmic oxyphilia, and peripheral basophilic cytoplasmic granules (based on H and E paraffin sections) were observed at dose levels of 5 ppm and above. The effect was minimal at 5 ppm (LOAEL) and more pronounced at higher doses. No effects were reported at 1 ppm, the NOEL level used as the basis for the RfD calculation. The authors believe the effect seen at 5 ppm "represents the smallest detectable morphologic effect, based on extensive observations of the rat liver as affected by a variety of chemicals."

DDT fed to rats for 2 years (Fitzhugh, 1948) caused liver lesions at all dose levels (10-800 ppm of diet). A LOAEL of 0.5 mg/kg bw/day was established. Application of a factor of 10 each for uncertainty of estimating a NOEL from a LOAEL, as well as for interspecies conversion and protection of sensitive human subpopulations (1000 total) results in the same RfD level as that calculated from the critical study. DDT-induced liver effects were observed in mice, hamsters and dogs as well.

The Laug et al. (1950) study was chosen for the RfD calculation because: 1) male rats appear to be the most sensitive animals to DDT exposure; 2) the study was of sufficient length to observe toxic effects; and 3) several doses were administered in the diet over the range of the dose-response curve. This study also established a LOAEL and a NOEL, with the LOAEL (0.25 mg/kg/day) being the lowest of any observed for this compound.

I.A.3. UNCERTAINTY AND MODIFYING FACTORS (ORAL RfD)

UF -- A factor of 10 each was applied for the uncertainty of interspecies conversion and to protect sensitive human subpopulations. An uncertainty factor for subchronic to chronic conversion was not included because of the corroborating chronic study in the data base.

MF --None

I.A.4. ADDITIONAL COMMENTS (ORAL RfD)

In one 3-generation rat reproduction study (Treon and Cleveland, 1955), offspring mortality increased at all dose levels, the lowest of which corresponds to about 0.2 mg/kg bw/day. Three other reproduction studies (rat and mouse) show no reproductive effects at much higher dose levels.

I.A.5. CONFIDENCE IN THE ORAL RfD

Study -- Medium
Data Base -- Medium
RfD -- Medium

The principal study appears to be adequate, but of shorter duration than that desired; therefore, confidence in the study can be considered medium to low. The data base is only moderately supportive of both the critical effect and the magnitude, and lacks a clear NOEL for reproductive effects; therefore, confidence in the data base can also be considered medium to low. Medium to low confidence in the RfD follows.

I.A.6. EPA DOCUMENTATION AND REVIEW OF THE ORAL RfD

Source Document -- This assessment is not presented in any existing US EPA document.

Other EPA Documentation -- None

Agency Work Group Review -- 12/18/85

Verification Date -- 12/18/85

I.A.7. EPA CONTACTS (ORAL RfD)

Joan S. Dollarhide / NCEA -- (513)569-7539

I.B. REFERENCE CONCENTRATION FOR CHRONIC INHALATION EXPOSURE (RfC)

Not available at this time.

II. CARCINOGENICITY ASSESSMENT FOR LIFETIME EXPOSURE

Section II provides information on three aspects of the carcinogenic risk assessment for the agent in question; the US EPA classification, and quantitative estimates of risk from oral exposure and from inhalation exposure. The classification reflects a weight-of-evidence judgment of the likelihood that the agent is a human carcinogen. The quantitative risk estimates are presented in three ways. The slope factor is the result of application of a low-dose extrapolation procedure and is presented as the risk per (mg/kg)/day. The unit risk is the quantitative estimate in terms of either risk per ug/L drinking water or risk per ug/cu.m air breathed. The third form in which risk is presented is a drinking water or air concentration providing cancer risks of 1 in 10,000, 1 in 100,000 or 1 in 1,000,000. Background Document 2 (Service Code 5) provides details on the rationale and methods used to derive the carcinogenicity values found in IRIS. Users are referred to Section I for information on long-term toxic effects other than carcinogenicity.

II.A. EVIDENCE FOR CLASSIFICATION AS TO HUMAN CARCINOGENICITY

II.A.1. WEIGHT-OF-EVIDENCE CLASSIFICATION

Classification -- B2; probable human carcinogen.

Basis -- Observation of tumors (generally of the liver) in seven studies in various mouse strains and three studies in rats. DDT is structurally similar to other probable carcinogens, such as DDD and DDE.

II.A.2. HUMAN CARCINOGENICITY DATA

Inadequate. The existing epidemiological data are inadequate. Autopsy studies relating tissue levels of DDT to cancer incidence have yielded conflicting results. Three studies reported that tissue levels of DDT and DDE were higher in cancer victims than in those dying of other diseases (Casarett et al., 1968; Dacre and Jennings, 1970; Wasserman et al., 1976). In other studies no such relationship was seen (Maier-Bode, 1960; Robinson et al., 1965; Hoffman et al., 1967). Studies of occupationally exposed workers and volunteers have been of insufficient duration to be useful in assessment of the carcinogenicity of DDT to humans.

II.A.3. ANIMAL CARCINOGENICITY DATA

Sufficient. Twenty-five animal carcinogenicity assays have been reviewed for DDT. Nine feeding studies, including two multigenerational studies, have been conducted in the following mouse strains: BALB/C, CF-1, A strain, Swiss/Bombay and (C57B1)x(C3HxAkR). Only one of these studies, conducted for 78 weeks, showed no indication of DDT tumorigenicity (NCI, 1978). Both hepatocellular adenomas and carcinomas were observed in six mouse liver tumor studies (Walker et al., 1973; Thorpe and Walker, 1973; Kashyap et al., 1977; Innes et al., 1969; Terracini et al., 1973; Turusov et al., 1973). Both benign and malignant lung tumors were observed in two studies wherein mice were exposed both in utero and throughout their lifetime (Shabad et al., 1973; Tarjan and Kemeny, 1969). Doses producing increased tumor incidence ranged from 0.15-37.5 mg/kg/day.

Three studies using Wistar, MRC Porton and Osborne-Mendel rats and doses from 25-40 mg/kg/day produced increased incidence of benign liver tumors (Rossi et al., 1977; Cabral et al., 1982; Fitzhugh and Nelson, 1946). Another study wherein Osborne-Mendel rats were exposed in this dietary dose range for 78 weeks was negative (NCI, 1978) as were three additional assays in which lower doses were given.

Tests of DDT in hamsters have not resulted in increased tumor incidence. Unlike mice and humans, hamsters accumulate DDT in tissue but do not metabolize it to DDD or DDE. Studies of DDT in dogs (Lehman, 1951, 1965) and monkeys (Adamson and Sieber, 1979, 1983) have not shown a carcinogenic effect. However, the length of these studies (approximately 30% of the animals' lifetimes) was insufficient to assess the carcinogenicity of DDT. DDT has been shown to produce hepatomas in trout (Halver, 1967).

II.A.4. SUPPORTING DATA FOR CARCINOGENICITY

DDT has been shown to act as a liver tumor promoter in rats initiated with 2-acetylaminofluorene, 2-acetamidophenanthrene or trans-4-acetylaminostilbene (Peraino et al., 1975; Scribner and Mottet, 1981; Hilpert et al., 1983).

DDT has produced both negative and positive responses in tests for genotoxicity. Positive responses have been noted in V79 mutation assays, for chromosome aberrations in cultured human lymphocytes, and for sister chromatid exchanges in V79 and CHO cells (Bradley et al., 1981; Rabello et al., 1975; Preston et al., 1981; Ray-Chaudhuri et al., 1982). In one study, DDT was reported to interact directly with DNA; this result was not confirmed in the absence of a metabolizing system (Kubinski et al., 1981; Griffin and Hill, 1978).

DDT is structurally related to the following chemicals which produce liver tumors in mice: DDE, DDD, dicofol and chlorobenzilate.

II.B.   QUANTITATIVE ESTIMATE OF CARCINOGENIC RISK FROM ORAL EXPOSURE

II.B.1. SUMMARY OF RISK ESTIMATES

Oral Slope Factor -- 3.4E-1 per (mg/kg)/day

Drinking Water Unit Risk -- 9.7E-6 per (ug/L)

Extrapolation Method -- Linearized multistage procedure, extra risk

Drinking Water Concentrations at Specified Risk Levels:

```
Risk Level                  Concentration
--------------------        -----------------
E-4 (1 in 10,000)           1E+1 ug/L
E-5 (1 in 100,000)          1E+0 ug/L
E-6 (1 in 1,000,000)        1E-1 ug/L
```

II.B.2. DOSE-RESPONSE DATA (CARCINOGENICITY, ORAL EXPOSURE)

Tumor Type -- Liver, benign and malignant (see table)
Test Animals -- mouse and rat (see table)
Route -- diet
Reference -- see table

Species/Strain Tumor Type	Slope Factor Male	Female	Reference
Mouse/CF-1, Benign	0.80	0.42	Turusov et al., 1973
Mouse/BALB/C, Benign	0.082		Terracini et al., 1973
Mouse/CF-1, Benign, Malignant	0.52	0.81	Thorpe and Walker, 1973
Mouse/CF-1, Benign	1.04	0.49	Tomatis and Turusov, 1975
Rat/MRC Porton		0.084	Cabral et al., 1982
Rat/Wistar, Benign	0.16	0.27	Rossi et al., 1977

II.B.3. ADDITIONAL COMMENTS (CARCINOGENICITY, ORAL EXPOSURE)

The estimate of the slope factor did not increase in the multigeneration feeding studies (Terracini et al., 1973; Turusov et al., 1973) but remained the same from generation to generation. A geometric mean of the above slope factors was used for the overall slope factor of 3.4E-1. This was done in order to avoid excluding relevant data (note that the appropriateness of this procedure is currently under study by US EPA). All tumors were of the liver; there were no metastases. A few malignancies were observed in the Turusov study; possible neoplasms were indicated in the Terracini and Tomatis studies. The Turusov study was carried out over six generations, the Terracini assay for two. The slope factor derived from data of Tarjan and Kemeny (1969) was not included in the calculation of the geometric mean because the tumors developed at different sites than in any other studies. In addition, there was a problem in this study with possible DDT contamination of the feed.

DDT is known to be absorbed by humans in direct proportion to dietary exposure; t(1/2) for clearance is 10-20 years.

The unit risk should not be used if the water concentration exceeds 1E+3 ug/L, since above this concentration the unit risk may not be appropriate.

II.B.4. DISCUSSION OF CONFIDENCE (CARCINOGENICITY, ORAL EXPOSURE)

Ten slope factors derived from six studies were within a 13-fold range. The slope factor derived from the mouse data alone was 4.8E-1 while that derived from the rat data alone was 1.5E-1. There was no apparent difference in slope factor as a function of sex of the animals. The geometric mean of the slope factors from the mouse and rat data combined was identical for the same tumor site as that for DDE [3.4E-1 per (mg/kg)/day], a structural analog.

II.C. QUANTITATIVE ESTIMATE OF CARCINOGENIC RISK FROM INHALATION EXPOSURE

II.C.1. SUMMARY OF RISK ESTIMATES

Inhalation Unit Risk -- 9.7E-5 (ug/cu.m)

Extrapolation Method -- Linear multistage procedure, extra risk

Air Concentrations at Specified Risk Levels:

Risk Level	Concentration
E-4 (1 in 10,000)	1E+0 ug/cu.m
E-5 (1 in 100,000)	1E-1 ug/cu.m
E-6 (1 in 1,000,000)	1E-2 ug/cu.m

II.C.2.   DOSE-RESPONSE DATA FOR CARCINOGENICITY, INHALATION
EXPOSURE

The inhalation risk estimates were calculated from the oral data presented in Section
II.B.2.

II.C.3.   ADDITIONAL   COMMENTS   (CARCINOGENICITY,   INHALATION
EXPOSURE)

The unit risk should not be used if the air concentration exceeds 1E+2 ug/cu.m, since
above this concentration the unit risk may not be appropriate.

II.C.4.   DISCUSSION OF CONFIDENCE (CARCINOGENICITY, INHALATION
EXPOSURE)

This inhalation risk estimate was calculated from the oral data presented in Section
II.B.2.

II.D. EPA DOCUMENTATION, REVIEW, AND CONTACTS (CARCINOGENICITY
ASSESSMENT)

II.D.1. EPA DOCUMENTATION

Source Document -- US EPA, 1985

The US EPA risk assessment document on DDT is an internal report and has not received
external review.

II.D.2. REVIEW (CARCINOGENICITY ASSESSMENT)

Agency Work Group Review -- 10/29/86, 11/12/86, 06/24/87

Verification Date -- 06/24/87

II.D.3. US EPA CONTACTS (CARCINOGENICITY ASSESSMENT)

James W. Holder / NCEA -- (202)260-5721

Chao W. Chen / NCEA -- (202)260-5898

III. HEALTH HAZARD ASSESSMENTS FOR VARIED EXPOSURE DURATIONS

        [Ed. Note: EPA does not have information yet in this section of IRIS].

IV. US EPA REGULATORY ACTIONS

EPA risk assessments may be updated as new data are published and as assessment
methodologies evolve. Regulatory actions are frequently not updated at the same time.
Compare the dates for the regulatory actions in this section with the verification dates for

the risk assessments in sections I and II, as this may explain inconsistencies. Also note that some regulatory actions consider factors not related to health risk, such as technical or economic feasibility. Such considerations are indicated for each action. In addition, not all of the regulatory actions listed in this section involve enforceable federal standards. Please direct any questions you may have concerning these regulatory actions to the US EPA contact listed for that particular action. Users are strongly urged to read the background information on each regulatory action in Background Document 4 in Service Code 5.

IV.A. CLEAN AIR ACT (CAA)

[Ed. Note: EPA does not have information yet in this section of IRIS].

IV.B. SAFE DRINKING WATER ACT (SDWA)

[Ed. Note: EPA does not have information yet in this section of IRIS].

IV.C. CLEAN WATER ACT (CWA)

IV.C.1. AMBIENT WATER QUALITY CRITERIA, Human Health

Water and Fish Consumption -- 2.4E-5 ug/L

Fish Consumption Only -- 2.4E-5 ug/L

Considers technological or economic feasibility? -- NO

Discussion -- For the maximum protection from the potential carcinogenic properties of this chemical, the ambient water concentration should be zero. However, zero may not be attainable at this time, so the recommended criteria represents a E-6 estimated incremental increase of cancer risk over a lifetime.

Reference -- 45 FR 79318 (11/28/80)

EPA Contact -- Criteria and Standards Division / OWRS (202)260-1315 / FTS 260-1315

IV.C.2. AMBIENT WATER QUALITY CRITERIA, Aquatic Organisms

Freshwater:
   Acute -- 1.1E+0 ug/L (at any time)
   Chronic -- 1.0E-3 ug/L (24-hour average)

Marine:
   Acute -- 1.3E-1 ug/L (at any time)
   Chronic -- 1.0E-3 ug/L (24-hour average)

Considers technological or economic feasibility? -- NO

Discussion -- Criteria were derived from a minimum data base consisting of acute and chronic tests on a variety of species. Requirements and methods are covered in the reference to the Federal Register.

Reference -- 45 FR 79318 (11/28/80)

EPA Contact -- Criteria and Standards Division / OWRS (202)260-1315 / FTS 260-1315

IV.D. FEDERAL INSECTICIDE, FUNGICIDE, AND RODENTICIDE ACT (FIFRA)

IV.D.1. PESTICIDE ACTIVE INGREDIENT, Registration Standard

Status -- List "B" Pesticide (1989)

Reference -- 54 FR 22706 (05/25/89)

EPA Contact -- Registration Branch / OPP (703)557-7760 / FTS 557-7760

IV.D.2. PESTICIDE ACTIVE INGREDIENT, Special Review

Action -- Most uses canceled (1972)

Considers technological or economic feasibility? -- YES

Summary of regulatory action -- Canceled, all products, except the following list of uses: 1) the US Public Health Service and other Health Service Officials for control of vector diseases, 2) the USDA or military for health quarantine, 3) in drugs, for controlling body lice (to be dispensed only by a physician), 4) in the formulation of prescription drugs for controlling body lice. PR Notice 71-1 (January 15, 1971) and 37 FR 13369 (July 7, 1972). Criterion of concern: carcinogenicity, bio-accumulation, wildlife hazard and other chronic effects.

Reference -- 37 FR 13369 (07/07/72)

EPA Contact -- Special Review Branch / OPP (703)557-7400 / FTS 557-7400

IV.E. TOXIC SUBSTANCES CONTROL ACT (TSCA)

[Ed. Note: EPA does not have information yet in this section of IRIS].

IV.F. RESOURCE CONSERVATION AND RECOVERY ACT (RCRA)

IV.F.1. RCRA APPENDIX IX, for Ground Water Monitoring

Status -- Listed

Reference -- 52 FR 25942 (07/09/87)

EPA Contact --   RCRA/Superfund Hotline (800)424-9346 / (202)260-3000 / FTS 260-3000

IV.G. SUPERFUND (CERCLA)

IV.G.1. REPORTABLE QUANTITY (RQ) for Release into the Environment

Value (status) -- 1 pound (Final, 1989)

Considers technological or economic feasibility? -- NO

Discussion -- The RQ for DDT is 1 pound, based on the aquatic toxicity, as established under CWA Section 311 (40 CFR 117.3). The available data indicate the aquatic 96-hour Median Threshold Limit for DDT is less than 0.1 ppm. This corresponds to an RQ of 1 pound. DDT has also been found to bioaccumulate in the tissues of aquatic and marine organisms, and has the potential to concentrate in the food chain.

Reference -- 54 FR 33418 (08/14/89)

EPA Contact -- RCRA/Superfund Hotline (800)424-9346 / (202)260-3000 / FTS 260-3000

VI. BIBLIOGRAPHY

VI.A. ORAL RfD REFERENCES

Fitzhugh, O.G. 1948. Use of DDT insecticides on food products. Ind. Eng. Chem. 40(4): 704-705.

Laug, E.P., A.A. Nelson, O.G. Fitzhugh and F.M. Kunze. 1950. Liver cell alteration and DDT storage in the fat of the rat induced by dietary levels of 1-50 ppm DDT. J. Pharmacol. Exp. Therap. 98: 268-273.

Treon, J.F. and F.P. Cleveland. 1955. Toxicity of certain chlorinated hydrocarbon insecticides for laboratory animals, with special reference to aldrin and dieldrin. J. Agric. Food Chem. 3(5): 402-408.

VI.B. INHALATION RfC REFERENCES

None

VI.C. CARCINOGENICITY ASSESSMENT REFERENCES

Adamson, R.H. and S.M. Sieber. 1979. The use of nonhuman primates for chemical carcinogenisis studies. Ecotoxicol. Environ. Qual. 2: 275-296.

Adamson, R.H. and S.M. Sieber. 1983. Chemical carcinogenesis studies in nonhuman primates. Basic Life Sci. 24: 129-156.

Bradley, M.O., B. Bhuyan, M.C. Francis, R. Langenbach, A. Peterson and E. Huberman. 1981. Mutagenesis by chemical agents in V79 Chinese hamster cells: A review and analysis of the literature. Mutat. Res. 87: 81-142.

Cabral, J.R.P., R.K. Hall, L. Rossi, S.A. Bronczyk and P. Shubik. 1982. Effects of long-term intake of DDT on rats. Tumorigenesis. 68: 11-17.

Casarett, L.J., G.C. Fryer, W.L. Yauger, Jr. and H.W. Klemmer. 1968. Organochlorine pesticide residues in human tissue--Hawaii. Arch. Environ. Health. 17: 306-311.

Dacre, J.C. and R.W. Jennings. 1970. Organochlorine insecticides in normal and carcinogenic human lung tissues. Toxicol. Appl. Pharmacol. 17: 277.

Fitzhugh, O.G. and A.A. Nelson. 1946. The chronic oral toxicity of DDT [2,2-bis(p-chlorophenyl-1,1,1-trichloroethane)]. J. Pharmacol. 89: 18-30.

Griffin, D.E. and W.E. Hill. 1978. In vitro breakage of plasmid DNA by mutagens and pesticides. Mutat. Res. 52: 161-169.

Halver, J.E. 1967. Crystalline aflatoxin and other vectors for trout hepatoma. In: J.E. Halver and I.A. Mitchell, Ed. Trout Hepatoma Research Conference Papers. Bureau of Sport Fisheries and Wildlife Research Rep. No. 70. Dept. of the Interior, Washington, DC: p. 78-102.

Hilpert, D., W. Romen and H-G. Neumann. 1983. The role of partial hepatectomy and of promoters in the formation of tumors in non-target tissues of trans-4-acetylaminostilbene in rats. Carcinogenesis. 4(12): 1519-1525.

Hoffman, W.S., H. Adler, W.I. Fishbein and F.C. Bauer. 1967. Relation of pesticide concentrations in fat to pathological changes in tissues. Arch. Environ. Health. 15: 758-765.

Innes, J.R.M., B.M. Ulland, M.G. Valerio, et al. 1969. Bioassay of pesticides and industrial chemicals for tumorgenicity in mice: A preliminary note. J. Natl. Cancer Inst. 42(6): 1101-1114.

Kashyap, S.K., S.K. Nigam, A.B. Karnik, R.C. Gupta and S.K. Chatterjee. 1977. Carcinogenicity of DDT (dichlorodiphenyl trichloroethane) in pure inbred Swiss mice. Int. J. Cancer. 19: 725-729.

Kubinski, H., G.E. Gutzke and Z.O. Kubinski. 1981. DNA-cell-binding (DCB) assay for suspected carcinogens and mutagens. Mutat. Res. 89: 95-136.

Lehman, A.J. 1951. Chemicals in Foods: A Report to the Association of Food and Drug Officials on Current Developments. Part II, Pesticides. Section V. Pathology, Q. Bull. Assoc. Food Drug Office, US 15(4): 126-132.

Lehman, A.J. 1965. Summaries of pesticide toxicity. Association of Food and Drug Officials of the United States, Topeka, Kansas.

Maier-Bode, H. 1960. Zur Frage der Herkunft des DDT im Koperfett des Menschen. Med. Exp. 3: 284-286. (Ger.)

NCI (National Cancer Institute). 1978. Bioassays of DDT, TDE and p,p'-DDE for possible carcinogenicity (CAS No. 50-29-3, 72-54-8, 72-55-9). NCI Report No. 131. DHEW Publ. No. (NIH) 78-1386.

Peraino, C., R.J.M. Fry, E. Staffeldt and J. P. Christopher. 1975. Comparative enhancing effects of phenobarbital, amobarbital, diphenylhydantoin, and dichlorodiphenyltrichloroethane of 2-acetylaminofluorene-induced hepatic tumorgenesis in the rat. Cancer Res. 35: 2884-2890.

Preston, R.J., W. Au, M.A. Bender, et al. 1981. Mammalian in vivo and in vitro cytogenetic assays: A report of the US EPA's Gene-Tox Program. Mutat. Res. 87: 143-188.

Rabello, M.N., W. Becak, W.F. DeAlmeida, et al. 1975. Cytogenetic study on individuals occupationally exposed to DDT. Mutat. Res. 28: 449-454.

Ray-Chaudhuri, R., M. Currens and P.T. Iype. 1982. Enhancement of sister-chromatid exchanges by tumor promoters. Br. J. Cancer. 45: 769-777.

Robinson, J., A. Richardson, C.G. Hunter, A.N. Crabtree and H.S. Rees. 1965. Organo-chlorine insecticide content of human adipose tissue in south-eastern England. Br. J. Ind. Med. 22: 220-229.

Rossi, L., M. Ravera, G. Repetti and L. Santi. 1977. Long-term administration of DDT or phenobarbital-Na in Wistar rats. Int. J. Cancer. 19: 179-185.

Scribner, J.D. and N.K. Mottet. 1981. DDT acceleration of mammary gland tumors induced in the male Sprague-Dawley rat by 2-acetomidophenanthrene. Carcinogenesis. 2(12): 1235-1239.

Shabad, L.M., T.S. Kolesnichenko and T.V. Nikonova. 1973. Transplacental and combined long-term effect of DDT in five generations of A-strain mice. Int. J. Cancer. 11: 688-693.

Tarjan, R. and T. Kemeny. 1969. Multigeneration studies on DDT in mice. Food Cosmet. Toxicol. 7: 215-222.

Terracini, B., M.C. Testa, J.R. Cabral and N. Day. 1973. The effects of long-term feeding of DDT to BALB/c mice. Int. J. Cancer. 11: 747-764.

Thorpe, E. and A.I.T. Walker. 1973. The toxicology of dieldrin (HEOD). II. Comparative long-term oral toxicity studies in mice with dieldrin, DDT, phenobarbitone, beta-BHC and gamma-BHC. Food Cosmet. Toxicol. 11: 433-442.

Tomatis, L. and V. Turusov. 1975. Studies on the carcinogenicity of DDT. Gann Monograph Cancer Res. 17: 219-241.

Turusov, V.S., N.E. Day, L. Tomatis, E. Gati and R.T. Charles. 1973. Tumors in CF-1 mice exposed for six consecutive generations to DDT. J. Natl. Cancer Inst. 51: 983-998.

US EPA. 1985. The Carcinogenic Assessment Groups Calculation of the Carcinogenicity of Dicofol (Kelthane), DDT, DDE and DDD (TDE). Prepared by the Office of Health and Environmental Assessment, Carcinogen Assessment Group, Washington, DC for the Hazard Evaluation Division, Office of Pesticide Programs, Office of Pesticides and Toxic Substances, Washington, DC.

Walker, A.I.T., E. Thorpe and D.E. Stevenson. 1973. The toxicology of dieldrin (HEOD). I. Long-term oral toxicity studies in mice. Food Cosmet. Toxicol. 11: 415-432.

Wasserman, M., D.P. Nogueira, L. Tomatis, et al. 1976. Organochlorine compounds in neoplastic and adjacent apparently normal breast tissue. Bull. Environ. Contam. Toxicol. 15(4): 478-484.

VI.D. DRINKING WATER HA REFERENCES

None

VII. REVISION HISTORY

Date	Section	Description
09/30/87	I.A.6.	Documentation changed
08/22/88	II.	Carcinogen summary on-line
01/01/91	II.	Text edited
01/01/91	II.C.1.	Inhalation slope factor removed (global change)
05/01/91	II.A.3.	Change Lehman, 1952 to '1951'
05/01/91	VI.	Bibliography on-line
01/01/92	I.A.7.	Secondary contact changed
01/01/92	IV.	Regulatory actions updated
02/01/96	I.A.7.	Contact changed

SYNONYMS

AGRITAN
ANOFEX
ARKOTINE
AZOTOX
BENZENE, 1,1'-(2,2,2-TRICHLOROETHYLIDENE)BIS(4-CHLORO-)
alpha, alpha-BIS(p-CHLOROPHENYL)-beta, beta, beta-TRICHLORETHANE
1,1-BIS-(p-CHLOROPHENYL)-2,2,2-TRICHLOROETHANE
2,2-BIS(p-CHLOROPHENYL)-1,1,1-TRICHLOROETHANE
BOSAN SUPRA
BOVIDERMOL
CHLOROPHENOTHAN
CHLOROPHENOTHANE
CHLOROPHENOTOXUM
CITOX
CLOFENOTANE
DDT
p,p'-DDT
DEDELO
DEOVAL
DETOX
DETOXAN
DIBOVAN
DICHLORODIPHENYLTRICHLOROETHANE
4,4'-DICHLORODIPHENYLTRICHLOROETHANE
Dichlorodiphenyltrichloroethane, p,p'-
DICOPHANE
DIDIGAM
DIDIMAC
DIPHENYLTRICHLOROETHANE
DODAT
DYKOL
ENT 1,506
ESTONATE
ETHANE, 1,1,1-TRICHLORO-2,2-BIS(p-CHLOROPHENYL)-
GENITOX
GESAFID
GESAPON
GESAREX

PEB1
PENTACHLORIN
PENTECH
PPZEIDAN
R50
RCRA WASTE NUMBER U061
RUKSEAM
SANTOBANE
TECH DDT
1,1,1-TRICHLOOR-2,2-BIS(4-CHLOOR FENYL)-ETHAAN
1,1,1-TRICHLOR-2,2-BIS(4-CHLOR-PHENYL)-AETHAN
1,1,1-TRICHLORO-2,2-BIS(p-CHLOROPHENYL)ETHANE
TRICHLOROBIS(4-CHLOROPHENYL)ETHANE
1,1,1-TRICHLORO-2,2-DI(4-CHLOROPHENYL)-ETHANE
1,1,1-TRICLORO-2,2-BIS(4-CLORO-FENIL)-ETANO
ZEIDANE
ZERDANE
GESAROL
GUESAPON
GUESAROL
GYRON
HAVERO-EXTRA
HILDIT
IVORAN
IXODEX
KOPSOL
MICRO DDT 75
MUTOXIN
NA 2761
NCI-C00464
NEOCID
PARACHLOROCIDUM
PEB1
PENTACHLORIN
PENTECH
PPZEIDAN
R50
RCRA WASTE NUMBER U061
RUKSEAM
SANTOBANE
TECH DDT

GESAROL	1,1,1-TRICHLOOR-2,2-BIS(4-CHLOOR
GUESAPON	FENYL)-ETHAAN
GUESAROL	1,1,1-TRICHLOR-2,2-BIS(4-CHLOR-
GYRON	PHENYL)-AETHAN
HAVERO-EXTRA	1,1,1-TRICHLORO-2,2-BIS(p-
HILDIT	CHLOROPHENYL)ETHANE
IVORAN	TRICHLOROBIS(4-
IXODEX	CHLOROPHENYL)ETHANE
KOPSOL	1,1,1-TRICHLORO-2,2-DI(4-
MICRO DDT 75	CHLOROPHENYL)-ETHANE
MUTOXIN	1,1,1-TRICLORO-2,2-BIS(4-CLORO-
NA 2761	FENIL)-ETANO
NCI-C00464	ZEIDANE
NEOCID	ZERDANE
PARACHLOROCIDUM	

Source: *Instant EPA's IRIS*, 1997, Instant Reference Sources, Inc., 7605 Rockpoint Drive, Austin, TX 78731 (Fax: 512-345-2386; Internet URL, http://www.instantref.com/inst-ref.htm).

Production, Use, and Pesticide Labeling Information

DDT is not currently listed in EPA's Pesticide Factsheet database and there is no indication of whether it will be added or not in the future.

1,2-Dibromo-3-chloropropane

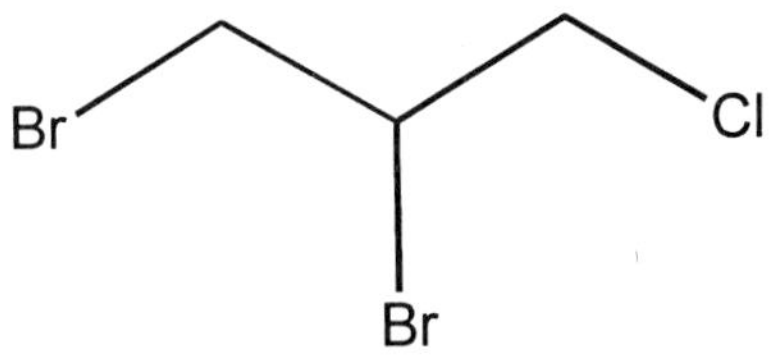

Introduction

1,2-Dibromo-3-chloropropane is used as a soil fumigant, nematicide, pesticide and intermediate in organic synthesis.

Identifying Information

CAS NUMBER: 96-12-8

NIOSH Registry Number: TX8750000

SYNONYMS: 3-Chloro-1,2-dibromopropane, Dibromochloropropane, 1-Chloro-2,3-dibromopropane

Endocrine Disruptor Information

1,2-Dibromo-3-chloropropane is a strongly suspected EED that it is listed by the Centers for Disease Control & Prevention [812], and the World Wildlife Fund [813] as a potential endocrine modifying chemical.

Chemical and Physical Properties

CHEMICAL FORMULA: $C_3H_5Br_2Cl$

MOLECULAR WEIGHT: 236.35

WLN: G1YE1E

PHYSICAL DESCRIPTION: Amber to colorless liquid

SPECIFIC GRAVITY: 2.08 @ 20/20°C [055], [395]

DENSITY: 2.05 g/mL @ 20°C [051], [062]

MELTING POINT: 6°C [107], [327]

BOILING POINT: 196°C [031], [047], [395], [430]

SOLUBILITY: Water: <0.1 mg/mL @ 18°C [700]; DMSO: >=100 mg/mL @ 20°C [700]; 95% Ethanol: >=100 mg/mL @ 20°C [700]; Acetone: >=100 mg/mL @ 20°C [700]; Methanol: Miscible [173], [395]

OTHER SOLVENTS: Dichloropropane: Miscible [031], [173]; Oils: Miscible [031], [051], [062], [205]; Isopropyl alcohol: Miscible [031], [173], [395]; Liquid hydrocarbons: Miscible [395]; Halogenated hydrocarbons: Miscible [395]; Alcohols: Soluble [430]

OTHER PHYSICAL DATA: Pungent odor [031], [173], [327], [395]; Refractive index: 1.5518 @ 25°C [395], [430] 1.553 @ 14°C [031], [047], [205]; Boiling point: 78°C @ 16 mm Hg [031], [047] 21°C @ 0.8 mm Hg [025,031]; Specific gravity: 2.09 @ 20/4°C [430]; Density: 2.093 g/mL @ 14°C [031], [047], [205]; Evaporation rate (butyl acetate = 1): Very much less than 1 [327]

Source: *Instant EPA's Air Toxics*, 1994, Instant Reference Sources, Inc., 7605 Rockpoint Drive, Austin, TX 78731 (Fax: 512-345-2386; Internet URL, http://www.instantref.com/inst-ref.htm).

Environmental Reference Materials

A source of EED environmental reference materials for 1,2-dibromo-3-chloropropane is Crescent Chemical Co., Hauppauge, NY (Fax: 516-348-0913); Catalog No. 36782.

Hazardous Properties

ACUTE/CHRONIC HAZARDS: 1,2-Dibromo-3-chloropropane is highly toxic via all routes of exposure [107]. It is a severe irritant [043], [031]. It is narcotic in high concentrations [031], [043]. When heated to decomposition it emits toxic fumes of hydrogen bromide, hydrogen chloride and carbon monoxide [043], [326].

HAP WEIGHTING FACTOR: 1 [713]

VOLATILITY:
         Vapor pressure: 0.8 mm Hg @ 21 °C [031], [055], [173], [395]

FIRE HAZARD: 1,2-Dibromo-3-chloropropane has a flash point of 76.6°C (170 °F) [043], [051], [062], [107]. It is combustible. Fires involving this material may be controlled with a dry chemical, carbon dioxide or Halon extinguisher.

REACTIVITY: 1,2-Dibromo-3-chloropropane reacts with chemically active metals such as aluminum, magnesium, tin and their alloys [173], [327]. It will attack some rubber materials and coatings [327], [395].

STABILITY: 1,2-Dibromo-3-chloropropane is stable under normal laboratory conditions. It is stable in neutral and acidic media but it is hydrolyzed in alkali [173], [395]. Solutions of it in water, DMSO, 95% ethanol or acetone should be stable for 24 hours under normal lab conditions [700].

Source: *Instant EPA's Air Toxics*, 1994, Instant Reference Sources, Inc., 7605 Rockpoint Drive, Austin, TX 78731 (Fax: 512-345-2386; Internet URL, http://www.instantref.com/inst-ref.htm).

Medical Symptoms of Exposure

Symptoms of exposure may included severe eye and skin irritation and irritation of the mucous membranes [031]. It is narcotic in high concentrations [031], [043] and has been implicated in sterility in males [043], [062], [186], [395]. It may also cause diminished renal function and degeneration and cirrhosis of the liver [186]. Other symptoms include nausea, conjunctivitis, respiratory irritation, pulmonary congestion, pulmonary edema, central nervous system depression, apathy, sluggishness and ataxia. Upon repeated exposure, erythema, inflammation and dermatitis may occur [327]. Exposure may also cause drowsiness, vomiting, liver and kidney damage, respiratory distress, testicular atrophy, spleen necrosis and sperm count depression [107].

Source: *Instant EPA's Air Toxics*, 1994, Instant Reference Sources, Inc., 7605 Rockpoint Drive, Austin, TX 78731 (Fax: 512-345-2386; Internet URL, http://www.instantref.com/inst-ref.htm).

Toxicological Information

The toxicological information below is gathered from several sources including two national databases: one from the *National Toxicology Program Chemical Repository Database* and another from *EPA's Integrated Risk Information System (IRIS)*.

Toxicology Information from the National Toxicology Program

CARCINOGENICITY:

Tumorigenic Data:

Type of Effect	Route/Animal	Amount Dosed (Notes)
TDLo	orl-rat	5475 mg/kg/73W-I
TCLo	ihl-rat	600 ppb/6H/2Y-I
TDLo	orl-mus	49 gm/kg/47W-I
TCLo	ihl-mus	600 ppb/6H/2Y-I
TDLo	skn-mus	100 gm/kg/74W-I
TD	orl-rat	9280 mg/kg/64W-I
TC	ihl-mus	3 ppm/6H/2Y-I
TC	ihl-rat	3 ppm/6H/2Y-I
TC	ihl-rat	600 ppb/6H/2Y-I
TC	ihl-rat	3 ppm/6H/84W-I
TC	ihl-mus	600 ppb/6H/2Y-I
TC	ihl-mus	3 ppm/6H/76W-I
TC	ihl-mus	600 ppb/6H/76W-I
TC	ihl-rat	600 ppb/6H/76W-I
TDLo	scu-rat	240 mg/kg/12W-I
TD	scu-rat	240 mg/kg/12W-I

Review:
> IARC Cancer Review: Animal Sufficient Evidence
> IARC Cancer Review: Human Inadequate Evidence
> IARC possible human carcinogen (Group 2B)

Standards and Regulations:
> OSHA cancer hazard

Status:
> NCI Carcinogenesis Bioassay (Gavage); Positive: Male and Female Rat, Male and Female Mouse
> NTP Carcinogenesis Bioassay (Inhalation); Positive: Male and Female Rat, Male and Female Mouse
> NTP Fourth Annual Report on Carcinogens, 1984
> NTP anticipated human carcinogen
> EPA Carcinogen Assessment Group

TERATOGENICITY:

Reproductive Effects Data:

Type of Effect	Route/Animal	Amount Dosed (Notes)
TDLo	orl-rat	500 mg/kg (6-15D preg)
TDLo	orl-rat	250 mg/kg (5D male)
TDLo	orl-rat	50 mg/kg (5D male)
TCLo	ihl-rat	10 ppm/6H (4W male)
TCLo	ihl-rat	1 ppm/6H (4W male)
TCLo	ihl-rat	10 ppm/7H (50D male)
TCLo	ihl-rbt	1 ppm/6H (70D male)
TCLo	ihl-rbt	10 ppm/6H (40D male)
TCLo	ihl-rbt	12 ppm/7H (66D male)
TCLo	ihl-gpg	12 ppm/7H (66D male)
TDLo	unr-rat	100 mg/kg (1D male)
TDLo	orl-rat	200 mg/kg (1D male)
TDLo	orl-rat	375 mg/kg (75D male)
TDLo	scu-rat	80 mg/kg (1D male)
TDLo	scu-rat	33 mg/kg (1D male)
TDLo	scu-rat	19 mg/kg (19D male)
TDLo	scu-rat	40 mg/kg (1D male)
TDLo	orl-rbt	92657 ug/kg (10W male)

MUTATION DATA:

Test	Lowest dose	Test	Lowest dose
mmo-sat	10 nL/plate	dlt-rat-orl	50 mg/kg/5D-C
mma-sat	500 ng/plate	dns-mus-ipr	100 mg/kg
dnr-esc	20900 ug/disc	sce-ham:lng	500 umol/L
mmo-bcs	1 nmol/plate	sln-dmg-ihl	17700 ppb/3H-C
dnd-rat:lvr	30 umol/L	dni-hmn:hla	10 mmol/L
cyt-rat-orl	3650 ug/kg/5D	spm-rat-orl	3650 ug/kg/5D
sln-dmg-orl	100 ppm/3D	trn-dmg-orl	200 ppm
dnd-rat-ipr	35 mg/kg	dnd-rat:tes	173 mg/L
spm-rbt-orl	375 mg/kg/10W-I	dns-mus-ipr	100 mg/kg
cyt-ham:lng	200 umol/L	dns-rat-ipr	150 mg/kg
msc-mus:lym	83700 ug/L	cyt-ham:ovr	50 mg/L
sce-ham:ovr	10 mg/L		

TOXICITY:

Typ. Dose	Mode	Specie	Amount	Units/Other
LD50	scu	rat	100	mg/kg
LD50	orl	rat	170	mg/kg
LC50	ihl	rat	103	ppm/8H
LD50	orl	mus	257	mg/kg
LD50	orl	rbt	180	mg/kg
LD50	skn	rbt	1400	mg/kg
LD50	orl	gpg	150	mg/kg
LD50	orl	ckn	60	mg/kg
LD50	ipr	mus	123	mg/kg

OTHER TOXICITY DATA:

> Skin and Eye Irritation Data:
>> skn-rbt 10 gm SEV
>> eye-rbt 1% MLD

> Review: Toxicology Review

> Standards and Regulations:
>> DOT-IMO: Poison B; Label: St. Andrew's Cross

Status:
> EPA TSCA Chemical Inventory, 1986
> EPA TSCA 8(a) Preliminary Assessment Information, Final Rule
> EPA TSCA Section 8(e) Status Report 8EHQ-0877-0003
> EPA TSCA Section 8(e) Status Report 8EHQ-0278-0056
> EPA TSCA Section 8(e) Status Report 8EHQ-0378-01138
> EPA TSCA Section 8(e) Status Report 8EHQ-0478-0123
> EPA TSCA Section 8(e) Status Report 8EHQ-0478-0128

EPA TSCA Section 8(e) Status Report 8EHQ-0678-0192 S
EPA TSCA Section 8(e) Status Report 8EHQ-0778-0219
Meets criteria for proposed OSHA Medical Records Rule
EPA Genetox Program 1988, Positive: Carcinogenicity-mouse/rat; E
coli polA without S9
EPA Genetox Program 1988, Positive: Histidine reversion-Ames test;
Sperm morphology-human
EPA TSCA Test Submission (TSCATS) Data Base, June 1988

SAX TOXICITY EVALUATION:

THR: Poison by ingestion, inhalation and subcutaneous routes. Moderately
toxic by skin contact. An eye and severe skin irritant. An experimental
carcinogen and teratogen. A suspected human carcinogen. Narcotic in high
concentrations. It has been implicated in causing human male sterility in factory
workers. Human mutagenic data. A soil fumigant.

Source: *Instant Tox-Base*, 1995, Instant Reference Sources, Inc., 7605 Rockpoint Drive,
Austin, TX 78731 (Fax: 512-345-2386; Internet URL, http://www.instantref.com/inst-
ref.htm).

Toxicology Information from EPA's Integrated Risk Information System (IRIS)

I. CHRONIC HEALTH HAZARD ASSESSMENTS FOR NONCARCINOGENIC
EFFECTS

I.A. REFERENCE DOSE FOR CHRONIC ORAL EXPOSURE (RfD)

Not available at this time.

I.B. REFERENCE CONCENTRATION FOR CHRONIC INHALATION EXPOSURE
(RfC)

The inhalation Reference Concentration (RfC) is analogous to the oral RfD and is
likewise based on the assumption that thresholds exist for certain toxic effects such as
cellular necrosis, but may not exist for other toxic effects such as carcinogenicity. The
inhalation RfC considers toxic effects for both the respiratory system (portal-of-entry)
and for effects peripheral to the respiratory system (extrarespiratory effects). It is
appropriately expressed in units of mg/cu.m. In general, the RfC is an estimate (with
uncertainty spanning perhaps an order of magnitude) of a daily inhalation exposure of the
human population (including sensitive subgroups) that is likely to be without an
appreciable risk of deleterious effects during a lifetime. Inhalation RfCs are derived
according to the Interim Methods for Development of Inhalation Reference Doses
(EPA/600/8-88/066F August 1989) developed by US EPA scientists and peer-reviewed.
For more information on the interim nature of these methods and future plans see the
INFOMORE Section of IRIS. RfCs can also be derived for the noncarcinogenic health
effects of compounds which are carcinogens. Therefore, it is essential to refer to other

sources of information concerning the carcinogenicity of this substance. If the US EPA has evaluated this substance for potential human carcinogenicity, a summary of that evaluation will be contained in Section II of this file when a review of that evaluation is completed.

I.B.1. INHALATION RfC SUMMARY

Critical Effect	Exposures*	UF	MF	RfC
Testicular effects	NOAEL: 0.94 mg/cu.m (0.1 ppm)	1000	1	2E-4
	NOAEL(ADJ): 0.17 mg/cu.m			mg/cu.m
	NOAEL(HEC): 0.17 mg/cu.m			
	LOAEL: 9.4 mg/cu.m (1 ppm)			
	LOAEL(ADJ): 1.7 mg/cu.m			
	LOAEL(HEC): 1.7 mg/cu.m			

13-Week Subchronic Rabbit Inhalation Study Rao et al., 1982

*Conversion Factors: MW = 236.3. Assuming 25C and 760 mm Hg,
NOAEL (mg/cu.m) = 0.1 ppm x 236.3/24.45 = 0.97; adjusted for compound purity of 0.973 = 0.94 mg/cu.m. NOAEL(ADJ) = NOAEL (mg/cu.m) x 6 hours/24 hours x 5 days/7 days = 0.17 mg/cu.m.

The NOAEL(HEC) was calculated for a gas: extrarespiratory effect in rabbits assuming periodicity was attained.

Since b:a lambda values are unknown for the experimental species (a) and humans (h), a default value of 1.0 is used for this ratio. NOAEL(HEC) = 0.17 mg/cu.m.

I.B.2. PRINCIPAL AND SUPPORTING STUDIES (INHALATION RfC)

Rao, K.S., J.D. Burek, F. Murray, et al. 1982. Toxicologic and reproductive effects of inhaled 1,2-dibromo-3-chloropropane in male rabbits. Fund. Appl. Toxicol. 2(5): 241-251.

Rao et al. (1982) exposed 6-month-old male New Zealand white rabbits (10/group) to 0, 0.1, 1 or 10 ppm (0, 0.94, 9.4 or 94 mg/cu.m) DBCP vapors (adjusted for 97.3% purity), 6 hours/day, 5 days/week (duration-adjusted to 0, 0.17, 1.7, and 17 mg/cu.m), for 14 weeks. The rabbits receiving the 10-ppm concentration were exposed for only 8 weeks due to high mortality (apparently from pneumonia). Body weights and hematological and clinical chemistry parameters were monitored, but no significant differences were found between the DBCP-exposed animals and the controls. Semen was collected and evaluated during the exposure and during a 32 to 38 week recovery period to assess sperm motility, viability, and count. The average sperm count of the rabbits exposed to the 10-ppm concentration was significantly less than that of the controls after 7 weeks of exposure, and remained decreased for the duration of the exposure period and through week 42 of postexposure. At 1 ppm DBCP, sperm counts were significantly reduced compared with controls from weeks 11 to 13 of exposure. At 0.1 ppm, sperm counts were sporadically lower than control values although this was statistically significant only once, during exposure week 12. The percentage of live sperm in the semen of the rabbits exposed to 10 ppm DBCP was also significantly decreased compared with control values during weeks 8 to 26. Rabbits exposed to 1 ppm DBCP, but not those exposed to 0.1 ppm, exhibited significant decreases in the percentage of live sperm during weeks 6, 12 and 13. From the 8th week of exposure onward, the rabbits exposed to 10 ppm DBCP had a marked decrease in the percentage of progressively motile sperm. No consistent statistically significant decreases in this parameter were noted in the two lower exposure groups. Abnormal spermatozoa within the seminiferous tubules of 3 to 4 rabbits from each exposure group were counted; the percentage of abnormal sperm at 14 weeks was 5% for controls, 10% for animals exposed to 0.1 ppm DBCP and 18% for animals exposed to 1 ppm DBCP.

To assess the effects of DBCP on fertility, exposed male rabbits were mated to unexposed female rabbits at weeks 14 and 41 of the study. DBCP did not affect the libido of the exposed male rabbits during week 14 based on the percentage of males (78-100%) that copulated with unexposed females. However, the five males exposed to 10 ppm DBCP were infertile since none of the females became pregnant. The mean number of implantations/litter in the 1 ppm group was significantly less than that of the control group. During week 41 (27 weeks post-exposure), all rabbits exposed to 0.1 and 1 ppm DBCP produced normal litters, and 2 of the 5 males exposed to 10 ppm regained fertility (i.e., increased sperm count) and produced normal litters. The follicle stimulating hormone (FSH) serum levels were also significantly elevated at 14 weeks in the males exposed to 1 ppm DBCP and at 46 weeks in the males exposed to 10 ppm DBCP. Increased FSH serum levels were consistent with a marked decrease in sperm count, whereas serum levels of testosterone were unchanged. The only gross lesion observed under macroscopic examination was the small size of the testes for rabbits exposed to 1 and 10 ppm DBCP. No gross lesions were observed in either the lungs or upper respiratory tract, and micropathology was not performed. Other histopathologic examination revealed changes in the reproductive system. These effects included atrophy of the testes, epididymides, and accessory sex glands including the prostate. The testes weight was significantly decreased to 50% of control values (week 14) in the group exposed to 1 ppm and to 75% of control values (week 8) in the group exposed to 10 ppm. Severe testicular atrophy was characterized by nearly complete or complete loss of spermatogenic elements in nearly all seminiferous tubules. Following the recovery period, tubular regeneration was observed in testes of some rabbits exposed to 10 ppm DBCP; 3 of 5 rabbits had regeneration such that 25% of the seminiferous tubules appeared normal. At the 1 ppm exposure, recovery was nearly complete in some rabbits although no incidences were given. Testes of rabbits exposed to 0.1 ppm DBCP appeared normal. Thus, rabbits exposed to 0.1 ppm DBCP showed no major treatment-related changes, and this level is designated as a no effect level. The NOAEL(HEC) is 0.17 mg/cu.m and the LOAEL(HEC) is 1.7 mg/cu.m. The respiratory tracts were not histologically examined in this study.

I.B.3. UNCERTAINTY AND MODIFYING FACTORS (INHALATION RfC)

UF -- An uncertainty factor of 10 is used for the protection of sensitive human subpopulations. A factor of 3 is used for interspecies extrapolation, as the concentration was dosimetrically adjusted to humans. A full factor of 10 is applied for the use of a subchronic study to reflect the marginal NOAEL in the principal study, as the minor testicular effects seen at the NOAEL were consistent with the effects seen at the higher LOAEL in this and other investigations. A study of chronic duration could result in these minor effects progressing into more delineated adverse effects. A factor of 3 is used for data base deficiency because of the lack of a multigenerational reproductive study, and inhalation development toxicity studies. The total uncertainty factor is therefore 1000.

MF -- None

I.B.4. ADDITIONAL STUDIES / COMMENTS (INHALATION RfC)

Whereas a number of occupational studies on exposure to DBCP demonstrate this compound to be a potent testicular toxicant in humans, none of these occupational studies to date have evaluated the possible respiratory tract effects of DBCP exposure. This is especially disturbing when the inhalation studies conducted by NTP (1982) established the efficacy of this compound to produce lesions and tumors in the nasal cavity of both rats and mice; Sax (1989) lists this compound as both an eye and skin irritant. Little reliable exposure data are available from any of these studies. Also, most are confounded as they have been conducted in pesticide manufacturing plants where workers are co-exposed to a number of other chemicals. Limited follow-up studies of effected worker populations indicate that paternal exposure to DBCP sufficient to produce oligospermia or azoospermia did not detectably increase the rate of congenital malformations or impair the health status of offspring conceived during or after DBCP exposure. DBCP is, however, a potential mutagen capable of inducing a dominant lethal effect in mice (Teramoto et al., 1980).

A cross-sectional study of 23 male workers at a DBCP production facility showed a 79% incidence (18 of 23) of azoospermia and oligospermia (Potashnik et al., 1979). No estimates of DBCP concentrations were given although exposure hours were presented. The exposure hours reported for a subgroup of 12 of these men diagnosed as azoospermic (sperm count = 0) ranged from 100 to 6726 hours. For another subgroup of six men diagnosed as oligospermic (designated as less than 10 million sperm/mL) the reported exposure hours ranged from 34 to 95 hours. The remaining 5 men had normal sperm counts and exposure hours ranging from 10 to 60. The azoospermic men had elevated FSH, but normal LH and testosterone levels. Testicular biopsy showed atrophy of seminiferous epithelium and tubules lined by Sertoli cells. In a 4-year follow-up study on 17 of these effected workers (plus three others not in the original study group), sperm count recovered in 4 of the 13 initially azoospermic men and 5 of the 7 oligospermic men (Potashnik, 1983). There was no improvement in the sperm count of the remaining 11 men who had prolonged exposure to DBCP. In another followup study (8-year) on 15 of the same group of effected workers, perinatal outcome as measured by birth defects, prematurity, mortality, or spontaneous abortions was not associated with paternal exposure to DBCP (Potashnik and Yanai-Inbar, 1987). In another subgroup of 11 of these effected workers, the rate of birth or health defects in their families following exposure was not different from that of a group of children from the same families conceived during the pre-exposure period (Potashnik and Phillip, 1988). In contrast, Kharrazi et al. (1980) report a statistically significant increase in the percentage of spontaneous abortions in wives of men working in Israeli banana plantations and in direct contact with DBCP. Levels of DBCP exposure were not given although the extent of exposure ranged from 1 season to 20 consecutive seasons. Sperm levels were not measured in this study, and there may have been selection bias among the participants as only 62 of 102 possible participants were interviewed for the study.

In a limited cross-sectional study, 11 of 25 men working in a DBCP-formulating plant were found to be azoospermic or oligospermic (designated as less than 1 million sperm/mL ejaculate) and had elevated serum levels of FSH and LH (Whorton et al., 1977). The average exposure of the 11 men with a very low sperm count was 8 years. DBCP levels were purported to have been measured early in 1977 with personal air-sampling devices and indicated an 8 hour average concentration of 0.4 ppm (3.9 mg/cu.m), although no further specifics were available. Testicular biopsy, performed in 10 of these 25 men by Biava et al. (1978) showed that the diminution of spermatogenesis was correlated with duration of exposure to DBCP. Men with 10-year exposures had ejaculate without sperm and seminiferous tubules devoid of germ cells. Exposure for 1 to 3 years produced marked diminution of sperm formation and spermatogenic activity limited to a few segments of the tubules. Spermatogenic activity in men exposed for less than a year was classified as normal. A followup study in a subgroup of these same workers conducted 7 years after the initial evaluation showed that 2 of 8 workers that were originally classified as azoospermic produced some sperm during the followup, although only one had normal sperm production (Eaton et al., 1986). These results suggest that damage to germinal tissue by DBCP exposure sufficient to produce sterility is permanent. Although the exposure is poorly characterized, 3.9 mg/cu.m DBCP appears to be a frank-effect-level (FEL) in humans based on cases of azoospermia.

Laboratory animal studies via other routes confirm the testicular, respiratory, and adrenal effects of DBCP. Studies with other species indicate the rabbit to be the most sensitive test species for testicular effects (Pease et al., 1991). The drinking water studies of Foote et al. (1986a,b) in rabbits were carefully and thoroughly executed to elucidate several aspects of DBCP on male reproductive function. In the 1986a study, dose-related decreases in the proportion of abnormal sperm as well as a biochemical indicator of impaired spermatogenesis (elevated FSH levels) were documented. The 1986b study demonstrated dose-related quantitative testicular histology including effects on testicular weight, alterations in seminiferous tubular diameter and a marked decrease of all germ cell types.

Sprague-Dawley rats (30/sex/group) were exposed to 0, 0.1, 1 or 10 ppm (0, 0.97, 9.7, or 97 mg/cu.m) DBCP vapor, 6 hours/day, 5 days/week (duration adjusted to 0, 0.17, 1.7 or 17 mg/cu.m), for 14 weeks, followed by a 32-week recovery period (Rao et al., 1983) for a total of 46 weeks. Body weight and clinical examinations were made throughout the study. At the 14-week sacrifice, absolute and relative testes and epididymides weights were significantly decreased compared with controls only in the group exposed to 10 ppm DBCP. At the 46-week sacrifice, only the relative testes weight in the males exposed to 10 ppm was significantly lower than controls. No significant differences were seen in organ weights of exposed female rats compared with their controls. There were no treatment-related gross lesions observed in animals after 4 weeks of exposure. At the 14-week sacrifice, histopathological changes occurred in testes (decreased size and dark color; decreased spermatogenesis in individual seminiferous tubules, lack of germinal cells in 5/5 males) and in the adrenal gland (foci of altered cells in cortex in 3/5 males and 3/5 females) of the animals exposed to 10 ppm DBCP. At the 46-week terminal sacrifice, testicular atrophy was observed in a concentration-related manner in all male groups (12/18 at 10 ppm, 5/20 at 1 ppm, 3/19 at 0.1 ppm) including controls (2/17). Adrenal cortical hyperplasia was noted in both sexes at the 10 ppm concentration (32/35) and in the females at the terminal kill at the 1 ppm concentration (7/19). Ovarian cysts were observed in females at the terminal sacrifice at the highest concentration (7/17). Another concentration-related effect observed only in the female animals at terminal (46 week) sacrifice was cortical hematocyst formation in 1 of 20 animals exposed to 0.1 ppm, in 4 of 19 animals exposed to 1 ppm, and in 16 of 17 animals exposed to 10 ppm. Mineralized deposits occurred in the cerebrum of the brain of both sexes (15/18 males and 6/17 females) in the high exposure animals at the terminal sacrifice. Although significance was not reported, DBCP markedly effected the animals exposed to 10 ppm and slightly effected the testes and the adrenals of the animals exposed to 1 ppm. Therefore, a mild LOAEL of 1 ppm DBCP (HEC = 1.73 mg/cu.m), based on testicular and adrenal effects, was determined from this subchronic study. The respiratory tract was not examined histopathologically in this study.

To assess fertility in male rats in this study (Rao et al., 1983), 20 males per exposure group were mated with unexposed females during weeks 2, 4, 6, 10, 12, 14, 16, 20, 24, 28, and 42. The percentage of males that impregnated at least one female was at least 85% for all the groups; no difference was seen between exposed and control males. In the group exposed to 10 ppm DBCP, a statistically significant increase ($p<0.05$) in post-implantation loss was observed during the fourth week of exposure and remained high through the remainder of the exposure period; this finding appears to be a treatment-related dominant lethal effect. By the tenth week of recovery, the average number of resorptions in the 10-ppm group was similar to that of the controls. No anomalies were observed in fetuses sired by exposed males during week 41 of the study. To assess fertility in exposed female rats, 20 exposed females per group were mated with unexposed males for a 5-day period during weeks 14, 18 and 20. Fertility of the exposed female rats was not significantly different from that of the controls except for a higher incidence of 10-ppm dams having litters of 4 or fewer pups. There were no significant differences between control and exposure groups and no major gross alterations in the pups.

The inhalation carcinogenesis bioassay conducted by NTP (1982) comprises four different studies involving rats and mice. These studies establish DBCP as a carcinogen, with high incidences of tumors appearing in the nasal cavity and on the tongues of rats and in the nasal cavity and lungs of mice. Focal hyperplasia in the nasal cavity was apparent in both species at the lower of the two exposure concentrations but may have been obscured by tumors at the higher concentration. Hyperplasia lower in the respiratory tract was also observed but at a less frequent incidence. Progression of these lesions, either into the lower respiratory tract or onto cancer, is implied but not proven by these observations. Consequently, HECs for respiratory effects in these studies are given for both extrathoracic (ET) and total pulmonary (TOT) surface areas. Though several HECs for effects in the ET area are as low as the proposed NOAEL(HEC), the RfC is based on testicular effects because of the human correlate and the confounding of nasal cavity lesions with cancerous lesions in the same anatomical area.

In a chronic carcinogenesis bioassay, F344 rats (50/sex/group) were exposed to 0, 0.6 or 3 ppm (0, 5.8 or 29 mg/cu.m) DBCP vapors, 6 hours/day, 5 days/week (duration adjusted to 0, 1.04 or 5.2 mg/cu.m). Results from this study are reported both in NTP (1982) and in Reznik et al. (1980b). The low-exposed rats were exposed for 103 weeks, and the high-exposed rats were exposed for 84 weeks due to excessive mortality. Body weights and clinical signs were recorded. Gross and microscopic examinations were performed on all major tissues including the testes and the nasal cavity where step cuts were made from the nostril to the cranium (the number of sections was not specified). A concentration-related respiratory effect observed in the male rats was focal hyperplasia of the nasal cavity in low-exposed and high-exposed (31/50 and 1/49) animals, respectively, that was not accompanied by an increased incidence in hyperplasia in either the bronchioles or the alveolar epithelium. In the female rats, the incidences of nasal cavity abscesses in the low- and high-exposed animals was 5/50 and 12/50, respectively, and 1/50 in the controls. Focal hyperplasia of the nasal cavity was noted in 24/50 of the low-exposure animals and in 23/50 of the high-exposure animals.   This nasal cavity hyperplasia was not accompanied by increased incidences of hyperplasia in either the bronchioles or the alveolar epithelium. It should be noted that the decrease in focal hyperplasia of the nasal cavity of both male and female rats at the highest exposure level was concomitant with an increase in neoplastic lesions at this exposure. The nasal cavity of the high-exposed females also showed chronic inflammation (6/50), hyperkeratosis (11/50), and squamous metaplasia (15/50). Other systemic effects in the high-exposed female rats include hyperkeratosis of the esophagus (22/49), stomach hyperkeratosis (15/48) and acanthosis (12/48), toxic nephropathy (46/49), and necrosis of cerebrum (8/49). Pigmentation of the spleen occurred in 10/50, 28/50, and 34/48 female rats in the 0-, 0.6- and 3-ppm DBCP groups, respectively. Degeneration of adrenal cortex occurred in 19/50 and 13/48 of low- and high-exposed females compared with 4/50 in the controls. The incidence of pathology of the testes was inversely related to the concentration, hyperplasia of interstitial cells occurring in 41/50 controls and in 18/50 of low- and 6/48 high-exposed animals. Likewise, testicular degeneration occurred in 8/50 low-exposed animals, but only in 4/48 high-exposed animals. Other systemic effects in the high-exposed males were splenic pigmentation (13/49) and atrophy (8/49), hyperkeratosis of esophagus (18/49), and toxic nephropathy (49/49). A LOAEL of 0.6 ppm was identified based on the splenic, and adrenal gland effects; LOAEL(HEC) = 1.04 mg/cu.m. The concentration of 0.6 ppm is also designated as a LOAEL for respiratory effects; for the total pulmonary area, LOAEL(HEC) = 2.3 mg/cu.m and for the extrathoracic area, LOAEL:(HEC) = 0.19 mg/cu.m.

B6C3F1 mice (50/sex/group) were exposed to 0, 0.6, and 3 ppm (0, 5.8 and 29 mg/cu.m) DBCP vapors, 6 hours/day, 5 days/week (duration adjusted to 0, 1.04 and 5.2 mg/cu.m). Results from this study are reported both in NTP (1982) and in Reznik et al. (1980a). The low-exposed female mice were exposed for 103 weeks and the low-exposed male mice and high-exposed animals were exposed for only 76 weeks due to excessive mortality. Body weights and clinical signs were recorded. Gross and microscopic examinations were performed on all major tissues including the testes and the nasal cavity where step cuts were made from the nostril to the cranium (number of sections not specified). Mean body weight gain was depressed by 17-28% in the high-exposed males after week 60 and by 25% in high-exposed females after week 76. The mortality in high-exposed females was significantly higher (p<0.001) than that of the other groups; 43 of 50 died during weeks 51 through 74, while mortality in the high-exposed males was comparable with other groups. Concentration-dependent respiratory effects observed in the male mice were focal hyperplasia of the nasal cavity in low-exposed and high-exposed (2/42 and 12/48) animals, respectively, as well as focal hyperplasia in the bronchioles (7/40 and 29/45) and hyperplasia of the alveolar epithelium (2/40 and 7/45). None of these effects were noted in any control animal. The high-exposed males also had a high incidence of supportive inflammation in the nasal cavity (21/48) and focal hyperplasia of bronchi (14/45). Splenic atrophy (16/45), toxic nephropathy (9/46), and leukocytosis of lungs (4/45) were also evident in high-exposed males. There was also a high incidence of hyperkeratosis (10/41 and 17/44) and acanthosis (6/41 and 11/44) in the stomach, kidney inflammation (9/42 and 7/46), and necrosis of prepuce (7/42 and 3/48) in the 0.6 and 3 ppm groups compared with very low or no incidences in the controls. No concentration-dependent effects were noted in the testes, seminal vesicles, or epididymides. In the female mice, the incidences in the low- and high-exposed animals of suppurative inflammation was 5/50 and 13/50; of focal hyperplasia of the nasal cavity, 17/50 and 3/50; of hyperplasia of bronchioles, 5/49 and 11/47; of hyperplasia of alveolar epithelium, 5/49 and 11/47. It should be noted that the decrease in focal hyperplasia of the nasal cavity of both male and female mice at the highest exposure level was concomitant with an increase in neoplastic lesions at this exposure. A greater incidence of splenic atrophy (19/43) and endometrium cyst (11/45) occurred in the high-exposed animals. A high incidence of hyperkeratosis (20/48 and 24/46) and acanthosis (12/48 and 18/46) of the stomach was evident in both exposure groups. A LOAEL of 0.6 ppm DBCP, due to gastrointestinal and kidney effects (HEC = 1.04 mg/cu.m) was determined for this chronic study. The concentration of 0.6 ppm is also designated a LOAEL for respiratory effects; for the total pulmonary area, LOAEL(HEC) = 6.7 mg/cu.m and for the extrathoracic area, LOAEL(HEC) = 0.19 mg/cu.m.

In the rat subchronic study (NTP, 1982; also reported in Reznik et al., 1980c), Fischer 344 rats (5/sex/group) inhaled 0 (filtered room air), 1, 5 or 25 ppm (0, 9.66, 48.3 or 241.6 mg/cu.m) DBCP (96% purity), 6 hours/day, 5 days/week (duration adjusted to 1.7, 8.6 or 43 mg/cu.m), for 13 weeks. Both sexes of rats in the highest group exhibited blood stains around the nasal orifice throughout the 13-week period. Two females died during weeks 10 and 11 of exposure while two females and one male were sacrificed during weeks 10, 11, and 12 due to moribund conditions. There was a 60% decrease in body weight compared with controls, and severe hair loss in the rats exposed to 25 ppm DBCP. The high exposure animals also had inflammation and severe necrosis of the respiratory and olfactory epithelium in the dorsal part of the nasal cavity. The incidence of these lesions is reported to be concentration-related and is reported in the narrative section of the report only. Necrosis of the tracheal epithelium was found in 7 of the 10 rats exposed to 25 ppm DBCP. In the lung, squamous metaplasia of the bronchial epithelium was present along with hyperplasia and partial regeneration of the bronchial and bronchiolar epithelium (data not presented). Atrophy with hypospermatogenesis was revealed in the testes of five male rats exposed to 25 ppm DBCP. With 1 and 5 ppm exposure levels, focal hepatic necrosis, hepatocytic hydropic changes, cytomegaly, and toxic tubular nephrosis were reported. A LOAEL of 1 ppm DBCP (HEC = 1.73 mg/cu.m) for liver and kidney alterations was determined. A LOAEL of 1 ppm is assumed for respiratory effects; for the total pulmonary area, LOAEL(HEC) = 2.23 mg/cu.m and for the extrathoracic area, LOAEL(HEC) = 0.19 mg/cu.m.

In the subchronic study in mice (NTP,1982; also reported in Reznik et al., 1980c) B6C3F1 mice (10/sex/group) were exposed to 0, 1, 5 or 25 ppm (9.66, 48.3 or 241.6 mg/cu.m) DBCP, 6 hours/day, 5 days/week (duration adjusted to 1.7, 8.6 or 43 mg/cu.m), for 13 weeks. Four of the males exposed to 25 ppm DBCP died before the end of the exposure period. Weight loss was 69% in males and 19% in females of the high-exposed group. Hydropic changes of the hepatocytes and nephrosis in the male mice exposed to 25 ppm and necrosis of the bronchiolar epithelium in the animals exposed to 25 ppm were reported. Regeneration and hyperplasia of the bronchiolar epithelium and megalocytic epithelial cells were found in all mice exposed to 5 ppm. Lesions in the epithelium of the nasal cavity (i.e., inflammation, necrosis, proliferative lesions) were also observed in mice exposed to 25 ppm DBCP. No data are presented for these effects; they are described in the narrative section of the report. Occurrence of lesions in other organs (e.g., testes, kidneys, or liver) are not discussed. A systemic NOAEL(HEC) for weight loss would be HEC = 1.73 mg/cu.m. A NOAEL of 1 ppm was assumed for respiratory effects; for the total pulmonary area, NOAEL(HEC) = 7.5 mg/cu.m and for the extrathoracic area, LOAEL(HEC) = 0.21 mg/cu.m.

Torkelson et al. (1961) conducted a series of inhalation exposures with DBCP in several species. In single exposures to rats at concentrations of 60 ppm (580 mg/cu.m) and higher, ocular and respiratory irritation was apparent. In preliminary range-finding studies, the authors reported mortality in rats subjected to a total of 15 7-hour exposures to 386 mg/cu.m (13/15 rats died), 48 such exposures to 193 mg/cu.m (10/15 died), and 50 such exposures at 97 mg/cu.m (2/15 died). Animals in the latter exposure group were described as having dulling of the corneas; weight loss; and hair loss, as well as gross lesions in the lungs, intestinal mucosa, kidneys, and testes. In a more extensive experiment, rats (20/sex), guinea pigs (10/sex), rabbits (3/sex), and monkeys (2 females) inhaled 12 ppm DBCP (116 mg/cu.m), 7 hours/day, 5 days/week, for a period of 70 to 92 days. Mortality was 40 to 50% in rats, and was attributed to lung infections. Much of the text is concerned with alterations in the genitalia, with severe atrophy and degeneration of the testes described in all species. Effects in rats included degenerative changes in the seminiferous tubules, increased Sertoli cells, reduced sperm count, and abnormal sperm. The respiratory tract was apparently not examined.

Although no inhalation developmental studies were located for this compound, Ruddick and Newsome (1979) performed a developmental study in which Wistar rats were gavaged with DBCP in corn oil. Pregnant rats (15/group) were randomized into four groups and gavaged doses of either 0 (vehicle), 12.5, 25, or 50 mg/kg of 97.5% DBCP on days 6 through 15 of gestation. Necropsies were carried out on day 22. DBCP was not teratogenic. No skeletal or visceral anomalies of significance were observed above those noted for control fetuses (data not shown). Mean fetal weights were significantly decreased in the highest dose group. Maternal weight gain was significantly decreased in the two highest dosed groups. The dose of 12.5 mg/kg was considered a NOAEL for maternal effects and 25 mg/kg was considered a NOAEL for fetal effects.

I.B.5. CONFIDENCE IN THE INHALATION RfC

Study -- Medium
Data Base -- Medium
RfC -- Medium

The subchronic inhalation study in rabbits of Rao et al (1982) is given a medium confidence rating due to the lack of reporting respiratory effects. The database is given medium confidence. Although chronic studies in 2 different species exist, the available reproductive studies were limited and there is uncertainty about occurrence of respiratory tract effects relative totesticular effects. A medium confidence in the RfC follows.

I.B.6. EPA DOCUMENTATION AND REVIEW OF THE INHALATION RfC

Source Document -- This assessment is not presented in any existing US EPA document.

Other EPA Documentation -- US EPA, 1988

Agency Work Group Review -- 08/15/91

Verification Date -- 08/15/91

I.B.7. EPA CONTACTS (INHALATION RfC)

Gary L. Foureman / NCEA -- (919)541-1183

Annie M. Jarabek / NCEA -- (919)541-4847

II. CARCINOGENICITY ASSESSMENT FOR LIFETIME EXPOSURE

This substance/agent has been evaluated by the US EPA for evidence of human carcinogenic potential. This does not imply that this agent is necessarily a carcinogen. The evaluation for this chemical is under review by an inter-office Agency work group. A risk assessment summary will be included on IRIS when the review has been completed.

III. HEALTH HAZARD ASSESSMENTS FOR VARIED EXPOSURE DURATIONS

[Ed. Note: EPA does not have information yet in this section of IRIS].

IV. US EPA REGULATORY ACTIONS

[Ed. Note: EPA does not have information yet in this section of IRIS].

VI. BIBLIOGRAPHY

VI.A. ORAL RfD REFERENCES

None

VI.B. INHALATION RfD REFERENCES

Biava, C., E. Smuckler and D. Whorton. 1978. The testicular morphology of individuals exposed to dibromochloropropane. Exp. Mol. Pathol. 29(3):448-458.

Eaton, M., M. Schenker, M. Whorton, S. Samuels, C. Perkins and J. Overstreet. 1986. Seven-year follow-up of workers exposed to 1,2-dibromo-3-chloropropane. J. Occup. Med. 28(11): 1145-1150.

Foote, R.H., E.C. Schermerhorn and M.E. Simkin. 1986a. Measurement of semenquality, fertility, and reproductive hormones to assess dibromochloropropane effects in live rabbits. Fund. Appl. Toxicol. 6: 628-637.

Foote, R.H., W.E. Berndtson and T.R. Rounsaville. 1986b. Use of quantitative testicular histology to assess the effect of dibromochloropropane on reproduction in rabbits. Fund. Appl. Toxicol. 6: 638-647.

Kharrazi, M., G. Potashnik and J.R. Goldsmith. 1980. Reproductive effects of dibromochloropropane. Isr. J. Med. Sci. 16(5): 403-406.

NTP (National Toxicology Program). 1982. Carcinogenesis bioassay of 1,2-dibromo-3-chloropropane (CAS No. 96-12-8) in F344 rats and B6C3F1 mice (inhalation study). Tech. Rep. Ser. No. 206. Pub. No. 82-1762. 188 p.

Pease, W., J. Vandenburg and K. Hooper. 1991. Comparing alternative approaches to establishing regulatory levels for reproductive toxicants: DBCP as a case study. Environ. Health Perspect. 91: 141-155.

Potashnik, G. 1983. A four-year assessment of workers with dibromochloropropane-induced testicular dysfunction. Andrologia. 15(2):164-170.

Potashnik, G. and M. Phillip. 1988. Lack of birth defects among offspring conceived during or after paternal exposure to dibromochloropropane (DBCP). Andrologica. 20(1): 90-94.

Potashnik, G. and I. Yanai-Inbar. 1987. Dibromochloropropane (DBCP): An 8-year reevaluation of testicular function and reproductive performance. Fertil. Steril. 47(2): 317-323.

Potashnik, G., I. Yanai-Inbar, M. Sacks and R. Israeli. 1979. Effect of dibromochloropropane on human testicular function. Isr. J. Med. Sci. 15(5): 438-442.

Rao, K.S., J.D. Burek, F. Murray, et al. 1982. Toxicologic and reproductive effects of inhaled 1,2-dibromo-3-chloropropane in male rabbits. Fund. Appl. Toxicol. 2(5): 241-251.

Rao, K.S., J. Burek, F. Murray, et al. 1983. Toxicologic and reproductive effects of inhaled 1,2-dibromo-3-chloropropane in rats. Fund. Appl. Toxicol. 3(2): 104-110.

Reznik, G., S. Stinson and J. Ward. 1980a. Lung tumors induced by chronic inhalation of 1,2-dibromo-3-chloropropane in B6C3F1 mice. Cancer Lett. 10(4): 339-342.

Reznik, G., H. Reznik-Schuller, J.M. Ward, and S.F. Stinson. 1980b. Morphology of nasal-cavity tumours in rats after chronic inhalation of 1,2-dibromo-3-chloropropane. Br. J. Cancer. 42(5): 772-781.

Reznik, G., S. Stinson and J. Ward. 1980c. Respiratory pathology in rats and mice after inhalation of 1,2-dibromo-3-chloropropane or 1,2-dibromoethane for 13 weeks. Arch. Toxicol. 46(3-4): 233-240.

Ruddick, J.A. and W.H. Newsome. 1979. A teratogenicity and tissue distribution study on dibromochloropropane in the rat. Bull. Environ. Contam. Toxicol. 21: 483-487.

Sax, N.I. 1987. Dangerous Properties of Industrial Materials. Van Nostrand Reinhold, Inc., New York.

Teramoto, S., R. Saito, H. Aoyama and Y. Shirasu. 1890. Dominant lethalmutation induced in male rats by 1,2-dibromo-3-chloropropane (DBCP). Mutat. Res. 77: 71-78.

Torkelson, T.R., S.E. Sadek, V.K. Rowe, et al. 1961. Toxicologic investigations of 1,2-dibromo-3-chloropropane. Toxicol. Appl. Pharmacol. 3: 545-559.

US EPA. 1988. Drinking Water Criteria Document for 1,2-dibromochloropropane(DBCP). Prepared by the Office of Health and Environmental Assessment,Environmental Criteria and Assessment Office, Cincinnati, OH for the Office of Drinking Water, Washington, DC. EPA 600/X-84/209-2.

Whorton, D., R. Krauss, S. Marshall and T. Milby. 1977. Infertility in malepesticide workers. Lancet. 2(8051): 1259-1261.

VI.C. CARCINOGENICITY ASSESSMENT REFERENCES

[Ed. Note: EPA does not have information yet in this section of IRIS].

VI.D. DRINKING WATER HA REFERENCES

[Ed. Note: EPA does not have information yet in this section of IRIS].

VII. REVISION HISTORY

```
--------  --------  ----------------------------------------------------------
Date      Section   Description
--------  --------  ----------------------------------------------------------
08/01/91  II.       Carcinogenicity assessment now under review
09/01/91  I.B.      Inhalation RfC now under review
10/01/91  I.B.      Inhalation RfC now on-line
10/01/91  VI.       Bibliography on-line
```

SYNONYMS

Propane, 1,2-dibromo-3-chloro-
Dibromochloropropane
1,2-dibromo-3-chloropropane
AI3-18445
BBC 12
Caswell No. 287
CCRIS 215
DBCP
Dibromchlorpropan [German]
Dibromochloropropane
EPA Pesticide Chemical Code 011301
Fumagon
Fumazone
Fumazone 86E
HSDB 1629
NCI-C00500
Nemabrom
Nemafume
Nemagon
NEMAGON SOIL FUMIGANT

Nemagon 20
Nemagon 20G
Nemagon 206
Nemagon 90
Nemanax
Nemanex
Nemapaz
Nemaset
Nemazon
OS 1897
OXY DBCP
PROPANE, 1-CHLORO-2,3-
DIBROMO-
RCRA WASTE NUMBER U066
SD 1897
1-CHLORO-2,3-DIBROMOPROPANE
1,2-DIBROM-3-CHLOR-PROPAN
[German]
1,2-DIBROMO-3-CHLOROPROPANE
1,2-DIBROMO-3-CLORO-PROPANO
[Italian]
1,2-DIBROOM-3-CHLOORPROPAAN
[Dutch]
3-CHLORO-1,2-DIBROMOPROPANE

Source: *Instant EPA's IRIS*, 1997, Instant Reference Sources, Inc., 7605 Rockpoint Drive, Austin, TX 78731 (Fax: 512-345-2386; Internet URL, http://www.instantref.com/inst-ref.htm).

Production, Use, and Pesticide Labeling Information

1,2-Dibromo-3-chloropropane is not currently listed in EPA's Pesticide Factsheet database and there is no indication of whether it will be added or not in the future.

2,4-Dichlorophenol

Introduction

2,4-Dichlorophenol is used in organic synthesis, and for the manufacture of 2,4-D. It also is used as a wood preservative, antiseptic and seed disinfectant. 2,4-Dichlorophenol may be released to the environment in effluents from its manufacture and use as a chemical intermediate and from chlorination processes involving water treatment and wood pulp bleaching. Releases can also occur from various incineration processes or from metabolism of various pesticides in soil. If released to the atmosphere, degradation can occur by reaction with photochemically formed hydroxyl radicals (estimated average half-life of 5.3 days). Physical removal from air may occur via rainfall. With a pKa of 7.8, 2,4-dichlorophenol can exist in both the non-dissociated and ionized forms in environmental soil and water depending upon the pH of the media. If released to soil, moderate to slow leaching is possible based on observed Koc values of 200-5000; the ionized form appears more susceptible to leaching than the non-dissociated form. Various biodegradation studies have demonstrated that 2,4-dichlorophenol is biodegradable under aerobic and anaerobic conditions in both soil and water. If released to water, adsorption to sediments may be important under various conditions determined, in part, by pH. Photodegradation in natural water can occur by direct photolysis or by reaction with sunlight-formed oxidants (singlet oxygen and peroxy radicals). The general population can be exposed to 2,4-dichlorophenol through consumption of contaminated tap water or by inhalation of contaminated air. [828]

Identifying Information

CAS NUMBER: 120-83-2

SYNONYMS: DCP, 2,4-DCP, o,p-dichlorophenol

Endocrine Disruptor Information

2,4-Dichlorophenol is a suspected EED that is listed by the World Wildlife Foundation, Canada [813].

Chemical and Physical Properties

CHEMICAL FORMULA: $C_6H_4Cl_2O$

MOLECULAR WEIGHT: 163.00

WLN: QR BG DG

PHYSICAL DESCRIPTION: White solid

SPECIFIC GRAVITY: 1.4 @ 15°C [051]

DENSITY: Not available

MELTING POINT: 45°C [017], [025], [042], [062]

BOILING POINT: 210°C [017], [025], [042], [062]

SOLUBILITY: Water: <0.1 mg/mL @ 18°C [700]; DMSO: >=100 mg/mL @ 18°C [700].; 95% Ethanol: >=100 mg/mL @ 18°C [700]; Acetone: >=100 mg/mL @ 18°C [700]

OTHER SOLVENTS: Carbon tetrachloride: Soluble [051], [062]; Chloroform: Very soluble [017], [047], [205]; Ether: Very soluble [017], [047], [205]; Benzene: Very soluble [017], [047,[205]; Alkaline solutions: Highly soluble [051]

OTHER PHYSICAL DATA: pKa = 7.69 @ 25°C (H$_2$O), Odor threshold: 0.21 ppm, Phenolic odor, Boiling point: 145-147°C @ 110 mm Hg [017], [047]

Source: *The National Toxicology Program's Chemical Database, Volume 2*, 1992, CRC Press, Inc./Lewis Publishers, 2000 Corporate Blvd., Boca Raton, FL 33431 (Fax: 407-998-9114).

Environmental Reference Materials

A source of EED environmental reference materials for 2,4-Dichlorophenol is Radian International LLC, Austin, TX (Fax 512-454-0268; Internet URL, http://www.radian.com/standards); Catalog No. ERD-077.

Hazardous Properties

ACUTE/CHRONIC HAZARDS: 2,4-Dichlorophenol is a severe irritant [062], [269]. When heated to decomposition or upon contact with acids or acid fumes it emits toxic fumes of chlorides, carbon monoxide and carbon dioxide [042], [269]. There is also limited evidence that 2,4-dichlorophenol causes cancer in humans [015].

VOLATILITY: Vapor pressure: 1 mm Hg @ 53°C [038], [042], [051]; Vapor density: 5.62 [042]

FIRE HAZARD: The flash point for 2,4-dichlorophenol is 113°C (237°F) [042], [062], [205], [269]. It is combustible. Fires involving this compound should be controlled using a dry chemical, carbon dioxide or Halon extinguisher.

REACTIVITY: This material can react vigorously with oxidizing agents [042], [051], [371]. It can also react with acids or acid fumes [042]. It is incompatible with acid chlorides and acid anhydrides [269].

STABILITY: 2,4-Dichlorophenol is stable under normal laboratory conditions. Solutions of 2,4-Dichlorophenol in water, DMSO, 95% ethanol or acetone should be stable for 24 hours under normal lab conditions [700].

Source: *The National Toxicology Program's Chemical Database, Volume 7*, 1992, CRC Press, Inc./Lewis Publishers, 2000 Corporate Blvd., Boca Raton, FL 33431 (Fax: 407-998-9114).

Medical Symptoms of Exposure

Symptoms of exposure may include severe irritation and burns of the skin, eyes, mucous membranes and the upper respiratory tract [269]. Other symptoms include tremors, convulsions, shortness of breath and inhibition of the respiratory system [051], [371].

Symptoms of exposure to this class of compounds include vomiting, collapse, coma, painless blanching or erythema, possible corrosion, profuse sweating, intense thirst, nausea, diarrhea, cyanosis from methemoglobinemia, hyperactivity, stupor, blood pressure fall, hyperpnea, abdominal pain, hemolysis, pulmonary edema followed by pneumonia, and occasional skin sensitivity reactions. If death from respiratory failure is not immediate, jaundice and oliguria or anuria may occur [301].

Source: *The National Toxicology Program's Chemical Database, Volume 4*, 1992, CRC Press, Inc./Lewis Publishers, 2000 Corporate Blvd., Boca Raton, FL 33431 (Fax: 407-998-9114).

Toxicological Information

The toxicological information below is gathered from several sources including two national databases: one from the *National Toxicology Program Chemical Repository Database* and another from *EPA's Integrated Risk Information System (IRIS)*.

Toxicology Information from the National Toxicology Program

CARCINOGENICITY:

Tumorigenic Data:

Type of Effect	Route/Animal	Amount Dosed (Notes)
TDLo:	skn-mus	16 gm/kg/39W-I

Status:
NTP Carcinogenesis Studies (Feed); No Evidence: Male and Female Rat, Male and Female Mouse

TERATOGENICITY:

Reproductive Effects Data:

Type of Effect	Route/Animal	Amount Dosed (Notes)
TDLo:	orl-rat	20 mg/kg (1-20D preg)
TDLo:	orl-rat	7500 mg/kg (6-15D preg)
TDLo:	scu-mus	666 mg/kg (6-14D preg)

MUTATION DATA:

Not available

TOXICITY:

Typ. Dose	Mode	Specie	Amount	Units	Other
LD50	orl	rat	580	mg/kg	
LD50	ipr	rat	430	mg/kg	
LD50	scu	rat	1730	mg/kg	
LD50	orl	mus	1276	mg/kg	
LD50	ipr	mus	153	mg/kg	
LD50	orl	mam	464	mg/kg	
LD50	skn	mam	790	mg/kg	

OTHER TOXICITY DATA:

Review:
Toxicology Review

Status:
EPA Genetox Program 1986, Inconclusive: Histidine reversion-Ames test
EPA TSCA Chemical Inventory, 1986
EPA TSCA Test Submission (TSCATS) Data Base, December 1986
Meets criteria for proposed OSHA Medical Records Rule

SAX TOXICITY EVALUATION:

THR = An experimental carcinogen. MODERATE via oral, intraperitoneal and subcutaneous routes.

Source: *Instant Tox-Base*, 1995, Instant Reference Sources, Inc., 7605 Rockpoint Drive, Austin, TX 78731 (Fax: 512-345-2386; Internet URL, http://www.instantref.com/inst-ref.htm).

Toxicology Information from EPA's Integrated Risk Information System (IRIS)

I. CHRONIC HEALTH HAZARD ASSESSMENTS FOR NONCARCINOGENIC EFFECTS

I.A. REFERENCE DOSE FOR CHRONIC ORAL EXPOSURE (RfD)

The Reference Dose (RfD) is based on the assumption that thresholds exist for certain toxic effects such as cellular necrosis, but may not exist for other toxic effects such as carcinogenicity. In general, the RfD is an estimate (with uncertainty spanning perhaps an order of magnitude) of a daily exposure to the human population (including sensitive subgroups) that is likely to be without an appreciable risk of deleterious effects during a lifetime. Please refer to Background Document 1 in Service Code 5 for an elaboration of these concepts. RfDs can also be derived for the noncarcinogenic health effects of compounds which are also carcinogens. Therefore, it is essential to refer to other sources of information concerning the carcinogenicity of this substance. If the US EPA has evaluated this substance for potential human carcinogenicity, a summary of that evaluation will be contained in Section II of this file when a review of that evaluation is completed.

I.A.1. ORAL RfD SUMMARY

Critical Effect	Experimental Doses*	UF	MF °	RfD
Decreased delayed hypersensitivity response	NOEL: 3 ppm (converted to (0.3 mg/kg/day)	100	1	3E-3 mg/kg/day
Rat, Subchronic to Chronic	LOAEL: 30 ppm (converted to 3.0 mg/kg/day)			

*Conversion Factors: Doses were estimated by the authors.

Exon and Koller, 1985

I.A.2. PRINCIPAL AND SUPPORTING STUDIES (ORAL RfD)

Exon, J.H. and L.D. Koller. 1985. Toxicity of 2-chlorophenol, 2,4-dichlorophenol and 2,4,6-trichlorophenol. In: Water Chlorination: Chemistry, Environmental Impact and Health Effects, Jolley et al., Ed. Vol. 5.

Female rats were exposed to 3, 30, or 300 ppm 2,4-dichlorophenol in drinking water from weaning age through breeding at 90 days, parturition, and weaning of pups. Ten randomly selected pups/group were weaned at 3 weeks and administered 2,4-dichlorophenol for an additional 15 weeks. The authors estimated the exposure to be approximately 0.3, 3.0, and 30.0 mg/kg bw/day for the low, medium, and high dose groups. Increases in serum antibody levels to keyhole limpet hemocyanin, as measured by an enzyme-linked immunosorbent assay (ELISA) were found to be treatment-related. The increase was statistically significant in the high-dose group, as were spleen and liver weights. Delayed-type hypersensitivity responses to bovine serum albumin in Freund's complete adjuvant were significantly decreased in those animals administered 3.0 mg/kg bw/day. The NOEL for 2,4-dichlorophenol was, therefore, determined to be 3 ppm or 0.3 mg/kg bw/day. This is substantially lower than the NOEL of 100 mg/kg bw/day reported by Kobayashi et al. (1972) for nonspecific liver changes in mice fed dichlorophenol in the diet for 180 days.

I.A.3. UNCERTAINTY AND MODIFYING FACTORS (ORAL RfD)

UF -- A factor of 10 each was employed for extrapolation from animal data to humans and for protection of sensitive human subpopulations. Since the test animals were exposed both in utero and through milk before the 15-week administration in drinking water, an additional factor for use of a subchronic study was not considered necessary.

MF -- None

I.A.4. ADDITIONAL COMMENTS (ORAL RfD)

Exon and Koller (1985) reported that exposure of dams to 300 ppm dichlorophenol resulted in a significant decrease in litter sizes.

I.A.5. CONFIDENCE IN THE ORAL RfD

Study -- Low Data Base -- Low RfD -- Low

The study (Exon and Koller, 1985) used an adequate number of animals and measured very sensitive endpoints (immunological functions) in an appropriate manner. As these endpoints are not commonly used in derivation of human health risk evaluations, confidence in the study is rated low. Additional published studies did not look for the critical effects and did not support the magnitude of the NOEL/LOAEL. Therefore, confidence in the data base is rated low. Low confidence in the RfD follows.

I.A.6. EPA DOCUMENTATION AND REVIEW OF THE ORAL RfD

Source Document -- US EPA, 1985

The Drinking Water Criteria Document is currently undergoing review.

Other EPA Documentation -- None

Agency Work Group Review -- 11/06/85, 01/22/86

Verification Date -- 01/22/86

I.A.7. EPA CONTACTS (ORAL RfD)

Julie Du / OST -- (202)260-7583

I.B. REFERENCE CONCENTRATION FOR CHRONIC INHALATION EXPOSURE (RfC)

[Ed. Note: EPA does not have information yet in this section of IRIS].

II. CARCINOGENICITY ASSESSMENT FOR LIFETIME EXPOSURE

This substance/agent has not been evaluated by the US EPA for evidence of human carcinogenic potential.

III. HEALTH HAZARD ASSESSMENTS FOR VARIED EXPOSURE DURATIONS

[Ed. Note: EPA does not have information yet in this section of IRIS].

IV. US EPA REGULATORY ACTIONS

EPA risk assessments may be updated as new data are published and as assessment methodologies evolve. Regulatory actions are frequently not updated at the same time. Compare the dates for the regulatory actions in this section with the verification dates for the risk assessments in sections I and II, as this may explain inconsistencies. Also note that some regulatory actions consider factors not related to health risk, such as technical or economic feasibility. Such considerations are indicated for each action. In addition, not all of the regulatory actions listed in this section involve enforceable federal standards. Please direct any questions you may have concerning these regulatory actions to the US EPA contact listed for that particular action. Users are strongly urged to read the background information on each regulatory action in Background Document 4 in Service Code 5.

IV.A. CLEAN AIR ACT (CAA)

[Ed. Note: EPA does not have information yet in this section of IRIS].

IV.B. SAFE DRINKING WATER ACT (SDWA)

[Ed. Note: EPA does not have information yet in this section of IRIS].

IV.C. CLEAN WATER ACT (CWA)

IV.C.1. AMBIENT WATER QUALITY CRITERIA, Human Health

Water and Fish Consumption: 3.09E+3 ug/L
Fish Consumption Only: 3.09E+3 ug/L

Considers technological or economic feasibility? -- NO

Discussion -- To control undesirable taste and odor qualities of ambient water, the estimated concentration is 3E-1 ug/L.

Reference -- 45 FR 79318 (11/28/80)

EPA Contact -- Criteria and Standards Division / OWRS(202)260-1315 / FTS 260-1315

IV.C.2. AMBIENT WATER QUALITY CRITERIA, Aquatic Organisms

Freshwater:
  Acute LEC -- 2.02E+3 ug/L
  Chronic LEC -- 3.65E+2 ug/L

Marine: None

Considers technological or economic feasibility? -- NO

Discussion -- The values that are indicated as "LEC" are not criteria, but are the lowest effect levels found in the literature. LECs are given when the minimum data required to derive water quality criteria are not available.

Reference -- 45 FR 79318 (11/28/80)

EPA Contact -- Criteria and Standards Division / OWRS(202)260-1315 / FTS 260-1315

IV.D. FEDERAL INSECTICIDE, FUNGICIDE, AND RODENTICIDE ACT (FIFRA)

        [Ed. Note: EPA does not have information yet in this section of IRIS].

IV.E. TOXIC SUBSTANCES CONTROL ACT (TSCA)

        [Ed. Note: EPA does not have information yet in this section of IRIS].

IV.F. RESOURCE CONSERVATION AND RECOVERY ACT (RCRA)

IV.F.1. RCRA APPENDIX IX, for Ground Water Monitoring

Status -- Listed

Reference -- 52 FR 25942 (07/09/87)

EPA Contact -- RCRA/Superfund Hotline (800)424-9346 / (202)260-3000 / FTS 260-3000

IV.G. SUPERFUND (CERCLA)

IV.G.1. REPORTABLE QUANTITY (RQ) for Release into the Environment

Value (status) -- 100 pounds (Final, 1985)

Considers technological or economic feasibility? -- NO

Discussion -- The final RQ is based on aquatic toxicity. The available data indicate that the aquatic 96-Hour Threshold Limit for 2,4-dichlorophenol is between 1 and 10 ppm.

Reference -- 50 FR 13456 (04/04/85); 54 FR 33418 (08/14/89)

EPA Contact -- RCRA/Superfund Hotline (800)424-9346 / (202)260-3000 / FTS 260-3000

VI. BIBLIOGRAPHY

VI.A. ORAL RfD REFERENCES

Exon, J.H. and L.D. Koller. 1985. Toxicity of 2-chlorophenol, 2,4-dichlorophenol and 2,4,6-trichlorophenol. In: Water Chlorination: Chemistry, Environmental Impact and Health Effects, Jolley et al., Ed., Vol.5. (Chap. 25). Lewis Publishers, Chelsea, MI. p. 307-330.

Kobayashi, S., T. Fukuda, K. Kawaguchi, H. Chang, S. Toda and H. Kawamura. 1972. Chronic toxicity of 2,4-dichlorophenol in mice: A simple design for checking the toxicity of residual metabolites of pesticides. J. Med. Soc., Toho Univ., Japan. 19: 356-362.

US EPA. 1985. Drinking Water Criteria Document for Chlorinated Phenols. Prepared by the Office of Health and Environmental Assessment, Environmental Criteria and Assessment Office, Cincinnati, OH for the Office of Drinking Water, Washington, DC.

VI.B. INHALATION RfC REFERENCES

None

VI.C. CARCINOGENICITY ASSESSMENT REFERENCES

None

VI.D. DRINKING WATER HA REFERENCES

None

VII. REVISION HISTORY

Date	Section	Description
09/30/87	I.A.2.	Citation corrected
03/01/88	I.A.1.	Dose conversion clarified
03/01/88	I.A.5.	Confidence levels revised
03/01/88	I.A.6.	Documentation corrected
06/30/88	I.A.7.	Primary contact changed
08/01/89	VI.	Bibliography on-line
01/01/92	IV.	Regulatory actions updated

SYNONYMS

DCP   NCI-C55345
2,4-DCP   PHENOL, 2,4-DICHLORO-
2,4-Dichlorophenol   RCRA WASTE NUMBER U081
Dichlorophenol, 2,4-

Source: *Instant EPA's IRIS*, 1997, Instant Reference Sources, Inc., 7605 Rockpoint Drive, Austin, TX 78731 (Fax: 512-345-2386; Internet URL, http://www.instantref.com/inst-ref.htm).

Dicofol

Introduction

Dicofol is an alcohol analog of DDT [053], [062]. It is a synthetic nonsystemic organochlorine acaricide which has been used primarily for the control of mites on field crops, vegetables, citrus and non-citrus fruits and in greenhouses.

Dicofol is released to the environment through its manufacture and use as a nonsystemic acaricide. If released to soil, it will be expected to bind to the soil strongly but under some circumstances it may reach groundwater, since it has been detected in groundwater. It is susceptible to hydrolysis in moist soils and evaporation from the surface of moist soils. It may be resistant to biodegradation. If it is released to water it will be expected to bind to the sediments and may bioconcentrate in aquatic organisms. It will be subject to hydrolysis and may directly photodegrade. It may be resistant to biodegradation but may be susceptible to evaporation. If it is released to the atmosphere it may be subject to direct photolysis. The estimated vapor phase half-life in the atmosphere is 2.92 days as a result of reaction with photochemically produced hydroxyl radicals. General exposure to dicofol occurs when kelthane is used and will occur mainly through consumption of contaminated foods. Dermal, ingestive, and inhalation exposure may also occur as a result of its manufacture and use as a non-systemic acaricide. [828]

Identifying Information

CAS NUMBER: 115-32-2

NIOSH Registry Number: DC8400000

SYNONYMS: KELTHANE, 1,1-BIS(4-CHLOROPHENYL)-2,2,2-TRICHLOROETHANOL, 4-CHLORO-ALPHA-(4-CHLOROPHENYL)-ALPHA-(TRICHLOROMETHYL)BENZENEMETHANOL, 4,4'-DICHLORO-ALPHA-(TRICHLOROMETHYL)BENZHYDROL, 1,1-BIS(CHLOROPHENYL)-2,2,2-TRICHLOROETHANOL, 2,2,2-TRICHLORO-1,1-BIS(P-CHLOROPHENYL)ETHANOL, ACARIN, ENT 23,648, CPCA, DI-(P-CHLOROPHENYL)TRICHLOROMETHYLCARBINOL, 2,2,2-TRICHLORO-1,1-DI-(4-CHLOROPHENYL)ETHANOL, HIFOL, 1,1-BIS(P-CHLOROPHENYL)-2,2,2-TRICHLOROETHANOL, CARBAX, DICHLOROKELTHANE, DTMC, FW 293,

KELTHANETHANOL, MITIGAN, NCI-C00486, CEKUDIFOL, DECOFOL, HILFOL 18.5 EC, KELTANE, PARA,PARA'-KELTHANE, KELTHANE A, KELTHANE DUST BASE, MILBOL, NA 2761

Endocrine Disruptor Information

Dicofol is a strongly suspected EED that it is listed by the Centers for Disease Control & Prevention [812], and the World Wildlife Fund [813] as a potential endocrine modifying chemical. It has been linked to sexual problems in alligators in Lake Apopka, FL. A spill of this chemical in 1980 resulted in male alligators with penises reduced to half or less of their normal size years later. Furthermore, the male alligators had skewed hormone ratios, showing a profile more typical of female alligators, i.e., elevated levels of estrogen and greatly reduced levels of testosterone. The female alligators also had elevated estrogen levels and an estrogen-to-testosterone ratio that was twice as high as it is in normal alligators. In addition, the ovaries of female alligators displayed abnormalities in their eggs and in the follicles, where the eggs matured before ovulation. Hormone disruption was also observed in turtles in Lake Apopka. Many red-eared turtles in the lake were females and many were found to be neither male or female but there were few male turtles. [810]

Chemical and Physical Properties

CHEMICAL FORMULA: $C_{14}H_9Cl_5O$

MOLECULAR WEIGHT: 370.50

WLN: GXGGXQR DG&R DG

PHYSICAL DESCRIPTION: White to gray powder

SPECIFIC GRAVITY: Not available

DENSITY: 1.45 g/mL @ 25°C (technical grade) [169], [172], [395]

MELTING POINT: 78.5-79.5°C [053], [169], [172], [395]

BOILING POINT: 225°C [051], [053], [072]

SOLUBILITY: Water: <0.1 mg/mL @ 22°C [700]; DMSO: >=100 mg/mL @ 21°C [700]; 95% Ethanol: >=100 mg/mL @ 21°C [700]; Acetone: >=100 mg/mL @ 20°C [700]

OTHER SOLVENTS: Distilled water: 0.8 mg/mL @ 20°C [055]; Most aliphatic solvents: Soluble [031], [169], [172], [395]; Most organic solvents: Soluble

OTHER PHYSICAL DATA: Brown, viscous oil (technical grade-~80% pure) [053], [169], [172], [395]; Boiling point: 180°C @ 0.1 mm Hg [169]; UV max (in ethanol): 226 nm, 258 nm, 266 nm, 276 nm (4.43, 2.82, 2.85, 2.60) [031]; Odorless [053], [395]

Source: *The National Toxicology Program's Chemical Database, Volume 2*, 1992, CRC Press, Inc./Lewis Publishers, 2000 Corporate Blvd., Boca Raton, FL 33431 (Fax: 407-998-9114).

Environmental Reference Materials

A source of EED environmental reference materials for dicofol is Crescent Chemical Co., Hauppauge, NY (Fax: 516-348-0913); Catalog No. 36677.

Hazardous Properties

ACUTE/CHRONIC HAZARDS: This compound is toxic by inhalation and ingestion [062]. It is an irritant [053], [371] and may be absorbed through the skin, respiratory tract and gastrointestinal tract. [295] When heated to decomposition it emits irritating gases and toxic fumes of chlorides. [043], [051], [072], [371]

VOLATILITY: Vapor pressure: Negligible @ room temperature [053], [169]

FIRE HAZARD: Literature sources indicate that dicofol is combustible [395]. Fires involving this material can be controlled with a dry chemical, carbon dioxide or Halon extinguisher.

REACTIVITY: Dicofol hydrolyzes in alkali [051], [169], [172], [395]. It is slightly corrosive to metals [053], [169]. Contact with steel at elevated temperatures causes formation of toxic gases [371].

STABILITY: Dicofol is stable under normal laboratory conditions. Solutions of dicofol in water, DMSO, 95% ethanol or acetone should be stable for 24 hours under normal lab conditions [700].

Source: *The National Toxicology Program's Chemical Database, Volume 7*, 1992, CRC Press, Inc./Lewis Publishers, 2000 Corporate Blvd., Boca Raton, FL 33431 (Fax: 407-998-9114).

Medical Symptoms of Exposure

SYMPTOMS: Symptoms of exposure to dicofol may include headache, dizziness, nausea, weight loss, convulsions, eye irritation and mild liver and kidney damage [371]. Allergic dermatitis may also occur [395].

Symptoms of exposure to this type of compound may include vomiting, paresthesia, giddiness, fatigue, tremors and coma [295], [455]. Other symptoms may include central nervous system stimulation, diarrhea, excitement, pulmonary edema, myocardial toxicity, hypothermia and respiration which is initially accelerated and later depressed [295]. Chronic exposure to this type of compound may cause loss of appetite, muscular weakness, apprehensive mental state and enhancement of hepatic microsomal enzyme activity [295].

Source: *The National Toxicology Program's Chemical Database, Volume 4*, 1992, CRC Press, Inc./Lewis Publishers, 2000 Corporate Blvd., Boca Raton, FL 33431 (Fax: 407-998-9114).

Toxicological Information

The toxicological information below is gathered from several sources including two national databases: one from the *National Toxicology Program Chemical Repository Database* and another from *EPA's Integrated Risk Information System (IRIS)*.

Toxicology Information from the National Toxicology Program

CARCINOGENICITY:

Tumorigenic Data:

Type of Effect	Route/Animal	Amount Dosed (Notes)
TDLo	orl-mus	17 gm/kg/78W-C
TD	orl-mus	35 gm/kg/78W-C

Review:

         IARC Cancer Review: Animal Limited Evidence

         IARC: Not classifiable as a human carcinogen (Group 3)

Status:

         NCI Carcinogenesis Bioassay (Feed); Negative: Male and Female Rat, Female Mouse

         NCI Carcinogenesis Bioassay (Feed); Positive: Male Mouse

TERATOGENICITY:

Reproductive Effects Data:

Type of Effect	Route/Animal	Amount Dosed (Notes)
TDLo	orl-rat	430 mg/kg (6-15D preg)

MUTATION DATA:

Test	Lwest dose		Test	Lowest dose
sce-hmn:lym	1 umol/L		cyt-hmn:leu	20 mg/L
mnt-mus-orl	215 mg/kg			

TOXICITY:

Typ. Dose	Mode	Specie	Amount	Units	Other
LD50	orl	rat	575	mg/kg	
LD50	skn	rat	100	mg/kg	
LD50	ipr	rat	1150	mg/kg	
LD50	orl	mus	420	mg/kg	
LD50	orl	rbt	1810	mg/kg	
LD50	skn	rbt	1870	mg/kg	
LD50	orl	gpg	1810	mg/kg	
LD50	orl	ckn	4365	mg/kg	

OTHER TOXICITY DATA:

Review: Toxicology Review-2
Standards and Regulations:
   DOT-Hazard: ORM-E; Label: None

Status:

EPA TSCA Test Submission (TSCATS) Data Base, January 1989
Meets criteria for proposed OSHA Medical Records Rule

SAX TOXICITY EVALUATION:

THR: Poison by ingestion and skin contact. Moderately toxic by intraperitoneal route. An experimental carcinogen. Human mutagenic data.

Source: *Instant Tox-Base*, 1995, Instant Reference Sources, Inc., 7605 Rockpoint Drive, Austin, TX 78731 (Fax: 512-345-2386; Internet URL, http://www.instantref.com/inst-ref.htm).

Toxicology Information from EPA's Integrated Risk Information System (IRIS)

I. CHRONIC HEALTH HAZARD ASSESSMENTS FOR NONCARCINOGENIC EFFECTS

I.A. REFERENCE DOSE FOR CHRONIC ORAL EXPOSURE (RfD)

[Ed. Note: EPA does not have information yet in this section of IRIS].

I.B. REFERENCE CONCENTRATION FOR CHRONIC INHALATION EXPOSURE (RfC)

[Ed. Note: EPA does not have information yet in this section of IRIS].

II. CARCINOGENICITY ASSESSMENT FOR LIFETIME EXPOSURE

The carcinogen assessment summary for this substance has been withdrawn pending further review by the CRAVE Agency Work Group.

Agency Work Group Review -- 06/03/87, 06/24/87, 08/05/87, 12/04/91, 06/02/92

EPA Contacts:

Jim Cogliano / NCEA / 202/260-3814

III. HEALTH HAZARD ASSESSMENTS FOR VARIED EXPOSURE DURATIONS

[Ed. Note: EPA does not have information yet in this section of IRIS].

IV. U.S. EPA REGULATORY ACTIONS

[Ed. Note: EPA does not have information yet in this section of IRIS].

VI. BIBLIOGRAPHY

[Ed. Note: EPA does not have information yet in this section of IRIS].

VII. REVISION HISTORY

```
--------  --------  --------------------------------------------------------
Date      Section   Description
--------  --------  --------------------------------------------------------
08/22/88  II.       Carcinogen summary on-line
01/01/90  VI.        Bibliography on-line
01/01/92  II.       Carcinogen assessment noted as pending change
01/01/92  IV.       Regulatory Action section on-line
04/01/92  II.       Carcinogen assessment withdrawn pending further review
04/01/92  IV.       Withdrawn; no assessments on-line
04/01/92  VI.       Bibliography withdrawn; no assessments on-line
07/01/92  II.       Work group review dates added
07/01/93  II.       EPA contact changed
```

SYNONYMS

acarin	FW293
Dicofol	kelthane
DTMC	mitigan

Source: *Instant EPA's IRIS*, 1997, Instant Reference Sources, Inc., 7605 Rockpoint Drive, Austin, TX 78731 (Fax: 512-345-2386; Internet URL, http://www.instantref.com/inst-ref.htm).

Production, Use, and Pesticide Labeling Information

1 . Description of Chemical

> Generic Name: 1,1-bis(chlorophenyl)-2,2,2-trichloroethanol

> Common Name: dicofol

> Trade Names: Acarin, Carbax, Decofol, Kelthane, Kelthane A, p,p'-Kelthane, Mibol, and Mitigan

> Chemical Abstracts Service (CAS) Number: 115-32-2

> EPA Shaughnessy Number: 010501

> Year of Initial Registration: 1957

> Pesticide Type: Acaricide

> U.S. and Foreign Producers: Rohm & Haas, Aceto, Makhteshim Beer- Shiva, Agan Chemical, Drexel Chemical, Tricon Chem International, and Ida, Inc.

2 . Use Patterns and Formulations

> - Foliar spray on agricultural crops and ornamentals, and in or around agricultural and domestic buildings for mite control.

> - Formulated as emulsifiable concentrates, wettable powders, dusts, ready-to-use liquids, and aerosol sprays.

3 . Science Findings

> - Chemical Characteristics

>> - Manufacturing-use dicofol products contain a number of DDT analogs as manufacturing impurities . These impurities include the o,p'- and p,p'- isomers of DDT, DDE, DDD, and a substance called extra-chlorine DDT or Cl-DDT.

>> - Dicofol is a nonflowable liquid (or waxy solid), ranging from dark to yellow-brown in color . It is stable under cool and dry conditions, is practically insoluble in water, but soluble in organic solvents. Its melting point ranges from 58 C to 78 °C . The chemical does not pose any unusual handling hazards.

- Toxicological Properties

    - Dicofol has a moderate acute oral toxicity (Toxicity Category III).

    - Dicofol has a relatively high degree of acute dermal toxicity (Toxicity Category II).

    - Results from an eye irritation test using a formulation intermediate indicate that the manufacturing-use product is probably a severe eye irritant (Toxicity Category I).

    - Toxicity studies on dicofol are as follows:

        - Oral LD50 in rats: 684-1495 mg/kg body weight

        - Oral LD50 in rabbits: 1810 mg/kg body weight

        - Dermal LD50 in rabbits: 2.1 gm/kg body weight

        - Eye Irritation: tests with a formulation intermediate showed corneal damage in some rabbits that persisted for seven days

    - Subchronic Oral Toxicity in Rats: NOEL is 20 ppm

    - Oncogenicity in Rats: Available data from the National Cancer Institute suggests that dicofol is a possible oncogen . However, the study is unacceptable due to the reported decomposition of the test material during the test.

    - Reproduction Study in Rats and Mice: The NOEL for both rats and mice is 100 ppm.

- Environmental Characteristics

    - Dicofol was stable for 350 days at 10 ppm in field plots under aerobic conditions.

    - Aged dicofol residues are negligibly mobile in sandy loam soil column leaching studies.

    - Field studies show that dicofol persists in soils for at least 4 years after application.

    - Dicofol residues may accumulate in fish as well as some rotational crops.

- Ecological Characteristics

    - The DDT analog contaminants in the dicofol products can cause reproductive impairment in various fish and flesh-eating birds (eggshell thinning).

    - Avian dietary LC50: 1237 to 3100 ppm for upland game birds (slightly toxic).

    - Prolonged exposure to low levels (5 to 10 ppm) had no significant effects on reproduction behavior of mallards.

    - Acute 96-hour LC50 for warm water fish: 0.31 to 0.51 ppm.

    - Acute 96-hour LC50 for cold water fish: 0.053 to 0.086 ppm.

    - Acute LD50 for marine grass shrimp: >0.439 ppm.

    - Findings show that dicofol is highly toxic to aquatic organisms.

    - Dicofol impairs the reproductive physiology of fish and aquatic invertebrates.

    - Dicofol does not seem to be phytotoxic to most plants for which it is registered.

    - Dicofol has been shown to be relatively nontoxic to honeybees and alfalfa leaf cutting bees.

- Tolerance Assessment

    - Current application rates on labeling exceed the rates associated with the established tolerances . Tolerances and/or labeling may need to be revised to correct this discrepancy.

    - Residues of the dicofol impurities will have to be examined for the tolerance reassessment.

- Summary of Science Findings

    - There are insufficient data available to characterize the environmental fate of dicofol; however, groundwater contamination is not expected due to dicofol's lack of mobility in soil and low leaching potential.

    - There are insufficient data available to determine the oncogenicity, teratogenicity, mutagenicity, and chronic feeding effects of dicofol.

- The data show that dicofol is highly toxic to aquatic organisms and only slightly toxic to birds, mammals, and beneficial insects.

- Dicofol bioaccumulates in some rotational crops and aquatic organisms.

- The DDT analog contaminants in dicofol may cause unreasonable adverse effects in certain bird and fish species.

4 . Summary of Regulatory Position and Rationale

- No use, formulation, or geographical restrictions are required . General use classification.

- A Special Review of all dicofol end-use products containing detectable amounts of DDT analogs will be initiated.

- No new registrations will be issued for new dicofol products intended for outdoor use which would further increase the amount of DDT in the environment.

- Registrants and applicants for registration must submit data regarding the composition of their products, and in particular the concentration of DDT contaminants in their manufacturing-use products . Registrants must indicate the lowest levels that could be achieved, the time frame needed, and the cost . The Agency is concerned about the presence of DDT contaminants at levels substantially below 1% of the technical product because a no-effect-level has not been determined for avian reproductive effects . Agency files show that currently registered manufacturing-use products contain from 9 to 15 percent DDT and DDT analog contamination

- Studies must be conducted by the registrants to measure whether DDT analog residues will be present on food or feed crops due to the use of dicofol products . A complete tolerance reassessment will be conducted after these data and other required residue data are submitted.

5 . Summary of Major Data Gaps

- Product Chemistry: product identity, certification of limits, physical and chemical properties, and special requirements as mentioned in Item 4, above . Due April, 1984.

- Residue Chemistry: metabolism in plants and animals, residue analytical methods, updated residue data (including residue data on DDT analogs) on most crops . Due December, 1986.

- Environmental Fate: degradation studies, soil metabolism studies, field volatility, field dissipation, accumulation in fish and rotational crops . Due December, 1986.

- Toxicology: inhalation studies, subchronic feeding, chronic toxicity, oncogenicity, teratogenicity, mutagenicity, metabolism . Due December, 1986.

- Ecological Effects: avian reproduction, aquatic organism testing . Due December, 1986.

- Special Testing For DDT Analog Contaminants: all environmental fate studies conducted on Cl-DDT, chronic testing with Cl-DDT on chronic toxicity on birds, fish, and aquatic invertebrates . Due December, 1986.

6 . Contact Person at EPA

Bruce A . Kapner (TS-767C)
Special Review Branch Office of Pesticide Programs
U. S. Environmental Protection Agency
401 M Street
SW Washington, DC 20460

Telephone: (703) 557-7400

Source: *Instant EPA's Pesticide Facts*, 1994, Instant Reference Sources, Inc., 7605 Rockpoint Drive, Austin, TX 78731 (Fax: 512-345-2386; Internet URL, http://www.instantref.com/inst-ref.htm).

Dieldrin

Introduction

Dieldrin is a nonsystemic, persistent insecticide with contact and stomach action. It was a broad spectrum insecticide used until 1974 when EPA restricted its use to termite control by direct soil injection and non-food seed and plant treatment. Dieldrin was used in tropical countries as a residual spray on the inside walls and ceilings of homes for the control of vectors of diseases, mainly malaria. Industrial uses include timber preservation, termite-proofing of plastic and rubber coverings of electrical and telecommunication cables, of plywood and building boards and as a termite barrier in building construction. [828]

Dieldrin is extremely persistent, but it is known to slowly photorearrange to photodieldrin (water half-life - 4 months). Dieldrin released to soil will persist for long periods (> 7 year). It can reach the air either through slow evaporation or adsorption on dust particles, will not leach, and can reach surface water with surface runoff. Once dieldrin reaches surface waters it will adsorb strongly to sediments, bioconcentrate in fish and slowly photodegrade. Biodegradation and hydrolysis are unimportant fate processes. The fate of dieldrin in the atmosphere is unknown but monitoring data have demonstrated that it can be carried long distances. Monitoring data demonstrates that dieldrin continues to be a contaminant in air, water, sediment, soil, fish, and other aquatic organisms, wildlife, foods, and humans. Human exposure appears to come mostly from food. [828]

Identifying Information

CAS NUMBER: 60-57-1

NIOSH Registry Number: IO1750000

SYNONYMS: HEXACHLOROEPOXYOCTAHYDRO-ENDO,EXO-DIMETHANONAPHTHALENE, 3,4,5,6,9,9-HEXACHLORO-1A,2,2A,3,6,6A,7,7A-OCTAHYDRO-2,7:3,6-, DIMETHANONAPHTH(2,3-B)OXIRENE, 1,2,3,4,10,10-HEXACHLORO-6,7-EPOXY-1,4,4A,5,6,7,8,8A-OCTAHYDRO-EXO-1,4-ENDO-, 5,8-DIMETHANONAPHTHALENE, 1,4:5,8-DIMETHANONAPHTHALENE, 1,2,3,4,10,10-HEXACHLORO-6,7-EPOXY-1,4,4a,, 5,6,7,8,8a-OCTAHYDRO, ENDO, EXO-, ENDO,EXO-1,2,3,4,10,10-HEXACHLORO-6,7-EPOXY-1,4,4A,5,6,7,8,8A-OCTAHYDRO-, 1,4:5,8-DIMETHANONAPHTHLENE, ALVIT, COMPOUND 497,

DIELDRIX, DIELDRITE, INSECTICIDE NO. 497, HEOD, ILLOXOL, ENT 16,225, QUINTOX, PANORAM D-31, OCTALOX, NCI-C00124, RCRA WASTE NUMBER P037, NA 2761

Endocrine Disruptor Information

Dieldrin is a strongly suspected EED that it is listed by EPA [811], the Centers for Disease Control & Prevention [812], and the World Wildlife Fund [813] as a potential endocrine modifying chemical. It also is listed as one of the commonly found pesticides in wildlife exhibiting symptoms of EDCs. [810]

Chemical and Physical Properties

CHEMICAL FORMULA: $C_{12}H_8Cl_6O$

MOLECULAR WEIGHT: 380.91

WLN: T E3 D5 C555 A D- FO KUTJ AG AG BG JG KG LG -ENDO EXO

PHYSICAL DESCRIPTION: White crystals to light tan flakes

SPECIFIC GRAVITY: 1.75 [016], [047], [055], [072]

DENSITY: 1.62 g/cm^3 @ 20°C [172]

MELTING POINT: 176-177°C [029], [033], [395], [421]

BOILING POINT: Not available

SOLUBILITY: Water: <1 mg/mL @ 24°C [700], DMSO: >=100 mg/mL @ 23°C [700], 95% Ethanol: <1 mg/mL @ 23°C [700], Acetone: >=100 mg/mL @ 23°C [700], Methanol: 10 g/L @ 20°C [169], Toluene: 410 g/L @ 20°C [169]

OTHER SOLVENTS: Carbon tetrachloride: 380 g/L @ 20°C [169], Aliphatic hydrocarbons: Sparingly soluble [421], [430], Petroleum oils: Slightly soluble [395], Aromatic solvents: Soluble [395], [421], [430], Halogenated solvents: Moderately soluble [421], [430], Most organic solvents: Moderately soluble [033], [043], Dichloromethane: 480 g/L @ 20°C [169], Benzene: 400 g/L @ 20°C [169], Alcohol: 1 in 4 [295], [455]

OTHER PHYSICAL DATA: Odor threshold: 0.041 mg/kg [055], [072], Mild chemical-like odor .

Source: *The National Toxicology Program's Chemical Database, Volume 2*, 1992, CRC Press, Inc./Lewis Publishers, 2000 Corporate Blvd., Boca Raton, FL 33431 (Fax: 407-998-9114).

Environmental Reference Materials

A source of EED environmental reference materials for Dieldrin is Radian International LLC, Austin, TX (Fax 512-454-0268; Internet URL, http://www.radian.com/standards); Catalog No. ERD-004.

Hazardous Properties

ACUTE/CHRONIC HAZARDS: Dieldrin is an irritant and may be toxic by skin contact, ingestion and inhalation [062], [346], [421], [430]. It is also toxic by eye contact [346]. It is readily absorbed through the skin [033], [295], [430] and it may accumulate in the body from chronic low dosages [043]. When heated to decomposition it emits toxic fumes of chlorides, hydrochloric acid and chlorinated hydrocarbon fumes. [043] Dieldrin may cause eye and skin irritation.

FIRE HAZARD: Literature sources indicate that dieldrin is nonflammable [173], [421], [430]. Fires involving this material can be controlled with a dry chemical, carbon dioxide or Halon extinguisher. A water spray may also be used.

REACTIVITY: Dieldrin is sensitive to mineral acids, acid catalysts, acid oxidizing agents and active metals. [169], [172], [173], [421] It reacts with phenols [072], [346], [421]. Dieldrin is also slightly corrosive to metals. [169] It may react vigorously with strong oxidizers such as chlorine and permanganates and strong acids such as sulfuric or nitric.

STABILITY: Dieldrin is stable under normal laboratory conditions. Solutions of dieldrin in water, DMSO, 95% ethanol or acetone should be stable for 24 hours under normal lab conditions. [700]

Source: *The National Toxicology Program's Chemical Database, Volume 7*, 1992, CRC Press, Inc./Lewis Publishers, 2000 Corporate Blvd., Boca Raton, FL 33431 (Fax: 407-998-9114).

Medical Symptoms of Exposure

Symptoms of exposure may include malaise, headache, nausea, vomiting and dizziness. [033], [151], [173], [346] Other symptoms may include tremors, clonic and tonic convulsions, coma and respiratory failure. [033], [151] Exposure may also lead to elevated temperature, dyspnea, cyanosis, tachycardia, aching muscles, fainting, disturbances of short term memory, severe convulsions, hematuria, total amnesia, hyperexcitability, hyperactivity, incoordination, nystagmus, quivering of outstretched fingers, sweating, dermatographia and death. [173] Severe depression, leukocytosis, blood pressure increase, metabolic acidosis and arrhythmias may also occur. [151] Other symptoms may include muscle spasms, drowsiness, loss of appetite, visual disturbances, insomnia, giddiness, eye irritation and redness, frothing at the mouth, enlarged liver and skin irritation.

Source: *The National Toxicology Program's Chemical Database, Volume 4*, 1992, CRC Press, Inc./Lewis Publishers, 2000 Corporate Blvd., Boca Raton, FL 33431 (Fax: 407-998-9114).

Toxicological Information

The toxicological information below is gathered from several sources including two national databases: one from the *National Toxicology Program Chemical Repository Database* and another from *EPA's Integrated Risk Information System (IRIS)*.

Toxicology Information from the National Toxicology Program

CARCINOGENICITY:

Tumorigenic Data:

Type of Effect	Route/Animal	Amount Dosed (Notes)
TDLo	orl-rat	200 mg/kg/2Y-C
TDLo	orl-mus	546 mg/kg/65W-C
TD	orl-mus	11 gm/kg/3Y-C
TD	orl-mus	610 mg/kg/73W-C
TD	orl-mus	714 mg/kg/85W-C
TD	orl-mus	8 mg/kg/2Y-C
TD	orl-mus	4550 mg/kg/65W-C

Review:
IARC Cancer Review; Animal Limited Evidence
IARC Cancer Review; Human Inadequate Evidence
IARC: Not classifiable as a human carcinogen (Group 3)

Status:
NCI Carcinogenesis Bioassay (Feed): Negative; Male and Female Rat, Female Mouse
NCI Carcinogenesis Bioassay (Feed): Equivocal; Male Mouse
EPA Carcinogen Assessment Group

TERATOGENICITY: See RTECS printout for data

MUTATION DATA: See RTECS printout for data

TOXICITY:

Typ. Dose	Mode	Species	Units
LDLo	orl	man	65mg/kg
LDLo	unr	hmn	28mg/kg
LD50	orl	rat	38300ug/kg
LC50	ihl	rat	13mg/m3/4H
LD50	skn	rat	56mg/kg
LD50	ipr	rat	35mg/kg
LD50	scu	rat	49mg/kg
LD50	ivn	rat	9mg/kg
LD50	orl	mus	38mg/kg
LDLo	ipr	mus	26mg/kg
LD50	ivn	mus	10500ug/kg
LD50	orl	dog	65mg/kg
LDL	ounr	dog	65mg/kg
LD50	orl	mky	3mg/kg
LDLo	orl	cat	500mg/kg
LC50	ihl	cat	80mg/m3/4H
LDLo	skn	cat	750mg/kg
LD50	orl	rbt	45mg/kg
LD50	skn	rbt	250mg/kg
LDLo	scu	rbt	150mg/kg
LD50	orl	pig	38mg/kg
LD50	orl	gpg	49mg/kg
LD50	orl	ham	60mg/kg
LD50	orl	pgn	23700ug/kg
LD50	ivn	pgn	1200mg/kg
LD50	orl	ckn	20mg/kg
LD50	orl	qal	10780ug/kg
LD50	orl	dck	381mg/kg
LD50	unr	mam	25mg/kg
LD50	orl	bwd	13300ug/kg

OTHER TOXICITY DATA:

Review: Toxicology Review-11

Standards and Regulations:
        DOT-Hazard: ORM-A; Label: None

Status:

        EPA Genetox Program 1988, Positive: V79 cell culture-gene mutation

        EPA Genetox Program 1988, Negative: Rodent dominant lethal; Host-mediated assay

        EPA Genetox Program 1988, Negative: Histidine reversion-Ames test; S cerevisiae-homozygosis

        EPA Genetox Program 1988, Inconclusive: D melanogaster sex-linked lethal

        EPA TSCA Test Submission (TSCATS) Data base, September 1989

        Lethal dose man: 65 mg/kg

        Lethal dose man: ~5 gm

SAX TOXICITY EVALUATION:

THR: A human poison by ingestion and possibly other routes. Poison experimentally by inhalation, ingestion, skin contact, intravenous, intraperitoneal and possibly other routes. An experimental carcinogen, neoplastigen, tumorigen, and teratogen. Experimental reproductive effects. Absorbed readily through the skin and other routes. It is a central nervous system stimulant. Human mutagenic data. Considerably more toxic than DDT by ingestion and skin contact. It may accumulate in the body from chronic low doses.

Source: *Instant Tox-Base*, 1995, Instant Reference Sources, Inc., 7605 Rockpoint Drive, Austin, TX 78731 (Fax: 512-345-2386; Internet URL, http://www.instantref.com/inst-ref.htm).

Toxicology Information from EPA's Integrated Risk Information System (IRIS)

I. CHRONIC HEALTH HAZARD ASSESSMENTS FOR NONCARCINOGENIC EFFECTS

I.A. REFERENCE DOSE FOR CHRONIC ORAL EXPOSURE (RfD)

The Reference Dose (RfD) is based on the assumption that thresholds exist for certain toxic effects such as cellular necrosis, but may not exist for other toxic effects such as carcinogenicity. In general, the RfD is an estimate (with uncertainty spanning perhaps an order of magnitude) of a daily exposure to the human population (including sensitive subgroups) that is likely to be without an appreciable risk of deleterious effects during a lifetime. Please refer to Background Document 1 in Service Code 5 for an elaboration of these concepts. RfDs can also be derived for the noncarcinogenic health effects of compounds which are also carcinogens. Therefore, it is essential to refer to other sources of information concerning the carcinogenicity of this substance. If the US EPA has evaluated this substance for potential human carcinogen-icity, a summary of that evaluation will be contained in Section II of this file when a review of that evaluation is completed.

I.A.1. ORAL RfD SUMMARY

Critical Effect	Experimental Doses*	UF	MF	RfD
Liver lesions	NOAEL: 0.1 ppm (0.005 mg/kg/day)	100	1	5E-5 mg/kg/day
	LOAEL: 1.0 ppm (0.05 mg/kg/day)			

*Conversion Factors: 1 ppm = 0.05 mg/kg/day (assumed rat food consumption)

2-Year Rat Feeding Study Walker et al., 1969

I.A.2. PRINCIPAL AND SUPPORTING STUDIES (ORAL RfD)

Walker, A.I.T., D.E. Stevenson, J. Robinson, R. Thorpe and M. Roberts. 1969. The toxicology and pharmacodynamics of dieldrin (HEOD): Two-year oral exposures of rats and dogs. Toxicol. Appl. Pharmacol. 15: 345-373.

Walker et al. (1969) administered dieldrin (recrystallized, 99% active ingredient) to Carworth Farm "E" rats (25/sex/dose; controls 45/sex) for 2 years at dietary concentrations of 0, 0.1, 1.0, or 10.0 ppm. Based on intake assumptions presented by the authors, these dietary levels are approximately equal to 0, 0.005, 0.05 and 0.5 mg/kg/day. Body weight, food intake, and general health remained unaffected throughout the 2-year period, although at 10.0 ppm (0.5 mg/kg/day) all animals became irritable and exhibited tremors and occasional convulsions. No effects were seen in various hematological and clinical chemistry parameters. At the end of 2 years, females fed 1.0 and 10.0 ppm (0.05 and 0.5 mg/kg/day) had increased liver weights and liver-to-body weight ratios (p<0.05). Histopathological examinations revealed liver parenchymal cell changes including focal proliferation and focal hyperplasia. These hepatic lesions were considered to be characteristic of exposure to an organochlorine insecticide. The LOAEL was identified as 1.0 ppm (0.005 mg/kg/day) and the NOAEL as 0.1 ppm (0.005 mg/kg/day).

I.A.3. UNCERTAINTY AND MODIFYING FACTORS (ORAL RfD)

UF -- The UF of 100 allows for uncertainty in the extrapolation of dose levels from laboratory animals to humans (10A) and uncertainty in the threshold for sensitive humans (10H).

MF -- None

I.A.4. ADDITIONAL COMMENTS (ORAL RfD)

Data considered for establishing the RfD:

1) 2-Year Feeding - rat: Principal study - see previous description

2) 2-Year Feeding (oncogenic) - dog: Systemic NOEL=0.005 mg/kg/day; LEL= 0.05 mg/kg/day (increased liver weight and liver/body weight ratios, increased plasma alkaline phosphatase, and decreased serum protein concentration) (Walker et al., 1969)

3) 2-Year Feeding - rat: Systemic LEL=0.5 ppm (approximately 0.025 mg/kg/day), (liver enlargement with histopathology); (Fitzhugh et al., 1964)

4) 2-Year Feeding (oncogenic) - mouse: Systemic LEL=0.1 ppm (0.015 mg/kg/day), (liver enlargement with histopathology); (Walker et al., 1972)

5) 25-Month Feeding - dog: Systemic NOEL=0.2 mg/kg/day; LEL=0.5 mg/kg/day, (weight loss and convulsions); (Fitzhugh et al., 1964)

6) Teratology - mouse: Teratogenic NOEL=6.0 mg/kg/day (HDT, gestational days 7-16); Maternal LEL=6.0 mg/kg/day (HDT, decrease in maternal weight gain); Fetotoxic LEL=6.0 mg/kg/day (HDT, decreased numbers of caudal ossification centers and increases in supernumerary ribs); (Chernoff et al., 1975). This study was not considered since 41% of the test dams died at the highest dose tested.

I.A.5. CONFIDENCE IN THE ORAL RfD

Study -- Low
Data Base -- Medium
RfD -- Medium

The principal study is an older study for which detailed data are not available and in which a wide range of doses was tested. The chronic toxicity evaluation is relatively complete and supports the critical effect, if not the magnitude of effects. Reproductive studies are lacking. The RfD is given a medium confidence rating because of the support for the critical effect from other dieldrin studies, and from studies on organochlorine insecticides in general.

I.A.6. EPA DOCUMENTATION AND REVIEW OF THE ORAL RfD

Source Document -- US EPA, 1987

Other EPA Documentation -- None

Agency Work Group Review -- 04/16/87

Verification Date -- 04/16/87

I.A.7. EPA CONTACTS (ORAL RfD)

Krishan Khanna / OST -- (202)260-7588

Henry Spencer / OST -- (202)557-4383

I.B. REFERENCE CONCENTRATION FOR CHRONIC INHALATION EXPOSURE (RfC)

[Ed. Note: EPA does not have information yet in this section of IRIS].

II. CARCINOGENICITY ASSESSMENT FOR LIFETIME EXPOSURE

Section II provides information on three aspects of the carcinogenic risk assessment for the agent in question; the US EPA classification, and quantitative estimates of risk from oral exposure and from inhalation exposure. The classification reflects a weight-of-evidence judgment of the likelihood that the agent is a human carcinogen. The quantitative risk estimates are presented in three ways. The slope factor is the result of application of a low-dose extrapolation procedure and is presented as the risk per (mg/kg)/day. The unit risk is the quantitative estimate in terms of either risk per ug/L drinking water or risk per ug/cu.m air breathed. The third form in which risk is presented is a drinking water or air concentration providing cancer risks of 1 in 10,000, 1 in 100,000 or 1 in 1,000,000. Background Document 2 (Service Code 5) provides details on the rationale and methods used to derive the carcinogenicity values found in IRIS. Users are referred to Section I for information on long-term toxic effects other than carcinogenicity.

II.A. EVIDENCE FOR CLASSIFICATION AS TO HUMAN CARCINOGENICITY

II.A.1. WEIGHT-OF-EVIDENCE CLASSIFICATION

Classification -- B2; probable human carcinogen

Basis -- Dieldrin is carcinogenic in seven strains of mice when administered orally. Dieldrin is structurally related to compounds (aldrin, chlordane, heptachlor, heptachlor epoxide, and chlorendic acid) which produce tumors in rodents.

II.A.2. HUMAN CARCINOGENICITY DATA

Inadequate. Two studies of workers exposed to aldrin and to dieldrin reported no increased incidence of cancer. Both studies were limited in their ability to detect an excess of cancer deaths. Van Raalte (1977) observed two cases of cancer (gastric and lymphosarcoma) among 166 pesticide manufacturing workers exposed 4-19 years and followed from 15-20 years. Exposure was not quantified, and workers were also exposed to other organochlorine pesticides (endrin and telodrin). The number of workers studied was small, the mean age of the cohort (47.7 years) was young, the number of expected deaths was not calculated, and the duration of exposure and of latency was relatively short.

In a retrospective mortality study, Ditraglia et al. (1981) reported no statistically significant excess in deaths from cancer among 1155 organochlorine pesticide manufacturing workers [31 observed vs. 37.8 expected, Standardized Mortality Ratio (SMR) = 82]. Workers were employed for 6 months or more and followed 13 years or more (24,939 person-years). Workers with no exposure (for example, office workers) were included in the cohort. Vital status was not known for 112 or 10% of the workers, and these workers were assumed to be alive; therefore additional deaths may have occurred but were not observed. Exposure was not quantified and workers were also exposed to other chemicals and pesticides (including endrin). Increased incidences of deaths from cancer were seen at several specific sites: esophagus (2 deaths observed, SMR = 235); rectum (3, SMR = 242); liver (2, SMR = 225); and lymphatic and hematopoietic system (6, SMR = 147), but these site-specific incidences were not statistically significantly increased.

II.A.3. ANIMAL CARCINOGENICITY DATA

Sufficient. Dieldrin has been shown to be carcinogenic in various strains of mice of both sexes. At different dose levels the effects range from benign liver tumors, to hepatocarcinomas with transplantation confirmation, to pulmonary metastases.

The Food and Drug Administration (FDA) conducted a long-term carcinogenesis bioassay for dieldrin (Davis and Fitzhugh, 1962). Ten ppm dieldrin was administered orally to 218 male and female C3HeB/Fe mice for 2 years. The study was compromised by the poor survival rate, lack of detailed pathology, loss of a large percentage of the animals to the study, and failure to treat the data for males and females separately. A statistically significant increase in incidence of hepatomas was observed in the treated groups versus the control groups in both males and females. In FDA follow-up study, Davis (1965) examined 100 male and 100 female C3H mice which had been orally administered 10 ppm dieldrin. The same limitations as the previous study were reported. The incidence of benign hepatomas and hepatic carcinomas was significantly increased in the dieldrin group. A reevaluation of the histological material of both studies was done by Reuber in 1974 (Epstein, 1975a,b; 1976). He concluded that the hepatomas were malignant and that dieldrin was hepatocarcinogenic for male and female C3HeB/Fe and C3H mice.

Walker et al. (1972) conducted several studies of dieldrin in CF1 mice of both sexes. Dieldrin was administered orally at concentrations of 0, 0.1, 1.0, and 10 ppm. Treatment groups varied from 87 to 288 animals of each sex. Surviving animals were sacrificed during weeks 132-140. Incidence of tumors was related to the number of dose levels and the dose administered. Effects were detected at the lowest dieldrin level tested (0.1 ppm) in both male and female mice. Dieldrin also produced significant increases (<0.05) in the incidence of pulmonary adenomas, pulmonary carcinomas, lymphoid tumors, and "other" tumors in female mice.

Diets containing 10 ppm dieldrin were fed to groups of 30 CF1 mice of both sexes for 110 weeks (Thorpe and Walker, 1973). The control group consisted of 45 mice of both sexes. A statistically significant increase (p<0.01) in incidence of liver tumors was found in both sexes of treated animals relative to controls. The liver tumors appeared much earlier in treated animals than controls.

Technical-grade dieldrin (>96%) was fed to B6C3F1 mice (50/sex/dose) at TWA doses of 0, 2.5, or 5 ppm for 80 weeks followed by an observation period of 10 to 13 weeks (NCI, 1978a). Matched control groups consisted of 20 untreated males and 10 untreated females. No significant difference in survival was noted. A significant dose-related increase in hepatocellular carcinoma was found in male mice when compared with pooled controls.

Tennekes et al. (1981) fed groups of 19 to 82 male CF1 mice control or dieldrin-supplemented (10 ppm) diets or control diets for 110 weeks. Dieldrin produced a statistically significant increased incidence of hepatocellular carcinomas in the treated group.

Dieldrin (>99%) was continuously fed in the diet for 85 weeks to 50 C3H/He, 62 B6C3Fl, and 71 C57Bl/6J male mice (Meierhenry et al., 1983). Controls were 50 to 76 males of each strain. Dieldrin produced a significant increase in the incidence of hepatocellular carcinomas compared with controls in all three strains.

Seven studies with four strains of rats fed 0.1 to 285 ppm dieldrin varying in duration of exposure from 80 weeks to 31 months did not produce positive results for carcinogenicity (Treon and Cleveland, 1955; Fitzhugh et al., 1964; Song and Harville, 1964; Walker et al., 1969; Deichmann et al., 1970; NCI, 1978a,b). Three of these studies used Osborne-Mendel rats, two studies used Carworth rats, and one each used Fischer 344 and Holtzman strains. Only three of the seven studies are considered adequate in design and conduct. The others used too few animals, had unacceptably high levels of mortality, were too short in duration, and/or had inadequate pathology examination or reporting.

II.A.4. SUPPORTING DATA FOR CARCINOGENICITY

Dieldrin causes chromosomal aberrations in mouse cells (Markaryan, 1966; Majumdar et al., 1976) and in human lymphoblastoid cells (Trepanier et al., 1977), forward mutation in Chinese hamster V79 cells (Ahmed et al., 1977), and unscheduled DNA synthesis in rat (Probst et al., 1981) and human cells (Rocchi et al., 1980). Dieldrin did not produce responses in 13 other mutagenicity tests. Negative responses were given in assays for gene conversion in S. cerevisiae, back-mutation in S. marcesans, forward mutation (Gal Rz2 in E. coli), and forward mutation to streptomycin resistance in E. coli (Fahrig, 1974). Negative responses were produced in reverse mutation assays with six strains of S. typhimurium with or without metabolic activation (Bidwell et al., 1975; Marshall et al., 1976; Shirasu et al., 1976; Wade et al., 1979; Haworth et al., 1983). Majumdar et al. (1977), however, reported that dieldrin was mutagenic for S. typhimurium with and without metabolic activation.

Five compounds structurally related to dieldrin - aldrin, chlordane, heptachlor, heptachlor epoxide, and chlorondic acid - have induced malignant liver tumors in mice. Chlorendic acid has also induced liver tumors in rats.

II.B.   QUANTITATIVE ESTIMATE OF CARCINOGENIC RISK FROM ORAL EXPOSURE

II.B.1. SUMMARY OF RISK ESTIMATES

Oral Slope Factor -- 1.6E+1 per (mg/kg)/day
Drinking Water Unit Risk -- 4.6E-4 per (ug/L)
Extrapolation Method -- Linearized multistage procedure, extra risk
Drinking Water Concentrations at Specified Risk Levels:

Risk Level	Concentration
E-4 (1 in 10,000)	2E-1 ug/L
E-5 (1 in 100,000)	2E-2 ug/L
E-6 (1 in 1,000,000)	2E-3 ug/L

II.B.2. DOSE-RESPONSE DATA (CARCINOGENICITY, ORAL EXPOSURE)

Tumor Type -- liver carcinoma
Test Animals -- mouse
Route -- diet
Reference -- see table

Sex/Strain	Slope Factor	Reference
Male, C3H	22	Davis (1965), reevaluated by Reuber, 1974 (cited in Epstein, 1975a)
Female, C3H	25	Davis (1965), reevaluated by Reuber, 1974 (cited in Epstein, 1975a)
Male, CF1	25	Walker et al. (1972)
Female, CF1	28	Walker et al. (1972)
Male, CF1	15	Walker et al. (1972)
Female, CF1	7.1	Walker et al. (1972)
Male, CF1	55	Thorpe and Walker (1973)
Female, CF1	26	Thorpe and Walker (1973)
Male, B6C3F1	9.8	NCI (1978a,b)
Male, CF1	18	Tennekes et al. (1981)
Male, C57B1/6J	7.4	Meierhenry et al. (1983)
Male, C3H/He	8.5	Meierhenry et al. (1983)
Male, B6C3F1	11	Meierhenry et al. (1983)

II.B.3. ADDITIONAL COMMENTS (CARCINOGENICITY, ORAL EXPOSURE)

The slope factor is the geometric mean of 13 slope factors calculated from liver carcinoma data in both sexes of several strains of mice. Inspection of the data indicated no strain or sex specificity of carcinogenic response.

The unit risk should not be used if the water concentration exceeds 20 ug/L, since above this concentration the unit risk may not be appropriate.

II.B.4. DISCUSSION OF CONFIDENCE (CARCINOGENICITY, ORAL EXPOSURE)

The individual slope factors calculated from 13 independent data sets range within a factor of 8.

II.C.   QUANTITATIVE   ESTIMATE   OF   CARCINOGENIC   RISK   FROM INHALATION EXPOSURE

II.C.1. SUMMARY OF RISK ESTIMATES

Inhalation Unit Risk -- 4.6E-3 per (ug/cu.m)
Extrapolation Method -- Linearized multistage procedure, extra risk
Air Concentrations at Specified Risk Levels:

Risk Level	Concentration
E-4 (1 in 10,000)	2E-2 ug/cu.m
E-5 (1 in 100,000)	2E-3 ug/cu.m
E-6 (1 in 1,000,000)	2E-4 ug/cu.m

II.C.2.   DOSE-RESPONSE   DATA   FOR   CARCINOGENICITY,   INHALATION EXPOSURE

Calculated from oral data in Section II.B.2.

II.C.3.   ADDITIONAL   COMMENTS   (CARCINOGENICITY,   INHALATION EXPOSURE)

The unit risk should not be used if air concentrations exceed 2 ug/cu.m, since above this concentration the unit risk may not be appropriate.

II.C.4.   DISCUSSION   OF   CONFIDENCE   (CARCINOGENICITY,   INHALATION EXPOSURE)

This inhalation risk estimate was based on oral data.

II.D. EPA DOCUMENTATION, REVIEW, AND CONTACTS (CARCINOGENICITY ASSESSMENT)

II.D.1. EPA DOCUMENTATION

Source Document -- US EPA, 1986

II.D.2. REVIEW (CARCINOGENICITY ASSESSMENT)

Agency Work Group Review -- 03/05/87

Verification Date -- 03/05/87

II.D.3. US EPA CONTACTS (CARCINOGENICITY ASSESSMENT)

Dharm Singh / NCEA -- (202)260-5958

Jim Cogliano / NCEA -- (202)260-3814

III. HEALTH HAZARD ASSESSMENTS FOR VARIED EXPOSURE DURATIONS

III.A. DRINKING WATER HEALTH ADVISORIES

The Office of Drinking Water provides Drinking Water Health Advisories (HAs) as technical guidance for the protection of public health. HAs are not enforceable Federal standards. HAs are concentrations of a substance in drinking water estimated to have negligible deleterious effects in humans, when ingested, for a specified period of time. Exposure to the substance from other media is considered only in the derivation of the lifetime HA. Given the absence of chemical-specific data, the assumed fraction of total intake from drinking water is 20%. The lifetime HA is calculated from the Drinking Water Equivalent Level (DWEL) which, in turn, is based on the Oral Chronic Reference Dose. Lifetime HAs are not derived for compounds which are potentially carcinogenic for humans because of the difference in assumptions concerning toxic threshold for carcinogenic and noncarcinogenic effects.   A more detailed description of the assumptions and methods used in the derivation of HAs is provided in Background Document 3 in Service Code 5.

III.A.1. ONE-DAY HEALTH ADVISORY FOR A CHILD

   Appropriate data for calculating a One-day HA are not available. It is recommended that the modified DWEL of 0.0005 mg/L be used as the One-day HA.

III.A.2. TEN-DAY HEALTH ADVISORY FOR A CHILD

   Appropriate data for calculating a Ten-day HA are not available. It is recommended that the modified DWEL of 0.0005 mg/L be used as the Ten-day HA.

III.A.3. LONGER-TERM HEALTH ADVISORY FOR A CHILD

Appropriate data for calculating Longer-term HAs for dieldrin are not available. It is recommended that the modified DWEL of 0.0005 mg/L be used as the Longer-term HA for the 10-kg child.

III.A.4. LONGER-TERM HEALTH ADVISORY FOR AN ADULT

Appropriate data for calculating Longer-term HAs for dieldrin are not available. It is recommended that the modified DWEL of 0.002 mg/L be used as the Longer-term HA for the 70-kg adult.

III.A.5. DRINKING WATER EQUIVALENT LEVEL / LIFETIME HEALTH ADVISORY

DWEL -- 2E-3 mg/L

Assumptions -- 2 L/day water consumption for a 70-kg adult

RfD Verification Date -- 04/16/87 (see Section I.A. in this file)

Lifetime HA -- None

Dieldrin is considered to be a probable human carcinogen. Lifetime HAs are not recommended for known or probable human carcinogens. The estimated excess cancer risk associated with lifetime exposure to drinking water containing dieldrin at the DWEL of 2 ug/L is approximately 8.05 x 10-4. Refer to Section II for the carcinogenicity assessment for dieldrin.

Principal Study -- Walker et al., 1969 (This study was used in the derivation of the chronic oral RfD; see Section I.A.2.)

III.A.6. ORGANOLEPTIC PROPERTIES

The odor threshold for dieldrin in water is reported as 0.04 mg/L.

III.A.7. ANALYTICAL METHODS FOR DETECTION IN DRINKING WATER

Determination of dieldrin is by a liquid-liquid extraction gas chromatographic procedure.

III.A.8. WATER TREATMENT

Available data indicate that reverse osmosis, granular activated carbon adsorption, ozonation, and conventional treatment will remove dieldrin from water.

III.A.9. DOCUMENTATION AND REVIEW OF HAs

US EPA. 1989. Drinking Water Health Advisories: Pesticides. Lewis Publishers, Chelsea, MI. p. 299-312.

EPA review of HAs in 1987.

Public review of HAs in January-March 1988.

Preparation date of this IRIS summary -- 08/20/90

III.A.10. EPA CONTACTS

Krishan Khanna / OST -- (202)260-7588

Edward V. Ohanian / OST -- (202)260-7571

III.B. OTHER ASSESSMENTS

[Ed. Note: EPA does not have information yet in this section of IRIS].

IV. US EPA REGULATORY ACTIONS

EPA risk assessments may be updated as new data are published and as assessment methodologies evolve. Regulatory actions are frequently not updated at the same time. Compare the dates for the regulatory actions in this section with the verification dates for the risk assessments in sections I and II, as this may explain inconsistencies. Also note that some regulatory actions consider factors not related to health risk, such as technical or economic feasibility. Such considerations are indicated for each action. In addition, not all of the regulatory actions listed in this section involve enforceable federal standards. Please direct any questions you may have concerning these regulatory actions to the US EPA contact listed for that particular action. Users are strongly urged to read the background information on each regulatory action in Background Document 4 in Service Code 5.

IV.A. CLEAN AIR ACT (CAA)

[Ed. Note: EPA does not have information yet in this section of IRIS].

IV.B. SAFE DRINKING WATER ACT (SDWA)

IV.B.1. MAXIMUM CONTAMINANT LEVEL GOAL (MCLG) for Drinking Water

[Ed. Note: EPA does not have information yet in this section of IRIS].

IV.B.2. MAXIMUM CONTAMINANT LEVEL (MCL) for Drinking Water

[Ed. Note: EPA does not have information yet in this section of IRIS].

IV.B.3. SECONDARY MAXIMUM CONTAMINANT LEVEL (SMCL) for Drinking Water

[Ed. Note: EPA does not have information yet in this section of IRIS].

IV.B.4. REQUIRED MONITORING OF "UNREGULATED" CONTAMINANTS

Status -- Listed (Proposed, 1991)

Discussion -- "Unregulated" contaminants are those contaminants for which EPA establishes a monitoring requirement but which do not have an associatedfinal MCLG, MCL, or treatment technique. EPA may regulate these contaminants in the future.

Monitoring requirement -- All systems to be monitored unless a vulnerability assessment determines the system is not vulnerable.

Analytical methodology -- Microextraction/gas chromatography (EPA 505); electron-capture/gas chromatography (EPA 508); GC/MS (EPA 525).

Reference -- 56 FR 3526 (01/30/91)

EPA Contact -- Drinking Water Standards Division / OGWDW /(202) 260-7575 / FTS 260-7575; or Safe Drinking Water Hotline / (800) 426-4791

IV.C. CLEAN WATER ACT (CWA)

IV.C.1. AMBIENT WATER QUALITY CRITERIA, Human Health

Water and Fish Consumption: 7.1E-5 ug/L

Fish Consumption Only: 7.6E-5 ug/L

Considers technological or economic feasibility? -- NO

Discussion -- For the maximum protection from the potential carcinogenic properties of this chemical, the ambient concentration should be zero. However, zero may not be attainable at this time, so the recommended criteria presents a E-6 estimated incremental increase of cancer risk over a lifetime. Reference -- 45 FR 79318 (11/28/80)

EPA Contact -- Criteria and Standards Division / OWRS (202)260-1315 / FTS 260-1315

IV.C.2. AMBIENT WATER QUALITY CRITERIA, Aquatic Organisms

Freshwater:
   Acute -- 1.0E+0 ug/L
   Chronic -- 1.9E-3 ug/L

Marine:
   Acute -- 7.1E-1 ug/L
   Chronic -- 1.9E-3 ug/L

Considers technological or economic feasibility? -- NO

Discussion -- Criteria were derived from a minimum data base consisting of acute tests on a variety of species. Requirements and methods are covered in the reference to the Federal Register.

Reference -- 45 FR 79318 (11/28/80)

EPA Contact -- Criteria and Standards Division / OWRS (202)260-1315 / FTS 260-1315

IV.D. FEDERAL INSECTICIDE, FUNGICIDE, AND RODENTICIDE ACT (FIFRA)

IV.D.1. PESTICIDE ACTIVE INGREDIENT, Registration Standard

[Ed. Note: EPA does not have information yet in this section of IRIS].

IV.D.2. PESTICIDE ACTIVE INGREDIENT, Special Review

Action -- Registration canceled (1974)

Considers technological or economic feasibility? -- NO

Summary of regulatory action -- Cancellation of all but termiticide and use. Criteria of concern: carcinogenicity, bio-accumulation, hazard to wildlife, and other chronic effects.

Reference -- 39 FR 37246 (10/18/74)

EPA Contact -- Special Review Branch / OPP (703)557-7400 / FTS 557-7400

IV.E. TOXIC SUBSTANCES CONTROL ACT (TSCA)

[Ed. Note: EPA does not have information yet in this section of IRIS].

IV.F. RESOURCE CONSERVATION AND RECOVERY ACT (RCRA)

IV.F.1. RCRA APPENDIX IX, for Ground Water Monitoring

Status -- Listed

Reference -- 52 FR 25942 (07/09/87)

EPA Contact -- RCRA/Superfund Hotline (800)424-9346 / (202)260-3000 / FTS 260-3000

IV.G. SUPERFUND (CERCLA)

IV.G.1. REPORTABLE QUANTITY (RQ) for Release into the Environment

Value (status) -- 1 pound (Final, 1989)

Considers technological or economic feasibility? -- NO

Discussion -- The RQ for dieldrin is based on aquatic toxicity as established under CWA Section 311 (40 CFR 117.3) and potential carcinogenicity. The available data indicate that the aquatic 96-Hour Median threshold Limit is less than 0.1 ppm, which corresponds to an RQ of 1 pound. Available data also indicate a hazard ranking of high and a weight of evidence classification of Group B2, which corresponds to an RQ of 1 pound.

Reference -- 54 FR 33418 (08/14/89)

EPA Contact -- RCRA/Superfund Hotline (800)424-9346 / (202)260-3000 / FTS 260-3000

VI. BIBLIOGRAPHY

VI.A. ORAL RfD REFERENCES

Chernoff, N., R.J. Kavlock, J.R. Kathrein, J.M. Dunn and J.K. Haseman. 1975. Prenatal effects of dieldrin and photodieldrin in mice and rats. Toxicol. Appl. Pharmacol. 31: 302-308.

Fitzhugh, O.G., A.A. Nelson and M.L. Quaife. 1964. Chronic oral toxicity of aldrin and dieldrin in rats and dogs. Food Cosmet. Toxicol. 2: 551-562.

US EPA. 1987. Dieldrin: Health Advisory. Office of Drinking Water, Washington, DC. NTIS PB 88-113543/AS.

Walker, A.I.T., D.E. Stevenson, J. Robinson, E. Thorpe and M. Roberts. 1969.The toxicology and pharmacodynamics of dieldrin (HEOD): Two-year oral exposures of rats and dogs. Toxicol. Appl. Pharmacol. 15: 345-373.

Walker, A.I.T., E. Thorpe and D.E. Stevenson. 1972. The toxicology of dieldrin (HEOD). I. Long-term oral toxicity studies in mice. Food Cosmet. Toxicol. 11: 415-432.

VI.B. INHALATION RfD REFERENCES

None

VI.C. CARCINOGENICITY ASSESSMENT REFERENCES

Ahmed, F.E., R.W. Hart and N.J. Lewis. 1977. Pesticide induced DNA damage and its repair in cultured human cells. Mutat. Res. 42: 161-174.

Bidwell, K., E. Weber, I. Neinhold, T. Connor and M.S. Legator. 1975. Comprehensive evaluation for mutagenic activity of dieldrin. Mutat. Res. 31: 314. (Abstract)

Davis, K.J. 1965. Pathology report on mice fed aldrin, dieldrin, heptachlor or heptachlor epoxide for two years. Internal FDA memorandum to Dr. A.J. Lehman. July 19. (Cited in: US EPA, 1986)

Davis, K.J. and O.G. Fitzhugh. 1962. Tumorigenic potential of aldrin and dieldrin for mice. Toxicol. Appl. Pharmacol. 4: 187-189.

Deichmann, W.B., W.E. MacDonald, E. Blum, et al. 1970. Tumorigenicity of aldrin, dieldrin and endrin in the albino rat. Ind. Med. Surg. 39: 426-434.

Ditraglia, D., D.P. Brown, T. Namekata and M. Iverson. 1981. Mortality study of workers employed at organochlorine pesticide manufacturing plants. Scand. J. Work. Env. Health. 7 (Suppl. 4): 140-146.

Epstein, S.S. 1975a. The carcinogenicity of dieldrin. Part 1. Sci. Total Environ. 4: 1-52.

Epstein, S.S. 1975b. The carcinogenicity of dieldrin. Part 2. Sci. Total Environ. 4: 205-217.

Epstein, S.S. 1976. Case study 5: Aldrin and dieldrin suspension based on experimental evidence and evaluation and societal needs. Ann. NY. Acad. Sci. 271: 187-195.

Fahrig, R. 1974. Comparative mutagenicity studies with pesticides. IARC Scientific Press No. 10.

Fitzhugh, O.G., A.A. Nelson and M.L. Quaife. 1964. Chronic oral toxicity of aldrin and dieldrin in rats and dogs. Food Cosmet. Toxicol. 2: 551-562.

Haworth, S., T. Lawlor, K. Mortelmans, W. Speck and E. Zeigler. 1983. Salmonella mutagenicity test results for 250 chemicals. Environ. Mutag. 5(Suppl. 1): 1-142.

Majumdar, S.K., H.A. Kopelman and M.J. Schnitman. 1976. Dieldrin-induced chromosome damage in mouse bone-marrow and WI-38 human lung cells. J. Hered. 67: 303-307.

Majumdar, S.K., L.G. Maharam and G.A. Viglianti. 1977. Mutagenicity of dieldrin in the Salmonella-microsome test. J. Hered. 68: 184-185.

Markaryan, D.S. 1966. Cytogenic effect of some chlorinated insecticides on mouse bone-marrow cell nuclei. Soviet Genetics. 2(1): 80-82.

Marshall, T.C., H.W. Dorough and H.E. Swim. 1976. Screening of pesticides for mutagenic potential using Salmonella typhimurium mutants. J. Agric. Chem. 24: 560-563.

Meierhenry, E.F., B.H. Reuber, M.E. Gershwin, L.S. Hsieh and S.W. French.   1983. Deildrin-induced mallory bodies in hepatic tumors of mice of different   strains. Hepatology. 3: 90-95.

NCI (National Cancer Institute). 1978a. Bioassays of aldrin and dieldrin for possible carcinogenicity. DHEW Publication No. (NIH) 78-821. National Cancer   Institute Carcinogenesis Technical Report Series, No. 21. NCI-CG-TR-21.

NCI (National Cancer Institute). 1978b. Bioassays of aldrin and dieldrin for possible carcinogenicity. DHEW Publication No. (NIH) 78-822. National Cancer   Institute Carcinogenesis Technical Report Series, No. 22. NCI-CG-TR-22.

Probst, G.S., R.E. McMahon, L.W. Hill, D.Z. Thompson, J.K. Epp and S.B. Neal.   1981. Chemically induced unscheduled DNA synthesis in primary rat hepatocyte cultures: A comparison with bacterial mutagenicity using 218 chemicals.   Environ. Mutagen. 3: 11-32.

Reuber, M.D. 1974. Exhibit 42. Testimony at hearings on aldrin/dieldrin.   (Cited in: Epstein, 1975a)

Rocchi, P., P. Perocco, W. Alberghini, A. Fini and G. Prodi. 1980. Effect of pesticides on scheduled and unscheduled DNA synthesis of rat thymocytes and   human lymphocytes. Arch. Toxicol. 45: 101-108.

Shirasu, Y., M. Moriya, K. Kato, A. Furuhashi and T. Kada. 1976.   Mutagenicity screening of pesticides in the microbial system. Mutat. Res. 40(1): 19-30.

Song, J. and W.E. Harville. 1964. Carcinogenicity of aldrin and dieldrin in mouse and rat liver. Fed. Proc. Fed. Am. Soc. Exp. Biol. 23: 336.

Tennekes, H.A., A.S. Wright, K.M. Dix and J.H. Koeman. 1981. Effects of dieldrin, diet, and bedding on enzyme function and tumor incidence in livers of male CF-1 mice. Cancer Res. 41: 3615-3620.

Thorpe, E. and A.I.T. Walker. 1973. The toxicology of dieldrin (HEOD). Part II. Comparative long-term oral toxicology studies in mice with dieldrin, DDT, phenobarbitone, beta-BHC and gamma-BHC. Food Cosmet. Toxicol. 11: 433-441.

Treon, J.F. and F.P. Cleveland. 1955. Toxicity of certain chlorinated   hydrocarbon insecticides for laboratory animals, with special reference to aldrin and dieldrin. Agric. Food Chem. 3: 402-408.

Trepanier, G., F. Marchessault, J. Bansal and A. Chagon. 1977. Cytological effects of insecticides on human lymphoblastoid cell line. In Vitro. 13: 201.

US EPA. 1986. Carcinogenicity Assessment of Aldrin and Dieldrin. Prepared by Carcinogen Assessment Group, Office of Health and Environmental Assessment,

Washington, DC for Hazard Evaluation Division, Office of Pesticide Programs, Office of Pesticides and Toxic Substances. OHEA-C-205.

Van Raalte, H.G.S. 1977. Human experience with dieldrin in perspective. Ecotox. Environ. Saf. 1: 203-210.

Wade, M.J., J.W. Moyer and C.H. Hine. 1979. Mutagenic action of a series of epoxides. Mutat. Res. 66(4): 367-371.

Walker, A.I.T., D.E. Stevenson, J. Robinson, E. Thorpe and M. Roberts. 1969. The toxicology and pharmacodynamics of dieldrin (HEOD): Two year oral exposures of rats and dogs. Toxicol. Appl. Pharmacol. 15: 345-373.

Walker, A.I.T., E. Thorpe and D.E. Stevenson. 1972. The toxicology of dieldrin (HEOD). I. Long-term oral toxicity studies in mice. Food Cosmet. Toxicol. 11: 415-432.

VI.D. DRINKING WATER HA REFERENCES

US EPA. 1989. Drinking Water Health Advisories: Pesticides. Lewis Publishers, Chelsea, MI. p. 299-312.

Walker, A.I.T., D.E. Stevenson, J. Robinson, E. Thorpe and M. Roberts. 1969. The toxicology and pharmacodynamics of dieldrin (HEOD). Two-year oral exposures of rats and dogs. Toxicol. Appl. Pharmacol. 15: 345-373.

VII. REVISION HISTORY

Date	Section	Description
09/07/88	I.A.	Oral RfD summary on-line
09/07/88	II.	Carcinogen summary on-line
03/01/90	II.A.2.	Ditraglia citation clarified
03/01/90	II.A.3.	Reuber citation year and Deichman spelling corrected
03/01/90	II.A.4.	Shirasu citation year corrected
03/01/90	II.B.2.	Reuber citation year corrected
03/01/90	VI.	Bibliography on-line
04/01/90	VI.C.	Treon and Cleveland, 1955 citation corrected
09/01/90	I.A.	Text edited
09/01/90	II.	Text edited
09/01/90	III.A.	Health Advisory on-line
09/01/90	VI.	Health Advisory references added
01/01/91	II.	Text edited
01/01/91	II.C.1.	Inhalation slope factor removed (global change)
01/01/92	IV.	Regulatory Action section on-line
07/01/93	II.D.3.	Secondary contact's phone number changed

SYNONYMS

ALVIT
COMPOUND 497
DIELDREX
Dieldrin
DIELDRINE
DIELDRITE
1,4:5,8-DIMETHANONAPHTHALENE,
1,2,3,4,10,10-HEXACHLORO-6,7-
EPOXY-1,4,4a,5,6,7,
  8,8a-OCTAHYDRO, endo, exo-
ENT 16,225
HEOD

HEXACHLOROEPOXYOCTAHYDRO-
endo,exo-DIMETHANONAPHTHALENE
3,4,5,6,9,9-HEXACHLORO-
1a,2,2a,3,6,6a,7,7a-OCTAHYDRO-
2,7:3,6-DIMETHANONAPHTH
  (2,3-b)OXIRENE
ILLOXOL
NA 2761
NCI-C00124
OCTALOX
PANORAM D-31
QUINTOX
RCRA WASTE NUMBER P037

Source: *Instant EPA's IRIS*, 1997, Instant Reference Sources, Inc., 7605 Rockpoint Drive, Austin, TX 78731 (Fax: 512-345-2386; Internet URL, http://www.instantref.com/inst-ref.htm).

Production, Use, and Pesticide Labeling Information

Dieldrin is not currently listed in EPA's Pesticide Factsheet database and there is no indication of whether it will be added or not in the future.

Di(2-ethylhexyl) phthalate

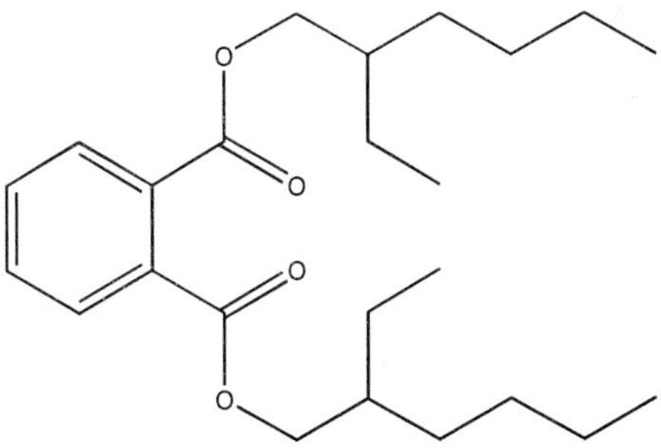

Introduction

Phthalates, best known for their ability to make plastics flexible are the most abundant industrial contaminants in the environment. [811] Di(2-ethylhexyl) phthalate is used as a plasticizer for polyvinyl chloride, especially in the manufacture of medical devices, and as a plasticizer for resins and elastomers. Some PVC products may contain up to 40% di(2-ethylhexyl) phthalate. It is also used in heat-seal coatings on metal foils such as those found in yoghurts, cream and indicudual portions of milk and in aluminum paper-foil laminates. [813] In addition, it is a solvent in erasable ink and dielectric fluid. It is also used in vacuum pumps, as an acaricide for use in orchards, an inert ingredient in pesticides, a detector for leaks in respirators, testing of air filtration systems and component in cosmetic products.

Di(2-ethylhexyl) phthalate in water will biodegrade (its half-life is 2-3 weeks), adsorb to sediments and bioconcentrate in aquatic organisms. Atmospheric di(2-ethylhexyl) phthalate will be carried long distances and be removed by rain. Human exposure will occur in occupational settings and from air, from consumption of drinking water, food (especially fish etc., where bioconcentration can occur) and food wrapped in PVC, as well as during blood transfusions from PVC blood bags. [828]

Identifying Information

CAS NUMBER: 117-81-7

NIOSH Registry Number: TI0350000

SYNONYMS: DEHP, Bis(2-ethylhexyl) phthalate, Dioctyl phthalate, 1,2-Benzenedicarboxylic acid, bis(2-ethylhexyl) ester

Endocrine Disruptor Information

Di(2-ethylhexyl) phthalate is a suspected EED that is listed by the World Wildlife Foundation, Canada [813].

527

Chemical and Physical Properties

CHEMICAL FORMULA: $C_{24}H_{38}O_4$

MOLECULAR WEIGHT: 390.54

WLN: 4Y2&1OVR BVO1Y4&2

PHYSICAL DESCRIPTION: colorless oily liquid

SPECIFIC GRAVITY: 0.9861 @ 20/20°C [051], [062], [395], [421]

DENSITY: 0.9732 g/mL @ 24°C [052]

MELTING POINT: -50°C [205], [269], [275], [430]

BOILING POINT: 384°C [047], [205], [269], [275]

SOLUBILITY: Water: <0.1 mg/mL @ 22°C; DMSO: 10-50 mg/mL @ 22°C 95%; Ethanol: >=100 mg/mL @ 22°C; Acetone: >=100 mg/mL @ 22°C

OTHER SOLVENTS: Mineral oil: Miscible [051], [062], [395], [421] Hexane: Miscible [421]

OTHER PHYSICAL DATA: Refractive index: 1.4853 @ 20°C [047], [205], [269], [275]; Pour point: -46°C [062], [395]; Mild odor [371] Boiling point: 222-230°C @ 4 mm Hg [047] 231°C @ 5 mm Hg [062], [395], [421]; Specific gravity: 0.9732 @ 24/21°C [052] 0.981 @ 20/4°C [269] 0.981 @ 25/25°C [205]; Viscosity: 81.4 centipoise @ 20°C [062,395,421]; Lambda max: 281(shoulder) nm, 275 nm, 225 nm (epsilon = 1130,1260,873) [052]; Liquid surface tension (estimated): 15 dynes/cm [371]; Liquid water interfacial tension (estimated): 30 dynes/cm [371]; Heat of combustion: -8410 cal/g [371].

Source: *Instant EPA's Air Toxics*, 1994, Instant Reference Sources, Inc., 7605 Rockpoint Drive, Austin, TX 78731 (Fax: 512-345-2386; Internet URL, http://www.instantref.com/inst-ref.htm).

Environmental Reference Materials

A source of EED environmental reference materials for Di(2-ethylhexyl) phthalate is Radian International LLC, Austin, TX (Fax 512-454-0268; Internet URL, http://www.radian.com/standards); Catalog No. ERB-048S

Hazardous Properties

ACUTE/CHRONIC HAZARDS: Di(2-ethylhexyl) phthalate is an irritant and may be harmful if ingested or inhaled [269]. It may be poorly absorbed through the skin and has low toxicity by all routes of exposure [421]. When heated to decomposition it emits acrid smoke and toxic fumes of carbon monoxide, carbon dioxide and traces of incompletely burned carbon products [043], [058], [269].

HAP WEIGHTING FACTOR: 1 [713]

VOLATILITY:
      Vapor pressure: 1.32 mm Hg @ 200°C [395], [421]
      Vapor density: 13.45 [055]

FIRE HAZARD: Di(2-ethylhexyl) phthalate has a flash point of 207°C (405°F); [205], [269], [275] it is combustible. Fires involving this material may be controlled with a dry chemical, carbon dioxide or Halon extinguisher.

The autoignition temperature of di(2-ethylhexyl) phthalate is 390°C (735 F) [451].

LEL: 0.3% @ 245°C [451]

REACTIVITY: Di(2-ethylhexyl) phthalate is incompatible with oxidizing materials [051], [058], [269], [346]. It is also incompatible with nitrates, strong acids and strong alkalis [346].

STABILITY: Di(2-ethylhexyl) phthalate is stable under normal laboratory conditions [051], [058], [371]. UV spectrophotometric stability screening indicates that solutions of it in 95% ethanol are stable for at least 24 hours [700].

Source: *Instant EPA's Air Toxics*, 1994, Instant Reference Sources, Inc., 7605 Rockpoint Drive, Austin, TX 78731 (Fax: 512-345-2386; Internet URL, http://www.instantref.com/inst-ref.htm).

Medical Symptoms of Exposure

Symptoms of exposure may include irritation of the eyes, mucous membranes [346], and skin [043], [058], [395]. It may also cause nausea, diarrhea [346], and mild gastric disturbances [151].

Source: *Instant EPA's Air Toxics*, 1994, Instant Reference Sources, Inc., 7605 Rockpoint Drive, Austin, TX 78731 (Fax: 512-345-2386; Internet URL, http://www.instantref.com/inst-ref.htm).

Toxicological Information

The toxicological information below is gathered from several sources including two national databases: one from the *National Toxicology Program Chemical Repository Database* and another from *EPA's Integrated Risk Information System (IRIS)*.

Toxicology Information from the National Toxicology Program

CARCINOGENICITY:

Tumorigenic Data:

Type of Effect	Route/Animal	Amount Dosed (Notes)
TDLo:	orl-rat	216 gm/kg/2Y-C
TDLo:	orl-mus	260 gm/kg/2Y-C
TD:	orl-mus	519 gm/kg/2Y-C
TD:	orl-mus	120 gm/kg/24W-C
TD:	orl-mus	262 gm/kg/2Y-C
TD:	orl-rat	433 gm/kg/2Y-C
TD:	orl-rat	524 gm/kg/2Y-C
TD:	orl-rat	438 gm/kg/2Y-C

Review:

IARC Cancer Review: Animal Sufficient Evidence
IARC possible human carcinogen (Group 2B) [015,395,610]
Status: NTP Carcinogenesis Bioassay (Feed); Positive: Male and Female Rat,
Male and Female Mouse [620]
NTP Fourth Annual Report on Carcinogens, 1984
NTP Carcinogenesis Studies; selected, July 1989
NTP anticipated human carcinogen [610]

TERATOGENICITY: See RTECS printout for data

MUTATION DATA: See RTECS printout for data

TOXICITY:

Typ.dose	Mode	Specie	Amount	Units	Other
LD50	ipr	mus	14	gm/kg	
LD50	ipr	rat	30700	mg/kg	
LD50	ivn	mus	1060	mg/kg	
LD50	ivn	rat	250	mg/kg	
LD50	orl	gpg	26	gm/kg	
TDLo	orl	man	143	mg/kg	
LD50	orl	mus	30	gm/kg	
LD50	orl	rat	30600	mg/kg	
LD50	orl	rbt	34	gm/kg	
LD50	skn	gpg	10	gm/kg	
LD50	skn	rbt	2 5	gm/kg	

OTHER TOXICITY DATA:

> Skin and Eye Irritation Data:
> skn-rbt 500 mg/24H MLD
> eye-rbt 500 mg
> eye-rbt 500 mg/24H MLD

> Review: Toxicology Review-8
> Status: EPA Genetox Program 1988, Inconclusive: In vitro SCE-nonhuman
> EPA Genetox Program 1988, Positive: Carcinogenicity-mouse/rat; Rodent
> dominant lethal
> EPA TSCA Chemical Inventory, 1986
> EPA TSCA 8(a) Preliminary Assessment Information, Final Rule
> EPA TSCA Section 8(e) Status Report 8EHQ-0982-0457
> EPA TSCA Test Submission (TSCATS) Data Base, September 1989
> NIOSH Analytical Methods: See Di(2-Ethylhexyl)phthalate, 5020

SAX TOXICITY EVALUATION:

> THR: Poison by intravenous route. Suspected human carcinogen and an
> experimental teratogen. Affects the human gastrointestinal tract. A mild skin
> and eye irritant.

Source: *Instant Tox-Base*, 1995, Instant Reference Sources, Inc., 7605 Rockpoint Drive, Austin, TX 78731 (Fax: 512-345-2386; Internet URL, http://www.instantref.com/inst-ref.htm).

Toxicology Information from EPA's Integrated Risk Information System (IRIS)

I. CHRONIC HEALTH HAZARD ASSESSMENTS FOR NONCARCINOGENIC EFFECTS

I.A. REFERENCE DOSE FOR CHRONIC ORAL EXPOSURE (RfD)

The Reference Dose (RfD) is based on the assumption that thresholds exist for certain toxic effects such as cellular necrosis, but may not exist for other toxic effects such as carcinogenicity. In general, the RfD is an estimate (with uncertainty spanning perhaps an order of magnitude) of a daily exposure to the human population (including sensitive subgroups) that is likely to be without an appreciable risk of deleterious effects during a lifetime. Please refer to Background Document 1 in Service Code 5 for an elaboration of these concepts. RfDs can also be derived for the noncarcinogenic health effects of compounds which are also carcinogens. Therefore, it is essential to refer to other sources of information concerning the carcinogenicity of this substance. If the US EPA has evaluated this substance for potential human carcinogenicity, a summary of that evaluation will be contained in Section II of this file when a review of that evaluation is completed.

I.A.1. ORAL RfD SUMMARY

Critical Effect	Experimental Doses*	UF	MF	RfD
Increased relative liver weight	NOAEL: none	1000	1	2E-2 mg/kg/day
	LOAEL: 0.04% of diet (19 mg/kg bw/day)			

*Conversion Factors: none

Guinea Pig Subchronic-to-Chronic Oral Bioassay Carpenter et al., 1953

I.A.2. PRINCIPAL AND SUPPORTING STUDIES (ORAL RfD)

Carpenter, C.P., C.S. Weil and H.F. Smyth. 1953. Chronic oral toxicity of di(2-ethylhexyl) phthalate for rats and guinea pigs. Arch. Indust. Hyg. Occup. Med. 8: 219-226.

The following numbers of guinea pigs were fed diets containing DEHP for a period of 1 year: 24 males and 23 females consumed feed containing 0.13% DEHP; 23 males and 23 females consumed feed containing 0.04% DEHP; and 24 males and 22 females were fed the control diet. These dietary levels corresponded to 64 or 19 mg/kg bw/day based on measured food consumption. No treatment-related effects were observed on mortality, body weight, kidney weight, or gross pathology and histopathology of kidney, liver, lung, spleen, or testes. Statistically significant increases in relative liver weights were observed in both groups of treated females (64 and 19 mg/kg bw/day).

Groups of 32 male and 32 female Sherman rats were maintained for 2 years on diets containing either 0.04, 0.13 or 0.4% DEHP (equivalent to 20, 60, and about 195 mg/kg bw/day based on measured food consumption). An F1 group of 80 animals was fed the 0.04% diet for 1 year. Mortality in the F1 treated and control groups was high; 46.2 and 42.7%, respectively, survived to 1 year. There was, however, no effect of treatment on either parental or F1 group mortality, life expectancy, hematology, or histopathology of organs. Both parental and F1 rats receiving the 0.4% DEHP diet were retarded in growth and had increased kidney and liver weights.

It appears that guinea pigs offer the more sensitive animal model for DEHP toxicity. A LOAEL in this species is determined to be 19 mg/kg/day.

I.A.3. UNCERTAINTY AND MODIFYING FACTORS (ORAL RfD)

UF -- Factors of 10 each were used for interspecies variation and for protection of sensitive human subpopulations. An additional factor of 10 was used since the guinea pig exposure was longer than subchronic but less than lifetime, and because, while the RfD is set on a LOAEL, the effect observed was considered to be minimally adverse.

MF -- None

I.A.4. ADDITIONAL COMMENTS (ORAL RfD)

Dietary levels of 0, 0.01, 0.1, and 0.3% DEHP (greater than 99% pure) were administered to male and female CD-1 mice that were examined for adverse fertility and reproductive effects using a continuous breeding protocol. DEHP was a reproductive toxicant in both sexes significantly decreasing fertility and the proportion of pups born alive per litter at the 0.3% level, and inducing damage to the seminiferous tubules. DEHP has been observed to be both fetotoxic and teratogenic (Singhe, 1972; Shiot and Nishimura, 1982).

I.A.5. CONFIDENCE IN THE ORAL RfD

The study by Carpenter et al. (1953) utilized sufficient numbers of guinea pigs and measured multiple endpoints. The fact that there were only two concentrations of DEHP tested precludes a rating higher than medium. Since there are corroborating chronic animal bioassays, the data base is likewise rated medium. Medium confidence in the RfD follows.

I.A.6. EPA DOCUMENTATION AND REVIEW OF THE ORAL RfD

The RfD has been reviewed by the RfD Work Group. Documentation may be found in the meeting notes of 01/22/86.

Other EPA Documentation -- None

Agency Work Group Review -- 01/22/86

Verification Date -- 01/22/86

I.A.7. EPA CONTACTS (ORAL RfD)

W. Bruce Peirano / NCEA -- (513)569-7540

I.B. REFERENCE CONCENTRATION FOR CHRONIC INHALATION EXPOSURE (RfC)

Not available at this time.

II. CARCINOGENICITY ASSESSMENT FOR LIFETIME EXPOSURE

Section II provides information on three aspects of the carcinogenic risk assessment for the agent in question; the US EPA classification, and quantitative estimates of risk from oral exposure and from inhalation exposure. The classification reflects a weight-of-evidence judgment of the likelihood that the agent is a human carcinogen. The quantitative risk estimates are presented in three ways. The slope factor is the result of application of a low-dose extrapolation procedure and is presented as the risk per (mg/kg)/day. The unit risk is the quantitative estimate in terms of either risk per ug/L drinking water or risk per ug/cu.m air breathed. The third form in which risk is presented is a drinking water or air concentration providing cancer risks of 1 in 10,000, 1 in

100,000 or 1 in 1,000,000. Background Document 2 (Service Code 5) provides details on the rationale and methods used to derive the carcinogenicity values found in IRIS. Users are referred to Section I for information on long-term toxic effects other than carcinogenicity.

II.A. EVIDENCE FOR CLASSIFICATION AS TO HUMAN CARCINOGENICITY

II.A.1. WEIGHT-OF-EVIDENCE CLASSIFICATION

Classification -- B2; probable human carcinogen.

Basis -- Orally administered DEHP produced significant dose-related increases in liver tumor responses in rats and mice of both sexes.

II.A.2. HUMAN CARCINOGENICITY DATA

Inadequate. Thiess et al. (1978) conducted a mortality study of 221 DEHP production workers exposed to unknown concentrations of DEHP for 3 months to 24 years. Workers were followed for a minimum of 5 to 10 years (mean follow-up time was 11.5 years). Eight deaths were reported in the exposed population. Deaths attributable to pancreatic carcinoma (1 case) and uremia (1 case in which the workers also had urethral and bladder papillomas) were significantly elevated in workers exposed for >15 years when compared to the corresponding age groups in the general population. The study is limited by a short follow-up period and unquantified worker exposure. Results are considered inadequate for evidence of a causal association.

II.A.3. ANIMAL CARCINOGENICITY DATA

Sufficient. In an NTP (1982) study, 50 male and 50 female fisher 344 rats per group were fed diets containing 0, 6000 or 12,000 ppm DEHP for 103 weeks. Similarly, groups of 50 male and 50 female B6C3F1 mice were given 0, 3000 or 6000 ppm DEHP in the diet for 103 weeks. Animals were killed and examined histologically when morbund or after 105 weeks. No clinical signs of toxicity were observed in either rats or mice. A statistically significant increase in the incidence of hepatocellular carcinomas and combined incidence of carcinomas and adenoma were observed in female rats and both sexes of mice.   The combined incidence of neoplastic nodules and hepatocellular carcinomas was statistically significantly increased in the high-dose male rats. A positive dose response trend was also noted.

Carpenter et al. (1953) found no malignant tumors in treated groups of 32 male and 32 female Sherman rats. Animals were given 400, 1300 or 4000 ppm DEHP in the diet for 1 year and reduced to a maximum of 8 males and 8 females and treated for another year. Controls, F1 and 4000 ppm groups were sacrificed after being maintained on control or 4000 ppm diets for 1 year. Only 40 to 47% of the animals in each group, including F1 animals, survived 1 year. Thus, an insufficient number of animals were available for a lifetime evaluation.

Carpenter et al. (1953) did not find a carcinogenic effect in guinea pigs and dogs exposed to 1300 or 4000 ppm DEHP. Both guinea pigs and dogs were terminated after 1 year of exposure. The treatment and survival periods for these animals were considerably below their lifetimes.

II.A.4. SUPPORTING DATA FOR CARCINOGENICITY

Studies indicate that DEHP is not a direct acting mutagen in either a forward mutation assay in Salmonella typhimurium (Seed, 1982) or the recassay in Bacillus subtilis (Tomita et al., 1982). DEHP did not induce mutations in a modified reverse mutation plate incorporation assay in Salmonella strains TA100 and TA98 at concentrations up to 1000 ug/plate in the presence or absence of S9 hepatic homogenate (Kozumbo et al., 1982). MEHP, the monoester form of DEHP and a metabolite is positive in the rec assay and in the reverse mutation assay in Salmonella. In the absence of exogenous metabolism MEHP produced chromosomal aberrations and sister chromatid exchanges in V79 cells. Both DEHP and MEHP induced chromosomal aberrations and morphological transformation in cultured fetal Syrian hamster cells exposed in utero (Tomita et al., 1982). Chromosomal effects were not found in CHO mammalian cells (Phillips et al., 1982) exposed to DEHP. DEHP was weakly positive with metabolic activation in only one of several studies testing for mutagenic activity at the thymidine kinase locus in L5178Y mouse lymphoma cells (Ashby et al., 1985). DEHP is a potent inducer of hepatic peroxisomal enzyme activity (Ganning et al., 1984).

II.B. QUANTITATIVE ESTIMATE OF CARCINOGENIC RISK FROM ORAL EXPOSURE

II.B.1. SUMMARY OF RISK ESTIMATES

Oral Slope Factor -- 1.4E-2/mg/kg/day

Drinking Water Unit Risk -- 4.0E-7 per (ug/L)

Extrapolation Method -- Linearized multistage procedure, extra risk

Drinking Water Concentrations at Specified Risk Levels:

Risk Level	Concentration
E-4 (1 in 10,000)	3E+2 ug/L
E-5 (1 in 100,000)	3E+1 ug/L
E-6 (1 in 1,000,000)	3E+0 ug/L

II.B.2. DOSE-RESPONSE DATA (CARCINOGENICITY, ORAL EXPOSURE)

Tumor Type -- Mouse/B6C3Fl, male
Test Animals -- hepatocellular carcinoma and adenoma
Route -- diet
Reference -- NTP, 1982

Dose Administered (ppm)	Human Equivalent (mg/kg)/day	Tumor Incidence
0	0	14/50
3000	32	25/48
6000	65	29/50

II.B.3. ADDITIONAL COMMENTS (CARCINOGENICITY, ORAL EXPOSURE)

In this study powdered rodent meal was provided in such a way that measured food consumption could include significant waste and spillage rather than true food intake. For this reason a standard food consumption rate of 13% mouse body weight was used in the dose conversion.

DEHP is hydrolyzed to monoesters including MEHP (Pollack et al., 1985; Lhuguenot et al., 1985; Kluwe, 1982). Although several species of animals have been determined to excrete glucuronide conjugates of monoethylhexyl phthalate (MEHP) upon exposure to DEHP, rats do not (Tanaka et al., 1975; Williams and Blanchfield, 1975; Albro et al., 1982).

Slope factors based on combined hepatocellular carcinoma and neoplastic nodule incidences were 4.5E-3/mg/kg/day for female rats, 3.2E-3/mg/kg/day for male rats. A slope factor based on hepatocellular adenomas or carcinomas in female mice is 1.0E-2/mg/kg/day.

The unit risk should not be used if the water concentration exceeds 4E+4 ug/L, since above this concentration the slope factor may differ from that stated.

II.B.4. DISCUSSION OF CONFIDENCE (CARCINOGENICITY, ORAL EXPOSURE)

An adequate number of animals was observed and a statistically significant increase in incidence of liver tumors was seen in both sexes and were dose dependent in both sexes of mice and female rats. A potential source of variability in the NTP study is the possibility of feed scattering. The above calculations are based on standard food consumption rates for mice (13% of body weight) and rats (5% of body weight).

II.C.   QUANTITATIVE ESTIMATE OF CARCINOGENIC RISK FROM INHALATION EXPOSURE

Not available.

II.D. EPA DOCUMENTATION, REVIEW, AND CONTACTS (CARCINOGENICITY ASSESSMENT)

II.D.1. EPA DOCUMENTATION

Source Document -- US EPA, 1988

The values in the 1988 Drinking Water Criteria Document for Phthalic Acid Esters (External Review Draft) have received Agency review.

II.D.2. REVIEW (CARCINOGENICITY ASSESSMENT)

Agency Work Group Review -- 08/26/87, 10/07/87

Verification Date -- 10/07/87

II.D.3. US EPA CONTACTS (CARCINOGENICITY ASSESSMENT)

Brian J. Commons / OST -- (202)260-7589

Linda R. Papa / NCEA -- (513)569-7587

III. HEALTH HAZARD ASSESSMENTS FOR VARIED EXPOSURE DURATIONS

[Ed. Note: EPA does not have information yet in this section of IRIS].

IV. US EPA REGULATORY ACTIONS

EPA risk assessments may be updated as new data are published and as assessment methodologies evolve. Regulatory actions are frequently not updated at the same time. Compare the dates for the regulatory actions in this section with the verification dates for the risk assessments in sections I and II, as this may explain inconsistencies. Also note that some regulatory actions consider factors not related to health risk, such as technical or economic feasibility. Such considerations are indicated for each action. In addition, not all of the regulatory actions listed in this section involve enforceable federal standards. Please direct any questions you may have concerning these regulatory actions to the US EPA contact listed for that particular action. Users are strongly urged to read the background information on each regulatory action in Background Document 4 in Service Code 5.

IV.A. CLEAN AIR ACT (CAA)

[Ed. Note: EPA does not have information yet in this section of IRIS].

IV.B. SAFE DRINKING WATER ACT (SDWA)

IV.B.1. MAXIMUM CONTAMINANT LEVEL GOAL (MCLG) for Drinking Water

Value -- 0 mg/L (Proposed, 1990)

Considers technological or economic feasibility? -- NO

Discussion -- The proposed MCLG for di(2-ethylhexyl) phthalate is zero based on the evidence of carcinogenic potential (B2)

Reference -- 55 FR 30370 (07/25/90)

EPA Contact -- Health and Ecological Criteria Division / OST / (202) 260-7571 / FTS 260-7571; or Safe Drinking Water Hotline / (800) 426-4791

IV.B.2. MAXIMUM CONTAMINANT LEVEL (MCL) for Drinking Water

Value -- 0.004 mg/L (Proposed, 1990)

Considers technological or economic feasibility? -- YES

Discussion -- MCL is based on 10x the MDL and is associated with a maximum lifetime individual risk of 1 E-6.

Monitoring requirements -- Community and non-transient water system monitoring based on state vulnerability assessment; vulnerable systems to be monitored quarterly for one year; repeat monitoring dependent upon detection and size of system.

Analytical methodology -- Photoionization/gas chromatography (EPA 502.2); gas chromatographic/mass spectrometry (EPA 524.1, 524.2): PQL= 0.004 mg/L.

Best available technology -- Granular activated carbon

Reference -- 55 FR 30370 (07/25/90)

EPA Contact -- Drinking Water Standards Division / OGWDW / (202) 260-7575 / FTS 260-7575; or Safe Drinking Water Hotline / (800) 426-4791

IV.B.3. SECONDARY MAXIMUM CONTAMINANT LEVEL (SMCL) for Drinking Water

[Ed. Note: EPA does not have information yet in this section of IRIS].

IV.B.4. REQUIRED MONITORING OF "UNREGULATED" CONTAMINANTS

Status -- Listed (Final, 1991)

Discussion -- "Unregulated" contaminants are those contaminants for which EPA establishes a monitoring requirement but which do not have an associated final MCLG, MCL, or treatment technique. EPA may regulate these contaminants in the future.

Monitoring requirement -- All systems to be monitored unless a vulnerability assessment determines the system is not vulnerable.

Analytical methodology -- Gas chromatography (EPA 506); gas chromatography/ mass spectrometry (EPA 525).

Reference -- 56 FR 3526 (01/30/91)

EPA Contact -- Drinking Water Standards Division / OGWDW / (202) 260-7575 / FTS 260-7575; or Safe Drinking Water Hotline / (800) 426-4791

IV.C. CLEAN WATER ACT (CWA)

IV.C.1. AMBIENT WATER QUALITY CRITERIA, Human Health

Water and Fish Consumption: 1.5E+4 ug/L

Fish Consumption Only: 5E+4 ug/L

Considers technological or economic feasibility? -- NO

Discussion -- The WQC of 1.5E+4 ug/L is based on consumption of contaminated aquatic organisms and water. A WQC of 5E+4 ug/L has also been established based on consumption of contaminated aquatic organisms alone.

Reference -- 45 FR 79318 (11/28/80)

EPA Contact -- Criteria and Standards Division / OWRS (202)260-1315 / FTS 260-1315

IV.C.2. AMBIENT WATER QUALITY CRITERIA, Aquatic Organisms

Freshwater:
  Acute -- 4.0E+2 ug/L
  Chronic -- 3.6E+2 ug/L

Marine:
  Acute -- 4.0E+2 ug/L
  Chronic -- 3.6E+2 ug/L

Considers technological or economic feasibility? -- NO

Discussion -- Criteria were derived from a minimum data base consisting of acute and chronic tests on a variety of species. EPA is currently considering withdrawing some or all the values.

Reference -- 45 FR 79318 (11/28/80)

EPA Contact -- Criteria and Standards Division / OWRS (202)260-1315 / FTS 260-1315

IV.D. FEDERAL INSECTICIDE, FUNGICIDE, AND RODENTICIDE ACT (FIFRA)

[Ed. Note: EPA does not have information yet in this section of IRIS].

IV.E. TOXIC SUBSTANCES CONTROL ACT (TSCA)

[Ed. Note: EPA does not have information yet in this section of IRIS].

IV.F. RESOURCE CONSERVATION AND RECOVERY ACT (RCRA)

IV.F.1. RCRA APPENDIX IX, for Ground Water Monitoring

Status -- Listed

Reference -- 52 FR 25942 (07/09/87)

EPA Contact -- RCRA/Superfund Hotline (800)424-9346 / (202)260-3000 / FTS 260-3000

IV.G. SUPERFUND (CERCLA)

IV.G.1. REPORTABLE QUANTITY (RQ) for Release into the Environment

Value (status) -- 100 pounds (Final, 1989)

Considers technological or economic feasibility? -- NO

Discussion -- The 100-pound RQ is based on assessment for potential carcinogenicity. Available data indicate a hazard ranking of low based on a potency factor of 0.015/mg/kg/day and weight-of-evidence group B2, which corresponds to an RQ of 100 pounds.

Reference -- 54 FR 33418 (08/14/89)

EPA Contact -- RCRA/Superfund Hotline (800)424-9346 / (202)260-3000 / FTS 260-3000

VI. BIBLIOGRAPHY

VI.A. ORAL RfD REFERENCES

Carpenter, C.P., C.S. Weil and H.F. Smyth. 1953. Chronic oral toxicity of di(2-ethylhexyl) phthalate for rats and guinea pigs. Arch. Indust. Hyg. Occup. Med. 8: 219-226.

NTP (National Toxicology Program). 1984. Di(2-ethylhexyl) phthalate: Reproduction and fertility assessment in CD-1 mice when administered by gavage. Final Report. NTP-84-079. NTP, Research Triangle Park, NC.

Shiota, K. and H. Nishimura. 1982. Teratogenicity of di-2-ethylhexyl phthalate and di-n-butyl phthalate in mice. Environ. Health Perspect. 45(0): 65-70.

Singhe, A.R., W.H. Lawrence and J. Autian. 1972. Teratogenicity of phthalate esters in rats. J. Pharmacol. Sci. 61: 51.

VI.B. INHALATION RₜC REFERENCES

None

VI.C. CARCINOGENICITY ASSESSMENT REFERENCES

Albro, P.W., J.T. Corbett, J.L. Schroeder, et al. 1982. Pharmacokinetics, interactions with macromolecules and species differences in metabolism of DEHP. Environ. Health Perspect. 45: 19-25.

Ashby, J., F.J. de Serres, M. Draper, et al. 1985. Evaluation of short-term tests for carcinogens. Report of the International Programme on Chemical Safety's Collaborative Study on In Vitro Assays. Elsevier Science Publishers, Amsterdam.

Carpenter, C.P., C.S. Weil and H.F. Smith, Jr. 1953. Chronic oral toxicity of di-(2-ethylhexyl) phthalate for rats, guinea pigs and dogs. AMA Arch. Ind. Hyg. Occup. Med. 8: 219-226.

Ganning, A.E., V. Brunk and G. Dallner. 1984. Phthalate esters and their effect on the liver. Hepatology. 4(3): 541-547.

Kluwe, W.M. 1982. Overview of phthalate ester pharmacokinetics in mammalian species. Environ. Health Perspect. 45: 3-10.

Kozumbo, W.J., R. Kroll and R.J. Rubin. 1982. Assessment of the mutagenicity of phthalate esters. Environ. Health Perspect. 45: 103-109.

Lhuguenot, J.C., A.M. Mitchell, G. Milner, E.A. Lock and C.R. Elcombe. 1985. The metabolism of di-(2-ethylhexyl) phthalate (DEHP) and mono-(2-ethylhexyl) phthalate (MEHP) in rats: In vivo and in vitro dose and time dependency of metabolism. Toxicol. Appl. Pharmacol. 80: 11-22.

NTP (National Toxicology Program). 1982. Carcinogenesis bioassay of di-(2-ethylhexyl) phthalate (CAS No. 117-81-7) in F344 rats and B6C3F, mice (feed study). NTP Tech. Rep. Ser. TR No. 217, NTP, Research Triangle Park, NC.

Phillips, B.J., T.E.B. James and S.D. Gangolli. 1982. Genotoxicity studies of di-(2-ethylhexyl) phthalate and its metabolites in CHO cells. Mutat. Res. 102: 297-304.

Pollack, G.M., R.C. Li, J.C. Ermer and D.D. Shen. 1985. Effects of route of administration and repetitive dosing on the disposition kinetics of di-(2-ethylhexyl) phthalate and its mono-de-esterified metabolite in rats. Toxicol. Appl. Pharmacol. 79: 246-256.

Seed, J.L. 1982. Mutagenic activity of phthalate esters in bacterial liquid suspension assays. Environ. Health Perspect. 45: 111-114.

Tanaka, A., T. Adachi, T. Takahashi and T. Yamaha. 1975. Biochemical studies on phthalic esters. I. Elimination, distribution and metabolism of di-(2-ethylhexyl) phthalate in rats. Toxicology. 4: 253-264.

Thiess, A.M., R. Frentzel-Beyme and R. Wieland. 1978. Mortality study in workers exposed to di-(2-ethylhexyl) phthalate (DOP). In: Moglichkerten und Grenzen des Biological Monitoring. Arbeitsmedizinische Probleme des Dienstleistungsqewerbes. Arbeitsmedizinische kolloquium [Possibilities and Limits of Biological Monitoring. Problems of Occupational Medicine in Small Industries. Colloquim in Occupational Medicine], Frankfurt/M., May 1978. Stuttgart, A.W. Gentner, p. 155-164. (Ger.)

Tomita, I., Y. Nakamura, N. Aoki and N. Inui. 1982. Mutagenic/carcinogenic potential of DEHP and MEHP. Environ. Health Perspect. 45: 119-125.

Williams, D.T. and B.J. Blanchfield. 1975. The retention, distribution, excretion and metabolism of dibutylphthalate-7-14C in the rat. J. Agric. Food Chem. 23: 854-857.

US EPA. 1988. Drinking Water Criteria Document for Phthalic Acid Esters. Prepared by the Office of Health and Environmental Assessment, Environmental Criteria and Assessment Office, Cincinnati, OH for the Office of Drinking Water, Washington, DC. (External Review Draft).

VI.D. DRINKING WATER HA REFERENCES

None

VII. REVISION HISTORY

Date	Section	Description
03/01/88	I.A.2.	Text added to paragraph 1
09/07/88	II.	Carcinogen summary on-line
02/01/89	II.A.2.	Study description revised
02/01/89	II.D.3.	Primary contact's phone number corrected
07/01/89	VI.	Bibliography on-line
08/01/89	I.A.4.	Text revised
05/01/90	II.A.4.	Text revised
05/01/90	VI.C.	Kozumbo et al., 1982 citation added
05/01/91	I.A.2.	Corrected principal study title
05/01/91	I.A.2.	2nd para, line 3 units corrected from g/kg to mg/kg
08/01/91	II.D.3.	Primary and secondary contacts changed
08/01/91	IV.F.1.	EPA contact changed
09/01/91	All	Primary name changed from Bis(2-ethylhexyl) phthalate
01/01/92	I.A.7.	Secondary contact changed
01/01/92	IV.	Regulatory actions updated
02/01/93	II.D.3.	Primary contact changed

SYNONYMS

BEHP
Bis(2-ethylhexyl)-1,2-benzene-dicarboxylate
Bis(2-ethylhexyl) phthalate
Bisoflex 81
Bisoflex DOP
Compound 889
DAF 68
DEHP
Di(2-ethylhexyl)orthophthalate
Di(2-ethylhexyl) phthalate
Dioctyl phthalate
Di-sec-octyl phthalate
DOP
Ergoplast FDO
Ethylhexyl phthalate
2-Ethylhexyl phthalate
Eviplast 80
Eviplast 81
Fleximel
Flexol DOP
Flexol plasticizer DOP
Good-Rite GP 264
Hatcol DOP

Hercoflex 260
Kodaflex DOP
Mollan O
NCI- C52733
Nuoplaz DOP
Octoil
Octyl phthalate
Palatinol AH
Phthalic acid, Bis(2-ethylhexyl) ester
Phthalic acid, dioctyl ester
Pittsburgh PX-138
Platinol DOP
RC Plasticizer DOP
RCRA waste number U028
Reomol D 79P
Reomol DOP
Sicol 150
Staflex DOP
Truflex DOP
Vestinol AH
Vinicizer 80
Witcizer 312

Source: *Instant EPA's IRIS*, 1997, Instant Reference Sources, Inc., 7605 Rockpoint Drive, Austin, TX 78731 (Fax: 512-345-2386; Internet URL, http://www.instantref.com/inst-ref.htm).

Di-n-butyl phthalate

Introduction

Phthalates, best known for their ability to make plastics flexible are the most abundant industrial contaminants in the environment. [811] Di-n-butyl phthalate is a ubiquitous pollutant due to its widespread use primarily as a plasticizer in plastics; for example it is widely used in PVC and nitrocellulose polyvinyl acetate. Other applications include carpet backing, hair spray, nail polish, and glue. It may contaminate food through its use as a plasticizer in coatings on cellophane and from its use in inks. [813] It also is used in cosmetics, safety glass, insecticides, paper coatings, adhesives, elastomers and explosives. It is used as a solvent in polysulfide dental impression materials, solvent for perfume oils, perfume fixative, textile lubricating agent and solid rocket propellent.

Di-n-butyl phthalate may be released into the environment as emissions and in wastewater during its production and use, incineration of plastics and migration of the plasticizer from materials containing it. If released into water it will adsorb moderately to sediment and particulates in the water column. The di-n-butyl phthalate will disappear in 3-5 days in moderately polluted waters and generally within 3 weeks in cleaner bodies of water. It will not bioconcentrate in fish since it is readily metabolized. If spilled on land it will adsorb moderately to soil and slowly biodegrade (66 and 98% degradation in 26 weeks from two soils). Di-n-butyl phthalate is found in groundwater under rapid infiltration sites and elsewhere. It has been suggested that its tendency to form complexes with water-soluble fulvic acids, a component of soils, may aid its transport into groundwater. Although it degrades under anaerobic conditions, its fate in groundwater is unknown. If released into air, di-n-butyl phthalate is generally associated with the particulate fraction and will be subject to gravitational settling. Vapor phase di-n-butyl phthalate will degrade by reaction with photochemically produced hydroxyl radicals (estimated half-life 18 hr). Human exposure is from air, drinking water and food in addition to in the workplace. [828]

Identifying Information

CAS NUMBER: 84-74-2

NIOSH Registry Number: TI0875000

544

SYNONYMS: PHTHALIC ACID, DIBUTYL ESTER, O-BENZENEDICARBOXYLIC ACID, DIBUTYL ESTER, BENZENE-O-DICARBOXYLIC ACID DI-N-BUTYL ESTER, N-BUTYL PHTHALATE, DIBUTYL 1,2-BENZENEDICARBOXYLATE, DIBUTYL PHTHALATE, CELLUFLEX DPB, DBP, ELAOL, HEXAPLAS M/B, PALATINOL C, POLYCIZER DBP, PX 104, STAFLEX DBP, WITCIZER 300, NA 9095, RCRA WASTE NUMBER U069

Endocrine Disruptor Information

Di-n-butyl phthalate is a strongly suspected EED that it is listed by EPA [811] and the World Wildlife Fund [813] as a potential endocrine modifying chemical. Susan Jobling of Brunel University in Uxbridge, England has studied butyl benzyl phthalate (BBP) and di-n-butyl phthalate (DBP) for their estrogenicity. It seems that not only do they appear estrogenic, but they also stimulate the growth of breast cancer cells in culture. [864], [811]

Chemical and Physical Properties

CHEMICAL FORMULA: $C_{16}H_{22}O_4$

MOLECULAR WEIGHT: 278.35

WLN: 4OVR BVO2

PHYSICAL DESCRIPTION: Clear, colorless, viscous liquid

SPECIFIC GRAVITY: 1.047-1.049 @ 20/20°C [043]

DENSITY: 1.05 g/mL @ 25°C [058], [102]

MELTING POINT: -35°C [025], [205], [275], [371]

BOILING POINT: 340°C [017], [058], [275], [451]

SOLUBILITY: Water: <1 mg/mL @ 20°C [700]; DMSO: >=100 mg/mL @ 20°C [700]; 95% Ethanol: >=100 mg/mL @ 20°C [700]; Acetone: >=100 mg/mL @ 20°C [700]

OTHER SOLVENTS: Alcohol: Soluble [017]; Ether: Soluble [017], [031]; Benzene: Soluble [017], [031]; Most organic solvents: Miscible [062]

OTHER PHYSICAL DATA: Very weak, ammoniacal odor [102], [346]; Viscosity: 0.203 poise @ 20°C [062]; Refractive index: 1.4911 @ 20°C [017], [047]; 1.4915 @ 25°C [062]; Critical temperature: 500°C [371]; Critical pressure: 17 atmospheres [371]; Liquid surface tension: 34 dynes/cm [371]; Liquid water interfacial tension: 27 dynes/cm [371]; Heat of combustion: -7400 cal/g [371]; Vapor pressure: 0.1 mm Hg @ 115°C [055]; 760 mm Hg @ 340°C [038]; Vapor pressure: 5 mm Hg @ 182.1°C; 10 mm Hg @ 198.2°C [038]; Vapor pressure: 20 mm Hg @ 216.2°C; 40 mm Hg @ 235.8°C [038]; Vapor pressure: 60 mm Hg @ 247.8°C; 100 mm Hg @ 263.7°C [038]

Source: *The National Toxicology Program's Chemical Database, Volume 2*, 1992, CRC Press, Inc./Lewis Publishers, 2000 Corporate Blvd., Boca Raton, FL 33431 (Fax: 407-998-9114).

Environmental Reference Materials

A source of EED environmental reference materials for Di-n-butyl phthalate is Radian International LLC, Austin, TX (Fax 512-454-0268; Internet URL, http://www.radian.com/standards); Catalog No. ERD-018.

Hazardous Properties

ACUTE/CHRONIC HAZARDS: This compound is an irritant and may be toxic at sufficient concentrations. It is also a lachrymator. When heated to decomposition it emits toxic fumes of carbon dioxide and carbon monoxide. [058]

VOLATILITY: Vapor pressure: 1 mm Hg @ 147°C; 1.1 mm Hg @ 150°C [058]. Vapor density: 9.58 [043], [071]

FIRE HAZARD: Di-n-butyl phthalate has a flash point of 157°C (315°F); [043], [058], [071], [102] it is combustible. Fires involving this material can be controlled with a dry chemical, carbon dioxide or Halon extinguisher.

The autoignition temperature of this compound is 403°C (757°F) [043], [071], [102], [451].

LEL: 0.5% [058], [102]   UEL: Not available

REACTIVITY: Di-n-butyl phthalate can react violently with chlorine [036], [043], [066], [451]. It is incompatible with nitrates, strong oxidizers, strong alkalis and strong acids [102].

STABILITY: Di-n-butyl phthalate is stable under normal laboratory conditions. Solutions of Di-n-butyl phthalate in water, DMSO, 95% ethanol or acetone should be stable for 24 hours under normal lab conditions [700].

Source: *The National Toxicology Program's Chemical Database, Volume 7*, 1992, CRC Press, Inc./Lewis Publishers, 2000 Corporate Blvd., Boca Raton, FL 33431 (Fax: 407-998-9114).

Medical Symptoms of Exposure

Symptoms of exposure may include irritation of the eyes, nasal passages, throat and upper respiratory tract. It may cause nausea, conjunctivitis, profuse lacrimation, vomiting, dizziness, headache, tearing of eyes and photophobia [058]. It may also cause liver and kidney damage [301].

Source: *The National Toxicology Program's Chemical Database, Volume 4*, 1992, CRC Press, Inc./Lewis Publishers, 2000 Corporate Blvd., Boca Raton, FL 33431 (Fax: 407-998-9114).

Toxicological Information

The toxicological information below is gathered from several sources including two national databases: one from the *National Toxicology Program Chemical Repository Database* and another from *EPA's Integrated Risk Information System (IRIS)*.

Toxicology Information from the National Toxicology Program

CARCINOGENICITY:

Status:
NTP Carcinogenesis Studies; on test (prechronic studies), October 1988

TERATOGENICITY:

Reproductive Effects Data:

Type of Effect	Route/Animal	Amount Dosed (Notes)
TDLo	orl-rat	2520 mg/kg (1-21D preg)
TDLo	orl-rat	12600 mg/kg (1-21D preg)
TDLo	ipr-rat	1017 mg/kg (5-15D preg)
TDLo	ipr-rat	305 mg/kg (5-15D preg)
TDLo	ipr-rat	6 gm/kg (3-9D preg)
TDLo	orl-mus	8640 mg/kg (1-18D preg)
TDLo	orl-mus	7200 mg/kg (1-18D preg)
TDLo	orl-mus	16800 mg/kg (7D male)
TDLo	orl-gpg	14 gm/kg (7D male)
TDLo	orl-rat	8400 mg/kg (7D male)
TDLo	orl-mus	20 gm/kg (6-13D preg)
TDLo	orl-rat	16800 mg/kg (7D male)

MUTATION DATA:

Test	Lowest dose		Test	Lowest dose
mmo-sat	100 ug/plate		cyt-ham:fbr	30 mg/L/24H

TOXICITY:

Typ. Dose	Mode	Specie	Amount	Units	Other
TDLo	orl	hmn	140	mg/kg	
LD50	orl	rat	8000	mg/kg	
LC50	ihl	mus	25	gm/m3/2H	
LD50	ipr	rat	3050	mg/kg	
LD50	orl	mus	5289	mg/kg	
LD50	ipr	mus	3570	mg/kg	
LD50	ivn	mus	720	mg/kg	
LD50	orl	gpg	10	gm/kg	
LDLo	skn	rat	6	gm/kg	

OTHER TOXICITY DATA:

Review: Toxicology Review-5

Standards and Regulations:
    DOT-Hazard: ORM-E; Label: None

Status:
    EPA Genetox Program 1988, Negative: S cerevisiae-reversion
    EPA Genetox Program 1988, Inconclusive: In vitro SCE-nonhuman
    EPA TSCA Chemical Inventory, 1986
    EPA TSCA 8(a) Preliminary Assessment Information, Final Rule
    EPA TSCA Section 8(e) Status Report 8EHQ-0886-0620
    EPA TSCA Test Submission (TSCATS) Data Base, January 1989
    NIOSH Analytical Methods: see Dibutyl phthalate, 5020
    Meets criteria for proposed OSHA Medical Records Rule.
    IDLH value: 9300 mg/m3

SAX TOXICITY EVALUATION:

THR: Moderately toxic by intraperitoneal and intravenous routes. Mildly toxic
by ingestion. An experimental teratogen. Experimental reproductive effects.
Mutagenic data.

Source: *Instant Tox-Base*, 1995, Instant Reference Sources, Inc., 7605 Rockpoint Drive,
Austin, TX 78731 (Fax: 512-345-2386; Internet URL, http://www.instantref.com/inst-
ref.htm).

Toxicology Information from EPA's Integrated Risk Information System (IRIS)

I. CHRONIC HEALTH HAZARD ASSESSMENTS FOR NONCARCINOGENIC
EFFECTS

I.A. REFERENCE DOSE FOR CHRONIC ORAL EXPOSURE (RfD)

The Reference Dose (RfD) is based on the assumption that thresholds exist for certain
toxic effects such as cellular necrosis, but may not exist for other toxic effects such as
carcinogenicity. In general, the RfD is an estimate (with uncertainty spanning perhaps an
order of magnitude) of a daily exposure to the human population (including sensitive
subgroups) that is likely to be without an appreciable risk of deleterious effects during a
lifetime. Please refer to Background Document 1 in Service Code 5 for an elaboration of
these concepts. RfDs can also be derived for the noncarcinogenic health effects of
compounds which are also carcinogens. Therefore, it is essential to refer to other sources
of information concerning the carcinogenicity of this substance. If the US EPA has
evaluated this substance for potential human carcinogenicity, a summary of that

evaluation will be contained in Section II of this file when a review of that evaluation is completed.

NOTE: The Oral RfD for dibutyl phthalate may change in the near future pending the outcome of a further review now being conducted by the Oral RfD Work Group.

I.A.1. ORAL RfD SUMMARY

Critical Effect	Experimental Doses*	UF	MF	RfD
Increased mortality	NOAEL: 0.25% of diet (125 mg/kg/day)	1000	1	1E-1 mg/kg/day
	LOAEL: 1.25% of diet (600 mg/kg bw/day)			

*Conversion Factors: The values of 125 mg/kg/day for 0.25% dibutyl phthalate in the diet and 600 mg/kg/day for 1.25% were estimated from a figure depicting daily intake in mg/kg in Smith (1953).

Rat Subchronic to Chronic, Oral Bioassay Smith, 1953

I.A.2. PRINCIPAL AND SUPPORTING STUDIES (ORAL RfD)

Smith, C.C. 1953. Toxicity of butyl sterate, dibutyl sebacate, dibutyl phthalate and methoxyethyl oleate. Arch. Hyg. Occup. Med. 7: 310-318.

Male Sprague-Dawley rats in groups of 10 were fed diets containing 0, 0.01, 0.05, 0.25, and 1.25% dibutyl phthalate for a period of 1 year. One-half of all rats receiving the highest dibutyl phthalate concentration died during the first week of exposure. The remaining animals survived the study with no apparent ill effects. There was no effect of treatment on gross pathology or hematology. While it was stated that several organs were sectioned and stained, no histopathologic evaluation was reported.

I.A.3. UNCERTAINTY AND MODIFYING FACTORS (ORAL RfD)

UF -- A factor of 10 was applied to account for interspecies variation, a factor of 10 for protection of sensitive human subpopulations, and an additional factor of 10 to account for both the less-than-chronic duration of the study and deficiencies in the study, such as the use of only male animals.

MF -- None

I.A.4. ADDITIONAL COMMENTS (ORAL RfD)

Fetotoxicity was observed when mice were fed 2100 mg/kg/day dibutyl phthalate throughout gestation (Shiota and Nishimura, 1982). An increase in terata of borderline statistical significance was observed in progeny of this treatment group. Dibutyl phthalate produces degeneration of the seminiferous tubules, probably as a result of increased urinary excretion of zinc (Gangolli, 1982).

I.A.5. CONFIDENCE IN THE ORAL RfD

Study -- Low
Data Base -- Low
RfD -- Low

The study by Smith (1953) used few animals of one sex only. It was not indicated in the paper whether the 50% mortality observed early in the study was considered treatment-related, nor was the cause of death indicated. This is the only subchronic bioassay of dibutyl phthalate reported in the literature. Confidence in the study, data base, and RfD are all rated low.

I.A.6. EPA DOCUMENTATION AND REVIEW OF THE ORAL RfD

Source Document -- US EPA, 1980

The RfD in the 1980 Ambient Water Quality Criteria document received extensive peer and public review.

Other EPA Documentation -- None

Agency Work Group Review -- 01/22/86

Verification Date -- 01/22/86

I.A.7. EPA CONTACTS (ORAL RfD)

Adib Tabri / NCEA -- (513)569-7505

I.B. REFERENCE CONCENTRATION FOR CHRONIC INHALATION EXPOSURE (RfC)

The health effects data for dibutyl phthalate were reviewed by the US EPA RfD/RfC Work Group and determined to be inadequate for derivation of an inhalation RfC. The verification status of this chemical is currently not verifiable. For additional information on health effects of this chemical interested parties are referred to the EPA documentation listed below.

US EPA. 1987. Drinking Water Criteria Document for Phthalic Acid Esters Prepared by the Office of Health and Environmental Assessment, Environmental Criteria and Assessment Office, Cincinnati, OH for the Office of Drinking Water, Washington, DC. (External Review Draft)

Agency Work Group Review -- 07/26/90

EPA Contacts:

Gary L. Foureman / NCEA -- (919)541-1183

Annie M. Jarabek / NCEA -- (919)541-4847

II. CARCINOGENICITY ASSESSMENT FOR LIFETIME EXPOSURE

Section II provides information on three aspects of the carcinogenic risk assessment for the agent in question; the US EPA classification, and quantitative estimates of risk from oral exposure and from inhalation exposure. The classification reflects a weight-of-evidence judgment of the likelihood that the agent is a human carcinogen. The quantitative risk estimates are presented in three ways. The slope factor is the result of application of a low-dose extrapolation procedure and is presented as the risk per (mg/kg)/day. The unit risk is the quantitative estimate in terms of either risk per ug/L drinking water or risk per ug/cu.m air breathed. The third form in which risk is presented is a drinking water or air concentration providing cancer risks of 1 in 10,000, 1 in 100,000 or 1 in 1,000,000. Background Document 2 (Service Code 5) provides details on the rationale and methods used to derive the carcinogenicity values found in IRIS. Users are referred to Section I for information on long-term toxic effects other than carcinogenicity.

II.A. EVIDENCE FOR CLASSIFICATION AS TO HUMAN CARCINOGENICITY

II.A.1. WEIGHT-OF-EVIDENCE CLASSIFICATION

Classification -- D; not classifiable.

Basis -- Pertinent data regarding carcinogenicity was not located in the available literature.

II.A.2. HUMAN CARCINOGENICITY DATA

None.

II.A.3. ANIMAL CARCINOGENICITY DATA

None.

II.A.4. SUPPORTING DATA FOR CARCINOGENICITY

DBP did not induce mutations in a modified reverse mutation plate incorporation assay in Salmonella strains TA100 and TA98 at concentrations up to 1000 ug/plate in the presence or the absence of S9 hepatic homogenate (Kozumbo et al., 1982). It was a weak direct-acting mutagen in a forward mutation assay in Salmonella typhimurium (Seed, 1982). DBP was mutagenic in the mouse lymphoma forward mutation assay only in the presence of metabolic activation (CMA, 1986). In addition, DBP showed some evidence of clastogenic activity in Chinese hamster fibroblasts (Ishidate and Odashima, 1977) but was negative in human leukocytes (Tsuchiya and Hattori, 1977). Research indicates that

DBP is hydrolyzed to monoesters (Kluwe, 1982; Rowland et al., 1977; Albro and Moore, 1974). There is evidence that DBP induces peroxisome proliferation (US EPA, 1987).

II.B.   QUANTITATIVE ESTIMATE OF CARCINOGENIC RISK FROM ORAL EXPOSURE

[Ed. Note: EPA does not have information yet in this section of IRIS].

II.C.   QUANTITATIVE ESTIMATE OF CARCINOGENIC RISK FROM INHALATION EXPOSURE

[Ed. Note: EPA does not have information yet in this section of IRIS].

II.D. EPA DOCUMENTATION, REVIEW, AND CONTACTS (CARCINOGENICITY ASSESSMENT)

II.D.1. EPA DOCUMENTATION

Source Document -- US EPA, 1987

The Drinking Water Criteria Document for Phthalic Acid Esters has received OHEA review.

II.D.2. REVIEW (CARCINOGENICITY ASSESSMENT)

Agency Work Group Review -- 08/26/87

Verification Date -- 08/26/87

II.D.3. US EPA CONTACTS (CARCINOGENICITY ASSESSMENT)

Brian J. Commons / OST -- (202)260-7589

Linda R. Papa / NCEA -- (513)569-7587

III. HEALTH HAZARD ASSESSMENTS FOR VARIED EXPOSURE DURATIONS

[Ed. Note: EPA does not have information yet in this section of IRIS].

IV. US EPA REGULATORY ACTIONS

EPA risk assessments may be updated as new data are published and as assessment methodologies evolve. Regulatory actions are frequently not updated at the same time. Compare the dates for the regulatory actions in this section with the verification dates for the risk assessments in sections I and II, as this may explain inconsistencies. Also note that some regulatory actions consider factors not related to health risk, such as technical or economic feasibility. Such considerations are indicated for each action. In addition, not all of the regulatory actions listed in this section involve enforceable federal standards. Please direct any questions you may have concerning these regulatory actions to the US EPA contact listed for that particular action. Users are strongly urged to read the background information on each regulatory action in Background Document 4 in Service Code 5.

IV.A. CLEAN AIR ACT (CAA)

[Ed. Note: EPA does not have information yet in this section of IRIS].

IV.B. SAFE DRINKING WATER ACT (SDWA)

IV.B.1. MAXIMUM CONTAMINANT LEVEL GOAL (MCLG) for Drinking Water

Value -- 0.8 mg/L (Proposed, 1990)

Considers technological or economic feasibility? -- NO

Discussion -- EPA is proposing to regulate dibutyl phthalate based on its potential adverse effects (increased mortality) reported in a one-year study in rats. The MCLG is based upon a DWEL of 4 mg/L and an assumed drinking water contribution of 20 percent.

Reference -- 55 FR 30370 (07/25/90)

EPA Contact -- Health and Ecological Criteria Division / OST / (202) 260-7571 / FTS 260-7571; or Safe Drinking Water Hotline / (800) 426-4791

IV.B.2. MAXIMUM CONTAMINANT LEVEL (MCL) for Drinking Water

[Ed. Note: EPA does not have information yet in this section of IRIS].

IV.B.3. SECONDARY MAXIMUM CONTAMINANT LEVEL (SMCL) for Drinking Water

[Ed. Note: EPA does not have information yet in this section of IRIS].

IV.B.4. REQUIRED MONITORING OF "UNREGULATED" CONTAMINANTS

[Ed. Note: EPA does not have information yet in this section of IRIS].

IV.C. CLEAN WATER ACT (CWA)

IV.C.1. AMBIENT WATER QUALITY CRITERIA, Human Health

Water and Fish Consumption: 3.4E+4 ug/L

Fish Consumption Only: 1.54E+5 ug/L

Considers technological or economic feasibility? -- NO

Discussion -- The WQC of 3.4E+4 ug/L is based on consumption of contaminated aquatic organisms and water. A WQC of 1.54E+5 ug/L has also been established based on consumption of contaminated aquatic organisms alone.

Reference -- 45 FR 79318 (11/28/80)

EPA Contact -- Criteria and Standards Division / OWRS (202)260-1315 / FTS 260-1315

IV.C.2. AMBIENT WATER QUALITY CRITERIA, Aquatic Organisms

Freshwater:
   Acute LEC -- 9.4E+2 ug/L
   Chronic LEC -- 3.0E+0 ug/L

Marine:
   Acute LEC -- 2.9E+3 ug/L
   Chronic LEC -- None

Considers technological or economic feasibility? -- NO

Discussion -- The values that are indicated as "LEC" are not criteria, but are the lowest effect levels found in the literature. LECs are given when the minimum data required to derive water quality criteria are not available. The values given are for the general class of phthalate esters and not specifically for dibutyl phthalate.

Reference -- 45 FR 79318 (11/28/80)

EPA Contact -- Criteria and Standards Division / OWRS (202)260-1315 / FTS 260-1315

IV.D. FEDERAL INSECTICIDE, FUNGICIDE, AND RODENTICIDE ACT (FIFRA)

IV.D.1. PESTICIDE ACTIVE INGREDIENT, Registration Standard

Status -- List "C" Pesticide (1989)

Reference -- 54 FR 30846 (07/24/89)

EPA Contact -- Registration Branch / OPP (703)557-7760 / FTS 557-7760

IV.D.2. PESTICIDE ACTIVE INGREDIENT, Special Review

        [Ed. Note: EPA does not have information yet in this section of IRIS].

IV.E. TOXIC SUBSTANCES CONTROL ACT (TSCA)

        [Ed. Note: EPA does not have information yet in this section of IRIS].

IV.F. RESOURCE CONSERVATION AND RECOVERY ACT (RCRA)

IV.F.1. RCRA APPENDIX IX, for Ground Water Monitoring

Status -- Listed

Reference -- 52 FR 25942 (07/09/87)

EPA Contact -- RCRA/Superfund Hotline (800)424-9346 / (202)260-3000 / FTS 260-3000

IV.G. SUPERFUND (CERCLA)

IV.G.1. REPORTABLE QUANTITY (RQ) for Release into the Environment

Value (status) -- 10 pounds (Final, 1985)

Considers technological or economic feasibility? -- NO

Discussion -- The final RQ is based on aquatic toxicity. The available data indicate that the aquatic 96-Hour Median Threshold Limit for dibutyl phthalate is between 0.1 and 1 ppm.

Reference -- 50 FR 13456 (04/04/85); 54 FR 33418 (08/14/89)

EPA Contact -- RCRA/Superfund Hotline (800)424-9346 / (202)260-3000 / FTS 260-3000

VI. BIBLIOGRAPHY

VI.A. ORAL RfD REFERENCES

Gangolli, S.D. 1982. Testicular effects of phthalate esters. Environ. Health Perspect. 45: 77-84.

Shiota, K. and H. Nishimura. 1982. Teratogenicity of di-2-ethylhexyl phthalate and di-n-butyl phthalate in mice. Environ. Health Perspect. 45(0): 65-70.

Smith C.C. 1953. Toxicity of butyl sterate, dibutyl sebacate, dibutyl phthalate and methoxyethyl oleate. Arch. Hyg. Occup. Med. 7: 310-318.

US EPA. 1980. Ambient Water Quality Criteria for Phthalate Esters. Prepared by the Office of Health and Environmental Assessment, Environmental Criteria and Assessment Office, Cincinnati, OH for the Office of Water Regulations and Standards, Washington, DC. EPA 440/5-80-067. NTIS PB 81- 117780.

VI.B. INHALATION RfC REFERENCES

US EPA. 1987. Drinking Water Criteria Document for Phthalic Acid Esters Prepared by the Office of Health and Environmental Assessment, Environmental Criteria and Assessment Office, Cincinnati, OH for the Office of Drinking Water, Washington, DC. (External Review Draft)

VI.C. CARCINOGENICITY ASSESSMENT REFERENCES

Albro, P.W. and B. Moore. 1974. Identification of the metabolites of simple phthalate diesters in rat urine. J. Chromatogr. 94: 209-218.

CMA (Chemical Manufacturers Association). 1986. Mutagenicity of 1C (di-n- butyl phthalate) in a mouse lymphoma mutation assay. Final report. Submitted to Hazleton Biotechnologies Company. HB Project No. 20989. September, 1986.

Ishidate, M., Jr. and S. Odashima. 1977. Chromosome tests with 134 compounds on Chinese hamster cells in vitro -- A screening test for chemical carcinogens. Mutat. Res. 48: 337-354.

Kluwe, W.M. 1982. Overview of phthalate ester pharmacokinetics in mammalian species. Environ. Health Perspect. 45: 3-10.

Kozumbo, W.J., R. Kroll and R.J. Rubin. 1982. Assessment of the mutagenicity of phthalate esters. Environ. Health Perspect. 45: 103-109.

Rowland, I.R., R.C. Cottrell and J.C. Phillips. 1977. Hydrolysis of phthalate esters by the gastro-intestinal contents of the rat. Food Cosmet. Toxicol. 15: 17-21.

Seed, J.L. 1982. Mutagenic activity of phthalate esters in bacterial liquid suspension assays. Environ. Health Perspect. 45: 111-114.

Tsuchiya, K. and K. Hattori. 1977. Chromosomal study on human leukocyte cultures treated with phthalic acid ester. Hokkaidoritus Eisei Kenkyusho Ho. 26: 114. (Abstract)

US EPA. 1987. Drinking Water Criteria Document for Phthalic Acid Esters. Prepared by the Office of Health and Environmental Assessment, Environmental Criteria and Assessment Office, Cincinnati, OH for the Office of Drinking Water, Washington, DC. External Review Draft.

VI.D. DRINKING WATER HA REFERENCES

None

VII. REVISION HISTORY

Date	Section	Description
09/07/88	II.	Carcinogen summary on-line
08/01/89	VI.	Bibliography on-line
03/01/90	I.A.4.	Text corrected
05/01/90	II.A.4.	First sentence revised
08/01/90	I.A.	Oral RfD summary noted as pending change
09/01/90	I.B.	Not verified; data inadequate
09/01/90	IV.F.1.	EPA contact changed
10/01/90	I.B.	Inhalation RfC message on-line
10/01/90	VI.B.	Inhalation RfC references added
08/01/91	II.D.3.	Primary and secondary contacts changed
01/01/92	I.A.7.	Secondary contact changed
01/01/92	IV.	Regulatory actions updated
02/01/93	II.D.3.	Primary contact changed

SYNONYMS

1,2-Benzenedicarboxylic Acid Dibutyl Ester	Elaol
o-Benzenedicarboxylic Acid, Dibutyl Ester	Ergoplast FDB
Benzene-o-Dicarboxylic Acid Di-n-Butyl Ester	Genoplast B
Butylphthalate	Hexaplast M/B
Celluflex DPB	N-Butylphthalate
Dibutyl 1,2-Benzene dicarboxylate	Palatinol C
Dibutyl phthalate	Phthalic Acid Dibutyl Ester
Di-n-Butylphthalate	Polycizer DBP
Dibutyl-o-Phthalate	PX 104
DPB	RC Plasticizer DBP

Source: *Instant EPA's IRIS*, 1997, Instant Reference Sources, Inc., 7605 Rockpoint Drive, Austin, TX 78731 (Fax: 512-345-2386; Internet URL, http://www.instantref.com/inst-ref.htm).

Endosulfan

Introduction

First introduced into the U.S. in 1954, endosulfan is an insecticide used widely on food crops. At least 20 million of pounds per year currently are used worldwide, including some 2 million pounds in the U.S. It is commonly sprayed on lettuce, tomatoes, artichokes, strawberries, pears, grapes, alfalfa, cotton, tea, tobacco and nuts. [882] It is a chlorinated hydrocarbon in the cyclodiene family and is a contact insecticide in use on a number of food crops It is marketed as a mixture of two isomers. This chemical is considered highly toxic in mammalian systems and is moderately toxic to birds and fish. [811] It is manufactured in several useful formulations including emulsifiable concentrates, wettable powders, granules, dustable powder, and smoke tablets. [817] There are two isomers of endosulfan: endosulfan-I (also called alpha-endosulfan) and endosulfan-II (also called beta-endosulfan). The commercial product is a mixture of these isomers and has its separate CAS Number which is the primary one below.

Identifying Information

CAS NUMBER: 115-29-7
       alpha-endosulfan CAS NUMBER: 959-98-8
        beta- endosulfan CAS NUMBER: 33213-65-9

NIOSH Registry Number: RB9275000

SYNONYMS: 1,4,5,6,7,7-HEXACHLORO-5-NORBORNENE-2,3-DIMETHANOL, CYCLIC SULFATE, ALPHA, BETA-1,2,3,4,7,7-HEXACHLOROBICYCLO)2.2.1)-2-HEPTENE-5,6-BISOXY-, METHYLENE SULFITE, 1,2,3,4,7,7-HEXACHLOROBICYCLO(2.2.1)HEPTEN-5,6-BIOXYMETHYLENE SULFITE, HEXACHLOROHEXAHYDROMETHANO 2,4,3-BENZODIOXATHIEPIN-3-OXIDE, 6,7,8,9,10,10-HEXACHLORO-1,5,5A,6,9,9A-HEXAHYDRO-6,9-METHANO-2,4,3-, BENZODIOXATHIEPIN-3-OXIDE, 1,4,5,6,7,7-HEXACHLORO-5-NORBORNENE-2,3-DIMETHANOL CYCLIC SULFITE, SULFUROUS ACID, CYCLIC ESTER WITH 1,4,5,6,7,7-HEXACHLORO-5-NORBORNENE-, 2-DIMETHANOL, CRISULFAN, DEVISULPHAN, ENDOCEL, ENDOSOL, ENDOSULPHAN, ENSURE, HILDAN, BENZOEPIN, BEOSIT, INSECTOPHENE, KOP-THIODAN, MALIX, THIFOR, THIMUL, THIODAN, THIOFOR, THIOMUL, THIONEX, THIOSULFAN,

TIONEL, TIOVEL, CHLORTHIEPIN, CYCLODAN, NCI-C00566, ENT 23,979, FMC 5462, BIO 5,642, HOE 2,671, NIA 5462, NIAGRA 5462, OMS 570, RCRA WASTE NUMBER P050

Endocrine Disruptor Information

Endosulfan is a strongly suspected EED that it is listed by EPA [811], the Centers for Disease Control & Prevention [812], and the World Wildlife Fund [813] as a potential endocrine modifying chemical. Endosulfan has presented mutagenic effects in bacteria and two mammalian species [863]. Endosulfan has also been shown to exhibit immunotoxic effects [826]

Chemical and Physical Properties

CHEMICAL FORMULA: $C_9H_6C_{16}O_3S$

MOLECULAR WEIGHT: 406.93

WLN: T C755 A EOSO KUTJ AG AG BG FO JG KG LG

PHYSICAL DESCRIPTION: Tan or white crystals with a sulfur dioxide odor

SPECIFIC GRAVITY: 1.745 g/cm^3 @ 20/20°C

DENSITY: Not available

MELTING POINT: 70-100°C (commercial technical mixture); 108-109°C (alpha-endosulfan); 205-208°C (beta-endosulfan)

BOILING POINT: Not available

SOLUBILITY: Water: <1 mg/mL @ 23°C; DMSO: 10-50 mg/mL @ 23°C; 95% Ethanol: 10-50 mg/mL @ 23°C; Acetone: >=100 mg/mL @ 23°C

OTHER SOLVENTS: Most organic solvents: Soluble; Xylene: Soluble; Kerosene: Soluble; Chloroform: Soluble

Source: *The National Toxicology Program's Chemical Database, Volume 2*, 1992, CRC Press, Inc./Lewis Publishers, 2000 Corporate Blvd., Boca Raton, FL 33431 (Fax: 407-998-9114).

Environmental Reference Materials

A source of EED environmental reference materials for Endosulfan is Radian International LLC, Austin, TX (Fax 512-454-0268; Internet URL, http://www.radian.com/standards); Catalog No. ERE-003 and ERE-004.

Hazardous Properties

ACUTE/CHRONIC HAZARDS: Endosulfan may be toxic by oral, skin absorption or inhalation routes. When heated to decomposition it emits toxic fumes.

VOLATILITY: Vapor pressure: 0.009 mm Hg @ 80°C

FIRE HAZARD: Flash point data for endosulfan are not available. It is probably combustible. Fires involving this material can be controlled with a dry chemical, carbon dioxide or Halon extinguisher.

REACTIVITY: This compound is hydrolyzed slowly by water and acids and rapidly by bases and alkalis to the alcohol and sulfur dioxide. The decomposition is catalyzed by iron which it corrodes.

STABILITY: Endosulfan is stable under normal laboratory conditions. Solutions of endosulfan in water, DMSO, 95% ethanol or acetone should be stable for 24 hours under normal lab conditions [700].

Source: *The National Toxicology Program's Chemical Database, Volume 7*, 1992, CRC Press, Inc./Lewis Publishers, 2000 Corporate Blvd., Boca Raton, FL 33431 (Fax: 407-998-9114).

Medical Symptoms of Exposure

Symptoms of exposure may include salivation, emesis, central nervous system stimulation or depression, generalized tonic and clonic convulsions, nervousness, agitation, tremors, hyperexcitability, ataxia, slight nausea, confusion, flushing and dry mouth.

Source: *The National Toxicology Program's Chemical Database, Volume 4*, 1992, CRC Press, Inc./Lewis Publishers, 2000 Corporate Blvd., Boca Raton, FL 33431 (Fax: 407-998-9114).

Toxicological Information

The toxicological information below is gathered from several sources including two national databases: one from the *National Toxicology Program Chemical Repository Database* and another from *EPA's Integrated Risk Information System (IRIS)*.

Toxicology Information from the National Toxicology Program

CARCINOGENICITY:

Tumorigenic Data:

Type of Effect	Route/Animal	Amount Dosed (Notes)
TDLo	orl-mus	330 mg/kg/78W-I
TDLo	scu-mus	2 mg/kg

Status:

> NCI Carcinogenesis Bioassay (Feed); Inadequate Study: Male Rat and
> Male Mouse
> NCI Carcinogenesis Bioassay (Feed); Negative: Female Rat and
> Female Mouse

TERATOGENICITY:

Reproductive Effects Data:

Type of Effect	Route/Animal	Amount Dosed (Notes)
TDLo	orl-rat	45 mg/kg (6-14D preg)
TDLo	orl-rat	600 mg/kg (60D male)

MUTATION DATA:

Test	Lowest dose		Test	Lowest dose
cyt-mus-unr	1 mg/kg		sln-dmg-orl	200 ppm/48H
sce-hmn:lym	1 umol/L			

TOXICITY:

Typ. Dose	Mode	Specie	Amount
LD50	orl	rat	18 mg/kg
LC50	ihl	rat	80 mg/m3/4H
LD50	skn	rat	74 mg/kg
LD50	ipr	rat	8 mg/kg
LD50	unr	rat	40 mg/kg
LD50	orl	mus	7360 ug/kg
LD50	ipr	mus	7 mg/kg
LD50	unr	mus	32 mg/kg
LD50	orl	cat	2 mg/kg
LD50	orl	rbt	28 mg/kg
LD50	skn	rbt	90 mg/kg
LD50	scu	rbt	360 mg/kg
LD50	orl	ham	118 mg/kg
LD50	orl	dck	33 mg/kg
LD50	orl	dom	26 mg/kg
LD50	skn	mam	147 mg/kg
LD50	orl	bwd	35 mg/kg

OTHER TOXICITY DATA:

    Review: Toxicology Review-3

    Standards and Regulations:
            DOT-Hazard: Poison B; Label: Poison
            DOT-Hazard: Poison B; Label: Poison, liquid

    Status:
            Meets criteria for proposed OSHA Medical Records Rule

SAX TOXICITY EVALUATION:

    THR: An experimental teratogen, neoplastigen and equivocal tumorigenic
    agent. VERY, VERY HIGH via oral route. VERY HIGH via dermal routes. A
    central nervous system stimulant producing convulsions.

Source: *Instant Tox-Base*, 1995, Instant Reference Sources, Inc., 7605 Rockpoint Drive,
Austin, TX 78731 (Fax: 512-345-2386; Internet URL, http://www.instantref.com/inst-
ref.htm).

Toxicology Information from EPA's Integrated Risk Information System (IRIS)

I.   CHRONIC HEALTH HAZARD ASSESSMENTS FOR NONCARCINOGENIC
EFFECTS

I.A. REFERENCE DOSE FOR CHRONIC ORAL EXPOSURE (RfD)

The Reference Dose (RfD) is based on the assumption that thresholds exist for certain
toxic effects such as cellular necrosis, but may not exist for other toxic effects such as
carcinogenicity. In general, the RfD is an estimate (with uncertainty spanning perhaps
an order of magnitude) of a daily exposure to the human population (including sensitive
subgroups) that is likely to be without an appreciable risk of deleterious effects during a
lifetime. Please refer to Background Document 1 in Service Code 5 for an elaboration
of these concepts. RfDs can also be derived for the noncarcinogenic health effects of
compounds which are also carcinogens.   Therefore, it is essential to refer to other
sources of information concerning the carcinogenicity of this substance. If the US EPA
has evaluated this substance for potential human carcinogenicity, a summary of that
evaluation will be contained in Section II of this file when a review of that evaluation is
completed.

I.A.1. ORAL RfD SUMMARY

Critical Effect	Experimental Doses*	UF	MF	RfD
Reduced body weight gain in males and females; increased incidence of marked progressive glomerulonephrosis and blood vessel aneurysms in males (1)	NOAEL: 15 ppm [0.6 mg/kg-day (male); 0.7 mg/kg-day (female)] LOAEL: 75 ppm [2.9 mg/kg-day (male); 3.8 mg/kg-day (female)]	100	1	6E3 MG--mg/kg/day
Decreased weight gain in males and neurologic findings in both sexes (2)	NOAEL: 10 ppm 0.57 mg/kg-day (female) LOAEL: 30 ppm [1.9 mg/kg-day (female); 2.1 mg/kg-day (male)]			

*Conversion Factors and Assumptions -- Actual dose tested

(1) 2-Year Rat Feeding Study Hoechst Celanese Corp., 1989a

(2) Hoechst Celanese Corp., 1989b

I.A.2. PRINCIPAL AND SUPPORTING STUDIES (ORAL RfD)

Hoechst Celanese Corporation. 1989a. MRID No. 40256502, 41099502. HED Doc. No. 007937. Available from EPA. Write to FOI, EPA, Washington, DC 20460.

Hoechst Celanese Corporation. 1989b. MRID No. 41099501. HED Doc. No. 007937. Available from EPA. Write to FOI, EPA, Washington, DC 20460.

Groups of Sprague-Dawley rats (50/sex/dose) were administered endosulfan in the diet for 2 years at dietary concentrations of 0, 3, 7.5, 15 and 75 ppm (Male: 0, 0.1, 0.3, 0.6 and 2.9 mg/kg-day; Female: 0, 0.1, 0.4, 0.7 and 3.8 mg/kg-day) (Hoechst Celanese Corp., 1989a). A satellite group of 20 animals/sex/dose were used for toxicity evaluation and sampled at intervals for hematology and clinic chemistry; survivors were sacrificed after 104 weeks. An additional group of 10 rats/sex were used for pretest hematology and health check. Animals received food and water ad libitum.

No effects of dosing on clinical signs, mortality, food and water consumption, ophthalmological examinations and urinalysis were observed. Mean body weight gains tended to be decreased in both males and females receiving 15 and 75 ppm. Weight gains were significantly depressed ($p<0.05$) during weeks 6-18 in males receiving 15 and 75 ppm when compared with controls; decreases did not achieve a level of significance in males or females receiving 15 ppm at other intervals. Between weeks 0-64, body weight gains were 9 and 13% lower in males and females of the 75 ppm group, respectively, than those of the controls. Overall weight gains (weeks 0-104) were 17% lower than the controls in both males and females receiving 75 ppm and 9% lower in rats receiving 15 ppm. The gains were significantly lower ($p<0.01$) in both sexes at the 75 ppm dose. No weight gain effects were seen in males and females receiving 3 or 7.5 ppm when compared with controls.

No toxicologically important changes in hematology and clinical chemistry parameters were observed. The incidence of bilaterally enlarged kidneys was increased in females of both the satellite (2 and 8 for the control and 75 ppm dose groups, respectively) and main (8 and 18 for the control and 75 ppm dose groups, respectively) groups receiving 75 ppm when compared with controls. Other findings in the kidneys (paleness, irregular or uniform cortical scarring and cysts) occurred at similar frequencies in control and dosed groups. The incidence of progressive glomerulonephrosis was high in all dose groups including controls which is not an uncommon finding in studies with chlorinated hydrocarbon pesticides. The severity appeared to be dose-related. The incidence of severe (marked) glomerulonephrosis was increased in both males and females receiving 75 ppm. In males, the increased incidence at 75 ppm was accounted for by rats that died. In descendants, the incidence combining both males in the main and satellite groups was 10/41, 10/43, 17/46 and 14/46 and 20/46 at 0, 3, 7.5, 15 and 75 ppm. The incidence in high-dose males (30/70, 43%) was reported to be higher than normally observed for historical controls. The laboratory control incidence in six studies was 70/300 (19.7%) with a range of 10 to 38%.

The incidence of aneurysms of the blood vessels was increased in the high-dose males of both the satellite (6/20) and the main study (13/50) when compared with controls (1/20 and 9/50). The percent incidence (27%) in the combined high-dose males was higher than normally found in historical controls (10%, range in five studies 4-18%). Other nonneoplastic findings were considered within the normal range of background.

Based on reduced body weight gain in males and females, and increased incidence of marked progressive glomerulonephrosis and blood vessel aneurysms in males, the LEL for systemic toxicity is 75 ppm (Male: 2.9 mg/kg-day; Female: 3.8 mg/kg-day). The NOEL for systemic toxicity is 15 ppm (Male: 0.6 mg/kg-day; Female: 0.7 mg/kg-day).

Endosulfan was fed in the diet to beagle dogs (6/sex/dose) for 1 year at dietary levels of 0, 3, 10 and 30 ppm (Male: 0, 0.2, 0.65 and 2.1 mg/kg-day; Female: 0, 0.18, 0.57 and 1.9 mg/kg-day) (Hoechst Celanese Corp., 1989b). An additional group of 6 dogs/sex received 30 ppm for 54 days, after which the dose was increased to 45 ppm (Male: 3.2 mg/kg-day; Female: 2.9 mg/kg-day) and continued at that level until a final increase to 60 ppm (Male: 4.1 mg/kg-day; Female: 3.8 mg/kg-day) was administered at 106 days. Animals received food and water ad libitum.

At the highest dose tested, severe nervous symptoms developed. A loss or weakening of placing and righting reactions was observed and substantial weight loss resulted (0.36 and 0.45 kg for males and females, respectively, between weeks 15 and 21). This group was sacrificed at 146-147 days owing to poor overall condition; one male had been sacrificed at 126 days. The overall weight gain in males receiving 30 ppm (to week 54) was 30% lower than the controls. Tonic contractions of the muscles of the abdomen and chaps a few hours after feeding was noted in both sexes receiving 30 ppm; one male was sacrificed at 39 weeks. The sacrificed dog at 30 ppm had gross and histologic changes in the lungs. The sacrificed dog receiving 60 ppm had pulmonary edema. The histologic findings in dosed and control groups for dogs sacrificed by design were generally unremarkable and incidental.

Based on decreased weight gain in males and neurologic findings in both sexes, the LEL for systemic toxicity is 30 ppm (Male: 2.1 mg/kg-day; Female: 1.9 mg/kg-day). The NOEL for systemic toxicity is 10 ppm (Male: 0.65 mg/kg-day; Female: 0.57 mg/kg-day).

I.A.3. UNCERTAINTY AND MODIFYING FACTORS (ORAL RfD)

UF -- The uncertainty factor of 100 reflects 10 for intraspecies variability and 10 for interspecies extrapolation.

MF -- None

I.A.4. ADDITIONAL STUDIES / COMMENTS (ORAL RfD)

The observation of a yellowish discoloration of the kidneys was found in a 30-day rat feeding study (Hoechst Celanese Corp., 1985), a 90-day rat feeding study (Hoechst Aktiengesellschaft, 1985), and a 2-generation reproduction study in rats (Hoechst Aktiengesellschaft, 1984a). This observation was not found in the rat chronic feeding/oncogenicity study (Hoechst Celanese Corp., 1989a), the 1-year dog feeding study (Hoechst Celanese Corp., 1989b), or the mouse carcinogenicity study (Hoechst Celanese Corp., 1988). Hoechst Celanese Corporation argues that data from chronic toxicity studies, metabolism, and special studies indicate that this is not an adverse hematopoietic effect, but is indicative of the physical presence and harmless process of elimination of endosulfan and its metabolites via the kidney. Electron microscopy and tissue residue analysis of the kidneys from the 30-day study indicated that the alpha-endosulfan and to a lesser extent beta-endosulfan, endosulfan sulfate, and endosulfan lactone were stored temporarily in the kidneys. Additionally, negative results were obtained from the staining of the kidneys with Prussian Blue to detect the presence of ferritin (evidence of hemosiderosis).

Data Considered for Establishing the RfD

1) 2-Year Feeding - rat: Principal study -- see previous description; Core grade minimum (Hoechst Celanese Corp., 1989a).

2) 1-Year Feeding - dog: Co-principal study -- see previous description; Core grade minimum (Hoechst Celanese Corp., 1989b).

3) 2-Generation Reproduction - rat: Core grade minimum (Hoechst Aktiengesellschaft, 1984a).

Four week old male and female rats of Crl:COBS CD(SD)BR strain were allowed to acclimate for 7 days and were then distributed randomly to groups of 32 of each sex for the F0 generation. F1b animals were distributed randomly to groups of 28 of each sex for the second generation. Animals were administered endosulfan in the diet at dose levels of 0, 3, 15 or 75 ppm (Male: 0, 0.2, 1.1 and 5.4 mg/kg-day; Female: 0, 0.25, 2.6 and 6.6 mg/kg-day).

Mortality, food/water consumption, and body weight gain were not affected in either generation, but a decrease in body weight gain ($p < 0.05$) was observed in the F0 females following the start of dosing. Pregnancy rate, gestation times, the ability to rear young to weaning, and precoital time were comparable among the groups at both matings in both generations. F0 males displayed increased heart weight at the mid- and high-dose levels (dose-related) and increased liver and kidney weights at the high-dose level. F0 females displayed increased brain and liver weights at the high-dose level. In the F1b adults, the high-dose males displayed increased kidney weights compared with the controls and females displayed increased liver weights at the mid- and high-dose levels. Although the changes in the heart and liver weights reported in the mid-dose F0 males and females were statistically significant, the US EPA noted that these effects were slight and limited to one sex of one litter and occurred only in one generation. In view of this and in the absence of histopathological changes in the liver and heart, the US EPA concluded that these statistically significant changes in organ weights should not be considered biologically significant. Therefore, based on a decrease in body weight gain in F0 females, the LEL for systemic toxicity is 75 ppm (Male: 5.4 mg/kg-day; Female: 6.6 mg/kg-day). The NOEL for systemic toxicity is 15 ppm (Male: 1.1 mg/kg-day; Female: 2.6 mg/kg-day).

No effect of treatment on litter size was observed throughout both matings of both generations. In the first mating of the F0 generation, an increase was noted in the cumulative litter loss (%) at the high-dose level. Litter and pup weights were comparable at birth among the groups in both generations, but a decrease in litter weight was observed during the lactation to weaning period in both matings in the F0 generation, which was significant at the high-dose level in the first mating and at the mid- and high-dose levels in the second mating (dose-related). Because there was no corroborative finding of a decrease in the number of pups per litter or in pup weight, this decrease in litter weight is not considered to be treatment-related. Increased pituitary weights (high-dose female pups of 1st mate of F0 generation) and increased uterine weights (high-dose female pup of 1st mate of F1b generation) were observed in the offspring. There were no histopathological findings observed in either the F1b adults of the selected pups from the second mate or the F1b generation that could be attributed to treatment. Based on increased pituitary and uterine weights, the LEL for offspring toxicity is 75 ppm (Male: 5.4 mg/kg-day; Female: 6.6 mg/kg-day). The NOEL for offspring toxicity is 15 ppm (Male: 1.1 mg/kg-day; Female: 2.6 mg/kg-day).

No evidence of reproductive toxicity was found at any of the dose levels tested. Therefore, the NOEL for reproductive toxicity is equal to or greater than 75 ppm (Male: 5.4 mg/kg-day; Female: 6.6 mg/kg-day).

4) Developmental toxicity - rat: Core grade supplementary (FMC Corp., 1980).

Groups of pregnant CD Sprague-Dawley rats were administered endosulfan by daily oral gavage on days 6 through 19 of gestation at dose levels of 0, 0.66, 2.0 or 6.0 mg/kg-day. Although the original protocol specified 25 animals per treatment group, 10 additional animals were added to the high-dose group (due to mortality among the original animals) and five additional animals were added to the control group (due to a loss of some tissues during the processing).

Maternal toxicity was apparent in the high-dose group in the form of significantly reduced body weights and body weight gain during gestation (p<0.01). Toxic signs observed in the high-dose group included face rubbing (20/35 animals), brown exudate (4/35), rough coat (5/35), flaccidity (8/35) and hyperactivity (11/35). Face rubbing was reported in 6/25 mid-dose animals and alopecia was reported in 2/25. No face rubbing was reported in low dose animals or controls. Mean fetal weight and crown-rump length were significantly reduced (p<0.05 and p<0.01, respectively) in the 6 mg/kg-day group. An increase in misaligned sternebrae was observed in all treated groups compared with concurrent controls. The increase in litters and fetuses affected at each dose level was above that reported in the historical control data base (18% of litters and 1.85% of fetuses based upon the examination of 65 litters and 863 fetuses). However, the variability between studies was not reported. An increased incidence of litters with extra ribs and poorly ossified and unossified sternebrae was observed at the high-dose level. More detailed historical control data may be useful in determining whether the apparent increase in misaligned sternebrae is due to an unusually low incidence in the concurrent control group and within the variability observed between studies. In the absence of such information, it is recommended that this finding be considered to be compound-related. An additional review of this study by the US EPA concluded that replacement of animals during or after the study made it difficult to interpret the data and derive a NOEL and LEL for this study. The US EPA has recommended a repeat of this study.

5) Developmental toxicity - rabbit: Core grade minimum (FMC Corp., 1981).

Groups of 20 pregnant New Zealand white rabbits were administered endosulfan by oral gavage on days 6 through 28 of gestation at dose levels of 0, 0.3, 0.7 or 1.8 mg/kg-day. When mortality was observed at the highest dose level, six more mated rabbits were added to this group.

Two animals in the control and one in the middle dose level showed nasal congestion. In the highest dose level, four animals showed a noisy and rapid breathing, hyperactivity and convulsions. Body weight gains during the days 19-29 and corrected for gravid uterine weights at sacrifice were less in the high-dose group than in controls. The former was also less than the control for the mid-dose group. However, these differences were not statistically significant. Based on these effects, the NOEL and LEL for maternal toxicity are 0.7 and 1.8 mg/kg-day, respectively. No developmental effects were observed at any dose tested. Therefore, the NOEL for developmental toxicity is equal to or greater than 1.8 mg/kg-day, the highest dose tested.

Other Data Reviewed:

6) 2-Year Feeding - mouse: Core grade minimum (Hoechst Celanese Corp., 1988).

Groups of Hoe:NMRKf mice (60/sex/dose) were fed endosulfan in the diet at dose levels
of 0, 2, 6, or 18 ppm (Male: 0, 0.28, 0.84, and 2.51 mg/kg-day; Female: 0, 0.32, 0.97,
and 2.86 mg/kg-day) for 2 years. A satellite group of 20 mice/sex/group was used for
interim sacrifices at 12 and 18 months. Animals were individually housed and received
food and water ad libitum.

No overt signs of toxicity or dose-related effects were noted on clinical observations,
food consumption, hematology, clinical chemistry, urinalysis, organ weights,
macroscopic pathology, or microscopic pathology. Decreased survival ($p<0.05$) in high-
dose females and body weight reduction ($p<0.05$) in high-dose males throughout the
study were considered to be compound-related effects. Based on these findings, the
NOEL and LEL for systemic toxicity are 6 ppm (Male: 0.84 mg/kg-day; Female: 0.97
mg/kg-day) and 18 ppm (Male: 2.51 mg/kg-day; Female: 2.86 mg/kg-day), respectively.

7) 90-Day Feeding - rat: Core grade minimum (Hoechst Aktiengesellschaft, 1985).

Groups of CD Sprague-Dawley rats (25/sex/dose) were fed endosulfan in the diet at dose
levels of 0, 10, 30, 60 or 360 ppm (0, 0.5, 1.5, 3 and 18 mg/kg-day) for 13 weeks.
Twenty of each group were sacrificed at 13 weeks and five were sacrificed after an
additional 4-week recovery period.

No significant mortality during the test period was reported.   Body weight was
marginally lowered in males and females at 360 ppm. Depressed RBC parameters were
observed in the 60 and 360 ppm groups (6 and 13 weeks, both sexes) and in males at 30
ppm (6 weeks). Several other statistically significant decreases from control were
observed but these were not dose-related. In general, the magnitude of the differences
observed at each time point is small ($<10\%$).   Kidney weight (relative) was increased in
both sexes at the high-dose level (360 ppm) and in males at 60 ppm. Two types of
histopathological findings in the kidney were noted at 13 weeks: (1) occasional cells o
proximal tubules showing yellowish discoloration of the cytoplasm (all dose levels), and
(2) darker and more particulate granular and/or clumped pigmentation, predominantly
in cells of the straight portions and, to a lesser extent in the proximal convoluted tubules
(both sexes at 360 ppm/males at 60 ppm). From the data following the 4-week recovery
period, the discoloration and/or pigmentation was not persistent after withdrawal of
treatment, which Hoechst Aktiengesellschaft states is indicative of the ongoing process
of excretion of test material via the kidneys. Hoechst Aktiengesellschaft concludes that,
although this process can be observed (due to the coloration of the kidney tissues and the
observation of "dark urine"), it is not indicative of an adverse effect on the kidney, as
confirmed by the histopathological findings.

8) 13-Week Feeding - mouse: Core grade minimum (Hoechst Aktiengesellschaft, 1984b).

Groups of CD-1 mice (20/sex/group) were fed endosulfan in the diet at dose levels of 0, 2, 6, 18 or 54 ppm (Male: 0, 0.24, 0.74, 2.13 and 7.3 mg/kg-day; Female: 0, 0.27, 0.8, 2.39 and 7.52 mg/kg-day) for 13 weeks. Ten animals of each sex were sacrificed after an approximately 20 day observation period prior to the test period and examined microscopically.

Increased mortality (12/20 males and 10/20 females) was observed at 54 ppm. Glucose levels in females were significantly ($p<0.01$) lowered at 6, 18 and 54 ppm. Hemoglobin levels were significantly ($p<0.05$) elevated at 2, 6 and 18 ppm in females and appeared elevated at 54 ppm; however, this value was not analyzed due to the few survivors. Mean corpuscular hemoglobin concentration was significantly ($p<0.05$) lowered at 2, 6, and 18 ppm in females and appeared lowered at 54 ppm, however this value was not analyzed for the reason cited above. Based on the effects observed in females at the lowest dose tested, the LEL for systemic toxicity was 2 ppm (0.27 mg/kg-day). A NOEL for systemic toxicity was not established.

9) 30-Day Feeding - rat: Core grade supplementary (Hoechst Celanese Corp., 1985).

Groups of male Wistar rats (10 animals for control, 50 animals for each test dose) were fed endosulfan in the diet at dose levels of 0, 360 and 720 ppm (0, 34, and 67.8 mg/kg-day) for 30 days.

No overt signs of toxicity or dose-related effects were observed on body weight, food or water consumption, clinical observations, or ophthalmology. Two dosed animals died during the study with no discernible signs of toxicity. Absolute and relative liver weights of males receiving doses of 360 and 760 ppm and kidney weights of males receiving 720 ppm were increased ($p<0.05$) following the dosing period; organ weights of dosed males were similar to controls following a 30-day recovery period. Macroscopic examination revealed discoloration of the kidneys following the dosing period; histopathologically, the number and size of the lysosomes of the proximal convoluted tubules of the kidneys were increased following the dosing period with this finding exhibited to a greater extent in high-dose males. The renal changes were found to be reversible following the recovery period without evidence of renal lesions. No evidence of comparable lysosmal activity in the brain or liver was reported. Electron microscopy and tissue analysis confirmed that alpha-endosulfan and to a lesser extent beta-endosulfan, endosulfan sulfate and endosulfan-lactone were stored temporarily in the kidneys during the dosing period; only negligible amounts of endosulfan metabolites were found in the liver. Based on kidney changes during the dosing period, the LEL for systemic toxicity was 360 ppm (34 mg/kg-day), the lowest dose tested. A NOEL for systemic toxicity was not established.

Data Gap(s): Rat Developmental Toxicity Study

I.A.5. CONFIDENCE IN THE ORAL RfD

Study -- Medium
Data Base -- Medium
RfD -- Medium

   The principal studies are of adequate quality and therefore the given a   medium
confidence rating. Additional studies are supportive of the principal studies. However,
due to the replacement of test animals during the available developmental toxicity study
(FMC Corp., 1980), the data are considered inadequate to address the requirement for
testing in a second species. Due to the lack of these developmental data in a second
species, the data base is given a medium-to-high confidence rating. Medium-to-high
confidence in the RfD follows.

I.A.6. EPA DOCUMENTATION AND REVIEW OF THE ORAL RfD

Source Document -- This assessment is not presented in any existing US EPA document.

Other EPA Documentation -- None

Agency Work Group Review -- 05/20/85, 05/31/85, 11/21/85, 02/05/86, 06/11/86,
03/18/87, 11/04/92, 03/31/93

Verification Date -- 03/31/93

I.A.7. EPA CONTACTS (ORAL RfD)

George Ghali / OPP -- (703)305-7490

William Burnam / OPP -- (703)305-7491

I.B. REFERENCE CONCENTRATION FOR CHRONIC INHALATION EXPOSURE
(RfC)

        [Ed. Note: EPA does not have information yet in this section of IRIS].

II. CARCINOGENICITY ASSESSMENT FOR LIFETIME EXPOSURE

This substance/agent has been evaluated by the US EPA for evidence of human
carcinogenic potential. This does not imply that this agent is necessarily a carcinogen.
The evaluation for this chemical is under review by an inter-office Agency work group.
A risk assessment summary will be included on IRIS when the review has been
completed.

III. HEALTH HAZARD ASSESSMENTS FOR VARIED EXPOSURE DURATIONS

        [Ed. Note: EPA does not have information yet in this section of IRIS].

IV. US EPA REGULATORY ACTIONS

EPA risk assessments may be updated as new data are published and as assessment methodologies evolve. Regulatory actions are frequently not updated at the same time. Compare the dates for the regulatory actions in this section with the verification dates for the risk assessments in sections I and II, as this may explain inconsistencies. Also note that some regulatory actions consider factors not related to health risk, such as technical or economic feasibility. Such considerations are indicated for each action. In addition, not all of the regulatory actions listed in this section involve enforceable federal standards. Please direct any questions you may have concerning these regulatory actions to the US EPA contact listed for that particular action. Users are strongly urged to read the background information on each regulatory action in Background Document 4 in Service Code 5.

IV.A. CLEAN AIR ACT (CAA)

[Ed. Note: EPA does not have information yet in this section of IRIS].

IV.B. SAFE DRINKING WATER ACT (SDWA)

[Ed. Note: EPA does not have information yet in this section of IRIS].

IV.C. CLEAN WATER ACT (CWA)

IV.C.1. AMBIENT WATER QUALITY CRITERIA, Human Health

Water and Fish Consumption -- 7.4E+1 ug/L

Fish Consumption Only -- 1.59E+2 ug/L

Considers technological or economic feasibility? -- NO

Discussion -- The WQC for the protection of human health is based on an ADI of 0.28 mg/day, which was derived from a 78-week mouse study from the National Cancer Institute. The bioconcentration factor was established to be 270.

Reference -- 45 FR 79318 (11/28/80)

EPA Contact -- Criteria and Standards Division / OWRS (202)260-1315 / FTS 260-1315

IV.C.2. AMBIENT WATER QUALITY CRITERIA, Aquatic Organisms

Freshwater:
    Acute -- 2.2E-1 ug/L (at any time)
    Chronic -- 5.6E-2 ug/L (24 hour average)

Marine:
    Acute -- 3.4E-2 ug/L (at any time)
    Chronic -- 8.7E-3 ug/L (24 hour average)

Considers technological or economic feasibility? -- NO

Discussion -- Criteria were derived from a minimum data base consisting of acute and chronic tests on a variety of species. Requirements and methods are covered in the reference to the Federal Register.

Reference -- 45 FR 79318 (11/28/80)

EPA Contact -- Criteria and Standards Division / OWRS (202)260-1315 / FTS 260-1315

IV.D. FEDERAL INSECTICIDE, FUNGICIDE, AND RODENTICIDE ACT (FIFRA)

IV.D.1. PESTICIDE ACTIVE INGREDIENT, Registration Standard

Status -- Issued (1982)

Reference -- Endosulfan Pesticide Registration Standard. April, 1982 (NTIS No. PB82-243999).

EPA Contact -- Registration Branch / OPP (703)557-7760 / FTS 557-7760

IV.D.2. PESTICIDE ACTIVE INGREDIENT, Special Review

[Ed. Note: EPA does not have information yet in this section of IRIS].

IV.E. TOXIC SUBSTANCES CONTROL ACT (TSCA)

[Ed. Note: EPA does not have information yet in this section of IRIS].

IV.F. RESOURCE CONSERVATION AND RECOVERY ACT (RCRA)

[Ed. Note: EPA does not have information yet in this section of IRIS].

IV.G. SUPERFUND (CERCLA)

IV.G.1. REPORTABLE QUANTITY (RQ) for Release into the Environment

Value (status) -- 1 pound (Final, 1985)

Considers technological or economic feasibility? -- NO

Discussion -- The final RQ is based on aquatic toxicity as established under CWA Section 311 (40 CFR 117.3). The available data indicate that the 96-Hour Median Threshold Limit for endosulfan is less than 0.1 ppm. Endosulfan is known to have a chronic effect but, since its RQ is set at the lowest possible level based on aquatic toxicity, no further evaluation has been carried out.

Reference -- 50 FR 13456 (04/04/85); 54 FR 33418 (08/14/89)

EPA Contact -- RCRA/Superfund Hotline (800)424-9346 / (202)260-3000 / FTS 260-3000

VI. BIBLIOGRAPHY

VI.A. ORAL RfD REFERENCES

FMC Corporation. 1980. MRID No. 00055544. HED Doc. No. 000416. Available from EPA. Write to FOI, EPA, Washington, DC 20460.

FMC Corporation. 1981. MRID No. 00094837. HED Doc. No. 001488. Available from EPA. Write to FOI, EPA, Washington, DC 20460.

Hoechst Aktiengesellschaft. 1984a. MRID No. 00148264. HED Doc. No. 004881, 008868, 009552. Available from EPA. Write to FOI, EPA, Washington, DC 20460.

Hoechst Aktiengesellschaft. 1984b. MRID No. 00147182. HED Doc. No. 004733. Available from EPA. Write to FOI, EPA, Washington, DC 20460.

Hoechst Aktiengesellschaft. 1985. MRID No. 00145668. HED Doc. No. 005115, 008868. Available from EPA. Write to FOI, EPA, Washington, DC 20460.

Hoechst Celanese Corporation. 1985. MRID No. 00147299, 40767601. HED Doc. No. 007163. Available from EPA. Write to FOI, EPA, Washington, DC 20460.

Hoechst Celanese Corporation. 1988. MRID No. 00162996, 40256501, 40792401. HED Doc. No. 007155. Available from EPA. Write to FOI, EPA, Washington, DC 20460.

Hoechst Celanese Corporation. 1989a. MRID No. 40256502, 41099502. HED Doc. No. 007937. Available from EPA. Write to FOI, EPA, Washington, DC 20460.

Hoechst Celanese Corporation. 1989b. MRID No. 41099501. HED Doc. No. 007937. Available from EPA. Write to FOI, EPA, Washington, DC 20460.

VI.B. INHALATION RfC REFERENCES

None

VI.C. CARCINOGENICITY ASSESSMENT REFERENCES

None

VI.D. DRINKING WATER HA REFERENCES

None

VII. REVISION HISTORY

```
-------- -----------  ----------------------------------------------------------
Date     Section      Description
-------- -----------  ----------------------------------------------------------
```

Date	Section	Description
07/01/89	I.A.2.	MRID numbers added to principal study
07/01/89	I.A.4.	Citations added
07/01/89	VI.	Bibliography on-line
01/01/92	IV.	Regulatory actions updated
12/01/92	I.A.	Withdrawn; new Oral RfD verified (in preparation)
12/01/92	VI.A.	Oral RfD references withdrawn
05/01/93	I.A.	Work group review date added
07/01/94	II.	Carcinogenicity assessment now under review
10/01/94	I.A.	Oral RfD summary replaced; new RfD
10/01/94	VI.A.	Oral RfD references replaced

SYNONYMS

BENZOEPIN
BEOSIT
BIO 5,462
CHLORTHIEPIN
CRISULFAN
CYCLODAN
DEVISULPHAN
ENDOCEL
ENDOSOL
Endosulfan
ENDOSULPHAN
ENSURE
ENT 23,979
FMC 5462
1,2,3,4,7,7-HEXACHLOROBICYCLO(2.2.1)HEPTEN-5,6-BIOXYMETHYLENESULFITE
alpha,beta-1,2,3,4,7,7-HEXACHLOROBICYCLO(2.2.1)-2-HEPTENE-5,6-BISOXYMETHYLENE SULFITE
HEXACHLOROHEXAHYDROMETHANO 2,4,3-BENZODIOXATHIEPIN-3-OXIDE
6,7,8,9,10,10-HEXACHLORO-1,5,5a,6,9,9a-HEXAHYDRO-6,9-METHANO-2,4,3-BENZODIOXATHIEPIN-3-OXIDE

1,4,5,6,7,7-HEXACHLORO-5-NORBORNENE-2,3-DIMETHANOL cyclic SULFITE
HILDAN
HOE 2,671
INSECTOPHENE
KOP-THIODAN
MALIX
NA 2761
NCI-C00566
NIA 5462
NIAGARA 5,462
5-NORBORNENE-2,3-DIMETHANOL, 1,4,5,6,7,7-HEXACHLORO-, CYCLIC SULFITE
OMS 570
RCRA WASTE NUMBER P050
THIFOR
THIMUL
THIODAN
THIOFOR
THIOMUL
THIONEX
THIOSULFAN
THIOSULFAN TIONEL
TIOVEL

Source: *Instant EPA's IRIS*, 1997, Instant Reference Sources, Inc., 7605 Rockpoint Drive, Austin, TX 78731 (Fax: 512-345-2386; Internet URL, http://www.instantref.com/inst-ref.htm).

Production, Use, and Pesticide Labeling Information

Endosulfan is not currently listed in EPA's Pesticide Factsheet database and there is no indication of whether it will be added or not in the future.

Endrin

Cl Cl Cl Cl Cl Cl O

Introduction

Endrin was used as an insecticide, avicide, rodenticide and pesticide and it is a stereoisomer of dieldrin. [025], [055], [151], [421]. It has been found as a contaminant throughout the environment, including foodstuffs, fish, human milk, etc. [025] and its manufacture and use is discontinued in the United States. [031] Endrin was used mainly on field crops such as cotton and grains. It was used to control the army cutworm, the pale western cutworm and it has also been used for grasshoppers in noncropland and to control voles and mice in orchards. [828]

Endrin is very persistent, but it is known to photodegrade to delta-ketoendrin. Endrin released to soil will persist for long periods (up to 14 years or more), will reach the air either through very slow evaporation or adsorption on dust particles, will not leach to groundwater, and will reach surface water from surface runoff. Once endrin reaches surface waters it will adsorb strongly to sediments, bioconcentrate in fish, and photodegrade. Biodegradation will not be an important process. Fate of endrin in the atmosphere is unknown, but it probably will be primarily associated with particulate matter and be removed mainly by rainout and dry deposition. Monitoring data demonstrates that endrin continues to be a contaminant in air, water, sediment, soil, fish, and other aquatic organisms. Human exposure appears to come mostly from food or occupational exposure. [828]

Identifying Information

CAS NUMBER: 72-20-8

NIOSH Registry Number: IO1575000

SYNONYMS: 1,2,3,4,10,10-HEXACHLORO-6,7-EPOXY-1,4,4A,5,6,7,8,8A-OCTAHYDRO-1,4-ENDO,ENDO-, 5,8-DIMETHANONAPHTHALENE, HEXACHLOROEPOXYOCTAHYDRO-ENDO,ENDODIMETHANO-NAPHTHALENE, 3,4,5,6,9,9-HEXACHLORO-1A,2,2A,3,6,6A,7,7A-OCTAHYDRO-2,7:3,6-DIMETHANO-, NAPHTH(2,3-B)OXIRENE, HEXADRIN, (1A ALPHA, 2 BETA, 2A BETA, 3 ALPHA, 6 ALPHA, 6A BETA, 7 BETA, 7A ALPHA)-, 3,4,5,6,9,9,-HEXACHLORO-1A,2,2A,3,6,6A,7,7A-OCTAHYDRO-2,7:3,6-

DIMETHANO-, NAPH[2,3-B]OXIRENE, COMPOUND 269, ENDREX, ENT 17,251, EXPERIMENTAL INSECTICIDE 269, MENDRIN, NA 2761, NCI-C00157, NENDRIN, OMS 197, RCRA WASTE NUMBER P051

Endocrine Disruptor Information

Endrin is on EPA's list of suspect EEDs [811] but little information has been located to determine the Agency's reason for its inclusion. More information should be forthcoming after EPA's studies are concluded.

Chemical and Physical Properties

CHEMICAL FORMULA: $C_{12}H_8Cl_6O$

MOLECULAR WEIGHT: 380.90

WLN: T E3 D5 C555 A D- FO KUTJ AG AG BG JG KG LG

PHYSICAL DESCRIPTION: White crystalline powder

SPECIFIC GRAVITY: 1.70 @ 20°C [173]

DENSITY: Not available

MELTING POINT: Decomposes @ 200°C [043], [051], [102], [395]

BOILING POINT: Decomposes [051], [102]

SOLUBILITY: Water: <1 mg/mL @ 20°C; DMSO: 5-10 mg/mL @ 20°C; 95% Ethanol: <1 mg/mL @ 20°C; Acetone: >=100 mg/mL @ 20°C; Methanol: Insoluble [051], [062], [421]

OTHER SOLVENTS: Benzene: 13.8 g/100 mL @ 25°C [031]; Carbon tetrachloride: 3.3 g/100 mL @ 25°C [031]; Hexane: 7.1 g/100 mL @ 25°C [031]; Xylene: 18.3 g/100 mL @ 25°C [031]; Alcohols: Sparingly soluble [169], [173]; Petroleum distillates: Sparingly soluble [173]; Mineral oils: Sparingly soluble [169]; Aromatic hydrocarbons: Soluble [451]; Esters: Soluble [451]; Ketones: Soluble [451]; Other common organic solvents: Moderately soluble [051], [062], [421]

OTHER PHYSICAL DATA: Technical grade endrin is a light tan powder [173], [451]; Specific gravity also reported as 1.65 @ 25°C [371]; Melting point also reported as 226-230°C (with decomposition) [169]; Mild chemical odor [102]; Odor threshold: 0.018-0.041 ppm [051]

Source: *The National Toxicology Program's Chemical Database, Volume 2*, 1992, CRC Press, Inc./Lewis Publishers, 2000 Corporate Blvd., Boca Raton, FL 33431 (Fax: 407-998-9114).

Environmental Reference Materials

A source of EED environmental reference materials for Endrin is Radian International LLC, Austin, TX (Fax 512-454-0268; Internet URL, http://www.radian.com/standards); Catalog No. ERE-007.

Hazardous Properties

ACUTE/CHRONIC HAZARDS: Endrin may be toxic by ingestion, inhalation and skin absorption. [025], [062], [371], [451] It is an irritant [051], [371] and, when heated to decomposition, it emits toxic fumes of carbon monoxide, chlorides, hydrogen chloride gas and phosgene. [051], [102], [371], [451]

FIRE HAZARD: Literature sources indicate that endrin is nonflammable. [102], [173], [371], [421] Fires involving this material can be controlled with a dry chemical, carbon dioxide or Halon extinguisher.

REACTIVITY: Endrin reacts with strong acids. [051], [169], [346], [395] It also reacts with certain metal salts and catalytically active carriers. [173] Mixtures with parathion dissolve exothermically in petroleum solvents and may cause an air-vapor explosion [043], [066], [451]. Endrin is incompatible with strong oxidizers. [051], [102], [346] It is slightly corrosive to metals. [169]

STABILITY: Endrin is sensitive to heat (will decompose at temperatures >200°C). [102], [173], [395] It is also sensitive to sunlight. [395]

Source: _The National Toxicology Program's Chemical Database, Volume 7_, 1992, CRC Press, Inc./Lewis Publishers, 2000 Corporate Blvd., Boca Raton, FL 33431 (Fax: 407-998-9114).

Medical Symptoms of Exposure

Symptoms of exposure may include epileptiform convulsions with violent muscular contractions, weakness and nausea. [151], [173], [371] Other symptoms may include twitching, tingling of the limbs, temporary deafness, mental confusion and unconsciousness. [151] Exposure may cause dizziness, abdominal discomfort, headache, lethargy, anorexia and death. [173], [346] It may also cause weakness of the legs, slight disorientation, slight aggression, hyperthermia (41 C or more), decerebrate rigidity in children, semiconsciousness and coma. [173] Facial congestion and frothing of the mouth may occur.[371] Abdominal pain may also occur. [451] Other symptoms may include vomiting, insomnia, aggressive confusion and respiratory failure [346]. Sleepiness and agitation have been reported. [102]

Symptoms of exposure to related compounds may include malaise, tremors, clonic and tonic convulsions, renal damage, ataxia, central nervous system depression and hepatic damage. [031] Other symptoms may include leukocytosis, rise in blood pressure, tachycardia, arrhythmias, metabolic acidosis, fever, disturbance of sleep, memory and behavior; generalized cerebral dysrhythmia, hematuria and albuminuria. [151] Hyperexcitability may occur. [301] Other symptoms reported in related compounds may include central nervous system stimulation, diarrhea, paresthesia, excitement, giddiness, fatigue, pulmonary edema, myocardial toxicity, hypothermia and an apprehensive mental state. Respiration may be accelerated initially, but then later depressed. [295]

Source: *The National Toxicology Program's Chemical Database, Volume 4*, 1992, CRC Press, Inc./Lewis Publishers, 2000 Corporate Blvd., Boca Raton, FL 33431 (Fax: 407-998-9114).

Toxicological Information

The toxicological information below is gathered from several sources including two national databases: one from the *National Toxicology Program Chemical Repository Database* and another from *EPA's Integrated Risk Information System (IRIS)*.

Toxicology Information from the National Toxicology Program

CARCINOGENICITY:

Review:

IARC Cancer Review: Animal Inadequate Evidence
IARC: Not classifiable as a human carcinogen (Group 3)

Status:

NCI Carcinogenesis Bioassay (Feed); Negative: Male and Female Rat, Male and Female Mouse

TERATOGENICITY:

Reproductive Effects Data:

Type of Effect	Route/Animal	Amount Dosed (Notes)
TDLo:	orl-rat	2320 ug/kg (4D pre)
TDLo:	itt-rat	10 mg/kg (10D male)
TDLo:	orl-mus	10 mg/kg (8-12D preg)
TDLo:	orl-mus	2500 ug/kg (9D preg)
TDLo:	orl-mus	2320 ug/kg (4D pre)
TDLo:	orl-mus	7 mg/kg (8D preg)
TDLo:	orl-mus	16500 ug/kg (7-17D preg)
TDLo:	orl-ham	15 mg/kg (5-14D preg)
TDLo:	orl-ham	5 mg/kg (8D preg)
TDLo:	orl-ham	7500 ug/kg (5-14D preg)

MUTATION DATA:

Test	Lowest dose		Test	Lowest dose
sce-ofs-mul	54 pmol/L		cyt-rat-par	1 mg/kg
spm-rat-par	10 mg/kg/10D-C			

TOXICITY:

Typ.Dose	Mode	Specie	Amount	Units	Other
LD50	orl	rat	3	mg/kg	
LD50	skn	rat	12	mg/kg	
LD50	unr	rat	17	mg/kg	
LD50	orl	mus	1370	ug/kg	
LD50	ivn	mus	2300	ug/kg	
LD50	orl	mky	3	mg/kg	
LDLo	orl	cat	5	mg/kg	
LD50	orl	rbt	7	mg/kg	
LD50	skn	rbt	60	mg/kg	
LD50	orl	gpg	16	mg/kg	
LD50	orl	ham	10	mg/kg	
LD50	orl	pgn	5600	ug/kg	
LD50	Ivn	pgn	1500	ug/kg	
LD50	orl	dck	5330	ug/kg	
LD50	orl	bwd	1780	ug/kg	
LDLo	orl	man	171	mg/kg	
LDLo	skn	cat	75	mg/kg	
LD50	unr	rbt	7	mg/kg	
LD50	orl	qal	4210	ug/kg	

OTHER TOXICITY DATA:

Review:
Toxicology Review-7

Standards and Regulations:
DOT-Hazard: Poison B; Label: Poison
DOT-Hazard: Poison B; Label: Poison liquid
EPA: Farm Worker Field Reentry

Status:
EPA Genetox Program 1988, Negative: Histidine reversion-Ames test
EPA Genetox Program 1988, Negative: In vitro UDS-human
fibroblast; TRP reversion
EPA Genetox Program 1988, Negative: S cerevisiae-homozygosis
EPA Genetox Program 1988, Negative/limited: Carcinogenicity-
mouse/rat
EPA Genetox Program 1988, Inconclusive: D melanogaster Sex-linked
lethal
EPA TSCA Test Submission (TSCATS) Database, Janaury 1989
Meets criteria for proposed OSHA Medical Records Rule IDLH value:
200 mg/m^3
Estimated lethal dose for adult: 0.5 tsp

SAX TOXICITY EVALUATION:

> THR: Poison by ingestion, skin contact, intravenous and possibly other routes. A suspected human carcinogen. An experimental teratogen. Experimental reproductive effects. Mutagenic data. A central nervous system stimulant. Highly toxic to birds, fish and humans. Many cases of fatal poisoning have been attributed to it. It does not accumulate in human tissue. In humans, ingestion of 1 mg/kg has caused symptoms. A dangerous fire hazard.

Source: *Instant Tox-Base*, 1995, Instant Reference Sources, Inc., 7605 Rockpoint Drive, Austin, TX 78731 (Fax: 512-345-2386; Internet URL, http://www.instantref.com/inst-ref.htm).

Toxicology Information from EPA's Integrated Risk Information System (IRIS)

I. CHRONIC HEALTH HAZARD ASSESSMENTS FOR NONCARCINOGENIC EFFECTS

I.A. REFERENCE DOSE FOR CHRONIC ORAL EXPOSURE (RfD)

The Reference Dose (RfD) is based on the assumption that thresholds exist for certain toxic effects such as cellular necrosis, but may not exist for other toxic effects such as carcinogenicity. In general, the RfD is an estimate (with uncertainty spanning perhaps an order of magnitude) of a daily exposure to the human population (including sensitive subgroups) that is likely to be without an appreciable risk of deleterious effects during a lifetime. Please refer to Background Document 1 in Service Code 5 for an elaboration of these concepts. RfDs can also be derived for the noncarcinogenic health effects of compounds which are also carcinogens. Therefore, it is essential to refer to other sources of information concerning the carcinogenicity of this substance. If the US EPA has evaluated this substance for potential human carcinogenicity, a summary of that evaluation will be contained in Section II of this file when a review of that evaluation is completed.

I.A.1. ORAL RfD SUMMARY

Critical Effect	Experimental Doses*	UF	MF	RfD
Mild histological lesions in liver, occasional convulsions	NOEL: 1 ppm in diet (0.025 mg/kg/day)	100	1	3E-4 mg/kg/day
	LOAEL: 2 ppm in diet (0.05 mg/kg/day)			

*Conversion Factors: 1 ppm = 0.025 mg/kg/day (assumed dog food consumption)

Dog Chronic Oral Bioassay, Velsicol Chemical, Corporation, 1969

I.A.2. PRINCIPAL AND SUPPORTING STUDIES (ORAL RfD)

Velsicol Chemical Corporation.   1969.   MRID. No. 00030198. Available from EPA. Write FOI, EPA, Washington, DC. 20460.

Groups of 3 to 7 dogs/sex were fed diets containing 0.1, 0.5, 1.0, 2.0 or 4.0 ppm endrin for 2 years. Dogs receiving 2 or 4 ppm experienced occasional convulsions, slightly increased relative liver weights, and mild histopathological effects in the liver (slight vacuolization of hepatic cells). No adverse effects on these parameters or on growth, food   consumption, behavior, serum chemistry, urine chemistry or histological appearance of major organs occurred at 1 ppm (NOEL) or less. The 2 ppm level is the LOAEL. The authors provided data concerning actual endrin consumptions as weekly averages, but no overall averages were calculated. Visual   inspection of these data indicated that application of the standard food factor of 2.5% bw/day would closely approximate actual consumption.   Therefore, the 1 ppm NOEL was equivalent to an endrin intake of 0.025 mg/kg/day.

An earlier study (Treon et al., 1955) established a dietary NOEL of 1 ppm for both dogs and rats for long-term feeding (18 months - 2 years). LOAELs of 3 ppm and 5 ppm were reported for dogs and rats, respectively. The primary target organs were the kidney and the liver. Dogs are judged to be more sensitive than rats to long-term exposure to endrin because of the lower food consumption of dogs (than rats) and because of the much shorter duration of exposure (in this study) relative to lifetime for dogs as compared to rats.

I.A.3. UNCERTAINTY AND MODIFYING FACTORS (ORAL RfD)

UF -- The UF of 100 allows for uncertainty in the extrapolation of dose levels from laboratory animals to humans (10A) and uncertainty in the   threshold for sensitive humans (10H).

MF -- None

I.A.4. ADDITIONAL COMMENTS (ORAL RfD)

Acute lethality data suggest that rabbits and monkeys are much more sensitive to endrin than rats (Treon et al., 1955). Long-term studies have not been conducted with species other than rats or dogs. Conflicting evidence exists as to the developmental toxicity of endrin. Developmental effects have been observed to occur at dose levels much greater than those associated with   chronic toxicity; these studies are discussed in US EPA (1987).

I.A.5. CONFIDENCE IN THE ORAL RfD

Study -- Medium Data Base -- Medium RfD -- Medium

The principal study was of average quality and is given medium confidence.   The data base is assigned medium confidence because, although the chronic data is supportive, information on reproductive effects is lacking. Medium confidence in the RfD follows.

I.A.6. EPA DOCUMENTATION AND REVIEW OF THE ORAL RfD

Source Document -- US EPA, 1987, 1985

Other EPA Documentation -- None

Agency Work Group Review -- 05/20/85, 08/13/87, 10/15/87, 04/20/88

Verification Date -- 04/20/88

I.A.7. EPA CONTACTS (ORAL RfD)

George Ghali / OPP -- (703)557-7490

Carolyn Smallwood / NCEA -- (513)569-7425

I.B. REFERENCE CONCENTRATION FOR CHRONIC INHALATION EXPOSURE (RfC)

Not available at this time.

II. CARCINOGENICITY ASSESSMENT FOR LIFETIME EXPOSURE

Section II provides information on three aspects of the carcinogenic risk assessment for the agent in question; the US EPA classification, and quantitative estimates of risk from oral exposure and from inhalation exposure.   The classification reflects a weight-of-evidence judgment of the likelihood   that the agent is a human carcinogen.   The quantitative risk estimates are presented in three ways. The slope factor is the result of application of a   low-dose extrapolation procedure and is presented as the risk per (mg/kg)/day.   The unit risk is the quantitative estimate in terms of either risk per ug/L drinking water or risk per ug/cu.m air breathed.   The third form in which risk is presented is a drinking water or air concentration providing cancer risks of 1 in 10,000, 1 in 100,000 or 1 in 1,000,000. Background Document 2 (Service Code 5) provides details on the rationale and methods used to derive the carcinogenicity values found in IRIS. Users are referred to Section I for information on long-term toxic effects other than carcinogenicity.

II.A. EVIDENCE FOR CLASSIFICATION AS TO HUMAN CARCINOGENICITY

II.A.1. WEIGHT-OF-EVIDENCE CLASSIFICATION

Classification -- D; not classifiable as to carcinogenicity for humans

Basis -- Oral administration of endrin did not produce carcinogenic effects in either sex of two strains of rats and three strains of mice. An NCI bioassay was suggestive of responses in male and female rats although NCI reported a no evidence conclusion. The inadequacies of several of the bioassays call into question the strength of the reported negative findings. These inadequacies and the suggestive responses in the NCI bioassay do not support a Group E classification; rather a Group D classification best reflects the equivocal data.

II.A.2. HUMAN CARCINOGENICITY DATA

Inadequate. Ditraglia et al. (1981) conducted a retrospective cohort study to examine the mortality of workers employed in the manufacture of organochlorine pesticides including endrin. No statistically significant excesses or deficits in mortality for any specific cancer site were noted.   Limited follow-up time (12 years), lack of exposure data, and few deaths give this study low power.

II.A.3. ANIMAL CARCINOGENICITY DATA

Inadequate. The potential carcinogenic effects of endrin have been evaluated following oral exposure to 1-100 ppm endrin in the diet of Carworth Farm rats, (Treon et al., 1955), Osborne-Mendel rats (Deichmann et al, 1970;   NCI, 1979), C57Bl/6J mice (Witherup et al., 1970), C3D2F1/J mice (Witherup et al., 1970) and B6C3F1 mice (NCI, 1979). There was no evidence of carcinogenicity in any of these studies. Treon et al., (1955) also failed to note any increase in tumorigenesis in dogs exposed up to 18.7 months at the   maximum tolerated dose. The length of this study was insufficient to provide for the expected latency period in dogs.

The NCI (1979) bioassay was done in Osborne-Mendel rats (50/sex/group) and B6C3F1 mice (50/sex/group); matched control groups included 10 animals/sex/ species. Since the number of animals in the matched-control groups was small, pooled-control groups from concurrent pesticide bioassays were used for statistical evaluation.

Endrin was administered daily in the diet for 80 weeks. Rats were observed for an additional 31 to 34 weeks and mice were observed for an additional 11 weeks. The initial doses for male rats and all mice were 2.5 or 5 ppm and for female rats were 5 or 10 ppm. Because of subsequent toxic effects, the doses for the female rats and male mice were reduced during the course of the studies. High-dose male mice were fed treatment and control diets on alternate weeks for 10 weeks. The resulting time-weighted average dose fed in the diets of treated animals was reported as follows: 2.5 or 5 ppm for male rats, 3 or 6 ppm for female rats, 1.6 or 3.2 ppm for male mice, and 2.5 or 5 ppm for female mice.

When compared with pooled controls, a statistically significant increase in hemangioma was observed in low-dose male rats (0/49, 5/46, 3/47), and a   significant increase in adrenal adenoma or carcinoma was seen in high-dose male rats (2/44, 4/46, 8/44). Islet-cell carcinoma incidence in male rats showed a significant positive trend but the pair wise comparisons were not significant. A statistically significant increase in pituitary adenoma was observed in the high-dose female rats (4/44, 11/47, 13/45) and a significant increase in adrenal adenoma or carcinoma was observed in the low-dose female rats (4/46, 14/49, 7/47).

Although NCI concluded from the bioassays that endrin was not carcinogenic, the responses noted above cannot be totally ignored. A primary reviewer for NCI noted that the negative findings could be a reflection of the high toxicity of endrin, which only permitted the administration of relatively low chronic doses. Furthermore, the reviewer observed that an accidental overdose among low-dose male mice resulted in the early death of several animals in this treatment group. The study was marred by a small number (10) of matched controls; however, this deficiency was compensated by the use of pooled controls.

Reuber (1978) reported positive carcinogenic effects of endrin had been observed in a FDA bioassay (Bierbower, 1965). Male and female Osborne-Mendel rats were exposed to 0.1 to 25 ppm endrin in the diet. At the 0.1 ppm dose, incidence of hyperplastic nodules and malignant tumors of the liver was significantly increased in female rats and in male and female rats combined. A variety of other tumors were observed including mammary gland, uterine, and thyroid tumors in females and thyroid and adrenal cortex tumors in males.

Reuber (1979) independently reevaluated several endrin carcinogenicity studies including the NCI (1979) study and the FDA (Bierbower, 1965) study and determined that a significant increase in tumor incidence was present. It is difficult to draw conclusions from Reuber's findings, however, since his criteria for classifying lesions as tumorigenic appear to differ from those of other investigators. Reuber did not provide slide by slide tabulation of his findings nor did he distinguish between primary and/or metastatic tumors in the liver (Albert, 1977).

II.A.4. SUPPORTING DATA FOR CARCINOGENICITY

Maslansky and Williams (1981) showed that endrin (10-3 and 10-4 M) was not genotoxic in the hepatocyte primary culture (HPC)/DNA repair assay using hepatocytes from male Fischer F344 rats, male CD-1 mice, and mal Syrian hamsters. DNA repair was observed in response to a positive control in all three systems. Endrin was not mutagenic in microbial systems with or without metabolic activation (Moriya et al., 1983; Probst et al., 1981; Glatt et al., 1983), and endrin exposure did not significantly affect sister-chromatid exchange frequencies in a human lymphoid cell line (Sobti et al., 1983). Endrin is also structurally related to aldrin, dieldrin, chlordane, chlorendic acid, and heptachlor which are known to be carcinogenic in animals.

II.B. QUANTITATIVE ESTIMATE OF CARCINOGENIC RISK FROM ORAL EXPOSURE

[Ed. Note: EPA does not have information yet in this section of IRIS].

II.C. QUANTITATIVE ESTIMATE OF CARCINOGENIC RISK FROM INHALATION EXPOSURE

[Ed. Note: EPA does not have information yet in this section of IRIS].

II.D. EPA DOCUMENTATION, REVIEW, AND CONTACTS (CARCINOGENICITY ASSESSMENT)

II.D.1. EPA DOCUMENTATION

Source Document -- US EPA, 1987a,b

The 1988 Carcinogenicity Assessment for Endrin has had Agency review.

II.D.2. REVIEW (CARCINOGENICITY ASSESSMENT)

Agency Work Group Review -- 03/23/88, 10/19/88

Verification Date -- 10/19/88

II.D.3. US EPA CONTACTS (CARCINOGENICITY ASSESSMENT)

Carolyn Smallwood / NCEA -- (513)569-7425

Dharm V. Singh / NCEA -- (202)260-5898

III. HEALTH HAZARD ASSESSMENTS FOR VARIED EXPOSURE DURATIONS

[Ed. Note: EPA does not have information yet in this section of IRIS].

IV. US EPA REGULATORY ACTIONS

EPA risk assessments may be updated as new data are published and as assessment
methodologies evolve. Regulatory actions are frequently not updated at the same time.
Compare the dates for the regulatory actions in this section with the verification dates for
the risk assessments in sections I and II, as this may explain inconsistencies. Also note
that some regulatory actions consider factors not related to health risk, such as technical
or economic feasibility. Such considerations are indicated for each action. In addition,
not all of the regulatory actions listed in this section involve enforceable federal
standards. Please direct any questions you may have concerning these regulatory actions
to the US EPA contact listed for that particular action. Users are strongly urged to read
the background information on each regulatory action in Background Document 4 in
Service Code 5.

IV.A. CLEAN AIR ACT (CAA)

[Ed. Note: EPA does not have information yet in this section of IRIS].

IV.B. SAFE DRINKING WATER ACT (SDWA)

IV.B.1. MAXIMUM CONTAMINANT LEVEL GOAL (MCLG) for Drinking Water

Value -- 0.002 mg/L (Proposed, 1990)

Considers technological or economic feasibility? -- NO

Discussion -- A MCLG of 0.002 mg/L is proposed based on potential adverse effects (clinical signs, mortality) reported in animal studies. The MCLG is based upon a DWEL of 0.009 mg/L and an assumed drinking water contribution of 20 percent.

Reference -- 55 FR 30370 (07/25/90)

EPA Contact -- Health and Ecological Criteria Division / OST / (202) 260-7571 / FTS 260-7571; or Safe Drinking Water Hotline / (800) 426-4791

IV.B.2. MAXIMUM CONTAMINANT LEVEL (MCL) for Drinking Water

Value -- 0.002 mg/L (Proposed, 1990)

Considers technological or economic feasibility? -- YES

Discussion -- EPA is proposing a MCL equal to the MCLG of 0.002 mg/L.

Monitoring requirements -- Community and non-transient water system monitoring based on state vulnerability assessment; vulnerable systems to be monitored quarterly for one year; repeat monitoring dependent upon detection and system size.

Analytical methodology -- Microextraction/gas chromatography (EPA 505); electron-capture/gas chromatography (EPA 508); gas chromatography/mass spectrometry (EPA 525).

Best available technology -- Granular activated carbon

Reference -- 55 FR 30370 (07/25/90)

EPA Contact -- Drinking Water Standards Division / OGWDW / (202) 260-7575 / FTS 260-7575; or Safe Drinking Water Hotline / (800) 426-4791

IV.B.3. SECONDARY MAXIMUM CONTAMINANT LEVEL (SMCL) for Drinking Water

[Ed. Note: EPA does not have information yet in this section of IRIS].

IV.B.4. REQUIRED MONITORING OF "UNREGULATED" CONTAMINANTS

[Ed. Note: EPA does not have information yet in this section of IRIS].

IV.C. CLEAN WATER ACT (CWA)

IV.C.1. AMBIENT WATER QUALITY CRITERIA, Human Health

Water and Fish Consumption: 1.0E+0 ug/L

Fish Consumption Only: None

Considers technological or economic feasibility? -- NO

Discussion -- The WQC of 1.0E+0 ug/L is based on consumption of contaminated aquatic organisms and water. A WQC based on consumption of contaminated aquatic organisms alone was not derived.

Reference -- 45 FR 79318 (11/28/80)

EPA Contact -- Criteria and Standards Division / OWRS (202)260-1315 / FTS 260-1315

IV.C.2. AMBIENT WATER QUALITY CRITERIA, Aquatic Organisms

Freshwater:
    Acute -- 1.8E-1 ug/L   Chronic -- 2.3E-3 ug/L (24-hour average)

Marine:
    Acute -- 3.7E-2 ug/L   Chronic -- 2.3E-3 ug/L (24-hour average)

Considers technological or economic feasibility? -- NO

Discussion -- Criteria were derived from a minimum data base consisting of acute tests on a variety of species. Requirements and methods are covered in the reference to the Federal Register.

Reference -- 45 FR 79318 (11/28/80)

EPA Contact -- Criteria and Standards Division / OWRS (202)260-1315 / FTS 260-1315

IV.D. FEDERAL INSECTICIDE, FUNGICIDE, AND RODENTICIDE ACT (FIFRA)

IV.D.1. PESTICIDE ACTIVE INGREDIENT, Registration Standard

Status -- Removed from list "B" pesticides/Registration canceled (1990)

Reference -- 55 FR 31166 (07/31/90)

EPA Contact -- Registration Branch / OPP (703)557-7760 / FTS 557-7760

IV.D.2. PESTICIDE ACTIVE INGREDIENT, Special Review

Action -- Final regulatory decision - PD4 (1979)

Considers technological or economic feasibility? -- NO

Summary of regulatory action -- All uses voluntarily canceled except one effective 09/12/85. Criteria of concern: oncogenicity, teratogenicity and reduction in endangered species and non-target species.

Reference -- 44 FR 43631 (07/25/79) [NTIS# PB81-109480]

EPA Contact -- Special Review Branch / OPP (703)557-7400 / FTS 557-7400

IV.E. TOXIC SUBSTANCES CONTROL ACT (TSCA)

[Ed. Note: EPA does not have information yet in this section of IRIS].

IV.F. RESOURCE CONSERVATION AND RECOVERY ACT (RCRA)

IV.F.1. RCRA APPENDIX IX, for Ground Water Monitoring

Status -- Listed

Reference -- 52 FR 25942 (07/09/87)

EPA Contact -- RCRA/Superfund Hotline (800)424-9346 / (202)260-3000 / FTS 260-3000

IV.G. SUPERFUND (CERCLA)

IV.G.1. REPORTABLE QUANTITY (RQ) for Release into the Environment

Value (status) -- 1 pound (Final, 1989)

Considers technological or economic feasibility? -- NO

Discussion -- The final RQ for endrin is based on aquatic toxicity as established under CWA Section 311 (40 CFR 117.3). The available data indicate that the aquatic 96-Hour Median Threshold Limit is less than 0.1 ppm, which corresponds to an RQ of 1 pound.

Reference -- 54 FR 33418 (08/14/89)

EPA Contact -- RCRA/Superfund Hotline (800)424-9346 / (202)260-3000 / FTS 260-3000

VI. BIBLIOGRAPHY

VI.A. ORAL RfD REFERENCES

Treon, J.F., F.P. Cleveland and J. Cappel. 1955. Toxicity of endrin for laboratory animals. Agric. Food Chem. 3(10): 842-848.

US EPA. 1987. Health Effects Assessment for Endrin. Final Draft. Environmental Criteria and Assessment Office, Cincinnati, OH. ECAO-CIN-H089.

US EPA. 1985. Drinking Water Criteria Document for Endrin. Final Draft. Environmental Criteria and Assessment Office, Cincinnati, OH. NTIS PB86- 117967.

Velsicol Chemical Corporation. 1969. MRID. No. 00030198. Available from EPA. Write FOI, EPA, Washington, DC. 20460.

VI.B. INHALATION RfD REFERENCES

None

VI.C. CARCINOGENICITY ASSESSMENT REFERENCES

Albert, R. 1977. Memorandum to Kyle Barbehenn, Endrin Project Manager, Office of Special Pesticide Review. Carcinogen Assessment Group, Washington, DC. June 20.

Bierbower, G.W. 1965. Final report on pathological study of rats fed endrin or dieldrin. Prepared as a memo to A.J. Lehman. Food and Drug Administration. (Cited in: Reuber, 1978)

Deichmann, W.B., W.E. MacDonald, E. Blum, et al. 1970. Tumorigenicity of aldrin, dieldrin and endrin in the albino rat. Ind. Med. 39: 426-434.

Ditraglia, D., D.P. Brown, T. Namekata and N. Iverson. 1981. Motality study of workers employed at organochlorine pesticide manufacturing plants. Scand. J. Work Environ. Health. 7: 140-146.

Glatt, H., R. Jung and F. Oesch. 1983. Bacterial mutagenicity investiga- tion of epoxides: Drugs, drug metabolites, steroids and pesticides. Mutat. Res. 11: 99-118.

Maslansky, C.J. and G.M. Williams. 1981. Evidence for an epigenetic mode of action in organochlorine pesticide hepatocarcinogenicity: A lack of geno- toxicity in rat, mouse, and hamster hepatocytes. J. Toxicol. Environ. Health. 8: 121-130.

Moriya, M., T. Ohta, K. Watanabe, T. Miyazawa, K. Kato and Y. Shirasu. 1983. Further mutagenicity studies on pesticides in bacterial reversion assay systems. Mutat. Res. 116: 185-216.

NCI (National Cancer Institute). 1979. Bioassay of endrin for possible carcinogenicity. Carcinogenesis Technical Report Series 12, NCR-CG-TR-12. Publ. No. (NIH) 79-812.

Probst, G.S., K.E. McMahon, L.E. Hill, et al. 1981. Chemically-induced unscheduled DNA synthesis in primary rat hepatocyte cultures: A comparison with bacterial mutagenicity using 218 compounds. Environ. Mutagen. 3: 11-32.

Reuber, M.D. 1978. Carcinomas, sarcomas and other lesions in Osborne-Mendel rats ingesting endrin. Exp. Cell. Biol. 46: 129-145.

Reuber, M.D. 1979. Carcinogenicity of endrin. Sci. Total Environ. 12: 101-135.

Sobti, R.C., A. Krishan and J. Davies. 1983. Cytokinetic and cytogenetic effect of agricultural chemicals on human lymphoid cells in vitro. II. Organochlorine pesticides. Arch. Toxicol. 52: 221-231.

Treon, J.F., F.P. Cleveland and J. Cappel. 1955. Toxicity of endrin for laboratory animals. J. Agric. Food Chem. 3: 842-848.

US EPA. 1987a. Drinking Water Criteria Document for Endrin. Prepared by the Office of Health and Environmental Assessment, Environmental Criteria and Assessment Office, Cincinnati, OH for the Office of Drinking Water, Washington, DC. (External Review Draft)

US EPA. 1987b. Health Effects Assessment for Endrin. Prepared by the Office of Health and Environmental Assessment, Environmental Criteria and Assessment Office, Cincinnati, OH for the Office of Emergency and Remedial Response, Washington, DC. (Final Draft)

Witherup, S., K.L. Stemmer, P. Taylor and P. Bietsch. 1970. The incidence of neoplasms in two strains of mice sustained on diets containing endrin. Kettering Lab., Univ. Cincinnati, Cincinnati, OH.

VI.D. DRINKING WATER HA REFERENCES

None

VII. REVISION HISTORY

Date	Section	Description
09/07/88	I.A.	Oral RfD summary on-line
04/01/89	V.	Supplementary data on-line
08/01/89	VI.	Bibliography on-line
10/01/89	II.	Carcinogen summary on-line
10/01/89	VI.C.	Carcinogen references added
04/01/91	I.A.	Secondary contact changed
04/01/91	II.	Primary contact changed
01/01/92	IV.	Regulatory Action section on-line
07/01/93	II.D.1.	EPA Documentation clarified
07/01/93	VI.C.	EPA references clarified

SYNONYMS

Endrin
mendrin, 1,2,3,4,10,10-hexachloro-6,7-epoxy-1,4,4(a)5,6,7,8,8a-octahydro-endo

Source: *Instant EPA's IRIS*, 1997, Instant Reference Sources, Inc., 7605 Rockpoint Drive, Austin, TX 78731 (Fax: 512-345-2386; Internet URL, http://www.instantref.com/inst-ref.htm).

Production, Use, and Pesticide Labeling Information

Endrin is not currently listed in EPA's Pesticide Factsheet database and there is no indication of whether it will be added or not in the future.

Heptachlor

Introduction

Heptachlor is an insecticide used for control of the cotton boll weevil, termites, ants, grasshoppers, cutworms, maggots, thrips, wireworms, flies, mosquitoes, soil insects, household insects and field insects. It has some fumigant action, and is applied as a soil treatment, a seed treatment or directly to foliage. However, the U. S. EPA has cancelled registration of pesticides containing Heptachlor with the exception of its use through subsurface ground insertion for termite control and dipping of roots or tops of nonfood plants [051], [072], [421]. The use of heptachlor in the United States was restricted to the control of fire ants in power transformers and its release to the environment may result from this use and past extensive pesticidal use prior to 1983. [828]

Release of heptachlor to soil surfaces will result in volatilization from the surface, especially in moist soils, but volatilization of heptachlor incorporated into soil will be slower. Hydrolysis in moist soils is expected to be significant. In soil, heptachlor will degrade to 1-hydroxychlordene, heptachlor epoxide and an unidentified metabolite less hydrophilic than heptachlor epoxide. Biodegradation may also be significant. Heptachlor is expected to adsorb strongly to soil and, therefore, to resist leaching to groundwater. Release of heptachlor to water will result in hydrolysis to 1-hydroxychlordene (half-life of about 1 day) and volatilization. Adsorption to sediments may occur. Biodegradation of heptachlor may occur, but is expected to be slow compared to hydrolysis. Bioconcentration of heptachlor may be significant. Direct and photosensitized photolysis may occur but are not expected to occur at a rate comparable to that of hydrolysis. In air, vapor phase heptachlor will react with photochemically generated hydroxyl radicals with an estimated half-life of 36 min. Direct photolysis may also occur. [828]

Identifying Information

CAS NUMBER: 76-44-8

NIOSH Registry Number: PC0700000

SYNONYMS: 1,4,5,6,7,8,8-Heptachloro-3a,4,7,7a-tetrahydro-4,7-methano-1H-indene

Endocrine Disruptor Information

Heptachlor is a strongly suspected EED that it is listed by EPA [811], the Centers for Disease Control & Prevention [812], and the World Wildlife Fund [813] as a potential endocrine modifying chemical.

Chemical and Physical Properties

CHEMICAL FORMULA: $C_{10}H_5Cl_7$

MOLECULAR WEIGHT: 373.32

WLN: L C555 A DU IUTJ AG AG BG FG HG IG JG

PHYSICAL DESCRIPTION: White crystals

SPECIFIC GRAVITY: 1.57-1.59 [062], [430]

DENSITY: 1.57-1.59 g/mL [051], [062], [072], [173]

MELTING POINT: 95-96°C [016], [025], [031], [062]

BOILING POINT: Decomposes [051], [058], [072]

SOLUBILITY: Water: <0.1 mg/mL @ 18°C [700]; DMSO: >=100 mg/mL @ 24°C [700]; 95% Ethanol: 10-50 mg/mL @ 24°C [700]; Acetone: >=100 mg/mL @ 24°C [700]

OTHER SOLVENTS: Ligroin: Soluble [016], [047]; Xylene: 102 g/100 mL @ 27°C [051], [072], [173], [395]; Carbon tetrachloride: 112 g/100 mL @ 27°C [051], [072], [173], [395]; Cyclohexane: 119 g/100 mL @ 27°C [051], [072], [173], [395]; Cyclohexanone: 119 g/100 mL [169]; Kerosene: 189 g/100 mL @ 27°C [173]; Hexane: Slightly soluble [421]; Ether: Soluble [016], [047]; Benzene: 106 g/100 mL @ 27°C [051], [072], [173], [395]; Most organic solvents: Soluble [043], [169];Petroleum distillates: Soluble [151]; Ketones: Soluble [151]; Alcohol: 4.5 g/100 mL @ 27°C [051,072,173]; Deodorized kerosene: 263 g/L [172]

OTHER PHYSICAL DATA: Density: 1.66 g/mL @ 20°C [051], [072]; Physical description (technical grade): Light tan waxy solid [169], [172]; Melting point (technical grade): 46-74° [058]; Boiling point also reported as 135-145°C [169], [395], [421]; Camphor-like odor [051], [058,072,173]

Source: *Instant EPA's Air Toxics*, 1994, Instant Reference Sources, Inc., 7605 Rockpoint Drive, Austin, TX 78731 (Fax: 512-345-2386; Internet URL, http://www.instantref.com/inst-ref.htm).

Environmental Reference Materials

A source of EED environmental reference materials for heptachlor is Radian International LLC, Austin, TX (Fax 512-454-0268; Internet URL, http://www.radian.com/standards); Catalog No. ERH-002.

Hazardous Properties

ACUTE/CHRONIC HAZARDS: Heptachlor may be toxic by ingestion, inhalation and skin absorption [033], [062]. It may be readily absorbed by the skin as well as by the lungs and gastrointestinal tract [173], [430]. When heated to decomposition it emits smoke, acrid fumes and toxic fumes of carbon monoxide [051]. Decomposition may also produce hydrogen chloride gas [058].

HAP WEIGHTING FACTOR: 10 [713]

VOLATILITY:
    Vapor pressure: 0.0003 mm Hg @ 25°C [033], [055], [395], [421]

FIRE HAZARD: Heptachlor is not combustible [058], [421]. Fires involving this material may be controlled with a dry chemical, carbon dioxide or Halon extinguisher.

REACTIVITY: Heptachlor is incompatible with strong alkali [169], [173]. It is corrosive to metals [169]. It can react with iron and rust to form toxic gases [058]. It can react vigorously with oxidizing materials [051]. It is susceptible to epoxidation [172].

STABILITY: Heptachlor is stable under normal lab conditions [169], [173], [186], [295]. Solutions of it in water, DMSO, 95% ethanol or acetone should be stable for 24 hours under normal lab conditions [700].

Source: *Instant EPA's Air Toxics*, 1994, Instant Reference Sources, Inc., 7605 Rockpoint Drive, Austin, TX 78731 (Fax: 512-345-2386; Internet URL, http://www.instantref.com/inst-ref.htm).

Medical Symptoms of Exposure

Symptoms of exposure may include irritation of the skin, central nervous system stimulation, ataxia, renal damage, tremors, convulsions, respiratory collapse, kidney damage and death [051], [072]. It can also cause congestion, edema, scattered petechial hemorrhages in the lungs, kidneys and brain, hyperexcitability, central nervous system depression and anuria [301]. Other symptoms may include paresthesia, excitement, giddiness, fatigue and coma [295]. If ingested, it can cause nausea, vomiting, diarrhea and irritation of the gastrointestinal tract [051], [072]. It may also cause liver necrosis and blood dyscrasias [033].

Source: *Instant EPA's Air Toxics*, 1994, Instant Reference Sources, Inc., 7605 Rockpoint Drive, Austin, TX 78731 (Fax: 512-345-2386; Internet URL, http://www.instantref.com/inst-ref.htm).

Toxicological Information

The toxicological information below is gathered from several sources including two national databases: one from the *National Toxicology Program Chemical Repository Database* and another from *EPA's Integrated Risk Information System (IRIS)*.

Toxicology Information from the National Toxicology Program

CARCINOGENICITY:

Tumorigenic Data:

Type of Effect	Route/Animal	Amount Dosed (Notes)
TDLo	orl-mus	403 mg/kg/80W-C
TD	orl-mus	930 mg/kg/80W-C
TD	orl-mus	876 mg/kg/2Y-C

Review:
> IARC Cancer Review: Animal Limited Evidence
> IARC Cancer Review: Human Inadequate Evidence
> IARC: Not classifiable as a human carcinogen (Group 3)

Status:
> NCI Carcinogenesis Bioassay (Feed); Negative: Male Rat
> NCI Carcinogenesis Bioassay (Feed); Equivocal: Female Rat
> NCI Carcinogenesis Bioassay (Feed); Positive: Male and Female Mouse
> EPA Carcinogen Assessment Group

TERATOGENICITY: Not available

MUTATION DATA: See RTECS printout for data

TOXICITY:

Typ. Dose	Mode	Specie	Amount	Units	Other
LD50	orl	rat	40	mg/kg	
LD50	skn	rat	119	mg/kg	
LD50	ipr	rat	27	mg/kg	
LD50	orl	mus	68	mg/kg	
LD50	ipr	mus	130	mg/kg	
LDLo	ivn	mus	20	mg/kg	
LD50	orl	gpg	116	mg/kg	
LD50	orl	ham	100	mg/kg	
LD50	unr	mam	60	mg/kg	
LDLo	orl	cat	50	mg/kg	
LCLo	ihl	cat	150	mg/m3/4H	
LDLo	skn	gpg	1	gm/kg	
LCLo	ihl	mam	200	mg/m3/4H	

OTHER TOXICITY DATA:

Review: Toxicology Review-6

Status:

EPA Genetox Program 1988, Positive/limited: Carcinogenicity-mouse/rat

EPA Genetox Program 1988, Inconclusive: Histidine reversion-Ames test

EPA Genetox Program 1988, Inconclusive: D melanogaster Sex-linked lethal

EPA TSCA Test Submission (TSCATS) Data Base, April 1990

SAX TOXICITY EVALUATION:

THR: An experimental carcinogen. Human mutagenic data.

Source: *Instant Tox-Base*, 1995, Instant Reference Sources, Inc., 7605 Rockpoint Drive, Austin, TX 78731 (Fax: 512-345-2386; Internet URL, http://www.instantref.com/inst-ref.htm).

Toxicology Information from EPA's Integrated Risk Information System (IRIS)

I. CHRONIC HEALTH HAZARD ASSESSMENTS FOR NONCARCINOGENIC EFFECTS

I.A. REFERENCE DOSE FOR CHRONIC ORAL EXPOSURE (RfD)

The Reference Dose (RfD) is based on the assumption that thresholds exist for certain toxic effects such as cellular necrosis, but may not exist for other toxic effects such as carcinogenicity. In general, the RfD is an estimate (with uncertainty spanning perhaps an order of magnitude) of a daily exposure to the human population (including sensitive subgroups) that is likely to be without an appreciable risk of deleterious effects during a lifetime. Please refer to Background Document 1 in Service Code 5 for an elaboration of these concepts. RfDs can also be derived for the noncarcinogenic health effects of compounds which are also carcinogens. Therefore, it is essential to refer to other sources of information concerning the carcinogenicity of this substance. If the US EPA has evaluated this substance for potential human carcinogenicity, a summary of that evaluation will be contained in Section II of this file when a review of that evaluation is completed.

I.A.1. ORAL RfD SUMMARY

Critical Effect	Experimental Doses*	UF	MF	RfD
Increased liver-to-body weight ratio in both males and females	NOEL: 3 ppm diet (0.15 mg/kg/day) LEL: 5 ppm diet (0.25 mg/kg/day)	300	1	5E-4 mg/kg/day

*Conversion Factors: 1 ppm = 0.05 mg/kg/day (assumed rat food consumption)

2-Year Rat Feeding Study Velsicol Chemical, 1955a

I.A.2. PRINCIPAL AND SUPPORTING STUDIES (ORAL RfD)

Velsicol Chemical Corporation. 1955a. MRID No. 00062599. Available from EPA. Write to FOI, EPA, Washington, DC 20460.

Six groups of CF strain white rats containing 20/sex were fed for 2 years with diets of 0, 1.5, 3, 5, 7, or 10 ppm of heptachlor. Lesions in the liver were limited to 7 ppm and above and were characteristic of chlorinated hydrocarbons (that is, hepatocellular swelling and peripheral arrangements of the cytoplasmic granules of cells of the central zone of the liver lobules). The NOEL for the lesions was 5 ppm and the LEL was 7 ppm. The NOEL for increased liver-to-body weight for males only was 3 ppm and the LEL was 5 ppm.

I.A.3. UNCERTAINTY AND MODIFYING FACTORS (ORAL RfD)

UF -- Based on a chronic exposure study, an uncertainty factor of 100 was used to account for inter- and intraspecies differences. An additional factor of 3 was considered appropriate because of the lack of chronic toxicity data in a second species, for a total uncertainty factor of 300. The serious deficiencies in the toxicologic data base would normally warrant a 10-fold factor for this area of uncertainty. However, toxicity data for other cyclodiene insecticides (aldrin, dieldrin, chlordane, and heptachlor epoxide) suggest that dogs and rats do not differ greatly in sensitivity to the effects of this class of compounds. Furthermore, liver toxicity has been fairly well established as the most sensitive endpoint for this class of compounds, which reduces the uncertainty attributable to the lack of information on other toxic effects.

MF -- None

I.A.4. ADDITIONAL COMMENTS (ORAL RfD)

Data Considered for Establishing the RfD:

1) 2-Year Feeding - rat: Principal study - see previous description; no core grade

2) 8-Month Feeding - rat: NOEL= none; LEL=5 ppm (0.25 mg/kg/day) (LDT) (swelling of cells); no core grade (Velsicol Chemical, 1964)

3) 1-Generation Reproduction - rat: NOEL=5 ppm (0.25 mg/kg/day); LEL=7 ppm (0.35 mg/kg/day) (increased pup death); no core grade (Velsicol Chemical, 1955b)

4) 3-Generation Reproduction - rat: NOEL=10 ppm (0.5 mg/kg/day) (HDT) (no adverse effects); no core grade (Velsicol Chemical, 1967)

Data Gap(s): Chronic Dog Feeding Study; Rat Teratology Study; Rabbit Teratology Study

I.A.5. CONFIDENCE IN THE ORAL RfD

Study -- Low
Data Base -- Low
RfD -- Low

The principal study is of low quality and is given a low confidence rating. Since ths data base on chronic toxicity is incomplete, the data base is given a low confidence rating. Low confidence in the RfD follows.

I.A.6. EPA DOCUMENTATION AND REVIEW OF THE ORAL RfD

Pesticide Registration Standard, August 1986

Pesticide Registration Files

Agency Work Group Review -- 05/20/85, 12/18/85, 02/26/86, 09/16/86, 04/16/87

Verification Date -- 04/16/87

I.A.7. EPA CONTACTS (ORAL RfD)

William Burnam / OPP -- (703)557-4791

George Ghali / OPP -- (703)557-7490

I.B. REFERENCE CONCENTRATION FOR CHRONIC INHALATION EXPOSURE (RfC)

[Ed. Note: EPA does not have information yet in this section of IRIS].

II. CARCINOGENICITY ASSESSMENT FOR LIFETIME EXPOSURE

Section II provides information on three aspects of the carcinogenic risk assessment for the agent in question; the US EPA classification, and quantitative estimates of risk from oral exposure and from inhalation exposure. The classification reflects a weight-of-evidence judgment of the likelihood that the agent is a human carcinogen. The quantitative risk estimates are presented in three ways. The slope factor is the result of application of a low-dose extrapolation procedure and is presented as the risk per (mg/kg)/day. The unit risk is the quantitative estimate in terms of either risk per ug/L drinking water or risk per ug/cu.m air breathed. The third form in which risk is presented is a drinking water or air concentration providing cancer risks of 1 in 10,000, 1 in 100,000 or 1 in 1,000,000. Background Document 2 (Service Code 5) provides details on the rationale and methods used to derive the carcinogenicity values found in IRIS. Users are referred to Section I for information on long-term toxic effects other than carcinogenicity.

II.A. EVIDENCE FOR CLASSIFICATION AS TO HUMAN CARCINOGENICITY

II.A.1. WEIGHT-OF-EVIDENCE CLASSIFICATION

Classification -- B2; probable human carcinogen

Basis -- Inadequate human data, but sufficient evidence exist from studies in which benign and malignant liver tumors were induced in three strains of mice of both sexes. Several structurally related compounds are liver carcinogens.

II.A.2. HUMAN CARCINOGENICITY DATA

Inadequate. There were 11 case reports involving central nervous system effects, blood dyscrasias, and neuroblastomas in children with pre- or postnatal exposure to chlordane and heptachlor (Infante et al., 1978). Since no other information was available, no conclusions can be drawn.

There were three epidemiologic studies of workers exposed to chlordane and/or heptachlor. One retrospective cohort study of pesticide applicators was considered inadequate in sample size and duration of follow-up. This study showed marginal statistically significant increased mortality from bladder cancer (3 observed) (Wang and McMahon, 1979a). The other two studies were retrospective cohort studies of pesticide manufacturing workers. Neither of them showed any statistically significant increased cancer mortality (Wang and McMahon, 1979b; Ditraglia et al., 1981). Both these populations also had confounding exposures from other chemicals.

II.A.3. ANIMAL CARCINOGENICITY DATA

Sufficient. Long-term carcinogenicity bioassays with heptachlor have been performed in rats and mice, with the latter showing a carcinogenic response. Davis (1965) fed groups of 100 male and 100 female C3H mice diets with 0 or 10 ppm heptachlor (purity not specified) for 2 years. Survival was low, with 50% of the controls and 30% of the treated mice surviving until the end of the experiment. A 2-fold increase in benign liver lesions over the controls was reported. After a histologic reevaluation, Reuber (as cited in Epstein, 1976), as well as four other pathologists, remarked a statistically significant increase in liver carcinomas in the treated male (64/87) and female (57/78) groups by comparison to controls (22/73 and 2/53 for males and females, respectively).

The NCI (1977) reported a significant dose-related increase of hepatocellular carcinomas in male and female B6C3F1 mice. Fifty male and 50 female mice were fed diets delivering technical-grade heptachlor at TWA concentrations of 6.1 and 13.8 ppm and 9 and 18 ppm, respectively. Treatment was for 80 weeks, followed by 10 weeks of observation. The authors also reported a statistically significant increase of hepatocellular carcinomas in high-dose males and females over the controls.

No indication of treatment-related increase of tumors has been reported in chronic studies with rats. In an early experiment, Witherup et al. (1955) fed 20 male and 20 female CFN rats each at 1.5, 3.5, 7.0, and 10.0 ppm in the diet for 110 weeks. Although no increase in tumors was found, liver lesions, described as the "chlorinated hydrocarbon" type, were observed at 7 and 10 ppm. Using 25 female CD rats, Jolley et al. (1966) also observed no malignant lesions of the liver but did find hepatocytomegaly when the rats were fed 7.5, 10, and 12.5 ppm heptachlor:heptachlor epoxide (mixture of 75:25). Over the 2 years of the experiment, a dose-related increase in mortality was observed. Two additional experiments, Cabral et al. (1972) and NCI (1977), found no increased incidence of hepatocellular carcinomas when the mixture was administered to Wistar rats by gavage or to Osborne-Mendel rats by diet.

II.A.4. SUPPORTING DATA FOR CARCINOGENICITY

Gene mutation assays indicate that heptachlor is not mutagenic in bacteria (Probst et al., 1981; Shirasu et al., 1976; Moriya et al., 1983) or mammalian liver cells (Telang et al., 1982). Negative results were reported in two dominant lethal assays using male germinal cells (Epstein et al., 1972; Arnold et al., 1977). DNA repair assays indicate that heptachlor is not genotoxic in rodent hepatocytes (Maslansky and Williams, 1981; Probst et al., 1981) but showed qualitative evidence of unscheduled DNA synthesis in human fibroblasts (Ahmed et al., 1977).

Five compounds structurally related to heptachlor (heptachlor epoxide, chlordane, aldrin, dieldrin, and chlorendic acid) have produced liver tumors in mice. Chlorendic acid has also produced liver tumors in rats.

II.B. QUANTITATIVE ESTIMATE OF CARCINOGENIC RISK FROM ORAL EXPOSURE

II.B.1. SUMMARY OF RISK ESTIMATES

Oral Slope Factor -- 4.5E+0 per (mg/kg)/day

Drinking Water Unit Risk -- 1.3E-4 per (ug/L)

Extrapolation Method -- Linearized multistage procedure, extra risk

Drinking Water Concentrations at Specified Risk Levels:

Risk Level	Concentration
E-4 (1 in 10,000)	8E-1 ug/L
E-5 (1 in 100,000)	8E-2 ug/L
E-6 (1 in 1,000,000)	8E-3 ug/L

II.B.2. DOSE-RESPONSE DATA (CARCINOGENICITY, ORAL EXPOSURE)

Tumor Type -- hepatocellular carcinomas
Test Animals -- mouse/C3H; mouse/B6C3F1
Route -- diet
Reference -- Davis, 1965; NCI, 1977

Administered Dose (ppm)	Human Equivalent Dose (mg/kg)/day	Tumor Incidence	Reference
Mouse/C3H, male			
0	0.000	22/73	Davis, 1965
10	0.108	64/87	as evaluated by Reuber, cited in Epstein, 1976
Mouse/C3H, female			
0	0.000	2/53	
10	0.108	57/78	
Mouse/B6C3F1, male (matched controls)			
0	0.000	5/19	NCI, 1977
6.1	0.063	11/46	
13.8	0.140	34/47	
Mouse/B6C3F1, female (matched controls)			
0.0	0.000	2/10	
9.0	0.094	3/47	

II.B.3. ADDITIONAL COMMENTS (CARCINOGENICITY, ORAL EXPOSURE)

Four data sets showed a significant increase in hepatocellular carcinomas in treatment groups compared with controls in mice. The quantitative estimate is the geometric mean of the slope factors from the four mouse data sets. The slope factors for each set are: 12.4 per (mg/kg)/day for C3H male mice, 14.9 per (mg/kg)/day for C3H female mice, 2.79 per (mg/kg)/day for B6C3F1 male mice, and 0.83 per (mg/kg)/day for B6C3F1 female mice. Although the magnitude of the responses differed somewhat, a combined risk estimate was chosen because the two strains are related and so that relevant data will not be discarded.

The above unit risk should not be used if the water concentration exceeds 80 ug/L, since above this concentration the unit risk may not be appropriate.

II.B.4. DISCUSSION OF CONFIDENCE (CARCINOGENICITY, ORAL EXPOSURE)

Adequate numbers of animals were treated and observed for the majority of their expected lifetime. The incidences of malignant lesions were significantly increased in all four data sets, and dose-response effects were observed in the NCI (1977) study.

II.C. QUANTITATIVE ESTIMATE OF CARCINOGENIC RISK FROM INHALATION EXPOSURE

II.C.1. SUMMARY OF RISK ESTIMATES

Inhalation Unit Risk -- 1.3E-3 per (ug/cu.m)

Extrapolation Method -- Linearized multistage procedure, extra risk

Air Concentrations at Specified Risk Levels:

Risk Level	Concentration
E-4 (1 in 10,000)	8E-2 ug/cu.m
E-5 (1 in 100,000)	8E-3 ug/cu.m
E-6 (1 in 1,000,000)	8E-4 ug/cu.m

II.C.2. DOSE-RESPONSE DATA FOR CARCINOGENICITY, INHALATION EXPOSURE

The risk estimates were calculated from the oral data presented in II.B.2.

II.C.3. ADDITIONAL COMMENTS (CARCINOGENICITY, INHALATION EXPOSURE)

The above unit risk should not be used if the air concentration exceeds 8 ug/cu.m, since above this concentration the unit risk may not be appropriate.

II.C.4. DISCUSSION OF CONFIDENCE (CARCINOGENICITY, INHALATION EXPOSURE)

See II.B.4.

II.D. EPA DOCUMENTATION, REVIEW, AND CONTACTS (CARCINOGENICITY ASSESSMENT)

II.D.1. EPA DOCUMENTATION

Source Document -- US EPA, 1986

The values in the 1986 Carcinogenicity Assessment for Chlordane and Heptachlor/Heptachlor Epoxide have been reviewed by the Carcinogen Assessment Group.

II.D.2. REVIEW (CARCINOGENICITY ASSESSMENT)

Agency Work Group Review -- 04/01/87

Verification Date -- 04/01/87

II.D.3. US EPA CONTACTS (CARCINOGENICITY ASSESSMENT)

Dharm V. Singh / NCEA -- (202)260-5958

Jim Cogliano / NCEA -- (202)260-3814

III. HEALTH HAZARD ASSESSMENTS FOR VARIED EXPOSURE DURATIONS

III.A. DRINKING WATER HEALTH ADVISORIES

The Office of Drinking Water provides Drinking Water Health Advisories (HAs) as technical guidance for the protection of public health. HAs are not enforceable Federal standards. HAs are concentrations of a substance in drinking water estimated to have negligible deleterious effects in humans, when ingested, for a specified period of time. Exposure to the substance from other media is considered only in the derivation of the lifetime HA. Given the absence of chemical-specific data, the assumed fraction of total intake from drinking water is 20%. The lifetime HA is calculated from the Drinking Water Equivalent Level (DWEL) which, in turn, is based on the Oral Chronic Reference Dose. Lifetime HAs are not derived for compounds which are potentially carcinogenic for humans because of the difference in assumptions concerning toxic threshold for carcinogenic and noncarcinogenic effects. A more detailed description of the assumptions and methods used in the derivation of HAs is provided in Background Document 3 in Service Code 5.

III.A.1. ONE-DAY HEALTH ADVISORY FOR A CHILD

Appropriate data for calculating a One-day HA for heptachlor are not available. It is recommended that the Ten-day HA of 0.010 mg/L be used as the One-day HA for heptachlor.

III.A.2. TEN-DAY HEALTH ADVISORY FOR A CHILD

Ten-day HA -- 1.0E-2 mg/L

LOAEL -- 1.0 mg/kg/day
UF -- 1000 (allows for interspecies and intra human variability with the use of a LOAEL from an animal study)
Assumptions -- 1 L/day water consumption for a 10-kg child

Principal Study -- Enan et al., 1982

Rats were administered heptachlor at 1.0 mg/kg/day in the feed for 14 days. Exposure resulted in evidence of liver damage and altered liver function: increased blood urea, increased blood glucose, decreased liver glycogen content, and increased acid and alkaline phosphatase levels when compared to controls. Thus, 1.0 mg/kg/day is the LOAEL in this study.

III.A.3. LONGER-TERM HEALTH ADVISORY FOR A CHILD

Appropriate data for calculating a Longer-term HA for heptachlor are not available. It is recommended that a modified DWEL (adjusted for a 10-kg child) of 0.005 mg/L be used as the Longer-term HA.

III.A.4. LONGER-TERM HEALTH ADVISORY FOR AN ADULT

Appropriate data for calculating a Longer-term HA for heptachlor are not available. It is recommended that the DWEL of 0.0175 mg/L be used as the Longer-term HA for the 70-kg adult.

III.A.5. DRINKING WATER EQUIVALENT LEVEL / LIFETIME HEALTH ADVISORY

DWEL -- 1.75E-2 mg/L

Assumptions -- 2 L/day water consumption for a 70-kg adult

RfD Verification Date = 04/16/87 (see Section I.A. of this file)

Lifetime HA -- None

Heptachlor is considered to be a probable human carcinogen. Refer to Section II of this file for information on the carcinogenicity of this substance.

Principal Study -- Velsicol Chemical Corporation, 1955 (This study was used in the derivation of the chronic oral RfD; see Section I.A.2.)

III.A.6. ORGANOLEPTIC PROPERTIES

[Ed. Note: EPA does not have information yet in this section of IRIS].

III.A.7. ANALYTICAL METHODS FOR DETECTION IN DRINKING WATER

Determination of heptachlor is by a liquid-liquid extraction gas chromatographic procedure.

III.A.8. WATER TREATMENT

Treatment techniques capable of removing heptachlor from drinking water include adsorption by granular activated carbon and ozone or ozone/ultra- violet oxidation.

III.A.9. DOCUMENTATION AND REVIEW OF HAs

US EPA. 1985. Final Draft of the Drinking Water Criteria Document on Heptachlor Epoxide. Office of Drinking Water, Washington, DC.

EPA review of HAs in 1985.

Public review of HAs following notification of availability in October, 1985.

Scientific Advisory Panel review of HAs in January, 1986.

Preparation date of this IRIS summary -- 06/17/87

III.A.10. EPA CONTACTS

Jennifer Orme Zavaleta / OST -- (202)260-7586

Edward V. Ohanian / OST -- (202)260-7571

III.B. OTHER ASSESSMENTS

Content to be determined.

IV. US EPA REGULATORY ACTIONS

EPA risk assessments may be updated as new data are published and as assessment methodologies evolve. Regulatory actions are frequently not updated at the same time. Compare the dates for the regulatory actions in this section with the verification dates for the risk assessments in sections I and II, as this may explain inconsistencies. Also note that some regulatory actions consider factors not related to health risk, such as technical or economic feasibility. Such considerations are indicated for each action. In addition, not all of the regulatory actions listed in this section involve enforceable federal standards. Please direct any questions you may have concerning these regulatory actions to the US EPA contact listed for that particular action. Users are strongly urged to read the background information on each regulatory action in Background Document 4 in Service Code 5.

IV.A. CLEAN AIR ACT (CAA)

[Ed. Note: EPA does not have information yet in this section of IRIS].

IV.B. SAFE DRINKING WATER ACT (SDWA)

IV.B.1. MAXIMUM CONTAMINANT LEVEL GOAL (MCLG) for Drinking Water

Value (status) -- 0 mg/L (Final, 1991)

Considers technological or economic feasibility? -- NO

Discussion -- An MCLG of 0 mg/L for heptachlor is promulgated based on evidence of carcinogenic effects (B2).

Reference -- 56 FR 3526 (01/30/91)

EPA Contact -- Health and Ecological Criteria Division / OST / (202) 260-7571 / FTS 260-7571; or Safe Drinking Water Hotline / (800) 426-4791

IV.B.2. MAXIMUM CONTAMINANT LEVEL (MCL) for Drinking Water

Value -- 0.0004 mg/L (Final, 1991)

Considers technological or economic feasibility? -- YES

Discussion -- EPA has set a MCL equal to the PQL of 0.0004 mg/L, which is associated with a lifetime individual risk 0.5 E-4.

Monitoring requirements -- All systems monitored for four consecutive quarters every 3 years; repeat monitoring dependent upon detection, vulnerability status and system size.

Analytical methodology -- Microextraction/gas chromatography (EPA 505); electron-capture/gas chromatography (EPA 508); gas chromatographic/mass spectrometry (EPA 525): PQL= 0.0004 mg/L.

Best available technology -- Granular activated carbon

Reference -- 56 FR 3526 (01/30/91)

EPA Contact -- Drinking Water Standards Division / OGWDW / (202) 260-7575 / FTS 260-7575; or Safe Drinking Water Hotline / (800) 426-4791

IV.B.3. SECONDARY MAXIMUM CONTAMINANT LEVEL (SMCL) for Drinking Water

[Ed. Note: EPA does not have information yet in this section of IRIS].

IV.B.4. REQUIRED MONITORING OF "UNREGULATED" CONTAMINANTS

[Ed. Note: EPA does not have information yet in this section of IRIS].

IV.C. CLEAN WATER ACT (CWA)

IV.C.1. AMBIENT WATER QUALITY CRITERIA, Human Health

Water and Fish Consumption -- 2.8E-4 ug/L

Fish Consumption Only -- 2.9E-4 ug/L

Considers technological or economic feasibility? -- NO

Discussion -- For the maximum protection from the potential carcinogenic properties of this chemical, the ambient water concentration should be zero. However, zero may not be attainable at this time, so the recommended criteria represents a E-6 estimated incremental increase of cancer risk over a lifetime.

Reference -- 45 FR 79318 (11/28/80)

EPA Contact -- Criteria and Standards Division / OWRS (202)260-1315 / FTS 260-1315

IV.C.2. AMBIENT WATER QUALITY CRITERIA, Aquatic Organisms

Freshwater:
  Acute -- 5.2E-1 ug/L (24-hour average)
  Chronic -- 3.8E-3 ug/L

Marine:
  Acute -- 5.3E-2 ug/L (24-hour average)
  Chronic -- 3.6E-3 ug/L

Considers technological or economic feasibility? -- NO

Discussion -- Criteria were derived from a minimum data base consisting of acute and chronic tests on a variety of species. Requirements and methods are covered in the reference to the Federal Register.

Reference -- 45 FR 79318 (11/28/80)

EPA Contact -- Criteria and Standards Division / OWRS (202)260-1315 / FTS 260-1315

IV.D. FEDERAL INSECTICIDE, FUNGICIDE, AND RODENTICIDE ACT (FIFRA)

IV.D.1. PESTICIDE ACTIVE INGREDIENT, Registration Standard

Status -- Issued (1986)

Reference -- Heptachlor Pesticide Registration Standard. December, 1986 (NTIS No. PB87-175808).

EPA Contact -- Registration Branch / OPP (703)557-7760 / FTS 557-7760

IV.D.2. PESTICIDE ACTIVE INGREDIENT, Special Review

Action -- Cancellation of many termiticide products (1988)

Considers technological or economic feasibility? -- NO

Summary of regulatory action -- 53 FR 11798 (04/08/88) [52 FR 42145 (11/03/870] announced cancellation of many termiticide products and termination of sale, distribution, and use (except homeowner products) as of 04/15/88. All remaining products are withdrawn or suspended under section 3(c)(2)(B). The agreement (order) accepted voluntary cancellations of the registrations of certain pesticide products and imposed limitations on the continued sale, distribution and use of existing stocks of such products. Under the provisions of the Administrator's acceptance of the settlement plan to phase out certain uses of chlordane, most registered products containing chlordane were effectively canceled or the applications for registration were denied by 12/31/80. A summary of those uses not affected by this settlement, or a previous suspension, follows: 1) subsurface ground insertion for termite control (clarified by 40 FR 30522, July 21, 1975, to apply to the use of emulsifiable or oil concentrate formulations for controlling subterranean termites on structural sites such as buildings, houses, barns, and sheds, using current control practices), 2) dipping of nonfood roots and tops. Criterion of concern: oncogenicity.

Reference -- 43 FR 12372 (03/24/78); 52 FR 42145 (11/03/87); 53 FR 11798 (04/08/88)

EPA Contact -- Special Review Branch / OPP (703)557-7400 / FTS 557-7400

IV.E. TOXIC SUBSTANCES CONTROL ACT (TSCA)

[Ed. Note: EPA does not have information yet in this section of IRIS].

IV.F. RESOURCE CONSERVATION AND RECOVERY ACT (RCRA)

IV.F.1. RCRA APPENDIX IX, for Ground Water Monitoring

Status -- Listed

Reference -- 52 FR 25942 (07/09/87)

EPA Contact -- RCRA/Superfund Hotline (800)424-9346 / (202)260-3000 / FTS 260-3000

IV.G. SUPERFUND (CERCLA)

IV.G.1. REPORTABLE QUANTITY (RQ) for Release into the Environment

Value -- 1 pound (Final, 1989)

Considers technological or economic feasibility? -- NO

Discussion -- The final RQ is based on carcinogenicity and aquatic toxicity as established under CWA Section 311 (40 CFR 117.3). Available data indicate a carcinogenic hazard rating of high and a weight of evidence classification of Group B2. The available data indicate that the aquatic 96-Hour Median Threshold Limit for heptachlor is <0.1 ppm, which corresponds to an RQ of 1 pound.

Reference -- 54 FR 33418 (08/14/89)

EPA Contact -- RCRA/Superfund Hotline (800) 424-9346 / (202) 260-3000 / FTS 260-3000

VI. BIBLIOGRAPHY

VI.A. ORAL RfD REFERENCES

Velsicol Chemical Corporation. 1955a. MRID No. 00062599. Available from EPA. Write to FOI, EPA, Washington, DC 20460.

Velsicol Chemical Corporation. 1955b. MRID No. 00062599. Available from EPA. Write to FOI, EPA, Washington, DC 20460.

Velsicol Chemical Corporation. 1964. MRID No. 00086210. Available from EPA. Write to FOI, EPA, Washington, DC 20460.

Velsicol Chemical Corporation. 1967. MRID No. 00147058. Available from EPA. Write to FOI, EPA, Washington, DC 20460.

VI.B. INHALATION RfC REFERENCES

None

VI.C. CARCINOGENICITY ASSESSMENT REFERENCES

Davis, K. 1965. Pathology Report on Mice Fed Aldrin, Dieldrin, Heptachlor and Heptachlor Epoxide for Two Years. Internal FDA memorandum to Dr. A.J. Lehman, July 19.

Epstein, S.S. 1976. Carcinogenicity of heptachlor and chlordane. Sci. Total Environ. 6: 103-154. NCI (National Cancer Institute). 1977. Bioassay of Heptachlor for Possible Carcinogenicity. NCI Carcinogenesis Tech. Rep. Ser. No. 9. (Also published as DHEW Publication No. [NIH] 77-809).

Reuber, M.D. 1977. Histopathology of Carcinomas of the Liver in Mice Ingesting Heptachlor or Heptachlor Epoxide. Exp. Cell Biol. 45: 147-157.

US EPA. 1986. Carcinogenicity Assessment of Chlordane and Hepta- chlor/Heptachlor Epoxide. Prepared by the Office of Health and Environmental Assessment, Carcinogen Assessment Group, Washington, DC. OHEA-C-204.

VI.D. DRINKING WATER HA REFERENCES

Enan, E.E., A.H. El-Sebae and O.H. Enan. 1982. Effects of some chlorinated hydrocarbon insecticides on liver function in white rats. Meded. Fac. Land- bouwwet., Rijksuniv. Gent. 47(1): 447-457.

US EPA. 1985. Final Draft of the Drinking Water Criteria Document on Heptachlor Epoxide. Office of Drinking Water, Washington, DC.

Velsicol Chemical Corporation. 1955. MRID No. 00062599. Available from EPA. Write to FOI, EPA, Washington, DC 20460.

VII. REVISION HISTORY

Date	Section	Description
03/01/88	II.A.1.	Basis clarified
03/01/88	II.B.4.	Confidence statement revised
03/01/88	III.A.	Health Advisory added
08/01/90	III.A.10	Primary contact changed
08/01/90	IV.F.1.	EPA contact changed
01/01/91	II.	Text edited
01/01/91	II.C.1.	Inhalation slope factor removed (global change)
03/01/91	I.A.4.	Citations added
03/01/91	VI.	Bibliography on-line
01/01/92	IV.	Regulatory actions updated
04/01/93	IV.C.2.	Freshwater and marine values corrected
07/01/93	II.D.3.	Secondary contact's phone number changed

SYNONYMS

AGROCERES
3-CHLOROCHLORDENE
DICYCLOPENTADIENE,
3,4,5,6,7,8,8a-HEPTACHLORO-
DRINOX
DRINOX H-34
E 3314
ENT 15,152
EPTACLORO
1,4,5,6,7,8,8-EPTACLORO-
3a,4,7,7a-TETRAIDRO-4,7-endo-
METANO-INDENE
GPKh
H
H-34
HEPTACHLOOR
1,4,5,6,7,8,8-HEPTACHLOOR-
3a,4,7,7a-TETRAHYDRO-4,7-
endo-METHANO-INDEEN
Heptachlor
HEPTACHLORE1(3a),4,5,6,7,8,8-
HEPTACHLORO-3a(1),4,7,7a-
TETRAHYDRO-4,7-
METHANOINDENE
3,4,5,6,7,8,8-
HEPTACHLORODICYCLOPENT
ADIENE

3,4,5,6,7,8,8a-
HEPTACHLORODICYCLOPENTADIENE
1,4,5,6,7,8,8-HEPTACHLORO-3a,4,7,7a-
TETRAHYDRO-4,7-ENDOMETHANOINDENE
1,4,5,6,7,10,10-HEPTACHLORO-4,7,8,9-
TETRAHYDRO-4,7-
ENDOMETHYLENEINDENE
1,4,5,6,7,8,8a-HEPTACHLORO-3a,4,7,7a-
TETRAHYDRO-4,7-METHANOINDANE
1,4,5,6,7,8,8-HEPTACHLORO-3a,4,7,7a-
TETRAHYDRO-4,7-METHANOINDENE
1,4,5,6,7,8,8-HEPTACHLORO-3a,4,7,7a-
TETRAHYDRO-4,7-METHANOL-1H-INDENE
1,4,5,6,7,10,10-HEPTACHLORO-4,7,8,9-
TETRAHYDRO-4,7-METHYLENEINDENE
1,4,5,6,7,8,8-HEPTACHLORO-3a,4,7,7,7a-
TETRAHYDRO-4,7-METHYLENE INDENE
1,4,5,6,7,8,8-HEPTACHLOR-3a,4,7,7,7a-
TETRAHYDRO-4,7-endo-METHANO-INDEN
HEPTAGRAN
IIEPTAMUL
4,7-METHANOINDENE, 1,4,5,6,7,8,8-
HEPTACHLORO-3a,4,7,7a-TETRAHYDRO-
NA 2761
NCI-C00180
RCRA WASTE NUMBER P059
RHODIACHLOR
VELSICOL 104

Source: *Instant EPA's IRIS*, 1997, Instant Reference Sources, Inc., 7605 Rockpoint Drive, Austin, TX 78731 (Fax: 512-345-2386; Internet URL, http://www.instantref.com/inst-ref.htm).

Production, Use, and Pesticide Labeling Information

1. Description of chemical

Generic Name: 1,4,5,6,7,8,8-heptachloro-3a,4,7,7a-tetrahydro-4,7-methano-1H-indene

Common Name: Heptachlor

Trade and Other Names: 1,4,5,6,7,8,8-heptachlor-3a,4,7,7a- tetrahydro-4,7-methanoindene; E-3314; Velsicol 104

EPA Shaughnessy Code: 044801

Chemical Abstracts Service (CAS) Number: 76-44-8

Year of Initial Registration: 1952

Pesticide Type: Insecticide

Chemical Family: Chlorinated cyclodiene

2. Use Pattern and Formulations

- ACTION: Notice of PROHIBITION OF CONTINUED SALE OR USE OF HEPTACHLOR PRODUCTS FOR SEED TREATMENT.

- The administrator has signed a Notice of Determination Pursuant to Section 6(a)(1) of FIFRA which will be published in the Federal Register. The Notice will prohibit any further sale or use of heptachlor products for seed treatment purposes. Any sale or use of heptachlor products for seed treatment will be a violation of the Federal Insecticide, Fungicide and Rodenticide Act (FIFRA).

3. [EDITOR'S NOTE: NPIRS advises that Section 3 has been omitted from this revision of the Fact Sheet.]

4. Summary of Regulatory Position and Rationale

- NOTICE OF INTENT TO CANCEL

- Prior to 1974, heptachlor (along with a related compound, chlordane) was registered for a wide variety of insecticide uses. On November 18, 1974, the Administrator issued a notice of intent to cancel registration for most uses of heptachlor (and chlordane). The basis for the notice of intent to cancel was evidence that heptachlor and chlordane had demonstrated toxic effects which may have significant adverse effects on human health, and evidence that both chemicals persist in the environment for many years after application, and as such, are subject to considerable movement from the site of actual application. The evidence on toxicity included a finding that heptachlor and its metabolite, heptachlor epoxide induce tumors in mice and that there was evidence of embryo toxicity in mice and rats. - Because of the persistence and wide application of heptachlor and chlordane products, heptachlor epoxide residues were routinely found in water, food sources, and human adult and fetal tissue. The Administrator therefore proposed to cancel all registered uses of chlordane and heptachlor, except those uses for subterranean termiticide control (see note) and dipping of non-food plants.

- NOTE: It should be noted that subsequently on October 1, 1987, EPA issued an Order accepting the voluntary cancellation of chlordane and heptachlor termiticide treatment products. A Notice signed on April 5, 1988, in response to a District Court ruling established limits on the sale and use of existing stock of termiticide products after April 15, 1988.

- On July 29, 1975, the administrator issued a notice of intent to suspend (pursuant to FIFRA Section 6(c)) the registrations of heptachlor and chlordane that were subject to the notice of intent to cancel. The grounds for the notice of intent to suspend were "new evidence which confirmed and heightened the human cancer hazard posed by chlordane and heptachlor" and the Administrator's determination that the cancellation proceeding resulting from the notice of intent to cancel would not be completed in time to "avert substantial additions of these persistent and ubiquitous compounds to an already serious human and environmental burden." The notice of intent to suspend applied to all uses covered by the notice to cancel.

- An evidentiary hearing on the proposed suspension took place between August and December of 1975. On December 2, 1975, the hearing examiner published a recommended decision dismissing the notice of intent to suspend. The basis for this recommendation was the hearing examiner's unwillingness to find "conclusively" that heptachlor and chlordane were (are) carcinogens in laboratory animals.

- Included in the recommended decision was a discussion of heptachlor for seed treatment. The document noted that inadequate alternatives for seed treatment existed at that time. The hearing examiner recommended that heptachlor for seed treatment not be suspended even if the Administrator were to disagree with the examiner on the question of the hazard posed by chlordane and heptachlor.

- On December 24, 1975, the Administrator issued his decision on the proposed suspension of chlordane and heptachlor products. The Administrator ordered a suspension of a number of chlordane and heptachlor uses during the pendency of the cancellation hearing.

- As the seed treatment, however, the Administrator found that no adequate alternatives to treatment with heptachlor existed at that time, and therefore found that the benefit from heptachlor for seed treatment exceeded the risks of such use during the time necessary to complete the cancellation hearing. Heptachlor for seed treatment was thus not one of the uses suspended by the Administrator.

- SETTLEMENT OF THE CANCELLATION PROCEEDING

- The cancellation proceeding continued until November of 1977, at which time the parties entered into settlement negotiations. The negotiations resulted in an agreement which was ratified in a Final

Order issued by the Administrator on March 6, 1978. The Final Order resulted in the eventual cancellation of all products subject to the original notice of intent to cancel notice. For seed treatment, the effective date of cancellation was September 1, 1982 for barley, oats, wheat, rye and corn, and July 1, 1983 for sorghum. The Order also contained production limitations; production of heptachlor for seed treatment was limited to 175,000 pounds annually from 1978 to 1982, and to 100,000 pounds in 1983. These production limitations were intentionally less than the use of heptachlor for seed treatment purposes in 1976 (which was 200,000 pounds).

- The purpose of the phased cancellations was to provide a "transition period" to allow users to make an orderly adjustment to alternative crops or pest control technologies where none then existed.

- EXISTING STOCKS DETERMINATION

- The sale and use existing stocks pesticide products canceled after a notice of intent to cancel is issued pursuant to Section 6(b) of FIFRA are controlled by Section 6(1) of FIFRA. It provides in part, "... the Administrator may permit the continued sale and use of existing stocks of a pesticide whose registration is canceled under Section 6(b) to such extent as he may specify if he determines that such sale or use is not inconsistent with the purposes of FIFRA and will not have unreasonable adverse effects on the environment."

- At the time the Agency issued the Final Order, it was expected based upon the use practices at that time that sale and use of existing stocks of canceled products would cease approximately within one year of the effective cancellation date. The existing stocks allowance and phased cancellation was to result in approximately a six year transition period for users of heptachlor treated seeds to adapt alternative management practices after 1978.

- The six year transitional period contemplated in 1984 ended over four years ago. The Agency believes that ten years is more than sufficient time for users to find alternatives to heptachlor seed treatment. Moreover, although some heptachlor continues to be used for seed treatment purposes, the transition away from heptachlor seed treatment has largely been completed (the amount of heptachlor used for seed treatment in 1987 was only 1% of the amount used in 1974).

- While the benefits associated with heptachlor seed treatment have greatly diminished in the past ten years, the Agency's general concerns with the use of heptachlor have not diminished.

- In addition, in late January and early February of 1986, the Food and Drug Administration (FDA) found very high levels of heptachlor and trans-chlordane in finished livestock feeds.

- A fermentation/distillation firm purchased and used obsolete pesticide treated seed grain in their fermentation process. The spent distillers mash was, in turn, used in the manufacture of finished animal feeds and fed to dairy cattle. When FDA tested the milk from dairy herds fed the contaminated feed, the levels of heptachlor epoxide (an animal metabolite of heptachlor) found exceeded, by as much as 75 times, the FDA action level of 0.1 ppm for heptachlor epoxide in the milk fat.

- As the result of this one incident, taxpayers have already incurred more than ten million dollars in investigative and indemnification costs. Total losses for all affected parties are expected to exceed sixteen million dollars.

- FDA and USDA subsequently carried out an extensive to determine how frequently obsolete pesticide treated seeds were being fed illegally to meat and/or milk producing animals or had entered the livestock feed markets. In over 1000 investigations, well over 100 violations were found. Feeding of obsolete heptachlor treated seed was involved in at least two of these additional violations.

- EPA subsequently has determined that sizable inventories of canceled heptachlor seed treatment products remain in the channels of trade. At the present levels of use, these products would be available for use for the next 70 years.

- As previously stated, under Section 6(a)(1), the Agency may permit the continued sale and use of existing stocks of a canceled pesticide only if the Agency determines that such sale and use is consistent with FIFRA and does not result in unreasonable adverse effects on the environment.

- Under the circumstances, the Agency can no longer find that continued sale or use of heptachlor for seed treatment will not have an unreasonable adverse effect on the environment. The Agency therefore no longer believes that such sale or use is consistent with Section 6(a)(1) of FIFRA.

- The Agency accordingly served notice in the Federal Register of _____ that sale or use of stocks of heptachlor for seed treatment is no longer permitted, and that any further sale or use shall be a violation of Section 12(a)(1)(A) and/or Section 12(a)(2)(K) of FIFRA.

- While any further use of heptachlor for seed treatment is not permitted, existing stocks of seed grain previously treated with heptachlor may be sold and planted in accordance with good agronomic practices.

- GUIDANCE ON THE STATUS OF HEPTACHLOR SEED TREATMENT PRODUCTS AS HAZARDOUS WASTES

- Unused quantities of canceled heptachlor seed treatment products can no longer be used as directed on their label. They, therefore, fir the definition of a solid waste as defined in 40 CFR 261.2 and 261.33 when they are discarded or held with the intent to discard.

- A hazardous waste is any solid waste which has been listed as a hazardous waste in 40 CFR Part 261 Subpart D or a solid waste which exhibits any of the characteristics of hazardous waste identified in 40 CFR Part 261 Subpart C ignitability, part 261.21; corrosivity, part 261.22; reactivity, part 261.23; and/or E.P. toxicity, part 261.24.

- Heptachlor is listed as an acutely hazardous waste (PO59) in 40 CFR part 261.33(e). Any unused heptachlor seed treatment products, rinsate or containers which have not been properly cleaned (triple rinsed as defined part 261.7) are therefore acutely hazardous wastes, as defined in 40 CFR part 261.33(e) if they are discarded or intended for discard.

- Any person by site who holds canceled heptachlor seed treatment products when they become wastes is a "generator" of hazardous wastes as defined in 40 CFR part 261. A generator must comply with the requirements of the Resource Conservation and Recovery Act (RCRA) and any other applicable Federal, State, and local laws and regulations.

- Those who hold canceled heptachlor seed treatment products at the time they become wastes are defined as "generators" and they fall into one of three categories of waste generators. They are:

- a) Conditionally Exempt Generator

- one who currently holds or generates no more than 1 kilogram (2.2 pounds) of acutely hazardous waste heptachlor seed treatment products, a listed acutely hazardous waste (PO59) and who generates no more than 100 kilograms (220 pounds) of other hazardous waste in any calendar month. - A conditionally exempt generator is not required to obtain a permit or interim status (40 CFR Part 261.5). He/she, however, is required to:

* Identify all hazardous waste held or generated, part 261.5(c)

* Send the hazardous waste to an authorized facility, part 261.5(f)(3)

* Never accumulate more than 1000 kilograms (2200 pounds) of hazardous waster and/or more than 1 kilogram (2.2 pounds) of acutely hazardous waste on his/her property, part 261.5(f)(2) and (g)(2).

- Acutely hazardous waste (PO59) may be held up to 1 kilogram (2.2 pounds) in containers which are in good condition (do not leak) and are compatible with the waste.

- b) Small Quantity Generator

- one who holds or generates no more than 1 kilogram (2.2 pounds) of acutely hazardous waste heptachlor seed treatment products, a listed acutely hazardous waste (PO59) and generates between 100 and 1000 kilograms (220 to 2200 pounds) of other hazardous waste in any calendar month.

- A small quantity generator must comply with the requirements of 40 CFR Part 262, Standards Applicable to Generators of Hazardous Waste including obtaining an EPA ID number, using the Uniform Hazardous Waste Manifest, accumulating waste in accordance with part 262.34(d) and complying with record keeping and reporting requirements of part 262.40(a), (c) and (d); part 262.42(b); and part 262.43.

- Small quantity generators who choose to store or treat beyond the allowances provided in 262.34(d)-(f) or to dispose of hazardous wastes or acutely hazardous wastes at their own facilities are subject to the full regulatory requirements of 40 CFR Parts 264 through 270 which pertain to the operation, maintenance and permitting of treatment, storage and disposal facilities.

- Generators must send heptachlor seed treatment products that are not treated or disposed of on site to a hazardous waste facility permitted to accept them.

- c) Generator

- one who holds or generates more than 1 kilogram (2.2 pounds) of acutely hazardous waste heptachlor seed treatment products, a listed acutely hazardous waste (PO59) or more than 1000 kilograms (2200 pounds) of hazardous waste in any calendar month.

- A hazardous waste "generator" as defined above must comply with all applicable hazardous waste management requirements set forth in 40 CFR Part 262, Standards Applicable to Generators of Hazardous Waste. Those who choose to transport their own hazardous waste must comply with 40 CFR Part 263, Standards Applicable to Transporters of Hazardous Waste.

- If a generator stores his/her waste for longer than 90 days, then he/she must obtain a RCRA hazardous waste storage permit and comply with the requirements of 40 CFR Part 264 and 40 CFR Part 265. An extension of 30 days may be granted by the Regional Administrator under certain emergency situations.

- Generators who choose to store or to treat beyond the allowances provided in Part 262.34(a) or to dispose of hazardous wastes or acutely hazardous wastes at their own facilities are subject to the full regulatory requirements of 40 CFR Parts 264 through 270 which pertain to the operation, maintenance and permitting of treatment, storage and disposal facilities.

- Generators must send heptachlor seed treatment products that are not treated or disposed of on-site to a hazardous waste facility permitted to accept them.

- Obsolete seed, which are no longer viable or suitable for planting and which have been treated with heptachlor are not "listed" hazardous in 40 CFR Subpart D. Their status as "characteristic" hazardous wastes under 40 CFR Subpart C 261.20 through 261.24 and 40 CFR Part 261 Appendix I, II and III must be determined by the generator under 40 CFR 261.11.

- It should be kept in mind, however, that some serious environmental impacts have resulted from the inappropriate disposal of obsolete heptachlor-treated seeds. Every effort should be made to plant existing stocks of heptachlor-treated seeds in accordance with good agronomic practices before they become obsolete.

- Should the generator find that obsolete heptachlor-treated seed is a "characteristic" hazardous waste under 40 CFR 262.11, then the seed may be stored, treated or disposed of only at a permitted hazardous waste facility. EPA recommends giving serious consideration to incineration.

- On the other hand, if after the aforementioned analysis, the obsolete heptachlor-treated seeds are determined to be non- hazardous, the obsolete heptachlor-treated seeds could be landfilled in accordance with the individual state and local requirements for disposal of solid waste. If landfill of the seed is not viable in your area, then consideration must again be given to incineration as the appropriate means of destruction.

5. CONTACT PERSON AT EPA, OFFICE OF PESTICIDE PROGRAMS:

[EDITOR'S NOTE: The usual Section 5, Summary of Major Data Gaps, has been omitted in this revision.]

James G. Touhey
Senior Agricultural Advisor (H-7506C)
Field Operations Division Office of Pesticide Programs
Environmental Protection Agency
401 M Street, SW
Washington, D.C. 20460

Office Location and Telephone Number:

Room 710 Crystal Mall, Building No. 2
1921 Jefferson Davis Highway
Arlington, VA 22202

Telephone: (703) 557-5664

6. CONTACT FOR ADDITIONAL INFORMATION REGARDING DISPOSAL

- For those states which have RCRA authorization, a concerned individual should contact the hazardous waste management agency of that state for additional information concerning the state disposal requirements.

- For non-authorized states the concerned individual should contact the hazardous waste management division of the EPA region in which his/her state falls.

- In addition, concerned parties may call the RCRA/Superfund Hotline toll free (1-800-424-9346) or may call commercially on (1-202-382- 3000) for more detailed information concerning RCRA requirements.

Source: *Instant EPA's Pesticide Facts*, 1994, Instant Reference Sources, Inc., 7605 Rockpoint Drive, Austin, TX 78731 (Fax: 512-345-2386; Internet URL, http://www.instantref.com/inst-ref.htm).

Heptachlor epoxide

Introduction

Heptachlor epoxide is a degradation product of the insecticide heptachlor. There are no commercial uses known for this compound. In 1978, the USEPA canceled the registration of heptachlor and chlordane, which contains 10% heptachlor and agricultural uses of these insecticides were phased out. After July 1, 1983 heptachlor and chlordane may only be used for underground termite control. [828]

Heptachlor epoxide adsorbs strongly to soil and is extremely resistant to biodegradation, persisting for many years in the upper soil layers. Some volatilization or photolysis loss may occur. If released into water, it will adsorb strongly to suspended and bottom sediment. Little biodegradation would be expected. Heptachlor epoxide is expected to exist in both the vapor and particulate phases in ambient air. Vapor phase reactions with photochemically produced hydroxyl radical may be an important fate process (it has an estimated half-life of 1.5 days). Heptachlor epoxide that associated with particulate matter and aerosols should be subject to gravitational settling and washout by rain. Due to its stability, long range dispersal occurs, resulting in the contamination of remote areas. Some photolysis loss probably occurs but there is no data to evaluate the rate of this process. Heptachlor epoxide is bioconcentrated extensively. It is taken up into the food chain by plants and bioconcentrates into fish, animals and milk. Residues in human milk primarily comes from eating meat and fish as is evident by the much lower concentration in the milk of vegetarians. However uptake in human via inhalation of vapors in houses treated with heptachlor and chlordane is also evident. Where monitoring at the top of the food chain was performed, such as residues in bald eagles and the Illinois Milk Survey, levels of heptachlor epoxide have not significantly changed three years after agricultural uses of the pesticide were being phased out. Exposure to heptachlor epoxide is primarily through the ingestion of food containing residues of the insecticide. The food classes most likely to be contaminated are dairy products and meat/poultry/fish. [828]

Identifying Information

CAS NUMBER: 1024-57-3

NIOSH Registry Number: PB9450000

SYNONYMS: HIPTACHLOR EPOXIDE, NCI-C06893, ENT 25,584, 1,4,5,6,7,8,8-HEPTACHLORO-2,3-EPOXY-3A,4,7,7A-TETRAHYDRO-4,7-METHANOINDAN, VELSICOL 53-CS-17, 2,5-METHANO-2H-OXIRENO(A)INDENE, 2,3,4,5,6,7,7-HEPTACHLORO-1A,1B,5,5A,6,6A-, HEXAHYDRO-, 2,3,4,5,6,7,7-HEPTACHLORO-1A,1B,5,5A,6,6A-HEXAHYDRO-2,5-METHANO-2H-, OXIRENO(A)INDENE

Endocrine Disruptor Information

Heptachlor epoxide is a strongly suspected EED that it is listed by the Centers for Disease Control & Prevention [812], and the World Wildlife Fund [813] as a potential endocrine modifying chemical.

Chemical and Physical Properties

CHEMICAL FORMULA: $C_{10}H_5Cl_7O$

MOLECULAR WEIGHT: 389.32

WLN: T D3 C555 A EO JUTJ AG AG BG GG IG JG KG

PHYSICAL DESCRIPTION: a liquid [816] (Ed. Note: the impure technical product is probably a liquid but the melting point below is characteristic of a crystalline solid.)

SPECIFIC GRAVITY: Not available

DENSITY: Not available

MELTING POINT: 157-160°C [816], [876]

BOILING POINT: Not available

SOLUBILITY: Water: 200 ug/L at 25°C [816], [875]

Environmental Reference Materials

A source of EED environmental reference materials for heptachlor epoxide is Radian International LLC, Austin, TX (Fax 512-454-0268; Internet URL, http://www.radian.com/standards); Catalog No. ERH-001.

Hazardous Properties

ACUTE/CHRONIC HAZARDS: Not available

STABILITY: Heptachlor epoxide is stable under normal laboratory conditions.

Source: *The National Toxicology Program's Chemical Database, Volume 7*, 1992, CRC Press, Inc./Lewis Publishers, 2000 Corporate Blvd., Boca Raton, FL 33431 (Fax: 407-998-9114).

Medical Symptoms of Exposure

This information is not available for heptachlor epoxide. See symptoms of exposure for heptachlor, a closely related compound.

Toxicological Information

The toxicological information below is gathered from several sources including two national databases: one from the *National Toxicology Program Chemical Repository Database* and another from *EPA's Integrated Risk Information System (IRIS)*.

Toxicology Information from the National Toxicology Program

CARCINOGENICITY: Not available

TERATOGENICITY: Not available

MUTAGENICITY: mma-hmn:fbr 10 umol/L

TOXICITY:

Typ. Dose	Mode	Specie	Amount	Unit	Other
LD50	orl	rat	62	mg/kg	
LDLo	orl	mus	40	mg/kg	
LDLo	ivn	mus	580mg/kg/69W-C		

OTHER TOXICITY DATA:

   Review: Toxicology Review

SAX TOXICITY EVALUATION: Not available

Source: *Instant Tox-Base*, 1995, Instant Reference Sources, Inc., 7605 Rockpoint Drive, Austin, TX 78731 (Fax: 512-345-2386; Internet URL, http://www.instantref.com/inst-ref.htm).

Toxicology Information from EPA's Integrated Risk Information System (IRIS)

I.   CHRONIC HEALTH HAZARD ASSESSMENTS FOR NONCARCINOGENIC EFFECTS

I.A. REFERENCE DOSE FOR CHRONIC ORAL EXPOSURE (RfD)

The Reference Dose (RfD) is based on the assumption that thresholds exist for certain toxic effects such as cellular necrosis, but may not exist for other toxic effects such as carcinogenicity. In general, the RfD is an estimate (with uncertainty spanning perhaps an order of magnitude) of a daily exposure to the human population (including sensitive subgroups) that is likely to be without an appreciable risk of deleterious effects during a lifetime. Please refer to Background Document 1 in Service Code 5 for an elaboration of these concepts. RfDs can also be derived for the noncarcinogenic health effects of compounds which are also carcinogens. Therefore, it is essential to refer to other sources of information concerning the carcinogenicity of this substance. If the US EPA has evaluated this substance for potential human carcinogenicity, a summary of that evaluation will be contained in Section II of this file when a review of that evaluation is completed.

I.A.1. ORAL RfD SUMMARY

Critical Effect	Experimental Doses*	UF	MF	RfD
Increased liver to body weight ratio in both males and females	NOEL: none LEL: 0.5 ppm (diet) (0.0125 mg/kg/day)	1000	1	1.3E-5 mg/kg/day

*Conversion Factors: 1 ppm = 0.025 mg/kg/day (assumed dog food consumption)

60-Week Dog Feeding Study Dow Chemical Co., 1958

I.A.2. PRINCIPAL AND SUPPORTING STUDIES (ORAL RfD)

Dow Chemical Company. 1958. MRID No. 00061912. Available from EPA. Write to FOI, EPA, Washington, DC 20460.

Beagle dogs from 23 to 27 weeks of age were divided into five groups (3 females and 2 males) and given diets containing 0, 0.5, 2.5, 5 or 7.5 ppm of heptachlor epoxide for 60 weeks. Liver-to-body weight ratios were significantly increased in a treatment-related fashion. Effects were noted for both males and females at the LEL of 0.5 ppm. A NOEL was not established.

I.A.3. UNCERTAINTY AND MODIFYING FACTORS (ORAL RfD)

UF -- Based on a chronic exposure study, an uncertainty factor of 1000 was used to account for inter- and intraspecies differences and to account for the fact that a NOEL was not attained.

MF -- None

I.A.4. ADDITIONAL COMMENTS (ORAL RfD)

None.

Data Considered for Establishing the RfD:

1) 60-Week Feeding - dog: Principal study - see previous description; no core grade

2) 2-Generation Reproduction - dog: NOEL=1 ppm (0.025 mg/kg/day); LEL=3 ppm (0.075 mg/kg/day) (liver lesions in pups); Reproductive NOEL=5 ppm (0.125 mg/kg/day); Reproductive LEL=7 ppm (0.175 mg/kg/day) (pup survival); no core grade (Velsicol Chemical, 1973a)

3) 3-Generation Reproduction - rat: NOEL=5 ppm (0.25 mg/kg/day); LEL=10 ppm (0.5 mg/kg/day) (pup mortality); no core grade (Velsicol Chemical, 1959a)

4) 2-Year Feeding - rat: LEL=0.5 ppm (0.025 mg/kg/day) (LDT) (females - vacuolar changes in central hepatic lobule); NOEL not established; no core grade (Velsicol Chemical, 1959b)

Other Data Reviewed:

1) Chronic Feeding Study - mouse: Heptachlor/Heptachlor Epoxide (1:3): NOEL= none; LEL=1 ppm (LDT) (vaculoation, enlarged nucleus, hepatocytomegaly); no core grade (Velsicol Chemical, 1973b)

2) Chronic Feeding Study - rat: Heptachlor/Heptachlor Epoxide (3:1): NOEL= none; LEL=5 ppm (LDT) (liver-to-body weight increase in females); no core grade (Velsicol Chemical, 1966)

3) 3-Generation Reproduction - rat: Heptachlor/Heptachlor Epoxide (3:1): NOEL=7 ppm (HDT); LEL= none; no core grade (Velsicol Chemical, 1967)

Data Gap(s): Rat Teratology Study; Rabbit Teratology

I.A.5. CONFIDENCE IN THE ORAL RfD

Study -- Low
Data Base -- Medium
RfD -- Low

The principal study is of low quality and is given a low confidence rating. Since the data base on chronic toxicity is complete but consists of low-quality studies, the data base is given a medium to low confidence rating. Low confidence in the RfD follows.

I.A.6. EPA DOCUMENTATION AND REVIEW OF THE ORAL RfD

Pesticide Registration Standard, August 1986

Agency Work Group Review -- 12/18/85, 09/16/86

Verification Date -- 09/16/86

I.A.7. EPA CONTACTS (ORAL RfD)

William Burnam / OPP -- (703)557-4791

George Ghali / OPP -- (703)557-7490

I.B. REFERENCE CONCENTRATION FOR CHRONIC INHALATION EXPOSURE (RfC)

Not available at this time.

II. CARCINOGENICITY ASSESSMENT FOR LIFETIME EXPOSURE

Section II provides information on three aspects of the carcinogenic risk assessment for the agent in question; the US EPA classification, and quantitative estimates of risk from oral exposure and from inhalation exposure. The classification reflects a weight-of-evidence judgment of the likelihood that the agent is a human carcinogen. The quantitative risk estimates are presented in three ways. The slope factor is the result of application of a low-dose extrapolation procedure and is presented as the risk per (mg/kg)/day. The unit risk is the quantitative estimate in terms of either risk per ug/L drinking water or risk per ug/cu.m air breathed. The third form in which risk is presented is a drinking water or air concentration providing cancer risks of 1 in 10,000, 1 in 100,000 or 1 in 1,000,000. Background Document 2 (Service Code 5) provides details on the rationale and methods used to derive the carcinogenicity values found in IRIS. Users are referred to Section I for information on long-term toxic effects other than carcinogenicity.

II.A. EVIDENCE FOR CLASSIFICATION AS TO HUMAN CARCINOGENICITY

II.A.1. WEIGHT-OF-EVIDENCE CLASSIFICATION

Classification -- B2; probable human carcinogen

Basis -- Sufficient evidence exists from rodent studies in which liver carcinomas were induced in two strains of mice of both sexes and in CFN female rats. Several structurally related compounds are liver carcinogens.

II.A.2. HUMAN CARCINOGENICITY DATA

Inadequate. There are no published epidemiological evaluations of heptachlor epoxide. It is not commercially available in the United States, but is a product of heptachlor oxidation.

There were 11 case reports involving central nervous system effects, blood dyscrasias and neuroblastomas in children with pre-/postnatal exposure to chlordane and heptachlor (Infante et al., 1978). Since no other information was available, no conclusions can be drawn.

There were three epidemiological studies of workers exposed to chlordane and/or heptachlor. One retrospective cohort study of pesticide applicators was considered inadequate in sample size and duration of follow-up. This study showed marginal statistically significant increased mortality from bladder cancer (3 observed) (Wang and McMahon, 1979a). Two other retrospective cohort studies were of pesticide manufacturing workers. Neither of them showed any statistically significant increased cancer mortality (Wang and McMahon, 1979b; Ditraglia et al., 1981). Both these populations also had confounding exposures from other chemicals.

II.A.3. ANIMAL CARCINOGENICITY DATA

Sufficient. Four long-term carcinogenesis bioassays of heptachlor epoxide have been reported. The major finding in mice has been an increased incidence of liver carcinomas. Davis (1965) fed groups of 100 male and 100 female C3H mice 0 or 10 ppm heptachlor epoxide for 2 years. Survival was generally low, with 50% of controls and 9.5% of treated mice living 2 years. A 2-fold increase in benign liver lesions (hepatic hyperplasia and benign tumors) over the controls was reported. Reevaluation by Reuber (1977b) revealed a significant increase in liver carcinomas in the dosed group (77/81 in females and 73/79 in males) over the controls (2/53 in females and 22/73 in males). The Velsicol Chemical Co. (1973) tested a 75:25 mixture of heptachlor epoxide:heptachlor in groups of 100 male and 100 female CD-1 mice. The mice were fed 0, 1, 5, and 10 ppm for 18 months. A statistically significant increase of hyperplasia was observed in the 5, and 10 ppm dose groups in both sexes; Reuber's reevaluation (US EPA, 1985) resulted in a change in diagnosis for benign to liver carcinomas, thereby increasing the incidence of hepatic carcinomas (p<0.01). Four independent pathologists concurred with Reuber's reevaluation.

The earliest bioassay with rats (Witherup et al., 1959) tested 25 male and 25 female CFN rats each at 0.5, 2.5, 5.0, 7.5, and 10 ppm for 108 weeks. The authors observed malignant and benign tumors randomly among test groups and controls. Reuber's reevaluation (1985) reported a significant increase of hepatic carcinomas above the controls at 5 and 10 ppm in the female rats. A reevaluation by Williams (1985) reported a significant increase of hepatic nodules at the 10 ppm level in the males over the controls. The Kettering Laboratory (Jolley et al., 1966) tested a mixture of 75:25 heptachlor:heptachlor epoxide in the diet of 25 female CD rats at 5, 7.5, 10, and 12.5 ppm for 2 years. Although no malignant lesions of the liver were observed, hepatocytomegaly was increased at 7.5, 10, and 12.5 ppm.

II.A.4. SUPPORTING DATA FOR CARCINOGENICITY

Gene mutation assays indicate that heptachlor epoxide is not mutagenic in bacteria (Moriya et al., 1983). In two mouse dominant lethal assays, heptachlor epoxide did not induce major chromosomal aberrations in male germinal cells (Arnold et al., 1977; Epstein et al., 1972). Ahmed et al. (1977) reported qualitative evidence of unscheduled DNA synthesis response in SV40 transformed human fibroblasts in the presence of hepatic homogenates and heptachlor epoxide.

Five compounds structurally related to heptachlor epoxide (chlordane, aldrin, dieldrin, heptachlor and chlorendic acid) have produced liver tumors in mice. Chlorendic acid has also produced liver tumors in rats.

II.B.   QUANTITATIVE ESTIMATE OF CARCINOGENIC RISK FROM ORAL EXPOSURE

II.B.1. SUMMARY OF RISK ESTIMATES

Oral Slope Factor -- 9.1E+0 per (mg/kg)/day
Drinking Water Unit Risk -- 2.6E-4 per (ug/L)
Extrapolation Method -- Linearized multistage procedure, extra risk
Drinking Water Concentrations at Specified Risk Levels:

Risk Level	Concentration
E-4 (1 in 10,000)	4E-1 ug/L
E-5 (1 in 100,000)	4E-2 ug/L
E-6 (1 in 1,000,000)	4E-3 ug/L

II.B.2. DOSE-RESPONSE DATA (CARCINOGENICITY, ORAL EXPOSURE)

Tumor Type -- hepatocellular carcinomas
Test Animals -- mouse/C3H (Davis); mouse/CD1 (Velsicol)
Route -- diet
Reference -- Davis, 1965; Velsicol, 1973 (see table)

Administered Dose (ppm)	Human Equivalent Dose (mg/kg/day)	Tumor Incidence	Reference
male			
0	0.0	22/73	Davis, 1965
10	0.108	73/79	as diagnosed
female			by Reuber, 1977
0	0.000	2/53	(cited in
10	0.108	77/81	Epstein, 1976)
female			
0	0.00	6/76	Velsicol, 1973
1	0.01	1/70	as evaluated
5	0.052	6/65	by Reuber, 1977
10	0.10	30/57	
male			
0	0.00	0/62	
1	0.01	2/68	
5	0.052	18/68	
10	0.10	52/80	

II.B.3. ADDITIONAL COMMENTS (CARCINOGENICITY, ORAL EXPOSURE)

The Davis (1965) study was designed to be for lifetime exposure. Thus, although survival was low, no correction for duration of experiment was made. Five data sets (four in mice and one in rats) show an increased incidence of hepatocellular carcinomas in treated groups compared with controls. There are four slope factors, 27.7 per (mg/kg)/day for C3H male mice, 36.2 per (mg/kg)/day for C3H female mice, 1.04 per (mg/kg)/day for CD-1 female mice, and 6.48 per (mg/kg)/day for CD-1 male mice. Since mice were the more sensitive species tested and to avoid discarding relevant data, the quantitative estimate is based on the geometric mean of 9.1 per (mg/kg)/day. This geometric mean is consistent with the potency estimate from rats of 5.8 per (mg/kg)/day (CFN females).

The above unit risk should not be used if the water concentration exceeds 40 ug/L, since above this concentration the unit risk may not be appropriate.

II.B.4. DISCUSSION OF CONFIDENCE (CARCINOGENICITY, ORAL EXPOSURE)

Adequate numbers of animals were treated in both studies, but survival in the Davis (1985) study was low. A dose-related increase in tumor incidence was observed in CD-1 mice. Slope factors were consistent in two species of rodents.

II.C. QUANTITATIVE ESTIMATE OF CARCINOGENIC RISK FROM INHALATION EXPOSURE

II.C.1. SUMMARY OF RISK ESTIMATES

Inhalation Unit Risk -- 2.6E-3 per (ug/cu.m)

Extrapolation Method -- Linearized multistage procedure, extra risk

Air Concentrations at Specified Risk Levels:

Risk Level	Concentration
E-4 (1 in 10,000)	4E-2 ug/cu.m
E-5 (1 in 100,000)	4E-3 ug/cu.m
E-6 (1 in 1,000,000)	4E-4 ug/cu.m

II.C.2. DOSE-RESPONSE DATA FOR CARCINOGENICITY, INHALATION EXPOSURE

The inhalation risk estimates were calculated from the oral data presented in II.B.2.

II.C.3. ADDITIONAL COMMENTS (CARCINOGENICITY, INHALATION EXPOSURE)

The above unit risk should not be used if the air concentration exceeds 4 ug/cu.m, since above this concentration the unit risk may not be appropriate.

II.C.4. DISCUSSION OF CONFIDENCE (CARCINOGENICITY, INHALATION EXPOSURE)

See II.B.4.

II.D. EPA DOCUMENTATION, REVIEW, AND CONTACTS (CARCINOGENICITY ASSESSMENT)

II.D.1. EPA DOCUMENTATION

Source Document -- US EPA, 1985, 1986

The values in the 1986 Carcinogenicity Assessment for Chlordane and Heptachlor/Heptachlor Epoxide have been reviewed by the Carcinogen Assessment Group.

II.D.2. REVIEW (CARCINOGENICITY ASSESSMENT)

Agency Work Group Review -- 04/01/87

Verification Date -- 04/01/87

II.D.3. US EPA CONTACTS (CARCINOGENICITY ASSESSMENT)

Dharm V. Singh / NCEA -- (202)260-5958

Jim Cogliano / NCEA -- (202)260-3814

III. HEALTH HAZARD ASSESSMENTS FOR VARIED EXPOSURE DURATIONS

III.A. DRINKING WATER HEALTH ADVISORIES

The Office of Drinking Water provides Drinking Water Health Advisories (HAs) as technical guidance for the protection of public health. HAs are not enforceable Federal standards. HAs are concentrations of a substance in drinking water estimated to have negligible deleterious effects in humans, when ingested, for a specified period of time. Exposure to the substance from other media is considered only in the derivation of the lifetime HA. Given the absence of chemical-specific data, the assumed fraction of total intake from drinking water is 20%. The lifetime HA is calculated from the Drinking Water Equivalent Level (DWEL) which, in turn, is based on the Oral Chronic Reference Dose. Lifetime HAs are not derived for compounds which are potentially carcinogenic for humans because of the difference in assumptions concerning toxic threshold for carcinogenic and noncarcinogenic effects. A more detailed description of the assumptions and methods used in the derivation of HAs is provided in Background Document 3 in Service Code 5.

III.A.1. ONE-DAY HEALTH ADVISORY FOR A CHILD

Appropriate data for calculating a One-day HA for heptachlor epoxide are not available. No recommendations are made for the One-day HA.

III.A.2. TEN-DAY HEALTH ADVISORY FOR A CHILD

Appropriate data for calculating a Ten-day HA for heptachlor epoxide are not available. No recommendations are made for the Ten-day HA.

III.A.3. LONGER-TERM HEALTH ADVISORY FOR A CHILD

Appropriate data for calculating a Longer-term HA for heptachlor epoxide are not available. It is recommended that a modified DWEL (adjusted for a 10-kg child) of 0.00013 mg/L (rounded to 0.00015 mg/L) be used as the Longer- term HA.

III.A.4. LONGER-TERM HEALTH ADVISORY FOR AN ADULT

Appropriate data for calculating a Longer-term HA for heptachlor epoxide are not available. It is recommended that the DWEL of 0.00044 mg/L (rounded to 0.0005 mg/L) be used as the Longer-term HA for the 70-kg adult.

III.A.5. DRINKING WATER EQUIVALENT LEVEL / LIFETIME HEALTH ADVISORY

DWEL = 4.4E-4 mg/L

Assumptions -- 2 L/day water consumption for a 70-kg adult

RfD Verification Date = 09/16/86 (see Section I.A. of this file)

Lifetime HA -- None

Heptachlor epoxide is considered to be a probable human carcinogen. Refer to Section II of this file for information on the carcinogenicity of this substance.

Principal Study -- Dow Chemical Co., 1958 (This study was used in the derivation of the chronic oral RfD; see Section I.A.2.)

III.A.6. ORGANOLEPTIC PROPERTIES

[Ed. Note: EPA does not have information yet in this section of IRIS].

III.A.7. ANALYTICAL METHODS FOR DETECTION IN DRINKING WATER

Determination of heptachlor epoxide is by a liquid-liquid extraction gas chromatographic procedure.

III.A.8. WATER TREATMENT

Treatment techniques capable of removing heptachlor epoxide from drinking water include adsorption by granular activated carbon and ozone or ozone/ultra-violet oxidation.

III.A.9. DOCUMENTATION AND REVIEW OF HAs

US EPA. 1985. Final Draft of the Drinking Water Criteria Document on Heptachlor Epoxide. Office of Drinking Water, Washington, DC.

EPA review of HAs in 1985.

Public review of HAs following notification of availability in October, 1985.

Scientific Advisory Panel review of HAs in January, 1986.

Preparation date of this IRIS summary -- 06/17/87

III.A.10. EPA CONTACTS

Jennifer Orme Zavaleta / OST -- (202)260-7586

Edward V. Ohanian / OST -- (202)260-7571

III.B. OTHER ASSESSMENTS

Content to be determined.

IV. US EPA REGULATORY ACTIONS

EPA risk assessments may be updated as new data are published and as assessment methodologies evolve. Regulatory actions are frequently not updated at the same time. Compare the dates for the regulatory actions in this section with the verification dates for the risk assessments in sections I and II, as this may explain inconsistencies. Also note that some regulatory actions consider factors not related to health risk, such as technical or economic feasibility. Such considerations are indicated for each action. In addition, not all of the regulatory actions listed in this section involve enforceable federal standards. Please direct any questions you may have concerning these regulatory actions to the US EPA contact listed for that particular action. Users are strongly urged to read the background information on each regulatory action in Background Document 4 in Service Code 5.

IV.A. CLEAN AIR ACT (CAA)

[Ed. Note: EPA does not have information yet in this section of IRIS].

IV.B. SAFE DRINKING WATER ACT (SDWA)

IV.B.1. MAXIMUM CONTAMINANT LEVEL GOAL (MCLG) for Drinking Water

Value -- 0 mg/L (Final, 1991)

Considers technological or economic feasibility? -- NO

Discussion -- EPA has set a MCLG of zero for heptachlor epoxide based on evidence of carcinogenic effects (B2).

Reference -- 56 FR 3526 (01/30/91)

EPA Contact -- Health and Ecological Criteria Division / OST / (202) 260-7571 / FTS 260-7571; or Safe Drinking Water Hotline / (800) 426-4791

IV.B.2. MAXIMUM CONTAMINANT LEVEL (MCL) for Drinking Water

Value -- 0.0002 mg/L (Final, 1991)

Considers technological or economic feasibility? -- YES

Discussion -- EPA has set a MCl equal to the PQL of 0.0002 mg/L, which is associated with a lifetime individual risk of 0.5 E-4.

Monitoring requirements -- All systems monitored for four consecutive quarters every 3 year; repeat monitoring dependent upon detection, vulnerability status and system size.

Analytical methodology -- Microextraction/gas chromatography (EPA 505); electron-capture/gas chromatography (EPA 508); gas chromatographic/mass spectrometry (EPA 525): PQL= 0.0002 mg/L.

Best available technology -- Granular activated carbon.

Reference -- 56 FR 3526 (01/30/91)

EPA Contact -- Drinking Water Standards Division / OGWDW / (202) 260-7575 / FTS 260-7575; or Safe Drinking Water Hotline / (800) 426-4791

IV.B.3. SECONDARY MAXIMUM CONTAMINANT LEVEL (SMCL) for Drinking Water

No data available

IV.B.4. REQUIRED MONITORING OF "UNREGULATED" CONTAMINANTS

[Ed. Note: EPA does not have information yet in this section of IRIS].

IV.C. CLEAN WATER ACT (CWA)

IV.C.1. AMBIENT WATER QUALITY CRITERIA, Human Health

Water and Fish Consumption: 2.8E-4 ug/L

Fish Consumption Only: 2.9E-4 ug/L

Considers technological or economic feasibility? -- NO

Discussion -- The WQC of 2.8E-4 ug/L represents a cancer risk level of 1E-6 based on consumption of contaminated aquatic organisms and water. A WQC of 2.9E-4 ug/L has also been established based on consumption of contaminated aquatic organisms alone. The heptachlor criteria for both aquatic life and human health serve as the bases for the heptachlor epoxide criteria. The Office of Water has not developed criteria specifically for heptachlor epoxide.

Reference -- 45 FR 79318 (11/28/80)

EPA Contact -- Criteria and Standards Division / OWRS (202)260-1315 / FTS 260-1315

IV.C.2. AMBIENT WATER QUALITY CRITERIA, Aquatic Organisms

Freshwater:
    Acute -- 5.2E-1 ug/L (24-hour average)
    Chronic -- 3.8E-3 ug/L

Marine:
    Acute -- 5.3E-2 ug/L (24-hour average)
    Chronic -- 3.6E-3 ug/L

Considers technological or economic feasibility? -- NO

Discussion -- Water quality criteria for the protection of aquatic life are derived from a minimum data base of acute and chronic tests on a variety of aquatic organisms. The data are assumed to be statistically representative and are used to calculate concentrations which will not have significant short- or long-term effects on 95% of the organisms exposed. Recent criteria (1985 and later) contain duration and frequency stipulations: the acute criteria maximum concentration is a 1-hour average and the chronic criteria continuous concentration is a 4-day average; these averages are not to be exceeded more than once every 3 years, on the average (Stephen et al., 1985). Earlier criteria (1980-1984) contained instantaneous acute and 24-hour average chronic concentrations which were not to be exceeded. These criteria are for heptachlor, rather than heptachlor epoxide.

Reference -- 45 FR 79318 (11/28/80)

EPA Contact -- Criteria and Standards Division / OWRS (202)260-1315 / FTS 260-1315

IV.D. FEDERAL INSECTICIDE, FUNGICIDE, AND RODENTICIDE ACT (FIFRA)

IV.D.1. PESTICIDE ACTIVE INGREDIENT, Registration Standard

Status -- Issued (1986)

Reference -- Heptachlor Pesticide Registration Standard. December, 1986 (NTIS No. PB87-175808).

EPA Contact -- Registration Branch / OPP (703)557-7760 / FTS 557-7760

IV.D.2. PESTICIDE ACTIVE INGREDIENT, Special Review

Action -- Cancellation of many uses (1988)

Considers technological or economic feasibility? -- YES

Summary of regulatory action -- Based on concern for oncogenicity.

Reference -- 43 FR 12372 (03/24/87; 52 FR 42145 (11/03/87); 53 FR 11798 (04/08/88)

EPA Contact -- Special Review Branch / OPP (703)557-7400 / FTS 557-7400

IV.E. TOXIC SUBSTANCES CONTROL ACT (TSCA)

[Ed. Note: EPA does not have information yet in this section of IRIS].

IV.F. RESOURCE CONSERVATION AND RECOVERY ACT (RCRA)

IV.F.1. RCRA APPENDIX IX, for Ground Water Monitoring

Status -- Listed

Reference -- 52 FR 25942 (07/09/87)

EPA Contact -- RCRA/Superfund Hotline (800)424-9346 / (202)260-3000 / FTS 260-3000

IV.G. SUPERFUND (CERCLA)

IV.G.1. REPORTABLE QUANTITY (RQ) for Release into the Environment

Value (status) -- 1 pound (Final, 1989)

Considers technological or economic feasibility? -- NO

Discussion -- The RQ for heptachlor epoxide is one pound based on potential carcinogenicity. Available data indicate a hazard ranking of high, based on a potency factor of 289.93/mg/kg/day and a weight-of-evidence group B2.

Reference -- 54 FR 33418 (08/14/89)

EPA Contact -- RCRA/Superfund Hotline (800)424-9346 / (202)260-3000 / FTS 260-3000

VI. BIBLIOGRAPHY

VI.A. ORAL RfD REFERENCES

Dow Chemical Company. 1958. MRID No. 00061912. Available from EPA. Write to FOI, EPA, Washington, DC 20460.

Dow Chemical Company. 1959a. MRID No. 00062676. Available from EPA. Write to FOI, EPA, Washington, DC 20460.

Dow Chemical Company. 1959b. MRID No. 00061911. Available from EPA. Write to FOI, EPA, Washington, DC 20460.

Dow Chemical Company. 1966. MRID No. 00086208. Available from EPA. Write to FOI, EPA, Washington, DC 20460.

Dow Chemical Company. 1967. MRID No. 00147057. Available from EPA. Write to FOI, EPA, Washington, DC 20460.

Dow Chemical Company. 1973a. MRID No. 00050058. Available from EPA. Write to FOI, EPA, Washington, DC 20460.

Dow Chemical Company. 1973b. MRID No. 000523262, 00062678, 00064943. Available from EPA. Write to FOI, EPA, Washington, DC 20460.

VI.B. INHALATION RfC REFERENCES

None

VI.C. CARCINOGENICITY ASSESSMENT REFERENCES

Davis, K.J. 1965. Pathology Report on Mice Fed Aldrin, Dieldrin, Heptachlor and Heptachlor Epoxide for Two Years. Internal FDA memorandum to Dr. A.J. Lehman, July 19.

Epstein, S.S. 1976. Carcinogenicity of heptachlor and chlordane. Sci. Total Environ. 6: 103-154.

Reuber, M.D. 1977. Histopathology of carcinomas of the liver in mice ingesting heptachlor or heptachlor epoxide. Exp. Cell Biol. 45: 147-157.

US EPA. 1985. Hearing Files on Chlordane, Heptachlor Suspension (unpublished draft). Available for inspection at: US EPA, Washington, DC.

US EPA. 1986. Carcinogenicity Assessment of Chlordane and Heptachlor/ Heptachlor Epoxide. Prepared by the Office of Health and Environmental Assessment, Carcinogen Assessment Group, Washington, DC. OHEA-C-204.

Velsicol Chemical Corporation. 1973. MRID No. 00062678. Available from EPA. Write to FOI, EPA, Washington, D.C. 20460.

VI.D. DRINKING WATER HA REFERENCES

Dow Chemical Company. 1958. MRID No. 00061912. Available from EPA. Write to FOI, EPA, Washington, DC 20460.

US EPA. 1985. Final Draft of the Drinking Water Criteria Document on Heptachlor Epoxide. Office of Drinking Water, Washington, DC.

VII. REVISION HISTORY

Date	Section	Description
09/30/87	II.	Carcinogen summary on-line
03/01/88	I.A.2.	Text clarified
03/01/88	I.A.5.	Confidence levels revised
03/01/88	II.B.4.	Confidence statement revised
03/01/88	III.A.	Health Advisory on-line
08/01/90	III.A.10	Primary contact changed
08/01/90	IV.F.1.	EPA contact changed
01/01/91	II.	Text edited
01/01/91	II.C.1.	Inhalation slope factor removed (global change)
03/01/91	I.A.4.	Citations added
03/01/91	VI.	Bibliography on-line
01/01/92	IV.	Regulatory actions updated
04/01/92	II.A.3.	Text revised
04/01/93	IV.C.2.	Freshwater and marine values corrected
07/01/93	II.D.3.	Secondary contact's phone number changed

SYNONYMS

ENT 25,584
EPOXYHEPTACHLOR
HCE
Heptachlor Epoxide
1,4,5,6,7,8,8-HEPTACHLORO-2,3-
EPOXY-2,3,3a,4,7,7a-HEXAHYDRO-4,7-
METHANOINDENE
1,4,5,6,7,8,8-HEPTACHLORO-2,3-
EPOXY-3a,4,7,7a-TETRAHYDRO-4,7-
METHANOINDAN
2,3,4,5,6,7,7-HEPTACHLORO-
1a,1b,5,5a,6,6a-HEXAHYDRO-2,5-
METHANO-2H-INDENO(1,2-
b)OXIRENE

HIPTACHLOR EPOXIDE
4,7-METHANOINDAN, 1,4,5,6,7,8,8-
HEPTACHLORO-2,3-EPOXY-3a,4,7,7a-
TETRAHYDRO-
2,5-METHANO-2H-
OXIRENO(a)INDENE, 2,3,4,5,6,7,7-
HEPTACHLORO-1a,1b,5,5a,6,6a-
 HEXAHYDRO-
VELSICOL 53-CS-17

Source: *Instant EPA's IRIS*, 1997, Instant Reference Sources, Inc., 7605 Rockpoint Drive, Austin, TX 78731 (Fax: 512-345-2386; Internet URL, http://www.instantref.com/inst-ref.htm).

Production, Use, and Pesticide Labeling Information

HEPTACHLOR EPOXIDE is not currently listed in EPA's Pesticide Factsheet database and there is no indication of whether it will be added or not in the future.

Hexachlorobenzene

Introduction

Hexachlorobenzene is used in organic synthesis, as a fungicide for seeds, as a wood preservative, in the manufacture of pentachlorophenol, in the production of aromatic fluorocarbons and in the impregnation of paper. As a fungicide for seeds it is used with sunflowers, safflowers and also for seedling diseases. It is also used for control of insects such as wireworms. In Europe, hexachlorobenzene has been used as the precursor for pentachlorophenol, though not in the USA. European pentachlorophenol made using alkaline hydrolysis of hexachlorobenzene has more polychlorinated dibenzo-p-dioxin and dibenzofuran impurities than the USA pentachlorophenol product. [828]

Hexachlorobenzene is formed as a waste product in the production of several chlorinated hydrocarbons and is a contaminant in some pesticides. It may enter the environment in air emissions and waste water in connection with the above and in flue gases and fly ash from waste incineration. Non-point source dispersal of hexachlorobenzene results from its presence as a contaminant in pesticides. It is a very persistent environmental chemical due to its chemical stability and resistance to biodegradation. If released to the atmosphere, it will exist primarily in the vapor phase and degradation will be extremely slow (the estimated half-life with hydroxyl radicals is 2 years). Long range global transport is possible. Physical removal from the atmosphere can occur via washout by rainfall and dry deposition. If released to water, it will significantly partition from the water colum to sediment and suspended matter. Volatilization from the water column is rapid; however, the strong adsorption to sediment can result in long periods of persistence. If released to soil, it will be strongly adsorbed and not generally susceptible to leaching. Hexachlorobenzene will bioconcentrate in fish and enter into the food chain (it has been detected in food during market basket surveys). Human exposure will be from ambient air, contaminated drinking water and food, as well as contact with contaminated soil or occupational atmospheres. [828]

Identifying Information

CAS NUMBER: 118-74-1

NIOSH Registry Number: DA2975000

SYNONYMS: Perchlorobenzene, Phenyl perchloryl

639

Endocrine Disruptor Information

Hexachlorobenzene is a strongly suspected EED that it is listed by the Centers for Disease Control & Prevention [812], and the World Wildlife Fund [813] as a potential endocrine modifying chemical.

Chemical and Physical Properties

CHEMICAL FORMULA: C_6Cl_6

MOLECULAR WEIGHT: 284.76

WLN: GR BG CG DG EG FG

PHYSICAL DESCRIPTION: White needles

SPECIFIC GRAVITY: 2.044 @ 24°C

DENSITY: Not available

MELTING POINT: 229°C

BOILING POINT: 322°C (sublimes)

SOLUBILITY: Water: <1 mg/mL @ 20°C [700]; DMSO: <1 mg/mL @ 20°C [700]; 95% Ethanol: <1 mg/mL @ 20°C [700]; Acetone: 1-5 mg/mL @ 23°C [700]

OTHER SOLVENTS: Chloroform: Soluble; Carbon disulfide: Soluble; Carbon tetrachloride: Sparingly soluble; Ether: Soluble; Benzene: Soluble

OTHER PHYSICAL DATA: Easily sublimable

Source: *Instant EPA's Air Toxics*, 1994, Instant Reference Sources, Inc., 7605 Rockpoint Drive, Austin, TX 78731 (Fax: 512-345-2386; Internet URL, http://www.instantref.com/inst-ref.htm).

Environmental Reference Materials

A source of EED environmental reference materials for hexachlorobenzene is Radian International LLC, Austin, TX (Fax 512-454-0268; Internet URL, http://www.radian.com/standards); Catalog No. ERH-024.

Hazardous Properties

ACUTE/CHRONIC HAZARDS: Hexachlorobenzene is a irritant and, when heated to decomposition, it emits toxic fumes of chlorides, carbon monoxide and carbon dioxide.

HAP WEIGHTING FACTOR: 10 [713]

VOLATILITY:
      Vapor pressure: 1 mm Hg @ 114.4°C

Vapor density: 9.8

FIRE HAZARD: Hexachlorobenzene has a flash point of 242°C (468°F). It is combustible. Fires involving this material can be controlled with a dry chemical, carbon dioxide or Halon extinguisher.

REACTIVITY: Hexachlorobenzene reacts violently with dimethylformamide.

STABILITY: Hexachlorobenzene is sensitive to moisture. Solutions of it in water, DMSO, 95% ethanol or acetone should be stable for 24 hours under normal lab conditions [700].

Source: *Instant EPA's Air Toxics*, 1994, Instant Reference Sources, Inc., 7605 Rockpoint Drive, Austin, TX 78731 (Fax: 512-345-2386; Internet URL, http://www.instantref.com/inst-ref.htm).

Medical Symptoms of Exposure

Symptoms of exposure may include irritation of the eyes, skin, mucous membranes and upper respiratory tract, corneal opacity, focal alopecia, atrophic hands, hypertrichosis, hepatomegaly, porphyria, anorexia, weight loss, enlargement of thyroid and lymph nodes, skin photosensitization and abnormal growth of hair.

Source: *Instant EPA's Air Toxics*, 1994, Instant Reference Sources, Inc., 7605 Rockpoint Drive, Austin, TX 78731 (Fax: 512-345-2386; Internet URL, http://www.instantref.com/inst-ref.htm).

Toxicological Information

The toxicological information below is gathered from several sources including two national databases: one from the *National Toxicology Program Chemical Repository Database* and another from *EPA's Integrated Risk Information System (IRIS)*. EPA regulatory actions.

Toxicology Information from the National Toxicology Program

CARCINOGENICITY:

Tumorigenic Data:

Type of Effect	Route/Animal	Amount Dosed (Notes)
TD	orl-rat	1050 mg/kg/30W-C
TDLo	orl-mus	6972 mg/kg/83W-C
TDLo	orl-ham	1000 mg/kg/18W-C
TD	orl-ham	3360 mg/kg/80W-C
TD	orl-ham	3360 mg/kg/80W-C
TDLo	orl-rat	2738 mg/kg/2Y-C
TD	orl-rat	5475 mg/kg/2Y-C
TD	orl-rat	6300 mg/kg/90W-C

Review:
            IARC Cancer Review: Human Inadequate Evidence
            IARC Cancer Review: Animal Sufficient Evidence
            IARC possible human carcinogen (Group 2B)

Status:
            NTP Fourth Annual Report on Carcinogens, 1984
            NTP anticipated human carcinogen
            EPA Carcinogen Assessment Group

TERATOGENICITY:

Reproductive Effects Data:

Type of Effect	Route/Animal	Amount Dosed (Notes)
TDLo	orl-rat	40 mg/kg (10-13D preg)
TDLo	orl-rat	6450 mg/kg (1-22D preg/21D post)
TDLo	orl-rat	88 mg/kg (70D male/70D pre-22D preg)
TDLo	orl-rat	812 mg/kg (MGN)
TDLo	orl-mus	1 gm/kg (7-16D preg)
TDLo	orl-rat	556 mg/kg (96D pre-21D post)
TDLo	orl-rat	212 mg/kg (14D pre-17D post)
TDLo	orl-mam	27562 ug/kg (66D pre-28D post)

MUTATION DATA:

Test	Lowest dose	Test	Lowest dose
dnd-esc	20 umol/L	mmo-smc	100 ppm

TOXICITY:

Typ. Dose	Mode	Specie	Amount	Units	Other
LDLo	unr	man	220	mg/kg	
LD50	orl	rat	10000	mg/kg	
LC50	ihl	rat	3600	mg/m3	
LD50	orl	mus	4	gm/kg	
LC50	ihl	mus	4	gm/m3	
LD50	orl	cat	1700	mg/kg	
LC50	ihl	cat	1600	mg/m3	
LD50	orl	rbt	2600	mg/kg	
LC50	ihl	rbt	1800	mg/m3	

OTHER TOXICITY DATA:

Review: Toxicology Review-2

Standards and Regulations:
DOT-IMO: Poison B; Label: St. Andrews Cross

Status:
"NIOSH Manual of Analytical Methods" Vol 7 343
"NIOSH Manual of Analytical Methods" to be revised by June, 1985
Reported in EPA TSCA Inventory, 1983
EPA Genetic Toxicology Program, January 1984
Meets criteria for proposed OSHA Medical Records Rule

SAX TOXICITY EVALUATION:

THR: An experimental neoplastigen, carcinogen, equivocal tumorigenic agent and teratogen. A suspected human carcinogen. Moderate via oral route.

Source: *Instant Tox-Base*, 1995, Instant Reference Sources, Inc., 7605 Rockpoint Drive, Austin, TX 78731 (Fax: 512-345-2386; Internet URL, http://www.instantref.com/inst-ref.htm).

Toxicology Information from EPA's Integrated Risk Information System (IRIS)

I. CHRONIC HEALTH HAZARD ASSESSMENTS FOR NONCARCINOGENIC EFFECTS

I.A. REFERENCE DOSE FOR CHRONIC ORAL EXPOSURE (RfD)

The Reference Dose (RfD) is based on the assumption that thresholds exist for certain toxic effects such as cellular necrosis, but may not exist for other toxic effects such as carcinogenicity. In general, the RfD is an estimate (with uncertainty spanning perhaps an order of magnitude) of a daily exposure to the human population (including sensitive subgroups) that is likely to be without an appreciable risk of deleterious effects during a lifetime. Please refer to Background Document 1 in Service Code 5 for an elaboration of these concepts. RfDs can also be derived for the noncarcinogenic health effects of compounds which are also carcinogens. Therefore, it is essential to refer to other sources of information concerning the carcinogenicity of this substance. If the US EPA has evaluated this substance for potential human carcinogenicity, a summary of that evaluation will be contained in Section II of this file when a review of that evaluation is completed.

I.A.1. ORAL RfD SUMMARY

Critical Effect	Experimental Doses*	UF	MF	RfD
Liver effects	NOAEL: 1.6 ppm (diet) (0.08 mg/kg/day)	100	1	8E-4 mg/kg/day
	LOAEL: 8.0 ppm (diet) (0.29 mg/kg/day)			

*Conversion Factors: doses were based on actual food consumption and body weights provided by Arnold at 30 weeks of exposure (US EPA, 1985, 1988).

Rat Chronic Feeding Study Arnold et al., 1985

I.A.2. PRINCIPAL AND SUPPORTING STUDIES (ORAL RfD)

Arnold, D.L., C.A. Moodie, S.M. Charbonneau, et al. 1985. Long-term toxicity of hexachlorobenzene in the rat and the effect of dietary Vitamin A. Fd. Chem. Toxic. 23(9): 779-793.

The derivation of the oral RfD is based on a 130-week study of Arnold et al. (1985). This study involved feeding male and female Sprague-Dawley rats, the F0 generation, diets containing 0, 0.32, 1.6, 8.0, or 40 ppm of hexachlorobenzene (analytical grade) for 90 days prior to mating and until 21 days after parturition (at weaning). The number of offspring (F1 generation) from these matings was reduced to 50 males and 50 females per dose group at 28 days of age and fed their respective parents' diets. Thus, the F1 animals were exposed to hexachlorobenzene and metabolites in utero, from maternal nursing and from their diets for the remainder of their lifetime (130 weeks). No hexachlorobenzene-induced adverse effects were reported in the 0.32 and 1.6 ppm hexachlorobenzene F1 groups, indicating that these levels are NOAELs. Although significant (p<0.05) increases were observed in the incidences of periportal glycogen depletion at 1.6 ppm, peribiliary lymphocytosis at 0.32, 1.6 and 40 ppm, and peribiliary fibrosis at 0.32 and 40 ppm in the F1 male rat groups, these effects are not being considered hexachlorobenzene-induced adverse effects because they were observed in a large number of F1 control males as well. The 8.0-ppm F1 groups were reported to have an increase (p<0.05) in hepatic centrilobular basophilic chromogenesis. The 40-ppm F1 groups showed increases (p<0.05) in pup mortality, hepatic centrilobular basophilic chromogenesis, and severe chronic nephrosis (males only).

I.A.3. UNCERTAINTY AND MODIFYING FACTORS (ORAL RfD)

UF -- An uncertainty factor of 100 was applied; 10 for interspecies and 10 for intraspecies variability.

MF -- None

I.A.4. ADDITIONAL COMMENTS (ORAL RfD)

The toxicity of long-term dietary exposure of humans to hexachlorobenzene was demonstrated by the epidemic of porphyria cutanea tarda (PCT) in Turkish citizens who accidentally consumed bread made from grain treated with hexachlorobenzene (Cam, 1963; Peters et al., 1966, 1982). In children less than 1 year of age, pink sore disease was observed along with 95% mortality. In addition to the PCT-associated symptoms of skin lesions, hypertrichosis, and hyperpigmentation, the exposure caused neurotoxicity and liver damage. Follow-up studies reported PCT symptoms, reduced growth and arthritic changes in the appendages of children who were directly or indirectly (i.e., through breast milk) exposed. These human data cannot be used for quantitative risk assessment purposes because accurate exposure data (dose and duration) are lacking.

An extensive number of animal research studies have been conducted on hexachlorobenzene including reproductive, teratology and carcinogenicity studies. These studies have been critiqued by US EPA (1985, 1988). Kuiper-Goodman et al. (1977) conducted another study that could be used to derive an oral RfD or the current RfD.

In a subchronic study, groups of 70 male and 70 female Charles River (COBS) rats were fed diets providing 0, 0.5, 2.0, 8.0, or 32.0 mg/kg/day of hexachlorobenzene, dissolved in corn oil, for up to 15 weeks (Kuiper-Goodman et al., 1977). Females were found to be more susceptible to hexachlorobenzene, as indicated by all parameters studied, and an "apparent" NOEL of 0.5 mg/kg/day was concluded by the authors. Increased liver porphyrin levels in females and increases in the size of centrilobular hepatocytes, along with the depletion of hepatocellular marker enzymes were noted with higher doses. In the two highest dose groups, there were increased liver-to-body weight ratios, as well as increased porphyrin levels in the kidney and spleen in both males and females. Exposure to the highest dose resulted in decreased survival, splenomegaly, and ataxia in females; increases in spleen-to-body weight and kidney-to-body weight ratios and intension tremors in males and females; and decreased body weight in males.

I.A.5. CONFIDENCE IN THE ORAL RfD

Study -- Medium
Data Base -- High
RfD -- Medium

The principal chronic study provided an unusual dosing scheme making it difficult to determine the true doses received by each experimental group. The study included extensive evaluation of systemic and neoplastic pathological endpoints and was critically reviewed before release and publication. The sensitive endpoint of porphyria was not evaluated in this study, otherwise a high confidence in the RfD could be assigned. The data base is rated high confidence due to the extensive number of quality research studies available.

I.A.6. EPA DOCUMENTATION AND REVIEW OF THE ORAL RfD

Source Document -- US EPA, 1985, 1988

Other EPA Documentation -- None

Agency Work Group Review -- 05/26/88

Verification Date -- 05/26/88

I.A.7. EPA CONTACTS (ORAL RfD)

W. Bruce Peirano / NCEA -- (513)569-7540

Charles Abernathy / OST -- (202)260-5374

I.B. REFERENCE CONCENTRATION FOR CHRONIC INHALATION EXPOSURE (RfC)

The health effects data for hexachlorobenzene were reviewed by the US EPA RfD/RfC Work Group and determined to be inadequate for derivation of an inhalation RfC. The verification status of this chemical is currently not verifiable. For additional information on health effects of this chemical, interested parties are referred to the EPA documentation listed below.

US EPA. 1986. Drinking Water Criteria Document for Hexachlorobenzene. Prepared by the Office of Health and Environmental Assessment, Environmental Criteria and Assessment Office, Cincinnati, OH for the Office of Drinking Water, Washington, DC. EPA-600/X-84-179-1. NTIS PB 86-117777.

US EPA. 1985. Health Assessment Document for Hexachlorobenzene. Prepared by the Office of Health and Environmental Assessment, Environmental Criteria and Assessment Office, Cincinnati, OH for the Office of Air Quality Planning and Standards, Research Triangle Park, NC. EPA/600/8-84/015F. NTIS PB 85-150332.

Agency Work Group Review -- 11/15/90

EPA Contacts:

Daniel J. Guth / NCEA -- (919)541-4930

Annie M. Jarabek / NCEA -- (919)541-4847

II. CARCINOGENICITY ASSESSMENT FOR LIFETIME EXPOSURE

Section II provides information on three aspects of the carcinogenic risk assessment for the agent in question; the US EPA classification, and quantitative estimates of risk from oral exposure and from inhalation exposure. The classification reflects a weight-of-evidence judgment of the likelihood that the agent is a human carcinogen. The quantitative risk estimates are presented in three ways. The slope factor is the result of application of a low-dose extrapolation procedure and is presented as the risk per (mg/kg)/day. The unit risk is the quantitative estimate in terms of either risk per ug/L drinking water or risk per ug/cu.m air breathed. The third form in which risk is presented is a drinking water or air concentration providing cancer risks of 1 in 10,000, 1 in 100,000 or 1 in 1,000,000. Background Document 2 (Service Code 5) provides details on the rationale and methods used to derive the carcinogenicity values found in IRIS. Users are referred to Section I for information on long-term toxic effects other than carcinogenicity.

II.A. EVIDENCE FOR CLASSIFICATION AS TO HUMAN CARCINOGENICITY

II.A.1. WEIGHT-OF-EVIDENCE CLASSIFICATION

Classification -- B2; probable human carcinogen.

Basis -- Hexachlorobenzene, when administered orally, has been shown to induce tumors in the liver, thyroid and kidney in three rodent species.

II.A.2. HUMAN CARCINOGENICITY DATA

Inadequate. The reported epidemiological studies of hexachlorobenzene have not been designed to measure increases in cancer incidence as an endpoint and are inadequate in this context.

Between 1954 and 1959 a large number of individuals in southeastern Turkey were exposed to hexachlorobenzene from the ingestion of seed grain treated with hexachlorobenzene as a fungicide. Approximately 5000 individuals developed adverse effects from this exposure with deaths of children under the age of 2 and about 4000 individuals developing porphyria. A follow-up of 204 patients 25-30 years after the onset of hexachlorobenzene-induced porphyria showed that a majority of the patients still showed symptoms of adverse effects (Cripps et al., 1984). One of these adverse effects was a 37% incidence of enlarged thyroids (thyromegaly), which is well above the average 5% incidence observed in southeastern Turkey. Two of the porphyria patients who underwent thyroidectomy showed no malignant changes. Clinical follow-up of these patients is continuing with emphasis in further evaluating the histopathology of thyromegaly patients.

II.A.3. ANIMAL CARCINOGENICITY DATA

Sufficient. The liver appears to be the primary target organ for hexachlorobenzene-induced cancer, although neoplasms of the thyroid and kidney have been observed as well.

Groups of 94 Sprague-Dawley rats/sex/dose were fed 0, 75, or 150 ppm hexachlorobenzene (purity >99.5%) in the diet for up to 2 years (Erturk et al., 1986). Interim kills of four rats/group were performed at weeks 0, 1, 2, 3, 4, 8, 16, 32, 48, 64, and 80. The remaining 50 animals/group were observed until natural death or until sacrifice at 2 years. Treated animals of both sexes surviving past 12 months showed significant increases in liver and renal tumors. Hemangiohepatomas, hepatocellular carcinomas and bile duct tumors were significantly increased in treated females; males and females in both dose groups had increased incidences of renal cell adenomas and hemangiohepatomas. Females were far more susceptible to hepatocarcinogenicity while males were generally more sensitive to renal carcinogenicity. The time-to-tumor onset in each dose group was generally longer than 1 year. The increase in hepatocellular carcinomas and bile duct tumors in males was not statistically significant. In this same study hepatomas were reported in Syrian golden hamsters that had been exposed for at least 90 days to 200 or 400 ppm hexachlorobenzene in the diet and killed after varying observation periods.

Groups of 30-60 Syrian golden hamsters/sex/dose were fed 0, 50, 100, or 200 ppm hexachlorobenzene (>99.5% pure) in the diet over lifetime (Cabral et al., 1977). After 50 weeks, survival in treated groups was comparable to controls; however, there was reduced lifespan among high-dose male and female animals after 70 weeks exposure. A significant dose-related increase in the incidence of hepatomas and liver hemangioendotheliomas was observed in males and in females. The incidence of hepatomas was statistically significantly increased in each treated group compared to controls while liver hemangioendothelioma incidence was statistically significantly elevated in the high-dose groups of both sexes and in middle-dose males. While thyroid alveolar adenomas were observed in all treated groups except low-dose males (none were observed in control groups), a significantly increased incidence was found only in high-dose males. There was a significant dose-related increase in the incidence of thyroid alveolar adenomas in males.

Smith and Cabral (1980) reported 100% incidence of liver tumors in small groups of female Agus (14) and Wistar (6) rats receiving 100 ppm hexachlorobenzene in arachis oil in the diet for 90 weeks compared to 0% in small groups of controls (12 Agus and 4 Wistar rats). In a 2-generation feeding study parental Sprague-Dawley rats were fed 0.32-40 ppm hexachlorobenzene in the diet for 3 months. Following mating, females were maintained on the diet through pregnancy and lactation. Pups received 0.32-40 ppm dietary hexachlorobenzene for 130 weeks. F1 females in the high-dose group had significant elevation in the incidence of neoplastic liver nodules (10/49 vs. 0/49 for controls) and adrenal pheochromocytomas (17/49 vs. 0/49 for controls), and F1 males showed increased parathyroid tumors (12/49 vs. 2/48 for controls) (Arnold et al., 1985).

Hepatomas were produced in a dose-related fashion in both male and female Swiss mice exposed through the diet to 50, 100, or 200 ppm hexachlorobenzene for up to 120 weeks (Cabral et al., 1979). The females in the high-dose group were observed to have a liver tumor incidence (14/41) significantly elevated over controls (0/49).

Short-term exposure to hexachlorobenzene did not significantly increase tumor incidence. Shorter exposures of Swiss mice (15 weeks) to 300 ppm hexachlorobenzene in the diet produced negligible incidences of liver tumors (1/26 female, 1/16 male) (Cabral et al., 1979). Lower doses of hexachlorobenzene (10 or 50 ppm) administered in the diet for 24 weeks did not result in increased liver tumor formation in ICR mice. There was hypertrophy of the centrilobular region, however, and 50 ppm hexachlorobenzene was found to enhance tumor induction and nodular hyperplasia in combination with 250 ppm polychlorinated terphenyl (Shirai et al., 1978).

II.A.4. SUPPORTING DATA FOR CARCINOGENICITY

Hexachlorobenzene is mutagenic for Saccharomyces cerevisiae (Guerzoni et al., 1976), but did not induce dominant lethal mutations in rats exposed by gavage (Simon et al., 1979); nor did it revert histidine auxotrophs of Salmonella typhimurium (Lawlor et al., 1979).

II.B. QUANTITATIVE ESTIMATE OF CARCINOGENIC RISK FROM ORAL EXPOSURE

II.B.1. SUMMARY OF RISK ESTIMATES

Oral Slope Factor -- 1.6 per (mg/kg)/day

Drinking Water Unit Risk -- 4.6E-5 per (ug/L)

Extrapolation Method -- Linearized multistage, extra risk

Drinking Water Concentrations at Specified Risk Levels:

Risk Level	Concentration
E-4 (1 in 10,000)	2 ug/L
E-5 (1 in 100,000)	2E-1 ug/L
E-6 (1 in 1,000,000)	2E-2 ug/L

II.B.2. DOSE-RESPONSE DATA (CARCINOGENICITY, ORAL EXPOSURE)

Tumor Type -- hepatocellular carcinoma
Test Animals -- rat/Sprague-Dawley, female
Route -- diet
Reference -- Erturk et al., 1986

Administered Dose (ppm)	Human Equivalent Dose (mg/kg)/day	Tumor Incidence
0	0	0/52
75	0.73	36/56
150	1.46	48/55

II.B.3. ADDITIONAL COMMENTS (CARCINOGENICITY, ORAL EXPOSURE)

Doses above are based on average food consumption and body weight data calculated by the study authors. Animals at risk (denominator) were those surviving until 12 months, the time of appearance of the first tumors. Note that this slope factor differs slightly from that reported in the Health Assessment Document (Lewis, 1989). This difference may result from rounding dose or body weight numbers differently and/or using a default rat body weight in the calculation of the human equivalent dose. A slope factor based on data from male hamsters was 1.7 per mg/kg/day (Cabral et al., 1977).

The unit risk should not be used if the water concentration exceeds 2E+2 ug/L, since above this concentration the slope factor may differ from that stated.

II.B.4. DISCUSSION OF CONFIDENCE (CARCINOGENICITY, ORAL EXPOSURE)

Significant increases in malignant tumors were observed among an adequate number of animals observed for their lifetime. Slope factors have been calculated from 14 different data sets encompassing 3 species, 4 studies and various endpoints. These fell within a range of approximately 1 order of magnitude (8.3E-2 to 1.7E+0).

II.C.   QUANTITATIVE ESTIMATE OF CARCINOGENIC RISK FROM INHALATION EXPOSURE

II.C.1. SUMMARY OF RISK ESTIMATES

Inhalation Unit Risk -- 4.6E-4 per (ug/cu.m)

Extrapolation Method -- Linearized multistage, extra risk

Air Concentrations at Specified Risk Levels:

Risk Level	Concentration
E-4 (1 in 10,000)	2E-1 ug/cu.m
E-5 (1 in 100,000)	2E-2 ug/cu.m
E-6 (1 in 1,000,000)	2E-3 ug/cu.m

II.C.2.   DOSE-RESPONSE DATA FOR CARCINOGENICITY, INHALATION EXPOSURE

Calculated from data in Section II.B.2.

II.C.3.   ADDITIONAL COMMENTS (CARCINOGENICITY, INHALATION EXPOSURE)

The unit risk should not be used if the air concentration exceeds 2E+1 ug/cu.m, since above this concentration the unit risk may not be appropriate.

II.C.4. DISCUSSION OF CONFIDENCE (CARCINOGENICITY, INHALATION EXPOSURE)

The inhalation risk estimate was based on data from oral exposure.

II.D. EPA DOCUMENTATION, REVIEW, AND CONTACTS (CARCINOGENICITY ASSESSMENT)

II.D.1. EPA DOCUMENTATION

Source Document -- US EPA, 1985, 1988

The values in the 1985 Health Assessment Document for Chlorinated Benzenes received extensive peer and public review.

The values in the 1988 Drinking Water Criteria Document for Hexachlorobenzene have received external and Agency review.

II.D.2. REVIEW (CARCINOGENICITY ASSESSMENT)

Agency Work Group Review -- 08/13/86, 02/03/88, 01/04/89, 03/01/89

Verification Date -- 03/01/89

II.D.3. US EPA CONTACTS (CARCINOGENICITY ASSESSMENT)

John Cicmanec / NCEA -- 513/569-7481

Robert E. McGaughy / NCEA -- 202/260-5889

III. HEALTH HAZARD ASSESSMENTS FOR VARIED EXPOSURE DURATIONS

III.A. DRINKING WATER HEALTH ADVISORIES

[Ed. Note: EPA does not have information yet in this section of IRIS].

IV. US EPA REGULATORY ACTIONS

EPA risk assessments may be updated as new data are published and as assessment methodologies evolve. Regulatory actions are frequently not updated at the same time. Compare the dates for the regulatory actions in this section with the verification dates for the risk assessments in sections I and II, as this may explain inconsistencies. Also note that some regulatory actions consider factors not related to health risk, such as technical or economic feasibility. Such considerations are indicated for each action. In addition, not all of the regulatory actions listed in this section involve enforceable federal standards. Please direct any questions you may have concerning these regulatory actions to the US EPA contact listed for that particular action. Users are strongly urged to read the background information on each regulatory action in Background Document 4 in Service Code 5.

IV.A. CLEAN AIR ACT (CAA)

[Ed. Note: EPA does not have information yet in this section of IRIS].

IV.B. SAFE DRINKING WATER ACT (SDWA)

IV.B.1. MAXIMUM CONTAMINANT LEVEL GOAL (MCLG) for Drinking Water

Value -- 0.0 mg/L (Proposed, 1990)

Considers technological or economic feasibility? -- NO

Discussion -- The proposed MCLG for hexachlorobenzene is zero based on the evidence of carcinogenic potential (classification B2).

Reference -- 55 FR 30370 (07/25/90)

EPA Contact -- Health and Ecological Criteria Division / OST / (202) 260-7571 / FTS 260-7571; or Safe Drinking Water Hotline / (800) 426-4791

IV.B.2. MAXIMUM CONTAMINANT LEVEL (MCL) for Drinking Water

Value -- 0.001 mg/L (Proposed, 1990)

Considers technological or economic feasibility? -- YES

Discussion -- The MCL is based on a PQL of 0.001 and is associated with a maximum lifetime individual risk of 0.5E-4.

Monitoring requirements -- Community and non-transient water system monitoring based on state vulnerability assessment; vulnerable systems to be monitored quarterly for one year; repeat monitoring dependent upon detection and system size.

Analytical methodology -- Microextraction/gas chromatography (EPA 505); electron-capture/gas chromatography (EPA 508); gas chromatography/mass spectrometry (EPA 525). PQL=0.001 mg/L.

Best available technology -- Granular activated carbon

Reference -- 55 FR 30370 (07/25/90)

EPA Contact -- Drinking Water Standards Division / OGWDW / (202) 260-7575 / FTS 260-7575; or Safe Drinking Water Hotline / (800) 426-4791

IV.B.3. SECONDARY MAXIMUM CONTAMINANT LEVEL (SMCL) for Drinking Water

[Ed. Note: EPA does not have information yet in this section of IRIS].

IV.B.4. REQUIRED MONITORING OF "UNREGULATED" CONTAMINANTS

Status -- Listed (Proposed, 1991)

Discussion -- "Unregulated" contaminants are those contaminants for which EPA establishes a monitoring requirement but which do not have an associated final MCLG, MCL, or treatment technique. EPA may regulate these contaminants in the future.

Monitoring requirement -- All systems to be monitored unless a vulnerability assessment determines the system is not vulnerable.

Analytical methodology -- Microextraction/gas chromatography (EPA 505); electron-capture/gas chromatography (EPA 508); gas chromatographic/mass spectrometry (EPA 525).

Reference -- 56 FR 3526 (01/30/91)

EPA Contact -- Drinking Water Standards Division / OGWDW / (202) 260-7575 / FTS 260-7575; or Safe Drinking Water Hotline / (800) 426-4791

IV.C. CLEAN WATER ACT (CWA)

IV.C.1. AMBIENT WATER QUALITY CRITERIA, Human Health

Water and Fish Consumption: 7.2E-4 ug/L

Fish Consumption Only: 7.4E-4 ug/L

Considers technological or economic feasibility? -- NO

Discussion -- For the maximum protection from the potential carcinogenic properties of this chemical, the ambient water concentration should be zero. Since zero, however, may not be attainable at this time, the recommended criteria represents an E-6 estimated incremental increase of cancer risk over a lifetime.

Reference -- 45 FR 79318 (11/28/80)

EPA Contact -- Criteria and Standards Division / OWRS (202)260-1315 / FTS 260-1315

IV.C.2. AMBIENT WATER QUALITY CRITERIA, Aquatic Organisms

Freshwater:
  Acute -- 6E+0 ug/L
  Chronic -- 3.68E+0 ug/L

Marine:
  Acute -- None
  Chronic -- None

Considers technological or economic feasibility? -- NO

Discussion -- Proposed criterion were derived from a minimum database consisting of acute and chronic tests on a variety of species. Requirements and methods are covered in the reference to the Federal Register.

Reference -- 55 FR 19986 (05/14/90)

EPA Contact -- Criteria and Standards Division / OWRS (202)260-1315 / FTS 260-1315

IV.D. FEDERAL INSECTICIDE, FUNGICIDE, AND RODENTICIDE ACT (FIFRA)

[Ed. Note: EPA does not have information yet in this section of IRIS].

IV.E. TOXIC SUBSTANCES CONTROL ACT (TSCA)

[Ed. Note: EPA does not have information yet in this section of IRIS].

IV.F. RESOURCE CONSERVATION AND RECOVERY ACT (RCRA)

IV.F.1. RCRA APPENDIX IX, for Ground Water Monitoring

Status -- Listed

Reference -- 52 FR 25942 (07/09/87)

EPA Contact -- RCRA/Superfund Hotline (800)424-9346 / (202)260-3000 / FTS 260-3000

IV.G. SUPERFUND (CERCLA)

IV.G.1. REPORTABLE QUANTITY (RQ) for Release into the Environment

Value (status) -- 10 pounds (Final, 1989)

Considers technological or economic feasibility? -- NO

Discussion -- The final RQ for hexachlorobenzene is based on potential carcinogenicity. Available data indicate a hazard ranking of medium and a weight of evidence classification of Group B2, which corresponds to an RQ of 10 pounds.

Reference -- 54 FR 33418 (08/14/89)

EPA Contact -- RCRA/Superfund Hotline (800)424-9346 / (202)260-3000 / FTS 260-3000

VI. BIBLIOGRAPHY

VI.A. ORAL RfD REFERENCES

Arnold, D.L., C.A. Moodie, S.M. Charbonneau, et al. 1985. Long-term toxicity of hexachlorobenzene in the rat and the effect of dietary Vitamin A. Food. Chem. Toxicol. 23(9): 779-793.

Cam, C. and C. Nigogosyan. 1963. Acquired toxic porphyria cutanea tarda due to hexachlorobenzene. Report of 348 cases caused by this fungicide. J. Am. Med. Assoc. 183: 88-91.

Kuiper-Goodman, T., D.L. Grant, C.A. Moodie, G.O. Korsrud and I.C. Munro. 1977. Subacute toxicity of hexachlorobenzene in the rat. Toxicol. Appl. Pharmacol. 40: 529-549.

Peters, H.A., S.A.M. Johnson, S. Cam, Y. M:uft:u, S. Oral and T. Ergene. 1966. Hexachlorobenzene-induced porphyria: Effect of chelation of the disease, porphyria, and metal metabolism. Am. J. Med. Sci. 251(3): 314-322.

Peters, H.A., A. Gocmen, D.J. Cripps, G.T. Bryan and I. Dobramaci. 1982. Epidemiology of hexachlorobenzene-induced porphyria in Turkey. Clinical and laboratory follow-up after 25 years. Arch. Neurol. 39(12): 744-749.

US EPA. 1985. Health Assessment Document for Chlorinated Benzenes. Prepared by the Environmental Criteria and Assessment Office, Cincinnati, OH for the Office of Air Quality, Planning and Standards, Washington, DC. EPA 600/8-84-015F. NTIS PB 85-150332.

US EPA. 1988. Drinking Water Criteria Document for Hexachlorobenzene. Prepared by the Environmental Criteria and Assessment Office, Cincinnati, OH for the Office of Drinking Water, Washington, DC.

VI.B. INHALATION RfD REFERENCES

US EPA. 1986. Drinking Water Criteria Document for Hexachlorobenzene. Prepared by the Office of Health and Environmental Assessment, Environmental Criteria and Assessment Office, Cincinnati, OH for the Office of Drinking Water, Washington, DC. EPA-600/X-84-179-1. NTIS PB 86-117777.

US EPA. 1985. Health Assessment Document for Chlorinated Benzenes. Prepared by the Office of Health and Environmental Assessment, Environmental Criteria and Assessment

Office, Cincinnati, OH for the Office of Air Quality Planning and Standards, Research Triangle Park, NC. EPA/600/8-84/015F. NTIS PB 85-150332

VI.C. CARCINOGENICITY ASSESSMENT REFERENCES

Arnold, D.L., C.A. Moodie, S.M. Charbonneau, et al. 1985. Long-term toxicity of hexachlorobenzene in the rat and the effect of dietary vitamin A. Food Chem. Toxicol. 23(9): 779-793.

Cabral, J.R.P., P. Shubik, T. Mollner and F. Raitano. 1977. Carcinogenic activity of hexachlorobenzene in hamsters. Nature. 269: 510-511.

Cabral, J.R.P., T. Mollner, F. Raitano and P. Shubik. 1979. Carcinogenesis of hexachlorobenzene in mice. Int. J. Cancer. 23(1): 47-51.

Cripps, D.J., H.A. Peters, A. Gocmen and I. Dogramaci. 1984. Porphyria turcica due to hexachlorobenzene: A 20- to 30-year follow-up study on 204 patients. Br. J. Dermatol. 111: 413-422.

Erturk, E., R.W. Lambrecht, H.A. Peters, D.J. Cripps, A. Gocmen, C.R. Morris and G.T. Bryan. 1986. Oncogenicity of hexachlorobenzene. In: Hexachlorobenzene: Proc. Int. Symp., C.R. Morris and J.R.P. Cabral, Ed. IARC Scientific Publ. No. 77, Oxford University Press, Oxford. p. 417-423.

Guerzoni, M.E., L. Del Cupolo and I. Ponti. 1976. Mutagenic activity of pesticides (attivita mutagenica delgi antiparassitari). Riv. Sci. Tecn. Alim. Nutri. Um. 6: 161-165.

Lawlor, T., S.R. Haworth and P. Voytek. 1979. Evaluation of the genetic activity of nine chlorinated phenols, seven chlorinated benzenes, and three chlorinated hexanes. Environ. Mutagen. 1: 143. (Abstr.)

Lewis, J. 1989. Syracuse Research Corperation, Cincinnati, OH. Letter to R. Schoeny, US EPA. March 22.

Shirai, T., Y. Miyata, K. Nakanishi, G. Murasaki and N. Ito. 1978. Hepatocarcinogenicity of polychlorinated terphenyl (PCT) in ICR mice and its enhancement by hexachlorobenzene (HCB). Cancer Lett. 4(5): 271-175.

Simon, G.S., R.G. Tardiff and J.F. Borzelleca. 1979. Failure of hexachlorobenzene to induce dominant lethal mutations in the rat. Toxicol. Appl. Pharmacol. 47: 415-419.

Smith, A.G. and J.R. Cabral. 1980. Liver-cell tumours in rats fed hexachlorobenzene. Cancer Lett. 11(2): 169-172.

US EPA. 1985. Health Assessment Document for Chlorinated Benzenes. Prepared by the Office of Health and Environmental Assessment, Environmental Criteria and Assessment Office, Cincinnati, OH for the Office of Air Quality Planning and Standards, Washington, DC. EPA/600/8-84-015F. NTIS PB 85-150332.

US EPA. 1988. Drinking Water Criteria Document for Hexachlorobenzene. Prepared by the Office of Health and Environmental Assessment, Environmental Criteria and Assessment Office, Cincinnati, OH for the Office of Drinking Water, Washington, DC. EPA-600/X-84-179-1. NTIS PB 86-117777.

VI.D. DRINKING WATER HA REFERENCES

None

VII. REVISION HISTORY

```
--------  --------  ----------------------------------------------------------
Date      Section   Description
--------  --------  ----------------------------------------------------------
09/26/88  I.A.      Oral RfD summary on-line
02/01/89  I.A.7.    Secondary contact's phone number corrected
05/01/89  II.       Carcinogen assessment now under review
06/01/90  VI.       Bibliography on-line
12/01/90  I.B.      Inhalation RfC now under review
03/01/91  I.A.7.    Secondary contact changed
03/01/91  I.B.      Inhalation RfC message on-line
03/01/91  II.       Carcinogenicity assessment on-line
03/01/91  VI.       Inhalation RfC and Carcinogenicity references added
04/01/91  I.A.      Text edited
01/01/92  IV.       Regulatory Action section on-line
04/01/92  II.D.2.   Review statement clarified
02/01/94  II.D.3.   Secondary contact's phone number changed
02/01/94  IV.C.2.   Freshwater: Acute value changed
```

SYNONYMS

granox	pentachlorophenyl chloride
Hexachlorobenzene	perchlorobenzene

Source: *Instant EPA's IRIS*, 1997, Instant Reference Sources, Inc., 7605 Rockpoint Drive, Austin, TX 78731 (Fax: 512-345-2386; Internet URL, http://www.instantref.com/inst-ref.htm).

Production, Use, and Pesticide Labeling Information

Hexachlorobenzene is not currently listed in EPA's Pesticide Factsheet database and there is no indication of whether it will be added or not in the future.

Indeno[1,2,3-c,d]pyrene

Introduction

Indeno[1,2,3-cd]pyrene is formed in most combustion or elevated temperature processes that involve compounds containing carbon and hydrogen. Known sources include coal, wood, and gasoline combustion, municipal waste incineration, coke ovens and cigarette smoke. It has also been found in gasoline, fresh and used motor oil, and road runoff. [828] Indeno[1,2,3-c,d]pyrene is a component of petroleum and coal products. It is a polycyclic aromatic hydrocarbon (PAH) found extensively in the environment, especially when associated with particulate matter from smoke and incomplete products of combustion.

When released to soil it will sorb strongly (estimated Koc = 20,146) and hence is not expected to leach. No information was found about volatilization from, hydrolysis in, or biodegradation in soil. When released to water it will sorb strongly to suspended particulate matter, biota and sediments. Although there is a high potential for indeno[1,2,3-cd]pyrene to bioconcentrate in most aquatic organisms, it may not bioconcentrate in fish since fish contain microsomal oxidase, which allows polyaromatic hydrocarbons to be metabolized. No information was found about indeno[1,2,3-cd]pyrene volatilization, photolysis, hydrolysis, or biodegradation in water. It will probably be persistent in the aquatic environment and concentrate in sediments. Almost all indeno[1,2,3-cd]pyrene released to the atmosphere will be sorbed to particulate matter; thus its atmospheric fate will primarily depend on physical processes such as dry and wet deposition. However, a computer-estimated half-life for indeno[1,2,3-cd]pyrene in the vapor phase is about 20 hours due to reaction with photochemically produced hydroxyl radicals. [828]

It has been found in rain, drinking water, groundwater, surface waters, treated industrial wastewaters, marine and freshwater sediments, suspended sediments, automobile exhaust, ambient air, foods (cereals, cooking oils, barley malt), powdered milk, infant formula, seafoods, and sewage sludge. The primary route of human exposure to indeno[1,2,3-cd]pyrene will probably be through ingestion of contaminated food. Other exposure to indeno[1,2,3-cd]pyrene may be from drinking water and breathing air that is contaminated with indeno[1,2,3-cd]pyrene. [828]

Identifying Information

CAS NUMBER: 193-39-5

NIOSH Registry Number: UR2625000

SYNONYMS: 2,3-O-PHENYLENEPYRENE, O-PHENYLENEPYRENE, INDENO(1,2,3-C,D)PYRENE, INDENO(1,2,3-C,D) PYRENE, ORTHO-PHENYLENEPYRENE, 2,3-ORTHO-PHENYLENEPYRENE

Endocrine Disruptor Information

Indeno[1,2,3-cd]pyrene is on EPA's list of suspected EEDs [811] and it is an aromatic coplanar compound. However, no additional information has been located with respect to its endocrine modifying properties.

Chemical and Physical Properties

CHEMICAL FORMULA: $C_{22}H_{12}$

MOLECULAR WEIGHT: 276.34

WLN: L E6 C5666 B6 3ABC VJ

PHYSICAL DESCRIPTION: white crystalline solid

SPECIFIC GRAVITY: Not available

DENSITY: Not available

MELTING POINT: 160-163°C [822]

BOILING POINT: 536°C [822]

SOLUBILITY: Water: 62 ug/L [816], [877]

Environmental Reference Materials

A source of EED environmental reference materials for indeno[1,2,3-c,d]pyrene is Radian International LLC, Austin, TX (Fax 512-454-0268; Internet URL, http://www.radian.com/standards); Catalog No. ERI-016.

Hazardous Properties

ACUTE/CHRONIC HAZARDS: Similar polynuclear aromatic hydrocarbons have produced tumors in animal tests and have caused mild irritation to skin and other exposed tissues [822].

STABILITY: This compound is stable under normal laboratory conditions [822].

Medical Symptoms of Exposure

Symptoms of exposure to indeno[1,2,3-cd]pyrene have not been found. They may be expected to be similar to symptoms recorded for some of the other PAHs such as anthracene, phenantharene and pyrene.

Toxicological Information

The toxicological information below is gathered from several sources including two national databases: one from the *National Toxicology Program Chemical Repository Database* and another from *EPA's Integrated Risk Information System (IRIS)*.

Toxicology Information from the National Toxicology Program

CARCINOGENICITY:

Tumorigenic Data:

Type of Effect	Route/Animal	Amount Dosed (Notes
TDLo:	scu-mus	72 mg/kg/9W-I

Review:
IARC Cancer Review: Animal Sufficient Evidence
IARC possible human carcinogen (Group 2B)

Status:
EPA Carcinogen Assessment Group
NTP anticipated human carcinogen

TERATOGENICITY: Not available

MUTAGENICITY:

Test	Lowest Dose
mma-sat	3 ug/plate/48H

TOXICITY: Not available

OTHER TOXICITY DATA:

Review:
Toxicology Review

Status:
Reported in EPA TSCA Inventory, 1980
EPA TSCA 8(a) Preliminary Assessment Information Proposed Rule

SAX TOXICITY EVALUATION:

   THR: An experimental (+) carcinogen.

Source: *Instant Tox-Base*, 1995, Instant Reference Sources, Inc., 7605 Rockpoint Drive, Austin, TX 78731 (Fax: 512-345-2386; Internet URL, http://www.instantref.com/inst-ref.htm).

Toxicology Information from EPA's Integrated Risk Information System (IRIS)

I. CHRONIC HEALTH HAZARD ASSESSMENTS FOR NONCARCINOGENIC EFFECTS

I.A. REFERENCE DOSE FOR CHRONIC ORAL EXPOSURE (RfD)

   [Ed. Note: EPA does not have information yet in this section of IRIS].

I.B. REFERENCE CONCENTRATION FOR CHRONIC INHALATION EXPOSURE (RfC)

   [Ed. Note: EPA does not have information yet in this section of IRIS].

II. CARCINOGENICITY ASSESSMENT FOR LIFETIME EXPOSURE

Section II provides information on three aspects of the carcinogenic risk assessment for the agent in question; the US EPA classification, and quantitative estimates of risk from oral exposure and from inhalation exposure. The classification reflects a weight-of-evidence judgment of the likelihood that the agent is a human carcinogen. The quantitative risk estimates are presented in three ways. The slope factor is the result of application of a low-dose extrapolation procedure and is presented as the risk per (mg/kg)/day. The unit risk is the quantitative estimate in terms of either risk per ug/L drinking water or risk per ug/cu.m air breathed. The third form in which risk is presented is a drinking water or air concentration providing cancer risks of 1 in 10,000, 1 in 100,000 or 1 in 1,000,000. Background Document 2 (Service Code 5) provides details on the rationale and methods used to derive the carcinogenicity values found in IRIS. Users are referred to Section I for information on long-term toxic effects other than carcinogenicity.

II.A. EVIDENCE FOR CLASSIFICATION AS TO HUMAN CARCINOGENICITY

II.A.1. WEIGHT-OF-EVIDENCE CLASSIFICATION

Classification -- B2, probable human carcinogen

Basis -- Based on no human data and sufficient data from animal bioassays. Indeno[1,2,3-cd]pyrene produced tumors in mice following lung implants, subcutaneous injection and dermal exposure. Indeno[1,2,3-cd]pyrene tested positive in bacterial gene mutation assays.

II.A.2. HUMAN CARCINOGENICITY DATA

None. Although there are no human data that specifically link exposure to indeno[1,2,3-cd]pyrene to human cancers, indeno[1,2,3-cd]pyrene is a component of mixtures that have been associated with human cancer. These include coal tar, soots, coke oven emissions and cigarette smoke (US EPA, 1984, 1990; IARC, 1984).

II.A.3. ANIMAL CARCINOGENICITY DATA

Sufficient. In carcinogen bioassays indeno[1,2,3-cd]pyrene exposure resulted in increased incidences of epidermoid carcinomas in a lung implantation study (Deutsch-Wenzel et al., 1983), injection site sarcomas in a subcutaneous injection assay (Lacassagne et al., 1963) and skin tumors in dermal application studies (Hoffman and Wynder, 1966; Rice et al., 1985a, 1986).

In a lifetime implant study, 3-month-old female Osborne-Mendel rats (35/group) received lung implants of indeno[1,2,3-cd]pyrene in 0.05 mL of a 1:1 (v:v) mixture of beeswax and trioctanoin (Deutsch-Wenzel et al., 1983). Rats received either 0.16 mg (0.65 mg/kg), 0.83 mg (3.4 mg/kg) or 4.15 mg (17 mg/kg) indeno[1,2,3-cd]pyrene. Controls consisted of an untreated group and a group receiving an implant of the vehicle. Median survival times in weeks were as follows: untreated controls, 118; vehicle controls, 104; low-dose, 116; mid-dose, 109; and high-dose, 92. Incidence of epidermoid carcinomas in the lung and thorax (combined) showed a statistically significant dose-related increase. The incidences were: untreated controls, 0/35; vehicle controls, 0/35; low-dose, 4/35 (11%); mid-dose, 8/35 (23%); and high-dose, 21/35 (60%).

Groups of male and female CD-1 mice (n=32) received intraperitoneal injections of indeno[1,2,3-cd]pyrene in dimethyl sulfoxide (DMSO) on days 1, 8 and 15 after birth (total dose = 580 ug/mouse) and were evaluated for tumors upon sacrifice at 52 weeks of age (LaVoie et al., 1987). One male mouse (1/11) developed a lung adenoma, no tumors occurred in female mice. Tumor incidence was not significantly different from vehicle controls. This test is considered to be a short-term lung tumor assay.

In mouse skin painting assays, indeno[1,2,3-cd]pyrene tested positive for cancer-initiating activity in several mouse strains (Hoffmann and Wynder, 1966; Rice et al., 1985a, 1986). In the Hoffmann and Wynder (1966) study female Swiss albino Ha/ICR/Mil mice (20/group) were given topical applications of indeno[1,2,3-cd]pyrene prepared as dioxane (at 0.05 and 0.1%) or in acetone solutions (at 0.01, 0.05 and 0.1%). Dioxane preparations did not induce skin tumors. By contrast, acetone solutions of indeno[1,2,3-cd]pyrene produced skin tumors in a dose-related fashion. No tumors were observed in animals painted with 0.01 or 0.05% indeno[1,2,3-cd]pyrene in acetone; 0.1% induced six papillomas and three carcinomas beginning at 9 months; and 0.5% resulted in seven papillomas and five carcinomas with the first tumor appearing at 3 months. The authors also reported that a total dose of 250 mg indeno[1,2,3-cd]pyrene delivered in 10 applications in 2 days was a sufficient initiating dose when followed by promotion with croton oil.

To examine the initiating capability of the compound's major metabolites in mouse skin, indeno[1,2,3-cd]pyrene was applied to the shaved backs of 20 Crl:CD-1(ICR)BR female mice (Rice et al., 1986). Acetone solutions were applied every other day for 10 days for a total initiating dose of 1 mg indeno[1,2,3-cd]pyrene. This was followed 10 days later by applications of the promotor tetradecanoylphorbol (TPA) (0.0025% in 100 mL acetone) 3 times/week for 20 weeks. Tumor incidence was essentially 100%. Indeno[1,2,3-cd]pyrene-1,2-diol and -1,2-oxide treatment both resulted in 80% tumor incidence in contrast to 8-hydroxy- and acetone-treated controls (approximately 25 and 5%, respectively).

An earlier initiation-promotion bioassay performed by Rice et al. (1985a) showed a pronounced dose-response relationship for tumors. Following the same protocol described above, an 80% tumor incidence was observed in mice receiving a total initiating dose of 1 mg indeno[1,2,3-cd]pyrene with an average of about four tumors/mouse after 22 weeks of promotion. However, when the total initiating dose was decreased to 100 or 300 mg/mouse, the number of tumor-bearing mice was not significantly increased.

Injection site sarcomas were reported in 10/14 male and 1/14 female XVIIc/Z mice administered 3 injections at 1-month intervals of 0.6 mg indeno[1,2,3-cd]pyrene. No concurrent controls appear to have been run in this experiment; the authors report, however, that in this mouse strain no spontaneous subcutaneous tumors have been reported (Lacassagne et al., 1963).

II.A.4. SUPPORTING DATA FOR CARCINOGENICITY

Indeno[1,2,3-cd]pyrene produced positive results in reverse mutation assays in Salmonella typhimurium strains TA100 and TA98 (2-3 ug/plate) (LaVoie et al., 1979; Hermann et al., 1980; Rice et al., 1985b).

II.B. QUANTITATIVE ESTIMATE OF CARCINOGENIC RISK FROM ORAL EXPOSURE

[Ed. Note: EPA does not have information yet in this section of IRIS].

II.C. QUANTITATIVE ESTIMATE OF CARCINOGENIC RISK FROM INHALATION EXPOSURE

[Ed. Note: EPA does not have information yet in this section of IRIS].

II.D. EPA DOCUMENTATION, REVIEW, AND CONTACTS (CARCINOGENICITY ASSESSMENT)

II.D.1. EPA DOCUMENTATION

Source Document -- US EPA, 1984, 1990

The 1990 Drinking Water Criteria Document for Polycyclic Aromatic Hydrocarbons has received Agency and external review.

II.D.2. REVIEW (CARCINOGENICITY ASSESSMENT)

Agency Work Group Review -- 02/07/90, 08/05/93, 09/21/93, 02/02/94

Verification Date -- 02/07/90

II.D.3. US EPA CONTACTS (CARCINOGENICITY ASSESSMENT)

Rita S. Schoeny / NCEA -- (513)569-7544

Robert E. McGaughy / NCEA -- (202)260-5889

III. HEALTH HAZARD ASSESSMENTS FOR VARIED EXPOSURE DURATIONS

III.A. DRINKING WATER HEALTH ADVISORIES

[Ed. Note: EPA does not have information yet in this section of IRIS].

IV. US EPA REGULATORY ACTIONS

EPA risk assessments may be updated as new data are published and as assessment methodologies evolve. Regulatory actions are frequently not updated at the same time. Compare the dates for the regulatory actions in this section with the verification dates for the risk assessments in sections I and II, as this may explain inconsistencies. Also note that some regulatory actions consider factors not related to health risk, such as technical or economic feasibility. Such considerations are indicated for each action. In addition, not all of the regulatory actions listed in this section involve enforceable federal standards. Please direct any questions you may have concerning these regulatory actions to the US EPA contact listed for that particular action. Users are strongly urged to read the background information on each regulatory action in Background Document 4 in Service Code 5.

IV.A. CLEAN AIR ACT (CAA)

[Ed. Note: EPA does not have information yet in this section of IRIS].

IV.B. SAFE DRINKING WATER ACT (SDWA)

IV.B.1. MAXIMUM CONTAMINANT LEVEL GOAL (MCLG) for Drinking Water

Value -- 0 mg/L (Proposed, 1990)

Considers technological or economic feasibility? -- NO

Discussion -- The proposed MCLG is zero. This value is based on carcinogenic PAH's as a class.

Reference -- 55 FR 30370 (07/25/90)

EPA Contact -- Health and Ecological Criteria Division / OST /(202) 260-7571 / FTS 260-7571; or Safe Drinking Water Hotline / (800) 426-4791

IV.B.2. MAXIMUM CONTAMINANT LEVEL (MCL) for Drinking Water

Value -- 0.0002 mg/L (Proposed, 1990)

Considers technological or economic feasibility? -- YES

Discussion -- The proposed MCL is equal to the PQL and is associated with a maximum lifetime individual risk of 1 E-4.

Monitoring requirements -- Community and non-transient water system monitoring based on state vulnerability assessment; vulnerable systems to be monitored quarterly for one year; repeat monitoring dependent upon detection and size of system.

Analytical methodology -- High pressure liquid chromatography (EPA 550, 550.1); gas chromatographic/mass spectrometry (EPA 525): PQL= 0.0002 mg/L.

Best available technology -- Granular activated carbon.

Reference -- 55 FR 30370 (07/25/90)

EPA Contact -- Drinking Water Standards Division / OGWDW /(202) 260-7575 / FTS 260-7575; or Safe Drinking Water Hotline / (800) 426-4791

IV.B.3. SECONDARY MAXIMUM CONTAMINANT LEVEL (SMCL) for Drinking Water

[Ed. Note: EPA does not have information yet in this section of IRIS].

IV.B.4. REQUIRED MONITORING OF "UNREGULATED" CONTAMINANTS

Status -- Listed (Proposed, 1991)

Discussion -- "Unregulated" contaminants are those contaminants for which EPA establishes a monitoring requirement but which do not have an associated final MCLG, MCL, or treatment technique. EPA may regulate these contaminants in the future.

Monitoring requirement -- All systems to be monitored unless a vulnerability assessment determines the system is not vulnerable.

Analytical methodology -- Gas chromatography/mass spectrometry (EPA 525); high pressure liquid chromatography (EPA 550, 550.1).

Reference -- 56 FR 3526 (01/30/91)

EPA Contact -- Drinking Water Standards Division / OGWDW /(202) 260-7575 / FTS 260-7575; or Safe Drinking Water Hotline / (800) 426-4791

IV.C. CLEAN WATER ACT (CWA)

IV.C.1. AMBIENT WATER QUALITY CRITERIA, Human Health

Water and Fish Consumption: 2.8E-3 ug/L

Fish Consumption Only: 3.11E-2 ug/L

Considers technological or economic feasibility? -- NO

Discussion -- For the maximum protection from the potential carcinogenic properties of this chemical, the ambient water concentration should be zero. However, zero may not be obtainable at this time, so the recommended criteria presents a E-6 estimated incremental increase of cancer over a lifetime. The values given represent polynuclear aromatic hydrocarbons as a class.   Reference -- 45 FR 79318 (11/28/80)

EPA Contact -- Criteria and Standards Division / OWRS(202)260-1315 / FTS 260-1315

IV.C.2. AMBIENT WATER QUALITY CRITERIA, Aquatic Organisms

Freshwater:
    Acute LEC --   none
    Chronic LEC --   none

Marine:
    Acute LEC --   3.0E+2 ug/L
    Chronic LEC -- none

Considers technological or economic feasibility? -- NO

Discussion -- The values that are indicated as "LEC" are not criteria, but are the lowest effect levels found in the literature. LEC's are given when the minimum data required to derive water quality criteria are not available. The values given represent polynuclear aromatic hydrocarbons as a class.

Reference -- 45 FR 79318 (11/28/80)

EPA Contact -- Criteria and Standards Division / OWRS(202)260-1315 / FTS 260-1315

IV.D. FEDERAL INSECTICIDE, FUNGICIDE, AND RODENTICIDE ACT (FIFRA)

[Ed. Note: EPA does not have information yet in this section of IRIS].

IV.E. TOXIC SUBSTANCES CONTROL ACT (TSCA)

[Ed. Note: EPA does not have information yet in this section of IRIS].

IV.F. RESOURCE CONSERVATION AND RECOVERY ACT (RCRA)

IV.F.1. RCRA APPENDIX IX, for Ground Water Monitoring

Status -- Listed

Reference -- 52 FR 25942 (07/09/87)

EPA Contact -- RCRA/Superfund Hotline(800)424-9346 / (202)260-3000 / FTS 260-3000

IV.G. SUPERFUND (CERCLA)

IV.G.1. REPORTABLE QUANTITY (RQ) for Release into the Environment

Value (status) -- 100 pounds (Final, 1989)

Considers technological or economic feasibility? -- NO

Discussion --   The final RQ for indeno[1,2,3-cd]pyrene is based on potential carcinogenicity. Available data indicate a hazard ranking of low and a weight of evidence classification of Group B2, which corresponds to an RQ of 100 pounds.

Reference -- 54 FR 33418 (08/14/89)

EPA Contact -- RCRA/Superfund Hotline(800)424-9346 / (202)260-3000 / FTS 260-3000

VI. BIBLIOGRAPHY

VI.A. ORAL RfD REFERENCES

None

VI.B. INHALATION RfC REFERENCES

None

VI.C. CARCINOGENICITY ASSESSMENT REFERENCES

Deutsch-Wenzel, R., H. Brune, G. Grimmer, G. Dettbarn and J. Misfeld. 1983. Experimental studies in rat lungs on the carcinogenicity and dose-response relationships of eight frequently occurring environmental polycyclic aromatic hydrocarbons. J. Natl. Cancer Inst. 71(3): 539-544.

Hermann, M., J.P. Durand, J.M. Charpentier, et al. 1980. Correlations of mutagenic activity with polynuclear aromatic hydrocarbon content of various mineral oils. In: Polynuclear Aromatic Hydrocarbons: Chemisty and Biological Effects, 4th Int. Symp., A. Bjorseth and A.J. Dennis, Ed. Battelle Press, Columbus, OH. p. 899-916.

Hoffmann, D. and E.L. Wynder, 1966. Beitrag zur carcinogen Wirkung von Dibenzopyrene. Z. Krebsforsch. 68(2): 137-149. (Ger.) Contribution on the carcinogenic effect of dibenzopyrenes.

IARC (International Agency for Research on Cancer). 1984. Monographs on the Evaluation of the Carcinogenic Risk of the Chemical to Man. Polynuclear Aromatic Hydrocarbons. Part 3. Industrial Exposures in Aluminum Production, Coal Gasification, Coke Production, and Iron and Steel Founding. Vol. 34. World Health Organization.

Lacassagne, A., N.P. Buu-Hoi, F. Zajdela, D. Lavit-Lamy and O. Chalvet. 1963. Activite cancerogene d'hyrocarbures aromatiques polycycliques a noyau flouranthene. Un. Int. Cancer Acta. 19(3-4): 490-496. (Fre.)

LaVoie, E.J., E.V. Bedenko, N. Hirota, S.S. Hecht and D. Hoffmann. 1979. A comparison of the mutagenicity, tumor-initiating activity and complete carcinogenicity of polynuclear aromatic hydrocarbons. In: Polynuclear Aromatic Hydrocarbons, P.W. Jones and P. Leber, Ed. Ann Arbor Science Publishers, Ann Arbor, MI. p. 705-721.

LaVoie, E.J., J. Braley, J.E. Rice and A. Rivenson. 1987. Tumorigenic activity for non-alternant polynuclear aromatic hydrocarbons in newborn mice. Cancer Lett. 34: 15-20.

Rice, J.E., D.T. Coleman, T.J. Hosted, E.J. LaVoie, D.J. McCaustland and J.C. Wiley. 1985a. On the metabolism, mutagenicity, and tumor-initiating activity of indeno[1,2,3-cd]pyrene. In: Polynuclear Aromatic Hydrocarbons: Mechanism, Methods and Metabolism, M. Cooke and A.J. Dennis, Ed. Batelle Press, Columbus, OH. p. 1097-1109.

Rice, J.E., D.T. Coleman, T.J. Hosted, Jr., E.J. LaVoie, D.J. McClausland and J.C. Wiley, Jr. 1985b. Identification of mutagenic metabolites in indeno[1,2,3-cd]pyrene formed in vitro with rat liver enzymes. Cancer Res. 45: 5421-5425.

Rice, J.E., T.J. Hosted, Jr., M.C. DeFloria, E.J. LaVoie, D.L. Fischer and J.C. Wiley, Jr. 1986. Tumor-initiating activity of major in vivo metabolites of indeno[1,2,3-cd]pyrene on mouse skin. Carcinogenesis. 7(10): 1761-1764.

US EPA. 1984. Carcinogen Assessment of Coke Oven Emissions. Office of Health and Environmental Assessment, Washington, DC. EPA 600/6-82-003F. NTIS PB 84-170181.

US EPA. 1990. Drinking Water Criteria Document for Polycyclic Aromatic Hydrocarbons (PAHs). Prepared by the Office of Health and Environmental Assessment, Environmental Criteria and Assessment Office, Cincinnati, OH for the Office of Drinking Water, Washington, DC. Final Draft. ECAO-CIN-D010, September, 1990.

VI.D. DRINKING WATER HA REFERENCES

None
VII. REVISION HISTORY

Date	Section	Description
12/01/90	II.	Carcinogen assessment on-line
12/01/90	VI.	Bibliography on-line
01/01/92	IV.	Regulatory Action section on-line
09/01/93	II.	Carcinogenicity assessment noted as pending change
09/01/93	II.D.2.	Work group review date added
11/01/93	II.D.2.	Work group review date added
03/01/94	II.	Pending change note removed; no change
03/01/94	II.D.2.	Work group review date added

SYNONYMS

Indeno(1,2,3-cd)pyrene
HSDB 5101
indeno(1,2,3-cd)pyrene
o-PHENYLENEPYRENE
RCRA WASTE NUMBER U137

1,10-(O-PHENYLENE)PYRENE
1,10-(1,2-Phenylene)pyrene
2,3-o-PHENYLENEPYRENE
2,3-PHENYLENEPYRENE

Source: *Instant EPA's IRIS*, 1997, Instant Reference Sources, Inc., 7605 Rockpoint Drive, Austin, TX 78731 (Fax: 512-345-2386; Internet URL, http://www.instantref.com/inst-ref.htm).

Lead

Pb

Introduction

Lead is very resistant to corrosion and is used in containers for corrosive liquids. Its alloys include solder, type metal, and various antifriction metals. Great quantities of lead and the dioxide are used in storage batteries. Much metal also goes into cable covering, plumbing, ammunition, and lead tetraethyl. The metal is used as a radiation and to absorb vibration. Lead oxide is used in producing fine "crystal glass" and "flint glass". The uses of lead in paints and gasoline, and the use of lead salts as insecticides are being drastically curtailed or eliminated due to health and environmental concerns. [706]

Identifying Information

CAS NUMBER: 7439-92-1

NIOSH Registry Number: OF7525000

SYNONYMS: PLUMBUM

Endocrine Disruptor Information

Lead is a strongly suspected EED that it is listed by EPA [811], the Centers for Disease Control & Prevention [812], and the World Wildlife Fund [813] as a potential endocrine modifying chemical. According to EPA [811] evidence has been obtained from rats, which includes Pb exposure *in utero*, "suggesting a dual site of lead action: (a) at the level of the hypothalamic pituitary unit, and (b) directly at the level of gonadal steroid biosynthesis." Blood-serum testosterone levels (in males) and estradiol levels (in females) were most severely suppressed in those animals where exposure began *in utero*. Both sexes in this group showed a 25% suppression in prepubertal growth (as measured by weight). [868]

Chemical and Physical Properties

ATOMIC FORMULA: Pb

ATOMIC WEIGHT: 207.19

PHYSICAL DESCRIPTION: Silver-bluish white soft solid

SPECIFIC GRAVITY: Not available

DENSITY: 11.3437 g/mL @ 16°C

MELTING POINT: 327.5°C

670

BOILING POINT: 1740°C

SOLUBILITY: Water: Insoluble; DMSO: Insoluble; 95% Ethanol: Insoluble; Acetone: Insoluble

Source: *Instant EPA's Air Toxics*, 1994, Instant Reference Sources, Inc., 7605 Rockpoint Drive, Austin, TX 78731 (Fax: 512-345-2386; Internet URL, http://www.instantref.com/inst-ref.htm).

Environmental Reference Materials

A source of EED environmental reference materials for lead is Radian International LLC, Austin, TX (Fax 512-454-0268; Internet URL, http://www.radian.com/standards); Catalog No. ICL-002L.

Hazardous Properties

ACUTE/CHRONIC HAZARDS: Lead powder may be toxic and the effects of exposure are cumulative. Care must be used in handling lead as it is a cumulative poison [706].

VOLATILITY: Vapor Pressure: 1 mm @ 973°C [703]

FIRE HAZARD: Flammable in the form of dust when exposed to heat or flame [703]. Metal is a noncombustible solid in bulk form [704].

REACTIVITY: Mixtures of hydrogen peroxide + trioxane explode on contact with lead. Rubber gloves containing lead may ignite in nitric acid. Violent reaction on ignition with chlorine trifluoride; concentrated hydrogen peroxide; ammonium nitrate (below 200°C with powdered lead); sodium acetylide (with powdered lead). Incompatible with NaN_3; Zr; disodium acetylide; oxidants. Can react vigorously with oxidizing materials. [703]

STABILITY: Moderately explosive in the form of dust when exposed to heat or flame. When heated to decomposition it emits highly toxic fumes of Pb. [703]

Source: *Instant EPA's Air Toxics*, 1994, Instant Reference Sources, Inc., 7605 Rockpoint Drive, Austin, TX 78731 (Fax: 512-345-2386; Internet URL, http://www.instantref.com/inst-ref.htm).

Medical Symptoms of Exposure

Symptoms of acute exposure may include abdominal pain, diarrhea, shock, muscular weakness and pain, headache, kidney damage, and coma. Symptoms of chronic exposure may include lead encephalopathy (especially in children), headache, vomiting, delirium/hallucinations, convulsions, coma, death from exhaustion and respiratory failure. Children typically show weight loss, weakness, anemia and exhibit GI and CNS complaints.

Source: *Instant EPA's Air Toxics*, 1994, Instant Reference Sources, Inc., 7605 Rockpoint Drive, Austin, TX 78731 (Fax: 512-345-2386; Internet URL, http://www.instantref.com/inst-ref.htm).

Toxicological Information

The toxicological information below is gathered from several sources including two national databases: one from the *National Toxicology Program Chemical Repository Database* and another from *EPA's Integrated Risk Information System (IRIS).*

Toxicology Information from the National Toxicology Program

CARCINOGENICITY:

> Review:
>> IARC Cancer Review: Human Inadequate Evidence
>> IARC Cancer Review: Animal Inadequate Evidence
>> IARC possible human carcinogen (Group 2B)

TERATOGENICITY:

Type of Effect	Route/Animal	Amount Dosed	Units (Notes)
TDLo	orl-rat	790	mg/kg (MGN)
TDLo	orl-rat	1140	mg/kg (14D pre-21D post)
TDLo	orl-rat	520	mg/kg (7-22D preg/10D post)
TDLo	orl-mus	1120	mg/kg (MGN)
TDLo	orl-mus	6300	mg/kg (1-21D preg)
TDLo	orl-mus	12600	mg/kg (1-21D preg)
TDLo	orl-mus	4800	mg/kg (1-16D preg)
TDLo	orl-dom	662	mg/kg (1-21W preg)

MUTAGENICITY: Not available

TOXICITY:

Typ. Dose	Mode	Specie	Amount	Unit	Other
TDLo	orl	wmn	450	mg/kg/6Y	TFX : CNS
LDLo	ipr	rat	1000	mg/kg	
LDLo	orl	pgn	160	mg/kg	

OTHER TOXICITY DATA:

> Review:
>> Toxicology Review-21

> Status:
>> "NIOSH Manual of Analytical Methods" Vol. 1: 102, 191, 195, 200, 214, 262; Vol. 3: S341
>> Reported in EPA TSCA Inventory, 1980

SAX TOXICITY EVALUATION:

>THR: A common air contaminant. It is a (S) carc of the lungs and kidney and an exper teratogen.

Source: *Instant Tox-Base*, 1995, Instant Reference Sources, Inc., 7605 Rockpoint Drive, Austin, TX 78731 (Fax: 512-345-2386; Internet URL, http://www.instantref.com/inst-ref.htm).

Toxicology Information from EPA's Integrated Risk Information System (IRIS)

I.A. REFERENCE DOSE FOR CHRONIC ORAL EXPOSURE (RfD)

A great deal of information on the health effects of lead has been obtained through decades of medical observation and scientific research. This information has been assessed in the development of air and water quality criteria by the Agency's Office of Health and Environmental Assessment (OHEA) in support of regulatory decision-making by the Office of Air Quality Planning and Standards (OAQPS) and by the Office of Drinking Water (ODW). By comparison to most other environmental toxicants, the degree of uncertainty about the health effects of lead is quite low. It appears that some of these effects, particularly changes in the levels of certain blood enzymes and in aspects of children's neurobehavioral development, may occur at blood lead levels so low as to be essentially without a threshold. The Agency's RfD Work Group discussed inorganic lead (and lead compounds) at two meetings (07/08/85 and 07/22/85) and considered it inappropriate to develop an RfD for inorganic lead. For additional information, interested parties are referred to the 1986 Air Quality Criteria for Lead (EPA-600/8-83/028a-dF) and its 1990 Supplement (EPA/600/8-89/049F) or the following Agency scientists: Harlal Choudhury / NCEA -- (513)569-7536

J. Michael Davis / NCEA -- (919)541-4162

Jeff Cohen / OST -- (202)260-5456

John Haines / OAQPS -- (919)541-5533

I.B. REFERENCE CONCENTRATION FOR CHRONIC INHALATION EXPOSURE (RfC)

>[Ed. Note: EPA does not have information yet in this section of IRIS].

II. CARCINOGENICITY ASSESSMENT FOR LIFETIME EXPOSURE

Section II provides information on three aspects of the carcinogenic risk assessment for the agent in question; the US EPA classification, and quantitative estimates of risk from oral exposure and from inhalation exposure.   The classification reflects a weight-of-evidence judgment of the likelihood that the agent is a human carcinogen. The quantitative risk estimates are presented in three ways. The slope factor is the result of application of a low-dose extrapolation procedure and is presented as the risk per (mg/kg)/day. The unit risk is the quantitative estimate in terms of either risk per ug/L drinking water or risk per ug/cu.m air breathed. The third form in which risk is presented is a drinking water or air concentration providing cancer risks of 1 in 10,000, 1 in 100,000 or 1 in 1,000,000. Background Document 2 (Service Code 5) provides details on the rationale and methods used to derive the carcinogenicity values found in IRIS. Users are referred to Section I for information on long-term toxic effects other than carcinogenicity.

II.A. EVIDENCE FOR CLASSIFICATION AS TO HUMAN CARCINOGENICITY

II.A.1. WEIGHT-OF-EVIDENCE CLASSIFICATION

Classification -- B2; probable human carcinogen

Basis -- Sufficient animal evidence. Ten rat bioassays and one mouse assay have shown statistically significant increases in renal tumors with dietary and subcutaneous exposure to several soluble lead salts. Animal assays provide reproducible results in several laboratories, in multiple rat strains with some evidence of multiple tumor sites. Short term studies show that lead affects gene expression. Human evidence is inadequate.

II.A.2. HUMAN CARCINOGENICITY DATA

Inadequate. There are four epidemiologic studies of occupational cohorts exposed to lead and lead compounds. Two studies (Dingwall-Fordyce and Lane, 1963; Nelson et al., 1982) did not find any association between exposure and cancer mortality. Selevan et al. (1985), in their retrospective cohort mortality study of primary lead smelter workers, found a slight decrease in the total cancer mortality (SMR=95). Apparent excesses were observed for respiratory cancer (SMR=111, obs=41, p>0.05) and kidney cancer (SMR=204,   obs=6, p>0.05). Cooper and Gaffey (1975) and Cooper (1985 update) performed a cohort mortality study of battery plant workers and lead smelter workers.   They found statistically significant excesses for total cancer mortality (SMR=113, obs=344), stomach cancer (SMR=168, obs=34), and lung cancer (SMR=124, obs=109) in the battery plant workers. Although similar excesses were observed in the smelter workers, they were not statistically significant.   Cooper and Gaffey (1975) felt it was possible that individual subjects were monitored primarily on the basis of obvious signs of lead exposure, while others who showed no symptoms of lead poisoning were not monitored.

All of the available studies lacked quantitative exposure information, as well as information on the possible contribution from smoking. All studies also included exposures to other metals such as arsenic, cadmium, and zinc for which no adjustment was done. The cancer excesses observed in the lung and stomach were relatively small (<200). There was no consistency of site among the various studies, and no study showed any dose-response relationship.   Thus, the available human evidence is considered to be inadequate to refute or demonstrate any potential carcinogenicity for humans from lead exposure.

II.A.3. ANIMAL CARCINOGENICITY DATA

Sufficient. The carcinogenic potential of lead salts (primarily phosphates and acetates) administered via the oral route or by injection has been demonstrated in rats and mice by more than 10 investigators. The most characteristic cancer response is bilateral renal carcinoma. Rats given lead acetate or subacetate orally have developed gliomas, and lead subacetate also produced lung adenomas in mice after i.p. administration. Most of these investigations found a carcinogenic response only at the highest dose. The lead compounds tested in animals are almost all soluble salts. Metallic lead, lead oxide and lead tetralkyls have not been tested adequately. Studies of inhalation exposure have not been located in the literature.

Azar et al. (1973) administered 10, 50, 100, and 500 ppm lead as lead acetate in dietary concentrations to 50 rats/sex/group for 2 years. Control rats (100/sex) received the basal laboratory diet. In a second 2-year feeding study, 20 rats/group were given diets containing 0, 1000, and 2000 ppm lead as lead acetate. No renal tumors were reported in the control groups or in treated animals of either sex receiving 10 to 100 ppm. Male rats fed 500, 1000, and 2000 ppm lead acetate had an increased renal tumor incidence of 5/50, 10/20, and 16/20, while 7/20 females in the 2000-ppm group developed renal tumors.

The Azar et al. (1973) study is limited by the lack of experimental detail. The possibility of environmental contamination from lead in the air or drinking water was not mentioned. The strains of rats used were not specified in the study, but the Health Effects Assessment for Lead (US EPA, 1984) indicates the rats were Wistar strain. The weight gain at 1000 and 2000 ppm was reported to be depressed, but details were not given.

Kasprzak et al. (1985), in investigating the interaction of dietary calcium on lead carcinogenicity, fed 1% lead subacetate (8500 ppm Pb) to male Sprague-Dawley rats in the diet for 79 weeks. Of the rats surviving (29/30) in this treatment group beyond 58 weeks, 44.8% had renal tumors. Four rats had adenocarcinomas; the remainaing nine had adenomas. Bilateral tumors were noted. No renal tumors were noted among the controls.

As part of a study to determine interactions between sodium nitrite, ethyl urea and lead, male Sprague-Dawley rats were given lead acetate in their drinking water for 76 weeks (Koller et al., 1986). The concentration of lead was 2600 ppm. No kidney tumors were detected among the 10 control rats. Thirteen of 16 (81%) lead-treated rats had renal tubular carcinoma; three tumors were detected at 72 weeks and the remainder detected at the termination of the study.

Van Esch and Kroes (1969) fed basic lead acetate at 0, 0.1%, and 1.0% in the diet to 25 Swiss mice/sex/group for 2 years. No renal tumors developed in the control group, but 6/25 male mice of 0.1% basic lead acetate group had renal tumors (adenomas and carcinomas combined). In the 1.0% group, one female had a renal tumor. The authors thought that the low incidence in the 1.0% group was due to early mortality.

Hamsters given lead subacetate at 0.5% and 1% in the diet had no significant renal tumor response (Van Esch and Kroes, 1969).

II.A.4. SUPPORTING DATA FOR CARCINOGENICITY

Lead acetate induces cell transformation in Syrian hamster embryo cells (DiPaolo et al., 1978) and also enhances the incidence of simian adenovirus induction. Lead oxide showed similar enhanced adenovirus induction (Casto et al., 1979).

Under certain conditions lead compounds are capable of inducing chromosomal aberrations in vivo and in tissue cultures. Grandjean et al. (1983) showed a relationship between SCE and lead exposure in exposed workers. Lead has been shown, in a number of DNA structure and function assays, to affect the molecular processes associated with the regulation of gene expression (US EPA, 1986).

II.B. QUANTITATIVE ESTIMATE OF CARCINOGENIC RISK FROM ORAL EXPOSURE

Not available.

Quantifying lead's cancer risk involves many uncertainties, some of which may be unique to lead. Age, health, nutritional state, body burden, and exposure duration influence the absorption, release, and excretion of lead.   In addition, current knowledge of lead pharmacokinetics indicates that an estimate derived by standard procedures would not truly describe the potential risk. Thus, the Carcinogen Assessment Group recommends that a numerical estimate not be used.

II.C.   QUANTITATIVE   ESTIMATE   OF   CARCINOGENIC   RISK   FROM INHALATION EXPOSURE

[Ed. Note: EPA does not have information yet in this section of IRIS].

II.D. EPA DOCUMENTATION, REVIEW, AND CONTACTS (CARCINOGENICITY ASSESSMENT)

II.D.1. EPA DOCUMENTATION

Source Document -- US EPA, 1984, 1986, 1989

US EPA, 1989 has received OHEA and SAB review.

The 1986 Air Quality Criteria Document for Lead has received Agency and External Review.

II.D.2. REVIEW (CARCINOGENICITY ASSESSMENT)

Agency Work Group Review -- 05/04/88

Verification Date -- 05/04/88

II.D.3. US EPA CONTACTS (CARCINOGENICITY ASSESSMENT)

William Pepelko / NCEA -- (202)260-5898

Jim Cogliano / NCEA -- (202)260-3814

III. HEALTH HAZARD ASSESSMENTS FOR VARIED EXPOSURE DURATIONS

III.A. DRINKING WATER HEALTH ADVISORIES

[Ed. Note: EPA does not have information yet in this section of IRIS].

IV. US EPA REGULATORY ACTIONS

EPA risk assessments may be updated as new data are published and as assessment methodologies evolve. Regulatory actions are frequently not updated at the same time. Compare the dates for the regulatory actions in this section with the verification dates for the risk assessments in sections I and II, as this may explain inconsistencies. Also note that some regulatory actions consider factors not related to health risk, such as technical or economic feasibility. Such considerations are indicated for each action. In addition, not all of the regulatory actions listed in this section involve enforceable federal standards. Please direct any questions you may have concerning these regulatory actions to the US EPA contact listed for that particular action. Users are strongly urged to read the background information on each regulatory action in Background Document 4 in Service Code 5.

IV.A. CLEAN AIR ACT (CAA)

IV.A.1. NATIONAL AMBIENT AIR QUALITY STANDARD (NAAQS)

Considers technological or economic feasibility? -- No

Discussion -- Under Section 109 of the CAA, EPA has set a primary (health-based) NAAQS for lead of 1.5 ug/cu.m, calendar quarter average not to be exceeded (43 FR 41258, 10/05/78). The secondary (welfare-based) NAAQS is identical to the primary standard. EPA is currently reviewing these standards to determine if changes are warranted.

Reference -- 40 CFR 50.12

US EPA Contact -- Air Quality Management Division / OAQPS / (919)541-5656 / FTS 629-5656

IV.B. SAFE DRINKING WATER ACT (SDWA)

IV.B.1. MAXIMUM CONTAMINANT LEVEL GOAL (MCLG) for Drinking Water

Value (status) -- 0 mg/L (Final, 1991)

Considers technological or economic feasibility? -- NO

Discussion --   The MCLG for lead is zero based on (1) occurrence of low level effects and difficulties in identifying clear threshold levels, (2) the overall Agency goal of reducing total lead exposures, and (3) the classification of lead as a group B2 carcinogen.

Reference -- 56 FR 26460 (06/07/91); 56 FR 32112 (07/15/91)

EPA Contact -- Health and Ecological Criteria Division / OST / (202) 260-7571 / FTS 260-7571; or Safe Drinking Water Hotline / (800) 426-4791

IV.B.2. MAXIMUM CONTAMINANT LEVEL (MCL) for Drinking Water

Value -- None (Final, 1991)

Considers technological or economic feasibility? -- YES

Discussion -- EPA concluded that setting an MCL for lead is not feasible and believes that the treatment approach contained in the final rule (corrosion control, source water reduction, public education and lead service line problems associated with establishing MCL's.

Monitoring requirements -- Tap water monitoring for lead and copper to determine whether a system is subject to the treatment technique requirements.   Water quality parameter sampling to determine the effectiveness of optional   corrosion control treatment. Source water monitoring for lead and copper to determine source water's contribution to total tap water lead and copper levels, and the need for treatment. Monitoring schedules vary by system size and type of monitoring.

Analytical methodology -- Atomic absorption/furnace technique (EPA 239.2; ASTM D-3559-85D; SM 3113); inductively-coupled plasma/mass spectrometry (EPA 200.8); atomic absorption/platform furnace technique (EPA 200.9).

Best available technology --

Optimal corrosion control treatment: pH/alkalinity adjustment, calcium adjustment; addition of corrosion inhibitor.

Source water treatment: Coagulation/filtration; ion exchange; lime softening; reverse osmosis.

Public education.

Lead service line replacement.

Reference -- 45 FR 57332 (08/27/80); 53 FR 31517 (08/18/88); 56 FR 26460 (06/07/91); 56 FR 32112 (07/15/91).

EPA Contact -- Drinking Water Standards Division / OGWDW / (202) 260-7575 / FTS 260-7575; or Safe Drinking Water Hotline / (800) 426-4791

IV.B.3. SECONDARY MAXIMUM CONTAMINANT LEVEL (SMCL) for Drinking Water

[Ed. Note: EPA does not have information yet in this section of IRIS].

IV.B.4. REQUIRED MONITORING OF "UNREGULATED" CONTAMINANTS

[Ed. Note: EPA does not have information yet in this section of IRIS].

IV.C. CLEAN WATER ACT (CWA)

IV.C.1. AMBIENT WATER QUALITY CRITERIA, Human Health

Water and Fish Consumption -- 5.0E+1 ug/L

Fish Consumption Only -- None

Considers technological or economic feasibility? -- NO

Discussion -- The criterion was set at the existing drinking water standard in 1980.

Reference -- 45 FR 79318 (11/28/80)

EPA Contact -- Criteria and Standards Division / OWRS (202)260-1315 / FTS 260-1315

IV.C.2. AMBIENT WATER QUALITY CRITERIA, Aquatic Organisms

Freshwater:
   Acute -- 8.2E+1 ug/L (1-hour average)
   Chronic -- 3.2E+0 ug/L (4-day average)

Marine:
   Acute -- 1.40E+2 ug/L (1-hour average)
   Chronic -- 5.6E+0 ug/L (4-day average)

Considers technological or economic feasibility? -- NO

Discussion -- Criteria were derived from a minimum data base consisting of acute and chronic tests on a variety of species. The toxicity of this compound in freshwater is hardness dependent. The values given are for a hardness of 100 mg/L CaCO3. For a more complete discussion, see the referenced notice.

Reference -- 50 FR 30784 (07/29/85)

EPA Contact -- Criteria and Standards Division / OWRS (202)260-1315 / FTS 260-1315

IV.D. FEDERAL INSECTICIDE, FUNGICIDE, AND RODENTICIDE ACT (FIFRA)

[Ed. Note: EPA does not have information yet in this section of IRIS].

IV.E. TOXIC SUBSTANCES CONTROL ACT (TSCA)

[Ed. Note: EPA does not have information yet in this section of IRIS].

IV.F. RESOURCE CONSERVATION AND RECOVERY ACT (RCRA)

IV.F.1. RCRA APPENDIX IX, for Ground Water Monitoring

Status -- Listed (total lead)

Reference -- 52 FR 25942 (07/09/87)

EPA Contact --   RCRA/Superfund Hotline (800)424-9346 / (202)260-3000 / FTS 260-3000

IV.G. SUPERFUND (CERCLA)

IV.G.1. REPORTABLE QUANTITY (RQ) for Release into the Environment

Value (status) -- 1 pound (Statutory, 1987)

Considers technological or economic feasibility? -- NO

Discussion -- The statutory 1-pound RQ for lead is retained pending assessment of its potential carcinogenicity and may be adjusted in a future notice of proposed rulemaking when the evaluation of available data is completed. Lead was evaluated for chronic toxicity, but was not ranked for toxicity because of insufficient data.

Reference -- 52 FR 8140 (03/16/87); 54 FR 33418 (08/14/89)

EPA Contact -- RCRA/Superfund Hotline (800)424-9346 / (202)260-3000 / FTS 260-3000

VI. BIBLIOGRAPHY

VI.A. ORAL RfD REFERENCES

None

VI.B. INHALATION RfD REFERENCES

None

VI.C. CARCINOGENICITY ASSESSMENT REFERENCES

Anderson, E.L., and CAG (Carcinogenic Assessment Group). 1983. Quantitative approaches in use to assess cancer risk. Risk Analysis. 3: 277-295.

Azar, A., H.J. Trochimowicz and M.E. Maxfield. 1973. Review of lead studies in animals carried out at Haskell Laboratory - Two year feeding study and response to hemorrhage study. In: Barth D., A. Berlin, R. Engel, P. Recht and J. Smeets, Ed. Environmental health aspects of lead: Proceedings International Symposium; October 1972; Amsterdam, The Netherlands. Commission of the European Communities, Luxemberg. p. 199-208.

Casto, B.C., J. Meyers and J.A. DiPaolo. 1979. Enhancement of viral transformation for evaluation of the carcinogenic or mutagenic potential of inorganic metal salts. Cancer Res. 39: 193-198.

Cooper, W.C. 1985. Mortality among employees of lead battery plants and lead producing plants, 1947-1980. Scand. J. Work Environ. Health. 11: 331-345.

Cooper, W.C. and W.R. Gaffey. 1975. Mortality of lead workers. In: Proceedings of the 1974 Conference on Standards of Occupational Lead Exposure, J.F. Cole, Ed., February, 1974. Washington, DC. J. Occup. Med. 17: 100-107.

Dingwall-Fordyce, I. and R.E. Lane. 1963. A follow-up study of lead workers. Br. J. Ind. Med. 20: 313-315.

DiPaolo, J.A., R.L. Nelson and B.C. Casto. 1978. In vitro neoplastic transformation of Syrian hamster cells by lead acetate and its relevance to environmental carcinogenesis. Br. J. Cancer. 38: 452-455.

Grandjean, P., H.C. Wulf and E. Niebuhr. 1983. Sister chromatid exchange in response to variations in occupational lead exposure. Environ. Res. 32: 199-204.

Kasprzak, K.S., K.L. Hoover and L.A. Poirier. 1985. Effects of dietary calcium acetate on lead subacetate carcinogenicity in kidneys of male Sprague- Dawley rats. Carcinogenesis. 6(2): 279-282.

Koller, L.D., N.I. Kerkvliet and J.H. Exon. 1986. Neoplasia induced in male rats fed lead acetate, ethyl urea and sodium nitrate. Toxicol. Pathol. 13: 50-57.

Nelson, D.J., L. Kiremidjian-Schumacher and G. Stotzky. 1982. Effects of cadmium, lead, and zinc on macrophage-mediated cytotoxicity toward tumor cells. Environ. Res. 28: 154-163.

Selevan, S.G., P.J. Landrigan, F.B. Stern and J.H. Jones. 1985. Mortality of lead smelter workers. Am. J. Epidemiol. 122: 673-683.

US EPA. 1984. Health Effects Assessment for Lead. Prepared by the Office of Health and Environmental Assessment, Environmental Criteria and Assessment Office, Cincinnati, OH, for the Office of Emergency and Remedial Response, Washington, DC. EPA/540/1-86/055. NTIS PB85-163996/AS.

US EPA. 1986. Air Quality Criteria Document for Lead. Volumes III, IV. Prepared by the Office of Health and Environmental Assessment, Environmental Criteria and Assessment Office, Research Triangle Park, NC, for the Office of Air Quality Planning and Standards. EPA-600/8-83/028dF.

US EPA. 1989. Evaluation of the potential carcinogenicity of lead and lead compounds: In support of reportable quantity adjustments pursuant to CERCLA Section 102. Prepared by the Office of Health and Environmental Assessment, Washington, DC. EPA/600/8-89/045A. (External Review Draft).

Van Esch, G.J. and R. Kroes. 1969. The induction of renal tumors by feeding of basic lead acetate to mice and hamsters. Br. J. Cancer. 23: 265-271.

VI.D. DRINKING WATER HA REFERENCES

None

VII. REVISION HISTORY

Date	Section	Description
09/26/88	II.	Carcinogen summary on-line
02/01/89	IV.B.1.	Effect level corrected in discussion
06/01/89	II.D.3.	Primary contact changed
06/01/89	IV.A.1.	Reference corrected - changed number for part in CFR
12/01/89	II.A.3.	Last paragraph - Correct Van Esch 1969 citation
12/01/89	VI.	Bibliography on-line
07/01/90	I.A.	Changed contact J. Cohen's office and telephone number
07/01/90	IV.F.1.	EPA contact changed
02/01/91	I.A.	Message revised to include new EPA document
02/01/91	I.A.	EPA contacts changed
05/01/91	II.A.	Text edited
01/01/92	IV.	Regulatory actions updated
06/01/92	IV.B.2.	MCL monitoring reqs. and BAT corrected
07/01/93	II.D.3.	Secondary contact's phone number changed
07/01/93	VI.C.	References alphabetized correctly
11/01/93	II.D.1.	US EPA 1987 replaced with 1989; rev. state. revised
11/01/93	VI.C.	US EPA 1987 deleted; US EPA 1989 added

SYNONYMS

Lead
Lead and compounds   plumbum

Source: ***Instant EPA's IRIS***, 1997, Instant Reference Sources, Inc., 7605 Rockpoint Drive, Austin, TX 78731 (Fax: 512-345-2386; Internet URL, http://www.instantref.com/inst-ref.htm).

Lindane

Introduction

Lindane is an organochlorine insecticide and fumigant and is a restricted use pesticide utilized in treating soil. It is persistent in the environment and highly toxic. Lindane is used as an insecticide, pediculicide, scabicide, ectoparasiticide and a pesticide. It is also used as a foliar spray and soil application for insecticidal control of a broad spectrum of phytophagous and soil dwelling insects, animal ectoparasites and public health pests. It is used on ornamentals, fruit trees, nut trees, vegetables, tobacco and timber. Lindane is found in baits and seed treatments for rodent control. It acts as a stomach and contact poison and has some fumigant action. It has been of value in the control of malaria and other vector-borne diseases and in the control of grasshoppers, cotton insects, rice insects, wireworms and other soil pests. It is used in pet shampoo to maintain the natural luster of the coat and aid in the prevention of ticks, lice and sarcoptic mange mites. It is also used on human patients in the treatment of head and crab lice and their ova. Lindane is stored in the fat of most animals and is highly toxic to birds and fish. There is only one isomer of lindane; it is the gamma isomer of 1,2,3,4,5,6-hexachlorocyclohexane. [818]

When released to water, lindane is not expected to volatilize significantly. Lindane released to acidic or neutral water is not expected to hydrolyze significantly, but in basic water, significant hydrolysis may occur (t1/2=95 hr at pH 9.3). Transport to sediment should be slow and result predominantly from diffusion rather than settling. Release of lindane to soil will most likely result in volatilization and slow leaching of it to groundwater. Lindane in the atmosphere is likely to be subject to rain-out and dry deposition. The estimated half-life for the reaction of vapor phase lindane with atmospheric hydroxyl radicals is 1.63 days. Lindane may slowly biodegrade in aerobic media and will rapidly degrade under anaerobic conditions. It has been reported to photodegrade in water in spite of the lack of a photoreactive center, but photolysis is not considered to be a major environmental fate process. Lindane will bioconcentrate slightly in fish. Monitoring data indicate that lindane is a contaminant in air, water, sediment, soil, fish and other aquatic organisms, wildlife, food, and humans. Human exposure results primarily from food. [828]

684

Identifying Information

CAS NUMBER: 58-89-9

NIOSH Registry Number: GV4900000

SYNONYMS: gamma-Benzene hexachloride, gamma-BHC, 1,2,3,4,5,6-
Hexachlorocyclohexane, gamma-isomer

Endocrine Disruptor Information

Lindane is a strongly suspected EED that it is listed by EPA [811], the Centers for
Disease Control & Prevention [812], and the World Wildlife Fund [813] as a potential
endocrine modifying chemical. There are reports of immunosuppression with lindane
[173] and EPA [811] reports that it presents some estrogenic effects in mice [844].
Lindane also is listed as one of the commonly found pesticides in wildlife exhibiting
symptoms of EDCs. [810] In animal studies, lindane has been shown to behave like
estrogen causing detrimental effects on the reproductive system of exposed males.
Lindane concentrates in breast milk and thus is of particular concern to nursing infants.
[882]

Chemical and Physical Properties

CHEMICAL FORMULA: $C_6H_6Cl_6$

MOLECULAR WEIGHT: 290.83

WLN: L6TJ AG BG CG DG EG FG GAMMA

PHYSICAL DESCRIPTION: Colorless or white crystalline solid

SPECIFIC GRAVITY: 1.87 @ 20/4°C [055]

DENSITY: 1.87 g/mL @ 20°C [055], [205]

MELTING POINT: 112-113°C [016], [169], [173]

BOILING POINT: 323.4°C [016], [395]

SOLUBILITY: Water: <1 mg/mL @ 24°C[700]; DMSO: 50-100 mg/mL @ 23°C[700];
95% Ethanol: 1-5 mg/mL @ 23°C [700]; Acetone: >=100 mg/mL @ 23°C [700];
Methanol: 7.4% [173], [169]; Toluene: >50 g/L [172]

OTHER SOLVENTS: Petroleum ether: 2.9% [169]; Chloroform: 24% [033], [169],
[173]; Chlorinated hydrocarbons: Moderately soluble [173]; Ethyl acetate: >50 g/L
[172]; Ether: >50 g/L [172]; Benzene: >50 g/L [172]; Acetic acid: 12.8% [169]; Carbon
tetrachloride: 6.7% [169]; Cyclohexane: 36.7% [169], [173]; Dioxane: 31.4% [169];
Kerosene: 2.0-3.2% [173]; Acid: Soluble [047]; Xylene: 24.7% [169], [173]; Rat fat:
12.55% [173]

OTHER PHYSICAL DATA: Vapor pressure is also reported as 0.0000094 mm Hg @ 20°C [033], [173]; Boiling point: 176.2°C @ 10 mm Hg [017] 288°C [051]; Slight, musty odor [033], [421]; Refractive index: 1.644 @ 20°C [033] Sublimes slowly [058]; Bitter taste [051]; Odor threshold: 12.0 mg/kg [055]

Source: *Instant EPA's Air Toxics*, 1994, Instant Reference Sources, Inc., 7605 Rockpoint Drive, Austin, TX 78731 (Fax: 512-345-2386; Internet URL, http://www.instantref.com/inst-ref.htm).

Environmental Reference Materials

A source of EED environmental reference materials for lindane is Radian International LLC, Austin, TX (Fax 512-454-0268; Internet URL, http://www.radian.com/standards); Catalog No. ERB-015.

Hazardous Properties

ACUTE/CHRONIC HAZARDS: Lindane is an irritant and may be toxic by ingestion, inhalation or skin absorption [033], [062], [269]. Oils may enhance skin absorption [158]. When heated to decomposition it emits toxic fumes of chlorine, hydrochloric acid and phosgene [043].

HAP WEIGHTING FACTOR: 1 [713]

VOLATILITY: Vapor pressure: 0.03 mm Hg @ 20°C [058], [395]

FIRE HAZARD: Lindane is nonflammable [102], [371].

REACTIVITY: Lindane is incompatible with strong bases [058]. It is incompatible with powdered metals such as iron, zinc and aluminum [058], [395]. It is also incompatible with oxidizing agents [269]. It can undergo oxidation when in contact with ozone [186]. When exposed to alkalis, lindane undergoes dehydrochlorination [172], [173], [395].

STABILITY: Lindane is stable under normal laboratory conditions. Solutions of it in water, DMSO, 95% ethanol or acetone should be stable for 24 hours under normal lab conditions [700]. Lindane is extremely stable to light, air and temperatures up to 180°C [169]. It decomposes at temperatures above 177 C [102].

Source: *Instant EPA's Air Toxics*, 1994, Instant Reference Sources, Inc., 7605 Rockpoint Drive, Austin, TX 78731 (Fax: 512-345-2386; Internet URL, http://www.instantref.com/inst-ref.htm).

Medical Symptoms of Exposure

Symptoms of exposure may include epileptic convulsions and serious EEG disturbances [421]. Other symptoms may include gastrointestinal disturbances, severe central nervous system involvement-cerebellar derangement, muscle spasms, blindness from optic nerve atrophy and diminution of vision [058]. Lindane also may cause malaise, faintness, dizziness followed by collapse and convulsions sometimes preceded by screaming and accompanied by foaming at the mouth and biting of the tongue, unconsciousness, retrograde amnesia, moderate rise in temperature, facial pallor, slight circumoral cyanosis, severe cyanosis of the face and extremities, slightly enlarged liver, depression and death from ingestion [173]. Effects of acute overexposure may be central nervous system stimulation, dyspnea, headache, nausea and irritation of the respiratory tract.

Effects of chronic overexposure may include irreversible renal changes, conjunctivitis, ecchymosis, staggering, fever, vomiting, mental confusion, pulmonary edema, dilation of the heart, extensive necrosis of blood vessels in the lungs, liver and kidney, fatty degeneration of the liver and kidneys and some cases of hypoplastic anemia.

Animal symptoms that have been observed are increased respiration, restlessness accompanied by frequency of micturition, intermittent muscular spasms of the whole body, salivation, grinding of the teeth and consequent bleeding from the mouth, backward movement with loss of balance and somersaulting, retraction of the head, convulsions, gasping and biting, collapse and death usually within a day. It may cause degenerative changes in the kidneys, pancreas, testes, nasal mucous membranes and liver (in extremely high doses). It may also cause immunosuppression [173]. It may cause respiratory failure [102].

Source: ***Instant EPA's Air Toxics***, 1994, Instant Reference Sources, Inc., 7605 Rockpoint Drive, Austin, TX 78731 (Fax: 512-345-2386; Internet URL, http://www.instantref.com/inst-ref.htm).

Toxicological Information

The toxicological information below is gathered from several sources including two national databases: one from the *National Toxicology Program Chemical Repository Database* and another from *EPA's Integrated Risk Information System (IRIS)*.

Toxicology Information from the National Toxicology Program

CARCINOGENICITY:

Tumorigenic Data:

Type of Effect	Route/Animal	Amount Dosed (Notes)
TD	orl-mus	25 gm/kg/73W-C
TDLo	orl-mus	14 gm/kg/2Y-C

Review:
    IARC Cancer Review: Human Inadequate Evidence
    IARC Cancer Review: Animal Limited Evidence
    IARC possible human carcinogen (Group 2B)
    IARC Note: Although IARC has assigned an overall evaluation to hexachlorocyclohexanes, it has not assigned an overall evaluation to all substances within this group

Status:

NCI Carcinogenesis Bioassay (Feed); Negative: Male and Female Rat, Male and Female Mouse

NTP Fifth Annual Report on Carcinogens, 1989

NTP anticipated human carcinogen

TERATOGENICITY: See RTECS printout for data

MUTATION DATA: See RTECS printout for data

TOXICITY:

Typ.Dose	Mode	Specie	Amount	Unit	Other
TDLo	orl	chd	180	mg/kg	
TDLo	orl	chd	111	mg/kg	
TDLo	skn	man	20	mg/kg/6W-I	
LD50	orl	rat	76	mg/kg	
LD50	skn	rat	500	mg/kg	
LD50	ipr	rat	35	mg/kg	
LD50	orl	mus	44	mg/kg	
LD50	ipr	mus	125	mg/kg	
LD50	orl	dog	40	mg/kg	
LDLo	ivn	dog	8	mg/kg	
LD50	orl	cat	25	mg/kg	
LDLo	unr	cat	50	mg/kg	
LD50	orl	rbt	60	mg/kg	
LD50	skn	rbt	50	mg/kg	
LDLo	ivn	rbt	4500	ug/kg	
LD50	orl	gpg	127	mg/kg	
LD50	orl	ham	360	mg/kg	

OTHER TOXICITY DATA:

Review: Toxicology Review-8

Standards and Regulations:
DOT-Hazard: ORM-A; Label: None

Status:

EPA Genetox Program 1988, Positive: S cerevisiae gene conversion

EPA Genetox Program 1988, Positive/limited: Carcinogenicity-mouse/rat

EPA Genetox Program 1988, Inconclusive: Host-mediated assay; D melanogaster Sex-linked   lethal

EPA TSCA Chemical Inventory, 1986

EPA TSCA Test Submission (TSCATS) Data Base, January 1990

NIOSH Analytical Methods: see Aldrin and Lindane, 5502

SAX TOXICITY EVALUATION:

> THR: A human systemic poison by ingestion. Also a poison by ingestion intraperitoneal, intravenous, skin contact and intramuscular routes. An experimental neoplastigen and teratogen. Other experimental animals reproductive effects. Mutagenic data.

Source: *Instant Tox-Base*, 1995, Instant Reference Sources, Inc., 7605 Rockpoint Drive, Austin, TX 78731 (Fax: 512-345-2386; Internet URL, http://www.instantref.com/inst-ref.htm).

Toxicology Information from EPA's Integrated Risk Information System (IRIS)

I.   CHRONIC HEALTH HAZARD ASSESSMENTS FOR NONCARCINOGENIC EFFECTS

I.A. REFERENCE DOSE FOR CHRONIC ORAL EXPOSURE (RfD)

The Reference Dose (RfD) is based on the assumption that thresholds exist for certain toxic effects such as cellular necrosis, but may not exist for other toxic effects such as carcinogenicity. In general, the RfD is an estimate (with uncertainty spanning perhaps an order of magnitude) of a daily exposure to the human population (including sensitive subgroups) that is likely to be without an appreciable risk of deleterious effects during a lifetime. Please refer to Background Document 1 in Service Code 5 for an elaboration of these concepts. RfDs can also be derived for the noncarcinogenic health effects of compounds which are also carcinogens. Therefore, it is essential to refer to other sources of information concerning the carcinogenicity of this substance. If the US EPA has evaluated this substance for potential human carcinogenicity, a summary of that evaluation will be contained in Section II of this file when a review of that evaluation is completed.

I.A.1. ORAL RfD SUMMARY

Critical Effect	Experimental Doses*	UF	MF	RfD
Liver and kidney toxicity	NOAEL: 4 ppm diet [0.33 mg/kg/day (females)]	1000	1	3E-4 mg/kg/day
	LOAEL: 20 ppm diet [1.55 mg/kg/day (males)]			

*Conversion Factor: Converted dose calculated from actual food consumption data.

Rat, Subchronic Oral Bioassay Zoecon Corp., 1983

I.A.2. PRINCIPAL AND SUPPORTING STUDIES (ORAL RfD)

Zoecon Corporation. 1983. MRID No. 00128356. Available from EPA. Write to FOI, EPA, Washington D.C. 20460.

Twenty male and 20 female Wistar KFM-Han (outbred) SPF rats/treatment group were administered 0, 0.2, 0.8, 4, 20, or 100 ppm lindane (99.85%) in the diet. After 12 weeks, 15 animals/sex/group were sacrificed. The remaining rats were fed the control diet for an additional 6 weeks before sacrifice. No treatment-related effects were noted on mortality, hematology, clinical chemistry, or urinalysis. Rats receiving 20 and 100 ppm lindane were observed to have greater-than-control incidence of the following: liver hypertrophy, kidney tubular degeneration, hyaline droplets, tubular distension, interstitial nephritis, and basophilic tubules. Since these effects were mild or rare in animals receiving 4 ppm, this represents a NOAEL. The reviewers of the study calculated the dose to be 0.29 mg/kg/day for males and 0.33 mg/kg/day for females, based on measured food intake.

In a 2-year feeding study (Fitzhugh, 1950), 10 Wistar rats/sex/group were exposed to 5, 10, 50, 100, 400, 800, or 1600 ppm lindane. Slight liver and kidney damage and increased liver weights were noted at the 100 ppm level. If a food intake equal to 5% body weight is assumed, a NOAEL of 2.5 mg/kg bw/day (50 ppm) can be determined from this assay. In a 2-year bioassay (Rivett et al., 1978), four beagle dogs/sex/group were administered 0, 25, 50, or 100 ppm lindane in the diet. Treatment-related effects noted in the animals of the 100 ppm group were increased serum alkaline phosphatase and enlarged dark friable livers. A NOAEL was determined to be 50 ppm (1.6 mg/kg bw/day).

I.A.3. UNCERTAINTY AND MODIFYING FACTORS (ORAL RfD)

UF -- A factor of 10 each was employed for use of a subchronic vs. a lifetime assay, to account for interspecies variation and to protect sensitive human subpopulations.

MF -- None

I.A.4. ADDITIONAL COMMENTS (ORAL RfD)

Data on reproductive effects of lindane are inconclusive. Most reports indicate that hexachlorocyclohexane isomers are nonteratogenic.

I.A.5. CONFIDENCE IN THE ORAL RfD

Study -- Medium
Data Base -- Medium
RfD -- Medium

The principal study used an adequate number of animals and measured multiple endpoints. Since there are other reported chronic and subchronic studies, confidence in the data base is medium. Medium confidence in the RfD follows.

I.A.6. EPA DOCUMENTATION AND REVIEW OF THE ORAL RfD

Source Document -- US EPA, 1985

The RfD in the Drinking Water Criteria Document has been extensively reviewed by US EPA scientists and selected outside experts.

Other EPA Documentation -- None

Agency Work Group Review -- 01/22/86

Verification Date -- 01/22/86

I.A.7. EPA CONTACTS (ORAL RfD)

W. Bruce Peirano / NCEA -- (513)569-7540

I.B. REFERENCE CONCENTRATION FOR CHRONIC INHALATION EXPOSURE (RfC)

A risk assessment for this substance/agent is under review by an EPA work group.

II. CARCINOGENICITY ASSESSMENT FOR LIFETIME EXPOSURE

This substance/agent has not been evaluated by the US EPA for evidence of human carcinogenic potential.

III. HEALTH HAZARD ASSESSMENTS FOR VARIED EXPOSURE DURATIONS

III.A. DRINKING WATER HEALTH ADVISORIES

The Office of Drinking Water provides Drinking Water Health Advisories (HAs) as technical guidance for the protection of public health. HAs are not enforceable Federal standards. HAs are concentrations of a substance in drinking water estimated to have negligible deleterious effects in humans, when ingested, for a specified period of time. Exposure to the substance from other media is considered only in the derivation of the lifetime HA. Given the absence of chemical-specific data, the assumed fraction of total intake from drinking water is 20%. The lifetime HA is calculated from the Drinking Water Equivalent Level (DWEL) which, in turn, is based on the Oral Chronic Reference Dose. Lifetime HAs are not derived for compounds which are potentially carcinogenic for humans because of the difference in assumptions concerning toxic threshold for carcinogenic and noncarcinogenic effects. A more detailed description of the assumptions and methods used in the derivation of HAs is provided in Background Document 3 in Service Code 5.

III.A.1. ONE-DAY HEALTH ADVISORY FOR A CHILD

Appropriate data for calculating a One-day HA are not available. It is recommended that the Ten-day HA of 1.2 mg/L (rounded to 1 mg/L) be used as the One-day HA.

III.A.2. TEN-DAY HEALTH ADVISORY FOR A CHILD

Ten-day HA -- 1.2E+0 mg/L

NOAEL -- 12.3 mg/kg/day
UF -- 100 (allows for interspecies and intrahuman variability)
Assumptions -- 1 L/day water consumption for a 10-kg child

Principal Study -- Muller et al., 1981

Rats were fed lindane at daily doses of 1.3, 12.3, or 25.4 mg/kg bw in the diet for 30 days. Nerve conduction delay was observed in the animals fed a daily dose of 25.4 mg/kg but was not observed at dose levels of 12.3 or 1.3 mg/kg. A NOAEL of 12.3 mg/kg/day was identified.

III.A.3. LONGER-TERM HEALTH ADVISORY FOR A CHILD

Longer-term (Child) HA -- 3.3E-2 mg/L

NOAEL -- 0.33 mg/kg/day
UF -- 100 (allows for interspecies and intrahuman variability with the use of a NOAEL from an animal study)
Assumptions -- 1 L/day water consumption for a 10-kg child

Principal Study -- Zoecon Corporation, 1983

Male and female rats were fed lindane at dietary levels of 0, 0.2, 0.8, 4, 20, or 100 ppm for 84 consecutive days. Liver hypertrophy, kidney tubular degeneration, hyaline droplets, tubular casts, tubular distension, interstitial nephritis, and basophilic tubules were observed in the 20 and 100 ppm groups. Effects were rare and very mild when noted at 4 ppm. The NOAEL was considered to be 4 ppm in this study. Based upon measured food consumption, the daily intake of lindane at 4 ppm in the diet was 0.29 mg/kg in males and 0.33 mg/kg in females. The dose of 0.33 mg/kg is identified as the NOAEL.

III.A.4. LONGER-TERM HEALTH ADVISORY FOR AN ADULT

Longer-term (Adult) HA -- 1.2E-1 mg/L

NOAEL -- 0.33 mg/kg/day
UF -- 100 (allows for interspecies and intrahuman variability with the use of a NOAEL from an animal study)
Assumptions -- 2 L/day water consumption for a 70-kg adult

Principal Study -- Zoecon Corporation, 1983 (study described in III.A.3.)

III.A.5. DRINKING WATER EQUIVALENT LEVEL / LIFETIME HEALTH ADVISORY

DWEL -- 1E-2 mg/L

RfD Verification Date -- 01/22/86 (see the RfD Section of this file)

Assumptions -- 2 L/day water consumption for a 70-kg adult

Lifetime HA -- 2E-4 mg/L

Assumptions -- 20% exposure by drinking water

Principal Study -- Zoecon Corporation, 1983

This study was used in the derivation of the oral chronic RfD; see the RfD Section for a description. NOTE: A safety factor of 10 was used in the derivation of this HA, in addition to the UF of 1000 for the RfD, to account for the possible carcinogenicity of this substance. The assessment for the potential human carcinogenicity of lindane is currently under review.

III.A.6. ORGANOLEPTIC PROPERTIES

No data available

III.A.7. ANALYTICAL METHODS FOR DETECTION IN DRINKING WATER

Determination of lindane is by a liquid-liquid extraction gas chromatographic procedure.

III.A.8. WATER TREATMENT

Treatment techniques capable of removing lindane from drinking water include adsorption on activated carbon, air stripping, reverse osmosis, and oxidation.

III.A.9. DOCUMENTATION AND REVIEW OF HAs

US EPA. 1985. Final Draft of the Drinking Water Criteria Document on Lindane. Office of Drinking Water, Washington, DC.

EPA review of HAs in 1985.

Public review of HAs following notification of availability in October, 1985.

Scientific Advisory Panel review of HAs in June, 1986.

Preparation date of this IRIS summary -- 06/17/87

III.A.10. EPA CONTACTS

Jennifer Orme Zavaleta / OST -- (202)260-7586

Edward V. Ohanian / OST -- (202)260-7571

III.B. OTHER ASSESSMENTS

Content to be determined.

IV. US EPA REGULATORY ACTIONS

EPA risk assessments may be updated as new data are published and as assessment methodologies evolve. Regulatory actions are frequently not updated at the same time. Compare the dates for the regulatory actions in this section with the verification dates for the risk assessments in sections I and II, as this may explain inconsistencies. Also note that some regulatory actions consider factors not related to health risk, such as technical or economic feasibility. Such considerations are indicated for each action. In addition, not all of the regulatory actions listed in this section involve enforceable federal standards. Please direct any questions you may have concerning these regulatory actions to the US EPA contact listed for that particular action. Users are strongly urged to read the background information on each regulatory action in Background Document 4 in Service Code 5.

IV.A. CLEAN AIR ACT (CAA)

[Ed. Note: EPA does not have information yet in this section of IRIS].

IV.B. SAFE DRINKING WATER ACT (SDWA)

IV.B.1. MAXIMUM CONTAMINANT LEVEL GOAL (MCLG) for Drinking Water

Value (status) -- 0.0002 mg/L (Final, 1991)

Considers technological or economic feasibility? -- NO

Discussion -- An MCLG of 0.0002 mg/L for lindane is promulgated based upon potential adverse effects reported in a dietary study in rats. The MCLG is based upon a DWEL of 0.01 mg/L and an assumed drinking water contribution of 20 percent. An additional uncertainty factor of 10 was applies since lindane was classed a category II contaminant (limited evidence of carcinogenicity via drinking water ingestion).

Reference -- 56 FR 3526 (01/30/91)

EPA Contact -- Health and Ecological Criteria Division / OST / (202) 260-7571 / FTS 260-7571; or Safe Drinking Water Hotline / (800) 426-4791

IV.B.2. MAXIMUM CONTAMINANT LEVEL (MCL) for Drinking Water

Value (status) -- 0.0002 mg/L (Final, 1991)

Considers technological or economic feasibility? -- NO

Discussion -- EPA has promulgated an MCL equal to the MCLG and PQL of 0.0002 mg/L.

Monitoring requirements -- All systems monitored for four consecutive quarters every three years; repeat monitoring dependent upon detection, vulnerability status and system size.

Analytical methodology -- Microextraction/gas chromatography (EPA 505); electron-capture/gas chromatography (EPA 508P; gas chromatographic/mass spectrometry (EPA 525): PQL= 0.0002 mg/L.

Best available technology -- Granular activated carbon

Reference -- 56 FR 3526 (01/30/91)

EPA Contact -- Drinking Water Standards Division / OGWDW / (202) 260-7575 / FTS 260-7575; or Safe Drinking Water Hotline / (800) 426-4791

IV.B.3. SECONDARY MAXIMUM CONTAMINANT LEVEL (SMCL) for Drinking Water

No data available

IV.B.4. REQUIRED MONITORING OF "UNREGULATED" CONTAMINANTS

[Ed. Note: EPA does not have information yet in this section of IRIS].

IV.C. CLEAN WATER ACT (CWA)

IV.C.1. AMBIENT WATER QUALITY CRITERIA, Human Health

Water and Fish Consumption: 1.86E-2 ug/L

Fish Consumption Only: 6.25E-2 ug/L

Considers technological or economic feasibility? -- NO

Discussion -- For the maximum protection from the potential carcinogenic properties of this chemical, the ambient concentration should be zero. However, zero may not be attainable at this time so the criteria given represents a E-6 incremental increase in cancer risk over a lifetime.

Reference -- 45 FR 79318 (11/28/80)

EPA Contact -- Criteria and Standards Division / OWRS (202)260-1315 / FTS 260-1315

IV.C.2. AMBIENT WATER QUALITY CRITERIA, Aquatic Organisms

Freshwater:
  Acute -- 2.0E+0 ug/L
  Chronic -- 8.0E-2 ug/L

Marine:
  Acute -- 1.6E-1 ug/L
  Chronic -- None

Considers technological or economic feasibility? -- NO

Discussion -- Water quality criteria for the protection of aquatic life are derived from a minimum data base of acute and chronic tests on a variety of aquatic organisms. The data are assumed to be statistically representative and are used to calculate concentrations which will not have significant short- or long-term effects on 95% of the organisms exposed. Recent criteria (1985 and later) contain duration and frequency stipulations: the acute criteria maximum concentration is a 1-hour average and the chronic criteria continuous concentration is a 4-day average which are not to be exceeded more than once every 3 years, on the average (see Stephen et al., 1985). Earlier criteria (1980-1984) contained instantaneous acute and 24-hour average chronic concentrations which were not to be exceeded. The freshwater chronic WQC is a 24-hour average.

Reference -- 45 FR 79318 (11/28/80)

EPA Contact -- Criteria and Standards Division / OWRS (202)260-1315 / FTS 260-1315

IV.D. FEDERAL INSECTICIDE, FUNGICIDE, AND RODENTICIDE ACT (FIFRA)

IV.D.1. PESTICIDE ACTIVE INGREDIENT, Registration Standard

Status -- Issued (1985)

Reference -- Lindane Pesticide Registration Standard. September, 1985 (NTIS No. PB86-175114).

EPA Contact -- Registration Branch / OPP (703)557-7760 / FTS 557-7760

IV.D.2. PESTICIDE ACTIVE INGREDIENT, Special Review

Action -- Final regulatory decision - PD4 (1984)
Considers technological or economic feasibility? -- YES

Summary of regulatory action -- Negotiated settlements have been made for Lindane in dog dips [49 FR 26282 (06/27/84)] and in smoke bombs [50 FR 5424 (02/08/85)].

Reference -- 45 FR 48513 (10/19/83); 49 FR 26282 (06/27/84)

EPA Contact -- Special Review Branch / OPP (703)557-7400 / FTS 557-7400

IV.E. TOXIC SUBSTANCES CONTROL ACT (TSCA)

[Ed. Note: EPA does not have information yet in this section of IRIS].

IV.F. RESOURCE CONSERVATION AND RECOVERY ACT (RCRA)

IV.F.1. RCRA APPENDIX IX, for Ground Water Monitoring

Status -- Listed
Reference -- 52 FR 25942 (07/09/87)

EPA Contact -- RCRA/Superfund Hotline (800)424-9346 / (202)260-3000 / FTS 260-3000

IV.G. SUPERFUND (CERCLA)

IV.G.1. REPORTABLE QUANTITY (RQ) for Release into the Environment

Value (status) -- 1 pound (Statutory, 1987)

Considers technological or economic feasibility? -- NO

Discussion -- The 1-pound RQ for lindane is based on aquatic toxicity as assigned by Section 311(b)(4) of the Clean Water Act (40 CFR 117.3). Available data indicate a 96-hour Median Threshold Limit of less than 0.1 ppm, which corresponds to an RQ of 1 pound.

Reference -- 52 FR 8140 (03/16/87); 54 FR 33418 (08/14/89)

EPA Contact -- RCRA/Superfund Hotline (800)424-9346 / (202)260-3000 / FTS 260-3000

VI. BIBLIOGRAPHY

VI.A. ORAL RfD REFERENCES

Fitzhugh, O.G., A.A. Nelson and J.P. Frawley. 1950. The chronic toxicities of technical benzene hexachloride and its alpha, beta and gamma isomers. J. Pharmacol. Exp. Ther. 100: 59-66.

Muller, D., H. Klepel, R.M. Macholz, H.J. Lewerenz and R. Engst. 1981. Electroneurophysiological studies on neurotoxic effects of hexachlorocyclo- hexane isomers and gamma-pentachlorocyclohexene. Bull. Environ. Contam. Toxicol. 27(5): 704-706.

Rivett, K.F., H. Chesterman, D.N. Kellett, A.J. Newman, and A.N. Worden. 1978. Effects of feeding lindane to dogs for periods of up to 2 years. Toxicology. 9: 273-289.

US EPA. 1985. Drinking Water Criteria Document for Lindane. Prepared by the Office of Health and Environmental Assessment, Environmental Criteria and Assessment Office, Cincinnati, OH for the Office of Drinking Water, Washington, DC.

Zoecon Corporation. 1983. MRID No. 00128356. Available from EPA. Write to FOI, EPA, Washington, DC 20460.

VI.B. INHALATION RfC REFERENCES

None

VI.C. CARCINOGENICITY ASSESSMENT REFERENCES

None

VI.D. DRINKING WATER HA REFERENCES

Muller, D., H. Klepel, R.M. Macholz, H.J. Lewerenz and R. Engst. 1981. Electroneurophysiological studies on neurotoxic effects of hexachlorocyclo- hexane isomers and gamma-pentachlorocyclohexene. Bull. Environ. Contam. Toxicol. 27(5): 704-706.

US EPA. 1985. Final Draft of the Drinking Water Criteria Document on Lindane. Office of Drinking Water, Washington, DC.

Zoecon Corporation. 1983. MRID No. 00128356. Available from EPA. Write to FOI, EPA, Washington, DC 20460.

VII. REVISION HISTORY

Date	Section	Description
03/31/87	IV.A.	Additional pesticide AI data added
03/01/88	I.A.1.	Principal study citation corrected
03/01/88	I.A.1.	Dose conversion corrected
03/01/88	I.A.2.	Text revised
03/01/88	III.A.	Health Advisory added
08/01/90	III.A.10	Primary contact changed
08/01/90	IV.F.1.	EPA contact changed
08/01/91	VI.	Bibliography on-line
01/01/92	I.A.7.	Secondary contact changed
01/01/92	IV.	Regulatory actions updated
07/01/92	I.B.	Inhalation RfC now under review
10/01/93	II.	Inadvertently listed as under review

SYNONYMS

AALINDAN
AFICIDE
AGRISOL G-20
AGRONEXIT
AMEISENATOD
AMEISENMITTEL MERCK
APARASIN
APHTIRIA
APLIDAL
ARBITEX
BBH
BEN-HEX
BENTOX 10
gamma-BENZENE HEXACHLORIDE
BENZENE HEXACHLORIDE-gamma-
isomer
BEXOL
BHC
gamma-BHC
CELANEX
CHLORESENE
CODECHINE
CYCLOHEXANE, 1,2,3,4,5,6-
HEXACHLORO-, gamma-isomer
DBH
DETMOL-EXTRAKT
DETOX 25
DEVORAN
DOL GRANULE
DRILL TOX-SPEZIAL AGLUKON
ENT 7,796
ENTOMOXAN
EXAGAMA
FORLIN
GALLOGAMA
GAMACARBATOX
GAMACID
GAMAPHEX
GAMENE
GAMISO
GAMMA-COL
GAMMAHEXA
GAMMAHEXANE
GAMMALIN
GAMMALIN 20
GAMMATERR
GAMMEX

gamma-HEXACHLOROBENZENE
1,2,3,4,5,6-
HEXACHLOROCYCLOHEXANE
1-alpha,2-alpha,3-beta,4-alpha,5-alpha,6-
beta-HEXACHLOROCYCLOHEXANE
Hexachlorocyclohexane, gamma-
gamma-1,2,3,4,5,6-
HEXACHLOROCYCLOHEXANE
1,2,3,4,5,6-
HEXACHLOROCYCLOHEXANE,
gamma-ISOMER
HEXACHLOROCYCLOHEXANE,
gamma-ISOMER
HEXATOX
HEXAVERM
HEXICIDE
HEXYCLAN
HGI
HORTEX
INEXIT
ISOTOX
JACUTIN
KOKOTINE
KWELL
LENDINE
LENTOX
LIDENAL
LINDAFOR
LINDAGAM
LINDAGRAIN
LINDAGRANOX
Lindane
gamma-LINDANE
LINDAPOUDRE
LINDATOX
LINDOSEP
LINTOX
LOREXANE
MILBOL 49
MSZYCOL
NA 2761
NCI-C00204
NEO-SCABICIDOL
NEXEN FB
NEXIT
NEXIT-STARK
NEXOL-E

GAMMEXANE   NICOCHLORAN
GAMMOPAZ   NOVIGAM
GEXANE   OMNITOX
HCCH   OWADZIAK
HCH   PEDRACZAK
gamma-HCH   PFLANZOL
HECLOTOX   QUELLADA
HEXA   RCRA WASTE NUMBER U129
HEXACHLORAN   SANG gamma
HEXACHLORANE   SILVANOL
gamma-HEXACHLORANE   SPRITZ-RAPIDIN
gamma-HEXACHLORAN   SPRUEHPFLANZOL
gamma-HEXACHLOR   STREUNEX
                                   TAP 85
                                   TRI-6
                                   VITON

Source: *Instant EPA's IRIS*, 1997, Instant Reference Sources, Inc., 7605 Rockpoint Drive, Austin, TX 78731 (Fax: 512-345-2386; Internet URL, http://www.instantref.com/inst-ref.htm).

Production, Use, and Pesticide Labeling Information

1. Description of Chemical

    Generic Name: gamma isomer of 1,2,3,4,5,6-hexachlorocyclohexane

    Common Name: lindane

    Trade Names: Exagamma, Forlin, Gallogamma, Gammaphex, Gammex, Gexane, Grammapoz, Grammexane, Inexit, Kwell, Lindafor, Lindagrain, Lindagram, Lindagranox, Lindalo, Lindamul, Lindapoudre, Lindaterra, Lindex, Lindust, Lintox, Novigram, and Silvanol

    EPA Shaughnessy Code: 009001

    Chemical Abstracts Service (CAS) Number: 58-89-9

    Year of Initial Registration: 1950

    Pesticide Type: insecticide/acaracide

    Chemical Family: chlorinated hydrocarbon

    US Producers: None

Foreign Producers: Celamerck GmbH KG, Ingelheim, Federal Republic of Germany; Rhone Poulenc Phytosanitaire, Lyon, France; Mitsui, Inc., Fukuoka, Japan; Tianjin International Trust & Investment Corp., Tianjin, China

2. Use Patterns and Formulations

- Application sites: field and vegetable crops (including seed treatment) and non-food crops (ornamentals and tobacco), greenhouse food crops (vegetables) and non-food crops (ornamentals), forestry (including Christmas tree plantations), domestic outdoor and indoor (pets and household), commercial indoor (food/feed storage areas and containers), animal premises (including manure), wood or wooden structures, and human skin/clothing (military use only).

- Percent of lindane used on various crops/sites:

  - Hardwood lumber 19%, seed treatment 48%, forestry <1%, livestock 20%, pineapple 2%, ornamentals 2%, pecans 3%, pets 3%, structures <1%, household 1%, and cucurbits 1%

- Types and methods of applications:

  Used as a dip tank solution (livestock, lumber, and pets), as a livestock spray, by ground equipment delivering a ground or foliar spray or dust, by soil incorporation, by soil injection in combination with a fumigant (for pineapple use only), as a smoke (for greenhouse fumigation only), and as a dust for human skin/clothing (military use only).

- Application rates:

  These range from 0.25 to 2.25 oz/100 lb. of seed for seed treatment; 0.1 to 2.06 lb/A for foliar and soil treatment; 0.8 to 1.5 oz/50,000 cubic feet of greenhouse; 0.006 to 0.11 lb/gal. for bark; 0.023 to 3% sprays, dips, and dusts for indoor and animal treatment; <0.01 lb/1,000 square feet for animal premises; <4 lb/ 1,000 square feet (14.64% solutions) for wood and wooden structures; and 1% dust for human skin/clothing treatment (military use only).

- Types of formulations:

  0.27-11.2% impregnated formulations, 0.5- 75% dusts, 3-73% wettable powders, 0.5-25% liquids, 0.25-3% pressurized liquids, 1-4% flowable concentrates, 0.45-40% emulsifiable concentrates.

3. Science Findings

- Chemical Characteristics

Technical lindane is a white crystalline solid. Its melting point is 112-113 C. It is soluble in most organic solvents and is relatively insoluble in water. Lindane is stable to light, heat, air, and strong acids, but decomposes to trichlorobenzenes and HCL in alkali.

- Toxicological Characteristics

- Acute oral: 88 mg/kg, Toxicity Category II
- Acute dermal: 300 mg/kg, Toxicity Category II
- Acute inhalation: data gap
- Primary eye irritation: data gap
- Primary skin irritation: Irritant, Toxicity Category I
- Skin sensitization: data gap

- Major routes of exposure: Human exposure from lindane is greatest during mixing, loading, and application. Dermal, ocular, and inhalation exposures to workers may occur during application. Exposure can be reduced by the use of approved respirators, protective clothing, and goggles.

- Oncogenicity: A 2-year mouse oncogenicity study demonstrated increased incidences of liver tumors (male and female) when dosed at 400 ppm. An 80-week mouse feeding study demonstrated increased incidences of liver tumors at the 80 ppm level, but not at the 160 ppm level. Two subchronic studies provide supportive evidence of oncogenicity. The mouse studies were referred to the Agency's Carcinogen Assessment Group (CAG) for evaluation. Based on the weight of the evidence, CAG classified lindane in the range B2-C. OPP believes that the classification C is appropriate at this time and, therefore, will regulate lindane as a class C carcinogen, pending receipt of the required rat oncogenicity study.

- Metabolism: Lindane does not appear to bioaccumulate in tissues.

- Teratology: Teratology studies in the rat, rabbit, and mouse were negative for teratogenic effects.

- Reproduction: A 3-generation rat reproduction study was negative at 100 ppm.

- Mutagenicity: Available data show lindane to be negative for gene mutation in bacterial Ames assays, host mediated, and dominant lethal assays. Lindane has been reported as negative in other in vitro assays for DNA damage/repair in bacteria, rat, and mouse hepatocytes, and mammalian cell transformation assays.

- Physiological and Biochemical Behavioral Characteristics

     - Mechanism of pesticidal action: Lindane acts in the nervous system through unknown mechanisms.

     - Metabolism and persistence in plants and animals: The metabolism of lindane in plants and livestock animals has not been adequately described. Additional data are being required.

- Environmental Characteristics

     - Available data are insufficient to assess the environmental fate of lindane. Data gaps exist for all required studies.

     - Preliminary adsorption data indicate that lindane has a low mobility in mineral soils and is relatively immobile in muck soils; however, the potential for lindane contamination of surface and groundwater exists, based on the results of a monitoring study conducted in certain southern states.

- Ecological Characteristics

     - Avian acute oral toxicity: data gap

     - Avian dietary toxicity: 882 ppm for bobwhite quail, 561 ppm for ring-necked pheasant (moderately toxic), and >5000 ppm for mallard duck (practically nontoxic).

     - Freshwater fish acute (LC50) toxicity: coldwater species (rainbow trout) 27 ppb for technical lindane (very highly toxic), warmwater species (bluegill) 68 ppb for technical lindane (very highly toxic).

     - Aquatic freshwater invertebrate toxicity: Daphnia 460 ppb (highly toxic). Additional data are required to fully characterize the ecological effects of lindane.

- Tolerance Assessment

     - The Agency is unable to complete a full tolerance assessment, because the metabolism of lindane in plants and livestock animals has not been adequately described. Also, seed treatment is now considered to be a food use requiring a tolerance, unless results of a radiolabeled study indicate that there is no translocation to edible parts of the plant following seed treatment. No new tolerances, except those required to support the existing seed treatment uses of lindane, will be considered until the toxicology and residue chemistry data gaps identified in the Standard have been filled.

- Established tolerances are published in 40 CFR 180.133.
- A listing of US tolerances includes the following:

Tolerance expressed Commodity as parts per million:

Commodity	Tolerance
Apples	1.0
Apricots	1.0
Asparagus	1.0
Avocados	1.0
Broccoli	1.0
Brussels sprouts	1.0
Cabbage	1.0
Cattle, fat	7.0
Cauliflower	1.0
Celery	1.0
Cherries	1.0
Collards	1.0
Cucumbers	3.0
Eggplant	1.0
Goats, fat	7.0
Grapes	1.0
Guavas	1.0
Hogs, fat	4.0
Horses, fat	7.0
Kale	1.0
Kohlrabi	1.0
Lettuce	3.0
Mangoes	1.0
Melons	3.0
Mushrooms	3.0
Mustard greens	1.0
Nectarines	1.0
Okra	1.0
Onions, dry bulb only	1.0
Peaches	1.0
Pears	1.0
Pecans	0.01
Peppers	1.0
Pineapples	1.0
Plums, inc. prunes	1.0
Pumpkins	3.0
Quinces	1.0
Sheep, fat	7.0
Spinach	1.0
Squash	3.0
Summer squash	3.0
Strawberries	1.0
Swiss chard	1.0
Tomatoes	3.0

- The best available data for determining an interim acceptable daily intake level of lindane is a subchronic feeding study in rats (1983) which demonstrated a No Observed Effect Level (NOEL) of 4 ppm. Based on dietary analysis, food intake, and body weight data from this particular study, the NOEL of 4 ppm is equivalent to 0.3 mg/kg/day. Using this latter value and a safety factor of 1000, the Provisional Acceptable Daily Intake (PADI) is 0.0003 mg/kg/day, and the Maximum Permissible Intake (MPI) for a 60-kg person is 0.018 mg/day. The Theoretical Maximum Residue Contribution (TMRC) for lindane, based on all established tolerances, is 1.4189 mg/day/ 1.5 kg of diet. The percent of the MPI used by the TMRC is 78 - 83%.

- Although the theoretical concentration from existing tolerances greatly exceeds the MPI, FDA market basket surveys indicate that actual residues of lindane are much lower. The Agency believes that the actual risk to consumers from the daily consumption of lindane, based on FDA market basket data for 1978-1982, is only 0.000002 mg/kg/day. Under this scenario, only 0.7% of the Maximum Permissible Intake is actually used.

- Summary Science Statement

Lindane is a chlorinated hydrocarbon of moderate mammalian acute toxicity. Lindane has been shown to be oncogenic in mice, but it is not genotoxic. The Agency has concluded that lindane is a possible human carcinogen. The Agency is requiring that another rat chronic/oncogenicity bioassay be performed. Lindane has been associated with possible induction of blood dyscrasia (aplastic anemia). The Agency is requiring a laboratory animal study to permit assessment of lindane's potential to cause blood dyscrasias. Other toxicology studies demonstrate systemic toxicity, targeting the liver and kidney. Lindane's behavior in the environment is not well-defined. The Agency is requiring a full complement of environmental fate studies. Lindane is slightly to moderately toxic to birds, and highly toxic to some aquatic organisms. Lindane is highly toxic to honeybees and certain beneficial parasites and predaceous insects. Additional studies on the ecological effects of lindane are required.

4. Summary of Regulatory Position and Rationale

- Required unique labeling summary:

- All manufacturing-use and end-use lindane products must bear appropriate labeling as specified in 40 CFR 162.10. In addition, the following information must appear on the labeling:

- All manufacturing-use products must state that they are intended for formulation into other manufacturing-use products or end-use products only for registered uses.

- All manufacturing-use products shall contain the following text in the Environmental Hazards section of the label: This pesticide is toxic to fish and aquatic invertebrates. Do not discharge effluent containing this product into lakes, streams, ponds, estuaries, oceans, or public waters unless this product is specifically identified and addressed in an NPDES permit. Do not discharge effluent containing this product into sewer systems without previously notifying the sewage treatment plant authority. For guidance, contact your State Water Board or Regional Office of the EPA.

- All end-use products containing lindane that were classified as restricted by the Final Notice of Determination concluding the RPAR shall continue to be classified for restricted use, and the restricted use label must include the cancer hazard warning statement.

- All end-use products shall continue to carry the applicator protection statements previously required by the Final Notice of Determination concluding the RPAR. Products with directions for foliar application to crops whose culture involves hand labor must bear the statements required under PR Notices 83-2 and 84-1 for farm worker safety, including a 24-hour reentry interval.

- End-use products with directions for spraying uninhabited buildings or empty storage bins must include protective clothing requirements, including the use of a respirator.

- All end-use products for indoor use shall indicate that lindane is not to be applied to edible product areas of food processing plants or to serving areas while food is exposed.

- All end-use products with uses on livestock or livestock premises must indicate not to contaminate food, feed, or water with the pesticide. Also, there must be a statement that indicates lindane is not to be applied to poultry houses, dairy barns, and milk rooms. All feed or water troughs must be covered, and all livestock should be removed from animal shelters (barns, sheds, etc.) prior to treatment of the structure.

- All end-use products for structural pest control must indicate that lindane is not to be applied in currently occupied areas (i.e. regular living or working areas, including finished basements or finished attics) of homes or other buildings. The characterization of a use site depends upon its intended function and not upon whether there are occupants in the area at the time of treatment.

- The Agency has determined that it should continue the registration of all currently registered uses of lindane. The Agency concluded in the RPAR that most uses of lindane would be continued, because a risk/benefit assessment demonstrated that the benefits from the uses outweighed the risks provided certain labeling restrictions, such as restricting some uses to certified pesticide applicators, requirements for protective clothing, and label statements describing necessary precautions, were added to all lindane labels. The Agency has reevaluated that decision and concludes that, except as described below, the risks and benefits are substantially the same as those described in the RPAR process.

- Based on FDA market basket residue levels, which the Agency believes in this case are more generally representative of actual residues than theoretical calculations, the estimate of the upper 95% confidence level for excess cancer risk is 2×10^{-6}. The estimate of the upper 95% confidence level for excess cancer risk to applicators for various uses is estimated to be from 10^{-4} to 10^{-7}, depending upon the site and method of application. The Agency has recalculated the exposures and margins of safety for applicators for 24 use patterns and has developed initial calculations for mixer/loaders or combination mixer/loader/applicators for 3 use patterns (forestry, cucurbits, and pecans). This reassessment is based on current Agency methods and models and consideration of a lower NOEL from the subchronic rat study for uses involving subchronic type exposure. Based on these calculations, the Agency will initiate a Special Review for the forestry and uninhabited buildings and empty storage bins spray uses on the basis of risks to applicators. Use of protective clothing, including an MSHA/NIOSH approved respirator, is now being required for spraying uninhabited buildings and empty storage bins while the Special Review is underway. Protective clothing requirements stipulated in the Final Notice of Determination of the lindane RPAR will continue for all other uses. No significant changes from the exposure values presented in PD-4 occurred for 12 of the uses. The calculated exposures for cucurbits, crawl spaces, dog dusts, dog dips, shelf paper, and commercial moth sprays increased by approximately an order of magnitude, but are still acceptable. Applicator exposure data is being required for seed treatment, structural treatment, livestock spraying, dog washes, dog shampoos, and dog dusts. Air monitoring data are required for the structural treatment and dog treatment uses. Exposure studies are required, in addition to toxicity studies, to support the registrations for application of lindane to human skin/clothing by the military.

- Lindane seed treatments were registered many years ago as non-food uses not requiring tolerances. The Agency now considers seed treatment to be a food use and requires data to support a tolerance, unless results of a radiolabeled study indicate that there is no translocation to edible parts of the plant following seed treatment. Data from one of these two alternatives must be submitted to support the seed treatment uses. No new tolerances, except those required to support the existing seed treatment uses, will be considered until the chronic feeding and residue chemistry data gaps identified in the Standard have been filled.

- Available data are insufficient to fully assess the environmental fate of lindane and its ecological effects. A full complement of such studies is being required. Precautionary label statements will continue to be required.

5. Summary of Major Data Gaps

Data Gap Due Date

An acute inhalation study	9 months
A 90-day inhalation study	15 months
A dermal sensitization study	9 months
A 21-day dermal toxicity study	12 months
A rat chronic/oncogenicity study	50 months
Laboratory animal blood dyscrasias	study pending
Full complement of environmental fate studies	39 months
Plant metabolism studies	24 months
Livestock animal metabolism studies	18 months
Residue chemistry studies on all crops, treatment uses	48 months including seed

6. Contact Person at EPA

George T. LaRocca
Product Manager (15)
Insecticide- Rodenticide Branch Registration Division (T-767C)
Office of Pesticide Programs
Environmental Protection Agency
401 M Street, SW
Washington, DC 20460

Office location and telephone number:

Room 204, Crystal Mall Bldg. 2
1921 Jefferson Davis Highway
Arlington, VA 22202

Telephone: (703)557-2400

Source: *Instant EPA's Pesticide Facts*, 1994, Instant Reference Sources, Inc., 7605 Rockpoint Drive, Austin, TX 78731 (Fax: 512-345-2386; Internet URL, http://www.instantref.com/inst-ref.htm).

Malathion

Introduction

Malathion is used as an insecticide for fruits, vegetables, ornamentals, household and livestock use. It is also used as an acaricide, in the control of flies and other insect pests in animal and poultry houses, in the control of adult mosquitoes in public health programs, in the control of human body and head lice and in flea and tick dips. It is used in veterinary medicine as an ectoparasiticide.

Release of malathion to the environment will occur as a result of its manufacture and use as a broad spectrum insecticide and acaricide. If malathion is released to soil, it should moderately bind to the soil, and will be subject to significant biodegradation and hydrolysis. Reported half lives in soil range from approximately 4 days to a reported literature avg of 6 days. If malathion is released to water it may moderately sorb to sediment, but will not be expected to bioconcentrate in aquatic organisms. It should be subject to biodegradation and may be subject to photodegradation at the surface of water. Hydrolysis in water may be an important fate process based on reported half life range of 0.2 weeks at pH 8.0 to 21 weeks at pH 6.0. Volatilization from water should not be an important fate process. If malathion is released to the atmosphere it may be subject to direct photolysis. The estimated vapor phase half life in the atmosphere is 1.50 days as a result of hydrogen abstraction by photochemically produced hydroxyl radicals. Occupational exposure to malathion will occur mainly from dermal contact and inhalation of contaminated air as a result of its use as an insecticide and acaricide. General exposure may result from the consumption of contaminated food. Minor exposure may occur from the consumption of contaminated water. [828]

Identifying Information

CAS NUMBER: 121-75-5

NIOSH Registry Number: WM8400000

SYNONYMS: MERCAPTOSUCCINIC ACID, DIETHYL ESTER, S-ESTER WITH O,O-DIMETHYL PHOSPHORO-, DITHIOATE, DIETHYL MERCAPTOSUCCINATE, O,O-DIMETHYL PHOSPHORODITHIOATE, O,O-DIMETHYL-S-(1,2-DICARBETHOXYETHYL)PHOSPHORODITHIOATE, 1,2-DI(ETHOXYCARBONYL)ETHYL O,O-DIMETHYL PHOSPHORODITHIOATE, S-

(1,2-DICARBETHOXYETHYL) O,O-DIMETHYLDITHIOPHOSPHATE,
CHEMATHION, CYTHION, CARBOPHOS, EMMATOS, FORMAL, FOSFOTHION,
FYFANON, KARBOFOS, KOP-THION, MALACIDE, MALAGRAN, MALAMAR,
MALATHON, MLT, PHOSPHOTHION, SADOFOS, NCI-C00215, ENT 17,034, S-
(1,2-BIS(CARBETHOXY)ETHYL) O,O-DIMETHYL DITHIOPHOSPHATE, S-(1,2-
BIS(ETHOXYCARBONYL)ETHYL O,O-DIMETHYLPHOSPHORODITHIOATE, S-
1,2-BIS(ETHOXYCARBONYL)ETHYL-O,O-DIMETHYL THIOPHOSPHATE,
CALMATHION, CARBETOX, CARBETHOXY MALATHION, CARBETOVUR,
CARBOFOS, CELTHION, CIMEXAN, COMPOUND 4049, DETMOL MA, DETMOL
MA 96%, DICARBOETHOXYETHYL O,O-DIMETHYL PHOSPHORODITHIOATE,
S-(1,2-DI(ETHOXYCARBONYL)ETHYL)
DIMETHYLPHOSPHOROTHIOLOTHIONATE, DIETHYL
(DIMETHOXYPHOSPHINOTHIOYLTHIO) BUTANEDIOATE, DIETHYL
(DIMETHOXYPHOSPHINOTHIOYLTHIO)SUCCINATE, DIETHYL
MERCAPTOSUCCINATE, O,O-DIMETHYL DITHIOPHOSPHATE, S-ESTER,
DIETHYL MERCAPTOSUCCINATE, O,O-DIMETHYL THIOPHOSPHATE,
DIETHYL MERCAPTOSUCCINATE S-ESTER WITH O,O-
DIMETHYLPHOSPHORODITHIOATE, DIETHYL MERCAPTOSUCCINIC ACID
O,O-DIMETHYL PHOSPHORODITHIOATE,
((DIMETHOXYPHOSPHINOTHIOYL)THIO)BUTANEDIOIC ACID DIETHYL
ESTER, O,O-DIMETHYL-S-(1,2-
BIS(ETHOXYCARBONYL)ETHYL)DITHIOPHOSPHATE, O,O-DIMETHYL-S-(1,2-
DICARBETHOXYETHYL) DITHIOPHOSPHATE, O,O-DIMETHYL-S-(1,2-
DICARBETHOXYETHYL)PHOSPHORODITHIOATE, O,O-DIMETHYL-S-(1,2-
DICARBETHOXYETHYL) THIOTHIONOPHOSPHATE, O,O-DIMETHYL S-1,2-
DI(ETHOXYCARBAMYL)ETHYL PHOSPHORODITHIOATE, O,O-
DIMETHYLDITHIOPHOSPHATE DIETHYLMERCAPTOSUCCINATE,
MERCAPTOSUCCINIC ACID DIETHYL ESTER, MERCAPTOTHION,
PHOSPHORODITHIOIC ACID, O,O-DIMETHYL ESTER, S-ESTER WITH
DIETHYL, MERCAPTOSUCCINATE

Endocrine Disruptor Information

Malathion is a suspected EED that is listed by the World Wildlife Foundation, Canada
[813].

Chemical and Physical Properties

CHEMICAL FORMULA: $C_{10}H_{19}O_6PS_2$

MOLECULAR WEIGHT: 330.36

WLN: 2OV1YVO2&SPS&O1&O1

PHYSICAL DESCRIPTION: Deep brown to yellow liquid

SPECIFIC GRAVITY: 1.23 @ 25/4°C [027], [031], [043], [172]

DENSITY: 1.2076 g/mL @ 20°C [017], [047]

MELTING POINT: 2.85°C [055], [169], [172], [395]

BOILING POINT: 156-157°C @ 0.7 mm Hg (decomposes) [017], [027], [047], [421]

SOLUBILITY: Water: <1 mg/mL @ 21.5°C [700]; DMSO: >=100 mg/mL @ 22°C [700]; 95% Ethanol: >=100 mg/mL @ 22°C [700]; Acetone: >=100 mg/mL @ 22°C [700]

OTHER SOLVENTS: Most organic solvents: Miscible [031], [062], [172], [395]; Esters: Miscible [031], [169], [421]; Ketones: Miscible [031], [169], [421]; Ethers: Miscible [031], [169], [421], [430]; Aromatic and alkylated aromatic hydrocarbons: Miscible [031]; Vegetable oils: Miscible [031], [430]; Petroleum ether: Soluble to 35% [031]; Benzene: Soluble [017], [047]; Chloroform: Miscible [455]; Some types of mineral oil: Slightly soluble [169]; Alcohols: Miscible [031], [169], [430], [455]; Certain paraffin hydrocarbons: Limited solubility [031]; Petroleum oils: Limited solubility [172], [173], [395]; Hexane: Miscible [421]

OTHER PHYSICAL DATA: Density: 1.2315 g/mL @ 25°C [062], [430]; Refractive index: 1.4960 @ 20°C [017], [047]; 1.4985 @ 25°C [025], [031], [169], [172]; Mercaptan odor [058]; Technical grade has a garlic-like odor [051]; Evaporation rate (butyl acetate = 1): Negligible [058]

Source: *The National Toxicology Program's Chemical Database, Volume 2*, 1992, CRC Press, Inc./Lewis Publishers, 2000 Corporate Blvd., Boca Raton, FL 33431 (Fax: 407-998-9114).

Environmental Reference Materials

A source of EED environmental reference materials for malathion is Crescent Chemical Co., Hauppauge, NY (Fax: 516-348-0913); Catalog No. 45551.

Hazardous Properties

ACUTE/CHRONIC HAZARDS: Malathion is an irritant [051], [102] and may be absorbed through the skin [062], [102], [451], [406]. It may also cause lacrimation [051], [421], and it is a cholinesterase inhibitor [151], [169], [301], [430]. When heated to decomposition it emits toxic fumes of carbon dioxide, carbon monoxide, sulfur oxides, hydrogen sulfide, phosphorous oxides and organic sulfide [043], [051], [058], [371]. Malaathion is less toxic to humans than other anticholinesterase agents because it is metabolized in the liver to an inactive form [102]. Humans may be more susceptible to the toxic effects of this compound than rats [151].

FIRE HAZARD: Malathion has a flash point of >163°C (>325°F) [051], [058], [371]. It is probably combustible. Fires involving this material can be controlled with a dry chemical, carbon dioxide or Halon extinguisher. A water spray may also be used [051], [058], [371].

REACTIVITY: Malathion is incompatible with strong oxidizers [058], [102], [346]. It is also incompatible with alkaline materials [058], [169], [173], [395]. It is decomposed by acids [169]. It may also be decomposed by prolonged contact with iron or iron-containing material, Raney nickel and sodium or lithium in liquid ammonia [051]. It is corrosive to copper, iron, lead and tin [421], [455]. It will also attack some forms of plastics, rubber and coatings [102].

STABILITY: Malathion hydrolyzes readily at a pH of greater than 7 or less than 5 [031], [173], [395], [421]. It is stable in aqueous solutions if buffered to a pH of 5.26 [031], [173]. It starts to decompose @ 49°C [051], [371].

Source: *The National Toxicology Program's Chemical Database, Volume 7*, 1992, CRC Press, Inc./Lewis Publishers, 2000 Corporate Blvd., Boca Raton, FL 33431 (Fax: 407-998-9114).

Medical Symptoms of Exposure

Symptoms may include tightness of the chest, wheezing, laryngeal spasms, excessive salivation, miosis, aching in and behind the eyes (attributed to ciliary spasm), blurring of distant vision, tearing, rhinorrhea and frontal headache following inhalation of extremely high concentrations. After ingestion, symptoms such as anorexia, nausea, vomiting, abdominal cramps, diarrhea, giddiness, confusion, ataxia and slurred speech may occur. [102]

Other effects on the eyes may include hyperemia of conjunctiva, constriction of the pupils and spasm of accomodation. Saku disease may also result from exposure to an anticholinesterase agent causing myopia, reduced vision from corneal astigmatism, narrowing of peripheral visual fields (with or without central scotoma), congestion or atrophy of optic nerves, difficulty with ocular pursuit movements and abnormality of ERG's [099]. Other symptoms of exposure to this compound may include irritation of the eyes and nose, bluish discoloration of the skin and runny nose. After skin absorption it may cause sweating and twitching in the area of absorption within 15 minutes to 4 hours. Severe intoxication by all routes may cause weakness, generalized twitching, paralysis, arrested breathing, dizziness, staggering, slurred speech, generalized sweating, irregular or slow heartbeat, convulsions and coma. Repeated exposure may result in increased susceptibility to this and other chemicals [102]. Other symptoms may include skin irritation, allergic sensitization of the skin, rhinitis, ataxia, lacrimation, dyspnea, slow pulse, tremors and muscular pain of the lower limbs [051]. It may also cause sensitivity to light, marked flaccidity of the limbs and unconsciousness [173]. It may cause a stinging sensation when in contact with the skin [371], [455]. It may also cause reddening of the skin [371]. Hypersecretion has also been reported [295]. Other symptoms may include bronchorrhea, bradycardia, bronchoconstriction, muscle fasiculations and possible death from respiratory failure. Arrhythmia, atrioventricular block and fatal aplastic anemia may occur [395]. Exposure may cause blood pressure depression [043]. It may also cause lightheadedness, loss of accomodation and sphinctor factor [058].

Source: *The National Toxicology Program's Chemical Database, Volume 4*, 1992, CRC Press, Inc./Lewis Publishers, 2000 Corporate Blvd., Boca Raton, FL 33431 (Fax: 407-998-9114).

Toxicological Information

The toxicological information below is gathered from several sources including two national databases: one from the *National Toxicology Program Chemical Repository Database* and another from *EPA's Integrated Risk Information System (IRIS)*.

Toxicology Information from the National Toxicology Program

CARCINOGENICITY:

    Review:

        IARC Cancer Review: Animal Inadequate Evidence
        IARC: Not classifiable as a human carcinogen (Group 3)

    Status:

        NCI Carcinogenesis Bioassay (Feed); Negative: Male and Female Rat, Male and Female Mouse

TERATOGENICITY:

    Reproductive Effects Data:

Type of Effect	Route/Animal	Amount Dosed (Notes)
TDLo:	orl-rat	5550 mg/kg (91D pre/1-20D preg)
TDLo:	orl-rat	43920 mg/kg (MGN)
TDLo:	orl-rat	191 mg/kg (9D preg)
TDLo:	orl-rat	283 mg/kg (9D preg)
TDLo:	unr-rat	80 mg/kg (2D male)

MUTATION DATA:

Test	Lowest dose		Test	Lowest dose
dnd-esc	100mg/L		mmo-sat	10mg/L
oms-hmn:leu	200mg/L		mmo-bcs	1nmol/plate
sce-hmn:fbr	5mg/L		mmo-omi	100mg/L
cyt-ham-ipr	240mg/kg		cyt-mus-ipr	230mg/kg
cyt-mus-orl	18gm/kg/50D-I		cyt-ham:lng	76mg/L
sce-ham:lng	40mg/L		sce-ham:ovr	300umol/L
cyt-ofs-mul	200nL/L		oms-hmn:lym	70mg/L
oms-mus-skn	1gm/kg		cyt-mus-skn	500mg/kg
cyt-ham:ovr	303mg/L			

TOXICITY:

Typ.Dose	Mode	Specie	Amount	Units	Other
LDLo	orl	man	471	mg/kg	
LDLo	orl	wmn	246	mg/kg	
LD50	skn	rat	4444	mg/kg	
LDLo	iat	cat	1820	ug/kg	
LD50	scu	rat	1000	mg/kg	
LD50	ivn	rat	50	mg/kg	
LC50	ihl	rat	84600	ug/m3/4H	
LD50	ipr	mus	193	mg/kg	
LD50	scu	mus	221	mg/kg	
LD50	ivn	mus	184	mg/kg	
LD50	ipr	ham	2400	mg/kg	
LD50	ipr	rat	250	mg/kg	
LD50	ipr	dog	1857	mg/kg	
LCLo	ihl	cat	10	mg/m3/4H	
LDLo	orl	rbt	1200	mg/kg	
LD50	orl	rbt	250	mg/kg	
LD50	orl	bwd	400	mg/kg	
LD50	orl	mus	190	mg/kg	
LD50	orl	rat	290	mg/kg	
LD50	skn	rbt	4100	mg/kg	
LD50	orl	gpg	570	mg/kg	
LD50	skn	gpg	6700	mg/kg	
LD50	ipr	gpg	550	mg/kg	
LD50	orl	ckn	600	mg/kg	
LD50	orl	dck	1485	ug/kg	
LD50	orl	dom	500	mg/kg	
LD50	orl	ctl	53	mg/kg	
LD50	unr	ctl	53	mg/kg	
LD50	unr	mam	500	mg/kg	
LD50	unr	mus	375	mg/kg	
LD50	unr	rat	450	mg/kg	

OTHER TOXICITY DATA:

Review:
Toxicology Review-6

Standard and Regulations:
DOT-Hazard: ORM-A; Label: None

Status:

> NIOSH Analytical Methods: see EPN, Malathion, and Parathion, 5012
> EPA Genetox Program 1988, Positive/dose response: In vitro SCE-human
> EPA Genetox Program 1988, Negative: D melanogaster Sex-linked lethal
> EPA Genetox Program 1988, Negative: In vitro UDS-human fibroblast; TRP reversion
> EPA Genetox Program 1988, Negative: S cerevisiae-homozygosis
> EPA Genetox Program 1988, Negative: Carcinogenicity-mouse/rat; Histidine reversion-Ames test
> EPA Genetox Program 1988, Inconclusive: B subtilis rec assay; E colipolA without S9
> EPA TSCA Test Submission (TSCATS) Data Base, January 1989
> Meets criteria for proposed OSHA Medical Records Rule

SAX TOXICITY EVALUATION:

> THR: A human poison by ingestion. An experimental poison by ingestion, inhalation, intraperitoneal, intravenous, intraarterial, subcutaneous and possibly other routes. Human systemic effects by ingestion. Human mutagenic data. An organic phosphate cholinesterase inhibitor-type insecticide.

Source: *Instant Tox-Base*, 1995, Instant Reference Sources, Inc., 7605 Rockpoint Drive, Austin, TX 78731 (Fax: 512-345-2386; Internet URL, http://www.instantref.com/inst-ref.htm).

Toxicology Information from EPA's Integrated Risk Information System (IRIS)

I. CHRONIC HEALTH HAZARD ASSESSMENTS FOR NONCARCINOGENIC EFFECTS

I.A. REFERENCE DOSE FOR CHRONIC ORAL EXPOSURE (RfD)

The Reference Dose (RfD) is based on the assumption that thresholds exist for certain toxic effects such as cellular necrosis, but may not exist for other toxic effects such as carcinogenicity. In general, the RfD is an estimate (with uncertainty spanning perhaps an order of magnitude) of a daily exposure to the human population (including sensitive subgroups) that is likely to be without an appreciable risk of deleterious effects during a lifetime. Please refer to Background Document 1 in Service Code 5 for an elaboration of these concepts. RfDs can also be derived for the noncarcinogenic health effects of compounds which are also carcinogens. Therefore, it is essential to refer to other sources of information concerning the carcinogenicity of this substance. If the US EPA has evaluated this substance for potential human carcinogenicity, a summary of that evaluation will be contained in Section II of this file when a review of that evaluation is completed.

I.A.1. ORAL RfD SUMMARY

Critical Effect	Experimental Doses*	UF	MF	RfD
RBC ChE depression	NOEL: 16 mg/day (0.23 mg/kg/day)	10	1	2E-2mg/kg/day
	LEL: 24 mg/day (0.34 mg/kg/day)			

*Dose Conversion Factors & Assumptions: Adult human male body weight = 70\kg

Subchronic Human Feeding Study Moeller and Rider, 1962

I.A.2. PRINCIPAL AND SUPPORTING STUDIES (ORAL RfD)

Moeller, H.C. and J.A. Rider. 1962. Plasma and red blood cell cholinesterase activity as indications of the threshold of incipient toxicity of ethyl-p-nitrophenyl thionobenzenephosphorate (EPN) and malathion in human beings.   Toxicol. Appl. Pharmacol. 4: 123-130.

Malathion was administered by gelatin capsules to groups of five healthy male volunteers ranging in age from 23-63 years at doses of either 8 mg/day for 32 days, 16 mg/day for 47 days, or 24 mg/day for 56 days. Cholinesterase activity was determined twice weekly before, during, and after administration of the chemical. The intermediate dose was a NOEL. The high dose was associated with a depression in plasma and RBC cholinesterase activity with no clinically manifested side effects.

I.A.3. UNCERTAINTY AND MODIFYING FACTORS (ORAL RfD)

UF -- An uncertainty factor of 10 was used to account for the range of sensitivity within the human population.

MF -- None

I.A.4. ADDITIONAL COMMENTS (ORAL RfD)

Data Considered for Establishing the RfD:

1) 47-Day Feeding - human: Principal study - see previous description; no core grade

2) 2-Year Feeding/Oncogenic - rat: Systemic NOEL=100 ppm (5 mg/kg/day); Systemic LEL=1000 ppm (50 mg/kg/day) (decreased brain cholinesterase and body weight); core grade minimum (American Cyanamid Co., 1980)

3) Reproduction - rat: Reproductive NOEL= none; LEL=240 mg/kg/day (only dose tested; reduced number of live pups and reduced pup body weight) (Kalow and Marton, 1961)

4) Teratology - rat: (i.p. injection) Reproductive NOEL and Terata NOEL=900 mg/kg/day; LEL= none (Kimbrough and Gaines, 1968)

5) 4-Week Inhalation - dog: NOEL= none; LEL=5 ppm (one dose and one dog tested) (Anonymous, 1965)

Data Gap(s): Chronic Dog Feeding Study; Rat Reproduction Study; Rat Teratology Study; Rabbit Teratology

I.A.5. CONFIDENCE IN THE ORAL RfD

Study -- Medium
Data Base -- Medium
RfD -- Medium

The principal study is of fair quality and is given a medium confidence rating. The data base supports the choice of the human cholinesterase NOEL as the basis of the RfD, but since the data base on chronic toxicity is incomplete, it is given a medium confidence rating. Medium confidence in the RfD follows.

I.A.6. EPA DOCUMENTATION AND REVIEW OF THE ORAL RfD

Pesticide Registration Files

Agency Work Group Review -- 07/22/85, 09/29/86, 03/18/87

Verification Date -- 03/18/87

I.A.7. EPA CONTACTS (ORAL RfD)

William Burnam / OPP -- (703)557-7491

George Ghali / OPP -- (703)557-7490

I.B. REFERENCE CONCENTRATION FOR CHRONIC INHALATION EXPOSURE (RfC)

A risk assessment for this substance/agent is under review by an EPA work group.

II. CARCINOGENICITY ASSESSMENT FOR LIFETIME EXPOSURE

This substance/agent has not been evaluated by the US EPA for evidence of human carcinogenic potential.

III. HEALTH HAZARD ASSESSMENTS FOR VARIED EXPOSURE DURATIONS

III.A. DRINKING WATER HEALTH ADVISORIES

[Ed. Note: EPA does not have information yet in this section of IRIS].

IV. US EPA REGULATORY ACTIONS

EPA risk assessments may be updated as new data are published and as assessment methodologies evolve. Regulatory actions are frequently not updated at the same time. Compare the dates for the regulatory actions in this section with the verification dates for the risk assessments in sections I and II, as this may explain inconsistencies. Also note that some regulatory actions consider factors not related to health risk, such as technical or economic feasibility. Such considerations are indicated for each action. In addition, not all of the regulatory actions listed in this section involve enforceable federal standards. Please direct any questions you may have concerning these regulatory actions to the US EPA contact listed for that particular action. Users are strongly urged to read the background information on each regulatory action in Background Document 4 in Service Code 5.

IV.A. CLEAN AIR ACT (CAA)

[Ed. Note: EPA does not have information yet in this section of IRIS].

IV.B. SAFE DRINKING WATER ACT (SDWA)

[Ed. Note: EPA does not have information yet in this section of IRIS].

IV.C. CLEAN WATER ACT (CWA)

IV.C.1. AMBIENT WATER QUALITY CRITERIA, Human Health

No data available

IV.C.2. AMBIENT WATER QUALITY CRITERIA, Aquatic Organisms

Freshwater:
    Acute -- none
    Chronic -- 1.0E-1 ug/L

Marine:
    Acute -- none
    Chronic -- 1.0E-1 ug/L

Considers technological or economic feasibility? -- NO

Discussion -- A criterion of 0.1 ug/L malathion for fresh water and marine aquatic life is recommended since it will not be expected to significantly inhibit AChE over a prolonged period of time and is in dose agreement with the criteria for other organophosphates.

Reference -- Quality Criteria for Water; July, 1976, PB-263943

EPA Contact -- Criteria and Standards Division / OWRS (202)260-1315 / FTS 260-1315

IV.D. FEDERAL INSECTICIDE, FUNGICIDE, AND RODENTICIDE ACT (FIFRA)

IV.D.1. PESTICIDE ACTIVE INGREDIENT, Registration Standard

Status -- Issued (1988)

Reference -- Malathion Pesticide Registration Standard. February, 1988 (NTIS No. PB88-184585). Deletion of certain uses and directions for use 56 FR 11420 (03/18/91).

EPA Contact -- Registration Branch / OPP (703)557-7760 / FTS 557-7760

IV.D.2. PESTICIDE ACTIVE INGREDIENT, Special Review

[Ed. Note: EPA does not have information yet in this section of IRIS].

IV.E. TOXIC SUBSTANCES CONTROL ACT (TSCA)

[Ed. Note: EPA does not have information yet in this section of IRIS].

IV.F. RESOURCE CONSERVATION AND RECOVERY ACT (RCRA)

[Ed. Note: EPA does not have information yet in this section of IRIS].

IV.G. SUPERFUND (CERCLA)

IV.G.1. REPORTABLE QUANTITY (RQ) for Release into the Environment

Value (status) -- 100 pounds (Final, 1985)

Considers technological or economic feasibility? -- NO

Discussion -- The final 100 pound RQ takes into consideration the hydrolysis of this hazardous substance. The lowest primary criteria RQ for malathion (10 pounds based on aquatic toxicity) has been adjusted upward one RQ level.

Reference -- 50 FR 13456 (04/04/85); 54 FR 33418 (08/14/89)

EPA Contact -- RCRA/Superfund Hotline (800)424-9346 / (202)260-3000 / FTS 260-3000

VI. BIBLIOGRAPHY

VI.A. ORAL RfD REFERENCES

American Cyanamid Company. 1980. MRID No. 00110562; HED Doc. No. 002504. Available from EPA. Write to FOI, EPA, Washington, DC 20460.

Anonymous. 1965. FAO/WHO Report. HED Doc. No. 000316, 000389. Available from EPA. Write to FOI, EPA, Washington, DC 20460.

Kalow, W. and A. Marton. 1961. Second-generation toxicity of malathion in rats. Nature (London). 192: 464-465.

Kimbrough, R. and T. Gaines. 1968. Effect of organic phosphorus compounds and alkylating agents on the rat fetus. Arch. Environ. Health. 16(Jun.): 805-808.

Moeller, H.C. and J.A. Rider. 1962. Plasma and red blood cell cholinesterase activity as indications of the threshold of incipient toxicity of ethyl-p-nitrophenyl thionobenzenephosphorate (EPN) and malathion in human beings. Toxicol. Appl. Pharmacol. 4: 123-130.

VI.B. INHALATION RfD REFERENCES

None

VI.C. CARCINOGENICITY ASSESSMENT REFERENCES

None

VI.D. DRINKING WATER HA REFERENCES

None

VII. REVISION HISTORY

| -------- | ----------- | --- |
Date	Section	Description
03/01/88	All	CASRN corrected
08/01/91	I.B.	Inhalation RfC now under review
01/01/92	I.A.4.	Citations added
01/01/92	IV.	Regulatory actions updated
01/01/92	VI.	Bibliography on-line

SYNONYMS

American Cyanamid 4,409
Cabofos
Calmathion
Carbethoxy Malathion
Carbetox
Carbophos
Chemathion
Cythion
Detmol MA 96%
Dicarboethoxyethyl O,O-Dimethyl
Phosphorodithioate
1,2-Di(Ethoxycarbonyl)Ethyl O,O-Dimethyl
Phosphorodithioate
Diethyl Mercaptosuccinate, O,O-Dimethyl
Dithiophosphate, S-Ester
Diethyl Mercaptosuccinate, O,O-Dimethyl
Phosphorodithioate
Diethyl Mercaptosuccinate, O,O-Dimethyl
Thiophosphate
Dithiophosphate de O,O-Dimethyle et de S-
(1,2-Dicarboethoxyethyle)
Emmatos Extra
ENT 17,034
Ethiolacar
Formal
Fosfothion
Fosfotion
Fyfanon
Karbofos
Kop-thion
Kypfos
Malacide
Malagran
Malakill
Malamar
Malamar 50
Malaphele
Malaphos
Malaspray
Malathion
Malathon
Malation
Malatol Malatox
Maldison

Malmed
Malphos
Maltox
Mercaptothion
Mercaptotion
MLT
NCl-C00215
Oleophosphothion
O,O-Dimethyldithiophosphate
Diethylmercaptosuccinate
O,O-Dimethyl S-(1,2-
Bis(Ethoxycarbonyl)Ethyl)Dithiophosph
ate
O,O-Dimethyl-S-(1,2-
Dicarbethoxyethyl) Dithiophosphate
O,O-Dimethyl S-(1,2-
Dicarbethoxyethyl)Phosphorodithioate
O,O-Dimethyl S-(1,2-
Dicarbethoxyethyl)Thiothionphosphate
O,O-Dimethyl S-1,2-
Di(Ethoxycarbamyl)Ethyl
Phosphorodithioate
O,O-Dimethyl-S-1,2-
Dikarbetoxylethyldithiofosfat
Phosphothion
Sadofos
Sadophos
S-(1,2-Bis(Aethoxy-Carbonyl)-Aethyl)-
O,O-Dimethyldithiophasphat
S-(1,2-Bis(Ethoxy-Carbonyl)-Ethyl)-
O,O-Dimethyldithiofosfaat
S-(1,2-Bis(Ethoxycarbonyl)Ethyl O,O-
Dimethyl Phosphorodithioate
S-1,2-Bis(Ethoxycarbonyl)Ethyl-O,O-
Dimethyl Thiophosphate
S(1,2-Bis(Etossi-Carbonil)-Etil)-O,O-
Dimetil-Ditiofosfato
S-(1,2-Di(Ethoxycarbonyl)Ethyl
Dimethyl Phosphorothiolothionate
SF 60
Siptox 1
Sumitox
Vegfru
Zithiol

Source: ***Instant EPA's IRIS***, 1997, Instant Reference Sources, Inc., 7605 Rockpoint Drive, Austin, TX 78731 (Fax: 512-345-2386; Internet URL, http://www.instantref.com/inst-ref.htm).

Production, Use, and Pesticide Labeling Information

1. Description of Chemical

Generic Name: O,O-dimethyl phosphorodithioate of diethyl mercaptosuccinate

Common Name: Malathion

Other Names: S-(1,2-bis-ethoxycarbonyl)ethyl O,O-dimethylphosphorodithioate; diethyl(dimethoxy-phosphinothioyl)thiobutanedioate; diethylmercaptosuccinate S-ether with O,O-dimethylphosphorothioate; O,O-dimethyl dithiophosphate of diethylmercaptosuccinate; S-(1,2-dicarbethoxyethyl) O,O-dimethy-phosphorodithioate; diethylmercaptosuccinic acid, S-ether with O,O-dimethylphosphorodithioate; American Cyanamid Co. (USP 2578 652) Code NO. E14049; Calmathion; Celethion; Cythion (deodorized grade); Chemathion; Malaspray; Detmol MA 96% (Albert & Co., Germany); Emmatos; Emmatos Extra; For-Mal (Forshaw Chemicals); Fyfanon; Hilthion; Karbofos; Kop-Thion; Kypfos; Malamar; Malaphele; Malathion ULV Concentrate; Malatol; Maltox (All-India Medical); Prentox Malathion 95% spray; Sumitox; Vegfru Malatox; Zithiol; Malmed.

EPA Pesticide Chemical Code (Shaughnessy Number): 057701

Chemical Abstract Service (CAS) Number: 121-75-5

Year of Initial Registration: 1956

Pesticide Type: Insecticide and Miticide

Chemical Family: Organophosphate

US and Foreign Producers: American Cyanamid Company; A/S Cheminova; McLaughlin Gormley King Company; Prentiss Drug and Chemical Corp.; Inc.; Carmel Chemical Corp.; Amvac Chemical Corp.; Prochimie International Inc.; Gowan Co.; Wesley Industries, Inc.; Trans Chemic Industries Inc.; Southern Mill Creek Products Co., Inc.; Octagon Process Inc.; FMC Corp.; and Aceto Chemical Co. Inc.

2. Use Patterns and Formulations

- Application Sites:

- Terrestrial food crop use on alfalfa, almond, anise, apple, apricot, asparagus, avocado, barley, beets, beets (seed crop), bermudagrass, blackberry, blueberry, boysenberry, broccoli, brussels sprouts, cabbage, cantaloupe, carrot, casaba melons, cauliflower, celery, cherry, chestnut, citrus fruits (nursery stock), clover, collards, corn, cotton, cowpeas (hay), crenshaw melons, cucumber, current, dandelion, date, dewberry, eggplant, endive, fig, filbert, flax, garlic, gooseberry, grapefruit, grapes, grass, grass hay, green beans, guava, honeydew melons, honey ball melons, horseradish, kale, kidney beans, kohlrabi, kumquat, leek, lemon, lespedeza, lettuce, lima beans, lime, loganberry, lupino, macadamia nut, mango, muskmelons, mustard greens, navy beans, nectarine, oats, okra, onion, onion (green), onion (seed crop), papaya, parsley, parsnip, passion fruit, pasture grasses, peach, peanuts, pear, peas, pecan, peppermint, peppers, persian melons, pineapple, rangeland grasses, raspberry, rutabaga, rye, safflower, salsify, shallot, snap beans, sorghum, soybeans, spearmint, spinach, squash, strawberry, sugar beets, sweet potato, swiss chard, tangelo, tangerine, tomato, turnips, vetch, walnut, watercress, watermelons, wax beans, and wheat.

- Terrestrial non-food crop use on tobacco, tobacco (transplant beds), ornamental flowering plants, ornamental lawns and turf, ornamental nursery stock, ornamental woody plants, pine seed orchards and uncultivated non-agricultural areas.

- Greenhouse food crop use on asparagus, beans, beets, celery, cole crops (including broccoli, cabbage, kale mustard greens, and turnips), corn cucumber, eggplant, endive, lettuce, melons, mushrooms, onion, peas, peppers, potato, radish, spinach, squash, summer squash, tomato, and watercress.

- Greenhouse non-food crop use on ornamental plants and Epcot display crops.

- Aquatic food crop uses on cranberry and rice.

- Aquatic non-food uses on intermittently flooded areas, irrigation systems, and sewage systems.

- Forestry uses on forest trees (including Douglas fir, eastern pine, hemlock, larch, pines, red pine, spruce, and true fir).

- Indoor uses on stored commodity treatment

- for almonds, barley, field corn, field or garden seeds, grapes (raisin), oats, peanuts, rice rye, sorghum, sunflower, wheat, bagged citrus pulp, and cattle feed concentrate blocks (non- medicated);

- pet and domestic animal uses for beef cattle, cats, chickens, dairy cattle (lactating and non-lactating), dogs, ducks, geese, goats, hogs, horses (including ponies), pigeons, sheep, and turkeys;

- animal premise uses for dairy and livestock barns, stables and pens, feed rooms, poultry houses, manure piles, garbage cans, garbage dumps, kennels, rabbits on wire, beef cattle feed lots and holding pens, cat sleeping quarters, dog sleeping quarters, poultry houses;

- agricultural premise uses for cull fruit and vegetable dumps; household uses for indoor domestic dwellings, human clothing (woolens and other fabrics), mattresses; - and commercial and industrial uses for bagged flour, cereal processing plants, edible and inedible commercial establishments, dry milk processing plants, edible and inedible eating establishments, edible and inedible food processing plants, packaged cereals, pet foods and feed stuff.

- Methods of Application: Sprays, aerosols and fogging equipment, ground and aerial equipment (including ULV), baits, paints, pet collars, dips, soil, bark and foliar application, dormant and delayed dormant application, animal dust bags and oilers, and cattle feed concentrate blocks.

- Formulations: Wettable powders, dusts, granules, emulsifiable concentrates, liquids, solids, impregnated materials, and pressurized sprays, pellets/tablets, liquids (ready to use).

Chemical/Physical Characteristics of the Technical Material

- Color: Colorless, yellow, amber, or brown

- Physical state: Liquid

- Odor: Mercaptan-like

- Specific gravity: 1.2315 at 25 °C

- Boiling Point: 156-157 °C at 0.7 mm Hg

- Solubility: 145 ppm in water at 25 °C; completely soluble in most alcohols, esters, high aromatic solvents, and ketones

- Vapor pressure: 0.00004 mm Hg at 30 °C

- Miscibility: Miscible with most organic solvents

- Stability: May gel in contact with iron, terreplate or tinplate

Toxicological Characteristics

- Acute Oral: Toxicity Category III (ranges from 1546 to 1945 mg/kg in female rats and 1522 to 1650 mg/kg in male rats).

- Acute Dermal: Toxicity Category III (>2000 mg/kg in female and male rats and rabbits).

- Acute Inhalation: Toxicity Category III based on toxicity values ranging from 1.7 to >4.0 mg/m^3 in rats.

- Primary Dermal Irritation: Toxicity Category IV based on mild dermal irritation reported in a rabbit study.

- Primary Eye Irritation: Toxicity Category III based on findings of mild conjunctival reactions 72 hours post application in rabbits' eyes.

- Skin Sensitization: Non-sensitizing

- Delayed Neurotoxicity: Data gap.

- Subchronic Inhalation: Data gap.

- Oncogenicity: Data gaps for mouse (using malathion) and rat (using malaoxon).

- Chronic Feeding: Data gaps for rodent and nonrodent (using malathion) and rodent (using malaoxon).

- Metabolism: Data gap.

- Teratogenicity: Data gap for rat. Data in rabbit indicated a NOEL = 25 mg/kg for developmental effects; it was not teratogenic in any dose group (Highest Dose Tested was 100 mg/kg).

- Reproduction: Data gap.

- Mutagenicity: Data gap.

Environmental Characteristics

- Data gaps exist for environmental fate. Data reviewed by the Agency indicate that malathion is very mobile in loamy sand and loam soils. Adsorption ratios reported (amount adsorbed/initial concentration) were 0.73 to 0.95. Data are needed before the Agency can assess the potential for malathion to contaminate groundwater.

Ecological Characteristics (technical grade)

- Avian oral toxicity 167 ppm for ring-necked pheasant and 1485 (8-day LD50) ppm for mallard.

- Avian dietary toxicity Acute toxicity value of 3497 ppm for (8-day LC50) bobwhite and >5000 ppm for mallard.

- Freshwater fish acute 200 ppm for rainbow trout and 40 to 103 ppm toxicity (96-hr LC50) for bluegill.

- Freshwater invertebrate 1 ppm for Daphnia magna toxicity (48-hr EC50)

- Estuarine invertebrate >1000 ppm for Eastern oyster toxicity

Tolerance Assessment

> - The available data pertaining to metabolism of malathion in plants are inadequate. Additional data are required on the uptake, distribution, and metabolism of malathion in alfalfa, cotton, soybeans, and either wheat or rice. The data pertaining to metabolism of malathion in animals are inadequate. Additional metabolism studies are required that utilize ruminants and poultry. Metabolism studies using cattle, poultry, and swine reflecting direct animal treatment are also required.

> - Analytical methodology for determining the levels of residues of malathion in plants and animals is adequate. Malathion is detected by the FDA-USDA multiresidue protocols.

> - Storage stability data demonstrate that residues of malathion in or on frozen plant commodities are stable up to 185 days after application and in milk stored at -10°C for 98 days after application. No data are currently available for animal tissues and are required. Additional storage stability data are also required in order to evaluate the adequacy of the malathion tolerances.

> - Insufficient data are available on the magnitude and levels of residues of malathion in or on all commodities listed in 40 CFR 180.111 except flax seed, hops, wild rice, and non-medicated cattle feed concentrate blocks. Processing studies are required.

> - Tolerances must be proposed and appropriate supporting residue data submitted for the following feed items: beanvines and hay; lentil forage and hay; cowpea seed; soybean straw; barley forage, hay and straw; corn forage and fodder; oat forage, hay and straw; rice straw; rye forage and straw; straw of wild rice; sorghum fodder; lespedeza forage; lupine forage; cotton forage; mint hay; peanut hulls, hay and vines; and pineapple forage.

> - Feed additive tolerances are required for residues of malathion in or on dried hops and spent hops. A tolerance for residues of malathion in or on anise must be proposed together with supporting residue data. Data are needed to support the use of malathion in food handling establishments. In addition, data reflecting the use of malathion on stored, unfinished tobacco are required.

- Based on a study in humans in which red blood cell and plasma cholinesterase activity were inhibited at a dose of 0.34 mg/kg (the lowest effect level or LEL, a NOEL has been extrapolated to 0.2 mg/kg/day. a provisional acceptable daily intake (PADI) of 0.02 mg/kg/day has been calculated using a 10-fold uncertainty factor. The PADI is provisional because the existing data base on malathion is lacking chronic toxicity studies, an acceptable teratology study in rats, an acceptable reproduction study, mutagenicity studies, and a metabolism study.

- The Theoretical Maximal Residue Contribution (TMRC) for the US population average is 0.1014 mg/kg/day, occupying 507% of the PADI. for children 1 to 6 years of age, the TMRC occupies 1133% of the PADI. The TMRC is based upon current tolerance levels and an assumption that 1005 of the sites are treated. Actual exposure levels are likely to be much lower. When the required data are submitted, the Agency will conduct a full tolerance reassessment.

Summary Science Statement

- Technical malathion is a mildly acutely toxic pesticide, which is placed in Toxicity Category III based on the oral, dermal and inhalation routes of exposure. Technical malathion is non-sensitizing and only mildly irritating to the eyes and skin (Toxicity Category III and IV, respectively). Additional data are required to assess the neurotoxic potential of malathion. Malathion is a cholinesterase inhibitor, reducing plasma and red blood cell cholinesterase.

- Although the Agency possesses a number of studies on the chronic effects of malathion and its principal metabolite malaoxon, several of these studies are deficient scientifically, and must be repeated.

- Of five studies concerning the oncogenicity of malathion and its metabolite, three are acceptable, and demonstrate that malathion is not carcinogenic in two species of rats, and that its metabolite malaoxon is not carcinogenic in mice. Because of questionable liver findings in the malathion mouse study and the malaoxon rat study, new studies must be conducted in these species.

- An acceptable rabbit teratology study demonstrated no teratogenicity at dosages up to 100 mg/kg/day. However, developmental and maternal toxicity were noted at dosages of 50 mg/kg/day. A similar study in rats was unacceptable and must be repeated. A 3-generation reproduction study was also unacceptable.

- Laboratory data show that technical malathion is potentially highly toxic to aquatic invertebrates, bees, and aquatic life stages of amphibians; moderately toxic to birds, and slightly toxic to fish. Based on theoretical calculations, both terrestrial and aquatic uses of malathion may pose significant risk to aquatic fauna. Reported fish kills and results of field studies suggest that adverse effects to both aquatic and terrestrial fauna may result from normal use of malathion. However, these studies are not adequately documented to enable EPA to propose restrictions on the use of malathion. EPA will reassess the impacts of malathion use on nontarget organisms after the required environmental fate and ecological effects data have been received and reviewed.

- The Agency is unable to assess the potential for malathion to contaminate groundwater because the environmental fate of malathion is largely uncharacterized. Preliminary data indicate that malathion is very mobile in loamy sand and loam soils. Additional data are needed in order for the Agency to assess its fate in the environment and potential for contaminating groundwater.

- A tolerance reassessment of malathion is not possible at this time, since most of the tolerances are not adequately supported, and because there are gaps in the chronic toxicology data base (chronic feeding studies, teratology study, reproduction study, mutagenicity studies, and a metabolism study). The Theoretical Maximal Residue Contribution

(TMRC) for the US population average is 0.1014 mg/kg/day and the Provisional Acceptable Daily Intake (PADI) is 0.02 mg/kg/day based on a human study in which plasma and red blood cell cholinesterase were monitored and a 10-fold uncertainty factor was used. The TMRC occupies 507% of the PADI.

4. Summary of Regulatory Position and Rationale

- No referral to Special Review is being made at this time.

- No new tolerances for raw agricultural commodities or significant new uses will be granted until the Agency has received data sufficient to perform a tolerance reassessment. Significant new uses will not be granted until the data gaps have been filled.

- The Agency is concerned about the potential hazards to aquatic organisms. However, no regulatory action is being considered at this time for fish and wildlife concerns. EPA will reassess the impacts of malathion use on nontarget organisms after the required environmental fate and ecological effects data have been received and reviewed.

- The Office of Endangered Species (OES) in the US Fish and Wildlife Service has determined that certain uses of malathion may jeopardize the continued existence of endangered species or critical habitat of certain endangered species. No additional labeling is required at this time; however, EPA is developing a program to reduce or eliminate exposure to these species, and may require labeling revisions when the program is developed.

- In order to meet the statutory standard for continued registration, the Agency has determined that malathion products must bear revised and updated fish and wildlife toxicity warnings.

- The Agency is deferring decisions concerning malathion's potential for contamination of groundwater until the environmental fate data have been submitted and reviewed.

- The Agency is not restricting the use of malathion products for retail sale only to certified applicators. Malathion does not meet any of the criteria of 40 CFR 162.11 and therefore products containing malathion do not warrant restricted use classification.

- The Agency is not establishing a longer reentry interval for agricultural uses of malathion beyond the minimum reentry interval for all agricultural uses of pesticides (sprays have dried, dusts have settled and vapors have dispersed). The Agency will reassess the need for reentry data/reentry intervals upon receipt of the required toxicology data.

5. Summary of Major Data Gaps

	Time Frame
- Toxicology	
- Delayed neurotoxicity	9 months
- 21-day dermal toxicity	9 months
- 90-day inhalation - rat	15 months
- Chronic toxicity (rodent and nonrodent--using malathion)	50 months
- Chronic toxicity (rodent) -- using malaoxon	50 months
- Oncogenicity (mouse -- using malathion	50 months
- Oncogenicity (rat) -- using malaoxon	50 months
- Teratogenicity - rat	15 months
- Reproductive effects - rat (2-generation)	39 months
- Mutagenicity	9-12 months
- Metabolism	24 months
- Domestic animal safety testing	15 months

- Environmental Fate/Exposure

- Hydrolysis	9 months
- Aerobic and anaerobic soil metabolism	27 months
- Aerobic and anaerobic aquatic metabolism	27 months
- Leaching and adsorption/desorption	12 months
- Terrestrial field dissipation	27 months
- Long-term field dissipation	50 months
- Forestry dissipation	27 months
- Aquatic (sediment) - field study	27 months
- Photodegradation in water, soil, air	9 months
- Volatility (lab)	12 months
- Rotational crops (confined)	39 months
- Accumulation in irrigated crops	39 months
- Accumulation in fish	12 months
- Accumulation in aquatic nontarget organisms	12 months
- Spray drift	18 months

- Residue Chemistry

- Storage stability data	18 months
- Plant and animal metabolism	18 months
- Residue data - raw agricultural commodities	18 months
- Processing studies	24 months
- Residue data on stored, unfinished tobacco	18 months
- Residues in water	15 months
- Residue data on food handling establishments	12 months

- Product Chemistry

- All	9-15 months

- Fish and Wildlife

- Acute toxicity to freshwater invertebrates	9 months
- Acute toxicity to estuarine and marine organisms	12 months
- Avian reproduction	24 months
- Fish early life stage	15 months
- Aquatic invertebrate life cycle	15 months
- Honeybee - toxicity of residues on foliate	15 months

6. Contact Person at EPA

> William H. Miller
> Product Manager 16
> Insecticide- Rodenticide Branch Registration Division (T-767C)
> Office of Pesticide Programs
> Environmental Protection Agency
> 401 M Street, SW
> Washington, DC 20460

> Office location and telephone number:

> Rm. 211, Crystal Mall #2
> 1921 Jefferson Davis Highway
> Arlington, VA 22202.

> Telephone: (703) 557-2600

Source: *Instant EPA's Pesticide Facts*, 1994, Instant Reference Sources, Inc., 7605 Rockpoint Drive, Austin, TX 78731 (Fax: 512-345-2386; Internet URL, http://www.instantref.com/inst-ref.htm).

Mancozeb

$$Mn^{++}$$

Introduction

Mancozeb is registered as a general use pesticide by the US Environmental Protection Agency (EPA). It is a combination of two other chemicals in this class: maneb and zineb. In July 1987, the Environmental Protection Agency announced the initiation of a special review of the ethylene bisdithiocarbamates (EBDCs), a class of chemicals to which mancozeb belongs. This Special Review was initiated because of concerns raised by laboratory tests on rats and mice. As part of the Special Review, EPA reviewed data from market basket surveys and concluded that actual levels of EBDC residues on produce purchased by consumers are too low to affect human health. [829]

Mancozeb is a fungicide used with crops such as apples, potatoes, and tomatoes. Mancozeb is also used on approximately 80 percent of the onion acreage in the United States. It is manufactured in numerous formulations including dusts, wettable powders, flowable concentrates, and granular end-use products. It is typically applied using either aerial or ground equipment. For ground equipment, mancozeb suspensions typically are made from a wettable powder or flowable concentrate that is applied by means of air blast sprayers or tractor-mounted boom sprayers. Dust formulations are typically applied by means of truck- or tractor-drawn duster or aerial equipment. It's mode of action is to inhibit enzyme activity by complexing with metal-containing enzymes including those involved with the production of adenosine triphosphate (ATP).

Mancozeb is slightly toxic to birds on an acute basis and it is generally toxic to fish, expecially to warm water fish. It is considered harmful to wildlife but not hazardous to honey bees. Because it is practically insoluble in water it is not likely to infiltrate to groundwater. [829]

Identifying Information

CAS NUMBER: 8018-01-7

SYNONYMS: Dithane® M-45, Manzate® 200, Fore®, Green-Daisen M, Karamate, Mancofol, Zimaneb,Manzeb, Policar, Dithane-Ultra Nemispot, Nemispor, Riozeb, Mancozin, Manzin

Endocrine Disruptor Information

Mancozeb is a strongly suspected EED that it is listed by the Centers for Disease Control & Prevention [812], and the World Wildlife Fund [813] as a potential endocrine modifying chemical. A toxicological concern from exposure to mancozeb is the hazard to the human thyroid from presence of ethylenethiourea (ETU), a contaminant, degradation product, and metabolite present in mancozeb and other EBDC products. [821] Mancozeb contains the element manganese, a metal similar in suspected EED properties to lead. [811] Evidence has been obtained from rats, which includes lead exposure *in utero*, "suggesting a dual site of lead action: (a) at the level of the hypothalamic pituitary unit, and (b) directly at the level of gonadal steroid biosynthesis." Blood-serum testosterone levels (in males) and estradiol levels (in females) were most severely suppressed in those animals where exposure began *in utero*. Both sexes in this group showed a 25% suppression in prepubertal growth (as measured by weight). [868], [811] Several studies of the effects of EBDCs on test animals have shown rapid reduction in the uptake of iodine and swelling of the thyroid (i.e. goiter). [829]

Chemical and Physical Properties

CHEMICAL FORMULA: $(C_4H_6MnN_2S_4)x$ $(Zn)y$

MOLECULAR WEIGHT: $(265.3)x + (65.4)y$

PHYSICAL DESCRIPTION: grayish-yellow powder [817]

SPECIFIC GRAVITY: Not available

DENSITY: Not available

MELTING POINT: 192-194°C (decomposes) [817]

BOILING POINT: Not available; it decomposes when heated

SOLUBILITY: insoluble in water and most organic solvents

Environmental Reference Materials

A source of EED environmental reference materials for mancozeb is Crescent Chemical Co., Hauppauge, NY (Fax: 516-348-0913); Catalog No. 45553.

Hazardous Properties

ACUTE/CHRONIC HAZARDS: Available data indicate that mancozeb itself is not a primary developmental toxicant or teratogen, however, an additional teratology study with mancozeb is required before its teratogenicity can be fully assessed. [821] Mancozeb has a very low acute toxicity to mammals. The major routes of exposure to mancozeb are through the skin or from inhalation. In spray or dust forms, the EBDCs are moderately irritating to the skin and respiratory mucous membranes. [829]

FIRE HAZARDS: Mancozeb does not readily ignite but it may burn and containers may explode in the heat of a fire. Suspensions of mancozeb dust in air, however, can ignite or explode. Thermal decomposition products include toxic oxides of nitrogen, sulfur, and oxygen. [829]

Medical Symptoms of Exposure

Symptons include decreased body weight, dilated ventricles, spinal cord hemorrhage, and delayed/ incomplete ossification of skull and ribs. [821]

Symptoms of poisoning from this class of chemicals include itching, scratchy throat, sneezing, coughing, inflammation of the nose or throat, and bronchitis. There is no evidence of 'neurotoxicity,' nerve tissue destruction or behavior change, from the ethylene bisdithiocarbamates (EDBCs). However, dithiocarbamates are partially chemically broken down, or metabolized, to carbon disulfide, a neurotoxin capable of damaging nerve tissue. [829]

Toxicological Information

Mancozeb is not currently listed in NTP's Chemical Repository Database or in EPA's IRIS database and there is no published schedule for its addition in the future.

Production, Use, and Pesticide Labeling Information

1. Description of Chemical

Generic Name: Zinc ion and manganese ethylenebisdithiocarbamate, coordination product

Common Name: Mancozeb

Trade Name: Dithane® M-45, Manzate® 200, Fore®

EPA Shaughnessy Code: 014504

Chemical Abstracts Service (CAS) Number: 8018-01-7

Year of Initial Registration: 1967

Pesticide Type: Fungicide (with minor insecticide use)

Chemical Family: Ethylene bisdithiocarbamate (EBDC)

US and Foreign Producers: Rohm and Haas produces a formulating intermediate; E.I. duPont de Nemours Co. produces mancozeb for its own end-use products.

2. Use Patterns and Formulations

- Application Sites: Terrestrial food and nonfood crops; aquatic (food); greenhouse (nonfood); forestry; outdoor domestic.

- Major Crops Treated: Apples, potatoes, and tomatoes. Mancozeb is also used on approximately 80 percent of the onion acreage in the United States.

- Formulation Types Registered: 80 percent active ingredient (ai) formulating intermediate; dust, wettable powder, flowable concentrate, and granular end-use products.

- Methods of Application: Foliar applications by aerial or ground equipment. For ground equipment, mancozeb suspensions typically are made from a wettable powder or flowable concentrate that is applied by means of air blast sprayers or tractor-mounted boom sprayers. Dust formulations are typically applied by means of truck- or tractor-drawn duster or aerial equipment.

- Treatment of the seeds may be accomplished by commercial seed treatment equipment by seed companies or by addition to the planter box at the farm site. Mancozeb may be used as a spray furrow treatment of soil at planting of onion sets. For treatment of surfaces of potato cut seed pieces or whole tuber seed pieces, mancozeb is applied by means of dip tanks or dusting equipment mounted over the seed pieces on a conveyor belt.

- Application Rates: Rates range from 0.4-3.3 lb/100 gal. or 0.6-19.06 lb ai/A for foliar application (high rates are for turf use rather than food crops); 0.04-4.32 lb/100 lb of seed; 0.78-27 lb/100 gal for pre-plant dip treatments and 2.4 lb/A for soil applications. Multiple applications may be made during the growing season.

- Usual Carrier: Water.

3. Science Findings

Chemical Characteristics

- Only limited product chemistry information is available as follows:

- Physical State: Solid at room temperature, decomposes when heated.

- Empirical Formula: $(C_4H_6MnN_2S_4)x$ $(Zn)y$

- Molecular Weight: $(265.3)x + (65.4)y$

Toxicology

    - Acute Toxicity: Note - Toxicity categories are discussed in 40 CFR 162.10

    - Oral Toxicity: 4500 mg/kg, Category III

    - Dermal Toxicity: > 5000 mg/kg, Category III

    - Inhalation Toxicity: > 5.14 mg/L, Category IV

    - Eye Irritation: Primary Irritation Score (PIS) - 2.3, Category III

    - Skin Irritation: PIS - 0.5, Category IV

    - Dermal Sensitization: Study required

- Subchronic Toxicity:

    - Oral (rats):
        No observed effect level (NOEL) for kidney effects - 60 ppm (3.5 mg/kg/day, males; 4.4 mg/kg/day females). NOEL for thyroid effects - 125 ppm (7.9 mg/kg/day, males; 9.2 mg/kg/day, females)

    - Oral (dogs):
        NOEL for systemic effects - 100 ppm (3.0 mg/kg/day, males; 3.4 mg/kg/day, females). NOEL for thyroid effects - 1000 ppm (29 mg/kg/day for both sexes).

    - Dermal: Studies required.

    - Inhalation: NOEL for systemic effects - 20 mg/m^3 NOEL for thyroid 80 mg/m^3

    - Major Routes of Exposure: Dermal and inhalation.

    - Chronic Toxicity: - Chronic Toxicity: Studies required.

    - Oncogenicity: Studies required.

- Teratology:

    - Maternal Toxicity: NOEL - 32 mg/kg/day; Lowest effects level (LEL) - 128 mg/kg/day (decreased body weight)

    - Fetal Toxicity: NOEL - 128 mg/kg/day; LEL - 512 mg/kg/day (increased resorptions).

- Teratogenic Effects: NOEL - 128 mg/kg/day; LEL - 512 mg/kg/day (dilated ventricles, spinal cord hemorrhage, delayed/ incomplete ossification of skull and ribs); A/D Ratio: 32/128 = 0.25. (Note: An A/D Ratio of less than 1 indicates that developmental toxicity may be ascribed to secondary affects of maternal toxicity.

- Reproduction: Study required.

- Mutagenicity: Negative - bacterial and in vitro mammalian cell systems, chromosome damage in vivo and in mammalian cell transformation. Positive - sister-chromatid exchanges in CHO cells in vitro.

Physiological and Behavioral Characteristics

- Mechanism of Pesticide Action: Mancozeb inhibits enzyme activity by complexing with metal-containing enzymes including those involved with the production of adenosine triphosphate (ATP).

- Metabolism: Mancozeb appears to be rapidly absorbed from the gastrointestinal tract, distributed to target organs and excreted almost totally by 96 hours. The major metabolite is ETU, comprising almost 24% of the bio-available dose in urine and bile. ETU residues in the thyroid and the liver were less than 1 ppm and were non- detectable after 24 hours.

Environmental Characteristics

- Hydrolysis: Mancozeb degrades with a half-life of 1 to 2 days at pH 5, 7, and 9. ETU is stable to hydrolysis at pH 5 and 7 and is very slowly hydrolyzed at pH 9.

- Anaerobic Aquatic Metabolism: Mancozeb and ETU declined with half-lives of 92 days and 29 to 35 days, respectively.

- Environmental Fate/Ground Water Concerns: Available studies indicate that ETU has the potential to leach. A complete assessment of the environmental fate, including the potential for groundwater contamination, from the use of mancozeb products will be undertaken when data are available.

- Exposure During Reentry Operations: Mancozeb is registered for use on crops which may involve substantial exposure to residues of the pesticide. Because ETU has demonstrated evidence of oncogenicity, mutagenicity, teratogenicity and thyroid effects and mancozeb has caused thyroid effects, reentry data are required. Until these data are received and evaluated, a 24-hour reentry interval is imposed.

Ecological Characteristics

- Avian Oral Toxicity:

  - Japanese quail - > 6400 mg/kg/day
  - European sparrow - 3000-6000 mg/kg/day
  - English house sparrow - 1500 mg/kg/day
  - Mallard duck - > 6400 mg/kg/day

- Freshwater Fish Toxicity:

  - Bluegill sunfish - 1.54 ppm
  - Rainbow trout - 0.46 ppm
  - Daphnia magna - 0.58 ppm

Tolerance Assessment

- Tolerances Established: Tolerances, expressed as zinc ethylene bisdithiocarbamate, have been established for residues of mancozeb in a variety of raw agricultural commodities and meat byproducts (40 CFR 180.176 and 40 CFR 180.319), and in processed food (21 CFR 193.460) and feed (21 CFR 561.410).

- Results of Tolerance Assessment: The toxicology data for mancozeb are insufficient to determine an Acceptable Daily Intake (ADI) or whether the toxicity observed in the studies is due to mancozeb or ETU.

- There are no acceptable chronic studies on which to calculate an ADI, therefore, a subchronic study has been used to calculate a Provisional

ADI (PADI). Because a subchronic study was used, an uncertainty factor of 1000 (rather than 100 used with chronic studies) was employed. The PADI for mancozeb is 0.003 mg/kg/day based on the 90-day dog feeding study with a NOEL of 3 mg/kg/day.

- The theoretical maximum residue contribution (TMRC), based on the assumption that 100 percent of each crop is treated and contains residues at the tolerance levels, is 0.028 mg/kg/day or approximately 900 percent of the PADI. Based on a more realistic dietary assessment, using average field trial residues and theoretical percent of crop treated, the estimated average consumption for the US population is 0.00097 mg/kg/day or 32.2 percent of the PADI.

Summary Science Statement

- The major toxicological concern from exposure to mancozeb is the hazard to the human thyroid from presence of ethylenethiourea (ETU), a contaminant, degradation product, and metabolite present in mancozeb and other EBDC products. ETU is an acknowledged goitrogen, teratogen and oncogen. Additional chronic studies of mancozeb are required for further evaluation. Available data indicate that mancozeb itself is not a primary developmental toxicant or teratogen, however, an additional teratology study with mancozeb is required before its teratogenicity can be fully assessed.

- Available data are not adequate to assess the environmental fate of mancozeb. However, studies do indicate that ETU has the potential to leach. A complete assessment of the environmental fate, including the potential for groundwater contamination, from the use of mancozeb products will be undertaken when data are available.

- Available data are insufficient to completely evaluate the ecological effects of mancozeb. Based on available data, mancozeb is no more than slightly toxic to avian wildlife on an acute basis; is highly toxic to warmwater fish; and appears to be at least moderately toxic to coldwater fish.

Summary of Regulatory Position and Rationale

- Summary of Agency Position: The Agency is currently evaluating the potential human health risks resulting from the use of mancozeb to determine whether additional regulatory action is warranted on mancozeb and the other EBDC pesticides containing the common contaminant, degradation product, and metabolite, ETU. At this time, the Agency will not establish any new food use tolerances or register any significant new uses. The Agency is also specifying precautionary labeling as set forth below.

- Unique Warning Statements Required on Labels:

- Manufacturing-Use Products

"This pesticide is toxic to fish. Do not discharge effluent containing this product into lakes, streams, ponds, estuaries, oceans, or public water unless this product is specifically identified and addressed in an NPDES permit. Do not discharge effluent containing this product to sewer systems without previously notifying the sewage treatment plant authority. For guidance, contact your State Water Board or Regional Office of the EPA."

- End-Use Products

> "This pesticide is toxic to fish. Drift and runoff from treated areas may be hazardous to aquatic organisms in neighboring areas. Do not apply directly to water or wetlands (swamps, bogs, marshes, and potholes). Do not contaminate water by cleaning of equipment or disposal of wastes."

- Seed Treatment Products

> "This pesticide is toxic to fish. Cover or incorporate spilled treated seed. Do not contaminate water by cleaning of equipment or disposal of wastes."

- Aquatic Food Use Products (cranberry, wild rice, taro)

> "This pesticide is toxic to fish. Drift and runoff from treated areas may be hazardous to aquatic organisms in neighboring areas. Do not contaminate water by cleaning of equipment or disposal of wastes."

- All Home Use Products

> "PROTECTIVE MEASURES: Always spray with your back to the wind. Wear long-sleeve shirt, long pants, and rubber gloves. Wash gloves thoroughly with soap and water before removing. Change your clothes immediately after using this product and launder separately from other laundry items before reuse. Shower immediately after use."

- Home Use Products with Food Uses

> "Pre-harvest intervals on this label are specified so that pesticide residues will be at an acceptable level when the crop is harvested."

- All Agricultural Products

> "After (sprays have dried/dusts have settled/vapors have dispersed, as applicable) do not enter or allow entry into treated areas until the 24-hour reentry interval has expired unless wearing the personal protective equipment listed on the label.

- WORKER SAFETY RULES

- "Keep all unprotected persons, children, livestock, and pets away from treated area or where there is danger of drift."

- "Do not rub eyes or mouth with hands. See First Aid (Practical Treatment Section).

- PERSONAL PROTECTIVE EQUIPMENT - For mixers, loaders, applicators and early reentry workers.

- "HANDLE THIS PRODUCT ONLY WHEN WEARING THE FOLLOWING PROTECTIVE CLOTHING AND EQUIPMENT: a long-sleeve shirt and long pants or a coverall that covers all parts of the body except the head, hands, and feet; chemical resistant gloves; shoes, socks, and goggles or a face shield. During mixing and loading, a chemical resistant apron must also be worn.

- "During application from a tractor with a completely enclosed cab with positive pressure filtration, or aerially with an enclosed cockpit, a long-sleeve shirt and long pants may be worn in place of the above protective clothing. Chemical resistant gloves must be available in the cab or cockpit and worn while exiting.

- "IMPORTANT! Before removing gloves, wash them with soap and water. Always wash hands, face, and arms with soap and water before eating, smoking or drinking. Always wash hands and arms with soap and water before using the toilet.
- "After work take off all clothes and shoes. Shower using soap and water. Wear only clean clothes. Do not use contaminated clothing. Wash protective clothing and protective equipment with soap and water after each use. Personal clothing worn during use must be laundered separately from household articles. Clothing and protective equipment heavily contaminated or drenched with mancozeb must be destroyed according to state and local regulations.

- "HEAVILY CONTAMINATED OR DRENCHED CLOTHING CANNOT BE ADEQUATELY DECONTAMINATED.

- "During aerial application, human flaggers are prohibited unless in totally enclosed vehicles."

5. Summary of Major Data Gaps

Study	Due Date - From Issuance of Standard
- Product Chemistry	6-15 Months
- Residue Chemistry:	18-24 Months

     - Plant and animal metabolism
     - Residue studies
     - NOTE: Dates are determined based on beginning of planting season after issuance of Standard.

- Toxicology:	9-50 Months

     - Dermal sensitization
     - Subchronic dermal
     - Chronic toxicity (rodent and nonrodent(mancozeb and ETU)
     - Oncogenicity (rat and mouse)
     - Teratology (rabbit)
     - Reproduction (rat) (mancozeb and ETU)
     - Mutagenicity (mancozeb and ETU)
     - Dermal (percutaneous) absorption (mancozeb and ETU)

- Ecological Effects:	9-18 Months

     - Avian dietary
     - Avian reproduction
     - Estuarine and marine organism
     - Fish early life stage
     - Aquatic invertebrate life cycle

- Environmental Fate:	9-50 Months

     - Hydrolysis (mancozeb and ETU)
     - Photodegradation (mancozeb and ETU)
     - Soil metabolism (mancozeb and ETU)
     - Aquatic metabolism (mancozeb and ETU)
     - Leaching and adsorption/desorption (mancozeb and ETU)
     - Volatility
     - Degradation (mancozeb and ETU)
     - Rotational crops
     - Irrigated crops
     - Fish accumulation

- Reentry	27 Months

6. Contact Person at EPA

Ms. Lois Rossi
Product Manager No. 21
Fungicide-Herbicide Branch Registration Division (TS-767C)
Office of Pesticide Programs
U. S. EPA
Washington, DC 20460

Telephone: (703)557-1900

Source: *Instant EPA's Pesticide Facts*, 1994, Instant Reference Sources, Inc., 7605 Rockpoint Drive, Austin, TX 78731 (Fax: 512-345-2386; Internet URL, http://www.instantref.com/inst-ref.htm).

Maneb

Mn^{++}

S$^-$ $^-$S

S=   =S

NH   NH

Introduction

Maneb is registered as a general use pesticide by the US Environmental Protection Agency (EPA). In July 1987, the Environmental Protection Agency announced the initiation of a special review of the ethylene bisdithiocarbamates (EBDCs), a class of chemicals to which mancozeb belongs. This Special Review was initiated because of concerns raised by laboratory tests on rats and mice. As part of the Special Review, EPA reviewed data from market basket surveys and concluded that actual levels of EBDC residues on produce purchased by consumers are too low to affect human health. Many home garden uses of EBDCs have been canceled because the EPA has assumed that home users of these pesticides do not wear protective clothing during application. [829]

Maneb is a fungicide used on crops such as fruits, vegetables, seed crops, nuts, flax, and grains and non-food crop including ornamentals, lawns, turf. It's predominate use is with apples, potatoes, tomatoes and sweet corn where it is effective for controlling foliar fungal diseases. It is manufactured in numerous formulations including dusts, granular, wettable powders, flowable concentrates and ready-to-use products. Maneb is typically applied in foliar applications to vegetable crops and apples by aerial equipment or ground equipment. Foliar treatment of tobacco or vegetable seed beds, application of sprays or dusts are typically made using hand held compressed air sprayers or dusting equipment. Potato and tomato foliage may be treated by means of solid set, wheel move, or center pivot sprinkler irrigation equipment. [821]

Maneb is one of the chemicals of a class called the ethylene-bisdithiocarbamates ('EBDCs') that are noted for their instability in the environment. The application of heat can break these chemicals down into a number of metabolites. In addition to natural environmental processes that break down EBDCs, cooking of vegetables that are contaminated with these fungicides can also change them into different metabolites. Ethylenethiourea, an EBDC metabolite that is considered to be cancer-producing, is formed when maneb-treated vegetables are cooked. [829]

Maneb is practically non-toxic to birds but it is highly toxic to fish although it is not very soluble in water. It adsorbs strongly to soil particles. Thus, despite its lengthy soil half life (60 days), maneb is not expected to contaminate groundwater. It may enter surface waters if erosion of soil with adsorbed maneb occurs. Maneb breaks down under both

744

aerobic and anaerobic soil conditions. Maneb's half-life in soil is four to eight weeks. [829]

Identifying Information

CAS NUMBER: 12427-38-2

SYNONYMS: manganese ethylene bisdithiocarbamate; Manesan; Manex; Manzate; Nereb; Newspor; Dithane M-22™; Manzate™

Endocrine Disruptor Information

Maneb is a strongly suspected EED that it is listed by the Centers for Disease Control & Prevention [812], and the World Wildlife Fund [813] as a potential endocrine modifying chemical. There are potential EDC effects to the thyroid gland from exposure to ETU, a decomposition product of maneb. Male monkeys exhibited increased weights of their thyroid glands when exposed to maneb. [821] Maneb contains the element manganese, a metal similar in suspected EED properties to lead. [811] Evidence has been obtained from rats, which includes lead exposure *in utero*, "suggesting a dual site of lead action: (a) at the level of the hypothalamic pituitary unit, and (b) directly at the level of gonadal steroid biosynthesis." Blood-serum testosterone levels (in males) and estradiol levels (in females) were most severely suppressed in those animals where exposure began *in utero*. Both sexes in this group showed a 25% suppression in prepubertal growth (as measured by weight). [868], [811]

Chemical and Physical Properties

CHEMICAL FORMULA: $(C_4H_6MnN_2S_4)x$

MOLECULAR WEIGHT: 265.26

PHYSICAL DESCRIPTION: yellow powder with a faint odor [821]

MELTING POINT: decomposes before melting [829]

SOLUBILITY: 0.5 ug/mL in water; soluble in chloroform and pyridine [829]

OTEHR PHYSICAL PROPERTIES: vapor pressure is $< 10^{-7}$ mbar at 20°C

Environmental Reference Materials

A source of EED environmental reference materials for maneb is Crescent Chemical Co., Hauppauge, NY (Fax: 516-348-0913); Catalog No. 45554.

Hazardous Properties

ACUTE/CHRONIC HAZARDS: Maneb is moderately toxic to humans and it is irritating to the skin and eyes. [829] Mutagenicity testing showed that maneb was positive for inducing chromosomal damage in an in vitro SCE (sister chromatid exchange) assay with metabolic activation. Evidence showed that Maneb is most likely not an initiating agent and the evidence in promotion capability was negative. Maneb has been found to be practically nontoxic to birds and mammals. [821]

FIRE HAZARDS: Maneb is a flammable, combustible material which may decompose, heat up, or self-ignite if it is exposed to air or moisture. Fire may produce irritating or poisonous gases such as toxic oxides of nitrogen and sulfur. Runoff from fire control or dilution may cause pollution. Maneb should be kept away from flames and sparks. [829]

Medical Symptoms of Exposure

Symptoms of exposure include increased thyroid gland weight. [821] Occasional signs of local irritation or inflammation of the skin, eyes, or respiratory tract have been experienced upon contact with maneb. Acute inhalation of large amounts of maneb dust or spray may cause irritation of the mucous membranes, resulting in a scratchy throat, sneezing, cough, and inflammation of the linings of the nose and upper respiratory tract. Signs of poisoning from large amounts of maneb may include nausea, vomiting, diarrhea, loss of appetite, weight loss, headache, confusion, drowsiness, coma, slowed reflexes, respiratory paralysis and death. [829]

Rats given 0.25% maneb in the diet for 2 years developed thyroid abnormalities and goiter. Other effects observed in studies of lab animals exposed to chronic doses of maneb include depression of reflexes, paralysis, impaired kidney function, and benign lung tumors. [829]

Symptoms of poisoning from this class of chemicals include itching, scratchy throat, sneezing, coughing, inflammation of the nose or throat, and bronchitis. There is no evidence of 'neurotoxicity,' nerve tissue destruction or behavior change, from the ethylene bisdithiocarbamates (EDBCs). However, dithiocarbamates are partially chemically broken down, or metabolized, to carbon disulfide, a neurotoxin capable of damaging nerve tissue. EBDC residues in or on foods convert readily to ethylenethiourea (ETU), a known teratogen, during commercial processing or home cooking. [829]

Toxicological Information

Maneb is not currently listed in NTP's Chemical Repository Database or in EPA's IRIS database and there is no published schedule for its addition in the future.

Production, Use, and Pesticide Labeling Information

1. Description of Chemical

> Chemical Name: Manganese ethylene bisdithiocarbamate

> Common Name: Maneb

> Trade Names: Dithane M-22™, Manzate™

EPA Shaughnessy Code: 014505

Chemical Abstracts Service (CAS) Number: 12427-38-2

Year of Initial Registration: Late 1940's

Pesticide Type: Fungicide

Chemical Family: Ethylene Bisdithiocarbamate (EBDC)

US and Foreign Producers: Pennwalt

2. Use Patterns and Formulations

- Registered Uses: Terrestrial food crop (fruits, vegetables, seed crops, nuts, flax, and grains); terrestrial non-food crop (ornamentals, lawns, turf); greenhouse food (tomatoes, rhubarb) and non-food crop (ornamentals).

- Predominant Uses: Apples, potatoes, tomatoes and sweet corn.

- Pests controlled: Foliar fungal diseases of selected fruit, nut, vegetable, grain, field and ornamental (including turf) crops.

- Types of Formulations: Technical, formulation intermediate, dust, granular, wettable powder, wettable powder/dust, flowable concentrate and ready-to-use.

- Types and Method of Application: Foliar application to vegetable crops and apples by aerial equipment or ground equipment. Foliar treatment of tobacco or vegetable seed beds, application of sprays or dusts might be by means of hand held compressed air sprayers or dusting equipment. Potato and tomato foliage may be treated by means of solid set, wheel move, or center pivot sprinkler irrigation equipment.

- Application Rates:

Terrestrial food crop: 0.01 - 8.4 lb. ai/A.
Terrestrial nonfood crop: 0.8 - 3.2 lb. ai/A.
Greenhouse food crop: 1.1 - 2.4 lb. ai/100 gal.
Greenhouse nonfood crop: 0.8 - 2.4 lb. ai/100 gal.

3. Science Findings

Chemical Characteristics

- Physical State: Powder
- Color: Yellow
- Odor: Faint
- Molecular Formula: $(C_4H_6MnN_2S_4)x$

Toxicological Characteristics

- Acute Oral: LD50 = 4,400 mg/kg bw (rat) (Toxicity Category III).

- Acute Dermal: LD50 > 2 gm/kg bw (rabbit) (Toxicity Category III).

- Acute Inhalation: LC50 > 2.22 +/= 0.26 mg/l (rat); (Toxicity Category III).

- Primary Dermal Irritation: Non-irritating (rabbit) (Toxicity Category IV).

- Primary Eye Irritation: Severe eye irritant (rabbit) (Toxicity Category I).

- Dermal Sensitization: Sensitizer (guinea pig).

- Major Routes of Exposure: Oral, dermal and inhalation.

- Subchronic Toxicity: No observed effect level (NOEL) = 100 ppm, LEL (increase in thyroid weight in males) = 300 ppm (Monkeys).

- Oncogenicity: Studies required.

- Chronic Feeding: Studies required.

- Metabolism: Studies in rate indicate that maneb is hydrolyzed, readily absorbed and excreted in the urine and feces. The major metabolite is ETU.

- Reproduction: Study required.

- Teratogenicity & Developmental Toxicity: Studies required.

- Mutagenicity: Mutagenicity testing showed that maneb was positive for inducing chromosomal damage in an in vitro SCE (sister chromatid exchange) assay with metabolic activation. Evidence showed that Maneb is most likely not an initiating agent and the evidence in promotion capability was negative. Additional data are required before the unscheduled DNA synthesis (UDS) assay can be upgraded to acceptable status. The following studies were negative: Sister chromatid exchange in CHO cells in the absence of a metabolic activation and host mediated assay in mice.

Physiological and Biochemical Characteristics

- Metabolism and Persistence in Plants and Animals: Metabolism of maneb is not completely understood. Additional data are being required in plants and livestock. ETU is a major metabolite of concern.

Environmental Characteristics

- Maneb degrades to ETU and other transient degradates in water and soil. ETU is stable in water at pH 5-9 and under sunlight and the degradation of ETU on soil is not enhanced by sunlight radiation.

- Maneb degrades very rapidly under anaerobic aquatic soil conditions but ETU is relatively stable under these conditions.

- ETU is stable in water at pH 5-9 and under sunlight and the degradation of ETU on soil is not enhanced by sunlight radiation. ETU is the degradate of major environmental concern. There are indications that ETU may leach and enter groundwater. However, additional data are required to complete the groundwater assessment.

Ecological Characteristics

- Maneb has been found to be practically nontoxic to birds and mammals.

- Avian dietary toxicity: LC50 > 9,000 ppm (bobwhite)

- The toxicity of a 80% product to warmwater fish is highly toxic.

- Tolerances, expressed as zinc ethylene bisdithiocarbamate, have been established for residues of maneb in a variety of raw agricultural commodities (40 CFR 180.110).

- The toxicology data for maneb are insufficient to determine an Acceptable Daily intake (ADI) or whether the toxicity observed in the studies is due to maneb or ETU. A subchronic study has been used to calculate a Provisional ADI (PADI). Because a subchronic study was used, an uncertainty factor of 1000 was employed. The PADI for maneb is 0.0005 mg/kg/day based on the six month feeding study with a NOEL of 5 mg/kg/day.

- The theoretical maximum residue contribution (TMRC), based on the assumption that 100 percent of each crop is treated and contains residues at the tolerance level, is 0.030 or approximately 600 percent of the PADI. Based on a more realistic dietary assessment, using anticipated field residues and estimate of percent crop treated, the estimated average consumption for the US population is 0.0036 mg/kg/day or 70 percent of the PADI.

4. Summary of Regulatory Positions and Rationale

- The Agency has initiated a Special Review for maneb along with the other EBDC's in June 1987 because of concern about the oncogenic risk to consumers from dietary exposure to ETU from food treated with these pesticides, and the risks of teratogenicity and adverse thyroid effects to applicators and mixer/loaders from exposure to ETU.

- ETU has been classified as a B2 oncogen (probable human carcinogen).

- The Agency will not consider establishment of new food use tolerances for maneb because the current residue chemistry and toxicology data are not sufficient to assess existing tolerances and the toxicology data base is insufficient to determine an ADI and also does not allow a decision as to whether observed toxicity is due to maneb or ETU.

- The Agency will consider the need for establishment of tolerances for ETU and any intermediate metabolites when data are sufficient to permit such decisions.

- The Agency will not establish any food/feed additive regulations pursuant to Section 409 of the Federal Food, Drug and Cosmetic Act (FFDC) and is deferring action on previously established food/feed additive regulations.

- Protective clothing labeling for maneb products, as specified as a result of the 1982 Decision Document should be updated.

- The Agency is requiring reentry data for maneb. In order to remain in compliance with FIFRA, an interim 24-hour reentry interval requirement must be placed on the label of all maneb end-use products registered for agricultural uses, until the required data are submitted and evaluated and any change in this reentry interval is announced.

- The Agency has screened and reviewed the environmental fate data to determine if maneb/ETU and/or its degradate(s) have the potential to leach into ground water. The Agency has decided that a small-scale retrospective groundwater monitoring study is required to further define the extent of the ground water problems.

- While the data gaps are being filled, currently registered manufacturing-use products (MP's) and end-use products (EP's) containing maneb as the sole active ingredient may be sold, distributed, formulated and used, subject to the terms and conditions specified in this Standard. However, new uses will not be registered. Registrants must provide or agree to develop additional data, as specified in the

Data Appendices of the Registration Standard, in order to maintain existing registrations.

- Labeling Requirements

   - All maneb products must bear appropriate labeling as specified in 40 CFR 156.10. Appendix II of the Registration Standard contains information on labeling.

   - The following are the major labeling specifications:

      - Environmental hazard statement
      - Protective clothing requirements
      - Preharvest interval
      - Worker safety rules
      - Grazing restrictions for almonds, apples, beans, corn, peanuts, potato, sugar beets, ornamental grasses, ornamental turf.

5. Summary of Major Data Gaps

   - Product Chemistry
      - All

   - Toxicology
      - Subchronic dermal (21-Day)
      - Subchronic inhalation (90-Day)
      - Chronic toxicity (rodent and nonrodent)
      - Oncogenicity (rat and mouse)
      - Teratology (rabbit and rat)
      - Reproduction (rat)
      - Mutagenicity (gene mutation) (other genotoxic effects)
      - Dermal absorption

   - Residue Chemistry
      - Nature of the Residue in Plants and Livestock
      - Analytical Methods
      - Magnitude of Residue for Variety of Commodities

- Environmental Fate
    - Hydrolysis
    - Photodegradation studies in water and soil
    - Aerobic soil studies
    - Aerobic aquatic
    - Leaching and absorption/desorption
    - Aquatic (sediment)
    - Dissipation Soil Studies
    - Small-scale retrospective monitoring study
    - Fish accumulation

- Reentry Protection
    - Reentry Studies on Foliar and Soil Dissipation

- Wildlife and Aquatic Organisms
    - Avian oral toxicity
    - Avian dietary toxicity
    - Avian reproduction
    - Freshwater fish toxicity
    - Acute freshwater invertebrates
    - Estuarine and marine organism toxicity
    - Fish early life stage and invertebrate life-cycle
    - Aquatic organism accumulation

- ETU Data Requirements

    - Toxicology
        - Chronic (rodent and non-rodent)
        - Reproduction

- Environmental Fate
    - Aerobic and anaerobic sol metabolism
    - Aerobic aquatic
    - Lab volatility
    - Degradation (soil)
    - Aquatic (sediment)
    - Degradation (soil long-term)
    - Fish Accumulation

6. Contact Person at EPA

Lois A. Rossi
Product Manager 21
Fungicide-Herbicide Branch Registration Division (TS-767C)
Office of Pesticide Programs
Environmental Protection Agency
401 M St., SW.
Washington, DC, 20460

Office Location and Phone Number

Room 227, Crystal Mall #2,
1921 Jefferson Davis Highway
Arlington, VA 22202.

Telephone: (703) 557-1900

Source: *Instant EPA's Pesticide Facts*, 1994, Instant Reference Sources, Inc., 7605 Rockpoint Drive, Austin, TX 78731 (Fax: 512-345-2386; Internet URL, http://www.instantref.com/inst-ref.htm).

Mercury

Hg

Introduction

Mercury is a virulent and cummulative poison that is readily absorbed through the respiratory tract, the gastrointestinal tract or through unbroken skin. It easily forms alloys with many metals, such as gold, silver, and tin, and is used in the recovery of gold from its ores. The metal is used for making thermometers, barometers, diffusion pumps, and many other instruments. It is also used widely in mercury-vapor lamps and advertising signs; in electrical apparatus, mercury cells for caustic soda and chlorine production; dental preparations; antifouling paint; batteries; and catalysts. [706] It formerly was used in pesticides but that use is now predominately discontinued.

Identifying Information

CAS NUMBER: 7439-97-6

NIOSH Registry Number: OV4550000

SYNONYMS: Mercure [French]; Mercurio [Italian and Spanish]; Quicksilber [German]; Quick silver

Endocrine Disruptor Information

Mercury is a strongly suspected EED that it is listed by EPA [811], the Centers for Disease Control & Prevention [812], and the World Wildlife Fund [813] as a potential endocrine modifying chemical. In support of this importance, mercury is considered a metal similar in suspected EED properties to lead. Evidence has been obtained from rats, which includes lead exposure *in utero*, "suggesting a dual site of lead action: (a) at the level of the hypothalamic pituitary unit, and (b) directly at the level of gonadal steroid biosynthesis." Blood-serum testosterone levels (in males) and estradiol levels (in females) were most severely suppressed in those animals where exposure began *in utero*. Both sexes in this group showed a 25% suppression in prepubertal growth (as measured by weight). [868], [811] In animals, methylmercury preferentially crosses the placental barrier and the fetal blood brain barrier, and is neuroteratological. Based on human and animal studies, inorganic and organic mercury may act on the endocrine system to alter hormonal levels, although the mechanisms of action are unproven. [882]

754

Chemical and Physical Properties

ATOMIC FORMULA: Hg

ATOMIC WEIGHT: 200.59 [702]

PHYSICAL DESCRIPTION: silver colored heavy liquid

SPECIFIC GRAVITY: 13.6 @ 20C/4°C [704]

MELTING POINT: -38.8°C [702]

BOILING POINT: 357°C [702]

SOLUBILITY: Water: 0.28 umoles/L [702]

OTHER PHYSICAL DATA: Freezing Point: -38°F [704] Vapor Pressure: 0.0012 mm [704] Solubility, water: Insoluble [704]

Environmental Reference Materials

A source of EED environmental reference materials for mercury is Radian International LLC, Austin, TX (Fax 512-454-0268; Internet URL, http://www.radian.com/standards); Catalog No. ICM-006L.

Hazardous Properties

ACUTE/CHRONIC HAZARDS: Poison by inhalation. Corrosive to skin, eyes, and mucous membranes. Mercury is an experimental tumorigen and teratogen. Experimental reproductive effects and human mutagenic data exist. [703] Elemental mercury is usually non-toxic when ingested unless GI fistula or other GI inflammatory disease is present or the mercury is retained for a prolonged period in the GI tract. Elemental mercury will readily vaporize if heated and form extremely toxic oxide fumes. The most consistent and pronounced effects of chronic exposure to elemental mercury vapor are on the CNS. Effects are neurological and psychiatric. Either acute or chronic exposure may produce permanent changes to affected organs and organ systems. Acute poisoning due to mercury vapors affects the lungs primarily, in the form of acute interstitial pneumonitis, bronchitis, and bronchiolitis. In general, chronic exposure produces four classical signs: gingivitis, sialorrhea, increased irritability, and muscular tremors. Rarely are all four seen together in an individual case. [723] It is still not known to what degree renal damage may occur in connection with chronic exposure to mercury vapor. Severe nephrotic changes have not been described in patients exposed only to mercury vapor. In patients exposed to a combination of mercury dust and vapor, such changes have been reported. The brain is the critical organ in humans for chronic mercury vapor exposure; in severe cases, spongeous degeneration of brain cortex can occur as a late sequela to past exposure. Renal proteinuria has been described following exposure to mercury vapor. Other reported effects from elemental mercury are contact dermatitis from mercury amalgam fillings and mercury sensitivity occurring among dental students. [723]

VOLATILITY: Vapor Pressure: 0.0012 mm [704]

FIRE HAZARD: Mercury is a noncombustible liquid [704].

REACTIVITY: Mercury may explode on contact with 3-bromopropyne; alkynes + silver perchlorate; ethylene oxide; lithium; methylsilane + oxygen (when shaken); peroxyformic acid; chlorine dioxide; and tetracarbonylnickel + oxygen. With methyl azide it explodes with shock or spark. Mercury vapor ignites on contact with boron diiodophosphide and it reacts violently with the following chemicals: acetylenic compounds; chlorine; chlorine dioxide, methyl azide; disodium acetylene; and nitromethane. [703] In addition, it may react violently with the following chemicals:

> **Ammonia**: Mercury may react with ammonia to form an explosive product. [703] The use of mercury manometers with ammonia should be avoided as intrinsically unsafe. Although pure dry ammonia and mercury do not react even under pressure at 340 kbar and 200°C, the presence of traces of water leads to the formation of an explosive compound, which may explode during depressurization of the system. Explosions in mercury-ammonia systems had been reported previously. [710]

> **Metals:** Mercury reacts violently with numerous metals (e.g., aluminum, calcium, potassium, sodium, and rubidium producing exothermic formation of amalgams). [703] The high mobility and tendency to dispersion exhibited by mercury, and the ease with which it forms alloys (amalgams) with many laboratory and electrical contact metals, can cause severe corrosion problems in laboratories. [710]

CORROSIVITY: The high mobility and tendency to dispersion exhibited by mercury, and the ease with which it forms alloys (amalgams) with many laboratory and electrical contact metals, can cause severe corrosion problems in laboratories [710].

Medical Symptoms of Exposure

Symptoms of exposure to mercury vapor may include coughing, chest pain, dyspnea, bronchial pneuitis; tremor, insomnia; irritability, indecision; headache, fatigue, weakness; stomatitis, salivation; gastrointestinal disturbance, anorexia, weight loss; proteinuria; irritation of eyes and skin. [704] Human systemic effects by inhalation include wakefulness, muscle weakness, anorexia, headache, tinnitus, hypermotility, diarrhea, liver changes, dermatitis and fever. [703] Mercury vapor will cause severe pulmonary damage if inhaled, as well as nephrotoxicity and gingivitis. Symptoms of chronic exposure to elemental mercury vapor may include depression, irritability, exaggerated response to stimulation (erethism), excessive shyness, insomnia, emotional instability, forgetfulness, confusion, and vasomotor disturbances such as excessive perspiration and uncontrolled blushing. Tremors are also common; these are exaggerated when task is required but minimal when patient is at rest or asleep. A fine trembling of fingers, eyelids, lips and tongue may be interrupted intermittently by coarse shaking movements. Erethism and tremors are reversible. [723] Acute intoxication from inhaling mercury vapor in high concentrations used to be common among those who extracted mercury from its ores. The condition is characterized by metallic taste, nausea, abdominal pain, vomiting, diarrhea, headache, and sometimes albuminuria. After a few days, salivary glands swell, stomatitis and gingivitis develop, and a dark line of mercury sulfide forms on inflamed gums. Teeth may loosen, and ulcers may form on lips and cheeks. In milder cases, recovery occurs within 10-14 days; but in others, poisoning of the chronic type may ensue. [723]

Toxicological Information

The toxicological information below is gathered from several sources including two national databases: one from the *National Toxicology Program Chemical Repository Database* and another from *EPA's Integrated Risk Information System (IRIS).*

Toxicology Information from the National Toxicology Program

Mercury is not currently listed in NTP's Chemical Repository database nor is it scheduled for addition in the future.

Toxicology Information from EPA's Integrated Risk Information System (IRIS)

I. CHRONIC HEALTH HAZARD ASSESSMENTS FOR NONCARCINOGENIC EFFECTS

I.A. REFERENCE DOSE FOR CHRONIC ORAL EXPOSURE (RfD)

[Ed. Note: EPA does not have information yet in this section of IRIS].

I.B. REFERENCE CONCENTRATION FOR CHRONIC INHALATION EXPOSURE (RfC)

The inhalation Reference Concentration (RfC) is analogous to the oral RfD and is likewise based on the assumption that thresholds exist for certain toxic effects such as cellular necrosis, but may not exist for other toxic effects such as carcinogenicity. The inhalation RfC considers toxic effects for both the respiratory system (portal-of-entry) and for effects peripheral to the respiratory system (extrarespiratory effects). It is appropriately expressed in units of mg/cu.m. In general, the RfC is an estimate (with uncertainty spanning perhaps an order of magnitude) of a daily inhalation exposure of the human population (including sensitive subgroups) that is likely to be without an appreciable risk of deleterious effects during a lifetime. Inhalation RfCs are derived according to the Interim Methods for Development of Inhalation Reference Doses (EPA/600/8-88/066F August 1989) developed by US EPA scientists and peer-reviewed. For more information on the interim nature of these methods and future plans see the INFOMORE Section of IRIS. RfCs can also be derived for the noncarcinogenic health effects of compounds which are carcinogens. Therefore, it is essential to refer to other sources of information concerning the carcinogenicity of this substance. If the US EPA has evaluated this substance for potential human carcinogenicity, a summary of that

evaluation will be contained in Section II of this file when a review of that evaluation is completed.

I.B.1. INHALATION RfC SUMMARY

Critical Effect	Exposures*	UF	MF	RfC
Hand tremor; increases in memory disturbances; slight subjective and objective evidence of autonomic dysfunction	NOAEL: None LOAEL: 0.025 mg/cu.m (converted to LOAEL [ADJ] of 0.009 mg/cu.m	30	1	3E-4 mg/cu.m

*Conversion Factors and Assumptions: This is an extrarespiratory effect of a vapor (gas). The LOAEL is based on an 8-hour TWA occupational exposure. MVho = 10 cu.m/day, MVh = 20 cu.m/day. LOAEL(HEC) = LOAEL(ADJ) = 0.025 mg/cu.m x MVho/MVh x 5 days/7 days = 0.009 mg/cu.m. Air concentrations (TWA) were measured in the Fawer et al. (1983), Ngim et al. (1992), and Liang et al. (1993) studies. Air concentrations were extrapolated from blood levels based on the conversion factor of Roels et al. (1987) as described in the Additional Comments section for the studies of Piikivi and Tolonen (1989), Piikivi and Hanninen (1989), and Piikivi (1989).

Human occupational inhalation studies Fawer et al., 1983; Piikivi and Tolonen, 1989; Piikivi and Hanninen, 1989; Piikivi, 1989; Ngim et al., 1992; Liang et al., 1993

I.B.2. PRINCIPAL AND SUPPORTING STUDIES (INHALATION RfC)

Fawer, R.F., U. DeRibaupierre, M.P. Guillemin, M. Berode and M. Lobe. 1983. Measurement of hand tremor induced by industrial exposure to metallic mercury. J. Ind. Med. 40: 204-208.

Piikivi, L. and U. Tolonen. 1989. EEG findings in chloralkali workers subjected to low long term exposure to mercury vapor. Br. J. Ind. Med. 46: 370-375.

Piikivi, L. and H. Hanninen. 1989. Subjective symptoms and psychological performance of chlorine-alkali workers. Scand. J. Work Environ. Health. 15: 69-74.

Piikivi, L. 1989. Cardiovascular reflexes and low long-term exposure to mercury vapor. Int. Arch. Occup. Environ. Health. 61: 391-395.

Ngim, C.H., S.C. Foo, K.W. Boey and J. Jeyaratnam. 1992. Chronic neurobehavioral effects of elemental mercury in dentists. Br. J. Ind. Med. 49: 782-790.

Liang, Y-X., R-K. Sun, Y. Sun, Z-Q. Chen and L-H. Li. 1993. Psychological effects of low exposure to mercury vapor: Application of a computer- administered neurobehavioral evaluation system. Environ. Res. 60: 320-327.

Fawer et al. (1983) used a sensitive objective electronic measure of intention tremor (tremors that occur at the initiation of voluntary movements) in 26 male workers (mean age of 44 years) exposed to low levels of mercury vapor in various occupations: fluorescent tube manufacture (n=7), chloralkali plants (n=12), and acetaldehyde production (n=7). Controls (n=25; mean age of 44.6 years) came from the same factories but were not exposed occupationally. Personal air samples (two per subject) were used to characterize an average exposure concentration of 0.026 mg/cu.m. It should be noted that it is likely that the levels of mercury in the air varied during the period of exposure and historical data indicate that previous exposures may have been higher. Exposure measurements for the control cohort were not performed. The average duration of exposure was 15.3 years. The measures of tremor were significantly increased in the exposed compared to control cohorts, and were shown to correspond to exposure and not to chronologic age. These findings are consistent with neurophysiological impairments that might result from accumulation of mercury in the cerebellum and basal ganglia. Thus, the TWA of 0.026 mg/cu.m was designated a LOAEL. Using the TWA and adjusting for occupational ventilation rates and workweek, the resultant LOAEL(HEC) is 0.009 mg/cu.m.

Piikivi and Tolonen (1989) used EEGs to study the effects of long-term exposure to mercury vapor in 41 chloralkali workers exposed for a mean of 15.6 +/- 8.9 years as compared with matched referent controls. They found that the exposed workers, who had mean blood Hg levels of 12 ug/L and mean urine Hg levels of 20 ug/L, tended to have an increased number of EEG abnormalities when analyzed by visual inspection only. When the EEGs were analyzed by computer, however, the exposed workers were found to have significantly slower and attenuated brain activity as compared with the referents. These changes were observed in 15% of the exposed workers. The frequency of these changes correlated with cortical Hg content (measured in other studies); the changes were most prominent in the occipital cortex less prominent in the parietal cortex, and almost absent in the frontal cortex. The authors extrapolated an exposure level associated with these EEG changes of 0.025 mg/cu.m from blood levels based on the conversion factor calculated by Roels et al. (1987).

Piikivi and Hanninen (1989) studied the subjective symptoms and psychological performances on a computer-administered test battery in 60 chloralkali workers exposed to mercury vapor for a mean of 13.7 +/- 5.5 years as compared with matched referent controls. The exposed workers had mean blood Hg levels of 10 ug/L and mean urine Hg levels of 17 ug/L. A statistically significant increase in subjective measures of memory disturbance and sleep disorders was found in the exposed workers. The exposed workers also reported more anger, fatigue and confusion. No objective disturbances in perceptual motor, memory or learning abilities were found in the exposed workers. The authors extrapolated an exposure level associated with these subjective measures of memory disturbance of 0.025 mg/cu.m from blood levels based on the conversion factor calculated by Roels et al. (1987).

Both subjective and objective symptoms of autonomic dysfunction were investigated in 41 chloralkali workers exposed to mercury vapor for a mean of 15.6 +/- 8.9 years as compared with matched referent controls (Piikivi, 1989). The quantitative non-invasive test battery consisted of measurements of pulse rate variation in normal and deep breathing, in the Valsalva maneuver and in vertical tilt, as well as blood pressure responses during standing and isometric work. The exposed workers had mean blood levels of 11.6 ug/L and mean urine levels of 19.3 ug/L. The exposed workers complained of more subjective symptoms of autonomic dysfunction than the controls, but the only statistically significant difference was an increased reporting of palpitations in the exposed workers. The quantitative tests revealed a slight decrease in pulse rate variations, indicative of autonomic reflex dysfunction, in the exposed workers. The authors extrapolated an exposure level associated with these subjective and objective measures of autonomic dysfunction of 0.030 mg/cu.m from blood levels based on the conversion factor calculated by Roels et al. (1987).

Two more recent studies in other working populations corroborate the neurobehavioral toxicity of low-level mercury exposures observed in the Fawer et al. (1983), Piikivi and Tolonen (1989), Piikivi and Hanninen (1989), and Piikivi (1989) studies.

Ngim et al. (1992) assessed neurobehavioral performance in a cross-sectional study of 98 dentists (38 female, 60 male; mean age 32, range 24-49 years) exposed to TWA concentrations of 0.014 mg/cu.m (range 0.0007 to 0.042 mg/cu.m) versus 54 controls (27 female, 27 male; mean age 34, range 23-50 years) with no history of occupational exposure to mercury. Air concentrations were measured with personal sampling badges over typical working hours (8-10 hours) and converted to an 8-hour TWA. No details on the number of exposure samples or exposure histories were provided. Blood samples from the exposed cohort were also taken and the data supported the correspondence calculated by Roels et al. (1987). Based on extrapolation of the average blood mercury concentration (9.8 ug/L), the average exposure concentration would be estimated at 0.023 mg/cu.m. The average duration of practice of the exposed dentists was 5.5 years. Exposure measurements of the control cohort were not performed. The exposed and control groups were adequately matched for age, amount of fish consumption, and number of amalgam dental fillings. The performance of the dentists was significantly worse than controls on a number of neurobehavioural tests measuring motor speed (finger tapping), visual scaning, visumotor coordination and concentration, visual memmory, and visuomotor coordination speed. These neurobehavioral effects are consistent with central and peripheral neurotoxicity and the TWA is considered a LOAEL. Using the TWA and adjusting for occupational ventilation rates and the reported 6-day workweek, the resultant LOAEL(HEC) is 0.006 mg/cu.m.

Liang et al. (1993) investigated workers in a fluorescent lamp factory with a computer-administered neurobehavioral evaluation system and a mood inventory profile. The exposed cohort (mean age 34.2 years) consisted of 19 females and 69 males exposed to ninterruptedly for at least 2 years prior to the study. Exposure was monitored with area samplers and ranged from 0.008 to 0.085 mg/cu.m across worksites. No details on how the exposure profiles to account for time spent in different worksites were constructed. The average exposure was estimated at 0.033 mg/cu.m. (range 0.005 to 0.19 mg/cu.m). The average duration was of working was 15.8 years for the exposed cohort. Urinary excretion was also monitored and reported to average 0.025 mg/L. The control cohort (mean age 35.1 years) consisted of 24 females and 46 males recruited from an embroidery factory. The controls were matched for age, education, smoking and drinking habits. Exposure measurements for the control cohort were not performed. The exposed cohort performed significantly worse than the control on tests of finger tapping, mental arithmetic, two-digit searches, switiching attention, and visual reaction time. The effect on performance persisted after the confounding factor of chronological age was controlled. Based on these neurobehavioral effects, the TWA of 0.033 mg/cu.m is designated as LOAEL. Using the TWA and adjusting for occupational ventilation rates and workweek, the resultant LOAEL(HEC) is 0.012 mg/cu.m.

The above studies were taken together as evidence for a LOAEL based on neurobehavioral effects of low-level mercury exposures. The LOAEL(HEC) levels calculated on measured air concentration levels of the Ngim et al. (1992) and the Liang et al. (1993) studies bracket that calculated based on the air concentrations measured by Fawer et al. (1983) as a median HEC level. Extrapolations of blood levels, used as biological monitoring that accounts for variability in exposure levels, also converge at 0.025 mg/cu.m as a TWA which results in the same HEC level. Thus, the TWA level of 0.025 mg/cu.m was used to represent the exposure for the synthesis of the studies described above. Using this TWA and taking occupational ventilation rates and workweek into account results in a LOAEL(HEC) of 0.009 mg/cu.m.

I.B.3. UNCERTAINTY AND MODIFYING FACTORS (INHALATION RfC)

UF -- An uncertainty factor of 10 was used for the protection of sensitive human subpopulations (including concern for acrodynia - see Additional Comments section) together with the use of a LOAEL. An uncertainty factor of 3 was used for lack of data base, particularly developmental and reproductive studies.

MF -- None

I.B.4. ADDITIONAL COMMENTS (INHALATION RfC)

Probably the most widely recognized form of hypersensitivity to mercury poisoning is the uncommon syndrome known as acrodynia, also called erythredema polyneuropathy or pink disease (Warkany and Hubbard, 1953). Infantile acrodynia was first described in 1828, but adult cases have also since been reported. While acrodynia has generally been associated with short-term exposures and with urine levels of 50 ug/L or more, there are some cases in the literature in which mercury exposure was known to have occurred, but no significant (above background) levels in urine were reported. There could be many reasons for this, but the most likely is that urine levels are not a simple measure of body burden or of target tissue (i.e., brain levels); however, they are the best means available for assessing the extent of exposure. It was felt that the RfC level estimated for mercury vapor based on neurotoxicity of chronic exposure in workers is adequate to protect children from risk of acrodynia because such exposures of long duration would be expected to raise urine levels by only 0.12 ug/L against a background level of up to 20 ug/L (i.e., such exposures would not add significantly to the background level of mercury in those exposed).

Roels et al. (1987) investigated the relationships between the concentrations of metallic mercury in air and levels monitored in blood or urine in workers exposed during manufacturing of dry alkaline batteries. Breathing zone personal samples were used to characterize airborne mercury vapors. Total mercury in blood and urine samples were analyzed using atomic absorption. The investigation controlled for several key factors including the use of reliable personal air monitoring, quality control for blood and urine analyses, standardization of the urinary mercury concentration for creatinine concentration, and stability of exposure conditions (examined subjects were exposed to mercury vapor for at least 1 year). Strong correlations were found between the daily intensity of exposure to mercury vapor and the end of workshift levels in blood (r=0.86; n=40) or urine (r=0.81; n=34). These relationships indicated a conversion factor of 1:4.5 (air:blood) and 1:1.22 (air:urine as ug/g creatinine). These factors were used to extrapolate blood or urine levels associated with effects in the reported studies to airborne mercury levels.

Sensory and motor nerve conduction velocities were studied in 18 workers from a mercury cell chlorine plant (Levine et al., 1982). Time-integrated urine Hg levels were used as an indicator of mercury exposure. Using linearized regression analysis, the authors found that motor and sensory nerve conduction velocity changes (i.e., prolonged distal latencies correlated with the time-integrated urinary Hg levels in asymptomatic exposed workers) occurred when urinary Hg levels exceeded 25 ug/L. This study demonstrates that mercury exposure can be associated with preclinical evidence of peripheral neurotoxicity.

Singer et al. (1987) studied nerve conduction velocity of the median motor, median sensor and sural nerves in 16 workers exposed to various inorganic mercury compounds (e.g., mercuric oxides, mercurial chlorides, and phenyl mercuric acid) for an average of 7.3 +/- 7.1 years as compared with an unexposed control group using t-tests. They found a slowing of nerve conduction velocity in motor, but not sensory, nerves that correlated with increased blood and urine Hg levels and an increased number of neurologic symptoms. The mean mercury levels in the exposed workers were 1.4 and 10 ug/L for blood and urine, respectively. These urine levels are 2-fold less than those associated with peripheral neurotoxicity in other studies (e.g., Levine et al., 1982). There was considerable variability in the data presented by Singer et al. (1987), however, and the statistical analyses (t-test) were not as rigorous as those employed by Levine et al. (1982) (linearized regression analysis). Furthermore, the subjects in the Levine et al. (1982) study were asymptomatic at higher urinary levels than those reported to be associated with subjective neurological complaints in the workers studied by Singer et al. (1987). Therefore, these results are not considered to be as reliable as those reported by Levine et al. (1982).

Miller et al. (1975) investigated several subclinical parameters of neurological dysfunction in 142 workers exposed to inorganic mercury in either the chloralkali industry or a factory for the manufacture of magnetic materials. They reported a significant increase in average forearm tremor frequency in workers whose urinary Hg concentrations exceeded 50 ug/L as compared with unexposed controls. Also observed were eyelid fasciculation, hyperactive deep-tendon reflexes and dermatographia, but there was no correlation between the incidence of these findings and urinary Hg levels.

Roels et al. (1985) examined 131 male and 54 female workers occupationally exposed to mercury vapor for an average duration of 4.8 years. Urinary mercury (52 and 37 ug/g creatinine for males and females, respectively) and blood mercury levels (14 and 9 ug/L for males and females, respectively) were recorded, but atmospheric mercury concentration was not provided. Symptoms indicative of CNS disorders were reported but not related to mercury exposure. Minor renal tubular effects were detected in mercury-exposed males and females and attributed to current exposure intensity rather (urinary Hg >50 ug/g creatinine) than exposure duration. Male subjects with urinary mercury levels of >50 ug/g creatinine exhibited preclinical signs of hand tremor. It was noted that females did not exhibit this effect and that their urinary mercury never reached the level of 50 ug/g creatinine. A companion study (Roels et al., 1987) related air mercury (Hg-air)levels to blood mercury (Hg-blood) and urinary mercury (Hg-U) values in 10 workers in a chloralkali battery plant. Duration of exposure was not specified. A high correlation was reported for Hg-air and Hg-U for preshift exposure (r=0.70, p<0.001) and post-shift (r=0.81, p<0.001) measurements. Based on these data and the results of their earlier (1985) study, the investigators suggested that some mercury-induced effects may occur when Hg-U levels exceed 50 ug/g creatinine, and that this value corresponds to a mercury TWA of about 40 ug/cu.m.

A survey of 567 workers at 21 chloralkali plants was conducted to ascertain the effects of mercury vapor inhalation (Smith et al., 1970). Mercury levels ranged from <0.01 to 0.27 mg/cu.m and chlorine concentrations ranged from 0.1 to 0.3 ppm at most of the working stations of these plants. Worker exposure to mercury levels (TWA) varied, with 10.2% of the workers being exposed to <0.01 mg/cu.m, 48.7% exposed to 0.01 to 0.05 mg/cu.m, 25.6% exposed to 0.06 to 0.10 mg/cu.m and 4.8% exposed to 0.24 to 0.27 mg/cu.m (approximately 85% were exposed to Hg levels less than or equal to 0.1 mg/cu.m). The duration of employment for the examined workers ranged from one year (13.3%) to >10 years (31%), with 55.7% of the workers being employed for 2 or 9 years. A group of 600 workers not exposed to chlorine served as a control group for assessment of chlorine effects, and a group of 382 workers not exposed to either chlorine or mercury vapor served as the reference control group. A strong positive correlation (p<0.001) was found between the mercury TWAs and the reporting of subjective neuropsychiatric symptoms (nervousness, insomnia), occurrence of objective tremors, and weight and appetite loss. A positive correlation (p<0.001) was also found between mercury exposure levels and urinary and blood mercury levels of test subjects. No adverse alterations in cardiorespiratory, gastrointestinal, renal or hepatic functions were attributed to the mercury vapor exposure. Additionally, biochemical (hematologic data, enzyme activities) and clinical measurements (EKG, chest X-rays) were no different between the mercury-exposed and non-exposed workers. No significant signs or symptoms were noted for individuals exposed to mercury vapor concentrations less than or equal to 0.1 mg/cu.m. This study provides data indicative of a NOAEL of 0.1 mg Hg/cu.m and a LOAEL of 0.18 mg Hg/cu.m. In a follow-up study conducted by Bunn et al. (1986), however, no significant differences in the frequency of objective or subjective findings such as weight loss and appetite loss were observed in workers exposed to mercury at levels that ranged between 50 and 100 ug/L. The study by Bunn et al. (1986) was limited, however, by the lack of information provided regarding several methodological questions such as quality assurance measures and control of possible confounding variables.

The mercury levels reported to be associated with preclinical and symptomatic neurological dysfunction are generally lower than those found to affect kidney function, as discussed below.

Piikivi and Ruokonen (1989) found no evidence of glomerular or tubular damage in 60 chloralkali workers exposed to mercury vapor for an average of 13.7 +/- 5.5 years as compared with their matched referent controls. Renal function was assessed by measuring urinary albumin and N-acetyl-beta-glucosaminidase (NAG) activity. The mean blood Hg level in the exposed workers was 14 ug/L and the mean urinary level was 17 ug/L. The authors extrapolated the NOAEL for kidney effects based on these results of 0.025 mg/cu.m from blood levels using the conversion factor calculated by Roels et al. (1987).

Stewart et al. (1977) studied urinary protein excretion in 21 laboratory workers exposed to 10-50 ug/cu.m of mercury. Their urinary level of mercury was about 35 ug/L. Increased proteinuria was found in the exposed workers as compared with unexposed controls. When preventive measure were instituted to limit exposure to mercury, proteinuria was no longer observed in the exposed technicians.

Lauwerys et al. (1983) found no change in several indices of renal function (e.g., proteinuria, albuminuria, urinary excretion of retinol-binding protein, aminoaciduria, creatinine in serum, beta-2-microglobulin in serum) in 62 workers exposed to mercury vapor for an average of 5.5 years. The mean urinary Hg excretion in the exposed workers was 56 ug/g creatinine, which corresponds to an exposure level of about 46 ug/cu.m according to a conversion factor of 1:1.22 (air:urine [ug/g creatinine]) (Roels et al., 1987). Despite the lack of observed renal effects, 8 workers were found to have an increased in serum anti-laminin antibodies, which can be indicative of immunological effects. In a followup study conducted by Bernard et al. (1987), however, there was no evidence of increased serum anti-laminin antibodies in 58 workers exposed to mercury vapor for an average of 7.9 years. These workers had a mean urinary Hg excretion of 72 ug/g creatinine, which corresponds to an exposure levels of about 0.059 mg/cu.m.

Stonard et al. (1983) studied renal function in 100 chloralkali workers exposed to inorganic mercury vapor for an average of 8 years. No changes in the following urinary parameters of renal function were observed at mean urinary Hg excretion rates of 67 ug/g creatinine: total protein, albumin, alpha-1-acid glycoprotein, beta-2-microglobulin, NAG, and gamma-glutamyl transferase. When urinary Hg excretion exceeded 100 ug/g creatinine, a small increase in the prevalence of higher activities of NAG and gamma-glutamyl transferase was observed.

The mercury levels reported to be associated with preclinical and symptomatic neurological dysfunction and kidney effects are lower than those found to pulmonary function, as discussed below.

McFarland and Reigel (1978) described the cases of 6 workers who were acutely exposed (4-8 hours) to calculated metallic mercury vapor levels of 1.1 to 44 mg/cu.m. These men exhibited a combination of chest pains, dyspnea, cough, hemoptysis, impairment of pulmonary function (reduced vital capacity), diffuse pulmonary infiltrates and evidence of interstitial pneumonitis. Although the respiratory symptoms resolved, all six cases exhibited chronic neurological dysfunction, presumably as a result of the acute, high-level exposure to mercury vapor.

Lilis et al. (1985) described the case of a 31-year-old male who was acutely exposed to high levels of mercury vapor in a gold-extracting facility. Upon admission to the hospital, the patient exhibited dyspnea, chest pain with deep inspiration, irregular infiltrates in the lungs and reduced pulmonary function (forced vital capacity [FVC]). The level of mercury to which he was exposed is not known, but a 24-hour urine collection contained 1900 ug Hg/L. Although the patient improved gradually over the next several days, 11 months after exposure he still showed signs of pulmonary function abnormalities (e.g., restriction and diffusion impairment).

Levin et al. (1988) described four cases of acute high-level mercury exposure during gold ore purification. The respiratory symptoms observed in these four cases ranged from minimal shortness of breath and cough to severe hypoxemia. The most severely affected patient exhibited mild interstitial lung disease both radiographically and on pulmonary function testing. One patient had a urinary Hg level of 245 ug/L upon hospital admission. The occurrence of long-term respiratory effects in these patients could not be evaluated since all but one refused follow-up treatment.

Ashe et al. (1953) reported that there was no histopathological evidence of respiratory damage in 24 rats exposed to 0.1 mg Hg/cu.m 7 hr/day, 5 days/week for 72 weeks. This is equivalent to a NOAEL[HEC] of 0.07 mg/cu.m.

Kishi et al. (1978) observed no histopathological changes in the lungs of rats exposed to 3 mg/cu.m of mercury vapor 3 hours/day, 5 days/week for 12-42 weeks.

Beliles et al. (1967) observed no histopathological changes in the lungs of pigeons exposed to 0.1 mg/cu.m of mercury vapor 6 hours/day, 5 days/week for 20 weeks.

Neurological signs and symptoms (i.e., tremors) were observed in 79 workers exposed to metallic mercury vapor whose urinary mercury levels exceeded 500 ug/L. Short-term memory deficits were reported in workers whose urine levels were less than 500 ug/L (Langolf et al., 1978).

Impaired performance in mechanical and visual memory tasks and psychomotor ability tests was reported by Forzi et al. (1978) in exposed workers whose urinary Hg levels exceeded 100 ug/L.

Decreased strength, decreased coordination, increased tremor, decreased sensation and increased prevalence of Babinski and snout reflexes were exhibited by 247 exposed workers whose urinary Hg levels exceeded 600 ug/L. Evidence of clinical neuropathy was observed at urinary Hg levels that exceeded 850 ug/L (Albers et al., 1988).

Preclinical psychomotor dysfunction was reported to occur at a higher incidence in 43 exposed workers (mean exposure duration of 5 years) whose mean urinary excretion of Hg was 50 ug/L. Workers in the same study whose mean urinary Hg excretion was 71 ug/L had a higher incidence of total proteinuria and albuminuria (Roels et al., 1982).

Postural and intention tremor was observed in 54 exposed workers (mean exposure duration of 7.7 years) whose mean urinary excretion of Hg was 63 ug/L (Roels et al., 1989).

Verbeck et al. (1986) observed an increase in tremor parameters with increasing urinary excretion of mercury in 21 workers exposed to mercury vapor for 0.5-19 years. The LOAEL for this effect was a mean urinary excretion of 35 ug/g creatinine.

Rosenman et al. (1986) evaluated routine clinical parameters (physical exams, blood chemistry, urinalysis), neuropsychological disorders, urinary NAG, motor nerve conduction velocities and occurrence of lenticular opacities in 42 workers of a chemical plant producing mercury compounds. A positive correlation (p<0.05 to p<0.001) was noted between urinary mercury (levels ranged from 100-250 ug/L) and the number of neuropsychological symptoms, and NAG excretions and the decrease in motor nerve conduction velocities.

Evidence of renal dysfunction (e.g., increased plasma and urinary concentrations of beta-galactosidase, increased urinary excretion of high molecular weight proteins and a slightly increased plasma beta-2-microglobulin concentration) was observed in 63 chloralkali workers. The incidence of these effects increased in workers whose urinary Hg excretion exceeded 50 ug/g creatinine (Buchet et al., 1980).

Increased urinary NAG levels were found in workers whose urinary Hg levels exceeded 50 ug/L (Langworth et al., 1992).

An increase in the concentration of urinary brush border proteins (BB-50) was observed in 20 workers whose mean urinary Hg excretion exceeded 50 ug/g creatinine (Mutti et al., 1985).

Foa et al. (1976) found that 15 out of 81 chloralkali workers exposed to 60-300 ug/cu.m mercury exhibited proteinuria.

An increased excretion of beta-glutamyl transpeptidase, indicative of renal dysfunction, was found in 509 infants dermally exposed to phenylmercury via contaminated diapers (Gotelli et al., 1985).

Berlin et al. (1969) exposed rats, rabbits and monkeys to 1 mg/cu.m of mercury vapor for 4 hours and measured the uptake and distribution of mercury in the brain as compared with animals injected intravenously with the same doses of mercury as mercuric salts. Mercury accumulated in the brain following inhalation exposure to metallic mercury vapor at levels that were 10 times higher than those observed following intravenous injection of the same dose of mercury as mercuric salts. These results demonstrate that mercury is taken up by the brain following inhalation of the vapor at higher levels than other forms of mercury and that this occurs in all species studied.

Limited animal studies concerning inhalation exposure to inorganic mercury are available. The results of a study conducted by Baranski and Szymczyk (1973) were reported in an English abstract. Adult female rats were exposed to metallic mercury vapor at 2.5 mg/cu.m for 3 weeks prior to fertilization and during gestation days 7-20. A decrease in the number of living fetuses was observed in the dams compared with unexposed controls, and all pups born to the exposed dams died by the sixth day after birth. However, no difference in the occurrence of developmental abnormalities was observed between exposed and control groups. The cause of death of the pups in the mercury-exposed group was unknown, although an unspecified percentage of the deaths was attributed by the authors to a failure of lactation in the dams. Death of pups was also observed in another experiment where dams were only exposed prior to fertilization (to 2.5 mg/cu.m), which supports the conclusion that the high mortality in the first experiment was due at least in part to poor health of the mothers. Without further information, this study must be considered inconclusive regarding developmental effects.

The only other study addressing the developmental toxicology of mercury is the one reported in abstract form by Steffek et al. (1987) and, as such, is included as a supporting study. Sprague-Dawley rats (number not specified) were exposed by inhalation to mercury vapor at concentrations of 0.1, 0.5 or 1.0 mg/cu.m throughout the period of gestation (days 1-20) or during the period of organogenesis (days 10-15). The authors indicated the exposure protocols to be chronic and acute exposure, respectively. At either exposure protocol, the lowest mercury level produced no detectable adverse effect. At 0.5 mg/cu.m, an increase in the number of resorptions (5/41) was noted for the acute group, and two of 115 fetuses exhibited gross cranial defects in the chronic group. At 1.0 mg/cu.m, the number of resorptions was increased in acute (7/71) and chronic (19/38) groups and a decrease in maternal and fetal weights also was detected in the chronic exposure group. No statistical analysis for these data was provided. A LOAEL of 0.5 mg/cu.m is provided based on these data.

Mishinova et al. (1980) investigated the course of pregnancy and parturition in 349 women exposed via inhalation to metallic mercury vapors (unspecified concentrations) in the workplace as compared to 215 unexposed women. The authors concluded that the rates of pregnancy and labor complication were high among women exposed to mercury and that the effects depended on "the length of service and concentration of mercury vapors." Lack of sufficient details preclude the evaluation of dose-response relationships.

In a questionnaire that assessed the fertility of male workers exposed to mercury vapor, Lauwerys et al. (1985) found no statistically significant change in the observed number of children born to the exposed group compared with a matched control group. The urinary excretion of mercury in the exposed workers ranged from 5.1 to 272.1 ug/g creatinine.

Another study found that exposure to metallic mercury vapor caused prolongation of estrus cycles in animals. Baranski and Szymczyk (1973) reported that female rats exposed via inhalation to mercury vapor at an average of 2.5 mg/cu.m, 6 hours/day, 5 days/week for 21 days experienced longer estrus cycles than unexposed animals. In addition, estrus cycles during mercury exposure were longer than normal estrus cycles in the same animals prior to exposure. Although the initial phase of the cycle was protracted, complete inhibition of the cycle did not occur. During the second and third weeks of exposure, these rats developed signs of mercury poisoning including restlessness, seizures and trembling of the entire body. The authors speculated that the effects on the estrus cycle were caused by the action of mercury on the CNS (i.e., damage to the hypothalamic regions involved in the control of estrus cycling).

Renal toxicity has been reported following oral exposure to inorganic mercury salts in animals, with the Brown-Norway rat appearing to be uniquely sensitive to this effect. These mercury-induced renal effects in the Brown-Norway rat are the basis for the oral

RfD for mercurial mercury. Several investigators have produced autoimmune glomerulonephritis by administering HgCl2 to Brown-Norway rats (Druet et al., 1978).

The current OSHA standard for mercury vapor is 0.05 mg/cu.m. NIOSH recommends a TWA Threshold Limit Value of 0.05 mg/cu.m for mercury vapor.

I.B.5. CONFIDENCE IN THE INHALATION RfC

Study -- Medium
Data Base -- Medium
RfC -- Medium

Due to the use of a sufficient number of human subjects, the inclusion of appropriate control groups, the exposure duration, the significance level of the reported results and the fact that exposure levels in a number of the studies had to be extrapolated from blood mercury levels, confidence in the key studies is medium. The LOAEL values derived from these studies can be corroborated by other human epidemiologic studies. The adverse effects reported in these studies are in accord with the well-documented effects of mercury poisoning. The lack of human or multispecies reproductive/ developmental studies precludes assigning a high confidence rating to the data base and inadequate quantification of exposure levels. Based on these considerations, the RfC for mercury is assigned a confidence rating of medium.

I.B.6. EPA DOCUMENTATION AND REVIEW OF THE INHALATION RfC

Source Document -- US EPA, 1995

This IRIS summary is included in The Mercury Study Report to Congress which was reviewed by OHEA and EPA's Mercury Work Group in November 1994. An interagency review by scientists from other federal agencies took place in January 1995. The report was also reviewed by a panel of non-federal external scientists in January 1995 who met in a public meeting on January 25-26. All reviewers comments have been carefully evaluated and considered in the revision and finalization of this IRIS summary. A record of these comments is summarized in the IRIS documentation files.

Other EPA Documentation -- None

Agency Work Group Review -- 11/16/89, 03/22/90, 04/19/90

Verification Date -- 04/19/90

I.B.7. EPA CONTACTS (INHALATION RfC)

Annie M. Jarabek / NCEA -- (919)541-4847

William F. Sette / OPP -- (703)305-6375

II. CARCINOGENICITY ASSESSMENT FOR LIFETIME EXPOSURE

Section II provides information on three aspects of the carcinogenic risk assessment for the agent in question; the US EPA classification, and quant- itative estimates of risk from oral exposure and from inhalation exposure. The classification reflects a weight-of-evidence judgment of the likelihood that the agent is a human carcinogen. The quantitative risk estimates are presented in three ways. The slope factor is the result of application of a low-dose extrapolation procedure and is presented as the risk per (mg/kg)/day. The unit risk is the quantitative estimate in terms of either risk per ug/L drinking water or risk per ug/cu.m air breathed. The third form in which risk is presented is a drinking water or air concentration providing cancer risks of 1 in 10,000, 1 in 100,000 or 1 in 1,000,000. Background Document 2 (Service Code 5) provides details on the rationale and methods used to derive the carcinogenicity values found in IRIS. Users are referred to Section I for information on long-term toxic effects other than carcinogenicity.

II. CARCINOGENICITY ASSESSMENT FOR LIFETIME EXPOSURE

II.A. EVIDENCE FOR CLASSIFICATION AS TO HUMAN CARCINOGENICITY

II.A.1. WEIGHT-OF-EVIDENCE CLASSIFICATION

Classification -- D; not classifiable as to human carcinogenicity

Basis -- Based on inadequate human and animal data. Epidemiologic studies failed to show a correlation between exposure to elemental mercury vapor and carcinogenicity; the findings in these studies were confounded by possible or known concurrent exposures to other chemicals, including human carcinogens, as well as lifestyle factors (e.g., smoking). Findings from genotoxicity tests are severely limited and provide equivocal evidence that mercury adversely affects the number or structure of chromosomes in human somatic cells.

II.A.2. HUMAN CARCINOGENICITY DATA

Inadequate. A number of epidemiological studies were conducted that examined mortality among elemental mercury vapor-exposed workers. Conflicting data regarding a correlation between mercury exposure and an increased incidence of cancer mortalities have been obtained. All of the studies have limitations that complicate interpretation of their results for associations between mercury exposure and induction of cancer; increased cancer rates were attributable to other concurrent exposures or lifestyle factors.

A retrospective cohort study examined mortality among 5663 white males who worked between 1953 and 1963 at a plant in Oak Ridge, Tennessee, where elemental mercury was used for lithium isotope separation (Cragle et al., 1984). The workers were divided into three cohorts: exposed workers who had been monitored on a quarterly basis for mercury levels in urine (n=2,133); workers exposed in the mercury process section for whom urinalysis monitoring data were not collected (n=270); and unexposed workers from other sections of the nuclear weapons production facility (n=3260). The study subjects worked at least 4 months during 1953-1958 (a period when mercury exposures were likely to be high); mortality data from death certificates were followed through the end of 1978. The mean age of the men at first employment at the facility was 33 years, and the average length of their employment was >16 years with a mean of 3.73 years of estimated mercury exposure. Air mercury levels were monitored beginning in 1955; during 1955 through the third quarter of 1956, air mercury levels were reportedly above 100 ug/cu.m in 30-80% of the samples. Thereafter, air mercury levels decreased to concentrations below 100 ug/cu.m. The mortality experience (i.e., the SMR) of each group was compared with the age-adjusted mortality experience of the US white male population.   Among exposed and monitored workers, no significant increases in mortality from cancer at any site were reported, even after the level or length of exposure was considered. A significantly lower mortality from all causes was observed. An excessive number of deaths was reportedly due to lung cancer in the exposed and monitored workers (42 observed, 31.36 expected), but also in the unexposed workers (71 observed, 52.93 expected). The SMR for each group was 1.34; the elevated incidence of lung cancer deaths was, therefore, attributed to some other factor at the plant and/or to lifestyle factors (e.g., smoking) common to both the exposed and unexposed groups. Study limitations include small cohort sizes for cancer mortality, which limited the statistical stability of many comparisons.

Barregard et al. (1990) studied mortality and cancer morbidity between 1958 and 1984 in 1190 workers from eight Swedish chloralkali plants that used the mercury cell process in the production of chlorine. The men included in the study had been monitored for urinary or blood mercury for more than one year between 1946 and 1984. Vital status and cause of death were ascertained from the National Population Register and the National Bureau of Statistics. The cancer incidence of the cohort was obtained from the Swedish Cancer Register. The observed total mortality and cancer incidences were compared with those of the general Swedish male population. Comparisons were not made between exposed and unexposed workers. Mean urinary mercury levels indicated a decrease in exposure between the 1950s and 1970s; the mean urinary mercury level was 200 ug/L during the 1950s, 150 ug/L during the 1960s and 50 ug/L in the 1970s. Mortality from all causes was not significantly increased in exposed workers. A significant increase in deaths from lung tumors was observed in exposed workers 10 years or more after first exposure (rate ratio, 2.0; 95% CI, 1.0-3.8). Nine of the 10 observed cases of lung cancer occurred among workers (457 of the 1190) possibly exposed to asbestos as well as to mercury. No dose response was observed with respect to mercury exposure and lung tumors. This study is limited because no quantitation was provided on smoking status, and results were confounded by exposure to asbestos.

Ahlbom et al. (1986) examined the cancer mortality during 1961-1979 of cohorts of Swedish dentists and dental nurses aged 20-64 and employed in 1960 (3454 male dentists, 1125 female dentists, 4662 female dental nurses). Observed incidences were compared with those expected based on cancer incidence during 1961-1979 among all Swedes employed during 1960 and the proportion of all Swedes employed as dentists and dental nurses. Data were stratified by sex, age (5-year age groups) and county. The incidence of glioblastomas among the dentists and dental nurses combined was significantly increased compared to survival rates (SMR, 2.1; 95% CI, 1.3-3.4); the individual groups had apparently elevated SMRs (2.0-2.5), but the 95% confidence intervals of these groups included unity. By contrast, physicians and nurses had SMRs of only 1.3 and 1.2, respectively. Exposure to mercury could not be established as the causative factor because exposure to other chemicals and X-rays was not ruled out.

Amandus and Costello (1991) examined the association between silicosis and lung cancer mortality between 1959 and 1975 in 9912 white male metal miners employed in the United States between 1959 and 1961. Mercury exposures were not monitored. Exposures to specific metals among the silicotic and nonsilicotic groups were analyzed separately. Lung cancer mortality in both silicotic and nonsilicotic groups was compared with rates in white males in the US population. Both silicotic (n=11) and nonsilicotic mercury miners (n=263) had significantly increased lung cancer mortality (SMR, 14.03; 95% CI, 2.89-40.99 for silicotics. SMR, 2.66; 95% CI, 1.15-5.24 for nonsilicotics). The analysis did not focus on mercury miners, and confounders such as smoking and radon exposure were not analyzed with respect to mercury exposure. This study is also limited by the small sample size for non-silicotic mercury miners.

A case-control study of persons admitted to a hospital in Florence, Italy, with lung cancer between 1981-1983 was performed to evaluate occupational risk factors (Buiatti et al., 1985). Cases were matched with one or two controls (persons admitted to the hospital with diagnoses other than lung cancer or suicide) with respect to sex, age, date of admission and smoking status. Women who had "ever worked" as hat makers had a significantly increased risk of lung cancer. The duration of employment as a hat maker averaged 22.2 years, and latency averaged 47.8 years. Workers in the Italian hat industry were known to be occupationally exposed to mercury; however, the design of this study did not allow evaluation of the relationship between cumulative exposure and cancer incidence. In addition, interpretation of the results of this study is limited by the small sample size (only 6/376 cases reported this occupation) and by exposure of hat makers to other pollutants including arsenic, a known lung carcinogen.

Ellingsen et al. (1992) examined the total mortality and cancer incidence among 799 workers employed for more than 1 year in two Norwegian chloralkali plants. Mortality incidence between 1953 and 1988 and cancer incidence between 1953 and 1989 were examined. Mortality and cancer incidence were compared with that of the age-adjusted general male Norwegian population. No increase in total cancer incidence was reported, but lung cancer was significantly elevated in the workers (rate ratio, 1.66; 95% CI, 1.0-2.6). No causal relationship can be drawn from the study between mercury exposure and lung cancer because no correlation existed between cumulative mercury dose, years of employment or latency time. Also, the prevalence of smoking was 10 20% higher in the exposed workers, and many workers were also exposed to asbestos.

II.A.3. ANIMAL CARCINOGENICITY DATA

Inadequate. Druckrey et al. (1957) administered 0.1 mL of metallic mercury to 39 male and female rats (BD III and BD IV strains) via intraperitoneal injection. Among the rats surviving longer than 22 months, 5/12 developed peritoneal sarcomas. The increase in the incidence of sarcomas was observed only in those tissues that had been in direct contact with the mercury. Although severe kidney damage was reported in all treated animals, no renal tumors or tumors at any site other than the peritoneal cavity were observed.

II.A.4. SUPPORTING DATA FOR CARCINOGENICITY

Cytogenetic monitoring studies of workers occupationally exposed to mercury by inhalation provide very limited evidence that mercury adversely affects the number or structure of chromosomes in human somatic cells. Popescu et al. (1979) compared four men exposed to elemental mercury vapor with an unexposed group and found a statistically significant increase in the incidence of chromosome aberrations in the WBCs from whole blood. Verschaeve et al. (1976) found an increase in aneuploidy after exposure to low concentrations of vapor, but results could not be repeated in later studies (Verschaeve et al., 1979). Mabille et al. (1984) did not find increases in structural chromosomal aberrations of lymphocytes of exposed workers. Similarly, Barregard et al. (1991) found no increase in the incidence or size of micronuclei and no correlation between micronuclei and blood or urinary mercury levels of chloralkali workers. A statistically significant correlation was observed between cumulative exposure to mercury and micronuclei induction in T lymphocytes in exposed workers, suggesting a genotoxic effect.

II.B. QUANTITATIVE ESTIMATE OF CARCINOGENIC RISK FROM ORAL EXPOSURE

None.

II.C. QUANTITATIVE ESTIMATE OF CARCINOGENIC RISK FROM INHALATION EXPOSURE

None.

II.D. EPA DOCUMENTATION, REVIEW, AND CONTACTS (CARCINOGENICITY ASSESSMENT)

II.D.1. EPA DOCUMENTATION

Source document -- US EPA, 1995

This IRIS summary is included in The Mercury Study Report to Congress which was reviewed by OHEA and EPA's Mercury Work Group in November 1994. An interagency review by scientists from other federal agencies took place in January 1995. The report was also reviewed by a panel of non-federal external scientists in January 1995 who met in a public meeting on January 25-26. All reviewers comments have been carefully evaluated and considered in the revision and finalization of this IRIS summary. A record of these comments is summarized in the IRIS documentation files.

II.D.2. REVIEW (CARCINOGENICITY ASSESSMENT)

Agency Work Group Review -- 01/13/88, 03/03/94

Verification Date -- 03/03/94

II.D.3. US EPA CONTACTS (CARCINOGENICITY ASSESSMENT)

Rita Schoeny / NCEA -- (513)569-7544

III. HEALTH HAZARD ASSESSMENTS FOR VARIED EXPOSURE DURATIONS

[Ed. Note: EPA does not have information yet in this section of IRIS].

IV. US EPA REGULATORY ACTIONS

EPA risk assessments may be updated as new data are published and as assessment methodologies evolve. Regulatory actions are frequently not updated at the same time. Compare the dates for the regulatory actions in this section with the verification dates for the risk assessments in sections I and II, as this may explain inconsistencies. Also note that some regulatory actions consider factors not related to health risk, such as technical or economic feasibility. Such considerations are indicated for each action. In addition, not all of the regulatory actions listed in this section involve enforceable federal standards. Please direct any questions you may have concerning these regulatory actions to the US EPA contact listed for that particular action. Users are strongly urged to read the background information on each regulatory action in Background Document 4 in Service Code 5.

IV.A. CLEAN AIR ACT (CAA)

[Ed. Note: EPA does not have information yet in this section of IRIS].

IV.B. SAFE DRINKING WATER ACT (SDWA)

IV.B.1. MAXIMUM CONTAMINANT LEVEL GOAL (MCLG) for Drinking Water

Value -- 0.002 mg/L (Final, 1991)

Considers technological or economic feasibility? -- NO

Discussion -- EPA has promulgated a MCLG of 0.002 mg/L based on potential adverse effects (renal toxicity) in three major studies. The MCLG is based upon a DWEL of 0.01 mg/L and an assumed drinking water contribution of 20 percent.

Reference -- 56 FR 3526 (01/30/91)

EPA Contact -- Health and Ecological Criteria Division / OST / (202) 260-7571 / FTS 260-7571; or Safe Drinking Water Hotline / (800) 426-4791

IV.B.2. MAXIMUM CONTAMINANT LEVEL (MCL) for Drinking Water

Value -- 0.002 mg/L (Final, 1991)

Considers technological or economic feasibility? -- YES

Discussion -- EPA has set an MCL equal to the MCLG of 0.002 mg/L.

Monitoring requirements -- Ground water systems monitored every three years; surface water systems monitored annually; systems out of compliance must begin monitoring quarterly until system is reliably and consistently below MCL.

Analytical methodology -- Manual cold vapor technique (EPA 245.1; ASTM D3223- 80; SM 303F); automated cold vapor technique (EPA 245.2): PQL=0.0005 mg/L.

Best available technology -- Coagulation/filtration; Lime softening; Reverse osmosis; Granular activated carbon.

Reference -- 56 FR 3526 (01/30/91)

EPA Contact -- Drinking Water Standards Division / OGWDW / (202) 260-7575 / FTS 260-7575; or Safe Drinking Water Hotline / (800) 426-4791

IV.B.3. SECONDARY MAXIMUM CONTAMINANT LEVEL (SMCL) for Drinking Water

[Ed. Note: EPA does not have information yet in this section of IRIS].

IV.B.4. REQUIRED MONITORING OF "UNREGULATED" CONTAMINANTS

[Ed. Note: EPA does not have information yet in this section of IRIS].

IV.C. CLEAN WATER ACT (CWA)

IV.C.1. AMBIENT WATER QUALITY CRITERIA, Human Health

Water and Fish Consumption: 1.44E-1 ug/L

Fish Consumption Only: 1.46E-1 ug/L

Considers technological or economic feasibility? -- NO

Discussion -- The WQC of 1.44E-1 ug/L is based on consumption of contaminated aquatic organisms and water. A WQC of 1.46E-1 ug/L has also been established based on consumption of contaminated aquatic organisms alone.

Reference -- 45 FR 79318 (11/28/80); 50 FR 30784 (07/29/85)

EPA Contact -- Criteria and Standards Division / OWRS (202)260-1315 / FTS 260-1315

IV.C.2. AMBIENT WATER QUALITY CRITERIA, Aquatic Organisms

Freshwater:
    Acute -- 2.4E+0 ug/L   (1-hour average)
    Chronic -- 1.2E-2 ug/L (4-day average)

Marine:
   Acute -- 2.1E+0 ug/L   (1-hour average)
   Chronic -- 2.5E-2 ug/L (4-day average)

Considers technological or economic feasibility? -- NO

Discussion -- Criteria were derived from a minimum data base consisting of acute tests on a variety of species. Requirements and methods are covered in the reference to the Federal Register. The Agency recommends an exceedence frequency of no more than 3 years.

Reference -- 45 FR 79318 (11/28/80); 50 FR 30784 (07/29/85)

EPA Contact -- Criteria and Standards Division / OWRS (202)260-1315 / FTS 260-1315

IV.D. FEDERAL INSECTICIDE, FUNGICIDE, AND RODENTICIDE ACT (FIFRA)

     [Ed. Note: EPA does not have information yet in this section of IRIS].

IV.E. TOXIC SUBSTANCES CONTROL ACT (TSCA)

     [Ed. Note: EPA does not have information yet in this section of IRIS].

IV.F. RESOURCE CONSERVATION AND RECOVERY ACT (RCRA)

IV.F.1. RCRA APPENDIX IX, for Ground Water Monitoring

Status -- Listed (total mercury)

Reference -- 52 FR 25942 (07/09/87)

EPA Contact -- RCRA/Superfund Hotline (800)424-9346 / (202)260-3000 / FTS 260-3000

IV.G. SUPERFUND (CERCLA)

IV.G.1. REPORTABLE QUANTITY (RQ) for Release into the Environment

Value (status) -- 1 pound (Final, 1989)

Considers technological or economic feasibility? -- NO

Discussion -- The final RQ for mercury is based on aquatic toxicity. The available data indicate that the aquatic 96-Hour Median Threshold Limit is less than 0.1 ppm, which corresponds to an RQ of 1 pound.

Reference -- 54 FR 33418 (08/14/89)

EPA Contact -- RCRA/Superfund Hotline (800)424-9346 / (202)260-3000 / FTS 260-3000

VI. BIBLIOGRAPHY

VI.A. ORAL RfD REFERENCES

None

VI.B. INHALATION RfC REFERENCES

Albers, J.W., L.R. Kallenbach, L.J. Fine, et al. 1988. Neurological abnormalities associated with remote occupational elemental mercury exposure. Ann. Neurol. 24(5): 651-659.

Ashe, W.F., E.J. Largent, F.R. Dutra, D.M. Hubbard and M. Blackstone. 1953. Behavior of mercury in the animal organism following inhalation. Ind. Hyg. Occup. Med. 17: 19-43.

Baranski, B. and I. Szymczyk. 1973. [Effects of mercury vapor upon reproductive functions of female white rats]. Med. Pr. 24(3): 249-261. (Czechoslovakian)

Beliles, R.P., R.S. Clark, P.R. Belluscio, C.L. Yuile and L.J. Leach. 1967. Behavioral effects in pigeons exposed to mercury vapor at a concentration of 0.1 mg/cu.m. Am. Ind. Hyg. J. 28(5): 482-484.

Berlin, M., J. Fazackerley and G. Nordberg. 1969. The uptake of mercury in the brains of mammals exposed to mercury vapor and to mercuric salts. Arch. Environ. Health. 18: 719-729.

Bernard, A.M., H.R. Roels, J.M. Foldart and R.L. Lauwerys. 1987. Search for anti-laminin antibodies in the serum of workers exposed to cadmium, mercury vapour or lead. Int. Arch. Occup. Environ. Health. 59: 303-309.

Buchet, J.P., H. Roels, A. Bernard and R. Lauwerys 1980. Assessment of renal function of workers exposed to inorganic lead, cadmium or mercury vapor. J. Occup. Med. 22(11): 741-750.

Bunn, W.B., C.M. McGill, T.E. Barber, J.W. Cromer and L.J. Goldwater. 1986. Mercury exposure in chloralkali plants. Am. Ind. Hyg. Assoc. J. 47(5): 249-254.

Druet, P., E. Druet, F. Potdevin, et al. 1978. Immune type glomerulonephritis induced by HgCl2 in the Brown-Norway rat. Ann. Immunol. 129C: 777-792.

Fawer, R.F., Y. DeRibaupierre, M.P. Guillemin, M. Berode and M. Lob. 1983. Measurement of hand tremor induced by industrial exposure to metallic mercury. J. Ind. Med. 40: 204-208.

Foa, V., L. Caimi, L. Amante, et al. 1976. Patterns of some lysosomal enzymes in the plasma and of proteins in urine of workers exposed to inorganic mercury. Int. Arch. Occup. Environ. Health. 37: 115-124.

Forzi, M., M.G. Cassitto, C. Bulgheroni and V. Foa. 1978. Psychological measures in workers occupationally exposed to mercury vapors: A validation study. In: Adverse Effects of Environmental Chemicals and Psychotropic Drugs: Neurophysiological and Behavioral Tests, Vol. 2, H.J. Zimmerman, Ed. Appleton-Century-Crofts, New York, NY. p. 165-171.

Gotelli, C.A., E. Astolfi, C. Cox, E. Cernichiari and T. Clarkson. 1985. Early biochemical effects of an organic mercury funcigicide on infants: "Dose makes the poison". Science. 277: 638-640.

Kishi, R., K. Hashimoto, S. Shimizu and M. Kobayashi. 1978. Behavioral changes and mercury concentrations in tissues of rats exposed to mercury vapor. Toxicol. Appl. Pharmacol. 46(3): 555-566.

Langolf, G.D., D.B. Chaffin, R. Henderson and H.P. Whittle. 1978. Evaluation of workers exposed to elemental mercury using quantitative tests of tremor and neuromuscular functions. Am. Ind. Hyg. Assoc. J. 39: 976-984.

Langworth, S., C.G. Elinder, K.G. Sundquist and O. Vesterberg. 1992. Renal and immunological effects of occupational exposure to inorganic mercury. Br. J. Ind. Med. 49: 394-401.

Lauwerys, R., A. Bernard, H. Roels, et al. 1983. Anti-laminin antibodies in workers exposed to mercury vapour. Toxicol. Lett. 17: 113-116.

Lauwerys, R., H. Roels, P. Genet, G. Toussaint, A. Bouckaert and S. De Cooman. 1985. Fertility of male workers exposed to mercury vapor or to manganese dust: A questionnaire study. Am. J. Ind. Med. 7(2): 171-176.

Levin, M., J. Jacobs and P.G. Polos. 1988. Acute mercury poisoning and mercurial pneumonitis from gold ore purification. Chest. 94(3): 554-558.

Levine, S.P., G.D. Cavender, G.D. Langolf and J.W. Albers. 1982. Elemental mercury exposure: Peripheral neurotoxicity. Br. J. Ind. Med. 39: 136-139.

Liang, Y-X., R-K. Sun, Y. Sun, Z-Q. Chen and L-H. Li. 1993. Psychological effects of low exposure to mercury vapor: Application of a computer- administered neurobehavioral evaluation system. Environ. Res. 60: 320-327.

Lilis, R., A. Miller and Y. Lerman. 1985. Acute mercury poisoning with severe chronic pulmonary manifestations. Chest. 88(2): 306-309.

McFarland, R.B. and H. Reigel. 1978. Chronic mercury poisoning from a single brief exposure. J. Occup. Med. 20(8): 532-534.

Miller, J.M., D.B. Chaffin and R.G. Smith. 1975. Subclinical psychomotor and neuromuscular changes in workers exposed to inorganic mercury. Am. Ind. Hyg. Assoc. J. 36: 725-733.

Mishonova, V.N., P.A. Stepanova and V.V. Zarudin. 1980. Characteristics of the course of pregnancy and labor in women coming in contact with low concentrations of metallic mercury vapors in manufacturing work places. Gig Tr Prof Zabol. Issue 2: 21-23.

Mutti, A., S. Lucertini, M. Fornari, et al. 1985. Urinary excretion of a brush-border antigen revealed by monoclonal antibodies in subjects occupationally exposed to heavy metals. Heavy Met Environ. International Conference 5th. Vol.1. p. 565-567.

Ngim, C.H., S.C. Foo, K.W. Boey and J. Jeyaratnam. 1992. Chronic neurobehavioral effects of elemental mercury in dentists. Br. J. Ind. Med. 49: 782-790.

Piikivi, L. 1989. Cardiovascular reflexes and low long-term exposure to mercury vapor. Int. Arch. Occup. Environ. Health. 61: 391-395.

Piikivi, L. and H. Hanninen. 1989. Subjective symptoms and psychological performance of chlorine-alkali workers. Scand. J. Work Environ. Health. 15: 69-74.

Piikivi, L. and A. Ruokonen. 1989. Renal function and long-term low mercury vapor exposure. Arch. Environ. Health. 44(3): 146-149.

Piikivi, L. and U. Tolonen. 1989. EEG findings in chlor-alkali workers subjected to low long term exposure to mercury vapor. Br. J. Ind. Med. 46: 370-375.

Roels, H., R. Lauwerys, J.P. Buchet, et al. 1982. Comparison of renal function and psychomotor performance in workers exposed to elemental mercury. Int. Arch. Occup. Environ. Health. 50: 77-93.

Roels, H., J.P. Gennart, R. Lauwreys, J.P. Buchet, J. Malchaire and A. Bernard. 1985. Surveillance of workers exposed to mercury vapor: validation of a previously proposed biological threshold limit value for mercury concentration in urine. Am. J. Ind. Med. 7: 45-71.

Roels, H., S. Abdeladim, E. Ceulemans and R. Lauwreys. 1987. Relationships between the concentrations of mercury in air and in blood or urine in workers exposed to mercury vapour. Ann. Occup. Hyg. 31(2): 135-145.

Roels, H., S. Abdeladim, M. Braun, J. Malchaire and R. Lauwerys. 1989. Detection of hand tremor in workers exposed to mercury vapor: A comparative study of three methods. Environ. Res. 49: 152-165.

Rosenman, K.D., J.A. Valciukas, L. Glickman, B.R. Meyers and A. Cinotti. 1986. Sensitive indicators of inorganic mercury toxicity. Arch. Environ. Health. 41(4): 208-215.

Singer, R., J.A. Valciukas and K.D. Rosenman. 1987. Peripheral neurotoxicity in workers exposed to inorganic mercury compounds. Arch. Environ. Health. 42(4): 181-184.

Smith, R.G., A.J. Vorwald, L.S. Patil and T.F. Mooney, Jr. 1970. Effects of exposure to mercury in the manufacture of chlorine. Am. Ind. Hyg. Assoc. J. 31(1): 687-700.

Steffek, A.J., R. Clayton, C. Siew and A.C. Verrusio. 1987. Effects of elemental mercury vapor exposure on pregnant Sprague-Dawley rats (abstract only). Teratology. 35: 59A.

Stewart, W.K., H.A. Guirgis, J. Sanderson and W. Taylor. 1977. Urinary mercury excretion and proteinuria in pathology laboratory staff. Br. J. Ind. Med. 34: 26-31.

Stonard, M.D., B.V. Chater, D.P. Duffield, A.L. Nevitt, J.J. O'Sullivan and G.T. Steel. 1983. An evaluation of renal function in workers occupationally exposed to mercury vapor. Int. Arch. Occup. Environ. Health. 52: 177-189.

US EPA. 1995. Mercury Study Report to Congress. Office of Research and Development, Washington DC 20460. EPA/600/P-94/002Ab. External Review Draft.

Verbeck, M.M., H.J.A. Salle and C.H. Kemper. 1986. Tremor in workers with low exposure to metallic mercury. Hyg. Assoc. J. 47(8): 559-562.

Warkany, J. and D.M. Hubbard. 1953. Acrodynia and mercury. J. Pediat. 42: 365-386.

VI.C. CARCINOGENICITY ASSESSMENT REFERENCES

Ahlbom, A., S. Norell, Y. Rodvall and M. Nylander. 1986. Dentists, dental nurses, and brain tumours. Br. Med. J. 292: 662.

Amandus, H. and J. Costello. 1991. Silicosis and lung cancer in US metal miners. Arch. Environ. Health. 46(2): 82-89.

Barregard, L., G. Sallsten and B. Jarvholm. 1990. Mortality and cancer incidence in chloralkali workers exposed to inorganic mercury. Br. J. Ind. Med. 47(2): 99-104.

Barregard, L., B. Hogstedt, A. Schutz, A. Karlsson, G. Sallsten and G. Thiringer. 1991. Effects of occupational exposure to mercury vapor on lymphocyte micronuclei. Scand. J. Work Environ. Health. 17: 263-268.

Buiatti, E., D. Kriebel, M. Geddes, M. Santucci and N. Pucci. 1985. A case control study of lung cancer in Florence, Italy. I. Occupational risk factors. J. Epidemiol. Comm. Health. 39: 244-250.

Cragle, D.L., D.R. Hollis, J.R. Qualters, W.G. Tankersley and S.A. Fry. 1984. A mortality study of men exposed to elemental mercury. J. Occup. Med. 26(11): 817-821.

Druckrey, H., H. Hamperl and D. Schmahl. 1957. Carcinogenic action of metallic mercury after intraperitoneal administration in rats. Z. Krebsforsch. 61: 511-519. (Cited in US EPA, 1985)

Ellingsen, D., A. Andersen, H.P. Nordhagen, J. Efskind and H. Kjuus. 1992. Cancer incidence and mortality among workers exposed to mercury in the Norwegian chloralkali industry. 8th International Symposium on Epidemiology in Occupational Health, Paris, France, September 10-12, 1991. Rev. Epidemiol. Sante Publique. 40(1): S93-S94.

Mabille, V., H. Roels, P. Jacquet, A. Leonard and R. Lauwerys. 1984. Cytogenetic examination of leucocytes of workers exposed to mercury vapor. Int. Arch. Occup. Environ. Health. 53: 257-260.

Popescu, H.I., L. Negru and I. Lancranjan. 1979. Chromosome aberrations induced by occupational exposure to mercury. Arch. Environ. Health. 34(6): 461-463.

US EPA. 1980. Ambient Water Quality Criteria Document for Mercury. Prepared by the Office of Health and Environmental Assessment, Environmental Criteria and Assessment Office, Cincinnati, OH for the Office of Water Regulation and Standards, Washington, DC. EPA/440/5-80/058. NTIS PB 81- 117699.

US EPA. 1984a. Mercury Health Effects Update: Health Issue Assessment. Final Report. Prepared by the Office of Health and Environmental Assessment, Environmental Criteria and Assessment Office, Cincinnati, OH for the Office of Air Quality Planning and Standards, Research Triangle Park, NC. EPA/600/8- 84/019F. NTIS PB81-85-123925.

US EPA. 1984b. Health Effects Assessment for Mercury. Prepared by the Office of Health and Environmental Assessment, Environmental Criteria and Assessment Office, Cincinnati, OH for the Office of Emergency and Remedial Response, Washington, DC. EPA/540/1086/042. NTIS PB86-134533/AS.

US EPA. 1985. Drinking Water Criteria Document for Mercury. Prepared by the Office of Health and Environmental Assessment Office, Cincinnati, OH for the Office of Drinking Water, Washington, DC. EPA/600/X-84/178. NTIS PB86- 117827.

US EPA. 1988. Drinking Water Criteria Document for Inorganic Mercury. Prepared by the Office of Health and Environmental Assessment, Environmental Criteria and Assessment Office, Cincinnati, OH for the Office of Drinking Water, Washington, DC. EPA/600/X-84/178. NTIS PB89-192207.

US EPA. 1993. Summary Review of Health Effects Associated with Mercuric Chloride: Health Issue Assessment (Draft). Prepared by the Office of Health and Environmental Assessment, Environmental Criteria and Assessment Office, Cincinnati, OH for the Office of Air Quality Planning and Standards, Research Triangle Park, NC. EPA/600/R-92/199.

US EPA. 1995. Mercury Study Report to Congress. Office of Research and Development, Washington, DC. External Review Draft. EPA/600/P-94/002Ab.

Verschaeve, L., M. Kirsch-Volders, C. Susanne et al. 1976. Genetic damage induced by occupationally low mercury exposure. Environ. Res. 12: 303-316.

Verschaeve, L., J.P. Tassignon, M. Lefevre, P. De Stoop and C. Susanne. 1979. Cytogenetic investigation on leukocytes of workers exposed to metallic mercury. Environ. Mutagen. 1: 259-268.

VI.D. DRINKING WATER HA REFERENCES

None

VII. REVISION HISTORY

```
--------  -----------  ---------------------------------------------------------
Date     Section   Description
--------  -----------  ---------------------------------------------------------
09/07/88 II.       Carcinogen summary on-line
09/01/89 VI.       Bibliography on-line
12/01/89 I.B.      Inhalation RfD now under review
05/01/91 II.A.3.   Text edited
01/01/92 IV.       Regulatory Action section on-line
04/01/94 II.       Carcinogenicity assessment noted as pending change
04/01/94 II.D.2.   Work group review date added
05/01/95 All       Name changed from mercury (inorganic)
05/01/95 II.       Carcinogen assessment replaced
05/01/95 VI.C.     Carcinogen assessment references replaced
06/01/95 I.B.      Inhalation RfC summary on-line
06/01/95 VI.B.     Inhalation RfC references on-line
```

SYNONYMS

hydragyrum	Liquid Silver
Mercury	Mercure [French]
Mercury, elemental	Mercurio [Italian]
Mercury, inorganic	Mercurio [Spanish]
Mercury, metallic	Mercury compounds
Mercury (organo) alkyl compounds	Mercury vapor
Caswell No. 546	NCI-C60399
COLLOIDAL MERCURY	Quecksilber [German]
EPA Pesticide Chemical Code 052301	Quicksilver
KWIK [Dutch]	

Source: *Instant EPA's IRIS*, 1997, Instant Reference Sources, Inc., 7605 Rockpoint Drive, Austin, TX 78731 (Fax: 512-345-2386; Internet URL, http://www.instantref.com/inst-ref.htm).

Methomyl

Introduction

Methomyl is an insecticide applied to food and non-food crops, aquatic food crops, forestry, and indoor human and animal premises. It is used as a nematocide type insecticide with tobacco, cotton, alfalfa, soybeans, and corn. It is also used as a foliar treatment for control of insects such as aphids, armyworms, cabbage looper, tobacco budworm, tomato fruitworm, cotton leaf perforator and the cotton bollworm. [828] It is typically applied on leaves and soil using gound and aircraft equipment to spray it. Methomyl is manufactured in several types of formulations including wettable powders, emulsifiable and soluble concentrates, granulars, baits, and dusts.

Methomyl is released into the environment primarily from its application to plants as an insecticide. If released to soil, methomyl will degrade primarily by microbial degradation with carbon dioxide as the principal end product; a lag period of one to two weeks may occur in unacclimated soils before biodegradation begins. A small degree of chemical hydrolysis may occur in moist soils. Methomyl may be susceptible to significant leaching. Field and greenhouse studies have shown that methomyl degrades rapidly in soil with half-lives of 14 days or less. If released to water, methomyl will hydrolyze at half-life rates of about 54, 38 and 20 weeks at pH's of 6.0, 7.0 and 8.0, respectively, at 25°C. Decomposition occurs more rapidly on aeration, in sunlight or with increased alkalinity. Methomyl may be susceptible to significant biodegradation in natural water as it has been shown to be readily biodegraded in soil. Aquatic volatilization, adsorption, and bioconcentration are not expected to be important. If released to the atmosphere, methomyl will react in the vapor-phase with photochemically produced hydroxyl radicals with an estimated half-life of 1.14 months. Direct photolysis may also contribute to its transformation in air. Methomyl adsorbed to particulates in air are subject to physical removal via wet and dry deposition. Major general population exposure to methomyl will occur through consumption of contaminated food. Occupational exposure by dermal and inhalation routes related to the use of methomyl as an insecticide may be significant.

Identifying Information

CAS NUMBER: 16752-77-5

SYNONYMS: methyl N-[[(methylamino)carbonyl]oxy]ethanimidothioate; methyl N-[(methylcarbamoxyl)oxy]thioacetimidate; Lannate, Lanox, and Nudrin

782

Endocrine Disruptor Information

Methomyl is a strongly suspected EED that it is listed by the Centers for Disease Control & Prevention [812], and the World Wildlife Fund [813] as a potential endocrine modifying chemical.

Chemical and Physical Properties

CHEMICAL FORMULA: $C_5H_{10}N_2O_2S$

MOLECULAR WEIGHT: 162.2.

PHYSICAL DESCRIPTION: white crystalline solid

DENSITY: 1.2946 at 24/4°C [817]

MELTING POINT: 78-79°C [817]

SOLUBILITY: Soluble in acetone, ethanol, methanol, isopropanol, and many other organic solvents. [817]

Environmental Reference Materials

A source of EED environmental reference materials for methomyl is Crescent Chemical Co., Hauppauge, NY (Fax: 516-348-0913); Catalog No. 45573.

Hazardous Properties

ACUTE/CHRONIC HAZARDS: A rat chronic/oncogenicity and a mouse oncogenicity study showed no oncogenic effects from exposure to methomyl. There has been some concern that acetamide, a suspected human carcinogen, may occur in plants following treatment with methomyl. However, recently available data on thiodicarb, a related insecticide that breaks down initially to methomyl in plants, reveal that acetamide will not occur in plants treated with methomyl. [821]

Medical Symptoms of Exposure

Dose-related histopathologic changes were observed in the kidneys and spleens of dogs fed methomyl at 400 and 1000 ppm and in the livers and bone marrow at 1000 ppm level.

Toxicological Information

The toxicological information below is gathered from several sources including two national databases: one from the *National Toxicology Program Chemical Repository Database* and another from *EPA's Integrated Risk Information System (IRIS)*. These two databases contain a wealth of information on "traditional" toxicological data which includes LD50, LC50, and LCLo data, data on carcinogenicity (e.g., tumorigenic data), teratogenicity (reproductive effects data), plus information on chronic health hazard assessments, carcinogenicity assessments, Drinking Water Health Advisories and other EPA regulatory actions.

Toxicology Information from the National Toxicology Program

Methomyl is not currently listed in NTP's Chemical Repository database nor is it scheduled for addition in the future.

Toxicology Informat*ion from EPA's Integrated Risk Information System (IRIS)

I. CHRONIC HEALTH HAZARD ASSESSMENTS FOR NONCARCINOGENIC EFFECTS

I.A. REFERENCE DOSE FOR CHRONIC ORAL EXPOSURE (RfD)

The Reference Dose (RfD) is based on the assumption that thresholds exist for certain toxic effects such as cellular necrosis, but may not exist for other toxic effects such as carcinogenicity. In general, the RfD is an estimate (with uncertainty spanning perhaps an order of magnitude) of a daily exposure to the human population (including sensitive subgroups) that is likely to be without an appreciable risk of deleterious effects during a lifetime. Please refer to Background Document 1 in Service Code 5 for an elaboration of these concepts. RfDs can also be derived for the noncarcinogenic health effects of compounds which are also carcinogens. Therefore, it is essential to refer to other sources of information concerning the carcinogenicity of this substance. If the US EPA has evaluated this substance for potential human carcinogenicity, a summary of that evaluation will be contained in Section II of this file when a review of that evaluation is completed.

I.A.1. ORAL RfD SUMMARY

Critical Effect	Experimental Doses*	UF	MF	RfD
Kidney and Spleen Pathology	NOEL: 100 ppm (2.5 mg/kg/day)	100	1	2.5E-2mg/kg/day
	LEL: 400 ppm (10 mg/kg/day)			

*Conversion Factors: 1 ppm = 0.025 mg/kg/day (assumed dog food consumption)

2-year Feeding Study Dogs, du Pont, 1968a

I.A.2. PRINCIPAL AND SUPPORTING STUDIES (ORAL RfD)

E.I. du Pont de Nemours & Company, Inc. 1968a. MRID No. 00007091, 00009012. Available from EPA. Write to FOI, EPA, Washington, DC 20460.

Beagle dogs (4/sex/dose) were fed methomyl in their ad libitum diets. The diets contained 0 (control), 50, 100, 400, and 1000 ppm methomyl. Dose-related histopathologic changes were observed in kidney and spleen at 400 and 1000 ppm and in the liver and bone marrow at 1000 ppm level. The enlarged prostate gland in one animal each of the 100 and 400 ppm dose group was not considered compound-related since the effect was not dose-related and since dogs tend to show prostate enlargement with age. The NOEL for systemic effects was 100 ppm (2.5 mg/kg/day). I.A.3. UNCERTAINTY AND MODIFYING FACTORS (ORAL RfD)

UF -- A UF of 100 was used to extrapolate animal data accounting for intra- and inter-species differences.

MF -- None

I.A.4. ADDITIONAL COMMENTS (ORAL RfD)

The NOEL (100 ppm) observed in the dog study is further supported by lifetime studies in rats and mice, and a reproduction study in rats. In converting ppm to mg/kg/day, the dog study yields the lowest NOEL of all species tested. The NOEL for maternal toxicity in the rabbit was 2 mg/kg/day. Although a fraction lower than the NOEL used to establish the RfD, this NOEL was not used since exposure in teratology studies is by gavage and the chronic study in dogs more closely reflects continuous dietary exposure.

Data Considered for Establishing the RfD:

1) 2-Year Feeding - dog: Principal study - see discussion above; core grade minimum

2) 22-Month Feeding - rat: NOEL 100 ppm (5 mg/kg/day); LEL 200 ppm (10 mg/kg/day) (effects on spleen) (females) (du Pont, 1968b)

3) 2-Year Feeding - rat: NOEL 100 ppm (5 mg/kg/day); LEL 400 ppm (10 mg/kg/day) (ChE inhibition, growth retardation) (1981); core grade minimum (du Pont, 1981a)

4) 3-Generation Reproduction - rat: NOEL 100 ppm (5 mg/kg/day); core grade minimum (du Pont, 1968c)

5) Teratology - rat: No teratogenic effects at highest dose, 400 ppm; maternal toxicity at 400 ppm (du Pont, 1978)

6) Teratology - rabbit: No teratogenic effects at highest dose 16 mg/kg/day; maternal toxicity at 6 mg/kg/day (death and CNS effects) (du Pont, 1983)

Data Gap(s): None

Other Data Reviewed:

1) 2-Year Feeding (oncogenic) - mice: Systemic NOEL=50 ppm (7.5 mg/kg/day); Systemic LEL=11 mg/kg/day (du Pont, 1981b)

2) Delayed Neurotoxicity - Hens: Not a neurotoxin - tested up to 200 mg/kg/day (du Pont, 1967)

I.A.5. CONFIDENCE IN THE ORAL RfD

Study -- Medium
Data Base -- High
RfD -- High

The 2-year dog study used for supporting the RfD is of adequate quality, but considering the study date (1968) not entirely in compliance with today's requirements. However, the rest of the data base is of very good quality and supports the finding in the dog study; therefore, confidence in the data base is high. High confidence in the RfD follows.

I.A.6. EPA DOCUMENTATION AND REVIEW OF THE ORAL RfD

Pesticide Registration Files

Agency Work Group Review -- 04/22/86

Verification Date -- 04/22/86

I.A.7. EPA CONTACTS (ORAL RfD)

William Burnam / OPP -- (703)557-7491

George Ghali / OPP -- (703)557-7490

I.B. REFERENCE CONCENTRATION FOR CHRONIC INHALATION EXPOSURE (RfC)

            [Ed. Note: EPA does not have information yet in this section of IRIS].

II. CARCINOGENICITY ASSESSMENT FOR LIFETIME EXPOSURE

This substance/agent has been evaluated by the US EPA for evidence of human carcinogenic potential. This does not imply that this agent is necessarily a carcinogen. The evaluation for this chemical is under review by an inter-office Agency work group. A risk assessment summary will be included on IRIS when the review has been completed.

III. HEALTH HAZARD ASSESSMENTS FOR VARIED EXPOSURE DURATIONS

III.A. DRINKING WATER HEALTH ADVISORIES

The Office of Drinking Water provides Drinking Water Health Advisories (HAs) as technical guidance for the protection of public health. HAs are not enforceable Federal standards. HAs are concentrations of a substance in drinking water estimated to have negligible deleterious effects in humans, when ingested, for a specified period of time.

Exposure to the substance from other media is considered only in the derivation of the lifetime HA. Given the absence of chemical-specific data, the assumed fraction of total intake from drinking water is 20%. The lifetime HA is calculated from the Drinking Water Equivalent Level (DWEL) which, in turn, is based on the Oral Chronic Reference Dose. Lifetime HAs are not derived for compounds which are potentially carcinogenic for humans because of the difference in assumptions concerning toxic threshold for carcinogenic and noncarcinogenic effects. A more detailed description of the assumptions and methods used in the derivation of HAs is provided in Background Document 3 in Service Code 5.

III.A.1. ONE-DAY HEALTH ADVISORY FOR A CHILD

Appropriate data for calculating a One-day HA are not available. It is recommended that the DWEL adjusted for a 10-kg child, 3E-1 mg/L, be used as the One-day HA.

III.A.2. TEN-DAY HEALTH ADVISORY FOR A CHILD

Appropriate data for calculating a Ten-day HA are not available. It is recommended that the DWEL adjusted for a 10-kg child, 3E-1 mg/L, be used as the Ten-day HA.

III.A.3. LONGER-TERM HEALTH ADVISORY FOR A CHILD

Appropriate data for calculating Longer-term HA for methomyl is not available. It is recommended that the DWEL adjusted for a 10-kg child, 3E-1 mg/L, be used as the longer-term HA for a 10-kg child.

III.A.4. LONGER-TERM HEALTH ADVISORY FOR AN ADULT

Appropriate data for calculating the Longer-term HA for methomyl is not available. It is recommended that the DWEL of 0.9 mg/L be used as the Longer- term HA for a 70-kg adult.

III.A.5. DRINKING WATER EQUIVALENT LEVEL / LIFETIME HEALTH ADVISORY

DWEL -- 9E-1 mg/L

Assumptions -- 2 L/day water consumption for a 70-kg adult

RfD Verification Date -- 04/22/86

Lifetime HA -- 2E-1 mg/L

Assumptions -- 20% exposure by drinking water

Principal Study -- Kaplan and Sherman, 1977 (This study was used in the derivation of the chronic oral RfD; see Section I.A.2.)

In a 2-year feeding study in beagle dogs (4/sex/dose), methomyl was administered at dietary levels of 0, 50, 100, 400, or 1000 ppm (approximately 1.25, 2.5, 10, or 25 mg/kg/day). Dogs receiving 1.25 or 2.5 mg/kg/day showed no evidence of toxic effects. Those receiving 10 mg/kg/day exhibited histopathological changes in the kidney and spleen. In addition to these effects, animals receiving the highest dose also exhibited symptoms of CNS toxicity, liver and bone marrow effects. A NOAEL of 2.5 mg/kg/day is identified for this study.

III.A.6. ORGANOLEPTIC PROPERTIES

No information is available on the organoleptic properties of methomyl.

III.A.7. ANALYTICAL METHODS FOR DETECTION IN DRINKING WATER

Analysis of methomyl is by a high performance liquid chromatographic procedure used for the determination of N-methyl carbamoyloximes and N-methylcarbamates in drinking water.

III.A.8. WATER TREATMENT

Available data indicate that granular activated carbon adsorption will remove methomyl from water.

III.A.9. DOCUMENTATION AND REVIEW OF HAs

EPA review of HAs in 1987.

Public review of HAs was January-March 1988.

Preparation date of this IRIS summary -- 09/30/88

III.A.10. EPA CONTACTS

Amal Mahfouz / OST -- (202)260-9568

Robert Cantilli / OST -- (202)260-5546

III.B. OTHER ASSESSMENTS

[Ed. Note: EPA does not have information yet in this section of IRIS].

IV. US EPA REGULATORY ACTIONS

EPA risk assessments may be updated as new data are published and as assessment methodologies evolve. Regulatory actions are frequently not updated at the same time. Compare the dates for the regulatory actions in this section with the verification dates for the risk assessments in sections I and II, as this may explain inconsistencies. Also note that some regulatory actions consider factors not related to health risk, such as technical or economic feasibility. Such considerations are indicated for each action. In addition, not all of the regulatory actions listed in this section involve enforceable federal standards. Please direct any questions you may have concerning these regulatory actions to the US EPA contact listed for that particular action. Users are strongly urged to read the background information on each regulatory action in Background Document 4 in Service Code 5.

IV.A. CLEAN AIR ACT (CAA)

[Ed. Note: EPA does not have information yet in this section of IRIS].

IV.B. SAFE DRINKING WATER ACT (SDWA)

Listed in the January 1991 Drinking Water Priority List and may be subject to future regulation (56 FR 1470, 01/14/91).

IV.B.1. MAXIMUM CONTAMINANT LEVEL GOAL (MCLG) for Drinking Water

[Ed. Note: EPA does not have information yet in this section of IRIS].

IV.B.2. MAXIMUM CONTAMINANT LEVEL (MCL) for Drinking Water

[Ed. Note: EPA does not have information yet in this section of IRIS].

IV.B.3. SECONDARY MAXIMUM CONTAMINANT LEVEL (SMCL) for Drinking Water

[Ed. Note: EPA does not have information yet in this section of IRIS].

IV.B.4. REQUIRED MONITORING OF "UNREGULATED" CONTAMINANTS

Status -- Listed (Final, 1991)

Discussion -- "Unregulated" contaminants are those contaminants for which EPA establishes a monitoring requirement but which do not have an associated final MCLG, MCL, or treatment technique. EPA may regulate these contaminants in the future.

Monitoring requirement -- All systems to be monitored unless a vulnerability assessment determines the system is not vulnerable.

Analytical methodology -- High pressure liquid gas chromatography derivatization (EPA 531.1).

Reference -- 56 FR 3526 (01/30/91)

EPA Contact -- Drinking Water Standards Division / OGWDW / (202) 260-7575 / FTS 260-7575; or Safe Drinking Water Hotline / (800) 426-4791

IV.C. CLEAN WATER ACT (CWA)

[Ed. Note: EPA does not have information yet in this section of IRIS].

IV.D. FEDERAL INSECTICIDE, FUNGICIDE, AND RODENTICIDE ACT (FIFRA)

IV.D.1. PESTICIDE ACTIVE INGREDIENT, Registration Standard

Status -- Issued (1981)

Reference -- Methomyl Pesticide Registration Standard. September, 1981 (NTIS No. PB82-180738) as amended on April, 1989 (NTIS No. PB89-193957).

EPA Contact -- Registration Branch / OPP (703)557-7760 / FTS 557-7760

IV.D.2. PESTICIDE ACTIVE INGREDIENT, Special Review

[Ed. Note: EPA does not have information yet in this section of IRIS].

IV.E. TOXIC SUBSTANCES CONTROL ACT (TSCA)

[Ed. Note: EPA does not have information yet in this section of IRIS].

IV.F. RESOURCE CONSERVATION AND RECOVERY ACT (RCRA)

[Ed. Note: EPA does not have information yet in this section of IRIS].

IV.G. SUPERFUND (CERCLA)

IV.G.1. REPORTABLE QUANTITY (RQ) for Release into the Environment

Value (status) -- 100 pounds (Final, 1985)

Considers technological or economic feasibility? -- NO

Discussion -- The final RQ was based on aquatic toxicity. The available data indicate that the aquatic 96-Hour Median Threshold Limit for methomyl is between 0.7 and 2.8 ppm.

Reference -- 50 FR 13456 (04/04/85); 54 FR 33418 (08/14/89)

EPA Contact -- RCRA/Superfund Hotline (800)424-9346 / (202)260-3000 / FTS 260-3000

VI. BIBLIOGRAPHY

VI.A. ORAL RfD REFERENCES

E.I. du Pont de Nemours & Company, Inc. 1967. MRID No. 00008827. Available from EPA. Write to FOI, EPA, Washington, DC 20460.

E.I. du Pont de Nemours & Company, Inc. 1968a. MRID No. 00007091, 00009012. Available from EPA. Write to FOI, EPA, Washington, DC 20460.

E.I. du Pont de Nemours & Company, Inc. 1968b. MRID No. 00007092, 00009011. Available from EPA. Write to FOI, EPA, Washington, DC 20460.

E.I. du Pont de Nemours & Company, Inc. 1968c. MRID No. 00007093. Available from EPA. Write to FOI, EPA, Washington, DC 20460.

E.I. du Pont de Nemours & Company, Inc. 1978. MRID No. 00008621. Available from EPA. Write to FOI, EPA, Washington, DC 20460.

E.I. du Pont de Nemours & Company, Inc. 1981a. MRID No. 00078361. Available from EPA. Write to FOI, EPA, Washington, DC 20460.

E.I. du Pont de Nemours & Company, Inc. 1981b. MRID No. 00078423. Available from EPA. Write to FOI, EPA, Washington, DC 20460.

E.I. du Pont de Nemours & Company, Inc. 1983. MRID No. 00131257. Available from EPA. Write to FOI, EPA, Washington, DC 20460.

VI.B. INHALATION RfC REFERENCES

None

VI.C. CARCINOGENICITY ASSESSMENT REFERENCES

None

VI.D. DRINKING WATER HA REFERENCES

Kaplan, M.A. and H. Sherman. 1977. Toxicity studies with methyl N-(((methylamino)carbonyl)oxy)-ethanimidothioate. Toxicol. Appl. Pharmacol. 40: 1-17.

VII. REVISION HISTORY

Date	Section	Description
03/31/87	IV.	Regulatory Action section on-line
12/01/88	I.A.4.	Core grades added to studies 1, 3 and 4
03/01/91	I.A.4.	Citations added
03/01/91	III.A.	Health Advisory on-line
03/01/91	VI.	Bibliography on-line
10/01/91	II.	Carcinogenicity assessment now under review
01/01/92	IV.	Regulatory actions updated

SYNONYMS

ACETIMIDIC ACID, N-
((METHYLCARBAMOYL)OXY)THIO-,
METHYL ESTER
ACETIMIDIC ACID, THIO-N-
((METHYLCARBAMOYL)OXY)-,
METHYL ESTER
ACETIMIDOTHIOIC ACID, METHYL-,
N-(METHYLCARBAMOYL) ESTER
DUPONT 1179
ENT 27,341
ETHANIMIDOTHIOIC ACID, N-
(((METHYLAMINO)CARBONYL)OXY)
-, METHYL ESTER
IN 1179
LANNATE
LANNATE L
MESOMILE

Methomyl
METHYL N-
((METHYLAMINO)CARBONYL)OXY)
ETHANIMIDO)THIOATE
2-METHYLTHIO-ACETALDEHYD-O-
(METHYLCARBAMOYL)-OXIM
2-METHYLTHIO-PROPIONALDEHYD-
O-(METHYLCARBAMOYL)-OXIM
METOMIL
N-
((((METHYLAMINO)CARBONYL)OXY)
ETHANIMIDOTHIOIC ACID METHYL
ESTER
NU-BAIT II
NUDRIN
RCRA WASTE NUMBER P066
SD 14999
3-THIABUTAN-2-ONE, O-
(METHYLCARBAMOYL)OXIME
WL 18236

Source: *Instant EPA's IRIS*, 1997, Instant Reference Sources, Inc., 7605 Rockpoint Drive, Austin, TX 78731 (Fax: 512-345-2386; Internet URL, http://www.instantref.com/inst-ref.htm).

Production, Use, and Pesticide Labeling Information

1. Description of Chemical

Generic Name: S-methyl N-[(methylcarbamoyl)oxy]thiocacetimidate

Common Name: Methomyl

Trade Name and other names: methyl N-[[(methylamino)carbonyl]oxy]ethanimidothioate; methyl N-[(methylcarbamoxyl)oxy]thioacetimidate; Lannate, Lanox, and Nudrin

EPA Shaughnessy Number: 090301

Chemical Abstracts Service (CAS) Number: 16752-77-5

Year of Initial Registration: 1968

Pesticide Type: Insecticide

Chemical Family: Carbamate

US Producer: E.I. duPont de Nemours and Company

2. Use Patterns and Formulations

- Application Sites: Terrestrial food and non-food crops, greenhouse food and non-food crops, aquatic food crops, forestry (ground only), and indoor- human and animal premise.

- Types and method of applications: Foliar and broadcast soil application by both ground and aircraft equipment.

- Types of formulations: Wettable powders, emulsifiable and soluble concentrates, granulars, baits, and dusts.

- Usual Carriers: Petroleum and clay carriers.

3. Science Findings

Physiochemical Characteristics

- Technical methomyl is a white crystalline solid with a melting point of 78-79 °C. Methomyl is soluble in water and most organic solvents. Methomyl's empirical formula is $C_5H_{10}N_2O_2S$ and its molecular weight is 162.2.

Toxicological Characteristics

- Acute Oral: 17 to 24 mg/kg (rat) Toxicity Category I.

- Acute Dermal: >5000 mg/kg (rabbit) Toxicity Category III.

- Primary Eye Irritation: Data gap.

- Acute Inhalation: 0.30 mg/liter/4 hours Toxicity Category III.

- Primary Skin Irritation: Data gap.

- Dermal Sensitization: Data gap.

- Acute Delayed Neurotoxicity: Data shows no potential for this effect.

- Subchronic Dermal: (21 day): Data gap.

- Oncogenicity: A rat chronic/oncogenicity study showed no oncogenic effects at the highest dose tested (HDT-400 ppm). A mouse oncogenicity study showed no oncogenic effects at HDT 200 ppm.

- Metabolism: Data gap. Preliminary data suggests that methomyl may be metabolized to the possible human oncogen acetamide. Additional metabolism data in the rat and monkey are required to detect the possible presence of acetamide in the tissues.

- Teratology: Not teratogenic or embryotoxic at HDT 400 ppm in rats. Not teratogenic or embryotoxic at HDT in rabbits (Maternal No Observed Effect Level NOEL of 2 mg/kg/day).

- Reproduction: No observed effects with a NOEL of 100 ppm.

- Mutagenicity: Not a mutagen in all the required tests.

Physiological and Biochemical Characteristics

- Mechanism of Pesticide Action: Methomyl kills by poisoning the insects' nervous system.

- Metabolism and persistence in plants and animals:

- Plants: Available data demonstrate that methomyl is relatively persistent on leaf surfaces and fruit. There has been some concern in the recent past that acetamide, a suspected human carcinogen, may occur in plants following treatment with methomyl. However, recently available data on thiodicarb, a related insecticide that breaks down initially to methomyl in plants, reveal that acetamide will not occur in plants treated with methomyl. Furthermore, any acetamide formed from acetonitrile, a known metabolite of methomyl, will be hydrolyzed to form acetic acid and ammonium ion.

- Animals: The available data are not adequate to assess the nature of methomyl in animals. Additional metabolism data (ruminants and poultry) are required to detect the possible presence of acetonitrile and acetamide. These data were requested in a FIFRA 3(c)(2)(B) letter dated March 23, 1987.

Environmental Characteristics

- Available data are insufficient to fully assess the environmental fate of methomyl. The Registration Standard issued in 1981 did not address the aquatic uses of methomyl. Data must now be submitted for the following: aquatic aerobic, aquatic anaerobic, aquatic field dissipation and irrigated crops. Vapor pressure data indicate the need for volatility data. Monitoring data are needed to assess the potential of this pesticide to contaminate groundwater.

Ecological Characteristics

- Acute avian oral toxicity: LD50 24.2 mg/kg for bobwhite quail (highly toxic).

- Avian dietary toxicity: LC50 of 1100, 1975, and 2883 ppm respectively for bobwhite quail, ring-necked pheasant, and mallard duck (slightly toxic).

- Freshwater fish acute toxicity: LC50 = 1.6 ppm for Rainbow trout. LC50 = 0.15 ppm for channel catfish (moderate to highly toxic for fish).

- Freshwater aquatic invertebrate toxicity: LC50 values of 0.0698 to 0.343 ppm suggest that is very highly toxic to freshwater invertebrates.

- Major Routes of Exposure:

- Dermal followed by inhalation. Human exposure occurs from mixing, loading and application. Exposure can be reduced by the use of goggles or face shield and protective clothing.

Tolerance Assessment

- Tolerances have been established for residues of methomyl in a variety of raw agricultural commodities (Refer to 40 CFR 180.253 for listing of tolerances). Methomyl's tolerances have been reassessed using the Tolerance Assessment System (TAS). The Acceptable Daily Intake (ADI) for this chemical is 0.025 mg/kg/day. The Theoretical

Maximum Residue Contribution (TMRC) for the US population is 0.016188 mg/kg/day, which occupies 65% of the ADI.

4. Summary of Regulatory Position and Rationale

- This review of methomyl is the second intensive evaluation of the compound. In its original Standard, issued in 1981, the Agency summarized the available data supporting the registration of methomyl and concluded that additional data were needed to fully evaluate the pesticide.

- The Agency has since received and reviewed the data and has revised its scientific and regulatory conclusions in relative to these data. Additionally, other information on the chemical (for example the acetamide issue) and the expanded data requirements promulgated in 1984 at 40 CFR Part 158, have added new data requirements. This Standard, which supersedes the 1981 document, is the Agency's updated assessment of the pesticide and the data needed to support its continued registration.

A. Methomyl is not being placed in Special Review at this time because none of the criteria for initiation of Special Review listed in 40 CFR 154.7 have been met. The Agency believes that the water soluble bag use restriction and the increased reentry intervals provide mixer/loader and field-worker protection. The Agency intends, however, to monitor State pesticide incidents monitoring systems to determine the effectiveness of labeling changes identified in the Standard. The Agency may impose further regulatory actions if these incidents reports indicate that these labeling changes are inadequate.

B. The Agency is requiring the submission of acute aquatic toxicity data and aquatic and non-aquatic field monitoring data on the end-use formulations and aquatic life stage data and avian reproduction data on the technical formulation in order to complete the wildlife risk assessment.

C. Based on methomyl's use pattern and toxicity data, the Agency has determined that methomyl may trigger the endangered species criteria for fish, aquatic organisms and insects. No endangered species labeling is required at this time. A program is being developed by the Agency to reduce or eliminate exposure of this chemical to these species. After this program is developed, the Agency will notify registrants of any additional labeling that may be required to remain in compliance with FIFRA. The labeling requirements affecting methomyl, e.g., those listed in PR Notices 87-4 and 87-5, have been withdrawn pending reissuance.

D. Various methomyl formulations were classified as restricted use products by regulation in 1978 (see 40 CFR 162.13). The Agency has now determined that the 90% water soluble bag formulation should also be classified as a restricted use pesticide. Labeling language for each of the restricted use products must specify that the restriction is based on high acute toxicity to humans.

E. The following reentry intervals are being imposed at this time based on the submitted reentry data: one day for beans, cabbages, roses grown outdoors and carnations, whether grown outdoors or in a greenhouse; three days for cotton, nectarines, and oranges/citrus; four days for peaches; and seven days for grapes. Because of the similarity in crops and in the work tasks performed in those crops, a three day reentry interval is being established for apples, and a one day reentry interval for alfalfa, asparagus, broccoli, brussel sprouts, carrots, cauliflower, celery, collards, cucumbers, lettuce, melons, onions, peanuts, peas, peppers, potatoes, sorghum, soybeans, summer squash, spinach, sugar beets, tobacco, and tomatoes. Additional data are being requested to set reentry intervals for mint, corn, roses grown in greenhouses, and chrysanthemums grown in greenhouses or outdoors. Until these data are received and evaluated, an interim seven day reentry interval is being established for corn, and a one day reentry interval is being established for these crops and all other crops and sites not specifically listed above.

F. The following labeling is required for all manufacturing use products:

- This pesticide is toxic to fish. do not discharge effluent containing this product into lakes, streams, ponds, estuaries, oceans, or public waters unless this product is specifically identified and addressed in the NPDES permit. Do not discharge effluent containing this product to sewer systems without previously notifying the sewage treatment plant authority. For guidance, contact your State Water Board or Regional Office of the EPA.

G. The following labeling is required for all end use products:

- PERSONAL PROTECTIVE EQUIPMENT:

USE ONLY WHEN WEARING THE FOLLOWING PERSONAL PROTECTIVE EQUIPMENT DURING MIXING/LOADING, APPLICATION, REPAIRING AND CLEANING OF MIXING, LOADING, AND APPLICATION EQUIPMENT, AND DISPOSAL OF THE PESTICIDE: long-sleeve shirt; long-legged pants; shoes and socks, chemical resistant gloves; face shield or goggles; NIOSH or MSHA approved respirator. During equipment repair and cleaning, the respirator need not be worn.

- IF APPLICATION IS PERFORMED USING AN ENCLOSED CAB OR COCKPIT, THE FOLLOWING

PROTECTIVE CLOTHING AND EQUIPMENT MAY BE WORN AS AN ALTERNATIVE: long-sleeved shirt and long-legged pants; shoes and socks. All other protective clothing and equipment required for use during application must be available in the cab and must be worn when exiting the cab into treated area. When used for this purpose, contaminated clothing may not be brought back into the cab unless in an enclosure such as a plastic bag.

- IMPORTANT! If pesticide comes in contact with skin, wash off with soap and water and contact a physician immediately. ALWAYS WASH HANDS, FACE, AND ARMS WITH SOAP AND WATER BEFORE USING TOBACCO PRODUCTS, EATING, DRINKING, OR TOILETING.

- AFTER WORK: Before removing gloves, wash them with soap and water. Take off all work clothes and shoes. Shower using soap and water. Wear only clean clothes when leaving job -- do not wear contaminated clothing. Personal clothing worn during work must be stored and laundered separately from protective clothing and household articles. Store protective clothing separately from personal clothing. Clean or launder protective clothing after each use. Respirators must be cleaned and filters replaced according to instructions included with the respirators. Protective clothing and protective equipment heavily contaminated or drenched with methomyl must be destroyed according to state and local regulations. HEAVILY CONTAMINATED OR DRENCHED CLOTHING CANNOT BE ADEQUATELY DECONTAMINATED.

- DURING AERIAL APPLICATION, HUMAN FLAGGERS ARE PROHIBITED.

- PROTECTIVE CLOTHING LABEL STATEMENTS FOR 2% BAITS:

> - Use only when wearing the following personal protective equipment during loading, application, repairing and cleaning of mixing, loading, and application equipment, and disposal of the pesticide: long-sleeve shirt; long-legged pants; shoes and socks; gloves.

- IMPORTANT! If pesticide comes in contact with skin, wash off with soap and water.

- ALWAYS WASH HANDS, FACE, AND ARMS WITH SOAP AND WATER BEFORE USING TOBACCO PRODUCTS, EATING, DRINKING, OR TOILETING.

- OTHER LABEL STATEMENTS FOR BAITS:

  - Do not contaminate feed and foodstuffs. Do not apply where poultry or other animals, especially dogs and young calves, can lick it.

  - Do not use in edible product areas of food processing plants, restaurants, or other areas where food is commercially prepared or processed. Do not use in serving areas while food is exposed.

- REENTRY INTERVALS:

  - The following reentry intervals are required in the directions for use section of all labels with terrestrial and greenhouse food and non-food uses: three days for cotton, nectarines, citrus and apples; four days for peaches; seven days for grapes and corn; all other crops, one day.

  - The following fish and wildlife statements are required to appear under the "Environmental Hazards" heading:

    - Granulars (including baits):

    - This pesticide is toxic to birds. Collect, cover or incorporate granules spilled on the soil surface. Do not apply directly to water or wetlands (swamps, bogs, marshes, and potholes). Do not contaminate water when disposing of equipment washwaters.

    - Non-Granular:

    - Aquatic (Watercress): This pesticide is toxic to fish. Drift and runoff from treated areas may be hazardous to aquatic organisms in neighboring areas. do not contaminate water when disposing of equipment washwaters.

    - Terrestrial: This pesticide is toxic to fish and wildlife. do not apply directly to water or wetlands (swamps, bogs, marshes, and potholes). Drift and runoff may be hazardous to aquatic organisms in neighboring areas. Do not contaminate water when disposing of equipment washwaters.

H. End use products (except granulars and baits) with outdoor crop uses must have the following bee caution:

- This product is highly toxic to bees exposed to direct treatment on blooming crops or weeds. Do not apply this product or allow it to drift to blooming crops or weeds while bees are actively visiting the treatment area.

- Products with lentil use must include a pregrazing interval of three days and a preharvest interval of seven days.

I. Pursuant to the data requirements in 40 CFR Part 158, the Agency has determined that the following data are essential to the Agency's assessment and should receive a priority review when they are received by the Agency:

- 40 CFR 158.240 Residue Chemistry
    - 171-4 Nature of the Residue (Metabolism-Livestock), Meat/Milk/Poultry and Eggs

- 40 CFR 158.390 Reentry Protection
    - 132-1 Foliar Dissipation (Reentry)
    - 201-1 Droplet Size Spectrum
    - 201-1 Drift Field Evaluation

- 40 CFR 158.340 Toxicology
    - 85-1 General Metabolism (Rat and Monkey)
    - 82-2 Subchronic Dermal (21-Day)

- 40 CFR 158.290 Environmental Fate
    - 162-3 Anaerobic Aquatic Metabolism
    - 162-4 Aerobic Aquatic Metabolism
    - 162-2 Laboratory Volatility
    - 164-2 Aquatic (Sediment)
    - 165-5 Accumulation in Non-Target Organisms
    - Groundwater Monitoring

- 40 CFR 158.490 Ecological Effects
    - 71-4 Avian Reproduction
    - 71-5 Simulated and Actual Field Testing - Birds
    - 72-1 Freshwater Fish Acute Toxicity
    - 72-2 Acute Toxicity - Freshwater Invertebrate

- 72-3 Acute Toxicity - Aquatic Estuarine and
  Marine Organism
- 72-5 Fish Life-Cycle
- 72-7 Field Testing for Aquatic Organism

5. Summary of Major Data Gaps

- The following studies are required to assess the toxicological characteristics of technical methomyl: Eye irritation, dermal sensitization, 21-day dermal toxicity, and general metabolism testing (rat and monkey).

- The following data are required to fully characterize methomyl's environmental fate: Reentry, volatility (lab), aquatic sediment dissipation, accumulation studies in irrigated crops and in fish, and groundwater monitoring.

- Additional residue and processing studies in certain commodities, are required to support existing tolerances.

- The following data are required to complete a wildlife risk assessment: Avian subacute dietary toxicity, avian reproduction, freshwater fish toxicity, acute toxicity to freshwater invertebrates, acute toxicity to estuarine and marine organisms, fish early life stage and aquatic invertebrate life cycle, and simulated or actual field testing for aquatic organisms and mammals and birds.

- Product chemistry and acute toxicity data are required.

6. Contact Person at EPA

Dennis Edwards
Product Manager
12 Insecticide- Rodenticide Branch Registration Division (H7505C)
Office of Pesticide Programs
Environmental Protection Agency
401 M. Street, SW
Washington, DC 20460

Office location and telephone number:

Room 202, Crystal Mall Building #2,
1921 Jefferson Davis Highway
Arlington, VA 22202

Telephone: (703) 557-2386

Source: *Instant EPA's Pesticide Facts*, 1994, Instant Reference Sources, Inc., 7605 Rockpoint Drive, Austin, TX 78731 (Fax: 512-345-2386; Internet URL, http://www.instantref.com/inst-ref.htm).

Methoxychlor

Introduction

Methoxychlor is an insecticide used to control a wide range of insect pests (particularly chewing insects) in field crops, forage crops, fruit, vines, flowers, vegetables and in forestry. It is also used for the control of insect pests in animal houses and dairies, and in household and industrial premises. It is used in veterinary medicine as an ectoparasiticide.

Release of methoxychlor to the environment is expected to occur primarily due to its use as an insecticide. Other sources of release may include loss during manufacturing, formulation, packaging, and disposal of methoxychlor. If released to soil, methoxychlor is expected to remain immobilized primarily in the upper layer of soil although a small percentage may migrate to lower depths, possibly into groundwater as suggested by the detection of methoxychlor in some groundwater samples. Under anaerobic conditions, biodegradation appears to be the dominant removal mechanism; however, under aerobic conditions, biodegradation is expected to be less rapid and possibly negligible. Rapid primary degradation of methoxychlor has been observed under anaerobic conditions in flooded soils (half-lives 1 week to < 2 months). Major degradation products under anaerobic conditions are dechlorinated methoxychlor (DMDD) and mono- and di-hydroxy derivatives of methoxychlor and DMDD. Methoxychlor may undergo indirect "sensitized" photolysis on the soil surfaces and it may undergo chemical hydrolysis in moist soils (half-life > 1 year). [828]

If released to water, methoxychlor may be removed or transported by several different mechanisms. Methoxychlor may adsorb to suspended solids and sediments or it may bioaccumulate in certain aquatic organisms, although fish are reported to metabolize methoxychlor fairly rapidly. Methoxychlor may undergo direct photolysis (the half-life is 4.5 months) or indirect "sensitized" photolysis (with a half-life <5 hours) depending upon the presence of photosensitizers. 1,1-Bis(p-methoxyphenyl)-2,2-dichloroethylene (DMDE) is a major photolysis product of methoxychlor. Volatilization of methoxychlor may be significant (the half-life is 4.5 days from a shallow river) and methoxychlor may also biodegrade under anaerobic conditions (the half-life is < 28 days in sediments) or aerobic conditions (where the half-life is >100 days in sediments). Oxidation and chemical hydrolysis are not expected to be significant fate processes. If released to the atmosphere, methoxychlor may exist in either vapor or particulate form. Methoxychlor may undergo reaction with photochemically generated hydroxyl radicals (estimated vapor phase half-life 3.7 hours) or physical removal by settling out or washing out in precipitation. The most probable route of exposure to methoxychlor would be inhalation or dermal contact during home use of this insecticide, inhalation of airborne particulate matter containing methoxychlor or ingestion of food or drinking water contaminated with methoxychlor. [828]

Identifying Information

CAS NUMBER: 72-43-5

NIOSH Registry Number: KJ3675000

SYNONYMS: 2,2-bis(p-Methoxyphenyl)-1,1,1-trichloroethane

Endocrine Disruptor Information

Methoxychlor is a strongly suspected EED that it is listed by the Centers for Disease Control & Prevention [812], and the World Wildlife Fund [813] as a potential endocrine modifying chemical. It also disrupts hormones but it does not leave telltale signs of exposure in body tissue. [810]

Chemical and Physical Properties

CHEMICAL FORMULA: $C_{16}H_{15}Cl_3O_2$

MOLECULAR WEIGHT: 345.66

WLN: GXGGYR DO1&R DO1

PHYSICAL DESCRIPTION: White to tan powder

SPECIFIC GRAVITY: 1.41 @ 25°C [055], [169], [173], [371]

DENSITY: Not available

MELTING POINT: 86-88°C [025], [031], [395], [421]

BOILING POINT: Decomposes [051], [102], [371], [421]

SOLUBILITY: Water: <1 mg/mL @ 23°C [700]; DMSO: >=100 mg/mL @ 23°C [700]; 95% Ethanol: 10-50 mg/mL @ 23°C [700]; Acetone: >=100 mg/mL @ 23°C [700]; Methanol: 50 g/kg @ 20°C [169], [172]

OTHER SOLVENTS: Aromatic solvents: Readily soluble [169], [173], [395]; Alcohol: Soluble [031], [062], [173], [395]; Petroleum oils: Moderately soluble [173], [395]; Xylene: 440 g/kg @ 20°C [169], [172] Chlorinated solvents: Readily soluble [169]; Vegetable oils: Readily soluble [169]; Chloroform: 440 g/kg @ 20°C [169], [172]; Ketonic solvents: Readily soluble [169]

OTHER PHYSICAL DATA: This compound exists as dimorphic crystals [025], [031], [047], [395]; Gray flaky powder (technical grade) [165], [169], [172], [395]; Melting point also reported as 78-78.2°C [025], [031], [395], [421]; Threshold odor concentration (in water): 4.7 mg/kg [055]; Slightly fruity odor [051], [346], [371], [421]; Spectroscopy data: lambda max (in benzene): 275 nm, 270 nm, 238 nm, 230 nm (E = 183, 241, 458, 575) [395]

Source: *Instant EPA's Air Toxics*, 1994, Instant Reference Sources, Inc., 7605 Rockpoint Drive, Austin, TX 78731 (Fax: 512-345-2386; Internet URL, http://www.instantref.com/inst-ref.htm).

Environmental Reference Materials

A source of EED environmental reference materials for methoxychlor is Radian International LLC, Austin, TX (Fax 512-454-0268; Internet URL, http://www.radian.com/standards); Catalog No. ERM-005.

Hazardous Properties

ACUTE/CHRONIC HAZARDS: Methoxychlor may be toxic by ingestion and inhalation [051], [371], [421] and it is an irritant [033], [051], [371]. When heated to decomposition it emits toxic fumes of chlorides, hydrogen chloride gas and carbon monoxide [043], [051], [102], [371].

HAP WEIGHTING FACTOR: 1 [713]

VOLATILITY:
   Vapor pressure: Very low [102], [169]
   Vapor density: 12 [043]

FIRE HAZARD: Methoxychlor is combustible. It burns only at high temperatures [051], [371]. Fires involving this material may be controlled with a dry chemical, carbon dioxide or Halon extinguisher. A water spray may also be used [051], [102], [371].

REACTIVITY: Methoxychlor is incompatible with alkaline materials, especially in the presence of catalytically-active metals [169], [173], [186], [395]. It is slightly corrosive to iron and aluminum [169]. It is decomposed by refluxing with sodium in isopropyl alcohol [186]. It is also incompatible with strong oxidizers [051], [102], [346]. It will attack some forms of plastics, rubber and coatings [102].

STABILITY: Methoxychlor turns pink or tan on exposure to light [169]. It is described as resistant to ultraviolet light, but other studies have shown it to break down rapidly under UV in hexane solution [186]. Solutions of it in water, DMSO, 95% ethanol or acetone should be stable for 24 hours under normal lab conditions [700].

Source: *Instant EPA's Air Toxics*, 1994, Instant Reference Sources, Inc., 7605 Rockpoint Drive, Austin, TX 78731 (Fax: 512-345-2386; Internet URL, http://www.instantref.com/inst-ref.htm).

Medical Symptoms of Exposure

Symptoms of exposure may include vomiting, tremors, convulsions and liver damage [301]. Somnolence may also occur [043]. Other symptoms may include generalized depression, headache, staggering, nausea and lethargy [051]. It may also cause diarrhea, numbness and partial paralysis [058]. Exposure over a prolonged period may cause kidney damage [033], [058].

Symptoms of exposure to this type of compound may include central nervous system stimulation, paresthesia, excitement, giddiness, fatigue, pulmonary edema, myocardial toxicity, hypothermia and coma. Respiration is initially accelerated and later depressed. Chronic exposure to this type of compound may cause loss of appetite, muscular weakness, apprehensive mental state and enhancement of hepatic microsomal enzyme activity [295].

Source: *Instant EPA's Air Toxics*, 1994, Instant Reference Sources, Inc., 7605 Rockpoint Drive, Austin, TX 78731 (Fax: 512-345-2386; Internet URL, http://www.instantref.com/inst-ref.htm).

Toxicological Information

The toxicological information below is gathered from several sources including two national databases: one from the *National Toxicology Program Chemical Repository Database* and another from *EPA's Integrated Risk Information System (IRIS)*.

Toxicology Information from the National Toxicology Program

CARCINOGENICITY:

Tumorigenic Data:

Type of Effect	Route/Animal	Amount Dosed (Notes)
TD	orl-rat	87360 mg/kg/2Y-C
TD	orl-rat	45500 mg/kg/1Y-C
TD	orl-rat	10920 mg/kg/1Y-C
TD	orl-rat	72800 mg/kg/2Y-C
TD	orl-mus	62622 mg/kg/2Y-C
TDLo	orl-dog	383 gm/kg/3Y-C
TDLo	orl-mus	56700 mg/kg/90W-C
TDLo	orl-rat	18200 mg/kg/2Y-C
TD	orl-rat	41 gm/kg/2Y-C
TD	orl-rat	80 gm/kg/2Y-C

Review:
IARC Cancer Review: Animal Inadequate Evidence
IARC: Not classifiable as a human carcinogen (Group 3)

Status:

> NCI Carcinogenesis Bioassay (Feed); Negative: Male and Female Rat,
> Male and Female Mouse

TERATOGENICITY: See RTECS printout for data

MUTATION DATA: See RTECS printout for data

TOXICITY:

Typ. Dose	Mode	Specie	Amount	Units	Other
LDLo	orl	hmn	6430	mg/kg	
TDLo	skn	hmn	2414	mg/kg	
LD50	orl	rat	5000	mg/kg	
LD50	ipr	ham	500	mg/kg	
LD50	orl	mus	1	gm/kg	

OTHER TOXICITY DATA:

> Review: Toxicology Review-3

Status:

> EPA TSCA Test Submission (TSCATS) Data Base, April 1990
> EPA Genetox Program 1988, Negative: Carcinogenicity-mouse/rat;
> SHE-clonal assay
> EPA Genetox Program 1988, Positive/dose response: Cell transform.-
> BALB/c-3T3
> EPA Genetox Program 1988, Negative: Cell transform.-RLV F344 rat
> embryo
> EPA Genetox Program 1988, Negative: Histidine reversion-Ames test
> EPA Genetox Program 1988, Negative: D melanogaster Sex-linked
> lethal
> EPA Genetox Program 1988, Negative: In vitro UDS-human
> fibroblast; TRP reversion
> EPA Genetox Program 1988, Negative: S cerevisiae-homozygosis
> EPA Genetox Program 1988, Inconclusive: B subtilis rec assay; E coli
> polA without S9
> Estimated fatal oral dose: 7.5 g/kg
> IDLH value: 7500 mg/m3

SAX TOXICITY EVALUATION:

THR: Moderately toxic by ingestion, intraperitoneal and skin contact.   An
experimental carcinogen, tumorigen and teratogen. Human systemic effects by
ingestion. Experimental reproductive effects. Mutagenic data.

Source: *Instant Tox-Base*, 1995, Instant Reference Sources, Inc., 7605 Rockpoint Drive, Austin, TX 78731 (Fax: 512-345-2386; Internet URL, http://www.instantref.com/inst-ref.htm).

Toxicology Information from EPA's Integrated Risk Information System (IRIS)

I. CHRONIC HEALTH HAZARD ASSESSMENTS FOR NONCARCINOGENIC EFFECTS

I.A. REFERENCE DOSE FOR CHRONIC ORAL EXPOSURE (RfD)

The Reference Dose (RfD) is based on the assumption that thresholds exist for certain toxic effects such as cellular necrosis, but may not exist for other toxic effects such as carcinogenicity. In general, the RfD is an estimate (with uncertainty spanning perhaps an order of magnitude) of a daily exposure to the human population (including sensitive subgroups) that is likely to be without an appreciable risk of deleterious effects during a lifetime. Please refer to Background Document 1 in Service Code 5 for an elaboration of these concepts. RfDs can also be derived for the noncarcinogenic health effects of compounds which are also carcinogens. Therefore, it is essential to refer to other sources of information concerning the carcinogenicity of this substance. If the US EPA has evaluated this substance for potential human carcinogenicity, a summary of that evaluation will be contained in Section II of this file when a review of that evaluation is completed.

I.A.1. ORAL RfD SUMMARY

Critical Effect	Experimental Doses*	UF	MF	RfD
Excessive loss of litters	NOEL: 5.01 mg/kg/day	1000	1	5E-3 mg/kg/day
	LEL: 35.5 mg/kg/day			

*Conversion Factors: Actual dose tested Rabbit Teratology Study, Kincaid Enterprises, 1986

I.A.2. PRINCIPAL AND SUPPORTING STUDIES (ORAL RfD)

Kincaid Enterprises, Inc. 1986. MRID No. 0015992. Available from EPA. Write to FOI, EPA, Washington, DC 20460.

Young adult female New Zealand White rabbits were randomized by a computerized process which assigned 17 animals each into 3 dose groups, 5.01, 35.5, and 251.0 mg/kg/day, and a control (a total of 68 animals). The females were artificially inseminated and the day of insemination considered as gestation day 0. All animals were dosed from days 7 through 19 of gestation. Animals were observed twice daily for mortality and moribundity, further they were observed once daily for clinical signs of toxicity. Individual body weights were taken on gestation days 0, 7, 10, 14, 20, 24, and 29. All surviving dams were sacrificed on gestation day 29.

Maternal toxicity was observed as excessive loss of litters (abortions) in the mid- and high-dose groups along with statistically significant decreases in body weight gain during the dosing period for both mid- and high-dose groups and in the mid dose following the dosing period and overall for the gestation period (the high dose was not analyzed due to total loss of litters). There also was an increase in clinical signs in both the mid- and high-dose groups; the deaths at the high dose were attributed to compound administration. The high incidence of lung agenesis noted in fetuses of all dose groups was unusual. No specific toxicity was noted in the low dose (5.01 mg/kg/day).

The tentative LEL for maternal toxicity is 35.5 mg/kg/day based on excessive loss of litters. The tentative NOEL for maternal toxicity is 5.01 mg/kg/day.

I.A.3. UNCERTAINTY AND MODIFYING FACTORS (ORAL RfD)

UF -- An uncertainty factor of 100 was used to account for the inter-and intraspecies differences. An additional UF of 10 was used to account for the poor quality of the critical study and for the incompleteness of the data base on chronic toxicity.

MF -- None

I.A.4. ADDITIONAL COMMENTS (ORAL RfD)

Methoxychlor is considered to have an estrogenic activity. Several recent papers in the open literature have addressed this action of methoxychlor. Kupfer and Bulger (1987) found that both methoxychlor and metabolites have estrogen-like activity with several metabolites having proestrogen activity. They used an in vitro system involving rat liver microsomes and NADPH for a metablizing system with estrogen receptors from immature rat uteri as a detection system.

Gray et al. (1989) investigated the effects of methoxychlor on the pubertal development and reproductive function in the male and female rat (Long-Evans hooded) by dosing rats from gestation, weaning, lactation, through puberty with either 25, 50, 100, or 200 mg/kg/day of methoxychlor. In females they found an acceleration of vaginal opening, abnormal estrus cycle, inhibition of luteal function and a blockage of implantation. In males they found an inhibition of somatic growth and accessory gland weight, elevated pituitary and serum prolactin levels, and a suppression of testicular Leydig cell function. Some of these effects occurred at levels as low as 25 mg/kg/day. These observations are consistent with the earlier reports that Methoxychlor mimics estrogen both in vivo and in vitro.

Goldman et al. (1986) investigated the subchronic effects of methoxychlor on the rat (Long-Evans hooded) reproductive system by dosing for 8 weeks with 25 or 50 mg/kg of methoxychlor by oral gavage. No effect was observed on the pituitary weight, serum LH, FSH, or prolactin levels and the pituitary LH of FSH concentrations. Pituitary prolactin levels were increased at both levels. There was an increase in GnRH levels in the mediobasal hypothalamus at the high-dose level. The authors determined that the reproductive effects of methoxychlor are mediated in part by an increase in prolaction release which   in turn influences the hypothalamic levels of GnRH.   This may be considered an early effect of methoxychlor on the rat reproductive system.

Cummings and Gray (1987) of the US EPA Health Effects Research Laboratory found that methoxychlor affects the decidual cell response of the rat uterus, suggesting a direct effect of the compound on the uterus with no effects on   uterine weight, serum progesterone levels, or corpora lutea maintenance.   Long-term exposure to methoxychlor reduced fertility and induced fetotoxicity.   The effects of reduced fertility and fetotoxicity were noted in a 3-generation reproduction study (see study #4).   Although the available data for these 3   studies were limited, it is apparent that methoxychlor at 1000 ppm produced reproductive effects in the form of reduced fertility index, reduced litter size, and reduced viability index.

Khera et al. (1978) on the teratogenicity of methoxychlor found that treatment   of pregnant rats with either technical grade or formulation of methoxychlor   produced maternal toxicity in the form of reduced body weight gain at all doses tested (50 to 400 mg/kg/day).   Developmental toxicity was noted as fetotoxicity at doses of 200 and 400 mg/kg/day and as a dose-related increase of wavy ribs at 100, 200, and 400 mg/kg/day.

A 2-year chronic rat study by du Pont de Nemours & Co. (1951) reported a systemic NOEL of 100 ppm (5 mg/kg/day); a 2-year chronic study by Hodge, et al. (1952) reported a systemic NOEL of 200 ppm (10 mg/kg/day).   Altough these studies are not definitive, they, along with the submitted studies from the registrant, support the NOEL of 5.01 mg/kg/day used for the calculation of the RfD for methoxychlor.

Data Considered for Establishing the RfD

1) Teratology - rabbit: Principal study - see previous description; core   grade supplementary (Kincaid Enterprises, Inc., 1986)

2) Teratology - rat: Dietary levels tested: 0, 200, 500, and 1250 ppm (10, 25, and 62.5 mg/kg/day); Female ChR-CD albino rats (animals were received   pregnant) were administered methoxychlor in the diet on gestation days 6   through 15.   There was maternal toxicity in the mid- and high-dose groups in the form of reduced body weight gain, reduced food consumption, increased   postimplantation loss, and a decreased number of liver fetuses per dam.   There was 1 and 2 dams in the mid- and high-dose groups, respectively, with total   resorptions of litters.   The increase in postimplantation loss resulted in a decrease in the litter size in the mid-and high-dose groups.   There was an   indication of 4 runts in one litter in the mid dose group, however, there was   no change in the mean fetal weight among dose groups.   The mid- and high-dose group had statistically significantly increased numbers of litters with wavy ribs.   Study deficiencies included the following: no individual animal data were provided; animals were received pregnant; and although dosing was by feed, the concentration analysis of the diet, diet preparation schedule, and stability of the test compound in the diet mixtures was not provided.   Therefore the tentative LEL is 500 ppm (25 mg/kg/day) based on the above effects.   The tentative NOEL is 200 ppm (10 mg/kg/day).; core grade supplementary (E.I. du Pont de Nemours and Co,, Inc., 1976a)

3) Teratology - rat: Dietary levels tested: 0, 34.6, 138.4, 242.2, and 346.0 mg/kg/day; Female Sprague-Dawley rats were dosed by gavage from gestation day 6 through 15. Control animals received corn oil in equivalent volumes to the test material which was administered at the high dose. There was evidence of reduced body weight gain at all doses tested. Further, at 138.4 mg/kg/day and above there was an increased number of resorptions, dead fetuses, and increased postimplantation loss. There was evidence of altered growth in the form of delayed ossification of skull bones and sternebrae and the reduced fetal body weight at the high dose. All doses tested had an increased incidence of hydronephrosis, and reduced or no ossification of skull bones, sternebrae and vertebrae along with wavy ribs. Study deficiencies include lack of stability and concentration analysis, dosing data, summary litter incidence, and maternal examination data. Based on the above effects observed at the lowest dose tested, the tentative LEL for maternal and developmental toxicity is 34.6 mg/kg/day. An NOEL for maternal and developmental toxicity could not be established.; core grade supplementary (Chemical Formulators, Inc. 1976b)

4) 3-Generation Reproduction - rat: Dietary levels tested: 0, 200, and 1000 ppm (0, 10, and 50 mg/kg/day); Male and female ChR-CD rats were administered methoxychlor in the diet for three generations. Three separate studies were conducted and reported in this study. The first reproduction study used dose levels of 0 and 200 ppm and the second reproduction study used dose levels of 0, 0 (2 control groups) and 1000 ppm. The third study was a pair feeding study with rats given 1000 ppm. The available data was limited for these 3 studies, however, it is apparent that methoxychlor at 1000 ppm produced reproductive effects in the form of reduced fertility index, reduced litter size, and reduced viability index. There was evidence of possible systemic toxicity at the 200 ppm dose, however, there was also evidence of reduced food consumption. Therefore the tentative NOEL and LEL are 200 ppm (10 mg/kg/day) and 1000 (50 mg/kg/day), respectively.; core grade supplementary (E.I. du Pont de Nemours and Co., Inc., 1966)

Other Data Reviewed:

1) Carcinogenicity Study - rat: Dietary levels tested: Male: 0, 360, 500, 720, and 1000 ppm (0, 18, 25, 36, and 50 mg/kg/day); Female: 0, 750, and 1500 ppm (0, 37.5, and 75 mg/kg/day); Male and female Osborne-Mendel rats were administered methoxychlor in the diet for 2 years. The initial dose levels for males were 360 and 720 ppm but were increased to 500 and 1000 ppm after week 30. Based on the data provided in this study, there is no substantial evidence that the MTD had been reached. The reduced male and female body weights noted in treated groups may be due to reduced food consumption (no food consumption data provided), also other studies with methoxychlor indicate that mixing the compound in the food tends to reduce food consumption and therefore weight.; core grade supplementary (US Department of Health, Education, and Welfare, 1977a)

2) Carcinogenicity Study - mouse: Dietary levels tested: Male: 0, 1400, 1750, 2800, and 3500 ppm (0, 210, 262.5, 420, and 525 mg/kg/day); Female: 0, 750, 1000, 1500, and 2000 ppm (0, 112.5, 150, 225, and 300 mg/kg/day); Male and female B6C3F1 were administered methoxychlor in the diet for 78 weeks. The initial dose levels for males were 1400 and 2800 ppm while females initially received 750 and 1500 ppm. After week two, doses were increased to 1750 and 3500 ppm for males and to 1000 and 2000

ppm for females. Based on the data provided in this study, there is no substantial evidence that the MTD had been reached. The reduced body weights noted in treated males (high dose only) and in treated females (all dose levels) may be due to reduced food consumption (no food consumption data provided). Other studies with methoxychlor indicate that mixing the compound in the food tends to reduced food consumption and therefore weight.; core grade supplementary (US Department of Health, Education, and Welfare, 1977b)

Data Gap(s): Chronic Rat Feeding/Carcinogenicity Study; Chronic Dog Feeding Study; Rat Reproduction Study; Rat Developmental toxicity Study; Rabbit Developmental toxicity Study; Chronic Mouse Feeding/Carcinogenicity Study

I.A.5. CONFIDENCE IN THE ORAL RfD

Study -- Low Data Base -- Low RfD -- Low

The critical study is given a low confidence rating since no conclusions could be made relative to the maternal or developmental toxicity of Methoxychlor due to the total loss of litters in the high-dose group and the small number of litters available for evaluation in the mid-dose group. The data base is given a low confidence rating because of the lack definitive chronic toxicity studies. Low confidence in the RfD follows.

I.A.6. EPA DOCUMENTATION AND REVIEW OF THE ORAL RfD

Source Document -- This assessment is not presented in any existing US EPA document.

Other EPA Documentation -- Pesticide Registration Standard, August 1988; Pesticide Registration Files

Agency Work Group Review -- 04/18/90, 05/17/90, 06/21/90

Verification Date -- 06/21/90

I.A.7. EPA CONTACTS (ORAL RfD)

George Ghali / OPP -- (703)557-7490

William Burnam / OPP -- (703)557-7491

I.B. REFERENCE CONCENTRATION FOR CHRONIC INHALATION EXPOSURE (RfC)

The health effects data for methoxychlor were reviewed by the US EPA RfD/RfC Work Group and determined to be inadequate for the derivation of an inhalation RfC. The verification status for this chemical is currently NOT VERIFIABLE. For additional information on the health effects of this chemical, interested parties are referred to the US EPA documentation listed below.

NOT VERIFIABLE status indicates that the US EPA RfD/RfC Work Group deemed the database at the time of review to be insufficient to derive an inhalation RfC according to the Interim Methods for Development of Inhalation Reference Concentrations (US EPA, 1990). This status does not preclude the use of information in cited references for assessment by others.

Derivation of an inhalation RfC for methoxychlor is not recommended at this time. No adequate long-term studies examining the effects of inhalation exposure to methoxychlor exist. No inhalation pharmacokinetic data exist for this compound. No data exist to definitively rule out portal-of-entry effects. The requirements for a minimal database have not been met (US EPA, 1990).

Methoxychlor [2,2-bis(4-methoxyphenyl)-1,1,1-trichloroethane], also known as methoxy-DDT, is a pale yellow, crystalline organochlorine insecticide. It is used principally as a larvacide. Vapor pressure data on methoxychlor are not available. Methoxychlor is the p-methoxy derivative of the insecticide dichlorodiphenyltrichloroethane (DDT). Technical grade methoxychlor contains approximately 88% methoxychlor, with the remaining 12% comprising at least 50 impurities.

The only study of methoxychlor using inhalation exposure is that of Haag et al. (1950) who exposed two dogs, two rabbits, and 10 rats to an atmosphere containing micronized dust in which 10% recrystallized methoxychlor was mixed with Pyrax (composition not described) plus 3% Santo-Cel (a dehydrated silica gel) for 2 hours/day, 5 days/week. Concentrations and duration of exposures for three replicate experiments were reported as 300 mg/cu.m for 4 weeks, 360 mg/cu.m for 4 weeks, and 430 mg/cu.m for 5 weeks. This diluent was itself toxic and caused death and weight changes in the control dogs and rats at about the same incidence as the group exposed to methoxychlor. Further, it is not clear from the report whether the amount of diluent was normalized for all the exposed groups. The toxicity of DDT was also investigated in this study. DDT and methoxychlor were of comparable toxicity in dogs and rabbits, but methoxychlor was less toxic than DDT in rats.

No reliable information is available on the effects of methoxychlor in humans, via inhalation or oral exposure. Ziem (1982) reported the case of a 49-year-old man who suffered from fatigue and bruising several weeks after he used a tomato dust pesticide containing methoxychlor. Two months after exposure he was diagnosed with aplastic anemia, and he died within 6 months. The man was well and had not been taking any drugs prior to exposure to methoxychlor. This is the only case of aplastic anemia reported in association with exposure to methoxychlor. Lehman (1949, as cited in US EPA, 1987a) estimated that the lethal oral dose of methoxychlor in humans is 450 g (6.4 g/kg for a 70 kg human). Stein (1970) reported the results of an experiment in which 16 human volunteers (prisoners) were orally administered either 0.5, 1.0, or 2.0 mg/kg methoxychlor for 5 to 8 weeks. Histopathological examination of biopsies of several tissues (liver, fat, bone marrow, and testicle) evidenced no abnormality. No weight disturbances or changes in clinical pathology (parameters measured not specified) were noted in the treated volunteers. Stein (1970) also reported the results of a study in which Sprague-Dawley rats (number not indicated) and Rhesus monkeys (3/group, sex not indicated) were administered 400-2500 mg/kg methoxychlor in 1% gum tragacanth by gavage for approximately 3 months (rats) or 6 months (monkeys). The rats demonstrated a dose-related depression in body weight gain after 4-6 weeks of treatment, but no weight disturbances were observed in the monkeys. No treatment-related effects on any of the clinical chemistry parameters measured were noted in either the rats or the monkeys. Similarly, no gross or microscopic evidence of treatment-related pathology was noted in either the rats or the monkeys. A decrease in hepatic triglycerides in both rats and monkeys was noted.

Several investigators have demonstrated that methoxychlor and its metabolites possess estrogenic properties (Bulger et al., 1978; Kupfer and Bulger, 1987). These estrogenic effects are manifested by changes in both male and female reproductive function and morphology in rodents. Administration of methoxychlor at rather high doses by gavage, in feed, or parenterally has been reported to stimulate the development of the reproductive tract in neonatal female rodents and their offspring, as evidenced by early vaginal opening, vaginal cornification, and an increase in the weight of reproductive organs (i.e., ovary and uterus) (Bulger et al., 1978; Eroschenko and Cooke, 1990; Gray et al., 1989; Harris et al., 1974). Methoxychlor administered to mature female rodents has been reported to inhibit reproductive function, as evidenced by inhibited folliculogenesis and atresia of follicles (Bal, 1984); decreased fertility; reduced implantations; and abnormal estrous cyclicity and/or persistent vaginal estrus (Gray et al., 1988, 1989; Martinez and Swartz, 1991). Atypical cell growth has also been noted in the uterus and oviducts (Eroschenko and Cooke, 1990; Gray et al., 1988). In a series of experiments conducted by Cummings and coworkers, it was demonstrated that the estrogenic, antifertility effects of methoxychlor are mediated in part by a direct effect on the uterus to suppress decidualization (Cummings and Gray, 1987), by suppression of serum progesterone levels (Cummings and Gray, 1989), and by accelerated transport of fertilized ova through the oviducts resulting in a loss of viable embryos that could account for the increase in preimplantation loss observed with methoxychlor (Cummings and Perreault, 1990). The estrogenic effects of methoxychlor have also been observed with regard to behavior. Behaviors thought to be mediated by estrogen (running wheel activity and sexual behavior) were enhanced in intact and ovariectomized female rats treated with methoxychlor, and the enhanced behaviors were suppressed by progesterone, which is known to block the effects of estrogen (Gray et al., 1988).

Effects on male reproductive function have also been reported following the administration of methoxychlor to rodents. Bal (1984) reported inhibited spermatogenesis, degeneration of spermatogonia and spermatocytes, and cytoplasmic vacuolation in the epithelium of the ductus epididymis in male rats following the administration of 100-200 mg/kg/day methoxychlor. A decrease in seminal vesicle and caudal epididymal weight and caudal sperm count as well as delayed puberty were observed in neonatal rats administered 25-200 mg/kg/day methoxychlor for one generation, indicating that the endocrine function of the testes and pituitary gland were affected (Gray et al., 1989). Cooke and Eroschenko (1990) also noted that the development of the neonatal male rat reproductive tract was inhibited by methoxychlor administration, as evidenced by a decrease in serum testosterone levels and decreased DNA content of the seminal vesicles, bulbourethral glands, and the ventral prostate. Rats fed 2000 ppm methoxychlor for 90 days exhibited decreased prostate size and cell content (Shain et al., 1977). Goldman et al. (1986) hypothesized that part of methoxychlor's effects on male reproductive function may be mediated by a prolactinemic effect since administration of 25 or 50 mg/kg/day methoxychlor to 21-day-old male rats caused an increase in serum prolactin levels and an increase in hypothalamic-gonadotropin-releasing hormone levels.

Methoxychlor has been demonstrated to be fetotoxic. Khera et al. (1978) studied the effects of oral administration of 50, 100, 200, or 400 mg/kg/day methoxychlor to pregnant Wistar rats on gestational days 6-15. Two formulations of methoxychlor were used: (1) technical grade and (2) a formulation that was 50% methoxychlor (the composition of the remaining 50% was unknown). Maternal body weight gain was depressed in all treatment groups and remained depressed after removal of the uterine contents, implying an adverse effect on the dam. Treatment with either formulation of methoxychlor at the two highest doses resulted in a reduced number of rats with live fetuses at term and a reduced number of live fetuses per pregnancy. Reduced fetal weight gain was observed at the two highest dose levels with both formulations. An increased incidence of fetal skeletal anomalies (mostly wavy ribs) was observed at the two highest dose levels with both formulations.

Several chronic oral carcinogenicity bioassays have been conducted with methoxychlor (see review by Reuber, 1980), the results of which have been equivocal such that methoxychlor has yet to be classified as a carcinogen by the US EPA. Aside from a depression in body weight gain observed in both rats fed at 1500 ppm and mice fed at 1994 ppm in a 2-year study conducted by NCI, no dose-related nonneoplastic effects were reported in these studies. Deichmann et al. (1967), however, fed 1000 ppm methoxychlor for 27 months to Osborne-Mendel rats and reported other nonneoplastic hepatic effects, including decreased absolute weight accompanied by hydropic swelling and some necrosis and congestion. Reuber (1980) reevaluated the slides from the carcinogenicity study of miniature swine and described the occurrence of moderate hyperplasia of the mammary gland with milk-like secretion, hyperplasia of the uterus, and chronic interstitial renal fibrosis. These lesions are similar to those observed in the subchronic study in swine reported by Stein (1970) and Tegeris et al. (1966) and may be interpreted to be due to the estrogenic properties of methoxychlor.

A series of studies conducted in dogs and swine indicates that the two species respond differently with respect to the toxicity of methoxychlor (Stein, 1970; Tegeris et al., 1966). Technical grade methoxychlor (1, 2, or 4 g/kg) was administered in the feed 7 days/week to groups of six animals each (with 12 animals serving as controls) for up to 6 months. Clinical examinations were conducted daily, weights were recorded weekly, and blood samples were taken for hematological and clinical chemical analyses at 6-week intervals throughout the experiment. Bone marrow morphology and complete necropsies, with histopathological evaluation of approximately 18 tissues, were conducted at study termination. All dogs that were fed methoxychlor lost weight throughout the experiment, but, after an initial 8-week weight loss, the swine receiving the two lower doses of methoxychlor began to gain weight, whereas the high-dose swine continued to lose weight. Most of the medium-dose (5/6) and all of the high-dose dogs (6/6) began exhibiting clinical signs of toxicity after 6 weeks of treatment. Symptoms included nervousness and apprehension, progressing to salivation, fasciculations, tremors, hyperesthesia, mydriasis, tonic seizures, and tetanic convulsions. Most of these dogs died 3 weeks thereafter. The swine exhibited no clinical signs of toxicity. No treatment-related changes in any of the hematological parameters studied were noted in either the dogs or the swine. The dogs exhibited dose- dependent elevations in SGOT, SGPT, and alkaline phosphatase (AP). At 24- weeks exposure, the enzyme values of the high-exposure group relative to control values were increased eightfold for SGOT, 30-fold for SGPT, and 30- fold for AP, whereas the swine exhibited only a two-fold increase in BUN. The only changes attributed to methoxychlor noted at gross and microscopic examination in dogs (including the liver) were a dose-dependent absence of adipose tissue from the normal depots and congestion of the small intestinal mucosa (without accompanying histopathology). In the swine, advanced chronic renal nephritis, hyperplastic and hypertrophic mammary glands, and hypertrophic uteri were noted in the treated animals. These latter effects on sex organs are most likely due to the estrogenic properties of methoxychlor.

Very little quantitative information is available on the toxicokinetics of methoxychlor, and the available information is for oral or parenteral routes of exposure only. Absorption of methoxychlor from the gastrointestinal tract can be inferred from the observation of toxic effects following oral administration. Kapoor et al. (1970) administered radiolabeled methoxychlor to mice and found that 98.3% of the administered radioactivity was eliminated within 24 hours, mostly in the feces. A number of studies show that methoxychlor does not accumulate in the body to any appreciable degree (e.g., Villeneuve et al., 1972), but accumulation of methoxychlor in fat has been observed following administration of very high dietary levels of methoxychlor (US EPA, 1987b). Methoxychlor is metabolized in the liver to readily excretable polar compounds (US EPA, 1987b). Methoxychlor and 26 metabolites were identified in the feces, urine, and bile of intact, colostomized, and bile-fistulated chickens orally administered methoxychlor (Davison et al., 1984). Lactating goats also eliminate methoxychlor and its metabolites primarily in the feces (Davison et al., 1982). The results of studies by Villeneuve et al. (1972) indicate that methoxychlor does not induce hepatic microsomal enzymes.

Bal, H.S. 1984. Effect of methoxychlor on reproductive systems of the rat (41861). Proc. Soc. Exp. Biol. Med. 176(2): 187-196.

Bulger, W.H., R.M. Muccitelli, and K. Kupfer. 1978. Studies on the in vivo and in vitro estrogenic activities of methoxychlor and its metabolites. Role of hepatic mono-oxygenase in methoxychlor activation. Biochem. Pharmacol. 27(20): 2417-2423.

Cooke, P.S. and V.P. Eroschenko. 1990. Inhibitory effects of technical grade methoxychlor on development of neonatal male mouse reproductive organs. Biol. Reprod. 42(3): 585-596.

Cummings, A.M. and L.E. Gray. 1987. Methoxychlor affects the decidual cell response of the uterus but not other progestational parameters in female rats. Toxicol. Appl. Pharmacol. 90(2): 330-336.

Cummings, A.M. and L.E. Gray. 1989. Antifertility effect of methoxychlor in female rats -- dose- and time-dependent blockade of pregnancy. Toxicol. Appl. Pharmacol. 97(3): 454-462.

Cummings, A.M. and S.D. Perreault. 1990. Methoxychlor accelerates embryo transport through the rat reproductive tract. Toxicol. Appl. Pharmacol. 102(1): 110-116.

Davison, K.L., V.J. Feil, and C.H. Lamoureux. 1982. Methoxychlor metabolism in goats. J. Agric. Food Chem. 30(1): 130-137.

Davison, K.L., C.H. Lamoureux, and V.J. Feil. 1984. Methoxychlor metabolism in chickens. J. Agric. Food Chem. 32(4): 900-908.

Deichmann, W.B., M. Keplinger, F. Sala, and E. Glass. 1967. Synergism among oral carcinogens. IV. The simultaneous feeding of four tumorigens to rats. Toxicol. Appl. Pharmacol. 11(1): 88-103.

Eroschenko, V.P. and P.S. Cooke. 1990. Morphological and biochemical alterations in reproductive tracts of neonatal female mice treated with the pesticide methoxychlor. Biol. Reprod. 42(3): 573-583.

Goldman, J.M., R.L. Cooper, G.L. Rehnberg, J.F. Hein, W.K. McElroy, and L.E. Gray Jr. 1986. Effects of low subchronic doses of methoxychlor on the rat hypothalamic-pituitary reproductive axis. Toxicol. Appl. Pharmacol. 86(3): 474-483.

Gray, L.E., Jr., J.S. Ostby, J.M. Ferrell, E.R. Sigmon, and J.M. Goldman. 1988. Methoxychlor induces estrogen-like alterations of behavior and the reproductive tract in the female rat and hamster: Effects on sex behavior, running wheel activity, and uterine morphology. Toxicol. Appl. Pharmacol. 96(3): 525-540.

Gray, L.E. Jr., J. Ostby, J. Ferrell et al. 1989. A dose-response analysis of methoxychlor-induced alterations of reproductive development and function in the rat. Fund. Appl. Toxicol. 12(1): 92-108.

Haag, H.B., J.K. Finnegan, P.S. Larson, W. Riese, and M.L. Dreyfuss. 1950. Comparative chronic toxicity for warm-blooded animals of DDT and DMDT (methoxychlor). Arch. Int. Pharmacodyn. 83(4): 491-504.

Harris, S.J., H.C. Cecil, and J. Bitman. 1974. Effect of several dietary levels of technical methoxychlor on reproduction in rats. J. Agric. Food Chem. 22(6): 969-973.

Kapoor, I.P., R.L. Metcalf, R.F. Nystrom, and G.K. Sangha. 1970. Comparative metabolism of methoxychlor, methiochlor, and DDT in mouse, insects, and in a model ecosystem. J. Agr. Food Chem. 18(6): 1145-1152.

Khera, K.S., C. Whalen, and G. Trivett. 1978. Teratogenicity studies on linuron, malathion, and methoxychlor in rats. Toxicol. Appl. Pharmacol. 45(2): 435-444.

Kupfer, D. and W.H. Bulger. 1987. Metabolic activation of pesticides with proestrogenic activity. Fed. Proc. 46(5): 1864-1869.

Martinez, E.M. and W.J. Swartz. 1991. Effects of methoxychlor on the reproductive system of the adult female mouse. I. Gross and histologic observations. Reprod. Toxicol. 5(2): 139-147.

Reuber, M.D. 1980. Carcinogenicity and toxicity of methoxychlor. Environ. Health Perspect. 36: 205-219.

Shain, S.A., J.C. Shaeffer, and R.W. Boesel. 1977. The effect of chronic ingestion of selected pesticides upon rat ventral prostate homeostasis. Toxicol. Appl. Pharmacol. 40(1): 115-130.

Stein, A.A. 1970. Comparative toxicology of methoxychlor. Pestic. Symp. Collect. Pap. Inter-Amer. Conf. Toxicol. Occup. Med., 6th. p. 225-229.

Tegeris, A.S., F.L. Earl, H.E. Smalley, Jr., and J.M. Curtis. 1966. Methoxychlor toxicity. Comparative studies in the dog and the swine. Arch. Environ. Health. 13(6): 776-787.

US EPA. 1987a. Drinking Water Criteria Document for Methoxychlor. Prepared by the Office of Health and Environmental Assessment, Environmental Criteria and Assessment Office, Cincinnati, OH, for the Office of Drinking Water, Washington, DC.

US EPA. 1987b. Health Advisories for 16 Pesticides (including Alachlor, Aldicarb, Carbofuran, Chlordane, DBCP, 1,2-Dichloropropane, 2,4-D, Endrin, Ethylene Dibromide, Heptachlor/Heptachlor Epoxide, Lindane, Methoxychlor, Oxamyl, Pentachlorophenol, Toxaphene, and 2,4,5-TP). Office of Drinking Water, Washington, DC.

US EPA. 1990. Interim Methods for Development of Inhalation Reference Concentrations, (Review Draft), Office of Research and Development, Washington, DC. EPA/600-8-90-066A. August, 1990.

Villeneuve, D.C., D.L. Grant, and W.E.J. Phillips. 1972. Modification of pentobarbital sleeping times in rats following chronic PCB ingestion. Bull. Environ. Contam. Toxicol. 7(5): 264-269.

Ziem, G. 1982. Aplastic anaemia after methoxychlor exposure [letter]. The Lancet. 2(8311): 1349.

Agency Work Group Review -- 11/07/91

EPA Contacts:

Gary L. Foureman / NCEA -- (919)541-1183

Annie M. Jarabek / NCEA -- (919)541-4847

II. CARCINOGENICITY ASSESSMENT FOR LIFETIME EXPOSURE

Section II provides information on three aspects of the carcinogenic risk assessment for the agent in question; the US EPA classification, and quant- itative estimates of risk from oral exposure and from inhalation exposure. The classification reflects a weight-of-evidence judgment of the likelihood that the agent is a human carcinogen. The quantitative risk estimates are presented in three ways. The slope factor is the result of application of a low-dose extrapolation procedure and is presented as the risk per (mg/kg)/day. The unit risk is the quantitative estimate in terms of either risk per ug/L drinking water or risk per ug/cu.m air breathed. The third form in which risk is presented is a drinking water or air concentration providing cancer risks of 1 in 10,000, 1 in 100,000 or 1 in 1,000,000. Background Document 2 (Service Code 5) provides details on the rationale and methods used to derive the carcinogenicity values found in IRIS. Users are referred to Section I for information on long-term toxic effects other than carcinogenicity.

II.A. EVIDENCE FOR CLASSIFICATION AS TO HUMAN CARCINOGENICITY

II.A.1. WEIGHT-OF-EVIDENCE CLASSIFICATION

Classification -- D; not classified as to human carcinogenicity

Basis -- Human data are unavailable, and animal evidence is inconclusive.

II.A.2. HUMAN CARCINOGENICITY DATA

None.

II.A.3. ANIMAL CARCINOGENICITY DATA

A number of chronic dietary studies have been done to test the carcinogenicity of methoxychlor in rats and mice (Nelson and Fitzhugh, 1951; Hodge et al., 1952, 1966; Radomski et al., 1965; Davis, 1969; Deichmann et al., 1967; NCI, 1978). In addition, two limited studies using mice (Hodge et al., 1966) have been performed by subcutaneous administration and skin application. Reuber (1980) reviewed these chronic studies, reevaluating raw data and the histological sections when possible.

In the Nelson and Fitzhugh (1951) study, Osborne-Mendel rats (12 rats/sex/group) ingested 0, 10, 25, 100, 200, 500 or 2000 ppm methoxychlor in the diet for 2 years. Animals were examined for gross lesions. Histological preparations were made only from the gross lesions found at autopsy. In the highest dose group four hepatic cell adenomas were observed, but this was not a statistically significant increase. No other changes or malignant lesions were noted in other organs. In his review of this study, Reuber (1980) concluded that the incidence of hepatic neoplasms in the treated animals was significantly greater than that in controls when hyperplastic nodules were included.

Groups of 25 male and 25 female rats (strain not specified) ingested 0, 25, 200, or 1600 ppm methoxychlor in the diet for 2 years (Hodge et al., 1952). At the end of 2 years, surviving animals were killed and many organs were examined grossly and histopathologically. In treated female rats, a greater number of total tumors was observed compared with controls. The authors considered this increase to be of no biological relevance because there was no significant increase in tumors of any one organ. Interpretation of these results is limited by the fact that many of the animals were not accounted for at the end of the study and that the liver was not routinely examined histologically.

Radomski et al. (1965) administered methoxychlor for 2 years in the diet at levels of 0 and 80 ppm to Osborne-Mendel rats (30 rats/sex/group). No increase in tumor incidence was found in the treated rats as compared with controls. Methoxychlor was also administered under the same regimen in a mixture with aramite, DDT, and thiourea at concentrations of 50 ppm each to 50 rats/sex/group. In this study an apparent increase in total tumors was observed in animals treated with the mixture as compared to controls.

Deichmann et al. (1967) administered methoxychlor in the diet to Osborne- Mendel rats (30/sex/dose) at levels of 0 and 1000 ppm for 27 months. The concentration was chosen to be 50% of the highest dose reported in the Nelson and Fitzhugh study (1951). An increase in the number of total tumors was observed in treated males as compared with controls, but the increase was not statistically significant.

NCI (1978) tested groups of 50 male and 50 female Osborne-Mendel rats and 50 male and 50 female B6C3F1 mice. Control groups of each species consisted of 20 males and 20 females. Rats were exposed to technical grade methoxychlor (95% pure) in the diet for 78 weeks, followed by a 33-week observation period without exposure to the test compound. Concentrations given the low-dose male rats were 360 ppm for the first 29 weeks followed by 500 ppm for the next 49 weeks. The high-dose group was given 720 ppm for 29 weeks, 1000 ppm for the following 29 weeks, then 1000 ppm administered in a cyclic pattern for 20 weeks of one dosage-free week followed by 4 weeks of treatment. The low-dose female rats were given 750 ppm for the entire 78 weeks. The high-dose group received 1500 ppm for 55 weeks followed by 23 weeks of the cyclic pattern of administration at the same concentration. The time-weighted average (TWA) concentration for the high- and low-dose groups, respectively, was 845 and 448 mg/kg for the male rats and 1385 and 750 mg/kg for female rats, respectively.

Male mice were given a concentration of 1400 ppm for 1 week, then 1750 ppm for 77 weeks or 2800 ppm for 1 week, then 3500 ppm for 77 weeks. Female mice were given concentrations of 750 ppm for 1 week, then 1000 ppm for 77 weeks or 1500 ppm for 1 week, then 2000 ppm for 77 weeks. The mice were observed for an additional 15 weeks with no methoxychlor treatment. The TWA concentration for high- and low-dose groups, respectively, was 3491 and 1746 mg/kg for the male mice and 1994 and 997 mg/kg for female mice. Necropsy was performed on all animals that died spontaneously or were killed when moribund or at the termination of the study. Histological examinations were performed on major organs and on any gross lesions of all animals, except where cannibalism or autolysis precluded such studies.

The only tumors observed at a higher incidence than in controls were hemangiosarcomas in male rats (1/20 control, 9/50 low-, 2/50 high-dose groups). Although historically this tumor type is not frequently observed in this strain or rats, the authors concluded that the increase was not a good indicator of the carcinogenicity of methoxychlor because the response was neither dose-related nor statistically different from control values. Other tumors observed in the treated rats also occurred in the controls at the same frequency. NCI concluded that under this experimental regimen, methoxychlor was not tumorigenic to Osborne-Mendel rats. In mice, a variety of tumors was observed, but the incidence was similar in both control and experimental groups. Recent reviews by Greiesemer and Cueto (1980) and Harper et al. (1982) indicated that the bioassays did not meet the current criteria for maximum tolerated doses and so were not powerful enough to detect carcinogenicity. The evidence of carcinogenicity was, therefore, judged to be inconclusive, rather than negative.

In the Davis (1969) study, male and female BALB/c and C3H mice (100/sex/strain) were fed diets containing 0 or 750 ppm methoxychlor for 2 years. Liver tumors were found in male and female BALB/c mice and in male C3H mice. Carcinomas of the testes were observed in male BALB/c mice. It was the author's preliminary judgment that the data did not show that methoxychlor was carcinogenic but suggested that a more complete statistical analysis was needed. In reviewing the original data, Reuber (1980) concluded that the increased incidences of liver carcinoma in C3H males and in BALB/c males and females were statistically significant, as well as increases in testicular carcinoma in BALB/c males and neoplasms at all sites in male and female BALB/c mice.

Nelson and Radomski (1953) fed methoxychlor at a dose of 300 mg/kg/day to four dogs. Two of the dogs died early in the study, but two female dogs survived the dosing period of 3.5 years. Liver foci were observed in one dog, and the other was described as exhibiting slight fibrosis in the liver. Reuber (1980) reexamined the histological sections and reported that one dog had developed liver carcinoma. The small number of animals used in this study precludes any definitive interpretation of these findings.

There is considerable disagreement between Reuber and the original authors in the interpretation of the histology and data from several of the chronic studies. NCI (1978), IARC (1979), and US EPA (1983) have concluded that the experimental evidence does not support the contention that methoxychlor is a carcinogen. US EPA (1987) has suggested that the differences in the conclusions may be due in part to the difficulty in distinguishing between regenerative hyperplasia, hyperplastic nodules, benign neoplasia, and malignant neoplasia, as well as the use of inappropriate control data in some of Reuber's statistical analyses.

II.A.4. SUPPORTING DATA FOR CARCINOGENICITY

In mutagenicity assays, negative results were obtained (with or without metabolic activation) in bacteria, yeast, in assays of methoxychlor-induced DNA damage, or in assays of unscheduled DNA synthesis in mammalian cell cultures (Probst et al., 1981). A weakly positive increase was observed in a transformation study using BALB/3T3 cell line (Dunkel et al., 1981). Methoxychlor is a structural analog of DDT.

II.B. QUANTITATIVE ESTIMATE OF CARCINOGENIC RISK FROM ORAL EXPOSURE

[Ed. Note: EPA does not have information yet in this section of IRIS].

II.C.   QUANTITATIVE   ESTIMATE   OF   CARCINOGENIC   RISK   FROM
INHALATION EXPOSURE

[Ed. Note: EPA does not have information yet in this section of IRIS].

II.D. EPA DOCUMENTATION, REVIEW, AND CONTACTS (CARCINOGENICITY
ASSESSMENT)

II.D.1. EPA DOCUMENTATION

Source Document -- US EPA, 1987, 1983

The 1987 Drinking Water Criteria document received OHEA review. The Multimedia
Risk Assessment received Agency review.

II.D.2. REVIEW (CARCINOGENICITY ASSESSMENT)

Agency Work Group Review -- 10/07/87

Verification Date -- 10/07/87

II.D.3. US EPA CONTACTS (CARCINOGENICITY ASSESSMENT)

Dharm V. Singh / NCEA -- (202)260-5889

III. HEALTH HAZARD ASSESSMENTS FOR VARIED EXPOSURE DURATIONS

III.A. DRINKING WATER HEALTH ADVISORIES

The Health Advisory for methoxychlor has been withdrawn on 12/01/93. A revised
Health Advisory is in preparation by the Office of Water. For further details contact
Amal Mahfouz / OST -- (202)260-9568.

III.B. OTHER ASSESSMENTS

Content to be determined.

IV. US EPA REGULATORY ACTIONS

EPA risk assessments may be updated as new data are published and as assessment
methodologies evolve. Regulatory actions are frequently not updated at the same time.
Compare the dates for the regulatory actions in this section with the verification dates for
the risk assessments in sections I and II, as this may explain inconsistencies. Also note
that some regulatory actions consider factors not related to health risk, such as technical
or economic feasibility. Such considerations are indicated for each action. In addition,
not all of the regulatory actions listed in this section involve enforceable federal
standards. Please direct any questions you may have concerning these regulatory actions
to the US EPA contact listed for that particular action. Users are strongly urged to read

the background information on each regulatory action in Background Document 4 in Service Code 5.

IV.A. CLEAN AIR ACT (CAA)

[Ed. Note: EPA does not have information yet in this section of IRIS].

IV.B. SAFE DRINKING WATER ACT (SDWA)

IV.B.1. MAXIMUM CONTAMINANT LEVEL GOAL (MCLG) for Drinking Water

Value -- 0.04 mg/L (Final, 1991)

Considers technological or economic feasibility? -- NO

Discussion -- A MCLG of 0.04 mg/L is promulgated based on potential adverse effects (developmental toxicity) reported in a rabbit study. The MCLG is based upon a DWEL of 2 mg/L and an assumed drinking water contribution of 20 percent.

Reference -- 56 FR 3526 (01/30/91)

EPA Contact -- Health and Ecological Criteria Division / OST / (202) 260-7571 / FTS 260-7571; or Safe Drinking Water Hotline / (800) 426-4791

IV.B.2. MAXIMUM CONTAMINANT LEVEL (MCL) for Drinking Water

Value -- 0.04 mg/L (Final, 1991)

Considers technological or economic feasibility? -- YES

Discussion -- EPA has promulgated a MCL equal to the MCLG of 0.04 mg/L.

Monitoring requirements -- All systems monitored initially for four consecutive quarters every three years; repeat monitoring dependent upon detection, vulnerability status and size.

Analytical methodology -- Microextraction/gas chromatography (EPA 505); gas chromatography/electron capture detector (EPA 508); liquid-solid extraction and column gas chromatography/mass spectrometry (EPA 525). PQL=0.001 mg/L.

Best available technology -- Granular activated carbon

Reference -- 56 FR 3526 (01/30/91)

EPA Contact -- Drinking Water Standards Division / OGWDW / (202) 260-7575 / FTS 260-7575; or Safe Drinking Water Hotline / (800) 426-4791

IV.B.3. SECONDARY MAXIMUM CONTAMINANT LEVEL (SMCL) for Drinking Water

[Ed. Note: EPA does not have information yet in this section of IRIS].

IV.B.4. REQUIRED MONITORING OF "UNREGULATED" CONTAMINANTS

[Ed. Note: EPA does not have information yet in this section of IRIS].

IV.C. CLEAN WATER ACT (CWA)

IV.C.1. AMBIENT WATER QUALITY CRITERIA, Human Health

Water and Fish Consumption: 1.0E+2 ug/L

Fish Consumption Only: None

Considers technological or economic feasibility? -- NO

Discussion -- This value is the same as the drinking water standard and approximates a safe level assuming consumption of contaminated organisms and water.

Reference -- Quality Criteria for Water, July 1976 (PB-263943).

EPA Contact -- Criteria and Standards Division / OWRS (202)260-1315 / FTS 260-1315

IV.C.2. AMBIENT WATER QUALITY CRITERIA, Aquatic Organisms

Freshwater:
  Acute -- None
  Chronic -- 3.0E-2 ug/L

Marine:
  Acute -- None
  Chronic -- 3.0E-1 ug/L

Considers technological or economic feasibility? -- NO

Discussion -- Criteria were derived from a minimum data base consisting of acute tests on a variety of species. Requirements and methods are covered in the reference to the Federal Register.

Reference -- Quality Criteria for Water, July 1976 (PB-263943).

EPA Contact -- Criteria and Standards Division / OWRS (202)260-1315 / FTS 260-1315

IV.D. FEDERAL INSECTICIDE, FUNGICIDE, AND RODENTICIDE ACT (FIFRA)

IV.D.1. PESTICIDE ACTIVE INGREDIENT, Registration Standard

Status -- Issued (1988)

Reference -- Methoxychlor Pesticide Registration Standard. December, 1988 (NTIS No. PB89-138523).

EPA Contact -- Registration Branch / OPP (703)557-7760 / FTS 557-7760

IV.D.2. PESTICIDE ACTIVE INGREDIENT, Special Review

   [Ed. Note: EPA does not have information yet in this section of IRIS].

IV.E. TOXIC SUBSTANCES CONTROL ACT (TSCA)

   [Ed. Note: EPA does not have information yet in this section of IRIS].

IV.F. RESOURCE CONSERVATION AND RECOVERY ACT (RCRA)

IV.F.1. RCRA APPENDIX IX, for Ground Water Monitoring

Status -- Listed

Reference -- 52 FR 25942 (07/09/87)

EPA Contact -- RCRA/Superfund Hotline (800)424-9346 / (202)260-3000 / FTS 260-3000

IV.G. SUPERFUND (CERCLA)

IV.G.1. REPORTABLE QUANTITY (RQ) for Release into the Environment

Value (status) -- 1 pound (Final, 1989)

Considers technological or economic feasibility? -- NO

Discussion -- The final RQ for methoxychlor is based on aquatic toxicity as established under CWA Section 311 (40 CFR 117.3). The available data indicate that the aquatic 96-Hour Median Threshold Limit is less than 0.1 ppm, which corresponds to an RQ of 1 pound.

Reference -- 54 FR 33418 (08/14/89)

EPA Contact -- RCRA/Superfund Hotline (800)424-9346 / (202)260-3000 / FTS 260-3000

VI. BIBLIOGRAPHY

VI.A. ORAL RfD REFERENCES

Chemical Formulators, Inc. 1976b. MRID No. 00070295. Available from EPA. Write to FOI, EPA, Washington, DC 20460.

Cummings, A.M. and L.E. Gray, Jr. 1987. Methoxychlor affects the decidual cell response of the uterus but not other progestational parameters in female rats. Toxicol. Appl. Pharmacol. 90: 330-336.

E.I. du Pont de Nemours and Company, Inc. 1951. MRID No. 00029282. Available from EPA. Write to FOI, EPA, Washington, DC 20460.

E.I. du Pont de Nemours and Company, Inc. 1966. MRID No. 00108732, 00113276. Available from EPA. Write to FOI, EPA, Washington, DC 20460.

E.I. du Pont de Nemours and Company, Inc. 1976a. MRID No. 00062704. Available from EPA. Write to FOI, EPA, Washington, DC 20460.

Goldman, J.M., R.L. Cooper, G.L. Rehnberg, J.F. Hein, W.K. McElroy and L.E. Gray, Jr. 1986. Effects of low subchronic doses of methoxychlor on the rat hypothalamic-pituitary reproductive axis. Toxicol. Appl. Pharmacol. 86: 474-483.

Gray, L.E., Jr., J. Ostby, J. Ferrell, et al. 1989. A dose-response analysis of methoxychlor-induced alterations of reproductive development and function in the rat. Fund. Appl. Toxicol. 12: 92-108.

Hodge, H.C., E.A. Maynard and H.J. Blanchet, Jr. 1952. Chronic oral toxicity tests of methoxychlor (2,2-Di-(P-methoxyphenyl)-1,1,1-trichloroethane) in rats and dogs. J. Pharmacol. Exp. Ther. 104: 60-66.

Khera, K.S., C. Whalen and G. Trivett. 1978. Teratogenicity studies on linuron, malathion, and methoxychlor in rats. Toxicol. Appl. Pharmacol. 45: 435-444.

Kincaid Enterprises, Inc. 1986. MRID No. 00159929. Available from EPA. Write to FOI, EPA, Washington, DC 20460.

Kupfer, D. and W.H. Bulger. 1987. Metabolic activation of pesticides with proestrogenic activity. Fed. Proceed. 48(5): 1864-1869.

US DHEW (US Department of Health, Education, and Welfare). 1977a. MRID No. 00026602. Available from EPA. Write to FOI, EPA, Washington, DC 20460.

US DHEW (US Department of Health, Education, and Welfare). 1977b. MRID No. 00026602. Available from EPA. Write to FOI, EPA, Washington, DC 20460.

VI.B. INHALATION RfC REFERENCES

Bal, H.S. 1984. Effect of methoxychlor on reproductive systems of the rat (41861). Proc. Soc. Exp. Biol. Med. 176(2): 187-196.

Bulger, W.H., R.M. Muccitelli, and K. Kupfer. 1978. Studies on the in vivo and in vitro estrogenic activities of methoxychlor and its metabolites. Role of hepatic mono-oxygenase in methoxychlor activation. Biochem. Pharmacol. 27(20): 2417-2423.

Cooke, P.S. and V.P. Eroschenko. 1990. Inhibitory effects of technical grade methoxychlor on development of neonatal male mouse reproductive organs. Biol. Reprod. 42(3): 585-596.

Cummings, A.M. and L.E. Gray. 1987. Methoxychlor affects the decidual cell response of the uterus but not other progestational parameters in female rats. Toxicol. Appl. Pharmacol. 90(2): 330-336.

Cummings, A.M. and L.E. Gray. 1989. Antifertility effect of methoxychlor in female rats -- dose- and time-dependent blockade of pregnancy. Toxicol. Appl. Pharmacol. 97(3): 454-462.

Cummings, A.M. and S.D. Perreault. 1990. Methoxychlor accelerates embryo transport through the rat reproductive tract. Toxicol. Appl. Pharmacol. 102(1): 110-116.

Davison, K.L., V.J. Feil, and C.H. Lamoureux. 1982. Methoxychlor metabolism in goats. J. Agric. Food Chem. 30(1): 130-137.

Davison, K.L., C.H. Lamoureux, and V.J. Feil. 1984. Methoxychlor metabolism in chickens. J. Agric. Food Chem. 32(4): 900-908.

Deichmann, W.B., M. Keplinger, F. Sala, and E. Glass. 1967. Synergism among oral carcinogens. IV. The simultaneous feeding of four tumorigens to rats. Toxicol. Appl. Pharmacol. 11(1): 88-103.

Eroschenko, V.P. and P.S. Cooke. 1990. Morphological and biochemical alterations in reproductive tracts of neonatal female mice treated with the pesticide methoxychlor. Biol. Reprod. 42(3): 573-583.

Goldman, J.M., R.L. Cooper, G.L. Rehnberg, J.F. Hein, W.K. McElroy, and L.E. Gray Jr. 1986. Effects of low subchronic doses of methoxychlor on the rat hypothalamic-pituitary reproductive axis. Toxicol. Appl. Pharmacol. 86(3): 474-483.

Gray, L.E., Jr., J.S. Ostby, J.M. Ferrell, E.R. Sigmon, and J.M. Goldman. 1988. Methoxychlor induces estrogen-like alterations of behavior and the reproductive tract in the female rat and hamster: Effects on sex behavior, running wheel activity, and uterine morphology. Toxicol. Appl. Pharmacol. 96(3): 525-540.

Gray, L.E. Jr., J. Ostby, J. Ferrell et al. 1989. A dose-response analysis of methoxychlor-induced alterations of reproductive development and function in the rat. Fund. Appl. Toxicol. 12(1): 92-108.

Haag, H.B., J.K. Finnegan, P.S. Larson, W. Riese, and M.L. Dreyfuss. 1950. Comparative chronic toxicity for warm-blooded animals of DDT and DMDT (methoxychlor). Arch. Int. Pharmacodyn. 83(4): 491-504.

Harris, S.J., H.C. Cecil, and J. Bitman. 1974. Effect of several dietary levels of technical methoxychlor on reproduction in rats. J. Agric. Food Chem. 22(6): 969-973.

Kapoor, I.P., R.L. Metcalf, R.F. Nystrom, and G.K. Sangha. 1970. Comparative metabolism of methoxychlor, methiochlor, and DDT in mouse, insects, and in a model ecosystem. J. Agr. Food Chem. 18(6): 1145-1152.

Khera, K.S., C. Whalen, and G. Trivett. 1978. Teratogenicity studies on linuron, malathion, and methoxychlor in rats. Toxicol. Appl. Pharmacol. 45(2): 435-444.

Kupfer, D. and W.H. Bulger. 1987. Metabolic activation of pesticides with proestrogenic activity. Fed. Proc. 46(5): 1864-1869.

Martinez, E.M. and W.J. Swartz. 1991. Effects of methoxychlor on the reproductive system of the adult female mouse. I. Gross and histologic observations. Reprod. Toxicol. 5(2): 139-147.

Reuber, M.D. 1980. Carcinogenicity and toxicity of methoxychlor. Environ. Health Perspect. 36: 205-219.

Shain, S.A., J.C. Shaeffer, and R.W. Boesel. 1977. The effect of chronic ingestion of selected pesticides upon rat ventral prostate homeostasis. Toxicol. Appl. Pharmacol. 40(1): 115-130.

Stein, A.A. 1970. Comparative toxicology of methoxychlor. Pestic. Symp. Collect. Pap. Inter-Amer. Conf. Toxicol. Occup. Med., 6th. p. 225-229.

Tegeris, A.S., F.L. Earl, H.E. Smalley, Jr., and J.M. Curtis. 1966. Methoxychlor toxicity. Comparative studies in the dog and the swine. Arch. Environ. Health. 13(6): 776-787.

US EPA. 1987a. Drinking Water Criteria Document for Methoxychlor. Prepared by the Office of Health and Environmental Assessment, Environmental Criteria and Assessment Office, Cincinnati, OH, for the Office of Drinking Water, Washington, DC.

US EPA. 1987b. Health Advisories for 16 Pesticides (including Alachlor, Aldicarb, Carbofuran, Chlordane, DBCP, 1,2-Dichloropropane, 2,4-D, Endrin, Ethylene Dibromide, Heptachlor/Heptachlor Epoxide, Lindane, Methoxychlor, Oxamyl, Pentachlorophenol, Toxaphene, and 2,4,5-TP). Office of Drinking Water, Washington, DC.

US EPA. 1990. Interim Methods for Development of Inhalation Reference Concentrations, (Review Draft), Office of Research and Development, Washington, DC. EPA/600-8-90-066A. August, 1990.

Villeneuve, D.C., D.L. Grant, and W.E.J. Phillips. 1972. Modification of pentobarbital sleeping times in rats following chronic PCB ingestion. Bull. Environ. Contam. Toxicol. 7(5): 264-269.

Ziem, G. 1982. Aplastic anaemia after methoxychlor exposure [letter]. The Lancet. 2(8311): 1349.

VI.C. CARCINOGENICITY ASSESSMENT REFERENCES

Davis, K.J. 1969. Histopathological diagnosis of lesions noted in mice fed DDT or methoxychlor. Memorandum to W. Hanson, Food and Drug Administration, Washington, DC, January 30. (Cited in Reuber, 1980)

Deichmann, W.B., M. Keplinger, F. Sala and E. Glass. 1967. Synergism among oral carcinogens. IV. The simultaneous feedings of four tumorigens to rats. Toxicol. Appl. Pharmacol. 11: 88-103.

Dunkel, V.C., R.J. Pienta, A. Sivak and K.A. Traul. 1981. Comparative neoplastic transformation responses of BALB/3T3 cells, Syrian hamster embryo cells and Rauscher murine leukemia virus-infected Fischer 344 rat embryo cells to chemical carcinogens. J. Natl. Cancer Inst. 67(6): 1303-1315.

Griesemer, R.A. and C. Cueto. 1980. Toward a classification scheme for degrees of experimental evidence for the carcinogenicity of chemicals for animals. In: Molecular and Cellular Aspects of Carcinogen Screening Tests, R. Montesano, Ed. IARC, Lyon, France. p. 259-281.

Harper, B.L., S.J. Rinkus, M. Scott, et al. 1982. Correlation of NCI and IARC carcinogens with their mutagenicity in Salmonella. In: Use of Mammalian Cells for Risk Assessment, a NATO Publication. Plenum Press.

Hodge, H.C., E.A. Maynard and H.J. Blanchet, Jr. 1952. Chronic oral toxicity tests of methoxychlor [2,2-di-(p-methoxyphenyl)-1,1,1- trichloroethane] in rats and dogs. J. Pharmacol. Exp. Ther. 104: 60-66.

Hodge, H.C., E.A. Maynard, W.L. Downs, J.K. Ashton and L.J. Salerno. 1966. Tests on mice for evaluating of the carcinogenic risk of chemicals to humans. Some halogenated hydrocarbons. WHO, IARC, Lyon, France. Vol. 20.

IARC (International Agency for Research on Cancer). 1979. IARC monographs on the evaluation of the carcinogenic risk of chemicals to humans. Some halogenated hydrocarbons. WHO, IARC, Lyon, France. Vol. 20.

NCI (National Cancer Institute). 1978. Bioassay of Methoxychlor for Possible Carcinogenicity. NCI-CG-TR-35. Carcinogenesis Program. p. 91.

Nelson, A.A. and O.G. Fitzhugh. 1951. Pathological changes produced in rats by feeding of methoxychlor at levels up to 0.2% of diet for 2 years. Prepared as a memorandum to A.J. Lehman, Food and Drug Administration, Washington, DC, June 9. (Cited in US EPA, 1983)

Nelson, A.A. and J.L. Radomski. 1953. Pathological changes produced in dogs for feeding of methoxychlor, 300 mg/kg/day for 3.5 years. Memorandum to A.J. Lehman, FDA, Washington, DC, June 9. (Cited in US EPA, 1983)

Probst, G.S., R.E. McMahon, L.E. Hill, C.Z. Thompson, J.K. Epp and S.B. Neal. 1981. Chemically-induced unscheduled DNA synthesis in primary rat hepatocyte cultures: A comparison with bacterial mutagenicity using 218 compounds. Environ. Mutat. 3: 11-32.

Radomski, J.L., W.B. Deichmann, W.E. MacDonald and E.M. Glass. 1965. Synergism among oral carcinogens. I. Results of the simultaneous feeding of four tumorigens to rats. Toxicol. Appl. Pharmacol. 7(5): 652-656.

Reuber, M.D. 1980. Carcinogenicity and toxicity of methoxychlor. Environ. Health Perspect. 36: 205-219.

US EPA. 1983. Multimedia Risk Assessment for Methoxychlor. Environmental Criteria and Assessment Office, Office of Water Regulation and Standards, Cincinnati, OH. (Draft: August, 1983).

US EPA. 1987. Drinking Water Criteria Document for Methoxychlor. Prepared by the Office of Health and Environmental Assessment, Environmental Criteria and Assessment Office, Cincinnati, OH for the Office of Drinking Water, Washington, DC.

VI.D. DRINKING WATER HA REFERENCES

Not available at this time

VII. REVISION HISTORY

```
--------  --------  ---------------------------------------------------------
Date      Section   Description
--------  --------  ---------------------------------------------------------
```

Date	Section	Description
09/07/88	II.	Carcinogen summary on-line
02/01/89	II.D.3.	Primary contact's phone number corrected
06/01/89	II.D.3.	Secondary contact deleted
12/01/89	VI.	Bibliography on-line
05/01/90	I.A.	Oral RfD now under review
09/01/90	I.A.	Oral RfD summary on-line
09/01/90	III.A.	Health Advisory on-line
09/01/90	VI.A.	Oral RfD references added
09/01/90	VI.D.	Health Advisory references added
10/01/90	II.	Text edited
08/01/91	I.A.4.	Khera citation year corrected
08/01/91	VI.A.	Khera reference year corrected
12/01/91	I.B.	Inhalation RfC now under review
01/01/92	IV.	Regulatory Action section on-line
04/01/92	I.B.	Inhalation RfC message on-line
04/01/92	VI.B.	Inhalation RfC references added
12/01/93	I.B.	Replaced with expanded assessment
12/01/93	III.A.	Health Advisory withdrawn
12/01/93	VI.B.	References revised
12/01/93	VI.D.	Health Advisory references withdrawn

SYNONYMS

2,2-di-p-anisyl-1,1,1-trichloroethane	Methoxychlor
DMDT	methoxy-DDT
marlate	moxie
methorcide	1,1,1-trichloro-2,2-bis(p-methoxyphenyl)ethane

Source: *Instant EPA's IRIS*, 1997, Instant Reference Sources, Inc., 7605 Rockpoint Drive, Austin, TX 78731 (Fax: 512-345-2386; Internet URL, http://www.instantref.com/inst-ref.htm).

Production, Use, and Pesticide Labeling Information

1. Description of Chemical

Generic Name: 2,2-bis(p-methoxyphenyl)-1,1,1-trichloroethane

Common Name: Methoxychlor

Trade Names: Marlate, Prentox, and Methoxcide

Other Chemical Nomenclature: 1,1,1-trichloro-2,2-bis(4-methoxyphenyl)ethane; 1,1,1-trichloro-2,2-di(4-methoxyphenyl)ethane; 1,1-(2,2,2-trichloroethylidene)-bis(4-methoxy)benzene; 1,1,1-trichloro-2,2-bis(p-methoxyphenyl)ethane; 2,2-bis(p-methoxyphenyl)-1,1,1-trichloroethane

CAS Registry No.: 72-43-5

EPA Pesticide Chemical Code (Shaughnessy Number): 034001

Empirical Formula: $C_{16}H_{15}C_{l3}O_2$

Molecular Weight: 345.7

Year of Initial Registration: 1948

Pesticide Type: Insecticide/Acaricide

Chemical Family: Chlorinated Hydrocarbon

US Registrants: Chemical Formulators: Prentiss Drug & Chemical Co., J.R. Simplot Co.; Dynachem Industries; Clover Chemical Co.; Drexel Chemical Co.; Kincaid Enterprises; and Wesley Industries.

2. Use Patterns and Formulations

- Registered Uses:

- TERRESTRIAL FOOD CROP Use: (1) seed treatment only use on grains and various vegetables; (2) foliar application (including seed treatment) use on vegetables and fruits; and (3) foliar application only use on vegetables and fruits.

- TERRESTRIAL NON-FOOD CROP use on grasses, ornamentals and trees.

- GREENHOUSE FOOD CROP use on mushrooms.

- DOMESTIC AND NON-DOMESTIC OUTDOOR use around dwellings and for garbage and sewer areas, general urban outdoor use.

- AQUATIC FOOD use on cranberry.

- AQUATIC NON-FOOD use for mosquito larvae control in aquatic sites, such as beaches, lakes, marshes and rivers.

- FORESTRY use on forest trees.

- INDOOR use on: (1) postharvest stored grain commodity and premise treatment; (2) direct animal treatment for dogs, cats, and farm animals; (3) agricultural premise use; (4) kennels, dog sleeping quarters and cat sleeping quarters; (5) indoor domestic dwellings for use on household contents such as human clothing (including woolens); (6) direct application to humans; (7) commercial and industrial use in food processing, storage transportation areas and equipment.

- Pests Controlled: Various nuisance species (some of public health significance) including cockroaches, mosquitoes, flies and chiggers; various arthropods attacking field crops, vegetables, fruits, ornamentals, stored grain, livestock and domestic pets.

- Method of Application: Sprays, fogs, paints, ground and aerial equipment, animal dust-bags, dips, sprays and back-rubbers.

- Formulations: Wettable powders, dusts, emulsifiable concentrates, flowable concentrates, liquid soluble concentrates, granules, ready- to-use products (liquids) and pressurized liquids.

3. Science Findings

Chemical/Physical Characteristics of the Technical Material

  - Color: Data Gap

  - Physical State: Crystalline solid (Farm Chemicals, 1987)

  - Odor: Data Gap

  - Melting Point: 89 °C (Farm Chemicals, 1987)

  - Specific Gravity: Data Gap

  - Solubility: Very soluble in aromatic chlorinated, or ketonic solvents, somewhat soluble in paraffinic types; essentially insoluble in water (Farm Chemicals, 1987).

  - Vapor Pressure: Data Gap

  - Flammability: Data Gap

  - pH: Data Gap

Toxicological Characteristics

  - With the exception of one mutagenicity study, there are no acceptable acute, subchronic or long-term toxicology/oncogenicity studies available to support technical methoxychlor. In the mutagenicity study, a mammalian cell in culture unscheduled DNA synthesis assay (UDS assay), no increase in abnormal DNA synthesis was noted.

Environmental Characteristics

- The Agency is unable to assess the environmental fate of methoxychlor because acceptable data are lacking. Preliminary data indicate that methoxychlor is stable to hydrolysis (half-life > 200 days); photodegradation in water (half-life of 4.5 months); and aerobic soil metabolism (half-life > 3 months in sandy loam soil). The half-life for anaerobic soil metabolism is reported at less than 1 month in sandy loam soil. Preliminary data also indicate that methoxychlor has a high adsorption rate to soil sediment (Kd value is 620).

Ecological Characteristics

- Based on acceptable laboratory data, technical methoxychlor is characterized as practically nontoxic to birds on both an acute oral and subacute dietary basis and very highly toxic to fish and aquatic invertebrates on an acute basis. There is fish and aquatic invertebrates on an acute basis. There is sufficient information to characterize methoxychlor as relatively nontoxic to honey bees. The acute toxicity value = 24 ug/bee.

  - Acute LD50 (bobwhite): > 2510 mg/kg

  - Dietary LC50: > 5620 ppm (upland gamebird).

  - Freshwater invertebrates toxicity (96-hr LC50) for daphnid: .78 ppb.

  - Fish acute toxicity (96-hr LC50) for rainbow trout: 1.31 ppm.

  - Fish acute toxicity (96-hr LC50) for brook trout: 0.009 ppm.

Tolerance Assessment

- Tolerances have been established for residues of methoxychlor in a variety of raw agricultural commodities, in meat, fat and meat byproducts (40 CFR 180.120). Tolerances are expressed in terms of methoxychlor per se.

- The nature of the residues of methoxychlor in plants and animals is not adequately understood. None of the tolerances for methoxychlor is adequately supported. Plant and animal metabolism studies, residue studies, analytical methodology, processing studies, and storage stability data are needed before the Agency can determine the adequacy of current tolerance levels.

- The Preliminary Limiting Dose (PLD) of methoxychlor is .005 mg/kg/day. This is based on a rabbit teratology study with a No Observed Effect Level (NOEL) of 5 mg/kg/day for increased loss of litters and an uncertainty factor of 1000 to account for inter- and intraspecies differences, poor quality of the study used and total incompleteness of the subchronic and chronic toxicity data base. The study is not considered to be adequate to define a NOEL for purposes of setting an Acceptable Daily Intake, since the experimental design was considered to be inadequate. It is being used on an interim basis for calculation of the PLD. The Agency is unable to complete a tolerance assessment of methoxychlor because of the incompleteness of the toxicology and residue chemistry data bases.

Summary Science Statement

- With the exception of one mutagenicity study, there are no acceptable acute, subchronic, or long-term toxicology/oncogenicity studies available to support technical methoxychlor. In the acceptable mutagenicity study, an unscheduled DNA synthesis assay in mammalian cells in culture, no abnormal DNA synthesis was noted at any of the dose levels tested.

- Based on acceptable laboratory data, technical methoxychlor is characterized as very highly toxic to fish and aquatic invertebrates, and practically nontoxic to birds and bees. Based on theoretical calculations, both terrestrial and aquatic uses of methoxychlor may pose a hazard to aquatic organisms, although there is no field evidence to support this. The impacts of methoxychlor use to nontarget organisms will be assessed upon receipt of ecological effects and environmental fate data.

- The environmental fate of methoxychlor cannot be characterized because acceptable data are lacking. Preliminary data suggest that methoxychlor is unlikely to contaminate groundwater because of its low solubility and high rate of adsorption to soil particles.

- The nature of the residues of methoxychlor in plants and animals is not adequately understood. None of the tolerances for methoxychlor is adequately supported. Plant and animal metabolism studies, residue studies, analytical methodology, processing studies, and storage stability data are needed before the Agency can determine the adequacy of current tolerance levels.

4. Summary of Regulatory Position and Rationale

- Methoxychlor is not being placed into Special Review at this time. Since there are so few acceptable studies available to support registration of products containing methoxychlor, the Agency is not yet able to make a determination as to whether any of the criteria of 40 CFR 154.7 have been met or exceeded.

- The Agency will not approve any new food uses, including minor uses for this chemical since none of the tolerances are adequately supported.

- The Agency is unable to assess methoxychlor's potential for contaminating groundwater. When data required in the Standard have been received and evaluated, the Agency will assess the potential for methoxychlor to contaminate groundwater.

- Updated worker safety rules are required for end-use product labels.

- The Agency is not establishing a longer reentry interval for agricultural uses of methoxychlor beyond the minimum reentry interval (sprays have dried, dusts have settled, and vapors have dispersed).

- Revised and updated fish and wildlife statements are required for end- use product labels. Since methoxychlor is practically nontoxic to bees, the bee statement imposed under FR Notice 68-19 is no longer appropriate. Registrants must remove the bee statement from the labeling. - The Agency is not classifying methoxychlor as a restricted use pesticide at this time, since it is unable to determine if this pesticide meets any of the risk criteria of 40 CFR 152.170. Upon receipt of data required under this Standard, the Agency will apply the criteria of 40 CFR 152.170 to determine if any uses of methoxychlor warrant restricted use classification.

- Since methoxychlor is an analogue of DDT, the Agency is requiring specific analysis of methoxychlor for the potential impurities 1,1,1-trichloro-2,2-bis(p-chlorophenyl)ethane (DDT) and other structurally similar compounds.

5. Summary of Major Data Gaps

- Toxicology   Time Frame

         - Acute Oral Toxicity   9 Months
         - Acute Dermal Toxicity   9 Months
         - Acute Inhalation Toxicity   9 Months
        - Eye Irritation   9 Months
        - Dermal Irritation   9 Months
        - Dermal Sensitization   9 Months
         - 21-Day Dermal Toxicity   9 Months
         - Chronic Testing (rodent)   50 Months
         - Chronic Testing (non-rodent)   50 Months
         - Oncogenicity (rat)   50 Months
         - Oncogenicity (mouse)   50 Months
         - Teratogenicity (rat)   15 Months
         - Teratogenicity (rabbit)   15 Months
         - Reproduction   39 Months
         - Gene Mutation   9 Months
         - Other Mechanisms of Mutagenicity   12 Months
         - Metabolism   24 Months

- Environmental Fate/Exposure

         - Hydrolysis   9 Months
         - Photodegradation in Water   9 Months
         - Photodegradation on Soil   9 Months
         - Aerobic Soil Metabolism   27 Months
         - Anaerobic Soil Metabolism   27 Months
         - Anaerobic Aquatic Metabolism   27 Months
         - Aerobic Aquatic Metabolism   27 Months
         - Leaching and Adsorption/Desorption   12 Months
         - Aquatic Dissipation   27 Months
         - Forestry   27 Months
         - Soil, Long-term   39 Months
         - Confined Rotational Crop   39 Months
         - Accumulation in Irrigated Crops   39 Months
         - Accumulation in Fish   12 Months
         - Accumulation in Aquatic Nontarget Organisms   12 Months

- Fish and Wildlife

         - Avian Reproduction   24 Months

- Freshwater Fish LC50 Studies (TEP)	9 Months
- Freshwater Invertebrate LC50 Studies (TEP)	9 Months
- Estuarine and Marine Organisms	
LC50 Studies (TEP)	12 Months
- Fish Early Life Stage and Invertebrate Life Cycle	15 Months
- Simulated or Actual Field Testing-	
Aquatic Organisms	24 Months
- Seed Germination/Seedling Emergence	9 Months
- Aquatic Plant Growth	9 Months

- Residue Chemistry

- Residue data - Raw Agricultural Commodities	18 Months
- Processing Studies	24 Months
- Plant and Animal Metabolism	18 Months
- Storage Stability	15 Months
- Residue Analytical Methods	15 Months

- Product Chemistry

- All Data	9-15 Months

6. Contact Person at EPA

Dennis H. Edwards, Jr.
Product Manager
12 Insecticide- Rodenticide Branch Registration Division (TS-767)
Environmental Protection Agency
Washington, DC 20460

Telephone (703) 557-2386

Source: *Instant EPA's Pesticide Facts*, 1994, Instant Reference Sources, Inc., 7605 Rockpoint Drive, Austin, TX 78731 (Fax: 512-345-2386; Internet URL, http://www.instantref.com/inst-ref.htm).

Metiram

No structure has been found for Metiram

Introduction

Metiram is registered as a general use pesticide by the US Environmental Protection Agency (EPA). It is a combination of two other chemicals in this class: maneb and zineb. In July 1987, the Environmental Protection Agency announced the initiation of a special review of the ethylene bisdithiocarbamates (EBDCs), a class of chemicals to which mancozeb belongs. This Special Review was initiated because of concerns raised by laboratory tests on rats and mice. As part of the Special Review, EPA reviewed data from market basket surveys and concluded that actual levels of EBDC residues on produce purchased by consumers are too low to affect human health. Many home garden uses of EBDCs have been canceled because the EPA has assumed that home users of these pesticides do not wear protective clothing during application. [829]

Metiram is a fungicide used on apples, asparagus, celery, corn, cotton, cucumber, peanuts, pecans, potatoes, sugar beets, and tomatoes; Terrestrial nonfood crops include tobacco and roses. However, its predominant use is with apples and potatoes. Metiram is used to control foliar fungal diseases of selected fruit, nut, vegetable, field and ornamental crops. It is slightly toxic to birds and fish but practically nontoxic to bees and it can be used around them with minimum injury to them. Being practically insoluble in water and adsorbing strongly to soil particles it is not expected to contaminate groundwater. However, it can enter surface waters if erosion of soil occurs. [829]

Metiram is manufactured in formulations that include dusts, wettable powders, and formulation intermediates. Typically, foliar application is made to fruits, vegetables and nuts by aerial equipment, as well as ground equipment. For ground equipment metiram suspensions are typically made from a wettable powder and applied using air blast sprayers. Metiram in water solution degrades primarily to ETU and other transient degradates. [821]

Identifying Information

CAS NUMBER: 9006-42-2

SYNONYMS: Polyram™, Polyram-Combi™

Endocrine Disruptor Information

Metiram is a strongly suspected EED that it is listed by the Centers for Disease Control & Prevention [812], and the World Wildlife Fund [813] as a potential endocrine modifying chemical. It belongs to ethylene bisdithiocarbamates (EBDC) class of chemicals and these have shown rapid reduction in the uptake of iodine and swelling of the thyroid. [829]

Chemical and Physical Properties

CHEMICAL FORMULA: $C_{16}H_{33}N_{11}S_{16}Zn_3$

MOLECULAR WEIGHT: 1088.65

MELTING POINT: decomposes above 120°C

PHYSICAL DESCRIPTION: light yellow powder with an odor typical of dithiocarbamates [829]

SOLUBILITY: water: 2.1 mg/L at 20°C; soluble in pyridine with decomposition; practically insoluble in organic solvents [829]

STABILITY: Metiram is unstable in strong acids and alkalis or a combination of heat and moisture [829]

OTHER PHYSICAL DATA: vapor pressure is $<7.5 \times 10^{-6}$ mbar at 20°C [829]

Environmental Reference Materials

A source of EED environmental reference materials for metiram is Crescent Chemical Co., Hauppauge, NY (Fax: 516-348-0913); Catalog No. 35621.

Hazardous Properties

ACUTE/CHRONIC HAZARDS: Metiram is moderately toxic by ingestion; it is not significantly absorbed through the skin. It is a moderate irritant. [829]

FIRE HAZARDS: Metiram may burn but it doesn't ignite readily. [829]

Medical Symptoms of Exposure

Metiram can cause skin and mucous membrane irritation. Prolonged or repeated exposure of the skin or eyes may cause dermatitis or conjunctivitis. [829]

Symptoms of poisoning from this class of chemicals include itching, scratchy throat, sneezing, coughing, inflammation of the nose or throat, and bronchitis. There is no evidence of 'neurotoxicity,' nerve tissue destruction or behavior change, from the ethylene bisdithiocarbamates (EDBCs). However, dithiocarbamates are partially chemically broken down, or metabolized, to carbon disulfide, a neurotoxin capable of damaging nerve tissue. EBDC residues in or on foods convert readily to ethylenethiourea (ETU), a known teratogen, during commercial processing or home cooking. [829]

Toxicological Information

Metiram is not currently listed in NTP's Chemical Repository Database or EPA's IRIS database and they are not scheduled for addition in the near future.

Production, Use, and Pesticide Labeling Information

1. Description of Chemical

   Chemical Name: Mixture of 5.2 parts by weight (83.9%) of ammoniates of ethylenebis(dithiocarbamate)-zinc with 1 part by weight (16.1%) ethylenebisdithiocarbamic acid, bimolecular and trimolecular cyclic anhydrosulfides and disulfides.

   Common Name: Metiram

   Trade Names: Polyram™, Polyram-Combi™

   EPA Shaughnessy Code: 014601

   Chemical Abstracts Service (CAS) Number: 9006-42-2

   Year of Initial Registration: late 1940's

   Pesticide Type: Fungicide

   Chemical Family: Ethylenebisdithiocarbamate (EBDC)

   US and Foreign Producers: FMC and BASF

2. Use Patterns and Formulations

   - Registered Uses: Terrestrial food crop uses on apples, asparagus, celery, corn (sweet), cotton, cucumber, peanuts, pecans, potatoes (including seed pieces), sugar beets, and tomatoes; Terrestrial nonfood crop uses on tobacco (field and transplants) and roses.

   - Predominant Uses: Apples and potatoes.

   - Pests Controlled: Foliar fungal diseases of selected fruit, nut, vegetable, field and ornamental crops.

   - Types of Formulations: Formulation intermediate, dust and wettable powder.

   - Types and Method of Application: Foliar application to fruits, vegetables and nuts by aerial equipment, as well as ground equipment. For ground equipment metiram suspensions typically made from a wettable powder, would be applied by means of air blast sprayers or in the case of some row crops by means of tractor mounted boom sprayers.

   - Application Rates: Terrestrial food crop: 0.3 - 6.4 lb ai/A Terrestrial nonfood crop: 1.2 - 2.4 lb ai/A

3. Science Findings

Chemical Characteristics

- Physical State: Solid

- Color: Light yellow

- Odor: Odorless

- Vapor Pressure: $<1 \times 10^{-7}$ mbar at 20 °C

- Molecular Formula: $(C_{16}H_{33}N_{11}S_{16}Zn_3)$

Toxicological Characteristics

- Acute Toxicity: All studies required.

- Major Routes of Exposure: Dermal, Inhalation and oral by ingestion of food residues.

- Subchronic Toxicity: Inhalation study is adequate, other studies required.

- Oncogenicity: Studies required.

- Chronic Feeding: Studies required.

- Metabolism: Studies in rats indicate that the polymer is hydrolyzed and readily absorbed and eliminated in the urine and feces. ETU was one of the metabolites in the urine and bile of rats.

- Reproduction: Study required.

- Teratogenicity & Developmental Toxicity: Studies required.

- Mutagenicity: Considering only the acceptable studies, the majority of mutagenicity studies on metiram were negative. However, the in vitro sister chromatid exchange assay in Chinese hamster ovary cells was positive and is considered a sensitive test for chromosomal effects. According to the present data, metiram is considered positive for chromosomal damage. A gene mutation assay is required.

Physiological and Biochemical Characteristics

- Metabolism and Persistence in Plants and Animals

- Metabolism of metiram is not completely understood. Additional data are being required in plants and livestock. ETU is a major metabolite of concern.

Environmental Characteristics

- Presently only the hydrolysis and photodegradation in soil and in water data requirements on both metiram and ETU were fully satisfied. Metiram has a very limited solubility in water. Metiram in water solution degrades primarily to ETU and other transient degradates. ETU is also a soil degradate of metiram and its formation on soil is enhanced by sunlight. ETU is stable in water at pH 5-9 and under sunlight and the degradation of ETU on soil is not enhanced by sunlight radiation. ETU is the degradate of major environmental concern. There are indications that ETU may leach and enter groundwater. However, additional data are required to complete the groundwater assessment.

Ecological Characteristics

- Metiram has been found to be slightly toxic to birds. Formulated metiram showed that LC50 values for mallard duck and bobwhite quail are both greater than 3712 ppm.

- Based on an acute contact honeybee toxicity study, there is sufficient information to characterize metiram as practically nontoxic to honeybees.

Tolerance Assessment

- Tolerances, expressed as zinc ethylene bisdithiocarbamate, have been established for residues of metiram in a variety of raw agricultural commodities (40 CFR 180.217 and 180.319).

- The toxicology data for metiram are insufficient to determine an Acceptable Daily Intake (ADI) or whether the toxicity observed in the studies is due to metiram or ETU. A three generation rat reproduction study has been used to calculate a Provisional ADI (PADI). Because a NOEL was not reached in the three generation study, an uncertainty factor of 1000 was employed. The PADI for metiram is 0.0003 mg/kg/day.

- The theoretical maximum residue contribution (TMRC), based on the assumption that 100 percent of each crop is treated and contains residues at the tolerance level, is 0.009 or approximately 3000 percent of the PADI. Based on a more realistic dietary assessment, using anticipated field residues and estimate of percent crop treated, the estimated average consumption for the US population is 0.00038 mg/kg/day or 122 percent of the PADI.

4. Summary of Regulatory Position and Rationale

- The Agency initiated a Special Review for metiram along with the other EBDC's in June 1987 because of concern about the oncogenic risk to consumers from dietary exposure to ETU from food treated with these pesticides, and the risks of teratogenicity and adverse thyroid effects to applicators and mixer/loaders from exposure to ETU.

- ETU has been classified as a B2 oncogen (probable human carcinogen).

- The Agency will not consider establishment of new food use tolerances for metiram because the current residue chemistry and toxicology data are not sufficient to assess existing tolerances and the toxicology data base is insufficient to determine an ADI and does not allow a decision as to whether observed toxicity is due to metiram or ETU.

- The Agency will consider the need for establishment of tolerances for ETU and any intermediate metabolites when data are sufficient to permit such decisions.

- The Agency will not establish any food/feed additive regulations pursuant to Section 409 of the Federal Food, Drug and Cosmetic Act (FFDCA) and is deferring action on previously established food/feed additive regulations.

- Protective clothing labeling for metiram products, as required as a result of the 1982 Decision Document, must be updated.

- The Agency is requiring reentry data for metiram. In order to remain in compliance with FIFRA, an interim 24-hour reentry interval requirement must be placed on the label of all metiram end-use products registered for agricultural uses, until the required data are submitted and evaluated and any change in this reentry interval is announced.

- The Agency has screened and reviewed the environmental fate data to determine if metiram/ETU and/or its degradate(s) have the potential to leach into ground water. The Agency has decided that in addition to environmental fate data requirements, a small-scale retrospective ground water monitoring study is also required to define the extent of the ground water problem.

- While the data gaps are being filled, currently registered manufacturing-use products (MP's) and end-use products (EP's) containing metiram as the sole active ingredient may be sold, distributed, formulated and used, subject to the terms and conditions specified in this Standard. However, new uses will not be registered. Registrants must provide or agree to develop additional data, as specified in the Data Appendices of the Registration Standard, in order to maintain existing registrations.

- Labeling Requirements:

    - All metiram products must bear appropriate labeling as specified in 40 CFR 156.10. Appendix II of the Registration Standard contains information on labeling requirements.

- In addition to the above, in order to remain in compliance with FIFRA, the Agency is requiring:

    - Protective clothing requirements
    - Environmental hazard precautions
    - Worker safety rules
    - Reentry interval
    - Grazing restrictions for apples, pecans, corn (sweet), cotton, peanuts, sugar beets and potato (seed pieces).

5. Summary of Data Gaps

    - Product Chemistry: All - due within 6 months
    - Technical Grade:
    - Preliminary analysis of product samples
    - MUP:
            - Analysis & certification of product ingredient
            - Oxidizing or reducing action
            - Flammability
            - Explodability
            - Storage stability
            - Corrosion characteristics

    - Toxicology: The last studies are due 12/90
            - Acute testing
            - Dermal sensitization
            - 90-Day feeding (rodent and nonrodent)
            - 21-Day subchronic dermal
            - Chronic toxicity (rodent and nonrodent)
            - Oncogenicity (rat and mouse)
            - Teratology (rabbit and rat)
            - Reproduction (rat)
            - Mutagenicity (point gene mutation)

    - Residue Chemistry: Data due 10/88 and 4/89
            - Nature of the Residue in Plants and Livestock
            - Analytical Methods
            - Magnitude of Residue for Variety of Commodities

- Environmental Fate: Last studies are due 7/90
    - Leaching and adsorption/desorption
    - Field dissipation
    - Degradation soil
    - Degradation (soil long-term)
    - Small-scale retrospective ground water monitoring
    - Confined rotational crops
    - Fish accumulation

- Reentry Protection : Data due 7/89
    - Reentry Studies on Foliar and Soil Dissipation

- Wildlife and Aquatic Organisms: Last data are due in 12 months
    - Avian oral toxicity
    - Freshwater fish toxicity
    - Acute freshwater invertebrates
    - Estuarine and marine organism toxicity

- ETU Data Requirements
    - Toxicology
        - Chronic (rodent and non-rodent) Data due 5/90
        - Reproduction Data due 12/90

- Environmental Fate: Last studies due 7/90
    - Aerobic and anaerobic soil metabolism
    - Aerobic aquatic
    - Lab volatility
    - Degradation (soil)
    - Aquatic (sediment)
    - Degradation (soil long-term)
    - Small-scale retrospective ground water monitoring study
    - Fish accumulation

6. Contact Person at EPA

Lois A. Rossi
Product Manager
21 Fungicide-Herbicide Branch Registration Division (TS-767C)
Office of Pesticide Programs
Environmental Protection Agency
401 M St., SW
Washington, DC 20460

Office location and phone number:

Room 227, Crystal Mall #2
1921 Jefferson Davis Highway
Arlington, VA 22202

Telephone: (703) 557-1900

Source: *Instant EPA's Pesticide Facts*, 1994, Instant Reference Sources, Inc., 7605 Rockpoint Drive, Austin, TX 78731 (Fax: 512-345-2386; Internet URL, http://www.instantref.com/inst-ref.htm).

Metolachlor

Introduction

Metolachlor is a preermengence herbicide used to control broadleaf and grassy weeds with corn, sorghum, cotton, potatoes, peanuts, soybeans, green beans, kidney and other beans, blackeye peas and other peas, stone fruits and tree nuts, ornamental plants, and along railroad and highway rights-of-way. [821] It is often used in combination with broad-leaved herbicides, in order to extend the spectrum of activity. It is a germination inhibitor active mainly on grasses at 1.0-2.5 kg ai/hectare Mixtures with other herbicides are also used in broad beans, carrots, hemp, lentils, paprika. [828] It is manufactured in several formulation types including granulars, emulsifiable concentrates, and flowable concentrates. It is typically applied using ground spray equipment, aircraft, or through center pivot irrigation systems. [821]

In the environment metolachlor has been found in several surface water surveys, some tapwater samples, and in ground water. [821] The release of metolachlor in the environment occurs during the manufacture and particularly from its use in the field. In soil, metolachlor is transformed to its metabolites primarily by biodegradation. The half-life of metolachlor disappearance from soil is about 90 days, although very little mineralization has been observed. Metolachlor is highly to moderately mobile. Although a slow process, biodegradation is a major pathway for the loss of metolachlor in water. The bioconcentration of metolachlor in aquatic organisms is not important. Reaction of metolachlor with photochemically produced hydroxyl radicals may be the most important loss process in the atmosphere. The half-life of metolachlor due to this reaction has been estimated to be 1.8 hrs. Partial removal of metolachlor will also occur as a result of dry and wet deposition. Metolachlor has been detected in surface water and groundwater around sprayed farmlands. It has rarely been found in community drinking water. The applicators of the herbicide are the most likely people for exposure to metolachlor by inhalation and dermal routes. [828]

Identifying Information

CAS NUMBER: 51218-45-2

SYNONYMS: BICEP; CGA-24705; alpha-CHLORO-2'-ETHYL-6'-METHYL-N-(1-METHYL-2-METHOXYETHYL)-ACETANILIDE; 2-CHLORO-6'-ETHYL-N-(2-METHOXY-1-METHYLETHYL)ACET-o-TOLUIDIDE; 2-CHLORO-N-(2-ETHYL-6-

METHYLPHENYL)-N-(2-METHOXY-1-METHYLETHYL)ACETAMIDE; CODAL;
COTORAN MULTI; DUAL; 2-ETHYL-6-METHYL-1-N-(2-METHOXY-1-
METHYLETHYL)CHLOROACETANILIDE; METELILACHLOR; MILOCEP;
ONTRACK 8E; PRIMAGRAM; PRIMEXTRA

Endocrine Disruptor Information

Metolachlor is on EPA's list of suspect EEDs [811] but little information has been
located to determine the Agency's reason for its inclusion. Testicular atrophy along with
increased incidence of neoplastic liver nodules and proliferative hepatic lesions has been
noted with rat studies. [821] In feeding studies with rats the thyroid-to-body weight ratio
and thyroid-to-brain weight ratio with males were significantly increased. [819]

Chemical and Physical Properties

CHEMICAL FORMULA: $C_{15}H_{22}ClNO_2$ [821]

MOLECULAR WEIGHT: 238.81

PHYSICAL DESCRIPTION: clear colorless liquid to tan liquid with no odor [821]

DENSITY: 1.12 at 20/4°C [817]

MELTING POINT: <25°C [817]

BOILING POINT: 100°C at 0.001 mm Hg [817]

SOLUBILITY: 530 mg/L in water at 20°C, miscible with xylene toluene, dimethyl
formamide, methyl cellosolve, butyl cellosolve, ethylene dichloride, and cyclohexanone.
Insoluble in ethylene glycol and propylene glycol. [821]

Environmental Reference Materials

A source of EED environmental reference materials for metolachlor is Crescent
Chemical Co., Hauppauge, NY (Fax: 516-348-0913); Catalog No. 45579.

Hazardous Properties

ACUTE/CHRONIC HAZARDS: Metolachlor is not considered to be teratogenic or cause
reproductive effects. It is not oncogenic in mice, but is considered an oncogen in rats, and
is tentatively classified as showing limited evidence of carcinogenicity in animals. [821]

Medical Symptoms of Exposure

Symptoms with dogs included decreased gain in body weight in males and females,
failure of the serum alkaline phosphatase to decrease with increased age, and possible
effects on blood clotting systems

Toxicological Information

The toxicological information below is gathered from several sources including two national databases: one from the *National Toxicology Program Chemical Repository Database* and another from *EPA's Integrated Risk Information System (IRIS)*.

Toxicology Information from the National Toxicology Program

Metolachlor is not currently listed in NTP's Chemical Repository database nor is it scheduled for addition in the future.

Toxicology Information from EPA's Integrated Risk Information System (IRIS)

I. CHRONIC HEALTH HAZARD ASSESSMENTS FOR NONCARCINOGENIC EFFECTS

I.A. REFERENCE DOSE FOR CHRONIC ORAL EXPOSURE (RfD)

The Reference Dose (RfD) is based on the assumption that thresholds exist for certain toxic effects such as cellular necrosis, but may not exist for other toxic effects such as carcinogenicity. In general, the RfD is an estimate (with uncertainty spanning perhaps an order of magnitude) of a daily exposure to the human population (including sensitive subgroups) that is likely to be without an appreciable risk of deleterious effects during a lifetime. Please refer to Background Document 1 in Service Code 5 for an elaboration of these concepts. RfDs can also be derived for the noncarcinogenic health effects of compounds which are also carcinogens. Therefore, it is essential to refer to other sources of information concerning the carcinogenicity of this substance. If the US EPA has evaluated this substance for potential human carcinogenicity, a summary of that evaluation will be contained in Section II of this file when a review of that evaluation is completed.

NOTE: The Oral RfD for metolachlor may change in the near future pending the outcome of a further review now being conducted by the RfD/RfC Work Group.

I.A.1. ORAL RfD SUMMARY

Critical Effect	Experimental Doses*	UF	MF	RfD
Decreased body weight gain	NOEL: 300 ppm (15 mg/kg/day)	100	1	1.5E-1mg/kg/day
	LEL: 3000 ppm (150 mg/kg/day)			
Reduced pup weights and parental food consumption	NOEL: 300 ppm (15 mg/kg/day)			
	LEL: 1000 ppm (50 mg/kg/day)			

*Conversion Factors: 1 ppm = 0.05 mg/kg/day (assumed rat food consumption)

2-Year Rat Feeding Study, Ciba-Geigy, 19832-Generation Rat Reproduction Study, Ciba-Geigy, 1981

I.A.2. PRINCIPAL AND SUPPORTING STUDIES (ORAL RfD)

Ciba-Geigy Corporation. 1983. MRID No. 00129377. Ciba-Geigy Corporation. 1981. MRID No. 00080897. Available from EPA. Write to FOI, EPA, Washington, DC 20460.

Albino CD rats were divided into four groups and fed diets containing 0 (70 animals/sex), 30 (60 animals/sex), 300 (60 animals/sex), and 3000 (70 animals/sex) ppm (0, 1.5, 15, and 150 mg/kg/day) of technical metolachlor for 2 years (Ciba-Geigy, 1983). The apparent increase in the incidence of "testicular atrophy" in male rats that died on test in this study is of doubtful toxicological significance. This finding was not present at final sacrifice, and historical control data demonstrate that this finding is relatively common in rats. Therefore, the NOEL for this study is 300 ppm, based on decreased body weight gain in rats fed 3000 ppm, the highest dose tested.

Metolachlor technical was fed in the diet at dose levels of 0, 30, 300, or 1000 ppm (0, 1.5, 15, and 50 mg/kg/day) to Charles River CD strain albino rats (15 males and 30 females/group) beginning at 32 days (Ciba-Geigy, 1981). Animals were mated after either 14 weeks (F0) or 17 weeks (F1) on test. Mating occurred once per generation. The F1 parental animals were randomly selected from the F1a litter after weaning of F1a. F0 males were sacrificed after 135 days on test and F0 females were sacrificed after 164 days on test. Gross examination was conducted on all F0 males and females that displayed "untoward developmental anomalies". After 157 to 167 days on test, F1 males were sacrificed and after 197 to 208 days, F1 females were sacrificed. Gross and histological examinations were performed on all F1 parents. Five randomly selected male and 5 female F1a progeny in each dose group were also examined histologically.

No compound related effect on parental body weight was observed. Food consumption was not effected by treatment in the F0 generation, but was significantly reduced for the F1 30 ppm females at week 16, 300 ppm females at weeks 6, 7, and 10 and the 1000 females at weeks 1, 6, 7, 8, 10, 13, 13, and 15, as compared to controls. Clinical observations of parental rats indicated no treatment-related effects. Pup survival was likewise not effected by treatment. Pup body weights of the 1000 ppm dose group were significantly reduced for F1a litters on days 14 and 21 and on days 4, 7, 14, and 21 for the F2a litters. Pup body weights of the 30 and 300 ppm dose groups did not appear to be effected in a compound-related manner. Liver-to-body weight ratios were significantly increased for both F1 parental males and females at 1000 ppm. The thyroid-to-body weight ratio and thyroid-to-brain weight ratio of 1000 ppm F1 males were significantly increased. Body weights of the weanling 1000 ppm F1a females and F2a males were reduced, though not significantly, and body weights of F2a weanling females were significantly reduced. The NOEL and LEL for reproductive toxicity are 300 and 1000 ppm (15 and 50 mg/kg/day), respectively, based on reduced pup weights and reduced parental food consumption.

I.A.3. UNCERTAINTY AND MODIFYING FACTORS (ORAL RfD)

UF -- The UF of 100 allows for uncertainty in the extrapolation of dose levels from laboratory animals to humans (10A) and uncertainty in the threshold for sensitive humans (10H).

MF -- None

I.A.4. ADDITIONAL COMMENTS (ORAL RfD)

Data Considered for Establishing the RfD

1) 2-Year Feeding (oncogenic) - rat: Principal study - see previous description; core grade minimum (Ciba-Geigy Corp., 1983)

2) 2-Generation Reproduction - rat: Co-Principal study - see previous description; core grade guideline (Ciba-Geigy Corp., 1981)

3) Teratology - rat: Maternal, Fetotoxic, and Teratogenic NOEL=360 mg/kg/day (HDT); Maternal, Fetotoxic, and Teratogenic LEL= none; core grade minimum (Ciba-Geigy Corp., 1976)

4) Teratology - rabbit: Maternal NOEL=120 mg/kg/day; Maternal LEL= none; Fetotoxic and Teratogenic NOEL=360 mg/kg/day; Fetotoxic and Teratogenic LEL= none; core grade minimum (Ciba-Geigy Corp., 1980)

Other Data Reviewed:

1) 2-Year Feeding (oncogenic) - mice: Systemic NOEL=1000 ppm (150 mg/kg/day); Systemic LEL=3000 ppm (450 mg/kg/day); core grade minimum (Ciba-Geigy Corp., 1982)

Data Gap(s): 6-Month Dog Study is under review

I.A.5. CONFIDENCE IN THE ORAL RfD

Study -- MediumData Base -- HighRfD -- High

Both co-critical studies are of good quality and are jointly given a medium confidence rating. Additional studies are supportive and of good quality; therefore, the data base is given a high confidence rating. High confidence in the RfD follows.

I.A.6. EPA DOCUMENTATION AND REVIEW OF THE ORAL RfD

Pesticide Registration Standard, September 1980

Pesticide Registration Files

Agency Work Group Review -- 04/22/86, 05/25/88, 06/22/88, 12/14/93

Verification Date -- 06/22/88

I.A.7. EPA CONTACTS (ORAL RfD)

William Burnam / OPP -- (703)557-7491

George Ghali / OPP -- (703)557-7490

I.B. REFERENCE CONCENTRATION FOR CHRONIC INHALATION EXPOSURE (RfC)

[Ed. Note: EPA does not have information yet in this section of IRIS].

II. CARCINOGENICITY ASSESSMENT FOR LIFETIME EXPOSURE

Section II provides information on three aspects of the carcinogenic risk assessment for the agent in question; the US EPA classification, and quantitative estimates of risk from oral exposure and from inhalation exposure. The classification reflects a weight-of-evidence judgment of the likelihood that the agent is a human carcinogen. The quantitative risk estimates are presented in three ways. The slope factor is the result of application of a low-dose extrapolation procedure and is presented as the risk per (mg/kg)/day. The unit risk is the quantitative estimate in terms of either risk per ug/L drinking water or risk per ug/cu.m air breathed. The third form in which risk is presented is a drinking water or air concentration providing cancer risks of 1 in 10,000, 1 in 100,000 or 1 in 1,000,000. Background Document 2 (Service Code 5) provides details on the rationale and methods used to derive the carcinogenicity values found in IRIS. Users are referred to Section I for information on long-term toxic effects other than carcinogenicity.

II.A. EVIDENCE FOR CLASSIFICATION AS TO HUMAN CARCINOGENICITY

II.A.1. WEIGHT-OF-EVIDENCE CLASSIFICATION

Classification -- C; possible human carcinogen.

Basis -- Classification is based on the appearance of proliferative liver lesions (combined neoplastic nodules and carcinomas) at highest dose tested (3000 ppm) in female rats.

II.A.2. HUMAN CARCINOGENICITY DATA

None.

II.A.3. ANIMAL CARCINOGENICITY DATA

Limited. Two chronic rat studies were conducted wherein metolachlor was incorporated in the diet for 2 years. Industrial Biotest Laboratories (IBT, 1979) fed 0, 30, 300, 1000, and 3000 ppm of metolachlor in the diet to 60 Charles River strain albino rats/sex/group. Proliferative hepatic lesions were significantly increased only in high-dose females when hyperplastic or neoplastic nodules were combined with angiosarcomas, cystic cholangiomas, cholangiomas and carcinomas. Inadequacies of this study, such as incomplete hematology, urinalysis, clinical chemistry, and dietary preparation records, prompted a repeat of the study.

Hazelton-Raltech, Inc. (1983) administered 0, 30, 300, or 3000 ppm metolachlor to 60-70 Charles River CD rats/sex/group for 104 weeks. There was a statistically significant increase in liver neoplastic nodules and carcinomas in the high-dose females when compared to controls. The increase was largely due to the occurrence of neoplastic nodules. No statistically significant increase in liver tumors was observed in male rats in either study.

Two chronic (2-year) mouse studies (IBT, 1977; Hazelton-Raltech, 1982) were conducted in which metolachlor was incorporated into the diet. IBT (1977) administered metolachlor at 0, 30, or 300 ppm; Halzelton-Raltech (1982) administered the compound at 300, 1000, or 3000 ppm. There were no oncogenic effects (p>0.05) noted in either study. The high dose produced weight reduction, thereby indicating that an MTD had been reached.

II.A.4. SUPPORTING DATA FOR CARCINOGENICITY

Metolachlor was not mutagenic in reverse mutation assays in Salmonella (US EPA, 1985). Its structure is similar to alachlor, which has been classified B2, but alachlor produces oncogenic response at different tumor sites (alachlor produces nasal turbinate, stomach and thyroid tumors). Available metabolic data indicate that both metolachlor and alachlor are metabolized to aniline derivatives (US EPA, 1985).

II.B. QUANTITATIVE ESTIMATE OF CARCINOGENIC RISK FROM ORAL EXPOSURE

[Ed. Note: EPA does not have information yet in this section of IRIS].

II.C. QUANTITATIVE ESTIMATE OF CARCINOGENIC RISK FROM INHALATION EXPOSURE

[Ed. Note: EPA does not have information yet in this section of IRIS].

II.D. EPA DOCUMENTATION, REVIEW, AND CONTACTS (CARCINOGENICITY ASSESSMENT)

II.D.1. EPA DOCUMENTATION

Source Document -- US EPA, 1985

The Toxicology Branch Peer Review Committee Office of Pesticide Programs, Office of Pesticides and Toxic Substances reviewed data on metolachlor.

II.D.2. REVIEW (CARCINOGENICITY ASSESSMENT)

Agency Work Group Review -- 11/10/87

Verification Date -- 11/10/87

II.D.3. US EPA CONTACTS (CARCINOGENICITY ASSESSMENT)

Elizabeth A. Doyle / OPP -- (703)308-2722

Reto Engler / OPP -- (703)308-2738

III. HEALTH HAZARD ASSESSMENTS FOR VARIED EXPOSURE DURATIONS

III.A. DRINKING WATER HEALTH ADVISORIES

The Office of Drinking Water provides Drinking Water Health Advisories (HAs) as technical guidance for the protection of public health. HAs are not enforceable Federal standards. HAs are concentrations of a substance in drinking water estimated to have negligible deleterious effects in humans, when ingested, for a specified period of time. Exposure to the substance from other media is considered only in the derivation of the lifetime HA. Given the absence of chemical-specific data, the assumed fraction of total intake from drinking water is 20%. The lifetime HA is calculated from the Drinking Water Equivalent Level (DWEL) which, in turn, is based on the Oral Chronic Reference Dose. Lifetime HAs are not derived for compounds which are potentially carcinogenic for humans because of the difference in assumptions concerning toxic threshold for carcinogenic and noncarcinogenic effects. A more detailed description of the assumptions and methods used in the derivation of HAs is provided in Background Document 3 in Service Code 5.

III.A.1. ONE-DAY HEALTH ADVISORY FOR A CHILD

Appropriate data for calculating a One-day HA are not available. It is recommended that the DWEL adjusted for a 10-kg child, 2E+0 mg/L, be used as the One-day HA.

III.A.2. TEN-DAY HEALTH ADVISORY FOR A CHILD

Appropriate data for calculating a Ten-day HA are not available. It is recommended that the DWEL adjusted for a 10-kg child, 2E+0 mg/L, be used as the Ten-day HA.

III.A.3. LONGER-TERM HEALTH ADVISORY FOR A CHILD

Appropriate data for calculating a Longer-term HA for a child are not available. It is recommended that the Drinking Water Equivalent Level (DWEL) adjusted for a 10-kg child, 2E+0 mg/L, be used for a Longer-term HA for a child.

III.A.4. LONGER-TERM HEALTH ADVISORY FOR AN ADULT

Appropriate data for calculating a Longer-term HA for an adult are not available. It is recommended that the Drinking Water Equivalent Level (DWEL) of 5E+0 mg/L be used for a Longer-term HA for an adult.

III.A.5. DRINKING WATER EQUIVALENT LEVEL / LIFETIME HEALTH ADVISORY

DWEL -- 5E+0 mg/L

Assumptions -- 2 L/day water consumption for a 70-kg adult

RfD Verification Date -- 06/22/88

Lifetime HA -- 1E-1 mg/L

Assumptions -- 20% exposure by drinking water

Principal Study -- Ciba-Geigy Corp., 1983 (This study was used in the derivation of the chronic oral RFD; see Section I.A.2.)

III.A.6. ORGANOLEPTIC PROPERTIES

No information is available on the organoleptic properties of metolachlor.

III.A.7. ANALYTICAL METHODS FOR DETECTION IN DRINKING WATER

Analysis of metolachlor is by a gas chromatographic method applicable to the determination of certain nitrogen- and phosphorous-containing pesticides in water samples.

III.A.8. WATER TREATMENT

Available data indicate that granular activated carbon adsorption will remove metolachlor from water.

III.A.9. DOCUMENTATION AND REVIEW OF HAs

EPA review of HAs in 1987.

Public review of HAs was January-March 1987.

Preparation date of this IRIS summary -- 09/30/88

III.A.10. EPA CONTACTS

Charles Abernathy / OST -- (202)260-5374

Edward Ohanian / OST -- (202)260-7571

III.B. OTHER ASSESSMENTS

Content to be determined.

IV. US EPA REGULATORY ACTIONS

EPA risk assessments may be updated as new data are published and as assessment methodologies evolve. Regulatory actions are frequently not updated at the same time. Compare the dates for the regulatory actions in this section with the verification dates for the risk assessments in sections I and II, as this may explain inconsistencies. Also note that some regulatory actions consider factors not related to health risk, such as technical or economic feasibility. Such considerations are indicated for each action. In addition, not all of the regulatory actions listed in this section involve enforceable federal standards. Please direct any questions you may have concerning these regulatory actions to the US EPA contact listed for that particular action. Users are strongly urged to read the background information on each regulatory action in Background Document 4 in Service Code 5.

IV.A. CLEAN AIR ACT (CAA)

[Ed. Note: EPA does not have information yet in this section of IRIS].

IV.B. SAFE DRINKING WATER ACT (SDWA)

Listed in the January 1991 Drinking Water Priority Listand may be subject to future regulation (56 FR 1470, 01/14/91).

IV.B.1. MAXIMUM CONTAMINANT LEVEL GOAL (MCLG) for Drinking Water

[Ed. Note: EPA does not have information yet in this section of IRIS].

IV.B.2. MAXIMUM CONTAMINANT LEVEL (MCL) for Drinking Water

[Ed. Note: EPA does not have information yet in this section of IRIS].

IV.B.3. SECONDARY MAXIMUM CONTAMINANT LEVEL (SMCL) for Drinking Water

[Ed. Note: EPA does not have information yet in this section of IRIS].

IV.B.4. REQUIRED MONITORING OF "UNREGULATED" CONTAMINANTS

Status -- Listed (Proposed, 1991)

Discussion --   "Unregulated" contaminants are those contaminants for which EPA establishes a monitoring requirement but which do not have an associated final MCLG, MCL, or treatment technique. EPA may regulate these contaminants in the future.

Monitoring requirement -- All systems to be monitored unless a vulnerability assessment determines the system is not vulnerable.

Analytical methodology -- Nitrogen-phosphorus detector/gas chromatography (EPA 507); gas chromatographic/mass spectrometry (EPA 525).

Reference -- 56 FR 3526 (01/30/91)

EPA Contact -- Drinking Water Standards Division / OGWDW /(202) 260-7575 / FTS 260-7575; or Safe Drinking Water Hotline / (800) 426-4791

IV.C. CLEAN WATER ACT (CWA)

[Ed. Note: EPA does not have information yet in this section of IRIS].

IV.D. FEDERAL INSECTICIDE, FUNGICIDE, AND RODENTICIDE ACT (FIFRA)

IV.D.1. PESTICIDE ACTIVE INGREDIENT, Registration Standard

Status -- Issued (1980)

Reference -- Metolachlor Pesticide Registration Standard. September, 1980(NTIS No. PB81-123280) as amended on January, 1987 (NTIS No. PB87-158853).

EPA Contact -- Registration Branch / OPP (703)557-7760 / FTS 557-7760

IV.D.2. PESTICIDE ACTIVE INGREDIENT, Special Review

[Ed. Note: EPA does not have information yet in this section of IRIS].

IV.E. TOXIC SUBSTANCES CONTROL ACT (TSCA)

[Ed. Note: EPA does not have information yet in this section of IRIS].

IV.F. RESOURCE CONSERVATION AND RECOVERY ACT (RCRA)

[Ed. Note: EPA does not have information yet in this section of IRIS].

IV.G. SUPERFUND (CERCLA)

[Ed. Note: EPA does not have information yet in this section of IRIS].

VI. BIBLIOGRAPHY

VI.A. ORAL RfD REFERENCES

Ciba-Geigy Corporation. 1976. MRID No. 00015396. Available from EPA. Write to FOI, EPA, Washington, DC 20460.

Ciba-Geigy Corporation. 1980. MRID No. 00041283. Available from EPA. Write to FOI, EPA, Washington, DC 20460.

Ciba-Geigy Corporation. 1981. MRID No. 00080897. Available from EPA. Write to FOI, EPA, Washington, DC 20460.

Ciba-Geigy Corporation. 1982. MRID No. 00039194, 00117597. Available from EPA. Write to FOI, EPA, Washington, DC 20460.

Ciba-Geigy Corporation. 1983. MRID No. 00063398, 00084005, 00129377, 00144364, 00158924. Available from EPA. Write to FOI, EPA, Washington, DC 20460.

VI.B. INHALATION RfC REFERENCES

None

VI.C. CARCINOGENICITY ASSESSMENT REFERENCES

Hazelton-Raltech, Inc. 1982. Carcinogenicity Study with Metolachlor in Albino Mice. Cited in US EPA, 1985.

Hazelton-Raltech, Inc. 1983. Chronic Rat Study of Metolachlor. Cited in US EPA, 1985.

IBT (Industrial Biotest Laboratories). 1977. Oncogenic Mice. Cited in US EPA, 1985.

IBT (Industrial Biotest Laboratories). 1979. Two-year Chronic Oncogenicity Oral Toxicity Study with Metolachlor in Albino Rats. Cited in US EPA, 1985.

US EPA. 1985. Toxicology Branch Peer Review Committee, Office of Pesticide Programs, Office of Pesticides and Toxic Substances memorandum on metolachlor. May 30.

VI.D. DRINKING WATER HA REFERENCES

Ciba-Geigy Corporation. 1983. MRID No. 00129377. Available from EPA. Write to FOI, EPA, Washington, DC. 20460.

VII. REVISION HISTORY

Date	Section	Description
03/31/87	IV.	Regulatory Action section on-line
03/01/88	I.A.5.	Confidence levels revised
06/30/88	I.A.	Withdrawn pending further review
08/22/88	II.	Carcinogen summary on-line
09/07/88	I.A.	Revised oral RfD summary added
12/01/88	I.A.1.	Clarified effect0
2/01/89	II.D.3.	Secondary contact's area code corrected
10/01/90	I.A.1.	Oral RfD corrected
02/01/91	I.A.	Text edited
02/01/91	II.	Text edited
02/01/91	VI.	Bibliography on-line
03/01/91	III.A.	Health Advisory on-line
03/01/91	VI.D.	Health Advisory references added
01/01/92	IV.	Regulatory actions updated
10/01/93	II.D.3.	Primary contact changed; secondary's phone no. changed
01/01/94	I.A.	Oral RfD noted as pending change
01/01/94	I.A.6.	Work group review date added

SYNONYMS

ACETAMIDE, 2-CHLORO-N-(6-ETHYL-
o-TOLYL)-N-(2-METHOXY-1-
METHYLETHYL)-
o-ACETOTOLUIDIDE, 2-CHLORO-6'-
ETHYL-N-(2-METHOXY-1-
METHYLETHYL)-
2-AETHYL-6-METHYL-N-(1-METHYL-
2-METHOXYAETHYL)-
CHLORACETANILID
BICEP
CGA-24705
alpha-CHLOR-6'-AETHYL-n-(2-
METHOXY-1-METHYLAETHYL)-
ACET-o-TOLUIDIN
alpha-CHLORO-2'-ETHYL-6'-METHYL-
N-(1-METHYL-2-METHOXYETHYL)-
ACETANILIDE
2-CHLORO-6'-ETHYL-N-(2-METHOXY-
1-METHYLETHYL)ACET-o-TOLUIDIDE

2-CHLORO-N-(2-ETHYL-6-
METHYLPHENYL)-N-(2-METHOXY-
1-METHYLETHYL)ACETAMIDE
CODAL
COTORAN MULTI
DUAL
2-ETHYL-6-METHYL-1-N-(2-
METHOXY-1-
METHYLETHYL)CHLOROACETANIL
IDE
METELILACHLOR
Metolachlor
MILOCEP
ONTRACK 8E
PRIMAGRAM
PRIMEXTRA

Source: *Instant EPA's IRIS*, 1997, Instant Reference Sources, Inc., 7605 Rockpoint Drive, Austin, TX 78731 (Fax: 512-345-2386; Internet URL, http://www.instantref.com/inst-ref.htm).

Production, Use, and Pesticide Labeling Information

1. Description of Chemical

Generic Name: 2-chloro-N-(2-ethyl-6-methylphenyl)N-(2-methoxy-1-methylethyl) acetamide

Common Name: Metolachlor

Trade Names: Dual, CGA-24705, Ontrack, Pennant

EPA Shaughnessy code: 108801

Chemical Abstracts Service (CAS) Number: 51218-45-2

Year of Initial Registration: 1976

Pesticide Type: Herbicide

Chemical Family: Chloracetanilide

US Producer: Ciba-Geigy Corporation, Agricultural Division (from active ingredient manufactured outside the US)

2. Use Patterns and Formulations

- Application sites: For pre-emergence control of certain broadleaf and grassy weeds in terrestrial crop areas (corn, sorghum, cotton, potatoes, peanuts, soybeans, green beans, kidney and other beans, blackeye peas and other peas, stone fruits and tree nuts) and terrestrial noncrop areas (ornamental plants, railroad, and highway rights-of-way).

- Types of formulations: 95% active ingredient (ai) technical grade manufacturing-use product. 5%, 15%, and 25% granulars, 86.4% emulsifiable concentrate, 27.5%, 31.8%, and 36.1% flowable concentrate with atrazine; 36.3% flowable concentrate with propazine; and 73.6% emulsifiable concentrate with metribuzin.

- Types and methods of application: End-use product is applied by ground spray equipment, aircraft, or through center pivot irrigation systems.

- Application rates: 1.25 to 4 lb metolachlor ai per acre (A) for terrestrial crop and non-crop areas.

- Usual carrier: Water, fluid fertilizers.

3. Science Findings

Chemical Characteristics

- Physical state: Liquid

- Color: White to tan

- Molecular weight: 238.8

- Solubility: 530 ppm in water at 20 °C, miscible with xylene toluene, dimethyl formamide, methyl cellosolve, butyl cellosolve, ethylene dichloride, and cyclohexanone. Insoluble in ethylene glycol and propylene glycol.

- Vapor pressure: About 10^{-5} mm Hg at 20 °C

Toxicological Characteristics

- Acute effects:

- Acute oral toxicity (rat): 2780 mg/kg (Toxicity Category III Moderately toxic)

- Acute dermal toxicity (Rabbit): > 10,000 mg/kg (Toxicity Category III - moderately toxic)

- Acute Inhalation Toxicity: >1.752 mg/L with 4-hour exposure (Toxicity Category IV - non toxic)

- Primary eye irritation: Non-irritating

- Primary dermal irritation: Non-irritating

- Dermal sensitization: Sensitizer in guinea pig

- Subchronic oral toxicity (dog): NOEL = 100 ppm (2.5 mg/kg) Decreased gain in body weight in males and females. Failure of the serum alkaline phosphatase to decrease with increased age, and possible effects on blood clotting systems at 300 and 1000 ppm.

- Chronic Effects:

- 3-generation reproduction (rat): NOEL = 300 ppm (15 mg/kg) Reduced pup weights and reduced parental food consumption at 1000 ppm (50 mg/kg)

- Teratogenicity:

- Rabbits - Not fetotoxic or teratogenic Maternal toxicity at high dose (360 mg/kg/day)

- Rats - Not fetotoxic or teratogenic. Decrease in food consumption at high dose (360 mg/kg/day) in first third of study.

- Chronic feeding/oncogenicity:

- Mice-Not oncogenic in two studies up to and including 3000 ppm (429 mg/kg)

- Rat - Systemic NOEL of 30 ppm (1.5 mg/kg). Systemic LEL of 300 ppm (testicular atrophy). In one study a statistically significant increase in primary liver neoplasms in females of high dose group (3000 ppm). In repeat study a statistically significant increased incidence of neoplastic liver nodules and proliferative hepatic lesions in females of the high dose group (3000 ppm)

- Mutagenicity: Negative in an Ames Test, and a mouse dominant lethal test.

- Major routes of exposure: Dermal, ocular, and inhalation from mixing concentrates and applying spray mixtures.

Physiological and Biochemical Behavior Characteristics

- Absorption: Generally applied prior to plant emergence. Absorbed through shoots just above seed and possibly roots.

- Mechanism of pesticidal action: Member of group of chloracetamide herbicides which are general growth inhibitors, especially of root elongation. Metolachlor may disrupt the integrity of plant cell membranes and inhibit lipid synthesis.

Environmental Characteristics

- Available data are insufficient to assess the environmental fate of metolachlor. There are indications that metolachlor is essentially stable in loamy sand soil over 64 days. Absorption constants in sandy clay loam, loam, and two sand soils indicate mobility in these soils.

Aged metolachlor ^{14}C residue was mobile in columns of loamy sand soil. Metolachlor ^{14}C residues were mobile in sandy loams, sand, and silt loam soil. Metolachlor ^{14}C residues were detected in plants grown in metolachlor-treated soil. Metolachlor has been found in several surface water surveys. Transient peaks of 1.2 to 4.4 ppb are reported in river water possibly as a result of runoff during spring and summer. Detectable levels of metolachlor were found in some tapwater samples. It has been found in ground water in two States.

Ecological Characteristics

- Avian oral toxicity: Mallard duck > 2510 mg/kg

- Avian dietary toxicity: Mallard duck > 10,000 ppm. Bobwhite
    quail >10,000 ppm

- Avian reproduction: Mallard duck - showed no impairment at any
test level - 1,000 or 10,000 ppm. Bobwhite quail - NOEL = 10 ppm,
impairment at 300 ppm but no effect at 1000 ppm

- Freshwater fish toxicity: Bluegill sunfish - 10.0 ppm, Rainbow trout -
    3.9 ppm

- Aquatic invertebrates: Daphnia Magna - 25.1 ppm

- Fish life cycle: Fathead minnow - Maximum acceptable toxicant
concentration > 0.78 < 1.6 ppm

Tolerance Assessment

- Tolerances have been established for residues of metolachlor and
its metabolites in raw agricultural commodities, milk, eggs, meat, fat
and meat byproducts 40 CFR 180.368(a) as follows:

Commodities	Parts Per Million
Almond hulls	0.3
Cattle, fat	0.02
Cattle, kidney	0.2
Cattle, liver	0.05
Cattle, meat	0.02
Cattle, meat byproducts (mbyp) (except kidney and liver)	0.02
Corn, fresh (inc. sweet, kernel plus cob with husk removed)	0.1
Corn, forage and fodder	8.0
Corn, grain	0.1
Cottonseed	0.1
Eggs	0.02
Goats, fat	0.02
Goats, kidney	0.2
Goats, liver	0.05

Table Continued)

Commodities	Parts Per Million
Goats, meat	0.02
Goats, mbyp (except kidney and liver)	0.02
Hogs, fat	0.02
Hogs, kidney	0.2
Hogs, liver	0.05
Hogs, meat	0.02
Hogs, mbyp (except kidney and liver)	0.02
Horses, fat	0.02
Horses, kidney	0.2
Horses, liver	0.05
Horses, meat	0.02
Horses, mbyp (except kidney and liver)	0.02
Legume vegetables group foliage (except soybean forage and soybean hay)	15.0
Milk	0.02
Peanuts	0.5
Peanut, forage and hay	30.0
Peanut, hulls	6.0
Peppers, chili	0.5
Potatoes	0.2
Poultry, fat	0.02
Poultry, liver	0.05
Poultry, meat	0.02
Poultry, mbyp (except liver)	0.02
Safflower seed	0.1
Seed and pod vegetables (except soybeans)	0.3
Sheep, fat	0.3
Sheep, kidney	0.2
Sheep, liver	0.05
Sheep, meat	0.02
Sheep, mbyp (except kidney and liver)	0.02
Sorghum, forage and fodder	2.0
Sorghum, grain	0.3
Soybeans	0.2
Soybeans, forage and hay	8.0
Stone fruits group	0.1
Tree nuts group	0.1

- Tolerances have been established {40 CFR 180.368(b) for indirect or inadvertent residues of metolachlor as a result of application of metolachlor to the growing crops listed in 180.368(a) as follows:

Commodities	Parts Per Million
Barley, fodder	0.5
Barley, forage	0.5
Barley, grain	0.1
Buckwheat, fodder	0.5
Buckwheat, forage	0.5
Buckwheat, grain	0.1
Millet, fodder	0.5
Millet, forage	0.5
Millet, grain	0.1
Milo, fodder	0.5
Milo, forage	0.5
Milo, grain	0.1
Oats, fodder	0.5
Oats, forage	0.5
Oats, grain	0.1
Rice, fodder	0.5
Rice, forage	0.5
Rice, grain	0.1
Rye, fodder	0.5
Rye, forage	0.5
Rye, grain	0.1
Wheat, fodder	0.5
Wheat, forage	0.5
Wheat, grain	0.1

- Canadian tolerances of 0.1 ppm have been established for residues of metolachlor in or on beans, corn, peas, potatoes, and soybeans. No Mexican tolerances or Codex Alimentarius Commission Maximum Residue Limits have been established for residues of metolachlor.

- Results of tolerance assessment:

- Using the NOEL of 30 ppm (1.5 mg/kg/day) from the rat chronic feeding study and a safety factor of 100, the acceptable daily intake (ADI) is 0.015 mg/kg/day, and the maximum permissible intake (MPI) is 0.9 mg/day for a 60 kg person. The established tolerances result in a total theoretical maximum residue concentration (TMRC) of 0.0755 mg/day (1.5 kg diet) which corresponds to 8.38 percent of the MPI for a 60 kg person.

Summary Science Statement

> - Metolachlor is not considered to be teratogenic or cause reproductive effects. It is not oncogenic in mice, but is considered an oncogen in rats, and is tentatively classified as showing limited evidence of carcinogenicity in animals. Metolachlor is not mutagenic in available studies, but mutagenicity and metabolism testing requirements not in effect at the time of issuance of the original Metolachlor Registration Standard in 1980 must be met. Metolachlor has been found in several surface water surveys, some tapwater samples, and in ground water in two States. Monitoring studies are required to determine the extent of contamination on a national scale. Metolachlor is slightly to moderately toxic to non-target organisms. Available data are insufficient to assess the environmental fate of metolachlor.

4. Summary of Regulatory Position and Rationale

Unique label precautionary statements:

> - Manufacturing-Use Products

> > Environmental Hazards

> > "Do not discharge effluent containing this product into lakes, streams, ponds, estuaries, oceans, or public water unless this product is specifically identified and addressed in an NPDES (National Pollution Discharge Elimination System) permit. Do not discharge effluent systems without previously notifying the sewage treatment plant authority. For guidance, contact your State Water Board or Regional Office of the EPA."

> - End-Use Products

> - a   Environmental hazards for emulsifiable concentrates and flowable concentrates

> > "Do not apply directly to water or wetlands (swamps, bogs, marshes, and potholes). Do not contaminate water by cleaning of equipment or disposal of wastes."

> - b   Environmental hazards for granules

> > "Cover or incorporate granules that are spilled during loading or are visible on soil surface in turn areas. Do not contaminate water by cleaning of equipment or disposal of wastes.

- c   Ground water and surface water advisory

"Metolachlor has been identified in limited sampling of ground water and there is the possibility that it may leach through soils to ground water, especially where soils are coarse and ground water is near the surface. Following application and during rainfall events that cause runoff, metolachlor may reach surface water bodies including streams, rivers and reservoirs."

"Care must be taken when using this product to prevent back siphoning into wells, spills or improper disposal of excess pesticide, spray mixtures or rinsates."

"Check valves or anti-siphoning devices must be used on all mixing and/or irrigation equipment."

- d   Crop rotation statement

"Crops other than beans (succulent or dry), fresh corn, grain corn, cotton, peanuts, peas (succulent and dry), chili peppers, potatoes, safflower, sorghum, soybeans, stone fruits, tree nuts, and barley, buckwheat, millet, milo, oats, rice, rye, and wheat may not be planted in metolachlor-treated soil until 12 months after application."

- Endangered species statements for products registered for crop use:

"It is a violation of Federal laws to use any pesticide in a manner that results in the death of an endangered species or adverse modification of their habitat."

"The use of this product may pose a hazard to certain Federally designated endangered species known to occur in specific areas within the CALIFORNIA counties of Merced, Sacramento, and Solano. Before using this product in these counties you must obtain the EPA Endangered Species Crop Bulletin. The bulletin is available from either your County Extension Agent, the Endangered Species Specialist in your State Wildlife Agency Headquarters, or the Regional Office of either the US Fish and Wildlife Service (Portland, Oregon) or the US Environmental Protection Agency (San Francisco, California). THIS BULLETIN MUST BE REVIEWED PRIOR TO PESTICIDE USE. THE USE OF THIS PRODUCT IS PROHIBITED IN THESE COUNTIES UNLESS SPECIFIED OTHERWISE IN THE BULLETIN."

5. Summary of Data Gaps

Toxicology

Mutagenicity studies	Oct. 1987
General metabolism	Oct. 1988
Effects on coagulation	Apr. 1987**

Environmental fate

Hydrolysis	July 1987
Photodegradation studies	July 1987
Metabolism studies	Jan. 1989
Mobility studies	Jan. 1989
Accumulation studies	Jan. 1990
Ground and surface water monitoring	Jan. 1987**

Product chemistry/residue chemistry

Product chemistry	Jan. 1988
Plant metabolism	Apr. 1988
Storage stability	Oct. 1988
Selected residue studies	Apr. 1989

- **NOTE: Date protocols are due. After acceptance, the Agency will provide time frame for submission of the reports.

6. Contact Person at EPA

Richard F. Mountfort
Product Manager No. 23
US Environmental Protection Agency, TS-767C
401 M Street SW
Washington, DC 20460

Telephone: (703) 557-1830

Source: *Instant EPA's Pesticide Facts*, 1994, Instant Reference Sources, Inc., 7605 Rockpoint Drive, Austin, TX 78731 (Fax: 512-345-2386; Internet URL, http://www.instantref.com/inst-ref.htm).

Metribuzin

Introduction

Metribuzin is a pre- and post-emergence triazone herbicide. [828] It is used to control of broadleaf weeds and grasses in soybeans, potatoes, barley, winter wheat, asparagus, sugarcane, tomatoes, lentils, and peas. It is manufactured in several types of formulations including wettable powders, flowable concentrates, and dry flowable concentrates. It may be applied using soil incorporation, on the surface of soils, applied to foliage, or broadcast with ground or aerial equipment or by using sprinkler irrigation. Metribuzin has been found in Ohio rivers and Iowa wells. Available data show that metribuzin has a potential to contaminate groundwater in soils low in organic and clay content. [821]

Metribuzin will be released to the environment primarily during agricultural spraying operations. If released to the atmosphere, degradation of vapor phase metribuzin by reaction with photochemically produced hydroxyl radicals (estimated half-life of 11 hrs) will be important. Metribuzin can be removed from air via rainfall and, in a particulate phase, metribuzin may be removed from air via dry deposition. If released to soil, biodegradation will be the primary fate process. Metribuzin is moderately adsorbed (Koc of 95) on soils with high clay and(or) organic content by a H-bonding mechanism and adsorption decreases with an increase in soil pH. Little leaching occurs on soils with high organic content, but metribuzin is readily leached in sandy soils. The soil half-life is in the range of 14-60 days. In water, biodegradation may be important based on studies in soil. Slow hydrolysis may aid in metribuzin degradation. Volatilization from water and bioconcentration in fish will not be important. Exposure of the general population to metribuzin may occur through ingestion of contaminated foods and drinking water as well as inhalation of dust and dermal contact resulting from its use. Workers may be exposed via dermal contact and inhalation of dust. [828]

Identifying Information

CAS NUMBER: 21097-64-9

SYNONYMS: 4-AMINO-6-(1,1-DIMETHYLETHYL)-3-(METHYLTHIO)-1,2,4-TRIAZIN-5(4H)-ONE; 4-AMINO-6-tert-BUTYL-3-(METHYLTHIO)-1,2,4-TRIAZIN-5-ON; 4-AMINO-6-tert-BUTYL-3-METHYLTHIO-as-TRIAZIN-5-ONE; BAY 61597; BAY DIC 1468; BAYER 6159H; BAYER 6443H; BAYER 94337; DIC 1468; LEXONE; SENCOR; SENCORAL; SENCORER; SENCOREX; 1,2,4-TRIAZIN-5(4H)-

ONE, 4-AMINO-6-(1,1-DIMETHYLETHYL)-3-(METHYLTHIO)-; 1,2,4-TRIAZIN-5-ONE, 4-AMINO-6-tert-BUTYL-3-(METHYLTHIO)-; as-TRIAZIN-5(4H)-ONE, 4-AMINO-6-tert-BUTYL-3-(METHYLTHIO)-

Endocrine Disruptor Information

Metribuzin is a strongly suspected EED that it is listed by the Centers for Disease Control & Prevention [812], and the World Wildlife Fund [813] as a potential endocrine modifying chemical. A chronic rat study indicated a statistically significant increase in the incidence of adenoma of liver bile duct and pituitary gland in females. [819]

Chemical and Physical Properties

CHEMICAL FORMULA: $C_8H_{14}N_4OS$ [817]

MOLECULAR WEIGHT: 214.28 [817]

PHYSICAL DESCRIPTION: white crystalline solid with a slight sulfur odor in the technical grade [817]

DENSITY: 1.31 at 20/4°C [817]

MELTING POINT: 125-126.5°C [817]

SOLUBILITY: soluble in aromatic and chlorinated hydrocarbon solvents, and in water (at 20 °C) to 1.220 g/L. [821]

Environmental Reference Materials

A source of EED environmental reference materials for metribuzin is Radian International LLC, Austin, TX (Fax 512-454-0268; Internet URL, http://www.radian.com/standards); Catalog No. ERM-033.

Hazardous Properties

ACUTE/CHRONIC HAZARDS: Metribuzin is not acutely toxic by oral, dermal, inhalation, or eye irritation routes of exposure. Available data indicate that metribuzin is not mutagenic. [821]

FIRE HAZARD: nonflammable [817]

Medical Symptoms of Exposure

Symptoms of exposure with dogs included reduced weight gain, increased mortality, hematological changes, and liver and kidney damage. [821]

Toxicological Information

The toxicological information below is gathered from several sources including two national databases: one from the *National Toxicology Program Chemical Repository Database* and another from *EPA's Integrated Risk Information System (IRIS)*.

Toxicology Information from the National Toxicology Program

Metribuzin is not currently listed in NTP's Chemical Repository database nor is it scheduled for addition in the future.

Toxicology Information from EPA's Integrated Risk Information System (IRIS)

I. CHRONIC HEALTH HAZARD ASSESSMENTS FOR NONCARCINOGENIC EFFECTS

I.A. REFERENCE DOSE FOR CHRONIC ORAL EXPOSURE (RfD)

The Reference Dose (RfD) is based on the assumption that thresholds exist for certain toxic effects such as cellular necrosis, but may not exist for other toxic effects such as carcinogenicity. In general, the RfD is an estimate (with uncertainty spanning perhaps an order of magnitude) of a daily exposure to the human population (including sensitive subgroups) that is likely to be without an appreciable risk of deleterious effects during a lifetime. Please refer to Background Document 1 in Service Code 5 for an elaboration of these concepts. RfDs can also be derived for the noncarcinogenic health effects of compounds which are also carcinogens. Therefore, it is essential to refer to other sources of information concerning the carcinogenicity of this substance. If the US EPA has evaluated this substance for potential human carcinogenicity, a summary of that evaluation will be contained in Section II of this file when a review of that evaluation is completed.

NOTE: The Oral RfD for metribuzin may change in the near future pending the outcome of a further review now being conducted by the RfD/RfC Work Group.

I.A.1. ORAL RfD SUMMARY

Critical Effect	Experimental Doses*	UF	MF	RfD
Liver and kidney effects, decreased body weight, mortality	NOEL: 100 ppm (2.5 mg/kg/day)	100	1	2.5E-2 mg/kg/day
	LEL: 1500 ppm (37.5 mg/kg/day)			

*Conversion Factors: 1 ppm = 0.025 mg/kg/day (assumed dog food consumption)

2-Year Feeding Study in Dogs, Mobay Chemical, 1974a

I.A.2. PRINCIPAL AND SUPPORTING STUDIES (ORAL RfD)

Mobay Chemical Corporation. 1974a. MRID No. 00061260, 00139397. Available from EPA. Write to FOI, EPA, Washington, DC 20460.

Sixteen male and 16 female Beagle dogs were fed 0 (control), 25, 100, and 1500 ppm of metribuzin in the diet for 2 years. Animals were examined daily for physical appearance and weighed weekly during the first year and biweekly thereafter. Clinical-chemical tests were performed at the beginning of the study and at 2, 4, 6, 12, and 24 months (an additional hematologic test was conducted at 23 months). High mortality, decreased body weight, increased relative liver weight along with related clinical tests, and histopathologic findings of liver and kidney damage were noted in the 1500 ppm test group. The two lower dose groups did not exhibit any treatment-related effects.

I.A.3. UNCERTAINTY AND MODIFYING FACTORS (ORAL RfD)

UF -- Based on a chronic exposure study, an uncertainty factor of 100 was used to account for the inter- and intra-species differences.

MF -- None

I.A.4. ADDITIONAL COMMENTS (ORAL RfD)

Data Considered for Establishing the RfD

1) 2-Year Feeding - dog: Principal study - see discussion above; core grade minimum

2) 2-Year Feeding (oncogenic) - rat: NOEL=100 ppm (5.0 mg/kg/day), LOEL=300 ppm (15.0 mg/kg/day) (decreased body weight gain and pathologic changes in the liver, kidneys, uterus and mammary glands); core grade minimum (Mobay Chemical, 1974b)

3) Teratology - rabbit: Developmental Toxicity NOEL=15 mg/kg/day, Developmental Toxicity LEL = 45 mg/kg/day (no malformations noted); Maternal NOEL=15 mg/kg/day, Maternal LEL=45 mg/kg/day; Teratogenic NOEL=135 mg/kg/day (HDT); A/D ratio = 1; core grade guideline (Mobay Chemical, 1981a)

4) Teratology - rat: Maternal Toxicity NOEL=100 mg/kg (HDT); Teratogenic NOEL=100 mg/kg (HDT); core grade supplementary (Mobay Chemical, 1972)

5) 3-Generation Reproduction - rat: Reproductive NOEL=300 ppm (15 mg/kg/day) (HDT); Maternal Toxicity NOEL=300 ppm (HDT); core grade supplementary (Mobay Chemical, 1974c)

Other Data Reviewed

1) 2-Year Oncogenic - mouse: Systemic NOEL=800 ppm (120 mg/kg/day), Systemic LEL=3200 ppm (480 mg/kg/day) (based on hematocrit, hemoglobin, and liver weight data); core grade guideline (Mobay Chemical, 1981b)

Data Gap(s): Rat Teratology Study; Rat Reproduction Study

I.A.5. CONFIDENCE IN THE ORAL RfD

Study -- High
Data Base -- Medium
RfD -- Medium

The principal study is of good quality; although there was a minor question of dosage selection, it is given a high rating. Additional studies are also of good quality, but reproductive data are missing; therefore, confidence in the data base can be considered medium to high. Confidence in the RfD can also be considered medium to high.

I.A.6. EPA DOCUMENTATION AND REVIEW OF THE ORAL RfD

Pesticide Registration Standard June, 1985

Agency Work Group Review -- 05/30/86, 02/16/94, 12/08/94

Verification Date -- 05/30/86

I.A.7. EPA CONTACTS (ORAL RfD)

William Burnam / OPP -- (703)557-7491

George Ghali / OPP -- (703)557-7490

I.B. REFERENCE CONCENTRATION FOR CHRONIC INHALATION EXPOSURE (RfC)

Not available at this time.

II. CARCINOGENICITY ASSESSMENT FOR LIFETIME EXPOSURE

Section II provides information on three aspects of the carcinogenic risk assessment for the agent in question; the US EPA classification, and quantitative estimates of risk from oral exposure and from inhalation exposure. The classification reflects a weight-of-evidence judgment of the likelihood that the agent is a human carcinogen. The quantitative risk estimates are presented in three ways. The slope factor is the result of application of a low-dose extrapolation procedure and is presented as the risk per (mg/kg)/day. The unit risk is the quantitative estimate in terms of either risk per ug/L drinking water or risk per ug/cu.m air breathed. The third form in which risk is presented is a drinking water or air concentration providing cancer risks of 1 in 10,000, 1 in 100,000 or 1 in 1,000,000. Background Document 2 (Service Code 5) provides details on the rationale and methods used to derive the carcinogenicity values found in IRIS. Users are referred to Section I for information on long-term toxic effects other than carcinogenicity.

II.A. EVIDENCE FOR CLASSIFICATION AS TO HUMAN CARCINOGENICITY

II.A.1. WEIGHT-OF-EVIDENCE CLASSIFICATION

Classification -- D; not classifiable as to human carcinogenicity

Basis -- No human data and inadequate evidence from animal bioassays. Metribuzin did not increase the incidence of tumors in a lifetime dietary study using CD-1 mice when compared with both concurrent and historic controls. In a 2-year feeding study in Wistar rats, no significant differences in neoplastic findings between the test and control groups were found. Short-term studies in bacteria and mammalian systems suggest that metribuzin is not mutagenic.

II.A.2. HUMAN CARCINOGENICITY DATA

None. There are no studies investigating an association between human cancer and exposure to metribuzin.

II.A.3. ANIMAL CARCINOGENICITY DATA

Inadequate. No evidence of carcinogenicity was seen in CD-1 mice or Wistar rats exposed to metribuzin in the diet for 2 years. The study in rats was determined to be an adequate cancer bioassay; the study in mice was adequate in females only, as it appears that the MTD may not have been achieved in male mice.

Mobay Chemical Corp. (1981) examined the carcinogenic properties of metribuzin in a 2-year feeding study in outbred CD-1 mice. In this study, metribuzin technical (92.9% pure) dissolved in corn oil was added to a commercial diet fed to the mice (50/sex/group) at levels of 0, 200, 800 or 3200 ppm for 104 weeks. Based on food consumption and analytical chemistry results, the investigators estimated that these dietary levels of metribuzin provided doses of 0, 25, 111 or 438 mg metribuzin/kg/day for males and 0, 35, 139 or 567 mg metribuzin/kg/day for females. The body weights of the treated males did not differ significantly from those of the control group. A statistically significant increase in body weight (approximately 7%) was noted in the 800-ppm female dose group twice during the first year of the bioassay; with this exception body weights of female dose groups did not differ significantly from controls. Treatment with the pesticide did not affect survival rates. All major organs were examined upon necropsy and the tissues were prepared for histology. Organ-specific tumor incidences were not reported; with the exception of the total number of tumor-bearing animals, data were reported only as percentages. Malignant lymphomas were found in 20- 26% of the female mice in each group, and in 6% (high dose) to 26% (control group) in males. Hepatocellular neoplasms (adenomas and adenocarcinomas) were reported almost exclusively in male mice (7% high-dose group, 25% control group). The incidence of pulmonary neoplasms ranged from 20% in control males to 6% in the high-dose males, and from 2% in high-dose females to 22% in the low- and middle-dose female groups. The total number of tumor-bearing males was inversely related to the dose (28/50, 26/50, 22/50 and 18/50 for controls, low-, middle- and high-dose groups, respectively). The total number of tumor-bearing females was increased in the low- and middle-dose groups (28/50 and 29/50, respectively), but the incidence in the high-dose groups (20/50) was comparable with that in control females (19/50). Early mortality did not depress tumor occurrences. The percentage of total tumors that was determined to be malignant was either lower than or comparable to that of controls: 91%, 81%, 85% and 83% for control, low-, middle- and high-dose males, respectively; and 92%, 94%, 81% and 91% for control, low-, middle- and high-dose females, respectively.

The US EPA Office of Pesticide Programs performed statistical analysis of these data using the Chi square test. A decrease in malignant and total tumor-bearing male mice in the high-dose group was statistically significant ($p=0.037$ and $p=0.045$, respectively). The number of tumor-bearing female mice appeared to be increased in the low-dose group (not statistically significant, $p=0.071$) and was statistically significantly increased in the middle-dose group ($p=0.45$ for malignant tumors and $p=0.0499$ for benign tumors). The tumor incidence in high-dose females was comparable to that in control female mice. The overall conclusion drawn was that under the test conditions in the Mobay Chemical Corp. (1981) study, metribuzin did not increase the incidences of tumors in mice. This study fulfills the criteria for an adequate cancer bioassay in female mice. Sufficient numbers of animals were used, the study was of sufficient duration, and survival was not affected. The MTD was achieved in the high-dose group females, as evidenced by a significant decrease in hemoglobin and hematocrit levels, and a statistically significant increase in liver weight (absolute and relative), kidney weight (absolute and relative), and spleen weight (relative only). In male mice, liver weight (absolute and relative) was increased in the middle-dose group only, suggesting that the MTD may not have been reached.

In a 2-year feeding study, 40 Wistar rats/sex/group received 25, 35, 100 and 300 ppm metribuzin (99.5% pure) in their diets; 80 rats/sex served as controls (Mobay Chemical Corp., 1974a). These doses corresponded to 0, 1.3, 1.9, 5.3 and 14.4 mg metribuzin/kg/day, respectively, in the males and 0, 1.7, 2.3, 6.5 and 20.4 mg metribuzin/kg/day, respectively, in females (Mobay Chemical Corp., 1974a). No significant differences in food consumption were reported between the treated animals and the respective control groups. The body weights of animals in the 25-, 35- and 100-ppm dose groups (both sexes) did not differ significantly from those of respective controls. Body weights in the male high-dose group were significantly decreased during weeks 70-80 and 90-100; high-dose female body weights were significantly decreased from weeks 20-100. At the end of study (104 weeks) there were no significant differences between the body weights of high-dose males and females and their respective controls.

All of the animals in the control and high-dose groups were subjected to a complete histopathologic evaluation; 10 rats/sex in the 25-, 35- and 100-ppm groups were subjected to a partial evaluation. The initial evaluation by Mobay Chemical Corp. (1974a) reported statistically significant increases in the incidence of liver bile duct adenomas and pituitary gland adenomas in the female high-dose groups by pair-wise comparison. The incidence of bile duct adenoma in the females was 13/71, 4/10, 5/10, 1/10 and 19/35 in the control, 25-, 35-, 100- and 300-ppm groups, respectively; the incidence in males was 19/66, 10/10, 8/10, 5/10 and 9/29, respectively. The incidence of pituitary gland adenomas in the female control and high-dose groups was 27/71 and 21/35, respectively; in males the incidences were 10/62 and 6/29, respectively. Pituitary glands were not subjected to histopathologic evaluation in other treated groups.

The US EPA Office of Pesticide Programs apparently reevaluated the original histopathology in the Mobay Chemical Corp. (1974) study; in this reevaluation all of the female liver bile duct adenomas were reclassified as bile duct proliferation. The pituitary glands from all animals were also histopathologically reevaluated; the incidences of pituitary adenoma in the female groups were 16/71 (23%), 6/34 (18%), 9/31 (29%), 11/33 (33%) and 14/35 (40%) in the control, 25-, 35-, 100- and 300-ppm groups, respectively. The increase in the high-dose group was statistically significantly increased by pair-wise comparison with the control. This increase was discounted because of high incidences reported in nine historical control studies; the average incidence reported in these studies was 22%. In four of these studies (two in each sex) the control incidences were 40%. The no-observed-effect level (NOEL) for systemic effects in this study was 100 ppm; the lowest-observed- effect level (LOEL) was 300 ppm, based on decreased weight gain and pathologic changes in the liver, kidneys, uterus and mammary glands. Based on these findings, this study was deemed adequate to determine that metribuzin was not oncogenic to SPF rats at dietary concentrations up to 300 ppm.

Pertinent data regarding the carcinogenic effects of metribuzin following inhalation exposure in animals were not located in the available literature.

II.A.4. SUPPORTING DATA FOR CARCINOGENICITY

Metribuzin was not mutagenic when tested in unspecified strains of Salmonella typhimurium or Escherichia coli (Mobay Chemical Corp., 1977; 1978). Metribuzin tested negative in the SOS Chromotest (DNA damage) conducted in Escherichia coli with or without metabolic activation (Xu and Schurr, 1990). Metribuzin did not induce reverse mutation in the D7 strain of Saccharomyces cerevisiae either in the presence or absence of metabolic activation (Mobay Chemical Corp., 1987).

Metribuzin was negative when tested for dominant lethal effects in male and female mice (unspecified strain) treated with doses of 300 mg metribuzin/kg (Mobay Chemical Corp., 1974b, 1975, 1976). Doses of 100 mg metribuzin/kg did not induce chromosomal aberrations in Chinese hamster spermatogonia (Mobay Chemical Corp., 1974c). Metribuzin also did not cause a significant increase in the unscheduled DNA synthesis when added to test cultures of rat primary hepatocytes (Mobay Chemical Corp., 1986a) and was found to be negative in the CHO/HGPRT mutation assay (Mobay Chemical Corp., 1986b). However, S-9 activated (but not nonactivated) metribuzin was found to be clastogenic in CHO cells (Mobay Chemical Corp., 1990).

Based on urinary excretion data, 36-51.9% of an orally administered dose of metribuzin was absorbed by Sprague-Dawley rats (Mobay Chemical Corp., 1972a). The half-life for elimination of radiolabeled metribuzin was 19.1- 30.4 hours for male rats and 22.4-33.6 hours for female rats (Mobay Chemical Corp., 1972a). In dogs, 52-60% of an administered oral dose was absorbed (Mobay Chemical Corp., 1972b). Within a 72- to 120-hour period, over 90% of the oral dose was excreted; about 52-60% was excreted in the urine as metabolites or conjugates and about 30% in the feces predominantly as unchanged metribuzin. In rats, about 90% of administered metribuzin was found to be excreted within 16 days by one investigator (Mobay Chemical Corp., 1972a) or within 5 days by another (Bleeke et al., 1985). Roughly equal amounts were excreted in the urine and feces. The major urinary metabolite was deamino metribuzin mercapturate (Bleeke et al., 1985). Mobay Chemical Corp. (1972a) also reported urinary metabolites of diketo metribuzin and deaminated diketo metribuzin. Conjugates of metribuzin were thought to account for the large water-soluble fraction of metabolites.

II.B. QUANTITATIVE ESTIMATE OF CARCINOGENIC RISK FROM ORAL EXPOSURE

[Ed. Note: EPA does not have information yet in this section of IRIS].

II.C. QUANTITATIVE ESTIMATE OF CARCINOGENIC RISK FROM INHALATION EXPOSURE

[Ed. Note: EPA does not have information yet in this section of IRIS].

II.D. EPA DOCUMENTATION, REVIEW, AND CONTACTS (CARCINOGENICITY ASSESSMENT)

II.D.1. EPA DOCUMENTATION

Source Document -- US EPA, 1992

The Drinking Water Quantification of Toxicologic Effects Document for Metribuzin has received Agency review.

II.D.2. REVIEW (CARCINOGENICITY ASSESSMENT)

Agency Work Group Review -- 02/03/93

Verification Date -- 02/03/93

II.D.3. US EPA CONTACTS (CARCINOGENICITY ASSESSMENT)

Susan Velazquez / NCEA -- (513)569-7571

Yogi Patel / OST -- (202)260-5849

III. HEALTH HAZARD ASSESSMENTS FOR VARIED EXPOSURE DURATIONS

III.A. DRINKING WATER HEALTH ADVISORIES

The Office of Drinking Water provides Drinking Water Health Advisories (HAs) as technical guidance for the protection of public health. HAs are not enforceable Federal standards. HAs are concentrations of a substance in drinking water estimated to have negligible deleterious effects in humans, when ingested, for a specified period of time. Exposure to the substance from other media is considered only in the derivation of the lifetime HA. Given the absence of chemical-specific data, the assumed fraction of total intake from drinking water is 20%. The lifetime HA is calculated from the Drinking Water Equivalent Level (DWEL) which, in turn, is based on the Oral Chronic Reference

Dose. Lifetime HAs are not derived for compounds which are potentially carcinogenic for humans because of the difference in assumptions concerning toxic threshold for carcinogenic and noncarcinogenic effects. A more detailed description of the assumptions and methods used in the derivation of HAs is provided in Background Document 3 in Service Code 5.

III.A.1. ONE-DAY HEALTH ADVISORY FOR A CHILD

Appropriate data for calculating a One-day HA are not available. It is recommended that the Ten-day HA of 5E+0 mg/L be used as the One-day HA.

III.A.2. TEN-DAY HEALTH ADVISORY FOR A CHILD

Ten-day HA -- 5E+0 mg/L

NOAEL -- 45 mg/kg/day
UF -- 100 (allows for interspecies and intrahuman variability with the use of
a NOAEL from an animal study)
Assumptions -- 1 L/day water consumption for a 10-kg child

Principal Study -- Mobay Chemical Corp. 1981

Metribuzin was administered by gavage to pregnant female rabbits (16 to 17/dose) on days 6 through 18 of gestation at doses of 15, 45, or 135 mg/kg/day. Following treatment, there was a statistically significant decrease ($p<0.05$) in body weight gain at the high dose (135 mg/kg/day). No maternal toxicity was reported in animals administered metribuzin at levels of 45 mg/kg/day or below. No treatment-related effects were reported at any dose level in fetuses based on gross, soft tissue, and skeletal examinations. A NOAEL of 45 mg/kg/day is identified in this study.

III.A.3. LONGER-TERM HEALTH ADVISORY FOR A CHILD

Longer-term (Child) HA -- 3E-1 mg/L

NOAEL -- 2.5 mg/kg/day
UF -- 100 (allows for interspecies and intrahuman variability with the use of
a NOAEL from an animal study)

Principal Study -- Mobay Chemical Corp., 1969

Metribuzan was administered to Wistar rats (15/sex/dose) for three months at dietary levels of 0, 50, 150, 500, or 1500 ppm (approximately 2.5, 7.5. 25, or 75 mg/kg/day). Following treatment, food consumption, growth, body weight, organ weight, clinical chemistry, hematology, urinalysis, and histopathology were measured. No significant effects on these parameters were observed in either sex at 50 ppm (2.5 mg/kg/day). Enlarged livers were found in females at 150 ppm and above ($p<0.05$) and thyroid glands were also enlarged in females at 500 ppm and 1,500 ppm ($p<0.05$ and $p<0.01$, respectively). In the males enlarged thyroids were reported at 500 ppm ($p<0.05$) and 1,500 ppm ($p<0.01$) while enlarged hearts were reported at 1500 ppm ($p<0.05$). Also, at

1,500 ppm (75 mg/kg/day), lower body weights (p<0.01) were reported in both sexes when compared to untreated controls. A NOAEL of 50 ppm (2.5 mg/kg/day) is identified for this study.

III.A.4. LONGER-TERM HEALTH ADVISORY FOR AN ADULT

Longer-term (Adult) HA -- 9E-1 mg/L

NOAEL -- 2.5 mg/kg/day
UF -- 100 (allows for interspecies and intrahuman variability with the use of a NOAEL from an animal study)
Assumptions -- 2 L/day water consumption for a 70-kg adult

Principal Study -- Mobay Chemical Corp., 1969 (study described in Section III.A.3.)

III.A.5.   DRINKING WATER EQUIVALENT LEVEL / LIFETIME HEALTH ADVISORY

DWEL -- 9E-1 mg/L

Assumptions -- 2 L/day water consumption for a 70-kg adult

RfD Verification Date -- 05/30/86

Lifetime HA -- 2E-1 mg/L

Assumptions -- 20% exposure by drinking water

Principal Study -- Mobay Chemical Corp., 1974 (This study was used in the derivation of the chronic oral RfD; see Section I.A.2.)

Four groups of beagle dogs (4/sex/dose) were administered metribuzin in the diet at dose levels of 0, 25, 100, or 1500 ppm (0, 0.625, 2.5, or 37.5 mg/kg/day, based on calculations in Lehman, 1959) for 24 months. Following treatment, food consumption, general behavior and appearance, clinical chemistry, hematology, urinalysis, body and organ weights, and histopathology were evaluated. No toxicologic effects were reported in animals administered 100 ppm (2.5 mg/kg/day) metribuzin or below for any of the parameters measured. Necrosis of the renal tubular cells, slight iron deposition as well as slight hyperglycemia and temporary hypercholesterolemia were noted in animals administered 1500 ppm (37.5 mg/kg/day) metribuzin. A NOAEL of 2.5 mg/kg/day is identified for this study.

III.A.6. ORGANOLEPTIC PROPERTIES

No information is available on the organoleptic properties of metribuzin.

III.A.7. ANALYTICAL METHODS FOR DETECTION IN DRINKING WATER

Analysis of metribuzin is by a gas chromatographic method applicable to the determination of certain organonitrogen pesticides in water samples.

III.A.8. WATER TREATMENT

Available data indicate that granular activated carbon adsorption and conventional treatment procedures will remove metribuzin from water.

III.A.9. DOCUMENTATION AND REVIEW OF HAs

EPA review of HAs in 1987.

Public review of HAs was January-March 1988.

Preparation date of this IRIS summary -- 09/30/88

III.A.10. EPA CONTACTS

Amal Mahfouz / OST -- (202)260-9568

Robert Cantilli / OST -- (202)260-5546

III.B. OTHER ASSESSMENTS

Content to be determined.

IV. US EPA REGULATORY ACTIONS

EPA risk assessments may be updated as new data are published and as assessment methodologies evolve. Regulatory actions are frequently not updated at the same time. Compare the dates for the regulatory actions in this section with the verification dates for the risk assessments in sections I and II, as this may explain inconsistencies. Also note that some regulatory actions consider factors not related to health risk, such as technical or economic feasibility. Such considerations are indicated for each action. In addition, not all of the regulatory actions listed in this section involve enforceable federal standards. Please direct any questions you may have concerning these regulatory actions to the US EPA contact listed for that particular action. Users are strongly urged to read the background information on each regulatory action in Background Document 4 in Service Code 5.

IV.A. CLEAN AIR ACT (CAA)

[Ed. Note: EPA does not have information yet in this section of IRIS].

IV.B. SAFE DRINKING WATER ACT (SDWA)

Listed in the January 1991 Drinking Water Priority List and may be subject to future regulation (56 FR 1470, 01/14/91).

IV.B.1. MAXIMUM CONTAMINANT LEVEL GOAL (MCLG) for Drinking Water

[Ed. Note: EPA does not have information yet in this section of IRIS].

IV.B.2. MAXIMUM CONTAMINANT LEVEL (MCL) for Drinking Water

[Ed. Note: EPA does not have information yet in this section of IRIS].

IV.B.3. SECONDARY MAXIMUM CONTAMINANT LEVEL (SMCL) for Drinking Water

[Ed. Note: EPA does not have information yet in this section of IRIS].

IV.B.4. REQUIRED MONITORING OF "UNREGULATED" CONTAMINANTS

Status -- Listed (Proposed, 1991)

Discussion -- "Unregulated" contaminants are those contaminants for which EPA establishes a monitoring requirement but which do not have an associated final MCLG, MCL, or treatment technique. EPA may regulate these contaminants in the future.

Monitoring requirement -- All systems to be monitored unless a vulnerability assessment determines the system is not vulnerable.

Analytical methodology -- Nitrogen-phosphorus detector/gas chromatography (EPA 507); electron-capture/gas chromatography (EPA 508); gas chromatographic/mass spectrometry (EPA 525).

Reference -- 56 FR 3526 (01/30/91)

EPA Contact -- Drinking Water Standards Division / OGWDW / (202)260-7575 or Safe Drinking Water Hotline / (800)426-4791

IV.C. CLEAN WATER ACT (CWA)

[Ed. Note: EPA does not have information yet in this section of IRIS].

IV.D. FEDERAL INSECTICIDE, FUNGICIDE, AND RODENTICIDE ACT (FIFRA)

IV.D.1. PESTICIDE ACTIVE INGREDIENT, Registration Standard

Status -- Issued (1985)

Reference -- Metribuzin Pesticide Registration Standard. June, 1985 (NTIS No. PB86-174216).

EPA Contact -- Registration Branch / OPP / (703)557-7760

IV.D.2. PESTICIDE ACTIVE INGREDIENT, Special Review

[Ed. Note: EPA does not have information yet in this section of IRIS].

IV.E. TOXIC SUBSTANCES CONTROL ACT (TSCA)

[Ed. Note: EPA does not have information yet in this section of IRIS].

IV.F. RESOURCE CONSERVATION AND RECOVERY ACT (RCRA)

[Ed. Note: EPA does not have information yet in this section of IRIS].

IV.G. SUPERFUND (CERCLA)

[Ed. Note: EPA does not have information yet in this section of IRIS].

VI. BIBLIOGRAPHY

VI.A. ORAL RfD REFERENCES

Mobay Chemical Corporation. 1972. MRID No. 00061257. Available from EPA. Write to FOI, EPA, Washington, DC 20460.

Mobay Chemical Corporation. 1974a. MRID No. 00061260, 00139397. Available from EPA. Write to FOI, EPA, Washington, DC 20460.

Mobay Chemical Corporation. 1974b. MRID No. 00061261, 00065136. Available from EPA. Write to FOI, EPA, Washington, DC 20460.

Mobay Chemical Corporation. 1974c. MRID No. 00061262, 00065135. Available from EPA. Write to FOI, EPA, Washington, DC 20460.

Mobay Chemical Corporation. 1981a. MRID No. 00087796. Available from EPA. Write to FOI, EPA, Washington, DC 20460.

Mobay Chemical Corporation. 1981b. MRID No. 00087795. Available from EPA. Write to FOI, EPA, Washington, DC 20460.

VI.B. INHALATION RfC REFERENCES

None

VI.C. CARCINOGENICITY ASSESSMENT REFERENCES

Bleeke, M.S., M.T. Smith and J.E. Casida. 1985. Metabolism and toxicity of metribuzin in mouse liver. Pestic. Biochem. Physiol. 23(1): 123-130.

Mobay Chemical Corporation. 1972a. MRID No. 00045265. HED Doc. No. 001146. Available from EPA. Write to FOI, EPA, Washington, DC 20460.

Mobay Chemical Corporation. 1972b. MRID No. 0004564. HED Doc. No. 001146. Available from EPA. Write to FOI, EPA, Washington, DC 20460.

Mobay Chemical Coporation. 1974a. MRID No. 00061261, 00065136, 00147942. HED Doc. No. 001147, 001148, 004767. Available from EPA. Write to FOI, EPA, Washington, DC 20460.

Mobay Chemical Corporation. 1974b. MRID No. 00086766. HED Doc. No. 001762. Available from EPA. Write to FOI, EPA, Washington, DC 20460.

Mobay Chemical Corporation. 1974c. MRID No. 00086765. HED Doc. No. 001762. Available from EPA. Write to FOI, EPA, Washington, DC 20460.

Mobay Chemical Corporation. 1975. MRID No. 00086767. HED Doc. No. 001762. Available from EPA. Write to FOI, EPA, Washington, DC 20460.

Mobay Chemical Corporation. 1976. MRID No. 00086768. HED Doc. No. 001762. Available from EPA. Write to FOI, EPA, Washington, DC 20460.

Mobay Chemical Corporation. 1977. MRID No. 00086770. HED Doc. No. 002778. Available from EPA. Write to FOI, EPA, Washington, DC 20460.

Mobay Chemical Corporation. 1978. MRID No. 00109254. HED Doc. No. 002778. Available from EPA. Write to FOI, EPA, Washington, DC 20460.

Mobay Chemical Corporation. 1981. MRID No. 00087795. HED Doc. No. 001761, 003911. Available from EPA. Write to FOI, EPA, Washington, DC 20460.

Mobay Chemical Corporation. 1986a. MRID No. 00157526. HED Doc. No. 005709. Available from EPA. Write to FOI, EPA, Washington, DC 20460.

Mobay Chemical Corporation. 1986b. MRID No. 00157527. HED Doc. No. 005709. Available from EPA. Write to FOI, EPA, Washington, DC 20460.

Mobay Chemical Corporation. 1987. MRID No. 40347701. HED Doc. No. 007369. Available from EPA. Write to FOI, EPA, Washington, DC 20460.

Mobay Chemical Corporation. 1990. MRID No. 41555102. HED Doc. No. 009520. Available from EPA. Write to FOI, EPA, Washington, DC 20460.

Xu, H.H. and K.M. Schurr. 1990. Genotoxicity of 22 pesticides in microtitration SOS chromotest. Tox. Assess. 5: 1-14.

US EPA. 1992. Revised and Updated Drinking Water Quantification of Toxicologic Effects for Metribuzin. Prepared by the Office of Health and Environmental Assessment, Environmental Criteria and Assessment Office, Cincinnati, OH for the Office of Drinking Water, Washington, DC. (Final Report)

VI.D. DRINKING WATER HA REFERENCES

Mobay Chemical Corporation. 1974. MRID No. 00061260. Available from EPA. Write to FOI, EPA, Washington, DC. 20460.

Mobay Chemical Corporation. 1969. MRID No. 00106161. Available from EPA. Write to FOI, EPA, Washington, DC. 20460.

Mobay Chemical Corporation. 1981. MRID No. 00087796. Available from EPA. Write to FOI, EPA, Washington, DC. 20460.

VII. REVISION HISTORY

Date	Section	Description
03/31/87	IV	Regulatory Action section on-line
03/01/88	I.A.5.	Confidence levels revised
03/01/91	I.A.4.	Citations added
03/01/91	III.A.	Health Advisory on-line
03/01/91	VI.	Bibliography on-line
01/01/92	IV.	Regulatory actions updated
03/01/93	II.	Carcinogenicity assessment now under review
12/01/93	II.	Carcinogenicity assessment on-line
12/01/93	VI.C.	Carcinogenicity assessment references on-line
03/01/94	I.A.	Oral RfD noted as pending change
03/01/94	I.A.6.	Work group review date added
01/01/95	I.A.6.	Work group review date added

SYNONYMS

4-AMINO-6-(1,1-DIMETHYLETHYL)-3-   DIC 1468
(METHYLTHIO)-1,2,4-TRIAZIN-5(4H)-   LEXONE
ONE   Metribuzin
4-AMINO-6-tert-BUTYL-3-   SENCOR
(METHYLTHIO)-1,2,4-TRIAZIN-5-ON   SENCORAL
4-AMINO-6-tert-BUTYL-3-   SENCORER
METHYLTHIO-as-TRIAZIN-5-ONE   SENCOREX
BAY 61597   1,2,4-TRIAZIN-5(4H)-ONE, 4-AMINO-6-
BAY DIC 1468   (1,1-DIMETHYLETHYL)-3-
BAYER 6159H   (METHYLTHIO)-
BAYER 6443H   1,2,4-TRIAZIN-5-ONE, 4-AMINO-6-tert-
BAYER 94337   BUTYL-3-(METHYLTHIO)-
                                           as-TRIAZIN-5(4H)-ONE, 4-AMINO-6-
                                           tert-BUTYL-3-(METHYLTHIO)-

Source: *Instant EPA's IRIS*, 1997, Instant Reference Sources, Inc., 7605 Rockpoint Drive, Austin, TX 78731 (Fax: 512-345-2386; Internet URL, http://www.instantref.com/inst-ref.htm).

Production, Use, and Pesticide Labeling Information

1. Description of the Chemical

>   Generic Name: 4-amino-6-(1,1-dimethylethyl)-3-(methylthio)-1,2,4- triazin-5(4H)-one

>   Common Name: metribuzin

>   Trade Names: Sencor, Lexone

>   EPA Shaughnessy Code: 101101

>   Chemical Abstracts Service (CAS) Number: 21097-64-9

>   Year of Initial Registration: 1973

>   Pesticide Type: Herbicide

>   Chemical Family: s-triazine

>   US Producer: Mobay Chemical Corporation

2. Use Patterns and Formulations

- Application sites: Metribuzin is registered for control of broadleaf weeds and grasses in soybeans, potatoes, barley, winter wheat, dormant established and sainfoin fields, asparagus, sugarcane, tomatoes, lentils, peas, and non-cropland.

- Types of formulations: Metribuzin is available as a 50% formulation intermediate, and 94% technical for formulation into end-use products, wettable powder, flowable concentrate, and dry flowable concentrate.

- Types and methods of application: Metribuzin may be soil-incorporated, surface applied, or applied foliarly, broadcast or band with ground equipment. It can be applied by aerial equipment or sprinkler irrigation (potatoes).

- Application rates: 0.25 to 4.0 a.i./A on crop sites; 6.0 to 8.0 a.i./ A on railroad rights-of-way.

- Usual carrier: Water

3. Science Findings

- Chemical Characteristics

Metribuzin is a solid at room temperature. Its molecular weight is 214.28. The melting point is 125.5 to 126.5 °C. Metribuzin is soluble in aromatic and chlorinated hydrocarbon solvents, and in water (at 20 °C) to 1220 ppm.

- Toxicological Characteristics

- Acute toxicity effects of metribuzin are as follows:

- Acute oral toxicity in rats: 2200 mg/kg body weight for males, 2345 mg/kg body weight for females. Toxicity Category III.

- Acute dermal toxicity in rats: 20,000 mg/kg body weight. Toxicity Category IV.

- Acute inhalation LC50 in rats: >20 mg/l/hr. Toxicity Category IV.

- Skin irritation in rabbits: PIS = 0.33/8.0. Toxicity Category IV.

- Eye irritation in rabbits: Not an irritant. Toxicity Category IV

- Dermal sensitization in guinea pig: Not a sensitizer. Toxicity Category IV.

- Subchronic and chronic effects: The 2-year dog study indicated dogs dosed with 1500 ppm (37.5 mg/kg) had reduced weight gain, increased mortality, hematological changes, and liver and kidney damage. The no-effect level is 100 ppm. The oncogenic potential of metribuzin is unclear at this time. The mouse oncogenicity study is negative for oncogenic effects. The chronic rat study indicates a statistically significant (p <0.05) increase in the incidence of adenoma of liver bile duct and pituitary gland in females at the 300 ppm dose level. Additional histopathology and historical control data on the incidence of these tumors in this particular strain of rats are needed before it can be determined if the increase is compound related. A teratology study in rabbits indicated no evidence of teratogenic effects at 135 mg/kg/day, the highest dose tested (HDT), and a NOEL of 15 mg/kg/day for maternal and fetal toxicity. Data gaps include rat chronic study, rat teratology study, and multi-generation reproduction study.

- Mutagenic effects: Available data indicate that metribuzin is not mutagenic. Data gaps exist in two categories of mutagenicity testing, specifically gene mutation studies in mammalian cells and tests for primary DNA damage, such as sister chromatid exchange or unscheduled DNA synthesis assay.

- N-nitroso contaminants: Available data do not provide grounds for concern at this time. The data are incomplete. The analysis for N-nitroso contaminants is requested.

- Major routes of human exposure: Primary nondietary exposure to the farmer is expected to be dermal and to occur during mixing, loading, and application. Exposure through ocular, inhalation, and ingestion routes is also expected.

- Physiological and Biochemical Behavioral Characteristics

> - Absorption characteristics: Metribuzin is absorbed through the leaves from surface treatment, but the major and significant route for uptake is via the root system.

> - Translocation characteristics: Uptake through the roots is best described as osmotic diffusion. Metribuzin is translocated upward in the xylem and moves distally when applied at the base of the leaves. It concentrates in the roots, stems, and leaves.

> - Mechanism of pesticidal action: Photosynthesis inhibitor.

> - Metabolism in plants: The major routes of detoxification are the action of oxidation and conversion to water soluble conjugated products.

- Environmental Characteristics

> - Adsorption and leaching in basic soil types: Metribuzin is moderately adsorbed on soils with high clay and/or organic matter content. Metribuzin is readily leached in sandy soils low in organic matter content.

- Microbial breakdown: Microbial breakdown appears to be the major mechanism by which metribuzin is lost from soils. Breakdown occurs fastest under aerobic conditions and at comparatively high temperatures.

- Loss from photodecomposition and/or volatilization: Slight loss.

- Average persistence at recommended rates: Half-life varies with soil type and climatic conditions. Half-life of metribuzin at normal use rates is one to two months.

- Potential groundwater problem: Metribuzin has been found in Ohio rivers and Iowa wells. Available data show that metribuzin has a potential to contaminate groundwater in soils low in organic and clay content. The Agency is requesting water monitoring studies on metribuzin and has determined that all uses of metribuzin should be classified for restricted use with appropriate labeling, including a groundwater advisory statement.

- Ecological Characteristics

   - Avian acute oral toxicity: 169.22 mg/kg (moderately toxic).

   - Subacute dietary toxicity: >4,000 ppm for mallard duck and bobwhite quail (slightly toxic).

   - Acute toxicity on freshwater invertebrates: 4.18 ppm (moderately toxic).

   - Acute toxicity on fish: 76.78 ppm for rainbow trout (slightly toxic), 75.96 ppm for bluegill sunfish (slightly toxic).

   - 96-hour LC50 on a marine/estuarine shrimp: 48.3 mg/l (slightly toxic).

   - Potential problem for endangered species:

      - The Agency evaluated metribuzin under the cluster/use pattern approach for use on corn, soybeans, and small grains. Available data indicate that metribuzin use on crops would probably not affect federally listed animal species.

      - Consultation with Office of Endangered Species (OES) on use of sulfometuron methyl indicated that several species of endangered plants which occur on or adjacent to rights-of-way would be jeopardized by exposure from its use. The Agency has concluded that these plants would be jeopardized by exposure to metribuzin. The Agency is imposing a statement concerning endangered plant species on all end-use products containing metribuzin and labeled for use on rights-of-way.

- Tolerance Assessment

- The Acceptable Daily Intake (ADI) is based on a no-observable-effect-level of 100 ppm (2.5 mg/kg) from the 2-year dog study. Using a 100-fold safety factor, the ADI is 0.025 mg/kg/day, with a Maximum Permissible Intake (MPI) of 1.5 mg/kg for a 60-kg adult human. Theoretical maximum residue contribution (TMRC) for metribuzin based on established tolerances is 0.3508 mg/day for a 1.5 kg diet. Currently, the permanent tolerances utilize 23.39% of the ADI.

- The Agency is unable to complete a full tolerance reassessment, because the available metribuzin toxicology and residue data do not fully support the established tolerances listed below. The metabolism of metribuzin in animals is not fully understood. Therefore, the Agency is requiring data on metabolism of metribuzin and related metabolites in ruminants, poultry, and several crops. The additional data will be used to assess dietary exposure to metribuzin and may lead to revisions in the existing tolerances.

Commodities	Parts per million
Alfalfa, green	2.0
Alfalfa, hay	7.0
Asparagus	0.05
Barley, grain	0.75
Barley, straw	1.0
Cattle, fat	0.7
Cattle, mbyp	0.7
Cattle, meat	0.7
Corn, fodder	0.1
Corn, forage	0.1
Corn, fresh (inc. sweet K + CWHR)	0.05
Corn, grain (inc. popcorn)	0.05
Eggs	0.01
Goats, fat	0.7
Goats, mbyp	0.7
Goats, meat	0.7
Grass	2.0
Grass, hay	7.0
Hogs, fat	0.7
Hogs, mbyp	0.7
Hogs, meat	0.7
Horses, fat	0.7
Horses, mbyp	0.7
Horses, meat	0.7
Lentils (dried)	0.05

(Table Continued)

Commodities	Parts per million
Lentils, forage	0.5
Lentils, vine hay	0.05
Milk	0.05
Peas	0.1
Peas (dried)	0.05
Peas, forage	0.5
Peas, vine hay	0.05
Potatoes	0.6
Poultry, fat	0.7
Poultry, mbyp	0.7
Poultry, meat	0.7
Sainfoin	2.0
Sainfoin, hay	7.0
Sheep, fat	0.7
Sheep, mbyp	0.7
Sheep, meat	0.7
Soybeans	0.1
Soybeans, forage	4.0
Soybeans, hay	4.0
Sugarcane	0.1
Tomatoes	0.1
Wheat, forage	2.0
Wheat, grain	0.75
Wheat, straw	1.0

Food

Commodities	Parts per million
Barley, milled fractions (except flour)	3.0
Potatoes, processed (inc. potato chips)	3.0
Sugarcane molasses	2.0
Wheat, milled fractions (except flour)	3.0

Feed

Commodities	Parts per million
Barley, milled fractions (except flour)	3.0
Potato waste, processed (dried)	3.0
Sugarcane bagasse	0.5
Sugarcane molasses	0.3
Tomato pomace, dried	2.0
Wheat, milled fractions (except flour)	3.0

- International tolerances: Canada

Commodities	Parts per million
Asparagus	0.1
Barley grain	0.1
Lentils	0.1
Peas	0.1
Potatoes	0.1
Soybeans	0.1
Tomatoes	0.1
Wheat grain	0.1

- Although the above Canadian tolerances differ from those in the United States, it is inappropriate for the Agency to harmonize these tolerances at the present time because of extensive toxicology and residue chemistry data gaps.

- There are no tolerances for residues of metribuzin in Mexico or <<Codex Alimentarius>>.

- Problems Known to have Occurred with Use

The Pesticide Incident Monitoring System (PIMS) does not indicate any incident involving agricultural uses of metribuzin.

- Summary Science Statement

- Metribuzin is not acutely toxic by oral, dermal, inhalation, or eye irritation routes of exposure. The available data do not indicate that any of the risk criteria listed in 162.11(a) of Title 40 of the US Code of Federal Regulations have been met or exceeded for the uses of metribuzin at the present time. Data gaps include rat chronic study, teratology study, multi-generation reproduction study, and two categories of mutagenicity testing. There are also extensive residue chemistry data gaps.

- Metribuzin has been found in Ohio rivers and Iowa wells. Although there are extensive data gaps in the area of environmental fate, available data indicate that metribuzin has a potential to contaminate groundwater in soils lower in organic matter and clay content.

- Available data indicate that metribuzin is moderately toxic to upland bird species on an acute oral basis, no more than slightly toxic to birds in the diet, and moderately toxic to freshwater fish and invertebrates. Metribuzin is slightly toxic to shrimp. A detailed ecological hazard assessment cannot be made until the acute dietary study on an upland game bird, acute toxicity studies on a marine/estuarine fish species and an oyster species, and appropriate environmental fate data are fulfilled.

4. Summary of Regulatory Position and Rationale

- Based on the review and evaluation of all available data and other relevant information on metribuzin, the Agency has made the following determinations:

- The available data do not indicate that any of the risk criteria listed in 162.11(a) of Title 40 of the US Code of Federal Regulations have been met or exceeded for the uses of metribuzin at the present time.

- The Agency will not allow any significant new uses to be established for metribuzin until the toxicology and residue chemistry deficien- cies identified have been satisfied.

- The Agency is requesting data on presence of nitroso-contaminants in metribuzin. Available data do not provide grounds for concern at this time.

- Based on concern for groundwater contamination, the Agency has determined that all uses of metribuzin should be classified as restricted use and carry appropriate labeling, including a ground- water advisory statement.

- The Agency is concerned about the exposure of endangered/threatened plant species occurring on or adjacent to rights-of-way from the use of metribuzin. An Endangered Species Statement is being required on the labeling.

- The Agency is imposing a rotational crop restriction. The extent of this restriction will be reconsidered when additional data are received.

- Specific label precautionary statements:

  - Hazard information: The human precautionary statements must appear on all MP labels as prescribed in 40 CFR 162.10.

- Environmental hazards statements:

  - All manufacturing-use products (MP's) intended for formulation into end-use products (EP's) must bear the following statements: Do not discharge effluent containing this product into lakes, streams, ponds, estuaries, oceans, or public water unless this product is specifically identified and addressed in an NPDES permit. Do not discharge effluent containing this product to sewer systems without previously notifying the sewage treatment plant authority. For guidance, contact your State Water Board or Regional Office of EPA.

  - All end-use products with outdoor uses must bear the following statement: Do not apply directly to water or wetlands. Do not contaminate water by cleaning of equipment or disposal of waste.

  - Groundwater statement: All end-use products (EP's) must be classified as Restricted Use (refer to 40 CFR 162.10(j)(2)(B)), and the labels must bear the following groundwater advisory: Metribuzin is a chemical which can travel (seep or leach) through soil and can contaminate groundwater which may be used as drinking water. Metribuzin has been found in groundwater as a result of agricultural use. Users are advised not to apply metribuzin where the water table (groundwater) is close to the surface and where soils are very permeable, i.e., well-drained soils such as loamy sands. Your local agricultural agencies can provide further information on the type of soil in your area and the location of groundwater.

- Endangered species: <<Notice>>: The use of this product on rights-of-way may pose a hazard to certain federally designated endangered plant species. They are known to be found in specific areas within the locations noted below. Prior to making applications, the user of this product must determine that no such species are located in or immediately adjacent to the area to be treated.

For information on protected species, contact the Endangered Species Specialist of the appropriate Regional Office of the US Fish and Wildlife Service listed below:

- Region 1, Portland, Oregon. California counties of Contra Costa, Solano, San Diego, Santa Barbara, Ventura, Los Angeles, and Orange. Idaho - Idaho County. Oregon - Harney County.

- Region 2, Albuquerque, New Mexico. Arizona counties of Coconino and Navajo. New Mexico counties of San Juan, Otero, Chaves, Lincoln, Eddy, and Dona Ana. Texas counties of El Paso, Pecos, and Runnels.

- Region 3, Twin Cities, Minnesota. Iowa counties of Allamakee, Clayton, and Jackson.

- Region 4, Atlanta, Georgia. Florida counties of Clay, Gulf, Gadsden, Franklin, and Liberty. Georgia counties of Wayne and Brantley. North Carolina, Henderson County. South Carolina, Greenville County.

- Region 5, Newton Corner, Massachusetts. New York, Ulster County.

- Region 6, Denver, Colorado. Utah counties of Emery, Piute, Garfield, Washington, Utah, and Wayne. Colorado counties of Montezuma, Delta, and Montrose.

- Restrictions on rotational crops: Do not plant food and feed crops other than those which are registered for use on metribuzin treated soils.

5. Summary of Major Data Gaps

- The following toxicological studies are required:
  - Acute inhalation toxicity (March 31, 1986).
  - Rat chronic/oncogenicity study (August 31, 1989).
  - Rat teratology study (September 30, 1986).
  - 2-generation rat reproduction study (September 30, 1988).
  - Mutagenicity testing (March to June, 1986).
  - General metabolism study (June 30, 1988).

- The following environmental fate data are required:
  - Hydrolysis (March 31, 1986).
  - Photodegradation in water or soil (March 31, 1986).
  - Metabolism studies in aerobic and anaerobic soils (September 30, 1987).
  - A mobility test involving leaching and adsorption/desorption (June 30, 1986).
  - Soil dissipation study (September 30, 1986).
  - Accumulation on rotational crops, confined (September 30, 1988).
  - Accumulation on rotational crops, field (February 28, 1989).
  - Accumulation study in fish (December, 1986).

- The following ecological effects data are required:
  - Avian subacute dietary toxicity, waterfowl (March 31, 1986).
  - Acute toxicity to estuarine and marine organisms, fish and mollusk (June 30, 1986).
  - Product chemistry data are required during 1986.

- The following residue chemistry data are required:
  - Plant metabolism data (June 30, 1987).
  - Metabolism studies utilizing ruminants and poultry (December 31, 1986).

6. Contact Person at EPA

Robert J. Taylor
Office of Pesticide Programs
US EPA Registration Division (TS-767C)
401 M Street S.W
Washington, DC 20460

Telephone: (703) 557-1800

Source: *Instant EPA's Pesticide Facts*, 1994, Instant Reference Sources, Inc., 7605 Rockpoint Drive, Austin, TX 78731 (Fax: 512-345-2386; Internet URL, http://www.instantref.com/inst-ref.htm).

Mirex

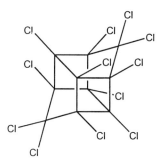

Introduction

Mirex is a highly stable insecticide formerly used for fire ant control in the southeastern US. Mirex was also employed as a flame-retardant. [828] Mirex was marketed under the trade name "Dechlorane" as a flame retardant coating for plastics, rubber, paint, paper and electrical goods. It was also used in antifouling paints, rodenticides, and additives for antioxidant and flame retardant mixtures for stabilized polymer compositions, ablative compositions, anthelmintic compositions and lubricant compositions. It was used in thermoplastic, thermosetting and elastomeric resin systems. It was also used as an insecticide with fire ants, yellow jackets, and western harvester ants. It was banned by the US EPA in 1978/

Release into the environment has occurred via effluents from manufacturing plants and sites where mirex was utilized as a fire resistant additive to polymers, and at points of application where it was used as a insecticide. Mirex is expected to persist in the environment despite the 1978 ban on its use in the US because it is resistant to biological and chemical degradation. Photolysis of mirex may occur. However, sorption is likely to be a more important fate process. Persistent compounds such as kepone, and monohydro- and dihydro- derivatives of mirex have been identified as products of extremely slow transformation of mirex. Mirex has been shown to bioconcentrate in aquatic organisms. A Koc value of 2.4×10^7 indicates mirex will strongly adsorb to organic materials in soils and sediments. Therefore mirex is expected to be immobile in soil and partition from the water column to sediments and suspended material. A Henry's Law Constant for mirex of 5.16×10^{-4} atm-cu m/mole at 22°C suggests rapid volatilization may occur from environmental waters and moist soils where absorption does not dominate. Based on this Henry's Law Constant, the volatilization half-life from a model river (22°C; 1 meter deep flowing 1 m/sec with a wind speed of 3 m/sec) has been estimated to be 10.7 hr; however, this estimation neglects the potentially important effect of adsorption. The volatilization half-life from an environmental pond, which considers the effect of adsorption, can be estimated to be about 1143 years. [828]

896

Identifying Information
CAS NUMBER: 2385-85-5

NIOSH Registry Number: PC8225000

SYNONYMS: 1,1A,2,2,3,3A,4,5,5,5A,5B,6-DODECACHLOROOCTAHYDRO-1,3,4-METHENO-1H-, CYCLOBUTA(CD)PENTALENE, HEXACHLOROCYCLOPENTADIENE DIMER, PERCHLOROPENTACYCLODECANE, DECHLORANE, DODECACHLOROOCTAHYDRO-1,3,4-METHENO-2H-CYCLOBUTA(C,D)PENTALENE, DODECACHLOROPENTACYCLODECANE, DODECACHLOROPENTACYCLO(3.2.2.0(SUP 2,6),0(SUP 3,9),0(SUP 5,10))DECANE, ENT 25,719, GC 1283, HRS 1276, PERCHLORODIHOMOCUBANE, PERCHLOROPENTACYCLO(5.2.1.0(SUP 2,6).0(SUP 3,9).0(SUP 5,8))DECANE, DODECACHLOROOCTAHYDRO-1,3,4-METHENO-1H-CYCLOBUTA(CD)PENTALENE, BICHLORENDO, CG-1283, DECHLORANE 4070, FERRIAMICIDE, 1,2,3,4,5,5-HEXACHLORO-1,3-CYCLOPENTADIENE DIMER, NCI-C06428

Endocrine Disruptor Information
Mirex is a strongly suspected EED that it is listed by the Centers for Disease Control & Prevention [812], and the World Wildlife Fund [813] as a potential endocrine modifying chemical.

Chemical and Physical Properties
CHEMICAL FORMULA: $C_{10}Cl_{12}$

MOLECULAR WEIGHT: 545.55

WLN: L545 B4 C5 D 4ABCE JTJ-/G 1 2

PHYSICAL DESCRIPTION: White crystals

SPECIFIC GRAVITY: Not available

DENSITY: Not available

MELTING POINT: 485°C (decomposes) [031], [043], [047], [346]

BOILING POINT: 485°C (Sublimes with decomposition) [371]

SOLUBILITY: Water: <1 mg/mL @ 24°C [700]; DMSO: 10-50 mg/mL @ 20.5°C [700]; 95% Ethanol: 1-10 mg/mL @ 24°C [700]; Acetone: 10-50 mg/mL @ 20.5° [700]

OTHER SOLVENTS: Benzene: 12.2% [031], [173], [395]; Dioxane: 15.3% [031], [051], [395]; Xylene: 14.3% [031], [173], [395]; Carbon tetrachloride: 7.2% [031], [173],

[395]; Methyl ethyl ketone: 5.6% [031], [051], [395]; Sea water: 0.2 ppb [051]; Aliphatic amines: Soluble.

OTHER PHYSICAL DATA: Odorless [043], [051], [072], [371]; UV absorption (in hexane): strong band @ 219.5 nm (log epsilon = 2.0); Vapor pressure: 0.000006 mm Hg @ 50°C [051], [072]

Source: *The National Toxicology Program's Chemical Database, Volume 2*, 1992, CRC Press, Inc./Lewis Publishers, 2000 Corporate Blvd., Boca Raton, FL 33431 (Fax: 407-998-9114).

Environmental Reference Materials

A source of EED environmental reference materials for mirex is Radian International LLC, Austin, TX (Fax 512-454-0268; Internet URL, http://www.radian.com/standards); Catalog No. ERM-001.

Hazardous Properties

ACUTE/CHRONIC HAZARDS: Mirex may be toxic and an irritant. When heated to decomposition mirex emits highly toxic fumes of carbon monoxide, carbon dioxide, hydrogen chloride gas, chlorine, carbon tetrachloride and phosgene. It can bioaccumulate [043].

FIRE HAZARD: Literature sources indicate that mirex is nonflammable [062], [173]. Fires involving this material can be controlled with a dry chemical, carbon dioxide or Halon extinguisher.

REACTIVITY: Mirex may react with strong oxidizers. It reacts with lithium and tertiary butyl alcohol [051].

STABILITY: Mirex is sensitive to exposure to sunlight. Solutions of mirex in water, DMSO, 95% ethanol or acetone should be stable for 24 hours under normal lab conditions [700].

Source: *The National Toxicology Program's Chemical Database, Volume 7*, 1992, CRC Press, Inc./Lewis Publishers, 2000 Corporate Blvd., Boca Raton, FL 33431 (Fax: 407-998-9114).

Medical Symptoms of Exposure

Symptoms of exposure may include an increase in central excitability, tremors and convulsions [151]. Central nervous system stimulation, hepatic injury, induction of the cytochrome P-450 system, neurotoxicity and hepatobiliary dysfunction, testicular atrophy, reduced sperm production, ocular flutter, hepatomegaly, splenomegaly, rashes, mental changes, and widened gaits may also occur [406]. Other symptoms may include congestion, edema, scattered petechial hemorrhages in the lungs, kidneys and brain, damage to tubular cells in the kidneys, hyperexcitability, ataxia, anuria and central nervous system depression that may terminate in respiratory failure [301]. In addition, gastrointestinal irritation, nausea, vomiting, diarrhea, malaise, headache, paresthesias, confusion and ventricular fibrillation may occur [371]. Other symptoms may include skin irritation, restlessness, weight loss, nervous system and liver abnormalities, skin rash and reproductive failure.

Source: *The National Toxicology Program's Chemical Database, Volume 4*, 1992, CRC Press, Inc./Lewis Publishers, 2000 Corporate Blvd., Boca Raton, FL 33431 (Fax: 407-998-9114).

Toxicological Information

The toxicological information below is gathered from several sources including two national databases: one from the *National Toxicology Program Chemical Repository Database* and another from *EPA's Integrated Risk Information System (IRIS)*.

Toxicology Information from the National Toxicology Program

CARCINOGENICITY:

Tumorigenic Data:

Type of Effect	Route/Animal	Amount Dosed (Notes)
TDLo	orl-mus	2222 mg/kg/58W-C
TDLo	orl-rat	1635 mg/kg/78W-C
TD	orl-rat	2340 mg/kg/56W-C
TD	orl-rat	2184 mg/kg/2Y-C

Review:
IARC Cancer Review: Animal Sufficient Evidence
IARC possible human carcinogen (Group 2B)

Status:
NTP Carcinogenesis Studies (Feed): Clear Evidence: Male and Female Rat
NTP Carcinogenesis Studies; test completed (camera copy in progress)
NTP Fifth Annual Report on Carcinogens, 1989; anticipated to be a carcinogen

TERATOGENICITY: See RTECS printout for most current data

Reproductive Effects Data:

Type of Effect	Route/Animal	Amount Dosed (Notes)
TDLo	orl-mus	22 mg/kg (1-21D preg)
TDLo	orl-mus	58 mg/kg (4W male/4W pre-2W post)
TDLo	orl-mus	37500 ug/kg (8-12D preg)
TDLo	orl-rat	50 mg/kg (5-9D preg)
TDLo	orl-rat	23750 ug/kg (4-22D preg)
TDLo	orl-rat	60 mg/kg (6-15D preg)
TDLo	orl-rat	56 mg/kg (8-15D preg)
TDLo	orl-rat	48 mg/kg (8-15D preg)
TDLo	orl-rat	60 mg/kg (10D male)
TDLo	orl-rat	18750 ug/kg (1-15D post)
TDLo	orl-rat	10 mg/kg (1-8D post)
TDLo	scu-rat	10 mg/kg (2D pre)
TDLo	scu-rat	2 mg/kg (1D pre)
TDLo	unr-mam	7 mg/kg (15-21D preg)
TDLo	unr-rat	8 mg/kg (8-15D preg)
TDLo	unr-rat	48 mg/kg (8-15D preg)
TDLo	orl-mus	1775 mg/kg (8-12D preg)

MUTATION DATA: See RTECS printout for most current data

Test	Lowest dose
oms-rat-orl	100 mg/kg

TOXICITY:

Typ. Dose	Mode	Specie	Amount	Units	Other
LC50	ihl	brd	1400	ppm	
LD50	orl	dck	2400	mg/kg	
LD50	orl	ham	125	mg/kg	
LD50	orl	rat	235	mg/kg	
LD50	skn	rbt	800	mg/kg	

OTHER TOXICITY DATA:

Review: Toxicology Review-2

Status:
EPA Genetox Program 1988, Negative: Rodent dominant lethal
EPA Genetox Program 1988, Positive: Carcinogenicity-mouse/rat
EPA TSCA Test Submission (TSCATS) Data Base, April 1990

SAX TOXICITY EVALUATION:

> THR: Poison by ingestion. Moderately toxic by inhalation and skin contact. A possible human carcinogen. An experimental carcinogen, tumorigen and teratogen. Experimental reproductive effects. Mutagenic data.

Source: *Instant Tox-Base*, 1995, Instant Reference Sources, Inc., 7605 Rockpoint Drive, Austin, TX 78731 (Fax: 512-345-2386; Internet URL, http://www.instantref.com/inst-ref.htm).

Toxicology Information from EPA's Integrated Risk Information System (IRIS)

I. CHRONIC HEALTH HAZARD ASSESSMENTS FOR NONCARCINOGENIC EFFECTS

I.A. REFERENCE DOSE FOR CHRONIC ORAL EXPOSURE (RfD)

The Reference Dose (RfD) is based on the assumption that thresholds exist for certain toxic effects such as cellular necrosis, but may not exist for other toxic effects such as carcinogenicity. In general, the RfD is an estimate (with uncertainty spanning perhaps an order of magnitude) of a daily exposure to the human population (including sensitive subgroups) that is likely to be without an appreciable risk of deleterious effects during a lifetime. Please refer to Background Document 1 in Service Code 5 for an elaboration of these concepts. RfDs can also be derived for the noncarcinogenic health effects of compounds which are also carcinogens. Therefore, it is essential to refer to other sources of information concerning the carcinogenicity of this substance. If the US EPA has evaluated this substance for potential human carcinogenicity, a summary of that evaluation will be contained in Section II of this file when a review of that evaluation is completed.

I.A.1. ORAL RfD SUMMARY

Critical Effect	Experimental Doses*	UF	MF	RfD
Liver cytomegaly, fatty metamorphosis, angiectasis, thyroid cystic follicles	NOAEL: 1 ppm (0.07 mg/kg/day)	300	1	2E-4 mg/kg/day
	LOAEL: 10 ppm (0.7 mg/kg/day)			

*Conversion Factors: None

Rat Chronic Dietary Feeding Study, NTP, 1990

I.A.2. PRINCIPAL AND SUPPORTING STUDIES (ORAL RfD)

NTP (National Toxicology Program). 1990. Toxicology and Carcinogenesis Studies of MIREX in F344/N Rats (Feed Studies). NTP TR 313.

Groups of 52 male and 52 female F344/N rats (initial body weight 120 and 100 g, respectively) were fed mirex (reported purity >96%) for 104 weeks. Reported mirex doses were in ppm and when converted to mg/kg-day were 0, 0.007, 0.07, 0.7, 1.8 and 3.8 mg/kg-day for males and 0, 0.007, 0.08, 0.7, 2.0 and 3.9 mg/kg-day for females. In a second study, 52 female F344/N rats were fed diets containing mirex at doses of 3.9 and 7.7 mg/kg-day. The following parameters were used to assess toxicity: clinical signs, body weight, survival and histologic examination of adrenal gland, bone marrow, brain, esophagus, heart, kidney, liver, lymph node (submandibular and/or mesenteric), lung and bronchi, mammary gland, pancreas, parathyroid gland, pituitary gland, prostrate/testis or ovary/uterus, salivary gland, skin, small and large intestine, spleen, stomach, thymus, thyroid gland, trachea and urinary bladder.

Survival of male rats in the 1.8 and 3.8 mg/kg-day groups was significantly less than controls (64% and 71% non-accidental deaths before termination vs. 15% in controls, p<0.001). Male rats in the 1.8 and 3.8 mg/kg-day dose groups gained less weight than controls during the first 70 weeks of exposure and lost weight between 70 and 104 weeks of exposure; body weights after 104 weeks of exposure were 11% (1.8 mg/kg-day) and 18% (3.8 mg/kg-day) less than controls. In the first study, female rats in the 3.9 mg/kg-day group gained less weight than controls; body weights after 104 weeks of exposure were 8% less than controls. In the second study, females in the 3.9 and 7.7 mg/kg-day groups gained less weight than controls; body weights after 104 weeks of exposure were 8% (3.9 mg/kg-day) and 18% (7.7 mg/kg-day) less than controls. No clinical signs of toxicity in male or female rats were reported.

Histologic examinations revealed dose-related changes in the parathyroid gland, kidney, liver, spleen and thyroid. In the parathyroid gland, dose- related increased incidence of hyperplasia was observed in male rats at and above 0.007 mg/kg-day. The incidences of hyperplasia were as follows: control, 6/32 (19%); 0.007 mg/kg-day, 12/39 (31%); 0.07 mg/kg-day, 13/39 (33%); 0.7 mg/kg-day, 18/40 (45%); 1.8 mg/kg-day, 22/50 (44%); and 3.8 mg/kg-day, 24/45 (53%). A dose-related increase in severity of nephropathy was observed in male rats at and above 0.7 mg/kg-day and in female rats at and above 2 mg/kg-day. Medullary hyperplasia was also seen in male rats at and above 0.7 mg/kg-day. Parathyroid hyperplasia and renal medullary hyperplasia are consistent with and may have been secondary to nephropathy, which was more severe in the rats exposed to mirex. In the liver, fatty metamorphosis, cytomegaly and angiectasis were detected in male rats at and above 0.7 mg/kg-day with necrosis at and above 1.8 mg/kg-day. Fatty metamorphosis, cytomegaly and necrosis were also observed in female rats at and above 0.7 mg/kg-day. Splenic fibrosis and cystic follicles of the thyroid were seen in male rats at and above 0.7 mg/kg-day. Based on liver and thyroid effects, this study defines a NOAEL of 0.07 mg/kg-day and a LOAEL of 0.7 mg/kg-day.

I.A.3. UNCERTAINTY AND MODIFYING FACTORS (ORAL RfD)

UF -- An uncertainty factor of 300 reflects 10 for intraspecies variability, 10 for interspecies extrapolation and 3 for lack of a complete data base, specifically lack of multi-generational data on reproductive effects and cardiovascular toxicity data.

MF -- 1.

I.A.4. ADDITIONAL STUDIES / COMMENTS (ORAL RfD)

The previously verified RfD was based on the Shannon (1976) study. Shannon (1976) performed studies in which prairie voles (Microtus ochrogaster) were exposed in the diet to mirex at 0, 0.1 and 0.5 ppm. The studies consisted of: single generation, 90-day exposure; single-generation, continuous exposure; and multi-generation, continuous exposure. Shannon chose M. ochrogaster to examine the effects of mirex on a wildlife species. Offspring (10/sex/dose group) from the first litter of the 0, 0.1 and 0.5 ppm groups of a single- generation, continuous exposure study were used as the first generation of the multi-generation, continuous exposure study. These animals were separated into 10 pairs (male and female) at approximately 60 days of age according to concentration of mirex exposure. The 0, 0.1 and 0.5 ppm mirex exposures continued through premating, mating, gestation and lactation for one litter. Shannon noted significant dose-related effects including decreased lactation index and increased percent mortality of pups in the first generation. In the second generation, significant differences were found in the percent survival of offspring to days 4 and 21 and in percent mortality of pups. Several difficulties were noted in this study; discrepancies in statistical analyses and lack of raw data to validate some of the results. The previous RfD was 2E-6 mg/kg-day.

Effects of mirex on the parathyroid and kidney reported in NTP (1990) have not been corroborated in chronic (Ulland et al., 1977) or subchronic (Larson et al., 1979) studies. However, NTP (1990) detected a dose-related increase in severity of nephropathy against a high but not unusual incidence of age- related nephropathy in rats (98% in controls). Thus, mirex may enhance the severity of age-related nephropathy, in which case, effects on the kidney would be detected only in a chronic study in which a rigorous severity assessment of the nephropathy was performed. The Ulland et al. (1977) study does not qualify as such a study; furthermore, histologic examinations were performed in the Ulland et al. (1977) study 6 months after exposure to mirex ended. Effects on the liver and thyroid reported in NTP (1990) have been corroborated (Gaines and Kimbrough, 1970; Fulfs et al., 1977; Ulland et al., 1977; Larson et al., 1979; Chu et al., 1981a; Yarbrough et al., 1981). Effects on the testis (hypocellularity and depressed spermatogenesis) reported in Yarbrough et al. (1981) were not detected in NTP (1990), however, these effects may have been masked in the NTP (1990) study by age-related degenerative changes in the testis. Reproductive and developmental effects (decreased fertility, fetal cataracts and edema) have been reported in several studies (Yarbrough et al., 1981; Gaines and Kimbrough, 1970; Chu et al., 1981b; Chernoff et al., 1979a,b; Chernoff and Kavlock, 1982; Grabowski and Payne, 1980; Kavlock et al., 1982; Scotti et al., 1981; Ware and Good, 1967). Effects on the fetal electrocardiogram have also been reported (Grabowski, 1983; Grabowski and Payne, 1980, 1983a,b).

Ulland et al. (1977) fed diets containing 0, 50 or 100 ppm mirex (reported purity 99%) to groups of 26 male and 26 female CD rats for 18 months. Reported mirex doses were 0, 1.5 and 3 mg/kg-day. Histologic evaluations made 6 months after exposure to mirex ended included adrenal glands, cerebrum, cerebellum, esophagus, heart, kidneys, liver, lungs, ovaries, pancreas, parathyroid gland, pituitary glands, spinal cord, small and large intestine, spleen, stomach, testis, thymus, thyroid, urinary bladder and uterus. Mortality of male rats in the dose groups at and above 50 ppm and females in the 100 ppm group was greater than rats in the control group. Histologic evaluation of the liver revealed cytomegaly, vacuolization, fatty metamorphosis and necrosis in rats exposed to mirex. No other treatment- related non-neoplastic lesions were reported in other tissues. This study identifies a FEL of 50 ppm (1.5 mg/kg-day).

Fulfs et al. (1977) conducted chronic bioassays in mice and monkeys. Groups of 100 male and 100 female CD-1 mice were fed diets containing 0, 1, 5, 15 or 30 ppm mirex (purity not specified) for 18 months. Estimated mirex doses based on US EPA (1987) and assumed average body weights of 0.036 kg were: 0, 0.17, 0.86, 2.6 and 5.2 mg/kg-day. Although survival was not specifically addressed, Fulfs et al. (1977) report that the 30 ppm group was removed from the study due to poor survival. This suggests that the 30 ppm dose is an FEL. Toxicity was assessed from biochemical, histochemical and histological (light and electron microscopy) evaluations of the liver. Female mice in the 1 ppm group had a significant increase in relative liver weight. Histologic evaluations of the liver revealed cellular hypertrophy, cellular or multicellular necrosis, and proliferation of the smooth endoplasmic reticulum (SER) at and above 5 ppm and nuclear inclusions at 30 ppm. Histochemical evaluations of the liver revealed centrilobular depletion of glucose-6- phosphatase activity at and above 5 ppm. Liver hypertrophy and SER proliferation are consistent with observed induction of mixed function oxidase over the same dose range in these animals (Byard et al., 1975). This study (Fulfs et al., 1977) identifies a NOAEL of 1 ppm (0.17 mg/kg-day) and a LOAEL of 5 ppm (0.86 mg/kg-day).

In the monkey study, Fulfs et al. (1977) administered oral gavage doses of 0 or 0.25 mg/kg-day mirex (purity not specified) in corn oil, 6 days/week for 36 months (0.21 mg/kg-day when multiplied by 6/7 to adjust to 7 days/week); or 1 mg/kg-day, 6 days/week for 16, 19 or 26 months (0.85 mg/kg-day when multiplied by 6/7 to adjust to 7 days/week) to groups of 2 male and 2 female rhesus monkeys (Macaca mulatta). Histochemical and histologic evaluations of the liver revealed "occasional" focal lymphocytic infiltration in treated monkeys (dose unspecified). This study identifies a 0.21-0.85 mg/kg-day LOAEL.

Yarbrough et al. (1981) conducted a 28-day study in which groups of 10 male Sprague-Dawley rats (60-85 g initial body weight) were fed diets containing 0, 0.5, 5.0, 50 or 75 ppm mirex (repurified from technical grade, reported purity 99.5%). Estimated mirex doses based on reported average food intake for controls of 14 g/day (which was not significantly different from mirex-treated groups) and reported average initial body weight (72 g) and weight gain (183 g) were 0, 0.05, 0.5, 5 and 7 mg/kg-day. Toxicity was assessed from measurements of serum enzymes, hematologic parameters, sperm counts, organ weights, and histologic evaluation of liver, thyroid and testis. There appeared to be a significantly decreased sperm count (53% decrease at 0.5 ppm, p<0.01), but this may have been due to the coincidence of the termination of the experiment and the inception of sexual maturity in all test groups. Thus, sexual maturity may have been delayed in dosed test groups due to mirex effects. Histologic evaluation of the liver revealed cytoplasmic alterations (increased density) that were "subtle and inconsistent" at 0.5 ppm, consistently observed at 5 ppm, and progressed to cytoplasmic inclusions with cell swelling at and above 50 ppm. Histologic evaluation of the testis revealed hypocellularity, decreased spermatogenesis, and luminal nucleated and giant cells, characteristic of testicular degeneration (dose not specified). Histologic evaluation of the thyroid revealed epithelial cell hypertrophy, colloid depletion, follicular atrophy and focal papillary formations present inconsistently at 5 ppm and more prominent at and above 50 ppm. This study identifies a NOAEL of 0.5 ppm (0.05 mg/kg-day) and a LOAEL of 5 ppm (0.5 mg/kg-day).

Groups of 40 male Sprague-Dawley rats (initial body weight 60-90 g) were fed diets containing 0, 5 or 50 ppm mirex (reported purity 98%) for 28 days and were evaluated for toxic effects at 28 days, or at 12, 24 or 48 weeks after exposure to mirex ended (Chu et al., 1981a). Estimated doses based on reported average food intakes and assumed average initial body weight of 75 g and reported final body weights were 0, 0.7 and 6.5 mg/kg-day. Toxicity was evaluated from complete hematologic analysis, serum chemistry, measurements of testicular sorbitol dehydrogenase and hepatic mixed function oxidase activity, serum T3 and T4 levels, and histologic examination of adrenal gland, bone marrow, brain, esophagus, heart, kidney, liver, lung, trachea and bronchi, lymph node (mesenteric and mediastinal), pancreas, parathyroid, pituitary, skeletal muscle, small and large intestine, spleen, stomach, testis and epididymis, thoracic aorta, thymus and thyroid. Histologic evaluations revealed lesions of the liver and thyroid in the 5 and 50 ppm dose groups. Liver lesions included fatty infiltration, cytoplasmic vacuolization, anisokaryosis and cellular necrosis. Thyroid lesions consisted of thickening of the follicular epithelium, loss of colloid and collapse of the follicles. Thyroid lesions regressed over the 48 week post-exposure observation period. Liver lesions persisted in the 50 ppm dose group. This study identifies a LOAEL of 5 ppm (0.7 mg/kg-day).

Groups of 10 male and 10 female Sherman rats (body weights not reported) were fed diets containing 0, 1, 5 or 25 ppm mirex (technical grade, reported purity 98%) for 166 days (Gaines and Kimbrough, 1970). Reported mirex doses were 0, 0.04-0.09, 0.21-0.48 and 1.3-3.1 mg/kg-day for males and 0, 0.06-0.10, 0.31-0.49 and 1.8-2.8 mg/kg-day for females. Toxicity was assessed from histologic examination (using light and electron microscopy) of the liver, which revealed dose-related hepatic cytomegaly in male and female rats. Incidences in male rats were: control, 0/10; 0.04-0.09 mg/kg-day, 2/10 (20%); 0.21-0.48 mg/kg-day, 5/10 (50%); and 1.3-3.1 mg/kg-day, 10/10 (100%). Incidences in females were: control, 0/10; 0.06-0.10 mg/kg-day, 0/10; 0.31- 0.49 mg/kg-day, 3/10 (30%); and 1.8-2.8 mg/kg-day, 5/10 (50%). This study identifies a LOAEL of 0.04-0.09 mg/kg-day.

In a series of subchronic studies, Larson et al. (1979) examined the effects of mirex in the diet of rats and beagle dogs. Groups of Charles River rats (10/sex/dose) were fed diets containing 0, 5, 20, 80, 320 or 1280 ppm mirex (reported purity 98%) for 13 weeks. Estimated mirex doses based on measured food consumption (the mean of food consumed at 4 and 13 weeks) were 0, 0.4, 1.3, 6.2, 24 and 96 mg/kg-day for males and 0, 0.3, 1.3, 5.8, 23 and 95 mg/kg-day for females. Toxicity was assessed from a hematologic analysis (unspecified), urinalysis (unspecified), and histologic examination of adrenal gland, bone marrow, brain, cecum, gonad, heart, kidney, liver, lung, pancreas, pituitary, small and large intestine, spleen, stomach, urinary bladder and thyroid. Male rats in the 80 ppm dose group and female rats in the 320 ppm dose group had significantly enlarged livers (increased liver/body weight ratio, p<0.05). Histologic examination revealed vacuolization and cytomegaly in livers of males in the 80 ppm dose group. Survival of male rats in the 1280 ppm dose group was decreased relative to controls (0 vs. 100% in controls) as was survival of female rats in the 1280 ppm group

(50% vs. 100% in controls). This study identifies a NOAEL of 20 ppm (1.3 mg/kg-day), a LOAEL of 80 ppm (6.2 mg/kg-day), and a FEL of 1280 ppm (95 mg/kg-day).

Larson et al. (1979) fed a diet containing 0, 4, 20 or 100 ppm mirex (reported purity 98%) to purebred beagle dogs (2/sex/dose) (7-12 kg initial body weight) for 13 weeks. Estimated mirex doses based on US EPA (1987) and reported average body weights of 11 kg for males and 9 kg for females were 0, 1, 5 and 27 mg/kg-day for males and 0, 1, 5 and 24 mg/kg-day for females. Toxicity was assessed from hematologic analyses (unspecified), blood chemistry (glucose, urea nitrogen, glutamic oxaloacetic transaminase, alkaline phosphatase, cholinesterase), sulfobromophthalein retention, urinalyses (unspecified), and histologic examination of adrenal gland, bone marrow, brain, cecum, gonad, heart, kidney, liver, lung, pancreas, pituitary, small and large intestine, spleen, stomach, urinary bladder and thyroid. Male and female dogs in the 100 ppm dose groups gained less weight than controls and had elevated serum alkaline phosphate levels. One male and one female dog in these groups died (13 and 10 weeks, respectively); the male had increased sulfobromophthalein retention. Results of histologic examinations were reported as unremarkable. This study identifies a FEL of 100 ppm (24 mg/kg-day).

Fulfs et al. (1977) fed Sprague-Dawley rats (number, sex and initial body weights not specified) diets containing 0, 5 or 30 ppm mirex (purity not specified) for 12 or 8 months, respectively. Mirex doses based on US EPA (1987) and assumed average body weights of 0.43 kg were 0, 0.4 and 2.2 mg/kg-day. Histochemical and histologic evaluations of the liver revealed "minimal" proliferation of smooth endoplasmic reticulum. This study identifies a NOAEL of 5 ppm (0.4 mg/kg-day).

Gaines and Kimbrough (1970) conducted a reproduction study of mirex (technical grade, reported purity 98%) in the diets of Sherman rats. Groups of 10 male rats (body weights not reported) were fed 0 or 25 ppm mirex for either 45 or 102 days. Groups of 10 female rats were fed 0 or 25 ppm mirex for 45 days or 0, 5 or 25 ppm mirex for 102 days. After the specified exposure period, rats fed mirex were mated with untreated rats. Females continued on their respective diets through gestation and lactation. Offspring from the rats exposed for 102 days were held after weaning and fed a nontreated diet until they were 90-100 days old. They were pair-mated within respective groups and their offspring held until weaning and checked for abnormalities. Reported mirex doses for males were 0 and 1.3-3.1 mg/kg-day. Reported doses for females were 0, 0.31-0.49 and 1.8-2.8 mg/kg-day.

Females exposed to 25 ppm for 45 days prior to mating and through gestation and lactation had significantly smaller litters (8.5 vs. 12.0 pups per litter, $p < 0.05$) and survival of pups to weaning was significantly decreased (53% vs. 89%, $p < 0.05$). Pups born to these females had a 33% incidence of cataracts compared to 0% in controls. Survival to weaning of pups born to females treated with 25 ppm for 102 days prior to mating and through gestation and lactation was significantly decreased (61% vs. 94% survival, $p < 0.05$) and pups born to these females had a 46.2% incidence of cataracts compared to 0% in controls. These parameters were not significantly affected in pups born to females treated with 5 ppm. Litters of pups born to females that had not been treated with mirex were transferred at birth to foster mothers that had been fed diets containing 0 or 5 ppm mirex for 73 days. Survival to weaning of pups transferred to treated females was significantly lower than pups transferred to untreated females (53.9% vs. 95.8%, p<0.05) and pups transferred to treated females had a significantly greater incidence of cataracts (37.5% vs. 0%, p<0.05). This study identifies a FEL of 5 ppm (0.31- 0.49 mg/kg-day).

Groups of 10 male and 20 female Sprague-Dawley rats (92 g initial body weight) were fed diets containing mirex (reported purity >98%) at 0, 5, 10, 20 or 40 ppm for 13 weeks prior to mating, during a 2-week mating period and through gestation and lactation (Chu et al., 1981b). Mated pairs were exposed to the same dietary levels; males were discarded from the study after mating. Estimated mirex doses for females based on US EPA (1987) and reported average initial body weight of 92 g and average weight gain of controls of 195 g were: 0.5, 1, 2 and 4 mg/kg-day. Toxicity in females was assessed from a complete hematologic analysis (hemoglobin, total and differential counts, bone marrow smear), serum chemistry (electrolytes, protein, enzymes, cholesterol, uric acid, bilirubin), measurements of liver aniline hydroxylase and aminopyrine demethylase activities; and histologic examination of adrenal gland, bone marrow, brain, bronchi, trachea and lungs, esophagus, eye, heart, kidney, liver, lymph nodes (mesenteric and mediastinal), ovaries, pancreas, parathyroid, peripheral nerves, pituitary, salivary glands, skeletal muscle, small and large intestine, skin, spleen, stomach, thoracic aorta, thymus, thyroid and uterus. The same histologic examinations (including prostrate, seminal vesicles and testes of males) and measurements of hepatic microsomal enzymes) were performed on pups that survived to 21 days. Histologic examinations revealed lesions of the liver and thyroid in adult females at all dose levels (at and above 5 ppm). Lesion incidences in the 5 ppm dose group were as follows: liver, 10/10 (100%) vs. 2/13 (15%) in controls; and thyroid, 6/10 (60%) vs. 2/13 (15%) in controls. Hepatic lesions included fatty infiltration and cytoplasmic vacuolization and anisokaryosis. Thyroid lesions included follicular epithelial thickening and collapse. Similar types of lesions of the liver and thyroid were observed in pups from treated females. Pups from treated females (at and above 5 ppm) had eye cataracts; incidences were 0/14 in controls and 4/10 (40%) in the 5 ppm group. Survival of pups to 21 days was significantly decreased (70% vs. 97% controls, p<0.001) in the 40 ppm dose group. This study identifies a LOAEL of 5 ppm (0.5 mg/kg-day) and a FEL of 40 ppm.

Chernoff et al. (1979a) fed diets containing 0 or 25 ppm mirex (reported purity >98%) to pregnant CD (Charles River) rats (numbers and initial weights not specified) from day 4 of gestation through day 34 post-parturition. Groups of 17-24 litters of pups were cross-fostered to yield four exposure groups: no exposure, prenatal exposure, postnatal exposure and perinatal exposure (prenatal and postnatal exposure). Estimated mirex doses based on US EPA (1987) and assumed average body weight of 250 g were 0 and 2.2 mg/kg-day. Exposure to mirex during gestation resulted in a significant increase in the incidence of fetal mortality (14% vs. 3% in controls, p<0.01). Perinatal and postnatal exposure resulted in significantly decreased pup survival to 8 days (77% and 74%, respectively, vs. 94% in controls, p<0.05) and significantly increased incidence of cataracts and other lens changes (31% and 43%, respectively, vs. 0% in controls, p<0.001). Other studies by the same investigators have demonstrated cataractogenicity of postnatal exposure to mirex in Sherman and Long-Evans rats and CD-1 mice (Chernoff et al., 1979b; Scotti et al., 1981). This study (Chernoff et al., 1979a) identifies a FEL of 25 ppm (2.2 mg/kg-day).

Groups of 10-37 pregnant female CD rats were administered mirex (technical grade, reported purity >98%) in corn oil by oral gavage; doses were 0, 5, 7, 9.5, 19 or 38 mg/kg-day on days 7-16 of gestation (Chernoff et al., 1979b). Animals were killed on day 21 of gestation, resorption sites and live fetuses were recorded, and live fetuses were examined for external and skeletal abnormalities. The incidence of edematous live fetuses was significantly elevated at doses at and above 7 mg/kg-day (p<0.05); incidences were as follows: control, 0%; 5 mg/kg-day, 5.8%; 7 mg/kg-day, 27.3%; 9.5 mg/kg-day, 22.9%; 19 mg/kg-day, 74.7%; and the number of sternal ossification centers was significantly decreased at doses at and above 7 mg/kg-day (p<0.05). Fetal mortality was significantly increased at doses at and above 19 mg/kg-day (p<0.001); control, 4%; 19 mg/kg-day, 65.1%; and 38 mg/kg-day, 100%. Maternal weight gain was significantly decreased at doses at and above 9.5 mg/kg-day, and maternal liver/body weight ratios were significantly increased at doses at and above 7 mg/kg-day. This study identifies a NOAEL of 5 mg/kg-day, a LOAEL of 7 mg/kg-day, and an FEL of 19 mg/kg-day.

Groups of 5 or 6 pregnant CD rats were administered 0, 6 or 12 mg/kg-day mirex (purity not specified) by oral gavage on days 7-16 of gestation (Kavlock et al., 1982). Dams were killed on day 21 of gestation and fetuses examined for external abnormalities. The following evaluations were made on fetuses that had no external abnormalities: body, brain, kidney, liver and lung weights; total DNA and protein in brain; total dipalmitoyl phosphatidylcholine and sphingomyelin in lung; total glycogen in liver; and total protein and alkaline phosphatase in kidney. External abnormalities observed at and above 6 mg/kg-day included edema, ectopic gonads and hydrocephaly. Brain and liver weights adjusted for fetal body weight were significantly decreased in the 6 and 12 mg/kg-day dose groups. Total liver glycogen, total kidney protein and total kidney alkaline phosphatase, all adjusted for fetal body weight, were significantly decreased in the 6 and 12 mg/kg-day groups (p<0.05). This study identifies a LOAEL of 6 mg/kg-day.

Groups of 20 mated female Wistar rats were administered mirex (reported purity 98%) in corn oil by oral gavage on days 6-15 of gestation; doses were 0, 1.5, 3, 6 or 12.5 mg/kg-day (Khera et al., 1976). Rats were killed on day 22 of gestation and fetuses were examined for external and skeletal abnormalities. Doses at and above 3 mg/kg-day increased the incidence of resorptions (7.4 vs. 3.7% in controls). Doses at and above 6 mg/kg-day significantly increased fetal mortality and the incidence of visceral anomalies (p<0.05). These included edema, scoliosis, runts and short tail at 6 mg/kg-day; and these anomalies in addition to cleft palate and heart defects at 12.5 mg/kg-day. Doses at and above 3 mg/kg-day decreased the incidence of pregnancy of survivors 22 days after mating. Doses at and above 6 mg/kg-day increased maternal mortality. This study identifies a NOAEL of 1.5 mg/kg-day, a LOAEL 3 mg/kg-day and an FEL of 6 mg/kg-day.

Groups of 20 male Wistar rats were administered mirex (reported purity 98%) in corn oil by oral gavage for 10 consecutive days; doses were 0, 1.5, 3 or 6 mg/kg-day (Khera et al., 1976). Following dosing, 14 mating trials were conducted in which each treated male was paired with two untreated virgin females for 5 days. Females were killed 13-15 days after pairing and viable embryos, deciduomas and corpora lutea were recorded. At a dose of 6 mg/kg-day, weight gain in the male rats was decreased and one male rat died, and the incidence of pregnancies (% of matings) resulting from the first trial (but not subsequent trials) was significantly decreased (p<0.05). This study identifies a 6 mg/kg-day LOAEL.

Groups of pregnant Long-Evans rats (group size not specified) were administered mirex (commercial grade, purity not specified) in peanut oil by oral gavage at doses of 0 or 0.25 mg/kg-day on days 15.5-21.5 of gestation (Grabowski, 1983). Pups from dams exposed to mirex had abnormal electrocardiograms, including significantly prolonged PR and QT intervals (p<0.03). In other studies with doeses at and above 5 mg/kg-day, prolongation of the PR interval correlated with edema and progressed to first and second degree heart block (missed ventricular beat) (Grabowski and Payne, 1980; 1983a,b). This study (Grabowski, 1983) identifies a LOAEL of 0.25 mg/kg-day.

Groups of 100-108 male and female Swiss Balb/c and CFW mice were fed diets containing 0 or 5 ppm technical grade mirex (reported purity 99%) for 30 days prior to pairing and for 90 days after pairing (Ware and Good, 1967). The study of Balb/c mice was repeated. Estimated mirex doses based on US EPA (1987) and assumed average body weight of 0.038 kg were 0 and 0.84 mg/kg-day. Numbers of litters, litter size, sex ratio and mortality were assessed. Mirex significantly increased parent mortality of Balb/c mice (p<0.05) but not CFW mice. Fecundity (young per producing pair) was significantly less in CFW mice exposed to mirex compared to controls (p<0.05). Fecundity and litter size were significantly less in Balb/c mice compared with controls (p<0.05) in one study; however, these effects were not corroborated in a duplication of the study. This study identifies an FEL of 0.84 mg/kg-day.

Groups of 24 or 25 pregnant CD-1 mice were administered 0 or 7.5 mg/kg-day mirex (in corn oil, purity not specified) by oral gavage on days 8-12 of gestation (Chernoff and Kavlock, 1982). Dams were allowed to give birth and litters were counted and weighed on postnatal days 1 and 3. Body weights and survival of pups in the mirex group were significantly decreased on postnatal days 1 and 3 compared to the control group (p<0.05). This study identifies an FEL of 7.5 mg/kg-day.

I.A.5. CONFIDENCE IN THE ORAL RfD

Study -- High
Data Base -- High
RfD -- High Confidence in the study can be considered high to medium. NTP (1990) used an adequate number of animals and examined all tissues identified as potential targets in other studies. Confidence in the data base can be considered is high to medium. Several good quality chronic, subchronic and developmental/reproductive studies have been conducted, although the data base is lacking a rodent multi-generation study. Reflecting high to medium confidence in the study and the data base, confidence in the RfD can be considered high to medium.

I.A.6. EPA DOCUMENTATION AND REVIEW OF THE ORAL RfD

Source Document -- This assessment is not presented in any existing US EPA documentation.

Other EPA Documentation -- US EPA, 1987

Agency Work Group Review -- 06/24/86, 04/15/87, 06/24/92
Verification Date -- 06/24/92

I.A.7. EPA CONTACTS (ORAL RfD)

Joan S. Dollarhide / NCEA -- (513)569-7539

Kenneth A. Poirier / OHEA -- (513)569-7553

I.B. REFERENCE CONCENTRATION FOR CHRONIC INHALATION EXPOSURE (RfC)

[Ed. Note: EPA does not have information yet in this section of IRIS].

II. CARCINOGENICITY ASSESSMENT FOR LIFETIME EXPOSURE

This substance/agent has been evaluated by the US EPA for evidence of human carcinogenic potential. This does not imply that this agent is necessarily a carcinogen. The evaluation for this chemical is under review by an inter-office Agency work group. A risk assessment summary will be included on IRIS when the review has been completed.

III. HEALTH HAZARD ASSESSMENTS FOR VARIED EXPOSURE DURATIONS

[Ed. Note: EPA does not have information yet in this section of IRIS].

IV. US EPA REGULATORY ACTIONS

EPA risk assessments may be updated as new data are published and as assessment methodologies evolve. Regulatory actions are frequently not updated at the same time. Compare the dates for the regulatory actions in this section with the verification dates for the risk assessments in sections I and II, as this may explain inconsistencies. Also note that some regulatory actions consider factors not related to health risk, such as technical or economic feasibility. Such considerations are indicated for each action. In addition, not all of the regulatory actions listed in this section involve enforceable federal standards. Please direct any questions you may have concerning these regulatory actions to the US EPA contact listed for that particular action. Users are strongly urged to read the background information on each regulatory action in Background Document 4 in Service Code 5.

IV.A. CLEAN AIR ACT (CAA)

[Ed. Note: EPA does not have information yet in this section of IRIS].

IV.B. SAFE DRINKING WATER ACT (SDWA)

[Ed. Note: EPA does not have information yet in this section of IRIS].

IV.C. CLEAN WATER ACT (CWA)

IV.C.1. AMBIENT WATER QUALITY CRITERIA, Human Health

[Ed. Note: EPA does not have information yet in this section of IRIS].

IV.C.2. AMBIENT WATER QUALITY CRITERIA, Aquatic Organisms

Freshwater: 1.0E-3 ug/L (at any time)

Marine: 1.0E-3 ug/L (at any time)

Considers technological or economic feasibility? -- NO

Discussion -- The criterion is based on the effects of mirex on fresh and salt water invertebrates, using an application factor of 0.01.

Reference -- Quality Criteria for Water, EPA 440/9-76-023 (07/76)

EPA Contact -- Criteria and Standards Division / OWRS (202)260-1315 / FTS 260-1315

IV.D. FEDERAL INSECTICIDE, FUNGICIDE, AND RODENTICIDE ACT (FIFRA)

IV.D.1. PESTICIDE ACTIVE INGREDIENT, Registration Standard

[Ed. Note: EPA does not have information yet in this section of IRIS].

IV.D.2. PESTICIDE ACTIVE INGREDIENT, Special Review

Action -- Registration canceled (1976)

Considers technological or economic feasibility? -- NO

Summary of regulatory action -- Administrative Law Hearing FIFRA Docket No. 293 (All registered products containing mirex were effectively canceled on December 1, 1977.) /criterion of concern: carcinogenicity, bio-accumulation, wildlife hazard and other chronic effects.

Reference -- 41 FR 56703 (12/29/76)

EPA Contact -- Special Review Branch / OPP (703)557-7400 / FTS 557-7400

IV.E. TOXIC SUBSTANCES CONTROL ACT (TSCA)

[Ed. Note: EPA does not have information yet in this section of IRIS].

IV.F. RESOURCE CONSERVATION AND RECOVERY ACT (RCRA)

[Ed. Note: EPA does not have information yet in this section of IRIS].

IV.G. SUPERFUND (CERCLA)

[Ed. Note: EPA does not have information yet in this section of IRIS].

VI. BIBLIOGRAPHY

VI.A. ORAL RfD REFERENCES

Byard, J.L., U.CH. Koepke, R. Abraham, L. Goldberg and F. Coulston. 1975.
Biochemical changes in the liver of mice fed mirex. Toxicol. Appl. Pharmacol. 33: 70-
77.

Chernoff, N., J.T. Stevens and E.H. Rogers. 1979a. Perinatal toxicology of mirex
administered in the diet: I. Viability, growth, cataractogenicity and   tissue levels.
Toxicol. Lett. 4: 263-268.

Chernoff, N., R.E. Linder, T.M. Scotti, E.H. Rogers, B.D. Carver and R.J. Kavlock.
1979b. Fetotoxicity and cataractogenicity of mirex in rats and mice with notes on
kepone. Environ. Res. 18: 257-269. Chernoff, N. and R.J. Kavlock. 1982. An in vivo
teratology screen utilizing pregnant mice. J. Toxicol. Environ. Health. 10: 541-550.

Chu, I., D.C. Villeneuve, B.L. MacDonald, V.E. Secours and V.E. Valli. 1981a.
Reversibility of the toxicological changes induced by photomirex and mirex. Toxicology.
21: 235-250.

Chu, I., D.C. Villeneuve, V.E. Secours, V.E. Valli and G.C. Becking. 1981b. Effects of
photomirex and mirex on reproduction in the rat. Toxicol. Appl. Pharmacol. 60: 549-
556.

Fulfs, J., R. Abraham, B. Drobeck, K. Pittman and F. Coulston. 1977. Species
differences in the hepatic response to mirex: Ultrastructural and histochemical studies.
Ecotoxicol. Environ. Saf. 1: 327-342.

Gaines, T.B. and R.D. Kimbrough. 1970. Oral toxicity of mirex in adult and suckling
rats. Arch. Environ. Health. 21: 7-14.

Grabowski, C.T. 1983. The electrocardiogram of fetal and newborn rats
anddysrhythmias induced by toxic exposure. In: Abnormal Functional Development of
the Heart, Lungs and Kidneys: Approaches to Functional Teratology. Alan R. Liss, Inc.,
NY. p. 185-206.

Grabowski, C.T. and D.B. Payne. 1980. An electrocardiographic study of
cardiovascular problems in mirex-fed rat fetuses. Teratology. 22: 167-177.

Grabowski, C.T. and D.B. Payne. 1983a. The causes of perinatal death induced by
prenatal exposure of rats to the pesticide, mirex. Part I: Pre-parturition observations of
the cardiovascular system. Teratology. 27: 7-11.

Grabowski, C.T. and D.B. Payne. 1983b. The causes of perinatal death induced by prenatal exposure of rats to the pesticide, mirex. Part II. Postnatal observations. J. Toxicol. Environ. Health. 11: 301-315.

Kavlock, R.J., N. Chernoff, E. Rogers et al. 1982. An analysis of fetotoxicity using biochemical endpoints of organ differentiation. Teratology. 26: 183-194.

Khera, K.S., D.C. Villeneuve, G. Terry, L. Panopio, L. Nash and G. Trivett. 1976. Mirex: A teratogenicity, dominant lethal and tissue distribution study in rats. Food Cosmet. Toxicol. 14: 25-29.

Larson, P.S., J.L. Egle, Jr., G.R. Hennigar and J.F. Borzelleca. 1979. Acute and subchronic toxicity of mirex in the rat, dog and rabbit. Toxicol. Appl. Pharmacol. 49: 271-277.

NTP (National Toxicology Program). 1990. Toxicology and Carcinogenesis Studies of MIREX (CAS No. 2385-85-5) in F344/N Rats (Feed Studies). NTP TR 313.

Scotti, T.M., N. Chernoff, R. Linder and W.K. McElroy. 1981. Histopathologic lens changes in mirex-exposed rats. Toxicol. Lett. 9: 289-294.

Shannon, V.C. 1976. The effects of mirex on the reproductive performance and behavioral development of the prairie vole Microtus ochrogaster. Ph.D. Thesis. University Microfilms International Dissertation Services, Ann Arbor, MI.

Ulland, B.M., N.P. Page, R.A. Squire, E.K. Weisburger and R.L. Cypher. 1977. A carcinogenicity assay of mirex in Charles River CD rats. J. Natl. Cancer Inst. 58: 133-140.

US EPA. 1987. Health Effects Assessment for Mirex. Prepared by the Office of Health and Environmental Assessment, Environmental Criteria and Assessment Office, Cincinnati, OH for the Office of Solid Waste and Emergency and Response, Washington, DC. EPA/600/8-88/046.

Yarbrough, J.D., J.E. Chambers, J.M. Grimley et al. 1981. Comparative study of 8-monohydromirex and mirex toxicity in male rats. Toxicol. Appl. Pharmacol. 58: 105-117.

Ware, G.W. and E.E. Good. 1967. Effects of insecticides on reproduction in the laboratory mouse. Toxicol. Appl. Pharmacol. 10: 54-61.

VI.B. INHALATION RfC REFERENCES

None

VI.C. CARCINOGENICITY ASSESSMENT REFERENCES

None

VI.D. DRINKING WATER HA REFERENCES

None

VII. REVISION HISTORY

Date	Section	Description
03/01/88	I.A.5.	Confidence levels revised
04/01/91	I.A.4.	Citations added
04/01/91	VI.	Bibliography on-line
01/01/92	IV.	Regulatory actions updated
08/01/92	I.A.	Withdrawn; new oral RfD verified (in preparation)
08/01/92	IV.	Regulatory actions withdrawn
08/01/92	VI.A.	Bibliography withdrawn
10/01/92	I.A.	Oral RfD summary replaced; RfD changed
10/01/92	IV.	Regulatory actions returned in conjunction with RfD
10/01/92	VI.A.	Bibliography replaced
07/01/93	II.	Carcinogenicity assessment now under review

SYNONYMS

BICHLORENDO
CG-1283
CYCLOPENTADIENE, HEXACHLORO-, DIMER
DECANE,PERCHLOROPENTACYCLO-DECHLORANE
DECHLORANE 4070
1,1a,2,2,3,3a,4,5,5,5a,5b,6-DODECACHLOROOCTAHYDRO-1,3,4-METHENO-1H-CYCLOBUTA (cd)PENTALENE
DODECACHLOROOCTAHYDRO-1,3,4-METHENO-2H-CYCLOBUTA (c,d)PENTALENE
DODECACHLOROPENTACYCLO(3.2.2.0(sup 2,6),0(sup 3,9),0(sup 5,10))DECANE
DODECACHLOROPENTACYCLODECANE
ENT 25,719

FERRIAMICIDE
GC 1283
HEXACHLOROCYCLOPENTADIENE DIMER
1,2,3,4,5,5-HEXACHLORO-1,3-CYCLOPENTADIENE DIMER
HRS 1276
1,3,4-METHENO-1H-CYCLOBUTA(cd)PENTALENE, DODECACHLOROOCTAHYDRO-
1,3,4-METHENO-1H-CYCLOBUTA(cd)PENTALENE, 1,1a,2,2,3,3a,4,5,5,5a,5b,6-DODECACHLOROOCTAHYDRO-
Mirex
NCI-C06428
PERCHLORODIHOMOCUBANE
PERCHLOROPENTACYCLO(5.2.1.0(sup 2,6).0(sup 3,9).0(sup 5,8))DECANE
PERCHLOROPENTACYCLODECANE

Source: *Instant EPA's IRIS*, 1997, Instant Reference Sources, Inc., 7605 Rockpoint Drive, Austin, TX 78731 (Fax: 512-345-2386; Internet URL, http://www.instantref.com/inst-ref.htm).

Production, Use, and Pesticide Labeling Information

MIREX is not currently listed in EPA's Pesticide Factsheet database and there is no indication of whether it will be added or not in the future.

Nitrofen

Introduction

Nitrofen is a contact herbicide for pre and post emergence control of annual grasses and broadleaf weeds on a variety of food ornamental crops and on rights-of-way. but it has not been used around homes and gardens. [828] Nitrofen is an herbicide used on many vegetables, a number of broad-leaved and grass weeds, cereals, rice, sugar beet, some ornamentals, broccoli, cauliflower, cabbage, brussel sprouts, onions, garlic and celery. It is also used in nurseries for roses and chrysanthemums.

Nitrofen may be released to the environment during its production and use as a pre- and post-emergence herbicide. However, nitrofen is no longer manufactured or sold in the US and Canada because of possible mutagenic and carcinogenic effects, although it has been used in other countries. When applied to soil, it will photolyze on the soil surface and biodegrade. It adsorbs strongly to soil and leaching will be negligible. Biodegradation is fairly rapid in flooded soil (with a half-life of 10 days at 30°C), but fairly slow in upland soil with substantial residues lasting through two growing seasons in cooler areas. If released in water, nitrofen would adsorb strongly to sediment and particulate matter in the water column, photolyze in surface layers (65% degradation in 1 week) and biodegrade (99% degradation in 50 days). Bioconcentration in fish and aquatic organisms will be appreciable. In the atmosphere, nitrofen would exist primarily adsorbed to particulate matter and in aerosols from spraying operations. It will be subject to gravitational settling and rapidly photolyze. Exposure to nitrofen would be primarily occupational via dermal contact, with agricultural workers who formulate and apply the herbicide or handle treated soil and crops being especially at risk. [828]

Identifying Information

CAS NUMBER:1836-75-5

NIOSH Registry Number: KN8400000

SYNONYMS: ETHER, 2,4-DICHLOROPHENYL P-NITROPHENYL, 2,4-DICHLOROPHENYL P-NITROPHENYL ETHER, BENZENE, 2,4-DICHLORO-1-(4-NITROPHENOXY)-, 2,4-DICHLORO-4'-NITRODIPHENYL ETHER, 2,4-DICHLORO-1-(4-NITROPHENOXY)BENZENE, 2,4-DECHLOROPHENYL P-

NITROPHENYL ETHER, 2',4'-DICHLORO-4-NITROBIPHENYL ETHER, 4-(2,4-DICHLOROPHENOXY)NITROBENZENE, 2,4-DICHLOROPHENYL 4-NITROPHENYL ETHER, 4'-NITRO-2,4-DICHLORODIPHENYL ETHER, FW 925, MEZOTOX, NCI-C00420, NICLOFEN, NIP, NITOFEN, NITRAFEN, NITRAPHEN, NITROCHLOR, NITROPHEN, NITROPHENE, PREPARATION 125, TOK, TOK-2, TOK E, TOK E-25, TOK E 40, TOKKORN, TOK WP-50, TRIZILIN

Endocrine Disruptor Information

Nitrofen is a strongly suspected EED that it is listed by the Centers for Disease Control & Prevention [812], and the World Wildlife Fund [813] as a potential endocrine modifying chemical.

Chemical and Physical Properties

CHEMICAL FORMULA: $C_{12}H_7Cl_2NO_3$

MOLECULAR WEIGHT: 284.10

WLN: WNR DOR BG DG

PHYSICAL DESCRIPTION: Colorless crystals

MELTING POINT: 70-71°C [031], [169], [172], [395]

BOILING POINT: 180-190°C @ 0.25 mm Hg [169]

SOLUBILITY: Water: <1 mg/mL @ 21° [700]; DMSO: >=100 mg/mL @ 21°C [700]; 95% Ethanol: 5-10 mg/mL @ 21°C [700]; Acetone: >=100 mg/mL @ 21°C [700]; Methanol: Soluble [395]

OTHER SOLVENTS: Benzene: 2000 g/kg @ 20°C [169]; Xylene: Soluble [395]; n-Hexane: 280 g/kg @ 20°C [169]

Source: *The National Toxicology Program's Chemical Database, Volume 2*, 1992, CRC Press, Inc./Lewis Publishers, 2000 Corporate Blvd., Boca Raton, FL 33431 (Fax: 407-998-9114).

Environmental Reference Materials

A source of EED environmental reference materials for nitrofen is Radian International LLC, Austin, TX (Fax 512-454-0268; Internet URL, http://www.radian.com/standards); Catalog No. ERN-020.

Hazardous Properties

ACUTE/CHRONIC HAZARDS: Nitrofen may be toxic by ingestion [062] and an irritant [043], [395]. When heated to decomposition it emits toxic fumes of nitrogen oxides and chlorine [043].

FIRE HAZARD: Flash point data for nitrofen are not available.   It is probably combustible. Fires involving this material can be controlled with a dry chemical, carbon dioxide or Halon extinguisher.

STABILITY: Nitrofen is sensitive to exposure to light (darkens in color) [169], [395]. Solutions of nitrofen in water, DMSO, 95% ethanol or acetone should be stable for 24 hours under normal lab conditions [700].

Source: *The National Toxicology Program's Chemical Database, Volume 7*, 1992, CRC Press, Inc./Lewis Publishers, 2000 Corporate Blvd., Boca Raton, FL 33431 (Fax: 407-998-9114).

Medical Symptoms of Exposure

Symptoms of exposure may include irritation of the skin and eyes [043], [395].

Source: *The National Toxicology Program's Chemical Database, Volume 4*, 1992, CRC Press, Inc./Lewis Publishers, 2000 Corporate Blvd., Boca Raton, FL 33431 (Fax: 407-998-9114).

Toxicological Information

The toxicological information below is gathered from several sources including two national databases: one from the *National Toxicology Program Chemical Repository Database* and another from *EPA's Integrated Risk Information System (IRIS)*.

Toxicology Information from the National Toxicology Program

CARCINOGENICITY:

Tumorigenic Data:

Type of Effect	Route/Animal	Amount Dosed (Notes)
TDLo	orl-rat	42 gm/kg/94W-C
TDLo	orl-mus	24 gm/kg/12W-C
TD	orl-mus	200 gm/kg/78W-C
TD	orl-mus	308 gm/kg/78W-C
TD	orl-mus	114 gm/kg/58W-C
TD	orl-mus	47 gm/kg/12W-C

Review:
IARC Cancer Review: Animal Sufficient Evidence
IARC possible human carcinogen (Group 2B)

Status:

NCI Carcinogenesis Bioassay (Feed); Negative: Male and Female Rat
NCI Carcinogenesis Bioassay (Feed): Positive: Male and Female
Mouse
NCI Carcinogenesis Bioassay (Feed); Inadequate Study: Male Rat
NCI Carcinogenesis Bioassay (Feed); Positive: Female Rat, Male and

Female Mouse

NTP Fourth Annual Report on Carcinogens, 1984
NTP anticipated human carcinogen

TERATOGENICITY: See RTECS printout for most current data

MUTATION DATA: See RTECS printout for most current data

TOXICITY:

Typ. Dose	Mode	Specie	Amount	Units	Other
LD50	orl	rat	740	mg/kg	
LD50	skn	rat	5000	mg/kg	
LD50	unr	rat	3000	mg/kg	
LD50	orl	mus	450	mg/kg	
LDLo	orl	cat	300	mg/kg	
LCLo	ihl	cat	620	mg/m3/4H	
LD50	orl	rbt	1620	mg/kg	

OTHER TOXICITY DATA:

Review: Toxicology Review

Status:

EPA Genetox Program 1988, Positive: Carcinogenicity-mouse/rat
EPA Genetox Program 1988, Inconclusive: Mammalian micronucleus
EPA TSCA Chemical Inventory, 1986
EPA TSCA Test Submission (TSCATS) Data Base, June 1989
Meets criteria for proposed OSHA Medical Records Rule

SAX TOXICITY EVALUATION:

THR: Poison by ingestion. Moderately toxic by inhalation and possibly other
routes. Mildly toxic by skin contact. A suspected human carcinogen. An
experimental carcinogen and teratogen. Experimental reproductive effects. A
skin and severe eye irritant. Mutagenic data.
NITROFEN is not currently listed in NTP's Chemical Repository database nor is it
scheduled for addition in the future.

Source: *Instant Tox-Base*, 1995, Instant Reference Sources, Inc., 7605 Rockpoint Drive, Austin, TX 78731 (Fax: 512-345-2386; Internet URL, http://www.instantref.com/inst-ref.htm).

Toxicology Information from EPA's Integrated Risk Information System (IRIS)

NITROFEN is not currently listed in EPA's IRIS database nor is it listed as scheduled for addition in the near future.

Production, Use, and Pesticide Labeling Information

NITROFEN is not currently listed in EPA's Pesticide Factsheet database and there is no indication of whether it will be added or not in the future.

Parathion

Introduction

Parathion is a broad spectrum insecticide used in agricultural applications. It is also used as an acaricide, a fumigant and a nematicide. There are two esters of parathion: the diethyl ester and the dimethyl ester. All of the information presented herein is for the diethyl ester derivative EXCEPT FOR THE PRODUCTION, USE AND PESTICIDE LABELING INFORMATION. The latter comes from EPA's Pesticide Fact Sheets database and it does not contain information on the diethyl ester and has very little information on the chemical, physical and hazardous properties of dimethyl parathion.

Parathion is released to the environment primarily in its use as a broad- spectrum insecticide in agricultural applications although more localized releases may occur in wastewater spills, and fugitive emissions during its production, transport, storage, and formulation. When sprayed on a field or orchard it will exist primarily as an aerosol in the source area and as a combination of vapor and aerosol downwind. The vapor will be rapidly photolyzed (with a half-life of 5 minutes in summer sunlight) to paraoxon. The parathion residue on foliage will decay with a half life of 1 day reaching low levels in a week or two. It will bind tightly to soil and decay by biological and chemical hydrolysis in several weeks forming p-nitrophenol, diethylthiophosphoric acid, and paraoxon. Accumulation from repeated doses is unlikely. Photolysis may occur on the soil surface. Degradation in flooded soil is much faster and is probably due to surface catalyzed hydrolysis; aminoparathion is formed under these low oxygen conditions. Residues will generally remain in the upper 6 inches of soil so leaching into groundwater is unlikely. [828]

While parathion residues resulting from spills may decrease markedly with time, especially during the first year, they may persist for many years. Parathion released into surface waters will be removed in approximately a week. The primary removal mechanism is adsorption to sediment and particulate matter where biodegradation or chemical hydrolysis will occur. While photooxidation in water occurs with a <1-10 days half life, the presence of radicals and sensitizers which are found in eutrophic waters may greatly accelerate this process. Bioconcentration will be low to moderate. Human exposure will result primarily during field application and formulation of the insecticide. Exposure will be primarily dermal and inhalation due to exposure to aerosols and vapors and from contact with sprayed surfaces as well as by inhalation. Exposure to the general public is by ingestion of produce containing parathion residues and from air in areas where spraying is taking place.[828]

Identifying Information

CAS NUMBER: 56-38-2

NIOSH Registry Number: TF4550000

SYNONYMS: Diethyl p-nitrophenylthiophosphate, O,O-Diethyl O-(p-nitrophenyl) phosphorothioate

Endocrine Disruptor Information

Parathion is a strongly suspected EED that it is listed by the Centers for Disease Control & Prevention [812], and the World Wildlife Fund [813] as a potential endocrine modifying chemical.

Chemical and Physical Properties

CHEMICAL FORMULA: $C_{10}H_{14}NO_5PS$

MOLECULAR WEIGHT: 291.27

WLN: WNR DOPS&O2&O2

PHYSICAL DESCRIPTION: Pale yellow to dark brown liquid

SPECIFIC GRAVITY: 1.265 @ 20/4°C [169], [172], [395], [430]

DENSITY: 1.270 g/cm³ @ 18.5°C [700]

MELTING POINT: 6°C [027], [031], [205], [371]

BOILING POINT: 375°C [017], [031], [205], [346]

SOLUBILITY: Water: <1 mg/mL @ 23°C [700]; DMSO: >=100 mg/mL @ 22°C [700]; 95% Ethanol: >=100 mg/mL @ 22°C [700]; Acetone: >=100 mg/mL @ 22°C [700]

OTHER SOLVENTS: Chloroform: Soluble [017], [047], [169]; Alcohol: Soluble [017], [031], [043], [169]; Esters: Soluble [031], [043], [173]; Ketones: Soluble [031], [043], [062], [173]; Petroleum ether: Insoluble [027], [031], [173], [205]; Kerosene: Insoluble [031], [043], [151], [173]; Ether: Soluble [017], [031], [047], [205]; Benzene: Soluble [169], [205]; Petroleum oils: Slightly soluble [172,395]; Usual spray oils: Insoluble [031], [062], [173]; Most organic solvents: Miscible [058], [172]; Animal and vegetable oils: Soluble [062]; Aromatic hydrocarbons: Soluble [031], [043], [062], [173]

OTHER PHYSICAL DATA: Specific gravity: 1.2704 @ 20/20°C [017], [047] 1.266 @ 25/15.6°C [058]; Boiling point: 157-162°C 0.6 mm Hg [027], [031], [055], [395]; Vapor pressure: 0.0000965 mm Hg @ 25°C [058] Refractive index: 1.5420 @ 20°C [027]

1.5370 @ 25°C [017], [031], [169], [205]; Surface tension: 39.2 dynes/cm @ 25°C [031]; Garlic-like odor [058], [102], [421]; Viscosity: 15.30 centipoise @ 25°C [031]; Heat of combustion: -5140 cal/g [371] log P octanol: 3.81 @ 20°C [055]

Source: *Instant EPA's Air Toxics*, 1994, Instant Reference Sources, Inc., 7605 Rockpoint Drive, Austin, TX 78731 (Fax: 512-345-2386; Internet URL, http://www.instantref.com/inst-ref.htm).

Environmental Reference Materials

A source of EED environmental reference materials for parathion is Radian International LLC, Austin, TX (Fax 512-454-0268; Internet URL, http://www.radian.com/standards); Catalog No. ERP-079.

Hazardous Properties

ACUTE/CHRONIC HAZARDS: Parathion is among the most poisonous materials commonly used for pest control. It is responsible for hundreds of human poisonings, many of which are fatal [395]. Parathion can be fatal by ingestion, inhalation and skin or eye contact [058]. It is a cholinesterase inhibitor [058], [102], [151], [295] and it can be absorbed through the skin [027], [052], [058], [173]. When heated to decomposition it emits toxic fumes of hydrogen sulfide, carbon monoxide, nitrogen oxides, phosphorus oxides and sulfur oxides [043], [058], [102].

HAP WEIGHTING FACTOR: 1 [713]

FIRE HAZARD: Parathion has a flash point of >93.3°C (>200°F) [058]. It is combustible [043]. Fires involving this material may be controlled with a dry chemical, carbon dioxide or Halon extinguisher. A water spray may also be used [058].

REACTIVITY: Parathion is incompatible with strong oxidizers [102], [346]. It reacts slowly with water and rapidly with bases or caustic solutions [058]. It is incompatible with substances having a pH higher than 7.5 [033]. It will attack some forms of plastics, rubber and coatings [102].

STABILITY: Parathion slowly decomposes in air [062]. It decomposes when exposed to temperatures >100°C [058], [102]. It is rapidly hydrolyzed under alkaline conditions [027], [151], [169], [172]. It is stable at a pH below 7.5 [173] and it is stable in distilled water and acid solutions [062], [395].

Source: *Instant EPA's Air Toxics*, 1994, Instant Reference Sources, Inc., 7605 Rockpoint Drive, Austin, TX 78731 (Fax: 512-345-2386; Internet URL, http://www.instantref.com/inst-ref.htm).

Medical Symptoms of Exposure

Symptoms of exposure to diethyl parathion may include nausea, vomiting, diarrhea, excessive salivation, muscle twitching, convulsions, respiratory failure, coma and cholinesterase inhibition [033], [058], [102], [295]. Anorexia, pinpoint pupils and bronchoconstriction also may occur [033]. Other symptoms may include tightness of the chest, wheezing, blurred vision, tearing, runny nose, headache, abdominal cramps, dizziness, staggering, slurred speech and irregular or slow heartbeat [058], [102], [151]. It may cause cyanosis (bluish discoloration of the skin), loss of appetite, sweating and confusion [058], [102]. It may also cause aching in and behind the eyes, weakness, loss of muscle coordination and incontinence [151]. Pulmonary edema may also occur [371]. Exposure may result in labored breathing, nervousness, drooling or frothing of the mouth and nose, excessive bronchial secretion, possible muscle paralysis (on exposure to large amounts), loss of reflexes and death [058]. It may also result in general anesthesia and effects of the pulmonary system, kidney, ureter and bladder [043]. In addition, parathion may cause reduced vision, narrowing of peripheral visual fields, congestion or atrophy of optic nerves, difficulty with ocular pursuit movements, abnormality of ERG, miosis and spasm of accommodation for near vision, wide and unreactive pupils, and scotoma [099].

Source: *Instant EPA's Air Toxics*, 1994, Instant Reference Sources, Inc., 7605 Rockpoint Drive, Austin, TX 78731 (Fax: 512-345-2386; Internet URL, http://www.instantref.com/inst-ref.htm).

Toxicological Information

The toxicological information below is gathered from several sources including two national databases: one from the *National Toxicology Program Chemical Repository Database* and another from *EPA's Integrated Risk Information System (IRIS)*.

Toxicology Information from the National Toxicology Program

CARCINOGENICITY:

Tumorigenic Data:

Type of Effect	Route/Animal	Amount Dosed (Notes)
TDLo	orl-rat	1260 mg/kg/80W-C

Review:
IARC Cancer Review: Animal Inadequate Evidence
IARC: Not classifiable as a human carcinogen (Group 3)

Status:
NCI Carcinogenesis Bioassay (Feed); Equivocal: Male and Female Rat
NCI Carcinogenesis Bioassay (Feed); Negative: Male and Female Mouse

TERATOGENICITY: See RTECS printout for data

MUTATION DATA: See RTECS printout for data

TOXICITY:

Typ. Dose	Mode	Specie	Amount	Units	other
LDLo	orl	hmn	171	ug/kg	
TDLo	orl	wmn	5670	ug/kg	
LDLo	unr	man	1471	ug/kg	
LD50	orl	rat	2	mg/kg	
LC50	ihl	rat	84	mg/m3/4H	
LD50	skn	rat	6800	ug/kg	
LD50	ipr	rat	2	mg/kg	
LD50	ivn	rat	3800	ug/kg	
LD50	orl	mus	5	mg/kg	
LD50	orl	dog	3	mg/kg	
LD50	orl	rbt	10	mg/kg	
LD50	orl	gpg	8	mg/kg	
LD50	ims	rat	6	mg/kg	
LCLo	ihl	mus	15	mg/m3	
LD50	ivn	mus	13	mg/kg	
LD50	ipr	mus	3	mg/kg	
LD50	ims	mus	7200	ug/kg	
LD50	scu	mus	10	mg/kg	
LD50	ivn	cat	3	mg/kg	
LD50	ipr	dog	12	mg/kg	
LD50	ivn	dog	12	mg/kg	
LD50	orl	cat	930	ug/kg	
LD50	ipr	cat	3	mg/kg	
LCLo	ihl	rbt	50	mg/m3/2H	
LD50	skn	rbt	15	mg/kg	
LCLo	ihl	gpg	14	mg/m3/2H	
LD50	skn	gpg	45	mg/kg	
LD50	ipr	gpg	12	mg/kg	
LD50	orl	pgn	1330	ug/kg	
LD50	orl	qal	4040	ug/kg	
LD50	orl	dck	2100	ug/kg	
LD50	skn	dck	28	mg/kg	
LD50	orl	hor	5	mg/kg	
LD50	unr	mam	6	mg/kg	
LD50	orl	bwd	1330	ug/kg	
LD50	skn	bwd	1800	ug/kg	
LD50	orl	hmn	3	mg/kg	
TDLo	orl	man	429	ug/kg/4D-I	
LDLo	skn	hmn	7143	ug/kg	
LDLo	itrh	mn	714	ug/kg	
LD50	orl	ckn	10	mg/kg	
LD50	ipr	ckn	2500	ug/kg	
LD50	par	frg	967	mg/kg	
LDLo	orl	dom	100mg/kg		
LDLo	ims	dom	20	mg/kg	
LD50	orl	mam	49	mg/kg	
LD50	skn	mus	19	mg/kg	

OTHER TOXICITY DATA:

Review: Toxicology Review-8

Standards and Regulations:
DOT-Hazard: Poison B; Label: Poison

Status:
EPA Genetox Program 1988, Negative: In vitro UDS-human fibroblast; TRP reversion
EPA Genetox Program 1988, Negative: S cerevisiae-homozygosis
EPA Genetox Program 1988, Inconclusive: B subtilis rec assay; E colipolA without S9
EPA Genetox Program 1988, Inconclusive: Histidine reversion-Ames test
EPA Genetox Program 1988, Inconclusive: D melanogaster Sex-linked lethal
EPA TSCA Test Submission (TSCATS) Data Base, April 1990
NIOSH Analytical Methods: see EPN, Malathion, and Parathion, 5012
OSHA Analytical Method #ID-62
IDLH value: 20 mg/m3
Acute lethal dose in man: 10-20 mg

SAX TOXICITY EVALUATION:

THR: A deadly human poison by an unspecified route. A deadly experimental poison by all routes. Human systemic effects by ingestion. An experimental carcinogen, tumorigen and teratogen. A possible human carcinogen. Experimental reproductive effects. Human mutagenic data. A cholinesterase inhibitor. Parathion, like the other organic phosphorus poisons, acts as an irreversible inhibitor of the enzyme cholinesterase and thus allows the accumulation of large amounts of acetylcholine. When a critical level of cholinesterase depletion is reached, grave symptoms appear. Whether death is actually caused entirely by cholinesterase depletion or by the disturbance of a number of enzymes is not yet known. Recovery is apparently complete if a poisoned animal or man has time to reform a critical amount of cholinesterase. The organism exposed remains susceptible to relatively low dosages of parathion until the cholinesterase level has regenerated. Small doses at frequent intervals are, therefore, more or less additive. There is not, however, at the present time, any indication that, when recovery from a given exposure is entirely complete, the exposed organism is prejudiced in any way.

Source: *Instant Tox-Base*, 1995, Instant Reference Sources, Inc., 7605 Rockpoint Drive, Austin, TX 78731 (Fax: 512-345-2386; Internet URL, http://www.instantref.com/inst-ref.htm).

Toxicology Information from EPA's Integrated Risk Information System (IRIS)

I. CHRONIC HEALTH HAZARD ASSESSMENTS FOR NONCARCINOGENIC EFFECTS

I.A. REFERENCE DOSE FOR CHRONIC ORAL EXPOSURE (RfD)

A risk assessment for this substance/agent is under review by an EPA work group.

I.B. REFERENCE CONCENTRATION FOR CHRONIC INHALATION EXPOSURE (RfC)

[Ed. Note: EPA does not have information yet in this section of IRIS].

II. CARCINOGENICITY ASSESSMENT FOR LIFETIME EXPOSURE

Section II provides information on three aspects of the carcinogenic risk assessment for the agent in question; the US EPA classification, and quantitative estimates of risk from oral exposure and from inhalation exposure. The classification reflects a weight-of-evidence judgment of the likelihood that the agent is a human carcinogen. The quantitative risk estimates are presented in three ways. The slope factor is the result of application of a low-dose extrapolation procedure and is presented as the risk per (mg/kg)/day. The unit risk is the quantitative estimate in terms of either risk per ug/L drinking water or risk per ug/cu.m air breathed. The third form in which risk is presented is a drinking water or air concentration providing cancer risks of 1 in 10,000, 1 in 100,000 or 1 in 1,000,000. Background Document 2 (Service Code 5) provides details on the rationale and methods used to derive the carcinogenicity values found in IRIS. Users are referred to Section I for information on long-term toxic effects other than carcinogenicity.

II.A. EVIDENCE FOR CLASSIFICATION AS TO HUMAN CARCINOGENICITY

II.A.1. WEIGHT-OF-EVIDENCE CLASSIFICATION

Classification -- C; possible human carcinogen.

Basis -- Increased adrenal cortical tumors in female and male Osborne-Mendel rats and positive trends for thyroid follicular adenomas and pancreatic islet-cell carcinomas in male rats in one study.

II.A.2. HUMAN CARCINOGENICITY DATA

None.

II.A.3. ANIMAL CARCINOGENICITY DATA

Limited. Osborne-Mendel rats, 50/sex/group were exposed to dietary levels of parathion; 10/sex served as controls. Due to toxicity, the original doses for males (40 and 80 ppm)

and females (20 and 40 ppm) were changed at 13 weeks to 30 and 60 ppm for both males and females. The female doses were returned to 20 and 40 ppm at week 46. Depressed body weight gain was observed while animals were on the test diets and demonstrated that the MTD was reached. Mortality was not affected. Adrenal cortical adenomas and combined adrenal adenomas and carcinomas were significantly increased in both sexes at the high dose and in the males at the low dose. Additionally, there were significant positive trends with respect to pooled controls for thyroid follicular adenomas and pancreatic islet cell carcinomas in the males. This study has the following limitations: the control group only contained 10 rats/sex; some tissues were not examined microscopically; some rats were only dosed for 80 weeks of the 112-week study; and lab audit of the study indicated instances of nonadherence to good laboratory practices by the test facility. Despite these limitations, the study was accepted, as a definitive tumor response at one site and marginal responses at two sites had been produced (NCI, 1979).

B6C3F1 mice (50/sex/dose, 10/sex for controls) were exposed to parathion (99.5% purity) in the diet at 0, 80, or 160 ppm for 62 to 80 weeks. No oncogenic effects were noted in either sex at either dose. Based on the data reported, the maximum tolerated dose (MTD) was reached in the male mice but was probably not achieved in the females. The study was flawed for reasons similar to those outlined for the NCI rat study (NCI, 1979).

In a well-conducted study, 60 male and 60 female Sprague-Dawley rats/sex/dose group were maintained on diets containing 0, 0.5, 5.0, and 50.0 ppm parathion for 110 (males) and 120 (females) weeks. The mortality in the 5-ppm males was increased between months 7 and 23, but mortality in all groups was comparable to controls at termination. The MTD was slightly exceeded at the high dose in this study, but no compound-induced oncogenic response was observed (Biodynamics, 1984).

II.A.4. SUPPORTING DATA FOR CARCINOGENICITY

A positive response was found with and without metabolic activation at 10E-5 and 10E-6 molar concentrations in an unscheduled DNA synthesis assay using human WI-38 cells (US EPA, 1986). Other mutagenicity assays were negative (reverse mutation in bacteria, mitotic recombination in yeast, and DNA repair in bacteria) but these tests were considered inadequate by the Office of Pesticide Programs Peer Review Panel.

II.B.   QUANTITATIVE ESTIMATE OF CARCINOGENIC RISK FROM ORAL EXPOSURE

[Ed. Note: EPA does not have information yet in this section of IRIS].

II.C.   QUANTITATIVE ESTIMATE OF CARCINOGENIC RISK FROM INHALATION EXPOSURE

[Ed. Note: EPA does not have information yet in this section of IRIS].

II.D. EPA DOCUMENTATION, REVIEW, AND CONTACTS (CARCINOGENICITY ASSESSMENT)

II.D.1. EPA DOCUMENTATION

Source Document -- US EPA, 1986

The Toxicology Branch Peer Review Committee, Office of Pesticide Programs, Office of Pesticides and Toxic Substances reviewed data on parathion.

II.D.2. REVIEW (CARCINOGENICITY ASSESSMENT)

Agency Work Group Review -- 08/05/87

Verification Date -- 08/05/87

II.D.3. US EPA CONTACTS (CARCINOGENICITY ASSESSMENT)

Elizabeth A. Doyle / OPP -- (703)308-2722

Reto Engler / OPP -- (703)308-2738

III. HEALTH HAZARD ASSESSMENTS FOR VARIED EXPOSURE DURATIONS

III.A. DRINKING WATER HEALTH ADVISORIES

[Ed. Note: EPA does not have information yet in this section of IRIS].

IV. US EPA REGULATORY ACTIONS

EPA risk assessments may be updated as new data are published and as assessment methodologies evolve. Regulatory actions are frequently not updated at the same time. Compare the dates for the regulatory actions in this section with the verification dates for the risk assessments in sections I and II, as this may explain inconsistencies. Also note that some regulatory actions consider factors not related to health risk, such as technical or economic feasibility. Such considerations are indicated for each action. In addition, not all of the regulatory actions listed in this section involve enforceable federal standards. Please direct any questions you may have concerning these regulatory actions to the US EPA contact listed for that particular action. Users are strongly urged to read the background information on each regulatory action in Background Document 4 in Service Code 5.

IV.A. CLEAN AIR ACT (CAA)

[Ed. Note: EPA does not have information yet in this section of IRIS].

IV.B. SAFE DRINKING WATER ACT (SDWA)

Listed in the January 1991 Drinking Water Priority List and may be subject to future regulation (56 FR 1470, 01/14/91).

IV.C. CLEAN WATER ACT (CWA)

IV.C.1. AMBIENT WATER QUALITY CRITERIA, Human Health

None available

IV.C.2. AMBIENT WATER QUALITY CRITERIA, Aquatic Organisms

Freshwater:
  Acute -- 6.5E-2 ug/L (1 hour average)
  Chronic -- 1.3E-2 ug/L (4 day average)

Marine: None.

Considers technological or economic feasibility? -- NO

Discussion -- Criteria were derived from a minimum data base consisting of acute tests on a variety of species. Requirements and methods are covered in the reference to the Federal Register.

Reference -- 51 FR 43665 (12/03/86)

EPA Contact -- Criteria and Standards Division / OWRS (202)260-1315 / FTS 260-1315

IV.D. FEDERAL INSECTICIDE, FUNGICIDE, AND RODENTICIDE ACT (FIFRA)

IV.D.1. PESTICIDE ACTIVE INGREDIENT, Registration Standard

Status -- Issued (1986)

Reference -- Ethyl Parathion Pesticide Registration Standard. December, 1986 (NTIS No. PB87-152120)

EPA Contact -- Registration Branch / OPP (703)557-7760 / FTS 557-7760

IV.D.2. PESTICIDE ACTIVE INGREDIENT, Special Review

        [Ed. Note: EPA does not have information yet in this section of IRIS].

IV.E. TOXIC SUBSTANCES CONTROL ACT (TSCA)

        [Ed. Note: EPA does not have information yet in this section of IRIS].

IV.F. RESOURCE CONSERVATION AND RECOVERY ACT (RCRA)

IV.F.1. RCRA APPENDIX IX, for Ground Water Monitoring

Status -- Listed

Reference -- 52 FR 25942 (07/09/87)

EPA Contact -- RCRA/Superfund Hotline (800)424-9346 / (202)260-3000 / FTS 260-3000

IV.G. SUPERFUND (CERCLA)

AIV.G.1. REPORTABLE QUANTITY (RQ) for Release into the Environment

Value (status) -- 10 pounds (Final, 1989)

Considers technological or economic feasibility? -- NO

Discussion -- The final RQ for parathion is based on aquatic toxicity as established under CWA Section 311 (40 CFR 117.3). The available data indicate that the aquatic 96-Hour Median Threshold Limit is 0.4 ppm, which corresponds to an RQ of 10 pounds.

Reference -- 54 FR 33418 (08/14/89)

EPA Contact -- RCRA/Superfund Hotline (800)424-9346 / (202)260-3000 / FTS 260-3000

VI. BIBLIOGRAPHY

VI.A. ORAL RfD REFERENCES

None

VI.B. INHALATION RfD REFERENCES

None

VI.C. CARCINOGENICITY ASSESSMENT REFERENCES

Biodynamics, Inc. 1984. MRID No. 00165344. Available from EPA. Write to FOI, EPA, Washington DC 20460.

NCI (National Cancer Institute). 1979. Bioassay of parathion for possible carcinogenicity. NCI-CG-TR-70. DHEW (NIH) 79-1320.

US EPA. 1986. Toxicology Branch Peer Review Committee, Office of Pesticide Programs, Office of Pesticides and Toxic Substances, memorandum on parathion, July 1.

VI.D. DRINKING WATER HA REFERENCES

None

VII. REVISION HISTORY

```
--------  -----------    ---------------------------------------------------------
Date      Section        Description
--------  -----------    ---------------------------------------------------------
```

08/22/88 II.   Carcinogen summary on-line
04/01/89 V.   Supplementary data on-line
12/01/89 VI.   Bibliography on-line
01/01/90 Synonyms   Added synonyms
01/01/91 II.   Text edited
01/01/92 IV.   Regulatory Action section on-line
10/01/93 II.D.3.   Primary contact changed; secondary's phone no. changed

SYNONYMS

AAT

AATP

AC 3422

ACC 3422

AI3-15108

ALKRON

ALLERON

AMERICAN CYANAMID 3422

APHAMITE

ARALO

B 404

BAY E-605

BAYER E-605

BLADAN

BLADAN F

CASWELL NO. 637

COMPOUND 3422

COROTHION

CORTHION

CORTHIONE

DANTHION

DIETHYL PARA-NITROPHENOL THIOPHOSPHATE

DIETHYL-P-NITROPHENYL MONOTHIOPHOSPHATE

O,O-DIETHYL O-(P-NITROPHENYL) PHOSPHOROTHIOATE

DIETHYL 4-NITROPHENYL PHOSPHOROTHIONATE

DIETHYL P-NITROPHENYL PHOSPHOROTHIONATE

O,O-DIETHYL-O,P-NITROPHENYL PHOSPHOROTHIOATE

O,O-DIETHYL-O-(4-NITRO-FENIL)-MONOTHIOFOSFAAT [DUTCH]

O,O-DIETHYL-O-P-NITROFENYLESTER KYSELINY THIOFOSFORECNE [CZECH]

O,O-DIETHYL O-4-NITROPHENYL PHOSPHOROTHIOATE

O,O-DIETHYL-O-(4-NITROPHENYL) PHOSPHOROTHIOATE

O,O-DIETHYL O-(4-NITROPHENYL) PHOSPHOROTHIOATE (9CI)

O,O-DIETHYL O-P-NITROPHENYLPHOSPHOROTHIOATE

O,O-DIETHYL-O-(P-NITROPHENYL)THIONOPHOSPHATE

O,O-DIETHYL O-4-NITROPHENYL THIOPHOSPHATE

O,O-DIETHYL O-P-NITROPHENYL THIOPHOSPHATE

O,O-DIETHYL O-(P-NITROPHENYL) PHOSPHOROTHIOATE

O,O-DIETIL-O-(4-NITRO-FENIL)-MONOTIOFOSFATO [ITALIAN]

O,O-DIETYL-O-P-NITROFENYLTIOFOSFAT [CZECH]

ORTHOPHOS

PAC

PACOL

DIETHYL P-NITROPHENYL
THIONOPHOSPHATE
O,O-DIETHYL O-P-NITROPHENYL
THIOPHOSPHATE
DIETHYL P-NITROPHENYL
THIOPHOSPHATE
DIETHYLPARATHION
DIETHYL PARATHION
DIETIL TIOFOSFATO DE P-
NITROFENILA [PORTUGUESE]
DNTP
DPP
DREXEL PARATHION 8E
E 605
E 605 F
E 605 FORTE
ECATOX
EKATIN WF   WF ULV
EKATOX
ENT 15,108
EPA PESTICIDE CHEMICAL CODE
057501
ETHLON
ETHYL PARATHION
ETILON
ETYLPARATION [CZECH]
FOLIDOL
FOLIDOL E
FOLIDOL E605
FOLIDOL E   E 605
FOLIDOL OIL
FOSFERNO
FOSFEX
FOSFIVE
FOSOVA
FOSTERN
FOSTOX
GEARPHOS
GENITHION
HSDB 197
KOLPHOS
KYPTHION
LETHALAIRE G-54
LIROTHION
MURFOS
NA 2783
NCI-C00226
NIRAN
NIRAN E-4

PANTHION
PARADUST
PARAFLOW
PARAMAR
PARAMAR 50
PARAPHOS
PARASPRAY
PARATHENE
PARATHION
PARATHION-ACETYL [GERMAN]
PARATHION-AETHYL [GERMAN]
PARATHION-ETHYL
PARATHION, LIQUID
PARATHION MIXTURE, DRY
PARATHION MIXTURE, LIQUID
PARAWET
PENNCAP E
PENPHOS
PESTOX PLUS
PETHION
PHENOL, P-NITRO-, O-ESTER WITH
O,O-DIETHYLPHOSPHOROTHIOATE
PHOSKIL
PHOSPHENOL
PHOSPHOROTHIOIC ACID, O,O-
DIETHYL O-(4-NITROPHENYL)
ESTER
PHOSPHOROTHIOIC ACID, O,O-
DIETHYL O-(P-NITROPHENYL)
ESTER
PHOSPHOSTIGMINE
RB
RCRA WASTE NUMBER P089
RHODIASOL
RHODIATOX
RHODIATROX
SELEPHOS
SIXTY-THREE SPECIAL E.C.
INSECTICIDE
SNP
SOPRATHION
STABILIZED ETHYL PARATHION
STATHION
STRATHION
SULPHOS
SUPER RODIATOX
T-47
THIOFOS
THIOMEX

NITROSTIGMIN (GERMAN)
NITROSTIGMINE
NITROSTYGMINE
NIUIF 100
NOURITHION
OLEOFOS 20
OLEOPARAPHENE
OLEOPARATHION
OMS 19
O,O-DIAETHYL-O-(4-NITRO-PHENYL)-
MONOTHIOPHOSPHAT [GERMAN]

THIOPHOS
THIOPHOS 3422
THIOPHOSPHATE DE O,O-
DIETHYLE ET DE O-(4-
NITROPHENYLE) [FRENCH]
TIOFOS
TOX 47
VAPOPHOR
VAPOPHOS
VITREX

Source: *Instant EPA's IRIS*, 1997, Instant Reference Sources, Inc., 7605 Rockpoint Drive, Austin, TX 78731 (Fax: 512-345-2386; Internet URL, http://www.instantref.com/inst-ref.htm).

Production, Use, and Pesticide Labeling Information

1. Description of Chemical

Generic Name: O,O-dimethyl O-(4-nitrophenyl) phosphorothioate

Common Name: Methyl Parathion

Trade Names: O,O-dimethyl O-(4-nitrophenyl) phosphorothioate; O,O-dimethyl O-(p-nitrophenyl) phosphorothioate; parathion-methyl; metaphos; cekumethion; Devithion; dimethyl parathion; E601; Folidol M; Fosferno M50; Parataf; Paratox; Partron M; Penncap M; Tekwaisa; Wofatox; Metacide; Bladan M; Metron; Dalf; Nitrox 80.

EPA Shaughnessy Code: 053501

Chemical Abstracts Service (CAS) Number: 298-00-0

Year of Initial Registration: 1954

Pesticide Type: Insecticide

Chemical Family: Organophosphate

US and Foreign Producers: Monsanto in the US, and Bayer, AG in West Germany, Chemiekombinat Bitterfield VEB in East Germany, and A/S Cheminova in Denmark.

2. Use Patterns and Formulations

- Application sites: Field, vegetable, tree fruit and nut crops, tobacco and ornamentals, forestry, aquatic food crops, mosquito abatement districts, terrestrial and non-crop sites.

- Pests controlled: A wide variety of insects and mites as well as tadpole shrimp.

- Types and methods of application: Usual application is foliar. May be applied by aircraft or ground equipment.

- Types of Formulations: Dusts, wettable powders, micro-encapsulated, emulsifiable concentrates, and ready-to-use liquid.

- Rates: 0.1 to 6.0 lbs. a.i. per acre
- Usual Carriers: Petroleum solvents, clay carriers.

3. Science Findings

Chemical Characteristics

- Little information is available. Technical methyl parathion has a vapor pressure of 0.14 mg/m^3 at 20 °C, and an octanol/water partition coefficient of 3300. Methyl parathion is soluble in most organic solvents and is slightly soluble in aliphatic hydrocarbons. This compound is practically insoluble in water.

Toxicology Characteristics

- Acute toxicity:

- Methyl parathion causes cholinesterase inhibition. It is highly toxic to mammals by all routes of exposure and is classified in Toxicity Category I (LD50 4.5 to 16 mg/kg).

- Major routes of exposure:

- The major route of exposure is acknowledged to be dermal with inhalation, ocular, and oral exposure being much smaller.

- Information from the California Department of Food and Agriculture reported incidents of worker poisonings and illnesses during mixing, loading and application of methyl parathion. EPA is requiring additional "Worker Safety Rules," including protective clothing, to reduce exposure.

- Delayed neurotoxicity:

  - Methyl parathion is not believed to cause delayed neurotoxicity.

- Chronic feeding/Oncogenicity studies:

  - The Agency has two 2 year chronic feeding/oncogenicity studies in the rat, one in the mouse and a one-year dog study.

  - The Agency is unable to definitively evaluate oncogenicity at this time; additional information is required in the Wistar rat strain and another mouse study is required. Chronic effects noted; retinal and sciatic nerve damage at high dose levels (50 ppm) was observed in the rat.

- Subchronic studies:

  - Subchronic feeding studies show cholinesterase as the primary target for the toxic action of methyl parathion. A NOEL was established in the rat at 2.5 ppm or 0.25 mg/kg/day. The NOEL in the dog was 0.3 mg/kg/day (this NOEL was used to establish the current PADI).

  However, additional subchronic studies in both the rat and dog are required to determine the NOEL for retinal and sciatic nerve damage in the rat and retinal damage in the dog.

- Teratogenicity:

  - Some evidence of embryotoxicity and fetotoxicity at 1.0 mg/kg in rats. However, maternal toxicity was not established. Additional data are required. No signs of developmental toxicity were noted in the rabbit.

- Reproduction:

  - No reproductive effects were observed in rats at dietary levels up to 25 ppm. No additional information is required.

- Mutagenicity:

  - The Agency has evaluated the reports of a number of assays which address the three major categories of alterations, i.e., 1) gene mutation, 2), structural chromosomal aberrations, and 3) other mechanisms of genotoxicity. Although results of several of the individual tests are negative, other tests in each of these major categories provide limited evidence that methyl parathion is genotoxic. No additional information is required.

Physiological and Biochemical Characteristics

- Methyl parathion acts by causing irreversible inhibition of cholinesterase enzyme, allowing accumulation of acetylcholine at cholinergic neuroeffector junctions and autonomic ganglia. Poisoning symptoms include headaches, nausea, vomiting, cramps, weakness, blurred vision, pinpoint pupils, tightness in the chest, drooling or frothing of mouth and nose, muscle spasms, coma, and death. The mechanism of pesticidal action is now known.

- Metabolism:

- Data gap; additional data are required.

Environmental Fate and Exposure

- Insufficient information is available for the analysis of the environmental fate and the exposure of humans and non-target organisms to methyl parathion. Additional data are required.

- Methyl parathion, it is believed, does not bioaccumulate.

- Dermal, ocular, and inhalation exposure can occur during mixing, loading, and application, cleaning and repair of equipment, and during early reentry. EPA is requiring additional "Worker Safety Rules," including protective clothing, to reduce exposure.

- Methyl Parathion, it is believed, has little or no potential to contaminate ground water. This chemical was not included on the list of potential ground water contaminators.

Ecological Characteristics

- Avian oral toxicity: 6.6 mg/kg for mallard duck and 7.6 mg/kg for bobwhite quail.

- Avian dietary toxicity: 336 ppm for mallard duck and 90 ppm for bobwhite quail.

- Small mammal oral toxicity: 75 to 379 mg/kg for microtine rodents.

- Avian reproduction: Laboratory studies showed no direct reproductive impairment; however, significant depression of brain cholinesterase activity was observed (These studies were conducted with the Penncap M formulation.) Field studies indicate the possibility of reproductive impairment. Effects on the survival of nestlings were also noted.

- Freshwater fish acute toxicity: 3.7 ppm for rainbow trout and 4.4 ppm for bluegill.

- Aquatic invertebrate acute toxicity: Daphnia magna 0.14 ppm.

- Marine and estuarine toxicity: Mysid shrimp 0.98 ppm Sheepshead minnow 12,000 ppb

- Endangered species: Previous consultations with the Office of Endangered Species have resulted in jeopardy opinions and labeling for crops (alfalfa, apples, barley, corn, cotton, pears, peanuts, sorghum, soybeans, and wheat), rangeland and pastureland, silvacultural sites, aquatic sites, and non-cropland use. Labeling is required in an effort to reduce the risk to endangered species.

Tolerance Assessment

- Present United States, Canadian, Mexican and Codex tolerances for methyl parathion in or on raw agricultural commodities are specified below. Established tolerances for residues of methyl parathion are also listed in 40 CFR Sections 180.121 (a) and (b). Because there are considerable gaps in both residue chemistry and toxicology, a tolerance assessment cannot be made at this time. The nature of the residue in plants and animals is not adequately understood because of inadequate metabolism data. When the required data are submitted to the Agency, the following will be evaluated: 1. the tolerance definition in plants; 2. the need for and nature of tolerances in or on animal commodities.

METHYL PARATHION

TABLE I

Summary of Present Tolerances

Commodity	Tolerances (ppm) United States	Canada	Mexico	(MRL) International (Codex)
Garden Beets	1.0	--	--	--
Carrots	1.0	--	--	--
*Parsnips	1.0	--	--	0.7
Potatoes	0.1	--	0.1	--
Radishes	1.0	--	1.0	--
*Rutabagas	1.0	--	--	--
Sugar Beets	0.1	--	--	0.5
Sweet Potatoes	0.1	--	0.1	--
Turnip	1.0	--	--	--

(Table Continued)

| Commodity | Tolerances (ppm) | | | (MRL) |
	United States	Canada	Mexico	International (Codex)
*Garlic	1.0	--	1.0	--
Onions	1.0	--	1.0	--
Celery	1.0	--	1.0	--
*Endive	1.0	--	--	--
Lettuce	1.0	--	1.0	--
*Parsley	1.0	--	--	--
Spinach	1.0	--	1.0	--
*Swiss Chard	1.0	--	--	--
Broccoli	1.0	--	1.0	0.2
Brussels sprouts	1.0	--	--	0.2
Cabbage	1.0	--	1.0	0.2
Cauliflower	1.0	--	--	0.2
Collards	1.0	--	--	0.2
Kale	1.0	--	--	0.2
Kohlrabi	1.0	--	--	0/2
Mustard Greens	1.0	--	--	0.2
Beans	1.0	--	1.0	0.2
*Guar Beans	0.2	--	1.0	--
Peas	1.0	0.7	1.0	0.7
*Lentils	1.0	--	1.0	--
Soybeans	0.1	--	0.1	--
Eggplant	1.0	--	1.0	--
Peppers	1.0	--	1.0	--
Tomatoes	1.0	--	1.0	0.2
Cucumbers	1.0	--	1.0	0.2
Melons	1.0	--	1.0	0.2
Pumpkins	1.0	--	1.0	0.2
Squash	1.0	--	1.0	--
*Summer Squash	1.0	--	1.0	0.2
*Citrus Fruits	1.0	--	1.0	1.2
Apples	1.0	--	1.0	--
Pears	1.0	--	1.0	--
*Quince	1.0	--	--	--
Apricots	1.0	--	--	0.2
Cherries	1.0	--	--	0.2
Nectarines	1.0	--	--	0.2
Peaches	1.0	--	1.0	0.2
Plums	1.0	--	--	0.2
*Blackberries	1.0	--	--	--
Blueberries	1.0	--	--	0.2
*Boysenberries	1.0	--	--	0.2
*Cranberries	1.0	--	--	0.2

(Table Continued)

Commodity	Tolerances (ppm) United States	Canada	Mexico	(MRL) International (Codex)
*Currants	1.0	--	--	--
*Dewberries	1.0	--	--	0.2
Gooseberries	1.0	--	--	0.2
Grapes	1.0	--	--	0.2
*Loganberries	1.0	--	--	0.2
*Raspberries	1.0	--	--	0.2
Strawberries	1.0	--	1.0	0.2
*Youngberries	1.0	--	--	0.2
Almonds	0.1	--	--	--
*Filberts	0.1	--	--	--
Pecans	0.1	--	--	--
*Walnuts	0.1	--	--	--
Barley	1.0	--	--	--
Corn	1.0	--	1.0	--
Oats	1.0	--	--	--
Rice	1.0	--	1.0	--
Rye	0.5	--	--	--
Sorghum	0.1	--	0.1	--
Wheat	1.0	--	--	--
Forage Grass	1.0	--	--	--
Alfalfa Forage	1.25	--	1.25	--
Alfalfa Hay	5.0	--	5.00	--
Clover Forage & Hay	1.0	--	--	--
*Trefoil	1.25	--	--	--
*Trefoil Hay	5.0	--	--	--
Vetch Forage & Hay	1.0	--	--	--
Miscellaneous Crops				
Artichokes	1.0	--	1.0	0.2
Avocados	1.0	--	1.0	0.2
Cottonseed	0.75	--	0.75	--
*Dates	1.0	--	--	0.2
*Figs	1.0	--	1.0	0.2
*Guavas	1.0	--	1.0	0.2
Hops	1.0	--	--	0.05
*Mangos	1.0	--	1.0	0.2
Mustard Seed	0.2	--	--	--
*Okra	1.0	--	1.0	--
*Olives	1.0	--	1.0	0.2
Peanuts	1.0	--	1.0	--
*Pineapple	1.0	--	1.0	0.2

(Table Continued)

| Commodity | Tolerances (ppm) | | | (MRL) |
	United States	Canada	Mexico	International (Codex)
*Rape Seed	0.2	--	--	--
*Sugarcane	0.1	--	0.1	--
Sunflower Seed	0.2	--	--	--

- The US, Canadian, and Mexican tolerances expressed in terms of residues of methyl parathion per se.

-The Codex Maximum Residue Levels expressed as residues of methyl parathion and its oxygen analog, methyl paraoxon.

- * These commodities have tolerances but no Federal Registrations.

- Because data gaps prevent the formulation of an acceptable daily intake, a provisional acceptable daily intake has been established. The PADI for methyl parathion is 0.0015 mg/kg/day with a safety factor of 200. This figure will be retained until additional data are received. The theoretical maximum residue contribution for methyl parathion is approximately 800% of the provisional acceptable daily intake.

Reported Pesticide Incidents

> - Most of the pesticide incidents reported involve illnesses during mixer/loading, application, and drift from target areas.

Science Summary Statement

> - Methyl parathion is a Toxicity Category I organophosphate compound which is highly toxic to laboratory mammals, humans, aquatic invertebrates, and birds. There is some evidence that methyl parathion may effect reproductive success in birds. Methyl parathion poses a hazard to many endangered species. In laboratory rats of the Wistar strain, oncogenicity could not be determined as the data were insufficient; the Agency is requiring additional information on this study as well as a repeat of the mouse study. Chronic toxicity data indicate that methyl parathion causes retinal and sciatic nerve damage in rats at high dose levels (50 ppm in diet). Because data are not available, the Agency is unable to determine a no observed effect level (NOEL) for sciatic nerve damage.

4. Summary of Regulatory Position and Rationale
> - A review of the data available indicates that no risk criteria listed in 40 CFR 154.7 have been met or exceeded for methyl parathion.

- The Agency is requiring avian reproduction and terrestrial full field testing and simulated or full field aquatic testing to better define the extent of exposure and hazard to wildlife.

- No new tolerances or new food uses will be considered until the Agency has received data sufficient to assess existing tolerances for methyl parathion.

- The Agency is concerned about the potential for human poisonings (cholinesterase inhibition) from the use of methyl parathion. The Agency will continue to classify for restricted use (due to very high acute toxicity). The Certified applicator must be physically present during mixing, loading, application, equipment repair, and equipment cleaning. Information from the California Department of Food and Agriculture reported incidents of worker poisonings and illnesses during mixing, loading, and application. EPA is requiring more stringent "Worker Safety Rules", including protective clothing, to reduce exposure.

- A 48 hour re-entry interval, previously established under 40 CFR 170.3 (b) (2) will remain in effect. - The Agency has concluded that data are not adequate to determine the oncogenic potential of methyl parathion, and is requiring another mouse study and additional information on the Wistar rat.

- The Agency is requiring glove permeability and drift studies because of the high acute toxicity of methyl parathion.

- All manufacturing-use products and end-use products must bear appropriate labeling as specified in 40 CFR 162.10. Additionally, the following information must appear on the labeling:

> - Labeling requirements have been imposed to protect fish and wildlife (including endangered species).

> - Methyl parathion will continue to be classified Restricted Use and the labeling must state the reason, "Due to very high acute toxicity". Certified applicator must be physically present during mixing, loading, application, repair and cleaning of equipment.

- Effluent containing methyl parathion may not be discharged into lakes, streams, ponds, estuaries, oceans or public waters unless this product is specifically identified in an NPDES permit. Discharge of effluent containing this product is forbidden without prior notice to the sewage treatment plant authority.

- Protective clothing requirements are mandatory in order to protect applicators, field workers, mixer/loaders, and persons who clean and repair application equipment.

- During aerial application, human flaggers are strictly prohibited.

5. Summary of Major Data Gaps

- Animal and plant metabolism studies
- Magnitude of residue in almost all crops
- Full battery of Environmental Fate data
- Additional subchronic toxicity testing to determine a NOEL for cholinesterase inhibition and other systemic effects (retinal degeneration, sciatic nerve effects, abnormal gait)
- Additional oncogenicity and teratogenicity information
- Glove permeability and drift studies.
- Aquatic accumulation studies
- Avian reproduction and terrestrial full field testing
- Simulated or full field aquatic testing
- Early life stage and fish life cycle studies
- Reentry studies
- Applicator Exposure Monitoring studies

6. Contact Person at EPA

Dennis Edwards
Acting Product Manager No. 12
Insecticide- Rodenticide Branch Registration Division (TS-767C)
Office of Pesticide Programs
Environmental Protection Agency
401 M Street, SW
Washington, DC 20460

Office location and telephone number:

Room 202, Crystal Mall #2
1921 Jefferson Davis Highway
Arlington, VA 22202

Telephone: (703) 557-2386

Source: *Instant EPA's Pesticide Facts*, 1994, Instant Reference Sources, Inc., 7605 Rockpoint Drive, Austin, TX 78731 (Fax: 512-345-2386; Internet URL, http://www.instantref.com/inst-ref.htm).

PCBs

Introduction

PCBs are mixtures of chlorinated biphenyl isomers with commercial mixtures named according to the approximate percentage of chlorine in the mixture. The commercial mixtures are known as Aroclor 12XX where the last two digits represent the percent chlorine. Thus, Aroclor 1221 has about 21 percent chlorine in the mixture. The common products are the Aroclors above with CAS numbers. The information herein on the chemical, physical and hazardous properties of PCBs are based on Aroclor 1254, one of the most commonly used of the Aroclors and an excellent representative of the rest of them. Aroclor 1254 contains biphenyls with 54% chlorine. It is composed of 11% tetra-, 49% penta-, 34% hexa-, and 6% heptachlorobiphenyls.

Because of certain properties such as non-flammability except at extremely high temperatures, low electrical conductivity, and stability to chemical and biological breakdown, PCBs have been suited for usage in electrical equipment, hydraulic equipment, and heat transfer systems. [716] PCBs have been used as dielectric fluids, fire retardants, heat transfer agents, hydraulic fluids, plasticizers and in other applications [717]. PCBs have also been used in railroad transformers, mining equipment, carbonless copy paper, pigments, electromagnets, as a microscopy mounting medium and immersion oil, as optical liquids, and in compressors and natural gas pipeline liquids [718]. Aroclor 1254 has been used in electrical capacitors, electrical transformers, vacuum pumps, gas-transmission turbines, high- temperature dielectrics for electric wires and electrical equipment, heat-exchange fluids, coatings, inks, insecticides, fillers, adhesives, paints and in duplicating papers. It has also been used as a plasticizer for cellulosics, vinyl resins and chlorinated rubbers and as hydraulic fluids, fire retardants, wax extenders, dedusting agents, pesticide extenders, lubricants, cutting oils, sealants and caulking compounds. However, PCBs are now banned for all of these uses and generally are found only in old applications of the above uses; they have been banned from use in the US since 1977.

Current evidence suggests that the major source of Aroclor 1254 release to the environment is an environmental cycling process of Aroclor 1254 previously introduced into the environment; this cycling process involves volatilization from ground surfaces (water, soil) into the atmosphere with subsequent removal from the atmosphere via wet/dry deposition and then revolatilization. PCBs, such as Aroclor 1254, are also currently released to the environment from landfills containing PCB waste materials and products, incineration of municipal refuse and sewage sludge, and improper (or illegal) disposal of PCB materials, such as waste transformer fluid, to open areas. [828]

944

In general, the persistence of the PCB congeners increase with an increase in the degree of chlorination. Screening studies have shown that Aroclor 1254 is generally resistant to biodegradation. Although biodegradation of Aroclor 1254 may occur slowly in the environmental, no other degradation mechanism have been shown to be important in natural water and soil systems; therefore, biodegradation may be the ultimate degradation process in water and soil. The PCB composition of the biodegraded Aroclor is different from the original Aroclor. If released to soil, the PCB congeners present in Aroclor 1254 will become tightly adsorbed to the soil particles. In the presence of organic solvents, PCBs may have a tendency to leach through soil. Although the volatilization rate of Aroclor 1254 may be low from soil surfaces, the total loss by volatilizaiton over time may be significant because of the persistence and stability of Aroclor 1254. Enrichment of the low chlorine-content PCBs occurs in the vapor phase relative to the original Aroclor; the residue will be enriched in the PCBs containing high chlorine content. [828]

If released to water, adsorption to sediment and suspended matter will be an important fate process. Although adsorption can immobilize Aroclor 1254 for relatively long periods of time, eventual resolution into the water column has been shown to occur. The PCB composition in water will be enriched in the lower chlorinated PCBs because of their greater water solubility, and the least water soluble PCBs (highest chlorine content) will remain adsorbed. In the absence of adsorption, Aroclor 1254 volatilizes relatively rapidly from water. However, strong PCB adsorption competes with volatilization which may have a half-life in excess of 4 years in typical bodies of water. Although the resulting volatilization rate may be low, the total loss by volatilization over time may be significant because of the persistence and stability of Aroclor 1254. Aroclor 1254 has been shown to bioconcentrate significantly in aquatic organisms. [828]

If released to the atmosphere, the PCB congeners in Aroclor 1254 will primarily exist in the vapor-phase with enrichment of the most volatile PCBs although a relatively small percentage will partition to the particulate phase. The dominant atmospheric transformation process for these congeners is probably the vapor-phase reaction with hydroxyl radicals which has estimated half-lives ranging for 3.1 months to 1.3 years. Physical removal of Aroclor 1254 from the atmosphere, which is very important environmentally due to the chemical stability of Aroclor 1254, is accomplished by wet and dry deposition. [828]

The major Aroclor 1254 exposure routes to humans are through food and drinking water, and by inhalation of contaminated air. Dermal exposure is important for workers involved with handling PCB-containing electrical equipment, spills or waste-site materials and for swimmers in polluted water. Exposure through consumption of contaminated fish may be especially important. [828]

Identifying Information

The CAS Numbers below include that for PCBs in general and also specific CAS Numbers for each of the Aroclors.

CAS NUMBER: 1336-36-3
        Aroclor 1221 CAS NUMBER: 11104-28-2
        Aroclor 1232 CAS NUMBER: 11141-16-5
        Aroclor 1242 CAS NUMBER: 53469-21-9
        Aroclor 1248 CAS NUMBER: 12672-29-6
        Aroclor 1254 CAS NUMBER: 11097-69-1
        Aroclor 1260 CAS NUMBER: 11096-82-5
        Aroclor 1262 CAS NUMBER: 37324-23-5
        Aroclor 1268 CAS NUMBER: 11100-14-4

NIOSH Registry Number: TQ1360000 for Aroclor 1254

SYNONYMS: AROCLORS; POLYCHLOROBIPHENYLS

Endocrine Disruptor Information

PCBs are a strongly suspected EED that it is listed by EPA [811], the Centers for Disease Control & Prevention [812], and the World Wildlife Fund [813] as a potential endocrine modifying chemical. Information from EPA [812] indicates that PCBs have been implicated in numerous pathological conditions ranging from breast cancer [847] to lowered IQS and memory disorders [848] and that there is a convincing body of evidence in the literature demonstrating profound estrogenic effects [849], [850], [851] and also effects on the thyroid hormones. [852] PCBs also are listed as one of the commonly found pollutants in wildlife exhibiting symptoms of EDCs. Poor reproduction of bald eagles has been linked to levels of PCBs found in their body fat. Deformaties in birds including clubbed feet, crossed bills, and missing eyes are also suspected of links with exposure to PCBs.[810]

Chemical and Physical Properties

CHEMICAL FORMULA: Variable

MOLECULAR WEIGHT: Variable

PHYSICAL DESCRIPTION: Clear, viscous liquids ranging in color from none to light yellow.

SPECIFIC GRAVITY: 1.4 for Aroclor 1254 (variable)

DENSITY: Variable

MELTING POINT: Not available

BOILING POINT: 275-420°C for Aroclor 1254

SOLUBILITY: Water: Insoluble (all Aroclors)

Source: *Instant EPA's Air Toxics*, 1994, Instant Reference Sources, Inc., 7605 Rockpoint Drive, Austin, TX 78731 (Fax: 512-345-2386; Internet URL, http://www.instantref.com/inst-ref.htm).

Environmental Reference Materials

A source of EED environmental reference materials for PCBs is Radian International LLC, Austin, TX (Fax 512-454-0268; Internet URL, http://www.radian.com/standards); Catalog No. ERA-001, ERA-002, and ERA-003.

Hazardous Properties

ACUTE/CHRONIC HAZARDS: PCBs in general, and Aroclor 1254 in particular, are moderately toxic by ingestion. PCBs are suspected human carcinogens; experimental carcinogens and tumorigens and have shown experimental reproductive effects. As with chlorinated naphthalenes, there are effects on skin and toxicity to liver. Severe liver damage may be fatal. [703] Hepatotoxicity appears to be increased with concurrent exposure to carbon tetrachloride. Toxicity increases with chlorine content. PCB oxides are more toxic than unoxidized materials. In 1968, an outbreak of poisoning occurred in Yusho, Japan, involving some 1000 people of all ages (15,000 victims referred to by Umeda), who ingested for several months rice bran oil that had been contaminated with PCBs to the extent of 1500 to 2000 ppm. [722] High exposure to PCBs was associated with reduced birth weight. This group also showed shortened gestational age after adjustment for these same variables. PCBs produce liver tumors in rats. Although human studies have not been conclusive, PCBs should be considered potential human carcinogens. [723] When heated to decomposition PCBs emit highly toxic fumes of Cl^-. [703]

FIRE HAZARD: Non-flammable except at extremely high temperatures. [716] For Aroclor 1254 the flash point is 222°C (432°F) [102], [107], [421]

REACTIVITY: Though chemically stable, PCBs are very reactive when exposed to ultraviolet light. Over the past several decades the photoreductive dehalogenation reactions of PCBs have been examined in organic solvents, neat thin films and solid phase, and as surface deposits on silica. [717] Aroclor 1254 is referenced as being incompatible with strong oxidizers [102], [107], [346]. It will attack some forms of plastics, rubber, and coatings [102].

STABILITY: Polychlorinated biphenyls are noted for their stability. They are resistant to heat, oxidation and attack by strong acids and bases. [717] PCBs decompose to highly toxic substances including dioxin when subjected to high temperatures [716]. Aroclor 1254 is sensitive to heat [102]. Solutions of Aroclor 1254 in water, DMSO, 95% ethanol, or acetone should be stable for 24 hours under normal laboratory conditions [700].

Source: *Instant EPA's Air Toxics*, 1994, Instant Reference Sources, Inc., 7605 Rockpoint Drive, Austin, TX 78731 (Fax: 512-345-2386; Internet URL, http://www.instantref.com/inst-ref.htm).

Medical Symptoms of Exposure

In persons who have suffered systemic intoxication, the usual signs and symptoms are nausea, vomiting, loss of weight, jaundice, edema and abdominal pain. Where the liver damage has been severe the patient may pass into coma and die. [703] After a latent period of 5 to 6 months people in Yuso, Japan that were exposed to PCB poisoning from ingesting contaminated rice bran oil in 1968 suffered from nausea, lethargy, chloracne, brown pigmentation of skin areas and nails, subcutaneous edema of the face, distinctive hair follicles, a cheese-like discharge from the eyes, swelling of eyelids and transient visual disturbances. Gastroenteric distress, and jaundice also were noted. Infants born to poisoned mothers had decreased birth weight and skin discoloration, and two stillbirths were reported. [722] The following symptoms have also been reported: conjunctival hyperemia, visual and hearing disturbances, increases in diastolic and systolic blood pressure, weakness and numbness of extremities, neurobehavioral and psychomotor impairment after occupational and in utero exposure, gastrointestinal upset and diarrhea, liver damage, clinical hepatitis, and asymptomatic hyperthyroxinemia. [723]

Source: *Instant EPA's Air Toxics*, 1994, Instant Reference Sources, Inc., 7605 Rockpoint Drive, Austin, TX 78731 (Fax: 512-345-2386; Internet URL, http://www.instantref.com/inst-ref.htm).

Toxicological Information

The toxicological information below is gathered from several sources including two national databases: one from the *National Toxicology Program Chemical Repository Database* and another from *EPA's Integrated Risk Information System (IRIS)*.

Toxicology Information from the National Toxicology Program

AROCHLOR 1254

CARCINOGENICITY:

    Tumorigenic Data:

Type of Effect	Route/Animal	Amount Dosed (Notes)
TDLo:	orl-rat	73500 mg/kg/2Y-C
TDLo:	orl-mus	17 gm/kg/48W-C
TDLo:	skn-mus	4 mg/kg
TDLo:	ipr-mus	500 mg/kg (19D preg)
TD :	orl-rat	1 mg/kg/D-C
TD :	orl-rat	3 mg/kg/D-C
TD :	orl-rat	4 gm/kg/2Y-I

    Review:
        IARC Cancer Review: Human Limited Evidence
        IARC Cancer Review: Animal Sufficient Evidence
        IARC probable human carcinogen (Group 2A)

Status:

NCI Carcinogenesis Bioassay (Feed); Equivocal: Male and Female Rat
NTP anticipated human carcinogen
EPA Carcinogen Assessment Group

TERATOGENICITY:

Reproductive Effects Data:

Type of Effect	Route/Animal	Amount Dosed (Notes)
TDLo:	orl-rat	192 mg/kg (6D post)
TDLo:	orl-rat	188 mg/kg (MGN)
TDLo:	orl-rat	645 mg/kg (MGN)
TDLo:	orl-rat	90 mg/kg (7-15D preg)
TDLo:	orl-rat	375 mg/kg (5D male)
TDLo:	orl-rat	750 mg/kg (5D male)
TDLo:	orl-mus	59400 ug/kg (3D pre-21D post)
TDLo:	orl-rbt	350 mg/kg (1-28D preg)
TDLo:	orl-rbt	280 mg/kg (1-28D preg)
TDLo:	orl-mam	14 mg/kg (30D pre)

MUTATION DATA:

Test	Lowestdose	Test	Lowestdose
spm-oin-orl	808 mg	otr-rat-orl	25ppm/2Y-C
dnd-rat-orl	1295 mg/kg	dns-rat:lvr	20mg/L
dnd-rat-ipr	500 mg/kg	dnd-rat:lvr	300umol/L
cyt-ofs-ipr	50 mg/kg	mnt-ofs-ipr	50mg/kg

TOXICITY:

Typ.dose	Mode	Specie	Amount	Units	Other
LD50	orl	rat	1010	mg/kg	
LD50	ivn	rat	358	mg/kg	
LD50	ipr	mus	2840	mg/kg	

OTHER TOXICITY DATA:

      Review: Toxicology Review-6

    Status:
           EPA Genetox Program 1988, Negative: SHE-clonal assay; Rodent
           dominant lethal
           EPA Genetox Program 1988, Negative: Sperm morphology-mouse
           EPA Genetox Program 1988, Inconclusive: Mammalian micronucleus
           EPA TSCA Test Submission (TSCATS) Data Base, June 1988
           NIOSH Current Intelligence Bulletin 7, 1976 and Bulletin 45, 1986
           NIOSH Analytical Methods: see Polychlorobiphenyls
           Meets criteria for proposed OSHA Medical Records Rule
           IDLH level: 5 mg/m3 [346]

SAX TOXICITY EVALUATION:

      THR: Poison by intravenous route. Moderately toxic by ingestion and
      intraperitoneal routes. A suspected human carcinogen. An experimental
      carcinogen and neoplastigen. Experimental teratogenic and reproductive
      effects. Mutagenic data.

Source: *Instant Tox-Base*, 1995, Instant Reference Sources, Inc., 7605 Rockpoint Drive, Austin, TX 78731 (Fax: 512-345-2386; Internet URL, http://www.instantref.com/inst-ref.htm).

Toxicology Information from EPA's Integrated Risk Information System (IRIS)

I. CHRONIC HEALTH HAZARD ASSESSMENTS FOR NONCARCINOGENIC EFFECTS

I.A. REFERENCE DOSE FOR CHRONIC ORAL EXPOSURE (RfD)

I.A.1. ORAL RfD SUMMARY

Please check the following individual Aroclor files for RfD assessments: Aroclor 1016, Aroclor 1248, Aroclor 1254, Aroclor 1260.

I.B. REFERENCE CONCENTRATION FOR CHRONIC INHALATION EXPOSURE (RfC)

      [Ed. Note: EPA does not have information yet in this section of IRIS].

II. CARCINOGENICITY ASSESSMENT FOR LIFETIME EXPOSURE

Section II provides information on three aspects of the carcinogenic risk assessment for the agent in question; the US EPA classification, and quantitative estimates of risk from oral exposure and from inhalation exposure. The classification reflects a weight-of-evidence judgment of the likelihood that the agent is a human carcinogen. The quantitative risk estimates are presented in three ways. The slope factor is the result of application of a low-dose extrapolation procedure and is presented as the risk per (mg/kg)/day. The unit risk is the quantitative estimate in terms of either risk per ug/L drinking water or risk per ug/cu.m air breathed. The third form in which risk is presented is a drinking water or air concentration providing cancer risks of 1 in 10,000, 1 in 100,000 or 1 in 1,000,000. Background Document 2 (Service Code 5) provides details on the rationale and methods used to derive the carcinogenicity values found in IRIS. Users are referred to Section I for information on long-term toxic effects other than carcinogenicity.

NOTE: EPA is currently reassessing the cancer potency factor for PCBs. Following a scientific external peer review and public review, EPA will revise the IRIS summary for PCBs to reflect the results of the reassessment. Requests for EPA's memorandum describing the reassessment of the IRIS cancer potency factor for PCBs and questions concerning the reassessment can be directed to the National Center for Environmental Assessment (202-260-3814).

II.A. EVIDENCE FOR CLASSIFICATION AS TO HUMAN CARCINOGENICITY

II.A.1. WEIGHT-OF-EVIDENCE CLASSIFICATION

Classification -- B2; probable human carcinogen

Basis -- A 1996 study found liver tumors in female rats exposed to Aroclors 1260, 1254, 1242, and 1016, and in male rats exposed to 1260. These mixtures contain overlapping groups of congeners that, together, span the range of congeners most often found in environmental mixtures. Earlier studies found high, statistically significant incidences of liver tumors in rats ingesting Aroclor 1260 or Clophen A 60 (Kimbrough et al., 1975; Norback and Weltman, 1985; Schaeffer et al., 1984). Mechanistic studies are beginning to identify several congeners that have dioxin-like activity and may promote tumors by different modes of action. PCBs are absorbed through ingestion, inhalation, and dermal exposure, after which they are transported similarly through the circulation. This provides a reasonable basis for expecting similar internal effects from different routes of environmental exposure. Information on relative absorption rates suggests that differences in toxicity across exposure routes are small. The human studies are being updated; currently available evidence is inadequate, but suggestive.

II.A.2. HUMAN CARCINOGENICITY DATA

Inadequate. A cohort study by Bertazzi et al. (1987) analyzed cancer mortality among workers at a capacitor manufacturing plant in Italy. PCB mixtures with 54%, then 42% chlorine were used through 1980. The cohort included 2100 workers (544 males and 1556 females) employed at least 1 week. At the end of follow-up in 1982, there were 64 deaths reported, 26 from cancer. In males, a statistically significant increase in death from gastrointestinal tract cancer was reported, compared with national and local rates (6 observed, 1.7 expected using national rates, SMR=346, CI=141-721; 2.2 expected using local rates, SMR=274, CI=112-572). In females, a statistically significant excess risk of death from hematologic cancer was reported, compared with local, but not national, rates (4 observed, 1.1 expected, SMR=377, CI=115- 877). Analyses by exposure duration, latency, and year of first exposure revealed no trend; however, the numbers are small.

A cohort study by Brown (1987) analyzed cancer mortality among workers at two capacitor manufacturing plants in New York and Massachusetts. At both plants the Aroclor mixture being used changed twice, from 1254 to 1242 to 1016. The cohort included 2588 workers (1270 males and 1318 females) employed at least 3 months in areas of the plants considered to have potential for heavy exposure to PCBs. At the end of followup in 1982, there were 295 deaths reported, 62 from cancer. Compared with national rates, a statistically significant increase in death from cancer of the liver, gall bladder, and biliary tract was reported (5 observed, 1.9 expected, SMR=263, p<0.05). Four of these five occurred among females employed at the Massachusetts plant. Analyses by time since first employment or length of employment revealed no trend; however, the numbers are small.

A cohort study by Sinks et al. (1992) analyzed cancer mortality among workers at a capacitor manufacturing plant in Indiana. Aroclor 1242, then 1016, had been used. The cohort included 3588 workers (2742 white males and 846 white females) employed at least 1 day. At the end of follow-up in 1986, there were 192 deaths reported, 54 from cancer. Workers were classified into five exposure zones based on distance from the impregnation ovens. Compared with national rates, a statistically significant excess risk of death from skin cancer was reported (8 observed, 2.0 expected, SMR=410, CI=180-800); all were malignant melanomas. A proportional hazards analysis revealed no pattern of association with exposure zone; however, the numbers are small.

Other occupational studies by NIOSH (1977), Gustavsson et al. (1986) and Shalat et al. (1989) looked for an association between occupational PCB exposure and cancer mortality. Because of small sample sizes, brief follow-up periods, and confounding exposures to other potential carcinogens, these studies are inconclusive.

Accidental ingestion: Serious adverse health effects, including liver cancer and skin disorders, have been observed in humans who consumed rice oil contaminated with PCBs in the "Yusho" incident in Japan or the "Yu-Cheng" incident in Taiwan. These effects have been attributed, at least in part, to heating of the PCBs and rice oil, causing formation of chlorinated dibenzofurans, which have the same mode of action as some PCB congeners (ATSDR, 1993; Safe, 1994).

II.A.3. ANIMAL CARCINOGENICITY DATA

Sufficient. Brunner et al. (1996) compared carcinogenicity across different Aroclors, dose levels, and sexes. Groups of 50 male or female Sprague-Dawley rats were fed diets with 25, 50, or 100 ppm Aroclor 1260 or 1254; 50 or 100 ppm Aroclor 1242; or 50, 100, or 200 ppm Aroclor 1016. There were 100 controls of each sex. The animals were killed at 104 weeks, after which a complete histopathologic evaluation was performed for control and high-dose groups; histopathologic evaluations of liver, brain, mammary gland, and male thyroid gland were also performed for low- and mid-dose groups.

Statistically significant increased incidences of liver adenomas or carcinomas were found in female rats for all Aroclors and in male rats for Aroclor 1260. Some of these tumors were hepatocholangiomas, a rare bile duct tumor seldom seen in control rats.

To investigate tumor progression after exposure has stopped, groups of 24 female rats were exposed for 52 weeks, then exposure was discontinued for an additional 52 weeks before the rats were killed. For Aroclors 1254 and 1242, tumor incidences from the stop study were approximately half those of the lifetime study; that is, nearly proportional to exposure duration. In contrast, stop-study tumor incidences were zero for Aroclor 1016, while for Aroclor 1260 they were generally greater than half those of the lifetime study. For 100 ppm Aroclor 1260, the stop study incidence was greater than that of the lifetime study, 71 vs. 48 percent.

Thyroid gland follicular cell adenomas or carcinomas were increased in males for all Aroclors; significant dose trends were noted for Aroclors 1254 and 1242. The increases did not continue proportionately above the lowest dose. No trends were apparent in females.

In female rats, the incidence of mammary tumors was decreased with lifetime exposure to Aroclor 1254 and, to a lesser extent, to 1260 or 1242; this result was not observed for Aroclor 1016. Decreases did not occur for any Aroclor in the stop study. The first mammary tumor was observed at a later age in the dosed groups.

Kimbrough et al. (1975) fed groups of 200 female Sherman rats diets with 0 or 100 ppm Aroclor 1260 for about 21 months. Six weeks later the rats were killed and their tissues were examined. Hepatocellular carcinomas and neoplastic nodules were significantly increased in rats fed Aroclor 1260.

The National Cancer Institute (NCI, 1978) fed groups of 24 male or female Fischer 344 rats diets with 0, 25, 50, or 100 ppm Aroclor 1254 for 104-105 weeks (24 months). Then the rats were killed and their tissues were examined. The combined incidence of leukemia and lymphoma in males was significantly increased by the Cochran-Armitage trend test; however, since Fisher exact tests were not also significant, NCI did not consider this result clearly related to Aroclor 1254. Hepatocellular adenomas and carcinomas were increased. Morgan et al. (1981) and Ward (1985) reevaluated gastric lesions from this study and found 6 adenocarcinomas in 144 exposed rats. This result is statistically significant, as gastric adenocarcinomas had occurred in only 1 of 3548 control male and female Fischer 344 rats in the NCI testing program. Intestinal metaplasia in exposed rats differed morphologically from controls, suggesting Aroclor 1254 can act as a tumor initiator.

Schaeffer et al. (1984) fed male weanling Wistar rats a standard diet for 8 weeks, then divided them into three groups. One group was fed the basic diet; for the other groups 100 ppm Clophen A 30 or A 60 was added. Rats were killed at 801 832 days (26.3 27.3 months) and were examined for lesions in the liver and some other tissues. For both mixtures, preneoplastic liver lesions were observed after 500 days (16.4 months) and hepatocellular carcinomas after 700 days (23 months) in rats dying before the end of the study. The investigators concluded, "Clophen A 60 had a definite, and Clophen A 30 a weak, carcinogenic effect on rat liver."

Norback and Weltman (1985) fed groups of male and female Sprague-Dawley rats diets of 0 or 100 ppm Aroclor 1260 for 16 months; the latter dose was reduced to 50 ppm for 8 more months. After 5 additional months on the control diet, the rats were killed and their livers were examined. Partial hepatectomy was performed on some rats at 1, 3, 6, 9, 12, 15, 18, and 24 months to evaluate sequential morphologic changes. In males and females fed Aroclor 1260, liver foci appeared at 3 months, area lesions at 6 months, neoplastic nodules at 12 months, trabecular carcinomas at 15 months, and adenocarcinomas at 24 months, demonstrating progression of liver lesions to carcinomas. By 29 months, 91% of females had liver carcinomas and 95% had carcinomas or neoplastic nodules; incidences in males were smaller, 4% and 15%, respectively. Vater et al. (1995) obtained individual animal results to determine whether the partial hepatectomies, which exert a strong proliferative effect on the remaining tissue, affected the incidence of liver tumors. They reported that the hepatectomies did not increase the tumor incidence. Among females fed Aroclor 1260, liver tumors developed in 4 of 7 animals with hepatectomies and 37 of 39 without hepatectomies; no liver tumors developed in controls or males with hepatectomies.

Moore et al. (1994) reevaluated the preceding rat liver findings (Kimbrough et al., 1975; NCI, 1978; Schaeffer et al., 1984; Norback and Weltman, 1985) using criteria and nomenclature that had changed to reflect new understanding of mechanisms of toxicity and carcinogenesis. The reevaluation found somewhat fewer tumors than did the original investigators. The apparent increase for Clophen A 30 (Schaeffer et al., 1984) is no longer statistically significant.

II.A.4. SUPPORTING DATA FOR CARCINOGENICITY

Several studies of less-than-lifetime exposure are supportive of a carcinogenic response (Kimbrough et al., 1972; Kimbrough and Linder, 1974; Kimura and Baba, 1973; Ito et al., 1973, 1974; Rao and Banerji, 1988).

PCBs give generally negative results in tests of genetic activity (ATSDR, 1993), implying that PCBs induce tumors primarily through modes of action that do not involve gene mutation. Initiation-promotion studies for several commercial PCB mixtures and congeners show tumor promoting activity in liver and lung; these studies are beginning to identify a subset of mixture components that may be significant contributors to cancer induction (Silberhorn et al., 1990). Toxicity of some PCB congeners is correlated with induction of mixed-function oxidases; some congeners are phenobarbital-type inducers, others are 3-methylcholanthrene-type inducers, and some have mixed inducing properties (McFarland and Clarke, 1989). The latter two groups most resemble 2,3,7,8-tetrachlorodibenzo-p-dioxin in structure and toxicity.

Studies of structurally related agents: Studies of 2,3,7,8- tetrachlorodibenzo-p-dioxin and a polybrominated biphenyl (PBB) mixture are summarized here because the pattern of tumors found by Brunner et al. (1996) mimics the tumors induced in rats by these structurally related agents. The National Toxicology Program (NTP, 1982) exposed groups of 50 male or female Osborne-Mendel rats by gavage to 0, 1.4, 7.1, or 71 ng/kg-day 2,3,7,8- tetrachlorodibenzo-p-dioxin for 2 years. Similar to the Brunner et al. (1996) study, liver tumors were increased in female rats and thyroid gland follicular cell tumors were increased in male rats. Mammary tumors were not, however, decreased in dosed female rats. In another study, NTP (1983) exposed groups of 51 male or female Fischer 344/N rats by gavage to 0, 0.1, 0.3, 1, 3, or 10 mg/kg-day of a PBB mixture ("Firemaster FF 1") for 6 months, then exposure was discontinued for 23 months before the animals were killed. Statistically significant increased incidences of liver tumors were found in male and female rats. Dose-related increased incidences of cholangiocarcinomas were found in male and female rats.

II.B. QUANTITATIVE ESTIMATE OF CARCINOGENIC RISK FROM ORAL EXPOSURE

II.B.1. SUMMARY OF RISK ESTIMATES

Oral Slope Factor -- See txt

Drinking Water Unit Risk -- See txt

Extrapolation Method -- Linear extrapolation below LED10s (US EPA, 1996b)

Drinking Water Concentrations at Specified Risk Levels:

Risk Level	Concentration
E-4 (1 in 10,000)	See txt
E-5 (1 in 100,000)	See txt
E-6 (1 in 1,000,000)	See txt

II.B.2. DOSE-RESPONSE DATA (CARCINOGENICITY, ORAL EXPOSURE)

Tumor Type -- Liver hepatocellular adenomas, carcinomas, cholangiomas, or cholangiocarcinomas
Test Animals -- Female Sprague-Dawley rats
Route -- Diet
Reference -- Brunner et al., 1996; Norback and Weltman, 1985

Administered Dose (mg/kg)/day (TWA)	Human Equivalent Dose (mg/kg)/day	Tumor Incidence
Aroclor 1260		
0	0	1/85
25	0.35	10/49
50	0.72	11/45
100	1.52	24/50
Aroclor 1254		
0	0	1/85
25	0.36	19/45
50	0.76	28/49
100	1.59	28/49
Aroclor 1242		
0	0	1/85
50	0.75	11/49
100	1.53	15/45
Aroclor 1016		
0	0	1/85
50	0.72	1/48
100	1.43	7/45
200	2.99	6/50
Aroclor 1260 (Norback and Weltman, 1985)		
0	0.75	1/45
100/50/0	1.3	41/46

II.B.3. ADDITIONAL COMMENTS (CARCINOGENICITY, ORAL EXPOSURE)

The cancer potency of PCB mixtures is determined using a tiered approach that depends on the information available. The following tier descriptions discuss all environmental exposure routes:

TIERS OF HUMAN SLOPE FACTORS FOR ENVIRONMENTAL PCBs

HIGH RISK AND PERSISTENCE

Upper-bound slope factor: 2.0 per (mg/kg)/day
Central-estimate slope factor: 1.0 per (mg/kg)/day

Criteria for use: - Food chain exposure - Sediment or soil ingestion - Dust or aerosol inhalation - Dermal exposure, if an absorption factor has been applied - Presence of dioxin-like, tumor-promoting, or persistent congeners - Early-life exposure (all pathways and mixtures)

LOW RISK AND PERSISTENCE

Upper-bound slope factor: 0.4 per (mg/kg)/day
Central-estimate slope factor: 0.3 per (mg/kg)/day

Criteria for use: - Ingestion of water-soluble congeners - Inhalation of evaporated congeners - Dermal exposure, if no absorption factor has been applied

LOWEST RISK AND PERSISTENCE

Upper-bound slope factor: 0.07 per (mg/kg)/day
Central-estimate slope factor: 0.04 per (mg/kg)/day

Criteria for use: Congener or isomer analyses verify that congeners with more than 4 chlorines comprise less than 1/2% of total PCBs.

Slope factors are multiplied by lifetime average daily doses to estimate the cancer risk. SAMPLE CALCULATIONS ARE GIVEN IN US EPA (1996a). Although PCB exposures are often characterized in terms of Aroclors, this can be both imprecise and inappropriate. Total PCBs or congener or isomer analyses are recommended.

When congener concentrations are available, the slope-factor approach can be supplemented by analysis of dioxin TEQs to evaluate dioxin-like toxicity. Risks from dioxin-like congeners (evaluated using dioxin TEQs) would be added to risks from the rest of the mixture (evaluated using slope factors applied to total PCBs reduced by the amount of dioxin-like congeners). SAMPLE CALCULATIONS ARE GIVEN IN US EPA (1996a).

Depending on the specific application, either central estimates or upper bounds can be appropriate. Central estimates describe a typical individual's risk, while upper bounds provide assurance that this risk is not likely to be underestimated if the underlying model is correct. The upper bounds calculated in this assessment reflect study design and provide no information about sensitive individuals or groups. Central estimates are useful for estimating aggregate risk across a population. Central estimates are used for comparing or ranking environmental hazards, while upper bounds provide information about the precision of the comparison or ranking.

Some PCBs persist in the body and retain biological activity after exposure stops (Anderson et al., 1991a). Compared with the current default practice of assuming that less-than-lifetime effects are proportional to exposure duration, rats exposed to a persistent mixture (Aroclor 1260) had more tumors, while rats exposed to a less persistent mixture (Aroclor 1016) had fewer tumors (Brunner et al., 1996). Thus there may be greater-than- proportional effects from less-than-lifetime exposure, especially for persistent mixtures and for early-life exposures.

Highly exposed populations include some nursing infants and consumers of game fish, game animals, or products of animals contaminated through the food chain. Highly sensitive populations include people with decreased liver function and infants (Calabrese and Sorenson, 1977).

Because of the potential magnitude of early-life exposures (ATSDR, 1993; Dewailly et al., 1991, 1994), the possibility of greater perinatal sensitivity (Calabrese and Sorenson, 1977; Rao and Banerji, 1988), and the likelihood of interactions among thyroid and hormonal development, it is reasonable to conclude that early-life exposures may be associated with increased risks. Due to this potential for higher sensitivity early in life, the "high risk" tier is used for all early-life exposure.

It is crucial to recognize that commercial PCBs tested in laboratory animals were not subject to prior selective retention of persistent congeners through the food chain (that is, the rats were fed Aroclor mixtures, not environmental mixtures that had been bioaccumulated). Bioaccumulated PCBs appear to be more toxic than commercial PCBs (Aulerich et al., 1986; Hornshaw et al., 1983) and appear to be more persistent in the body (Hovinga et al., 1992). For exposure through the food chain, risks can be higher than those estimated in this assessment.

In calculating these estimates, administered doses were expressed as a lifetime daily average calculated from weekly body weight measurements and food consumption estimates (Keenan and Stickney, 1996). Doses were scaled from rats to humans using a factor based on the 3/4 power of relative body weight.

UNIT RISK ESTIMATE AND DRINKING WATER CONCENTRATIONS

For ingestion of water-soluble congeners, the middle-tier slope factor can be converted to a unit risk estimate and drinking water concentrations associated with specified risk levels.

Upper-bound slope factor: 0.4 per (mg/kg)/day
Upper-bound unit risk: $1 \times 10\text{-}5$ per ug/L

Drinking water concentration associated with a risk of:

1 in 10,000   10 ug/L
1 in 100,000   1 ug/L
1 in 1,000,000   0.1 ug/L

These estimates should not be used if drinking water concentrations exceed 1000 ug/L, since above this concentration the dose-response curve in the experimental range may provide better estimates.

For food chain exposure or ingestion that includes contaminated sediment or soil, the slope factor for "high risk and persistence" should be used instead.

II.B.4. DISCUSSION OF CONFIDENCE (CARCINOGENICITY, ORAL EXPOSURE)

Joint consideration of cancer studies and environmental processes leads to a conclusion that environmental PCB mixtures are highly likely to pose a risk of cancer to humans. Although environmental mixtures have not been tested in cancer assays, this conclusion is supported by several complementary sources of information. Statistically significant, dose-related, increased incidences of liver tumors were induced in female rats by

Aroclors 1260, 1254, 1242, and 1016 (Brunner et al., 1996). These mixtures contain overlapping groups of congeners that, together, span the range of congeners most frequently found in environmental mixtures. Several congeners have dioxin-like activity (Safe, 1994) and may promote tumors by different modes of action (Silberhorn et al., 1990); these congeners are found in environmental samples and in a variety of organisms, including humans (McFarland and Clarke, 1989).

The range of potency observed for commercial mixtures is used to represent the potency of environmental mixtures. The range reflects experimental uncertainty and variability of commercial mixtures, but not human heterogeneity or differences between commercial and environmental mixtures. Environmental processes alter mixtures through partitioning, transformation, and bioaccumulation, thereby decreasing or increasing toxicity. The overall effect can be considerable, and the range observed for commercial mixtures may underestimate the true range for environmental mixtures (Hutzinger et al., 1974; Callahan et al., 1979). Limiting the potency of environmental mixtures to the range observed for commercial mixtures reflects a decision to base potency estimates on experimental results, however uncertain, rather than apply safety factors to compensate for lack of information.

A tiered approach allows use of different kinds of information in estimating the potency of environmental mixtures. When congener information is limited, exposure pathway is used to indicate whether environmental processes have decreased or increased a mixture's potency. Partitioning, transformation, and bioaccumulation have been extensively studied (Hutzinger et al., 1974; Callahan et al., 1979) and can be associated with exposure pathway, thus the use of exposure pathway to represent environmental processes increases confidence in the risks inferred for environmental mixtures. For example, evaporated or dissolved congeners tend to be lower in chlorine content than the original mixture; they tend also to be more inclined to metabolism and elimination and lower in persistence and toxicity. On the other hand, congeners adsorbed to sediment or soil tend to be higher in chlorine content and persistence, and bioaccumulated congeners ingested through the food chain tend to be highest of all. Rates of these processes vary over several orders of magnitude (Hutzinger et al., 1974; Callahan et al., 1979). When available, congener information is an important tool for refining a potency estimate that was based on exposure pathway.

Extrapolation to environmental levels is based on models that are linear at low doses. Low-dose-linear models are appropriate when a carcinogen acts in concert with other exposures and processes that cause a background incidence of cancer (Crump et al, 1976; Lutz, 1990). Even when the mode of action indicates a nonlinear dose-response curve in homogeneous animal populations, the presence of genetic and lifestyle factors in a heterogeneous human population tends to make the dose-response curve more linear (Lutz, 1990). This is because genetic and lifestyle factors contribute to a wider spread of human sensitivity, which extends and straightens the dose-response curve over a wider range.

Uncertainty around these estimates extends in both directions. The slope factor ranges primarily reflect mixture variability, and so are not necessarily appropriate for probabilistic analyses that attempt to describe model uncertainty and parameter uncertainty. Estimates based on animal studies benefit from controlled exposures and absence of confounding factors; however, there is uncertainty in extrapolating dose and response rates across species. Information is lacking to evaluate high-to-low-dose differences. PCBs are absorbed through ingestion, inhalation, and dermal exposure, after which they are transported similarly through the circulation (ATSDR, 1993). This provides a reasonable basis for expecting similar internal effects from different routes of environmental exposure. Information on relative absorption rates suggests that differences in toxicity across exposure routes are small. The principal uncertainty, though, is using commercial mixtures to make inferences about environmental mixtures.

When exposure involves the food chain, uncertainty extends principally in one direction: through the food chain, living organisms selectively bioaccumulate persistent congeners, but commercial mixtures tested in laboratory animals were not subject to prior selective retention of persistent congeners. Bioaccumulated PCBs appear to be more toxic than commercial PCBs (Aulerich et al., 1986; Hornshaw et al., 1983) and appear to be more persistent in the body (Hovinga et al., 1992). For exposure through the food chain, risks can be higher than those estimated in this assessment. Two highly exposed populations, nursing infants and consumers of contaminated game animals, are exposed through the food chain.

The dioxin-like nature of some PCBs raises a concern for cumulative exposure, as dioxin-like congeners add to background exposure of other dioxin-like compounds and augment processes associated with dioxin toxicity. This weighs against considering PCB exposure in isolation or as an increment to a background exposure of zero. Confidence in this assessment's use of low-dose-linear models is enhanced when there is additivity to background exposures and processes (Crump et al, 1976; Lutz, 1990).

II.C. QUANTITATIVE ESTIMATE OF CARCINOGENIC RISK FROM INHALATION EXPOSURE

II.C.1. SUMMARY OF RISK ESTIMATES

Inhalation Unit Risk -- See txt

Extrapolation Method -- Linear extrapolation below LED10s (US EPA, 1996b)

Air Concentrations at Specified Risk Levels:

Risk Level	Concentration
E-4 (1 in 10,000)	See txt
E-5 (1 in 100,000)	See txt
E-6 (1 in 1,000,000)	See txt

II.C.2. DOSE-RESPONSE DATA (CARCINOGENICITY, INHALATION EXPOSURE)

See Dose-Response Data for oral exposure.

II.C.3. ADDITIONAL COMMENTS (CARCINOGENICITY, INHALATION EXPOSURE)

See Additional Comments for oral exposure.

For inhalation of evaporated congeners, the middle-tier slope factor can be converted to a unit risk estimate and ambient air concentrations associated with specified risk levels.

Upper-bound slope factor: 0.4 per (mg/kg)/day
Upper-bound unit risk: $1 \times 10\text{-}4$ per ug/L

Ambient air concentration associated with a risk of:

1 in 10,000   1 ug/m3
1 in 100,000   0.1 ug/m3
1 in 1,000,000   0.01 ug/m3

These estimates should not be used if ambient air concentrations exceed 100 ug/m3, since above this concentration the dose-response curve in the experimental range may provide better estimates.

For inhalation of an aerosol or dust contaminated with PCBs, the slope factor for "high risk and persistence" should be used instead.

II.C.4. DISCUSSION OF CONFIDENCE (CARCINOGENICITY, INHALATION EXPOSURE)

See Discussion of Confidence for oral exposure. Information on relative absorption rates suggests that differences in toxicity across exposure routes are small.

II.D. EPA DOCUMENTATION, REVIEW, AND CONTACTS (CARCINOGENICITY ASSESSMENT)

II.D.1. EPA DOCUMENTATION

Source Document -- US EPA, 1996a [Available from the Risk Information Hotline, Cincinnati OH. Telephone: (513)569-7254; FAX (513)569-7159)].

The source document and IRIS summary were considered at a public, external peer review workshop in May 1996. A workshop report was written by the review panel (US EPA, 1996c). All comments have been carefully evaluated and considered in this IRIS summary. A record of these comments is summarized in the IRIS documentation files.

Other EPA Documentation -- US EPA, 1988

II.D.2. REVIEW (CARCINOGENICITY ASSESSMENT)

Agency Work Group Review -- 08/22/96

Verification Date -- 08/22/96

II.D.3. US EPA CONTACTS (CARCINOGENICITY ASSESSMENT)

Jim Cogliano / NCEA -- (202)260-3814

Charli Hiremath / NCEA -- (202)260-5725

III. HEALTH HAZARD ASSESSMENTS FOR VARIED EXPOSURE DURATIONS

[Ed. Note: EPA does not have information yet in this section of IRIS].

IV. US EPA REGULATORY ACTIONS

EPA risk assessments may be updated as new data are published and as assessment methodologies evolve. Regulatory actions are frequently not updated at the same time. Compare the dates for the regulatory actions in this section with the verification dates for the risk assessments in sections I and II, as this may explain inconsistencies. Also note that some regulatory actions consider factors not related to health risk, such as technical or economic feasibility. Such considerations are indicated for each action. In addition, not all of the regulatory actions listed in this section involve enforceable federal standards. Please direct any questions you may have concerning these regulatory actions to the US EPA contact listed for that particular action. Users are strongly urged to read the background information on each regulatory action in Background Document 4 in Service Code 5.

IV.A. CLEAN AIR ACT (CAA)

[Ed. Note: EPA does not have information yet in this section of IRIS].

IV.B. SAFE DRINKING WATER ACT (SDWA)

IV.B.1. MAXIMUM CONTAMINANT LEVEL GOAL (MCLG) for Drinking Water

Value -- 0.0 mg/L (Final, 1991)

Considers technological or economic feasibility? -- NO

Discussion -- The MCLG for polychlorinated biphenyls is zero based on the evidence of carcinogenic potential (classification B2).

Reference -- 56 FR 3526 (01/30/91)

EPA Contact -- Health and Ecological Criteria Division / OST /(202) 260-7571 / FTS 260-7571; or Safe Drinking Water Hotline / (800) 426-4791

IV.B.2. MAXIMUM CONTAMINANT LEVEL (MCL) for Drinking Water

Value -- 0.0005 mg/L (Final, 1991)

Considers technological or economic feasibility? -- YES

Discussion -- The MCL is based on a PQL of 0.0005 mg/L and is associated with a maximum lifetime individual risk of E-4.

Monitoring requirements -- All systems monitored initially for four consecutive quarters every three years; repeat monitoring dependent upon detection, vulnerability status and system size.

Analytical methodology -- Microextraction/gas chromatography (EPA 505); electron capture detector (EPA 508); perchlorination/gas chromatography (EPA 508A). PQL= 0.0005 mg/L.

Best available technology -- Granular activated carbon

Reference -- 56 FR 3526 (01/30/91)

EPA Contact -- Drinking Water Standards Division / OGWDW /(202) 260-7575 / FTS 260-7575; or Safe Drinking Water Hotline / (800) 426-4791

IV.B.3. SECONDARY MAXIMUM CONTAMINANT LEVEL (SMCL) for Drinking Water

[Ed. Note: EPA does not have information yet in this section of IRIS].

IV.B.4. REQUIRED MONITORING OF "UNREGULATED" CONTAMINANTS

[Ed. Note: EPA does not have information yet in this section of IRIS].

IV.C. CLEAN WATER ACT (CWA)

IV.C.1. AMBIENT WATER QUALITY CRITERIA, Human Health

Water and Fish Consumption: 7.9E-5 ug/L

Fish Consumption Only: 7.9E-6 ug/L

Considers technological or economic feasibility? -- NO

Discussion -- For the maximum protection from the potential carcinogenic properties of this chemical, the ambient water concentration should be zero. Since zero, however, may not be attainable at this time, the recommended criteria represents an E-6 estimated incremental increase of cancer risk over a lifetime.

Reference -- 45 FR 79318 (11/28/80)

EPA Contact -- Criteria and Standards Division / OWRS (202)260-1315 / FTS 260-1315

IV.C.2. AMBIENT WATER QUALITY CRITERIA, Aquatic Organisms

Freshwater:
  Acute -- 2.0E+0 ug/L
  Chronic -- 1.4E-2 ug/L

Marine:
  Acute -- 1.0E+1 ug/L
  Chronic -- 3.0E-2 ug/L

Considers technological or economic feasibility? -- NO

Discussion -- Criteria were derived from a minimum data base consisting of acute tests on a variety of species. Requirements and methods are covered in the reference to the Federal Register.

Reference -- 45 FR 79318 (11/28/80)

EPA Contact -- Criteria and Standards Division / OWRS (202)260-1315 / FTS 260-1315

IV.D. FEDERAL INSECTICIDE, FUNGICIDE, AND RODENTICIDE ACT (FIFRA)

        [Ed. Note: EPA does not have information yet in this section of IRIS].

IV.E. TOXIC SUBSTANCES CONTROL ACT (TSCA)

IV.E.1. TSCA, SECTION 6

Status -- Final Rule (1988)

Discussion -- Prohibits the manufacture, processing, distribution in commerce, or the use of PCBs other than in a "totally enclosed manner" unless specifically exempted by the EPA. Reporting, disposal and record- keeping requirements. Advance notice of proposed rulemaking [56 FR 26738, (06/10/91)] to amend TSCA PCB disposal regulations.

Reference -- 52 FR 27322 (07/19/88); 55 FR 21033 (05/22/90)

EPA Contact -- Chemical Control Division / OTS (202) 260-3749 / FTS 260-3749

IV.F. RESOURCE CONSERVATION AND RECOVERY ACT (RCRA)

IV.F.1. RCRA APPENDIX IX, for Ground Water Monitoring

Status -- Listed

Reference -- 52 FR 25942 (07/09/87)

EPA Contact -- RCRA/Superfund Hotline (800)424-9346 / (202)260-3000 / FTS 260-3000

IV.G. SUPERFUND (CERCLA)

IV.G.1. REPORTABLE QUANTITY (RQ) for Release into the Environment

Value (status) -- 1 pound (Final, 1989)

Considers technological or economic feasibility? -- NO

Discussion -- The final RQ for polychlorinated biphenyls is based on aquatic toxicity. The available data indicate that the 96-Hour Median Threshold Limit is less than 0.1 ppm, which corresponds to an RQ of 1 pound.

Reference -- 54 FR 33418 (08/14/89)

EPA Contact -- RCRA/Superfund Hotline (800)424-9346 / (202)260-3000 / FTS 260-3000

VI. BIBLIOGRAPHY

VI.A. ORAL RfD REFERENCES

None

VI.B. INHALATION RfD REFERENCES

None

VI.C. CARCINOGENICITY ASSESSMENT REFERENCES

Amano, M., K. Yagi, H. Nakajima, R. Takehara, H. Sakai and G. Umeda. 1984. Statistical observations about the causes of the death of patients with oil poisoning. Japan Hygiene. 39(1): 1-5.

Bahn, A.K., I. Rosenwaike, N. Herrmann, P. Grover, J. Stellman and K. O'Leary. 1976. Melanoma after exposure to PCB's. New Engl. J. Med. 295: 450.

Bahn, A.K., P. Grover, I. Rosenwaike, K. O'Leary and J. Stellman. 1977. Reply to letter from C. Lawrence entitled, "PCB? and melanoma". New Engl. J. Med. 296: 108.

Bertazzi, P.A., L. Riboldi, A. Pesatori, L. Radice and C. Zacchetti. 1987. Cancer mortality of capacitor manufacturng workers. Am. J. Ind. Med. 11(2): 165-176.

Brown, D.P. 1987. Mortality of workers exposed to polychlorinated biphenyls -- An update. Arch. Environ. Health. 42(6): 333-339.

Brown, D.P. and M. Jones. 1981. Mortality and industrial hygiene study of workers exposed to polychlorinated biphenyls. Arch. Environ. Health. 36(3): 120-129.

Green, S., J.V. Carr, K.A. Palmer and E.J. Oswald. 1975. Lack of cytogenetic effects in bone marrow and spermatogonial[sic] cells in rats treated with polychlorinated biphenyls (Aroclors 1242 and 1254). Bull. Environ. Contam. Toxicol. 13(1): 14-22.

Heddle, J.A. and W.R. Bruce. 1977. Comparison of tests for mutagenicity or carcinogenicity using assays for sperm abnormalities, formation of micronuclei and mutations in Salmonella. In: Origins of Human Cancer, H.H. Hiatt et al., Ed. Cold Spring Harbor Conf. Cell Prolif., Cold Spring Harbor Lab., Cold Spring Harbor, NY. 4: 1549-1557.

Ito, N., H. Nagasaki, M. Arai, S. Makiura, S. Sugihara and K. Hirao. 1973. Histopathologic studies on liver tumorigenesis induced in mice by technical polychlorinated biphenyls and its promoting effect on liver tumors induced by benzene hexachloride. J. Natl. Cancer Inst. 51(5): 1637-1646.

Ito, N., H. Nagasaki, S. Makiura and M. Arai. 1974. Histopathological studies on liver tumorigenesis in rats treated with polychlorinated biphenyls. Gann. 65: 545-549.

Kimbrough, R.D. 1987. Human health effects of polychlorinated biphenyls (PCBs) and polybrominated biphenyls (PBBs). Ann. Rev. Pharmacol. Toxicol. 27: 87-111.

Kimbrough, R.D. and R.E. Linder. 1974. Induction of adenofibrosis and hepatomas in the liver of BALB/cJ mice by polychlorinated biphenyls (Aroclor 1254). J. Natl. Cancer Inst. 53(2): 547-552.

Kimbrough, R.D., R.A. Squire, R.E. Linder, J.D. Strandberg, R.J. Montali and V.W. Burse. 1975. Induction of liver tumors in Sherman strain female rats by polychlorinated biphenyl Aroclor 1260. J. Natl. Cancer Inst. 55(6): 1453-1459.

Kimura, N.T. and T. Baba. 1973. Neoplastic changes in the rat liver induced by polychlorinated biphenyl. Gann. 64: 105-108.

NCI (National Cancer Institute). 1978. Bioassay of Aroclor (trademark) 1254 for possible carcinogenicity. CAS No. 27323-18-8. NCI Carcinogenesis Tech. Rep. Ser. No. 38.

NIOSH (National Institute for Occupational Safety and Health). 1977. Criteria for a Recommended Standard . . . Occupational Exposure to Polychlorinated Biphenyls (PCBs). US DHEW, PHS, CDC, Rockville, Md. Publ. No. 77-225.

Norback, D.H. and R.H. Weltman. 1985. Polychlorinated biphenyl induction of hepatocellular carcinoma in the Sprague-Dawley rat. Environ. Health Perspect. 60: 97-105.

Peakall, D.B., J.L. Lincer and S.E. Bloom. 1972. Embryonic mortality and chromosomal alterations caused by Aroclor 1254 in ring doves. Environ. Health Perspect. 1: 103-104.

Schaeffer, E., H. Greim and W. Goessner. 1984. Pathology of chronic polychlorinated biphenyl (PCB) feeding in rats. Toxicol. Appl. Pharmacol. 75: 278-288.

Schoeny, R. 1982. Mutagenicity testing of chlorinated biphenyls and chlorinated dibenzofurans. Mutat. Res. 101: 45-56.

Schoeny, R.S., C.C. Smith and J.C. Loper. 1979. Non-mutagenicity for Salmonella of the chlorinated hydrocarbons Aroclor 1254, 1,2,4-trichlorobenzene, mirex and kepone. Mutat. Res. 68: 125-132.

US EPA. 1988. Drinking Water Criteria Document for Polychlorinated Biphenyls (PCBs). Prepared by the Office of Health and Environmental Assessment, Environmental Criteria and Assessment Office, Cincinnati, OH for the Office of Drinking Water, Washington, DC.

Wyndham, C., J. Devenish and S. Safe. 1976. The in vitro metabolism, macromolecular binding and bacterial mutagenicity of 4-chlorobiphenyl, a model PCB substrate. Res. Commun. Chem. Pathol. Pharmacol. 15: 563-570.

VI.D. DRINKING WATER HA REFERENCES

None

VII. REVISION HISTORY

Date	Section	Description
05/01/89	II.	Carcinogen summary on-line
01/01/90	II.	Text edited
01/01/90	VI.	Bibliography on-line
01/01/92	IV.	Regulatory Action section on-line
06/01/94	I.A.	Message only
01/01/96	II.	Note added to assessment
10/01/96	II.	File replaced; cancer potency of mixtures addressed

SYNONYMS

AROCLOR	FENCLOR
AROCLOR 1221	INERTEEN
AROCLOR 1232	KANECHLOR
AROCLOR 1242	KANECHLOR 300
AROCLOR 1248	KANECHLOR 400
AROCLOR 1254	MONTAR
AROCLOR 1260	NOFLAMOL
AROCLOR 1262	PCB
AROCLOR 1268	PCBs
AROCLOR 2565	PHENOCHLOR
AROCLOR 4465	PHENOCLOR
AROCLOR 5442	POLYCHLORINATED BIPHENYL
BIPHENYL, POLYCHLORO-	Polychlorinated Biphenyls
CHLOPHEN	POLYCHLOROBIPHENYL
CHLOREXTOL	PYRALENE
CHLORINATED BIPHENYL	PYRANOL
CHLORINATED DIPHENYL	SANTOTHERM
CHLORINATED DIPHENYLENE	SANTOTHERM FR
CHLORO BIPHENYL	SOVOL
CHLORO 1,1-BIPHENYL	THERMINOL FR-1
CLOPHEN	UN 2315
DYKANOL	

Source: ***Instant EPA's IRIS***, 1997, Instant Reference Sources, Inc., 7605 Rockpoint Drive, Austin, TX 78731 (Fax: 512-345-2386; Internet URL, http://www.instantref.com/inst-ref.htm).

Pentachlorophenol

Introduction

Pentachlorophenol is used in large quantities as a wood preservative for utility poles, crossarms, and fenceposts. These uses may result in some environmental releases from the wood and during spills. [828] It is also used an insecticide for termite control, pre-harvest defoliant, general herbicide, wood preservative, synthesis of pentachlorophenyl esters, molluscide, fungicide, bactericide, antimildew agent, slimicide and algaecide. The technical material finds extensive use in cooling towers of electric plants, as additives to adhesives based on starch and vegetable and animal protein, in shingles, roof tiles, brick walls, concrete blocks, insulation, pipe sealant compounds, photographic solutions, and textiles and in drilling mud in the petroleum industry.

Releases to soil can decrease in concentrations due to slow biodegradation and leaching into groundwater. If released in water, pentachlorophenol will adsorb to sediment, photodegrade (especially at higher pHs) and slowly biodegrade. Bioconcentration in fish will be moderate. In air, pentachlorophenol will be lost due to photolysis and reaction with photochemically produced hydroxyl radicals. Human's will be occupationally exposed to pentachlorophenol via inhalation and dermal contact primarily in situations where they use this preservative or are in contact with treated wood product. The general population will be exposed primarily from ingesting food contaminated with pentachlorophenol. [828]

Identifying Information

CAS NUMBER: 87-86-5

NIOSH Registry Number: SM6300000

SYNONYMS: Chem-Tol; Chlorophen; Cryptogil OL; Dowcide 7; Dowicide EC-7; DP-2, technical; Durotox; EP 30; Fungifen; Glazd penta; Grundier arbezol; 1-Hydroxy-2,3,4,5,6-pentachlorobenzene; Lauxtol; Lauxtol A; Liroprem; PCP; Penchlorol; Penta; Pentachloorfenol; Pentachlorofenol; Pentachlorofenolo; Pentachlorophenate; Pentaclorofenolo; Pentacon; Penta-Kil; Pentasol; Penwar; Peratox; Permacide; Permagard; Permasan; Permatox; Permatox dp-2; Permatox penta; Permite; Phenol, pentachloro-; Preventol P; Priltox; Santobrite; Santophen; Santophen 20; Sinituho; Term-i-trol

Endocrine Disruptor Information

Pentachlorophenol is a strongly suspected EED that it is listed by EPA [811], the Centers for Disease Control & Prevention [812], and the World Wildlife Fund [813] as a potential endocrine modifying chemical. Information from EPA [811] indicates that in addition to being a teratogen, carcinogen and a mutagen,[856] pentachlorophenol also is a potent endocrine disruptor impacting adrenal , thyroid and pituitary functions [857], [858], and that it has also been implicated as an immuno-suppressant. [859]

Chemical and Physical Properties

CHEMICAL FORMULA: $C_6C_{l5}OH$

MOLECULAR WEIGHT: 266.34

WLN: QR BG CG DG EG FG

PHYSICAL DESCRIPTION: White to light brown crystalline solid

SPECIFIC GRAVITY: 1.978 @ 22/4°C [017], [031], [062], [205]

MELTING POINT: 191°C [043], [169], [173], [295]

BOILING POINT: 309-310°C (with decomposition) [031], [169], [421]

SOLUBILITY: Water: <1 mg/mL @ 20°C [700]; DMSO: 50-100 mg/mL @ 20°C [700]; 95% Ethanol: >=100 mg/mL @ 20°C [700]; Acetone: >=100 mg/mL @ 20°C [700]; Methanol: Soluble [051], [421]

OTHER SOLVENTS: Benzene: Soluble [017], [031], [062], [395]; Dilute alkali: Soluble [062]; Paraffins: Slightly soluble [169], [172]; Carbon tetrachloride: Slightly soluble [169], [172]; Cold petroleum ether: Slightly soluble [031], [043], [205], [395]; Alkanes: Slightly soluble [421]; Most organic solvents: Soluble [169], [172]; Alcohol: Very soluble [031], [043], [205], [295]; Sodium hydroxide solution: Freely soluble [295]; Ether: Soluble [017], [043], [295], [395]; Paraffinic petroleum oils: Moderately soluble [173]

OTHER PHYSICAL DATA: Weak acid (pKa 4.71) [172] pKa: 8.2 (in methanol) [025]; Melting point (hydrous): 174°C [017], [025], [047], [395]; Phenolic (carbolic) odor [036]; Pungent taste [421]; Molar refraction: 53.5 [051] Log octanol/water partition coefficient: 5.01 [051], [055]; Odor threshold, medium: 0.857 ppm [051]; Taste threshold, upper: 0.857 ppm [051]; Vapor pressure: 0.12 mm Hg @ 100 C [051] 20 mm Hg @ 192.2°C [038]; Boiling point also given as 293.08°C [051]; Sublimes in needles [051]; Lambda maximum (in methanol): 326 nm (shoulder), 304 nm, 293 nm (shoulder), 254 nm (shoulder); epsilon 0.0651, 0.2831, 0.2181, 0.4416 [052]

Source: *Instant EPA's Air Toxics*, 1994, Instant Reference Sources, Inc., 7605 Rockpoint Drive, Austin, TX 78731 (Fax: 512-345-2386; Internet URL, http://www.instantref.com/inst-ref.htm).

Environmental Reference Materials

A source of EED environmental reference materials for pentachlorophenol is Radian International LLC, Austin, TX (Fax 512-454-0268; Internet URL, http://www.radian.com/standards); Catalog No. ERP-080.

Hazardous Properties

ACUTE/CHRONIC HAZARDS: Pentachlorophenol may be fatal if swallowed, inhaled or absorbed through the skin [036], [051]. It is an irritant and may be readily absorbed through the skin [058], [102], [421]. When heated to decomposition it may emit toxic fumes of carbon monoxide, carbon dioxide [269], hydrogen chloride and chlorinated hydrocarbons [451].

HAP WEIGHTING FACTOR: 1 [713]

VOLATILITY:
          Vapor pressure: 40 mm Hg @ 211.2°C [043], [058]
          Vapor density: 9.2 [055], [058]

FIRE HAZARD: Pentachlorophenol is nonflammable [058], [172], [371], [421].

REACTIVITY: Pentachlorophenol may react with strong oxidizing agents [058], [102], [269], [346]. It is also incompatible with strong bases, acid chlorides and acid anhydrides [269]. It forms salts with alkaline metals [395]. Solutions of pentachlorophenol in oil cause natural rubber to deteriorate, but synthetic rubber may be used in equipment and for protective clothing [172].

STABILITY: Pentachlorophenol is stable under normal laboratory conditions [058], [169], [395]. Irradiation of a dilute aqueous solution (100 ppm) by sunlight or ultraviolet light results in the formation of photodegradation products [430]. Solutions of it in DMSO, 95% ethanol or acetone should be stable for 24 hours under normal lab conditions [700].

Source: *Instant EPA's Air Toxics*, 1994, Instant Reference Sources, Inc., 7605 Rockpoint Drive, Austin, TX 78731 (Fax: 512-345-2386; Internet URL, http://www.instantref.com/inst-ref.htm).

Medical Symptoms of Exposure

Exposure may cause muscular weakness and unconsciousness [036]. It may cause irritation of the throat, sneezing, coughing, anorexia, weight loss, profuse sweating, headaches, dizziness, nausea, vomiting, dyspnea, and chest pains [346]. It may also cause loss of appetite, shortness of breath and irritation of the respiratory tract. Risk of serious or fatal intoxication is increased in hot weather [058]. Other symptoms may include abdominal pain, intense thirst and death [051]. Ingestion causes an increase then a decrease in respiration, blood pressure and urinary output; fever, increased bowel action, motor weakness, collapse with convulsions and lung, kidney and liver damage [031]. It may cause skin and conjunctiva irritation, exfoliation of the epidermal layer, painful and red skin, impairment of the autonomic function and circulation, visual damage, including an arcuate type of scotoma, dehydration, coma, terminal spasm and chloracne. Children are more susceptible than adults. Symptoms in children may include intermittent delirium, rigors, flushing and excitement. Babies nurse avidly. Tachycardia, hepatomegaly, progressive metabolic acidosis, proteinuria, azotemia, irritability followed by lethargy, pneumonia or bronchitis and aplastic anemia may occur [173]. It may also cause acneform dermatitis, allergic skin response, eye inflammation, permanent corneal injury and irritation to the nose and pharynx [151]. Other symptoms may include inflammation of the conjunctiva, characteristically shaped corneal opacity, corneal numbness, slight mydriasis, tachypnea and hepatic enlargement [421]. It may cause irritation of the eyes, burns to the skin and eyes, irritation of the mucous membranes and anesthesia [371]. Other symptoms may include systemic intoxication, hyperpyrexia, gastrointestinal upset, congestion of the lungs, edema in the brain, cardiac dilatation, centrilobular degeneration in the liver and mild degeneration of the renal tubules [430]. It may also cause profuse diaphoresis and Hodgkin's disease [395].

Source: *Instant EPA's Air Toxics*, 1994, Instant Reference Sources, Inc., 7605 Rockpoint Drive, Austin, TX 78731 (Fax: 512-345-2386; Internet URL, http://www.instantref.com/inst-ref.htm).

Toxicological Information

The toxicological information below is gathered from several sources including two national databases: one from the *National Toxicology Program Chemical Repository Database* and another from *EPA's Integrated Risk Information System (IRIS)*.

Toxicology Information from the National Toxicology Program

CARCINOGENICITY:

Tumorigenic Data:

Type of Effect	Route/Animal	Amount Dosed (Notes)
TDLo	scu-mus	46 mg/kg

Review:
         IARC Cancer Review: Human Limited Evidence
         IARC Cancer Review: Animal Inadequate Evidence
         IARC possible human carcinogen (Group 2B)
         IARC Note: Although IARC has assigned an overall evaluation to
         chlorophenols, it has not assigned an overall evaluation to all
         substances within this group

Status:

> NTP Carcinogenesis Studies (Feed); Clear Evidence: Male and Female Mouse (Dowicide EC-7)
> NTP Carcinogenesis Studies (Feed); Clear Evidence: Male Mouse (for technical grade)
> NTP Carcinogenesis Studies (Feed); Some Evidence: Female Mouse (for technical grade)

TERATOGENICITY:

Reproductive Effects Data:

Type of Effect	Route/Animal	Amount Dosed (Notes)
TDLo	orl-rat	60 mg/kg (9D preg)
TDLo	orl-rat	50 mg/kg (6-15D preg)
TDLo	orl-rat	4 gm/kg (77D pre-28D post)
TDLo	orl-rat	120 mg/kg (8-11D preg)
TDLo	scu-mus	450 mg/kg (6-14D preg)

MUTATION DATA:

Test	Lowest dose	Test	Lowest dose
mma-sat	40 nmol/plate	mrc-smc	190 umol/L
mmo-smc	400 mg/L	slt-mus-ipr	50 mg/kg
otr-ham:emb	100 mg/L	cyt-ham:ovr	80 mg/L
sce-ham:ovr	3 mg/L		

TOXICITY:

Typ. Dose	Mode	Specie	Amount	Units	Other
LDLo	orl	man	401	mg/kg	
LD50	orl	rat	27	mg/kg	
LC50	ihl	rat	355	mg/m3	
LD50	skn	rat	96	mg/kg	
LD50	ipr	rat	56	mg/kg	
LD50	scu	rat	100	mg/kg	
LDLo	scu	dog	135	mg/kg	
LDLo	orl	rbt	70	mg/kg	
LDLo	skn	rbt	40	mg/kg	
LDLo	ipr	rbt	135	mg/kg	
LD50	orl	mus	117	mg/kg	
LC50	ihl	mus	225	mg/m3	
LD50	ipr	mus	58	mg/kg	
LD50	unr	dog	70	mg/kg	
LDLo	scu	rbt	70	mg/kg	
LD50	unr	gpg	100	mg/kg	
LD50	orl	ham	168	mg/kg	
LD50	orl	dck	380	mg/kg	
LD50	unr	frg	36	mg/kg	

OTHER TOXICITY DATA:

       Skin and Eye Irritation Data:
           skn-rbt 10 mg/24H open MLD

       Review: Toxicology Review-2

       Standards and Regulations:
           DOT-Hazard: ORM-E; Label: None

      Status:
           NIOSH Analytical Methods: see Pentachlorophenol in blood, 8001, in urine, see 8303
           EPA TSCA Chemical Inventory, 1986
           EPA TSCA Test Submission (TSCATS) Data Base, January 1989
           EPA Genetox Program 1986, Positive: Cell transform.-SA7/SHE; S cerevisiae gene conversion
           EPA Genetox Program 1986, Positive: S cerevisiae-forward mutation
           EPA Genetox Program 1986, Negative: Host-mediated assay; Mouse spot test
           EPA Genetox Program 1986, Negative: Histidine reversion-Ames test; S cerevisiae-homozygosis
           Meets criteria for proposed OSHA Medical Records Rule
           IDLH Level: 150 mg/m3 in air

SAX TOXICITY EVALUATION:

       THR: Human poison by ingestion. Poison experimentally by ingestion, skin contact, intraperitoneal and subcutaneous routes. A carcinogen. An experimental teratogen. A skin irritant. Mutagenic data.

Source: *Instant Tox-Base*, 1995, Instant Reference Sources, Inc., 7605 Rockpoint Drive, Austin, TX 78731 (Fax: 512-345-2386; Internet URL, http://www.instantref.com/inst-ref.htm).

Toxicology Information from EPA's Integrated Risk Information System (IRIS)

I. CHRONIC HEALTH HAZARD ASSESSMENTS FOR NONCARCINOGENIC EFFECTS

I.A. REFERENCE DOSE FOR CHRONIC ORAL EXPOSURE (RfD)

The Reference Dose (RfD) is based on the assumption that thresholds exist for certain toxic effects such as cellular necrosis, but may not exist for other toxic effects such as carcinogenicity. In general, the RfD is an estimate (with uncertainty spanning perhaps an order of magnitude) of a daily exposure to the human population (including sensitive subgroups) that is likely to be without an appreciable risk of deleterious effects during a lifetime. Please refer to Background Document 1 in Service Code 5 for an elaboration of these concepts. RfDs can also be derived for the noncarcinogenic health effects of compounds which are also carcinogens. Therefore, it is essential to refer to other sources of information concerning the carcinogenicity of this substance. If the US EPA has evaluated this substance for potential human carcinogenicity, a summary of that evaluation will be contained in Section II of this file when a review of that evaluation is completed.

I.A.1. ORAL RfD SUMMARY

Critical Effect	Experimental Doses*	UF	MF	RfD
Liver and kidney pathology	NOAEL: 3 mg/kg/day	100	1	3E-2 mg/kg/day
	LOAEL: 10 mg/kg/day			

*Conversion Factors: none

Rat Oral Chronic Study, Schwetz et al., 1978

I.A.2. PRINCIPAL AND SUPPORTING STUDIES (ORAL RfD)

Schwetz, B.A., J.F. Quast, P.A. Keelev, C.G. Humiston and R.J. Kociba. 1978. Results of 2-year toxicity and reproduction studies on pentachlorophenol in rats. In: Pentachlorophenol: Chemistry, Pharmacology and Environmental Toxicology, K.R. Rao, Ed. Plenum Press, NY. p. 301. Only one chronic study regarding oral exposure (Schwetz et al., 1978) was located in the available literature. Twenty-five rats/sex were administered 1 of 3 doses in the diet. At the 30 mg/kg/day level of treatment, a reduced rate of body weight gain and increased specific gravity of the urine were observed in females. Pigmentation of the liver and kidneys was observed in females exposed at 10 mg/kg/day or higher levels and in males exposed to 30 mg/kg/day. The 3 mg/kg/day level of exposure was reported as a chronic NOAEL.

A number of studies that have investigated the teratogenicity of orally administered pentachlorophenol in rodents are available in the literature. Although these studies (Larsen et al., 1975; Schwetz and Gehring, 1973; Schwetz et al., 1978; Hinkle, 1973) did not reveal teratogenic effects, feto- maternal toxicity was seen at 30 mg/kg/day (Schwetz and Gehring, 1973). Since pentachlorophenol apparently does not cross the placental barrier, the observed fetotoxicity may be a reflection of maternal toxicity (Larsen et al., 1975). The NOAEL in these studies was concluded to be 3.0 mg/kg/day (US EPA, 1984), which is the same as for the chronic study reported earlier.

I.A.3. UNCERTAINTY AND MODIFYING FACTORS (ORAL RfD)

UF -- The 100-fold factor accounts for the expected intra- and inter- species variability to the toxicity of this chemical in lieu of specific data.

MF -- None

I.A.4. ADDITIONAL COMMENTS (ORAL RfD)

None.

I.A.5. CONFIDENCE IN THE ORAL RfD

Study -- High
Data Base -- Medium
RfD -- Medium

The confidence in the chosen study is rated high because a moderate number of animals/sex were used in each of three doses, a comprehensive analysis of parameters was conducted, and a reproductive study was also run. Confidence in the supporting data base is rated medium because only one chronic study is available. Other subchronic studies provide adequate but weaker supporting data. The confidence in the RfD is medium. More chronic/reproductive studies are needed to provide a higher confidence in the RfD.

I.A.6. EPA DOCUMENTATION AND REVIEW OF THE ORAL RfD

US EPA. 1984. Health Effects Assessment for Pentachlorophenol. Prepared by the Office of Health and Environmental Assessment, Environmental Criteria and Assessment Office, Cincinnati, OH for the Office of Emergency and Remedial Response, Washington, DC.

Limited Peer Review and Agency-wide Internal Review, 1984.

US EPA. 1985. Drinking Water Criteria Document for Pentachlorophenol. Prepared by the Office of Health and Environmental Assessment, Environmental Criteria and Assessment Office, Cincinnati, OH for the Office of Drinking Water, Washington, DC.

Two external peer reviews and an Agency internal review.

Agency Work Group Review -- 05/20/85

Verification Date -- 05/20/85

I.A.7. EPA CONTACTS (ORAL RfD)

Linda R. Papa / NCEA -- (513)569-7587

I.B. REFERENCE CONCENTRATION FOR CHRONIC INHALATION EXPOSURE (RfC)

A risk assessment for this substance/agent is under review by an EPA work group.

II. CARCINOGENICITY ASSESSMENT FOR LIFETIME EXPOSURE

Section II provides information on three aspects of the carcinogenic risk assessment for the agent in question; the US EPA classification, and quantitative estimates of risk from oral exposure and from inhalation exposure. The classification reflects a weight-of-evidence judgment of the likelihood that the agent is a human carcinogen. The quantitative risk estimates are presented in three ways. The slope factor is the result of application of a low-dose extrapolation procedure and is presented as the risk per (mg/kg)/day. The unit risk is the quantitative estimate in terms of either risk per ug/L drinking water or risk per ug/cu.m air breathed. The third form in which risk is presented is a drinking water or air concentration providing cancer risks of 1 in 10,000, 1 in 100,000 or 1 in 1,000,000. Background Document 2 (Service Code 5) provides details on the rationale and methods used to derive the carcinogenicity values found in IRIS. Users are referred to Section I for information on long-term toxic effects other than carcinogenicity.

II.A. EVIDENCE FOR CLASSIFICATION AS TO HUMAN CARCINOGENICITY

II.A.1. WEIGHT-OF-EVIDENCE CLASSIFICATION

Classification -- B2; probable human carcinogen

Basis -- The classification is based on inadequate human data and sufficient evidence of carcinogenicity in animals: statistically significant increases in the incidences of multiple biologically significant tumor types (hepatocellular adenomas and carcinomas, adrenal medulla pheochromocytomas and malignant pheochromocytomas, and/or hemangiosarcomas and hemangiomas) in one or both sexes of B6C3F1 mice using two different preparations of pentachlorophenol (PeCP). In addition, a high incidence of two uncommon tumors (adrenal medulla pheochromocytomas and hemangiomas/hemangiosarcomas) was observed with both preparations. This classification is supported by mutagenicity data, which provides some indication that PeCP has clastogenic potential.

II.A.2. HUMAN CARCINOGENICITY DATA

Inadequate. Gilbert et al. (1990) attempted to study the effects of exposure to PeCP and other chemical preservatives among a cohort of 182 men employed in the wood treating industry in Hawaii. The study included both current and former workers who had experienced a minimum of 3 months of continuous employment treating wood between 1960 and November 1981. The first part of the study consisted of a cross-sectional clinical assessment of 88 workers (66 current, 22 former) and 58 nonexposed men employed in other occupations. Significantly elevated levels of urinary PeCP were found among the wood treaters but this was not related to any morbidity or mortality endpoint.

In part two of the study, the authors attempted to compare the mortality experience of the cohort with that expected in Hawaiian males of the same age. Only deaths that occurred in Hawaii were ascertained. Six deaths were observed compared with eight expected. Overall, the authors concluded that their results do not suggest any clinically significant adverse health effects nor any increased cancer morbidity or mortality from exposure to

PeCP and other wood preserving chemicals. These conclusions must be seriously questioned based on the following: inadequate detail of selection for participation, particularly among the 58 unexposed "controls"; only 50% of eligible workers participated in the clinical portion which creates the potential for selection bias; employment eligibility criteria were different for current versus former workers; the clinical examiner was not blinded as to the exposure status of participants which raises questions about the presence of observation bias; the clinical data were presented and analyzed in a nonstandard way; no details are given about methods used to compute mortality "rates"; and, failure to ascertain deaths occurring outside of Hawaii. With over 30% of the original cohort apparently lost to follow-up, the study is of questionable validity. It cannot be used as evidence of no effect of the exposures but instead must be viewed as uninformative.

II.A.3. ANIMAL CARCINOGENICITY DATA

Sufficient. Two different 90% pure preparations of PeCP were tested in 2-year bioassays in B6C3F1 mice (NTP, 1989). Typical impurities present in both preparations included tri- and tetrachlorophenol, hexachlorobenzene, polychlorinated dibenzo-p-dioxins, and polychlorinated dibenzofurans. Technical grade PeCP (TG-PeCP) is a composite that consisted of equal proportions of product from Monsanto, Reichold and Vulcan. These specific products are no longer being produced. The second 90% pure preparation of PeCP, EC-7 PeCP, differed from TG-PeCP in the level and nature of impurities present (e.g., EC-7 PeCP contained lower levels of dioxins and dibenzofurans). TG-PeCP was administered daily in the feed at dose levels of 0, 100, and 200 ppm to groups of 50 male and 50 female B6C3F1 mice for 2 years. The average doses of TG-PeCP were approximately 17-18 or 35 mg/kg for males and females, respectively. Two groups of control mice (35/sex) were fed basal diets. Survival of the mice did not appear to be affected by exposure to TG-PeCP at any dose level tested. However, it should be noted that survival of the male control mice (12/35) was low compared with historical control values. The early deaths were found to be due to urinary tract infections resulting from injuries sustained during fighting among the group-housed control male mice. After month 16 of the study, the male mice were singly housed to reduce the incidence of fighting and consequent high mortality. The incidences of hepatocellular adenomas and/or carcinomas were significantly increased in male mice exposed to TG-PeCP when compared with controls; the incidences were 7/32, 26/47 and 37/48 in control, low-dose and high-dose male mice, respectively. The incidences of benign and malignant pheochromocytomas of the adrenal medulla were also significantly greater in dosed male mice than in controls; the incidences were 0/31 in controls, 10/45 in low-dose animals and 23/45 in the high-dose animals. There was no significant increase in the numbers of liver tumors or pheochromocytomas in female mice exposed to TG-PeCP. However, the nonsignificant increase in liver tumors in TG-PeCP exposed females was considered biologically significant. TG-PeCP- and EC-7 PeCP-exposed females showed comparable responses in the 100- and 200-ppm dose groups with a marked increase observed only at 600 ppm in the EC-7 PeCP females. The liver tumor incidences for TG-PeCP exposed females was 3/33, 9/49 and 9/50, respectively. Vascular tumors (hemangiomas and/or hemangiosarcomas) were observed in female mice but not in male mice. Incidences of the hemangiosarcoma tumors were statistically significantly increased when compared to controls and all were malignant (0/30, 3/48, 6/46 in the control, low-dose and high-dose females, respectively).

EC-7 PeCP was administered daily in the feed at dose levels of 0, 100, 200, and 600 ppm (NTP, 1989). The average daily doses of EC-7 PeCP were approximately 17-18, 34-37, and 114-118 mg/kg, for the low-, mid-, and high- dose groups, respectively. Two groups of control mice (35/sex) were fed basal diets. Survival did not appear to be affected by exposure to EC-7 PeCP at any of the dose levels tested. The incidences of hepatocellular adenomas and/or carcinomas were significantly increased in dosed male mice exposed to EC-7 PeCP when compared with controls (6/35, 19/48, 21/48, 34/49 in the control, low-, mid-, and high-dose males, respectively). The incidences of benign and malignant pheochromocytomas of the adrenal medulla in males were also significantly greater in treated males than in the controls (1/34, 4/48, 21/48, 45/49 in the control, low-, mid-, and high-dose males, respectively). There was a significant increase in liver tumors (adenomas and/or carcinomas) (1/34, 4/50, 6/49 and 31/48 in the control, low-, mid-, and high-dose females, respectively) and benign and malignant pheochromocytomas in female mice exposed to EC-7 PeCP at the high-dose only (0/35, 2/49, 2/46, 38/49 in the control, low-, mid-, and high-dose females, respectively). Vascular tumors (hemangiomas and/or hemangiosarcomas) were observed in female mice but not in male mice. The incidence of these latter tumors was statistically significantly elevated in the high-dose group when compared with controls and all but one of the tumors was malignant (0/34, 1/50, 3/48, 9/47 in the control, low-, mid-, and high-dose females, respectively).

In a study reported by BRL (1968) and Innes et al. (1969), 18 male and 18 female crossbred mice were administered 46.4 mg/kg EC-7 PeCP in gelatin by gavage on days 7 through 28 after birth, followed by administration of 130 ppm (17 mg/kg/day) EC-7 PeCP in the diet for 18 months. It is not possible to ascertain whether the EC-7 PeCP used in this study is the same as the EC-7 used in the NTP (1989) study, since the level and nature of the impurities present in the preparation were not reported by Innes or BRL. Groups of mice from each strain served as negative or vehicle controls. Results indicated that there was no difference between the incidence of tumors in the PeCP-treated group and the control groups. Only tumor incidences were reported, so it is not known what other toxic effects (if any) may have occurred. This study is limited for drawing conclusions concerning the carcinogenicity of PeCP, however, because only one dose level was used. Furthermore, an insufficient number of animals (according to current guidelines) was studied.

In a chronic oral study on a different species conducted by Schwetz et al. (1978), groups of 25 Sprague-Dawley rats/sex were fed diets of 0, 8, 23, 77, or 231 ppm PeCP for 22 (for male) or 24 (for female) months (equivalent to 1, 3, 10, or 30 mg PeCP/kg/day). The PeCP preparation used in this study was reported to be 90% pure, and representative of the commercially available Dowicide EC-7 PeCP used in the NTP (1989) study. Results from the experiment indicated that in the high-dose group a reduced rate of body weight gain (i.e., a 12% lower mean monthly body weight during the last 12 months of the study) and an increased specific gravity of the urine were observed in females. Pigmentation of the liver and kidneys was observed in females exposed at 10 mg/kg/day or higher levels and in males exposed to 30 mg/kg/day. There was no significant increase in tumor incidence as compared with controls. A slight increase in pheochromocytomas of the adrenal medulla was noted at the lower dose levels. Survival was reported to be unaffected by treatment. Since the high dose (30 mg/kg/day) elicited signs of only mild toxicity, NTP suggested that the MTD had been reached but not exceeded in this study.

Catilina et al. (1981) also found no evidence of carcinogenicity in Wistar rats following subcutaneous administration of purified and technical grades of PeCP (6 mg/kg/dose). Test compounds were administered 3 times/week for 40 weeks followed by a 3-month post-treatment observation period. The use of only one dose, the use of an inappropriate route of administration, the relatively short exposure time, and excessive mortality limit the usefulness of this study for drawing conclusions concerning the carcinogenicity of PeCP.

In another study, Boutwell and Bosch (1959) applied a 20% solution of commercial grade PeCP in benzene to the shaved skin of Sutter mice twice weekly for 13 weeks following an initial exposure with 0.3% DMBA in benzene. Because of the dose level, frequency and duration of exposure in this study, only limited conclusions concerning the effectiveness of PeCP as a complete carcinogen can be made; these results, however, are sufficient to conclude that PeCP was not a tumor promoter in this assay.

II.A.4. SUPPORTING DATA FOR CARCINOGENICITY

Results from cytogenetic studies provide evidence for the clastogenic potential of PeCP. In cytogenicity studies with cultured CHO cells, TG-PeCP produced an increase in chromosomal aberrations in the presence but not the absence of S9 hepatic homogenate activation. Conversely, SCEs were induced only in the absence of S9 hepatic homogenate (NTP, 1989).

II.B. QUANTITATIVE ESTIMATE OF CARCINOGENIC RISK FROM ORAL EXPOSURE

II.B.1. SUMMARY OF RISK ESTIMATES

Oral Slope Factor -- 1.2E-1 per (mg/kg)/day

Drinking Water Unit Risk -- 3E-6 per (ug/L)

Extrapolation Method -- Linearized multistage procedure

Drinking Water Concentrations at Specified Risk Levels:

Risk Level	Concentration
E-4 (1 in 10,000)	3E+l ug/L
E-5 (1 in 100,000)	3E+O ug/L
E-6 (1 in 1,000,000)	3E-l ug/L

II.B.2. DOSE-RESPONSE DATA (CARCINOGENICITY, ORAL EXPOSURE)

Tumor Type -- hepatocellular adenoma/carcinoma, pheochromocytoma/malignant pheochromocytoma, hemangiosarcoma/hemangioma (pooled incidence Test Animals -- mouse/B6C3F1, female

Route -- diet

Reference -- NTP, 1989

Administered Dose		Human Equivalent Dose (mg/kg)/day	Pooled Hepatocellular and Hemangiosarcoma Tumor Incidence
ppm	(mg/kg)/day		
Technical grade pentachlorophenol			
0	0	0	5/31
100	17	1.4	12/48
200	35	2.7	15/46
Dowicide EC-7 pentachlorophenol			
0	0	0	1/34
100	17	1.3	6/49
200	34	2.7	9/46
600	114	8.7	42/49

II.B.3. ADDITIONAL COMMENTS (CARCINOGENICITY, ORAL EXPOSURE)

Two different pentachlorophenol preparations induced liver tumors, pheochromocytomas and hemangiosarcomas in female mice and liver tumors and pheochromocytomas in male mice. All three tumor types are considered related to the administration of pentachlorophenol. The hemangiosarcomas, however, are considered to be the tumor of greatest concern; the EPA Science Advisory Board found that "these tumors were related to the administration of the pentachlorophenol formulations tested, occurred in a dose-response manner in the treated animals, and are morphologically related to known fatal human cancers that are induced by xenobiotics." Hemangiosarcomas were found only in female mice. To give preference to the data on hemangiosarcomas and because some male groups experienced significant early loss, only the female mice are used in the quantitative risk assessment.

In developing these estimates, benign and malignant tumors are combined; the liver tumors and pheochromocytomas were mostly benign. The pooled incidence counts animals with any of the three tumor types. Animals dying before the first tumor was observed are not considered to be at risk and are not included in the totals. Equivalent human doses are calculated using a surface-area adjustment. There are no pharmacokinetic data on pentachlorophenol. The slope factor is calculated as the geometric mean of the slope factors for each pentachlorophenol preparation.

II.B.4. DISCUSSION OF CONFIDENCE (CARCINOGENICITY, ORAL EXPOSURE)

For purposes of comparison, a slope factor of 0.05 can be derived from the incidence of hemangiosarcomas alone. Also for comparison, a slope factor of 0.5 can be derived from the pooled incidence of liver tumors and pheochromocytomas in male B6C3F1 mice.

The carcinogenicity assessment is based on results in a single animal species.

II.C.   QUANTITATIVE   ESTIMATE   OF   CARCINOGENIC   RISK   FROM INHALATION EXPOSURE

Not available.

II.D. EPA DOCUMENTATION, REVIEW, AND CONTACTS (CARCINOGENICITY ASSESSMENT)

II.D.1. EPA DOCUMENTATION

Source Document -- This assessment is not presented in any existing US EPA document.

II.D.2. REVIEW (CARCINOGENICITY ASSESSMENT)

Agency Work Group Review -- 11/10/87, 09/22/88, 10/19/88, 12/06/89, 02/08/90, 08/02/90

Verification Date -- 08/02/90

II.D.3. US EPA CONTACTS (CARCINOGENICITY ASSESSMENT)

Jim Cogliano / NCEA -- (202)260-3814

Rita Schoeny / NCEA -- (513)569-7544

III. HEALTH HAZARD ASSESSMENTS FOR VARIED EXPOSURE DURATIONS

III.A. DRINKING WATER HEALTH ADVISORIES

The Office of Drinking Water provides Drinking Water Health Advisories (HAs) as technical guidance for the protection of public health. HAs are not enforceable Federal standards. HAs are concentrations of a substance in drinking water estimated to have negligible deleterious effects in humans, when ingested, for a specified period of time. Exposure to the substance from other media is considered only in the derivation of the lifetime HA. Given the absence of chemical-specific data, the assumed fraction of total intake from drinking water is 20%. The lifetime HA is calculated from the Drinking Water Equivalent Level (DWEL) which, in turn, is based on the Oral Chronic Reference Dose. Lifetime HAs are not derived for compounds which are potentially carcinogenic for humans because of the difference in assumptions concerning toxic threshold for carcinogenic and noncarcinogenic effects. A more detailed description of the assumptions and methods used in the derivation of HAs is provided in Background Document 3 in Service Code 5.

III.A.1. ONE-DAY HEALTH ADVISORY FOR A CHILD

One-day HA -- 1E+0 mg/L

NOAEL -- 10 mg/kg/day
UF -- 100 (allows for interspecies and intrahuman variability with the use of a NOAEL from an animal study)
Assumptions -- 1 L/day water consumption for a 10-kg child

Principal Study -- Nishimura et al., 1982

Increased liver/body weight ratios were observed in male rats after single oral doses of sodium pentachlorophenate at levels greater than 10 mg/kg. The authors described the doses in terms of pentachlorophenol content.

III.A.2. TEN-DAY HEALTH ADVISORY FOR A CHILD

Appropriate data for calculating a Ten-day HA are not available. It is recommended that the Longer-term HA for the 10-kg child of 0.30 mg/L be used as the Ten-day HA.

III.A.3. LONGER-TERM HEALTH ADVISORY FOR A CHILD

Longer-term (Child) HA -- 3E-1 mg/L

NOAEL -- 3 mg/kg/day
UF -- 100 (allows for interspecies and intrahuman variability with the use of a NOAEL from an animal study)
Assumptions -- 1 L/day water consumption for a 10-kg child

Principal Study -- Johnson et al., 1973

Pentachlorophenol was fed to rats by diet at levels of 3, 10, or 30 mg/kg/day for 90 days. Increased liver and kidney weights were induced at the two higher doses, whereas increased liver and kidney weights were not evident at the 3-mg/kg/day feeding level.

III.A.4. LONGER-TERM HEALTH ADVISORY FOR AN ADULT

Longer-term (Adult) HA -- 1.05E+0 mg/L

NOAEL -- 3 mg/kg/day
UF -- 100 (allows for interspecies and intrahuman variability with the use of a NOAEL from an animal study)
Assumptions -- 2 L/day water consumption for a 70-kg adult
Principal Study -- Johnson et al., 1973 (study described in III.A.3.)

III.A.5. DRINKING WATER EQUIVALENT LEVEL / LIFETIME HEALTH ADVISORY

DWEL -- 1.05E+0 mg/L

Basis -- Derived from an oral chronic RfD of 3.0E-2 (unrounded); verification date - 05/20/85; Refer to Section I.A. for a discussion of the RfD.

Assumptions -- 2 L/day water consumption for a 70-kg adult

Lifetime Health Advisory -- 2E-1 mg/L
Assumptions -- 20% exposure by drinking water

Principal Study -- Schwetz et al., 1978 (This study was used in the derivation of the oral chronic RfD; see Section I.A.2.)

III.A.6. ORGANOLEPTIC PROPERTIES

Odor perception threshold (water) -- 1600 ug/L.

Taste perception threshold (water) -- 30 ug/L.

III.A.7. ANALYTICAL METHODS FOR DETECTION IN DRINKING WATER

Determination of pentachlorophenol is by a liquid-liquid extraction gas chromatographic procedure.

III.A.8. WATER TREATMENT

Treatment techniques for removal of pentachlorophenol from drinking water pertain predominantly to adsorption. The use of air stripping also has been considered.

III.A.9. DOCUMENTATION AND REVIEW OF HAs

US EPA. 1985. Final Draft of the Drinking Water Criteria Document on Endrin. Office of Drinking Water, Washington, DC.

EPA review of HAs in 1985.

Public review of HAs following notification of availability in October, 1985.

Preparation date of this IRIS summary -- 06/17/87

III.A.10. EPA CONTACTS

Jennifer Orme Zavaleta / OST -- (202)260-7586

Edward V. Ohanian / OST -- (202)260-7571

III.B. OTHER ASSESSMENTS

Content to be determined.

IV. US EPA REGULATORY ACTIONS

EPA risk assessments may be updated as new data are published and as assessment methodologies evolve. Regulatory actions are frequently not updated at the same time. Compare the dates for the regulatory actions in this section with the verification dates for the risk assessments in sections I and II, as this may explain inconsistencies. Also note that some regulatory actions consider factors not related to health risk, such as technical or economic feasibility. Such considerations are indicated for each action. In addition, not all of the regulatory actions listed in this section involve enforceable federal standards. Please direct any questions you may have concerning these regulatory actions to the US EPA contact listed for that particular action. Users are strongly urged to read the background information on each regulatory action in Background Document 4 in Service Code 5.

IV.A. CLEAN AIR ACT (CAA)

  [Ed. Note: EPA does not have information yet in this section of IRIS].

IV.B. SAFE DRINKING WATER ACT (SDWA)

IV.B.1. MAXIMUM CONTAMINANT LEVEL GOAL (MCLG) for Drinking Water

Value (status) -- 0 mg/L (Final, 1991)

Considers technological or economic feasibility? -- NO

Discussion -- The MCLG of 0 mg/L for pentachlorophenol is based upon potential carcinogenic effects (B2).

Reference -- 56 FR 3600 (01/30/91); 56 FR 30266 (07/01/91)

EPA Contact -- Health and Ecological Criteria Division / OST /(202) 260-7571 / FTS 260-7571; or Safe Drinking Water Hotline / (800) 426-4791

IV.B.2. MAXIMUM CONTAMINANT LEVEL (MCL) for Drinking Water

Value -- 0.001 mg/L (Final, 1991)

Considers technological or economic feasibility? -- YES

Discussion -- EPA has set an MCL equal to the PQL of 0.001 mg/L, which is associated with a lifetime individual risk of less than E-6.

Monitoring requirements -- All systems monitored for four consecutive quarters every three years; repeat monitoring dependent upon detection, vulnerability status and system size.

Analytical methodology -- Gas chromatographic/mass spectrometry (EPA 525); electron-capture/gas chromatography (EPA 515.1): PQL = 0.001 mg/L.

Best available technology -- Granular activated carbon.

Reference -- 56 FR 3600 (01/30/91); 56 FR 3526 (01/30/91); 56 FR 30266 (07/01/91).

EPA Contact -- Drinking Water Standards Division / OGWDW /(202) 260-7575 / FTS 260-7575; or Safe Drinking Water Hotline / (800) 426-4791

IV.B.3. SECONDARY MAXIMUM CONTAMINANT LEVEL (SMCL) for Drinking Water

Value -- 0.03 mg/L (Proposed, 1989)

Considers technological or economic feasibility? -- NO

Discussion -- SMCLs are non-enforceable and establish limits for contaminants which may affect the aesthetic qualities (e.g. taste and odor) of drinking water. It is recommended that systems monitor for these contaminants every three years. More frequent monitoring for contaminants such as pH, color, odor or others may be appropriate under certain circumstances. The SMCL for pentachlorophenol is based on adverse taste. Promulgation has been deferred following public comment (56 FR 3526).

Reference -- 54 FR 22062 (05/22/89); 56 FR 3526 (01/30/91)

EPA Contact -- Drinking Water Standards Division / OGWDW /(202) 260-7575 / FTS 260-7575; or Safe Drinking Water Hotline / (800) 426-4791

IV.B.4. REQUIRED MONITORING OF "UNREGULATED" CONTAMINANTS

[Ed. Note: EPA does not have information yet in this section of IRIS].

IV.C. CLEAN WATER ACT (CWA)

IV.C.1. AMBIENT WATER QUALITY CRITERIA, Human Health

Water and Fish Consumption: 1.01E+3 ug/L

Fish Consumption Only: None

Considers technological or economic feasibility? -- NO

Discussion -- The WQC necessary for the protection of public health is 1.01E+3 ug/L. Its basis is a NOAEL of 3 mg/kg in a mammalian study, a safety factor of 100, and an assumption of daily ingestion of 2 L of water and 6.5 g of fish. A WQC of 30.0 ug/L based upon organoleptic effects has also been derived. However, organoleptic endpoints have limited value in setting water quality standards, since there is no demonstrated relationship between taste/odor effect and adverse health effects.

Reference -- 45 FR 79318 (11/28/80)

EPA Contact -- Criteria and Standards Division / OWRS (202)260-1315 / FTS 260-1315

IV.C.2. AMBIENT WATER QUALITY CRITERIA, Aquatic Organisms

Freshwater:
    Acute -- 2.2E+1 ug/L (1 hour average)
    Chronic -- 1.3E+1 ug/L (4 day average)

Marine:
    Acute -- 1.3E+1 ug/L (1 hour average)
    Chronic -- 7.9E+0 ug/L (4 day average)

Considers technological or economic feasibility? -- NO

Discussion -- Criteria were derived from a minimum data base consisting of acute and chronic tests on a variety of species. The toxicity of penta- chlorophenol is dependent on the pH of the ambient water. The value given is for a pH of 7.8. A more complete discussion can be found in the reference document.

Reference -- 51 FR 43665 (12/03/86)

EPA Contact -- Criteria and Standards Division / OWRS (202)260-1315 / FTS 260-1315

IV.D. FEDERAL INSECTICIDE, FUNGICIDE, AND RODENTICIDE ACT (FIFRA)

IV.D.1. PESTICIDE ACTIVE INGREDIENT, Registration Standard

Status -- List "B" Pesticide (1989)

Reference -- 54 FR 22706 (05/25/89)

EPA Contact -- Registration Branch / OPP (703)557-7760 / FTS 557-7760

IV.D.2. PESTICIDE ACTIVE INGREDIENT, Special Review

Action -- Final regulatory decision - PD4 (1987)

Considers technological or economic feasibility? -- YES

Summary of regulatory action -- Non-deferred uses canceled for non-wood use; amendment to PD4 (clarifying uses) 53 FR 5524. Previous decision (1984) required label changes including a restricted use classification; two amendments to PD4 [51FR 1334 (01/10/86) and 52 FR 140 (01/02/87)]. Comprehensive DCI issued 08/04/89.

Reference -- 52 FR 2282 (01/21/87); 51FR 1334 (01/10/86); 52 FR 140 (01/02/87); 53 FR 5524

EPA Contact -- Special Review Branch / OPP (703)557-7400 / FTS 557-7400

IV.E. TOXIC SUBSTANCES CONTROL ACT (TSCA)

[Ed. Note: EPA does not have information yet in this section of IRIS].

IV.F. RESOURCE CONSERVATION AND RECOVERY ACT (RCRA)

IV.F.1. RCRA APPENDIX IX, for Ground Water Monitoring

Status -- Listed
Reference -- 52 FR 25942 (07/09/87)
EPA Contact --   RCRA/Superfund Hotline (800)424-9346 / (202)260-3000 / FTS 260-3000

IV.G. SUPERFUND (CERCLA)

IV.G.1. REPORTABLE QUANTITY (RQ) for Release into the Environment

Value (status) -- 10 pounds (Final, 1989)
Considers technological or economic feasibility? -- NO

Discussion -- The RQ is based on aquatic toxicity as assigned by Section 311(b)(4) of the Clean Water Act (40 CFR 117.3). Available data indicate a 96-hour Median Threshold Limit between 0.2 and 0.6 ppm, corresponding to an RQ of 10 pounds.

Reference -- 54 FR 33418 (08/14/89)
EPA Contact -- RCRA/Superfund Hotline (800)424-9346 / (202)260-3000 / FTS 260-3000

VI. BIBLIOGRAPHY

VI.A. ORAL RfD REFERENCES

Hinkle, D.K. 1973. Fetotoxic effects of pentachlorophenol in the Golden Syrian Hamster. Toxicol. Appl. Pharmacol. 25: 445.

Larsen, R.V., G.S. Born, W.V. Kessler, S.M. Shaw and D.C. Van Sickle. 1975. Placental transfer and teratology of pentachlorophenol in rats. Environ. Lett. 10: 121-128.

Schwetz, B.A. and P.J. Gehring. 1973. The effect of tetrachlorophenol and pentachlorophenol on rat embryonal and fetal development. Toxicol. Appl. Pharmacol. 25: 455.

Schwetz, B.A., J.F. Quast, P.A. Keelev, C.G. Humiston and R.J. Kociba. 1978. Results of 2-year toxicity and reproduction studies on pentachlorophenol in rats. In: Pentachlorophenol: Chemistry, Pharmacology and Environmental Toxicology, K.R. Rao, Ed. Plenum Press, NY. p. 301.

US EPA. 1984. Health Effects Assessment for Pentachlorophenol. Prepared by the Office of Health and Environmental Assessment, Environmental Criteria and Assessment Office, Cincinnati, OH for the Office of Emergency and Remedial Response, Washington, DC.

US EPA. 1985. Drinking Water Criteria Document for Pentachlorophenol. Prepared by the Office of Health and Environmental Assessment, Environmental Criteria and Assessment Office, Cincinnati, OH for the Office of Drinking Water, Washington, DC.

VI.B. INHALATION RfC REFERENCES

None

VI.C. CARCINOGENICITY ASSESSMENT REFERENCES

Boutwell, R.K. and D.K. Bosch. 1959. The tumor-promoting action of phenol and related compounds for mouse skin. Cancer Res. 19: 413-424.

BRL (Bionetics Research Laboratories). 1968. Evaluation of the carcinogenic, teratogenic and mutagenic activities of selected pesticides and industrial chemicals, Vol. 1. Carcinogenic Study. Prepared for the National Cancer Institute, Bethesda, MD. NTIS PB-223-159. p. 393.

Catilina, P., A. Chamoux, M.J. Catilina and J. Champeix. 1981. Study of the pathogenic properties of substances used as wood protectives: Pentachlorophenol. Arch. Mal. Prof. Med. Trav. Secur. Soc. 42(6): 334-337. (Fre.)

Gilbert, F., C. Minn, R. Duncan and J. Wilkinson. 1990. Effects of pentachlorophenol and other chemical preservatives on the health of wood- treating workers in Hawaii. Arch. Environ. Contam. Toxicol. 19(4): 603-609.

Innes, J.R.M., B.M. Ulland, M.G. Valerio, et al. 1969. Bioassay of pesticides and industrial chemicals for tumorigenicity in mice. A preliminary note. J. Natl. Cancer Inst. 42: 1101-1114.

NTP (National Toxicology Program). 1989. Technical Report on the Toxicology and Carcinogenesis Studies of Pentachlorophenol (CAS No. 87-86-5) in B6C3F1 mice (Feed Studies). NTP Tech. Report No. 349. NIH Publ. No. 89-2804.

Schwetz, B.A., J.F. Quast, P.A. Keeler, C.G. Humiston and R.J Kociba. 1978. Results of two-year toxicity and reproduction studies on pentachlorophenol in rats. In:

Pentachlorophenol: Chemistry, Pharmacology and Environmental Toxicology, K.R. Rao, Ed. Plenum Press, NY. p. 301-309.

VI.D. DRINKING WATER HA REFERENCES

Nishimura, H., N. Nishimura and H. Oshima. 1982. Effects of pentachlorophenol on the levels of hepatic glycogen. Sangyo Isaku. 24(4):398-399.

Johnson, R.L., P.J. Gehring, R.J. Kociba and B.A. Schwetz. 1973. Chlorinated dibenzodioxins and pentachlorophenol. Environ. Health Perspect., Exp. Issue No. 5, September, 1973. p. 171.

Schwetz, B.A., J.F. Quast, P.A. Keelev, C.G. Humiston and R.J. Kociba. 1978. Results of 2-year toxicity and reproduction studies on pentachlorophenol in rats. In: Pentachlorophenol: Chemistry, Pharmacology and Environmental Toxicology, K.R. Rao, Ed. Plenum Press, NY. p. 301.

US EPA. 1985. Drinking Water Criteria Document for Endrin. Office of Drinking Water, Washington, DC. (Final Draft)

VII. REVISION HISTORY

Date	Section	Description
03/01/88	III.A.	Health Advisory added
06/30/88	I.A.6.	Documentation year corrected
01/01/90	II.	Carcinogen assessment now under review
01/01/90	VI.	Bibliography on-line
04/01/90	I.A.2.	NOEL corrected to NOAEL in last sentence, 1st paragraph
07/01/90	I.B.	Inhalation RfC now under review
07/01/90	IV.F.1.	EPA contact changed
08/01/90	III.A.10	Primary contact changed
03/01/91	II.	Carcinogenicity assessment on-line
03/01/91	VI.C.	Carcinogenicity references added
01/01/92	I.A.7.	Primary contact changed
01/01/92	IV.	Regulatory actions updated
02/01/93	I.A.7.	Minor text change
07/01/93	II.D.3.	Primary contact's phone number changed

SYNONYMS

Chem-Tol
Chlorophen
Cryptogil OL
Dowcide 7
Dowicide EC-7
DP-2, technical
Durotox
EP 30
Fungifen
Glazd penta
Grundier arbezol
1-Hydroxy- 2,3,4,5,6-pentachlorobenzene
Lauxtol
Lauxtol A
Liroprem
NCI-C54933
NCI-C55378
NCI-C55389
NCI-C56655
PCP
Penchlorol
Penta
Pentachloorfenol
Pentachlorofenol
Pentachlorofenolo
Pentachlorophenate
Pentachlorophenol
2,3,4,5,6-Pentachlorophenol.
Pentachlorphenol
Pentaclorofenolo
Pentacon
Penta-Kil
Pentasol
Penwar
Peratox
Permacide
Permagard
Permasan
Permatox
Permatox dp-2
Permatox penta
Permite
Phenol, pentachloro-
Preventol P
Priltox
Santobrite
Santophen
Santophen 20
Sinituho
Term-i-trol
WLN: QR BG CG DG EG FG

Source: *Instant EPA's IRIS*, 1997, Instant Reference Sources, Inc., 7605 Rockpoint Drive, Austin, TX 78731 (Fax: 512-345-2386; Internet URL, http://www.instantref.com/inst-ref.htm).

Production, Use, and Pesticide Labeling Information

Pentachlorophenol is not currently listed in EPA's Pesticide Factsheet database and there is no indication of whether it will be added or not in the future.

Pentachloronitrobenzene

Introduction

Pentachloronitrobenzene is used as a an herbicide and also as a fungicide for seed and soil treatment. It is used for damping off of cotton; black root and club root of cabbage, cauliflower, brussels sprouts, and broccoli potatoes; southern stem and root rot of peanuts; southern blight of tomatoes and peppers; root and stem rot and white mold of beans; white rot of garlic; bunt of wheat; botrytis storage rot of roses; brown patch and snow mold of turf; petal blight of azaleas; root rot of easter lilies; flower blight of camellia; stem rot of various ornamentals; and crown and black rot of bulbous ornamentals It is used on cabbage during soil treatment against club root and on avocados during soil and seed treatment against rhizoctonia. It is used on ornamental turf and lawns during foliar treatment against snow mold and as an intermediate, herbicide, fungicide for seed and soil treatment, and as a slime inhibitor in industrial waters. [828]

Pentachloronitrobenzene released to soil is not expected to leach extensively. Field and laboratory half-lives for it in soil vary from several weeks to almost 2 years. Volatilization may be the most significant loss mechanism for pentachloronitrobenzene from aerobic soils, followed by biodegradation. In a anaerobic soil, its loss was principally by conversion to pentachloroaniline (PCA). [828]

Pentachloronitrobenzene released to water will sorb to sediments, suspended sediments, and biota. An estimated Henry's Law Constant for pentachloronitrobenzene indicates that volatilization from water may be significant; however, sorption of it to organic particulate matter in water will decrease the significance of volatilization. Pentachloronitrobenzene will have a low to moderate tendency to bioconcentrate. Photolysis and hydrolysis of pentachloronitrobenzene are probably not significant degradative processes. Biodegradation appears to be relatively slow; however, it may be a significant degradative process for it in water. [828]

No information was found on the fate of pentachloronitrobenzene in the atmosphere; however, it will probably adsorb to particulate matter and thus may be removed from the atmosphere by wet and dry deposition. Pentachloronitrobenzene has been found in drinking water, well water, crop land and nursery soils, spinach leaves, cheese, fruits, ground grains, leaf and stem vegetables, nuts, and oilseed by-products. The most probable route of human exposure to PCNB will probably be through the ingestion of contaminated food. [828]

Identifying Information
CAS NUMBER: 82-68-8

NIOSH Registry Number: DA6650000

SYNONYMS: Nitropentachlorobenzene, Quintocene, Quintozene, Quintobenzene

Endocrine Disruptor Information
Penetachloronitrobenzene is on EPA's list of suspect EEDs [811] but little information has been located to determine the Agency's reason for its inclusion. More information should be forthcoming after EPA's studies are concluded.

Chemical and Physical Properties
CHEMICAL FORMULA: $C_6Cl_5NO_2$

MOLECULAR WEIGHT: 295.34

WLN: WNR BG CG DG EG FG

PHYSICAL DESCRIPTION: Crystalline pale yellow to white solid

SPECIFIC GRAVITY: 1.718 @ 25/4°C [017], [031], [062], [205]

DENSITY: 1.718 g/mL @ 25°C [051], [172]

MELTING POINT: 146°C [043], [172], [346], [395]

BOILING POINT: 328°C @ 760 mm Hg (decomposes) [031], [051], [062], [169]

SOLUBILITY: Water: <1 mg/mL @ 22°C [700]; DMSO: 50-100 mg/mL @ 20°C [700]; 95% Ethanol: 10-50 mg/mL @ 20°C [700]; Acetone: >=100 mg/mL @ 20°C [700]

OTHER SOLVENTS: Alcohol: Practically insoluble cold [031]; Benzene: Freely soluble [031], [051,395]; Chloroform: Freely soluble [031], [051], [395]; Carbon disulfide: Freely soluble [031], [051], [169], [395]; Ethanol: ~2.0 g/kg @ 25°C [169], [172], [173]; Aromatic solvents: Readily soluble [051], [169]; Ketones: Readily soluble [051], [169]; Chlorinated hydrocarbons: Readily soluble [051], [169]; Alkanols: Slightly soluble [051]

OTHER PHYSICAL DATA: Melting point (technical grade): 142-145°C [058], [062]; Vapor pressure: 0.00001161 mm Hg @ 10 C; 0.00005 mm Hg @ 20°C [051]; pH (saturated solution): 2.5-4.0 [058]; Musty odor [051], [062]; Odor is similar to moth balls [058]

Source: *Instant EPA's Air Toxics*, 1994, Instant Reference Sources, Inc., 7605 Rockpoint Drive, Austin, TX 78731 (Fax: 512-345-2386; Internet URL, http://www.instantref.com/inst-ref.htm).

Environmental Reference Materials

A source of EED environmental reference materials for pentachloronitrobenzene is Radian International LLC, Austin, TX (Fax 512-454-0268; Internet URL, http://www.radian.com/standards); Catalog No. ERP-081.

Hazardous Properties

ACUTE/CHRONIC HAZARDS: Pentachloronitrobenzene may be harmful if swallowed, inhaled or absorbed through the skin. It is an irritant [269] and when heated to decomposition it emits highly toxic fumes of chlorine, carbon monoxide, carbon dioxide, nitrogen oxides, hydrogen chloride gas and phosgene [043], [058], [269].

HAP WEIGHTING FACTOR: 1 [713]

VOLATILITY:
Vapor pressure: 0.013 mm Hg @ 25°C [043], [173], [395]
Vapor density: 10.2 [051]

FIRE HAZARD: Flash point data for pentachloronitrobenzene are not available. It is probably combustible. Fires involving this material may be controlled with a dry chemical, carbon dioxide or Halon extinguisher. A water spray may also be used [058], [269].

REACTIVITY: Pentachloronitrobenzene is incompatible with strong oxidizers [058], [269]. It is also incompatible with strong bases [269]. It is corrosive to unlined metal containers [058].

STABILITY: Pentachloronitrobenzene is hydrolyzed by alkalis [169]. In one study, ultraviolet radiation has resulted in decomposition [395]. Solutions of it in water, DMSO, 95% ethanol or acetone should be stable for 24 hours under normal lab conditions [700].

Source: *Instant EPA's Air Toxics*, 1994, Instant Reference Sources, Inc., 7605 Rockpoint Drive, Austin, TX 78731 (Fax: 512-345-2386; Internet URL, http://www.instantref.com/inst-ref.htm).

Medical Symptoms of Exposure

Symptoms of exposure may include irritation of the skin and eyes [269]. Skin contact may result in erythema, itching, edema and formation of small vesicles [051], [173]. Skin sensitization may also occur [151]. Eye contact may result in conjunctivitis and corneal injury [173]. Kidney and liver damage may occur [301]. Vomiting may also occur [058].

Exposure to this type of compound can cause central nervous system stimulation, vomiting, diarrhea, paresthesia, excitement, giddiness, fatigue, tremors, convulsions, coma, pulmonary edema, hypothermia and liver, kidney and myocardial toxicity.

Respiration may be initially accelerated and then later depressed. Chronic exposure to this type of compound leads to headache, loss of appetite, muscular weakness, fine tremors and apprehensive mental state [295].

Source: *Instant EPA's Air Toxics*, 1994, Instant Reference Sources, Inc., 7605 Rockpoint Drive, Austin, TX 78731 (Fax: 512-345-2386; Internet URL, http://www.instantref.com/inst-ref.htm).

Toxicological Information

The toxicological information below is gathered from several sources including two national databases: one from the *National Toxicology Program Chemical Repository Database* and another from *EPA's Integrated Risk Information System (IRIS)*.

Toxicology Information from the National Toxicology Program

CARCINOGENICITY:

Tumorigenic Data:

Type of Effect	Route/Animal	Amount Dosed (Notes)
TDLo:	orl-mus	135 gm/kg/77W-C

Review:
IARC Cancer Review: Animal Limited Evidence
IARC: Not classifiable as a human carcinogen (Group 3)

Status:
NCI Carcinogenesis Bioassay (Feed); Negative: Male and Female Rat, Male and Female Mouse
NTP Carcinogenesis Studies (Feed); No Evidence: Male and Female Mouse
EPA Carcinogen Assessment Group

TERATOGENICITY:

Reproductive Effects Data:

Type of Effect	Route/Animal	Amount Dosed (Notes)
TDLo:	orl-mus	5 gm/kg (7-16D preg)
TDLo:	orl-mus	4176 mg/kg (6-14D preg)
TDLo:	orl-mus	4176 mg/kg (6-10D preg)
TDLo:	orl-mus	1935 mg/kg (6-14D preg)

MUTATION DATA:

Test	Lowestdose		Test	Lowestdose
dnd-esc	20umol/L		mmo-asn	5 umol/L
mmo-esc	10mg/plate		sln-asn	17 umol/L
mrc-asn	40umol/L		dnr-sat	1 mg/disc
cyt-ham:ovr	7500ug/L			

TOXICITY:

Typ.dose	Mode	Specie	Amount	Units	Other
LD50	orl	rat	1100	mg/kg	
LC50	ihl	rat	1400	mg/m3	
LD50	orl	mus	1400	mg/kg	
LC50	ihl	mus	2	gm/m3	
LD50	ipr	mus	4500	mg/kg	
LD50	orl	rbt	800	mg/kg	

OTHER TOXICITY DATA:

Review: Toxicology Review-2

Status:
EPA Genetox Program 1988, Negative: Host-mediated assay; In vitro UDS-human fibroblast
EPA Genetox Program 1988, Negative: TRP reversion; S cerevisiae-homozygosis
EPA Genetox Program 1988, Negative/limited: Carcinogenicity-mouse/rat
EPA Genetox Program 1988, Inconclusive: B subtilis rec assay; E colipolA without S9
EPA Genetox Program 1988, Inconclusive: Histidine reversion-Ames test
EPA Genetox Program 1988, Inconclusive: D melanogaster Sex-linked lethal
EPA TSCA Chemical Inventory, 1986
EPA TSCA Test Submission (TSCATS) Data Base, January 1989
Meets criteria for proposed OSHA Medical Records Rule

SAX TOXICITY EVALUATION:

THR: An experimental carcinogen and neoplastigen. Moderately toxic by ingestion and possibly other routes. Experimental reproductive effects. Mutagenic data.

Source: *Instant Tox-Base*, 1995, Instant Reference Sources, Inc., 7605 Rockpoint Drive, Austin, TX 78731 (Fax: 512-345-2386; Internet URL, http://www.instantref.com/inst-ref.htm).

Toxicology Information from EPA's Integrated Risk Information System (IRIS)

I. CHRONIC HEALTH HAZARD ASSESSMENTS FOR NONCARCINOGENIC EFFECTS

I.A. REFERENCE DOSE FOR CHRONIC ORAL EXPOSURE (RfD)

The Reference Dose (RfD) is based on the assumption that thresholds exist for certain toxic effects such as cellular necrosis, but may not exist for other toxic effects such as carcinogenicity. In general, the RfD is an estimate (with uncertainty spanning perhaps an order of magnitude) of a daily exposure to the human population (including sensitive subgroups) that is likely to be without an appreciable risk of deleterious effects during a lifetime. Please refer to Background Document 1 in Service Code 5 for an elaboration of these concepts. RfDs can also be derived for the noncarcinogenic health effects of compounds which are also carcinogens. Therefore, it is essential to refer to other sources of information concerning the carcinogenicity of this substance. If the US EPA has evaluated this substance for potential human carcinogenicity, a summary of that evaluation will be contained in Section II of this file when a review of that evaluation is completed.

I.A.1. ORAL RfD SUMMARY

Critical Effect	Experimental Doses*	UF	MF	RfD
Liver toxicity	NOEL: 30 ppm (0.75 mg/kg/day)	300	1	3E-3mg/kg/day
	LEL: 180 ppm (4.5 mg/kg/day)			

*Conversion Factors: 1 ppm = 0.025 mg/kg/day (assumed dog food consumption)

2-Year Dog Feeding Study Olin Mathieson Corp., 1968a

I.A.2. PRINCIPAL AND SUPPORTING STUDIES (ORAL RfD)

Olin Mathieson Chemical Corporation. 1968a. MRID No. 00001667, 00083328, 00114201; HED Doc. No. 005188. Available from EPA. Write to FOI, EPA, Washington, DC 20460.

A 2-year feeding study with dogs (four males and four females/group) given diets containing 0, 30, 180, or 1080 ppm indicated that PCNB (1.4% hexachlorobenzene) caused liver weight increases, increased liver-to-body weight ratios, elevated serum alkaline phosphatase levels, and microscopically observed cholestatic hepatosis with secondary bile nephrosis at 1080 ppm (the highest dose tested). An interim sacrifice at 1 year occurred with one dog/sex/group; the remaining animals were sacrificed at 2 years. The cholestatic changes were observed in all animals given diets containing 180 and 1080 ppm PCNB, and one of three male dogs in the 30 ppm dose group exhibited the microscopic changes (no female dogs were affected). The authors noted that these histopathologic changes were moderate in the 1080 ppm group and minimal in the 180 ppm group. Based on these results, 30 ppm was the NOEL and 180 ppm was the LEL in dogs.

I.A.3. UNCERTAINTY AND MODIFYING FACTORS (ORAL RfD)

UF -- An uncertainty factor of 100 was used to account for the inter- and intraspecies differences. An additional UF of 3 was used since the data base for chronic toxicity is incomplete.

MF -- None

I.A.4. ADDITIONAL COMMENTS (ORAL RfD)

Data Considered for Establishing the RfD:

1) 2-Year Feeding - dog: Principal study - see previous description; core grade minimum

2) 3-Generation Reproduction - rat: NOEL=500 ppm (25 mg/kg/day) (highest level tested); LEL=none; core grade minimum (Olin Mathieson Corp., 1968b)

3) 3-Month feeding - mouse: NOEL=1250 ppm (187.5 mg/kg/day) for males and 2500 ppm (375 mg/kg/day) for females; core grade minimum (NTP, 1986)

Data Gap(s): Chronic Rat Feeding Study; Rat Teratology; Rabbit Teratology

I.A.5. CONFIDENCE IN THE ORAL RfD

Study -- Medium Data Base -- Medium RfD -- Medium

The principal study appears to be of fair quality and is given a medium confidence rating. Because of the lack of a complete data base on chronic toxicity, the data base is given a medium confidence rating. Medium confidence in the RfD follows.

I.A.6. EPA DOCUMENTATION AND REVIEW OF THE ORAL RfD

Pesticide Registration Standard, July 1986

Pesticide Registration Files

Agency Work Group Review -- 05/20/85, 08/19/86, 04/15/87

Verification Date -- 04/15/87

I.A.7. EPA CONTACTS (ORAL RfD)

William Burnam / OPP -- (703)557-7491

George Ghali / OPP -- (703)557-7490

I.B. REFERENCE CONCENTRATION FOR CHRONIC INHALATION EXPOSURE (RfC)

[Ed. Note: EPA does not have information yet in this section of IRIS].

II. CARCINOGENICITY ASSESSMENT FOR LIFETIME EXPOSURE

This substance/agent has been evaluated by the US EPA for evidence of human carcinogenic potential. This does not imply that this agent is necessarily a carcinogen. The evaluation for this chemical is under review by an inter-office Agency work group. A risk assessment summary will be included on IRIS when the review has been completed.

III. HEALTH HAZARD ASSESSMENTS FOR VARIED EXPOSURE DURATIONS

III.A. DRINKING WATER HEALTH ADVISORIES

[Ed. Note: EPA does not have information yet in this section of IRIS].

IV. US EPA REGULATORY ACTIONS

EPA risk assessments may be updated as new data are published and as assessment methodologies evolve. Regulatory actions are frequently not updated at the same time. Compare the dates for the regulatory actions in this section with the verification dates for the risk assessments in sections I and II, as this may explain inconsistencies. Also note that some regulatory actions consider factors not related to health risk, such as technical or economic feasibility. Such considerations are indicated for each action. In addition, not all of the regulatory actions listed in this section involve enforceable federal standards. Please direct any questions you may have concerning these regulatory actions to the US EPA contact listed for that particular action. Users are strongly urged to read the background information on each regulatory action in Background Document 4 in Service Code 5.

IV.A. CLEAN AIR ACT (CAA)

[Ed. Note: EPA does not have information yet in this section of IRIS].

IV.B. SAFE DRINKING WATER ACT (SDWA)

[Ed. Note: EPA does not have information yet in this section of IRIS].

IV.C. CLEAN WATER ACT (CWA)

IV.C.1. AMBIENT WATER QUALITY CRITERIA, Human Health

[Ed. Note: EPA does not have information yet in this section of IRIS].

IV.C.2. AMBIENT WATER QUALITY CRITERIA, Aquatic Organisms

Freshwater:
    Acute LEC -- 2.5E+2 ug/L
    Chronic LEC -- 5.0E+1 ug/L

Marine:
    Acute LEC -- 1.6E+2 ug/L
    Chronic LEC -- 1.29E+2 ug/L

Considers technological or economic feasibility? -- NO

Discussion -- The values that are indicated as "LEC" are not criteria, but are the lowest effect levels found in the literature. LECs are given when the minimum data required to derive water quality criteria are not available. The values given represent chlorinated benzenes as a class.

Reference -- 45 FR 79318 (11/28/80)

EPA Contact -- Criteria and Standards Division / OWRS(202)260-1315 / FTS 260-1315

IV.D. FEDERAL INSECTICIDE, FUNGICIDE, AND RODENTICIDE ACT (FIFRA)

IV.D.1. PESTICIDE ACTIVE INGREDIENT, Registration Standard

Status -- Issued (1987)

Reference -- Pentachloronitrobenzene Pesticide Registration Standard. January, 1987 (NTIS No. PB88-101126).

EPA Contact -- Registration Branch / OPP(703)557-7760 / FTS 557-7760

IV.D.2. PESTICIDE ACTIVE INGREDIENT, Special Review

Action -- Special Review Terminated (1982)

Considers technological or economic feasibility? -- NO

Summary of regulatory action -- Special review was terminated through negotiated agreement with registrants to reduce levels of the HCB contaminant and make label changes to reduce exposure. Criterion of concern: oncogenicity.

Reference -- 47 FR 18177 (04/19/82)

EPA Contact -- Special Review Branch / OPP(703)557-7400 / FTS 557-7400

IV.E. TOXIC SUBSTANCES CONTROL ACT (TSCA)

[Ed. Note: EPA does not have information yet in this section of IRIS].

IV.F. RESOURCE CONSERVATION AND RECOVERY ACT (RCRA)

IV.F.1. RCRA APPENDIX IX, for Ground Water Monitoring

Status -- Listed

Reference -- 52 FR 25942 (07/09/87)

EPA Contact -- RCRA/Superfund Hotline(800)424-9346 / (202)260-3000 / FTS 260-3000

IV.G. SUPERFUND (CERCLA)

IV.G.1. REPORTABLE QUANTITY (RQ) for Release into the Environment

Value (status) -- 100 pounds (Final, 1989)

Considers technological or economic feasibility? -- NO

Discussion -- The RQ is based on potential carcinogenicity. Available data indicate a hazard ranking of low based on a potency factor of 1.42/mg/kg/day and assignment to weight-of-evidence group C. This corresponds to an RQ of 100 pounds.

Reference -- 54 FR 33418 (08/14/89)

EPA Contact -- RCRA/Superfund Hotline (800)424-9346 / (202)260-3000 / FTS 260-3000

VI. BIBLIOGRAPHY

VI.A. ORAL RfD REFERENCES

NTP (National Toxicology Program). 1986. Technical Report on the Toxicology and Carcinogenesis Studies of Pentachloronitrobenzene in B6C3F1 Mice (feed studies). Report No. NIH 86-2581.

Olin Mathieson Chemical Corporation. 1968a. MRID No. 00001667, 00083328, 00114201; HED Doc. No. 005188. Available from EPA. Write to FOI, EPA, Washington, DC 20460.

Olin Mathieson Chemical Corporation. 1968b. MRID No. 00001666; HED Doc. No. 000654. Available from EPA. Write to FOI, EPA, Washington, DC 20460.

VI.B. INHALATION RfD REFERENCES

None

VI.C. CARCINOGENICITY ASSESSMENT REFERENCES

None

VI.D. DRINKING WATER HA REFERENCES

None

VII. REVISION HISTORY

```
--------  --------  ---------------------------------------------------------
Date      Section   Description
--------  --------  ---------------------------------------------------------
01/01/92  I.A.4.    Citations added
01/01/92  IV.       Regulatory actions updated
01/01/92  VI.       Bibliography on-line
04/01/92  VI.A.     NTP reference year corrected
```

SYNONYMS

AVICOL	PENTACHLORNITROBENZOL
BATRILEX	Pentachloronitrobenzene
BOTRILEX	PENTAGEN
BRASSICOL	PKhNB
EARTHCIDE	QUINTOCENE
FARTOX	QUINTOZEN
FOLOSAN	QUINTOZENE
FOMAC 2	RCRA WASTE NUMBER U185
FUNGICLOR	SANICLOR 30
GC 3944-3-4	TERRACHLOR
KOBU	TERRACLOR
KOBUTOL	TERRAFUN
KP 2	TILCAREX
NCI-C00419	TRI-PCNB
OLPISAN	UN 1282
PCNB	

Source: *Instant EPA's IRIS*, 1997, Instant Reference Sources, Inc., 7605 Rockpoint Drive, Austin, TX 78731 (Fax: 512-345-2386; Internet URL, http://www.instantref.com/inst-ref.htm).

Production, Use, and Pesticide Labeling Information

Pentachloronitrobenzene is not currently listed in EPA's Pesticide Factsheet database and there is no indication of whether it will be added or not in the future.

Phenantharene

Introduction

Phenanthrene is an ingredient in coke oven emissions. It is used in the synthesis of dyestuffs, explosives, and as an intermediate in many organic synthetic reactions.

Release of phenanthrene most likely results from the incomplete combustion of a variety of organic compounds including wood and fossil fuels. Release to the soil will likely result in biodegradation. Volatilization is not expected to be significant. Phenanthrene is expected to bind strongly to soil and not leach extensively to groundwater. When released to water, adsorption of phenanthrene to suspended sediments is expected to remove most of the compound from solution. Photolysis is expected to occur near the water surface and biodegradation of phenanthrene in the water column is expected. Oxidation, volatilization and bioconcentration are not expected to be significant. Phenanthrene released to the atmosphere is expected to rapidly adsorb to particulate matter. Phenanthrene adsorbed on fly ash has been shown to photolyze rapidly (half-life 49 hr) and phenanthrene adsorbed on particulate matter will be subject to wet and dry deposition. Vapor phase phenanthrene will react with photochemically generated, atmospheric hydroxyl radicals with an estimated half-life of 1.67 days. Phenanthrene is a contaminant in air, water, sediment, soil, fish and other aquatic organisms and food. Human exposure results primarily from ingestion of food contaminated with phenanthrene. [828]

Identifying Information

CAS NUMBER: 85-01-8

NIOSH Registry Number: SF7175000

SYNONYNMS: Phenanthren

Endocrine Disruptor Information

Phenanthrene is on EPA's list of suspected EEDs [811] and it is an aromatic coplanar compound. However, no additional information has been located with respect to its endocrine modifying properties.

Chemical and Physical Properties

CHEMICAL FORMULA: $C_{14}H_{10}$

1004

MOLECULAR WEIGHT: 178.24 [703]

PHYSICAL DESCRIPTION: Solid or monoclinic crystals [703]

DENSITY: 1.179 @ 25°C [703]

MELTING POINT: 100°C [703]

BOILING POINT: 339°C [703]

SOLUBILITY: Water: Insoluble [703]; Hot alcohol: Soluble [703]; Ether: Very soluble [703]; Carbon Disulfide: Soluble [703]; Benzene: Soluble [703]

OTHER PHYSICAL DATA: Vapor Pressure: 1 mm @ 118.3°C [703]; Vapor Density: 6.14 [703]

Source: *Instant EPA's Air Toxics*, 1994, Instant Reference Sources, Inc., 7605 Rockpoint Drive, Austin, TX 78731 (Fax: 512-345-2386; Internet URL, http://www.instantref.com/inst-ref.htm).

Environmental Reference Materials

A source of EED environmental reference materials for phenantharene is Radian International LLC, Austin, TX (Fax 512-454-0268; Internet URL, http://www.radian.com/standards); Catalog No. ERP-003.

Hazardous Properties

ACUTE/CHRONIC HAZARDS: Phenanthrene is an irritant and, when heated to decomposition it, emits acrid smoke and fumes.

VOLATILITY:
>   Vapor pressure: 1 mm Hg @ 20°C
>   Vapor density: 6.14

FIRE HAZARD: Phenanthrene has a flash point of 171°C (340°F); it is combustible. Fires involving this compound can be controlled using a dry chemical carbon dioxide or Halon extinguisher; a water spray can also be used.

REACTIVITY: Phenanthrene may react with oxidizing materials.

STABILITY: Phenanthrene is stable under normal laboratory conditions.

Source: *Instant EPA's Air Toxics*, 1994, Instant Reference Sources, Inc., 7605 Rockpoint Drive, Austin, TX 78731 (Fax: 512-345-2386; Internet URL, http://www.instantref.com/inst-ref.htm).

Medical Symptoms of Exposure

Symptoms of exposure may include skin sensitization, dermatitis, bronchitis, cough, dyspnea, respiratory neoplasm, kidney neoplasm, skin irritation, and respiratory irritation.

Source: *Instant EPA's Air Toxics*, 1994, Instant Reference Sources, Inc., 7605 Rockpoint Drive, Austin, TX 78731 (Fax: 512-345-2386; Internet URL, http://www.instantref.com/inst-ref.htm).

Toxicological Information

The toxicological information below is gathered from several sources including two national databases: one from the *National Toxicology Program Chemical Repository Database* and another from *EPA's Integrated Risk Information System (IRIS)*.

Toxicology Information from the National Toxicology Program

CARCINOGENICITY:

Tumorigenic Data:

Type of Effect	Route/Animal	Amount Dosed (Notes)
TDLo:	skn-mus	71 mg/kg
TD:	skn-mus	22 gm/kg/10W-I

Review:
  IARC Cancer Review: Animal Inadequate Evidence
  IARC: Not classifiable as a human carcinogen

TERATOGENICITY: Not available

MUTAGENICITY:

Mutation Data:

Test	Lowest Dose	Test	Lowest Dose
mma-sat	100 ug/plate	cyt-ham:lng	40 mg/L/27H
dnd-sal:spr	3 gm/L	sce-ham-ipr	900 mg/kg/24H
dnd-sal:tes	5 ug/1H-C	sce-ham:fbr	10 umol/L
dnd-ham:fbr	5 mg/L/24H	msc-hmn:lym	100 umol/L
dnd-ham:kdy	5 mg/L	dnd-rat:lvr	3 mmol/L

TOXICITY:

Typ.Dose	Mode	Specie	Amount	Unit	Other
LD50	orl	mus	700	mg/kg	
LD50	ivn	mus	56	mg/kg	

OTHER TOXICITY DATA:

    Status:

        "NIOSH Manual of Analytical Methods" Vol 1 206
        Reported in EPA TSCA Inventory, 1980
        EPA TSCA 8(a) Preliminary Assessment Information Proposed Rule
        EPA Genetic Toxicology Program, January 1984
        Meets criteria for proposed OSHA Medical Records Rule

SAX TOXICITY EVALUATION:

    THR: MUTATION data. An experimental neoplastigen and equivocal tumorigenic agent. HIGH via intravenous routes. MODERATE via oral routes. A human skin photosensitizer. A slight fire hazard.

Source: *Instant Tox-Base*, 1995, Instant Reference Sources, Inc., 7605 Rockpoint Drive, Austin, TX 78731 (Fax: 512-345-2386; Internet URL, http://www.instantref.com/inst-ref.htm).

Toxicology Information from EPA's Integrated Risk Information System (IRIS)

I. CHRONIC HEALTH HAZARD ASSESSMENTS FOR NONCARCINOGENIC EFFECTS

I.A. REFERENCE DOSE FOR CHRONIC ORAL EXPOSURE (RfD)

[Ed. Note: EPA does not have information yet in this section of IRIS].

I.B. REFERENCE CONCENTRATION FOR CHRONIC INHALATION EXPOSURE (RfC)

A risk assessment for this substance/agent is under review by an EPA work group.

II. CARCINOGENICITY ASSESSMENT FOR LIFETIME EXPOSURE Section II provides information on three aspects of the carcinogenic risk assessment for the agent in question; the US EPA classification, and quantitative estimates of risk from oral exposure and from inhalation exposure. The classification reflects a weight-of-evidence judgment of the likelihood that the agent is a human carcinogen. The quantitative risk estimates are presented in three ways. The slope factor is the result of application of a low-dose extrapolation procedure and is presented as the risk per (mg/kg)/day. The unit risk is the quantitative estimate in terms of either risk per ug/L drinking water or risk per ug/cu.m air breathed. The third form in which risk is presented is a drinking water or air concentration providing cancer risks of 1 in 10,000, 1 in 100,000 or 1 in 1,000,000. Background Document 2 (Service Code 5) provides details on the rationale and methods used to derive the carcinogenicity values found in IRIS. Users are referred to Section I for information on long-term toxic effects other than carcinogenicity.

II.A. EVIDENCE FOR CLASSIFICATION AS TO HUMAN CARCINOGENICITY

II.A.1. WEIGHT-OF-EVIDENCE CLASSIFICATION

Classification -- D, not classifiable as to human carcinogenicity

Basis -- Based on no human data and inadequate data from a single gavage study in rats and skin painting and injection studies in mice.

II.A.2. HUMAN CARCINOGENICITY DATA

None.

II.A.3. ANIMAL CARCINOGENICITY DATA

Inadequate. Data from a rat gavage study and mouse skin application and injection studies are not adequate to assess the carcinogenicity of phenanthrene. Ten female Sprague-Dawley rats received a single oral dose of 200 mg phenanthrene in sesame oil (Huggins and Yang, 1962). No mammary tumors occurred. The observation period was not specified; however, based on the discussion of other experiments in the report it was probably at least 60 days. Controls were not reported.

Complete carcinogenic activity was not shown in two skin painting assays. Kennaway (1924) reported no tumors in 100 mice (strain and sex not specified) treated with phenanthrene (purity not specified) in 90% benzene (dose not reported) for 9 months. Roe and Grant (1964) reported in an abstract that mice (number, sex and strain not specified) did not develop tumors after dermal exposure to 5% phenanthrene (purity not specified, vehicle not specified) 3 times/week for 1 year.

Five studies of cancer-initiating activity in skin painting assays in mice have yielded one positive result. Groups of 30 female CD-1 mice received a single dermal application of 1.8 mg phenanthrene in benzene, followed by twice-weekly applications of tetradecanoylphorbol acetate (TPA, 3 mg), a promoter, for 35 weeks (Scribner, 1973). Phenanthrene used in the study was purified by preparative thin-layer chromatography (TLC) and determined to be homogeneous on TLC. It is stated in the report that the dose of TPA was 3 mg (5 umol); however, it is not clear whether this refers to the twice weekly or total dose. Controls were treated with TPA (6 mg); it is not clear whether controls received benzene (vehicle). The tumor incidence (skin papilloma) at 35 weeks was 12/30 (40%) in treated mice and 0/30 in TPA controls.

Tumor-initiating activity was not shown in the four other mouse skin painting studies. In the first study, male Swiss albino (Ha/ICR) mice (15 to 20/group) received 10 applications of a 0.1% solution of phenanthrene in acetone (total dose 1 mg) or acetone alone, followed by repeated applications of TPA (2.5 ug in acetone) 3 times/week for 20 weeks (LaVoie et al., 1981). Phenanthrene was >99.5% pure as determined by high pressure liquid chromatography (HPLC). No tumors occurred in treated or control mice. Wood et al. (1979) exposed female CD-1 mice (30/group) to a single application of 1.8 mg phenanthrene in acetone:ammonium hydroxide (1000:1) or vehicle alone, followed by TPA (10 ug) twice weekly for 35 weeks. Phenanthrene used in this study was >98% pure and homogeneous on HPLC. Tumor incidence (skin papillomas) out of 27-29 survivors in each group was 17% in treated mice and 7% in vehicle controls (not statistically different). In another study, albino mice (10/sex/dose, strain not specified) received four dermal applications of phenanthrene (total dose 1.2 mg, purity not specified) in acetone or to acetone alone, followed by croton oil once each week for 20 weeks (Roe, 1962). Tumor incidence (skin papillomas) was 4/19 (21%) in treated mice and 2/20 (10%) in vehicle controls. In the last study (Salaman and Roe, 1956), groups of 20 "S" strain mice (sex unspecified) received 10 dermal applications (3 times/week) of 18% phenanthrene (total dose 0.54 g, purity not specified) in acetone, followed by 18 weekly applications of croton oil. Controls were treated with 18 applications of croton oil; 10 controls survived until termination. The tumor incidence (skin papillomas) was 5/20 (25%) in treated mice and 4/10 (40%) in croton oil controls.

Parenterally administered phenanthrene was not shown to have tumorigenic activity in three studies. In the first (Buening et al., 1979), groups of Swiss Webster BLU:Ha ICR mice (100/group, approximately 50% of each sex) received intraperitoneal injections of phenanthrene (total dose 0.25 mg) in dimethyl sulfoxide (DMSO) or DMSO alone on days 1, 8, and 15 after birth. Phenanthrene was >98% pure and homogeneous on HPLC. Incidence of pulmonary tumors (adenomas) at 38 to 42 weeks was 1/18 (6%) and 5/17 (30%) in female and male treated mice and 7/38 (18%) and 2/10 (19%) in female and male controls; the apparent differences were not statistically significant. No hepatic tumors occurred in treated or control mice. One treated female mouse developed malignant lymphoma. In the second study (Grant and Roe, 1963), albino mice (sex, strain and group size not specified) received single subcutaneous injections of phenanthrene (40 ug, purity not specified) in an acetone/gelatin vehicle or only the vehicle. Incidence of pulmonary adenomas after 52-62 weeks was 3/39 (6%) in treated mice and 8/34 (24%) in vehicle controls. Other tumors reported were 4 hepatomas and 2 skin papillomas in treated mice, and 1 mammary adenocarcinoma, 1 hepatoma and 1 hemangioma in control mice. Finally in the Steiner (1955) study, groups of 40 to 50 male and female C57BL mice (numbers per sex not specified) received single subcutaneous injections of 5 mg phenanthrene (purity not specified) in tricaprylin. No tumors were reported in 27 surviving mice after 4 months. Vehicle controls were not reported.

II.A.4. SUPPORTING DATA FOR CARCINOGENICITY

Phenanthrene has not yielded positive results in assays for DNA damage in Bacillus subtilis and Escherichia coli (Rosenkrantz and Poirier, 1979; McCarroll et al., 1981). Tests for mutagenicity in Salmonella typhimurium have yielded positive (Oesch et al., 1981; Sakai et al., 1985; Bos et al., 1988) and negative results (Wood et al., 1979; McCann et al., 1975; LaVoie et al., 1981; Kaden et al., 1979; Bos et al., 1988). The results of phenanthrene in a fungi recombination assay (Simmon, 1979) and in tests for DNA damage in several mammalian cell cultures were not positive (Lake et al., 1978;

Probst et al., 1981; Rice et al., 1984). A test for forward mutation in Chinese hamster ovary cells exposed to 1 ug/mL was not positive (Huberman and Sachs, 1976), whereas a test in human lymphoblast TK6 cells incubated with rat liver S9 (Arochlor) and 9 ug/mL phenanthrene yielded positive results (Barfknecht et al., 1981). Phenanthrene did not yield positive results in sister chromatid exchange and chromosome aberration assays in mammalian cell cultures (Popescu et al., 1977) or in cell transformation assays in several types of mammalian cells (5-40 ug/mL) (Marquardt and Heidelberger, 1972; Kakunaga, 1973; Evans and DiPaolo, 1975; Pienta et al., 1977).

Current theories regarding the mechanisms of metabolic activation of polycyclic aromatic hydrocarbons lead to predictions of a carcinogenic potential for phenanthrene. Jerina et al. (1978) considered phenanthrene to have a "bay-region" structure. It is metabolized by mixed function oxidases to reactive diol epoxides (Nordqvist et al., 1981; Vyas et al., 1982) that have been shown to be weakly mutagenic in some bacterial and mammalian cell assays (Wood et al., 1979). Evidence from in vivo assays indicates, however, that phenanthrene metabolites have a relatively low tumorigenic potential. The 1,2-, 3,4- and 9,10-dihydrodiol metabolites of phenanthrene did not show tumor initiating activity in mouse skin painting assays (Wood et al., 1979).   The 1,2-diol-3,4-epoxides of phenanthrene did not produce lung tumors when injected into newborn mice (Buening et al., 1979).   The relatively weak mutagenic and tumorigenic activity of phenanthrene diol epoxides is inconsistent with the "bay region theory" of PAH carcinogenesis. The reason for the inconsistency has not been elucidated. Phenanthrene epoxides have a relatively small molecular size (relative to other more active PAH epoxides such as chrysene diol epoxides) and as a result may have a lower affinity for DNA or may be transported less efficiently into the mammalian nucleus (Wood et al., 1979).   While some studies have considered phenanthrene to have a "bay-region" structure, it may not clearly fall into this category.

II.B.   QUANTITATIVE ESTIMATE OF CARCINOGENIC RISK FROM ORAL EXPOSURE

None.

II.C. QUANTITATIVE ESTIMATE OF RISK FROM INHALATION EXPOSURE

None.

II.D. EPA DOCUMENTATION, REVIEW, AND CONTACTS (CARCINOGENICITY ASSESSMENT)

The 1990 Drinking Water Criteria Document for Polycyclic Aromatic Hydrocarbons (PAHs) has received Agency and external review.

II.D.1. EPA DOCUMENTATION

Source Document -- US EPA, 1990

The 1990 Drinking Water Criteria Document for Polycyclic Aromatic Hydrocarbons has received Agency and external review.

II.D.2. REVIEW (CARCINOGENICITY ASSESSMENT)

Agency Work Group Review -- 02/07/90, 05/03/90

Verification Date -- 05/03/90

II.D.3. US EPA CONTACTS (CARCINOGENICITY ASSESSMENT)

Rita Schoeny / NCEA -- (513)569-7544

Robert McGaughy / NCEA -- (202)260-5889

III. HEALTH HAZARD ASSESSMENTS FOR VARIED EXPOSURE DURATIONS

[Ed. Note: EPA does not have information yet in this section of IRIS].

IV. US EPA REGULATORY ACTIONS

EPA risk assessments may be updated as new data are published and as assessment methodologies evolve. Regulatory actions are frequently not updated at the same time. Compare the dates for the regulatory actions in this section with the verification dates for the risk assessments in sections I and II, as this may explain inconsistencies. Also note that some regulatory actions consider factors not related to health risk, such as technical or economic feasibility. Such considerations are indicated for each action. In addition, not all of the regulatory actions listed in this section involve enforceable federal standards. Please direct any questions you may have concerning these regulatory actions to the US EPA contact listed for that particular action. Users are strongly urged to read the background information on each regulatory action in Background Document 4 in Service Code 5.

IV.A. CLEAN AIR ACT (CAA)

[Ed. Note: EPA does not have information yet in this section of IRIS].

IV.B. SAFE DRINKING WATER ACT (SDWA)

[Ed. Note: EPA does not have information yet in this section of IRIS].

IV.C. CLEAN WATER ACT (CWA)

IV.C.1. AMBIENT WATER QUALITY CRITERIA, Human Health

Water and Fish Consumption: 2.8E-3 ug/L

Fish Consumption Only: 3.11E-2 ug/L

Considers technological or economic feasibility? -- NO

Discussion -- For the maximum protection from the potential carcinogenic properties of this chemical, the ambient water concentration should be zero. However, zero may not be obtainable at this time, so the recommended criteria represents a E-6 estimated incremental increase of cancer over a lifetime. The values given represent polynuclear aromatic hydrocarbons as a class.   Reference -- 45 FR 79318 (11/28/80)

EPA Contact -- Criteria and Standards Division / OWRS(202)260-1315 / FTS 260-1315

IV.C.2. AMBIENT WATER QUALITY CRITERIA, Aquatic Organisms

Freshwater:
      Acute --   3.0E+1 ug/L
      Chronic -- 6.3E+0 ug/L

Marine:
      Acute --   7.7E+0 ug/L
      Chronic -- 4.6E+0 ug/L

Considers technological or economic feasibility? -- NO

Discussion -- Proposed criterion were derived from a minimum database consisting of acute and chronic tests in a variety of species. Requirements and methods are covered in . the reference to the Federal Register.

Reference -- 55 FR 19986 (05/14/90)

EPA Contact -- Criteria and Standards Division / OWRS(202)260-1315 / FTS 260-1315

IV.D. FEDERAL INSECTICIDE, FUNGICIDE, AND RODENTICIDE ACT (FIFRA)

No data available

IV.E. TOXIC SUBSTANCES CONTROL ACT (TSCA)

      [Ed. Note: EPA does not have information yet in this section of IRIS].

IV.F. RESOURCE CONSERVATION AND RECOVERY ACT (RCRA)

IV.F.1. RCRA APPENDIX IX, for Ground Water Monitoring

Status -- Listed
Reference -- 52 FR 25942 (07/09/87)

EPA Contact -- RCRA/Superfund Hotline(800)424-9346 / (202)260-3000 / FTS 260-3000

IV.G. SUPERFUND (CERCLA)

IV.G.1. REPORTABLE QUANTITY (RQ) for Release into the Environment

Value (status) --   5000 pounds (Final, 1989)
Considers technological or economic feasibility? -- NO

Discussion --   The available data for acute hazard (oral LD50 in mice of 700 mg/kg) lies above the upper limit for the 5000-pound RQ, but since it is a designated hazardous substance, the largest assignable RQ is 5000 pounds.

Reference -- 54 FR 33418 (08/14/89)

EPA Contact -- RCRA/Superfund Hotline(800)424-9346 / (202)260-3000 / FTS 260-3000

VI. BIBLIOGRAPHY

VI.A. ORAL RfD REFERENCES

None

VI.B. INHALATION RfC REFERENCES

None

VI.C. CARCINOGENICITY ASSESSMENT REFERENCES

Barfknecht, T.R., B.M. Andon, W.G. Thilly and R.A. Hites. 1981. Soot and mutation in bacteria and human cells. In: Chemical Analysis and Biological Fate: Polynuclear Aromatic Hydrocarbons. 5th Int. Symp., M. Cooke and A.J. Dennis, Ed. Battelle Press, Columbus, OH. p. 231-242.

Bos, R.P., J.L.G. Theuws, F.J. Jongeneelen and P.Th. Henderson. 1988. Mutagenicity of bi-, tri- and tetra-cyclic aromatic hydrocarbons in the "taped-plate assay" and in the conventional Salmonella mutagenicity assay. Mutat. Res. 204: 203-206.

Buening, M.K., W. Levin, J.M. Karle, et al. 1979. Tumorigenicity of bay-region epoxides and other derivatives of chrysene and phenanthrene in newborn mice. Cancer Res. 39: 5063-5068.

Evans, C.H. and J.A. DiPaolo. 1975. Neoplastic transformation of guinea pig fetal cells in culture induced by chemical carcinogens. Cancer Res. 35: 1035-1044.

Grant, G.A. and F.J.C. Roe. 1963. The effect of phenanthrene on tumor induction by 3,4-benzopyrene administered to newly born mice. Br. J. Cancer. 17: 261-265.

Huberman, E. and L. Sachs. 1976. Mutability of different genetic loci in mammalian cells by metabolically activated carcinogenic polycyclic hydrocarbons. Proc. Natl. Acad. Sci. USA. 73(1): 188-192.

Huggins, C. and N.C. Yang. 1962. Induction and extinction of mammarycancer. Science 137(3562): 257-262.

Jerina, D.M., H. Yagi, R.E. Lehr, et al. 1978. The bay-region theory of carcinogenesis by polycyclic aromatic hydrocarbons. In: Polycyclic Hydrocarbons and Cancer, Vol. 1, Environment, Chemistry and Metabolism, H.V. Gelboin and P.O.P. Ts'o, Ed. Academic Press, NY. p. 173-188.

Kaden, D.A., R.A. Hites and W.G. Thilly. 1979. Mutagenicity of soot and associated polycyclic aromatic hydrocarbons to Salmonella typhimurium. Cancer Res. 39: 4152-4159.

Kakunaga, T. 1973. A quantitative system for assay of malignant transformation by chemical carcinogens using a clone derived from BALB/3T3. Int. J. Cancer. 12: 463-473.

Kennaway, E.L. 1924. On cancer-producing tars and tar-fractions. J. Ind. Hyg. 5(12): 462-488.

Lake, R.S., M.L. Kropko, M.R. Pezzutti, R.H. Shoemaker and H.J. Igel. 1978. Chemical induction of unscheduled DNA synthesis in human skin epithelial cell cultures. Cancer Res. 38: 2091-2098.

LaVoie, E.J., L. Tulley-Freiler, V. Bedenko and D. Hoffmann. 1981. Mutagenicity, tumor-initiating activity, and metabolism of methylphenanthrenes. Cancer Res. 41: 3441-3447.

Marquardt, H. and C. Heidelberger. 1972. Influence of "feeder cells" and inducers and inhibitors of microsomal mixed-function oxidases on hydrocarbon-induced malignant transformation of cells derived from C3H mouse prostate. Cancer Res. 32: 721-725.

McCann, J.E., E. Choi, E. Yamasaki and B.N. Ames. 1975. Detection of carcinogens as mutagens in the Salmonella/microsome test: Assay of 300 materials. Proc. Natl. Acad. Sci. USA. 72(12): 5135-5139.

McCarroll, N.E., B.H. Keech and C.E. Piper. 1981. A microsuspension adaptation of the Bacillus subtilis 'rec' assay. Environ. Mutagen. 3:607-616.

Nordqvist, M., D.R. Thakker, K.P. Vyas, et al. 1981. Metabolism of chrysene and phenanthrene to bay-region diol epoxides by rat liver enzymes. Mol. Pharmacol. 19: 168-178.

Oesch, F., M. Bucker and H.R. Glatt. 1981. Activation of phenanthrene to mutagenic metabolites and evidence for at least two different activation pathways. Mutat. Res. 81: 1-10.

Pienta, R.J., J.A. Poiley and W.B. Lebherz, III. 1977. Morphological transformation of early passage golden Syrian hamster embryo cells derived from cryopreserved primary cultures as a reliable in vitro bioassay for identifying diverse carcinogens. Int. J. Cancer. 19: 642-655.

Popescu, N.C., D. Turnbull and J.A. DiPaolo. 1977. Sister chromatid exchange and chromosome aberration analysis with the use of several carcinogens and noncarcinogens: Brief communication. J. Natl. Cancer Inst. 59(1): 289-293.

Probst, G.S., R.E. McMahon, L.E. Hill, C.Z. Thompson, J.K. Epp and S.B. Neal. 1981. Chemically-induced unscheduled DNA synthesis in primary rat hepatocyte cultures: A comparison with bacterial mutagenicity using 218 compounds. Environ. Mutagen. 3: 11-32.

Rice, J.E., T.J. Hosted, Jr. and E.L. LaVoie. 1984. Fluoranthene and pyrene enhance benzo[a]pyrene-DNA adduct formation in vivo in mouse skin. Cancer Lett. 24: 327-333.

Roe, F.J.C. 1962. Effect of phenanthrene on tumor-initiation by 3,4-benzopyrene. Br. J. Cancer. 16: 503-506

Roe, F.J.C. and G.A. Grant. 1964. Tests of pyrene and phenanthrene forincomplete carcinogenic and anticarcinogenic activity. Br. Emp. Cancer Campaign. 41: 59-69. (Abstract)

Rosenkrantz, H.S. and L.A. Poirier. 1979. Evaluation of the mutagenicity and DNA-modifying activity of carcinogens and noncarcinogens in microbial systems. J. Natl. Cancer Inst. 62(4): 873-892.

Sakai, M., D. Yoshida and S. Mizusaki. 1985. Mutagenicity of polycyclic aromatic hydrocarbons and quinones on Salmonella typhimurium TA97. Mutat. Res. 156: 61-67.

Salaman, M.H. and F.J.C. Roe. 1956. Further tests for tumor-initiating activity: N,N-di-(2-chloroethyl)-p-aminophenylbutyric acid (CB1348) as an initiator of skin tumor formation in the mouse. Br. J. Cancer. 10: 363-378.

Scribner, J.D. 1973. Brief Communication: Tumor initiation by apparently noncarcinogenic polycyclic aromatic hydrocarbons. J. Natl. Cancer Inst. 50(6): 1717-19.

Simmon, V.F. 1979. In vitro assays of recombinogenic activity of chemical carcinogens and related compounds with Saccharomyces cerevisiae D3. J. Natl. Cancer Inst. 62(4): 901-909.

Steiner, P.E. 1955. Carcinogenicity of multiple chemicals simultaneously administered. Cancer Res. 15: 632-635.

Vyas, K.P., D.R. Thakker, W. Levin, H. Yagi, A.H. Conney and D.M. Jerina. 1982. Stereoselective metabolism of the optical isomers of trans- 1,2-dihydroxy-1,2-dihydrophenanthrene to bay-region diol epoxides by rat liver microsomes. Chem-biol. Interactions. 38: 203-213.

US EPA. 1990. Drinking Water Criteria Document for Polycyclic Aromatic Hydrocarbons (PAHs). Prepared by the Office of Health and Environmental Assessment, Environmental Criteria and Assessment Office, Cincinnati, OH for the Office of Drinking Water, Washington, DC. Final Draft. ECAO-CIN-D010, September, 1990.

Wood, A.W., R.L. Chang, W. Levin, et al. 1979. Mutagenicity and tumorigenicity of phenanthrene and chrysene epoxides and diol epoxides. Cancer Res. 39: 4069-4077.

VI.D. DRINKING WATER HA REFERENCES

None

VII. REVISION HISTORY

```
--------  --------  ---------------------------------------------------------
Date      Section   Description
--------  --------  ---------------------------------------------------------
12/01/90  II.       Carcinogen assessment on-line
12/01/90  VI.       Bibliography on-line
01/01/92  IV.       Regulatory Action section on-line
07/01/93  VI.C.     References clarified
09/01/94  I.B.      Inhalation RfC now under review
```

SYNONYMS

Phenanthrene	Phenanthren [German]
HSDB 2166	Phenanthrene
NSC 26256	

Source: *Instant EPA's IRIS*, 1997, Instant Reference Sources, Inc., 7605 Rockpoint Drive, Austin, TX 78731 (Fax: 512-345-2386; Internet URL, http://www.instantref.com/inst-ref.htm).

Pyrene

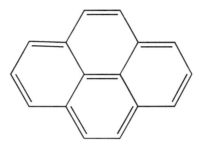

Introduction

Pyrene occurs in coal tar [031]. It is used in biochemical research and as an intermediate in the synthesis of 3,4-benzpyrene and many other polyaromatic hydrocarbons.

Pyrene's release to the environment is ubiquitous since it is a ubiquitous product of incomplete combustion. It is largely associated with particulate matter, soils and sediments. Although environmental concentrations are highest near sources, its presence in places distant from primary sources indicates that it is reasonably stable in the atmosphere and capable of long distance transport. When released to air it may be subject to direct photolysis, although adsorption to particulates apparently can retard this process. [828]

If released to water, it will adsorb very strongly to sediments and particulate matter, bioconcentrate in aquatic organisms slightly to moderately, but will not hydrolyze. It may be subject to significant biodegradation, and direct photolysis may be important near the surface of waters. Evaporation may be important with a half-life of 4.8 to 39.2 days predicted for evaporation from a river 1 m deep, flowing at 1 m/sec with a wind velocity of 3 m/sec; half-life for evaporation from a model pond was 1176 days. Adsorption to sediments and particulates will limit evaporation. [828]

If released to soil it will be expected to adsorb very strongly to the soil and will not be expected to appreciably leach to the groundwater, although its presence in groundwater illustrates that it can be transported there. It will not be expected to hydrolyze or significantly evaporate from soils and surfaces. It may be subject to appreciable biodegradation in soils. Human exposure will be from inhalation of contaminated air and consumption of contaminated food and water. Especially high exposure will occur through the smoking of cigarettes and the ingestion of certain foods (eg, smoked and charcoal-broiled meats and fish). [828]

Identifying Information

CAS NUMBER: 129-00-0

NIOSH Registry Number: UR2450000

SYNONYMS: Benzo(def)phenanthrene

Endocrine Disruptor Information

Pyrene is on EPA's list of suspected EEDs [811] and it is an aromatic coplanar compound. However, no additional information has been located with respect to its endocrine modifying properties.

Chemical and Physical Properties

CHEMICAL FORMULA: $C_{16}H_{10}$

MOLECULAR WEIGHT: 202.26 [703]

PHYSICAL DESCRIPTION: Colorless solid, solutions have a slight blue color [703]; Tetracene impurities sometimes cause pyrene crystals to have a yellow color.

DENSITY: 1.271 @ 23°C [703]

MELTING POINT: 156°C [703]

BOILING POINT: 404°C [703]

SOLUBILITY: Water: Insoluble [703]; Organic solvents: Fairly soluble [703]

OTHER PHYSICAL DATA: Melting Point: 151.2°C [706]

Source: *Instant EPA's Air Toxics*, 1994, Instant Reference Sources, Inc., 7605 Rockpoint Drive, Austin, TX 78731 (Fax: 512-345-2386; Internet URL, http://www.instantref.com/inst-ref.htm).

Environmental Reference Materials

A source of EED environmental reference materials for pyrene is Radian International LLC, Austin, TX (Fax 512-454-0268; Internet URL, http://www.radian.com/standards); Catalog No. ERP-004.

Hazardous Properties

ACUTE/CHRONIC HAZARDS: Pyrene can be absorbed through the skin [062]. When heated to decomposition it emits acrid smoke and fumes [042].

VOLATILITY:
        Vapor pressure: 2.60 mm Hg @ 200.4°C; 6.90 mm Hg @ 220.8°C [039]

FIRE HAZARD: Flash point data for pyrene are not available. It is probably combustible. Fires involving this material may be controlled with a dry chemical, carbon dioxide or Halon extinguisher.

REACTIVITY: Pyrene reacts with nitrogen oxides to form nitro derivatives. It also reacts with 70% nitric acid [395].

STABILITY: Pyrene is stable under normal laboratory conditions. Solutions of it in water, DMSO, 95% ethanol or acetone should be stable for 24 hours under normal lab conditions [700].

Source: *Instant EPA's Air Toxics*, 1994, Instant Reference Sources, Inc., 7605 Rockpoint Drive, Austin, TX 78731 (Fax: 512-345-2386; Internet URL, http://www.instantref.com/inst-ref.htm).

Medical Symptoms of Exposure

In rats, inhalation has caused hepatic, pulmonary and intragastric pathologic changes. Cutaneous absorption for 10 days has caused hyperemia, weight loss and hematopoietic changes. Applications for 30 days have produced dermatitis.

Source: *Instant EPA's Air Toxics*, 1994, Instant Reference Sources, Inc., 7605 Rockpoint Drive, Austin, TX 78731 (Fax: 512-345-2386; Internet URL, http://www.instantref.com/inst-ref.htm).

Toxicological Information

The toxicological information below is gathered from several sources including two national databases: one from the *National Toxicology Program Chemical Repository Database* and another from *EPA's Integrated Risk Information System (IRIS)*.

Toxicology Information from the National Toxicology Program

CARCINOGENICITY:

Tumorigenic Data:

Type of Effect	Route/Animal	Amount Dosed (Notes)
TDLo:	skn-mus	10 gm/kg/3 W-I

Review:
IARC Cancer Review: Animal Inadequate Evidence
IARC: Not classifiable as a human carcinogen (Group 3)

TERATOGENICITY: Not available

MUTATION DATA:

Test	Lowest dose		Test	Lowest dose
mmo-sat	5 ug/plate	|	msc-hmn:oth	12 umol/L
mma-sat	300 ng/plate	|	dns-rat:lvr	2500 umol/L
mma-esc	1 ug/plate	|	msc-rat:emb	10 mg/L
dnd-esc	10 umol/L	|	otr-ham:emb	10 mg/L
dnd-nml	8600 nmol/L	|	otr-ham:kdy	62500 ug/L
dns-hmn:fbr	100 mg/L	|	dns-ham:lvr	100 umol/L
dnd-sal:tes	5 ug/1H-C	|	cyt-ham:lng	10 mg/L
oms-esc	50 nmol/tube	|		

TOXICITY:

Typ.Dose	Mode	specie	amount	units	other
LD50	orl	rat	2700	mg/kg	
LC50	ihl	rat	170	mg/m3	
LD50	orl	mus	800	mg/kg	
LD50	ipr	mus	514	mg/kg	

OTHER TOXICITY DATA:

Skin and Eye Irritation Data: skn-rbt 500 mg/24H MILD
Status:

EPA TSCA Chemical Inventory, 1986

EPA Genetox Program 1986, Inconclusive: Carcinogenicity-mouse/rat; Mammalian micronucleus

EPA Genetox Program 1986, Negative: Cell transform.-BALB/c-3T3; SHE-clonal assay

EPA Genetox Program 1986, Negative: Cell transform.-mouse embryo

EPA Genetox Program 1986, Negative: Cell transform.-RLV F344 rat embryo

EPA Genetox Program 1986, Negative: In vitro cytogenetics-nonhuman; Host-mediated assay

EPA Genetox Program 1986, Negative: Histidine reversion-Ames test; Invivo SCE-nonhuman

EPA Genetox Program 1986, Negative: Sperm morphology-mouse; In vitro UDS-human   fibroblast

EPA Genetox Program 1986, Negative: V79 cell culture-gene mutation

EPA Genetox Program 1986, Inconclusive: In vitro SCE-nonhuman

EPA TSCA Test Submission (TSCATS) Data Base, March 1988

NIOSH Analytical Methods: see PAHs (HPLC)

Meets criteria for proposed OSHA Medical Records Rule

SAX TOXICITY EVALUATION:

> THR: MUTATION data. A skin irritant. An experimental equivocal
> tumorigenic agent.

Source: ***Instant Tox-Base***, 1995, Instant Reference Sources, Inc., 7605 Rockpoint Drive, Austin, TX 78731 (Fax: 512-345-2386; Internet URL, http://www.instantref.com/inst-ref.htm).

Toxicology Information from EPA's Integrated Risk Information System (IRIS)

I. CHRONIC HEALTH HAZARD ASSESSMENTS FOR NONCARCINOGENIC EFFECTS

I.A. REFERENCE DOSE FOR CHRONIC ORAL EXPOSURE (RfD)

The Reference Dose (RfD) is based on the assumption that thresholds exist for certain toxic effects such as cellular necrosis, but may not exist for other toxic effects such as carcinogenicity. In general, the RfD is an estimate (with uncertainty spanning perhaps an order of magnitude) of a daily exposure to the human population (including sensitive subgroups) that is likely to be without an appreciable risk of deleterious effects during a lifetime. Please refer to Background Document 1 in Service Code 5 for an elaboration of these concepts. RfDs can also be derived for the noncarcinogenic health effects of compounds which are also carcinogens. Therefore, it is essential to refer to other sources of information concerning the carcinogenicity of this substance. If the US EPA has evaluated this substance for potential human carcinogenicity, a summary of that evaluation will be contained in Section II of this file when a review of that evaluation is completed.

I.A.1. ORAL RfD SUMMARY

Critical Effect	Experimental Doses*	UF	MF	RfD
Kidney effects (renal tubular pathology, decreased kidney weights)	NOAEL: 75 mg/kg/day	3000	1	3E-2 mg/kg/day
	LOAEL: 125 mg/kg/day			

*Conversion Factors: None

Mouse Subchronic Oral Bioassay, US EPA, 1989

I.A.2. PRINCIPAL AND SUPPORTING STUDIES (ORAL RfD)

US EPA. 1989. Mouse Oral Subchronic Toxicity of Pyrene. Study conducted by Toxicity Research Laboratories, Muskegon, MI for the Office of Solid Waste, Washington, DC.

Male and female CD-1 mice (20/sex/group) were gavaged with 0, 75, 125, or 250 mg/kg/day pyrene in corn oil for 13 weeks. The toxicological parameters examined in this study included body weight changes, food consumption, mortality, clinical pathological evaluations of major organs and tissues, and hematology and serum chemistry. Nephropathy, characterized by the presence of multiple foci of renal tubular regeneration, often accompanied by interstitial lymphocytic infiltrates and/or foci of interstitial fibrosis, was present in 4, 1, 1, and 9 male mice in the control, low-, medium-, and high-dose groups, respectively. Similar lesions were seen in 2, 3, 7, and 10 female mice in the 0, 75, 125, and 250 mg/kg treatment groups. The kidney lesions were described as minimal or mild in all dose groups. Relative and absolute kidney weights were reduced in the two higher dosage groups. Based on the results of this study, the low dose (75 mg/kg/day) was considered the NOAEL and 125 mg/kg/day the LOAEL for nephropathy and decreased kidney weights.

I.A.3. UNCERTAINTY AND MODIFYING FACTORS (ORAL RfD)

UF -- An uncertainty factor of 3000 reflects 10 each for intra- and interspecies variability, 10 for the use of a subchronic study for chronic RfD derivation, and an additional 3 to account for the lack of both toxicity studies in a second species and developmental/reproductive studies.

MF -- None

I.A.4. ADDITIONAL COMMENTS (ORAL RfD)

White and White (1939) fed six male rats (unspecified strain) a diet containing 2000 mg pyrene/kg for 40 days. The average reported food intake for two animals was 6.1 g/day, and the average body weight for these two animals was 94.3 g. A decrease in body weight gain was observed in two animals. The authors stated that this body weight gain was representative of the whole group; although there was no change in food intake. White and White (1939) also observed enlarged livers and increased hepatic lipid content in animals treated with pyrene, benzpyrene or methylcholanthrene in the diet; however, incidence data were not reported and it is unclear whether this effect occurred in the pyrene treated rats. Interpretation of this study is further complicated by the lack of experimental controls and statistical analysis, small sample size, and incomplete reporting of histopathology results.

I.A.5. CONFIDENCE IN THE ORAL RfD

Confidence in the principal study is medium, as it was a well-designed experiment that examined a variety of toxicological endpoints and identified both a NOAEL and LOAEL for the critical effect. Confidence in the data base is low, due to the lack of supporting subchronic, chronic, and developmental/reproductive studies. Accordingly, confidence in the RfD is low.

I.A.6. EPA DOCUMENTATION AND REVIEW OF THE ORAL RfD

Source Document -- This assessment is not presented in any existing US EPA document.

Other EPA Documentation -- US EPA, 1989

Agency Work Group Review -- 11/15/89

Verification Date -- 11/15/89

I.A.7. EPA CONTACTS (ORAL RfD)

Harlal Choudhury / NCEA -- (513)569-7536

Kenneth A. Poirier / OHEA -- (513)569-7553

I.B. REFERENCE CONCENTRATION FOR CHRONIC INHALATION EXPOSURE (RfC)

A risk assessment for this substance/agent is under review by an EPA work group.

II. CARCINOGENICITY ASSESSMENT FOR LIFETIME EXPOSURE

Section II provides information on three aspects of the carcinogenic risk assessment for the agent in question; the US EPA classification, and quantitative estimates of risk from oral exposure and from inhalation exposure. The classification reflects a weight-of-evidence judgment of the likelihood that the agent is a human carcinogen. The quantitative risk estimates are presented in three ways. The slope factor is the result of application of a low-dose extrapolation procedure and is presented as the risk per (mg/kg)/day. The unit risk is the quantitative estimate in terms of either risk per ug/L drinking water or risk per ug/cu.m air breathed. The third form in which risk is presented is a drinking water or air concentration providing cancer risks of 1 in 10,000, 1 in 100,000 or 1 in 1,000,000. Background Document 2 (Service Code 5) provides details on the rationale and methods used to derive the carcinogenicity values found in IRIS. Users are referred to Section I for information on long-term toxic effects other than carcinogenicity.

II.A. EVIDENCE FOR CLASSIFICATION AS TO HUMAN CARCINOGENICITY

II.A.1. WEIGHT-OF-EVIDENCE CLASSIFICATION

Classification -- D, not classifiable as to human carcinogenicity

Basis -- Based on no human data and inadequate data from animal bioassays.

II.A.2. HUMAN CARCINOGENICITY DATA

None.

II.A.3. ANIMAL CARCINOGENICITY DATA

Inadequate. Groups of 14-29 newborn male and 18-49 newborn female CD-1 mice on 1, 8, and 15 days of age received intraperitoneal injections of pyrene (purity unknown) in dimethyl sulfoxide (DMSO) (total dose = 40, 141 or 466 ug/mouse), or DMSO alone (Wislocki et al., 1986). Tumors were evaluated in animals that died spontaneously after weaning and in all remaining animals at 1 year after exposure. The mid-dose group was initiated 10 weeks after the other groups and had a separate vehicle control. The survival rate in the high-dose groups (male and female) was 25 to 35%; most of the mice died between the last injection and weaning. This high mortality was not observed in the control, low- or mid-dose groups (the survival rates were not stated). A statistically significant increase in the incidence of liver carcinomas occurred in the mid-dose males (3/25) relative to their vehicle control group (0/45), but not in the high-dose males (1/14) or low-dose males (0/29) or in female mice, when compared with their respective controls. The incidences of total liver tumors (adenomas and carcinomas), lung tumors or malignant lymphomas were not statistically significantly elevated in treated animals. The results of this 1-year experiment were not considered to be positive because of the overall lack of tumorigenic response in the short-term.

Mouse skin-painting assays of pyrene as a complete skin carcinogen or as an initiator of carcinogenicity were either not positive or inconclusive (Badger et al., 1940; Horton and Christian, 1974; Van Duuren and Goldschmidt, 1976; Salaman and Roe, 1956; Scribner, 1973).

A subcutaneous pyrene injection did not produce tumors in Jackson A mice; the mice were observed for 18 months after injection (Shear and Leiter, 1941).

II.A.4. SUPPORTING DATA FOR CARCINOGENICITY

In DNA damage assays in Escherichia coli and Bacillus subtilis pyrene was not mutagenic (Ashby and Kilbey, 1981). In bacterial gene mutation tests both positive (Kinae et al., 1981; Bridges et al., 1981; Matijasevic and Zeiger, 1985; Sakai et al., 1985; Kaden et al., 1979; Bos et al., 1988) and negative (McCann et al., 1975; LaVoie et al., 1979; Ho et al., 1981; Bos et al., 1988) results have been reported. The consensus conclusion on the international collaborative study (which involved 20 bacterial test sets) was that protocol or evaluation criteria were critical factors in individual test verdicts. Pyrene induced increased incidence of mitotic gene conversion but not other genetic endpoints in yeast (de Serres and Hoffman, 1981). Pyrene did not induce an increase in sex-linked recessive lethals in Drosophila (Valencia and Houtchens, 1981).

Mixed results have also been observed in mammalian assays in vitro, again with protocol and evaluation criteria being a factor in at least some of the cases. In the collaborative study Evans and Mitchell (1981) concluded pyrene was positive for SCE induction in CHO cells when all concentrations were different from controls, but no apparent increase when the concentration was increased 10-fold. In the same volume, two other laboratories reported pyrene negative both for SCE and for chromosome aberrations in CHO cells (Brookes and Preston, 1981). Tong et al. (1981) also reported that pyrene did not induce SCE in a rat liver epithelial cell system. Jotz and Mitchell (1981) reported pyrene was positive in the L5178Y mouse lymphoma gene mutation assay.

Pyrene did not induce chromosome aberrations (as detected by micronuclei) or SCE in bone marrow of several mouse strains receiving i.p. injections of pyrene (Purchase and Ray, 1981). Results of mammalian cell transformation assays in a variety of cell types have not been positive (DiPaolo et al., 1969; Pienta et al., 1977; Casto, 1979; Chen and Heidelberger, 1969; DiPaolo et al., 1972; Kakunaga, 1973; Evans and DiPaolo, 1975).

II.B. QUANTITATIVE ESTIMATE OF CARCINOGENIC RISK FROM ORAL EXPOSURE

None.

II.C. QUANTITATIVE ESTIMATE OF CARCINOGENIC RISK FROM INHALATION EXPOSURE

None.

II.D. EPA DOCUMENTATION, REVIEW, AND CONTACTS (CARCINOGENICITY ASSESSMENT)

II.D.1. EPA DOCUMENTATION

Source Document -- US EPA, 1990

The 1990 Drinking Water Criteria Document for Polycyclic Aromatic Hydrocarbons has undergone Agency and external review.

II.D.2. REVIEW (CARCINOGENICITY ASSESSMENT)

Agency Work Group Review -- 02/07/90

Verification Date -- 02/07/90

II.D.3. US EPA CONTACTS (CARCINOGENICITY ASSESSMENT)

Rita S. Schoeny / NCEA -- (513)569-7544

Robert E. McGaughy / NCEA -- (202)260-5889

III. HEALTH HAZARD ASSESSMENTS FOR VARIED EXPOSURE DURATIONS

[Ed. Note: EPA does not have information yet in this section of IRIS].

IV. US EPA REGULATORY ACTIONS

EPA risk assessments may be updated as new data are published and as assessment methodologies evolve. Regulatory actions are frequently not updated at the same time. Compare the dates for the regulatory actions in this section with the verification dates for the risk assessments in sections I and II, as this may explain inconsistencies. Also note that some regulatory actions consider factors not related to health risk, such as technical or economic feasibility. Such considerations are indicated for each action. In addition, not all of the regulatory actions listed in this section involve enforceable federal standards. Please direct any questions you may have concerning these regulatory actions to the US EPA contact listed for that particular action. Users are strongly urged to read the background information on each regulatory action in Background Document 4 in Service Code 5.

IV.A. CLEAN AIR ACT (CAA)

[Ed. Note: EPA does not have information yet in this section of IRIS].

IV.B. SAFE DRINKING WATER ACT (SDWA)

[Ed. Note: EPA does not have information yet in this section of IRIS].

IV.C. CLEAN WATER ACT (CWA)

IV.C.1. AMBIENT WATER QUALITY CRITERIA, Human Health

Water and Fish Consumption: 2.8E-3 ug/L

Fish Consumption Only: 3.11E-2 ug/L

Considers technological or economic feasibility? -- NO

Discussion -- For the maximum protection from the potential carcinogenic properties of this chemical, the ambient water concentration should be zero. However, zero may not be obtainable at this time, so the recommended criteria represents a E-6 estimated incremental increase of cancer over a lifetime. The values given represent polynuclear aromatic hydrocarbons as a class.   Reference -- 45 FR 79318 (11/28/80)

EPA Contact -- Criteria and Standards Division / OWRS(202)260-1315 / FTS 260-1315

IV.C.2. AMBIENT WATER QUALITY CRITERIA, Aquatic Organisms

Freshwater:
   Acute LEC -- none
   Chronic LEC -- none

Marine:
   Acute LEC --   3.0E+2 ug/L
   Chronic LEC -- none

Considers technological or economic feasibility? -- NO

Discussion -- The values that are indicated as "LEC" are not criteria, but are the lowest effect levels found in the literature. LEC's are given when the minimum data required to derive water quality criteria are not available. The values given represent polynuclear aromatic hydrocarbons as a class.

Reference -- 45 FR 79318 (11/28/80)

EPA Contact -- Criteria and Standards Division / OWRS(202)260-1315 / FTS 260-1315

IV.D. FEDERAL INSECTICIDE, FUNGICIDE, AND RODENTICIDE ACT (FIFRA)

[Ed. Note: EPA does not have information yet in this section of IRIS].

IV.E. TOXIC SUBSTANCES CONTROL ACT (TSCA)

[Ed. Note: EPA does not have information yet in this section of IRIS].

IV.F. RESOURCE CONSERVATION AND RECOVERY ACT (RCRA)

IV.F.1. RCRA APPENDIX IX, for Ground Water Monitoring

Status -- Listed

Reference -- 52 FR 25942 (07/09/87)

EPA Contact -- RCRA/Superfund Hotline(800)424-9346 / (202)260-3000 / FTS 260-3000

IV.G. SUPERFUND (CERCLA)

IV.G.1. REPORTABLE QUANTITY (RQ) for Release into the Environment

Value (status) -- 5000 pounds (Final, 1989)

Considers technological or economic feasibility? -- NO

Discussion -- No data have been found to permit the ranking of this hazardous substance. The available data for acute hazards may lie above the upper limit for the 5000-pound RQ, but since it is a designated hazardous substance, the largest assignable RQ is 5000 pounds. This chemical has been assessed for carcinogenicity and chronic toxicity and is not, at this time, considered to be a chronic toxicant nor a potential carcinogen.

Reference -- 54 FR 33418 (08/14/89)

EPA Contact -- RCRA/Superfund Hotline(800)424-9346 / (202)260-3000 / FTS 260-3000

VI. BIBLIOGRAPHY

VI.A. ORAL RfD REFERENCES

US EPA. 1989. 13-Week Mouse Oral Subchronic Toxicity with Pyrene. TRL Study #042-012. Study conducted by Toxicity Research Laboratories, Muskegon, MI for the Office of Solid Waste, Washington, DC.

White, J. and A. White. 1939. Inhibition of growth of the rat by oral administration of methylcholanthrene, benzpyrene, or pyrene and the effects of various dietary supplements. J. Biol. Chem. 131: 149-161.

VI.B. INHALATION RfC REFERENCES

None

VI.C. CARCINOGENICITY ASSESSMENT REFERENCES

Ashby, J. and B. Kilbey. 1981. Summary report on the performance of bacterial repair, phase induction, degranulation, and nuclear enlargement assays. In: Evaluation of Short-Term Tests for Carcinogens. Report of the International Collaborative Program. Progress in Mutation Research, Vol. 1, F.J. de Serres and J. Ashby, Ed. Elsevier, North Holland, New York.p. 33-48.

Badger, G.M., J.W. Cook, C.L. Hewett, et al. 1940. The production of cancer by pure hydrocarbons. V. Proc. R. Soc. London Ser. B. 129: 439-467.

Bos, R.P., J.L.G. Theuws, F.J. Jongeneelen and P.Th. Henderson. 1988. Mutagenicity of bi-, tri- and tetra-cyclic aromatic hydrocarbons in the "taped-plate assay" and in the conventional Salmonella mutagenicity assay. Mutat. Res. 204: 203-206.

Bridges, B.A., D.B. McGregor and E. Zeiger, et al. 1981. Summary report on the performance of bacterial mutation assays. In: Evaluation of Short-term Tests for Carcinogens. Report of the International Collaborative Program. Progress in Mutation Research, Vol. 1, F.J. de Serres and J. Ashby, Ed. Amsterdam, Elsevier, North Holland. p. 49-67.

Brookes, P. and R.J. Preston. 1981. Summary report on the performance of in vitro mammalian assays. In: Evaluation of Short-term Tests for Carcinogens. Report of the International Collaborative Program. Progress in Mutation Research, Vol. 1, F.J. de Serres and J. Ashby, Ed. Amsterdam, Elsevier, North Holland. p. 77-85.

Casto, B.C. 1979. Polycyclic hydrocarbons and Syrian hamster embryo cells: Cell transformation, enhancement of viral transformation and analysis of DNA damage. In: Polynuclear Aromatic Hydrocarbons, P.W. Jones and P. Leber, Ed. Ann Arbor Science Publ., Ann Arbor, MI. p. 51-66.

Chen, T.T. and C. Heidelberger. 1969. Quantitative studies on the malignant transformation of mouse prostate cells by carcinogenic hydrocarbons in vitro. Int. J. Cancer. 4: 166-178.

Dean, B.J. 1981. Activity of 27 coded compounds in the RL1 chromosome assay. In: Evaluation of Short-term Tests for Carcinogens. Report of the International Collaborative Program. Progress in Mutation Research, Vol. 1, F.J. de Serres and J. Ashby, Ed. Amsterdam, Elsevier, North Holland. p. 570-579.

de Serres, F.J., G.R. Hoffman, J. Von Borstel, et al. 1981. Summary report on the performance of yeast assays. In: Evaluation of Short-term Tests for Carcinogens. Report of the International Collaborative Program. Progress in Mutation Research, Vol. 1, F.J. de Serres and J. Ashby, Ed. Amsterdam, Elsevier, North Holland. p. 68-76.

DiPaolo, J.A., P. Donovan and R. Nelson. 1969. Quantitative studies of in vitro transformation by chemical carcinogens. J. Natl. Cancer Inst. 42(5):867-874.

DiPaolo, J.A., K. Takano and N.C. Popescu. 1972. Quantitation of chemically induced neoplastic transformation of BALB/3T3 cloned cell lines. Cancer Res. 32: 2686-2695.

Evans, C.H. and J.A. DiPaolo. 1975. Neoplastic transformation of guinea pig fetal cells in culture induced by chemical carcinogens. Cancer Res. 35: 1035-1044.

Evans, E.L. and A.D. Mitchell. 1981. Effect of 20 coded chemicals on sister chromatid exchange frequencies in cultured Chinese hamster cells. In: Evaluation of Short-term Tests for Carcinogens. Report of the International Collaborative Program. Progress in Mutation Research, Vol. 1, F.J. de Serres and J. Ashby, Ed. Amsterdam, Elsevier, North Holland. p. 538-550.

Ho, C-H., B.R. Clark, M.R. Guerin, B.D. Barkenhus, T.K. Rao and J.L. Epler. 1981. Analytical and biological analyses of test materials from the synthetic fuel technologies. IV. Studies of chemical structure-mutagenic activity relationships of aromatic nitrogen compounds relevant to synfuels. Mutat. Res. 85: 335-345.

Horton, A.W. and G.M. Christian. 1974. Carcinogenic versus incomplete carcinogenic activity among aromatic hydrocarbons: Contrast between chrysenes and benzo[b]triphenylene. J. Natl. Cancer Inst. 53(4): 1017-1020.

Jotz, M.M. and A.D. Mitchell. 1981. Effects of 20 coded chemicals on the forward mutation frequency at the thymidine kinase locus in L5178Y mouse lymphoma cells. In: Evaluation of Short-term Tests for Carcinogens. Report of the International Collaborative Program. Progress in Mutation Research, Vol. 1, F.J. de Serres and J. Ashby, Ed. Amsterdam, Elsevier, North Holland. p. 580-593.

Kaden, D.A., R.A. Hites and W.G. Thilly. 1979. Mutagenicity of soot and associated polycyclic aromatic hydrocarbons to Salmonella typhimurium. Cancer Res. 39: 4152-4159.

Kakunaga, T. 1973. A quantitative system for assay of malignant transformation by chemical carcinogens using a clone derived from BALB/3T3. Int. J. Cancer. 12: 463-473.

Kinae, N., T. Hashizume, T. Makita, I. Tomita, I. Kimura and H. Kanamori. 1981. Studies on the toxicity of pulp and paper mill effluents - 1. Mutagenicity of the sediment samples derived from Kraft paper mills. Water Res. 15: 17-24.

Lake, R.S., M.L. Kropko, M.R. Pezzutti, R.H. Shoemaker and H.J. Igel. 1978. Chemical induction of unscheduled DNA synthesis in human skin epithelial cell cultures. Cancer Res. 38: 2091-2098.

LaVoie, E.J., E.V. Bedenko, N. Hirota, S.S. Hecht and D. Hoffmann. 1979. A comparison of the mutagenicity, tumor-initiating activity and complete carcinogenicity of polynuclear aromatic hydrocarbons. In: Polynuclear Aromatic Hydrocarbons, P.W. Jones and P. Leber, Ed. Ann Arbor Science Publishers, Ann Arbor, MI. p. 705-721.

Martin, C.N., A.C. McDermid and R.C. Garner. 1978. Testing of known carcinogens and noncarcinogens for their ability to induce unscheduled DNA synthesis in HeLa cells. Cancer Res. 38: 2621-2627.

Matijasevic, Z. and E. Zeiger. 1985. Mutagenicity of pyrene in Salmonella. Mutat. Res. 142: 149-152.

McCann, J.E., E. Choi, E. Yamasaki and B.N. Ames. 1975. Detection of carcinogens as mutagens in the Salmonella/microsome test: Assay of 300 chemicals. Natl. Acad. Sci. 72(12): 5135-5139.

Perry, P.E. and E.J. Thompson. 1981. Evaluation of the sister chromatid exchange method in mammalian cells as a screening system for carcinogens. In: Evaluation of Short-term Tests for Carcinogens. Report of the International Collaborative Program. Progress in Mutation Research, F.J. de Serres and J. Ashby, Ed. Elsevier/North Holland, NY. p. 560-569.

Pienta, R.J., J.A. Poiley and W.B. Libherz, III. 1977. Morphological transformation of early passage golden Syrian hamster embryo cells derived from cryopreserved primary cultures as a reliable in vitro bioassay for identifying diverse carcinogens. Int. J. Cancer. 19: 642-655.

Popescu, N.C., D. Turnbull and J.A. DiPaolo. 1977. Sister chromatid exchange and chromosome aberration analysis with the use of several carcinogens and noncarcinogens: Brief communication. J. Natl. Cancer Inst. 59(1): 289-293.

Probst, G.S., R.E. McMahon, L.E. Hill, C.Z. Thompson, J.K. Epp and S.B. Neal. 1981. Chemically-induced unscheduled DNA synthesis in primary rat hepatocyte cultures: A comparison with bacterial mutagenicity using 218 compounds. Environ. Mutagen. 3: 11-32.

Purchase, I.F.H. and V. Ray. 1981. Summary report on the performance of in vivo assays. In: Evaluation of Short-term Tests for Carcinogens. Report of the International

Collaborative Program. Progress in Mutation Research, Vol. 1, F.J. de Serres and J. Ashby, Ed. Amsterdam, Elsevier, North Holland. p. 86-95.

Rice, J.E., T.J. Hosted, Jr. and E.J. LaVoie. 1984. Fluoranthene and pyrene enhance benzo[a]pyrene -- DNA adduct formation in vivo in mouse skin. Cancer Lett. 24: 327-333.

Robinson, D.E. and A.D. Mitchell. 1981. Unscheduled DNA synthesis response of human fibroblasts, WI-38 cells, to 20 coded chemicals. In: Evaluation of Short-term Tests for Carcinogens. Report of the International Collaborative Program. Progress in Mutation Research, Vol. 1, F.J. de Serres and J. Ashby, Ed. Amsterdam, Elsevier, North Holland. p. 517-527.

Sakai, M., D. Yoshida and S. Mizusdki. 1985. Mutagenicity of polycyclic aromatic hydrocarbons and quinones on Salmonella typhimurium TA97. Mutat. Res. 156: 61-67.

Salaman, M.H. and F.J.C. Roe. 1956. Further tests for tumor-initiating activity: N,N-di-(2-chloroethyl)-p-aminophenylbutyric acid (CB1348) as an initiator of skin tumor formation in the mouse. Br. J. Cancer. 10: 363-378.

Scribner, J.D. 1973. Brief Communication: Tumor initiation by apparently noncarcinogenic polycyclic aromatic hydrocarbons. J. Natl. Cancer Inst. 50: 1717-1719.

Shear, M.J. and J. Leiter. 1941. Studies in carcinogenesis. XVI. Production of subcutaneous tumors in mice by miscellaneous polycyclic compounds. J. Natl. Cancer Inst. 2: 241-258.

Tong, C., S.V. Brat and G.M. Williams. 1981. Sister-chromatid exchange induction by polycyclic aromatic hydrocarbons in an intact cell system of adult rat-liver epithelial cells. Mutat. Res. 91: 467-473.

US EPA. 1990. Drinking Water Criteria Document for Polycyclic Aromatic Hydrocarbons (PAHs). Prepared by the Office of Health and Environmental Assessment, Environmental Criteria and Assessment Office, Cincinnati, OH for the Office of Drinking Water, Washington, DC. ECAO-CIN-D010, September, 1990. (Final Draft)

Valencia, R. and K. Houtchens. 1981. Mutagenic activity of 10 coded compounds in the Drosophila sex-linked recessive lethal test. In: Evaluation of Short-term Tests for Carcinogens. Report of the International Collaborative Program. Progress in Mutation Research, Vol. 1, F.J. de Serres and J. Ashby, Ed. Amsterdam, Elsevier, North Holland, NY. p. 651-659.

Van Duuren, B.L. and B.M. Goldschmidt. 1976. Cocarcinogenic and tumor-promoting agents in tobacco carcinogenesis. J. Natl. Cancer Inst. 51(6): 1237-1242.

Wislocki, P.G., E.S. Bagan, A.Y.H. Lu et al. 1986. Tumorigenicity of nitrated derivatives of pyrene, benz[a]anthracene, chrysene and benzo[a]pyrene in the newborn mouse assay. Carcinogenesis. 7(8): 1317-1322.

VI.D. DRINKING WATER HA REFERENCES

None
VII. REVISION HISTORY

```
--------  --------   -------------------------------------------------------
Date      Section  Description
--------  --------   -------------------------------------------------------
09/01/90  I.A.     Oral RfD summary on-line
09/01/90  VI.      Bibliography on-line
01/01/91  II.      Carcinogen assessment on-line
01/01/91  VI.C.    Carcinogen assessment references added
07/01/91  I.A.7.   Primary and secondary contacts changed
08/01/91  VI.A.    US EPA, 1989 citation clarified
01/01/92  IV.      Regulatory Action section on-line
07/01/93  I.A.6.   Other EPA Documentation added
09/01/94  I.B.     Inhalation RfC now under review
```

SYNONYMS

BENZO(DEF)PHENANTHRENE	PYREN [GERMAN]
HSDB 4023	PYRENE
NSC 17534	BETA-PYRENE

Source: *Instant EPA's IRIS*, 1997, Instant Reference Sources, Inc., 7605 Rockpoint Drive, Austin, TX 78731 (Fax: 512-345-2386; Internet URL, http://www.instantref.com/inst-ref.htm).

Simazine

Introduction

Simazine is recommended for the control of broad-leaved and grass weeds in deep rooted crops, pre-emergence herbicide and soil sterilant. A major use is on maize which can convert simazine to the herbicidally inactive 6-hydroxy analogue. In the USA it is also used to control algae in farm ponds, fish hatcheries, etc.

Simazine may be released to the environment via effluents at manufacturing sites and at points of application where it is employed as a herbicide. If released to water, simazine is not expected to bioconcentrate in aquatic organisms, adsorb to sediment and suspended particulate matter, or to volatilize. Slow biodegradation of simazine may occur in water based upon the slow biodegradation observed in soil. Simazine is fairly resistant to hydrolysis with reported half-lives for hydrolysis in aqueous buffer solutions at 25°C at pH 5, 7, and 9 of 70, >200, and >200 days, respectively. The product of simazine hydrolysis is 2-hydroxy-4,6-bis(ethylamino)-1,3,5-triazine. [828]

If released to soil, the mobility of simazine will be expected to vary from slight to high in soil-types ranging from clay soils to sandy loams soils, respectively, based upon soil column, soil thin-layer chromatography, and Koc experiments. Therefore it may leach to groundwater; adsorption of simazine in soil has been observed to increase as titratable acidity, organic matter and, to a lesser extent, clay content of the soil increased. Simazine may be susceptible to slow hydrolysis in soil based upon reported half-lives for degradation (purportedly mainly soil catalyzed hydrolysis) of simazine in two soils at 45 and 100 days. Simazine can be utilized by certain soil microorganisms as a source of energy and mineralization. No degradation of simazine was detected in a soil suspension test without the addition of glucose as an energy source suggesting that degradation of simazine in these soil experiments was due to cometabolism. 2-Chloro-4-amino-6-ethylamino-1,3,5-triazine and 2,4-dihydroxy-6-amino-1,3,5-triazine have been identified as microbial transformation products of simazine in soil. Reported persistence of simazine in soil varies from a half-life of <1 month to no degradation being observed in 3.5 months. Simazine is not expected to volatilize from near surface soils or surfaces under normal environmental conditions. [828]

If released to the atmosphere, simazine is expected to exist almost entirely in the particulate phase. Vapor phase reactions with photochemically produced hydroxyl

1033

radicals in the atmosphere may be important (with an estimated half-life of about 2.8 hours). Photolysis may be an important removal mechanism in the atmosphere. The most probable exposure should be occupational exposure which may occur through dermal contact or inhalation at places where simazine is produced or used as a herbicide. [828]

Identifying Information

CAS NUMBER: 122-34-9

NIOSH Registry Number: XY5250000

SYNONYMS: S-TRIAZINE, 2-CHLORO-4,6-BIS(ETHYLAMINO)-, A 2079, AKTINIT S, AQUAZINE, BATAZINA, 2,4-BIS(ETHYLAMINO)-6-CHLORO-s-TRIAZINE, BITEMOL, BITEMOL S 50, CAT, CAT(HERBICIDE), CDT, CEKUSAN, CEKUZINAS, CET, 1-CHLORO,3,5-BISETHYLAMINO-2,4,6-TRIAZINE, 2-CHLORO-4,6-BIS(ETHYLAMINO)-S-TRIAZINE, 2-CHLORO-4,6-BIS(ETHYLAMINO)-1,3,5-TRIAZINE, FRAMED, G 27692, GEIGY 27,692, GESARAN, GESATOP, GESATOP 50, H 1803, HERBAZIN, HERBAZIN 50, HERBEX, HERBOXY, HUNGAZIN DT, PERMAZINE, PRIMATOL S, PRINCEP, PRINTOP, RADOCON, RADOKOR, SIMADEX, SIMANEX, SIMAZIN, SIMAZINE 80W, SYMAZINE, TAFAZINE, TAFAZINE 50-W, TAPHAZINE, TRIAZINE A 384, W 6658, ZEAPUR

Endocrine Disruptor Information

Simazine is on EPA's list of suspect EEDs [811] but little information has been located to determine the Agency's reason for its inclusion. More information should be forthcoming after EPA's studies are concluded.

Chemical and Physical Properties

CHEMICAL FORMULA: $C_7H_{12}ClN_5$

MOLECULAR WEIGHT: 201.69

WLN: T6N CN ENJ BM2 DM2 FG

PHYSICAL DESCRIPTION: White crystalline powder

SPECIFIC GRAVITY: Not available

DENSITY: 1.302 g/cm^3 @ 20°C [169], [172]

MELTING POINT: 225-227°C [025], [169], [172], [430]

BOILING POINT: Not available

SOLUBILITY: Water: <1 mg/mL @ 22°C [700]; DMSO: 10-50 mg/mL @ 22°C [700]; 95% Ethanol: <1 mg/mL @ 22°C [700]; Acetone: <1 mg/mL @ 22°C [700]; Methanol: 400 mg/L @ 20° [169], [172]

OTHER SOLVENTS: Chloroform: 900 mg/L @ 20°C [169], [172]; Light petroleum: 2 mg/L @ 20°C [172]; Diethyl ether: 300 mg/L @ 20°C [172]; Dioxane: Slighty soluble [031]; Ethyl Cellosolve: Slighty soluble [031]; Most organic solvents: Slightly soluble [062]; Petroleum ether: 2 mg/L @ 20°C [169]

OTHER PHYSICAL DATA: pK is 1.7 @ 21°C [172]

Source: *The National Toxicology Program's Chemical Database, Volume 2*, 1992, CRC Press, Inc./Lewis Publishers, 2000 Corporate Blvd., Boca Raton, FL 33431 (Fax: 407-998-9114).

Environmental Reference Materials

A source of EED environmental reference materials for simazine is Radian International LLC, Austin, TX (Fax 512-454-0268; Internet URL, http://www.radian.com/standards); Catalog No. ERS-057.

Hazardous Properties

ACUTE/CHRONIC HAZARDS: Simazine may be toxic and an irritant. When heated to decomposition it emits toxic fumes of hydrogen chloride and nitrogen oxides [043].

FIRE HAZARD: Literature sources indicate that simazine is nonflammable [051]. Fires involving this material can be controlled with a dry chemical, carbon dioxide or Halon extinguisher.

REACTIVITY: This compound is hydrolyzed by strong acids and alkalis [169].

STABILITY: Simazine is decomposed by UV radiation [169]. It is stable in neutral and slightly basic or acidic media, but it is hydrolyzed by stronger acids and bases [169], [173]. Solutions of simazine in water, DMSO, 95% ethanol or acetone should be stable for 24 hours under normal lab conditions [700].

Source: *The National Toxicology Program's Chemical Database, Volume 7*, 1992, CRC Press, Inc./Lewis Publishers, 2000 Corporate Blvd., Boca Raton, FL 33431 (Fax: 407-998-9114).

Medical Symptoms of Exposure

Symptoms of exposure may include contact dermatitis, slight edema, erythema, pruritus and burning [346]. It may cause skin and eye irritation [043]. It may also cause liver and kidney damage, coma or convulsions [301].

Source: *The National Toxicology Program's Chemical Database, Volume 4*, 1992, CRC Press, Inc./Lewis Publishers, 2000 Corporate Blvd., Boca Raton, FL 33431 (Fax: 407-998-9114).

Toxicological Information

The toxicological information below is gathered from several sources including two national databases: one from the *National Toxicology Program Chemical Repository Database* and another from *EPA's Integrated Risk Information System (IRIS).*

Toxicology Information from the National Toxicology Program

CARCINOGENICITY:

Tumorigenic Data:

Type of Effect	Route/Animal	Amount Dosed (Notes)
TDLo:	scu-rat	16 gm/kg/61W-I
TDLo:	scu-mus	35 gm/kg/87W-I

TERATOGENICITY:

Reproductive Effects Data:

Type of Effect	Route/Animal	Amount Dosed (Notes)
TCLo:	ihl-rat	17 mg/m3/2H (7-14D preg)
TDLo:	orl-rat	3120 mg/kg (6-15D preg)
TDLo:	orl-rat	780 mg/kg (6-15D preg)
TDLo:	orl-rat	25 gm/kg (6-15D preg)

MUTATION DATA:

Test	lowestdose		Test	Lowestdose
sln-dmg-par	396 umol/L		sln-dmg-orl	2000 ppm
dlt-dmg-par	396 umol/L		dlt-dmg-orl	6000 ppm
msc-mus:lym	400 mg/L		sce-hmn:lym	1mg/L

TOXICITY:

Typ.dose	Mode	Specie	Amount	Units	Other
LD50	orl	rat	971	mg/kg	
LD50	ivn	mus	100	mg/kg	
LD50	unr	mam	1400	mg/kg	
LCLo	ihl	rat	580	mg/m3	
LD50	unr	rat	1390	mg/kg	
LD50	orl	mam	2014	mg/kg	

OTHER TOXICITY DATA:

Skin and Eye Irritation Data:
skn-rbt 500 mg open MLD
eye-rbt   80 mg MOD

Review: Toxicology Review

Status:
EPA Genetox Program 1988, Positive: Cell transform.-
SA7/SHE
EPA Genetox Program 1988, Negative: D melanogaster-
whole sex chrom. loss
EPA Genetox Program 1988, Negative: In vitro UDS-human
fibroblast
EPA Genetox Program 1988, Negative: S cerevisiae gene
conversion; S cerevisiae-homozygosiis
EPA Genetox Program 1988, Inconclusive: D melanogaster-
nondisjunction
EPA Genetox Program 1988, Inconclusive: D melanogaster-
partial sex chrom. loss
EPA Genetox Program 1988, Inconclusive: Histidine
reversion-Ames test
EPA TSCA Chemical Inventory, 1986
Meets criteria for proposed OSHA Medical Records Rule

SAX TOXICITY EVALUATION:

THR: Poison intravenous route. Mutagenic data. An experimental tumorigen.
A skin and eye irritant.

Source: *Instant Tox-Base*, 1995, Instant Reference Sources, Inc., 7605 Rockpoint Drive,
Austin, TX 78731 (Fax: 512-345-2386; Internet URL, http://www.instantref.com/inst-
ref.htm).

Toxicology Information from EPA's Integrated Risk Information System (IRIS)

I.   CHRONIC HEALTH HAZARD ASSESSMENTS FOR NONCARCINOGENIC
EFFECTS

I.A. REFERENCE DOSE FOR CHRONIC ORAL EXPOSURE (RfD)

The Reference Dose (RfD) is based on the assumption that thresholds exist for certain
toxic effects such as cellular necrosis, but may not exist for other toxic effects such as
carcinogenicity. In general, the RfD is an estimate (with uncertainty spanning perhaps an
order of magnitude) of a daily exposure to the human population (including sensitive
subgroups) that is likely to be without an appreciable risk of deleterious effects during a

lifetime. Please refer to Background Document 1 in Service Code 5 for an elaboration of these concepts. RfDs can also be derived for the noncarcinogenic health effects of compounds which are also carcinogens. Therefore, it is essential to refer to other sources of information concerning the carcinogenicity of this substance. If the U.S. EPA has evaluated this substance for potential human carcinogenicity, a summary of that evaluation will be contained in Section II of this file when a review of that evaluation is completed.

I.A.1. ORAL RfD SUMMARY

Critical Effect	Experimental Doses*	UF	MF	RfD
Reduction in weight gains; hematological changes in females	NOAEL: 10 ppm (0.52 mg/kg-day)	100	1	5E-3 mg/kg-day
	LOAEL: 100 ppm (5.3 mg/kg-day)			

*Conversion Factors and Assumptions: Actual dose tested

2-Year Rat Feeding Study, Ciba-Geigy Corp., 1988a

I.A.2. PRINCIPAL AND SUPPORTING STUDIES (ORAL RfD)

Ciba-Geigy Corp. 1988a. MRID No. 40614405; HED Doc No. 007240, 007449 Available from EPA. Write to FOI, EPA, Washington, DC 20460.

Groups of Sprague-Dawley rats were fed technical simazine for 2 years at dietary levels of 0, 10, 100 or 1000 ppm (Male: 0, 0.41, 4.2, 45.8 mg/kg-day; Female: 0, 0.52, 5.3, 63.1 mg/kg-day). Animals tested for chronic toxicity effects consisted of 40 rats/sex in the control and high-dose groups and 30/sex in the low- and mid-dose groups, while 50/sex/dose were tested for carcinogenicity. After approximately 52 weeks of treatment in the chronic toxicity test, 10 rats/sex/group were sacrificed and an additional 10 rats/sex from the control and high-dose groups were maintained on an untreated diet for approximately 52 weeks at which time all the remaining animals were sacrificed. After 104 weeks of treatment, all remaining animals from the chronic toxicity and carcinogenicity test were sacrificed.

Mean body weights for high-dose male and female rats were statistically significantly lower than the control group beginning on day 7 of study and continuing to study termination. Mid-dose female rats had statistically significant lower mean body weights compared with controls both at different time intervals throughout the study and at study termination. Mean body weight gains were also statistically significantly lower in high-dose male and female rats as compared with controls throughout the study. For mid-dose male and female rats, statistically significantly lower body weight gains were seen occasionally at different time intervals but not at study termination. A statistically significant reduction in food consumption was observed in high-dose male rats beginning at day 7 through 700 of the study. Statistically significant depression of food intake was also reported for high-dose female rats on days 7 through 560 on study, but not during the final 6 months of the study. The reduced food consumption in high-dose animals was correlated with decreased body weight and body weight gains in the same groups throughout the study. Changes in food consumption of low- and mid-dose animals were seen only rarely during the study.

A number of hematology parameters appeared were affected by simazine treatment. These apparent treatment-related effects were pronounced mainly in the high-dose females on most sampling days. Statistically significant differences between the control and high-dose group values were seen in females as follows: the red blood cell (RBC) count was depressed at all sample times; hemoglobin (HGB) was depressed on days 361, 537, and 725 on study; hematocrit (HCT) was depressed on days 361, 537, and 725 of sampling; mean corpuscular hemoglobin (MCHB) was elevated on days 361, 537, and 725 of sampling; mean corpuscular hemoglobin concentration (MCHC) was elevated on days 174 of sampling; white blood cell (WBC) count was elevated on days 174, 361, 537, and 725 of sampling; the percent of neutrophils was elevated on day 316 of sampling; and lymphocytes were depressed on day 361 of sampling. Changes in these parameters, although only occasionally statistically significant, were also observed in the mid-dose females. Comparable changes between the control and high-dose groups were also seen in females of the recovery group. In males, the MCHC was statistically significantly higher in the high-dose group compared with the control group on day 361 of sampling (with an apparent dose-related trend); while the leukocyte count was statistically significantly lower than controls in the mid- and high-dose groups on day 537 of sampling. Other changes seen were not considered treatment-related. For males in the recovery group, hematology parameter values were comparable for the most part between the high-dose group and the control groups. Statistically significantly lower values were seen on day 537 for mean corpuscular volume (MCV) and on days 537 and 725 for MCHB.

Based on decreased body weight gains and hematologic parameters in females, the LEL for systemic toxicity of Simazine is 100 ppm (5.3 mg/kg-day). The NOEL for systemic toxicity is 10 ppm (0.52 mg/kg-day).

I.A.3. UNCERTAINTY AND MODIFYING FACTORS (ORAL RfD)

UF -- The uncertainty factor of 100 reflects 10 for interspecies extrapolation and 10 for intraspecies variability.

MF -- None

I.A.4. ADDITIONAL STUDIES / COMMENTS (ORAL RfD)

Reduction in body weight gains with accompanying hematologic deficits were the toxic endpoints for simazine. These effects are seen across all studies that lasted at least 1 year, including the 2-year rat feeding study, the 2-year mouse feeding study, and the 1-year dog feeding study. The effects are also seen in both the rat and rabbit gavage developmental toxicity studies. The data from these gavage studies support the findings of other studies which related the reduction in body weight to simazine and not to a palatability effect, as has been suggested in the feeding studies.

1) 2-Year Feeding/Oncogenicity - rat: Principal study -- see previous discussion; core grade minimum (Ciba-Geigy Corp., 1988a)

2) 2-Generation Reproduction - rat: Dietary levels tested: 0, 10, 100, and 500 ppm (Male: 0, 0.56, 5.61, and 28.89 mg/kg-day; Female: 0, 0.7, 7.04, and 34.96 mg/kg-day); Groups of Sprague-Dawley rats (30/sex/dose) were fed diets containing simazine technical over 2 generations. Compound-related parental toxicity was observed at 100 and 500 ppm. At 500 ppm, significant decreases were consistently present in food consumption, body weight, and body weight gain. The relative increases in testicular and ovarian weights were considered secondary effects because of the weight loss observed in the animals dose. At 100 ppm, a consistent, but not always significant, decrease was observed in body weight. Due to the fact that this decrease in body weight was present in both sexes and generations and occurred in a dose-related manner, the effects at 100 ppm are considered to be compound-related. The sporadically decreased food consumption noted at 100 ppm was not considered to be compound-related. Based on decreased body weight, the LEL for parental toxicity is 100 ppm (Male: 5.61 mg/kg-day; Female: 7.04 mg/kg-day). The NOEL for parental toxicity is 10 ppm (Male: 0.56 mg/kg-day; Female: 0.7 mg/kg-day). No compound-related effects were observed in this study. The mean numbers of stillborn pups per litter were slightly but insignificantly increased for the F1 pups at 10 and 100 ppm, and for F2a and F2b pups at 500 ppm. In each of these groups, an increased number of dead pups (9-16) in one single litter was responsible for this increase. These increases were not considered compound-related but rather normal variations, since they occurred in only one litter in each respective dose group; they did not occur in a dose-related manner; and no consistent pattern was noted across two generations. This conclusion is further supported by the fact that no changes were noted in the numbers of viable neonates per litter in the live birth indices. The percentage of males was slightly decreased at 500 ppm for F1 pups (46%) and F2b pups (44%). These decreases were not significant because they did not occur among F2a pups. When survival indices to day 21 were reported separately for males and females, no differences were noted between sexes, the dose groups, or the generations. Based on these results, the NOEL for reproductive toxicity is equal to or greater than 500 ppm (Male: 28.89 mg/kg-day; Female: 34.96 mg/kg-day). Core grade minimum (Ciba-Geigy Corp., 1991)

3) 1-Year Feeding - dog: Dietary levels tested: 0, 20, 100, and 1250 ppm (Male: 0, 0.68, 3.4, and 43 mg/kg-day; Female: 0, 0.76, 3.6, and 45 mg/kg-day); Groups of beagle dogs (4/sex/dose) were fed diets containing simazine for 1 year. An additional 4 dogs/sex/dose were used for a clinical study (pre-dose and weeks 14, 25, and 52). Toxicity was demonstrated in high-dose males by decrements in body weight gain; variable but reversible decrements in red blood cell counts, hemoglobin concentration and hematocrit; and statistically significant increases in platelet counts. Similar toxicity was demonstrated in high-dose females by statistically significant larger decrements in body weight gain, and in mid- and high-dose females by decrements in the red blood cell counts, hemoglobin concentration, and hematocrit. Slight increases occurred in platelet counts in high-dose females. Decreased body weight gain occurred in one mid-dose female; this decrement was considered compound-related, although no other effects were noted in this animal. The efficiency of food utilization was apparently decreased in high-dose females. In high-dose males the absolute organ weight, organ-to-brain weight, and organ-to-body-weight ratios were increased for the adrenals (130%), kidneys (111%), and liver (108%); and decreased for the spleen (69%) and thyroid/parathyroid (60%). In

high-dose females the weight of the adrenals (129%), liver (104%), and thyroid/parathyroid (114%) weights may have been increased. These and other organ weight effects were not accompanied by any findings at histological examination, and thus, they may be incidental to the study. Based on decreased body weight gain; decreased RBC, Hgb, and HCT; and nominally increased platelet count in females, the LEL for systemic toxicity is 100 ppm (3.6 mg/kg-day). The NOEL for systemic toxicity is 20 ppm (0.76 mg/kg-day). Core grade minimum (Ciba-Geigy Corp., 1988b)

4) Developmental Toxicity - rat: Dose levels tested: 0, 30, 300, and 600 mg/kg-day; Groups of pregnant CDL COBS CD SD BR rats (25/dose) were administered simazine in 2% carboxymethylcellulose by gavage on gestation days 6 through 15. Maternal toxicity was demonstrated by statistically significant decreases in body weight and body weight gain, and decreases in food consumption and relative efficiency of utilization at the mid- and high-dose levels during treatment. Based on the above effects, the NOEL and LEL for maternal toxicity are 30 and 300 mg/kg-day, respectively. Developmental toxicity was also demonstrated at the two highest dose levels. The incidence of additional centra/vertebrae was statistically significantly increased in litters at the mid- and high-dose levels. Several other skeletal parameters indicating dose-related toxicity at these doses were statistically significant on a fetal basis, but not on a litter basis, such as head not completely ossified, teeth not ossified, centra/vertebrae not ossified, rudimentary rib and sternebrae not ossified. These parameters were nominally elevated in litters. Most of the parameters affected frequently occur in association with maternal toxicity, and although some may disappear in fetuses after birth, these effects are considered indicative of developmental toxicity. Based on the above effects, the NOEL and LEL for developmental toxicity are 30 and 300 mg/kg-day, respectively. Core grade supplementary (additional data must be submitted) (Ciba-Geigy Corp., 1986)

5) Developmental Toxicity - rabbit: Dose levels tested: 0, 5, 75, and 200 mg/kg-day; Groups of pregnant New Zealand White rabbits (19/dose) were administered simazine with 3% corn starch by gavage on gestation days 7 through 19. A significant decrease in fetal weights and an increase in skeletal variations were observed at 200 mg/kg-day. Based on the above effects, the NOEL and LEL for developmental toxicity are 75 and 200 mg/kg-day, respectively. Significant decreases in body weight gain and food consumption, tremors, and abortions were observed in dams receiving 75 and 200 mg/kg-day. Based on these effects, the NOEL and LEL for maternal toxicity are 5 and 75 mg/kg-day, respectively. Core grade guideline (Ciba-Geigy Corp., 1984)

Other Data Reviewed:

1) 95-Week Feeding/Oncogenicity - mice: Dietary levels tested: 0, 40, 1000, and 4000 ppm (Male: 0, 5.3, 131.5, and 542 mg/kg-day; Female: 0, 6.2, 160.0, and 652.1 mg/kg-day); Groups of CD-1 mice (60/sex/dose) were fed diets containing simazine for 95 weeks. Mean body weight decreased in both males and females in the mid- and high-dose groups and food consumption decreased in mid- and high-dose males and in mid-dose females. Decreases in erythroid parameters may have been related to weight loss. Other hematologic parameters were not affected. Clinical chemistry values and urinary parameters were normal in all dosed groups. Organ-to-body-weight ratios were increased in high-dose females for several organs; however, there were no histologic correlates and the changes were accompanied by decreased body weights at termination.

The incidence of amyloidosis was high in all groups. The LEL for systemic toxicity is 1000 ppm (Male: 131.5 mg/kg-day; Female: 160.0 mg/kg-day) based on decreased weight gain. The NOEL for systemic toxicity is 40 ppm (Male: 5.3 mg/kg-day; Female: 6.2 mg/kg-day). Core grade guideline (Ciba-Geigy Corp., 1988c)

2) 3-Generation Reproduction - rat: Dietary levels tested: 0, 50, and 100 ppm (0, 2.5, and 5 mg/kg-day); Groups of albino rats (strain not specified) were fed diets containing simazine over two generations. Reduced weight gain in parental male and female animals was observed at both dose levels. The weight gain of male was also significantly reduced by approximately 11% in the F0 generation during their premating period when compared with controls. Based on the effects observed at the LDT, the LEL for parental toxicity is 50 ppm (2.5 mg/kg-day). A NOEL for parental toxicity was not established. Reproductive toxicity could not be determined based on lack of histologic evaluations in apparently sterile males in the F1b generation. Up to 33% of the potential paternal males in the 100 ppm group did not impregnate a female in 2 successive breeding sessions. The small sample size of the F3b pups examined, the length of gestation, and pup and litter weights at 14 and 21 days were not determined. The male and female parents were not examined histologically in any generation. Core grade supplementary (Ciba-Geigy Corp., 1965)

3) 13-Week Feeding - rat: Dietary levels tested: 0, 200, 2000, and 4000 ppm (Male: 0, 13.5, 128.2, and 249 mg/kg-day; Female: 0, 16, 171.3, and 291.1 mg/kg-day); Groups of Sprague-Dawley rats (10/sex/dose) were fed simazine in a powdered feed admixture ad libitum for 13 weeks. Reduced mean feed intake in treated rats is most likely due to the palatability of simazine in the diet. Lower individual body weights and reduced body weight gains paralleled mean food intake in treated rats. The majority of alterations in clinical chemistry values [reduced erythrocyte (males and females) and leucocyte counts (males only); elevated cholesterol and inorganic phosphate levels (males and females)] may be related to feed consumption in rats. Renal calculi and attending hyperplasia were the only dose-related lesions detected microscopically. Based on the effects observed at lowest dose tested, the LEL for systemic toxicity is 200 ppm (Male: 13.5 mg/kg-day; Female: 16 mg/kg-day). A NOEL for systemic toxicity was not established. Core grade supplementary (Ciba-Geigy Corp., 1985a)

4) 13-Week Feeding - dog: Dietary levels tested: 0, 200, 2000, and 4000 ppm (Male: 0, 6.9, 65.2, and 133.6 mg/kg-day; Female: 0, 8.2, 64.3, 136.7 mg/kg-day); Groups of beagle dogs (4/sex/dose) were fed a dietary admixture of simazine ad libitum for 13 weeks. It appears that reduced mean feed intake in treated dogs is most likely due to the palatability of simazine in the diet. Lower individual body weights and reduced body weight gains paralleled mean feed intake in treated dogs. The majority of the alterations in clinical chemistry values [reduced albumin levels, increased globulin levels, and elevated urinary specific gravity (males only) and increased ketone levels (males and females)] and organ weights observed in the mid- and high-dose groups may be related to feed consumption in treated dogs. Based on the effects observed at the mid-dose, the LEL for systemic toxicity is 2000 ppm (Male: 65.2 mg/kg-day; Female: 64.3 mg/kg-day). The NOEL for systemic toxicity is 200 ppm (Male: 6.9 mg/kg-day; Female: 8.2 mg/kg-day). Core grade minimum (Ciba-Geigy Corp., 1985b)

Data Gap(s): Rat Developmental Toxicity Study

I.A.5. CONFIDENCE IN THE ORAL RfD

Study -- MediumData Base -- HighRfD -- High

The principal study is of adequate quality and is given a medium confidence rating. Additional studies, in particular the chronic dog study and the 2-generation reproduction study, are supportive of the NOEL and LEL established in the principal. The quality of the supporting studies is adequate and therefore, the data base is given a high confidence rating. High confidence in the RfD follows.

I.A.6. EPA DOCUMENTATION AND REVIEW OF THE ORAL RfD

Source Document -- This assessment is not presented in any existing U.S. EPA document.

Other EPA Documentation -- OPP Registration Standard, 1985
Agency Work Group Review -- 06/24/86, 06/15/89, 08/14/91
Verification Date -- 08/14/91I.A.7. EPA CONTACTS (ORAL RfD)

George Ghali / OPP -- (703)305-7490
William Burnam / OPP -- (703)305-7491

I.B. REFERENCE CONCENTRATION FOR CHRONIC INHALATION EXPOSURE (RfC)

[Ed. Note: EPA does not have information yet in this section of IRIS].

II. CARCINOGENICITY ASSESSMENT FOR LIFETIME EXPOSURE

This substance/agent has been evaluated by the U.S. EPA for evidence of human carcinogenic potential. This does not imply that this agent is necessarily a carcinogen. The evaluation for this chemical is under review by an inter-office Agency work group. A risk assessment summary will be included on IRIS when the review has been completed.

III. HEALTH HAZARD ASSESSMENTS FOR VARIED EXPOSURE DURATIONS

[Ed. Note: EPA does not have information yet in this section of IRIS].

IV. U.S. EPA REGULATORY ACTIONS

EPA risk assessments may be updated as new data are published and as assessment methodologies evolve. Regulatory actions are frequently not updated at the same time. Compare the dates for the regulatory actions in this section with the verification dates for the risk assessments in sections I and II, as this may explain inconsistencies. Also note that some regulatory actions consider factors not related to health risk, such as technical or economic feasibility. Such considerations are indicated for each action. In addition, not all of the regulatory actions listed in this section involve enforceable federal standards. Please direct any questions you may have concerning these regulatory actions to the U.S. EPA contact listed for that particular action. Users are strongly urged to read the background inform-ation on each regulatory action in Background Document 4 in Service Code 5.

IV.A. CLEAN AIR ACT (CAA)

[Ed. Note: EPA does not have information yet in this section of IRIS].

IV.B. SAFE DRINKING WATER ACT (SDWA)

IV.B.1. MAXIMUM CONTAMINANT LEVEL GOAL (MCLG) for Drinking Water

Value -- 0.001 mg/L (Proposed, 1990)

Considers technological or economic feasibility? -- NO

Discussion -- EPA is proposing to regulate simazine based on its potential adverse effects (increased mortality, reduced body weight gain and tumor incidence) reported in a chronic/carcinogenicity study in rats. The MCLG is based upon a DWEL of 0.06 mg/L and an assumed drinking water contribution of 20 percent. The MCLG includes an additional uncertainty factor of 10 for Group C contaminants.

Reference -- 55 FR 30370 (07/25/90)

EPA Contact -- Health and Ecological Criteria Division / OST /(202) 260-7571 / FTS 260-7571; or Safe Drinking Water Hotline / (800) 426-4791IV.B.2. MAXIMUM CONTAMINANT LEVEL (MCL) for Drinking Water

Value -- 0.001 mg/L (Proposed, 1990)

Considers technological or economic feasibility? -- YES

Discussion -- EPA is proposing an MCL equal to the proposed MCLGof 0.001 mg/L.

Monitoring requirements -- Community and non-transient water system monitoring based on state vulnerability assessment; vulnerable systems must be monitored quarterly for one year; repeat monitoring dependent upon detection and size of system.

Analytical methodology -- Nitrogen-phosphorus detector/gas chromatography(EPA 507): PQL= 0.0007 mg/L.

Best available technology -- Granular activated carbon

Reference -- 55 FR 30370 (07/25/90)

EPA Contact -- Drinking Water Standards Division / OGWDW /(202) 260-7575 / FTS 260-7575; or Safe Drinking Water Hotline / (800) 426-4791

IV.B.3. SECONDARY MAXIMUM CONTAMINANT LEVEL (SMCL) for Drinking Water

[Ed. Note: EPA does not have information yet in this section of IRIS].

IV.B.4. REQUIRED MONITORING OF "UNREGULATED" CONTAMINANTS

Status -- Listed (Final, 1991)

Discussion -- "Unregulated" contaminants are those contaminants for which EPA establishes a monitoring requirement but which do not have an associated final MCLG, MCL, or treatment technique. EPA may regulate these contaminants in the future.

Monitoring requirement -- All systems to be monitored unless a vulnerability assessment determines the system is not vulnerable.

Analytical methodology -- Microextraction/gas chromatography (EPA 505);nitrogen-phosphorus detector/gas chromatography (EPA 507); gas chromatographic/mass spectrometry (EPA 525).

Reference -- 56 FR 3526 (01/30/91)

EPA Contact -- Drinking Water Standards Division / OGWDW /(202) 260-7575 / FTS 260-7575; or Safe Drinking Water Hotline / (800) 426-4791

IV.C. CLEAN WATER ACT (CWA)

[Ed. Note: EPA does not have information yet in this section of IRIS].

IV.D. FEDERAL INSECTICIDE, FUNGICIDE, AND RODENTICIDE ACT (FIFRA)

IV.D.1. PESTICIDE ACTIVE INGREDIENT, Registration Standard

Status -- Issued (1984)

Reference -- Simazine Pesticide Registration Standard. March, 1984(NTIS No. PB84-212349).

EPA Contact -- Registration Branch / OPP (703)557-7760 / FTS 557-7760

IV.D.2. PESTICIDE ACTIVE INGREDIENT, Special Review

[Ed. Note: EPA does not have information yet in this section of IRIS].

IV.E. TOXIC SUBSTANCES CONTROL ACT (TSCA)

[Ed. Note: EPA does not have information yet in this section of IRIS].

IV.F. RESOURCE CONSERVATION AND RECOVERY ACT (RCRA)

[Ed. Note: EPA does not have information yet in this section of IRIS].

IV.G. SUPERFUND (CERCLA)

[Ed. Note: EPA does not have information yet in this section of IRIS].

VI. BIBLIOGRAPHY

VI.A. ORAL RfD REFERENCES

Ciba-Geigy Corporation. 1965. MRID No. 00023365, 00080631; HED Doc No. 003689, 007240, 007449. Available from EPA. Write to FOI, EPA, Washington, DC 20460.

Ciba-Geigy Corporation. 1984. MRID No. 00161407; HED Doc No. 004535, 005127, 007240. Available from EPA. Write to FOI, EPA, Washington, DC 20460.

Ciba-Geigy Corporation. 1985a. MRID No. 00143265; HED Doc No. 004656, 007240. Available from EPA. Write to FOI, EPA, Washington, DC 20460.

Ciba-Geigy Corporation. 1985b. MRID No. 00146655; HED Doc No. 004656, 007240. Available from EPA. Write to FOI, EPA, Washington, DC 20460.

Ciba-Geigy Corporation. 1986. MRID No. 40614403; HED Doc No. 007240, 007449. Available from EPA. Write to FOI, EPA, Washington, DC 20460.

Ciba-Geigy Corporation. 1988a. MRID No. 40614405; HED Doc No. 007240, 007449. Available from EPA. Write to FOI, EPA, Washington, DC. 20460.

Ciba-Geigy Corporation. 1988b. MRID No. 40614402; HED Doc No. 007240, 007449. Available from EPA. Write to FOI, EPA, Washington, DC 20460.

Ciba-Geigy Corporation. 1988c. MRID No. 40614404; HED Doc No. 007240, 007449. Available from EPA. Write to FOI, EPA, Washington, DC 20460.

Ciba-Geigy Corporation. 1991. MRID No. 41803601; HED Doc No. 008461. Available from EPA. Write to FOI, EPA, Washington, DC. 20460.

VI.B. INHALATION RfD REFERENCES

None

VI.C. CARCINOGENICITY ASSESSMENT REFERENCES

None

VI.D. DRINKING WATER HA REFERENCES

None

VII. REVISION HISTORY

Date	Section	Description
03/01/88	I.A.4.	Data gaps revised
07/01/89	I.A.	Withdrawn; new Oral RfD verified (in preparation)
09/01/89	II.	Carcinogen assessment now under review
11/01/89	I.A.	Oral RfD summary replaced; RfD changed
11/01/89	VI.	Bibliography on-line
09/01/91	I.A.	Withdrawn; new Oral RfD verified (in preparation)
09/01/91	VI.A.	Bibliography withdrawn
01/01/92	IV.	Regulatory actions updated
09/01/93	I.A.	Oral RfD summary replaced; RfD changed
04/01/94	I.A.1.	Study type corrected to 2-year

SYNONYMS

A 2079
AKTINIT S
AQUAZINE
BATAZINA
2,4-BIS(AETHYLAMINO)-6-CHLOR-1,3,5-
TRIAZIN
2,4-BIS(ETHYLAMINO)-6-CHLORO-s-
TRIAZINE
BITEMOL
BITEMOL S 50
CAT
CDT
CEKUSAN
CEKUZINA-S
CET1-CHLORO, 3,5-BISETHYLAMINO-2,4,6-
TRIAZINE
2-CHLORO-4,6-BIS(ETHYLAMINO)-1,3,5-
TRIAZINE
2-CHLORO-4,6-BIS(ETHYLAMINO)-s-
TRIAZINE
6-CHLORO-N,N'-DIETHYL-1,3,5-TRIAZINE-
2,4-DIYLDIAMINE
FRAMED
G 27692
GEIGY 27,692
GESARAN

GESATOP 50
H 1803
HERBAZIN 50
HERBEX
HERBOXY
HUNGAZIN DT
PREMAZINE
PRIMATOL S
PRINCEP
PRINTOP
RADOCON
RADOKOR
SIMADEX
SIMANEX
SIMAZIN
Simazine
SYMAZINE
TAFAZINE 50-W
TAPHAZINE
TRIAZINE A 384
1,3,5-Triazine-2,4,6-triamine, N-
cyclopropyl-
W 6658
ZEAPUR

Source: *Instant EPA's IRIS*, 1997, Instant Reference Sources, Inc., 7605 Rockpoint Drive, Austin, TX 78731 (Fax: 512-345-2386; Internet URL, http://www.instantref.com/inst-ref.htm).

Production, Use, and Pesticide Labeling Information

1. Description of the Chemical

    Generic Name: 2-chloro-4,6-bis(ethylamino)-s-triazine

    Common Name: simazine

    Trade Names: Princep, Aquazine, Gespun, Amazine, Primatol S, Simtrol, Algigon, Algi-eater, Atomicide, Algaecide, Algidize, Algae-A-Way, Cimacide, Cekusan, Framed, Gesatop, Simadex, and Simanex.

    EPA Shaughnessy Code: 080807

    Chemical Abstracts Service (CAS) Registry Number: 122-34-9

    Year of Initial Registration: 1957

    Pesticide Type: Herbicide

    Chemical Family: Triazine

    U.S. & Foreign Producers: Ciba-Geigy Corporation, Griffin Corporation, Aceto Chemical Company, Inc., Drexel Chemical Company, and Vertac Chemical Company.

2. Use Patterns and Formulations

    - Application sites: Simazine is registered for use as a selective or nonselective herbicide and algaecide. It is registered for use on agricultural, noncrop, forest, and aquatic sites.

    - Types of formulations: Wettable powder, granular, liquid, flowable concentrate, soluble concentrate, dry flowable and liquid ready-to-use.

    - Types and methods of applications: Broadcast, band, soil incorporated and soil surface application using ground or aerial equipment. The specific method of application and type of equipment are determined by site, formulation, and equipment availability.

    - Application rates: 1.6 lbs. a.i./A to 9.6 lbs. a.i./A, generally 4.0 lbs. a.i./A

    - Usual carriers: Water, oil, and clay

3. Science Findings

    - Chemical Characteristics:

        Simazine is a white, odorless, crystalline solid. It is stable to heat, and the melting point is 225-227 °C. Simazine is nonflammable and does not present unusual handling characteristics. Storage stability is > three years at room temperature under dry conditions.

    - Toxicological Characteristics:

        - Simazine is a moderate eye and dermal irritant (Toxicity Category III) and has low oral and dermal toxicities (Toxicity Category IV).

        - There are no data available for inhalation toxicity (a data gap exists for this requirement).

        - Toxicology studies on simazine are as follows:

            - Oral LD50 in rats: > 15,380 mg/kg body weight

            - Dermal LD50 in rats: 10.2 mg/kg body weight

            - Inhalation in rats: No data available for review

            - Skin irritation in rabbits: Slight irritant

            - Eye irritation in rabbits: Five test animals showed moderate irritation after 72 hours which was reversible in days. No corneal opacity was observed.

            - Teratology in rats: No data available for review. A data gap exists for this requirement in two species.

            - 3-generation reproduction study in rats: The reproduction NOEL is > 100 parts per million (ppm). No adverse effects on reproductive performance at a dietary level of 100 ppm for three generations over a total study period of 93 weeks.

            - Chronic feeding/oncogenicity in rats: Chronic toxicity and oncogenic potential could not be determined in this study. A data gap exists for oncogenic or chronic toxicity in the rodent (rat)

            - Chronic feeding in dogs: Neither chronic toxicity nor oncogenic potential could be determined from this study. A data gap exists for the chronic toxicity in non-rodents (dogs).

- Oncogenicity in mice: An oncogenicity study in a second species is required.

- Mutagenicity: Mutagenicity studies were not available for review Data gaps exist for the entire category of mutagenicity testing required for registration.

- General metabolism: A data gap exists for a required general metabolism study which identifies and quantifies metabolites formed in the exposure of mammalian species.

- Physiological & Biochemical Behavioral Characteristics:

- Foliar absorption: Absorbed mostly through plant roots with little or no foliar penetration. It has low adhering ability and is readily washed from foliage by rain.

- Translocation: Following root absorption, it is translocated acropetally in the xylem, accumulating in the apical meristems and leaves of plants.

- Mechanism of pesticidal action: A photosynthetic inhibitor, but may have additional effects.

- Metabolism & persistence in plants: Simazine is readily metabolized by tolerant plants to hydroxysimazine and amino acid conjugates. The hydroxysimazine can be further degraded by dealkylation of the side chains and by hydrolysis of resulting amino groups on the ring and some CO_2 production. These alterations of simazine are major protective mechanisms in most tolerant crop and weed species. Unaltered simazine accumulates in sensitive plants, causing chlorosis and death.

- Environmental Characteristics:

- Adsorption and leaching in basic soil types: Simazine is more readily adsorbed on muck or clay soils than in soils of low clay and organic matter content. The downward movement or leaching of simazine is limited by its low water solubility and adsorption to certain soil constituents. Tests have shown that for several months after surface application the greatest portion will be found in the surface two inches of soil. It has little, if any, lateral movement in soil, but can be washed along with soil particles.

- Microbial breakdown: Microbial breakdown is one of several processes involved in the degradation of simazine. In soils, microbial activity possibly accounts for decomposition of a significant amount of simazine.

- Loss from photodecomposition and/or volatilization: Under normal climatic conditions, loss of simazine from soil by photodecomposition and/or volatilization is considered insignificant.

- Bioaccumulation: Simazine has a low potential to bioaccumulate in fish.

- Resultant average persistence: The average half-life of simazine under anaerobic soil conditions is greater than 12 weeks. The half-life of simazine under aerobic soil conditions is 8 to 12 weeks. The persistence of simazine in ponds is dependent upon many factors, including the level of algae and weed infestation. The average half-life for simazine in ponds is 30 days.

- Ecological Characteristics:

- Avian oral LD50: > 4640 mg/kg (practically non-toxic)

- Avian dietary LC50: 2000 ppm to 11,000 ppm (moderately to slightly toxic)

- Fish LC50: 6.4 to 70.5 ppm (moderately to slightly toxic)

- Aquatic invertebrate LC50: 3.2 to 100 ppm (moderately to slightly toxic)

- The use of simazine could affect endangered aquatic species only if there was a direct application to the water where they dwell. Terrestrial endangered species may be affected, particularly for such uses as ditchbanks and rights-of-way. Formal consultation with the Office of Endangered Species, USFWS, may be initiated.

- Tolerance Assessments:

Tolerances have been established for simazine in a variety of food and forage crops, meat and poultry, milk and dairy products, and shellfish.

The Agency has reevaluated the existing data base, which revealed significant deficiencies. Evaluation of acute toxicity data did not reveal any adverse acute effects of simazine. Data are either lacking or insufficient to determine long-term chronic effects, oncogenicity potentials, teratogenicity, and mutagenicity. These data are crucial and necessary for the continuation of existing tolerances and for the consideration of additional tolerances. The information specifying sameness or differences between metabolites formed in plants and animals is also necessary.

- Problems Which are Known to have Occurred with Use of Chemical:

The files of the Pesticide Incident Monitoring System (PIMS) indicate 71 incidents involving simazine during the period 1966 to June, 1981. Two groups of reports were distinguished in these incidents in which alleged adverse effects were reported. One group containing 18 reports cited the involvement of simazine alone. The other contained 53 reports and cited simazine in combination with other ingredients. Humans were involved in 13 incidents in which simazine alone was cited as causing the alleged adverse effects. One person was hospitalized, and 12 received medical attention in these incidents. Humans were involved in 43 incidents in which simazine was cited in combination with other ingredients. Nine people were hospitalized, more than 35 received medical attention, and 412 were affected or involved and did not seek medical advice in these incidents.

- Summary of Science Findings:

Simazine is a moderate eye and dermal irritant with low oral and dermal toxicities. The available toxicity data are insufficient to fully assess the long-term chronic effects or the oncogenic, teratogenic, and mutagenic potential of simazine. The key data gap for most treated agricultural commodities is the simazine metabolites. Most established residue tolerances for raw agricultural commodities are expressed in terms of the parent compound only. Available data are insufficient to fully assess the environmental fate of simazine and the exposure of humans and nontarget organisms to simazine. The available simazine product chemistry data are insufficient to totally assess the chemical's characteristics. Simazine is not very toxic to nontarget insects, birds, or estuarine and marine organisms.

4. Summary of Regulatory Position and Rationale

- All terrestrial uses are RESTRICTED; all other uses are classified GENERAL.

- No major use, formulation, or geographical restrictions are required, except for groundwater as addressed below.

- No unique warning statements, protective clothing requirements, nor reentry interval statements are required on the labeling; however, the labeling must bear the following groundwater contamination precautionary statements:

Simazine is a chemical which can travel (seep or leach) through soil and can contaminate groundwater which may be used as drinking water. Simazine has been found in groundwater as a result of agricultural use. Users are advised not to apply simazine where the water table (groundwater) is close to the surface and where the soils are very permeable, i.e. well-drained soils such as loamy sands. Your local agricultural agencies can provide further information on the type of soil in your area and the location of groundwater.

- No risk assessments were conducted.

5. Summary of Major Data Gaps

- The available product and residue chemistry data are insufficient to fully assess the chemical's characteristics. After receipt of guidance package, data must be submitted within 6 months for short-term studies and within 4 years for long-term studies. Long-term studies include simazine and its metabolites in meat, milk, poultry, eggs, and other commodities.

- The available toxicity data are insufficient to fully assess the long- term chronic effects and the oncogenic, teratogenic, and mutagenic potential of simazine. Data gap also exists for a general metabolism study in mammalian species. These studies must be submitted within four years after receipt of guidance package.

- The following data are required to fully assess the environmental fate and transport of, and the potential exposure to simazine: (a) photo- degradation studies on soil and in water; (b) aerobic soil metabolism; (c) anaerobic and aerobic aquatic metabolism; (d) leaching and adsorption; (e) field dissipation studies: aquatic, forestry and long- term; (f) accumulation studies on rotational and irrigated crops. These studies must be submitted within the time periods indicated in Table A, Generic Data Requirements for Simazine, 158.130 Environmental Fate.

6. Contact Person at EPA:

Richard F. Mountfort
Product Manager (23)
Environmental Protection Agency (TS-767C)
401 M Street, S.W.
Washington, D.C. 20460

Telephone: (703) 557-1830

Source: *Instant EPA's Pesticide Facts*, 1994, Instant Reference Sources, Inc., 7605 Rockpoint Drive, Austin, TX 78731 (Fax: 512-345-2386; Internet URL, http://www.instantref.com/inst-ref.htm).

Styrene

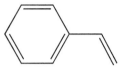

Introduction

Styrene is used in the manufacture of plastics, synthetic rubber, resins, and in insulators. It is also important in the manufacture of some paints and it is used in the preparation of acrylonitrile-butadiene-styrene and stryene-acrylonitrile polymer resins. It is an important ingredient in the manufacture of styrenated polyesters, rubber-modified polystyrene, and copolymer resin systems used as a diluent to reduce viscosity of uncured resin systems. It is used with glass fibers in the construction of boats and in the synthesis of styrene-divinylbenzene copolymers as a matrix for ion exchange resins. Styrene is used as a cross-linking agent in polyester resins. It is a FDA-approved flavoring agent for ice cream and candy. [828]

Significant amounts of styrene may be released to the environment from emissions generated by its production and use and from automobile exhaust. If released to the atmosphere, styrene will react rapidly with both hydroxyl radicals and ozone with a combined, calculated half-life of about 2.5 hours. If released to environmental bodies of water, styrene will volatilize relatively rapidly and may be subject to biodegradation, but is not expected to hydrolyze. If released to soil it will biodegrade and leach with a low-to-moderate soil mobility. While styrene has been detected in various US drinking waters, it was not detected in a groundwater supply survey of 945 US finished water supplies which use groundwater sources. Styrene has been detected in various US chemical, textile, latex, oil refinery and industrial wastewater effluents. Styrene has been frequently detected in the ambient air of source dominated locations and urban areas, has been detected in the air of a national forest in Alabama, and has been detected in the vicinity of oil fires. Food packaged in polystyrene containers has been found to contain small amounts of styrene. [828]

Identifying Information

CAS NUMBER: 100-42-5

NIOSH Registry Number: WL3675000

SYNONYMS: Ethenylbenzene, Phenylethylene, Vinylbenzene

Endocrine Disruptor Information

Styrene is a suspected EED that is listed by the Centers for Disease Control and Prevention [812].

Chemical and Physical Properties

CHEMICAL FORMULA: C_8H_8

MOLECULAR WEIGHT: 104.16

WLN: 1U1R

PHYSICAL DESCRIPTION: Colorless to yellowish, oily liquid

SPECIFIC GRAVITY: 0.9060 @ 20/4°C

DENSITY: 0.909 g/cc @ 20°C [700]

MELTING POINT: -31 to -30.6°C

BOILING POINT: 145-146°C

SOLUBILITY: Water: <1 mg/mL @ 19°C [700]; DMSO: >=100 mg/mL @ 19°C [700]; 95% Ethanol: >=100 mg/mL @ 19°C [700]; Acetone: >=100 mg/mL @ 19°C [700]

OTHER SOLVENTS: methanol: Soluble; carbon disulfide: Soluble; petroleum ether: Soluble; ether: Soluble; benzene: Soluble

OTHER PHYSICAL DATA: Liquid surface tension: 32.14 dynes/cm Liquid-water interfacial tension: 35.48 dynes/cm Critical temperature: 373°C Critical pressure: 39.46 atm Odor threshold: 0.148 ppm Latent heat of vaporization: 156 Btu/lb Heat of polymerization: -277 Btu/lb Penetrating odor

Source: *Instant EPA's Air Toxics*, 1994, Instant Reference Sources, Inc., 7605 Rockpoint Drive, Austin, TX 78731 (Fax: 512-345-2386; Internet URL, http://www.instantref.com/inst-ref.htm).

Environmental Reference Materials

A source of EED environmental reference materials for styrene is Radian International LLC, Austin, TX (Fax 512-454-0268; Internet URL, http://www.radian.com/standards); Catalog No. ERS-058.

Hazardous Properties

ACUTE/CHRONIC HAZARDS: Styrene is an irritant and it may be narcotic in high concentrations. When heated to decomposition it emits acrid fumes.

HAP WEIGHTING FACTOR: 1 [713]

VOLATILITY:
   Vapor pressure: 4.3 mm Hg @ 15°C; 9.5 mm Hg @ 30°C; 10 mm Hg @ 35°C
   Vapor density: 1.1

FIRE HAZARD: The flash point of styrene is 31.1°C (88°F) and it is flammable. Fires involving this chemical should be controlled using a dry chemical, carbon dioxide or Halon extinguisher.

The autoignition temperature of styrene is 490°C (914°F).

LEL: 1.1%   UEL: 6.1%

REACTIVITY: Styrene reacts violently with chlorosulfonic acid, oleum, sulfuric acid, and alkali metal graphite. It also reacts vigorously with oxidizing materials. Styrene readily undergoes polymerization when heated or exposed to light or a peroxide catalyst. The polymerization releases heat and it may react violently. It is incompatible with aluminum chloride. It can corrode copper and copper alloys.

STABILITY: Styrene is sensitive to heat, light and may be sensitive to air. UV spectrophotometric stability screening indicates that solutions of it in ethanol are stable for 24 hours [700].

Source: *Instant EPA's Air Toxics*, 1994, Instant Reference Sources, Inc., 7605 Rockpoint Drive, Austin, TX 78731 (Fax: 512-345-2386; Internet URL, http://www.instantref.com/inst-ref.htm).

Medical Symptoms of Exposure

Symptoms of exposure may include weakness, irritation to the eyes and mucous membranes, conjunctivitis, central nervous system depression, loss of coordination, drowsiness, lack of appetite, nausea and vomiting and dermatitis. It is also narcotic in high concentrations.

Source: *Instant EPA's Air Toxics*, 1994, Instant Reference Sources, Inc., 7605 Rockpoint Drive, Austin, TX 78731 (Fax: 512-345-2386; Internet URL, http://www.instantref.com/inst-ref.htm).

Toxicological Information

The toxicological information below is gathered from several sources including two national databases: one from the *National Toxicology Program Chemical Repository Database* and another from *EPA's Integrated Risk Information System (IRIS)*.

Toxicology Information from the National Toxicology Program

CARCINOGENICITY:

Tumorigenic Data:

Type of Effect	Route/Animal	· Amount Dosed (Notes)
TDLo:	scu-rat	16 gm/kg/61W-I
TDLo:	scu-mus	35 gm/kg/87W-I

TERATOGENICITY:

Reproductive Effects Data:

Type of Effect	Route/Animal	Amount Dosed (Notes)
TCLo:	ihl-rat	17 mg/m3/2H (7-14D preg)
TDLo:	orl-rat	3120 mg/kg (6-15D preg)
TDLo:	orl-rat	780 mg/kg (6-15D preg)
TDLo:	orl-rat	25 gm/kg (6-15D preg)

MUTATION DATA:

Test	lowestdose		Test	Lowestdose
sln-dmg-par	396 umol/L	|	sln-dmg-orl	2000 ppm
dlt-dmg-par	396 umol/L	|	dlt-dmg-orl	6000 ppm
msc-mus:lym	400 mg/L	|	sce-hmn:lym	1mg/L

TOXICITY:

Typ.dose	Mode	Specie	Amount	Units	Other
LD50	orl	rat	971	mg/kg	
LD50	ivn	mus	100	mg/kg	
LD50	unr	mam	1400	mg/kg	
LCLo	ihl	rat	580	mg/m3	
LD50	unr	rat	1390	mg/kg	
LD50	orl	mam	2014	mg/kg	

OTHER TOXICITY DATA:

Skin and Eye Irritation Data:
skn-rbt 500 mg open MLD
eye-rbt 80 mg MOD

Review: Toxicology Review

Status:
EPA Genetox Program 1988, Positive: Cell transform.-
SA7/SHE
EPA Genetox Program 1988, Negative: D melanogaster-
whole sex chrom. loss
EPA Genetox Program 1988, Negative: In vitro UDS-human
fibroblast
EPA Genetox Program 1988, Negative: S cerevisiae gene
conversion; S cerevisiae-homozygosiis
EPA Genetox Program 1988, Inconclusive: D melanogaster-
nondisjunction
EPA Genetox Program 1988, Inconclusive: D melanogaster-
partial sex chrom. loss
EPA Genetox Program 1988, Inconclusive: Histidine
reversion-Ames test
EPA TSCA Chemical Inventory, 1986
Meets criteria for proposed OSHA Medical Records Rule

SAX TOXICITY EVALUATION:

THR: Poison intravenous route. Mutagenic data. An experimental tumorigen.
A skin and eye irritant.

Source: *Instant Tox-Base*, 1995, Instant Reference Sources, Inc., 7605 Rockpoint Drive, Austin, TX 78731 (Fax: 512-345-2386; Internet URL, http://www.instantref.com/inst-ref.htm).

Toxicology Information from EPA's Integrated Risk Information System (IRIS)

I. CHRONIC HEALTH HAZARD ASSESSMENTS FOR NONCARCINOGENIC EFFECTS

I.A. REFERENCE DOSE FOR CHRONIC ORAL EXPOSURE (RfD)

The Reference Dose (RfD) is based on the assumption that thresholds exist for certain toxic effects such as cellular necrosis, but may not exist for other toxic effects such as carcinogenicity. In general, the RfD is an estimate (with uncertainty spanning perhaps an order of magnitude) of a daily exposure to the human population (including sensitive subgroups) that is likely to be without an appreciable risk of deleterious effects during a lifetime. Please refer to Background Document 1 in Service Code 5 for an elaboration of these concepts. RfDs can also be derived for the noncarcinogenic health effects of compounds which are also carcinogens. Therefore, it is essential to refer to other sources of information concerning the carcinogenicity of this substance. If the US EPA has evaluated this substance for potential human carcinogenicity, a summary of that

evaluation will be contained in Section II of this file when a review of that evaluation is completed.

NOTE: The Oral RfD for styrene may change in the near future pending the outcome of a further review now being conducted by the Oral RfD Work Group.

I.A.1. ORAL RfD SUMMARY

Critical Effect	Experimental Doses*	UF	MF	RfD
Red blood cell and liver effects	NOAEL: 200 mg/kg-day	1000	1	2E-1 mg/kg-day
	LOAEL: 400 mg/kg-day			

*Conversion Factors: none

Dog Subchronic Oral Study, Quast et al., 1979

I.A.2. PRINCIPAL AND SUPPORTING STUDIES (ORAL RfD)

Quast, J.F., C.G. Humiston, R.Y. Kalnins, et al. 1979. Results of a toxicity study of monomeric styrene administered to beagle dogs by oral intubation for 19 months. Toxicology Research Laboratory, Health and Environmental Sciences, DOW Chemical Co., Midland, MI. Final Report.

Four beagle dogs/sex were gavaged with doses of 0, 200, 400, or 600 mg styrene/kg bw/day in peanut oil for 560 days. No adverse effects were observed for dogs administered styrene at 200 mg/kg-day. In the higher dose groups, increased numbers of Heinz bodies in the RBCs, decreased packed cell volume, and sporadic decreases in hemoglobin and RBC counts were observed. In addition, increased iron deposits and elevated numbers of Heinz bodies were found in the livers. Marked individual variations in blood cell parameters were noted for animals at the same dose level. Other parameters examined were body weight, organ weights, urinalyses, and clinical chemistry. The NOAEL in this study is 200 mg/kg-day and the LOAEL is 400 mg/kg-day.

Long-term studies (120 weeks) in rats and mice (Ponomarkov and Tomatis, 1978) showed liver, kidney, and stomach lesions for rats (dosed weekly with styrene at 500 mg/kg) and no significant effects for mice (dosed weekly with 300 mg/kg). Rats receiving an average daily oral dose of 95 mg styrene/kg bw for 185 days showed no adverse effects, while those receiving 285 or 475 mg/kg-day showed reduced growth and increased liver and kidney weights (Wolf et al., 1956). Other subchronic rat feeding studies found LOAELs in the 350-500 mg/kg-day range and NOAELs in the range of 100-400 mg/kg-day.

The lifetime studies in rats and mice (Ponomarkov and Tomatis, 1978) are not appropriate for risk assessment of chronic toxicity because of the dosing schedule employed. The Wolf et al. (1956) study is of insufficient duration (185 days) to be considered chronic.

I.A.3. UNCERTAINTY AND MODIFYING FACTORS (ORAL RfD)

UF -- The uncertainty factor of 1000 reflects 10 for both intraspecies and interspecies variability to the toxicity of this chemical in lieu of specific data, and 10 for extrapolation of subchronic effects to chronic effects.

MF -- None

I.A.4. ADDITIONAL COMMENTS (ORAL RfD)

None.

I.A.5. CONFIDENCE IN THE ORAL RfD

Study -- Medium
Data Base -- Medium
RfD -- Medium

The principal study is well done and the effect levels seem reasonable, but the small number of animals/sex/dose prevents a higher confidence than medium at this time. The data base offers strong support, but lacks a bona fide full-term chronic study; thus, it is also considered to have medium confidence. Medium confidence in the RfD follows.

I.A.6. EPA DOCUMENTATION AND REVIEW OF THE ORAL RfD

US EPA. 1984. Health and Environmental Effects Profile for Styrene. Prepared by the Office of Health and Environmental Assessment, Environmental Criteria and Assessment Office, Cincinnati, OH for the Office of Solid Waste, Washington, DC.

US EPA. 1985. Drinking Water Criteria Document for Styrene. Prepared by the Office of Health and Environmental Assessment, Environmental Criteria and Assessment Office, Cincinnati, OH for the Office of Drinking Water, Washington, DC.

The ADI in the 1984 Health and Environmental Effects Profile document has received an Agency review with the help of two external scientists.

Agency Work Group Review -- 10/09/85, 11/06/85, 10/09/85

Verification Date -- 10/09/85

I.A.7. EPA CONTACTS (ORAL RfD)

Julie Du / OST -- (202)260-7583

Edward V. Ohanian / OST -- (202)260-7571

I.B. REFERENCE CONCENTRATION FOR CHRONIC INHALATION EXPOSURE (RfC)

The inhalation Reference Concentration (RfC) is analogous to the oral RfD and is likewise based on the assumption that thresholds exist for certain toxic effects such as cellular necrosis, but may not exist for other toxic effects such as carcinogenicity. The inhalation RfC considers toxic effects for both the respiratory system (portal-of-entry) and for effects peripheral to the respiratory system (extrarespiratory effects). It is appropriately expressed in units of mg/cu.m. In general, the RfC is an estimate (with uncertainty spanning perhaps an order of magnitude) of a daily inhalation exposure of the human population (including sensitive subgroups) that is likely to be without an appreciable risk of deleterious effects during a lifetime. Inhalation RfCs are derived according to the Interim Methods for Development of Inhalation Reference Doses (EPA/600/8-88/066F August 1989) developed by US EPA scientists and peer-reviewed. For more information on the interim nature of these methods and future plans see the INFOMORE Section of IRIS. RfCs can also be derived for the noncarcinogenic health effects of compounds which are carcinogens. Therefore, it is essential to refer to other sources of information concerning the carcinogenicity of this substance. If the US EPA has evaluated this substance for potential human carcinogenicity, a summary of that evaluation will be contained in Section II of this file when a review of that evaluation is completed.

I.B.1. INHALATION RfC SUMMARY

Critical Effect	Exposures*	UF	MF	RfC
CNS effects	NOAEL: 94 mg/cu.m (25 ppm = 150 mmole urinary styrene metabolites/mole creatinine adjusted to lower 95% confidence limit = 22 ppm) NOAEL(HEC): 34 mg/cu.m LOAEL: >94 mg/cu.m (>22 ppm derived as in NOAEL listing)	30	1	1E+0 mg/cu.m

*Conversion Factors: MW = 104.15. Assuming 25 C and 760 mmHg, NOAEL (mg/cu.m) = NOAEL (ppm) x MW/24.45 = 94 mg/cu.m. The NOAEL exposure level is based on a back extrapolation from worker urinary concentration of styrene metabolites reported in the principal study and adjusted to the lower 95% confidence limit listed in Guillemin et al. (1982), which was 88%, 25 ppm x 0.88 = 22 ppm. The NOAEL(HEC) is calculated using an 8-hour TWA occupational exposure. MVho = 10 cu.m/day, MVh = 20 cu.m/day. NOAEL(HEC) = 94 mg/cu.m x MVho/MVh x 5 days/7 days = 34 mg/cu.m. The feasibility of applying the exposure model of Perbellini et al. (1988) for extrapolation of the values in the principal study is currently being investigated.

Application of this model may result in changes in the NOAEL(HEC) value and, therefore, the RfC.

Occupational Study, Mutti et al. (1984)

I.B.2. PRINCIPAL AND SUPPORTING STUDIES (INHALATION RfC)

Mutti, A., A. Mazzucchi, P. Rusticelli, G. Frigeri, G. Arfini, and I. Franchini. 1984. Exposure-effect and exposure-response relationships between occupational exposure to styrene and neuropsychological functions. Am. J. Ind. Med. 5: 275-286.

In a cross-sectional study, Mutti et al. (1984) examined the neuropsychological function in 50 workers whose mean duration of styrene exposure was 8.6 (SD of 4.5) years. Styrene exposure was assessed by the authors to correspond to air concentrations ranging from 10-300 ppm as a mean daily exposure. These concentrations were estimated from the summation of the principal urinary metabolites of styrene, mandelic acid (MA) and phenylglyoxylic acid (PGA). Urinary metabolite levels are considered as reliable biological indicators of styrene exposure (ACGIH, 1986; WHO, 1983), and several laboratories have determined collectively that the specific method used in this study, the summation of the principal metabolites collected in next-morning urine, is the most reliable and representative of actual air exposure concentrations (Guillemin et al., 1978, 1982; Ikeda et al., 1982; Franchini et al., 1983). Workers with absence of metabolic and neurologic disorders, smoking habits of <20 cigarettes/day, and an alcohol intake of <80 mL of ethanol/day were chosen. These same eligibility criteria were used to select a control group of 50 workers that was matched for age, sex, and educational level. The exposed workers were further segregated into four subgroups (n = 9-14) according to increasing levels of urinary styrene metabolites. A battery of neuropsychological tests was conducted on the same day as the urine collection and included exams evaluating visuo-motor speed, memory, and intellectual function. No other endpoints were considered. Correlation analysis of the test results and urinary metabolite levels showed a clear concentration response in at least three of eight tests, including block design (intellectual function), digit-symbol (memory), and reaction times (visuo-motor speed). Evidence of a concentration-response relationship was also present for short- and long-term logical memory and embedded figures (impaired visual perception). When the results were analyzed using duration of exposure as a covariate, increases in reaction times and a decrease in digit symbol (memory, concentration) were apparent. The only test showing results in the lowest exposure group, short-term verbal memory loss, exhibited no concentration-response relationship. The neuropsychological results from this study are from established tests for CNS dysfunction, are present when compared against a stringently matched control population, and show concentration-response relationships. Also, the deficiencies noted in the reaction-times corroborate the results presented by Moller et al. (1990) and others discussed below.

The concentration-response relationship between urinary metabolite concentration (mandelic acid and phenylglyoxylic acid levels normalized to creatinine in "morning-after" urine) and test results indicated a significant effect level in the subgroup whose urine contained 150-299 mmole urinary metabolites/mole creatinine. Workers with metabolite concentrations of up to 150 mmoles/mole appeared to have no significant effects, and this level is therefore designated as the NOAEL in this study. The authors state that this level of urinary metabolites corresponds to a mean daily 8-hour exposure to air styrene of 25 ppm (106 mg/cu.m). Derivation of this air level is from the creatinine-normalized, combined concentration of the styrene metabolites, MA and PGA, in urine collected from the workers on Saturday mornings. Guillemin et al. (1982) demonstrated a logarithmic relationship (r = 0.871) between the summation of urinary metabolites (MA + PGA, next morning) and air concentrations of styrene (ppm x hours). Guillemin calculated the mean combined urinary metabolite concentration (next morning) for an 8-hour exposure to 100 ppm. This relationship was used by both Mutti et al. (1984) and Guillemin and Berode (1988) in a proportional manner to obtain styrene air levels at lower urinary metabolite concentrations. The 95% confidence interval was also calculated for an 8-hour exposure at 100 ppm, the lower limit of the confidence calculation being 88% of the mean styrene exposure. This factor was applied directly to the NOAEL of 25 ppm [25 ppm x 0.88 = 22 ppm (94 mg/cu.m)]. Due to the construction of the subgroups, designation of a LOAEL was the lower limit of the subgroup in which adverse effects were observed [i.e., greater than the NOAEL of 22 ppm (94 mg/cu.m)].

I.B.3. UNCERTAINTY AND MODIFYING FACTORS (INHALATION RfC)

UF -- A partial UF of 3 was used for data base inadequacy, including the lack of concentration-response information on respiratory tract effects. A partial UF of 3 instead of 10 was used for intraspecies variability since the lower 95% confidence limit of the exposure extrapolation was used and because Perbellini et al. (1988) demonstrated that this biological exposure index (i.e., urinary metabolites) accounts for differences in pharmacokinetic/ physiologic parameters such as alveolar ventilation rate. A partial UF of 3 instead of 10 was also evoked for lack of information on chronic studies as the average exposure duration of the principal study of Mutti et al. (1984) was not long enough (8.6 years) to be considered chronic. The total uncertainty is therefore 30 (three times the one-half logarithm of 10).

MF -- None

I.B.4. ADDITIONAL STUDIES / COMMENTS (INHALATION RfC)

Central nervous system effects caused by exposure to styrene have been reported by several investigators in studies since the early 1970s. Gotell et al. (1972) noted significantly increased reaction times in a group of six polyester plant workers exposed to >150 ppm styrene as compared with unexposed controls and another group exposed to <150 ppm. Gamberle et al. (1975) noted longer and more irregular reaction times in 106 exposed workers (estimated at 13-101 ppm styrene for an average of 2.7 years) as compared with a control population. Cherry et al. (1980) found no alteration in reaction times among 27 workers exposed to a mean of 92 ppm styrene (duration unspecified) and no difference in performance on four behavioral tests of memory and vigilance. However, a follow-up study conducted 21 months later by these authors on 8 of the same 27 individuals (Cherry et al., 1981) found that the three men with the highest urinary mandelic acid concentrations all improved reaction times significantly (p < 0.05) as compared with their earlier values. These alterations were correlated with those individuals having slow clearance of mandelic acid. Among 98 male laminating workers (mean styrene exposure of 5.1 years), Lindstrom et al. (1976) reported weak correlations of CNS-behavioral testing (poor psychomotor performance and visuomotor inaccuracy) with levels of styrene exposure estimated to be between 25 and 75 ppm based on urinary mandelic acid levels. The study of Mutti et al. (1984) is one of few in which extensive CNS-behavioral testing was carried out. The recent studies of Moller et al. (1990) and Flodin et al. (1989) couple functional decrements with adverse behavioral effects in a chronically exposed worker population. The effect of styrene on peripheral nerves has also been reported, although not to the extent of central effects. These studies are briefly described in the recent study of Murata et al. (1991) in which data on a small number of workers suggest that styrene may differentially affect sensory nerves.

The studies of Moller et al. (1990) and Flodin et al. (1989) both examined central effects of styrene in the same worker population. The Flodin et al. (1989) study on neuropsychiatric effects of styrene exposure provides documentation of the styrene levels to which these workers were exposed.

In a cross-sectional occupational study, Moller et al. (1990) studied 18 male Swedish boatbuilders exposed to styrene for an average of 10.8 years (range of 6-15 years). Personal sampling (8-hour TWA concentrations) available for 7/10 years showed that the workers had been exposed to 50-140 mg/cu.m styrene. The exposure data is discussed at length in the text of the Flodin et al. (1989) study below. These workers were not further subgrouped into high-and low-exposure groups as in the Flodin et al. (1989) study. Two reference groups were used for evaluation of the tests; both were unexposed to industrial solvents and matched to the exposure group with respect to smoking and alcohol consumption. These workers were subsequently given a thorough otoneurological examination, evaluating auditory, visual, and vestibular systems, as well as coordination of vestibular sensations with compensatory eye movements manifest in the vestibuloocular reflex (VOR). The test showing the greatest and most consistent deviation in performance between the exposed workers and the reference population was the visual suppression of the VOR. Execution of this test required the subjects to fixate on a target that moved with movement of the subjects' chair. The normal VOR functions to maintain steady gaze of the eyes despite movements of the head. However, when the target moves with the subject, this reflex must be suppressed in order to follow the target. Quantitation of a subject's capacity to suppress VOR is measured as the ratio between eye and target velocity (gain) and in the temporal relationship between eye and target velocity (phase). Abnormal functioning would be manifest by an increase in gain and a decrease in phase, if the values exceeded the mean +/- 2 SD units of the corresponding measurement in a reference population. Abnormal phase shifts were recorded for 4/18 workers; in each of the four cases, the gain was also abnormal. These differences were reflected in the group values that showed a higher mean gain (p < 0.001) and a decrease in angle of phase lead (p < 0.001) in examinations carried out with predictable (sinusoidal frequency sweep) and unpredictable (pseudorandomized) conditions. The deficits shown in these tests suggest lesions in the brainstem or cerebellar regions, because these findings are in accordance with findings in patients with known brainstem or cerebellar disorders (Odkvist et al., 1982, 1987). Other tests reported in this study also gave indications of neurological deficits. The posturography test demonstrated increases in sway area of subjects with their eyes closed, which are consistent with vestibular deficits. The overall results of this test are, however, internally inconsistent as more of these workers showed abnormal scores with their eyes open than with their eyes closed. Although examination of saccadic eye movements (brief, fast movements occurring with change in fixation point) did reveal abnormal scores in latency for 7/18 subjects, none of the 18 were found to have abnormal velocity. This also is an internal inconsistency within the exam, making these results equivocal. A similar internal inconsistency was noted in the results of the smooth-pursuit eye movement during which the subject is asked to visually track a slowly moving stimulus. Seven of the 18 subjects had abnormal results in lag time (phase), whereas none of the 18 tested abnormal for gain. Disturbances were found in the central auditory pathways of seven workers but the significance of these effects were not evaluated (see discussion of Pryor et al., 1987). This study provides credible toxicological information on the possible long- term consequences of human exposure to styrene. No effect levels are assigned based on these data as the number of subjects is small and no indication is given of a dose-response relationship among these chronically exposed workers.

Twenty-one male Swedish workers were exposed to styrene for an average of 11.6 years (range of 6-21 years) as a consequence of their occupation, boat building (Flodin et al., 1989). Personal sampling (8-hour average concentrations) had been carried out in 9 of the last 12 years, with the number of sampling days for each year ranging from 3 to 27; the data presented indicate an exposure range of around 40-140 mg/cu.m during this time. In one year, 30-minute sampling times were performed on most tasks and showed that peak exposures were exceptional, with only about 1% of all measurements of styrene >300 mg/cu.m; about 44% of the samples were <25 mg/cu.m, and about 65% were <50 mg/cu.m. Based on these sampling data, the workers were classified into two groups: those exposed to about 50 mg/cu.m styrene for the past 7 years and those exposed to about 25 mg/cu.m. The higher exposed workers were primarily involved in lamination processes. Workers were examined by a clinical interview that included a detailed inquiry about neuropsychiatric symptoms; psychometrical tests and some clinical chemistry were also included. The workers were examined twice, the first time (21 workers examined) occurring when they had not been exposed for 1 week. The second examination (17 workers examined; nine in the higher exposure group and eight in the lower exposure group) occurred after the workers had not been exposed for a minimum of 3 months because the factory had gone bankrupt. Two of the remaining four workers refused participation in the second interview, and results from two others were disallowed as one was diagnosed as having pulmonary thrombosis and the other psycho-organic syndrome (POS) (see discussion of Moller et al., 1989). No concurrent control group was used. According to the questionnaire, the symptoms at the first examination were distributed in a exposure-related manner, with the higher exposure group having an average of 10.4 symptoms/worker and the lower exposure group having an average of 5.3 symptoms/worker. On the second examination, the average symptoms/worker decreased to 1.9 in both exposure groups. The claim that the higher exposure group performed significantly worse in one psychometrical test (manual dexterity) is not interpretable due to lack of specifics in the text. Internal inconsistencies of the results and the lack of a concurrent matched control group cause the results from this study to be equivocal. No effect levels were designated in this study.

The levels of styrene reported in the study of Mutti et al. (1984) were estimated from levels of urinary styrene metabolites. As already discussed, this method is considered as a biological indicator of exposure and has been shown to reliably reflect the total concentration of styrene to which individuals have been exposed. The relationships between styrene exposure and urinary metabolite concentrations have been established by a number of studies including those of Guillemin et al. (1982) and Ikeda et al. (1982)

and reviewed by the WHO (1983). The work of Perbellini et al. (1988) has shown that the variability of urinary metabolite levels observed in styrene-exposed worker populations may reflect physiologic and metabolic variability inherent in humans. These characteristics are not shared by the extensive studies on styrene levels reported by Lemasters et al. (1985a) and of Jensen et al. (1990), both of which are historical and are based on area and personal sampling procedures.

A diagnosis of psycho-organic syndrome has been used to describe the symptomatology observed in workers heavily exposed to a variety of organic solvents. Examples of specific effects include neurasthenia, personality alterations, unsteadiness, dizziness, and vertigo. Moller et al. (1989) examined nine men who had been diagnosed as having POS, subjecting them to a battery of audiological and vestibular-oculomotor tests, the latter of which measure the capacity to transform signals from the inner ear to compensatory eye movements for purposes such as maintenance of equilibrium. The men were described as having been exposed to various mixtures of alcoholic, aromatic, and aliphatic industrial solvents during their working careers of 8-30 years (mean = 21 years). Seven of them had been granted disability pensions at the time of the examinations. Abnormal static posture was noted in 4/9 members of the POS group (p < 0.001). Voluntary saccades (rapid intermittent movements of the eye) were abnormal (both prolonged latency and decreased maximum speed) in 5/9 of the POS group, but in only 2/9 matched controls. As these functions are considered to be controlled centrally in the area of the cerebellum, these findings indicate that cerebellar lesions may occur from chronic exposures to a variety of organic solvents. The results presented by Moller et al. (1990) for these tests in workers exposed to styrene indicate this chemical to be capable of eliciting this toxicity.

A duration-response between length of exposure to solvents and incidence of altered cerebellar functioning (as evidenced by effects on hearing and the vestibular-oculomotor system) is indicated in the study of Odkvist et al. (1987). This study examined 23 workers, all of whom had been extensively exposed to aliphatic and aromatic solvents. Sixteen of the 23 workers were diagnosed with POS. The average length of exposure of these 16 workers was 27 years (range 9-40 years), considerably more than the average of 21 years (range 5-30 years) for the remaining seven individuals. In the battery of 11 vestibular-oculomotor tests, five (including saccade, visual-suppression, Romberg's test, and electronystagmography) showed a dose-response relationship with the percentage incidences being higher in the group diagnosed with POS; the other six tests either were not conducted in one or the other group or exhibited no change. Thus, both the number of workers diagnosed with POS and incidence of cerebellar dysfunctioning were correlated with duration of exposure to organic solvents.

The specific capacity of styrene to cause alterations in cerebellar function in humans under short-term acute exposure conditions was experimentally shown by Odkvist et al. (1982). Ten people (5/sex) were exposed to 370-591 mg/cu.m styrene for 80 minutes. A battery of six vestibular-oculomotor tests was administered before, during, and after the exposure. Visual suppression and saccade tests both showed statistically significant alterations in 8/10 subjects. Results between exposed subjects and controls did not differ for the optovestibular, optokinetic, and slow pursuit movement test or the sinusoidal swing test. These results indicate that acute exposures to high concentrations of styrene may affect processes within the cerebellum. It should be noted, however, that the results obtained by Moller et al. (1990) and Flodin et al. (1989) were obtained in workers that had not been exposed to styrene for a minimum of 3 months.

Larsby et al. (1978) investigated the relationship between vestibulo- oculomotor function and styrene levels in arterial blood and cerebrospinal fluid. Rabbits (n = 16) were cannulated and infused with a 10% solution of styrene introduced at 3.1-12.6 mg/minute. Vestibular function was evaluated using electronystagmography in response to rotary acceleration. Styrene concentration was monitored in both arterial (ear) and cerebrospinal fluid. During the exposure, positional nystagmus (i.e., involuntary eye movement response to rotary movement evident only when the animal is lying on one side or the other, not prone) was observed in 10/11 rabbits tested. In addition, a paradoxical rotary response was observed in 6/16 rabbits. This phenomenon is described as involuntary eye movement opposite to the direction of rotation (left-beating eye movements when rotated clockwise and vice versa). Both observations indicate that vestibular function was affected. Results also showed that, at an infusion rate of 4.9 mg/minute, arterial blood levels reached a constant level only after about 2 hours. Styrene concentrations in cerebrospinal fluid had the same shape as the arterial curve and was constantly 5-10% of the corresponding arterial concentration.

A number of other occupational studies investigating the effects of styrene on workers are available in the literature. Most of these reports examine either central or peripheral nervous function, although blood effects (Stengel et al., 1990), nephrotoxicity (Viau et al., 1987), and liver effects (Hotz et al., 1980) have also been examined. Nearly all of these reports suffer from one or more deficiencies, the most common being lack of exposure information. Almost all, however, indicate that styrene affects central processes in humans. An overview of several of these studies follows.

In addition to the auditory data listed in the Moller et al. (1990) study, two other studies have reported hearing loss, one in humans and the other in laboratory animals. Muijser et al. (1988) evaluated hearing thresholds up to 16 kHz in 59 workers exposed to airborne styrene. Air samples (4-hour mean averages, 6-16 samples/group work area) were taken in the breathing zone during 3 consecutive days in work areas where 31 workers were directly exposed to styrene (mean = 138 mg/cu.m) and 28 workers were indirectly exposed to styrene (mean = 61 mg/cu.m). The duration of employment for the combined group was 8.6 +/- 6.5 years. The control population consisted of 88 individuals not exposed to styrene or other chemicals but who were comparable in age and socio-economic status to the exposed workers. Audiometric analyses were corrected for age. Comparison of hearing thresholds between the controls and both exposed groups revealed no differences in hearing thresholds at frequencies up to and including 16 kHz. Comparison between the experimental groups, however, did reveal that the directly exposed workers had higher thresholds for frequencies at 8-16 kHz than did the indirectly exposed workers (p value from multivariate analysis of covariance = 0.012). Quantitative and qualitative differences in the background noise between the control and study plants may have compromised the control population in this study. Although difficulties with the control population prevent definite assignment of effect levels, human exposure to 138 mg/cu.m styrene appears to have resulted in high-frequency hearing loss [LOAEL(HEC) = 69 mg/cu.m]. Pryor et al. (1987) exposed male Fisher 344/N rats (12/group) to 0 (clean air), 800, 1000, or 1200 ppm (0, 3408, 4260, or 5112 mg/cu.m, respectively) styrene for 14 hours/day, 7 days/week for 3 weeks. The duration-adjusted values are 0, 1988, 2485, or 2982 mg/cu.m, respectively. Auditory response thresholds were determined by both behavioral and electrophysiologic methods, apparently some weeks after the last exposure. Increases in auditory thresholds were recorded at 8-20 kHz with both methods at the lowest concentration used [LOAEL(HEC) = 1988 mg/cu.m].

Several occupational studies have reported adverse health effects on workers whose exposure was at or near 25 ppm (110 mg/cu.m) styrene. Lindstrom et al. (1976) reported that the visuomotor accuracy of styrene-exposed workers (n = 98, average exposure 4.9 years, range 0.5-14 years) was significantly poorer (p < 0.05) than that of the nonexposed workers. Moreover, this deficit was shown to be related to concentration in two subgroups whose exposures (estimated from urinary mandelic acid) were 25 and 75 ppm. Seppalainen and Harkonen (1976) conducted a cross-sectional study on 96 styrene workers, all of whom received EEG examinations 24 hours after termination of exposure. Mean exposure to styrene was estimated from urinary mandelic acid to be 808 mg/L, approximately 36 ppm in air. Twenty-three of 96 (24%) EEG's were abnormal in the exposed group as compared with a normal population (indicated as about 10% in the text), a difference that is statistically significant (p < 0.05).

Human irritation from styrene exposure has been characterized in a limited study by Stewart et al. (1968). Nine male volunteers were exposed to air concentrations of styrene of 50-375 ppm (213-1597 mg/cu.m) for periods of 1-7 hours. Urinalysis, hematology, and blood chemistry studies were conducted prior to exposure and at 16 and 72 hours postexposure; subjective symptoms were also recorded. Within 15 minutes following a 60-minute exposure to 375 ppm styrene, 4/5 volunteers complained of mild eye and nasal irritation. (Throughout the remainder of the exposure, all subjects noted a progressive loss of their ability to perceive the odor of styrene.) It is not clear from the text whether or not the irritation remained throughout the exposure. After 45 minutes of exposure, one of the subjects reported being nauseated and two others complained of feeling slightly inebriated. At 216 ppm styrene, 1/3 subjects noted nasal irritation 20 minutes into a 60-minute exposure. No adverse symptoms were reported by the three subjects exposed to 52 ppm styrene for 1 hour. Six subjects were then exposed to 99 ppm styrene for a total of 7 hours. Three of the six subjects complained of mild eye and throat irritation 20 minutes after the start of the exposure; in two of these subjects, the eye irritation persisted for 30 minutes before subsiding. At the end of the exposure, none of the subjects reported nausea, headache, or eye, nose, or throat irritation. The clinical studies were all normal. None of the six individuals exposed to 99 ppm (422 mg/cu.m) styrene for 7 hours or 216 ppm for 1 hour experienced any subjective symptoms of significant consequence. Although these results suggest no symptoms of consequence in humans exposed to styrene concentrations as high as 216 ppm (920 mg/cu.m), the population tested was small (only three subjects), and the duration was minimal.

Considerable success has been attained in modeling levels of inhaled styrene in biological systems. The physiologically based pharmacokinetic model for styrene of Ramsey and Andersen (1984) allows simulation over a wide range of concentrations on the time course of styrene distributed to four main tissue groups: (1) highly perfused organs, (2) moderately perfused organs (predominately muscle), (3) slowly perfused tissue (predominately fat), and (4) liver. When applied to actual rat data, this model accurately predicted blood styrene levels of 80-1200 ppm (340-5112 mg/cu.m). Then the behavior of inhaled styrene in humans was simulated successfully by substitution of human physiological parameters. These authors were able to demonstrate that blood concentrations of inhaled styrene in rats, mice, and humans were nearly identical at air concentrations of less than or equal to 200 ppm (852 mg/cu.m) but differed widely at higher concentrations. Perbellini et al. (1988) developed a physiologically based mathematical model for human exposure to airborne styrene that accounts for metabolism, subsequent synthesis, transfer, and urinary excretion of the principal metabolites MA and PGA. The model comprises eight compartments: (1) lung; (2) the richly perfused tissues of heart, brain, and kidney; (3) muscle; (4) fat; (5) liver tissue for catabolism of styrene; (6) liver tissue for transfer of metabolites; (7) body water in which the metabolites are distributed; and (8) urine in which the metabolites are excreted. Simulation results using this model were in agreement with reported urinary metabolite concentrations measured in various studies of worker populations, including that of Guillemin et al. (1982). Further simulations demonstrated that the use of urinary metabolites as a biological exposure index can accurately account for variability in pharmacokinetic/ physiologic parameters such as the alveolar ventilation rate. Simulations using an alveolar ventilation rate of 6-12 L/minute resulted in less than a three-fold range in model output (urinary metabolite concentration).

Jersey et al. (1978) exposed Sprague-Dawley rats (96/sex/group) to 0, 600, or 1200 ppm (0, 2556, or 5112 mg/cu.m, respectively) of 99.5% styrene for 6 hours/day, 5 days/week for up to 20 months. The exposure concentration of the 1200 ppm exposure group was reduced to 1000 ppm (4260 mg/cu.m) after 2 months because the males showed signs of toxicity (narcosis leading to anesthesia and excessive weight loss) with death coming to three animals. Exposures were terminated when mortality reached 50% for one exposure group of each sex; this was at 18.3 months for males and 20.7 months for females. All surviving rats were euthanized at the end of 2 years with interim group samplings at 6 and 12 months. Hematology, clinical chemistry, body weights, gross anatomical and histopathological analysis, and cage-side observations were performed for evaluation of toxicity. The respiratory tract (including the lungs, trachea, and nasal turbinates) was not examined in all animals; the nasal turbinates, for example, were examined only in a portion of the controls and high-exposure animals (14/28 animals, sexes combined). The number of sections examined in the nasal turbinates, trachea, or lungs is not indicated in the text. No exposure-related increase in mortality was noted in either sex. An inverse relationship between mortality and exposure was noted in male rats. A high incidence of murine pneumonia was associated with an increased mortality, but only in the control and high-exposure animals; no dose-response relationship was apparent from the data. Average body weights of both females and males were decreased at both dose levels at various times throughout the experiment. However, only the body weights of the males exposed to the highest concentration were decreased more than 10% (14% maximum), consistently only during treatment days 82-263. The only concentration-dependent alteration in organ weights observed was in absolute and relative liver weights in females sacrificed at 6 months. At the terminal sacrifice, an increase in absolute liver weights was observed only in the females exposed to the highest concentration; no histopathology accompanied this alteration. The only histological result considered to be concentration-dependent was an increase in incidence of alveolar histiocytosis (areas containing lipid-laden alveolar macrophages) that corresponded to grossly visible subpleural pale foci in the lungs of the females exposed to the highest concentration. No concentration- related effects were reported in any groups for hematology, clinical chemistry, or urinalysis. Deficiencies in this study preclude assigning effect levels.

Conti et al. (1988) exposed Sprague-Dawley rats (30/sex/dose) to 0, 25, 50, 100, 200, or 300 ppm (0, 106, 213, 426, 852, or 1278 mg/cu.m, respectively) styrene for 4 hours/day, 5 days/week for 52 weeks. The animals were kept under observation until spontaneous death. Histopathologic examinations were performed on each animal; tissues examined included brain, liver, kidneys, gonads, spleen, and pancreas. The lungs were apparently the only portion of the respiratory tract examined in this study. No noncancer results were reported or discussed in this study. No effect levels could be assigned in this study.

Effects of styrene on the respiratory tract have been addressed in mouse subchronic studies by the NTP (NTP, 1991a). B6C3F1 mice (10/sex/group) were exposed to 0, 62.5, 125, 250, or 500 ppm (0, 266, 532, 1065, or 2130 mg/cu.m, respectively) styrene for 6 hours/day, 5 days/week for 13 weeks. The duration-adjusted values for this exposure regime are 0, 47.5, 95, 190, or 380 mg/cu.m. Body weight changes, hematology, serum chemistry, sperm morphology, vaginal cytology, gross pathology, and histopathology (including the entire respiratory tract) were monitored for toxicity. Death occurred in the first week of exposure but only in the males exposed to 250 ppm; no deaths at the highest concentration were noted. Histopathology of these animals showed evidence of thymic and renal cortical necrosis. In exposed female mice, the average liver to body weight ratio was increased in the animals at the two highest concentrations. Histopathology revealed concentration-related increases in centrilobular liver cytomegaly, karyomegaly, and necrosis at these same concentrations with no effects being recorded at 125 ppm styrene or below. The lung to body weight ratios were increased at all levels relative to the control values. Histopathology of the respiratory tract revealed that the incidence of metaplasia and degeneration of the olfactory epithelium of the nasal cavity were already total (10/10 mice) in females at the lowest concentration, with necrosis being observed at higher concentrations. Likewise, bronchiolar regeneration was present in all female animals at all concentrations. The incidence of epithelial hyperplasia of the forestomach was also maximal at the lowest concentration. Similar results were noted for the males. No NOAEL for either respiratory or extrarespiratory effects was achieved in this study. The LOAEL(ADJ) for this study would be 47.5 mg/cu.m. The effects described occurred in both the nasal cavity [extrathoracic (ET)] and bronchiolar region [tracheobronchiolar (TB)]. The LOAEL(HEC) would therefore be based on effects in these regions with an RGDR of 1.55 = 74 mg.cu.m.

In a rat subchronic study (NTP, 1991b), F344/N rats (10/sex/group) were exposed to 0, 125, 250, 500, 1000, or 1500 ppm (0, 532, 1065, 2130, 4260, or 6390 mg/cu.m, respectively) styrene for 6 hours/day, 5 days/week for 13 weeks. The duration-adjusted values for this exposure regime are 0, 95, 190, 380, 761, or 1141 mg/cu.m. The same toxicological endpoints as in the mouse study were monitored. No deaths were recorded. Liver to body weight ratios were elevated at the three highest exposure levels in males and the two highest exposure levels in females, although no histopathology accompanied this alteration. Concentration-related goblet-cell hypertrophy was noted in the nasopharyngeal duct starting at 125 ppm styrene in males and at 250 ppm in females. Concentration-related degeneration of the olfactory epithelium (described as minimal to mild) was noted starting at 1000 ppm styrene in females and at 1500 ppm in males. Degeneration of the olfactory epithelium was noted also at the two highest exposure levels in both sexes. A NOAEL of 500 ppm is designated for extrathoracic effects [NOAEL(ADJ) = 380 mg/cu.m x RGDR of 0.107 = NOAEL(HEC) = 41 mg/cu.m].

The effect of styrene on the trachea of rats was addressed in two studies conducted by Ohashi et al. (1985, 1986). Both ciliary activity and histopathology were evaluated. Male Sprague-Dawley rats (10/exposure group and 10 controls) were exposed to styrene at 30 or 800 ppm for 8 consecutive weeks or at 150 or 1000 ppm for 3 consecutive weeks. Nasal and tracheal mucosa were examined by electron microscopy, immediately and several weeks after cessation of exposure. The 8-week study (Ohashi et al., 1985) found that there was a strong increase in mucus secretion, with an increase in number of dense bodies in the nasal, but not tracheal, mucosal cells after exposure to 30 ppm styrene. There was an increase in secretory granules in goblet cells at 3-weeks postexposure. At the higher concentration in the 8- week exposure, there was a marked increase in mucus secretion, vacuolation, and sloughing of epithelial cells in both the nasal and tracheal epithelium.   Compound cilia were observed, together with nuclear pyknosis, vacuolization of epithelial cells, and changes in electron density in goblet cells, which also exhibited cores with high electron density. Sloughing of epithelial cells from the basement membrane also persisted at 3-weeks postexposure. Severe ciliary denudation was observed in the high-exposure group (1000 ppm styrene) in the 3-week exposure, with ciliary activity in the nose disabled and decreased to 18% of control values in the trachea. No effect levels were designated from these studies as the quantitative relationship between ciliary activity and mucus transport is not clear. In his review of mucociliary transport, Wanner (1977) suggests considerable functional reserve of this system; in chicken trachea 30-50% of particle transport activity was present at a time when only 10% of the epithelium was ciliated.

In two studies, Lemasters et al. (1985b, 1989) examined the reproductive outcomes of female workers involved in plastics manufacturing. In the 1985 study, data from a total of 174 styrene-exposed and 449 unexposed women were collected and analyzed. No increased prevalence in menstrual disorders was observed in subgroups of the workers exposed to either 13 or 52 ppm styrene. In the 1989 study, the authors examined the relationship between styrene exposure and lowered birth weights. During the study, the authors collected and analyzed data from 819 no-, 154 low- (2-29 ppm), and 75 high- (30-82 ppm) exposed pregnancies. There was not a statistically significant concentration-response relationship in decreasing average birth weights. In women who worked at the most highly exposed jobs (estimated at 82 ppm), however, a 4% reduction in average birth weight that approached statistical significance (p = 0.08), despite the small sample size (n = 50), was detected.

Murray et al. (1978) exposed pregnant Sprague-Dawley rats and New Zealand rabbits to inhaled styrene at concentrations of 0, 300, or 600 ppm (0, 1278, or 2556 mg/cu.m., respectively) for 7 hours/day from gestation days 6-15 (rats) and 6-18 (rabbits). No concentration-related developmental toxicity was evident in either species by either route. Adverse maternal effects (decreased food consumption and a p < 0.05 decrease in weight gain only during the first 3 days of exposure) were noted. This study identifies a freestanding NOAEL for developmental effects of 2556 mg/cu.m.

Beliles et al. (1985) conducted a three-generation reproductive study concomitantly with a 2-year chronic study of exposure of rats to styrene in their drinking water. Sprague-

Dawley rats were treated with monomeric styrene in their drinking water at 0, 125, or 250 ppm. These doses corresponded to 8- 14 mg/kg/day for males and 12-21 mg/kg/day for females. After animals were dosed for 90 days, 20 females and 10 males from the styrene groups and 30 females and 15 males from the controls were used for the F0 generation and then returned to the chronic study. Representatives of these pups, the F1 generation, were then exposed until they were 110 days of age, at which time they were mated to produce the F2 generation. The F3 generation was produced in the same manner. Each generation was evaluated for fertility (male and female), litter size, pup viability, pup survival, sex ratio, pup body weight, weanling liver and kidney weight, physical and behavioral abnormalities on each day of lactation, and marrow cytogenetics. Reduction in the gestation and 1-, 7-, and 14-day survival indices of the high dose F2 pups was observed. A reduction in survival was also noted among the high dose F1 pups, but only at 21 days. No other evidences of fetotoxicity were noted. Although the authors claim these effects to be due to extensive losses in only 1 or 2 litters, only data on individual fetuses is presented. The high-dose level is designated as a NOAEL for reproductive effects.

Kankaanpaa et al. (1980) exposed pregnant BMR/T6T6 mice (15 controls and 13 exposed) to 250 ppm (1065 mg/cu.m) >99% pure styrene on gestation days 6-16 for 6 hours/day. Parameters monitored included number of litters and fetuses (total, live, dead, and malformed). No description of maternal toxicity is given, although narrative is provided on two preliminary experiments, one conducted at 500 ppm (2130 mg/cu.m) in which 2/6 pregnant mice and the surviving females had a fetal death rate of 47%. The other experiment was conducted at 700 ppm (2982 mg/cu.m) in which 3/5 pregnant animals died, with the surviving dams having a 95% fetal death rate. No dams died during the 250 ppm experiment, and the difference in the fetal death rate between controls and exposed dams was not statistically significant (27% vs. 18% in the controls; $p < 0.10$). The number of malformed fetuses was also increased in the exposed vs. the control mice (2.9% vs. 0.9%), but no statistical analysis was performed. A steep concentration response is indicated by this study: 500 ppm bringing death to both dams and fetuses, whereas 250 ppm appears to be without effect [NOAEL(HEC) = 1065 mg/cu.m]. These authors also exposed pregnant Chinese hamsters (2-7/treatment group and 15 controls) to 0, 300, 500, 750, or 1000 ppm (1278, 2130, 3195, or 4260 mg/cu.m, respectively) styrene for 6 hours/day on gestation days 6-18. Although the small number of animals limits the interpretation of this study, the highest concentration appears to be an effect level [LOAEL(HEC) = 4260 mg/cu.m], as the number of dead or resorbed fetuses was 66% as compared with 26% in the controls. There were no incidences of malformed fetuses in any treatment group or in the controls.

I.B.5. CONFIDENCE IN THE INHALATION RfC

Study -- Medium
Data Base -- Medium
RfC -- Medium

The study of Mutti et al. (1984) documents concentration-response relationships of CNS effects in a relatively small worker population. However, the results of this study are consistent with a number of other studies showing central effects in chronically exposed worker populations, most notably that of Moller et al. (1990). The urinary metabolites,

MA and PGA, are direct biological indicators of exposure to styrene. Numerous studies have demonstrated the relationship between urinary metabolites and air levels of styrene to be reliable and quantitative. Physiologically based pharmacological modeling of this exposure methodology demonstrates that it reflects and incorporates at least a portion of intrahuman variability related to pharmacokinetics. The study is therefore assigned a medium confidence level. The data base can be considered medium to high as chronic laboratory animal studies addressing noncancer endpoints are not yet available, but a number of human exposure studies support the choice of critical effect. Preliminary information in mice indicate that styrene is a respiratory tract irritant in mice at concentrations lower than 47.5 mg/cu.m. The RfC is assigned an overall confidence rating of medium.

I.B.6. EPA DOCUMENTATION AND REVIEW OF THE INHALATION RfC

Source Document -- The assessment is not presented in any existing US EPA document.

Other EPA Documentation -- US EPA 1984a,b, 1985, 1989, 1991

Agency Work Group Review -- 09/20/89, 03/26/92

Verification Date -- 03/26/92

I.B.7. EPA CONTACTS (INHALATION RfC)

Gary L. Foureman / NCEA -- (919)541-1183

Annie M. Jarabek / NCEA -- (919)541-4847

II. CARCINOGENICITY ASSESSMENT FOR LIFETIME EXPOSURE

This substance/agent has been evaluated by the US EPA for evidence of human carcinogenic potential. This does not imply that this agent is necessarily a carcinogen. The evaluation for this chemical is under review by an inter-office Agency work group. A risk assessment summary will be included on IRIS when the review has been completed.

III. HEALTH HAZARD ASSESSMENTS FOR VARIED EXPOSURE DURATIONS

III.A. DRINKING WATER HEALTH ADVISORIES

The Office of Drinking Water provides Drinking Water Health Advisories (HAs) as technical guidance for the protection of public health. HAs are not enforceable Federal standards. HAs are concentrations of a substance in drinking water estimated to have negligible deleterious effects in humans, when ingested, for a specified period of time. Exposure to the substance from other media is considered only in the derivation of the lifetime HA. Given the absence of chemical-specific data, the assumed fraction of total intake from drinking water is 20%. The lifetime HA is calculated from the Drinking Water Equivalent Level (DWEL) which, in turn, is based on the Oral Chronic Reference Dose. Lifetime HAs are not derived for compounds which are potentially carcinogenic for humans because of the difference in assumptions concerning toxic threshold for carcinogenic and noncarcinogenic effects. A more detailed description of the assumptions and methods used in the derivation of HAs is provided in Background Document 3 in Service Code 5.

III.A.1. ONE-DAY HEALTH ADVISORY FOR A CHILD

One-day HA -- 2E+1 mg/L

NOAEL -- 22.5 mg/kg/day
UF -- 10 (allows for intrahuman variability with the use of a NOAEL from a human study)
Assumptions -- 1 L/day water consumption for a 10-kg child

Principal Study -- Stewart et al., 1968

Human volunteers exposed to styrene by inhalation at 217 mg/cu.m (51 ppm) and 499 mg/cu.m (117 ppm) for 1 and 2 hours, respectively, showed no signs of toxicity. The moderately strong initial styrene odor diminished after 5 minutes. At 921 mg/cu.m (216 ppm) nasal irritation resulted after 20 minutes. Eye and nose irritation, strong odor, and altered neurological function were reported for volunteers exposed to styrene at 1600 mg/cu.m (376 ppm) for 1 hour. Most volunteers exposed to this level exhibited reduced performance in the Crawford Manual Dexterity Collar and Pin Test, the modified Romberg Test and the Flannagan Coordination Test. Six subjects were exposed to 422 mg/cu.m (99 ppm) styrene vapor for 7 hours and no serious adverse effects were noted; this level was identified as the NOAEL. Based on the conditions of exposure and an assumed absorption rate of 64%, this level is equivalent to a dose of 22.5 mg/kg/day.

III.A.2. TEN-DAY HEALTH ADVISORY FOR A CHILD

Appropriate data for calculating a Ten-day HA for styrene are not available. It is recommended that the Longer-term HA for a 10-kg child of 2 mg/L be used as the Ten-day HA.

III.A.3. LONGER-TERM HEALTH ADVISORY FOR A CHILD

Longer-term (Child) HA -- 2E+0 mg/L

NOAEL -- 200 mg/kg/day
UF -- 1000 (allows for interspecies and intrahuman variability with the use of a NOAEL from an animal study with a small number of animals per treatment group)
Assumptions -- 1 L/day water consumption for a 10-kg child

Principal Study -- Quast et al., 1978

Beagle dogs (four/dosage group) were given styrene by gavage 7 days/week for 560 days at 200, 400, or 600 mg/kg bw/day. At the two higher dose levels, minimal histopathologic effects were noted in the liver (increased iron deposits within the reticuloendothelial cells) as well as hematologic effects that included increased Heinz bodies in erythrocytes and a decreased packed cell volume. At the lowest dose level, these effects were not noted. Therefore, the NOAEL for this study was 200 mg/kg/day.

III.A.4. LONGER-TERM HEALTH ADVISORY FOR AN ADULT

Longer-term (Adult) HA -- 7E+0 mg/L

NOAEL -- 200 mg/kg/day
UF -- 1000 (allows for interspecies and intrahuman variability with the use of a NOAEL from an animal study with a small number of animals per treatment group)
Assumptions -- 2 L/day water consumption for a 70-kg adult

Principal Study -- Quast et al., 1978 (study described in III.A.3.)

III.A.5. DRINKING WATER EQUIVALENT LEVEL / LIFETIME HEALTH ADVISORY

DWEL -- 7E+0 mg/L

Assumptions -- 2 L/day water consumption for a 70-kg adult

RfD Verification Date -- 10/09/85

Lifetime HA -- 1E-1 mg/L

Assumptions -- 20% exposure through drinking water

NOTE: For this chemical the quantitative cancer risk assessment indicates that an additional safety factor of 10 is necessary to account for possible cancer risk.

Principal Study -- Quast et al., 1978

This study is described for the longer-term (child) HA.

III.A.6. ORGANOLEPTIC PROPERTIES

No information is available on the organoleptic properties of styrene.

III.A.7. ANALYTICAL METHODS FOR DETECTION IN DRINKING WATER

Styrene content is determined by a purge-and-trap gas chromatographic procedure used for the determination of volatile aromatic and unsaturated organic compounds in water and unsaturated organic compounds.

III.A.8. WATER TREATMENT

Methods available for the removal of styrene from water include air stripping, granular activated carbon, and oxidation.

III.A.9. DOCUMENTATION AND REVIEW OF HAs

US EPA. 1989. Final Draft of the Drinking Water Criteria Document on Styrene. Office for Drinking Water, Washington, DC.

US EPA. 1988. Final Health Advisory for Styrene. Rev. Environ. Contam. Toxicol. 107: 131-146.

EPA review of HAs in 1985.

Public review of HAs following notification of availability in October, 1985.

Science Advisory Board review of HAs in January, 1986.

Preparation date of this IRIS summary -- 08/20/90

III.A.10. EPA CONTACTS

Julie Du / OST -- (202)260-7583

Edward V. Ohanian / OST -- (202)260-7571

III.B. OTHER ASSESSMENTS

Content to be determined.

IV. US EPA REGULATORY ACTIONS

EPA risk assessments may be updated as new data are published and as assessment methodologies evolve. Regulatory actions are frequently not updated at the same time. Compare the dates for the regulatory actions in this section with the verification dates for the risk assessments in sections I and II, as this may explain inconsistencies. Also note that some regulatory actions consider factors not related to health risk, such as technical or economic feasibility. Such considerations are indicated for each action. In addition, not all of the regulatory actions listed in this section involve enforceable federal standards. Please direct any questions you may have concerning these regulatory actions to the US EPA contact listed for that particular action. Users are strongly urged to read the background information on each regulatory action in Background Document 4 in Service Code 5.

IV.A. CLEAN AIR ACT (CAA)

[Ed. Note: EPA does not have information yet in this section of IRIS].

IV.B. SAFE DRINKING WATER ACT (SDWA)

IV.B.1. MAXIMUM CONTAMINANT LEVEL GOAL (MCLG) for Drinking Water

Value (status) -- 0.1 mg/L (Final, 1991)

Considers technological or economic feasibility? -- NO

Discussion -- EPA has identified styrene as a Category II chemical and has promulgated an MCLG based on hepatic and hematologic effects reported in a chronic oral study in dogs. The MCLG is based upon a DWEL of 7 mg/L and an assumed drinking water contribution of 20 percent. In addition, the agency carefully examined the overall weight of evidence of cancer, especially: (1) the comparatively low estimated cancer potency; and (2) the lack of a carcinogenic response in an adequately conducted drinking water study. Moreover, styrene is not likely to be widespread in drinking water based on occurrence information available to the agency.

Reference -- 56 FR 3526 (01/30/91)

EPA Contact -- Health and Ecological Criteria Division / OST /(202) 260-7571 / FTS 260-7571; or Safe Drinking Water Hotline / (800) 426-4791

IV.B.2. MAXIMUM CONTAMINANT LEVEL (MCL) for Drinking Water

Value -- 0.1 mg/L (Final, 1991)

Considers technological or economic feasibility? -- YES

Discussion -- The EPA has promulgated a MCL equal to the MCLG of 0.1 mg/L.

Monitoring requirements -- All systems initially monitored for four consecutive quarters; repeat monitoring dependent upon detection, vulnerability status and system size.

Analytical methodology -- Gas chromatographic/mass spectrometry (EPA 524.2); purge and trap capillary gas chromatography (EPA 502.2); gas chromatography/ mass spectrometry (EPA 524.1); purge and trap gas chromatography (EPA 503.1); PQL= 0.005 mg/L.

Best available technology -- Granular activated carbon; packed tower aeration.

Reference -- 56 FR 3526 (01/30/91); 56 FR 30266 (07/01/91)

EPA Contact -- Drinking Water Standards Division / OGWDW /(202) 260-7575 / FTS 260-7575; or Safe Drinking Water Hotline / (800) 426-4791

IV.B.3. SECONDARY MAXIMUM CONTAMINANT LEVEL (SMCL) for Drinking Water

Value -- 0.01 mg/L (Proposed, 1989)

Considers technological or economic feasibility? -- NO

Discussion -- SMCLs are non-enforceable and establish limits for contaminants which may affect the aesthetic qualities (e.g. taste and odor) of drinking water. It is recommended that systems monitor for these contaminants every three years. More frequent monitoring for contaminants such as pH, color, odor or others may be appropriate under certain circumstances. The SMCL for styrene is based on odor qualities. Promulgation has been deferred following public comment (54 FR 22062).

Reference -- 54 FR 22062 (05/22/89); 56 FR 3526 (01/30/91)

EPA Contact -- Drinking Water Standards Division / OGWDW /(202) 260-7575 / FTS 260-7575; or Safe Drinking Water Hotline / (800) 426-4791

IV.B.4. REQUIRED MONITORING OF "UNREGULATED" CONTAMINANTS

[Ed. Note: EPA does not have information yet in this section of IRIS].

IV.C. CLEAN WATER ACT (CWA)

[Ed. Note: EPA does not have information yet in this section of IRIS].

IV.D. FEDERAL INSECTICIDE, FUNGICIDE, AND RODENTICIDE ACT (FIFRA)

[Ed. Note: EPA does not have information yet in this section of IRIS].

IV.E. TOXIC SUBSTANCES CONTROL ACT (TSCA)

[Ed. Note: EPA does not have information yet in this section of IRIS].

IV.F. RESOURCE CONSERVATION AND RECOVERY ACT (RCRA)

IV.F.1. RCRA APPENDIX IX, for Ground Water Monitoring

Status -- Listed

Reference -- 52 FR 25942 (07/09/87)

EPA Contact -- RCRA/Superfund Hotline (800)424-9346 / (202)260-3000 / FTS 260-3000

IV.G. SUPERFUND (CERCLA)

IV.G.1. REPORTABLE QUANTITY (RQ) for Release into the Environment

Value (status) -- 1000 pounds (Final, 1985)

Considers technological or economic feasibility? -- NO

Discussion -- The final RQ is based on aquatic toxicity [as established under Section 311(b)(4) of the Clean Water Act, 40 CFR 117.3], ignitability, and reactivity. The available data indicate that the aquatic 96-Hour Median Threshold Limit for styrene is between 10 and 100 ppm. In addition, styrene is easily combustible when exposed to heat or flame and can react vigorously with oxidizing materials.

Reference -- 50 FR 13456 (04/04/85); 54 FR 33418 (08/14/89)

EPA Contact -- RCRA/Superfund Hotline (800)424-9346 / (202)260-3000 / FTS 260-3000

VI. BIBLIOGRAPHY

VI.A. ORAL RfD REFERENCES

Ponomarkov, V. and L. Tomatis. 1978. Effects of long-term oral administration of styrene to mice and rats. J. Work Environ. Health. 4(suppl. 2): 127-135.

Quast, J.F., C.G. Humiston, R.Y. Kalnins, et al. 1979. Results of a toxicity study of monomeric styrene administered to beagle dogs by oral intubation for 19 months. Toxicology Research Laboratory, Health and Environmental Sciences, DOW Chemical Co., Midland, MI. Final Report.

US EPA. 1984. Health and Environmental Effects Profile for Styrene. Prepared by the Office of Health and Environmental Assessment, Environmental Criteria and Assessment Office, Cincinnati, OH for the Office of Solid Waste, Washington, DC.

US EPA. 1985. Drinking Water Criteria Document for Styrene. Prepared by the Office of Health and Environmental Assessment, Environmental Criteria and Assessment Office, Cincinnati, OH for the Office of Drinking Water, Washington, DC.

VI.B. INHALATION RfC REFERENCES

ACGIH (American Conference of Governmental Industrial Hygienists). 1986. Documentation of the Threshold Limit Values and Biological Exposure Indices, styrene monomer BEI. Cincinnati, OH.

Beliles, R.P., J.H. Butala, C.R. Stack, and S. Makris. 1985. Chronic toxicity and 3-generation reproduction study of styrene monomer in the drinking water of rats. Fund. Appl. Toxicol. 5: 855-868.

Cherry, N., H.A. Waldron, G.G. Wells, R.T. Wilkinson, H.K. Wilson, and S. Jones. 1980. An investigation of the acute behavioral effects of styrene on factory workers. Br. J. Ind. Med. 37: 234-240.

Cherry, N., B. Rodgers, H. Venables, H.A. Waldron, and G.G. Wells. 1981. Acute behavioral effects of styrene exposure: A further analysis. Br. J. Ind. Med. 38: 346-350.

Conti, B., C. Maltoni, G. Perion, and A. Ciliberti. 1988. Long-term carcinogenicity bioassays on styrene administered by inhalation, ingestion and injection and styrene oxide administered by ingestion in Sprague-Dawley rats, and para-methylstyrene administered by ingestion in Sprague-Dawley rats and Swiss mice. Ann. N.Y. Acad. Sci. 534: 203-234.

Flodin, U., K. Ekberg, and L. Andersson. 1989. Neuropsychiatric effects of low exposure to styrene. Br. J. Ind. Med. 46 (11): 805-808.

Franchini, I., A. Angiolini, C. Arcari et al. 1983. Mandelic acid and phenylglyoxylic acid excretion in workers exposed to styrene under model conditions. In: Developments in the Science and Practice of Toxicology: Proceedings of the Third International Congress of Toxicology, A.W. Hayes, R.C. Schnell, and T.S. Miya, ed. Elsevier Science Publishers, New York, NY.

Gamberale, F., H.O. Lisper, and B. Anshelm-Olson. 1975. Effect of styrene gases on reaction time among workers in plastic boat industry. Arbete och Halsa. 8: 23. (Swedish; cited in World Health Organization, 1983)

Gotell, P., O. Axelson, and B. Lindelof. 1972. Field studies on human styrene exposure. Work Environ. Health. 9(2): 76-83. (Swedish; cited in World Health Organization, 1983)

Guillemin, M.P., and D. Bauer. 1978. Biological monitoring of exposure to styrene by analysis of combined urinary mandelic and phenylglyoxylic acids. Am. Ind. Hyg. Assoc. J. 39(11): 873-879.

Guillemin, M.P., D. Bauer, B. Martin, and A. Marazzi. 1982. Human exposure to styrene. IV. Industrial hygiene investigations and biological monitoring in the polyester industry. Int. Arch. Occup. Environ. Health. 51(2): 139-150.

Guillemin, M.P., and M. Berode. 1988. Biological monitoring of styrene: A review. Am. Ind. Hyg. Assoc. J. 49(10): 497-505.

Hotz, P., M.P. Guillemin, and M. Lob. 1980. Study of some hepatic effects (induction and toxicity) caused by occupational exposure to styrene in the polyester industry. Scand. J. Work Environ. Health. 6(3): 206-215.

Ikeda, M., A. Koizumi, M. Miyasaka, and T. Watanabe. 1982. Styrene exposure and biologic monitoring in FRP boat production plants. Int. Arch. Occup. Environ. Health. 49(3-4): 325-339.

Jensen, A.A., N.O. Breum, J. Bacher, and E. Lynge. 1990. Occupational exposures to styrene in Denmark, 1955-1988. Am. J. Ind. Med. 17: 593-606.

Jersey, G.C., M.F. Balmer, J.F. Quast et al. 1978. Two-year chronic inhalation toxicity and carcinogenicity study on monomeric styrene in rats. Dow Chemical Study for the Chemical Manufacturing Association, December 6, 1978.

Kankaanpaa, J.T.J., E. Elovaara, K. Hemminki, and H. Vainio. 1980. The effect of maternally inhaled styrene on embryonal and fetal development in mice and Chinese hamsters. Acta Pharmacol. Toxicol. 47(2): 127-129.

Larsby, B., R. Tham, L.M. Odkvist, D. Hyden, I. Bunnfors, and G. Aschan. 1978. Exposure of rabbits to styrene. Scand. J. Work Environ. Health.4(1): 60-65.

Lemasters, G.K., A. Carson, and S.J. Samuels. 1985a. Occupational styrene exposure for 12 product categories in the reinforced-plastics industry. Am. Ind. Hyg. Assoc. J. 46(8): 434-441.

Lemasters, G.K., A. Hagen, and S.J. Samuels. 1985b. Reproductive outcomes inwomen exposed to solvents in 36 reinforced plastics companies. I. Menstrual dysfunction. J. Occup. Med. 27(7): 490-494.

Lemasters, G.K., S.J. Samuels, J.A. Morrison, and S.M. Brooks. 1989. Reproductive outcomes of pregnant workers employed at 36 reinforced plastics companies. II. Lowered birth weight. J. Occup. Med. 31(2): 115-120.

Lindstrom, K., H. Harkonen, and S. Hernberg. 1976. Disturbances in psychological functions of workers occupationally exposed to styrene. Scand. J. Work Environ. Health. 3: 129-139.

Moller, C., L.M. Odkvist, J. Thell et al. 1989. Otoneurological findings in Psycho-organic Syndrome caused by industrial solvent exposure. Acta Otolaryngol. 107(1-2): 5-12.

Moller, C., L. Odkvist, B. Larsby et al. 1990. Otoneurological findings in workers exposed to styrene. Scand. J. Work Environ. Health. 16(3): 189-194.

Muijser, H., E.M.G. Hoogendijk, and J. Hooisma. 1988. The effects of occupational exposure to styrene on high-frequency hearing thresholds. Toxicology. 49(2-3): 331-340.

Murata, K., S. Araki, and K. Yokoyama. 1991. Assessment of the peripheral, central, and autonomic nervous system function in styrene workers. Am. J. Ind. Med. 20: 775-784.

Murray, F.J., J.A. John, M.F. Balmer, and B.A. Schwetz. 1978. Teratologic evaluation of styrene given to rats and rabbits by inhalation or by gavage. Toxicology. 11(4): 335-343.

Mutti, A., A. Mazzucchi, P. Rustichelli, G. Frigeri, G. Arfini, and I. Franchini. 1984. Exposure-effect and exposure-response relationships between occupational exposure to styrene and neuropsychological functions. Am. J. Ind. Med. 5(4): 275-286.

NTP (National Toxicology Program). 1991a. Thirteen-week subchronic inhalation toxicity study in mice. Prepared by Battelle, Pacific Northwest Laboratories for the National Toxicology Program under Contract No. N01-ES- 95281. This report underwent review by NTP, but the final report is not yet available.

NTP (National Toxicology Program). 1991b. Thirteen-week subchronic inhalation toxicity study in rats. Prepared by Battelle, Pacific Northwest Laboratories for the National Toxicology Program under Contract No. N01-ES- 95281. This report underwent review by NTP, but the final report is not yet available.

Odkvist, L.M., B. Larsby, R. Tham et al. 1982. Vestibulo-oculomotor disturbances in humans exposed to styrene. Acta Otolaryngol. 94(5-6):487-493.

Odkvist, L.M., S.D. Arlinger, C. Edling, B. Larsby, and L.M. Bergholtz. 1987. Audiological and vestibulo-oculomotor findings in workers exposed to solvents and jet fuel. Scand. Audiol. 16(2): 75-81.

Ohashi, Y., Y. Nakai, H. Ikeoka et al. 1985. Electron microscopic study of the respiratory toxicity of styrene. Osaka City Med. J. 31(1): 11-21.

Ohashi, Y., Y. Nakai, H. Ikeoka et al. 1986. Degeneration and regeneration of respiratory mucosa of rats after exposure to styrene. J. Appl. Toxicol. 6(6): 405-412.

Perbellini, L., P. Mozzo, P.V. Turri, A. Zedde, and F. Brugnone. 1988. Biological exposure index of styrene suggested by a physiologico-mathematical model. Int. Arch. Occup. Environ. Health. 60(3): 187-193.

Pryor, G.T., C.S. Rebert, and R.A. Howd. 1987. Hearing loss in rats caused by inhalation of mixed xylenes and styrene. J. Appl. Toxicol. 7(1): 55-61.

Ramsey, J.C. and M.E. Andersen. 1984. A physiologically based description of the inhalation pharmacokinetics of styrene in rats and humans. Toxicol. Appl. Pharmacol. 73(1): 159-175.

Seppalainen, A.M. and H. Harkonen. 1976. Neurophysiological findings among workers occupationally exposed to styrene. Scand. J. Environ. Health. 3: 140-146.

Stengel, B., A. Touranchet, H.L. Boiteau, H. Harousseau, L. Mandereau, and D. Hemon. 1990. Hematological findings among styrene-exposed workers in the reinforced plastics industry. Int. Arch. Occup. Environ. Health. 62(1):11-18.

Stewart, R.D., H.C. Dodd, E.D. Baretta, and A.W. Schaffer. 1968. Human exposure to styrene vapor. Arch. Environ. Health. 16(5): 656-662.

US EPA. 1984a. Health and Environmental Effects Profile for Styrene. Prepared by the Office of Health and Environmental Assessment, Environmental Criteria and Assessment Office, Cincinnati, OH, for the Office of Solid Waste and Emergency Response, Washington, DC. EPA-600/X-84/325. NTIS Pub. No. PB88-182175/AS.

US EPA. 1984b. Drinking Water Criteria Document for Styrene. Prepared by the Office of Health and Environmental Assessment, Environmental Criteria and Assessment Office, Cincinnati, OH, for the Office of Drinking Water, Washington, DC. EPA-600/X-84/195. NTIS Pub. No. PB86-118056/AS.

US EPA. 1985. Reportable Quantity Document for Styrene. Prepared by the Office of Health and Environmental Assessment, Environmental Criteria and Assessment Office, Cincinnati, OH, for the Office of Solid Waste and Emergency Response, Washington, DC. EPA-600/X-85/221.

US EPA. 1989. Health Effects Assessment Document for Styrene. Prepared by the Office of Health and Environmental Assessment, Environmental Criteria and Assessment Office, Cincinnati, OH, for the Office of Solid Waste and Emergency Response, Washington, DC. EPA-600/8-88/054. NTIS Pub. No. PB90- 142357/AS.

US EPA. 1991. Drinking Water Criteria Document for Styrene. Prepared by the Office of Health and Environmental Assessment, Environmental Criteria and Assessment Office, Cincinnati, OH, for the Office of Drinking Water, Washington, DC.

Viau, C., A. Bernard, R. De Russis, A. Ouled, P. Maldague, and R. Lauwerys. 1987. Evaluation of the nephrotoxic potential of styrene in man and rat. J. Appl. Toxicol. 7(5): 313-316.

Wanner, A. 1977. Clinical aspects of mucociliary transport. Am. Rev. Resp. Dis. 116: 73-125.

WHO (World Health Organization). 1983. Environmental Health Criteria 26 Document on Styrene. International Program on Chemical Safety, Institute of Occupational Health, Helsinki, Finland.

VI.C. CARCINOGENICITY ASSESSMENT REFERENCES

None

VI.D. DRINKING WATER HA REFERENCES

Quast, J.F., R.P. Kalnins, K.J. Olson, et al. 1978. Results of a toxicity study in dogs and rats administered monomeric styrene. Toxicol. Appl. Pharmacol. 45: 293-294.

Stewart, R.D., H.C. Dodd, E.D. Baretta and A.W. Schaffer. 1968. Human exposure to styrene vapor. Arch. Environ. Health. 16(5): 656-662.

US EPA. 1988. Final Health Advisory for Styrene. Rev. Environ. Contam. Toxicol. 107: 131-146.

US EPA. 1989. Final Draft of the Drinking Water Criteria Document on Styrene. Office for Drinking Water, Washington, DC.

VII. REVISION HISTORY

```
-------- ----------- ----------------------------------------------------------
Date     Section     Description
-------- ----------- ----------------------------------------------------------
```

Date	Section	Description
06/30/88	I.A.7.	Contacts changed
10/01/89	I.B.	Inhalation RfD now under review
02/01/90	VI.	Bibliography on-line
05/01/90	I.A.	Oral RfD summary noted as pending change
05/01/90	II.	Carcinogen assessment now under review
09/01/90	I.A.	Text edited
09/01/90	III.A.	Health Advisory on-line
09/01/90	VI.D.	Health Advisory references added
01/01/92	IV.	Regulatory actions updated
11/01/92	I.B.	Inhalation RfC summary on-line
11/01/92	VI.B.	Inhalation RfC references on-line
07/01/93	I.B.1.	'E' notation added

SYNONYMS

BENZENE, VINYL-	STYREEN
CINNAMENE	STYREN
CINNAMENOL	Styrene
CINNAMOL	STYRENE, MONOMER
DIAREX HF 77	STYROL
ETHENYLBENZENE	STYROLE
ETHYLENE, PHENYL-	STYROLENE
NCI-C02200	STYRON
PHENETHYLENE	STYROPOL
PHENYLETHENE	STYROPOR
PHENYLETHYLENE	UN 2055
STIROLO	

Source: *Instant EPA's IRIS*, 1997, Instant Reference Sources, Inc., 7605 Rockpoint Drive, Austin, TX 78731 (Fax: 512-345-2386; Internet URL, http://www.instantref.com/inst-ref.htm).

2,4,5-T

Introduction

2,4,5-T is used as a foliar translocated herbicide with some residual action for the control of woody weeds. It is a selective weed killer and is used for the control of shrubs and trees and is applied as a foliage, dormant shoot or basal bark spray. It is also used for girdling, injection or cut-stump treatment. Commercial samples may be contaminated with toxic tetrachloro-dioxins and 2,4-D [025]. 2,3,7,8-TCDD is usually present as a contaminant [043]. 2,4,5-T was used as a defoliating agent in the Vietnam conflict [295].

2,4,5-T has been used as a growth regulator to increase size of citrus fruits and reduce excessive drop of deciduous fruit. The use of 2,4,5-T in the United States has been canceled since 1985. Some or all applications may be classified by the US EPA as Restricted Use Pesticides. [828]

Release of 2,4,5-T to the environment occurred during its past use as a herbicide and it can form in the environment as a hydrolysis product of its herbicide esters. Other sources of release may include losses during formulation, packaging or disposal of 2,4,5-T, its esters and the acaricide, tetradifon. Since 2,4,5-T has a pKa of 2.88 it will be found in the dissociated form in all environmental media. If released in soil, 2,4,5-T can biodegrade and its mobility is expected to vary from highly mobile in sandy soil to slightly mobile in muck (due to adsorption to humic acids and other organic matter). Removal by biodegradation apparently limits the extent of leaching, however, and groundwater contamination is likely only by rapid flow through large channels and deep soil cracks. 2,4,5-trichlorophenol and 2,4,5-trichloroanisole are the primary microbial degradation products of 2,4,5-T. Chemical hydrolysis in moist soils and volatilization from dry and moist surfaces should not be significant. The persistence of 2,4,5-T in soil is reported to vary between 14 to 300 days, but usually does not exceed one full growing season regardless of the application rate. Degradation under anaerobic conditions in flooded soils is much slower (half-life less than or equal to 48 weeks) than in field moist soils. [828]

If released to water, photochemical decomposition, volatilization and biodegradation of 2,4,5-T appear to be the dominant removal mechanisms. The primary degradation product of 2,4,5-T in water is 2,4,5-trichlorophenol. The aquatic near surface half-life for direct photolysis has been calculated to be 15 days during summer at latitude 40°. Humic substances can photosensitize 2,4,5-T and humic induced photoreactions may dominate photodegradation processes when humic substance concentrations exceed 15 mg/L of organic carbon per liter. Primary photodegradation products are 2,4,5-trichlorophenol and 2-hydroxy-4,5-dichlorophenoxyacetic acid. Adsorption of 2,4,5-T to humic acids in suspended solids and sediments may be significant. Oxidation, chemical hydrolysis, volatilization and bioaccumulation should not be significant. [828]

If released to the atmosphere, 2,4,5-T should exist as fine droplets and adsorbed on airborne particulates. 2,4,5-T has the potential to undergo (a) direct photolysis due to ultraviolet absorption at >290 nm, (b) a reaction with photochemically generated hydroxyl radicals (the estimated vapor phase half-life is 1.12 days) or (c) be physically removed by settling out or washout in rainfall. The most probable route of exposure to 2,4,5-T would be inhalation and dermal exposure of workers involved in the manufacture, handling or application of 2,4,5-T, related ester compounds or certain tetradifon formulations which contain 2,4,5-T. The general public could potentially be exposed by inhalation of particulate matter or ingestion of fruit, milk or drinking water contaminated with 2,4,5-T. [828]

Identifying Information
CAS NUMBER: 93-76-5

NIOSH Registry Number: AJ8400000

SYNONYMS: 2,4,5-TRICHLOROPHENOXYACETIC ACID; 2,4,5-TRICHLORO PHENOXYACETIQUE; ACIDO (2,4,5-TRICLORO-FENOSSI)-ACETICO; BCF-BUSHKILLER; BRUSH RHAP; BRUSHTOX; DACAMINE; DEBROUSSAILLANT CONCENTRE; DECAMINE 4T; DED-WEED BRUSH KILLER; DINOXOL; ENVERT-T; ESTERON 245; ESTERON 245 BE; FENCE RIDER; FORRON; FORST U 46; FORTEX; FRUITONE A; INVERTON 245; LINE RIDER; NA 2765; PHORTOX; REDDON; REDDOX; SPONTOX; SUPER D WEEDONE; TIPPON; TORMONA; TRANSAMINE; TRIBUTON; (2,4,5-TRICHLOOR-FENOXY)-AZIJNZUUR; (2,4,5-TRICHLOR-PHENOXY)-ESSIGSAEURE; TRINOXOL; TRIOXON; TRIOXONE; VEON; VEON 245; VERTON 2T; VISKO RHAP LOW VOLATILE ESTER; WEEDAR; WEEDONE

Endocrine Disruptor Information
2,4,5-T is a strongly suspected EED that it is listed by the Centers for Disease Control & Prevention [812], and the World Wildlife Fund [813] as a potential endocrine modifying chemical.

Chemical and Physical Properties
CHEMICAL FORMULA: $C_8H_5Cl_3O_3$

MOLECULAR WEIGHT: 255.49

WLN: QV1OR BG DG EG

PHYSICAL DESCRIPTION: White to light tan crystals

SPECIFIC GRAVITY: 1.80 @ 20/20°C [031], [172]

DENSITY: 1.803 g/mL @ 20°C [051], [371]

MELTING POINT: 158°C [051], [055], [371], [395]

BOILING POINT: Decomposes [051], [102], [371], [421]

SOLUBILITY: Water: <0.1 mg/mL @ 21.0°C [700]; DMSO: >=100 mg/mL @ 23.5°C [700]; 95% Ethanol: >=100 mg/mL @ 23.5°C [700]; Acetone: >=100 mg/mL @ 23.5°C [700]; Methanol: 496 g/L @ 25°C [169]; Toluene: 7.32 g/L @ 25°C [169];

OTHER SOLVENTS: Benzene: >10% [047]; Ether: 234.3 g/L @ 25°C [169]; Polar solvents: Soluble [455]; Alcohol: Soluble [016], [031], [062], [205]; Ethanol: 548.2 g/L @ 20°C [169]; Heptane: 400 mg/L @ 25°C [172]; Xylene: 6.08 g/L @ 25°C [169]; Isopropyl alcohol: Soluble [395]; Petroleum oils: Insoluble [169], [172]

OTHER PHYSICAL DATA: Odorless [051], [346], [371], [421]; Odor threshold concentration (detection): 2.92 mg/kg [055]; Spectroscopy data: Lambda max: 206 nm, 289 nm, 297 nm (E = 1872,99,91) [395]; Heat of combustion: 3600 cal/g [051]; Colorless crystals (technical grade) [172]; Melting point (technical grade): 150-151°C [173]; Metallic taste [195]

Source: *The National Toxicology Program's Chemical Database, Volume 2*, 1992, CRC Press, Inc./Lewis Publishers, 2000 Corporate Blvd., Boca Raton, FL 33431 (Fax: 407-998-9114).

Environmental Reference Materials

A source of EED environmental reference materials for 2,4,5-T is Radian International LLC, Austin, TX (Fax 512-454-0268; Internet URL, http://www.radian.com/standards); Catalog No. ERT-047.

Hazardous Properties

VOLATILITY: Vapor pressure: ~0.0 mm Hg @ 20°C [102], [421]

FIRE HAZARD: Literature sources indicate that 2,4,5-T is nonflammable [051], [102, [421]. Fires involving this material can be controlled with a dry chemical, carbon dioxide or Halon extinguisher. A water spray may also be used [051], [371].

LEL: Not available   UEL: Not available

REACTIVITY: 2,4,5-T forms water-soluble salts with alkali metals and amines [025], [172], [173]. It reacts with organic and inorganic bases to form salts and with alcohols to form esters [395]. It is incompatible with strong oxidizing agents and strong bases [269]. It can be corrosive to common metals [051], [371]. It may be deleterious to painted surfaces [051]. Precipitation occurs with hard water in the absence of sequestering agents [172].

STABILITY: 2,4,5-T is stable under normal laboratory conditions [173], [371], [395]. It is stable in aqueous solution at pH 5-9 [169], [172]. Solutions of 2,4,5-T in water, DMSO, 95% ethanol or acetone should be stable for 24 hours under normal laboratory conditions [700].

ACUTE/CHRONIC HAZARDS: 2,4,5-T may be harmful if ingested, inhaled or absorbed through the skin. It is also an irritant [269] and, when heated to decomposition, it emits toxic fumes of carbon monoxide, carbon dioxide, hydrogen chloride gas and phosgene [043], [102], [269], [371].

Source: *The National Toxicology Program's Chemical Database, Volume 7*, 1992, CRC Press, Inc./Lewis Publishers, 2000 Corporate Blvd., Boca Raton, FL 33431 (Fax: 407-998-9114).

Medical Symptoms of Exposure

Symptoms of exposure may include weakness, lethargy, anorexia, diarrhea, ventricular fibrillation, cardiac arrest and death [043], [051]. Other symptoms may include nausea, vomiting, fatigue, lowered blood pressure, convulsions and coma [371]. It can cause irritation of the skin, eyes [051], [371] mucous membranes and upper respiratory tract [269]. Other symptoms may include porphyria cutanea tarda, chloracne, abdominal pains, weakness in the lower extremities, nervousness, irritability, headache, neurological and behavioral changes, fat metabolism disorders, decreased auditory acuity and blood in the stools. In pregnant women, it may cause stillbirths and miscarriages [395]. Exposure can cause contact dermatitis [406]. It can also cause a metallic taste in the mouth [173]. Prolonged contact with high concentrations may cause skin burns [151]. Chronic exposure may cause liver damage [102], [269]. It may cause increased pulse rate and respiration, falling blood pressure, respiratory alkylosis, hemoconcentration, oliguria, rising blood urea nitrogen, restlessness, progressive shock, submucosal hemorrhage, moderate congestion, edema of the mucosa of the small intestine, congestion of the lungs, some congestion of other organs, consolidation of an area of the right middle lobe, acute necrosis of the mucousa of the large and small intestine, eczema and slight fatty infiltration of the liver [195].

Source: *The National Toxicology Program's Chemical Database, Volume 4*, 1992, CRC Press, Inc./Lewis Publishers, 2000 Corporate Blvd., Boca Raton, FL 33431 (Fax: 407-998-9114).

Toxicological Information

The toxicological information below is gathered from several sources including two national databases: one from the *National Toxicology Program Chemical Repository Database* and another from *EPA's Integrated Risk Information System (IRIS)*.

Toxicology Information from the National Toxicology Program

CARCINOGENICITY:

Tumorigenic Data:

Type of Effect	Route/Animal	Amount Dosed (Notes)
TDLo:	orl-mus	3379 mg/kg/33W-C
TDLo:	scu-mus	215 mg/kg

Review: IARC Cancer Review: Human Limited Evidence
IARC Cancer Review: Animal Inadequate Evidence
IARC possible human carcinogen (Group 2B)
IARC Note: Although IARC has assigned an overall evaluation to chlorophenoxy herbicides, it has not assigned an overall evaluation to all substances within this group

TERATOGENICITY:

Reproductive Effects Data:

TypeofEffect	Route/Animal	Amount Dosed (Notes)
TDLo:	orl-rat	100 mg/kg (8D preg)
TDLo:	orl-rat	6 mg/kg (8D preg)
TDLo:	orl-rat	714 mg/kg (MGN)
TDLo:	orl-rat	27600 ug/kg (10-15D preg)
TDLo:	orl-rat	100 mg/kg (11D preg)
TDLo:	orl-rat	500 mg/kg (6-10D preg)
TDLo:	orl-mus	200 mg/kg (6-15D preg)
TDLo:	orl-mus	6250 ug/kg (10D male)
TDLo:	orl-mus	300 mg/kg (12D preg)
TDLo:	orl-mus	900 mg/kg (6-15D preg)
TDLo:	orl-mus	150 mg/kg (6-15D preg)
TDLo:	orl-mus	1 gm/kg (8-12D preg)
TDLo:	orl-ham	500 mg/kg (7-11D preg)
TDLo:	ipr-ham	200 mg/kg (7-11D preg)
TDLo:	scu-rat	300 mg/kg (13-14D preg)
TDLo:	unr-rat	27600 ug/kg (10-15D preg)
TDLo:	unr-rat	278 mg/kg (10-15D preg)
TDLo:	unr-rat	60 mg/kg (10-15D preg)
TDLo:	scu-mus	1 gm/kg (6-15D preg)
TDLo:	scu-mus	450 mg/kg (6-14D preg)

MUTATION DATA:

Test	Lowestdose		Test	Lowestdose
cyt-dmg-orl	250 ppm		dnd-sal:spr	100 umol/L
sln-dmg-orl	1000 ppm/15D		dnd-mam:lym	100 umol/L
cyt-grb-ipr	250 mg/kg		cyt-grb-par	250 mg/kg
sln-dmg-unr	1000 ppm/15D		mmo-bcs	1 nmol/plate
mmo-smc	35 mg/L		cyt-ham:ovr	1750 mg/L
sce-ham:ovr	300 mg/L			

TOXICITY:

typ.dose	mode	specie	amount	units	other
LD50	orl	ckn	310	mg/kg	
LD50	orl	dog	100	mg/kg	
LD50	orl	gpg	381	mg/kg	
LD50	orl	ham	425	mg/kg	
LD50	orl	mam	500	mg/kg	
LD50	orl	mus	242	mg/kg	
LD50	orl	rat	300	mg/kg	
LD50	skn	rat	1535	mg/kg	
LD50	unr	rat	500	mg/kg	

OTHER TOXICITY DATA:

    Review: Toxicology Review-9

    Standards and Regulations:
        DOT-Hazard: ORM-A; Label: None

    Status:
        EPA Genetox Program 1988, Positive: D melanogaster Sex-linked
        lethal; S cerevisiae-reversion
        EPA Genetox Program 1988, Positive/dose response: In vivo
        cytogenetics-nonhuman bone   marrow
        EPA Genetox Program 1988, Negative: D melanogaster-whole sex
        chrom. loss
        EPA Genetox Program 1988, Negative: D melanogaster-nondisjunction
        EPA Genetox Program 1988, Inconclusive: Host-mediated assay;
        Mammalian micronucleus
        EPA TSCA Chemical Inventory, 1986
        EPA TSCA Test Submission (TSCATS) Data Base, January 1989
        NIOSH Analytical Methods: see 2,4-D and 2,4,5-T, 5001
        Meets criteria for proposed OSHA Medical Records Rule
        IDLH value: 5000 mg/m3
        Approximate lethal dose to 150 pound man: 4 tsps

SAX TOXICITY EVALUATION:

> THR: Poison by ingestion. Moderately toxic by unspecified route. An experimental neoplastigen, tumorigen and teratogen. May be a human and experimental carcinogen. Experimental reproductive effects. Mutagenic data. A highly toxic chlorinated phenoxy acid herbicide which is rapidly excreted after ingestion. Readily absorbed by inhalation and ingestion routes, slowly by skin contact.

Source: *Instant Tox-Base*, 1995, Instant Reference Sources, Inc., 7605 Rockpoint Drive, Austin, TX 78731 (Fax: 512-345-2386; Internet URL, http://www.instantref.com/inst-ref.htm).

Toxicology Information from EPA's Integrated Risk Information System (IRIS)

I. CHRONIC HEALTH HAZARD ASSESSMENTS FOR NONCARCINOGENIC EFFECTS

I.A. REFERENCE DOSE FOR CHRONIC ORAL EXPOSURE (RfD)

The Reference Dose (RfD) is based on the assumption that thresholds exist for certain toxic effects such as cellular necrosis, but may not exist for other toxic effects such as carcinogenicity. In general, the RfD is an estimate (with uncertainty spanning perhaps an order of magnitude) of a daily exposure to the human population (including sensitive subgroups) that is likely to be without an appreciable risk of deleterious effects during a lifetime. Please refer to Background Document 1 in Service Code 5 for an elaboration of these concepts. RfDs can also be derived for the noncarcinogenic health effects of compounds which are also carcinogens. Therefore, it is essential to refer to other sources of information concerning the carcinogenicity of this substance. If the US EPA has evaluated this substance for potential human carcinogenicity, a summary of that evaluation will be contained in Section II of this file when a review of that evaluation is completed.

I.A.1. ORAL RfD SUMMARY

Critical Effect	Experimental Doses*	UF	MF	RfD
Increased urinary coproporphyrins	NOAEL: 3 mg/kg/day	300	1	E-2 mg/kg/day
Reduced neonatal survival	LOAEL: 10 mg/kg/day			

*Conversion Factors: None
2-Year Rat Feeding StudyKociba et al., 1979
3-Generation Rat Feeding Study Smith et al., 1981

I.A.2. PRINCIPAL AND SUPPORTING STUDIES (ORAL RfD)

Kociba, R.J., D.G. Keyes, R.W. Lisowe, et al. 1979. Results of a two-year chronic toxicity and oncogenic study of rats ingesting diets containing 2,4,5-trichlorophenoxyacetic acid (2,4,5-T). Food Cosmet. Toxicol. 17: 205-221.

Smith, F.A., F.J. Murray, J.A. John, et al. 1981. Three-generation reproduction study of rats ingesting 2,4,5-trichloropenoxyacetic acid in the diet. Toxicol. Res. Lab., Dow Chemical, Midland, MI.

Groups of Sprague-Dawley rats (50/sex) were maintained on diets supplying 0, 3, 10, or 30 mg 2,4,5-T/kg bw/day for 2 years (Kociba et al., 1979). An additional 10 animals/sex/group were sacrificed after 4 months. Toxicological endpoints measured were body weight, food consumption, tumorigenicity, hematology, urinalysis, serum chemistry and histopathology. No effects were observed at 3 mg/kg/day. An increase in urinary excretion of coproporphyrins (at 4 months only) was reported for males at 10 and 30 mg/kg/day, and for females only at the high-dose level. A mild dose-related increase in the incidence of mineralized deposits in the renal pelvis was reported for females at 10 and 30 mg/kg/day after 2 years.

In a 3-generation reproduction study (Smith et al.,1981), rats were fed levels of 2,4,5-T corresponding to 0, 3, 10, or 30 mg/kg bw/day. No effects were observed at the low dose. Reduced neonatal survival was observed at both higher doses.

The reproductive endpoint is well documented. Other studies have shown effect on reproduction in mice, rats, hamsters and monkeys. A LOAEL of 15 mg/kg with a NOEL of 8 mg/kg for reduced fetal body weight was reported for mice. Fetal mortality was observed following administration of 2,4,5-T at 40 mg/kg (highest dose) to pregnant hamsters. Cleft palate was induced in A/JAX mice at 15 mg/kg; lower doses were not tested (Cranmer, 1978). Other strains of mice were less sensitive. Higher doses (about 200 mg/kg) induce frank teratogenic effects in rats. A qualitative association between 2,4,5-T exposure and human birth defects has been suggested. Terata and fetotoxic effects have not been observed in monkeys up to a dose of 40 mg/kg.

I.A.3. UNCERTAINTY AND MODIFYING FACTORS (ORAL RfD)

UF -- The UF of 300 includes uncertainty in the extrapolation of dose levels from laboratory animals to humans (10A), uncertainty in the threshold for sensitive humans (10H), and uncertainty because of deficiencies in the chronic toxicity data base (3D).

MF -- None

I.A.4. ADDITIONAL COMMENTS (ORAL RfD)

None.

I.A.5. CONFIDENCE IN THE ORAL RfD

Study -- HighData Base -- MediumRfD -- Medium

The Smith et al. (1981) study appears to be adequate for the assessment of long-term reproductive effects; a clear NOEL is established. The data base is highly supportive of both the nature and the magnitude of the reproductive effect. The Kociba et al. (1979) study rates a high confidence rating because of its relative completeness. Confidence in the RfD is medium (tending towards high) because of the mutual support of the two studies and strength of the reproductive data base. The relative weakness of the chronic toxicity data base precludes a higher confidence level.

I.A.6. EPA DOCUMENTATION AND REVIEW OF THE ORAL RfD

Source Docment -- US EPA, 1982

Other EPA Documentation -- None

Agency Work Group Review -- 12/18/85, 05/15/86, 08/13/87, 10/15/87, 01/20/88

Verification Date -- 01/20/88

I.A.7. EPA CONTACTS (ORAL RfD)

Joan Dollarhide / NCEA -- (513)569-7539

I.B. REFERENCE CONCENTRATION FOR CHRONIC INHALATION EXPOSURE (RfC)

[Ed. Note: EPA does not have information yet in this section of IRIS].

II. CARCINOGENICITY ASSESSMENT FOR LIFETIME EXPOSURE

This substance/agent has not been evaluated by the US EPA for evidence of human carcinogenic potential.

III. HEALTH HAZARD ASSESSMENTS FOR VARIED EXPOSURE DURATIONS

III.A. DRINKING WATER HEALTH ADVISORIES

The Office of Drinking Water provides Drinking Water Health Advisories (HAs) as technical guidance for the protection of public health. HAs are not enforceable Federal standards. HAs are concentrations of a substance in drinking water estimated to have negligible deleterious effects in humans, when ingested, for a specified period of time. Exposure to the substance from other media is considered only in the derivation of the lifetime HA. Given the absence of chemical-specific data, the assumed fraction of total intake from drinking water is 20%. The lifetime HA is calculated from the Drinking Water Equivalent Level (DWEL) which, in turn, is based on the Oral Chronic Reference Dose. Lifetime HAs are not derived for compounds which are potentially carcinogenic for humans because of the difference in assumptions concerning toxic threshold for carcinogenic and noncarcinogenic effects. A more detailed description of the assumptions and methods used in the derivation of HAs is provided in Background Document 3 in Service Code 5.

III.A.1. ONE-DAY HEALTH ADVISORY FOR A CHILD

Appropriate data for calculating a One-day HA are not available. It is recommended that the Ten-day HA of 0.8 mg/L be used as the One-day HA.

III.A.2. TEN-DAY HEALTH ADVISORY FOR A CHILD

Ten-day HA -- 8E-1 mg/L

NOAEL -- 8 mg/kg/dayUF -- 100 (allows for interspecies and intrahuman variability with the use of a NOAEL from an animal study)Assumptions -- 1 L/day water consumption for a 10-kg childPrincipal Study -- Neubert and Dillman, 1972

The effects of 2,4,5-T were studied in pregnant NMRI mice using three samples of 2,4,5-T: one had <0.02 ppm dioxin and was considered "dioxin-free"; a second sample had a dioxin content of 0.05 plus/minus 0.02 ppm; the third sample had an undetermined dioxin content. Each sample containing 2,4,5-T was administered by gavage on days 6 through 15 at doses from 8 to 120 mg/kg. Fetuses were removed on day 18, examined grossly and histologically. Even with the purest sample of 2,4,5-T, cleft palate frequency exceeding ($p<0.05$) that of controls was observed with doses higher than 30 mg/kg/day. Reductions ($p<0.05$) in fetal weight were observed with all samples tested at doses as low as 10 to 15 mg/kg/day. There was no clear increase in embryo lethality over that of controls at these lower doses. With the purest sample of 2,4,5-T, single oral doses of 150 to 300 mg/kg were capable of producing significant ($p<0.05$) incidences of cleft palate. The maximal teratogenic effect was seen when the 2,4,5-T was administered on day 12 to 13 of gestation. Based on the data obtained with the purest sample of 2,4,5-T, the teratogenic NOAEL is 15 mg/kg/day, and the fetotoxic NOAEL is 8 mg/kg/day.

III.A.3. LONGER-TERM HEALTH ADVISORY FOR A CHILD

Longer-Term (Child) HA -- 3E-1 mg/L

NOAEL -- 3 mg/kg/dayUF -- 100 (allows for interspecies and intrahuman variability with the use ofa NOAEL from an animal study)Assumptions -- 1 L/day water consumption for a 10-kg child

Principal Study -- Smith et al., 1981

Male and female Spraque-Dawley rats (F0) were fed a diet containing2,4,5-T (<0.03 ppb TCDD) to provide dosage levels of 0, 3, 10, or 30 mg/kg/day for 90 days and then were mated. At day 21 of lactation, the resulting pups were randomly selected for the F1 generation and the rest were necropsied. Subsequent matings were conducted to produce F2, F3a and F3b litters, successive generations being fed from weaning on the appropriate test or control diet. Fertility was decreased (p<0.05) in the matings of the 3b litters in the group given 10 mg/kg/day. Postnatal survival was significantly decreased (p<0.05) in the F2 litters of the 10 mg/kg/day group and the F1, F2 and F3 litters of the 30 mg/kg/day group. A significant decrease (p<0.05) in relative thymus weight was seen only in the F3b generations of the 30 mg/kg group, but the relative liver weight of weanlings was significantly increased (p<0.05) in the F2, F3a and F3b litters. The authors concluded that dose levels of 2,4,5-T sufficiently high to cause signs of toxicity in neonates had no effect on the reproductive capacity of the rats, except for a tendency toward a reduction of postnatal survival at doses of 10 and 30 mg/kg/day. Reproduction was not impaired at the lowest dose level of 3 mg/kg; this apparent NOAEL with respect to reproductive capacity and fetotoxic effects in this study is 3 mg/kg/day. The authors noted a significant decrease (p<0.05) in F1 (10 and 30 mg/kg on days 14 and 21) and F3 (3 mg/kg on day 14, and 10 and 30 mg/kg on day 21) litters, and they concluded that there was no effect of 2,4,5-T on rat reproduction except for a tendency toward a reduction in neonatal survival at 10 and 30 mg/kg/day.

III.A.4. LONGER-TERM HEALTH ADVISORY FOR AN ADULT

Longer-term (Adult) HA -- 1E+0 mg/L

NOAEL -- 3 mg/kg/dayUF -- 100 (allows for interspecies and intrahuman variability with the use ofa NOAEL from an animal study)Assumptions -- 2 L/day water consumption for a 70-kg adult

Principal Study -- Smith et al., 1981

This study is described in the Longer-term (Child) HA discussion.

III.A.5. DRINKING WATER EQUIVALENT LEVEL / LIFETIME HEALTH ADVISORY

DWEL -- 3.5E-1 mg/L

Assumptions -- 2 L/day water consumption for a 70-kg adult

RfD Verification Date -- 01/20/88 (see the oral RfD)

Lifetime HA -- 7E-2 mg/L

Assumptions -- 20% exposure by drinking water
Principal Study -- Kociba et al., 1979

Groups of Spraque-Dawley rats (50/sex/level) were maintained on diets containing 2,4,5-T at 3, 10, or 30 mg/kg/day for two years. The 2,4,5-T was approximately 99% pure; dioxins were not present at the detection limit of 0.33 ppb. Control groups included 86 animals of each sex. The highest dose level was associated with some degree of toxicity including a decrease in body weight gain (p<0.05 in females) and increases in relative kidney weight (p<0.05). Increases in the volume of urine excreted and in the urinary excretion of coproporphyrin and uroporphyrin were also observed at this dose level. Increased incidence (p<0.05) of morphological changes were observed in the kidney, liver and lungs of animals administered 30 mg/kg/day. The kidney changes involved primarily the presence of mineralized deposits in the renal pelvis in females. Effects noted at the 10 mg/kg/day level were primarily an increased incidence (p<0.05) of mineralized deposits in the renal pelvis in females and, in males andonly during the early phase of the study, an increase (p<0.05) in urinary excretion of coproporphyrin. At the lowest dose level (3 mg/kg/day), there were no changes were considered to be related to treatment throughout the 2-year period. For this study, the NOAEL is identified as 3 mg/kg/day.

III.A.6. ORGANOLEPTIC PROPERTIES

No information is available on the organoleptic properties of 2,4,5-T.

III.A.7. ANALYTICAL METHODS FOR DETECTION IN DRINKING WATER

Determination of 2,4,5-T is by liquid-liquid extraction gas chromatographic procedure.

III.A.8. WATER TREATMENT

Available data indicate that granular activated carbon and powdered activated carbon adsorption will effectively remove 2,4,5-T from water.

III.A.9. DOCUMENTATION AND REVIEW OF HAs

US EPA. 1988. Drinking Water Health Advisory for 2,4,5-T. Office of Water, Washington, DC.

EPA review of HAs was in 1987
Public review of the HAs was in January-March 1988
Preparation date of this IRIS summary -- 08/30/92

III.A.10. EPA CONTACTS

Amal Mahfouz / OST -- (202)260-9568
Edward V. Ohanian / OST -- (202)260-7571

III.B. OTHER ASSESSMENTS

[Ed. Note: EPA does not have information yet in this section of IRIS].

IV. US EPA REGULATORY ACTIONS

EPA risk assessments may be updated as new data are published and as assessment methodologies evolve. Regulatory actions are frequently not updated at the same time. Compare the dates for the regulatory actions in this section with the verification dates for the risk assessments in sections I and II, as this may explain inconsistencies. Also note that some regulatory actions consider factors not related to health risk, such as technical or economic feasibility. Such considerations are indicated for each action. In addition, not all of the regulatory actions listed in this section involve enforceable federal standards. Please direct any questions you may have concerning these regulatory actions to the US EPA contact listed for that particular action. Users are strongly urged to read the background information on each regulatory action in Background Document 4 in Service Code 5.

IV.A. CLEAN AIR ACT (CAA)

[Ed. Note: EPA does not have information yet in this section of IRIS].

IV.B. SAFE DRINKING WATER ACT (SDWA)

Listed in the January 1991 Drinking Water Priority Listand may be subject to future regulation (56 FR 1470, 01/14/91).

IV.C. CLEAN WATER ACT (CWA)

[Ed. Note: EPA does not have information yet in this section of IRIS].

IV.D. FEDERAL INSECTICIDE, FUNGICIDE, AND RODENTICIDE ACT (FIFRA)

[Ed. Note: EPA does not have information yet in this section of IRIS].

IV.E. TOXIC SUBSTANCES CONTROL ACT (TSCA)

[Ed. Note: EPA does not have information yet in this section of IRIS].

IV.F. RESOURCE CONSERVATION AND RECOVERY ACT (RCRA)

IV.F.1. RCRA APPENDIX IX, for Ground Water Monitoring

Status -- Listed

Reference -- 52 FR 25942 (07/09/87)

EPA Contact -- RCRA/Superfund Hotline(800)424-9346 / (202)260-3000 / FTS 260-3000

IV.G. SUPERFUND (CERCLA)

IV.G.1. REPORTABLE QUANTITY (RQ) for Release into the Environment

Value (status) -- 1000 pounds (Final, 1989)

Considers technological or economic feasibility? -- NO

Discussion -- The final RQ for 2,4,5-trichlorophenoxyacetic acid is based on aquatic toxicity. The available data indicates that the aquatic 96-Hour Median Threshold Limit is 10.4 ppm, which corresponds to an RQ of 1000 pounds.

Reference -- 54 FR 33418 (08/14/89)

EPA Contact -- RCRA/Superfund Hotline(800)424-9346 / (202)260-3000 / FTS 260-3000

VI. BIBLIOGRAPHY

VI.A. ORAL RfD REFERENCES

Cranmer, M. 1978. Use of herbicides in forestry. USDA/EPA Symposium, Arlington, VA, February 21-22,.

Kociba, R.J., D.G. Keyes, R.W. Lisowe, et al. 1979. Results of a two-year chronic toxicity and oncogenic study of rats ingesting diets containing 2,4,5-trichlorophenoxyacetic acid (2,4,5-T). Food Cosmet. Toxicol. 17: 205-221.

Smith, F.A., F.J. Murray, J.A. John, K.D. Nitschke, R.J. Kociba and B.A. Schwetz. 1981. Three-generation reproduction study of rats ingesting 2,4,5-trichlorophenoxyacetic acid in the diet. Toxicol. Res. Lab., Dow Chemical, Midland, MI.

US EPA. 1982. Draft Interim Criterion Statement: Chlorophenoxy Herbicides. Ambient Water Quality Criterion for the Protection of Human Health. ECAO-CIN-82-D005. (Internal Review Draft). This document was developed from a multimedia document on the Chlorophenoxy Herbicides which during 1982 had fairly comprehensive external review with a limited internal review.

VI.B. INHALATION RfD REFERENCES

None

VI.C. CARCINOGENICITY ASSESSMENT REFERENCES

None

VI.D. DRINKING WATER HA REFERENCES

Kociba, R.J., D.J. Keyes, R.W. Lisowe et al. 1979. Results of a two-year chronic toxicity and oncogenic study of rats ingesting diets containing 2,4,5-trichlorophenoxyacetic acid (2,4,5-T). Food Cosmet. Toxicol. 17: 205-221.

Neubert, D. and I. Dillman. 1972. Embryotoxic effects in mice treated with 2,4,5-trichlorophenoxyacetic acid and 2,3,7,8-tetrachlorodibenzo-p-dioxin. Naunyn-Schmiedeberg's Arch. Pharmacol. 272: 243-264.

Smith, F.A., F.J. Murray, J.A. John, K.D. Nitschke, R.J. Kociba and B.A. Schwetz. 1981. Three-generation reproduction study of rats ingesting 2,4,5-trichlorophenoxyacetic acid in the diet. Food Cosmet. Toxicol. 19: 41-45.

US EPA. 1988. Drinking Water Health Advisory for 2,4,5-T. Office of Water, Washington, DC.

VII. REVISION HISTORY

Date	Section	Description
09/07/88	I.A.	Oral RfD summary on-line
01/01/89	I.A.7.	Contacts phone numbers corrected
06/01/89	I.A.6.	Work group review dates added
08/01/89	VI.	Bibliography on-line
08/01/89	Synonyms	Silvex deleted from synonyms
01/01/92	I.A.7.	Secondary contact changed
01/01/92	IV.	Regulatory Action section on-line
02/01/93	III.A.	Health Advisory on-line
02/01/93	VI.D.	Health Advisory references on-line

SYNONYMS

2,4,5-T	REDDON
ACIDE 2,4,5-TRICHLORO	REDDOX
PHENOXYACETIQUE	SPONTOX
ACIDO (2,4,5-TRICLORO-FENOSSI)-	SUPER D WEEDONE
ACETICO	TIPPON
BCF-BUSHKILLER	TORMONA
BRUSH RHAP	TRANSAMINE
BRUSHTOX	TRIBUTON
DACAMINE	(2,4,5-TRICHLOOR-FENOXY)-
DEBROUSSAILLANT CONCENTRE	AZIJNZUUR
DECAMINE 4T	2,4,5-Trichlorophenoxyacetic acid
DED-WEED BRUSH KILLER	Trichlorophenoxyacetic acid, 2,4,5-
DINOXOL	(2,4,5-TRICHLOR-PHENOXY)-
ENVERT-T	ESSIGSAEURE
ESTERON 245	TRINOXOL
ESTERON 245 BE	TRIOXON
FENCE RIDER	TRIOXONE
FORRON	U 46
FORST U 46	VEON
FORTEX	VEON 245
FRUITONE A	VERTON 2T
INVERTON 245	VISKO RHAP LOW VOLATILE
LINE RIDER	ESTER
NA 2765	WEEDAR
PHORTOX	WEEDONE
RCRA WASTE NUMBER U232	WEEDONE 2,4,5-T

Source: *Instant EPA's IRIS*, 1997, Instant Reference Sources, Inc., 7605 Rockpoint Drive, Austin, TX 78731 (Fax: 512-345-2386; Internet URL, http://www.instantref.com/inst-ref.htm).

Production, Use, and Pesticide Labeling Information

2,4,5-T is not currently listed in EPA's Pesticide Factsheet database and there is no indication of whether it will be added or not in the future.

Source: *Instant EPA's Pesticide Facts*, 1994, Instant Reference Sources, Inc., 7605 Rockpoint Drive, Austin, TX 78731 (Fax: 512-345-2386; Internet URL, http://www.instantref.com/inst-ref.htm).

2,3,7,8-TCDD

Introduction

2,3,7,8-Tetrachlorodibenzo-p-dioxin is a contaminant created in the manufacture of Agent Orange, a widely used defoliant in Vietnam [031]. 2,3,7,8-TCDD is also present in certain herbicide and fungicide formulations, such as 2,4,5-T and pentachlorophenol. It has been tested for use in flame proofing polyesters and against insects and wood-destroying fungi. It is implicated as the causative agent of various symptoms described by veterans exposed in the war [031], [051]. It is considered to be the most toxic chemical manufactured [406]. 2,3,7,8-Tetrachlorodibenzodioxin is currently released to the environment primarily through emissions from the incineration of municipal and chemical wastes, in exhaust from automobiles using leaded gasoline, and from the improper disposal of certain chlorinated chemical wastes. [828]

If released to the atmosphere, gas-phase 2,3,7,8-tetrachlorodibenzodioxin may be degraded by reaction with hydroxyl radicals and direct photolysis. Particulate-phase 2,3,7,8-tetrachlorodibenzodioxin may be physically removed from air by wet and dry deposition. If released to water, it will predominantly be associated with sediments and suspended material. 2,3,7,8-Tetrachlorodibenzodioxin near the water's surface may experience significant photodegradation. Volatilization from the water column may be important, but adsorption to sediment will limit the overall rate by which it is removed from water. The persistence half-life of 2,3,7,8-tetrachlorodibenzodioxin in lakes has been estimated to be in excess of 1.5 yr. Bioconcentration in aquatic organisms has been demonstrated. If released to soil, it is not expected to leach. Photodegradation on terrestrial surfaces may be an important transformation process. Volatilization from soil surfaces during warm conditions may be a major removal mechanism. The persistence half-life of 2,3,7,8-tetrachlorodibenzodioxin on soil surfaces may vary from less than a year to three years, but half-lives in soil interiors may be as long as 12 years. Screening studies have shown that 2,3,7,8-tetrachlorodibenzodioxin is generally resistant to biodegradation. A major route of exposure to the general population results from incineration processes and exhausts from leaded gasoline engines. [828] The greatest exposure pathway to the general population appears to be from ingestion of contaminated food. [882]

Identifying Information

CAS NUMBER: 1746-01-6

NIOSH Registry Number: HP3500000

SYNONYMS: 2,3,7,8-Tetrachlorodibenzodioxin; Dioxin

Endocrine Disruptor Information

2,3,7,8-TCDD is a suspected EED that is listed by the World Wildlife Foundation, Canada [813]. It is one of 75 isomers of "dioxin," an impurity in Agent Orange and a byproduct of volcanoes, forest fires, and the production of bleached paper, wood preservatives and other products. It is linked to three types of cancer: soft-tissue sarcoma, non-Hodgkin's lymphoma, and Hodgkin's disease. Dioxin acts almost exclusively through an aryl hydrocarbon (Ah) receptor. Once it occupies the receptor in a human cell it binds to DNA in the cell nucleus and prompts many of the sae changes in gene expression (such as developmental disruption) seen in animal experiments. It is still uncertain, however, whether dioxin interferes with the development of the brain and disrupting sexual behavior. [810]

Chemical and Physical Properties

CHEMICAL FORMULA: $C_{12}H_4Cl_4O_2$

MOLECULAR WEIGHT: 321.96

WLN: T C666 BO IOJ EG FG LG MG

PHYSICAL DESCRIPTION: Colorless to white crystals

SPECIFIC GRAVITY: Not available

DENSITY: Not available

MELTING POINT: 295°C [029], [031]

BOILING POINT: Decomposes @ 500°C [051], [072], [346]

SOLUBILITY: Water: <1 mg/mL @ 25°C [700]; DMSO: <1 mg/mL @ 25°C [700]; 95% Ethanol: <1 mg/mL @ 25°C [700]; Acetone: <1 mg/mL @ 25°C [700]; Methanol: 0.01 mg/mL @ 25°C [072], [395], [900]; Toluene: <1 mg/mL @ 20°C [700]

OTHER SOLVENTS: Benzene: 0.57 mg/mL [072], [395]; Chloroform: 0.37 mg/mL [072], [395]; Perchloroethylene: 0.68 mg/mL @ 25°C [900]; Chlorobenzene: 0.72 mg/mL [072], [395]; o-Dichlorobenzene: 1.4 mg/mL [072], [395]; n-Octanol: 0.05 mg/mL [051], [072], [395]; Lard oil: 0.04 mg/mL [395]; Hexane: 0.28 mg/mL @ 25°C [900]

OTHER PHYSICAL DATA: Crystals from anisole (melting point: 320-325°C) [031], [072]; Decomposition begins @ 500°C and is virtually complete within 21 seconds @ a temperature of 800°C [051], [072], [346]; Partition coefficient in water (hexane system): 1000 [072]; Lambda max (in chloroform): 248 nm, 310 nm (E = 92.2, 173.6) [395]

Source: *Instant EPA's Air Toxics*, 1994, Instant Reference Sources, Inc., 7605 Rockpoint Drive, Austin, TX 78731 (Fax: 512-345-2386; Internet URL, http://www.instantref.com/inst-ref.htm).

Environmental Reference Materials

A source of EED environmental reference materials for 2,3,7,8-TCDD is Cambridge Isotope Laboratories, 50 Frontage Rd., Andover, MA 01810-5413 (Fax: 508-749-2768; Internet URL, http://www.isotope.com); Catalog No. ED-901C.

Hazardous Properties

ACUTE/CHRONIC HAZARDS: 2,3,7,8-Tetrachlorodibenzo-p-dioxin may be very toxic [026], [031], an irritant, and an allergin. It also may cause eye irritation [043], [301].

HAP WEIGHTING FACTOR: 100,000 [713]

VOLATILITY:
   Vapor pressure: 0.00000000064 mm Hg @ 20°C;
               0.0000000014 mm Hg @ 25°C [901]

FIRE HAZARD: 2,3,7,8-Tetrachlorodibenzo-p-dioxin is nonflammable [051], [072].

REACTIVITY: 2,3,7,8-Tetrachlorodibenzo-p-dioxin is in general an unreactive compound. It is changed chemically when exposed as solutions in iso-octane or n-octanol to ultraviolet light [395]. It undergoes catalytic perchlorination [051], [072].

STABILITY: 2,3,7,8-Tetrachlorodibenzo-p-dioxin undergoes slow photochemical degradation and slow bacterial degradation. It is extremely stable but is chemically degraded by temperatures in excess of 500°C or by irradiation with ultraviolet light under certain conditions [051], [072]. Photodecomposition is negligible in aqueous solutions [055]. Solutions of it in water, DMSO, 95% ethanol or acetone should be stable for 24 hours when protected from light [700].

Source: *Instant EPA's Air Toxics*, 1994, Instant Reference Sources, Inc., 7605 Rockpoint Drive, Austin, TX 78731 (Fax: 512-345-2386; Internet URL, http://www.instantref.com/inst-ref.htm).

Medical Symptoms of Exposure

Symptoms of exposure may include eye irritation, allergic dermatitis, wasting, hepatic necrosis, thymic atrophy, hemorrhage, lymphoid depletion and chloracne [043]. It may cause hypercholesterolemia and psychiatric disturbances [406]. It may also cause hyperpigmentation, liver damage, Hodgkin's lymphoma, raised serum hepatic enzyme levels, disorders of fat metabolism, disorders of carbohydrate metabolism, cardiovascular disorders, urinary tract disorders, respiratory disorders, pancreatic disorders, polyneuropathies, lower extremity weakness, sensorial impairments and neurasthenic or depressive syndromes [395]. Exposure may lead to excessive oiliness of the skin, abdominal pains, excessive flatulence, loss of body weight, oppressive headaches, excessive fatigue, unusual loss of vigor, porphyria cutanea tarda, porphyrinuria, uncharacteristic irritability and high blood cholesterol [173]. It may cause hepatotoxicity, thrombocytopenia, suppression of cellular immunity and death [346]. It may also cause a burning sensation to the eyes, nose and throat, headache, dizziness, nausea, vomiting, itching, redness, swelling of the face, nodules on the face, forearms, shoulders, neck and throat which may progress to come domes and cysts, acneform eruptions, aching muscles (mainly thighs and chest), insomnia, hirsutism, loss of libido, pain, hepatic dysfunction, hyperlipidemia and emotional disorders [301]. Other symptoms may include hemorrhagic cystitis, focal pyelonephritis, arthralgias, diabetes mellitus, burn-like sores, spontaneous abortions, liver damage (fatty changes, mild fibrosis, hemofuscin deposition and degeneration) and death [072].

Source: *Instant EPA's Air Toxics*, 1994, Instant Reference Sources, Inc., 7605 Rockpoint Drive, Austin, TX 78731 (Fax: 512-345-2386; Internet URL, http://www.instantref.com/inst-ref.htm).

Toxicological Information

The toxicological information below is gathered from several sources including two national databases: one from the *National Toxicology Program Chemical Repository Database* and another from *EPA's Integrated Risk Information System (IRIS)*.

Toxicology Information from the National Toxicology Program

CARCINOGENICITY:

Tumorigenic Data:

Type of Effect	Route/Animal	Amount Dosed (Notes)
TD	orl-rat	73 ug/kg/2Y-C
TDLo	orl-rat	52 ug/kg/2Y-I
TD	orl-rat	328 ug/kg/78W-C
TDLo	orl-mus	52 ug/kg/2Y-I
TD	skn-mus	80 ug/kg
TD	orl-rat	137 ug/kg/65W-C
TD	orl-rat	1 ug/kg/2Y-I
TD	orl-mus	1 ug/kg/2Y-I
TDLo	skn-mus	62 ug/kg/2Y-I
TD	orl-rat	27 ug/kg/65W-C
TD	orl-mus	36 ug/kg/52W-I

Review:
IARC Cancer Review: Animal Sufficient Evidence
IARC Cancer Review: Human Inadequate Evidence
IARC possible human carcinogen (Group 2B)

Status:

NCI Carcinogenesis Bioassay (Gavage); Positive: Male and Female
Rat, Male and Female Mouse
NCI Carcinogenesis Bioassay (Dermal); Equivocal: Male Mouse
NCI Carcinogenesis Bioassay (Dermal); Positive: Female Mouse
NTP Fourth Annual Report on Carcinogens, 1984
NTP anticipated human carcinogen
EPA Carcinogen Assessment Group

TERATOGENICITY:

Reproductive Effects Data:

Type of Effect	Route/Animal	Amount Dosed (Notes)
TDLo	orl-rat	1250 ng/kg (6-15D preg)
TDLo	orl-rat	20 ug/kg (1D pre)
TDLo	orl-rat1	27 ng/kg (MGN)
TDLo	ipr-rat	6 ug/kg (17D preg)
TDLo	scu-rat	5 mg/kg (6-15D preg)
TDLo	orl-mus	30 ug/kg (6-15D preg)
TDLo	orl-mus	23 ug/kg (11D preg)
TDLo	orl-mus	235 ug/kg (28D pre-21D post)
TDLo	orl-mus	20 ug/kg (14D preg/3D post)
TDLo	ipr-mus	25 ug/kg (7-11D preg)
TDLo	ipr-mus	20 ug/kg (11D preg)
TDLo	scu-mus	250 ug/kg (7-16D preg)
TDLo	scu-mus	100 ug/kg (2D preg)
TDLo	scu-mus	100 ug/kg (10D preg)
TDLo	scu-mus	30 ug/kg (10D preg)
TDLo	orl-mus	1 ug/kg (10D preg)
TDLo	orl-rbt	1 ug/kg (6-15D preg)
TDLo	orl-rbt	10 ug/kg (6-15D preg)
TDLo	orl-rat	1500 ng/kg (1-3D preg)
TDLo	orl-rat	1270 ng/kg (MGN)
TDLo	orl-rbt	2500 ng/kg (6-15D preg)
TDLo	orl-mus	12 ug/kg (10-13D preg)
TDLo	unr-rat	1500 mg/kg (1D male)

MUTATION DATA:

Test	Lowest dose		Test	Lowest dose
dni-mus-ipr	400 ug/kg		msc-mus:lym	50 mg/L
dni-rat-orl	200 ug/kg		mmo-sat	2 mg/L
dni-rat-ipr	10 ug/kg		cyt-rat-orl	100 ug/kg
pic-esc	500 ug/L		cyt-rat-ipr	10 mg/kg
mmo-esc	2 mg/L		cyt-mus-orl	100 ug/kg
mmo-smc	10 mg/L		mrc-smc	10 mg/L
dns-rat-orl	5 ug/kg		hma-mus/smc	25 ug/kg
mma-smc	2 mg/L		cyt-mus-ipr	10 ug/kg
otr-mus:fbr	200 nmol/L		dni-hmn:oth	10 nmol/L
oms-mus:oth1 nmo	l/L		dnd-rat-orl	100 ug/kg
dns-mus-orl	800 pmol/kg			

TOXICITY:

Typ. Dose	Mode	Specie	Amount	Units	Other
LD50	orl	rat	20	ug/kg	
LD50	orl	mus	114	ug/kg	
LDLo	skn	mus	80	ug/kg	
LDLo	unr	mus	200	ug/kg	
LD50	ipr	ham	3	mg/kg	
LD50	orl	mky	2	ug/kg	
LD50	skn	rbt	275	ug/kg	
LD50	orl	gpg	500	ng/kg	
LD50	orl	ham	1157	ug/kg	
LDLo	orl	ckn	25	ug/kg	
LD50	ipr	rat	60	ug/kg	
TDLo	skn	hmn	107	ug/kg	
LD50	ipr	mus	120	ug/kg	
LD50	orl	dog	1	ug/kg	
LD50	ipr	rbt	252	ug/kg	
LD50	orl	frg	1	mg/kg	

OTHER TOXICITY DATA:

Skin and Eye Irritation Data:
eye-rbt 2 mg MOD

Review: Toxicology Review-21

Status:

> Meets criteria for proposed OSHA Medical Records Rule
> EPA TSCA Section 8(e) Status Report 8EHQ-0381-0390
> EPA TSCA Section 8(e) Status Report 8EHQ-0778-0209
> EPA Genetox Program 1988, Negative: Rodent dominant lethal
> EPA TSCA Test Submission (TSCATS) Data Base, January 1989
> NIOSH Current Intelligence Bulletin 40, 1984
> EPA Genetox Program 1988, Positive: Carcinogenicity-mouse/rat

SAX TOXICITY EVALUATION:

> THR: One of the most toxic synthetic chemicals. A deadly experimental poison by ingestion, skin contact, intraperitoneal and possibly other routes.May be a human carcinogen. An experimental carcinogen, neoplastigen, tumorigen and teratogen. Experimental reproductive effects. Humanmutagenic data. TCDD is the most toxic member of the 75 dioxins. Itcauses death in rats by hepatic cell necrosis. Death can follow a lethaldose by weeks. A by product of the manufacture of polychlorinatedphenols. It is found at low levels in 2,4,5-T; 2,4,5-trichlorophenol andhexachlorophene. It is also formed during various combustion processes.Incineration of chemical wastes, including chlorophenols, chlorinatedbenzenes and biphenyl ethers, may result in the presence of TCDD in fluegases, fly ash and soot particles. It is immobile in contaminated soiland may be retained for years. TCDD has potential for bio-accumulationin animals. An accident in Seveso, Italy and inadvertant soil contamination in Missouri have resulted in abandonment of the contaminated areas.

Source: *Instant Tox-Base*, 1995, Instant Reference Sources, Inc., 7605 Rockpoint Drive, Austin, TX 78731 (Fax: 512-345-2386; Internet URL, http://www.instantref.com/inst-ref.htm).

Toxicology Information from EPA's Integrated Risk Information System (IRIS)

2,3,7,8-TCDD is not currently listed in EPA's IRIS database nor is it listed as scheduled for addition in the near future.

Toxaphene

Toxaphene is a mixture of chlorinated terpenes so there is no representative sturcture.

Introduction

Toxaphene is a mixture of more than 175-179 components produced by chlorination of camphene. It has been used extensively as a pesticide on cotton as well as other crops. [828] Toxaphene is an insecticide and pesticide. It is used on cotton crops, cattle, swine, soybeans, corn, wheat, peanuts, lettuce, tomatoes, grains, vegetables, fruit and other food crops. It is used in the control of animal ectoparasites, grasshoppers, army-worms, cutworms and all major cotton pests. It controls livestock pests such as flies, lice, ticks, scab mites and mange. It also controls mosquito larvae, leaf miners, bagworms, church bugs, yellow jackets and caterpillars.

Toxaphene has conditional and restricted use as an insecticide and as a miticide in foliar treatment of: cranberries, strawberries, apples, pears, quinces, nectarines, peaches, bananas, pineapple, eggplant, peppers, pimentos, tomatoes, broccoli, brussel sprouts, cabbage, cauliflower, collards, kale, kohlrabi, spinach, lettuce (head and leaf), parsnips, rutabagas, beans (lima, green and snap), corn (sweet), cowpeas, okra, alfalfa, barley, oats, rice, rye, wheat, celery, cotton, horseradish, peanuts, peas, sunflowers, soybeans, ornamental plants, birch, elm, hickory, maple oak, and noncrop areas. It is also used in seed crop foliar treatment of clover and trefoil; in soil treatment of corn; in back rubber of beef cattle; in animal treatment of goats, sheep, beef cattle, and hogs; and aerial application and tank mixtures. [828]

Toxaphene is very persistent. When released to soil it will persist for long periods (1 to 14 yr), is not expected to leach to groundwater or be removed significantly by runoff unless adsorbed to clay particles which are removed by runoff. Biodegradation may be enhanced by anaerobic conditions such as flooded soil. Evaporation from soils and surfaces will be a significant process for toxaphene. Toxaphene released in water will not appreciably hydrolyze, photolyze, or significantly biodegrade. It will strongly sorb to sediments and bioconcentrate in aquatic organisms. Field tudies have shown it to be detoxified rapidly in shallow and very slowly in deep bodies of water. [828]

Toxaphene may undergo very slow direct photolysis in the atmosphere. However vapor phase reactions with photochemically produced hydroxyl radicals should be more important fate process (estimated half-life 4-5 days). Toxaphene can be transported long distances in the air (1200 km) probably adsorbed to particular matter. Monitoring data demonstrates that toxaphene is a contaminant in some air, water, sediment, soil, fish and other aquatic organisms, foods and birds. Human exposure appears to come mostly from food or occupational exposure. [828]

Identifying Information

CAS NUMBER: 8001-35-2

NIOSH Registry Number: XW5250000

SYNONYMS: Chlorinated camphene, Chlorocamphene, Octachlorocamphene, Polychlorocamphene

Endocrine Disruptor Information

Toxaphene is a strongly suspected EED that it is listed by the Centers for Disease Control & Prevention [812], and the World Wildlife Fund [813] as a potential endocrine modifying chemical.

Chemical and Physical Properties

CHEMICAL FORMULA: $C_{10}H_{10}Cl_8$

MOLECULAR WEIGHT: 413.80

WLN: L55 A CYTJ CU1 D1 D1 XG XG XG

PHYSICAL DESCRIPTION: Yellow or amber waxy solid

SPECIFIC GRAVITY: 1.63 [051], [072]

DENSITY: 1.66 g/mL @ 27°C [051], [062], [072]

MELTING POINT: 65-90°C [031], [043], [062], [395]

BOILING POINT: Decomposes [051], [072]

SOLUBILITY: Water: <1 mg/mL @ 19°C [700]; DMSO: >=100 mg/mL @ 19°C [700]; 95% Ethanol: 5-10 mg/mL @ 19°C [700]; Acetone: >=100 mg/mL @ 19°C [700]; Toluene: Soluble [430]

OTHER SOLVENTS: Chloroform: Soluble [051], [072]; Petroleum oils: Soluble [072], [173]; Aromatic hydrocarbons: Very soluble [031], [043], [051], [072]; Most organic solvents: Soluble [051], [062], [072], [395]; Hexane: Soluble [051], [072], [430]; Aliphatic hydrocarbons: Soluble [051], [072]; Xylene: Soluble [430]; Deodorized kerosene: >280 [430]; Mineral oil: 55-60 [430]

OTHER PHYSICAL DATA: Density: 1.65 g/mL @ 25°C [051], [072]; The flash point for a solution of this compound in 10% xylene is 28.9°C (84°F) [051], [072]; The autoignition temperature for a solution of this compound in 10% xylene is 530°C (986°) [051], [072]; Slight pine odor [031], [043], [151], [173]; Odor threshold (detection): 0.14 mg/kg [055]; Odor threshold in water: 0.0052 mg/L [051]; Burning rate: 5.8 mm/min

[051]; Viscosity: 89 centipoise @ 110°C, 57 centipoise @ 120°C [051]; Viscosity: 39.1 centipoise @ 130°C [051]; Specific heat: 0.258 cal/g @ 41 C° [051]

Source: *Instant EPA's Air Toxics*, 1994, Instant Reference Sources, Inc., 7605 Rockpoint Drive, Austin, TX 78731 (Fax: 512-345-2386; Internet URL, http://www.instantref.com/inst-ref.htm).

Environmental Reference Materials

A source of EED environmental reference materials for toxaphene is Radian International LLC, Austin, TX (Fax 512-454-0268; Internet URL, http://www.radian.com/standards); Catalog No. ERT-002.

Hazardous Properties

ACUTE/CHRONIC HAZARDS: Toxaphene may be toxic by ingestion, inhalation and skin absorption [062]. It also is an irritant [031], [072] and a lachrymator [173]. When heated to decomposition it emits toxic fumes of chlorides [072].

HAP WEIGHTING FACTOR: 1 [713]

VOLATILITY: Vapor pressure: 0.2-0.4 mm Hg @ 25°C [173]

FIRE HAZARD: Flash point data for toxaphene are not available. It is probably combustible. Fires involving this material may be controlled with a dry chemical, carbon dioxide or Halon extinguisher.

REACTIVITY: Toxaphene is decomposed in the presence of alkali [033], [051], [395]. It is corrosive to iron [033] and is incompatible with strong oxidizers [346]. It is noncorrosive in the absence of moisture [395].

STABILITY: Toxaphene is decomposed by sunlight and heat [033], [051], [173], [395].

Source: *Instant EPA's Air Toxics*, 1994, Instant Reference Sources, Inc., 7605 Rockpoint Drive, Austin, TX 78731 (Fax: 512-345-2386; Internet URL, http://www.instantref.com/inst-ref.htm).

Medical Symptoms of Exposure

Symptoms of exposure may include nausea, mental confusion, jerking of the arms and legs, convulsions, cyanosis, lacrimation, eye pain, headache, vertigo, abdominal pain, liquid stools and weakness [173]. Other symptoms may include somnolence, effect on seizure threshold, coma, allergic dermatitis, skin irritation and liver injury [043]. It may cause central nervous system stimulation, tremors and death [031]. It may also cause salivation, leg and back muscle spasms, vomiting, hyper excitability, shivering, tetanic contractions of all skeletal muscles and respiratory failure [072]. Auditory reflex excitability may occur [301]. Exposure may cause agitation, dry and red skin, and unconsciousness [346].

Source: *Instant EPA's Air Toxics*, 1994, Instant Reference Sources, Inc., 7605 Rockpoint Drive, Austin, TX 78731 (Fax: 512-345-2386;   Internet URL, http://www.instantref.com/inst-ref.htm).

Toxicological Information

The toxicological information below is gathered from several sources including two national databases: one from the *National Toxicology Program Chemical Repository Database* and another from *EPA's Integrated Risk Information System (IRIS)*.

Toxicology Information from the National Toxicology Program

CARCINOGENICITY:

Tumorigenic Data:

Type of Effect	Route/Animal	Amount Dosed (Notes)
TDLo	orl-rat	30 gm/kg/80W-C
TDLo	orl-mus	6600 mg/kg/80W-C
TD	orl-mus	13 gm/kg/80W-C

Review:
   IARC Cancer Review: Animal Sufficient Evidence
   IARC possible human carcinogen (Group 2B)

Status:
   EPA Carcinogen Assessment Group
   NTP anticipated human carcinogen
   NTP Fourth Annual Report on Carcinogens, 1984
   NCI Carcinogenesis Bioassay (Feed); Equivocal: Male and Female Rat
   NCI Carcinogenesis Bioassay (Feed); Positive: Male and Female
Mouse

TERATOGENICITY: See RTECS printout for most current data

MUTATION DATA: See RTECS printout for most current data

TOXICITY:

Typ. Dose	Mode	Specie	Amount	Units	Other
LDLo	orl	man	29	mg/kg	
TDLo	skn	hmn	657	mg/kg	
LDLo	orl	hmn	28	mg/kg	
LDLo	unr	man	44	mg/kg	
LD50	orl	rat	50	mg/kg	
LD50	skn	rat	600	mg/kg	
LDLo	ipr	rat	70	mg/kg	
LD50	unr	rat	240	mg/kg	
LD50	orl	mus	112	mg/kg	
LCLo	ihl	mus	2000	mg/m3/2H	
LD50	ipr	mus	42	mg/kg	
LD50	unr	mus	45	mg/kg	
LD50	orl	dog	15	mg/kg	
LD50	orl	rbt	75	mg/kg	
LD50	skn	rbt	1025	mg/kg	
LD50	orl	gpg	250	mg/kg	
LD50	orl	ham	200	mg/kg	
LD50	orl	dck	31	mg/kg	

OTHER TOXICITY DATA:

Skin and Eye Irritation Data:
skn-mam 500 mg MOD

Review: Toxicology Review-3

Standards and Regulations:
DOT-Hazard: ORM-A; Label: None

Status:

EPA Genetox Program 1988, Positive: Carcinogenicity-mouse/rat
EPA TSCA Test Submission (TSCATS) Data Base, June 1989
Meets criteria for proposed OSHA Medical Records Rule
Lethal oral dose for adult: ~4-7 g
IDLH value: 200 mg/m3

SAX TOXICITY EVALUATION:

THR: Human poison by ingestion and possibly other routes. Experimental poison by ingestion, intraperitoneal and possibly other routes. Moderately toxic experimentally by inhalation and skin contact. An experimental carcinogen, tumorigen and teratogen. It may be a human carcinogen.

Source: *Instant Tox-Base*, 1995, Instant Reference Sources, Inc., 7605 Rockpoint Drive, Austin, TX 78731 (Fax: 512-345-2386; Internet URL, http://www.instantref.com/instref.htm).

Toxicology Information from EPA's Integrated Risk Information System (IRIS)

I. CHRONIC HEALTH HAZARD ASSESSMENTS FOR NONCARCINOGENIC EFFECTS

I.A. REFERENCE DOSE FOR CHRONIC ORAL EXPOSURE (RfD)

[Ed. Note: EPA does not have information yet in this section of IRIS].

I.B. REFERENCE CONCENTRATION FOR CHRONIC INHALATION EXPOSURE (RfC)

[Ed. Note: EPA does not have information yet in this section of IRIS].

II. CARCINOGENICITY ASSESSMENT FOR LIFETIME EXPOSURE

Section II provides information on three aspects of the carcinogenic risk assessment for the agent in question; the US EPA classification, and quantitative estimates of risk from oral exposure and from inhalation exposure. The classification reflects a weight-of-evidence judgment of the likelihood that the agent is a human carcinogen. The quantitative risk estimates are presented in three ways. The slope factor is the result of application of a low-dose extrapolation procedure and is presented as the risk per (mg/kg)/day. The unit risk is the quantitative estimate in terms of either risk per ug/L drinking water or risk per ug/cu.m air breathed. The third form in which risk is presented is a drinking water or air concentration providing cancer risks of 1 in 10,000, 1 in 100,000 or 1 in 1,000,000. Background Document 2 (Service Code 5) provides details on the rationale and methods used to derive the carcinogenicity values found in IRIS. Users are referred to Section I for information on long-term toxic effects other than carcinogenicity.

II.A. EVIDENCE FOR CLASSIFICATION AS TO HUMAN CARCINOGENICITY

II.A.1. WEIGHT-OF-EVIDENCE CLASSIFICATION

Classification -- B2; probable human carcinogen.

Basis -- The classification is based on increased incidence of hepatocellular tumors in mice and thyroid tumors in rats and is supported by mutagenicity in Salmonella.

II.A.2. HUMAN CARCINOGENICITY DATA

None.

II.A.3. ANIMAL CARCINOGENICITY DATA

Sufficient. Two long-term carcinogenicity bioassays with toxaphene have been performed in rats and mice with both species showing a carcinogenic response. Dietary toxaphene was administered for 18 months at doses of 0, 7, 20 and 50 ppm to 54 B6C3F1 mice/sex/group. Animals were observed 6 months post-treatment. An increased incidence of hepatocellular carcinomas and neoplastic nodules (adenomas) was seen in both sexes and was statistically significant in males administered 50 ppm (Litton Bionetics, 1978).

In a second study (NCI, 1979), dietary toxaphene was administered to 50 Osborne-Mendel rats/sex/group and 50 B6C3F1 mice/sex/group for 80 weeks. Rats received TWA doses of 556 and 1112 ppm for males and 540 and 1080 ppm for females. The animals were observed for 28-30 weeks post-treatment. Controls consisted of 10 matched controls/sex and 45 additional pooled controls/sex. A statistically significant dose-related increased incidence of thyroid tumors (adenomas and carcinomas) was seen in both male and female rats.

Mice received TWA doses of 99 and 198 ppm for both sexes. Controls consisted of 10 matched controls/sex and 40 additional pooled controls/sex. A statistically significantly increased incidence of liver cancer in treated animals was observed and was dose-related (NCI, 1979).

II.A.4. SUPPORTING DATA FOR CARCINOGENICITY

Toxaphene is mutagenic to Salmonella (Hill, 1977). It was negative in a modified dominant lethal assay of male ICR/Ha Swiss mice (Epstein, 1972). No significant differences were found between rates of chromosomal aberrations in leukocytes of workers occupationally exposed to toxaphene and of unexposed workers (US EPA, 1978).

II.B. QUANTITATIVE ESTIMATE OF CARCINOGENIC RISK FROM ORAL EXPOSURE

II.B.1. SUMMARY OF RISK ESTIMATES

Oral Slope Factor -- 1.1E+0 per (mg/kg)/day

Drinking Water Unit Risk -- 3.2E-5 per (ug/L)

Extrapolation Method -- linearized multistage procedure, extra risk

Drinking Water Concentrations at Specified Risk Levels:

Risk Level	Concentration
E-4 (1 in 10,000)	3E+0 ug/L
E-5 (1 in 100,000)	3E-1 ug/L
E-6 (1 in 1,000,000)	3E-2 ug/L

II.B.2. DOSE-RESPONSE DATA (CARCINOGENICITY, ORAL EXPOSURE)

Tumor Type -- hepatocellular carcinomas and neoplastic nodules
Test Animals -- mouse/B6C3F1, males
Route -- diet
Reference -- Litton Bionetics, 1978

Administered Dose		Human Equivalent	Tumor
(ppm)	(mg/kg)/day	Dose (mg/kg)/day	Incidence
0	0.0	0	10/53
7	0.91	0.051	10/54
20	2.6	0.144	12/53
50	6.5	0.361	18/51

II.B.3. ADDITIONAL COMMENTS (CARCINOGENICITY, ORAL EXPOSURE)

The Litton Bionetics (1978) study was used for derivation of a slope factor because more dose levels were used, and a positive carcinogenic response was found at a lower dose than in the NCI study (1979). Weight of the animals was assumed to be 0.03 kg, and animal lifetime was taken as 735 days, the duration of the experiment.

The unit risk should not be used if the water concentration exceeds 3E+2 ug/L, since above this concentration the unit risk may not be appropriate.

II.B.4. DISCUSSION OF CONFIDENCE (CARCINOGENICITY, ORAL EXPOSURE)

An adequate number of animals was observed. A dose-response effect was seen in a study with 3 non-zero dose levels.

II.C. QUANTITATIVE ESTIMATE OF CARCINOGENIC RISK FROM INHALATION EXPOSURE

II.C.1. SUMMARY OF RISK ESTIMATES

Inhalation Unit Risk -- 3.2E-4 per (ug/cu.m)

Extrapolation Method -- linearized multistage procedure, extra risk

Air Concentrations at Specified Risk Levels:

Risk Level	Concentration
E-4 (1 in 10,000)	3E-1 ug/cu.m
E-5 (1 in 100,000)	3E-2 ug/cu.m
E-6 (1 in 1,000,000)	3E-3 ug/cu.m

II.C.2. DOSE-RESPONSE DATA FOR CARCINOGENICITY, INHALATION EXPOSURE

The unit risk was calculated from the oral data presented in II.B.2.

II.C.3. ADDITIONAL COMMENTS (CARCINOGENICITY, INHALATION EXPOSURE)

The unit risk should not be used if the air concentration exceeds 3.1E+1 ug/cu.m, since above this concentration the unit risk may not be appropriate.

II.C.4. DISCUSSION OF CONFIDENCE (CARCINOGENICITY, INHALATION EXPOSURE)

This inhalation risk estimate was based on oral data.

II.D. EPA DOCUMENTATION, REVIEW, AND CONTACTS (CARCINOGENICITY ASSESSMENT)

II.D.1. EPA DOCUMENTATION

Source Document -- US EPA, 1978, 1980

The values in the 1980 Ambient Water Quality Criteria document have received both Agency and outside review.

II.D.2. REVIEW (CARCINOGENICITY ASSESSMENT)

Agency Work Group Review -- 03/05/87

Verification Date -- 03/05/87

II.D.3. US EPA CONTACTS (CARCINOGENICITY ASSESSMENT)

Charlie Hiremath / NCEA -- (202)260-5725

William E. Pepelko / NCEA -- (202)260-5904

III. HEALTH HAZARD ASSESSMENTS FOR VARIED EXPOSURE DURATIONS

[Ed. Note: EPA does not have information yet in this section of IRIS].

IV. US EPA REGULATORY ACTIONS

EPA risk assessments may be updated as new data are published and as assessment methodologies evolve. Regulatory actions are frequently not updated at the same time. Compare the dates for the regulatory actions in this section with the verification dates for the risk assessments in sections I and II, as this may explain inconsistencies. Also note that some regulatory actions consider factors not related to health risk, such as technical or economic feasibility. Such considerations are indicated for each action. In addition, not all of the regulatory actions listed in this section involve enforceable federal standards. Please direct any questions you may have concerning these regulatory actions to the US EPA contact listed for that particular action. Users are strongly urged to read the background information on each regulatory action in Background Document 4 in Service Code 5.

IV.A. CLEAN AIR ACT (CAA)

[Ed. Note: EPA does not have information yet in this section of IRIS].

IV.B. SAFE DRINKING WATER ACT (SDWA)

IV.B.1. MAXIMUM CONTAMINANT LEVEL GOAL (MCLG) for Drinking Water

Value -- 0.00 mg/L (Final, 1991)

Considers technological or economic feasibility? -- NO

Discussion -- The final MCLG for toxaphene is zero based on the evidence of carcinogenic potential (classification B2).

Reference -- 56 FR 3526 (01/30/91)

EPA Contact -- Health and Ecological Criteria Division / OST /(202) 260-7571 / FTS 260-7571; or Safe Drinking Water Hotline / (800) 426-4791

IV.B.2. MAXIMUM CONTAMINANT LEVEL (MCL) for Drinking Water

Value -- 0.003 mg/L (Final, 1991)

Considers technological or economic feasibility? -- YES

Discussion -- The MCL is based on a PQL of 0.003 mg/L and is associated with a maximum lifetime individual risk of E-4.

Monitoring requirements -- All systems initially monitored for four consecutive quarters every three years; repeat monitoring dependent upon detection, vulnerability status and system size.

Analytical methodology -- Microextraction/gas chromatography (EPA 505). PQL=0.003 mg/L.

Best available technology -- Granular activated carbon

Reference -- 56 FR 3526 (01/30/91)

EPA Contact -- Drinking Water Standards Division / OGWDW /(202) 260-7575 / FTS 260-7575; or Safe Drinking Water Hotline / (800) 426-4791

IV.B.3. SECONDARY MAXIMUM CONTAMINANT LEVEL (SMCL) for Drinking Water

   [Ed. Note: EPA does not have information yet in this section of IRIS].

IV.B.4. REQUIRED MONITORING OF "UNREGULATED" CONTAMINANTS

   [Ed. Note: EPA does not have information yet in this section of IRIS].

IV.C. CLEAN WATER ACT (CWA)

IV.C.1. AMBIENT WATER QUALITY CRITERIA, Human Health

   [Ed. Note: EPA does not have information yet in this section of IRIS].

IV.C.2. AMBIENT WATER QUALITY CRITERIA, Aquatic Organisms

Freshwater:
   Acute -- 7.3E-1 ug/L (1 hour average)
   Chronic -- 2E-4 ug/L (4 day average)

Marine:
   Acute -- 2.1E-1 ug/L (1 hour average)
   Chronic -- 2E-4 ug/L (4 day average)

Considers technological or economic feasibility? -- NO

Discussion -- Criteria were derived from a minimum data base consisting of acute tests on a variety of species. Requirements and methods are covered in the reference to the Federal Register.

Reference -- 51 FR 43665 (12/03/86)

EPA Contact -- Criteria and Standards Division / OWRS (202)260-1315 / FTS 260-1315

IV.D. FEDERAL INSECTICIDE, FUNGICIDE, AND RODENTICIDE ACT (FIFRA)

IV.D.1. PESTICIDE ACTIVE INGREDIENT, Registration Standard

Status -- Removed from list "B" pesticides/Registration canceled (1990)
Reference -- 55 FR 31166 (07/31/90)
EPA Contact -- Registration Branch / OPP (703)557-7760 / FTS 557-7760

IV.D.2. PESTICIDE ACTIVE INGREDIENT, Special Review

Action -- Final Regulatory Decision - PD 4 (1982)
Considers technological or economic feasibility? -- NO

Summary of regulatory action -- Cancellation of registrations for most uses and
continued registration of certain uses under specific terms and conditions.

Reference -- 47 FR 53784 (11/29/82) [NTIS# PB83-144204]

EPA Contact -- Special Review Branch / OPP (703)557-7400 / FTS 557-7400

IV.E. TOXIC SUBSTANCES CONTROL ACT (TSCA)

        [Ed. Note: EPA does not have information yet in this section of IRIS].

IV.F. RESOURCE CONSERVATION AND RECOVERY ACT (RCRA)

IV.F.1. RCRA APPENDIX IX, for Ground Water Monitoring

Status -- Listed

Reference -- 52 FR 25942 (07/09/87)

EPA Contact -- RCRA/Superfund Hotline (800)424-9346 / (202)260-3000 / FTS 260-
3000

IV.G. SUPERFUND (CERCLA)

IV.G.1. REPORTABLE QUANTITY (RQ) for Release into the Environment

Value (status) -- 1 pound (Final, 1989)

Considers technological or economic feasibility? -- NO

Discussion -- The final RQ for toxaphene is based on aquatic toxicity as established
under CWA Section 311 (40 CFR 117.3). The available data indicate that the aquatic 96-
Hour Median Threshold Limit is less than 0.1 ppm, which corresponds to an RQ of 1
pound.

Reference -- 54 FR 33418 (08/14/89)

EPA Contact -- RCRA/Superfund Hotline (800)424-9346 / (202)260-3000 / FTS 260-3000

VI. BIBLIOGRAPHY

VI.A. ORAL RfD REFERENCES

None

VI.B. INHALATION RfD REFERENCES

None

VI.C. CARCINOGENICITY ASSESSMENT REFERENCES

Epstein, S.S. E. Arnold, J. Andrea, W. Bass and Y. Bishop. 1972. Detection of chemical mutagens by the dominant lethal assay in the mouse. Toxicol. Appl. Pharmacol. 23(2): 288-325.

Hill, R.N. 1977. Memorandum to Fred Hageman. Off. Spec. Pestic. Rev., US EPA. December 15.

Litton Bionetics. 1978. Carcinogenic evaluation in mice: Toxaphene. Final report. Prepared by Litton Bionetics, Inc., Kensington, MD for Hercules, Inc., Wilmington, DE. LBI Project No. 20602.

NCI (National Cancer Institute). 1979. Bioassay of Toxaphene for Possible Carcinogenicity. Carcinogenesis Testing Program. Division of Cancer Cause and Prevention. NCI, National Institute of Health, Bethesda, Maryland, 20014. US Department of Health, Education and Welfare. DHEW Publication No. (NIH) 79-837.

US EPA. 1978. Occupational Exposure to Toxaphene. A Final Report by the Epidemiologic Studies Program, Human Effects Monitoring Branch, Benefits and Field Studies Division, OPP, OTS, EPA.

US EPA. 1980. Ambient Water Quality Criteria for Toxaphene. Prepared by the Office of Health and Environmental Assessment, Environmental Criteria and Assessment Office, Cincinnati, OH for the Office of Water Regulations and Standards. Washington, DC. EPA 440/5-80-076. NTIS PB 81-117863.

VI.D. DRINKING WATER HA REFERENCES

None

VII. REVISION HISTORY

```
--------   -----------   ---------------------------------------------------------
Date       Section       Description
--------   -----------   ---------------------------------------------------------
```

Date	Section	Description
08/22/88	II.	Carcinogen summary on-line
04/01/89	V.	Supplementary data on-line
06/01/90	VI.	Bibliography on-line
01/01/91	II.	Text edited
01/01/91	II.C.1.	Inhalation slope factor removed (global change)
01/01/92	IV.	Regulatory Actions section on-line

SYNONYMS

alltox	phenacide
chlorinated-camphene	toxadust
geniphene	toxakil
penphene	Toxaphene

Source: *Instant EPA's IRIS*, 1997, Instant Reference Sources, Inc., 7605 Rockpoint Drive, Austin, TX 78731 (Fax: 512-345-2386; Internet URL, http://www.instantref.com/inst-ref.htm).

Production, Use, and Pesticide Labeling Information

Toxaphene is not currently listed in EPA's Pesticide Factsheet database and there is no indication of whether it will be added or not in the future.

Trifluralin

Introduction

Trifluralin is an anthropogenic compound used as a pre-emergence herbicide. It is a selective herbicide for grasses and broadleaf weeds in soybean, cotton and other types of field crops. It is used with alfalfa and pasture lands. It is also used with vegetables, fruits and nuts. [828] In August, 1979, trifluralin was brought under Special Review by the EPA because of the presence of a N-nitrosamine contaminant which had been shown to cause tumors and to have mutagenic effects in animals. The principle manufacturer of trifluralin had already instituted manufacturing methods to reduce N-nitrosamine contaminant levels. The Special Review was concluded in 1982, with the requirement that N-nitrosamine contaminant levels in trifluralin not exceed 0.5 ppm, a level which EPA believes will have no toxic effects. [829]

Trifluralin may be released to the environment during its production and will be released during its application to agricultural fields. If released to soil, trifluralin is expected to biodegrade and volatilize to the atmosphere. The persistence of trifluralin in soil has been estimated at approximately 6 months although it is more persistent in northern climates. Trifluralin is expected to biodegrade under both aerobic and anaerobic conditions in soil. It is also expected to rapidly volatilize from both moist and dry soils to the atmosphere. Trifluralin will strongly adsorb to soil. [828]

If released to water, trifluralin is expected to biodegrade under both aerobic and anaerobic conditions and undergo direct photolytic degradation. It is expected to bioconcentrate in fish and aquatic organisms and adsorb strongly to sediment and suspended organic matter. It may also volatilize from water to the atmosphere. If released to the atmosphere, trifluralin is expected to undergo a rapid gas-phase photolysis. The rate of this process increases in the presence of ozone. Trifluralin may also undergo atmospheric removal by a gas-phase reaction with photochemically produced hydroxyl radicals with an estimated half-life of 4.6 hours. Occupational exposure to trifluralin may occur by inhalation or dermal contact. The general population may be exposed to trifluralin from lawn products or by ingestion of contaminated agricultural products or fish. [828]

Trifluralin is not hazardous to birds but it is toxic to fish and other aquatic organisms. However, since it is strongly adsorbed on soils and very insoluble in water it shouldn't be a large contributor to groundwater contamination. [829]

Identifying Information

CAS NUMBER: 1582-09-8

NIOSH Registry Number: XU9275000

SYNONYMS: 2,6-Dinitro-N,N-dipropyl-4-(trifluoromethyl)benzenamine, Flurene SE, Treflan, Tri-4, Trust, M.T.F., Trifluralina 600, Elancolan, Su Seguro Carpidor, Trefanocide, Treficon, Trim, L-36352, Crisalin, TR-10, Triflurex, Ipersan.

Endocrine Disruptor Information

Trifluralin is a strongly suspected EED that it is listed by EPA [811], the Centers for Disease Control & Prevention [812], and the World Wildlife Fund [813] as a potential endocrine modifying chemical.

Chemical and Physical Properties

CHEMICAL FORMULA: $C_{13}H_{16}F_3N_3O_4$

MOLECULAR WEIGHT: 335.32

WLN: FXFFR CNW ENW DN3&3

PHYSICAL DESCRIPTION: Yellowish-orange solid

SPECIFIC GRAVITY: Not available

DENSITY: 1.294 g/mL @ 25°C

MELTING POINT: 46-47°C

BOILING POINT: Decomposes

SOLUBILITY: Water: <0.1 mg/mL @ 22.5°C [700]; DMSO: >=100 mg/mL @ 18°C [700]; 95% Ethanol: 10-50 mg/mL @ 18°C [700]; Acetone: >=100 mg/mL @ 18°C [700]

OTHER SOLVENTS: Xylene: Very soluble; Stoddard solvent: Very soluble

OTHER PHYSICAL DATA: Heat of combustion: -5020 cal/g; Boiling point: 139-140°C @ 4.2 mm Hg

Source: *Instant EPA's Air Toxics*, 1994, Instant Reference Sources, Inc., 7605 Rockpoint Drive, Austin, TX 78731 (Fax: 512-345-2386; Internet URL, http://www.instantref.com/inst-ref.htm).

Environmental Reference Materials

A source of EED environmental reference materials for trifluralin is Crescent Chemical Co., Hauppauge, NY (Fax: 516-348-0913); Catalog No. 45700.

Hazardous Properties

ACUTE/CHRONIC HAZARDS: Trifluralin may be toxic and an irritant. It is also an experimental carcinogen. When heated to decomposition it emits toxic fumes of hydrogen fluoride and nitrogen oxides. It can also be irritating to the skin and eyes. However, it was not acutely toxic to animals by oral, dermal or inhalation routes of exposure. Repeated skin contact may cause allergic dermatitis. [829]

HAP WEIGHTING FACTOR: 1 [713]

VOLATILITY: Vapor pressure: 0.000199 mm Hg @ 29.5°C [055]

FIRE HAZARD: Trifluralin has a flash point of >85°C (>185°F) [355], [371]. It is combustible. Fires involving this material may be controlled with a dry chemical, carbon dioxide or Halon extinguisher. A water spray may also be used [371]. It is not flamable but emulsifiable formulations of it may be flammable. It can pose a fire and explosion hazard in the presence of strong oidizers. [829]

STABILITY: Trifluralin is sensitive to exposure to light [168]. Solutions of it in water, DMSO, 95% ethanol or acetone should be stable for 24 hours when protected from light [700].

Source: *Instant EPA's Air Toxics*, 1994, Instant Reference Sources, Inc., 7605 Rockpoint Drive, Austin, TX 78731 (Fax: 512-345-2386; Internet URL, http://www.instantref.com/inst-ref.htm).

Medical Symptoms of Exposure

Symptoms of exposure may include irritation of the skin, eyes, gastrointestinal tract, and respiratory tract, convulsions, and coma. Nausea and severe gastrointestinal discomfort may occur from ingestion of it. When applied to the eyes of rabbits, trifluralin produced slight irritation which cleared within 7 days. Skin sensitization (allergies) may occur in some individuals. Inhalation may cause irritation of the lining of the mouth, throat or lungs. The solvent in emulsifiable concentrates of trifluralin may cause irritation to the skin. Most cases of poisoning result from the carrier or solvent in formulated trifluralin products, rather than from the trifluralin itself. Consumption at high levels for prolonged periods of time caused liver and kidney damage in animals. [829]

Toxicological Information

The toxicological information below is gathered from several sources including two national databases: one from the *National Toxicology Program Chemical Repository Database* and another from *EPA's Integrated Risk Information System (IRIS)*.

Toxicology Information from the National Toxicology Program

CARCINOGENICITY:

Tumorigenic Data:

Type of Effect	Route/Animal	Amount Dosed (Notes)
TDLo	orl-mus	180 gm/kg/78W-C
TDLo	ipr-mus	2600 ug/kg/39D-I
TDLo	scu-mus	2600 ug/kg/39D-I
TD	orl-mus	340 gm/kg/78W-C

Status:
NCI Carcinogenesis Bioassay (Feed); Positive: Female Mouse
NCI Carcinogenesis Bioassay (Feed); Negative: Male and Female Rat,
Male Mouse

TERATOGENICITY: See RTECS printout for most current data

Reproductive Effects Data:

Type of Effect	Route/Animal	Amount Dosed (Notes)
TDLo	orl-mus	10 mg/kg (6-15D preg)
TDLo	orl-mus	10 gm/kg (6-15D preg)
TDLo	ipr-mus	200 mg/kg (1D male)

MUTATION DATA: See RTECS printout for most current data

Test	Lowest dose	Test	Lowest dose
sln-nsc	1 mg/L	mrc-asn	100 ug/plate
cyt-hmn:lym	2 ppm	cyt-mus-ipr	200 mg/kg
dlt-mus-ipr	200 mg/kg	sce-ham:ovr	160 ug/L
sce-hmn:lym	1 mg/L		

TOXICITY:

Typ. Dose	Mode	Specie	Amount	Units	Other
LD50	orl	mus	5000	mg/kg	
LDLo	ipr	mus	1500	mg/kg	
LD50	orl	mam	3700	mg/kg	
LDLo	unr	mam	5	gm/kg	

OTHER TOXICITY DATA:

Standards and Regulations:
EPA FIFRA 1988 Pesticide subject to registration or re-registration

Status:
EPA Genetox Program 1988, Positive: N crassa-aneuploidy
EPA Genetox Program 1988, Weakly Positive: S cerevisiae-homozygosis
EPA Genetox Program 1988, Positive/limited: Carcinogenicity-mouse/rat
EPA Genetox Program 1988, Negative: D melanogaster Sex-linked lethal
EPA Genetox Program 1988, Negative: In vitro UDS-human fibroblast; TRP reversion
EPA Genetox Program 1988, Inconclusive: B subtilis rec assay; E colipolA without S9
EPA TSCA Test Submission (TSCATS) Data Base, April 1990

SAX TOXICITY EVALUATION:

THR: An experimental teratogen and carcinogen. MODERATE via intraperitoneal and oral routes. LOW via oral route. MUTATION data.

Source: *Instant Tox-Base*, 1995, Instant Reference Sources, Inc., 7605 Rockpoint Drive, Austin, TX 78731 (Fax: 512-345-2386; Internet URL, http://www.instantref.com/inst-ref.htm).

Toxicology Information from EPA's Integrated Risk Information System (IRIS)

I. CHRONIC HEALTH HAZARD ASSESSMENTS FOR NONCARCINOGENIC EFFECTS

I.A. REFERENCE DOSE FOR CHRONIC ORAL EXPOSURE (RfD)

The Reference Dose (RfD) is based on the assumption that thresholds exist for certain toxic effects such as cellular necrosis, but may not exist for other toxic effects such as carcinogenicity. In general, the RfD is an estimate (with uncertainty spanning perhaps an order of magnitude) of a daily exposure to the human population (including sensitive subgroups) that is likely to be without an appreciable risk of deleterious effects during a lifetime. Please refer to Background Document 1 in Service Code 5 for an elaboration of these concepts. RfDs can also be derived for the noncarcinogenic health effects of compounds which are also carcinogens. Therefore, it is essential to refer to other sources of information concerning the carcinogenicity of this substance. If the US EPA has evaluated this substance for potential human carcinogen- icity, a summary of that evaluation will be contained in Section II of this file when a review of that evaluation is completed.

I.A.1. ORAL RfD SUMMARY

Critical Effect	Experimental Doses*	UF	MF	RfD
Increased liver weights; increase in methemoglobin	NOEL: 30 ppm (0.75 mg/kg/day) LEL: 150 ppm (3.75 mg/kg/day)	100	1	7.5E-3 mg/kg/day

*Conversion Factors: 1 ppm = 0.025 mg/kg/day (assumed dog food consumption)

12-Month Dog Feeding Study, Hoechst Aktiengesellschaft, 1984a

I.A.2. PRINCIPAL AND SUPPORTING STUDIES (ORAL RfD)

Hoechst Aktiengesellschaft. 1984a. MRID No. 00151908. Available from EPA. Write to FOI, EPA, Washington, DC 20460.

Beagle dogs (6/sex/dose) were fed diets containing 0, 30, 150, or 750 ppm (0, 0.75, 3.75, and 18.75 mg/kg/day) of trifluralin for 12 months. At 750 ppm (HDT; 18.75 mg/kg/day) there was a decreased weight gain in males and females. There were some significant decreases in red blood cell parameters in high- dose males and females. There was an increase in methemoglobin in mid- and high-dose males and females. Total serum lipids, triglycerides, and cholesterol were increased in high-dose males and females when compared with controls. There were increases in liver weight in males receiving 150 and 750 ppm (3.75 and 18.75 mg/kg/day) and females receiving 750 ppm trifluralin and increases in mean spleen weight in females receiving 750 ppm. There was no histologic findings that correlated with organ weight changes. Based on the increases in liver weights and methemoglobin, the LEL is 150 ppm (3.75 mg/kg/day) and the NOEL is 30 ppm (0.75 mg/kg/day).

I.A.3. UNCERTAINTY AND MODIFYING FACTORS (ORAL RfD)

UF -- An uncertainty factor of 100 was used to account for the inter- and intraspecies differences.

MF -- None

I.A.4. ADDITIONAL COMMENTS (ORAL RfD)

The previous RfD for trifluralin was established using a 3-month rat feeding study (Eli Lilly & Co., 1985) with a Systemic LEL of 2.5 mg/kg/day (lowest dose tested) based on increased alpha 1, alpha 2 and beta globulins in the urine. The original data from this study was re-examined with regard to total protein, alpha 1, alpha 2, and beta and gamma globulin. This reexamination concluded that an NOEL was established at 50 ppm (2.5 mg/kg/day) and an LEL at 200 ppm (10 mg/kg/day) based on evidence of protein excretion (TP, alpha 1, alpha 2, and beta globulins). Therefore, when the complete database for trifluralin is considered, the chronic dog study is the appropriate study to establish the RfD.

Data Considered for Establishing the RfD

1) 1-Year Feeding - dog: Principal study - see previous description; core grade guideline

2) 2-Year Feeding (oncogenic) - rat: Systemic NOEL=200 ppm (10 mg/kg/day); Systemic LEL=800 ppm (40 mg/kg/day) (body weight changes); core grade guideline (Hoechst Aktiengesellschaft, 1986a)

3) 2-Generation Reproduction - rat: Systemic NOEL=200 ppm (10 mg/kg/day); Systemic LEL=630 ppm (31.5 mg/kg/day) (decreased body weight); Reproductive NOEL=2000 ppm (100 mg/kg/day) (HDT); Reproductive LEL=none; core grade minimum (Elanco Product Co., 1986)

4) 2-Generation Reproduction - rat: Reproductive NOEL=650 ppm (32.5 mg/kg/day); Reproductive LEL=2000 ppm (100 mg/kg/day) (HDT; reduced litter size); Developmental NOEL=200 ppm (10 mg/kg/day); Developmental LEL=650 ppm 32.5 mg/kg/day) (increased weanling body weight); Parental NOEL=none; Parental LEL=200 ppm (10 mg/kg/day) (LDT; increased kidney weights); At 650 ppm renal lesions of the proximal tubules and increased relative kidney weights; core grade minimum (Hoechst Aktiengesellschaft, 1984b)

5) Teratology - rat: Maternal NOEL=225 mg/kg/day; Maternal LEL=475 mg/kg/day (decreased body weight and food consumption); Fetotoxic NOEL=475 mg/kg/day; Fetotoxic LEL=1000 mg/kg/day (decreased mean fetal body weight); Teratogenic NOEL=1000 mg/kg/day (HDT); Teratogenic LEL=none; core grade minimum (Elanco Product Co., 1984a)

6) Teratology - rabbit: Maternal NOEL=100 mg/kg/day; Maternal LEL=225 mg/kg/day (body weight loss); Fetotoxic NOEL=225 mg/kg/day; Fetotoxic LEL=500 mg/kg/day (HDT; decreased fetal weight and increased number of fetal runts); Teratogenic NOEL=500 mg/kg/day (HDT); Teratogenic LEL=none; core grade minimum (Elanco Product Co., 1984b)
Other Data Reviewed:

1) Oncogenicity - mouse: Systemic NOEL=50 ppm (7.5 mg/kg/day); Systemic LEL=200 ppm (30 mg/kg/day) (increased liver weight in males); At 800 ppm (120 mg/kg/day) (HDT) an increase in liver weight in males and females was observed; core grade supplementary (pending submission of historical control data) (Hoechst Aktiengesellschaft, 1986b)

2) 6-Month Feeding - dog: NOEL=none; LEL=400 ppm (10 mg/kg/day) (LDT; enlarged livers, discolored kidneys, corneal vascularization, hemolytic anemia and increase alkaline phosphatase); core grade supplementary (Hoechst Aktiengesellschaft, 1981)

3) 3-Month Feeding - rat: NOEL=none; LEL=800 ppm (40 mg/kg/day) (LDT; liver/body weight increases and pituitary/body weight decreases in all doses); core grade minimum (Hoechst Aktiengesellschaft, 1980)

4) 3-Month Special Urinalysis Study - rat: NOEL=50 ppm (2.5 mg/kg/day); LEL=200 ppm (10 mg/kg/day) [evidence of protein excretion (TP, alpha 1, alpha 2, and beta globulins)]; core grade minimum (Eli Lilly & Co., 1985)

5) Teratology - rat: Maternal NOEL=100 mg/kg/day; Maternal LEL=500 mg/kg/day (decreased food consumption and increased liver and spleen weights); Developmental NOEL=none; LEL=20 mg/kg/day (reduced skeletal maturity and increased vascular fragility); core grade supplementary (Hoechst Aktiengesellschaft, 1983)

6) Teratology - rabbit: Maternal and Developmental NOEL=60 mg/kg/day (HDT); Maternal and Developmental LEL=none; core grade supplementary (Hoechst Aktiengesellschaft, 1984e)

Data Gap(s): None

I.A.5. CONFIDENCE IN THE ORAL RfD

Study -- High
Data Base -- High
RfD -- High

The critical study is of good quality and is given a high confidence rating. Additional studies are supportive and of good quality; therefore, the data base is given a high confidence rating. High confidence in the RfD follows.

I.A.6. EPA DOCUMENTATION AND REVIEW OF THE ORAL RfD

Source Document -- This assessment is not presented in any existing US EPA document.

Other EPA Documentation -- Pesticide Registration Standard, June 1985; Position Document 1/2/3, August 1979; Position Document 4, July 1982; Pesticide Registration Files

Agency Work Group Review -- 05/30/86, 02/18/87. 04/20/89

Verification Date -- 04/20/89

I.A.7. EPA CONTACTS (ORAL RfD)

George Ghali / OPP -- (703)557-7490

William Burnam / OPP -- (703)557-7491

I.B. REFERENCE CONCENTRATION FOR CHRONIC INHALATION EXPOSURE (RfC)

[Ed. Note: EPA does not have information yet in this section of IRIS].

II. CARCINOGENICITY ASSESSMENT FOR LIFETIME EXPOSURE

Section II provides information on three aspects of the carcinogenic risk assessment for the agent in question; the US EPA classification, and quant- itative estimates of risk from oral exposure and from inhalation exposure. The classification reflects a weight-of-evidence judgment of the likelihood that the agent is a human carcinogen. The quantitative risk estimates are presented in three ways. The slope factor is the result of application of a low-dose extrapolation procedure and is presented as the risk per (mg/kg)/day. The unit risk is the quantitative estimate in terms of either risk per ug/L drinking water or risk per ug/cu.m air breathed. The third form in which risk is presented is a drinking water or air concentration providing cancer risks of 1 in 10,000, 1 in 100,000 or 1 in 1,000,000. Background Document 2 (Service Code 5) provides details on the rationale and methods used to derive the carcinogenicity values found in IRIS. Users are referred to Section I for information on long-term toxic effects other than carcinogenicity.

II.A. EVIDENCE FOR CLASSIFICATION AS TO HUMAN CARCINOGENICITY

II.A.1. WEIGHT-OF-EVIDENCE CLASSIFICATION

Classification -- C; possible human carcinogen.

Basis -- Classification is based on the induction of urinary tract tumors (renal pelvis carcinomas and urinary bladder papillomas) and thyroid tumors (adenomas/carcinomas combined) in one animal species (F344 rats) in one study. Trifluralin is structurally similar to ethalfluralin, a carcinogen in the rat.

II.A.2. HUMAN CARCINOGENICITY DATA

None.

II.A.3. ANIMAL CARCINOGENICITY DATA

Limited. A chronic bioassay of trifluralin was performed in F344 rats in which 60 animals/sex received dietary doses of 0, 813, 3250 and 6500 ppm for 2 years (Emmerson et al., 1980). Statistically significant (p<0.05) increases in the incidences of bladder papillomas and renal pelvis carcinomas were found at the highest dose level tested in female and male rats, respectively. In addition, a significant (p<0.05) increase in the incidence of follicular cell tumors of the thyroid gland (adenomas plus carcinomas combined) occurred at the highest dose tested in male rats. All of the previous increased tumor incidences exceeded historical incidences for similar tumors in other studies performed at the test laboratory.

Four other rodent chronic bioassays of trifluralin in the diet have been performed. These included a 2-year study in Sprague-Dawley rats (0, 200, 1000 and 2000 ppm) (Eli Lilly, 1966), a 78-week study in Osborne-Mendel rats (0, 3250 and 6500 ppm) (NCI, 1978a), a

78-week study in B6C3F1 mice (0, 2375 and 5000 ppm) (NCI, 1978b) and a 2-year study in B6C3F1 mice (0, 563, 2250 and 4500 ppm) (Eli Lilly, 1980). Trifluralin did not produce statistically significant increases in tumors in any of these studies.

II.A.4. SUPPORTING DATA FOR CARCINOGENICITY

Trifluralin is structurally related to ethalfluralin, which is oncogenic, producing mammary gland fibroadenomas in female F344 rats. In addition, both trifluralin and ethalfluralin produce a common urinary metabolite in rats that produces nonneoplastic renal pathology, including bladder calculi.

There was no evidence of mutagenicity for trifluralin in rat dominant lethal, L5178Y mouse lymphoma, Salmonella typhimurium, Saccharomyces cerevisiae, and DNA repair assays, nor did it induce sister chromatid exchange in Chinese hamster ovary cells.

II.B. QUANTITATIVE ESTIMATE OF CARCINOGENIC RISK FROM ORAL EXPOSURE

II.B.1. SUMMARY OF RISK ESTIMATES

Oral Slope Factor -- 7.7E-3/mg/kg/day

Drinking Water Unit Risk -- 2.2E-7/ug/L

Extrapolation Method -- linearized multistage procedure, extra risk

Drinking Water Concentrations at Specified Risk Levels:

Risk Level	Concentration
E-4 (1 in 10,000)	5E+2 ug/L
E-5 (1 in 100,000)	5E+1 ug/L
E-6 (1 in 1,000,000)	5 ug/L

II.B.2. DOSE-RESPONSE DATA (CARCINOGENICITY, ORAL EXPOSURE)

Tumor Type -- combined renal pelvis carcinomas, urinary bladder papillomas and/or thyroid adenomas and carcinomas
Test Animals -- rat/F344, male
Route -- diet
Reference -- Emmerson et al., 1980

Administered Dose (ppm)	Human Equivalent Dose (mg/kg)/day	Tumor Incidence
0	0	5/60
813	5.1	5/60
3250	21.9	9/60
6500	46.5	7/60

II.B.3. ADDITIONAL COMMENTS (CARCINOGENICITY, ORAL EXPOSURE)

Incidence data were based on observation of at least one tumor at any of the indicated sites.

The unit risk should not be used if the water concentration exceeds 5E+4 ug/L, since above this concentration the slope factor may differ from that stated.

II.B.4. DISCUSSION OF CONFIDENCE (CARCINOGENICITY, ORAL EXPOSURE)

Tumors were induced at different sites in F344 rats of one or both sexes. An adequate number of animals was observed in a lifetime study.

II.C. QUANTITATIVE ESTIMATE OF CARCINOGENIC RISK FROM INHALATION EXPOSURE

[Ed. Note: EPA does not have information yet in this section of IRIS].

II.D. EPA DOCUMENTATION, REVIEW, AND CONTACTS (CARCINOGENICITY ASSESSMENT)

II.D.1. EPA DOCUMENTATION

Source Document -- US EPA, 1982, 1986

The Toxicology Branch Peer Review Committee reviewed data on trifluralin.

II.D.2. REVIEW (CARCINOGENICITY ASSESSMENT)

Agency Work Group Review -- 05/13/87, 06/03/87, 06/24/87

Verification Date -- 06/24/87

II.D.3. US EPA CONTACTS (CARCINOGENICITY ASSESSMENT)

Elizabeth A. Doyle / OHEA -- (703)308-2722

Reto Engler / OPP -- (703)308-2738

III. HEALTH HAZARD ASSESSMENTS FOR VARIED EXPOSURE DURATIONS

[Ed. Note: EPA does not have information yet in this section of IRIS].

IV. US EPA REGULATORY ACTIONS

EPA risk assessments may be updated as new data are published and as assessment methodologies evolve. Regulatory actions are frequently not updated at the same time. Compare the dates for the regulatory actions in this section with the verification dates for the risk assessments in sections I and II, as this may explain inconsistencies. Also note that some regulatory actions consider factors not related to health risk, such as technical or economic feasibility. Such considerations are indicated for each action. In addition, not all of the regulatory actions listed in this section involve enforceable federal standards. Please direct any questions you may have concerning these regulatory actions to the US EPA contact listed for that particular action. Users are strongly urged to read the background information on each regulatory action in Background Document 4 in Service Code 5.

IV.A. CLEAN AIR ACT (CAA)

[Ed. Note: EPA does not have information yet in this section of IRIS].

IV.B. SAFE DRINKING WATER ACT (SDWA)

[Ed. Note: EPA does not have information yet in this section of IRIS].

IV.C. CLEAN WATER ACT (CWA)

[Ed. Note: EPA does not have information yet in this section of IRIS].

IV.D. FEDERAL INSECTICIDE, FUNGICIDE, AND RODENTICIDE ACT (FIFRA)

IV.D.1. PESTICIDE ACTIVE INGREDIENT, Registration Standard

Status -- Issued (1987)

Reference -- Trifluralin Pesticide Registration Standard. April, 1987 (NTIS No. PB87-201935).

EPA Contact -- Registration Branch / OPP (703)557-7760 / FTS 557-7760

IV.D.2. PESTICIDE ACTIVE INGREDIENT, Special Review

Action -- Final regulatory decision - PD4 (1982)

Considers technological or economic feasibility -- NO

Summary of regulatory action -- Registration allowed to continue if total N-nitrosamine contamination is kept below 0.5 ppm for technical products, and below a figure based on trifluralin content for formulated products. Criterion of concern: oncogenicity and mutagenicity.

Reference -- 47 FR 33777 (08/04/82) [NTIS# PB82-263252]

EPA Contact -- Special Review Branch / OPP (703)557-7400 / FTS 557-7400

IV.E. TOXIC SUBSTANCES CONTROL ACT (TSCA)

[Ed. Note: EPA does not have information yet in this section of IRIS].

IV.F. RESOURCE CONSERVATION AND RECOVERY ACT (RCRA)

[Ed. Note: EPA does not have information yet in this section of IRIS].

IV.G. SUPERFUND (CERCLA)

[Ed. Note: EPA does not have information yet in this section of IRIS].

VI. BIBLIOGRAPHY

VI.A. ORAL RfD REFERENCES

Elanco Product Company. 1984a. MRID No. 00152419. Available from EPA. Write to FOI, EPA, Washington D.C. 20460.

Elanco Product Company. 1984b. MRID No. 00152421. Available from EPA. Write to FOI, EPA, Washington D.C. 20460.

Elanco Product Company. 1986. MRID No. 00162543. Available from EPA. Write to FOI, EPA, Washington D.C. 20460.

Eli Lilly and Company. 1985. MRID No. 00157156, 40138301. Available from EPA. Write to FOI, EPA, Washington D.C. 20460.

Hoechst Aktiengesellschaft. 1980. MRID No. 00151906. Available from EPA. Write to FOI, EPA, Washington D.C. 20460.

Hoechst Aktiengesellschaft. 1981. MRID No. 00151907. Available from EPA. Write to FOI, EPA, Washington D.C. 20460.

Hoechst Aktiengesellschaft. 1983. MRID No. 00151899. Available from EPA. Write to FOI, EPA, Washington D.C. 20460.

Hoechst Aktiengesellschaft. 1984a. MRID No. 00151908. Available from EPA. Write to FOI, EPA, Washington D.C. 20460.

Hoechst Aktiengesellschaft. 1984b. MRID No. 00151901, 00151903. Available from EPA. Write to FOI, EPA, Washington D.C. 20460.

Hoechst Aktiengesellschaft. 1984c. MRID No. 00151900. Available from EPA. Write to FOI, EPA, Washington D.C. 20460.

Hoechst Aktiengesellschaft. 1986a. MRID No. 00153496, 00162456, 00162458. Available from EPA. Write to FOI, EPA, Washington D.C. 20460.

Hoechst Aktiengesellschaft. 1986b. MRID No. 00158935. Available from EPA. Write to FOI, EPA, Washington D.C. 20460.

VI.B. INHALATION RfD REFERENCES

None

VI.C. CARCINOGENICITY ASSESSMENT REFERENCES

Eli Lilly Company. 1966. Rat oncogenicity study of trifluralin. (Cited in US EPA, 1986)

Eli Lilly Company. 1980. Mouse oncogenicity study of trifluralin. (Cited in US EPA, 1986)

Emmerson, J.L., E.C. Pierce, J.P. McGrath, et al. 1980. The chronic toxicity of compound 36352 (trifluralin) given as a compound of the diet to the Fischer 344 rats for two years. Studies R-87 and R-97 (unpublished study received September 18, 1980 under 1471-35; submitted by Elanco Products Co., Division of Eli Lilly and Co., Indianapolis, IN).

NCI (National Cancer Institute). 1978a. Rat oncogenicity study of trifluralin. (Cited in US EPA, 1986)

NCI (National Cancer Institute). 1978b. Mouse oncogenicity study of trifluralin. (Cited in US EPA, 1986)

US EPA. 1982. Trifluralin (TREFLAN) Position Document 4. Office of Pesticides and Toxic Substances, Washington, DC, July.

US EPA. 1986. Toxicology Branch Peer Review Committee Memorandum on Trifluralin, April 11.

VI.D. DRINKING WATER HA REFERENCES

None

VII. REVISION HISTORY

Date	Section	Description
08/22/88	II.	Carcinogen summary on-line
02/01/89	I.A.	Oral RfD summary noted as pending change
05/01/89	I.A.	Withdrawn; new RfD verified (in preparation)
07/01/89	I.A.	Oral RfD summary replaced; RfD changed
07/01/89	VI.	Bibliography on-line
01/01/92	IV.	Regulatory Action section on-line
10/01/93	II.D.3.	Primary contact changed; secondary's phone no. changed

SYNONYMS

AGREFLAN
AGRIFLAN 24
BENZENAMINE, 2,6-DINITRO-N,N-DIPROPYL-4-(TRIFLUOROMETHYL)-
CRISALIN
DIGERMIN
2,6-DINITRO-N,N-DIPROPYL-4-(TRIFLUOROMETHYL)BENZENAMINE
2,6-DINITRO-N,N-DI-n-PROPYL-alpha,alpha,alpha-TRIFLUORO-p-TOLUIDINE
2,6-DINITRO-4-TRIFLUORMETHYL-N,N-DIPROPYLANILIN
4-(DI-n-PROPYLAMINO)-3,5-DINITRO-1-TRIFLUOROMETHYLBENZENE
ELANCOLAN
L-36352
LILLY 36,352
NCI-C00442
NITRAN
N,N-DIPROPYL-2,6-DINITRO-4-TRIFLUORMETHYLANILIN
N,N-DI-n-PROPYL-2,6-DINITRO-4-TRIFLUOROMETHYLANILINE
N,N-DIPROPYL-4-TRIFLUOROMETHYL-2,6-DINITROANILINE
OLITREF
SU SEGURO CARPIDOR
TREFANOCIDE
TREFICON
TREFLAM
TREFLAN
TREFLANOCIDE ELANCOLAN
s-TRIAZINE, 2-CHLORO-4,6-BIS(ETHYLAMINO)-
TRIFLUORALIN
alpha,alpha,alpha-TRIFLUORO-2,6-DINITRO-N,N-DIPROPYL-p-TOLUIDINE
Trifluralin
TRIFLURALINE
TRIFUREX
TRIKEPIN
TRIM

Source: *Instant EPA's IRIS*, 1997, Instant Reference Sources, Inc., 7605 Rockpoint Drive, Austin, TX 78731 (Fax: 512-345-2386; Internet URL, http://www.instantref.com/inst-ref.htm).

Production, Use, and Pesticide Labeling Information

TRIFLURALIN is not currently listed in EPA's Pesticide Factsheet database and there is no indication of whether it will be added or not in the future.

Vinclozolin

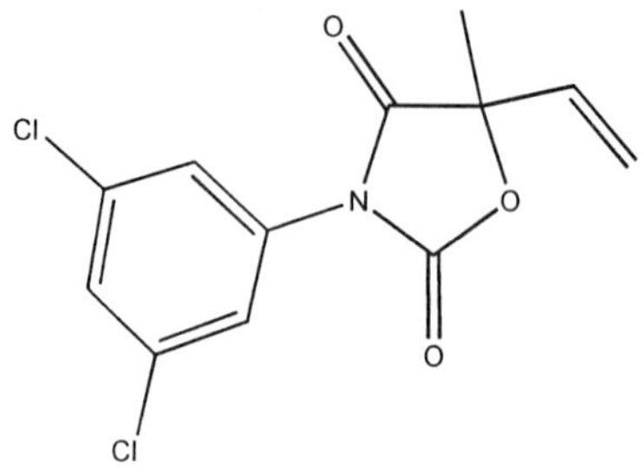

Introduction

Vinclozolin is a fungicide widely used on fruits in the US [879]

Identifying Information

CAS NUMBER: 50471-44-8

SYNONYMS: Ronilan; 3-(3,5-dichlorophenyl)-5-ethenyl-5-methyl-2,4-oxazolidinedione

Endocrine Disruptor Information

Vinclozolin is a strongly suspected EED that it is listed by EPA [811] and the World Wildlife Fund [813] as a potential endocrine modifying chemical. It also is reported to target the androgen receptor and occupies it. It blocks the receptor's function and doesn't let testosterone get produced at normal levels during development of males. The result is that males may become hermaphrodites, an ambiguous state where the person cannot function as either male or female. [810] In the US, female rats exposed to vinclozolin, a synthetic chemical widely used to kill fungus on fruit, gave birth to males with female traits. [879]

Chemical and Physical Properties

CHEMICAL FORMULA: $C_{12}H_9NO_3Cl_2$ [814]

MOLECULAR WEIGHT: 286.11 [814]

PHYSICAL DESCRIPTION:

MELTING POINT: 108°C [814]

BOILING POINT: 131°C [814]

Environmental Reference Materials

A source of EED environmental reference materials for vinclozolin is Radian International LLC, Austin, TX (Fax 512-454-0268; Internet URL, http://www.radian.com/standards); Catalog No. ERV-009.

Hazardous Properties

ACUTE/CHRONIC HAZARDS: Tests reported in EPA's IRIS database do not indicate significant acute or chronic hazards. [819]

Medical Symptoms of Exposure

Symptoms of exposure with dogs included significant increases in absolute and relative adrenal weights in male and females. In female dogs there was vacuolation of the zona fasciculata and birefringence of the cortex in the adrenal while in males the adrenal showed vacuolation of the zona fasciculata. Also in male dogs a decrease in absolute kidney weights was noted and with higher concentrations the kidneys showed fat droplets in the distal tubules. [819]

Toxicological Information

The toxicological information below is gathered from several sources including two national databases: one from the *National Toxicology Program Chemical Repository Database* and another from *EPA's Integrated Risk Information System (IRIS)*.

Toxicology Information from the National Toxicology Program

Vinclozolin is not currently listed in NTP's Chemical Repository database nor is it scheduled for addition in the future.

Toxicology Information from EPA's Integrated Risk Information System (IRIS)

I. CHRONIC HEALTH HAZARD ASSESSMENTS FOR NONCARCINOGENIC EFFECTS

I.A. REFERENCE DOSE FOR CHRONIC ORAL EXPOSURE (RfD)

The Reference Dose (RfD) is based on the assumption that thresholds exist for certain toxic effects such as cellular necrosis, but may not exist for other toxic effects such as carcinogenicity. In general, the RfD is an estimate (with uncertainty spanning perhaps an order of magnitude) of a daily exposure to the human population (including sensitive subgroups) that is likely to be without an appreciable risk of deleterious effects during a lifetime. Please refer to Background Document 1 in Service Code 5 for an elaboration of these concepts. RfDs can also be derived for the noncarcinogenic health effects of compounds which are also carcinogens. Therefore, it is essential to refer to other sources of information concerning the carcinogenicity of this substance. If the US EPA has evaluated this substance for potential human carcinogenicity, a summary of that evaluation will be contained in Section II of this file when a review of that evaluation is completed.

I.A.1. ORAL RfD SUMMARY

Critical Effect	Experimental Doses*	UF	MF	RfD
Organ weight changes	NOEL: 100 ppm (2.5 mg/kg/day) LEL: 300 ppm (7.5 mg/kg/day)	100	1	2.5E-2 mg/kg/day

*Dose Conversion Factors & Assumptions: 1 ppm = 0.025 mg/kg/day (assumed dog food consumption)

6-Month Feeding Dog Study, BASF Corp., 1982

I.A.2. PRINCIPAL AND SUPPORTING STUDIES (ORAL RfD)

BASF Wyandotte Corporation. 1982. MRID No. 00110446; HED Doc. No. 002214. Available from EPA. Write to FOI, EPA, Washington D.C. 20460.

Vinclozolin was administered in the diet to 6 dogs/sex/group at levels of 0 (controls), 100, 300, 600, and 2000 ppm for 6 months. Toxic effects were noted at 300 ppm and consisted of significant increases in absolute and relative adrenal weights in male and females. This diagnosis was supported by the absolute and relative weight increases of the adrenals at 600 ppm and 2000 ppm and accompanied by histological findings. In females at 600 ppm and above there was vacuolation of the zona fasciculata and birefringence of the cortex in the adrenal. In males at 2000 ppm, the adrenal showed vacuolation of the zona fasciculata. In male dogs at 300 ppm and above, a decrease in absolute kidney weights was noted, and at 600 ppm and above the kidneys showed fat droplets in the distal tubules.

I.A.3. UNCERTAINTY AND MODIFYING FACTORS (ORAL RfD)

UF -- An uncertainty factor of 100 was used to account for the inter- and intraspecies differences. The 6-month dog study is of marginal duration to qualify as a long-term study. An additional UF of 10 to account for the lack of a "chronic" dog study (at least 1 year) was not considered necessary, since 1) the other toxicologic studies on vinclozolin generally did not indicate that the toxicologic endpoints would be seen at lower doses with extended exposure, 2) the effects seen at the LEL in the 6-month dog study are of a minor toxicologic nature, and 3) the dog is the most sensitive species tested by far.

MF -- None

I.A.4. ADDITIONAL COMMENTS (ORAL RfD)

Data Considered for Establishing the RfD

1) 6-Month Feeding - dog: Principal study - see discussion above; core grade minimum

2) 103-Week Feeding (Oncogenic) - rat: Systemic NOEL=486 ppm (24.3 mg/kg/day), Systemic LEL=1458 ppm (72.9 mg/kg/day) (body weight reduction, reduced serum bilirubin); no core grade (BASF Wyandotte Corp., 1977a)

3) 3-Generation Reproduction - rat: NOEL=1458 ppm (72.9 mg/kg/day) (HDT); Teratogenic NOEL=1458 ppm (HDT); no core grade (BASF Wyandotte Corp., 1977b)

4) Teratology - rabbit: NOEL=300 mg/kg/day (HDT); Fetotoxic NOEL=80 mg/kg/day; Teratogenic NOEL=300 mg/kg/day (HDT); core grade minimum (BASF Wyandotte Corp., 1981)

5) Teratology - mice: Teratogenic NOEL=6000 ppm (900 mg/kg/day); Fetotoxic LEL=6000 ppm (resorptions); no core grade (BASF Wyandotte Corp., 1975)

Other Data Reviewed

1) 26-Month Feeding (Oncogenic) - mouse: Systemic NOEL=486 ppm (72.9 mg/kg/day); Systemic LEL=1458 ppm (7219 mg/kg/day) (decrease body weight gain in males); core grade minimum (BASF Wyandotte Corp., 1977)

Data Gap(s): None

I.A.5. CONFIDENCE IN THE ORAL RfD

Study -- High Data Base -- HighRfD -- High

The critical study is of high quality and is given a high confidence rating. The other studies are also of good quality; thus, the data base is given a high confidence rating. High confidence in the RfD follows.

I.A.6. EPA DOCUMENTATION AND REVIEW OF THE ORAL RfD

Pesticide Registration Files

Agency Work Group Review -- 07/08/86

Verification Date -- 07/08/86

I.A.7. EPA CONTACTS (ORAL RfD)

William Burnam / OPP -- (703)557-7491

George Ghali / OPP -- (703)557-7490

I.B. REFERENCE CONCENTRATION FOR CHRONIC INHALATION EXPOSURE (RfC)

Not available at this time.

II. CARCINOGENICITY ASSESSMENT FOR LIFETIME EXPOSURE

This substance/agent has not been evaluated by the US EPA for evidence of human carcinogenic potential.

III. HEALTH HAZARD ASSESSMENTS FOR VARIED EXPOSURE DURATIONS

[Ed. Note: EPA does not have information yet in this section of IRIS].

IV. US EPA REGULATORY ACTIONS

EPA risk assessments may be updated as new data are published and as assessment methodologies evolve. Regulatory actions are frequently not updated at the same time. Compare the dates for the regulatory actions in this section with the verification dates for the risk assessments in sections I and II, as this may explain inconsistencies. Also note that some regulatory actions consider factors not related to health risk, such as technical or economic feasibility. Such considerations are indicated for each action. In addition, not all of the regulatory actions listed in this section involve enforceable federal standards. Please direct any questions you may have concerning these regulatory actions to the US EPA contact listed for that particular action. Users are strongly urged to read the background inform-ation on each regulatory action in Background Document 4 in Service Code 5.

IV.A. CLEAN AIR ACT (CAA)

[Ed. Note: EPA does not have information yet in this section of IRIS].

IV.B. SAFE DRINKING WATER ACT (SDWA)

[Ed. Note: EPA does not have information yet in this section of IRIS].

IV.C. CLEAN WATER ACT (CWA)

[Ed. Note: EPA does not have information yet in this section of IRIS].

IV.D. FEDERAL INSECTICIDE, FUNGICIDE, AND RODENTICIDE ACT (FIFRA)

IV.D.1. PESTICIDE ACTIVE INGREDIENT, Registration Standard

Status -- List "B" Pesticide (1989)

Reference -- 54 FR 22706 (05/25/89); Notice of receipt of request to amend registration to delete all plum and prune use; 56 FR 24190 (05/29/91).

EPA Contact -- Registration Branch / OPP(703)557-7760 / FTS 557-7760

IV.D.2. PESTICIDE ACTIVE INGREDIENT, Special Review

[Ed. Note: EPA does not have information yet in this section of IRIS].

IV.E. TOXIC SUBSTANCES CONTROL ACT (TSCA)

[Ed. Note: EPA does not have information yet in this section of IRIS].

IV.F. RESOURCE CONSERVATION AND RECOVERY ACT (RCRA)

[Ed. Note: EPA does not have information yet in this section of IRIS].

IV.G. SUPERFUND (CERCLA)

[Ed. Note: EPA does not have information yet in this section of IRIS].

VI. BIBLIOGRAPHY

VI.A. ORAL RfD REFERENCES

BASF Wyandotte Corporation. 1975. MRID No. 00062644; HED Doc. No. 000244. Available from EPA. Write to FOI, EPA, Washington, DC 20460.

BASF Wyandotte Corporation. 1977a. MRID No. 00070036; HED Doc. No. 000244. Available from EPA. Write to FOI, EPA, Washington, DC 20460.

BASF Wyandotte Corporation. 1977b. MRID No. 00062645; HED Doc. No. 000244. Available from EPA. Write to FOI, EPA, Washington, DC 20460.

BASF Wyandotte Corporation. 1977c. MRID No. 00070037; HED Doc. No. 000244, 001885, 002717, 002669. Available from EPA. Write to FOI, EPA, Washington, DC 20460.

BASF Wyandotte Corporation. 1981. MRID No. 00085079; HED Doc. No. 002409. Available from EPA. Write to FOI, EPA, Washington, DC 20460.

BASF Wyandotte Corporation. 1982. MRID No. 00110446; HED Doc. No. 002214. Available from EPA. Write to FOI, EPA, Washington, DC 20460.

VI.B. INHALATION RfD REFERENCES

None

VI.C. CARCINOGENICITY ASSESSMENT REFERENCES

None

VI.D. DRINKING WATER HA REFERENCES

None

VII. REVISION HISTORY

```
--------  --------   -----------------------------------------------------------
Date      Section     Description
--------  --------   -----------------------------------------------------------
03/01/88  I.A.4.  Text added
03/01/88  I.A.5.  Confidence levels revised
01/01/92  I.A.4.  Citations added
01/01/92  IV.     Regulatory Action section on-line
01/01/92  VI.     Bibliography on-line
```

SYNONYMS

BAS 35202F   Ronilan
BAS 35204   Vinclozolin
BAS352-04F

Source: *Instant EPA's IRIS*, 1997, Instant Reference Sources, Inc., 7605 Rockpoint Drive, Austin, TX 78731 (Fax: 512-345-2386; Internet URL, http://www.instantref.com/inst-ref.htm).

Production, Use, and Pesticide Labeling Information

Vinclozolin is not currently listed in EPA's Pesticide Factsheet database and there is no indication of whether it will be added or not in the future.

Source: *Instant EPA's Pesticide Facts*, 1994, Instant Reference Sources, Inc., 7605 Rockpoint Drive, Austin, TX 78731 (Fax: 512-345-2386; Internet URL, http://www.instantref.com/inst-ref.htm).

Zineb

Introduction

Zineb is registered as a general use pesticide by the US EPA. It is used as a fungicide and an insecticide to protect fruit and vegetable crops from a wide range of foliar and other diseases. In July 1987, the Environmental Protection Agency announced the initiation of a special review of the ethylene bisdithiocarbamates (EBDCs), a class of chemicals to which mancozeb belongs. This Special Review was initiated because of concerns raised by laboratory tests on rats and mice. As part of the Special Review, EPA reviewed data from market basket surveys and concluded that actual levels of EBDC residues on produce purchased by consumers are too low to affect human health. Many home garden uses of EBDCs have been canceled because the EPA has assumed that home users of these pesticides do not wear protective clothing during application. [829] It is used It is available in the US as wettable powder and dust formulations.

Identifying Information

CAS NUMBER: 12122-67-7

NIOSH Registry Number: ZH3325000

SYNONYMS: ASPOR, ASPORUM, BLIGHTOX, CARBADINE, CHEM ZINEB, CINEB, DITHANE Z, DITHIANE Z-78, ENT 14,874, 1,2-ETHANEDIYLBISCARBAMODITHIOIC ACID, ZINC COMPLEX, 1,2-ETHANEDIYLBISCARBAMOTHIOIC ACID, ZINC SALT, ETHYLENEBIS(DITHIOCARBAMATO)ZINC, ETHYL ZIMATE, HEXATHANE, MILTOX, NOVOSIR N, NOVOZIR N, PARZATE, PARZATE ZINEB, PEROSIN, PEROZINE, Z-78, ZEBENIDE, ZEBTOX, ZIMATE, ZINC ETHYLENEBISDITHIOCARBAMATE, ZINC ETHYLENE-1,2-BISDITHIOCARBAMATE, ZINEB 80, ZINOSAN

Endocrine Disruptor Information

Zineb is a strongly suspected EED that it is listed by the Centers for Disease Control & Prevention [812], and the World Wildlife Fund [813] as a potential endocrine modifying chemical. Mice subjected to zineb exhibited sterility, resorption of fetuses, abnormal tails in offspring, abortions and weak pups. [829]

Chemical and Physical Properties

CHEMICAL FORMULA: $C_4H_6N_2S_4Zn$

MOLECULAR WEIGHT: 275.73

WLN: T9MYS-Zn-SYMTJ BUS FUS

PHYSICAL DESCRIPTION: Off-white powder

SPECIFIC GRAVITY: Not available

DENSITY: Not available

MELTING POINT: Not available

SOLUBILITY: Water: <1 mg/mL @ 25°C [700]; DMSO: <1 mg/mL @ 25°C [700]; 95% Ethanol: <1 mg/mL @ 25°C [700]; Acetone: <1 mg/mL @ 19°C [700]

Source: *The National Toxicology Program's Chemical Database, Volume 2*, 1992, CRC Press, Inc./Lewis Publishers, 2000 Corporate Blvd., Boca Raton, FL 33431 (Fax: 407-998-9114).

Environmental Reference Materials

A source of EED environmental reference materials for zineb is Crescent Chemical Co., Hauppauge, NY (Fax: 516-348-0913); Catalog No. 45707.

Hazardous Properties

ACUTE/CHRONIC HAZARDS: Zineb may be toxic by ingestion and inhalation and it is also an irritant and a dermal sensitizer. Cross sensitization with maneb and mancozeb may occur. [829] Toxic fumes are emitted when it is heated to decomposition.

FIRE HAZARD: Flash point data for zineb are not available. It is probably combustible. Fires involving this compound should be controlled with a dry chemical, carbon dioxide or Halon extinguisher.

STABILITY: Zineb will decomposes if exposed to heat.

Source: *The National Toxicology Program's Chemical Database, Volume 7*, 1992, CRC Press, Inc./Lewis Publishers, 2000 Corporate Blvd., Boca Raton, FL 33431 (Fax: 407-998-9114).

Medical Symptoms of Exposure

Early symptoms from exposure of humans to inhalation of zineb include tiredness, dizziness and weakness. More severe symptoms include headache, nausea, fatigue, slurred speech, convulsions and unconsciousness Mucous membrane irritation has also been reported in humans. Occupational exposure to inhalation of zineb can lead to changes in liver enzymes, moderate anemia and other blood changes, increased incidence of poisoning symptoms during pregnancy, and chromosomal changes in the lymphocytes. Liver functioning was affected in workers exposed to zineb. Moderate anemia and other blood changes were also reported in 150 workers exposed to zineb in a chemical plant. A five-month study of zineb showed that concentrations of 20 and 200 mg/m3 caused a decrease in the activity of cholinesterase. Repeated or prolonged dermal exposure may cause dermatitis or conjunctivitis. Farm workers who were repeatedly exposed to zineb, in fields sprayed with 0.5% suspension of the fungicide, reported severe and extensive contact dermatitis. Poisoning from this class of chemicals include itching, scratchy throat, sneezing, coughing, inflammation of the nose or throat and bronchitis. [829]

Toxicological Information

The toxicological information below is gathered from several sources including two national databases: one from the *National Toxicology Program Chemical Repository Database* and another from *EPA's Integrated Risk Information System (IRIS)*.

Toxicology Information from the National Toxicology Program

CARCINOGENICITY:

> Review:
> > IARC Cancer Review: Animal Inadequate Evidence
> > IARC: Not classifiable as a human carcinogen

TERATOGENICITY:

> Reproductive effects data:

Type of Effect	Route/Animal	Amount Dosed (Notes)
TDLo	orl-rat	4 gm/kg(13D preg)
TDLo	orl-rat	4 gm/kg(11D preg)
TDLo	orl-rat	8 gm/kg(9D preg)
TDLo	orl-rat	4 gm/kg(13D preg)

MUTAGENICITY: Mutation data: cyt-mus-orl 10 mg/kg

TOXICITY:

Typ. Dose	Mode	Specie	Amount	Unit	Other
TDLo	orl	rat	54gm/kg/95W-I	TFX:CAR	
TDLo	imp	rat	80	mg/kg	TFX:ETA
TDLo	orl	mus	7800 mg/kg/5W-I	TFX:ETA	
TDLo	ipr	mus	1760mg/kg(11-21D preg)	TFX:ETA	
LD50	orl	rat	5200	mg/kg	
LD50	ihl	rat	1850	mg/kg	
LD50	unk	rat	1850	mg/kg	
LD50	orl	mus	7600	mg/kg	
LDLo	orl	rbt	600	mg/kg	
LD50	unk	mam	1350	mg/kg	

OTHER TOXICITY DATA:

Review: Toxicology Review-2

SAX TOXICITY EVALUATION:

THR: MOD via unknown route. An irritant via oral and inhalation routes. THR=A fungicide. Irritant to the skin and mucous membranes.

Source: *Instant Tox-Base*, 1995, Instant Reference Sources, Inc., 7605 Rockpoint Drive, Austin, TX 78731 (Fax: 512-345-2386; Internet URL, http://www.instantref.com/inst-ref.htm).

Toxicology Information from EPA's Integrated Risk Information System (IRIS)

I. CHRONIC HEALTH HAZARD ASSESSMENTS FOR NONCARCINOGENIC EFFECTS

I.A. REFERENCE DOSE FOR CHRONIC ORAL EXPOSURE (RfD)

The Reference Dose (RfD) is based on the assumption that thresholds exist for certain toxic effects such as cellular necrosis, but may not exist for other toxic effects such as carcinogenicity. In general, the RfD is an estimate (with uncertainty spanning perhaps an order of magnitude) of a daily exposure to the human population (including sensitive subgroups) that is likely to be without an appreciable risk of deleterious effects during a lifetime. Please refer to Background Document 1 in Service Code 5 for an elaboration of these concepts. RfDs can also be derived for the noncarcinogenic health effects of compounds which are also carcinogens. Therefore, it is essential to refer to other sources of information concerning the carcinogenicity of this substance. If the US EPA has evaluated this substance for potential human carcinogenicity, a summary of that evaluation will be contained in Section II of this file when a review of that evaluation is completed.

I.A.1. ORAL RfD SUMMARY

Critical Effect	Experimental Doses*	UF	MF	RfD
Thyroid hyperplasia	NOEL: none LOAEL: 500 ppm (diet) (converted to 25 mg/kg/day)	500	1	5E-2 mg/kg/day

*Dose Conversion Factors & Assumptions: Assumed rat food consumption = 5% bw/day

Rat, Chronic Oral Bioassay Blackwell-Smith et al., 1953

I.A.2. PRINCIPAL AND SUPPORTING STUDIES (ORAL RfD)

Blackwell-Smith, Jr., R.J., J.K. Finnegan, P.S. Larson, P.F. Sahyoun, M.L. Dreyfuss and H.B. Haag. 1953. Toxicologic studies on zinc and disodium ethylene bisdithiocarbamates. J. Pharmacol. Exp. Therap. 109: 159-166.

Groups of 10 male and 10 female rats received 0, 500, 1000, 2500, 5000 or 10,000 ppm zineb in the diet for 2 years. Blackwell-Smith et al. (1953) used the criteria for grading thyroid hyperplasia established by Seifter and Ehrich (1948), who had published an earlier report on goitrogenic effects of zineb feeding in weanling rats. This system grades hyperplastic responses on a scale of 1-5, with a response of 2+ being regarded as within normal limits. In order to ensure consistency in grading of slides, Blackwell-Smith et al. (1953) submitted sections from 10 rats used in a short-term study to Dr. Seifter for evaluation. These readings were used as a guide for subsequent classification of responses. Thyroid hyperplasia was observed in rats of all treated groups, establishing the LOEL of 25 mg/kg/day (500 ppm). At this dose 6/16 rats exhibited hyperplasia graded 3+ or above as compared with the control group, for which no response greater than 2+ was observed. At higher dosages, rats developed more severe thyroid hyperplasia in addition to renal congestion, nephritis and nephrosis; increased mortality was noted for rats consuming the two highest dietary levels.

Other data summarized in US EPA (1984) attest to the development of thyroid hyperplasia as a consequence of zineb consumption, and collectively support the choice of 25 mg/kg/day as the proper effect level from which to derive an RfD.

I.A.3. UNCERTAINTY AND MODIFYING FACTORS (ORAL RfD)

UF -- A composite uncertainty factor of 500 was used consisting of 2 factors of 10 to account for inter- and intraspecies variability and a 5-fold factor for use of a LOAEL. The intermediate uncertainty factor of 5 was used as the observed effects were graded on a subjective, albeit controlled scale, with the resulting impression that these effects were only minimally adverse. Thus, a full 10-fold factor for the extrapolation of a LOAEL to a NOAEL was not deemed necessary.

MF -- None

I.A.4. ADDITIONAL COMMENTS (ORAL RfD)

Several studies in rats and dogs support the choice of a LOAEL. Terata have been observed in rats administered amounts of zineb in excess of 2000 mg/kg/day (Short et al., 1980).

I.A.5. CONFIDENCE IN THE ORAL RfD

Study -- Medium
Data Base -- Medium
RfD -- Medium

The critical study was of chronic duration, measured multiple endpoints, and included five dose levels. The numbers of animals, however, were relatively small, leading to a medium confidence rating. Supportive studies exist in several species, however, adequate reproductive data are not available. Thus, confidence in the data base is also medium. Medium confidence in the RfD follows.

I.A.6. EPA DOCUMENTATION AND REVIEW OF THE ORAL RfD

Source Document -- US EPA, 1984

The ADI in the 1984 Health and Environmental Effects Profile has received an Agency review with the help of two external scientists.

Other EPA Documentation -- None

Agency Work Group Review -- 11/06/85, 02/05/86, 05/15/86

Verification Date -- 02/05/86

I.A.7. EPA CONTACTS (ORAL RfD)

Adib Tabri / NCEA -- (513)569-7505

I.B. REFERENCE CONCENTRATION FOR CHRONIC INHALATION EXPOSURE (RfC)

[Ed. Note: EPA does not have information yet in this section of IRIS].

II. CARCINOGENICITY ASSESSMENT FOR LIFETIME EXPOSURE

This substance/agent has been evaluated by the US EPA for evidence of human carcinogenic potential. This does not imply that this agent is necessarily a carcinogen. The evaluation for this chemical will be reviewed at a later date by an inter-office

Agency work group. A risk assessment summary will be included on IRIS when the review has been completed.

III. HEALTH HAZARD ASSESSMENTS FOR VARIED EXPOSURE DURATIONS

[Ed. Note: EPA does not have information yet in this section of IRIS].

IV. US EPA REGULATORY ACTIONS

EPA risk assessments may be updated as new data are published and as assessment methodologies evolve. Regulatory actions are frequently not updated at the same time. Compare the dates for the regulatory actions in this section with the verification dates for the risk assessments in sections I and II, as this may explain inconsistencies. Also note that some regulatory actions consider factors not related to health risk, such as technical or economic feasibility. Such considerations are indicated for each action. In addition, not all of the regulatory actions listed in this section involve enforceable federal standards. Please direct any questions you may have concerning these regulatory actions to the US EPA contact listed for that particular action. Users are strongly urged to read the background information on each regulatory action in Background Document 4 in Service Code 5.

IV.A. CLEAN AIR ACT (CAA)

[Ed. Note: EPA does not have information yet in this section of IRIS].

IV.B. SAFE DRINKING WATER ACT (SDWA)

[Ed. Note: EPA does not have information yet in this section of IRIS].

IV.C. CLEAN WATER ACT (CWA)

[Ed. Note: EPA does not have information yet in this section of IRIS].

IV.D. FEDERAL INSECTICIDE, FUNGICIDE, AND RODENTICIDE ACT (FIFRA)

IV.D.1. PESTICIDE ACTIVE INGREDIENT, Registration Standard

Status -- Removed from list "B" pesticides/Registration cancelled (1990)

Reference -- 54 FR 22706/55 FR 31166 (07/31/90)

EPA Contact -- Registration Branch / OPP (703)557-7760 / FTS 557-7760

IV.D.2.a. PESTICIDE ACTIVE INGREDIENT, Special Review

Action -- Final Regulatory Decision - PD4 (1982)

Considers technical or economic feasibility? -- No

Summary of regulatory action -- Requires modified labeling and additional toxicological data. Criterion of concern: oncogenicity.

Reference -- 47 FR 47669 (10/27/82) [NTIS# PB87-181350]

EPA Contact -- Special Review Branch / OPP (703)557-7400 / FTS 557-7400

IV.D.2.b. PESTICIDE ACTIVE INGREDIENT, Special Review

Action -- Notice of Preliminary Determination - PD 2/3 (1989)

Considers technological or economic feasibility? -- NO

Summary of regulatory action -- Preliminary determination to cancel all uses of zineb. Comment period reopened 55 FR 40206 (10/02/90). PD 4 to be issued in FY '91. Criterion of concern: thyroid effects.

Reference -- 54 FR 52158 (12/20/89).

EPA Contact -- Special Review Branch / OPP (703)557-7400 / FTS 557-7400

IV.E. TOXIC SUBSTANCES CONTROL ACT (TSCA)

> [Ed. Note: EPA does not have information yet in this section of IRIS].

IV.F. RESOURCE CONSERVATION AND RECOVERY ACT (RCRA)

> [Ed. Note: EPA does not have information yet in this section of IRIS].

IV.G. SUPERFUND (CERCLA)

> [Ed. Note: EPA does not have information yet in this section of IRIS].

VI. BIBLIOGRAPHY

VI.A. ORAL RfD REFERENCES

Blackwell-Smith, R., J.K. Finnegan, P.S. Larson, P.F. Sahyoun, M.L. Dreyfuss and H.B. Haag. 1953. Toxicologic studies on zineb and disodium ethylene bis(dithiocarbamates). J. Pharmacol. Accol. Exp. Therap. 109: 159-166.

Seifter, J. and W.E. Ehrich. 1948. Goitrogenic compounds: Pharmacological and pathological effects. J. Pharmacol. Exp. Therp. 92: 303-314. (Cited in: Blackwell-Smith, R., et al. 1953)

Short, R.D., J.L. Minor and T.M. Unger. 1980. Teratology of a zineb formulation. US EPA Report No. 600/1-80-017. NTIS PB80-181175.

US EPA. 1984. Health and Environmental Effects Profile for Zineb. Prepared by the Office of Health and Environmental Assessment, Environmental Criteria and Assessment Office, Cincinnati, OH for the Office of Solid Waste, Washington, DC.

VI.B. INHALATION RfD REFERENCES

None

VI.C. CARCINOGENICITY ASSESSMENT REFERENCES

None

VI.D. DRINKING WATER HA REFERENCES

None

VII. REVISION HISTORY

```
--------  --------  -----------------------------------------------------------
Date      Section   Description
--------  --------  -----------------------------------------------------------
03/01/88  I.A.2.    Text revised
03/01/88  I.A.4.    Text added
03/01/90  VI.       Bibliography on-line
01/01/92  I.A.7.    Secondary contact changed
01/01/92  IV.       Regulatory Action section on-line
```

SYNONYMS

ASPOR	MICIDE
ASPORUM	MILTOX
BERCEMA	MILTOX SPECIAL
BLIGHTOX	NOVOSIR N
BLITEX	NOVOZIN N 50
BLIZENE	NOVOZIR
CARBADINE	NOVOZIR N
CHEM ZINEB	NOVOZIR N 50
CINEB	PAMOSOL 2 FORTE
CRITTOX	PARZATE
CYNKOTOX	PARZATE C
DAISEN	PARZATE ZINEB
DIPHER	PEROSIN
DITHANE 65	PEROSIN 75B
DITHANE Z	PEROZIN
DITHANE Z-78	PEROZINE
DITHIANE Z-78	POLYRAM Z
DITIAMINA	SPERLOX-Z
ENT 14,874	THIODOW

1,2-
ETHANEDIYLBIS(CARBAMODITHIO
ATO) (2-)-S,S'-ZINC
((1,2-
ETHANEDIYLBIS(CARBAMODITHIO
ATO))(2-)ZINC
1,2-
ETHANEDIYLBISCARBAMODITHIOI
C ACID, ZINC COMPLEX
1,2-
ETHANEDIYLBISCARBAMOTHIOIC
ACID, ZINC SALT
ETHYLENEBIS(DITHIOCARBAMATO
)ZINC
ETHYLENEBIS(DITHIOCARBAMIC
ACID), ZINC SALT
ETHYL ZIMATE
HEXATHANE
KUPRATSIN
KYPZIN
LIROTAN
LONACOL

TIEZENE
TRITOFTOROL
TSINEB
Z-78
ZEBENIDE
ZEBTOX
ZIDAN
ZIMATE
ZINC
ETHYLENEBISDITHIOCARBAMATE
ZINC ETHYLENE-1,2-
BISDITHIOCARBAMATE
ZINC,
(ETHYLENEBIS(DITHIOCARBAMATO)
)-
Zineb
ZINEB 75
ZINEB 75 WP
ZINEB 80
ZINK-(N,N'-AETHYLEN-
BIS(DITHIOCARBAMAT))
ZINOSAN

Source: *Instant EPA's IRIS*, 1997, Instant Reference Sources, Inc., 7605 Rockpoint Drive, Austin, TX 78731 (Fax: 512-345-2386; Internet URL, http://www.instantref.com/inst-ref.htm).

Production, Use, and Pesticide Labeling Information

ZINEB is not currently listed in EPA's Pesticide Factsheet database and there is no indication of whether it will be added or not in the future.

Ziram

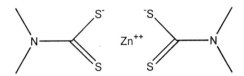

Introduction

Ziram is a carbamate agricultural pesticide. It is applied to foliage as well as being used for soil and seed treatment. It is primarily used on almonds and stone fruit. Ziram is also used as a rubber vulcanization accelerator. It is used in adhesives including those used in food packaging, paper coats for non-food contact, industrial cooling water, latex-coated articles, neoprene, paper and paperboard, plastics (polyethylene and polystyrene) and textiles. It is also used as an agricultural fungicide and as a repellent to birds and rodents.

Identifying Information

CAS NUMBER: 137-30-4

SYNONYMS: ZINC DIMETHYLDITHIOCARBAMATE, METHYL ZIMATE, MILBAM, DIMETHYLDITHIOCARBAMIC ACID ZINC SALT, COROZATE, FUCLASIN, KARBAM WHITE, METHYL CYMATE, METHASAN, ZERLATE, ZIMATE, ZIRBERK, ZINC BIS(DIMETHYLTHIOCARBAMOYL) DISULFIDE, CARBAZINC, CUMAN, DRUPINA 90, FUNGOSTOP, HEXAZIR, MEZENE, PRODARAM, TRICARBAMIX Z, TRISCABOL, VANCIDE MZ-96, ZINCMATE, ZIRAMVIS, ZIRASAN 90, ZIREX 90, ZIRIDE, ZITOX, FUCLASIN ULTRA, METHAZATE, ACETO ZDED, ACETO ZDMD, METHYL ZIRAM, MEXENE, MILBAN, MOLURAME, POMARSOL Z FORTE, CORONA COROZATE, CYMATE, DIMETHYLCARBAMODITHIOIC ACID ZINC COMPLEX, RHODIACID, SOXINAL PZ, VULCACURE, VULKACITER 1, ZINC N,N-DIMETHYLDITHIOCARBAMATE, ZIRAME, ZIRASAN, AAVOLEX, AAZIRA, ANTENE, AMYL ZIMATE, CIRAM, MYCRONIL, AAPROTECT, ZIRTHANE

Endocrine Disruptor Information

Ziram is a strongly suspected EED that it is listed by the Centers for Disease Control & Prevention [812], and the World Wildlife Fund [813] as a potential endocrine modifying chemical. The primary target organ with exposed humans is the thyroid. [880], [829] Rats dosed with ziram prior to pregnancy had marked reductions in fertility and litter size [878], [829] and became largely sterile [879], [829]. Male mice exhibited wasting away of testes when dosed with ziram. [880], [829] When chickens were exposed to ziram the hens became infertile and roosters exhibited retarded testicular development. [881], [829] Rats fet low doses of ziram for two years showed buildup of zinc in the male reproductive system and specifically in the prostate. High concentrations of zinc were also found in bone, liver, kidney, pancreas and endocrine glands. [881], [829]

Chemical and Physical Properties

CHEMICAL FORMULA: $C_6H_{12}N_2S_4.Zn$

MOLECULAR WEIGHT: 305.81

WLN: 1N1&YUS&S-ZN-SYUS&N1&1

PHYSICAL DESCRIPTION: White powder

SPECIFIC GRAVITY: 1.66 @ 25/4°C [031], [169]

DENSITY: 1.66 g/mL @ 23°C [173]

MELTING POINT: 250°C [031], [173]

BOILING POINT: Not available

SOLUBILITY: Water: <1 mg/mL @ 21°C [700]; DMSO: <1 mg/mL @ 21°C [700]; 95% Ethanol: <1 mg/mL @ 21°C [700]; Acetone: <1 mg/mL @ 21°C [700]; Methanol: <1 mg/mL @ 18°C [700]; Toluene: <1 mg/mL @ 18°C [700]

OTHER SOLVENTS: 5% Ammonium hydroxide: <1 mg/mL @ 22°C [700]; Benzene: <0.5 g/100 mL @ 25°C [031], [395]; Ether: <0.2 g/100 mL @ 25°C [031]; Carbon tetrachloride: <0.2 g/100 mL @ 25°C [031]; Chloroform: Soluble [031], [169], [172], [173]; Naphtha: 0.5 g/100 mL @ 25°C [031]; Dilute alkalies: Soluble [062], [169], [172], [173]; Carbon disulfide: Soluble [062], [169], [172], [173]; Concentrated hydrochloric acid: Soluble [062]; Alcohol: <0.2 g/100 mL @ 25°C [031]; Most organic solvents: Slightly soluble [058]; Dilute caustic solutions: Soluble [031]

OTHER PHYSICAL DATA: Melting point (dust): 148°C; (crystals): 250°C [395]; Odorless [173], [346]; Specific gravity: 1.65 @ 20/20 C°[043]

Source: *The National Toxicology Program's Chemical Database, Volume 2*, 1992, CRC Press, Inc./Lewis Publishers, 2000 Corporate Blvd., Boca Raton, FL 33431 (Fax: 407-998-9114).

Environmental Reference Materials

A source of EED environmental reference materials for ziram is Crescent Chemical Co., Hauppauge, NY (Fax: 516-348-0913); Catalog No. 45708.

Hazardous Properties

ACUTE/CHRONIC HAZARDS: This compound may be toxic, an irritant [031], [043], [062], and it can be absorbed through intact skin [151]. When heated to decomposition it may emit toxic fumes of carbon monoxide, carbon dioxide, nitrogen oxides, sulfur oxides and zinc oxides [058].

VOLATILITY: Vapor pressure: Negligible @ room temperature [169], [172], [173]

FIRE HAZARD: Ziram has a flash point of 93°C (200°F) [058]; it is combustible. Fires involving this material can be controlled with a dry chemical, carbon dioxide or Halon extinguisher. A water spray may also be used [058]. This compound may form explosive dust-air mixtures [033], [058].

REACTIVITY: Ziram is corrosive to iron and copper [169]. It is incompatible with strong oxidizing agents and acids [058]. It is also incompatible with mercury [172], [173].

STABILITY: Ziram is stable under normal laboratory conditions [058]. UV spectrophotometric stability screening indicates that solutions of this chemical in 95% ethanol are stable for at least 24 hours [700].

Source: *The National Toxicology Program's Chemical Database, Volume 7*, 1992, CRC Press, Inc./Lewis Publishers, 2000 Corporate Blvd., Boca Raton, FL 33431 (Fax: 407-998-9114).

Medical Symptoms of Exposure

Symptoms of exposure may include irritation of the skin, nose, throat and eyes. It may also cause gastritis, reduced hemoglobin and vegetodystonia [173]. Other symptoms may include brain edema and hemorrhage. In vivo, it may be corrosive to the eyes and cause hemolysis, dystrophy of the muscle, liver and kidney damage, emphysema, local necrosis of the intestine, neural and visual disturbances, and dermatitis [151]. It can also cause headache, tightness of the chest and irritation of the respiratory tract [058].

Source: *The National Toxicology Program's Chemical Database, Volume 4*, 1992, CRC Press, Inc./Lewis Publishers, 2000 Corporate Blvd., Boca Raton, FL 33431 (Fax: 407-998-9114).

Toxicological Information

The toxicological information below is gathered from several sources including two national databases: one from the *National Toxicology Program Chemical Repository Database* and another from *EPA's Integrated Risk Information System (IRIS)*.

Toxicology Information from the National Toxicology Program

CARCINOGENICITY:

Tumorigenic Data:

Type of Effect	Route/Animal	Amount Dosed (Notes)
TD	orl-rat	13160 mg/kg/94W-I
TDLo	imp-rat	60 mg/kg
TDLo	orl-mus	840 mg/kg/13W-I
TDLo	orl-rat	12978 mg/kg/2Y-I
TD	orl-rat	25956 mg/kg/2Y-I

Review:
> IARC Cancer Review: Animal Inadequate Evidence
> IARC: Not classifiable as a human carcinogen (Group 3)

Status:
> NTP Carcinogenesis Bioassay (Feed); Positive: Male Rat
> NTP Carcinogenesis Bioassay (Feed); Negative: Female Rat and Male
Mouse
> NTP Carcinogenesis Bioassay (Feed); Equivocal: Female Mouse

TERATOGENICITY: See RTECS printout for most current data

Reproductive Effects Data:

Type of Effect	Route/Animal	Amount Dosed (Notes)
TDLo	orl-rat	500 mg/kg (6-15D preg)
TDLo	orl-rat	250 mg/kg (6-15D preg)
TDLo	orl-rat	1 gm/kg (6-15D preg)

MUTATION DATA: See RTECS printout for most current data

Test	Lowest dose	Test	Lowest dose
mmo-sat	5 ug/plate	mma-sat	5 ug/plate
dnd-esc	1 umol/L	dnr-bcs	600 ng/disc
cyt-hmn:lym	10 nmol/L	sln-dmg-orl	500 mg/L
cyt-mus-orl	1750 mg/kg/5D-C	msc-mus:lym	62500 ng/L
cyt-ham:ovr	25 ug/L	mnt-mus-unr	350 mg/kg/24H

TOXICITY:

Typ. Dose	Mode	Specie	Amount	Units	Other
LD50	orl	rat	1400	mg/kg	
LD50	unr	rat	1230	mg/kg	
LD50	ipr	rat	23	mg/kg	
LD50	orl	mus	480	mg/kg	
LD50	ipr	mus	73	mg/kg	
LD50	ivn	mus	18	mg/kg	
LD50	orl	rbt	400	mg/kg	
LDLo	ipr	rbt	50	mg/kg	
LD50	orl	gpg	200	mg/kg	
LDLo	ipr	gpg	30	mg/kg	
LD50	orl	bwd	100	mg/kg	
LD50	unr	mam	1400	mg/kg	
LD50	scu	rbt	400	mg/kg	
LD50	scu	rat	1340	mg/kg	
LD50	scu	mus	800	mg/kg	

OTHER TOXICITY DATA:

Status:

EPA Genetox Program 1988, Negative: S cerevisiae gene conversion
EPA Genetox Program 1988, Inconclusive: B subtilis rec assay; D melanogaster Sex-linked lethal
EPA TSCA Chemical Inventory, 1986
EPA TSCA Test Submission (TSCATS) Data Base, January 1989
Meets criteria for proposed OSHA Medical Records Rule

SAX TOXICITY EVALUATION:

THR: Poison by ingestion, intraperitoneal and intravenous routes. Moderately toxic by inhalation. Mutagenic data. An experimental carcinogen and tumorigen.

Source: *Instant Tox-Base*, 1995, Instant Reference Sources, Inc., 7605 Rockpoint Drive, Austin, TX 78731 (Fax: 512-345-2386; Internet URL, http://www.instantref.com/inst-ref.htm).

Toxicology Information from EPA's Integrated Risk Information System (IRIS)

ZIRAM is not currently listed in EPA's IRIS database nor is it listed as scheduled for addition in the near future.

Production, Use, and Pesticide Labeling Information

ZIRAM is not currently listed in EPA's Pesticide Factsheet database and there is no indication of whether it will be added or not in the future.

References

[001] Chemical Abstracts Service. STN International: CA and REGISTRY On-line. Columbus, OH.

[005] Christensen, H.E. and T.T. Luginbyhl, Eds. Registry of Toxic Effects of Chemical Substances. National Institute for Occupational Safety and Health. Rockville, MD. 1975.

[007] Fairchild, E.J., R.J. Lewis, Sr. and R.L. Tatken, Eds. Registry of Toxic Effects of Chemical Substances. DHEW (NIOSH) Publication No. 78-104-A. National Institute for Occupational Safety and Health. Cincinnati, OH. 1977.

[010] Lewis, R.J., Sr. and R.L. Tatken, Eds. Registry of Toxic Effects of Chemical Substances. DHEW (NIOSH) Publication No. 79-100. National Institute for Occupational Safety and Health. Cincinnati, OH. 1979.

[012] Lewis, R.J., Sr. and R.L. Tatken, Eds. Registry of Toxic Effects of Chemical Substances. DHEW (NIOSH) Publication No. 81-116. National Institute for Occupational Safety and Health. Cincinnati, OH. 1980.

[015] Lewis, R.J., Sr. and R.L. Tatken, Eds. Registry of Toxic Effects of Chemical Substances. On-line Ed. National Institute for Occupational Safety and Health. Cincinnati, OH, April 1991.

[016] Weast, R.C., D.R. Lide, M.J. Astle, and W.H. Beyer, Eds. CRC Handbook of Chemistry and Physics. 70th Ed. CRC Press, Inc. Boca Raton, FL. 1989.

[017] Weast, R.C., M.J. Astle, and W.H. Beyer, Eds. CRC Handbook of Chemistry and Physics. 67th Ed. CRC Press, Inc. Boca Raton, FL. 1986.

[018] Weast, R.C., M.J. Astle, and W.H. Beyer, Eds. CRC Handbook of Chemistry and Physics. 65th Ed. CRC Press, Inc. Boca Raton, FL. 1984.

[019] Weast, R.C. and M.J. Astle, Eds. CRC Handbook of Chemistry and Physics. 63rd Ed. CRC Press, Inc. Boca Raton, FL. 1982.

[020] Weast, R.C. and M.J. Astle, Eds. CRC Handbook of Chemistry and Physics. 60th Ed. CRC Press, Inc. Boca Raton, FL. 1979.

[021] Weast, R.C. and M.J. Astle. Eds. CRC Handbook of Chemistry and Physics. 57th Ed. CRC Press, Inc. Boca Raton, FL. 1977.

[022] Weast, R.C. and M.J. Astle, Eds. CRC Handbook of Chemistry and Physics. 56th Ed. CRC Press, Inc. Boca Raton, FL. 1976.

[025] Buckingham, J., Ed. Dictionary of Organic Compounds. 5th Ed. Chapman and Hall. New York. 1982.

[027] Edmundson, R.S. Ed. Dictionary of Organophosphorus Compounds. Chapman and Hall. New York. 1988.

[028] Buckingham, J., Ed. Dictionary of Organic Compounds. 5th Ed. Chapman and Hall. New York. 1988. Supplement 6.

[029] Buckingham, J., Ed. Dictionary of Organic Compounds. 5th Ed. and Supplements. Chapman and Hall. New York. 1988.

[030] Windholz, M., Ed. The Merck Index. 9th Ed. Merck and Co. Rahway, NJ. 1976.

[031] Windholz, M., Ed. The Merck Index. 10th Ed. Merck and Co. Rahway, NJ. 1983.

[032] Merck & Company. MRCK: The Merck Index On-line. BRS Information Technologies. Latham, New York 1991.

[033] Budavari, Susan, Ed. The Merck Index. 11th Ed. Merck and Co., Inc. Rahway, NJ. 1989.

[035] Bretherick, L., Ed. Hazards in the Chemical Laboratory. 3rd Ed. The Royal Society of Chemistry. London. 1981.

[036] Bretherick, L., Ed. Hazards in the Chemical Laboratory. 4th Ed. The Royal Society of Chemistry. London. 1986.

[038] Stull, D.R. Vapor pressure of pure substances: Organic Compounds. Industrial and Engineering Chem. 39(4):517-550. 1947.

[039] Boublik, T., V. Fried and E. Hala. The Vapor Pressures of Pure Substances. New York, Elsevier Scientific Pub. Co., 1973.

[040] Sax, N.I. Dangerous Properties of Industrial Materials. 4th Ed. Van Nostrand Reinhold. New York. 1975.

[041] Sax, N.I. Dangerous Properties of Industrial Materials. 5th Ed. Van Nostrand Reinhold. New York. 1979.

[042] Sax, N.I. Dangerous Properties of Industrial Materials. 6th Ed. Van Nostrand Reinhold. New York. 1984.

[043] Sax, N.I. and Richard J. Lewis, Sr. Dangerous Properties of Industrial Materials. 7th Ed. Van Nostrand Reinhold. New York. 1989.

[045] Sax, N.I., Ed. Industrial Pollution. Van Nostrand Reinhold. New York. 1974.

[046] Sax, N.I., Ed. Cancer Causing Chemicals. Van Nostrand Reinhold. New York. 1981.

[047] Weast, R.C. and M.J. Astle, Eds. CRC Handbook of Data on Organic Compounds. CRC Press, Inc. Boca Raton, FL. 1985.

[050] International Technical Information Institute. Toxic and Hazardous Industrial Chemicals Safety Manual for Handling and Disposal with Toxicity and Hazard Data. International Technical Information Institute. 1978.

[051] Sax, N. Irving, Ed. Dangerous Properties of Industrial Materials Report. Bimonthly Updates. Van Nostrand Reinhold Company, Inc. New York.

[052] Midwest Research Institute. NTP Analytical Chemistry Reports. Kansas City, MO.

[053] Arthur D. Little, Inc. NTP Health and Safety Packages. Arthur D. Little, Inc. Cambridge, MA.

[054] Tracor Jitco. NTP Safety and Toxicity Packages. Tracor Jitco. Bethesda, MD.

[055] Verschueren, K. Handbook of Environmental Data on Organic Chemicals. 2nd Ed. Van Nostrand Reinhold. New York. 1983.

[058] Information Handling Services. Material Safety Data Sheets Service. Microfiche Ed. Bimonthly Updates.

[060] Hawley, G.G., Ed. The Condensed Chemical Dictionary. 9th Ed. Van Nostrand Reinhold. New York. 1977.

[061] Hawley, G.G., Ed. The Condensed Chemical Dictionary. 10th Ed. Van Nostrand Reinhold. New York. 1981.

[062] Sax, N.I. and R.J. Lewis Sr., Eds. Hawley's Condensed Chemical Dictionary. 11th Ed. Van Nostrand Reinhold. New York. 1987.

[065] Bretherick, L. Handbook of Reactive Chemical Hazards. 2nd Ed. Butterworths. London. 1984.

[066] Bretherick, L. Handbook of Reactive Chemical Hazards. 3rd Ed. Butterworths. London. 1985.

[070] Proctor, N.H. and J.P. Hughes. Chemical Hazards of the Workplace. J.B. Lippincott. Philadelphia. 1978.

[071] Sax, N. Irving, Ed. Hazardous Chemicals Information Annual, No. 1. Van Nostrand Reinhold Information Services. New York. 1986.

[072] Sax, N. Irving, Ed. Hazardous Chemicals Information Annual, No. 2. Van Nostrand Reinhold Information Services. New York. 1987.

[075] Nielson, J.M., Ed. Material Safety Data Sheets. Technology Marketing Operation, General Electric Co. Schenectady, NY. 1980.

[076] Nielson, J.M., Ed. Material Safety Data Sheets. Volume II: Tradename Materials. Technology Marketing Operation, General Electric Co. Schenectady, NY. 1980.

[080] U.S. Environmental Protection Agency, Office of Toxic Substances. Toxic Substances Control Act Chemical Substances Inventory, Initial Inventory. 6 Vols. U.S. Environmental Protection Agency. Washington, D.C. 1979.

[081] U.S. Environmental Protection Agency, Office of Toxic Substances. Toxic Substances Control Act Chemical Substances Inventory, Cumulative Supplement II to the Initial Inventory. U.S. Environmental Protection Agency. Washington, DC. 1982.

[082] U.S. Environmental Protection Agency, Office of Toxic Substances. Toxic Substances Control Act Chemical Substance Inventory: 1985 Edition. 5 Vols. U.S. Environmental Protection Agency. Washington, D.C. January 1986.

[084] Commission of the European Communities. Constructing EINECS: Basic Documents. Official Journal of the European Communities. Office for Official Publications of the European Communities. Luxembourg. 1981.

[085] Commission of the European Communities. Constructing EINECS: Basic Documents, Compendium of Known Substances. Vols. 1-3. Office for Official Publications of the European Communities. Luxembourg. 1981.

[086] Commission of the European Communities. Constructing EINECS: Basic Documents, European Core Inventory. Vols. 1-4. Office for Official Publications of the European Communities. Luxembourg. 1981.

[090] Steere, N.V., Ed. Handbook of Laboratory Safety. 2nd Ed. CRC Press, Inc. Cleveland, OH. 1971.

[092] Wilhoit, Randolph C. and Bruno J. Zwolinski. Handbook of Vapor Pressures and Heats of Vaporization of Hydrocarbons and Related Compounds. Evans Press. Fort Worth, TX. 1971.

[095] Gardner, W., E.I. Cooke and R.W.I. Cooke. Handbook of Chemical Synonyms and Trade Names. 8th Ed. CRC Press. Boca Raton, FL. 1978.

[099] Grant, W. Morton, M.D. Toxicology of the Eye. 3rd Ed. Charles C. Thomas, Publisher. Springfield, IL. 1986.

[100] Grasselli, J.G., Ed. CRC Atlas of Spectral Data and Physical Constants for Organic Compounds. CRC Press, Inc. Cleveland, OH. 1973.

[102] U.S. Department of Health and Human Services and U.S. Department of Labor. NIOSH/OSHA Occupational Health Guidelines for Chemical Hazards. 3 Vols. DHHS (NIOSH) Publication No. 81-123. January, 1981.

[105] The Society of Dyers and Colourists. Colour Index. Vols. 1-8. The Society of Dyers and Colourists. Yorkshire, England. American Association of Textile Chemists and Colorists. Research Triangle Park, NC. 1971-1987.

[107] Occupational Health Services, Inc. Hazardline. Occupational Health Services, Inc. New York.

[110] Oak Ridge National Laboratory. Environmental Mutagen Information Center (EMIC), Bibliographic Data Base. Oak Ridge National Laboratory. Oak Ridge, TN.

[120] Oak Ridge National Laboratory. Environmental Teratogen Information Center (ETIC), Bibliographic Data Base. Oak Ridge National Laboratory. Oak Ridge, TN.

[125] Shepard, T.H. Catalog of Teratogenic Agents. 3rd Ed. The Johns Hopkins University Press. Baltimore. 1980.

[130] Strauss, H.J. Handbook for Chemical Technicians. McGraw-Hill. New York. 1979.

[135] Braker, W. and A.L. Mossman. Matheson Gas Data Book. Matheson Gas Products. Secaucus, NJ. 1980.

[137] Braker, W. and A.L. Mossman The Matheson Unabridged Gas Data Book. Matheson Gas Products. Secaucus, NJ. 1974.

[140] Manufacturing Chemists Association. Guide for Safety in the Chemical Laboratory. 2nd Ed. Van Nostrand Reinhold. New York. 1972.

[141] Veterinary Medicine Publishing Co. Veterinary Pharmaceuticals and Biologicals. 6th Ed. Veterinary Medicine Publishing Co. Lenexa, KS. 1988.

[142] U.S. Department of the Interior, Bureau of Mines. Dust Explosibility of Chemicals, Drugs, Dyes and Pesticides. National Technical Information Service, U.S. Department of Commerce. Springfield, VA. May 1968.

[143] U.S. Department of the Interior, Bureau of Mines. Explosibility of Agricultural Dusts. National Technical Information Service, U.S. Department of Commerce. Springfield, VA. 1961.

[144] U.S. Department of the Interior, Bureau of Mines. Explosibility of Dusts used in the Plastics Industry. National Technical Information Service, U.S. Department of Commerce. Springfield, VA. 1962.

[145] International Working Group on the Toxicology of Rubber Additives. Rubber Chemicals Safety Data and Handling Precautions. International Working Group on the Toxicology of Rubber Additives. Belgium. 1984.

[149] Wagner, Sheldon L. Clinical Toxicology of Agricultural Chemicals. Noyes Data Corporation. Park Ridge, NJ. 1983.

[150] Gosselin, R.E., H.C. Hodge, R.P. Smith and M.N. Gleason. Clinical Toxicology of Commercial Products. 4th Ed. Williams and Wilkins, Co. Baltimore. 1976.

[151] Gosselin, R.E., H.C. Hodge, and R.P. Smith. Clinical Toxicology of Commercial Products. 5th Ed. Williams and Wilkins, Co. Baltimore. 1984.

[155] Thomas, C.L., Ed. Taber's Cyclopedic Medical Dictionary. 14th Ed. F.A. Davis Co. Philadelphia. 1981.

[157] Huff, B.B., Ed. Physicians' Desk Reference. 44th Ed. Medical Economics Co. Oradell, NJ. 1990.

[158] Huff, B.B., Ed. Physicians' Desk Reference. 43rd Ed. Medical Economics Co. Oradell, NJ. 1989.

[159] Huff, B.B., Ed. Physicians' Desk Reference. 41st Ed. Medical Economics Co. Oradell, NJ. 1987.

[160] Huff, B.B., Ed. Physicians' Desk Reference. 33rd Ed. Medical Economics Co. Oradell, NJ. 1979.

[161] Huff, B.B., Ed. Physicians' Desk Reference. 36th Ed. Medical Economics Co. Oradell, NJ. 1982.

[162] Huff, B.B., Ed. Physicians' Desk Reference. 38th Ed. Medical Economics Co. Oradell, NJ. 1984.

[163] Thompson, J.F., Ed. Analytical Reference Standards and Supplemental Data for Pesticides and Other Organic Compounds. EPA Publ. No. 600/9-76-012. U.S. Environmental Protection Agency, Office of Research and Development, Health Effects Research Laboratory. Research Triangle Park, NC. 1976.

[164] Huff, B.B., Ed. Physicians' Desk Reference for Nonprescription Drugs. 11th Ed. Medical Economics Co. Oradell, NJ. 1990.

[165] Wiswesser, W.J., Ed. Pesticide Index. Entomological Society of America. College Park, MD. 1976.

[168] Hartley, Douglas B.Sc., Ph.D., M.I.Inf.Sc. and Hamish Kidd B.Sc., Eds. The Agrochemicals Handbook. The Royal Society of Chemistry. Nottingham, England. 1983.

[169] Hartley, Douglas B.Sc., Ph.D., M.I.Inf.Sc. and Hamish Kidd B.Sc., Eds. The Agrochemicals Handbook. 2nd Ed. The Royal Society of Chemistry. Nottingham, England. 1987.

[170] Worthing, C.R., Ed. The Pesticide Manual, A World Compendium. 6th Ed. British Crop Protection Council. London, England. 1979.

[171] Worthing, C.R., Ed. The Pesticide Manual, A World Compendium. 7th Ed. British Crop Protection Council. London, England. 1983.

[172] Worthing, C.R., Ed. The Pesticide Manual, A World Compendium. 8th Ed. British Crop Protection Council. London, England. 1987.

[173] Hayes, W.J., Jr. Pesticides Studied in Man. Williams and Wilkins. Baltimore. 1982.

[175] Council for Agricultural Science and Technology. The Phenoxy Herbicides. 2nd Ed. CAST Report No. 77. Council for Agricultural Science and Technology. Iowa State University, Ames, Iowa. 1978.

[176] Advisory Committee on Pesticides. Further Review of the Safety for Use in the U.K. of the Herbicide 2,4,5-T. Advisory Committee on Pesticides. 1980.

[180] Ross, S.S., Ed. Toxic Substances Sourcebook. Series 1. Environmental Information Center, Inc., Toxic Substances Reference Department. New York. 1978.

[185] Esposito, M.P., T.O. Tiernan and F.E. Dryden. Dioxins. EPA Report No. 600/2-80-197. Industrial Environmental Research Laboratory. Environmental Protection Agency. Cincinnati, Ohio. 1980.

[186] Sittig, Marshall, Ed. Pesticide Manufacturing and Toxic Materials Control Encyclopedia. Noyes Data Corporation. Park Ridge, NJ. 1980.

[190] Packer, K., Ed. Nanogen Index, A Dictionary of Pesticides and Chemical Pollutants. Nanogens International. Freedom, CA. 1975 (Updated 1979).

[195] Estrin, F.E., P.A. Crosley and C.R. Haynes, Eds. CFTA Cosmetic Ingredient Dictionary. 3rd Ed and Supplement. The Cosmetic, Toiletry and Fragrance Assn. Inc. Washington. 1982.

[200] International Air Transport Association. Restricted Articles Regulations. 23nd Ed. International Air Transport Assn. Geneva. 1980.

[201] International Air Transport Association. Dangerous Goods Regulations and Supplement. 28th Ed. International Air Transport Association. Montreal, Quebec. 1987.

[205] Dean, John A., Ed. Lange's Handbook of Chemistry. 13th Ed. McGraw-Hill Book Company. New York. 1985.

[210] U.S. Coast Guard, Department of Transportation. Chemical Data Guide for Bulk Shipment by Water. 5th Ed. U.S. Coast Guard Publication No. CG-388. U.S. Coast Guard. Washington, DC. 1976.

[215] Rom, William N., Ed. Environmental and Occupational Medicine. Little, Brown and Company. Boston. 1983.

[220] Airline Tariff Publishing Co. Official Air Transport Restricted Articles Tariff No. 6-D Governing the Transportation of Restricted Articles by Air. Airline Tariff Publishing Co. Washington, DC. 19 April 1980 Revision.

[225] Armour, Margaret-Ann, Lois M. Browne and Gordon L. Weir. Hazardous Chemicals Information and Disposal Guide. University of Alberta. Alberta, Canada. 1984.

[230] Cross, R.J. and Mingos, D.M.P., Eds. Organometallic Compounds of Nickel, Palladium, Platinum, Copper, Silver and Gold. Chapman and Hall. London. 1985.

[231] Harrison, P.G., Ed. Organometallic Compounds of Germanium, Tin and Lead. Chapman and Hall. London. 1985.

[232] Knox, G.R., Ed. Organometallic Compounds of Iron. Chapman and Hall. London. 1985.

[233] Wardell, J.L., Ed. Organometallic Compounds of Zinc, Cadmium and Mercury. Chapman and Hall. London. 1985.

[234] White, C., Ed. Organometallic Compounds of Cobalt, Rhodium and Iridium. Chapman and Hall. London. 1985.

[240] Grayson, Martin, Ed. Kirk-Othmer Encyclopedia of Chemical Technology. Volumes 1-24 and Supplement. 3rd Ed. John Wiley & Sons. New York. 1978-1984.

[245] Kidd, Hamish and Douglas Hartley, Eds. UK Pesticides for Farmers and Growers. The Royal Society of Chemistry. Nottingham, England. 1987.

[250] Office of the Federal Register, National Archives and Records Service, General Services Administration. Code of Federal Regulations, Title 49, Transportation, Parts 100 to 199 (Revised as of October 1, 1979). Government Printing Office. Washington, DC. 1979.

[251] Office of the Federal Register, National Archives and Records Service, General Services Administration. Code of Federal Regulations, Title 49, Transportation, Parts 100 to 199 (Revised as of October 1, 1982). Government Printing Office. Washington, DC. 1982.

[252] Office of the Federal Register, National Archives and Records Service, General Services Administration. Code of Federal Regulations, Title 49, Transportation, Parts 100 to 199 (Revised as of October 1, 1983). Government Printing Office. Washington, DC. 1983.

[253] Office of the Federal Register, National Archives and Records Service, General Services Administration. Code of Federal Regulations, Title 49, Transportation, Parts 100 to 199 (Revised as of October 1, 1986). Government Printing Office. Washington, DC. 1986.

[255] Office of the Federal Register, National Archives and Records Service, General Services Administration. Code of Federal Regulations, Title 40, Protection of Environment, Parts 100 to 399 (Revised as of July 1, 1980). Government Printing Office. Washington, DC. 1979.

[260] Slein, M.W. and E.B. Sansone. Degradation of Chemical Carcinogens. Van Nostrand Reinhold. New York. 1980.

[269] Lenga, Robert E. The Sigma-Aldrich Library of Chemical Safety Data. Edition 1. Sigma-Aldrich Corporation. Milwaukee, WI. 1985.

[270] Aldrich Chemical Company. Aldrich Catalog/Handbook of Fine Chemicals. Aldrich Chemical Co., Inc. Milwaukee, WI. 1978.

[271] Aldrich Chemical Company. Aldrich Catalog/Handbook of Fine Chemicals. Aldrich Chemical Co., Inc. Milwaukee, WI. 1980.

[272] Aldrich Chemical Company. Aldrich Catalog/Handbook of Fine Chemicals. Aldrich Chemical Co., Inc. Milwaukee, WI. 1982.

[273] Aldrich Chemical Company. Aldrich Catalog/Handbook of Fine Chemicals. Aldrich Chemical Co., Inc. Milwaukee, WI. 1984.

[274] Aldrich Chemical Company. Aldrich Catalog/Handbook of Fine Chemicals. Aldrich Chemical Co., Inc. Milwaukee, WI. 1986.

[275] Aldrich Chemical Company. Aldrich Catalog/Handbook of Fine Chemicals. Aldrich Chemical Co., Inc. Milwaukee, WI. 1988.

[276] Aldrich Chemical Company. Aldrich Catalog/Handbook of Fine Chemicals. Aldrich Chemical Co., Inc. Milwaukee, WI. Unspecified edition.

[280] Aldrich Chemical Company. The Library of Rare Chemicals. 4th Ed. Aldrich Chemical Co., Inc. Milwaukee, WI. 1978.

[290] Deichmann, W.B. and H.W. Gerarde. Toxicology of Drugs and Chemicals. Academic Press. New York. 1969.

[295] Reynolds, James E.F., Ed. Martindale The Extra Pharmacopoeia. 28th Ed. The Pharmaceutical Press. London. 1982.

[300] Dreisbach, R.H. Handbook of Poisoning: Prevention, Diagnosis and Treatment. 10th Ed. Lange Medical Publications. Los Altos, CA. 1980.

[301] Dreisbach, R.H. Handbook of Poisoning: Prevention, Diagnosis and Treatment. 11th Ed. Lange Medical Publications. Los Altos, CA. 1983.

[305] Block, J.B. The Signs and Symptoms of Chemical Exposure. Charles H. Thomas. Springfield, IL. 1980.

[310] Haque, R., Ed. Dynamics, Exposure and Hazard Assessment of Toxic Chemicals, Ann Arbor Science Publishers. Ann Arbor, MI. 1980.

[315] Florey, Klaus Ed. Analytical Profiles of Drug Substances. Academic Press, Inc. Orlando, FL. 1973-1987.

[320] Occupational Safety and Health Administration. Tentative OSHA Listing of Confirmed and Suspected Carcinogens by Category. Occupational Safety and Health Administration. Washington, DC. 1979.

[321] Soderman, Jean V. Ed. CRC Handbook of Identified Carcinogens and Noncarcinogens: Carcinogenicity-Mutagenicity Database. Vols. 1-2. CRC Press, Inc. Boca Raton, FL. 1982.

[325] Office of the Federal Register National Archives and Records Administration. Code of Federal Regulations, Title 29, Labor, Parts 1900 to 1910. U.S. Government Printing Office. Washington. 1986.

[326] Office of the Federal Register National Archives and Records Administration. Code of Federal Regulations, Title 29, Labor, Parts 1900 to 1910. U.S. Government Printing Office. Washington. 1987.

[327] Office of the Federal Register National Archives and Records Administration. Code of Federal Regulations, Title 29, Labor, Parts 1900 to 1910. U.S. Government Printing Office. Washington. 1988.

[329] Office of the Federal Register National Archives and Records Administration. Code of Federal Regulations, Title 29, Labor, Part 1900.1200: Hazard Communication. U.S. Government Printing Office. Washington. 1986.

[330] Rappoport, Z.V.I., Ed. CRC Handbook of Organic Compound Identification. 3rd Ed. CRC Press, Inc. Cleveland, OH. 1975.

[335] United States Environmental Protection Agency. EPA Toxicology Handbook. Government Institutes, Inc. Rockville, Maryland. September 1986.

[340] Sittig, M. Hazardous and Toxic Effects of Industrial Chemicals. Noyes Data Corporation. Park Ridge, NJ. 1979.

[345] Sittig, M. Handbook of Toxic and Hazardous Chemicals. Noyes Publications. Park Ridge, NJ. 1981.

[346] Sittig, M. Handbook of Toxic and Hazardous Chemicals and Carcinogens. 2nd Ed. Noyes, Publications. Park Ridge, NJ. 1985.

[350] Sittig, M., Ed. Priority Toxic Pollutants, Health Impacts and Allowable Limits. Noyes Data Corporation. Park Ridge, NJ. 1980.

[355] Weiss, G., Ed. Hazardous Chemicals Data Book. Noyes Data Corporation. Park Ridge, NJ. 1980.

[360] Sunshine, I., Ed. CRC Handbook Series in Analytical Toxicology, Section A: General Data, Vol. 1. CRC Press, Inc. Boca Raton, FL. 1969.

[365] Connors, Kenneth A., Gordon L. Amidon and Valentino J. Stella. Chemical Stability of Pharmaceuticals: A Handbook for Pharmacists. 2nd Ed. John Wiley & Sons. New York. 1986.

[370] U.S. Coast Guard, Department of Transportation. Chemical Hazards Response Information System (CHRIS), A Condensed Guide to Chemical Hazards. U.S. Coast Guard Publication No. CG-446-1. U.S. Coast Guard. Washington, DC. 1974.

[371] U.S. Coast Guard, Department of Transportation. CHRIS Hazardous Chemical Data. U.S. Coast Guard. Washington, D.C. 1985.

[375] Barton, A.F.M. Solubility Data Series, Volume 15: Alcohols with Water. Pergamon Press. New York. 1984.

[380] U.S. Coast Guard, Department of Transportation. Chemical Hazards Response Information System (CHRIS), Hazardous Chemical Data. U.S. Coast Guard Publication No. CG-446-2. U.S. Coast Guard. Washington, DC. 1978.

[385] Williams, L.R., E. Calliga and R. Thomas. Hazardous Materials Spill Monitoring: Safety Handbook and Chemical Hazard Guide, Part A. U.S. EPA Report No. EPA-600/4-79-008a. U.S. Environmental Protection Agency. Washington, DC. 1979.

[386] Williams, L.R., E. Calliga and R. Thomas. Hazardous Materials Spill Monitoring: Safety Handbook and Chemical Hazard Guide, Part B - Chemical Data. U.S. EPA Report No. EPA-600/4-79-008b. U.S. Environmental Protection Agency. Washington, DC. 1979.

[390] Bureau of Explosives, Association of American Railroads. Hazardous Materials Regulations of the Department of Transportation by Air, Rail, Highway, Water and Military Explosives by Water, Including Specifications for Shipping Containers. Bureau of Explosives Tariff No. BOE-6000-A. Bureau of Explosives. Washington, DC. 1981.

[393] Student, P.J. Emergency Handling of Hazardous Materials in Surface Transportation. Bureau of Explosives. Washington, D.C. 1981.

[395] International Agency for Research on Cancer, World Health Organization. IARC Monographs on the Evaluation of Carcinogenic Risk of Chemicals to Man. International Agency for Research on Cancer. Geneva, Switzerland.

[400] U.S. Department of Transportation, Materials Transportation Bureau, Office of Hazardous Materials Operations. An Index to the Hazardous Materials Regulations, Title 49, Code of Federal Regulations, Parts 100-199 (January 3, 1977 Revision). U.S. Department of Transportation. Washington, DC. 1977.

[401] Nutt, A. R. Toxic Hazards of Rubber Chemicals. Elsevier Applied Science Publishers. New York. 1984.

[405] Goodman, L.S. and A. Gilman. The Pharmacological Basis of Therapeutics. 5th Ed. Macmillan Publishing Co. New York. 1975.

[406] Goodman, L.S., A. Gilman, F. Murad and T.W. Rall, Eds. The Pharmacological Basis of Therapeutics. 7th Ed. Macmillan Publishing Co. New York. 1985.

[410] American Conference of Governmental Industrial Hygenists. Threshold Limit Values for Chemical Substances and Physical Agents in the Work Environment with Intended Changes for 1982. American Conference of Governmental Industrial Hygenists. Cincinnati, OH. 1982.

[411] American Conference of Governmental Industrial Hygenists. Threshold Limit Values for Chemical Substances and Physical Agents in the Work Environment with Intended Changes for 1984-85. American Conference of Governmental Industrial Hygenists. Cincinnati, OH. 1984.

[412] American Conference of Governmental Industrial Hygienists. Threshold Limit Values and Biological Exposure Indices for 1985-1986. American Conference of Governmental Industrial Hygienists. Cincinnati, OH. 1985.

[413] American Conference of Governmental Industrial Hygienists. Threshold Limit Values and Biological Exposure Indices for 1986-1987. American Conference of Governmental Industrial Hygienists. Cincinnati, OH. 1986.

[414] American Conference of Governmental Industrial Hygienists. Threshold Limit Values and Biological Exposure Indices for 1987-1988. American Conference of Governmental Industrial Hygienists. Cincinnati, OH. 1987.

[415] American Conference of Governmental Industrial Hygienists. Threshold Limit Values and Biological Exposure Indices for 1988-1989. American Conference of Governmental Industrial Hygienists. Cincinnati, OH. 1988.

[416] American Conference of Governmental Industrial Hygienists. Threshold Limit Values and Biological Exposure Indices for 1990-1991. American Conference of Governmental Industrial Hygienists. Cincinnati, OH. 1990.

[420] American Conference of Governmental Industrial Hygenists. Documentation of the Threshold Limit Values. 4th ed. American Conference of Governmental Industrial Hygenists. Cincinnati, OH. 1980.

[421] American Conference of Governmental Industrial Hygienists. Documentation of the Threshold Limit Values. 5th Ed. American Conference of Governmental Industrial Hygienists. Cincinnati, OH. 1986.

[425] Schwope, A.D., P.P. Costas, J.O. Jackson and D.J. Weitzman. Guidelines for the Selection of Chemical Protective Clothing. Vol. I: Field Guide; Vol. II: Technical and Reference Manual. American Conference of Governmental Industrial Hygenists, Inc. Cincinnati, OH. 1983.

[426] Schwope, A.D., P.P. Costas, J.O. Jackson and D.J. Weitzman. Guidelines for the Selection of Chemical Protective Clothing. Vol. I: Field Guide; Vol. II: Technical and Reference Manual. American Conference of Governmental Industrial Hygenists, Inc. Cincinnati, OH. 1985.

[430] Clayton, G.D. and F.E. Clayton, Eds. Patty's Industrial Hygiene and Toxicology. Vol. 2. Third Revised Edition. John Wiley and Sons. New York. 1981.

[435] Lyman, Warren J., Ph.D., William F. Reehl and David H. Rosenblatt, Ph.D. Eds. Handbook of Chemical Property Estimation Methods. McGraw-Hill Book Company. New York. 1982.

[440] National Fire Protection Association. Flash Point Index of Trade Name Liquids. 9th Ed. National Fire Protection Association. Boston, MA. 1978.

[445] Baker, Charles J. The Firefighter's Handbook of Hazardous Materials. 4th Ed. Maltese Enterprises, Inc. Indianapolis, Indiana. 1984.

[450] National Fire Protection Association. Fire Protection Guide on Hazardous Chemicals. 7th Ed. National Fire Protection Association. Boston. 1978.

[451] National Fire Protection Association. Fire Protection Guide on Hazardous Materials. 9th Ed. National Fire Protection Association. Quincy, MA. 1986.

[455] The Pharmaceutical Society of Great Britain. The Pharmaceutical Codex. 11th Edition. The Pharmaceutical Press. London. 1979.

[460] National Fire Protection Association. Hazardous Chemicals Data. National Fire Protection Association. Boston. 1975.

[465] National Toxicology Program. Chemical Registry Handbook, Parts I and II. National Toxicology Program. Research Triangle Park, NC. 1981.

[470] Meyer, E. Chemistry of Hazardous Materials. Prentiss-Hall. Englewood Cliffs, NJ. 1977.

[475] United States Environmental Protection Agency. Good Laboratory Practice Compliance Inspection Manual. Government Institutes, Inc. August 1985.

[480] Fletcher, J.H., O.C. Dermer and R.B. Fox, Eds. Nomenclature of Organic Compounds, Principles and Practice. American Chemical Society. Washington, DC. 1974.

[485] Rigaudy, J. and S.P. Klesney. Nomenclature of Organic Chemistry, Sections A, B, C, D, E, F and H. Pergamon Press. New York. 1979.

[490] Meidl, J.H. Explosive and Toxic Hazardous Materials. Glencoe Publishing Co. Encino, CA. 1970.

[495] Williams, Phillip L. and James L. Burson Eds. Industrial Toxicology. Van Nostrand Reinhold Company. New York. 1985.

[500] Mackison, F.W., R.S. Stricoff and L.J. Partridge, Eds. Occupational Health Guidelines for Chemical Hazards. DHHS (NIOSH) Publication No. 81-123. National Institute for Occupational Safety and Health. Cincinnati, OH. 1981.

[501] Mackison, F.W., R.S. Stricoff and L.J. Partridge, Eds. NIOSH/OSHA Pocket Guide to Chemical Hazards. DHEW (NIOSH) Publication No. 78-210. National Institute for Occupational Safety and Health. Cincinnati, OH. 1978.

[505] American Lung Association of Western New York. Chemical Emergency Action Manual. 2nd Ed. C.V. Mosby Co. St. Louis, MO. 1983.

[510] Alliance of American Insurers. Handbook of Organic Industrial Solvents. 5th Ed. Alliance of American Insurers. Chicago. 1980.

[515] Committee on Specifications and Criteria for Biochemical Compounds Division of Chemistry and Chemical Technology National Research Council. Specifications and Criteria for Biochemical Compounds. 3rd Ed. National Academy of Sciences. Washington, D.C. 1984.

[520] Lyman, W.J., W.F. Reehl and D.H. Rosenblatt. Handbook of Chemical Property Estimation Methods, Environmental Behavior of Organic Compounds. McGraw-Hill. New York. 1982.

[525] Hartwell, J.L. Survey of Compounds Which Have Been Tested for Carcinogenic Activity. 2nd Ed. Public Health Service Publication No. 149. National Cancer Institute, National Institutes of Heath. Bethesda, MD. 1963.

[527] Shubik, P. and J.L. Hartwell. Survey of Compounds Which Have Been Tested for Carcinogenic Activity. Supplement 2. Public Health Service Publication No. 149. National Cancer Institute, National Institutes of Heath. Bethesda, MD. 1969.

[528] National Cancer Institute, National Institutes of Health, Public Health Service, U.S. Department of Health, Education and Welfare. Survey of Compounds Which Have Been Tested for Carcinogenic Activity. 1961-1967 Vol. Sections I and II. DHEW (NIH) Publication No. 73-35. Public Health Service Publication No. 149. National Cancer Institute. Bethesda, MD. 1973.

[529] National Cancer Institute, National Institutes of Health, Public Health Service, U.S. Department of Health, Education and Welfare. Survey of Compounds Which Have Been Tested for Carcinogenic Activity. 1968-1969 Vol. DHEW (NIH) Publication No. 72-35,

Public Health Service Publication No. 149. National Cancer Institute. Bethesda, MD. 1972.

[530] National Cancer Institute, National Institutes of Health, Public Health Service, U.S. Department of Health, Education and Welfare. Survey of Compounds Which Have Been Tested for Carcinogenic Activity. 1970-1971 Vol. DHEW (NIH) Publication No. 73-453, Public Health Service Publication No. 149. National Cancer Institute. Bethesda, MD. 1973.

[540] National Cancer Institute, National Institutes of Health, Public Health Service, U.S. Department of Health, Education and Welfare. Survey of Compounds Which Have Been Tested for Carcinogenic Activity. 1978 Vol. DHEW (NIH) Publication No. 80-453 (Formerly Public Health Service Publication No. 149). National Cancer Institute. Bethesda, MD. 1973.

[545] Office of the Federal Register National Archives and Records Administration. Federal Register, Dept. of Labor, Part III. U.S. Government Printing Office. Washington. January 19, 1989.

[550] U.S. Environmental Protection Agency, Office of Pesticides and Toxic Substances. TSCA Chemical Assessment Series, Chemical Hazard Information Profiles (CHIPs), August 1976 - August 1978. U.S. EPA Publication No. EPA-560/11-80-011. U.S. Environmental Protection Agency. Washington, DC. 1980.

[560] U.S. Environmental Protection Agency, Office of Pesticides and Toxic Substances. TSCA Chemical Assessment Series, Chemical Screening: Initial Evaluations of Substantial Risk Notices, Section 8(e), January 1, 1977 - June 30, 1979. U.S. EPA Publication No. EPA-560/11-80-008. U.S. Environmental Protection Agency. Washington, DC. 1980.

[565] Chemical and Pharmaceutical Press. MSDS Reference for Crop Protection Chemicals. Chemical and Pharmaceutical Press. New York, NY. 1989.

[566] Chemical and Pharmaceutical Press. MSDS Reference for Crop Protection Chemicals. Updates. Chemical and Pharmaceutical Press. New York, NY. 1989.

[570] Cone, M.V., M.F. Baldauf, F.M. Martin and J.T. Ensminger, Eds Chemicals Identified in Human Biological Media, A Data Base, First Annual Report, Vol. I, Parts 1 and 2, Records 1 - 1580. Interagency Collaborative Group on Environmental Carcinogenesis, National Cancer Institute, National Institutes of Health. Bethesda, MD. 1980.

[571] Cone, M.V., M.F. Baldauf, F.M. Martin and J.T. Ensminger, Eds. Chemicals Identified in Human Biological Media, A Data Base, Second Annual Report, Vol. II, Parts 1 and 2, Records 1581-3500. Interagency Collaborative Group on Environmental Carcinogenesis, National Cancer Institute, National Institutes of Health. Bethesda, MD. 1981.

[575] Consolidated Midland Corporation. Technical Bulletin no. 1002, Phorbol, TPA and Derivatives. Brewster, New York, 1982.

[580] Frick, G.W., Ed. Environmental Glossary, 2nd Ed. Government Institutes, Inc. Rockville, MD. 1982.

[582] Arbuckle, J.G., G.W. Frick, R.M. Hall Jr., M.L. Miller, T.F.P. Sullivan and T.A. Vanderver, Jr. Environmental Law Handbook. 7th Ed. Government Institutes, Inc. Rockville, MD. 1980.

[584] Government Institutes, Inc. Environmental Statutes. 1983 Ed. Government Institutes, Inc. Rockville, MD. 1983.

[590] Stever, Donald W. Law of Chemical Regulation and Hazardous Waste. Clark Boardman Company, Ltd. New York. 1986.

[600] Hazards Research Corporation. Vapor Pressure Determinations. HRC Report. Hazards Research Corporation. Rockaway, NJ.

[601] Safety Consulting Engineers Inc. Vapor Pressure Test Results. Safety Consulting Engineers Inc. Rosemont, IL.

[610] Clansky, Kenneth B., Ed. Suspect Chemicals Sourcebook: A Guide to Industrial Chemicals Covered Under Major Federal Regulatory and Advisory Programs. Roytech Publications, Inc. Burlingame, CA. 1990. Section 3.

[620] United States National Toxicology Program. Chemical Status Report. NTP Chemtrack System. Research Triangle Park, NC. January 4, 1991.

[650] 3M Company. 1989 Respirator Selection Guide. 3M Occupational Health and Environmental Safety Division. St. Paul, MN. 1989.

[651] 3M Company. 1990 Respirator Selection Guide. 3M Occupational Health and Environmental Safety Division. St. Paul, MN. 1990.

[700] Experimentally determined by Radian Corporation, Austin, Texas.

[701] EPA's IRIS Chemical Information Database. Adapted for publication by Lawrence H. Keith; Lewis Publishers, Inc., 1992.

[702] Screening Methods for the Development of Air Toxics Emission Factors, EPA-450/4- 91-021, September 1991.

[703] Dangerous Properties of Industrial Materials, Seventh Edition, N. Irving Sax and Richard J. Lewis, Sr., Van Nostrand Reinhold, New York, 1989.

[704] NIOSH Pocket Guide to Chemical Hazards, U.S. Department of Health and Human Services, Public Health Service, Centers for Disease Control, National Institute for Occupational Safety and Health, DHHS (NIOSH) Publication No. 90-117, June 1990.

[705] EPA's Pesticide Fact Sheet Database, Mary M.Walker and Lawrence H. Keith, Editors, Lewis Publishers, Inc., 1992.

[706] Handbook of Chemistry and Physics, Seventy-second Edition, David R. Lide, Editor in Chief, CRC Press, 1992.

[707] "Asbestos: Scientific Developments and Implications for Public Policy," B.T. Mossman, J. Bignon, M. Corn, A. Seaton, J.B.L. Gee, SCIENCE, Vol. 247, pp 294 - 301, 19 January 1990.

[708] Clean Air Act Amendments, Public Law #101-549, Section 301, November 15, 1990.

[709] Pesticide Handbook (Entoma), Twenty-ninth Edition, Robert L. Caswell, Kathleen J. DeBold, Lorraine S. Gilbert, Editors, Entomological Society of America, 1981.

[710] Bretherick, L. Handbook of Reactive Chemical Hazards. 3rd Ed. Butterworths. London. 1985.

[711] Kant, James A., Ed. Riegel's Handbook of Industrial Chemistry. 8th Ed. Van Nostrand Reinhold. New York. 1983.

[712] Gross, Paul and Daniel Brown. Toxic and Biomedical Effects of Fibers. Noyes Publications. Parkridge, NJ. 1984.

[713] 56 Federal Register 27338-27356. June 13, 1991.

[714] National Research Council, Committee on Hazardous Substances in the Laboratory, Assembly of Mathematical and Physical Sciences. Prudent Practices for Handling Hazardous Chemicals in Laboratories. National Academy Press. Washington, D.C. 1981.

[715] Budavari, Susan, Ed. The Merck Index: An Encyclopedia of Chemicals, Drugs, and Biologicals. 11th Ed. Merck & Co., Inc. 1989.

[716] McClanahan, James L. and Mary M. Walker. Guidebook on Polychlorinated Biphenyls. March, 1987.

[717] Exner, Jurgen H., Ed. Solving Hazardous Waste Problems. Learning from Dioxins. American Chemical Society. Washington, D.C. 1987.

[718] 40 CFR Part 761.

[719] Mobay Corporation, Agricultural Chemicals Division. Material Safety Data Sheet, Baygon 1.5. February 1, 1991.

[720] National Fire Protection Association. Fire Protection Guide on Hazardous Materials. Ninth Ed. National Fire Protection Association. Quincy, MA. 1986.

[801] "Screening Methods for the Development of Air Toxics Emission Factors," EPA 450/4/91 021, Inventory Guidance and Evalustion Section, Emission Inventory Branch, TSD, U. S. EPA, Research Triangle Park, NC, 27711, September 1991

[802] Keith, L. H., Chapter 4, "Sampling Air Matrices," in Environmental Sampling and Analysis - A Practical Guide, pp 41-50, Lewis Publishers, Inc., 1991

[803] Lewis, R.G., "Problems Associated With Sampling for Semivolatile Organic Chemicals in air," Proceedings, 1986 EPA/APCA Symposium on Measurement of Toxic Air Pollutants," APCA Special Publication VIP-7, Air and Waste Management Association, Pittsburgh, PA, pp. 134-145, 1986.

[804] Clemments, J. B. and R. G. Lewis, "Sampling for Organic Compounds," in Principles of Environmental Sampling, L. H. Keith, Ed., American Chemical Society, p. 287, 1988.

[805] Hicks, B. B., T. P. Meyers, and D. D. Baldocchi, "Aerometric Measurement Requirements for Quantifying Dry Deposition," in Principles of Environmental Sampling, L. H. Keith, Ed., American Chemical Society, p. 297, 1988.

[806] Keith, L. H. "Practical QC," Instant Reference Sources, Inc., 7605 Rockpoint Dr., Austin, TX, 78731 1995.

[810] Colborn, T., D. Dumanoski, and Meyers, J. P., "Our Stolen Future," Penguin Books, New York, NY, 1996.

[811] Jones, Tammy, U. S. EPA, Las Vegas, NV, Private Communication, December 12, 1996.

[812] Needham, Larry, National Center for Environmental Health, Centers for Disease Control & Prevention, Public Health Service, DHHS, Atlanta, GA, Private Communication, August 25, 1996.

[813] World Wildlife Fund Canada [Online] Available http://www.wwfcanada.org/hormone-disruptors/, December 15, 1996.

[814] Chemfinder [Online] Available http://chemfinder.camsoft.com, January 5, 1997.

[815] Keith, L. H., and D. B. Walters, The National Toxicology Program's Chemical Data Compendium, Lewis Publishers, 1992.

[816] Montgomery, John H., Groundwater Chemicals Desk Reference, Second Edition, CRC Press, 1996.

[817] Montgomery, John H., Agrochemicals Desk Reference, Environmental Data, Lewis Publishers, 1993.

[818] Keith, L. H. and Mary M. Walker, Instant EPA's Air Toxics, Instant Reference Sources, Inc., Austin, TX, 1994.

[819] Keith, L. H., Instant EPA's IRIS, Instant Reference Sources, Inc., Austin, TX, 1997.

[820] Keith, L. H., Instant Tox-Base, Instant Reference Sources, Inc., Austin, TX, 1995.

[821] Walker, Mary M. And L. H. Keith, Instant Pesticide Facts, Instant Reference Sources, Inc., Austin, TX, 1994.

[822] Keith, L. H. and D. B. Walters, The National Toxicology Program's Chemical Database, Volume 2, CRC Press, Inc./Lewis Publishers, Boca Raton, FL, 1992.

[823] Keith, L. H. and D. B. Walters, The National Toxicology Program's Chemical Database, Volume 4, CRC Press, Inc./Lewis Publishers, Boca Raton, FL, 1992.

[824] Keith, L. H. and D. B. Walters, The National Toxicology Program's Chemical Database, Volume 7, CRC Press, Inc./Lewis Publishers, Boca Raton, FL, 1992.

[825] Hileman, Bette, Chemical & Engineering News, p. 23, March 18, 1996.

[826] Hileman, Bette, Chemical & Engineering News, p. 28, May 13, 1996.

[827] Guillette, L. J., T. S. Gross, G. R. Masson, J. M. Matter, H. J. Percival, and A. R. Woodware, Environ. Health Perspect., 102, 680, 1994.

[828] Spectrum Chemical Fact Sheets [Online] Available http://www.speclab.com/ January 5, 1997.

[829] EXTOXNET [Online} Available http://ace.ace.orst.edu/info/extonet/ January 5, 1997.

[830] EPA Chemical Factsheets [Online] Available gopher://ecosys.drdr.virginia.edu:70/11/library/gen/toxics December 29, 1996.

[831] Science Advisory Board, Environmental Futures Committee, U.S. Environmental Protection Agency, Beyond The Horizon: Using Foresight to Protect the Environmental Future, January 1995.

[832] Office of Research and Development, U.S. Environmental Protection Agency, Strategic Plan for the Office of Research and Development, May 1996.

[833] Colborn, T.; F.S. von Saal, A. M. Soto, Environ. Health Perspect., 101(5), 378-384 October 1993.

[834] Leavenworth, S. "Sold Down the River", News & Observer, Raleigh, North Carolina, Available http://WWW2.NANDO.NET/NAO/NEUSE/NEUSE.HTML, March 3, 1995.

[835] Thayer, John S., Environmental Chemistry of the Heavy Elements: Hydrido and Organo Compounds, VCH Publishers, New York, 1995.

[836] Huggett, R.J. et.al., Environ. Sci. Technol.. 26 (2), 1992.

[837] ATSDR Information Sheet, Tin -,Agency for Toxic Substances and Disease Registry, September 1995.

[838] TR-183, Bioassay of Dibutyltin Diacetate for Possible Carcinogenicity (CAS No.1067-33-0), National Cancer Institute, September 1995.

[839] Bradlow, H.L. et.al., Environ Health Persp. 103: 147-150, 1995.

[840] Hays, W.J. ed., Handbook of Pesticide Toxicology Vol.3 Academic Press Inc. New York, 1990.

[841] Repetto, Robert, et.al., World Resources Inst. (Pesticides and the Immune System) ISBN 1-56973-087-3, 1996.

[842] Barron, M.G. et.al., Rev. Environ Contam.Toxicol. 144: 1-93, 1995.

[843] Thrasher, J.D. et.al., Arch. Environ. Health 48: (2) 89-93, 1993.

[844]Smith, A.G., Chlorinated Hydrocarbon Pesticides Vol.3 Academic Press Inc. New York, 1991.

[845] Fry D.M. et.al., Science 231: 919-924, 1989.

[846] Bulger, W.H., Am. J. Ind. Med. 4: 163-173, 1983.

[847] Falck, F. et.al., Arch. .Environ. Health 47:143-146, 1992.

[848] Jackson, J.L. et.al., New Eng. J. of Med. Vol.355 (11) :783-789, 1996.

[849] McKinney, J.D. et.al., Environ. Health Persp. 102: 290-297, 1994.

[850] Biessmann,A. et.al., Environ. Pollut. (Sec. A) Eco. Biol. 27: 15-30, 1982.

[851] Bergeron, J.M. et.al., Environ. Health Persp. 102: 780-781, 1994.

[852] Pluim, H.J. et.al., Environ. Health Persp. 101: 504-508, 1993.

[853] Barrass, N. et.al., Environ. Health Persp. 101 (Supp.5):219-223, 1993.

[854] Buckley. J. et.al., Cancer. Resh. 49: 4030-4037, 1990.

[855] Blyler, G. et.al., Fund. Appl. Toxicol 23 (2): 188-193, 1994.

[856] Reigner, B.G. et.al., Hum. Exp. Toxicol. 12 (3): 215-225, 1993.

[857] Jekat, F.W. et.al., Toxicol. Let. 71(1): 9-25, 1994.

[858] Frigters, Raaij J.A. et.al., Toxicol. 94(1-3): 197-208, 1994.

[859] Exotoxnet (Pesticide Information Project of the Cooperative Extension Service) Online files Oregon State Univ., Revision 9/93.

[860] Quevauviller, P.H. et.al., Appl. Organometallic Chem. 5: 125-129, 1991.

[861] Raloff, J., Science News 145: 56-59, 1994.

[862] Science News, 145: 142, 1994.

[863] Ecotoxnet (Pesticide Information Project of the Cooperative Extension Service) Online files Oregon State Univ., Revision 9/93

[864] Jobling, S.,Environ. Health Persp. (As reported in Science News 148, July 15,1995.)

[865] Greco, D.S. et.al., Vet. Clin. North Am. Small Anim. Pract. 24(4): 765-782, 1994.

[866] Perez-Martinez,C. et.al., Am. J. Vet. Resh. 56(12): 1615-1619, 1995.

[867] Miller, M.M. et.al., Am. J. Obstet. Gynecol. 166: 1535-1541, 1992.

[868] Ronis, M.J., Badger, T.M., Shema, S.J., Roberson, P.K., and Shaikh, F., Toxicol. Appl. Pharmacol. 136:361-371, 1996.

[869] U. S. EPA, Health Advisory on Aldicarb, Office of Drinking Water, 1987.

[870] Meister, R. T. (Ed.), Farm Chemicals Handbook '92, Meister Publishing Co., Willoughby, OH, 1992.

[871] U. S. EPA, Pesticide Fact Sheet No. 158: Allethrin Stereoisomers, U. S. EPA, Office of Pesticide Progras, Registration Div., Washington, DC, March 24, 1988.

[872] Occupational Health Services, Inc., MSDS for Allethrin, OHS Inc., Secaucus, NJ, Nov 17, 1992.

[873] Environmental Health Criteria [Online] Available gopher://gopher.who.ch:70/OO/.anonymousftp/programme/pcs/ehc158.asc December 28, 1996.

[874] Weiss, G. Hazardous Chemicals Data Book, Noyes Data Corp., p. 1069, Park Ridge, NJ, 1986.

[875] Bigger, J. W., and I. R. Riggs, Hilgardia, 42 (10) pp 383-391, 1974.

[876] Singh, J., Bull. Environ. Contam. Toxicol, 4(2), pp 77-79, 1969.

[877] Sims, R. C., W. C. Doucette, J. E. McLean, W. J. Grenney, and R. R. DuPont, U. S. EPA Report-600/6-88-001, 1988.

[878] National Research Council. Drinking Water and Health, Advisory Center on Toxicology, Assembly of Life Sciences. Safe Drinking Water Committee, National Academy of Sciences, Washington, DC. 1977.

[879] Edwards, I.R., D.G. Ferry and W.A. Temple. Fungicides and Related Compounds. In Handbook of Pesticide Toxicology, Volume 3, Classes of Pesticides. Wayland J. Hayes and Edward R. Laws (eds.) Academic Press, NY. 1991.

[880] National Toxicology Program. Carcinogenesis Bioassay of Ziram (CAS No. 137-30-4) in F344/N Rats and B6CF1 Mice (Feed Study). U.S. epartment of Health and Human Services, Public Health Service, National Institutes of Health, Technical Report Series No. 238, 1983.

[881] National Library of Medicine. Hazardous Substances Databank. TOXNET, Medlars Management Section, Bethesda, MD. 1993.

[882] Schmidt, Wayne A., "Hormone Copycats" [Online] Available http://www.greatlakes.nwf.org:80/toxics/hcc1-int.htm, Feb.11, 1997.

[883] Birnbaum, L.S., Re-evaluation of dioxin (presentation). Great Lakes Water Quality Board, International Joint Commission. Windsor, Ontario. 1993.

[884] Porter, W.P., S.M. Green, N.L. Debbink and I. Carlson. "Groundwater Pesticides: Interactive Effects of Low Concentrations of Carbamates, Aldicarb and Methomyl and the Triazine Metribuzin on Thyroxine and Somatotropin Levels in White Rats." J. Toxicol. & Environ. Health 40:15-34, 1993.

[885] vom Saal, F.S., M.M. Montano and M.H. Wang. "Sexual Differentiation in Mammals," in: Chemically induced Alterations in Sexual and Functional Development:The wildlife/human Connection. Colborn, T. and C. Clement, eds. Princeton, NJ: Princeton Scientific Publishing Co., Inc., pp. 17-83, 1992.

[886] Bern, H.A., "The Fragile Fetus." in: Chemically Induced Alterations in Sexual and Functional Development:The wildlife/human Cnnection. Colborn, T. and C. Clement, eds. Princeton, NJ: Princeton Scientific Publishing Co., Inc., pp. 9-15, 1992.

[887]. Kalland, T. "Long-term Effects on the Immune System of an Early Life Exposure to Diethylstilbestrol," in: Environmental Factors in Human Growth and Development, Cold Spring Harbor Laboratory, Cold Spring Harbor, NY, pp. 217-242, 1982.

[888] Hines, M., "Surrounded by Estrogens? Considerations for Neurobehavioral Development in Human Beings," in: Chemically Induced Alterations in Sexual and

Functional Development:The wildlife/human Connection. Colborn, T. & C. Clement, eds. Princeton, NJ: Princeton Scientific Publishing Co., Inc., pp. 261-281, 1991.

[889] Jansen, H.T., P.S. Cooke, J. Porcelli, T-C. Liu and L.G. Hansen. "Estrogenic and Antiestrogenic Actions of PCBs in the Female Rat, In Vitro and in Vivo Studies. Reproductive Toxicol. 7:237-248, 1993.

[890]. Leatherland, J., "Endocrine and Reproductive Function in Great Lakes Salmon," in: Chemically Induced Alterations in Sexual and Functional Development: The Wildlife/Human Connection. Colborn, T. & C. Clement, eds. Princeton, NJ: Princeton Scientific Publishing Co., Inc., pp. 129-145, 1992.

[891] Carlsen, E. et al. "Evidence for Decreasing Quality of Semen During Past 50 Years," British Med. J. 305:609-613, 1992.

[892]. Sharpe, R.M. "Are Environmental Chemicals a Threat to Male Fertility?" Chemistry & Industry 3:87-94, 1992.

[893] Guo, Y.L., T.J. Lai, S.H. Ju, Y.C. Chen and C.C. Hsu., Sexual Developments and Biological Fndings in Yucheng Children," Dioxin '93: 13th International Symposium on Chlorinated and Related Compounds, Sept. 1993, Vienna, Vol. 14:235-238, 1993.

[894] Mably, T.A., D.L. Bjerke, R.S. Moore, A. Gendron-Fitzpatrick and R.E. Peterson. "In Utero and Lactational Exposure of Male Rates to 2,3,7,8-Tetrachlorodibenzo-p-dioxin (series of three papers)." Toxicol. & Appl. Pharmacol. 114:97-126, 1992.

[895] Osterlind, A., "Diverging Trends in Incidence and Mortality of Testicular Cancer in Denmark," British J. Cancer 53:501-505, 1986.

[896] Giwercman, A. and N.E. Skakkebaek. "The human testis--an organ at risk?" Intern. J. Andrology 15:373-375, 1992.

[897] Beardsley, T. A war not won," Scientific Am. 270:130-138, 1994."

[898]. Pollner, F. "A Holistic Approach to Breast Cancer Research," Environ. Health Perspectives 101:116-120, 1993.

[899]. Davis, D.L., H.L. Bradlow, M. Wolff, T. Woodruff, D.G. Hoel and H. Anton-Culver. "Medical Hypothesis: Xenoestrogens as Preventable Causes of Breast Cancer," Environ. Health Perspectives 101:372-377, 1993.

[900] Wolff, M.S., P.G. Toniolo, E.W. Lee, M. Rivera and N. Dubin, "Blood Levels of Organohlorine Residues and Risk of Breast Cancer," J. Nat. Cancer Institute. 85:648-652, 1993.

[901] Rier, S.E., D.C. Martin, R.E. Bowman, W.P. Dmowski and J.L. Becker. "Endometriosis in Rhesus Monkeys (Macaca mulatta) Following Chronic Exposure to 2,3,7,8-Tetrachlorodibenzo-p-dioxin," Fund. & Applied Toxicol. 21:433-441, 1993.

[902] Gerhard, I. and B. Runnebaum. "Fertility Disorders may Result from Heavy Metal and Pesticide Contamination Which Limits Effectiveness of Hormone Therapy," Zentralblatt fü r Gynä kalogie 114:593-602, 1992.

[903] Klein, P.A., "Immunology and Biotechnology for the Study and Control of Infectious Diseases in Wildlife Populations." J. Zoo & Wildl. Medicine 24:346-351, 1993.

[904] Dewailly, E., P. Ayotte, S. Bruneau, C. Laliberté , D.C.G. Muir and R.J. Norstrom. "Inuit Exposure to Organochlorines Through the Aquatic Food Chain in Arctic Qué bec." Environ. Health Perspectives 101:618-620, 1993.

[905] Fein, G.G., J.L. Jacobson, S.W. Jacobson, P.M. Schwartz and J.K. Dowler. "Prenatal Exposure to Polychlorinated Biphenyls: Effects on Birth Size and Gestational Age," J. Pediatrics 105:315-320, 1984.

[906] Jacobson, S.W., G.G. Fein, J.L. Jacobson, P.M. Schwartz, J.K. Dowler, "The Effects of PCB Exposure on Visual Recognition Memory." Child Development 56:853-860, 1985.

[907] Dodds, E. C. and Lawson, W., Molecular structure in relation to oestrogenic activity. Compounds without a phenanthrene nucleus. Proc. Royal Soc. Lon. B. 125: 222-232, 1938.

[908] Krishnan, A. V., Starhis, P., Permuth, S. F., Tokes, L. and Feldman, D., Bisphenol-A: an estrogenic substance is released from polycarbonate flasks during autoclaving. Endocrin. 132: 2279-2286, 1993.

[909] Brotons, J. A., Olea-Serrano, M. F., Villalobos, M., Pedraza, V. and Olea, N., Xenoestrogens released from lacquer coatings in food cans. Environ. Health Persp. 103: 608-612, 1995.

Appendix 1 - Abbreviations

Abbreviation	Definition
AADI	Adjusted Acceptable Daily Intake
ACGIH	American Conference of Governmental Industrial Hygienists
AChE	acetylcholinesterase
Action Level	the exposure concentration at which certain provisions of the NIOSH recommended standard must be initiated.
ADI	Acceptable Daily Intake
AHFS	American Hospital Formulary Service
AIHA	American Industrial Hygiene Associaton
AMA	American Medical Association
AOAC	Association of Official Analytical Chemists
APA	American Pharmaceutical Association
ASCII	American Standard Code for Information Exchange
ASHP	American Society of Hospital Pharmacists
ASTM	American Society for Testing and Materials
AUR	Air Unit Risk
BHP	biodegradation, hydrolysis, and photolysis
BOD5	biochemical oxygen demand as measured in the standard 5-day test
brd	domestic or laboratory bird
BUN	blood urea nitrogen
Bw	body weight
BWa	Body weight (kg) for experimental animal species used in the HEC derivation of an RfC
bwd	wild bird species
BWh	Body weight (kg) for human used in the HEC derivation of an RfC
Ca	potential human carcinogen
CAG	Carcinogen Assessment Group, US EPA
cal/g	calories per gram
CAS	Chemical Abstracts Service
CAS	Chemical Abstract Service
CASRN	Chemical Abstract Service Registry Number
cat	adult cat
CBI	Confidential business information
cc	(1) cubic centimeter, (2) closed cup
CCINFO	Canadian Centre for Occupational Health and Safety, Toronto, Canada
CDC	Centers for Disease Control
CDC	Center for Disease Control & Prevention (in Atlanta, Georgia)
CERCLA	Comprehensive Environmental Response, Compensation, and Liability Act of 1980
CF	Conversion factor based on PBPK modeling used in the HEC derivation of an RfC for gases. Note: The CF is specific for the experimental animal species and the exposure regimen and concentration simulated.

Abbreviation	Definition
CFR	Code of Federal Regulations
chd	child
ChE	cholinesterase
CIB	NIOSH Current Intelligence Bulletin
CIIT	Chemical Industry Institute of Toxicology
CIS	Fein-Marquardt and Co., Chemical Information System
ckn	chicken, adult (male or female)
CL	ceiling limit - the concentration that should not be exceeded even instantaneously.
Clear Evidence	when carcinogenicity is demonstrated by studies that are interpreted as showing a chemically related increased incidence of malignant neoplasms, studies that exhibit a substantially increased incidence of benign neoplasms, or studies that exhibit an increased incidence of a combination of malignant and benign neoplasms where each increases with dose.
CNS	central nervous system
CODEN	a unique six-letter character code derived from the American Society for Testing and Materials CODEN for Periodical Titles and the CAS Source Index.
CPK	creatine phosphokinase
CRAVE	Carcinogen Risk Assessment Verification Endeavor
ctl	cattle or horse
CWA	Clean Water Act
DASE	Dutch Association of Safety Experts
dBA	decibel, weighted according to the A scale, which approximates the response of the human ear.
dck	duck
DEA	US Drug Enforcement Administration
DHEW	Department of Health and Human Services)
DMSO	dimethyl sulfoxide
DNA	deoxyribonucleic acid
dog	adult dog
dom	domestic animals such as goat or sheep
DOT	US Department of Transportation
DW	drinking water
DWEL	Drinking Water Equivalent Level
ECG	electrocardiogram
EEG	electroencephalogram
EKG	electrocardiogram
ELISA	enzyme-linked immunosorbent assay
EMTD	estimated maximum tolerated dose
EP	Extraction Procedure
EPA	US Environmental Protection Agency
EPUB	Electronic Publishing computer system (part of EPA's EMail system)
Equivocal Evidence	when carcinogenicity is demonstrated by studies that are interpreted as showing a chemically related marginal increase of neoplasms.

Abbreviation	Definition
ER	Extrarespiratory. Refers to effects peripheral to the respiratory system as the portal-of-entry, or systemic effects.
ET	Extrathoracic region of the respiratory tract
F1	first filial generation (in experimental animals)
fbr	fiber
FEL	frank-effect level
FIFRA	Federal Insecticide, Fungicide, and Rodenticide Act
FLP	Flash point
FOI	Freedom of Information
FR	Federal Register
FRC	functional reserve capacity
frg	adult frog
FTS	Federal Telecommunications System
FWS	US Fish and Wildlife Service
g/mL	grams per milliliter
GC	gas chromatography (a technique used to separate mixtures of volatile chemicals from each other)
GI	gastrointestinal
gpg	guinea pig
GPT	glutamic-pyruvic transaminase
grb	gerbil
HA	Health Advisory
ham	hamster
HAPPS	Hazardous Air Pollution Prioritization System
HAS	Health Assessment Summary
HCT	hematocrit
HDT	highest dose tested
HEC	human equivalent concentration
HEEP	Health and Environmental Effects Profile
Hg	mercury
Hgb	hemoglobin
HHS	US Department of Health and Human Services
hmn	human
hor	horse or donkey
HPLC	high performance liquid chromatography (a technique used to separate mixtures of nonvolatile chemicals from each other)
HSDB	Hazardous Substance Data Base
ial	intraaural (ear)
IARC	United Nations International Agency for Research on Cancer.
iat	intraarterial (artery)
IATA	International Air Transportation Association
ice	intracerebral (cerebrum)
ICR	Institute of Cancer Research
ICRP	International Commission for Radiological Protection
icv	intracervical (cervix)
idr	intradermal (dermis (skin))
idu	intraduodenal (duodenum)

Abbreviation	Definition
ihl	inhalation (route of exposure used for a chemical)
IMO	International Maritime Organization
imp	surgical implant
ims	intramuscular (muscle)
Inadequate Evidence	when, because of major qualitative or quantitative limitatons, the studies cannot be interpreted as showing eiher the presence or absence of a carcinogenic effect. This indicates that one of two conditions prevailed: (a) there are few pertinent data; or (b) the available studies, while showing evidence of association, do not exclude chance, bias, or confounding.
Inadequate Study	when carcinogenicity is not demonstrated because of major qualitative or quantitative limitations, and the studies cannot be interpreted as valid for showing either the presence or absence of a carcinogenic effect.
inf	human infant
ipc	intraplacental (placenta)
ipl	intrapleural (pleural cavity)
ipr	intraperitoneal (peritoneal cavity)
IRIS	Integrated Risk Information System
irn	intrarenal (kidney)
isp	intraspinal (spinal canal)
ITII	International Technical Information Institute
itr	intratracheal (trachea)
itt	intratesticular (testes)
iut	intrauterine (uterus)
ivg	intravaginal (vagina)
ivn	intravenous (vein)
kbar	kilobar
kdy	kidney
kg	killigram (one thousand grams)
L	liter (one thousand milliliters)
LC	lethal concentration
LC50	Lethal Concentration 50 - a calculated concentration of a substance in air, exposure to which for a specified length of time, is expected to cause the death of 50% of an entire defined experimental population. It is determined from the exposure to the substance of a significant number from that population.
LCLo	Lethal Concentration Low - the lowest concentration of a substance in air, other than LC50, which has been reported to have caused death in humans or animals. The reported concentrations may be entered for periods of exposure which are less than 24 hours (acute) or greater than 24 hours (subacute and chronic).
LD	lethal dose
LD50	Lethal Dose 50 - a calculated dose of a substance which is expected to cause the death of 50% of an experimental animal population. It is determined from the exposure to the substance by any route other than inhalation of a significant number from that population.

Abbreviation	Definition
LDH	lactic-acid dehydrogenase
LDLo	Lethal Dose Low - the lowest dose (lower than LD50) of a substance introduced by any route, other than inhalation, over any given period of time, in one or more divided portions and reported to have caused death in humans or animals.
LDT	lowest dose tested
LEL	(1) lower explosive limit, (2) lowest-effect level
leu	leukemia
LFL	lower flammability limit
Limited Evidence	evidence of carcinogenicity when data suggest a carcinogenic effect but are limited because: (a) the studies involve a single species, strain or experiment; or (b) the experiments are restricted by inadequate dosage levels, inadequate duration of exposure to the agent, inadequate duration of exposure to the agent, inadequate period of follow-up, poor survival, too few animals, or inadequate reporting; or (c) the neoplasms produced often occur spontaneously and, in the past, have been difficult to classify as malignant by histological criteria alone. This indicates that a causal interpretation is credible, but that alternative explanations, such as chance, bias, or confounding, could not adequately be excluded.
LOAEL	lowest-observed-adverse-effect level
LOAEL(ADJ)	LOAEL adjusted to continuous exposure duration from an intermittent regimen by hour/day and days/7 days.
LOAEL(HEC)	LOAEL adjusted for dosimetric differences across species to a human equivalent concentration.
lym	lymphatic (pertaining to lymph glands, cells or the lymphatic system)
mam	mammal of an unidentified species
man	adult man
MCL	maximum contaminant level
MCLG	maximum contaminant level goal
MED	minimum effective dose
MeV	million electron volt
MEV	minimum effective dose
MF	modifying factor
mg	milligram
mg/kg	milligrams per kilogram
mg/L	milligrams per liter
mky	monkey
MLD	mild - a well defined erythema and slight edema on the skin where the dose was applied.
MMAD	mass median aerodynamic diameter
mmHg	millimeters of mercury; a measure of pressure
mmol	millimole
MOD	moderate - moderate to severe erythema and severe edema on the skin where the dose was applied resulting in a raised area of about 1 mm.

Abbreviation	Definition
MOE	margin of exposure
MOS	margin of safety
MSHA	Mine Safety and Health Administration
MTD	maximum tolerated dose
MTL	median threshold limit
mul	multiple
mus	mouse
MVa	Minute ventilatory volume for experimental animal species (composite value expressed in cu.m/day) used in the HEC derivation of an RfC.
MVh	Minute ventilatory volume for human (composite value expressed in cu.m/day) used in the HEC derivation of an RfC.
MVho	Minute ventilatory volume for human in an occupational environment, assuming 8 hour/day exposure (composite value expressed in cu.m/day), used in the HEC derivation of an RfC.
NAAQS	National Ambient Air Quality Standards
NAS	National Academy of Sciences
NCI	National Cancer Institute
NESHAP	National Emission Standards for Hazardous Air Pollutants
NFPA	National Fire Protection Association
ng	nanogram
NIH	National Institutes of Health
NIOSH	National Institute for Occupational Safety and Health
NLM	National Library of Medicine
nm	nanometer
nml	non-mammalian species
nmol	nanomole
No Evidence	when several adequate studies are available which show that, within the limits of the tests used, the chemical is not carcinogenic.
NOAEL	no-observed-adverse-effect level
NOAEL(ADJ)	NOAEL adjusted to continuous exposure duration from an intermittent regimen by hour/day and days/7 days.
NOAEL(HEC)	NOAEL adjusted for dosimetric differences across species to a human equivalent concentration.
NOEL	no observed effect level
NRC	National Research Council
nse	non-standard exposure (for example, a spill or accidental exposure)
NSPS	New Source Performance Standards
NTIS	National Technical Information Service
NTP	National Toxicology Program
OAQPS	Office of Air Quality Planning and Standards, US EPA
OAR	Office of Air and Radiation, US EPA
OARM	Office of Administration and Resources Management, US EPA
OC	open cup
ocu	ocular (eye)
OHEA	Office of Health and Environmental Assessment, US EPA
OHM/TADS	Oil and Hazardous Materials Technical Assistance Data Systems

Abbreviation	Definition
OPP	Office of Pesticide Programs, US EPA
OPPE	Office of Policy Planning and Evaluation, US EPA
OPTS	Office of Pesticides and Toxic Substances, US EPA
ORD	US EPA Office of Research and Development
orl	oral (through the mouth via feeding or drinking)
OSHA	US Occupational Safety and Health Administration
OST	Office of Science and Technology, US EPA
OSWER	Office of Solid Waste and Emergency Response, US EPA
otr	other
OTS	US EPA Office of Toxic Substances
OW	Office of Water, US EPA
OWRS	Office of Water Regulations and Standards, US EPA
par	parenteral (skin)
PBPK	physiologically based pharmacokinetic
PCB	polychlorinated biphenyl
pCi	picocurie
PD	Position Document
PEL	permissible exposure limit
PEL	OSHA permissible exposure level
pg	picogram
pgn	pigeon
PHS	US Public Health Service
pig	adult pig
pmol	picomole
ppb	parts per billion
pph	parts per hundred
ppm	parts per million
ppt	parts per trillion
PU	pulmonary region of the respiratory tract
qal	laboratory quail
rat	adult male, adult female or unspecifided sex of rat
RBC	red blood cell(s)
rbt	adult rabbit
RCRA	Resource Conservation and Recovery Act
RDDR	Regional deposited dose ratio used in derivation of an HEC for particles.
RDDR(ER)	Regional deposited dose ratio used in the HEC derivation of an RfC for an observed extrarespiratory effect of particles.
RDDR(ET)	Regional deposited dose ratio used in the HEC derivation of an RfC for an observed effect of particles in the extrathoracic region of the respiratory tract.
RDDR(PU)	Regional deposited dose ratio used in the HEC derivation of an RfC for an observed effect of particles in the pulmonary region of the respiratory tract.
RDDR(TB)	Regional deposited dose ratio used in the HEC derivation of an RfC for an observed effect of particles in the tracheobronchial region of the respiratory tract.

Abbreviation	Definition
RDDR(TH)	Regional deposited dose ratio used in the HEC derivation of an RfC for an observed effect of particles in the thoracic region of the respiratory tract.
RDDR(TOTAL)	Regional deposited dose ratio used the HEC derivation of an RfC for an observed effect of particles in the total respiratory tract.
rec	rectal (rectum or colon)
REL	NIOSH recommended exposure limi
RfC	Inhalation Reference Concentration
RfD	Oral Reference Dose
RgD	Regulatory Dose
RGDR	Regional gas dose ratio used in derivation of an HEC for gases.
RGDR(ET)	Regional gas dose ratio used in the HEC derivation of an RfC for an observed effect of a gas in the extrathoracic region of the respiratory tract.
RGDR(PU)	Regional gas dose ratio used in the HEC derivation of an RfC for an observed effect of a gas in the pulmonary region of the respiratory tract.
RGDR(TB)	Regional gas dose ratio used in the HEC derivation of an RfC for an observed effect of a gas in the tracheobronchial region of the respiratory tract.
RGDR(TH)	Regional gas dose ratio used in the HEC derivation of an RfC for an observed effect of a gas in the thoracic region of the respiratory tract.
RGDR(TOTAL)	Regional gas dose ratio used in the HEC derivation of an RfC for an observed effect of a gas in the total respiratory tract.
RM	risk management
rns	rinsed
RPAR	rebuttable presumption against registration
RQ	Reportable Quantity
RTECS	Registry of Toxic Effects of Chemical Substances
RV	residual volume
Sa	Surface area (in sq.cm) for respiratory tract region for experimental animal species used in the HEC derivation of an RfC.
Sa(ET)	Surface area (in sq.cm) of extrathoracic region for experimental animal species used in the HEC derivation of an RfC.
Sa(PU)	animal species used in the HEC derivation of an RfC.
Sa(TB)	Surface area (in sq.cm) of tracheobronchial region for experimental animal species used in the HEC derivation of an RfC.
Sa(TH)	animal species used in the HEC derivation of an RfC.
Sa(TOTAL)	Surface area (in sq.cm) of total respiratory system for experimental animal species used in the HEC derivation of an RfC.
SAB	Science Advisory Board
SANSS	Structure and Nomenclature Search System
SAP	serum alkaline phosphatase
SARA	Superfund Amendments and Reauthorization Act of 1986
sat	saturated
scu	subcutaneous (under the skin)

Abbreviation	Definition
SDWA	Safe Drinking Water Act
SEV	severe - severe erythema (beet redness) to slight eschar formation (injuries in depth) and severe edema (raised more than 1 mm and extending beyond area of exposure) on the skin where the dose was applied.
SGOT	serum glutamic-oxaloacetic transaminase
SGPT	serum glutamic-pyruvic transaminase
Sh	Surface area (in sq.cm) of respiratory tract for human used in the HEC derivation of an RfC.
Sh(ET)	Surface area (in sq.cm) of extrathoracic region for human used in the HEC derivation of an RfC.
Sh(PU)	Surface area (in sq.cm) of pulmonary region for human used in the HEC derivation of an RfC.
Sh(TB)	Surface area (in sq.cm) of tracheobronchial region for human used in the HEC derivation of an RfC.
Sh(TH)	Surface area (in sq.cm) of thoracic region for human used in the HEC derivation of an RfC.
Sh(TOTAL)	Surface area (in sq.cm) of total respiratory system for human used in the HEC derivation of an RfC.
sigma g	geometric standard deviation
(skin)	potential contribution to overall exposure by the cutaneous route including mucous membranes and eyes.
skn	skin
SMCL	secondary maximum contaminant level
SMR	standard mortality rate
Some Evidence	when carcinogenicity is demonstrated by studies that are interpreted as showing a chemically related increased incidence of benign neoplasms, studies that exhibit marginal increases in neoplasms of several organs/tissues, or studies that exhibit a slight increase in uncommon malignant or benign neoplasms.
specie	species of animal tested (e.g., rat, rabbit, human, etc.)
sql	squirrel
SRI	Stanford Research Institute
STEL	short term exposure limit
Sufficient Evidence	evidence of carcinogenicity when there is an increased incidence of malignant tumors: (a) in multiple species or strains; or (b) in multiple experiments (preferably with different routes of administration or using different dose levels); or (c) to an unusual degree with regard to incidence, site or type of tumor, or age at onset. Additional evidence may be provided by data on dose-response effects. This indicates that there is a causal relationship between the exposure and human cancer.
T 1/2	half-life (time)
TB	tracheobronchial region of the respiratory tract
TC	toxic concentration
TCC	Tagliabue closed cup, a standard method of determining flash points

Abbreviation	Definition
TCLo	Toxic Concentration Low - the lowest concentration of substance in air to which humans or animals have been exposed for any given period of time, that has produced any toxic effect in humans, or has produced a tumorigenic or reproductive effect in animals or humans.
TD	toxic dose
TDB	Toxicology Data Base
TDLo	Toxic Dose Low - the lowest dose of a substance introduced by any route other than inhalation, over any given period of time, to which humans or animals have been exposed and reported to produce any non-significant toxic effects in humans or to produce non-significant tumorigenic or reproductive effects in animals or humans.
TEC	Toxic Effects Code
TERT	tertiary
TH	thoracic (TB + PU) region of the respiratory tract
TLV	Threshold Limit Values - recommended limits proposed by the American Conference of Governmental Industrial Hygienists (ACGIH) to which most workers can be exposed wihout adverse effect. TLVs may be expressed as a time-weighted average (TWA), as a short term exposure limit (STEL), or as a ceiling value (CL).
TOC	Tagliabue open cup, a standard method of determining flash points
tod	toad
TOTAL	total respiratory tract
TOXNET	Toxicology Data Network
trk	turkey
TSCA	Toxic Substances Control Act
TSCATS	Toxic Substance Control Act Test Submission
TWA	time weighted average - The concentration of a compound to which nearly all workers may be repeatedly exposed, for a normal 8-hour workday and a 40-hour workweek, without adverse effect
UCL	upper confidence limit
UEL	Upper Explosive Limits
UF	uncertainty factor
UFL	Upper Flammability Limit
ug	microgram
umol	micromole
UN/ID	United Nations Identification Number
unr	unreported
UV	ultraviolet light
v/v	volume for volume
VAa	Alveolar ventilation rate (cu.m/day) for experimental animal species used in HEC derivation of an RfC.
VAh	Alveolar ventilation rate (cu.m/day) for human used in HEC derivation of an RfC.
VOC	volatile organic compound
WBC	white blood cell(s)
WHO	United Nations World Health Organization.

Abbreviation	Definition
WLN	Wiswesser Line Notation
wmn	woman
WQC	Water Quality Criteria

Appendix 2 - Glossary of Medical Terms

This glossary is intended to provide a brief description of a limited number of medically related terms. For additional and more complete information consult a medical dictionary or a medical professional.

Symptom of Exposure	Description of the Symptom
Abortifacient	An agent inducing expulsion of the fetus
Abscess	Localized collection of pus in a cavity
Acaricide	A chemical that destroys ticks or mites
Adiadochokinesia	Inability to do perform alternating movements
Adrenomegaly	Enlargement of one or both of the adrenal glands
Agitatio	Restlessness
Agranulocytosis	A decrease in granulocytic leukocytes in the blood
Akathisis	Motor restlessness
Albuminuria	The presence of readily detectable amounts of albumin protein in the urine
Allergin	The antibody responsible for anaphylaxis
Allergy	An abnormal response of a hypersensitive person to chemical and physical stimuli
Alopecia	Baldness or deficiency of hair
Alveoli	Air spaces within the lungs
Alveolitis	Inflammation of a small saclike dilation
Amblyopia	Vision dimness without apparent eye lesion
Amenorrhea	Absence of menstruation
Amyotrophy	Atrophy of muscle
Anaphylactic shock	A syndrome occurring after reintroduction of an antigen into an animal previously sensitized to the antigen
Anemia	Deficiency in the hemoglobin and/or red blood cells
Angioedema	Swelling of subcutaneous tissue, submucosa or viscera
Angina	Any disease characterized by attacks of choking or suffocation
Angular	Sharply bent
Annulare	Ring shaped
Anogenital	Area around the anal-genital region
Anorexia	Loss of appetite
Anthelmintic	A remedy for the destruction or elimination of intestinal worms
Antiketogenic	Limiting formation of ketones
Antimicrobial	A compound that destroys or suppresses the growth of microbes
Antinarcotic	A compound that prevents narcosis
Antipyresis	Therapeutic use of agents to lower fever
Anuria	Urinary suppression or failure of kidney function
Apathy	Lack of feeling or emotion

1197

Symptom of Exposure	Description of the Symptom
Aphthous stomatis	Small mouth ulcers
Aplastic Anemia	Failure of bone marrow to produce red blood cells
Apnea	Suspension of breathing
Arachnoiditis	Inflammation of the membrane surrounding the brain
Arcuate	Arc shaped
Areflexia	Absence of reflexes
Argyria	Poisoning by silver
Arrhythmias	Change of normal cardiac rhythm
Arthralgia	Pain in a joint
Arthus phenomenon	Development of an inflammatory lesion, e.g., ulcer
Ascites	Accumulation of serous fluid in the abdominal cavity
Aspermia	Nonemission of semen
Asphyxia	Suffocation from lack of oxygen
Asthenia	Lack or loss of strength
Asterixis	A motor disturbance marked by lack of assumed posture
Asthenovegetative syndrome	Characterized by extreme weakness
Asthma	Wheezing due to bronchial contractions
Asymptomatic	Without any symptoms
Ataxia	Muscular incoordination
Atelectasis	Incomplete expansion of lungs at birth
Athetosis	Slow involuntary movements
Atrophy	Reduction in size
Axonal	Pertaining to the vertebral column
Azoospermia	Absence of live sperm in semen
Azotemia	Excess of urea in blood
Basophilia	Reaction of immature erythrocytes to basic dyes
Behcet's syndrome	Severe uveitis and retinal vasculitis with mouth and genitalia lesions
Benign	Harmless
Bell's palsy	Facial paralysis resulting in facial distortion
Bilirubin	A breakdown product of heme
Blanching	To clean, declororize or remove the skin
Blepharitis	Inflammation of the eyelid
Blepharospasm	Eye spasm
Blood Dyscrasia	Any abnormal condition of the formed elements of the blood or of the constituents required for clotting
Bradycardia	Slowness of the heartbeat
Bronchiogenic	Originating in the bronchi of the lungs
Bullae	A large blister beneath the skin
Bulbar	Bulb-like, formerly pertaining to the medulla oblongata
Cachexia	General ill health and malnutrition
Calculus	Abnormal concentrations in the body, often of mineral salts
Candidiasis	A candida fungal infection

Symptom of Exposure	Description of the Symptom
Carboxyhemoglobin	Compound formed when hemoglobin is exposed to carbon monoxide
Catamenia	Mensuration
Catatonic	A particular type of psychological schizophrenic condition
Cephalalgia	Headache
Cerebral dysrhythmia	Irregular brain waves
Cerebellar	Pertaining to the cerebella part of the brain
Cerebri	Pertaining to the brain
Cheilitis	Inflammation of the lips
Cheyne Stokes breathing	Alternating periods of breathing and not breathing
Chemosis	Edema of the ocular conjunctiva
Chemotic	An agent that increases production of lymph in the conjunctiva
Cholangitis	Inflammation of the bile duct
Chloasma	Melasma
Cholestasia	Stoppage or suppression of bile flow
Chloracne	An acne eruption caused by chlorinated hydrocarbons
Chorea	Rapid involuntary jerky movements which appear well coordinated
Choroid	Vascular coating of the eye
Choroidopathy	Disease of the vascular coat of the eye
Chromodacryorrhea	Bloody tears
Cirrhosis	Progressive fibrosis of the liver
Clonic	Rapid alternating muscular contractions and relaxation
Clonus	Rapidly alternating successive muscle contractions and relaxations
Colic	Acute abdominal pain
Colitis	Inflammation of the colon
Coma	Unconsciousness from which patient can't be aroused
Comedo	A blockage in the excretory duct of the skin
Conjunctivitis	Inflammation of the membrane lining the eyelids and eyeball
Coomb's test	A test usually for presence of proteins
Cor pulmonale	Acute or chronic right heart disfunction
Coryza	Profuse nasal discharge
Crepitation	Rasping sound of rubbing things together
Cubital	Pertaining to the elbow or forearm
Cushing's syndrome	A condition, more common to females, resulting from neoplasms of the adrenal cortex or pituitary
Cutaneous	Pertaining to, or affecting the skin
Cyanosis	Bluish skin and mucous membrane color due to reduced hemoglobin in blood
Cylindruria	Presence of tube casts in urine
Cystitis	Inflammation of the urinary bladder
Decerebrate	Elimination or decrease of brain function

Symptom of Exposure	Description of the Symptom
Dementia	Mental deterioration
Demyelination	Remove or destroy the myelin sheath of a nerve
Dental Caries	A localized, progressive and molecular disintegration of the teeth
Depilation	Removal of hair
Dermatitis	Inflammation of the skin
Desquamation	Peeling of skin
Diabetes mellitus	A pancreatic metabolism disorder
Diaphoresis	Profuse perspiration
Diarrhea	A common symptom of gastrointestinal disease characterized by increased frequency and fluid consistency of the stools
Diplopia	A sight disorder where one object appears as two
Diverticulum	A pouch in the lining of the mucous membrane
Dysarthria	Impaired speech
Dyscrasia	A morbid condition involving an imbalance of components
Dyskinesia	An abnormal movement
Dysmenorrhea	Painful mensuration
Dysphagia	Immediately painful swelling
Dyspnea	Difficulty in breathing
Dystonia	Lack of normal tone
Dystrophy	Disorder arising from faulty nutrition
Dysuria	Difficult or painful urination
Ecchymosis	A small hemorrhagic spot in skin or mucous membrane
Eccrine	Outward secretion from a sweat gland
Ectopia	A displacement or misposition
Ectropion	Outward turning, as of an eyelid
Ectrodactylia	Absence of all or part of the fingers or toes
Eczema	Itching or inflammation of the skin
Edema	A swelling of tissues
Emesis	Vomiting
Encephalopathy	Degenerative brain disease
Endometrium	Mucous membrane of the uterus
Endothelium	Squamous lining of any closed body cavity
Enteritis	Inflammation of the intestines
Enterocolitis	Inflammation of both large and small intestines
Enuresis	Urinary incontinence; also bed wetting
Eosinophilia	An abnormal increase in the number of blood eosinophils
Epigastalgia	Pain in the upper middle abdomen
Epiglottis	Cartilage overhanging the larynx
Epileptiform	Resembling epilepsy
Epiphora	Abnormal overflow of tears
Epistaxis	Nosebleed
Epithelioma	A tumor of the internal or external covering of the body
Erythema	Reddening of the skin

Symptom of Exposure	Description of the Symptom
Erythematous	Characterized by reddening of the skin
Euphoria	An exaggerated feeling of well-being
Exanthemata	An eruptive disease or fever
Excoriation	Skin abrasion
Exfoliation	Separation of bone or other tissues into layers
Exfoliative dermatitis	Falling off of layers or scaling of skin
Exophthalmos	Abnormal protrusion of the eyeball
Extrasystole	Premature beat
Fasciculation	A small local contraction of muscles visible through skin
Fibrosis	A thickening, associated with growth of fibrous tissue
Fibrotic	Abnormal formation of fibrous tissue
Flare	A reddened or flushed skin area
Flatus	Gas or air in the intestinal tract
Folliculitis	Inflammation of a follicle, i.e., sac or cavity as in hair follicle
Fontanel	A soft spot in an infants skull
Fossae	Trench, cannel or depressed area
Fungicide	A chemical that destroys fungi
Furunculosis	Persistent reoccurring painful skin nodules; also called boils
Galactorrhea	Excessive or spontaneous milk flow
Gangrenous	Death and decomposition of body tissue due to failure of blood supply, injury or disease
Gastritis	Inflammation of the lining of the stomach
Gingivae	Pertaining to the gums
Gingivitis	Inflammation of the gums with swelling and bleeding
Gingivostomatitis	Inflammation of gums and oral mucosa
Gluconeogenesis	Formation of glucose by breakdown of glycogen
Glomerulonephritis	An inflammatory kidney disease
Glossitis	Inflammation of the tongue
Glottis	Vocal part of the larynx
Glycosuria	The presence of glucose in the urine
Gonadotropin	A gonad stimulating hormone
Gout	Disorder characterized by excess uric acid in blood
Grand mal seizure	Generalized seizure
Granulocytopenia	Abnormal number of granulocytes in blood
Granulomas	A tumor composed of granulation tissue
Gynecomastia	Excessive development of male mammary glands
Heinz bodies	Coccoid inclusion bodies resulting from oxidative injury to and precipitation of hemoglobin
Hematemesis	Vomiting of blood
Hematocrit	The volume of red blood cells
Hematopoietic	Pertaining to the formation of blood in the body
Hematuria	Blood in the urine
Hemoglobin	Red coloring matter of the blood that caries oxygen
Hemofuscin	Pigmentation from degeneration of hemoglobin

Symptom of Exposure	Description of the Symptom
Hemolysis	Liberation of hemoglobin
Hemoptysis	Spitting blood
Hemorrhage	Profuse bleeding
Hemosiderosis	A condition characterized by the deposition of iron-containing pigment from the disintegration of hemoglobin into the liver and spleen
Hepatic	Pertaining to the liver
Hepatic porphyria	Increased quantities of phorphyrins in the liver
Hepatitis	Inflammation of the liver
Hepatomegaly	Enlargement of the liver
Hepatorenal	Pertaining to the liver or kidneys
Herpes zoster	Acute inflammation of the cerebral ganglia
Hilarity	Exhilaration of spirits
Hirsutism	Abnormal growth of hair in unusual places
Hydrocephalus	Abnormal accumulation of fluid in the cranium
Hyperaminoacidemia	An abnormal amount of amino acids in the urine
Hyperchloremia acidosis	A metabolic disorder common with renal disease
Hyperemia	Engorgement with blood, excess blood
Hyperhidrosis	Excessive sweating
Hyperesthesia	Abnormally increased skin sensitivity
Hyperestrogenism	Excess secretion of estrogen
Hyperglycemia	Abnormal increase of sugar in the blood
Hyperkalemia	Abnormal elevation of potassium in blood
Hyperkeratosis	Hypertrophy of the cornea
Hyperlipidemia	Excess of blood lipids
Hypernea	Excessive mental activity
Hyperosmolar	Abnormally high concentration due to osmosis
Hyperpigmentation	Development of increased skin pigmentation
Hyperplasia	Increase in size of tissue or organ due to increase in number of cells
Hyperplastic	Excessive proliferation of cells
Hyperpyrexia	Highly elevated body temperature
Hypertrophy	Enlargement of an organ due to increased size of its cells
Hypofibrinogenemia	Abnormally low fibrinogen content in blood
Hypothalamic syndrome	Disfunction of the hypothalamus
Hypertrichosis	Excessive growth of hair
Hyperuricemia	Excess uric acid in blood
Hypocalceia	Deficiency of calcium
Hypokalemia	Abnormally low potassium in blood
Hypomagnesemia	Abnormally low concentration of magnesium in blood
Hypotonia	Diminished tone of skeletal muscles
Hypoxemia	Deficient oxygenation of the blood
Ichthyosis	Skin disorder characterized by dryness
Icterus	Jaundice
Incontinence	Inability to control extrectory functions

Symptom of Exposure	Description of the Symptom
Inflammation	The reaction of body tissue to injury
Infiltration	Accumulation in cells or tissue of substances not normal to it
Inotropic	Affecting force of muscle contractions
Iridocyclitis	Irritation of the iris and the cilary body
Iritis	Inflammation of the iris
Insecticide	A compound that kills insects
Ischemia	Deficiency of blood due to constrictionor obstruction
Insoluble	Incapable of dissolving in a given liquid
Interossel	Situated between bones
Jaundice	Yellowness of skin, eyes mucous membranes and body fluids
Keratoma	A horny growth, e.g., wart or callus
Keratoses	A skin growth
Keratitis	Inflammation of the cornea of the eye
Ketosis	Abnormally high concentrations of ketones
Kallikrein	One of a specific group of enzymes present in blood, certain glands, organs, urine or lymph system
Lachrymator	Substances which increase tear flow
Lacrimation	Secretion of tears from the eyes
Lacrimator	See lachrymator
Laryngeal	Pertaining to the larynx
Lassitude	Weariness or debility
Latent Period	The time elapsed between exposure and the first manifestation of symptoms
Lateral nystagmus	Oscillation of the eyes from side to side
Lathyrism	Morbid condition from ingestion of seeds of the lathyrus plant family
Lesch-Nyhan syndrome	Disorder of purine metabolism, characterized by physical and mental retardation
Lesion	Abnormal change, injury or damage to a tissue or organ
Lethargy	Drowsiness or indifference
Leukemia	A blood disease distinguished by a marked increase of white blood cells
Leukemogen	Any substance the produces or incites leukemia
Leukoblastosis	Proliferation of leukocytes
Leukocyte	A white blood cell
Leukocytosis	A temporary increase in the number of leukocytes in the blood
Leukopenia	A decrease in the number of white blood cells
Libido	Sexual desire
Lichenification	Thickening of the epidermis
Lividity	Discoloration
Lobule	Small lobe
Lower nephron syndrome	Retrogressive kidney change associated with shock
Lumen	A cavity within a tubular organ; also a unit of light

Symptom of Exposure	Description of the Symptom
Lupus erythematosus	Inflammation of the skin
Lyell's disease	Toxic epidermal necrolysis
Lymphadenopathy	Disease of the lymph nodes
Lymphocytosis	Excess of lymphocytes in blood
Lymphocytopenia	Reduction of lymphocytes in peripheral blood
Lymphopenia	Decrease in the number of blood lymphocytes
Lymphosarcoma	Malignant neoplastic disorder of lymphoid system
Maceration	Softening of a solid by soaking
Macrocytosis	Erythrocytes are larger than usual
Macula	Corneal spot
Maculopapular eruptions	A small discolored elevation of the skin
Malaise	A feeing of illness or depression
Malabsorption	Decreased intestinal absorption of nutrients
Malignancy	A neoplasm or tumor that is cancerous
Mediastinitis	Substernal pain with fever
Medullary	Pertaining to the marrow
Melena	Dark, blood containing, stool or vomit
Melanoma	Tumor of melanin pigmented cells
Melasma	Discolorations, typically during pregnancy on cheeks
Megaloblast	The originator of abnormal blood cells
Meibomian Gland	A gland of the eye
Meningismus	Meningeal inflammation associated with febrile illness or dehydration
Menorrhagia	Excessive uterine bleeding
Mentation	Mental activity
Mesothelioma	A malignant tumor of the membrane which surrounds the internal organs of the body
Metastasis	Spreading of cancer cells from one part of the body to another
Micrognathia	Smallness of the jaw
Micturition	Urination
Miliaria	Changes associated with sweat retention
Miosis	Constriction of the eye
Monilial	Former name for Candida fungus
Monoclinic	Along a single axis
Monocytosis	Increased proportion of monocytes in blood
Morbilliform	Resembling measles
Mucin	Chief constituent of mucus
Mucopurulent	Consisting of mucous and pus
Myoclonus	Shocklike muscle contractions
Mucositis	Inflammation of a mucous membrane
Myalgia	Tenderness or pain in the muscles
Mydriasis	Dilation of the pupil of the eyeball
Myeloblastic	A condition where the bone marrow cell develops into a large cell in bone marrow from which blood cells are formed

Symptom of Exposure	Description of the Symptom
Myelofibrosis	Replacement of bone marrow by fibrous tissue
Myelopathy	Disorder of the spinal cord
Myelopoieses	The development of bone marrow or formation of cells derived from bone marrow
Myeloproliferative syndrome	Medullary and extramedullary proliferation of bone marrow constituents
Myocardial	Concerning heart muscle
Myocardial infarction	Heart attack
Myopathy	Any disease of a muscle
Myotonia	Increased muscular irritability or tonic muscle spasms
Myxedema	Dry, waxy swelling, associated with hypothyroidism
Narcotic	Producing stupor or sleep
Nasopharyngitis	Inflammation of nasal passages and pharynx
Necrotic	Death of a portion of tissue
Neoplasm	A new and abnormal formation of tissue, as a tumor or growth
Neonate	Newly born
Nephritis	Inflammation in the kidneys
Neuropathies	Any disease of the nerves
Neutropenia	Decrease in number of blood neutrophilic leukocytes
Nociceptive avoidance behavior	Avoidance of pain
Nocturia	Urination at night
Node	A small round or oval mass of lymphoid tissue
Nodule	A small node
Nuchal	Back of the neck
Nystagmus	Oscillatory movement of the eyeball
Oligospermia	Reduction of the number of spermatozoa in the semen
Oliguria	Excretion of a reduced amount of urine relative to fluid intake
Onycholysis	Separation of a nail from its bed
Oogenesis	Formation of ova
Opacities	Areas or spots that are not transparent
Opisthotonos	Spasm which bends head and heels backward and body forward
Osmiophilic	Easily stained by osmium
Osseous	Bone quality
Osteomalacia	Softening of the bones
Ototoxicity	Deleterious to organs of hearing and balance
Oxalosis	Presence of oxalic acid or an oxalate
Oxypurine	Purine compound containing oxygen
Pallor	Absence of skin coloration
Palpebral	Pertaining to the eyelid
Palpitation	Abnormal rhythm of the heart of which a person is acutely aware

Symptom of Exposure	Description of the Symptom
Pancreatitis	Inflammation of the pancreas
Pancytopenia	A reduction in all cellular elements of the blood
Papilledema	Swelling of the optic disk
Papillomas	Benign epithelial or endothelial tumors
Papular	A small superficial elevation of the skin
Parenteral	Injection by route other than the alimentary canal
Paresthesia	Abnormal sensation such as numbness, prickling, or tingling
Parkinson's disease	Paralysis agitans, characterized by muscle rigidity
Pathological	Abnormal or diseased
Patella	Bone in front of knee
Paresis	Incomplete paralysis
Parotid	Near the ear
Paroxysmal	A sudden recurrence of symptoms
Pectoral	Pertaining to the chest or breast
Pediculicide	A compound that kills lice
Peliosis hepatitis	Mottled blue liver
Pellagra	Disorder due to deficiency of niacin
Pemphigoid	A group of dermatological syndromes
Percutaneous	Effected through the skin
Periarteritis nodosa	Inflammation of the small and medium sized arteries
Peribronchial	Located around the bronchus
Pericardium	The fibroserous sac surrounding the heart
Pericholangitis	Inflammation of tissue surrounding the bile duct Peripheral
Neuritis	Inflammation of peripheral nerves
Pernicious Anemia	Severe form of blood disease marked by progressive decrease in red blood corpuscles, muscular weakness, and gastrointestinal and neural disturbances
Peristalsis	A progressive contraction of waves in tubes
Peritonitis	Inflammation of the peritoneum
Perivascular	Near or around the blood vessels
Petechia	A pinpoint, nonraised skin hemorrhage
Peyronie's disease	Disease resulting in the downward bowing of the penis
Pharyngitis	Inflammation of the pharynx
Pharynx	The musculomembranous sac between the mouth and the esophagus
Phlegm	Thick mucous from the respiratory passages
Phlebitis	Inflammation of a vein
Photophobia	Abnormal visual intolerance of light
Pleurisy	Inflammation of the lining of the lungs or chest cavity
Pneumonitis	Inflammation of the lungs
Polyarthritis	Inflammation of several joints
Polycythemia	Increase in total body red cells
Polydipsia	Excessive thirst over extended time period
Polyneuropathy	Disease of several nerves
Polyneuritis	Inflammation of many nerves

Symptom of Exposure	Description of the Symptom
Polyuria	The passage of a large volume of urine in a given period of time
Porphyria	Disturbances of porphyrin metabolism
Porphyrea cutanea tarda	Porphyria appearing in late adulthood
Proctitis	Inflammation of the rectum
Proptosis	Bulging of the eye
Prostration	Absolute exhaustion
Proteinuria	Protein in the urine
Pruritus	Severe itching
Pruritus ani	Chronic itching in anal region
Psoriasis	Chronic particular type of skin lesions
Ptosis	Prolapse of an organ
Ptyalism	Excess secretion of saliva
Punctiform	Point like shape
Purgative	Evacuation of the bowel
Purpura	Hemorrhage into the skin or mucous membranes
Pustular	Characterized by small elevations of the skin filled with pus
Pyelonephritis	Inflammation of the kidney and the pelvis
Pyorrhea	Discharge of pus
Pyrexia	A fever
Rale	Abnormal respiratory sound
Raynaud's disease	Vascular disorder
Reticulocytosis	Increase in the number of red blood cells containing a network of granules or filaments in circulating blood
Retinopathy	Any noninflammatory disease of the retina
Retrosternal	Behind the sternum
Retrobulbar	Behind the eyeball
Rhinitis	Inflammation of the nasal mucosa
Rhinorrhea	Thin watery discharge from the nose
Rhonchus	Rale or rattling in throat
Rickets	Caused by deficiency of vitamin D marked by bending of the bones, especially in infants and children
Rodenticide	A compound that is poisonous to rats and mice
Romberg's sign	Swaying or falling of the body when standing with feet close together and the eyes closed
Sanguinolent	Containing, or tinged with blood
Salicylism	Toxic effects from salicylic acid or its salts
Sarcoidoisis	A chronic, progressive granulomatous reticulosis of unknown origin of almost any organ
Scarlatiniform	Resembling scarlet fever
Sclerosis	Hardening from inflammation
Scotoma	An area of suppressed vision in the visual field
Seborrhea	Excessive secretion of sebum
Seminiferous	Producing or conveying semen

Symptom of Exposure	Description of the Symptom
Sepsis	Presence in blood or tissue of a toxin or pathogenic microorganism
Septicemia	Persistence of pathogenic microorganisms in blood
Sialoadenitis	Inflammation of the salivary gland
Siderosis	Iron pigment in blood, tissue, or body fluid
Sloughing	Casting off of necrotic tissue
Somnolence	Sleepiness or unusual drowsiness
Spasticity	Increase over normal muscle tone
Spermatogenesis	Formation of spermatozoa
Splenomegaly	Enlargement of the spleen
Sphincter	Ringlike band of muscles that constrict a passage or close an orifice
Stasis	Stoppage of decrease in blood or other body fluid flow
Steatorrhea	Fatty stools
Stertorous	Snoring
Stevens - Johnson syndrome	Severe form of erythema multiforme
Stomatitis	Inflammation of the mouth
Strabismus	The visual axes of the eyes do not meet as desired
Striate	Striped
Stroma	The supporting framework of an organ
Subcutaneous	Beneath the skin
Substernal	Beneath the sternum
Suppuration	Formation of pus
Symblepharon	Adhesion between tarsal and bulbar conjunctiva
Syncope	Fainting
Systemic	Spread throughout the body
Tachycardia	Abnormally increased heart rate
Tachypnea	Abnormally increased respiration
Telangiectasis	Vascular lesion
Tenesmus	Straining to empty bowels
Tetany	Sharp twitching or flexing of muscles or joints
Threshold Limit Value (TLV)	An atmospheric exposure level under which most people can work without harmful effects
Thrombocytopenia	Decrease in number of blood platelets
Thrombosis	Formation of an aggregation of blood factors, primarily blood platelets and fibrin, causing vascular obstruction
Time-Weighed Average	The concentration of a compound to which nearly all workers may be repeatedly exposed, for a normal 8-hour workday and a 40-hour workweek, without adverse effect
Tonic	Pertaining to tone
Tinnitus	A ringing sound in the ears
Tourette's syndrome	Facial or vocal tics progressing to other body parts
Toxemia	Blood containing poisonous products

Symptom of Exposure	**Description of the Symptom**
Toxic Nephrosis	Kidney failure due to toxic degeneration of the kidney or renal tubules
Tracheitis	Inflammation of the trachea
Tremulous	Quivering
Tuberculostatic	Something that inhibits the growth of tubercle bacilli
Tumor	A swelling or enlargement; may also refer to a spontaneous growth of new tissue
Ulcerative	Causing ulcers
Urobilinuria	Excess urobilin in the urine
Urticaria	Hives or a rash
Uveitis	Inflammation of the uvea or iris and surrounding areas
Vacuoles	Space in a cell
Vaginitis	Inflammation of the vagina
Varices	Enlargement of veins, arteries or lymphatic vessels
Vasodilation	Relaxation of the blood vessels
Vascular	Blood vessels
Verruca	Wart
Vertigo	Dizziness
Vesication	Blistering
Vesicle	A sac containing liquids
Vesiculation	The presence or formation of vesicles
Virilism	Masculinity
Viscera	Internal organs enclosed within a cavity such as the abdominal or thoracic cavities
Wheal	A raised body area which is redder or paler then surrounding skin and may itch
Wilson's disease	Hepatolenticular degeneration
Xerosis	Abnormal dryness
Yusho disease	Disease outbreak in Japan due to exposure of cooking oil contaminated with chlorinated dibenzofurans

Appendix 3 - Glossary of Risk Assessment Related Terms

This glossary contains definitions for risk assessment-related terms used in IRIS. It is meant to assist the user in understanding terms used by the US EPA in risk assessment activities. These definitions are not all-encompassing, nor should they be construed to be "official" definitions. It is assumed that the user has some familiarity with risk assessment and health science. For terms that are not included in this glossary, the user should refer to standard health science, biostatistical, and medical textbooks and dictionaries.

Acceptable Daily Intake -- An estimate of the daily exposure dose that is likely to be without deleterious effect even if continued exposure occurs over a lifetime.

Acute exposure -- One dose or multiple doses occurring within a short time (24 hours or less).

Acute hazard or toxicity -- see Health hazard.

Added risk -- The difference between the cancer incidence under the exposure condition and the background incidence in the absence of exposure; AR = P(d) - P(O).

Aerodynamic diameter -- Term used to describe particles with common inertial properties to avoid the complications associated with the effects of particle size, shape and physical density.

Anecdotal data -- Data based on descriptions of individual cases rather than on controlled studies.

Attributable risk -- The difference between risk of exhibiting a certain adverse effect in the presence of a toxic substance and that risk in the absence of the substance.

Benign -- Not malignant; remaining localized.

Bioassay -- The determination of the potency (bioactivity) or concentration of a test substance by noting its effects in live animals or in isolated organ preparations, as compared with the effect of a standard preparation.

Bioavailability -- The degree to which a drug or other substance becomes available to the target tissue after administration or exposure.

Blood-to-air partition coefficient -- Ratio of concentrations for a given chemical achieved between blood and air at equilibrium.

Carcinogen -- An agent capable of inducing a cancer response.

1211

Carcinogenesis -- The origin or production of cancer, very likely a series of steps. The carcinogenic event so modifies the genome and/or other molecular control mechanisms in the target cells that these can give rise to a population of altered cells.

Case-control study -- An epidemiologic study that looks back in time at the exposure history of individuals who have the health effect (cases) and at a group who do not (controls), to ascertain whether they differ in proportion exposed to the chemical under investigation.

Chronic effect -- An effect that is manifest after some time has elapsed from initial exposure. See also Health Hazard.

Chronic exposure -- Multiple exposures occurring over an extended period of time, or a significant fraction of the animal's or the individual's lifetime.

Chronic hazard or toxicity -- see Health hazard.

Chronic study -- A toxicity study designed to measure the (toxic) effects of chronic exposure to a chemical.

Cohort study -- An epidemiologic study that observes subjects in differently exposed groups and compares the incidence of symptoms. Although ordinarily prospective in nature, such a study is sometimes carried out retrospectively, using historical data.

Confounder -- A condition or variable that may be a factor in producing the same response as the agent under study. The effects of such factors may be discerned through careful design and analysis.

Control group -- A group of subjects observed in the absence of agent exposure or, in the instance of a case/control study, in the absence of an adverse response.

Core grade(s) -- Quality ratings, based on standard evaluation criteria established by the Office of Pesticide Programs, given to toxicological studies after submission by registrants.

Critical effect -- The first adverse effect, or its known precursor, that occurs as the dose rate increases.

Developmental toxicity -- The study of adverse effects on the developing organism (including death, structural abnormality, altered growth, or functional deficiency) resulting from exposure prior to conception (in either parent), during prenatal development, or postnatally up to the time of sexual maturation.

Dose-response relationship -- A relationship between the amount of an agent (either administered, absorbed, or believed to be effective) and changes in certain aspects of the biological system (usually toxic effects), apparently in response to that agent.

Endpoint -- A response measure in a toxicity study.

Estimated exposure dose (EED) -- The measured or calculated dose to which humans are likely to be exposed considering exposure by all sources and routes.

Excess lifetime risk -- The additional or extra risk incurred over the lifetime of an individual by exposure to a toxic substance.

Extra risk -- The added risk to that portion of the population that is not included in measurement of background tumor rate; $ER(d) = [P(d) - P(O)]/ [1-P(O)]$.

Extrapolation -- An estimation of a numerical value of an empirical (measured) function at a point outside the range of data which were used to calibrate the function. The quantitative risk estimates for carcinogens are generally low-dose extrapolations based on observations made at higher doses. Generally one has a measured dose and measured effect.

Frank-effect level (FEL) -- Exposure level which produces unmistakable adverse effects, such as irreversible functional impairment or mortality, at a statistically or biologically significant increase in frequency or severity between an exposed population and its appropriate control.

Gamma multi-hit model -- A dose-response model of the form

$$P(d) = \text{Integral from 0 to d of } \{[a^{**}k][s^{**}(k-1)][\exp(-as)]/G(u)\}ds$$

where: $G(u)$ = integral from 0 to infinity of $[s^{**}(u-1)][\exp(-s)]ds$
  $P(d)$ = the probability of cancer from a dose rate d
  k = the number of hits necessary to induce the tumor
  a = a constant
when $k = 1$, see the one-hit model.

Guidelines for Carcinogen Risk Assessment -- US EPA guidelines intended to guide Agency evaluation of suspect carcinogens in line with statutory policies and procedures. See FR 33992-34003, September 24, 1986.

Guidelines for Exposure Assessment -- US EPA guidelines intended to guide Agency analysis of exposure assessment data in line with statutory policies and procedures. See 51 FR 34042-34054, September 24, 1986.

Guidelines for Health Assessment of Suspect Developmental Toxicants -- US EPA guidelines intended to guide Agency analysis of developmental toxicity data in line with statutory policies and procedures. See 51 FR 34028-34040, September 24, 1986.

Guidelines for the Health Risk Assessment of Chemical Mixtures -- US EPA guidelines intended to guide Agency analysis of information relating to health effects data on chemical mixtures in line with statutory policies and procedures. See 51 FR 34014-34025, September 24, 1986.

Guidelines for Mutagenicity Risk Assessment -- US EPA guidelines intended to guide Agency analysis of mutagenicity data as related to heritable mutagenic risks, in line with statutory policies and procedures. See 51 FR 34006-34012, September 24, 1986.

Health Advisory -- An estimate of acceptable drinking water levels for a chemical substance based on health effects information; a Health Advisory is not a legally enforceable Federal standard, but serves as technical guidance to assist Federal, state, and local officials.

Health hazard (types of) --

1. *Acute toxicity* -- The older term used to describe immediate toxicty. Its former use was associated with toxic effects that were severe (e.g., mortality) in contrast to the term "subacute toxicity" that was associated with toxic effects that were less severe. The term "acute toxicity" is often confused with that of acute exposure.

2. *Allergic reaction* -- Adverse reaction to a chemical resulting from previous sensitization to that chemical or to a structurally similar one.

3. *Chronic toxicity* -- The older term used to describe delayed toxicity. However, the term "chronic toxicity" also refers to effects that persist over a long period of time whether or not they occur immediately or are delayed. The term "chronic toxicity" is often confused with that of chronic exposure.

4. *Idiosyncratic reaction* -- A genetically determined abnormal reactivity to a chemical.

5. *Immediate versus delayed toxicity* -- Immediate effects occur or develop rapidly after a single administration of a substance, while delayed effects are those that occur after the lapse of some time. These effects have also been referred to as acute and chronic, respectively.

6. *Reversible versus irreversible toxicity* -- Reversible toxic effects are those that can be repaired, usually by a specific tissue's ability to regenerate or mend itself after chemical exposure, while irreversible toxic effects are those that cannot be repaired.

7. *Local versus systemic toxicity* -- Local effects refer to those that occur at the site of first contact between the biological system and the toxicant; systemic effects are those that are elicited after absorption and distribution of the toxicant from its entry point to a distant site.

Human equivalent concentration -- Exposure concentration for humans that has been adjusted for dosimetric differences between experimental animal species and humans to be equivalent to the exposure concentration associated with observed effects in the experimental animal species. If occupational human exposures are used for extrapolation, the human equivalent concentration represents the equivalent human exposure concentration adjusted to a continuous basis.

Human equivalent dose -- The human dose of an agent that is believed to induce the same magnitude of toxic effect as that which the known animal dose has induced.

Incidence -- The number of new cases of a disease within a specified period of time.

Incidence rate -- The ratio of the number of new cases over a period of time to the population at risk.

Individual risk -- The probability that an individual person will experience an adverse effect. This is identical to population risk unless specific population subgroups can be identified that have different (higher or lower) risks.

Initiation -- The ability of an agent to induce a change in a tissue which leads to the induction of tumors after a second agent, called a promoter, is administered to the tissue repeatedly. See also Promoter.

Interspecies dose conversion -- The process of extrapolating from animal doses to equivalent human doses.

Latency period -- The time between the initial induction of a health effect and the manifestation (or detection) of the health effect; crudely estimated as the time (or some fraction of the time) from first exposure to detection of the effect.

Limited evidence -- According to the US EPA's Guidelines for Carcinogen Risk Assessment, limited evidence is a collection of facts and accepted scientific inferences which suggests that the agent may be causing an effect, but this suggestion is not strong enough to be considered established fact.

Linearized multistage procedure -- The modified form of the multistage model (see Multistage Model). The constant q1 is forced to be positive (>0) in the estimation algorithm and is also the slope of the dose-response curve at low doses. The upper confidence limit of q1 (called q1*) is called the slope factor.

Logit model -- A dose-response model of the form

$$P(d) = 1/[1 + \exp -(a + b \log d)]$$

where P(d) is the probability of toxic effects from a continuous dose rate d, and a and b are constants.

Lowest-observed-adverse-effect level (LOAEL) -- The lowest exposure level at which there are statistically or biologically significant increases in frequency or severity of adverse effects between the exposed population and its appropriate control group.

Lowest-effect level (LEL) -- Same as LOAEL.

Malignant -- Tending to become progressively worse and to result in death if not treated; having the properties of anaplasia, invasiveness, and metastasis.

Margin of Exposure (MOE) -- The ratio of the no observed adverse effect level (NOAEL) to the estimated exposure dose (EED).

Margin of Safety (MOS) -- The older term used to describe the margin of exposure.

Mass median aerodynamic diameter (MMAD) -- Mass median of the distribution of mass with respect to aerodynamic diameter.

Metastasis -- The transfer of disease from one organ or part to another not directly connected with it; adj., metastatic.

Model -- A mathematical function with parameters which can be adjusted so that the function closely describes a set of empirical data. A "mathematical" or "mechanistic" model is usually based on biological or physical mechanisms, and has model parameters that have real world interpretation. In contrast, "statistical" or "empirical" models are curve-fitting to data where the math function used is selected for its numerical properties. Extrapolation from mechanistic models (e.g., pharmacokinetic equations) usually carries higher confidence than extrapolation using empirical models (e.g., logit).

Modifying factor (MF) -- An uncertainty factor which is greater than zero and less than or equal to 10; the magnitude of the MF depends upon the professional assessment of scientific uncertainties of the study and database not explicitly treated with the standard uncertainty factors (e.g., the completeness of the overall data base and the number of species tested); the default value for the MF is 1.

Multistage model -- A dose-response model often expressed in the form

$$P(d) = 1 - \exp \{-[q(0) + q(1)d + q(2)d^{**}2 + \ldots + q(k)d^{**}k]\}$$

where P(d) is the probability of cancer from a continuous dose rate d, the q(i) are the constants, and k is the number of dose groups (or, if less, k is the number of biological stages believed to be required in the carcinogenesis process). Under the multistage model, it is assumed that cancer is initiated by cell mutations in a finite series of steps. A one-stage model is equivalent to a one-hit model.

No data -- According to the US EPA Guidelines for Carcinogen Risk Assessment, "no data" describes a category of human and animal evidence in which no studies are available to permit one to draw conclusions as to the induction of a carcinogenic effect.

No evidence of carcinogenicity -- According to the US EPA Guidelines for Carcinogen Risk Assessment, a situation in which there is no increased incidence of neoplasms in at least two well-designed and well-conducted animal studies of adequate power and dose in different species.

No-observed-adverse-effect level (NOAEL) -- An exposure level at which there are no statistically or biologically significant increases in the frequency or severity of adverse effects between the exposed population and its appropriate control; some effects may be produced at this level, but they are not considered as adverse, nor precursors to adverse effects. In an experiment with several NOAELs, the regulatory focus is primarily on the highest one, leading to the common usage of the term NOAEL as the highest exposure without adverse effect.

No-observed-effect level (NOEL) -- An exposure level at which there are no statistically or biologically significant increases in the frequency or severity of any effect between the exposed population and its appropriate control.

One-hit model -- A dose-response model of the form

$$P(d) = a - \exp(-b\,d)$$

where P(d) is the probability of cancer from a continuous dose rate d, and b is a constant. The one-hit model is based on the concept that a tumor can be induced after a single susceptible target or receptor has been exposed to a single effective dose unit of a substance.

Organoleptic -- Affecting or involving a sense organ as of taste, smell, or sight.

Physiologically based pharmacokinetic (PBPK) model -- Physiologically based compartmental model used to quantitatively describe pharmacokinetic behavior.

Principal study -- The study that contributes most significantly to the qualitative and quantitative risk assessment.

Probit model -- A dose-response model of the form

$$P(d) = 0.4\{\text{integral from minus infinity to } [\log(d - u)]/s \text{ of } [\exp-(y^{**}2)/2]dy\}$$

where P(d) is the probability of cancer from a continuous dose rate d, and u and s are constants.

Promoter -- In studies of skin cancer in mice, an agent which results in an increase in cancer induction when administered after the animal has been exposed to an initiator, which is generally given at a dose which would not result in tumor induction if given alone. A cocarcinogen differs from a promoter in that it is administered at the same time as the initiator. Cocarcinogens and promoters do not usually induce tumors when administered separately. Complete carcinogens act as both initiator and promoter. Some known promoters also have weak tumorigenic activity, and some also are initiators. Carcinogens may act as promoters in some tissue sites and as initiators in others.

Proportionate mortality ratio (PMR) -- The number of deaths from a specific cause and in a specific period of time per 100 deaths in the same time period.

Prospective study -- A study in which subjects are followed forward in time from initiation of the study. This is often called a longitudinal or cohort study.

q1* -- Upper bound on the slope of the low-dose linearized multistage procedure.

Reference Concentration (RfC) -- An estimate (with uncertainty spanning perhaps an order of magnitude) of a continuous inhalation exposure to the human population (including sensitive subgroups) that is likely to be without an appreciable risk of deleterious noncancer effects during a lifetime.

Reference Dose (RfD) -- An estimate (with uncertainty spanning perhaps an order of magnitude) of a daily exposure to the human population (including sensitive subgroups) that is likely to be without an appreciable risk of deleterious effects during a lifetime.

Regional deposited dose (RDD) -- The deposited dose of particles calculated for the region of interest as related to the observed effect. For respiratory effects of particles, the deposited dose is adjusted for ventilatory volumes and the surface area of the respiratory region effected (mg/min-sq.cm). For extrarespiratory effects of particles, the deposited dose in the total respiratory system is adjusted for ventilatory volumes and body weight (mg/min-kg).

Regional deposited dose ratio (RDDR) -- The ratio of the regional deposited dose calculated for a given exposure in the animal species of interest to the regional deposited dose of the same exposure in a human. This ratio is used to adjust the exposure effect level for interspecies dosimetric differences to derive a human equivalent concentration for particles.

Regional gas dose (RGD) -- The gas dose calculated for the region of interest as related to the observed effect for respiratory effects. The deposited dose is adjusted for ventilatory volumes and the surface area of the respiratory region effected (mg/min-sq.cm).

Regional gas dose ratio (RGDR) -- The ratio of the regional gas dose calculated for a given exposure in the animal species of interest to the regional gas dose of the same exposure in humans. This ratio is used to adjust the exposure effect level for interspecies dosimetric differences to derive a human equivalent concentration for gases with respiratory effects.

Registration (of a pesticide) -- Under FIFRA and its amendments, new pesticide products cannot be sold unless they are registered with the US EPA. Registration involves a comprehensive evaluation of risks and benefits based on all relevant data.

Regulatory dose (RgD) -- The daily exposure to the human population reflected in the final risk management decision; it is entirely possible and appropriate that a chemical with a specific RfD may be regulated under different statutes and situations through the use of different RgDs.

Relative risk (sometimes referred to as risk ratio) -- The ratio of incidence or risk among exposed individuals to incidence or risk among nonexposed individuals.

Reportable quantity -- The quantity of a hazardous substance that is considered reportable under CERCLA. Reportable quantities are: (1) one pound, or (2) for selected substances, an amount established by regulation either under CERCLA or under Section 311 of the Clean Water Act. Quantities are measured over a 24-hour period.

Risk -- The probability of injury, disease, or death under specific circumstances. In quantitative terms, risk is expressed in values ranging from zero (representing the certainty that harm will not occur) to one (representing the certainty that harm will occur). The following are examples showing the manner in which risk is expressed in IRIS: E-4 = a risk of 1/10,000; E-5 = a risk of 1/100,000; E-6 = a risk of 1/1,000,000. Similarly, 1.3E-3 = a risk of 1.3/1000 = 1/770; 8E-3 = a risk of 1/125; and 1.2E-5 = a risk of 1/83,000.

Risk assessment -- The determination of the kind and degree of hazard posed by an agent, the extent to which a particular group of people has been or may be exposed to the agent, and the present or potential health risk that exists due to the agent.

Risk management -- A decisionmaking process that entails considerations of political, social, economic, and engineering information with risk-related information to develop, analyze, and compare regulatory options and to select the appropriate regulatory response to a potential chronic health hazard.

Safety Factor -- See Uncertainty Factor.

Short-term exposure -- Multiple or continuous exposures occurring over a week or so.

Slope Factor -- The slope of the dose-response curve in the low-dose region. When low-dose linearity cannot be assumed, the slope factor is the slope of the straight line from 0 dose (and 0 excess risk) to the dose at 1% excess risk. An upper bound on this slope is usually used instead of the slope itself. The units of the slope factor are usually expressed as 1/(mg/kg-day).

Standardized mortality ratio (SMR) -- The ratio of observed deaths to expected deaths.

Subchronic exposure -- Multiple or continuous exposures occurring usually over 3 months.

Subchronic study -- A toxicity study designed to measure effects from subchronic exposure to a chemical.

Sufficient evidence -- According to the US EPA's Guidelines for Carcinogen Risk Assessment, sufficient evidence is a collection of facts and scientific references which is definitive enough to establish that the adverse effect is caused by the agent in question.

Superfund -- Federal authority, established by the Comprehensive Environmental Response, Compensation, and Liability Act (CERCLA) in 1980, to respond directly to releases or threatened releases of hazardous substances that may endanger health or welfare.

Supporting studies -- Those studies that contain information that is useful for providing insight and support for the conclusions.

Systemic effects -- Systemic effects are those that require absorption and distribution of the toxicant to a site distant from its entry point, at which point effects are produced. Most chemicals that produce systemic toxicity do not cause a similar degree of toxicity in all organs, but usually demonstrate major toxicity to one or two organs. These are referred to as the target organs of toxicity for that chemical.

Systemic toxicity -- See Systemic effects.

Target organ of toxicity -- See Systemic effects.

Threshold -- The dose or exposure below which a significant adverse effect is not expected. Carcinogens are thought to be non-threshold chemicals, to which no exposure can be presumed to be without some risk of adverse effect.

Threshold Limit Values (TLVs) -- Recommended guidelines for occupational exposure to airborne contaminants published by the American Conference of Governmental Industrial Hygienists (ACGIH). The TLVs represent the average concentration (in mg/cu.m) for an 8-hour workday and a 40-hour work week to which nearly all workers may be repeatedly exposed, day after day, without adverse effect.

Tumor progression -- The sequence of changes in which a tumor develops from a microscopic lesion to a malignant stage.

Uncertainty factor -- One of several, generally 10-fold factors, used in operationally deriving the Reference Dose (RfD) from experimental data. UFs are intended to account for (1) the variation in sensitivity among the members of the human population; (2) the uncertainty in extrapolating animal data to the case of humans; (3) the uncertainty in extrapolating from data obtained in a study that is of less-than-lifetime exposure; and (4) the uncertainty in using LOAEL data rather than NOAEL data.

Unit Risk -- The upper-bound excess lifetime cancer risk estimated to result from continuous exposure to an agent at a concentration of 1 ug/L in water, or 1 ug/cu.m in air.

Upper bound -- An estimate of the plausible upper limit to the true value of the quantity. This is usually not a statistical confidence limit.

Weibull model -- A dose-response model of the form

$$P(d) = 1 - \exp[-b(d^{**}m)]$$

where P(d) is the probability of cancer due to continuous dose rate d, and b and m are constants.

Weight-of-evidence for carcinogenicity -- The extent to which the available biomedical data support the hypothesis that a substance causes cancer in humans.

Index

I

J

R

U

V

W

Y

Z